# A Consistent and Effective Approach to Problem Solving

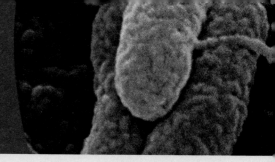

*Genetic Analysis* provides students with a unique, integrated problem-solving strategy.

The biggest challenge in a genetics course is teaching students how to become effective problem solvers. We employ a unique problem-solving feature, **Genetic Analysis,** that gives students a consistent, repeatable, three-step approach to solving problems. Genetic Analysis is integrated throughout each chapter, following discussions of important content, to help students immediately apply concepts in a problem-solving context.

Every Genetic Analysis example presents a consistent three-step approach that trains students how to **Evaluate, Deduce,** and then **Solve** problems.

Each Genetic Analysis is presented in a **clear, two-column format** that helps students see the Solution Strategy in one column and its corresponding execution in a separate Solution Step column.

## GENETIC ANALYSIS 20.2

You are interested in the development of the body plan of kelp, a common brown alga found along many coastlines. Would reverse or forward genetics approaches—as represented, respectively, by looking for gene homology and by the use of mutagenesis—be more suited to identifying the genes required for early kelp development?

| Solution Strategies | Solution Steps |
|---|---|
| **Evaluate** | |
| 1. Identify the topic this problem addresses, and explain the nature of the required answer. | 1. This problem concerns the investigation of genes determining development of kelp. Devising an answer requires evaluating the relative potential of reverse genetic analysis versus forward genetic analysis (see Chapters 16 and 17). |
| 2. Identify the critical information given in the problem. | 2. Kelp is identified as brown algae, a form of life distinct from land plants and animals. |
| **Deduce** | |
| 3. Determine if looking for gene homology (a reverse genetic approach) has a high probability of successfully identifying developmental genes in kelp. | 3. Examination of Figure 18.12 indicates that kelp is only distantly related to either land plants or animals. Therefore, searching for brown algal genes based on the sequences of plant or animal developmental genes is something of a fishing expedition. |

TIP: Review Figure 18.12, to find the relationship of kelp to the other organisms you have been studying.

PITFALL: Distantly related organisms are likely to have evolved substantially since they last shared a common ancestor, and the extent of gene homology decreases as evolutionary distance between species increases.

| Solution Strategies | Solution Steps |
|---|---|
| **Solve** | |
| 4. Determine whether the use of mutagenesis (a forward genetic approach) is likely to help identify kelp developmental genes. | 4. A good approach to finding developmental genes is to perform a mutagenesis experiment that will identify mutants in which pattern formation is perturbed. Mutagenesis can potentially affect any gene; thus, the forward genetic approach is not biased or restricted to genes that share homology with genes in other species. Mutants displaying abnormalities of wild-type pattern formation are likely to carry mutations of pattern-forming genes. |

TIP: How were genes that regulate development in *Drosophila* originally identified?

For more practice, see Problems 17, 19, 23, 25, and 28.

For **additional practice,** students are directed to similar problems at the end of the chapter.

Genetic Analysis examples include helpful **Tips** to highlight critical steps and **Pitfalls** to avoid.

The accompanying **Student Solutions Manual and Study Guide** (ISBN 0-13-174167-5) provides additional worked problems along with tips for solving problems. It also presents solutions to all of the textbook problems in a consistent *Evaluate, Deduce,* and *Solve* format to complement the approach modeled in the Genetic Analysis examples.

# MasteringGenetics™ helps students master key concepts and provides additional problem-solving practice

## An Engaging Experience

Dynamic, engaging experiences personalize and activate learning for each student. To learn more, go to www.masteringgenetics.com

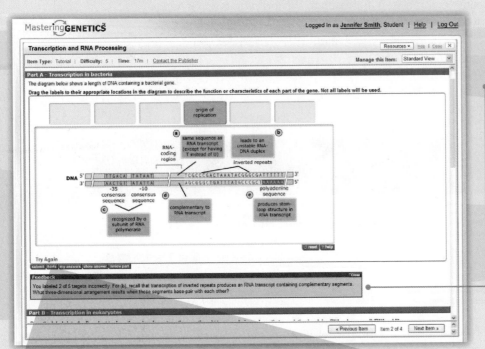

**Guided tutorials**

**In-depth tutorials**, focused on key genetics concepts, reinforce problem-solving skills by coaching students with **hints and feedback** specific to their misconceptions. Tutorial topics include pedigree analysis, sex linkage, gene interactions, DNA replication, and more.

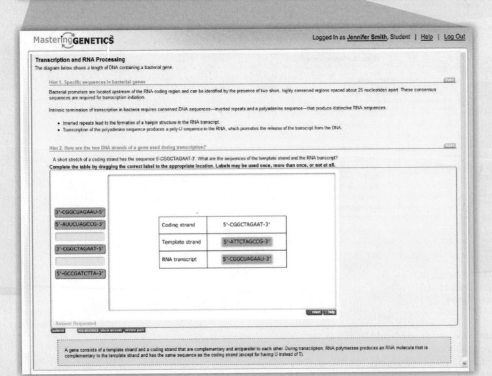

**24/7 Coaching in Solving Genetics Problems**

If students working on a tutorial get stuck, they can access **hints** to get back on track. If an incorrect answer is submitted, MasteringGenetics gives instant feedback specific to the error made, helping students overcome misconceptions and strengthen problem-solving skills.

## A Trusted Partner

Used by over 1.5 million science students, Mastering is the **most effective** and **widely used online tutorial, homework, and assessment system** for the sciences.

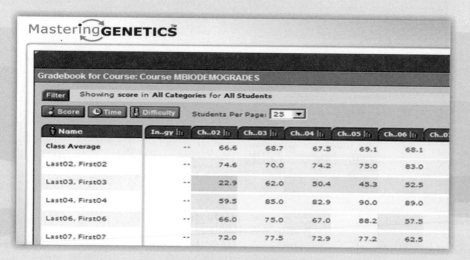

### Gradebook

In addition to the MasteringGenetics tutorials, instructors may also assign **end-of-chapter problems** and **test bank questions that are automatically scored** and entered into the Mastering gradebook.

The **color-coded gradebook,** which instantly identifies students in trouble and challenging topics for your class as a whole, provides insight into how well your students are doing throughout the course.

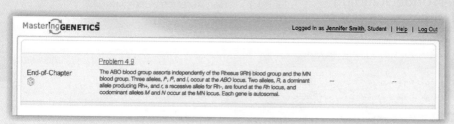

### Ensure that students do their own work

Selected questions with randomized values automatically provide individual students with different values for a given question, thereby ensuring that students do their own work. These questions are identified with an icon in the MasteringGenetics item library.

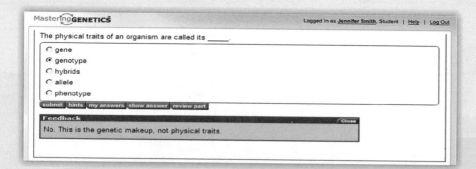

### Easy Customization

Any question or answer choice can be easily edited to match the precise language you use.

## Proven Results

The Mastering platform is the only online homework system with research showing that it improves student learning. A wide variety of published papers based on NSF-sponsored research and tests illustrate the benefits of the Mastering program.

Results documented in **scientifically valid efficacy papers** are available at www.masteringgenetics.com/site/results/

# An integrated approach to evolution and to Mendelian and molecular genetics

## The text presents an integrated approach to evolution and to Mendelian and molecular genetics.

### 1.4 Evolution Has a Molecular Basis

As biologists survey varieties of life, assess the genetic similarities and differences between species, and explore the relationship of modern organ[...] to their extinct ancestors, it beco[...] is connected through DNA. Rich[...] and author of several books on[...] this molecular connection when[...]

Chapter 1

### 5.6 Genetic Linkage Analysis Traces Genome Evolution

Evolution changes genomes as populat[...] species diverge from common ancesto[...] include alterations of allele frequenc[...] changes in the number and structure[...]

Chapter 5

**An integrated evolutionary perspective** is demonstrated throughout the book, helping students keep sight of important evolutionary principles as they are learning the core genetics concepts.

Chapter 8

Natural selection has operated to retain strong sequence similarity in consensus regions and to retain the position of the consensus regions relative to the start of transcription. The effectiveness of evolution in maintaining promoter consensus sequences is illustrated by comparison with the sequences between and around −10 and −35, which are not conserved and which exhibit considerable variation. In addition, the spacing between [...] and their placement relative to the +1 nucl[...] RNA polymerase is a large molecule that bi[...]

Chapter 8

These similarities of RNA polymerase structure and function are a direct result of the shared evolutionary history of bacteria, archaea, and eukaryotes.

## The Integration of Genetic Approaches: Understanding Sickle Cell Disease

**Unique Chapter 10: The Integration of Genetic Approaches** explores the hereditary and molecular basis of sickle cell disease in humans, integrating discussions of many research techniques.

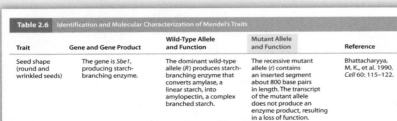

| Table 2.6 | Identification and Molecular Characterization of Mendel's Traits | | | | |
|---|---|---|---|---|---|
| **Trait** | **Gene and Gene Product** | **Wild-Type Allele and Function** | **Mutant Allele and Function** | **Reference** | |
| Seed shape (round and wrinkled seeds) | The gene is *Sbe1*, producing starch-branching enzyme. | The dominant wild-type allele (*R*) produces starch-branching enzyme that converts amylase, a linear starch, into amylopectin, a complex branched starch. | The recessive mutant allele (*r*) contains an inserted segment about 800 base pairs in length. The transcript of the mutant allele does not produce an enzyme product, resulting in a loss of function. | Bhattacharyya, M. K., et al. 1990. *Cell* 60: 115–122. | |

Within a traditional chapter organization, Sanders and Bowman **integrate transmission genetics and molecular genetics in tables, figures, and text.** This approach helps students understand how today's geneticists think.

### The Molecular Basis of Dominance

A character is called dominant if it is seen in the homozygous and heterozygous genotypes, and it is called recessive if it is observed only in a single homozygous genotype. In this sense, dominance and recessiveness have a phenotypic basis. The phenotypes are, however, a consequence of the activities of proteins produced by the alleles of a gene. In this sense, dominance and recessiveness have a molecular basis. The dominance of one allele over another is determined by the activity of the protein products of the allele—by the manner in which the protein products of alleles work to produce the phenotype.

Let's compare two examples to illustrate the molecular basis of dominance and recessiveness. In both examples, a wild-type allele produces an active enzyme and a mutant allele produces either very little enzyme or none at all. In the first example the mutant allele is recessive,

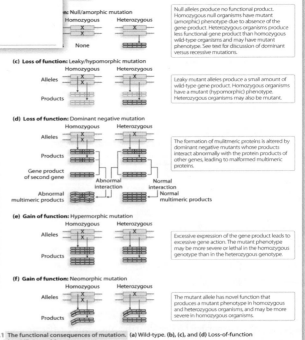

The expression of the products of wild-type alleles produces wild-type phenotype.

Null alleles produce no functional product. Homozygous null organisms have mutant (amorphic) phenotype due to absence of the gene product. Heterozygous organisms produce less functional gene product than homozygous wild-type organisms and may have mutant phenotype. See text for discussion of dominant versus recessive mutations.

Leaky mutant alleles produce a small amount of wild-type gene product. Homozygous organisms have a mutant (hypomorphic) phenotype. Heterozygous organisms may also be mutant.

The formation of multimeric proteins is altered by dominant negative mutants whose products interact abnormally with the protein products of other genes, leading to malformed multimeric proteins.

Excessive expression of the gene product leads to excessive gene action. The mutant phenotype may be more severe or lethal in the homozygous genotype than in the heterozygous genotype.

The mutant allele has novel function that produces a mutant phenotype in homozygous and heterozygous organisms, and may be more severe in homozygous organisms.

**Figure 4.1** The functional consequences of mutation. **(a)** Wild-type. **(b)**, **(c)**, and **(d)** Loss-of-function mutations. **(e)** and **(f)** Gain-of-function mutations.

# Carefully crafted figures teach complex genetics processes

**The figures in *Genetic Analysis* help students understand the most important—and often challenging—lessons in each chapter.**

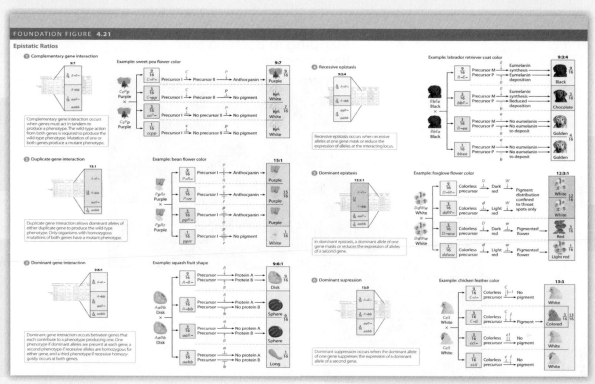

**Foundation Figures** integrate text and art to illustrate pivotal genetics concepts in a concise, easy-to-follow format.

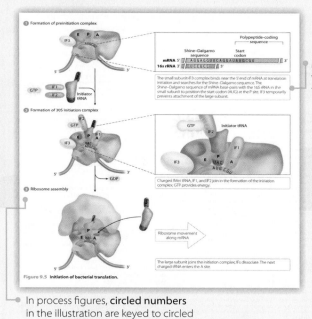

Figure 9.5  Initiation of bacterial translation.

**Many key figures include annotations,** which focus students on key aspects of the concept being illustrated.

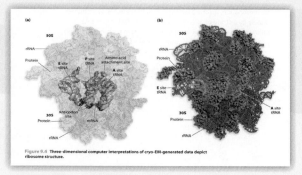

Figure 9.4  Three-dimensional computer interpretations of cryo-EM-generated data depict ribosome structure.

In process figures, **circled numbers** in the illustration are keyed to circled numbers in the figure legend or text.

Selected figures incorporate a **realistic style of art,** including 3-D molecular models, as appropriate.

# Clear organization provides a learning path and focuses students on important lessons

## *Genetic Analysis* provides a clear learning path.

For a quick overview or review, each chapter opens with an **outline** of the chapter and a list of **Essential Ideas.**

---

6

## Genetic Analysis and Mapping in Bacteria and Bacteriophage

### CHAPTER OUTLINE

6.1 Bacteria Transfer Genes by Conjugation
6.2 Interrupted Mating Analysis Produces Time-of-Entry Maps
6.3 Conjugation with F′ Strains Produces Partial Diploids
6.4 Bacterial Transformation Produces Genetic Recombination
6.5 Bacterial Transduction Is Mediated by Bacteriophages
6.6 Bacteriophage Chromosomes Are Mapped by Fine-Structure Analysis

### ESSENTIAL IDEAS

■ Bacterial conjugation is a one-way transfer of genetic material from a donor cell to a recipient cell. Three types of donor cells can conjugate with recipient cells to transfer donor DNA.
■ Donor bacterial genetic maps are derived from conjugation analysis.
■ A particular type of bacteria conjugation can produce bacteria with genomes that are partially diploid.
■ Transformation is the absorption of extracellular DNA across the cell wall and membrane of a recipient bacterial cell, and its analysis leads to mapping of donor bacterial genes.
■ Transduction, mediated by bacteriophages, is the transfer of DNA from a donor bacterial cell to a recipient cell, and its analysis leads to mapping of donor bacterial genes.
■ Fine-structure genetic analysis of a bacteriophage genome demonstrated that DNA nucleotide base pairs are the fundamental units of mutation and recombination.

184

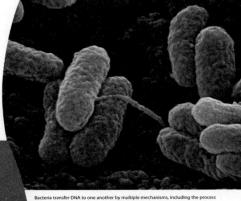

Bacteria transfer DNA to one another by multiple mechanisms, including the process of gene transfer called conjugation shown here. The bacterial "donor" (center left) transfers DNA through a tube that connects it to a bacterial "recipient" (lower right).

Here's a disturbing little secret of human life: Your body contains approximately 100 trillion cells, but only about 10 trillion of them are yours! The other 90% of the cells you carry around are bacteria, fungi, and other forms of microscopic life. Many of these biological hitchhikers perform useful, even essential, functions. For example, you carry hundreds of species of bacteria in your gut that collectively have a mass of more than 3 pounds. Without these intestinal bacteria, your digestion of carbohydrates would be impaired, and your ability to manufacture essential nutrients such as vitamin B$_{12}$ and vitamin K would be disabled. The bacteria teeming in your digestive tract also help keep potentially harmful bacteria at bay by vigorously

---

Genetic Insight  Bacterial transcription termination is dependent on the presence of certain DNA sequences. In intrinsic termination, inverted repeat DNA sequences followed immediately by a string of adenines produce an mRNA stem-loop followed by a poly-U string. In rho-dependent termination, a rho protein binds to mRNA and catalyzes the release of the transcript from RNA polymerase.

**Genetic Insights** appear several times within each chapter and help students recognize and understand key concepts at a glance.

The chapter outline and essential ideas are synthesized and expanded at the end of the chapter, in the **Summary.**

---

SUMMARY     Mastering**GENETICS**     *For activities, animations, and review quizzes, go to the study area at www.masteringgenetics.com.*

### 2.1 Gregor Mendel Discovered the Basic Principles of Genetic Transmission

■ A broad education in science and mathematics prepared Gregor Mendel to design hybridization experiments that could reveal the principles of hereditary transmission.

### 2.2 Monohybrid Crosses Reveal the Segregation of Alleles

■ Mendel's experimental design had five important features: controlled crosses, use of pure-breeding parental strains, examination of discreet traits, quantification of results, and the use of replicate and reciprocal crosses.
■ Crosses between pure-breeding parental plants with different phenotypes produce monohybrid F$_1$ progeny with the dominant phenotype.
■ Monohybrid crosses produce a 3:1 ratio of the dominant to the recessive phenotype among F$_2$ progeny and demonstrate the operation of the law of segregation.
■ The law of segregation states that two alleles at a locus will separate from one another during gamete formation, each allele has an equal probability of inclusion in a gamete, and gametes unite at random during reproduction.
■ Mendel used test-cross analysis to demonstrate that F$_1$ plants are monohybrid, and he used the self-fertilization of F$_2$ plants with the dominant phenotype to demonstrate that the latter have a 2:1 ratio of heterozygotes to homozygotes.

### 2.3 Dihybrid and Trihybrid Crosses Reveal the Independent Assortment of Alleles

■ The F$_2$ progeny of dihybrid F$_1$ plants display a 9:3:3:1 phenotype ratio that demonstrates the operation of the law of independent assortment.
■ Mendel used trihybrid-cross analysis to demonstrate that alleles of multiple genes are transmitted in accordance with the predictions of the law of independent assortment.

### 2.4 Probability Theory Predicts Mendelian Ratios

■ The product rule of probability is used to determine the likelihood of two or more independent events occurring simultaneously or consecutively. The joint probability is determined by multiplying the probabilities of the independent events.
■ The sum rule of probability is applied when two or more outcomes are possible. The individual probabilities of the outcomes are added together to determine the joint probability.
■ Conditional probability is the probability of outcomes that are contingent on particular conditions.
■ Binomial probability theory describes the distribution of outcomes of an experiment in terms of the number of outcome classes and the frequency of each class. Pascal's triangle is a convenient tool for determining the distribution of binomial outcomes.

### 2.5 Chi-Square Analysis Tests the Fit between Observed Values and Expected Outcomes

■ The chi-square test ($\chi^2$) compares observed results with the results predicted by a genetic hypothesis that is based on chance.
■ The result of the chi-square test determines how closely predictions match results.
■ The significance of a chi-square value is determined by the *P* (probability) value corresponding to the number of degrees of freedom in the experiment.

### 2.6 Autosomal Inheritance and Molecular Genetics Parallel the Predictions of Mendel's Hereditary Principles

■ Traits transmitted by autosomal inheritance are equally likely in males and females.
■ Autosomal dominant inheritance produces a vertical pattern of transmission in which each organism with the dominant trait has at least one parent with the trait.

# Engage students in learning about today's research

*Genetic Analysis* engages students in the process of science with thought-provoking discussions of experiments, research techniques, and practical applications of research.

---

**Experimental Insight 5.1**

**Mapping a Gene for Breast and Ovarian Cancer Susceptibility**

Most cases of cancer develop through the acquisition of multiple mutations in somatic cells, meaning that there is no inherited mutation that increases the likelihood of cancer development. In some families, however, the frequent occurrence of a particular kind of cancer in a pattern consistent with single-gene inh can suggest the hereditary transmission of a mutant al

families in which multiple cases appeared at young ages, and in lateral cancer occurred (affecting in a single patient) in patterns o

**Experimental Insight 17.1**

**Plant-Derived Antimalarial Drugs Produced in *E. coli***

The production of amorphadiene in *E. coli* exemplifies the use of genetic engineering to produce a high-value pharmaceutical product. Amorphadiene is the immediate precursor to artemisinin, a potent antimalarial drug. Artemisinin has been touted as the next-generation antimalarial drug since it is effective at treating multiple stages of malarial infection and

produces IPP and DMAPP, the pa inhibition, preventing large q from accumulating.
  This obstacle to producing larg in *E. coli* is circumvented by use of from *Saccharomyces cerevisiae* an

**Experimental Insight essays** discuss influential experiments, summarize real data derived from the experiments, and explain conclusions drawn from the analysis of results.

**Research Technique 9.1**

**Two-Dimensional Gel Electrophoresis and the Identification of Ribosomal Proteins**

PURPOSE All ribosomes are composed of two subunits that are each a complex mixture of rRNA and dozens of proteins. One approach to determining the number of proteins contained in each ribosomal subunit uses a method of electrophoresis known as two-dimensional

Gel S    Gel L

**Research Technique 10.1**

**The Production and Detection of DNA Restriction Fragment Length Polymorphisms**

PURPOSE Restriction digestion followed by DNA gel electrophoresis is one method for detecting variation of DNA sequence that alters the number or the relative positions of RFLP sequences. Variation in the number or length of restriction fragments can result from DNA sequence changes that alter a restriction sequence, making it unrecognizable, or that create a new restriction sequence. RFLP changes can also result from the insertion or deletion of DNA between restriction sequences that increase or decrease the length of restriction fragments.

MATERIALS AND PROCEDURES DNA is isolated from cells and treated with one or more restriction enzymes to produce DNA restriction fragments. The restriction fragments are then separated by DNA gel electrophoresis, causing the fragments to be visualized as "bands" on the gel. Laboratory methods described in Research Technique 10.2 can also aid in the identification of specific restriction fragments.

DESCRIPTION DNA sequence variation altering the number or length of restriction fragments (RFLPs) produces distinctive restriction fragments for each allele. Organisms that are homozygous for DNA sequence at a restriction site produce the same restriction fragment from homologous chromosomes and produce one DNA band on a gel. Heterozygous organisms have different DNA sequences, produce a different restriction fragment from each chromosome, and have two DNA bands on a gel. Some examples of these processes are shown to the right.

CONCLUSION DNA base substitution changes that alter a restriction sequence and the insertion or deletion of DNA between two restriction sequences are the principal ways DNA sequence alterations can produce RFLPs. RFLP alleles form genotypes whose DNA restriction fragments produce distinctive patterns in gel electrophoresis. Each genotype has a distinctive combination of band number and band size on the gel.

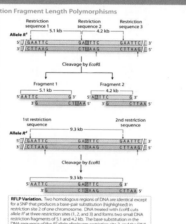

**RFLP Variation.** Two homologous regions of DNA are identical except for a SNP that produces a base-pair substitution (highlighted in restriction site 2 of one chromosome. DNA treated with *Eco*RI cuts allele $R^1$ at three restriction sites (1, 2, and 3) and forms two small DNA restriction fragments of 5.1 and 4.2 kb. The base substitution in the DNA sequence of the $R^2$ allele eliminates restriction site 2, and the DNA is cut only at sites 1 and 3, resulting in a single DNA restriction fragment of 9.3 kb.

**Research Technique boxes** explore important research methods and visually illustrate the results and interpretations of the techniques.

---

**CASE STUDY**

**The Evolution of Antibiotic Resistance**

Alexander Fleming got a little sloppy with his sterile technique one day in 1929 and made a mistake that has since saved millions of lives. Fleming was working with *Staphylococcus*, a common bacterial strain that causes a serious and potentially fatal "staph" infection when it enters the body through a cut or abrasion. O
nated his *Stap*

They infect people in hospitals, medical clinics, and locker rooms, killing thousands around the world each year. The strains are so common that you may have MRSA on your skin at this moment!
  What happened to bring about this shift? Where once

**Case Studies** are short, real-world examples that appear at the end of every chapter and highlight central ideas or concepts of the chapter to remind students of some of the practical applications of genetics.

**CASE STUDY**

**Environmental Epigenetics**

Here's a simple question: How are traits passed from one generation to the next? The first answer that came to your mind was probably (and not incorrectly) that traits are passed by the transmission of ge
decade or so,
unexpected d
tain cases, pa

controlled modifications of gene expression and that in a few select instances, the affected genes can be transmitted to offspring in their epigenetically modified form. More current

**CASE STUDY**

**Positional Cloning in Humans: The Huntington Disease Gene**

Huntington disease, an inevitably fatal, late-onset neurodegenerative disorder, is named for George Huntington, the physician who published the classic description of the disease and its inheritance in 1872. His description specified the symptoms of movement disorder, personality change, and cognitive decline and, notably, outlined the autosomal dominant pattern of inheritance, a feature that went unappreciated until

one-third of the families, suggesting the *HD* gene was likely to reside within 500 kb of the shared haplotype (see Chapter 5 for a discussion of haplotypes).

CANDIDATE GENE IDENTIFICATION Before 2001, the year a draft of the human genome sequence was published, identification of genes in large stretches of human genomic

# G E N E T I C
## AN INTEGRATED APPROACH
# A N A L Y S I S

Mark F. Sanders
University of California at Davis

John L. Bowman
Monash University, Melbourne,
Australia
University of California at Davis

**Benjamin Cummings**

Boston  Columbus  Indianapolis  New York  San Francisco  Upper Saddle River
Amsterdam  Cape Town  Dubai  London  Madrid  Milan  Munich  Paris  Montréal  Toronto
Delhi  Mexico City  São Paulo  Sydney  Hong Kong  Seoul  Singapore  Taipei  Tokyo

Editor-in-Chief: *Beth Wilbur*
Executive Director of Development: *Deborah Gale*
Senior Acquisitions Editor: *Michael Gillespie*
Project Editor: *Anna Amato*
Development Editor: *Moira Lerner*
Art Development Editor: *Adam Steinberg*
Associate Editor: *Brady Golden*
Assistant Editor: *Leslie Allen*
Executive Managing Editor: *Erin Gregg*
Managing Editor: *Michael Early*
Production Project Manager: *Camille Herrera*
Supplements Production Project Manager: *Jane Brundage*
Production Management and Compositor: *Progressive Publishing Alternatives*
Interior Designer: *Tani Hasegawa*
Cover Designer: *Jodi Notowitz*
Illustrators: *Precision Graphics*
Photo Researcher: *Eric Schrader*
Image Lead: *Donna Kalal*
Manufacturing Buyer: *Michael Penne*
Printer and Binder: *RR Donnelly Willard*
Cover Printer: *Lehigh Phoenix*
Director of Editorial Content: *Tania Mlawer*
Senior Media Producer: *Laura Tommasi*
Director of Marketing: *Christy Lawrence*
Executive Marketing Manager: *Lauren Harp*
Cover Photo Credit: *Eye of Science / Photo Researchers, Inc.*

Credits and acknowledgments borrowed from other sources and reproduced, with permission, in this textbook appear on p. **C-1**.

If you purchased this book within the United States or Canada, you should be aware that it has been imported without the approval of the Publisher or the Author.

ISBN 10: 0-321-81846-6; ISBN 13: 978-0-321-81846-1 (International)
1 2 3 4 5 6 7 8 9 10—DOW—15 14 13 12 11

# Table of Contents

# 7 DNA Structure and Replication  222

# 8 Molecular Biology of Transcription and RNA Processing  260

# 13 Chromosome Aberrations and Transposition   422

# 14 Regulation of Gene Expression in Bacteria and Bacteriophage   459

# 16 Forward Genetics and Recombinant DNA Technology 522

# 17 Applications of Recombinant DNA Technology and Reverse Genetics 562

# 18 Genomics: Genetics from a Whole-Genome Perspective 602

# 19 Cytoplasmic Inheritance and the Evolution of Organelle Genomes 641

# 20 Developmental Genetics 674

# 21 Genetic Analysis of Quantitative Traits 708

## 22  Population Genetics and Evolution  737

# About the Authors

 **Mark F. Sanders** has been a faculty member in the Department of Molecular and Cellular Biology at the University of California, Davis for 27 years. In that time, he has taught more than 150 genetics courses to nearly 35,000 undergraduate students. Specializing in teaching the genetics course for which this book is written, he also teaches a genetics laboratory course, an advanced human genetics course for biology majors, and a human heredity course for nonscience majors. His teaching experience also includes introductory biology as well as courses in population genetics and evolution.

Dr. Sanders received his B.A. degree in Anthropology from San Francisco State University and his M.A. and Ph.D. degrees in Biological Anthropology from the University of California, Los Angeles. Following graduation, he spent four years at the University of California, Berkeley as a postdoctoral researcher studying inherited susceptibility to human breast and ovarian cancer. At UC Berkeley, he also taught his first genetics courses. Since coming to the University of California, Davis, he has maintained a full-time teaching schedule and promotes academic achievement by undergraduate students in numerous ways, including as an active student advisor, through his direction of several undergraduate student programs, and service as the associate dean for Undergraduate Academic Programs in the College of Biological Sciences.

 **John L. Bowman** is a professor in the School of Biological Sciences at Monash University in Melbourne, Australia, and an adjunct professor in the Department of Plant Biology at the University of California, Davis in the United States. He received a B.S. in Biochemistry at the University of Illinois at Urbana-Champaign in 1986 and a Ph.D. in Biology from the California Institute of Technology in Pasadena, California. His Ph.D. research focused on how the identities of floral organs are specified in *Arabidopsis* (described in Chapter 20). He conducted postdoctoral research at Monash University on the regulation of floral development. From 1996 to 2006, his laboratory at UC Davis focused on developmental genetics of plants, focusing on how leaves are patterned. From 2006 to 2011, he was a Federation Fellow at Monash University, where his laboratory is studying land plant evolution using a developmental genetics approach. At UC Davis he taught genetics, "from Mendel to cancer," to undergraduate students, and he continues to teach genetics courses at Monash University.

# Dedication

To my extraordinary wife and partner Ita whose support, patience, and encouragement throughout this project make me very fortunate. She is a treasure. To my wonderful children Jana and Nick who have lived with Dad working on this project so long they might be forgiven for thinking they have another, more problematic, sibling named "the book". To John and Lincoln who complete our family circle for now. And to all my students from whom I have learned as much as I have taught.

*Mark F. Sanders*

For my parents, Lois and Noel, who taught me to love and revere nature. And to all my genetics students who have inspired us over the years, I hope that the inspiration was mutual.

*John L. Bowman*

# Preface

Modern genetics is almost 150 years old. Its origins are traced to the mid-1860s, when Gregor Mendel described the results and analysis of his experiments on pea plants that, for the first time, correctly described the basic rules by which genes are transmitted from one generation to the next. Mendel's work languished largely unknown and unappreciated until 1900, when it was rediscovered. Innumerable experimental analyses during the first half of the 20th century clarified and expanded Mendel's principles and their analysis. The second half of the century brought the recognition and rapid development of principles and methods of molecular genetic analysis. Today, new information accumulates at an unprecedented rate, yet new discoveries in genetics remain rooted in the principles of transmission, molecular, and evolutionary genetics.

At its core, genetics remains a way of thinking about the analysis of organisms. It is an experimental discipline in which analysis and problem solving, through the application of principles, are the keys to understanding. In *Genetic Analysis: An Integrated Approach*, our foremost goals are to give students of genetics a solid foundation in principles of transmission and molecular genetics, help them develop strong problem-solving skills, provide exposure to both the experimental nature of genetics and the evolutionary context of genetics, and prepare them to process and organize new information as it is acquired.

Through the many classes we have taught over the years, we continue to see one learning challenge emerge again and again as a roadblock for understanding genetics—problem solving. While students read solutions provided in textbooks and are generally knowledgeable enough to understand the problem and the logic of the answers presented, they all too often have no clear idea of how they might personally devise the solutions on their own. Similarly, in the classroom, as an instructor solves a problem, students nod affirmatively as they follow the steps of the analysis to its conclusion. But when faced with similar homework or exam problems, students discover that they don't know how to solve or even approach the problems. Thus all too often, students are left to sink or swim when it comes to solving genetics problems. Swimmers manage to devise their own de novo approach to problem solving to gain the required level of understanding, whereas many others are unable to do so and fail to develop a deep comprehension of genetics. To address this issue head-on, we have designed an innovative problem-solving strategy that can help students attain the level of proficiency required to truly understand genetics and its applications.

## A Problem-Solving Approach

Although providing clear explanations of genetic principles is essential to a student's ability to comprehend the discipline, learning requires more than understanding the facts of genetics. Students must actively engage with the material through the practice of solving problems. The idea that students of genetics must solve problems to effectively learn the discipline of genetics is not a new one. However, we offer a unique approach to learning problem solving: it was developed in the classroom and provides a logical, repeatable framework for students to use when encountering genetics problems of any type. Having a strategy enables students to more effectively apply what they learn to new situations, such as problems in the ends of chapters, on MasteringGenetics, and on exams.

Research and our own classroom experience shows that effective problem solvers follow a general strategy for tackling complex problems. So to help train students to become more effective problem solvers, we employ a unique problem-solving feature called Genetic Analysis that gives students a consistent, repeatable method to help them know how to start thinking about a problem, what the end goal is, and the kind of analysis required to get there. The three stages of this problem-solving framework are *evaluate*, *deduce*, and *solve*.

**Evaluate:** Students learn to identify the topic of the problem, specify the nature or format of the answer, and identify critical information given in the problem.

**Deduce:** Students learn how to use conceptual knowledge to analyze data, make connections, and infer additional information or next steps.

**Solve:** Students learn how to accurately apply analytical tools and to execute their plan to solve a given problem.

Irrespective of the type of problem a student faces, this framework guides students through the stages of problem solving and gives them the confidence to undertake new problems.

Each Genetic Analysis is organized in a two-column format to help students easily follow each step of the Solution Strategy in the left-hand column along with its corresponding execution in the Solution Step in the right-hand column. We also include problem-solving Tips to highlight critical steps and Pitfalls to avoid, gathered from our teaching experience. Genetic Analysis is therefore integrated throughout each chapter, right after discussions of important content, to help students immediately apply concepts. Each chapter includes two or three Genetic Analysis features, and the book contains 50 in all.

We pair Genetic Analysis with strong end-of-chapter problems that are divided into two groups. Chapter Concept problems come first and review the critical information, principles, and analytical tools discussed in the chapter. These are followed by Application and Integration problems that are more challenging and give students practice in solving problems that are broader in scope.

All solutions to the end-of-chapter problems in the *Study Guide and Solutions Manual* use the evaluate-deduce-solve model to reinforce the approach. At the end of the book we provide answers to selected even-numbered end-of-chapter problems, and we have included a few example solutions from the *Study Guide and Solutions Manual* to reinforce the importance of our problem-solving strategy. We highly recommend, however, that students purchase the *Study Guide and Solutions Manual* to accompany the textbook. It is a robust and informative supplement that provides valuable support for learning how to solve genetics problems.

## An Evolutionary Perspective

In his 1973 essay, "Nothing in Biology Makes Sense Except in the Light of Evolution," the evolutionary biologist Theodosius Dobzhansky clearly stated his concept of the central role of evolution. Today his viewpoint is more apt than ever, particularly in the field of genetics. DNA connects all life and is the reason genetic principles are applicable to all forms of life, past and present. Geneticists, then, are acutely aware of evolutionary relationships between genes, genomes, and organisms. Evolutionary processes at the organismal level discovered through comparative biology can also shed light on the function of genes and organization of genomes at the molecular level. Likewise, the function of genes and organization of genomes informs the evolutionary model.

To set the tone, Chapter 1 introduces evolutionary principles and then connects them to similarities and variations of organisms. Evolution is a strong theme of Chapter 10, and it is the central focus of Chapter 22. Evolutionary ideas are frequently included to remind students of its connection to genetics. For example, in Chapter 4 (page 112), the evolutionary implications of shared ABO blood groups by humans, great apes, and many monkey species are discussed. In Chapter 9 (Section 9.4), we describe the genetic code in the context of the single origin of life.

## Connecting Transmission and Molecular Genetics

Experiments that shed light on principles of transmission genetics preceded the discovery of the structure and function of DNA and its role in inherited molecular variation by several decades. Yet biologists recognize that DNA variation is the basis of inherited morphological variation observed in transmission genetics. Understanding how these two approaches to genetics are connected is vital to thinking like a geneticist. We have highlighted this connection in multiple chapters. In Chapter 2, for example, we identify and describe the wild-type and mutant activities of the four genes known to have been studied by Mendel. In Chapter 4, dominant and recessive alleles are given a molecular context. And, in Chapter 10, we explore both the hereditary and molecular basis of sickle cell disease in humans.

## Textbook Organization

This textbook contains 22 chapters organized into five units that span the breadth of genetics. We follow the commonly used Mendel-first organization that begins with gene transmission and function and ends with genetic analysis of populations.

### Unit I Gene Transmission and Function

We start by reviewing basic concepts of DNA replication, transcription, translation, evolutionary concepts, and the connection between inherited variation and evolutionary processes at the level covered in introductory biology courses (Chapter 1). Mendelian genetics is then introduced, featuring a discussion of the identification of four of Mendel's genes (Chapter 2). Cell division and sex-linked inheritance (Chapter 3) and gene interactions (Chapter 4) follow. Finally, identification and analysis of genetic linkage in eukaryotic genomes, and the analysis of gene transfer in bacterial genomes wrap up the unit (Chapters 5 and 6).

### Unit II Central Dogma

Building on the introduction to molecular processes in Unit I, this unit describes the details of DNA replication (Chapter 7), transcription and RNA processing (Chapter 8), and translation (Chapter 9). As we explore these key concepts in greater detail, we also introduce important laboratory techniques such as polymerase chain reaction and DNA sequencing. The unit concludes with an exploration of the human autosomal recessive trait sickle cell disease that features integration of transmission genetics, molecular analysis, molecular techniques, and evolution (Chapter 10).

## Unit III Genome Structure and Function

We begin Unit III by describing bacterial and eukaryotic chromosome structure (Chapter 11). We then move to gene mutation, DNA repair, and homologous recombination (Chapter 12), which is followed by chromosome mutations and transposition (Chapter 13). Discussion of the regulation of gene expression in bacteria and bacteriophage follows (Chapter 14), and the last chapter emphasizes the role of epigenetic phenomena in the regulation of gene expression in eukaryotes (Chapter 15).

## Unit IV Genome Expression and Analysis

We discuss recombinant DNA technology and its use in forward and reverse genetic approaches (Chapters 16 and 17), followed by genomic analyses and applications (Chapter 18) and inheritance of cytoplasmic genomes, and evolution of cytoplasmic organelles (Chapter 19). The final chapter in the unit describes the use of genetic approaches to understand the development of organisms (Chapter 20).

## Unit V Genetic Analysis of Populations

We tie together the course concepts by exploring quantitative genetics (Chapter 21)—which includes discussion of historically significant experiments that contributed to development of the field as well as modern applications of genome-wide association analysis—and by refocusing on evolution in the context of population genetics (Chapter 22).

# Pathways Through the Book

This book is written with a Mendel-first approach that many instructors find offers the most effective pedagogical approach for teaching genetics. We are cognizant, however, that the scope of information covered in genetics courses varies and that instructor preferences differ. We have kept differences and alternative approaches in mind while writing the book. Thus we provide *five pathways* through the book that instructors can use to meet their varying course goals and objectives. Each pathway features integration of problem solving through the inclusion of Genetic Analysis features in each chapter.

## 1. Mendel-First Approach

**Ch 1–22**

This pathway provides a traditional approach that begins with Mendelian genetics and integrates it with evolutionary concepts and connects it to molecular genetics. As examples, we discuss genes responsible for four of Mendel's traits (Chapter 2) as well as gene structure in relation to dominance and functional level (Chapter 5). We draw together hereditary variation, molecular variation, and evolution in the discussion of sickle cell disease (Chapter 10).

## 2. Molecular-First Approach

**Ch 1 → Ch 7–10 → Ch 2–6 → Ch 11–22**

This pathway provides a molecular-first approach to develop a clear understanding of the molecular basis of heredity and variation before delving into the analysis of hereditary transmission.

## 3. Integration of Molecular Analysis

**Ch 1 → Ch 10 → Ch 2–15 → Ch 16–22**

This pathway focuses on the parallels of transmission and molecular genetic analyses right from the start, and it best reflects the way a geneticist would approach study of the field. We recommend this pathway for students who already have a strong genetics background and are familiar with some common molecular techniques.

## 4. Quantitative Genetics Focus

**Ch 1–2 → Ch 21 → Ch 3–20 → Ch 22**

This pathway incorporates quantitative genetics early in the course by introducing polygenic inheritance (Chapter 2) and following it up with a comprehensive discussion of quantitative genetics (Chapter 21).

## 5. Population Genetics Focus

**Ch 1–2 → Ch 22 → Ch 3–21**

This pathway incorporates population genetics early in the course. Instructors can use the introduction to evolutionary principles and processes (Chapter 1) and the role of genes and alleles in transmission (Chapter 2) and then address evolution at the population level and at higher levels (Chapter 22).

# Chapter Features

A principal goal of our writing style and chapter organization is to engage the reader both intellectually and visually to invite continuous reading, all the while clearly explaining complex and difficult ideas. Our conversational tone encourages student reading and comprehension, and our attractive design and realistic art program visually engages students and puts them at ease. Experienced instructors of genetics know that students are more engaged when they can relate concepts to the real world. To that end, we use real experimental data to illustrate genetic principles and analysis as well as to familiarize students with exciting research and creative researchers in the field. We also discuss a broad array of organisms—such as humans, bacteria, yeast, plants, fruit flies, nematodes, vertebrates, and viruses—to exemplify genetic principles. We have found that this approach helps students see the diversity of genetics and relate better to the concepts.

Careful thought has been given to our chapter features; each one serves to improve student learning and bring students closer to thinking like a geneticist. Many features are included in every chapter or appear more occasionally, but the following features illustrate how we highlight central ideas, problems, and methods that are important for understanding genetics.

- **Genetic Analysis:** This is our key problem-solving feature that guides students through the problem-solving process by using the *evaluate-deduce-solve* framework.

- **Foundation Figures:** Highly detailed illustrations of pivotal concepts in genetics.

- **Experimental Insights:** Discusses critical or illustrative experiments, the data derived from the experiments, and the conclusions drawn from analysis of experimental results.

- **Research Techniques:** Explores important research methods and visually illustrates the results and interpretations.

- **Genetic Insights:** Short in-chapter summaries of critical principles, ideas, and conclusions.

- **Case Studies:** Short, real-world examples, at the end of every chapter, highlight central ideas or concepts of the chapter with interesting examples that remind students of some practical applications of genetics.

## Reviewing and Testing Understanding

At the end of every chapter, we have a Summary of key concepts organized by chapter section, Keywords to help students self-test during chapter review, and diverse Problem Sets to help students prepare for exams. Problem sets are separated into conceptual problems and application/integration problems. The Chapter Concept problems help students assess their understanding of the key concepts in the chapter, while Application and Integration problems give students practice working with real data sets or synthesizing concepts across chapters. Both problem sets can be approached using the *evaluate-deduce-solve* strategy, as demonstrated in the worked solutions provided in the *Study Guide and Solutions Manual*.

## NEW MasteringGenetics

Another key reviewing and testing tool is MasteringGenetics, the most powerful online homework and assessment system available. Tutorials follow the Socratic method, coaching students to the correct answer by offering feedback specific to a student's misconceptions as well as providing hints students can access if they get stuck.

The interactive approach of the tutorials provides a unique way for students to learn genetics concepts while developing and honing their problem-solving skills. In addition to tutorials, MasteringGenetics includes animations, quizzes, and end-of-chapter problems from the textbook. This exclusive product of Pearson greatly enhances learning genetics through problem solving. We've identified 14 of the most challenging topics in genetics and have built tutorials to help students think through their application:

- Mendel's Law of Segregation
- Mendel's Law of Independent Assortment
- Pedigree Analysis
- Recombination and Linkage Mapping
- Sex Linkage
- Incomplete Dominance and Codominance
- Gene Interactions
- Bacterial Genetics: Conjugation
- Bacterial Genetics: Transduction
- Quantitative Genetics
- Gel-Based Analysis
- DNA Replication
- Transcription and RNA Processing
- Translation

## Student Supplements

### MasteringGenetics

(ISBN: 0-321-70706-0 / 978-0-321-70706-2)

### Study Guide and Solutions Manual

(ISBN: 0-13-174167-5 / 978-0-13-174167-6)

Written by Peter Mirabito from the University of Kentucky, the *Study Guide and Solutions Manual* is divided into five sections: Genetics Problem-Solving Toolkit, Types of Genetics Problems, Solutions to End-of-Chapter Problems, and Test Yourself. In the "toolkit," students are reminded of key terms and concepts and key relationships that are needed to solve the types of problems in a chapter. This is followed by a breakdown of the types of problems students will encounter in the end-of-chapter problems for a particular chapter: they learn the key strategies to solve each type, variations on a problem type that they may encounter, and a worked example modeled after the Genetic Analysis feature of the main textbook. The solutions also reflect the *evaluate-deduce-solve* strategy of the Genetic Analysis feature. Finally, for more practice, we've included 5–10 Test Yourself problems and accompanying solutions.

## Instructor Supplements

### MasteringGenetics

(ISBN: 0-321-70706-0 / 978-0-321-70706-2)

MasteringGenetics engages and motivates students to learn and allows you to easily assign automatically graded activities. Tutorials provide students with personalized coaching and feedback. Using the gradebook, you can quickly monitor and display student results. MasteringGenetics easily captures data to demonstrate assessment outcomes. Resources include:

▌ In-depth tutorials that coach students with hints and feedback specific to their misconceptions

▌ An item library of thousands of assignable questions including reading quizzes and end-of-chapter problems. You can use publisher-created prebuilt assignments to get started quickly. Each question can be easily edited to match the precise language you use.

▌ A gradebook that provides you with quick results and easy-to-interpret insights into student performance.

### TestGen TestBank

(ISBN: 0-321-70684-6 / 978-0-321-70684-3

Test questions are available as part of the TestGen EQ Testing Software, a text-specific testing program that is networkable for administering tests. It also allows instructors to view and edit questions, export the questions as tests, and print them out in a variety of formats.

### Instructor Resource DVD

(ISBN: 0-131-74163-2 / 978-0-131-74163-8)

The Instructor Resource DVD offers adopters of the text convenient access to the most comprehensive and innovative set of lecture presentation and teaching tools offered by any genetics textbook. Developed to meet the needs of veteran and newer instructors alike, these resources include:

▌ The JPEG files of all text line drawings with labels individually enhanced for optimal projection results (as well as unlabeled versions) and all text tables.

▌ Most of the text photos, including all photos with pedagogical significance, as JPEG files.

▌ A set of PowerPoint® presentations consisting of a thorough lecture outline for each chapter augmented by key text illustrations, and animations.

▌ PowerPoint® presentations containing a comprehensive set of in-class Classroom Response System (CRS) questions for each chapter.

▌ In Word files, a complete set of the assessment materials and study questions and answers from the testbank.

## We Welcome Your Comments and Suggestions

Genetics is continuously changing and textbooks must also change continuously to keep pace with the field and to meet the needs of instructors and students. Communication with our talented and dedicated users is a critical driver of change. We welcome all suggestions and comments and invite you to communicate with us directly through a dedicated email address for the book. Please send comments about the book to us at sbgenetics@uc-davis.edu.

## Acknowledgments

The adage that begins with the words "It takes a village . . ." applies more accurately to the development and assembly of a first-edition textbook than to any other enterprise we know. We can attest that this book would not exist without the invaluable contributions of numerous professionals whose talents and single-minded dedication to producing the best possible genetics textbook are evident on every page. This has been a true team effort, and we are grateful to all of our teammates.

Gary Carlson has been a personal friend for many years and has been a mentor, advocate, and facilitator for this book since it was barely a glimmer of an idea. We cannot thank him enough for his unwavering support and friendship. This book would not exist without him. Michael Gillespie took over the reins as acquisitions editor toward the end of this project and brought with him both considerable skill and unrelenting enthusiasm for the book. We greatly value his work in bringing this project to its conclusion and look forward to working with him as we move forward.

Adam Steinberg served as our art developmental editor and is a professional beyond compare. His superb skills as an artist are matched only by his intimate knowledge of the subject matter. He produced many of the dynamic and intellectually engaging pieces of art in the book, and his artistic concept adds immeasurably to the book. The fine artists of Precision have converted Adam's directions into many of the superb pieces of art in the book. Our developmental editor, Moira Lerner Nelson, brought her exceptional editorial skills to bear on each chapter. Her guidance and dedication has been invaluable. We also want to thank Carol Pritchard-Martinez, who provided us important editorial guidance early in the project.

Development on our project has been in the very capable hands of Deborah Gale. We deeply appreciate her involvement, both for her expertise and for her guidance during development and writing. Anna Amato has undertaken the difficult and often thankless job of managing the day-to-day and week-to-week activities of book development and writing with grace, competence, and humor.

She has been the glue that sometimes held things together. Camille Herrera and her staff have composed and produced an exceptionally attractive book that makes our writing look better and make better sense. We also thank Crystal Clifton and her staff of copy editors for their assistance.

Lauren Harp, Michelle Cadden, and Brooke Suchomel provided us with critically important support and information through marketing surveys, class testing, and student focus groups. Their work helped shape and improve the book you hold. Beth Wilbur, Michael Young, and Paul Corey have each played important roles in making this book a reality, and we thank them for all they have done on our behalf and on behalf of this project.

Peter Mirabito, who has undertaken the Herculean task of writing the solutions to end-of-chapter problems and the *Study Guide and Solutions Manual*, has been invaluable. From the first time we met, Peter has been an enthusiastic supporter of this book. We couldn't be more pleased to have Peter as part of our team.

Finally, we thank the many reviewers and accuracy checkers who have taken the time to read and comment on our manuscript at various stages. By availing us of their expertise and perspectives, they are responsible for more improvements to this book than we can possibly count.

## Reviewers

Bert Abbott, *Clemson University*
Mary Alleman, *Duquesne University*
Ancha Baranova, *George Mason University*
Timothy Bloom, *Campbell University*
James Bradley, *Auburn University*
Carol Burdsal, *Tulane University*
Patrick Burton, *Wabash College*
Pat Calie, *Eastern Kentucky University*
Vicki Cameron, *Ithaca College*
Kimberly Carlson, *University of Nebraska at Kearney*
Steven M. Carr, *Memorial University of Newfoundland*
Aaron Cassill, *University of Texas–San Antonio*
Clarissa Cheney, *Pomona College*
Francis Choy, *University of Victoria*
Beth Conway, *Lipscomb University*
Kirsten Crossgrove, *University of Wisconsin–Whitewater*
Kenneth Curr, *California State University, East Bay*
Kim Dej, *McMaster University*
Chunguang Du, *Montclair State University*
John Elder, *Valdosta State University*
Victoria Finnerty, *Emory University*
Robert Fowler, *San Jose State University*
Rick Gaber, *Northwestern University*
Anne Galbraith, *University of Wisconsin–La Crosse*
Susan Godfrey, *Penn State University*
Michael Goodisman, *Georgia Tech University*
Nels Granholm, *South Dakota State University*
Jody Hall, *Brown University*
Pam Hanratty, *Indiana University*
Mike Harrington, *University of Alberta*

Patrick Hayes, *Oregon State University*
Jutta Heller, *Loyola University*
Jerald Hendrix, *Kennesaw State University*
Kathleen Hill, *University of Western Ontario*
Nancy Huang, *Colorado College*
Erik Johnson, *Wake Forest University*
Cheryl Jorcyk, *Boise State University*
Cliff Keil, *University of Delaware*
Steven Kempf, *Auburn University*
Oliver Kerscher, *College of William & Mary*
Elliot Krause, *Seton Hall University*
Jocelyn Krebs, *University of Alaska*
Alan Leonard, *Florida International University*
Min-Ken Liao, *Furman University*
Kirill Lobachev, *Georgia Tech University*
Heather Lorimer, *Youngstown State University*
Fordyce Lux, *Metropolitan State College of Denver*
Clint Magill, *Texas A&M University*
Jeffrey Marcus, *Western Kentucky University*
Phillip McClean, *North Dakota State University*
Philip Meneely, *Haverford College*
Scott Michaels, *Indiana University*
Peter Mirabito, *University of Kentucky*
Margaret A. Olney, *St. Martin's University*
Greg Orloff, *Emory University*
John C. Osterman, *University of Nebraska–Lincoln*
John N. Owens, *retired*
J. S. Parkinson, *University of Utah*
Bernie Possidente, *Skidmore College*
Chara J. Ragland, *Texas A&M University*
Dennis Ray, *University of Arizona*
Rosie Redfield, *University of British Columbia*
John Rinehart, *Eastern Oregon University*
Malcolm Schug, *University of North Carolina at Greensboro*
Rodney Scott, *Wheaton College*
Linda Sigismondi, *University of Rio Grande*
Tin Tin Su, *University of Colorado–Boulder*
Martin Tracey, *Florida International University*
Tara Turley-Stoulig, *Louisiana State University*
Fyodor Umov, *University of California, Berkeley*
Sarah VanVickle-Chavez, *Washington University in St. Louis*
Dennis Venema, *Trinity Western University*
David Waddell, *University of North Florida*
Dunkan Walker, *private business*
Clifford Weil, *Purdue University*
Karen Weiler, *West Virginia University*
Dan Wells, *University of Houston*
Bruce Wightman, *Muhlenberg College*
Diana Wolf, *University of Alaska Fairbanks*
Andrew J. Wood, *Southern Illinois University*
Joanna Wysocka-Diller, *Auburn University*
Lev Yampolsky, *East Tennessee State University*
Ann Yezerski, *King's College*
Janey Youngblom, *California State University, Stanislaus*
Chaoyang Zeng, *University of Wisconsin–Milwaukee*

## Class Testers

Daron Barnard, *Worcester University*
Mary Bedell, *University of Georgia*
Indrani Bose, *Western Carolina University*

Mirjana Brockett, *Georgia Institute of Technology*
Gerald Buldak, *Loyola University Chicago*
Hui Min Chung, *University of West Florida*
Craig Coleman, *Brigham Young University*
Cynthia Cooper, *Washington State University Vancouver*
Kenyon Daniel, *University of South Florida*
John Hamlin, *Louisiana State University, Eunice*
Kelly Hogan, *University of North Carolina at Chapel Hill*
Barbara Hollar, *University of Detroit Mercy*
Rick Jellen, *Brigham Young University*
David Johnson, *Samford University*
Diana Johnson, *George Washington University*
Hope Johnson, *California State University, Fullerton*
Christopher Jones, *Moravian College*
Todd Kelson, *Brigham Young University, Idaho*
Joomyeong Kim, *Louisiana State University*
Melanie Lee-Brown, *Guilford College*
Alan Lloyd, *University of Texas at Austin*
Heather Lorimer, *Youngstown State University*
Peter Mirabito, *University of Kentucky*
Paul Morris, *Bowling Green State University*
Marlene Murray-Nsuela, *Andrews University*
Nikolas Nikolaidis, *California State University, Fullerton*
Kavita Oomen, *Georgia State University*
Rebekah Rampey, *Harding University*
Mike Robinson, *Miami University, Ohio*
Melissa Rowland-Goldsmith, *Chapman University*
John Scales, *Midwestern State University*
Lillie Searles, *University of North Carolina*
Marty Shankland, *University of Texas, Austin*
Patricia Shields, *University of Maryland*
Bin Shuai, *Wichita State University*
Leslie Slusher, *West Chester University of Pennsylvania*
Tom Snyder, *Michigan Technical College*
Jeff Stuart, *Purdue University*
Susan Sullivan, *Louisiana State University, Alexandria*
Christine Terry, *Augusta State University*
Jimmy Triplett, *Jacksonville State University*
Virginia Vandergon, *California State University, Northridge*
David Westenberg, *Missouri University of Science & Technology*
Bruce Wightman, *Muhlenberg College*
Craig Woodard, *Mt. Holyoke College*
Roger Young, *Drury University*

## Participants in Student Focus Groups

### California State Polytechnic University, Pomona

**Professor Glenn Kageyama**

Jacqueline Aguayo, Angela Beal, Rosslyn K. Beard, Tiffany Chau, Jacklyn Chen, Joyce Chen, Nancy Chen, Stephen Chun, Jacquelyn Coello, Sarah Cunningham, Elizabeth Demarest, Daniela Dorantes, Andrew Esterson, Erick Estrada, Chirag Fain, Janae Ferguson, Kamaljit Gill, Josiah Gin, Scion Gomez, Israel Gonzales, Molly Gouin, Leslie Hernandez, Phu Hoang, Leticia Jaramillo, Amera Kchech, Daniel Lee, Jonathan Lee, Fabiola Lugo, Marisa Martinez, Neha Mohindroo, King Moon, Jared Nojima, Justin Padiernos, Gricel Plascencia, Bernice Ramirez, Sara Villa, Lu Wang, Brittany Ward, Ebony White

### University of Texas, Austin

**Professor Inder Saxena**

Matthew Bourneuf, Kayla Bradfield, Shelley C. Bradley, Kiersten DeHart, Nathan Fu, Jered Heinrich, Grant Hogan, Glenda Hope, Garcia, Ramu Kharel, Nafeeshathul S. Riyaj, Jarrett Scott, Elaina Spriggs, Diana Weng

### University of California, Los Angeles

**Professor John Merriam**

Cameron Arakaki, Jasmine Chen, Michaela Ching, Darwin Dirks, April Gamboa, Zhyliana Garcia Valdez, Elizabeth Geledzhyan, Victoria Hsu, Sophianne Kajouke, Michael Kashiktchian, Jonathan Liem, Danielle Lingnau, Tram Loi, Michael Opene, Michaela Scott, Mital Shah, Eve Tang

## Supplements Contributors

Pat Calie, *Eastern Kentucky University*
Kathleen Fitzpatrick, *Simon Fraser University*
Jutta Heller, *Loyola University*
Peter Mirabito, *University of Kentucky*
Louise Paquin, *McDaniel College*
Sarah Van Vickle-Chavez, *Washington University in St. Louis*
Dennis Venema, *Trinity Western University*
Andrew J. Wood, *Southern Illinois University*

# The Molecular Basis of Heredity, Variation, and Evolution

# 1

A three-story sculpture of DNA hanging in the atrium of the Life Sciences building at the University of California at Davis.

## ESSENTIAL IDEAS

■ Modern genetics developed during the 20th century and now is a prominent discipline of the biological sciences.

■ DNA replication produces exact copies of the original molecule.

■ The "central dogma of biology" describing the relationship between DNA, RNA, and protein is a foundation of molecular biology.

■ Gene expression is a two-step process that first produces an RNA transcript of a gene, and then synthesizes an amino acid string by translation of RNA.

■ Evolution is a foundation of modern genetics that occurs through four processes.

L ife is astounding, both in the richness of its history and in its diversity. From the first simple, single-celled organisms that arose billions of years ago have descended millions of species of microorganisms, plants, and animals. All of these species are connected by a shared evolutionary past that is revealed by the study of genetics, the science that explores the organization, transmission, expression, variation, and evolution of hereditary characteristics of organisms.

Genetics is a dynamic discipline that finds applications everywhere humans interact with one another and with other organisms. In research laboratories, farms, grocery stores, medical offices, courtrooms, and many other settings, genetics plays a large and expanding role in our lives. Modern genetics is an increasingly gene- and genome-based discipline—that is, it is increasingly focused on events at the molecular level—that nevertheless remains strongly focused on the traditional areas of heredity and variation at the level of physical appearance. Welcome to the fascinating discipline of genetics; you are in for an exciting and rewarding journey.

In this chapter, we survey the scope of modern genetics and present some basic information about deoxyribonucleic acid—DNA, the carrier of genetic information. We begin with a brief overview of the origins of genetic science. Next we retrace some of the fundamentals of *DNA replication,* and of *transcription* and *translation* (the two main components of gene expression), by reviewing what you learned about these processes in previous biology courses. In the final section, we describe the central position of evolution in genetics and discuss the roles of heredity and variation in evolution.

## 1.1 Modern Genetics Has Entered Its Second Century

Humans have been implicitly aware of genetics for more than 10,000 years (**Figure 1.1**). From the time of the domestication of rice in Asia, maize in Central America, and wheat in the Middle East, humans have recognized that desirable traits found in plants and animals can be reproduced and enhanced in succeeding generations through selective mating. On the other hand, explicit exploration and understanding of the hereditary principles of genetics—what we might think of as the science of modern genetics—is a much more recent development.

### The First Century of Modern Genetics

In 1900, three botanists working independently of one another—Carl Correns in Germany, Hugo de Vries in Holland, and Erich von Tschermak in Austria—reached strikingly similar conclusions about the pattern of transmission of hereditary traits in plants (**Figure 1.2**). Each reported that his results mirrored those published in 1866 by an obscure amateur botanist and Augustinian monk named Gregor Mendel. (Mendel's work is discussed in Chapter 2.) Although Correns, de Vries, and Tschermak had actually *rediscovered* an explanation of hereditary

transmission that Mendel had published 34 years earlier, their announcement of the identification of principles of hereditary transmission gave modern genetics its start.

Biologists immediately began testing, verifying, and expanding on the newly appreciated explanation of heredity. In 1901, William Bateson, an early and vigorous proponent of "Mendelism," read a publication by a British physician-scientist named Archibald Garrod describing the appearance of the hereditary disease alkaptonuria in multiple members of unrelated families. Bateson immediately realized that Garrod's description depicted "exactly the conditions most likely to enable a rare, usually recessive character to show itself." Garrod, with Bateson's interpretive assistance, had produced the first documented example of a human hereditary disorder.

Shortly thereafter, Walter Sutton and Theodore Boveri independently used microscopy to observe chromosome movement during cell division in reproductive cells. They each noted that the patterns of chromosome movement mirrored the transmission of the newly rediscovered Mendelian hereditary units. This work suggested that the hereditary units, or *genes*, posited by Mendel are located on *chromosomes*. **Genes** are the physical units of heredity. We now know they are composed of defined DNA sequences that collectively control gene transcription and contain the information to produce RNA molecules or proteins. **Chromosomes** consist of single long molecules of double-stranded DNA and may be associated with many different kinds of protein. Chromosomes of bacteria usually occur singly, whereas the chromosomes of sexually reproducing organisms typically occur in pairs known as **homologous pairs** or, more simply, as **homologs.** Each chromosome carries many genes, and homologs carry genes for the same traits in the same order on each member of the pair.

Bacteria are single-celled organisms that do not have a true nucleus. In almost all cases, bacteria have a single chromosome that occupies a cell space known as the nucleoid. The chromosome replicates in conjunction with bacterial cell division. In eukaryotes—a classification that includes all single-celled and multicellular plants and animals—a true nucleus contains multiple pairs of homologous chromosomes. The chromosomes remain permanently in the nucleus. A complete set of chromosomes are transmitted during a cell division process called **mitosis** to produce identical daughter cells. Sexual reproduction to produce offspring occurs by way of the cell division process called **meiosis** that produces reproductive cells or **gametes**—sperm and egg in animals and pollen and egg in plants.

Predictable patterns of gene transmission during sexual reproduction are a focus of later chapters, including hereditary transmission and the analysis of transmission ratios (Chapter 2), cell division and chromosome heredity (Chapter 3), gene action and interaction of genes in producing variation of physical appearance (Chapter 4), and the analysis of genetic linkage between genes (Chapter 5).

**(a)**

**(b)**

**Figure 1.1  Ancient applications of genetics.  (a)** An early record of human genetic manipulation is this Assyrian relief from 882–859 BCE. It shows priests in bird masks artificially pollinating date palms. **(b)** Modern maize (left) is thought to have developed through human domestication of its wild ancestor teosinte (right).

Genetic experiments taking place in roughly the first half of the 20th century developed the concept of the gene as the physical unit of heredity and revealed the relationship between **phenotype,** meaning the observable traits of an organism, and **genotype,** meaning the genetic constitution of an organism. Biologists also described how hereditary variation is attributable to alternative forms of a gene, called **alleles.** During this period, the study of gene transmission was established as a foundation of genetics. The concepts of gene action and gene interaction in producing phenotype variation were described, as was the concept of mapping genes. It was also during this period that evolutionary biologists developed gene-based models of evolution. These ideas form the foundation of genetic principles and analysis, and their use continues to the present day.

An experiment conducted in 1944 by Oswald Avery, Colin MacLeod, and Maclyn McCarty, identified *deoxyribonucleic acid (DNA)* as the hereditary material and is commonly credited with inaugurating the "molecular era"

**(a)** Carl Correns

**(b)** Hugo de Vries

**(c)** Erich von Tschermak

**Figure 1.2  Early 20th-century genetics.  (a)** Carl Correns, **(b)** Hugo de Vries, and **(c)** Erich von Tschermak simultaneously rediscovered the experiments and principles of Gregor Mendel in 1900.

in genetics (see Chapter 7). This new era, which spanned the second half of the 20th century and continues to the present day, began as an effort to discover the molecular structure of DNA. This research reached a milestone in 1953, when the efforts of many biologists, including, most famously, James Watson, Francis Crick, Maurice Wilkins, and Rosalind Franklin, contributed to the identification of the double-helical structure of DNA. A few years later, in 1958, the mechanism of DNA replication was ascertained. By the mid-1960s, the basic mechanisms of transcription and translation were laid out, and the genetic code by which mRNA is translated into proteins was deciphered. On the basis of these achievements, genetic analysis at the molecular level emerged as a principal field of genetic inquiry during the second half of the 20th century.

Cloning and the development of recombinant DNA technology progressed rapidly during the 1970s. By the early 1980s, biologists realized that to properly understand the unity and complexity of life, they would have to study and compare entire genomes. This realization launched the "genomics era" in genetics. The **genome** is the complete set of genetic information carried by a species. Genome biologists initiated hundreds of genome sequencing projects to decipher the complete DNA sequences of organisms ranging from bacteria to humans. Fittingly, in 2001, a century after Garrod and Bateson's historic identification of alkaptonuria as a human hereditary disease, scientific groups from around the world published the completed "first draft" of the human genome. The remarkable information provided by the Human Genome Project, and by hundreds of other genome sequencing projects that have been undertaken and completed in the past 25 years, promises that the second century of genetics will be every bit as remarkable as its first century.

## Genetics—Central to Modern Biology

One of the foundations of modern biology is the demonstration that all life on Earth shares a common origin, or *progenote*, a term meaning "progenitor form" (**Figure 1.3**). All life is descended from this common ancestor and is commonly divided into three major domains of life. These three domains are **Eukarya, Bacteria,** and **Archaea.** Eukarya are also called eukaryotes; they consist of single-celled and multicellular organisms with a true nucleus, multiple chromosomes, and other distinguishing characteristics identified in the figure. Bacteria and archaea are single-celled organisms that lack a true nucleus and generally have single chromosomes, among other distinguishing characteristics. Plant and animal cells contain specialized organelles called *mitochondria,* and plant cells also contain organelles called *chloroplasts* that perform essential functions. Mitochondria and chloroplasts carry their own DNA, and they are descended from ancient parasitic bacterial invaders that evolved an endosymbiotic relationship with their eukaryotic hosts (see Chapter 19).

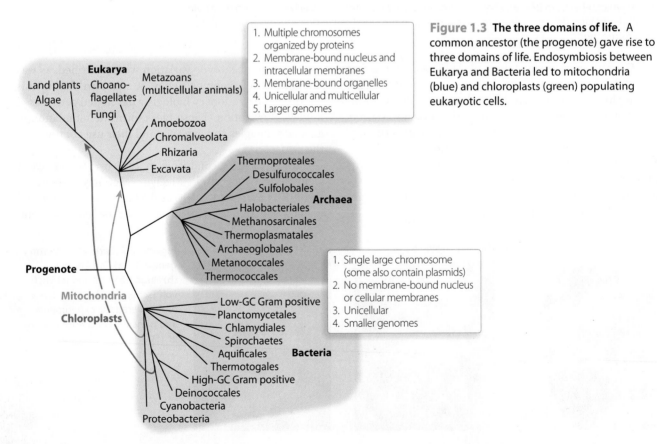

**Figure 1.3  The three domains of life.** A common ancestor (the progenote) gave rise to three domains of life. Endosymbiosis between Eukarya and Bacteria led to mitochondria (blue) and chloroplasts (green) populating eukaryotic cells.

A second foundation of biology is the recognition that the hereditary material—the molecular substance that conveys and stores genetic information—is **deoxyribonucleic acid (DNA)** in all organisms. A few viruses use **ribonucleic acid (RNA)** as their hereditary material, but viruses are noncellular biological entities that must invade host cells to reproduce. DNA has a double-stranded structure described as a **DNA double helix,** or as a **DNA duplex,** consisting of two strands joined together in accordance with specific biochemical rules.

Eukarya, Bacteria, and Archaea share general mechanisms of **DNA replication,** the process that precisely duplicates the DNA duplex prior to cell division, and they also share general mechanisms of gene expression, the processes that express the genetic information in the genome. All organisms express their genetic information by **transcription,** a process in which one strand of DNA is used to direct synthesis of a single strand of RNA. Transcription produces various forms of RNA, including **messenger RNA (mRNA),** which in all organisms undergoes **translation** to produce proteins at nucleoprotein structures called **ribosomes.**

As the biological discipline devoted to the examination of all aspects of heredity and variation between generations and through evolutionary time, genetics is central to modern biology. Modern genetics has three major branches. **Transmission genetics,** also known as **Mendelian genetics,** is the study of the transmission of traits and characteristics in successive generations. **Evolutionary genetics** studies the origins of and genetic relationships between organisms and examines the evolution of genes and genomes. **Molecular genetics** studies inheritance and variation of nucleic acids and proteins, and tries to connect them to inherited variation and evolution in organisms. These branches of genetics are not rigid. There is substantial cross-communication among the branches, and it is rare to find a geneticist today who doesn't use analytical approaches from all three. Similarly, not only are most biological scientists, to a greater or lesser extent, also geneticists, but many of the methods and techniques of genetic experimentation and analysis are shared by all biological scientists. After all, genetic analysis interprets the common language of life by integrating information from all three branches.

## 1.2 The Structure of DNA Suggests a Mechanism for Replication

At its core, hereditary transmission is the process of dispersing genetic information from parents to offspring. In sexually reproducing organisms, this process is accomplished by creation of male gametes (sperm or pollen) and female gametes (eggs), followed by uniting of gametes to form a fertilized egg that goes on to develop into an organism. DNA is the hereditary molecule in gametes. Similarly, in somatic (body) cells of plants and animals, and in organisms that reproduce by asexual processes, DNA is the hereditary molecule that ensures that successive generations of cells are identical.

Experiments and research on cells taking place over 75 years, from the late-1800s through the mid-1900s, culminated in the identification of DNA as the hereditary material (see Section 7.1). The identification of DNA was of monumental importance to biologists and biochemists, and was the foundation of new molecular-focused approaches in biological science research. With the identification of DNA, they realized that understanding the molecular structure of DNA was key to two fundamental questions: (1) how DNA could carry the diverse array of genetic information present in the various genomes of animals and plants, and (2) how the molecule replicated. In this section, we introduce basic concepts of DNA structure and DNA replication, with molecular details of DNA replication provided later (Chapter 7).

### The DNA Double Helix

The immediate impact of the identification of DNA as the hereditary material was to encourage many biologists to turn their attention to discovering the structure of DNA. In the early 1950s, James Watson, an American in his mid-twenties who had recently completed a doctoral degree, and Francis Crick, a British biochemist in his mid-thirties, began working together at Cambridge University in England to solve the puzzle of DNA structure. Their now legendary collaboration culminated in a 1953 publication that shook the scientific world.

Watson and Crick's paper accurately described the molecular structure of DNA as a double helix composed of two strands of DNA with an invariant sugar-phosphate backbone on the outside and nucleotide bases—adenine, thymine, guanine, and cytosine—arrayed in complementary pairs, oriented toward the center of the molecule. This was a discovery of enormous importance that formed a cornerstone of modern biology. With the structure of DNA in hand, the "gene" was no longer just a conceptual entity. Rather, it had a physical form that could be quantified and sequenced; a gene could be compared with other genes in its genome and with genes in the genomes of other species.

Watson and Crick's description of DNA structure was not the product of their work exclusively. In fact, unlike others who made significant contributions to the discovery of DNA structure, Watson and Crick were not actively engaged in laboratory research. Outside of their salaries, they had very little financial support available to conduct research. In lieu of laboratory research, Watson and Crick put their efforts into DNA-model building, basing their interpretations on experimental data gathered by others.

**(a)**                    **(b)**

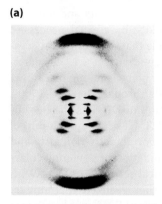

**Figure 1.4  X-ray diffraction evidence of DNA structure.**
(a) This X-shaped pattern is consistent with the diffraction of X-ray beams by a helical molecule composed of two strands. (b) Rosalind Franklin obtained this X-ray diffraction result.

Rosalind Franklin, a biophysicist working with Maurice Wilkins at King's College London, was one of the principal sources of information used by Watson and Crick. Franklin used an early form of X-ray diffraction imagery to examine the crystal structure of DNA. In Franklin's method, X-rays passing through crystal preparations of DNA are diffracted as they encounter the atoms in the crystals. The diffracted X-rays are recorded on X-ray film, and the crystal structure is interpreted from the visual pattern the X-rays produce (Figure 1.4). Franklin's most famous X-ray diffraction photograph clearly shows (to the well-trained eye) that DNA is a duplex, consisting of two strands twisted around one another in a double helix.

Watson and Crick combined Franklin's X-ray diffraction data with information published a few years earlier by Erwin Chargaff, who found that in the DNA of most organisms the percentages of adenine and thymine are approximately equal to one another and that the percentages of cytosine and guanine are equal to one another as well (Table 1.1). Known as **Chargaff's rule,** this information helped Watson and Crick formulate the hypothesis that DNA nucleotides are arranged in **complementary base pairs.** Adenine, on one strand of the double helix, pairs only with thymine on the other DNA strand, and cytosine pairs only with guanine to form the other base pair. With these data, their own knowledge of biochemistry, and their analysis of incorrect models of DNA structure, Watson and Crick built a table-top model of DNA out of implements and materials scattered around their largely inactive research laboratory space—wire, tin, tape, and paper, supported by ring stands and clamps (Figure 1.5).

## DNA Nucleotides

Each strand of the double helix is composed of **DNA nucleotides** that have three principal components: a five-carbon deoxyribose sugar, a phosphate group, and one of four nitrogen-containing nucleotide bases, designated

| Source Genome | Percentage of Nucleotide Bases | | | | Ratios | |
|---|---|---|---|---|---|---|
| | Adenine (A) | Guanine (G) | Cytosine (C) | Thymine (T) | G+C | G/C |
| *Bacteria* | | | | | | |
| *E. coli* (B) | 23.8 | 26.8 | 26.3 | 23.1 | 53.1 | 1.02 |
| *Yeast* | | | | | | |
| *S. cerevisiae* | 31.3 | 18.7 | 17.1 | 32.9 | 35.8 | 1.09 |
| *Fungi* | | | | | | |
| *N. crassa* | 23.0 | 27.1 | 26.6 | 23.3 | 53.7 | 1.02 |
| *Invertebrate* | | | | | | |
| *C. elegans* | 31.2 | 19.3 | 20.5 | 29.1 | 39.8 | 0.94 |
| *D. melanogaster* | 27.3 | 22.5 | 22.5 | 27.6 | 45.0 | 1.00 |
| *Plant* | | | | | | |
| *A. thaliana* | 29.1 | 20.5 | 20.7 | 29.7 | 41.2 | 0.99 |
| *Vertebrate* | | | | | | |
| *M. musculus* | 29.2 | 21.7 | 19.7 | 29.4 | 41.4 | 1.10 |
| *H. sapiens* | 30.6 | 19.7 | 19.8 | 30.3 | 39.5 | 0.99 |

**Table 1.1**    Nucleotide-Base Composition of Various Genomes

**Figure 1.5**  James Watson (left) and Francis Crick (right) in 1953 with their cardboard-and-wire model of DNA.

gives the strand a sugar-phosphate backbone; the nucleotide bases project away from the backbone.

Nucleotides can occur in any order along one strand of the molecule, but complementary base pairing between the strands matches an A on one strand with a T on the complementary strand, and a G on one strand with a C on the other. Complementary base pairing is the basis of Chargaff's rule and produces equal percentages of A and T, and of C and G, in double-stranded DNA molecules. **Hydrogen bonds,** noncovalent bonds consisting of weak electrostatic attractions, form between complementary base pairs to join the two DNA strands into a double helix. Each strand of DNA has a 5′ end and a 3′ end. These designations refer to the phosphate group (5′) and hydroxyl group (3′) at the ends of each strand of DNA and establish **strand polarity,** that is, the 5′-to-3′ orientation of each strand. Complementary strands of DNA are **antiparallel,** meaning that the polarities of the complementary strands run in opposite directions—one strand is oriented 5′ to 3′ and the complementary strand is oriented 3′ to 5′.

If you are like many biology students, you have probably wondered from time to time what DNA actually looks like, both on the macroscopic and microscopic level. Even today's best microscopes have difficulty capturing high-resolution images of DNA, although computer-aided techniques for analyzing molecular structure can produce an interpretation of its microscopic appearance. However, you do not need sophisticated instrumentation

**adenine (A), guanine (G), thymine (T),** and **cytosine (C)** (**Figure 1.6**). The nucleotides forming a strand are linked together by a covalent **phosphodiester bond** between the 5′ phosphate group of one nucleotide and the 3′ hydroxyl (OH) group of the adjacent nucleotide. This bonding

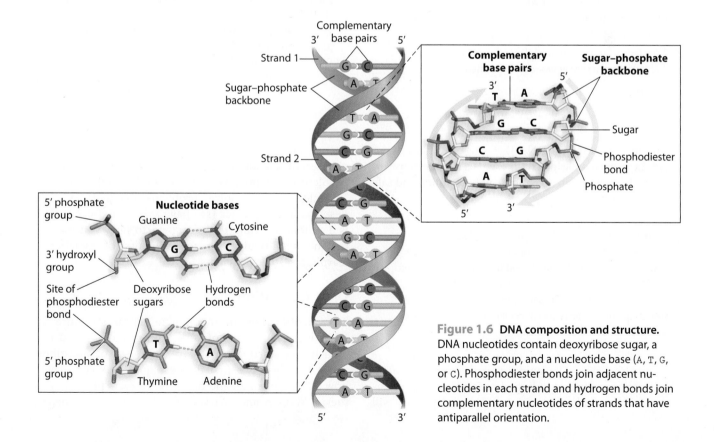

**Figure 1.6  DNA composition and structure.** DNA nucleotides contain deoxyribose sugar, a phosphate group, and a nucleotide base (A, T, G, or C). Phosphodiester bonds join adjacent nucleotides in each strand and hydrogen bonds join complementary nucleotides of strands that have antiparallel orientation.

## Experimental Insight **1.1**

### Countertop DNA Isolation—Try This at Home!

For all the abundance of DNA in cells, its molecular structure is too small to see without the aid of the most powerful electron microscopes. However, that doesn't mean DNA must remain invisible to the naked eye. The key to seeing it is simply a question of volume. If enough DNA is collected together, it can be seen—although not, of course, in its molecular detail. Using a rich source of DNA (such as onions, that are available year-round, or strawberries, whose nuclei contain eight copies of each chromosome) and a few familiar household items, you can collect a visible sample of DNA in your own home in about 30 minutes.

#### INGREDIENTS

1 small peeled onion (about 1 cup) or about 1 cup strawberries with leaves removed

1 to 2 cups water with 1 teaspoon salt per cup

2 tablespoons dishwashing liquid

1 tablespoon meat tenderizer (containing "papain" from papaya)

4 to 6 ounces isopropyl ("rubbing") alcohol (95% is best, but 70% is sufficient)

#### EQUIPMENT

Food processor (for onion) or a potato masher or ricer (for strawberries)

Small bowl

Clear glass jar or container with vertical sides

Cheesecloth to layer over the top of the glass container with a few inches to spare all around

1 rubber band to go around the glass container

1 chopstick or a similar wooden implement

#### DIRECTIONS

1. Peel onion and finely chop in food processor. Alternatively, thoroughly mash strawberries in bowl.

2. Add 1 to 2 cups water to onion and process into a fine slurry. Pour slurry into small bowl. If using strawberries, add about 1 cup water and mash into a fine slurry.

3. Add 2 tablespoons liquid dishwashing soap to slurry and stir gently. Be careful not to let the soap get foamy. Let mixture stand several minutes to let the soap break down the cell and nuclear membranes.

4. Add 1 tablespoon meat tenderizer to mixture and stir gently. Once again, avoid foaming the soap. The papain in the tenderizer will digest much of the protein released by the ruptured cells as well as the proteins attached to DNA. Let mixture stand several minutes to allow papain time to work.

5. Place 2 to 3 layers cheesecloth loosely over the opening of the glass container, allowing the cloth to form a small "bowl" inside the opening. Use the rubber band to hold the cheesecloth in place. Pour the slurry mixture through the cheesecloth, scooping out the onion or strawberry debris as it fills the cheesecloth bowl. Approximately 8 to 12 ounces of "juice" will collect at the bottom of the container. Discard the cheesecloth and its contents.

6. Pour the alcohol into the juice and stir very briefly. Let the juice mixture rest for 5 to 10 minutes. The juice settles to the bottom of the mixture, the alcohol rises to the top, and the large mass of floating cottony material is DNA.

7. When the alcohol has completely separated from the juice, you can "spool" the DNA onto a chopstick by slowly twirling the stick in the cottony DNA.

---

to produce a sample of DNA that you can hold in your hand. **Experimental Insight 1.1** presents a simple recipe for DNA isolation you can do at home with common and safe household compounds.

---

Genetic Insight  Each DNA strand in a double-helix has a 5′ and a 3′ end. The strands consist of nucleotides linked by covalent phosphodiester bonds. Hydrogen bonding between the strands requires complementary nucleotide base pairing, in which A pairs with T and C pairs with G. The complementary strands of a double helix are antiparallel, so that the 5′ end of one faces the 3′ end of the other.

---

Genetic Analysis 1.1 guides you through a problem that tests your understanding of base-pair complementation and complementary strand polarity. You will encounter multiple Genetic Analysis problems in each chapter of the book. Each one presents you with a problem and guides you to the solution following the "Evaluate–Deduce–Solve" problem-solving strategy we describe in the preface (see page xvi). Genetic Analysis is a unique learning tool designed to help you master the most difficult task for students of genetics—learning to solve problems.

## DNA Replication

The identification of the double-helical structure of DNA was a milestone of 20th-century biology that established a starting point for a new set of questions about heredity. The first of these questions concerned how DNA replicates, and how DNA directs protein synthesis. Watson and Crick concluded their 1953 paper

Write the sequence and polarity of the DNA strand complementary to the strand shown below.

$$5'-…ACGGATCCTCCCTAGTGCGTAATTCG…-3'$$

| Solution Strategies | Solution Steps |
|---|---|
| **Evaluate** | |
| 1. Identify the topic this problem addresses and explain the nature of the requested answer. | 1. This problem concerns the complementarity and polarity of DNA strands in a duplex. You are asked to give the nucleotide sequence and polarity of a complementary strand. |
| 2. Identify the critical information given in the problem. | 2. The problem provides the nucleotide sequence and polarity of one strand of a DNA duplex. |
| **Deduce** | |
| 3. Review the relationship of nucleotides in the second strand to those in the strand given. <br><br> TIP: Complementary DNA strands are antiparallel. <br><br> TIP: Strands in a DNA duplex are joined by hydrogen bonding between complementary base pairs. | 3. DNA complementary base pairing joins adenine with thymine and guanine with cytosine to form a DNA duplex. |
| 4. Describe the polarity relationship of the complementary DNA strands. | 4. The second strand of this duplex will be oriented with its 3′ end to the left and its 5′ end to the right. |
| **Solve** | |
| 5. Give the sequence and polarity of the complementary DNA strand. | 5. By the rules of complementary base pairing and antiparallel strand orientation, the second DNA strand is <br><br> $3'-TGCCTAGGAGGGATCACGCATTAAGC-5'$ |

For more practice, see Problems 11, 12, and 14.

---

with a directive for future research on the question of DNA replication. They wrote, "It has not escaped our notice that the specific base-pairing we have proposed immediately suggests a possible copying mechanism for the genetic material."

Indeed, it was evident that each single strand of DNA contains the information necessary to generate the companion strand of DNA through complementary base pairing, and that DNA replication must generate two identical DNA duplexes from the original parental duplex during each replication cycle. At the time Watson and Crick described the structure of DNA, however, the mechanism of replication was not known. It would take another 5 years for Matthew Meselson and Franklin Stahl, in an ingenious experiment of simple design, to prove that DNA replicates by a *semiconservative* mechanism (Chapter 7).

As you may recall from previous courses, **semiconservative replication** requires separation of the two complementary strands of original DNA and the use of each strand as a template for synthesis of a new, complementary strand of DNA. The mechanism is termed "semiconservative" because each new duplex is composed of one **parental strand** (which is conserved from the original DNA) and one newly synthesized **daughter strand** (Figure 1.7).

DNA replication begins with the parental strands separating from one another by breaking of the hydrogen bonds that hold the strands together. (This process is much like what happens when a zipper comes undone.) DNA polymerases are the enzymes active in DNA replication. These enzymes use nucleotides on parental DNA strands as a template to direct the assembly of new nucleotides to form daughter strands. DNA polymerase first identifies the nucleotide that is complementary to the parental strand nucleotide; the enzyme then catalyzes formation of a phosphodiester bond to join the new nucleotide to the previous nucleotide in the nascent (growing) daughter strand.

The biochemistry of nucleic acids and DNA polymerases dictates that DNA strands elongate only in the 5′-to-3′ direction. In other words, nucleotides are added exclusively to the 3′ end of the nascent strand, leading to 5′-to-3′ growth. Like the parental duplex, each new DNA duplex contains antiparallel strands. Each parental strand–daughter strand combination forms a new double helix of DNA that is an exact replica of the original parental duplex.

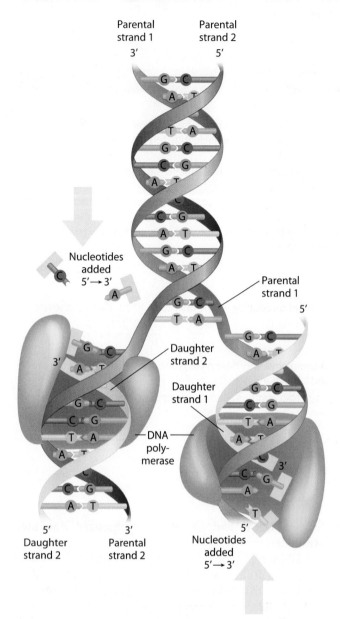

**Figure 1.7  Semiconservative DNA replication.** Each parental DNA strand serves as the template for synthesis of its daughter strand. DNA polymerase synthesizes daughter strands one nucleotide at a time.

## 1.3 Transcription and Translation Express Genes

The **central dogma of biology** is a statement describing the flow of hereditary information. It summarizes the critical relationships between DNA, RNA, and protein; the functional role that DNA plays in maintaining, directing, and regulating the expression of genetic information; and the roles played by RNA and proteins in gene function. Francis Crick proposed the original version of the central dogma, shown in **Figure 1.8a**, in 1956 to encapsulate the role DNA plays in directing transcription of RNA and, in turn, the role messenger RNA plays in translation of proteins. As Crick told the story years later, he wrote this concept as "DNA → RNA → protein" (spoken as "DNA to RNA to protein") on a slip of paper and taped it to the wall above his desk to remind himself of the direction of information transfer during the expression of genetic information. The most important idea it conveys is that DNA does not code directly for protein. Rather, DNA makes up the genome of an organism and is a permanent repository of genetic information in each cell that directs gene expression by the transcription of DNA to RNA and the translation of mRNA to proteins.

Over the decades since Crick first introduced the central dogma, biologists have developed a clear understanding of the role of DNA in maintaining and expressing genetic information. Most of the details of the two-stage process by which genetic information in DNA sequences of genes is transcribed to RNA and then translated to protein are known. Later chapters discuss transcription (Chapter 8) and translation (Chapter 9). Today, however, biologists also know that several forms of RNA are found in cells, and all these RNA molecules are transcribed and play a variety of roles in cells, but only mRNA is translated. Two important categories of RNA that are not translated but nonetheless play critical roles in translation are ribosomal RNA and transfer RNA. **Ribosomal RNA (rRNA)** forms part of the ribosomes, the plentiful cellular structures where protein assembly takes place. **Transfer RNA (tRNA)**

**Figure 1.8  The central dogma of biology.** **(a)** Francis Crick's original central dogma of biology. **(b)** The updated central dogma of biology.

**(a)**

*Replication*

DNA —— *Transcription* —→ RNA —— *Translation* —→ **Protein**

**(b)**

*Replication*

DNA —— *Transcription* —→
- Messenger RNA (mRNA) —— *Translation* —→ **Protein**
- Ribosomal RNA (rRNA)
- Transfer RNA (tRNA) ⟶ To the ribosome
- Micro RNA (miRNA)
- Other RNA
- Retrovirus RNA

*Reverse transcription*

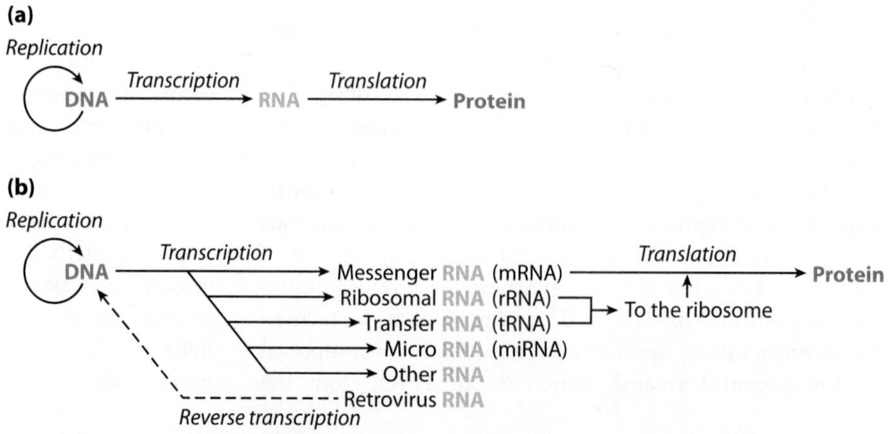

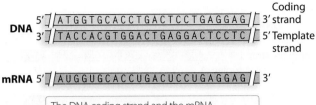

The DNA coding strand and the mRNA transcript have the same polarity and sequence, substituting U in mRNA for T in DNA.

**Figure 1.9 The correspondence of RNA to DNA template and coding strands.**

carries **amino acids,** the building blocks of proteins, to ribosomes. An updated central dogma of biology is shown in **Figure 1.8b.** In addition to mRNA, rRNA, and tRNA, the figure identifies **reverse transcription,** a form of information flow that synthesizes DNA from an RNA template in RNA-containing viruses (retroviruses) by using an enzyme called reverse transcriptase. It also identifies micro-RNA (miRNA), the focus of a rapidly emerging new area of RNA investigation that studies the role of these small RNA molecules in the regulation of gene expression in plants and animals (see Chapter 15).

## Transcription

Transcription uses one strand of the DNA making up a gene to direct synthesis of a single-stranded RNA transcript. The DNA strand from which the transcript is synthesized is called the **template strand.** The RNA-synthesizing enzyme RNA polymerase pairs template-strand nucleotides with complementary RNA nucleotides to synthesize new transcript in the 5′-to-3′ direction; the transcript is antiparallel to the DNA template strand (**Figure 1.9**).

The complementary partner of the DNA template strand is known as the **coding strand.** Because the coding strand is both complementary and antiparallel to the DNA template strand, it has the same 5′ → 3′ polarity as

the RNA transcript synthesized from the template strand; moreover, the RNA transcript and the DNA coding strand are identical in nucleotide sequence, except for the appearance of U in the place of T.

RNA is composed of four nucleotides that are chemically very similar to DNA. RNA nucleotides consist of a sugar (ribose, in this case), a phosphate group, and one of four nitrogenous bases. Three of the RNA nucleotide bases are adenine, cytosine, and guanine, the same as those found in DNA. The fourth RNA base is **uracil** (U). In DNA–RNA and RNA–RNA complementary base pairing, C pairs with G, and A pairs with U.

Transcription producing RNA is a controlled process in cells. RNA polymerase is the enzyme responsible for synthesizing RNA transcripts. To begin transcription, RNA polymerase and other transcriptionally active proteins must locate a gene and gain access to the template DNA strand. Once the coding sequence of the gene has been transcribed, the RNA polymerase must stop transcription and disconnect from DNA.

**Promoters** are the most common type of DNA sequences recognized by RNA polymerase as indicating the presence of a nearby gene. Promoters help regulate the initiation of transcription by controlling RNA polymerase access to DNA. Promoters are not transcribed. Instead, the transcription of a gene begins near the promoter at the **start of transcription.** Transcription ends at the **termination sequence,** a DNA sequence that facilitates the cessation of transcription (**Figure 1.10a**). In bacteria, protein-producing genes are transcribed into mRNA that is quickly translated to produce the protein. Eukaryotic genes have a different structure than do bacterial genes. They are subdivided into **exons,** which contain the coding information that will be used during translation, and **introns,** which intervene between exons and are removed from the transcript before translation (**Figure 1.10b**). The removal of introns from eukaryotic mRNA and other modifications before translation occurs in the nucleus (see Chapter 8).

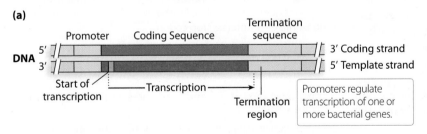

Promoters regulate transcription of one or more bacterial genes.

**Figure 1.10 Gene structure in bacteria and eukaryotes.** Coding sequences contain information to be transcribed into RNA. Promoter sequences regulate the initiation of transcription, and termination sequences control the cessation of transcription. **(a)** Bacterial genes contain a single coding sequence that carries the information of the gene. **(b)** The coding sequence of eukaryotic genes is split up into exons, which are separated by introns.

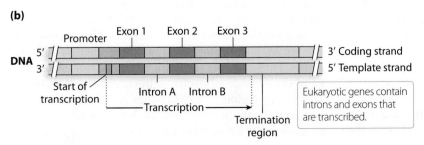

Eukaryotic genes contain introns and exons that are transcribed.

Genetic Insight  RNA molecules are the single-stranded transcripts of genes. They are synthesized by the enzyme RNA polymerase using one DNA strand as a template. Each messenger RNA from bacterial genes carries a single coding sequence, but eukaryotic mRNAs contain intron and exon sequences and require processing prior to translation.

**(a)**

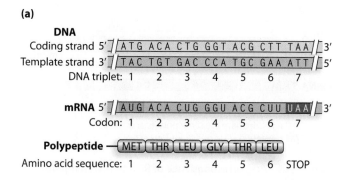

## Translation

The DNA sequences of genes are transcribed into mRNA, and mRNA is translated into proteins. Translation converts the genetic message carried by mRNA nucleotides into sequences of amino acids using the *genetic code.* The amino acids are joined to one another by a covalent bond called a **peptide bond.** The resulting string of amino acids is a **polypeptide,** which upon folding makes up all or part of a **protein.**

Translation of mRNA occurs at the ribosome, where sets of three consecutive nucleotides, each set called a **codon,** specify the sequence of amino acids making up the polypeptide. Each mRNA codon, then, is a triplet of RNA nucleotides coded by three complementary DNA nucleotides on the template strand that are identified as the DNA triplet (**Figure 1.11a**). Translation begins with mRNA attaching to a ribosome in a manner that places the **start codon,** the codon specifying the first amino acid of a polypeptide, in the necessary location (**Figure 1.11b**). The start codon is most commonly AUG and is the codon at which translation begins. From the start codon, ribosomes move in the 5′ → 3′ direction along mRNA to assemble the amino acid string, with each successive set of three mRNA nucleotides forming a codon.

Amino acids are transported to ribosomes by tRNAs. At each codon, complementary base pairing occurs between codon nucleotides and a three-nucleotide sequence of tRNA called an **anticodon.** This interaction assembles amino acids in the order dictated by mRNA sequence. Ribosomal proteins power the continuous progression of the ribosome along mRNA and catalyze peptide bond formation in the growing polypeptide chain. Translation continues until the ribosome encounters one of three **stop codons,** thus bringing translation to a halt.

The **genetic code,** through which mRNA codons specify amino acids, was deciphered by a series of experiments that took place during the early 1960s. The experiments revealed that the genetic code contains 64 codons; every codon consists of three positions that are each filled by one of the four RNA nucleotides. A series of codons in mRNA specifies the amino acid sequence in a polypeptide. An mRNA codon is read in the 5′-to-3′ direction: The first base of the codon is at its 5′ end, the

**(b)**

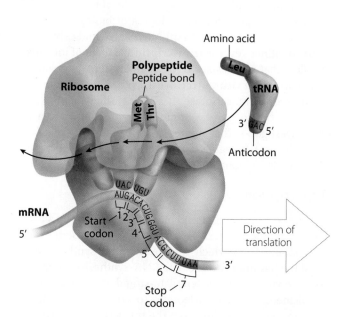

**Figure 1.11  Overview of translation.  (a)** Messenger RNA codons are complementary and antiparallel to DNA triplets of the template strand. **(b)** Ribosomes initiate translation of mRNA at the start codon and move along the mRNA in the 3′ direction, adding a new amino acid to the nascent polypeptide by reading each codon. Transfer RNA molecules carry amino acids to ribosomes, where the tRNA anticodon sequences interact with codon sequences of mRNA. Translation terminates when the ribosome encounters a stop codon.

third base is at its 3′ end, and the second base is in the middle.

Because there are 64 codons (4³) and only 20 common kinds of amino acids in proteins, the genetic code is redundant: Most amino acids can be specified by two or more different codons. For two amino acids—methionine (Met or M) and tryptophan (Trp or W)—there is only one codon, but for others there are as many as six. A total of 61 of the 64 codons specify amino acids, and the other 3 are the stop codons. The 64 codons are displayed in **Table 1.2** using the three-letter and one-letter abbreviations for amino acids listed in **Table 1.3.**

**Table 1.2**    The Genetic Code

Second Position

| First Position (5' end) | U | C | A | G | Third Position (3' end) |
|---|---|---|---|---|---|
| **U** | UUU ⎤ Phe (F) <br> UUC ⎦ <br> UUA ⎤ Leu (L) <br> UUG ⎦ | UCU ⎤ <br> UCC ⎥ Ser (S) <br> UCA ⎥ <br> UCG ⎦ | UAU ⎤ Tyr (Y) <br> UAC ⎦ <br> UAA – stop <br> UAG – stop | UGU ⎤ Cys (C) <br> UGC ⎦ <br> UGA – stop <br> UGG – Trp (W) | U <br> C <br> A <br> G |
| **C** | CUU ⎤ <br> CUC ⎥ Leu (L) <br> CUA ⎥ <br> CUG ⎦ | CCU ⎤ <br> CCC ⎥ Pro (P) <br> CCA ⎥ <br> CCG ⎦ | CAU ⎤ His (H) <br> CAC ⎦ <br> CAA ⎤ Gln (Q) <br> CAG ⎦ | CGU ⎤ <br> CGC ⎥ Arg (R) <br> CGA ⎥ <br> CGG ⎦ | U <br> C <br> A <br> G |
| **A** | AUU ⎤ <br> AUC ⎥ Ile (I) <br> AUA ⎦ <br> AUG – Met (M) | ACU ⎤ <br> ACC ⎥ Thr (T) <br> ACA ⎥ <br> ACG ⎦ | AAU ⎤ Asn (N) <br> AAC ⎦ <br> AAA ⎤ Lys (K) <br> AAG ⎦ | AGU ⎤ Ser (S) <br> AGC ⎦ <br> AGA ⎤ Arg (R) <br> AGG ⎦ | U <br> C <br> A <br> G |
| **G** | GUU ⎤ <br> GUC ⎥ Val (V) <br> GUA ⎥ <br> GUG ⎦ | GCU ⎤ <br> GCC ⎥ Ala (A) <br> GCA ⎥ <br> GCG ⎦ | GAU ⎤ Asp (D) <br> GAC ⎦ <br> GAA ⎤ Glu (E) <br> GAG ⎦ | GGU ⎤ <br> GGC ⎥ Gly (G) <br> GGA ⎥ <br> GGG ⎦ | U <br> C <br> A <br> G |

**Genetic Analysis 1.2** allows you to work through the transcription and translation of the DNA sequence assessed in Genetic Analysis 1.1.

Genetic Insight    Translation converts the genetic information contained in mRNA sequences into polypeptides using a triplet genetic code. Translation begins at the start codon; successive amino acids are linked by peptide bonds; and translation ceases at the stop codon.

## 1.4 Evolution Has a Molecular Basis

As biologists survey varieties of life, assess the genetic similarities and differences between species, and explore the relationship of modern organisms to one another and to their extinct ancestors, it becomes apparent that all life is connected through DNA. Richard Dawkins, a biologist and author of several books on evolution, made note of this molecular connection when observing that life "is a river of DNA, flowing and branching through geologic time." Dawkins's "river through time" that connects all organisms past and present is DNA. This shared DNA is a basis for identifying and studying relationships between organisms and tracing their evolutionary histories.

Life is not static or uniform, of course; it evolves as DNA diverges, through acquired mutational changes, into separate "branches" whose metaphorical forking leads to new species. The Dawkins quote suggests that for heredity to maintain genetic continuity across generations, and for variation to develop between organisms and evolve new species, the biochemical processes that replicate DNA and express the genetic information must also be universal. From this perspective the universality of DNA as the hereditary molecule of life, and the shared processes of DNA replication and gene expression, are consistent with the idea of a single origin of life that has evolved into the millions of species inhabiting Earth today as well as countless millions that preceded them but are now extinct.

Life on Earth most likely originated from a single source between 3.5 and 4.0 billion years ago. The oldest

| Table 1.3 | Amino Acid Abbreviations | |
|---|---|---|
| **Amino Acid** | **Three-Letter Abbreviation** | **One-Letter Abbreviation** |
| Alanine | Ala | A |
| Arginine | Arg | R |
| Asparagine | Asn | N |
| Aspartic acid | Asp | D |
| Cysteine | Cys | C |
| Glutamine | Gln | Q |
| Glutamic acid | Glu | E |
| Glycine | Gly | G |
| Histidine | His | H |
| Isoleucine | Ile | I |
| Leucine | Leu | L |
| Lysine | Lys | K |
| Methionine | Met | M |
| Phenylalanine | Phe | F |
| Proline | Pro | P |
| Serine | Ser | S |
| Threonine | Thr | T |
| Tryptophan | Trp | W |
| Tyrosine | Tyr | Y |
| Valine | Val | V |

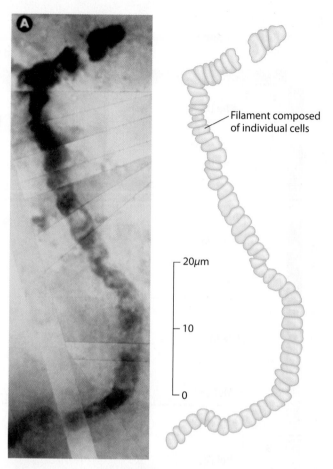

**Figure 1.12  Ancient life-forms.**  Living bacteria of the genus *Lyngbya* (left) resemble *Paleolyngbya* fossilized remains from the 950-million-year-old Lakhanda Formation in Siberia (right).

fossilized life-forms are spherical and filament-shaped organisms from rock formations in Western Australia that are 2.5 billion years old. Some of the earliest fossils have an appearance similar to bacteria that exist today (**Figure 1.12**). These early life-forms have given rise to a dazzling array of species, most now extinct. They are ancestral, however, to modern species that inhabit every conceivable ecological niche on Earth, from the most temperate to the most extreme.

## Darwin's Theory of Evolution

Over the millennia since life originated, untold millions of species have come and gone. These changes have occurred through **evolution,** the theory that all organisms are related by common ancestry and have diversified over time, primarily through the process of *natural selection.* The contemporary notion of evolution was proposed, independently, by Charles Darwin and Alfred Wallace in the late 1850s. Both these authors based their proposals on firsthand observations of the distribution and diversity of life across the globe. Each author described higher rates of survival and reproduction of certain forms of a species over alternative forms through the process of natural selection. Natural selection works at the phenotypic level but is based on underlying genetic variation: As natural selection increases the frequency of one morphological form over another in the population, the frequencies of the alleles controlling each form also change. Over many generations, morphological forms that produce more offspring also leave more copies of the alleles that control the phenotype, creating a genetic change in the population.

Charles Darwin's theory of evolution by natural selection is now a firmly established scientific fact incorporating three principles of population genetics that were obvious to many naturalists in Darwin's day but were not assembled into a coherent model until Darwin articulated their connection in his 1859 publication *On the Origin of Species.* Darwin's union of these principles into an evolutionary theory had a revolutionary effect on biology that laid the foundation of modern biological science. Here are Darwin's principles of populations:

Returning to the DNA strand presented in Genetic Analysis 1.1 (reproduced below), and also using the complementary sequence determined in that problem, answer the following questions.

3'-...ACGGATCCTCCCTAGTGCGTAATTCG...-5'

a. One strand of the double-stranded DNA sequence you produced in Genetic Analysis 1.1 is used as a template to produce mRNA containing five amino acid codons and a stop codon. The mRNA is translated into a polypeptide containing five amino acids, the first of which is methionine (Met), coded by the start codon. Identify the segment of the DNA template strand that contains the sequences encoding the start codon, stop codon, and amino acids.

b. Write the sequence and polarity of the mRNA transcript, specifying the codons for the five amino acids and the stop codon.

c. Write the amino acid sequence of the protein produced by translation. Use both the three-letter and one-letter codes for the sequence.

| Solution Strategies | Solution Steps |
|---|---|
| **Evaluate** | |
| 1. Identify the topic this problem addresses and explain the nature of the requested answer. | 1. The problem concerns identification of a DNA segment encoding a sequence of five amino acids in a protein. In addition to the coding segment of DNA, the answer requires identification of the corresponding mRNA transcript and the resulting amino acid sequence. |
| 2. Identify the critical information given in the problem. | 2. The necessary information is the double-stranded sequence produced as the answer to Genetic Analysis 1.1. |
| **Deduce** | |
| 3. Write out the double-stranded DNA sequence and identify DNA triplets that might be a <u>start codon</u>.<br><br>TIP: The most common start codon in mRNA is 5'-AUG-3' (methionine), coded by the template–DNA strand triplet 3'-TAC-5'. | 3. The double-stranded DNA sequence (from Genetic Analysis 1.1) is<br><br>5'-TGCCTAGGAGGGATCACGCATTAAGC-3'<br>3'-ACGGATCCTCCCTAGTGCGTAATTCG-5'<br><br>The single DNA triplet encoding a potential start codon is highlighted in bold type. |
| 4. Identify DNA triplets corresponding to possible <u>stop codons</u>.<br><br>TIP: There are three stop codons, UAA, UAG, and UGA, coded by DNA triplets ATT, ATC, and ACT, respectively. | 4. Five DNA triplets potentially encoding stop codons are shown in bold type below.<br><br>5'-TGCCTAGGAGGGATCACGCATTAAGC-3'<br>3'-ACGGATCCTCCCTAGTGCGTAATTCG-5' |
| **Solve** | Answer a |
| 5. Scan both strands of DNA looking for a 3'-TAC-5' (start codon) sequence that is followed by 12 nucleotides (four codons) encoding amino acids and then by a stop codon.<br><br>TIP: Messenger RNA is complementary and antiparallel to the DNA template strand. | 5. The one potential start codon is in the upper strand to the right (3'-TAC-5'), making the upper strand the template strand. The triplet encoding the stop codon is to the left (3'-ATC-5'). Four codons separate the start and stop that are highlighted below.<br><br>5'-TGCCTAGGAGGGATCACGCATTAAGC-3'<br>3'-ACGGATCCTCCCTAGTGCGTAATTCG-5' |
| 6. Determine the mRNA <u>sequence and polarity</u>.<br><br>TIP: Use the genetic code to translate mRNA. | Answer b<br>6. The mRNA sequence is<br><br>5'-AUG CGU GAU CCC UCC UAG-3' |
| 7. Determine the <u>amino acid sequence</u> of the polypeptide encoded by this mRNA. | Answer c<br>7. The polypeptide encoded by mRNA is Met-Tyr-Asp-Asn-Ala, or M-Y-D-N-A. |

For more practice, see Problems 15, 16, and 19.

1. Variation exists among the individual members of populations with regard to the expression of traits.

2. Hereditary transmission allows the variation in traits to be passed from one generation to the next.

3. Certain variant forms of traits give the individuals that carry them a higher rate of survival and reproduction in particular environmental conditions. These organisms leave more offspring and increase the frequency of the variant form in the population.

Darwin laid out the general process by which species evolved, but he never understood the underlying hereditary mechanisms that allowed the process to occur. Today, more than 150 years after Darwin introduced his revolutionary proposal, biologists fully understand the role of genetics in evolution. With regard to Darwin's evolutionary principles, biology has established these findings:

1. Phenotypic variation of expressed traits reflects inherited genetic variation. DNA-sequence differences (allelic variation) must be the cause of phenotypic variation if evolution is to occur.

2. Hereditary transmission of phenotypic variation requires that offspring inherit and express the alleles that were responsible for the variation in parental organisms.

3. Organisms carrying alleles that are favored by natural selection have a reproductive advantage over organisms that do not carry favored alleles. The former group therefore leaves more copies of their alleles in the next generation, causing the population to evolve through a change in allele frequency.

In other words, progressive phenotypic change in a population is paralleled by genetic changes.

Evolution by natural selection implies that one form reproduces in greater numbers than others in a population because of being better adapted to the conditions driving natural selection. This process, known as adaptive evolution, is common; but many examples of so-called nonadaptive evolution, the evolution of characteristics that are reproductively equivalent to other forms in the population, are also observed. Nonadaptive traits are neutral with respect to natural selection, conferring neither a selective advantage nor a selective disadvantage to their bearer. Processes of evolution other than natural selection help explain the evolution of nonadaptive variation.

## Four Evolutionary Processes

Evolutionary genetics studies and compares genetic changes in populations and species over time. Its foundations were established in the first four decades of the 20th century by several notable evolutionary biologists and innumerable lesser-known individuals. Interestingly, this work took place before DNA was identified as the hereditary material, and before the chemical structure of genes was defined and understood. Ronald Fisher, Sewall Wright, J. B. S. Haldane, and many others devised mathematical and statistical models of gene distribution and gene evolution in populations and species. This work produced evolutionary hypotheses that have been tested and verified countless times in laboratory and natural populations.

Through this massive body of work, evolutionary biology has confirmed Darwin's model of the evolution of species by natural selection and expanded the description of evolutionary processes to account for the evolution of individual genes and nonadaptive traits. Biologists identify four processes of evolution. Each one leads to *changes in the frequency of alleles in a population over time*, a characteristic that is the hallmark of evolutionary change. The four evolutionary processes are as follows:

1. **Natural selection**—the differential reproductive rates of members of a species owing to possession of different forms of an adaptive character. Population members with the best-adapted form are most successful at reproducing and leave more offspring than those possessing less adaptive forms. Over time, the frequency of the best-adapted form, and the alleles that produce it, increase in the population.

2. **Migration**—the movement of members of a species from one population to another. This migratory movement transfers alleles from one population to another, and if the allele frequencies are different enough or the size of the migrating population is sufficiently large, migration can rapidly alter allele frequencies.

3. **Mutation**—the slow addition of allelic variation that increases the hereditary diversity of populations and serves as the "raw material" of evolutionary change. Evolution is impossible without inherited variation for adaptive traits. Mutation provides the hereditary diversity required for evolutionary change.

4. **Random genetic drift**—the random change of allele frequencies due to chance in randomly mating populations. Genetic drift occurs in all populations, but it is most pronounced in very small populations, where statistical fluctuations in allele frequencies can be significant from one generation to the next.

By the middle of the 20th century, the **modern synthesis of evolution**—the name given to the merging of evolutionary theory with the results of experimental and molecular population biology—emerged as a unifying view of evolution. The modern synthesis tells the story of morphologic and molecular evolution of plant and animal species using experimentally verified processes and mechanisms.

Among the best-known principal architects of the modern synthesis are Theodosius Dobzhansky and Ernst Mayr, who drew together ideas from Darwin, Fisher, Wright, Haldane, and others to demonstrate how evolution operates in real populations. Dobzhansky and Mayr profoundly influenced the thinking and research of generations of biologists by demonstrating that hereditary mechanisms revealed by laboratory investigations, and the structure and evolution of natural populations, are consistent with the predictions made by Fisher, Wright, and Haldane. In simple terms, Dobzhansky and Mayr showed that evolution in populations and evolution in species occur as predicted by evolutionary theory. Today, having been fleshed out by the work of countless researchers, the modern synthesis gives a clear and virtually

---

Genetic Insight  Four evolutionary processes—natural selection, migration, mutation, and random genetic drift—shape populations and species. The modern synthesis of evolution unites Darwin's ideas of evolution with those of contemporary evolutionary biologists to provide a comprehensive view of the mechanisms and processes of evolutionary change.

---

complete picture of the factors that produce the evolutionary changes in populations and of the mechanisms that produce the evolution of species. We incorporate evolutionary examples into many chapters and specifically discuss evolution in species and in populations (see Chapter 22).

## Tracing Evolutionary Relationships

Evolutionary biologists investigate evolution by studying the morphologic (physical) evolution and molecular (DNA, RNA, and protein) evolution of organisms. Both morphologic and molecular comparisons can identify relationships between living species, as well as revealing ancestor–descendant relationships. Morphological or molecular similarities and differences identifying the evolutionary relationships of ancestors and descendants can be depicted in a diagram called a **phylogenetic tree.** These trees summarize the evolutionary histories of populations or species using branching points in the tree to indicate the common ancestors of descendant organisms.

The most commonly used approach to phylogenetic tree construction is the **cladistic** approach, which reconstructs evolutionary relationships and assorts them into groups called **clades** based on the identification of **shared derived characteristics** that can be either morphological or molecular. Shared derived characteristics develop through evolution from more primitive characteristics and display a range of differences among the

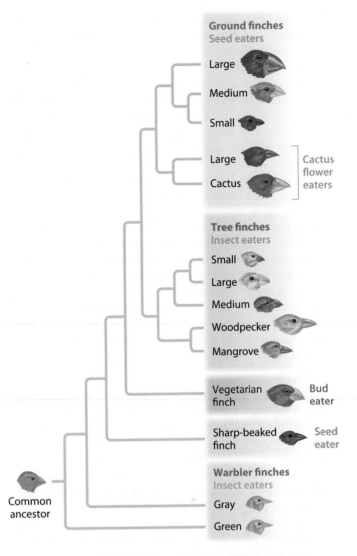

**Figure 1.13  Morphologic evolution.** A phylogenetic tree based on morphological characteristics shows the apparent relationships between 14 species of finches inhabiting the Galápagos Islands. (Adapted by permission from N. A. Campbell and J. B. Reece, *Biology*, 8th ed., Fig 1.22, p. 17. © 2008.)

groups being studied. The cladistic approach to phylogenetic tree construction often uses the concept of parsimony, meaning the use of the simplest or most economical choices, to construct a phylogenetic tree positing the fewest number of changes necessary to account for the known differences between groups. The idea behind the parsimony concept is that the simplest explanation for the known differences among groups has the greatest likelihood of being correct.

**Figure 1.13** shows a phylogenetic tree proposed for 14 finch species that inhabit the Galápagos Islands. These finch species were one of the groups studied by Darwin as he formulated his evolutionary theory. The tree shown here is based on a variety of morphological and behavioral characteristics, including the beak shape, beak size, feeding habits, and habitat of each species, as well as its degree of isolation or separation from other species in the Galápagos Islands.

Knowledge of the evolutionary relationships of a group of organisms allows a better understanding of their biology. For example, among Darwin's finches, the tree finches form a **monophyletic group**—a set of organisms that includes a common ancestor and all of its descendants. The ground finches also form a monophyletic group. In contrast, insect-eating finches form a **paraphyletic group**—a set of organisms that includes a common ancestor but does not include all of its descendants. Knowledge of whether a group of organisms is monophyletic or paraphyletic allows inferences about the evolution of traits. For example, since the tree finches are monophyletic, this trait, arboreal living, likely evolved once in the common ancestor of the tree finches. Conversely, since insect-eating finches form a paraphyletic group, the trait of preferring insects as a food source either evolved multiple times or, alternatively, has been lost multiple times in the evolution of Darwin's finches.

**Constructing Phylogenetic Trees Using Morphology and Anatomy**    Consider the features shared by various groupings of the animals listed in **Figure 1.14**: salmon, crocodile, platypus, kangaroo, wolf, gorilla, and human. One common morphologic feature of all these animals is the presence of a backbone, which unites this group into a clade we know as vertebrates that all share a common ancestor—in this case, an ancestral vertebrate. A second morphologic feature, the presence of four legs, unites all the tetrapod animals and excludes salmon. Thus, all the

animals except the salmon can be united into a clade we call tetrapods. Because fish are not within the clade of tetrapods, they form an *outgroup* to tetrapods. An **outgroup** is a taxon or group of taxa that is related to, but not included within, the clade in question. The species within the clade of interest are called the **ingroup.** In our example, each successive clade is identified by grouping species based on other shared characteristics.

After a phylogenetic tree has been constructed, it may be used to infer the characters of ancestral species. For example, we can infer that the common ancestor of all the taxa in Figure 1.14 had a backbone, which would therefore be an ancestral character; but it did not have four legs, which in this case would be a derived character that evolved later, in the common ancestry of tetrapods.

**Constructing Phylogenetic Trees Using Molecules**
Phylogenetic trees based on molecular characteristics are constructed in the same manner as those based on morphologic characteristics, except the shared features are DNA or protein sequences. Descendant groups have nucleic acid or amino acid sequences that are derived from ancient sequences possessed by their common ancestors. The concept of parsimony says that the most closely related molecular sequences are those that have the smallest number of differences between them.

**Figure 1.15** considers the DNA sequences derived from the first 15 nucleotides of the β-globin gene from seven species (**a** to **g**). In the figure, the sequences have been aligned vertically and the number of differences between the top sequence and each of the other sequences is noted in the first step of the figure.

A common method of constructing a phylogenetic tree begins with pairwise comparisons, grouping the most similar sequences closest together (on the assumption that they are the most closely related), and subsequently brings in the more distantly related sequences to obtain the most parsimonious tree. Analysis begins with sequences **a** and **b**, since they are identical, and successively adds in more distantly related sequences to construct a tree. Sequence information from **c,** which is most similar to **a** and **b,** is added next, followed by the other sequences. A completed phylogenetic tree constructed following the enumerated steps results in a tree that recapitulates the known phylogeny of vertebrates.

Care must be taken in the construction of phylogenetic trees to ensure that the morphologic or molecular characteristics they are based upon did not evolve independently, since this would violate the assumption that they are evidence of a common ancestor. For example, we know that wings in birds and bats have independent origins, because many other shared derived features unite bats with mammals and unite birds with reptiles. Shared characteristics that have independent

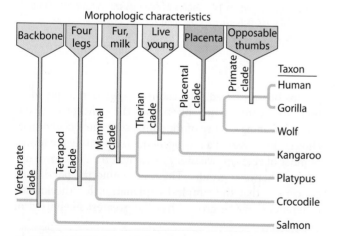

**Figure 1.14 The generation of a morphologic phylogenetic tree.** Organisms are assessed for the presence or absence of a series of morphological characters, and those that share characteristics form clades. The origins of specific traits can be traced on the phylogenetic tree.

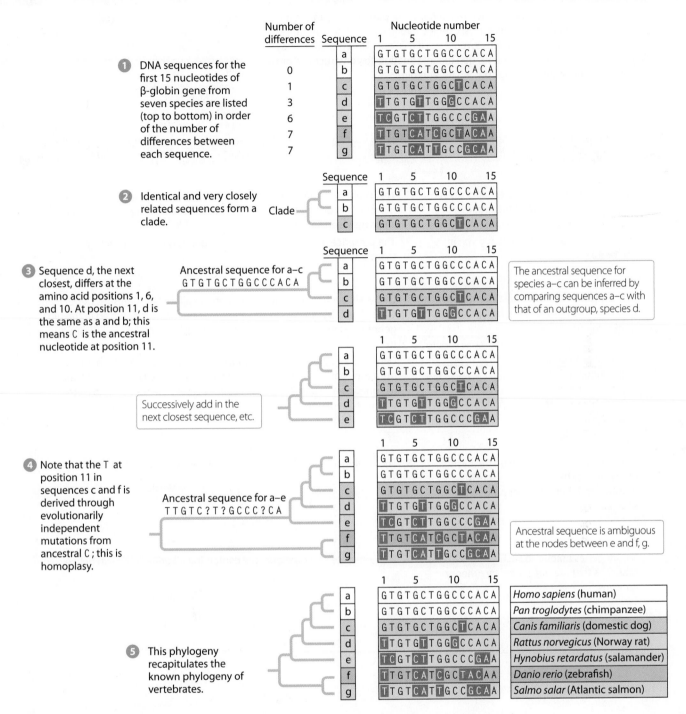

**Figure 1.15 Construction of a phylogenetic tree based on molecular characters, using the principle of parsimony.**

evolutionary origins arise by **convergent evolution,** or **homoplasy.** While the frequency of convergent evolution to the same character state is generally low, high levels may sometimes confound the construction of trees. **Genetic Analysis 1.3** guides you through phylogenetic tree construction.

The availability of DNA sequence data for most lineages of vertebrates has revolutionized how we view animal phylogeny. Some groups that were traditionally grouped together, such as mammals, birds, and amphibians, do prove, from sequence data, to form monophyletic groups. However, analyses have indicated that reptiles

Evolutionary biologists have searched the genomes of pigs, whales, and cows to identify the presence or absence of six genes (labeled *A* to *F* in the table at right). A gene is marked with a plus symbol (+) if it is found in a genome, or by a minus symbol (−) if it is not found. Use the information to construct the most likely phylogenetic tree relating cow, whale, and pig.

| Organism | Gene | | | | | |
|---|---|---|---|---|---|---|
| | *A* | *B* | *C* | *D* | *E* | *F* |
| Pig | + | − | − | + | − | − |
| Whale | + | + | + | − | + | − |
| Cow | + | + | + | − | − | + |

| Solution Strategies | Solution Steps |
|---|---|

### Evaluate

1. Identify the topic this problem addresses and explain the nature of the requested answer.

2. Identify the critical information given in the problem.

1. This problem requires evaluation of gene presence and absence in three mammals in order to construct a phylogenetic tree depicting the relationships between the groups.

2. The presence or absence of each of six genes is given for each type of mammal.

### Deduce

3. Identify genes shared by all three groups, genes shared by two of the groups, and genes unique to one group.

TIP: Genes shared by organisms are likely to have been present in their common ancestor.

4. Assign shared genes to phylogenetic branches that in completed tree will be shared by the corresponding organisms.

3. Of the six genes tested, gene *A* is found in all three organisms. Genes *B* and *C* are shared by whale and cow genomes but are not detected in the pig genome. Gene *D* is unique to pigs, *E* is unique to whales, and *F* is unique to cows.

4. Gene *A* is assigned to the base of the phylogenetic tree, which ascends from the common ancestor of the three organisms. Genes *B* and *C* are assigned to a branch shared by whale and cow. Genes *D*, *E*, and *F* are unique to separate groups and therefore are placed on separate branches.

### Solve

5. Begin to construct the tree on the basis of the genes shared by the greatest number and then by lesser numbers of organisms.

5. The phylogenetic tree based on shared genes is shown below.

6. Assign genes unique to each genome to branches that are not shared by other organisms.

6. The complete phylogenetic tree containing all genes is shown below.

For more practice, see Problem 18.

---

and fish do not form monophyletic groups and are paraphyletic. For example, crocodiles are more closely related to birds than to other reptiles. Morphologic and molecular analyses of dinosaurs suggest they are the sister group of birds, implying that extant birds are modern-day dinosaurs, as discussed in the Case Study at the end of the chapter.

**Genetic Insight** Morphological and molecular differences between contemporary taxonomic groups are used to construct phylogenetic trees that depict the evolutionary relationships between the groups. In tree construction, primitive characters or ancestral sequences are found in common ancestors, and shared derived characters are present in descendant groups.

## CASE STUDY

### Back in Time—Genetic and Genomic Analysis of Dinosaurs

The reptiles commonly known as dinosaurs first appeared approximately 250 million years ago. They are still alive today—at least in the form of their direct descendants, modern birds. The hypothesis that birds evolved from dinosaurs has prevailed for many decades and rests on multiple sources of information, including the observation that feathers first evolved in dinosaurs (**Figure 1.16**). Two studies published in 2007, and subsequent supporting data, lend additional credence to the connection and bring dinosaurs face to face with modern genetic and genomic analysis.

*Tyrannosaurus rex*, with its large stature, huge jaws and teeth, and fearsome reputation as a voracious carnivore, is perhaps the most familiar of all dinosaurs. The species was widely distributed, and its fossils have been found in relative abundance in North America. A *T. rex* leg bone found in the Hell Creek Formation in Montana in 2003 provided a sample from which Mary Schweitzer and colleagues extracted a minute amount of the bone protein collagen.

Collagen protein forms a highly ordered triple helix in bones that is conserved across taxonomic groups. The amino acid glycine, the smallest of the amino acids, is conserved at each turn of the protein helix and constitutes approximately 33% of the amino acid sequence of collagen. Alanine is also relatively common in collagen, as are the amino acids proline and lysine.

From their triple helix structure and amino acid profile, collagen molecules are distinctive and readily identifiable. Schweitzer and her colleagues detected such distinctive structures when they examined small fragments of the Hell Creek Formation *T. rex* leg bones with electron microscopy. A series of antibody tests confirmed the presence of collagen in the samples, after which Schweitzer and John Asara used a highly sensitive type of mass spectrometry to do an amino acid analysis of the presumed *T. rex* collagen. Mass spectrometry analysis indicated an abundance of glycine and relatively high levels of alanine and proline, consistent with what would be expected for collagen. Numerous controls and repeated tests confirm the detection of collagen in the *T. rex* samples.

Although the total amount of collagen detected in the *T. rex* bones was tiny, it provided enough mass spectrometry data to determine the ratio of the amounts of glycine to alanine. The Gly : Ala ratio for *T. rex* collagen is 2.6:1. In comparison, collagen from chickens has a Gly : Ala ratio of 2.5:1. These ratios are consistent with a close relationship between dinosaur collagen and chicken collagen. This ongoing research project—using the oldest-known protein ever extracted from a fossil specimen—adds support to the prevailing view that dinosaurs and birds share a close evolutionary relationship.

Additional support comes from a 2007 study of dinosaur bones by Chris Organ and colleagues that took a very different direction. Organ is interested in the evolution of genome size, and specifically in why birds have the smallest genomes of all vertebrates. Many theories have been proposed to explain this small genome size, but they take two general forms: (1) that birds evolved small genome size as an adaptive response to their high metabolic rate (suggesting that large genome size is too costly to maintain in metabolically active organisms) or (2) that small genome size in birds is an ancient ancestral characteristic of the bird lineage.

Organ developed an analytical method based on a well-established general correlation between genome size and cell size. In general, organisms with larger genomes also have larger cells. In examining the fossilized remains from 31 extinct species of dinosaurs, his team estimated the size of bone cells, called osteocytes, by measuring the volume of cavities in fossilized bone where the cells were once located. Previously, they had computed the correlation between osteocyte size and genome size in modern species; they then applied the same method to estimating the size of dinosaur genomes. Their results indicate that dinosaurs had small genomes of about the same size as those of modern bird species. These small genomes date back approximately 250 million years, thus predating the appearance of the first birds by about 100 million years. The evolutionary explanation for small dinosaur genomes remains an open question, but Organ's results lend credence to the view that small genome size is an ancestral characteristic of birds and not a more recent adaptation.

The significance of these dinosaur studies is twofold. First, they suggest the possibility of performing genetic and genomic analyses on the remnants of organisms that lived

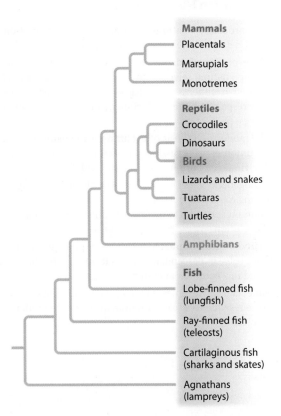

**Figure 1.16** **An animal phylogeny, based on morphologic and molecular data, showing that birds are the closest living relatives of dinosaurs.**

hundreds of millions of years ago. Sophisticated laboratory methods for the isolation, extraction, and measurement of trace amounts of nucleic acid and protein from extinct species can be combined with sensitive detection procedures and computational methods to perform these studies. Second, they provide new support for existing theories of the close evolutionary relationship between dinosaurs and modern birds.

---

SUMMARY      *For activities, animations, and review quizzes, go to the study area at www.masteringgenetics.com.*

## 1.1 Modern Genetics Has Entered Its Second Century

■ Genetic principles first outlined by Gregor Mendel in 1865 were "rediscovered" in 1900 and so made modern genetics a 20th-century scientific discipline.

■ Study of the transmission of morphological variation during the first half of the 20th century established transmission genetics as a central focus of genetic analysis.

■ The analysis of DNA, RNA, and protein beginning in the second half of the 20th century established genetics as a molecular discipline.

■ Life on Earth has three domains—Bacteria, Archaea, and Eukarya—that share a common evolutionary history.

## 1.2 The Structure of DNA Suggests a Mechanism for Replication

■ Deoxyribonucleic acid (DNA) is the genetic material. DNA is a double helix composed of four nucleotides that are composed of a five-carbon deoxyribose sugar, a phosphate group, and one of four nucleotide bases: adenine (A), thymine (T), cytosine (C), or guanine (G).

■ Nucleotides in a DNA strand are joined by covalent phosphodiester bonds between the 5′ phosphate of one nucleotide and the 3′ OH of the adjoining nucleotide.

■ DNA strands are joined by hydrogen bonds that form between complementary base pairs. A pairs with T and C pairs with G.

■ Strands of the DNA duplex are antiparallel; one strand is oriented 5′ → 3′ and the complementary strand is oriented 3′ → 5′.

■ DNA replicates by a semiconservative process that produces exact copies of the original DNA double helix.

■ DNA polymerase uses one strand of DNA as a template to synthesize a complementary daughter strand one nucleotide at a time in the 5′-to-3′ direction.

## 1.3 Transcription and Translation Express Genes

■ The central dogma of biology (DNA → RNA → protein) identifies DNA as an information repository and describes how DNA dictates protein structure through a messenger RNA intermediary that in turn directs polypeptide synthesis.

■ Transcription is the process that synthesizes single-stranded RNA from a template DNA strand.

■ RNA transcripts have the same 5′ → 3′ polarity and sequence as the coding strand of DNA; they differ only in the presence of U rather than T.

■ Certain DNA sequences, most commonly promoters, bind RNA polymerase and other transcriptional proteins.

■ Translation is the process that uses messenger RNA (mRNA) sequences to synthesize proteins.

■ Messenger RNA codons base-pair with tRNA anticodons at the ribosome.

■ Each tRNA carries a specific amino acid that is added to the growing polypeptide chain.

■ The genetic code contains 61 codons that specify amino acids and 3 that are stop codons.

## 1.4 Evolution Has a Molecular Basis

■ Four processes—natural selection, mutation, migration, and genetic drift—drive the evolution of populations and species.

■ The evolution of adaptive morphological characters occurs through natural selection pressures exerted on species by their environments. Nonadaptive characters that are neutral with respect to natural selection evolve by other evolutionary processes.

■ The modern synthesis of evolution is the name applied to the union of transmission genetics, molecular genetics, Darwinian evolution, and modern evolutionary genetics.

■ Phylogenetic trees describe the evolutionary relationships among modern species and trace their descent from common ancestors to identify the most likely pattern of evolution.

■ Shared derived characteristics are morphologic attributes that evolve in descendant species from ancient characters found in a common ancestor.

■ Molecular phylogenies trace the evolution of nucleic acid or protein sequences from common ancestors to modern species.

---

KEYWORDS

allele *(p. 3)*
amino acid *(p. 11)*
anticodon *(p. 12)*
antiparallel *(p. 7)*

Archaea *(p. 4)*
Bacteria *(p. 4)*
central dogma of biology *(p. 10)*
Chargaff's rule *(p. 6)*

chromosome *(p. 2)*
cladistics (clade) *(p. 17)*
coding strand *(p. 11)*
codon *(p. 12)*

complementary base pairs *(p. 6)*
convergent evolution (homoplasy) *(p. 19)*
daughter strand *(p. 9)*
deoxyribonucleic acid (DNA) *(p. 5)*
DNA double helix (DNA duplex) *(p. 5)*
DNA nucleotides: adenine (A), guanine (G), thymine (T), cytosine (C) *(p. 6)*
DNA replication (semiconservative replication) *(pp. 5, 9)*
Eukarya (eukaryote) *(p. 4)*
evolution *(p. 14)*
evolutionary genetics *(p. 5)*
exon *(p. 11)*
gametes *(p. 2)*
gene *(p. 2)*
genetic code *(p. 12)*
genome *(p. 4)*
genotype *(p. 3)*
homologous chromosomes (homologous pair, homologs) *(p. 2)*

hydrogen bond *(p. 7)*
ingroup *(p. 18)*
intron *(p. 11)*
meiosis *(p. 2)*
messenger RNA (mRNA) *(p. 5)*
migration *(p. 16)*
mitosis *(p. 2)*
modern synthesis of evolution *(p. 16)*
molecular genetics *(p. 5)*
monophyletic group *(p. 18)*
mutation *(p. 16)*
natural selection *(p. 16)*
outgroup *(p. 18)*
paraphyletic group *(p. 18)*
parental strand *(p. 9)*
peptide bond *(p. 12)*
phenotype *(p. 3)*
phosphodiester bond *(p. 7)*
phylogenetic tree *(p. 17)*
promoter *(p. 11)*

protein (polypeptide) *(p. 12)*
random genetic drift *(p. 16)*
reverse transcription *(p. 11)*
ribonucleic acid (RNA) *(p. 5)*
ribosomal RNA (rRNA) *(p. 10)*
ribosome *(p. 5)*
shared derived characteristics *(p. 17)*
start codon *(p. 12)*
start of transcription *(p. 11)*
stop codon *(p. 12)*
strand polarity (5′ and 3′) *(p. 7)*
template strand *(p. 11)*
termination sequence (transcription termination) *(p. 11)*
transcription *(p. 5)*
transfer RNA (tRNA) *(p. 10)*
translation *(p. 5)*
transmission genetics (Mendelian genetics) *(p. 5)*
uracil (U) *(p. 11)*

---

PROBLEMS        *For instructor-assigned tutorials and problems, go to www.masteringgenetics.com.*

## Chapter Concepts

*For answers to selected even-numbered problems, see Appendix: Answers.*

1. Genetics affects many aspects of our lives. Identify three ways genetics affects your life or the life of a family member or friend. The effects can be regularly encountered or can be one time only or occasional.

2. How do you think the determination that DNA is the hereditary material affected the direction of biological research?

3. A commentator once described genetics as "the queen of the biological sciences." The statement was meant to imply that genetics is of overarching importance in the biological sciences. Do you agree with this statement? In what ways do you think the statement is accurate?

4. All life shares DNA as the hereditary material. From an evolutionary perspective, why do you think this is the case?

5. Define the terms *allele, chromosome,* and *gene* and explain how they relate to one another. Develop an analogy between these terms and the process of using a street map to locate a new apartment to live in next year, i.e., consider which term is analogous to a street, which to a type of building, and which to an apartment floor plan.

6. Define the terms *genotype* and *phenotype,* and relate them to one another.

7. Define *natural selection,* and describe how natural selection operates as a mechanism of evolutionary change.

8. Describe the modern synthesis of evolution, and explain how it connects Darwinian evolution to molecular evolution.

9. What are the four processes of evolution? Briefly describe each process.

10. Define each of the following terms:
    a. transcription
    b. allele
    c. central dogma of biology
    d. translation
    e. DNA replication
    f. gene
    g. chromosome
    h. antiparallel
    i. phenotype
    j. complementary base pair
    k. nucleic acid strand polarity
    l. genotype
    m. natural selection
    n. mutation
    o. modern synthesis of evolution

## Application and Integration

*For answers to selected even-numbered problems, see Appendix: Answers.*

11. If thymine makes up 21% of the DNA nucleotides in the genome of a plant species, what are the percentages of the other nucleotides in the genome?

12. What reactive chemical groups are found at the 5′ and 3′ carbons of nucleotides? What is the name of the bond formed when nucleotides are joined in a single strand? Is this bond covalent or noncovalent?

13. Identify two differences in chemical composition that distinguish DNA from RNA.

14. What is the central dogma of biology? Identify and describe the molecular processes that accomplish the flow of genetic information described in the central dogma.

15. A portion of a polypeptide contains the amino acids Trp-Lys-Met-Ala-Val. Write the possible mRNA and template strand DNA sequences. (Hint: Use A / G and T / C to indicate that either adenine/guanine or thymine/cytosine could occur in a particular position, and use N to indicate that any DNA nucleotide could appear.)

16. The following segment of DNA is the template strand transcribed into mRNA:

    5'-…GACATGGAA…-3'

    a. What is the sequence of mRNA created from this sequence?
    b. What is the amino acid sequence produced by translation?

17. Consider the following segment of DNA:

    5'-…ATGCCAGTCACTGACTTG…-3'
    3'-…TACGGTCAGTGACTGAAC…-5'

    a. How many phosphodiester bonds are required to form this segment of double-stranded DNA?
    b. How many hydrogen bonds are present in this DNA segment?
    c. If the lower strand of DNA serves as the template transcribed into mRNA, how many peptide bonds are present in the polypeptide fragment into which the mRNA is translated?

18. Examine Figure 1.14 and answer the following questions.
    a. How many clades are shown in the figure?
    b. What characteristic is shared by all clades in the figure?
    c. What characteristics are shared by the mammalian clade and the human clade? What characteristics distinguish these two clades?

19. Fill in the missing nucleotides so there are three per block, and the missing amino acid abbreviations in the graphic below.

20. Four nucleic acid samples are analyzed to determine the percentages of the nucleotides they contain. Survey the data in the table below and determine which samples are DNA and which are RNA and specify whether each sample is double-stranded or single-stranded. Justify each answer.

| | A | G | T | U | C |
|---|---|---|---|---|---|
| Sample 1 | 22% | 28% | 22% | 0 | 28% |
| Sample 2 | 30% | 30% | 0 | 20% | 20% |
| Sample 3 | 18% | 32% | 0 | 18% | 32% |
| Sample 4 | 29% | 29% | 21% | 0 | 21% |

21. Are seed-eating finches among Darwin's finches monophyletic or paraphyletic? What about cactus flower–eating finches?

22. If one is constructing a phylogeny of reptiles using DNA sequence data, which taxa (birds, mammals, amphibians, or fish) might be suitable to use as an outgroup?

23. Consider the following amino acid sequences obtained from different species of apes. Applying parsimony, construct a phylogenetic tree of the apes.

| | | | | | | | | | | | | |
|---|---|---|---|---|---|---|---|---|---|---|---|---|
| *Pongo pygmaeus* | G | G | P | H | Y | R | L | I | A | V | E | D |
| *Pongo abelii* | G | G | P | H | Y | R | L | I | A | V | E | D |
| *Pan paniscus* | G | A | P | H | F | R | L | L | A | V | E | E |
| *Pan troglodytes* | G | A | P | H | F | R | L | L | A | V | E | E |
| *Gorilla gorilla* | G | A | P | H | F | R | L | I | A | V | E | E |
| *Gorilla beringei* | G | A | P | H | F | R | L | I | A | V | E | E |
| *Homo sapiens* | G | A | P | H | F | N | L | L | A | V | E | E |
| *Hylobates lar* | G | G | P | H | Y | R | L | I | S | V | E | D |
| *Hoolock hoolock* | G | G | P | H | Y | R | L | I | S | V | D | D |
| *Common Ancestor* | G | G | P | H | Y | R | L | I | S | V | D | D |

**DNA**

Coding 5' ... GGC GA ... T ... 3'
Template 3' ... C ... G ... 5'

**mRNA codon**

5' ... UAC ... A  A ... 3'

**tRNA anticodon**

3' ... UUA ... 5'

**Amino acid**

3-letter  MET ...
1-letter  ... E  S ...

# Transmission Genetics

The friars of St. Thomas were dedicated teachers of the arts and sciences. Gregor Mendel (standing, far right) ponders a flower while he poses with his fellow teachers.

## ESSENTIAL IDEAS

■ Mendel's hereditary experiments with pea plants identified two laws of heredity known as segregation and independent assortment.

■ Consistent and predictable phenotype ratios in generations descending from two parents differing for a single trait support the law of segregation.

■ The inheritance of two or more traits is predicted by the law of independent assortment.

■ The rules of probability predict genetic inheritance.

■ The statistical method known as chi-square analysis is used to evaluate how closely the predicted outcomes of genetic experiments match experimental observations.

■ The inheritance of traits in human families follows the hereditary laws of segregation and independent assortment.

■ The inheritance of DNA, RNA, and protein variation parallels the inheritance of phenotypic variation.

When Gregor Mendel identified and described two fundamental laws of hereditary transmission, he ushered in a new era of understanding in biology. The terms *Mendelian genetics* and *Mendelism* were coined to recognize this contribution, and they are used as synonyms for **transmission genetics,** the field that describes the transfer of genes from parents to offspring. Like his contemporary Charles Darwin, who elegantly described the process of evolution by natural selection, Mendel helped articulate a new way to view the world through his discoveries.

Mendel was by no stretch of the imagination the first person to examine the transmission of hereditary traits in plants. Many amateur botanists of the 18th and early

19th centuries conducted what were then called studies of "plant hybridization" on many species, including the edible pea plant (*Pisum sativum*) that was the subject of Mendel's experiments. Others before him had even carried out crosses similar to Mendel's, some made observations like those on which Mendel based his two principles of heredity, and some even came close to articulating a description of the hereditary principles Mendel described. But no one described hereditary transmission as precisely as Mendel did. Mendel succeeded because of his superior experimental design and his quantification of results. His approach allowed him to formulate and test genetic hypotheses with a level of rigor that no one had achieved before him, or would achieve for 35 years after Mendel articulated his hypothesis.

In this chapter, we examine how Mendel used experimental results to identify two pivotal principles of hereditary transmission. We see (1) how Mendel's unprecedented experimental designs enabled him to detect genetic phenomena that escaped identification by his predecessors and (2) how the transmission of traits can be predicted using random probability theory. The chapter concludes with a discussion that extends transmission genetic principles to the examination of inherited traits and DNA-sequence variants in humans and other organisms. We begin, however, with a short biography of Gregor Mendel that reveals how his educational experiences profoundly influenced his approach to scientific exploration.

## 2.1 Gregor Mendel Discovered the Basic Principles of Genetic Transmission

Born in 1822 to a farming family of modest means in what is now part of the Czech Republic, Johann (later known by his clerical name, Gregor) Mendel completed the equivalent of high school at age 18 with a certificate attesting to exceptional academic abilities. He began his higher education at the Olmutz Philosophical Institute in 1840, but these studies took a severe toll on his mental and physical health, and he soon gave them up. In 1843, after attempting unsuccessfully to restart his education at

Olmutz, he decided to pursue higher learning by entering the priesthood instead. Based on its strong reputation in teacher training and a recommendation from a former teacher at Olmutz, he selected St. Thomas monastery in the Czech city of Brno. Mendel's duties at St. Thomas included temporary teaching of natural science at a middle school in Brno. His keen interest in teaching science and his desire to become a permanent teacher led monastery administrators to send Mendel to the University of Vienna in 1851 to study natural science as preparation for a teaching examination.

In Vienna, Mendel studied plant physiology and plant biology with Professor Franz Unger and physics with Professor Christian Doppler as well as Doppler's successor Professor Andreas von Ettinghausen. From Professor Unger, Mendel learned to think critically about prevailing theories of plant reproduction and hybridization. Doppler, an experimental physicist famous for the Doppler effect, espoused a "particulate" view of physics and taught Mendel how to separate individual characteristics from one another in experiments. Professor Ettinghausen taught Mendel the mathematics of combinatorial analysis. Mendel would apply each of these lessons to his later research. In 1853, Mendel returned to Brno to teach natural science. He took and passed the written portion of the permanent teachers' examination but apparently never completed the oral portion, remaining a "temporary" teacher at the school in Brno until he became abbot of the monastery in 1868.

In the summer of 1856, after a 3-year period during which he pondered how he might pursue his interest in natural science, Mendel began his work on trait heredity in the edible pea plant *Pisum sativum*. Mendel began his studies by gathering 34 different varieties of peas collected from local suppliers. Over the next 2 years, he tested each variety for its ability to uniformly reproduce identical characteristics from one generation to the next. Ultimately, he settled on 14 strains of *Pisum* representing seven individual traits, each of which had two easily distinguished forms of expression in a seed or plant (**Figure 2.1**). Mendel worked with these 14 strains for the next 5 years, concluding his experiments in 1863.

On February 8 and March 8, 1865, Mendel discussed his work on peas at two meetings of the Brno Society for the Study of Natural Science. The society published his report in its *Proceedings* the following year, 1866. After publication of his work, Mendel corresponded with several prominent botanists in Europe, most notably Karl Naegeli. Mendel's letters to Naegeli have scientific significance because they clearly lay out his experiments, his results, and his conclusions. Unfortunately, neither Naegeli nor any of his contemporaries seemed to grasp the importance of Mendel's work.

After becoming abbot of the monastery, Mendel gave up his scientific work and faithfully served the monastery until his death in 1884. Mendel died in obscurity, never having had the importance of his experiments understood or appreciated. Sixteen years later, in the year 1900,

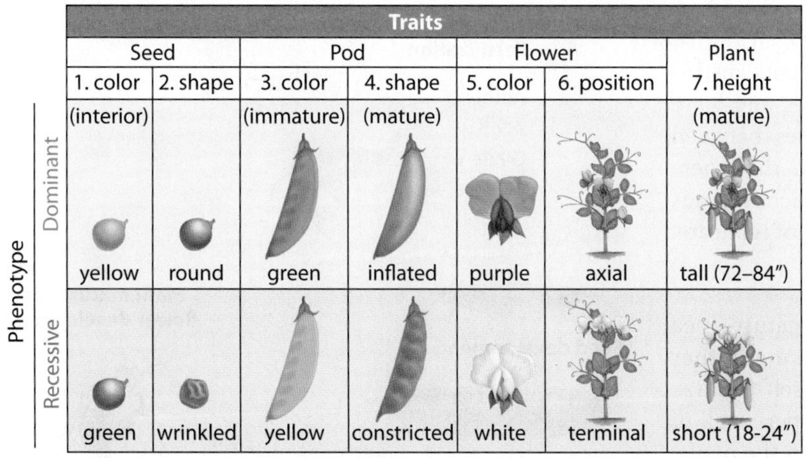

| Traits | | | | | | |
|---|---|---|---|---|---|---|
| Seed | | Pod | | Flower | | Plant |
| 1. color | 2. shape | 3. color | 4. shape | 5. color | 6. position | 7. height |
| **Dominant** (interior) yellow | round | (immature) green | (mature) inflated | purple | axial | (mature) tall (72–84") |
| **Recessive** green | wrinkled | yellow | constricted | white | terminal | short (18-24") |

**Figure 2.1 The seven dichotomous traits of *Pisum sativum* studied by Mendel.** Each trait has a dominant phenotype and a recessive phenotype that are easily distinguished.

biologists would replicate and rediscover his experiments and launch a revolution in biology. Today, nearly 150 years after the work was carried out, the former St. Thomas Monastery in Brno is the location of the Mendel Museum, which houses displays honoring Gregor Mendel. You can visit the museum online to see some of the displays and enjoy a number of interactive features at www.mendel-museum.com.

## Mendel's Modern Experimental Approach

Mendel successfully identified principles of hereditary transmission that eluded investigators who preceded him and continued to elude investigators for many years after his death. Was Mendel more insightful? Did he make fortuitous choices by selecting *Pisum sativum* as his experimental organism and in selecting his seven characteristics? Did he have a superior approach to genetic experimentation and analysis? The answer to each of these questions is yes.

Mendel's superior insight came principally from his familiarity with quantitative thinking and his understanding of the particulate nature of matter, learned through the study of physics with Doppler. Central to Mendel's experimental success was counting the number of progeny with specific phenotypes. This logical and now routine component of data gathering was the key to Mendel's ability to formulate the hypotheses that explained his results. Under Doppler and Ettinghausen, Mendel had learned to study individual properties of matter separately and to think in quantitative terms about combinations of outcomes.

Mendel made a fortuitous choice in selecting the pea plant as his experimental organism. Peas were commonly used for hybridization studies in Mendel's time, so a large number of strains displaying different phenotypic characteristics were available. The pea plant is hearty and was easy for a skilled botanist like Mendel to manipulate and crossbreed.

In choosing to study individual traits of the pea plant, Mendel designed his experiments to test the **blending theory** of heredity that was the predominant hereditary

theory at the time. The blending theory viewed the traits of progeny as a mixture of the characteristics possessed by the two parental forms. Under this theory, progeny were believed to display characteristics that were approximately intermediate between those of the parents. For example, the blending theory would predict that crossing a black cat and a white cat would produce gray kittens, and that the original black or white colors would never reappear if the gray kittens were bred to one another. Mendel reasoned that if the blending theory were true, he would see evidence of it in each trait. If no blending were seen in individual traits, the blending theory would be disproved.

As crucial as his quantitative approach and choice of *Pisum* were to his ultimate success, Mendel's radically new experimental design was his most important innovation. Mendel was ahead of his time in that his scientific experiments were hypothesis driven. In other words, following an initial observation, he devised a hypothesis to explain the observation and then carried out an independent experiment to test the hypothesis. An experimenter employing this approach, the basis of the modern *scientific method,* will follow these steps:

1. Make initial observations about a phenomenon or process.

2. Formulate a testable hypothesis to explain observations.

3. Design a controlled experiment to test the hypothesis.

4. Collect data from the controlled experiment.

5. Interpret experimental results, comparing the observed results to those expected under assumptions of the hypothesis.

6. Draw reasonable conclusions, reformulating or retesting the hypothesis if necessary.

In the following sections, we discuss how careful planning of his experimental design allowed Mendel to collect data on individual traits of the pea plant, formulate hypotheses to explain his phenotypic observations, and conduct independent experiments to test his predictions.

## Five Critical Experimental Innovations

Five features of Mendel's breeding experiments distinguish them from those of his contemporaries and were critical to his success: (1) controlled crosses between plants; (2) use of pure-breeding strains to begin the experimental controlled crosses; (3) selection of dichotomous traits; (4) quantification of results; and (5) use of replicate, reciprocal, and test crosses.

**Controlled Crosses between Plants**  In nature, pea plant flowers contain both a pollen-producing anther and an egg-containing ovule and usually self-fertilize (**Figure 2.2**). Self-fertilization occurs when a sperm cell of pollen from the anther fertilizes an egg within the ovule. Fertilized ovules develop in the ovary, which matures into fruit (seed pod) as seeds (peas) develop inside. A mature seed pod usually contains five to seven peas, each of which results from a different fertilization event. In genetic experiments, peas can be collected and scored for their phenotypes or can be planted to produce pea plants that are scored for their traits.

Pea plants are also capable of cross-pollination, if pollen from one plant is used to fertilize the ovules of another. In nature, plants are cross-pollinated by insects, birds, mammals, and wind. Mendel used his familiarity with plants to carry out **artificial cross-fertilization** (**Figure 2.3**). First, he emasculated developing pea flowers by cutting off the nascent anthers. This modification made the plants incapable of self-pollination, but the

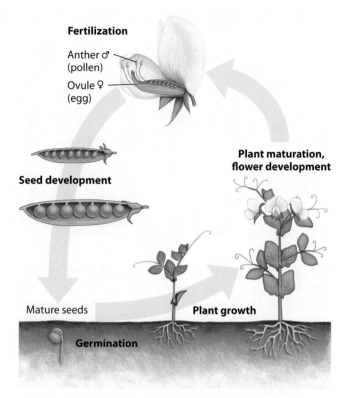

**Figure 2.2  Life cycle of *Pisum sativum*.** Seeds (peas) are planted and germinate, growing into mature flowering plants. Eggs in the flower ovule are fertilized by pollen produced from anthers. Immature seeds arise from individual fertilized eggs in the pod that forms as seeds develop. After seeds mature, they are dispersed to renew the cycle.

**Figure 2.3  Artificial cross-fertilization of pea plants.**

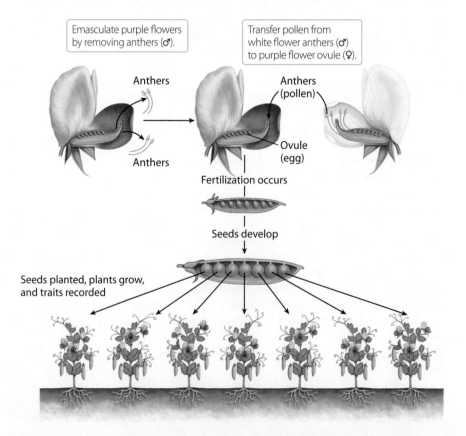

ovules could still be fertilized by cross-fertilization with pollen from another plant. Mendel carried out artificial cross-fertilization by using a small paintbrush to lift mature pollen from a non-emasculated flower and brush it onto an emasculated flower. With this manipulation Mendel restricted reproduction to those plants he identified beforehand as likely to yield informative results, thus performing what is now known as a **controlled genetic cross** between selected organisms.

**Pure-Breeding Strains to Begin Experimental Crosses**
During the 2 years before beginning his hereditary experiments, Mendel performed numerous controlled genetic crosses to obtain strains that consistently produced a single phenotype without variation. Strains of this kind that consistently produce the same phenotype are called **pure-breeding strains,** also known as **true-breeding strains.** The self-fertilization of a pure-breeding purple-flowered plant will yield only purple flowers among progeny plants. Two plants from a pure-breeding line can be crossed to one another and will produce progeny with the same phenotype.

Mendel spent almost 2 years generating 14 pure-breeding strains for his 7 traits. He studied the inheritance of each trait the same way, beginning each experiment by crossing pure-breeding parental plants that expressed different phenotypes of the selected trait. For example, Mendel crossed pure-breeding purple-flowered plants with pure-breeding white-flowered plants. By artificial cross-fertilization of these **parental generation (P generation)** plants, Mendel produced seeds that were grown into the **first filial generation ($F_1$ generation)** of plants (**Figure 2.4**). The $F_1$ plants were then used as the sources of pollen and ovum to produce the seeds that were grown into the **second filial generation ($F_2$ generation)**. The **third filial generation ($F_3$ generation)** was produced by crossing plants from the $F_2$ generation, and so on for as many generations as needed.

**Selection of Single Traits with Dichotomous Phenotypes**    Each of the seven traits that Mendel chose is found in just two dichotomous forms. The two phenotypes are readily distinguished from one another, so there can be no ambiguity of assignment, and there are no intermediate phenotypes. For example, one trait was seed color; every seed was either yellow or green.

The alternative forms of the seven traits Mendel studied are illustrated in Figure 2.1. The 14 pure-breeding strains were bred for (1) seed color (yellow or green), (2) seed shape (round or wrinkled), (3) pod color (green or yellow), (4) pod shape (inflated or constricted), (5) flower color (purple or white), (6) flower position (axial or terminal), and (7) plant height (tall or short).

It is interesting to note that Mendel initially had selected an eighth trait producing either gray or white exterior seed coats. Early in his analysis, however, he found that plants with purple flowers *always* had gray seed coats and that those with white flowers *always* had white seed

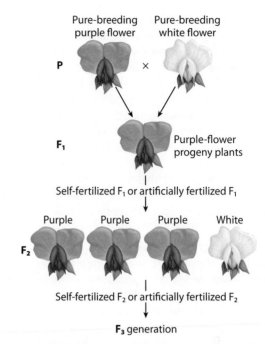

**Figure 2.4  Production of three generations of pea plants.** Plants of the P generation are artificially cross-fertilized to produce the $F_1$ generation. Self-fertilization, or crossing of $F_1$-generation plants, produces the $F_2$ generation. $F_2$ plants either self-fertilize or are crossed to one another to produce the $F_3$ generation.

coats. He speculated that flower color and seed-coat color were determined by the same genetic mechanism, and he was correct. The pigment anthocyanin is produced by plants that have purple flower color and gray seed coats, but a mutation eliminates anthocyanin production in plants with white flowers and white seed coats.

**Quantification of Results**    Each time Mendel made a controlled cross, he carefully counted the number of progeny plants of each phenotype. This seemingly simple act—now standard in scientific data gathering—was revolutionary in Mendel's day. By expressing his results numerically, Mendel could easily analyze them for revealing patterns such as the occurrence of consistent ratios between phenotypes. These ratios were critically important to Mendel's discovery of the rules by which he could predict transmission of alleles during reproduction, and they are the foundation of Mendel's two laws of heredity.

**Replicate-, Reciprocal-, and Test-Cross Analysis**    The final feature that distinguished Mendel's experiments from those of his predecessors was his repetition of crosses. Rather than simply counting the results of a single cross, Mendel made many **replicate crosses,** producing hundreds of $F_1$ plants and several thousand $F_2$ plants by repeating the same cross several times.

Mendel also performed **reciprocal crosses,** in which the same genotypes are crossed but the sexes of the donating parents are switched. The plant providing the ovule in the first cross is used as a source of pollen in the reciprocal cross. An example of a reciprocal cross is shown in **Figure 2.5a**, where pollen from a strain producing yellow peas (*G–*) is used to fertilize the ova of plants from a strain producing green peas (*g–*). The reciprocal cross is performed using pollen from the green-pea–producing plant to fertilize ovules of the yellow-pea–producing plant. Note that these reciprocal crosses produce F$_1$ with yellow peas. We discuss the importance of this result in the following section.

Finally, Mendel performed **test crosses** (**Figure 2.5b**). Here, *R* and *r* represent alleles of the rugose gene, meaning "full of wrinkles". We examine the results and significance of this kind of controlled genetic cross below.

---

Genetic Insight  Mendel's experiments on heredity introduced the use of controlled crosses; the use of pure-breeding parental strains; the selection of dichotomous traits; the careful quantification of cross results; and the performance of replicate crosses, reciprocal crosses, and test crosses as central elements of a careful experimental design.

---

**(a) Reciprocal crosses**

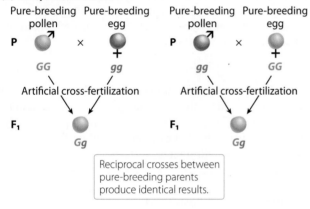

Reciprocal crosses between pure-breeding parents produce identical results.

**(b) Test cross**

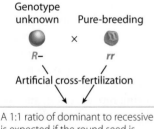

A 1:1 ratio of dominant to recessive is expected if the round seed is heterozygous (*Rr*); all progeny are dominant if the round seed is homozygous (*RR*).

**Figure 2.5  Reciprocal crosses and test cross.  (a)** Two reciprocal crosses between different pure-breeding yellow (*GG*) and green (*gg*) parents produce F$_1$ plants with yellow seeds (*Gg*). **(b)** A test cross is made between an F$_1$ with the dominant phenotype suspected to be heterozygous (as indicated by *R–*) and a pure-breeding (*rr*) plant with the recessive phenotype. See Section 2.2 for definitions of these terms.

## 2.2 Monohybrid Crosses Reveal the Segregation of Alleles

In Mendel's time, genetic experiments did not generally follow modern scientific method, nor did hereditary studies generally quantify the results of crosses. Mendel owes his success at proposing laws of heredity to his use of the scientific method and to the quantitative approach he brought to analysis of his crosses. Mendel's carefully executed experiments and their equally precise analysis disproved the blending theory of heredity and produced a new hereditary theory.

### Identifying Dominant and Recessive Traits

In this section we illustrate the results and interpretation of Mendel's crosses by studying the transmission of two of Mendel's traits, pea color (yellow or green) and, in separate crosses, the transmission of pea shape (round or wrinkled). The results and interpretations we describe apply equally well to the five other traits Mendel examined. The uniformity of the experimental results and interpretations are due to Mendel's decision to conduct experiments on each trait in the same way. He began hereditary experiments on each trait by artificial cross-fertilization of pure-breeding parental plants to produce an F$_1$ generation, and he then self-fertilized or intercrossed F$_1$ plants to produce the F$_2$ generation.

Crossing pure-breeding yellow-pea–producing plants and pure-breeding green-pea producers in replicate and reciprocal crosses, for example, Mendel consistently found that all of the F$_1$ plants produced yellow peas and none produced green peas (**Figure 2.6**). Mendel identified yellow as the **dominant phenotype** on the basis of its presence in the F$_1$, and he identified green as the **recessive phenotype** since it is not seen among F$_1$ progeny. Mendel next crossed F$_1$ yellow plants to produce the F$_2$ and observed reemergence of the recessive green phenotype. Among the F$_2$, Mendel found that approximately three-fourths (75%) of the peas were yellow and the remaining one-fourth (25%) were green. The yellow : green ratio in the F$_2$ is $\frac{3}{4}:\frac{1}{4}$, or roughly 3:1. Mendel made similar observations for his experiments testing inheritance of pea shape. Replicate and reciprocal crosses of pure-breeding round-pea–producing plants with pure-breeding wrinkled-pea–producing plants produced F$_1$ plants bearing exclusively round peas. This result identifies round as the dominant phenotype and wrinkled as the recessive phenotype. His F$_1$ cross produced F$_2$ peas in the ratio 75% round to 25% wrinkled—once again a roughly 3:1 ratio.

Tabulating results over several growing seasons for all seven traits, Mendel counted more than 20,000 F$_2$ peas or plants. **Table 2.1** displays Mendel's results revealing three consistent features: (1) dominance of one phenotype over the other in the F$_1$ generation, (2) reemergence of the recessive phenotype in the F$_2$ generation, and (3) a ratio of

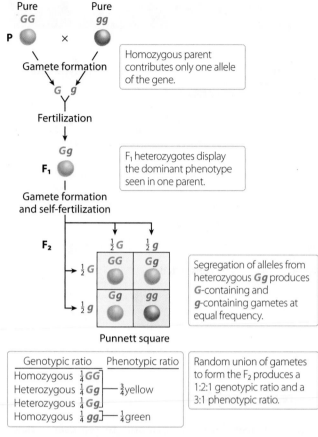

**Figure 2.6 Segregation of alleles for seed color.** In the cross between yellow-seeded and green-seeded pure-breeding parental plants, $F_1$ progeny display the dominant yellow phenotype. Note that the 3:1 phenotypic ratio and 1:2:1 genotypic ratio displayed in the $F_2$ generation result from crossing the $F_1$.

approximately 3:1 (dominant : recessive) among $F_2$ phenotypes. Mendel determined that yellow is dominant to green and round is dominant to wrinkled based on $F_1$ results. Green pea color and wrinkled pea shape reemerge in the $F_2$, which displays a consistent 3:1 ratio between the dominant and recessive phenotypes. For example, Mendel classified 8023 $F_2$ peas by their color and 6324 $F_2$ peas by their shape. Among the $F_2$ peas classified by color, he found 6022 yellow seeds and 2001 green seeds, a ratio of almost exactly three to one. Of the $F_2$ seeds classified for pea shape, 5474 were round and 1850 were wrinkled, again a ratio of very nearly three to one. Data for each of the other five characteristics revealed the same 3:1 ratio of dominant to recessive in the $F_2$.

## Evidence of Particulate Inheritance and Rejection of the Blending Theory

Mendel's $F_1$ experimental results reject the blending theory of heredity. Specifically, the observation that all $F_1$ progeny have the same phenotype (i.e., the dominant phenotype) that is indistinguishable from the phenotype of one of the

pure-breeding parents contradicts the blending theory prediction that the $F_1$ would display a mixture of the parental phenotypes. The persistence of the dominant phenotype and the reemergence of the recessive phenotype in the $F_2$ also run counter to the predictions of the blending theory.

Having rejected the blending theory, however, Mendel went on to propose a new hereditary hypothesis. Taking advantage of the analytical superiority of his quantitative approach to data analysis, Mendel proposed that each trait is determined by two "particles of heredity." Mendel used the German word *elementen*, a term meaning "unit or element," to describe the two discrete units of hereditary information for each trait. This idea is the basis of Mendel's theory of **particulate inheritance,** which proposes that each plant carries two particles of heredity for each trait. A plant receives one unit of heredity in the egg and the second unit in pollen. Each parental plant passes one of its two particles to offspring during reproduction.

The hereditary particles that are passed from one generation to the next are called *alleles* in modern terminology. This term had not been invented in Mendel's time (nor had the term *gene*, for that matter), but he correctly surmised that two *elemente* (alleles) were present for each trait in a plant and together determined the phenotype of the trait. Mendel used letters as symbols to represent the alleles for each trait, and he proposed a pattern of allele transmission from parents to offspring that explained his phenotypic observations in the $F_1$ and the $F_2$. Mendel proposed that pure-breeding lines contain two identical copies of the same allele.

Pure-breeding organisms have a **homozygous genotype,** a term meaning that the two alleles (i.e., the two copies of the gene) carried by an organism are identical. If a homozygous plant is self-fertilized or, if two organisms pure-breeding for the same trait are crossed, the progeny receive identical alleles from each parent and have the same homozygous genotype as the parents as well as the same homozygous genotype. In contrast, if a genetic cross is made between pure-breeding parents with different traits, each parent is homozygous for a different allele. The progeny receive a distinct allele from each parent and have a **heterozygous genotype,** a term meaning that two different alleles make up the genotype. Heterozygous organisms can have a dominant phenotype if they carry a copy of the dominant allele.

Geneticists now know that inheritance of the seven traits Mendel described is controlled by pairs of alleles of seven different genes. Thus, while Mendel did not use the words *gene* or *allele*, he understood the concept embodied by each term. Contemporary genetics describes inheritance of Mendel's traits in terms of genes and alleles, and continues to use letters to represent alleles. Different notational schemes and gene-naming conventions have been adopted for different species. (A table describing gene naming, gene nomenclature, and other information about the genes and genomes of model genetic organisms is located on the endsheets.)

| Table 2.1 | Mendel's Observations for Seven Monohybrid Traits in the $F_1$ and $F_2$ Generations | | | | |

| Crosses between Pure-Breeding Parental Phenotypes | $F_1$ Phenotype | $F_2$ Phenotypes | | $F_2$ Phenotype Ratio |
|---|---|---|---|---|
| | | Dominant | Recessive | |
| Round × wrinkled seeds[a] | All round seeds | 5474 round | 1850 wrinkled | 2.96:1 |
| Yellow × green seeds (interior seed color) | All yellow seeds | 6022 yellow | 2001 green | 3.01:1 |
| Purple × white flowers[b] (gray × white seed coat, or exterior seed color) | All purple flowers (gray seed coat) | 705 purple | 224 white | 3.15:1 |
| Axial × terminal flowers | All axial flowers | 651 axial | 207 terminal | 3.14:1 |
| Green × yellow pods | All green pods | 428 green | 152 yellow | 2.82:1 |
| Inflated × constricted pods | All inflated pods | 882 inflated | 299 constricted | 2.95:1 |
| Tall × short plants | All tall plants | 787 tall | 277 short | 2.84:1 |
| TOTAL | | 14,949 | 5010 | 2.98:1 |

[a] The dominant phenotype is written first and always appears as the $F_1$ phenotype.
[b] A single gene controls both flower color and seed-coat color. Mendel discussed both traits but recognized they were controlled by the same gene.

Central to understanding the inheritance of the seven traits Mendel studied is the concept that pure-breeding organisms have homozygous genotypes. In Figure 2.6, for example, the pure-breeding yellow parent has the *GG* homozygous genotype, and the pure-breeding green parent has the *gg* homozygous genotype. Crosses of pure-breeding parents of different homozygous genotypes produce heterozygous (*Gg*) $F_1$ progeny that all have the dominant yellow phenotype. According to Mendel's hypothesis, each pure-breeding parent passes one allele to the $F_1$, making it heterozygous. One allele, *G* in this case, is dominant and produces the dominant phenotype in all the $F_1$.

The heterozygous $F_1$ are then crossed with one another or are self-fertilized in a **monohybrid cross,** a term referring to a cross between two organisms that have the same heterozygous genotype for one gene. With a dominant and a recessive allele in the heterozygous genotype of plants undergoing a monohybrid cross, a **3:1 phenotypic ratio** is predicted for the $F_2$. At the same time, $F_2$ organisms are predicted to have three genotypes: The two homozygous genotypes (the same genotypes present in the original pure-breeding parents) are each expected to occur in one-fourth of the $F_2$ progeny, and the heterozygous genotype is predicted in the remaining one-half of the $F_2$ progeny. Therefore, among the $F_2$, a **1:2:1 genotypic ratio** is predicted. The one-fourth of the $F_2$ that are homozygous *GG* plus the one-half of $F_2$ progeny that are heterozygous *Gg* are the three-fourths of the $F_2$ with the dominant (yellow) phenotype. The remaining one-fourth of the $F_2$ contain the homozygous *gg* genotype and have the recessive (green) phenotype. The same inheritance pattern occurs for all the other traits studied by Mendel.

## Segregation of Alleles

Figure 2.6 uses letters as symbols to represent alleles and genotypes in parental, $F_1$, and $F_2$ organisms and introduces a simple and functional tool of genetic analysis—the Punnett square. The **Punnett square** method of diagramming the genetic content of gametes and their union to form offspring is named in honor of Sir Reginald Punnett, a famous geneticist of the early 20th century. The Punnett square separates the two alleles carried by each reproducing organism, placing those from one parent along the vertical margin of the square and those from the other parent along the horizontal margin. These separated alleles represent the **gametes** of reproducing organisms, the sperm and egg cells, each of which carries only one copy of each gene. The squares in the body of the Punnett diagram show the results expected from random uniting of the gametes, identifying the genotype of offspring produced by each possible combination of parental gametes. In Figure 2.6, the gametes of the $F_1$ parents are placed at the margins of the Punnett square, and gamete union produces the $F_2$ generation in the genotype proportions shown in the body of the Punnett square.

Mendel used the concept of particulate inheritance to analyze his experiments and to formulate a hypothesis to explain his results. Mendel's first hypothesis is known as the **law of segregation,** sometimes also known as **Mendel's first law.** This hypothesis describes the particulate nature of inheritance, identifies the segregation (separation) of alleles during gamete formation, and proposes the random union of gametes to produce progeny in predictable proportions:

**The law of segregation** *The two alleles for each trait will separate (segregate) from one another during gamete formation, and each allele will have an equal probability ($\frac{1}{2}$) of inclusion in a gamete. Random union of gametes at fertilization will unite one gamete from each parent to produce progeny in ratios that are determined by chance.*

The law of segregation applies to each of the seven traits Mendel examined, and each experiment produces similar results. We can take flower color as an example and use the law of segregation to explain the events shown in Figure 2.4, from the parental cross through the production of $F_2$ progeny. Gametes formed by pure-breeding purple (*PP*) parents all contain *P*. Similarly, gametes from pure-breeding white (*pp*) parents all contain *p*. The $F_1$ all have the dominant purple phenotype and have a heterozygous (*Pp*) genotype. Segregation of alleles is more easily visualized among gametes produced by the heterozygous $F_1$ plants: One-half of the gametes from those plants are expected to contain *P* and one-half to contain *p*. The random union of gametes from the heterozygous $F_1$ plants leads to the combinations and frequencies shown in the Punnett square of Figure 2.6, leading to the 1:2:1 genotypic ratio and the 3:1 phenotypic ratio.

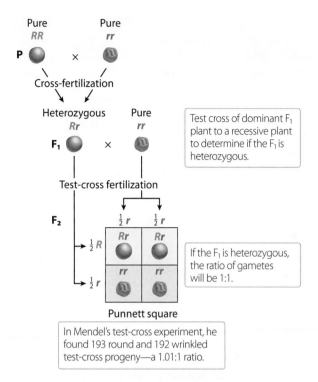

Figure 2.7 is performed between a plant grown from a round $F_1$ seed and a pure-breeding wrinkled-seed plant.

---

Genetic Insight   Mendel's law of segregation predicts a 3:1 phenotypic ratio among the $F_2$ progeny of $F_1$ heterozygous plants. Mendel recognized that a 1:2:1 genotypic ratio underlies the $F_2$ phenotypic ratio, correctly surmising that alleles segregate during gamete formation and unite at random to produce zygotes.

---

## Hypothesis Testing by Test-Cross Analysis

Mendel proposed the law of segregation to explain the phenotype proportions he observed in the $F_1$ and $F_2$ generations of his breeding experiments. Consistent with good scientific method, he considered the law of segregation to be a hypothesis that made testable predictions about cross progeny. Mendel's proposal that $F_1$ progeny are heterozygous is critical to the proposal that the gametes that produce the $F_2$ will have an equal chance of containing one or the other of the alleles. Based on his segregation hypothesis, Mendel expected one-half of the gametes derived from the heterozygous $F_1$ to carry the dominant allele and the remaining one-half to carry the recessive allele.

To test this prediction, Mendel performed test-cross analysis, by matting $F_{1s}$ suspected to be heterozygous with a pure-breeding recessive plant (**Figure 2.7**). Mendel's test crosses were identical in structure to those used countless times in genetic analysis since Mendel. Based on the segregation hypothesis, Mendel predicted that test-cross progeny phenotypes would be 50% dominant and 50% recessive. The test cross diagrammed in

In this test, the wrinkled-seed plant is homozygous *rr* and produces only *r*-containing gametes. Therefore, if the $F_1$ plant is heterozygous, the cross should produce *R* gametes and *r* gametes at a frequency of $\frac{1}{2}$ each. Consequently, the progeny of the cross would be $\frac{1}{2}$ *Rr* and $\frac{1}{2}$ *rr*, resulting in a 1:1 ratio of round : wrinkled. As the figure indicates, Mendel performed this cross and observed 193 round peas and 192 wrinkled peas in test-cross progeny. Mendel performed this kind of test-cross analysis for several of his traits and consistently observed a 1:1 ratio in test-cross progeny (**Table 2.2**).

Mendel's test-cross results validate two components of his segregation hypothesis. First, the results show that $F_1$ plants with the dominant phenotype have a heterozygous genotype. Second, the results validate the proposal that chance determines the frequency of gametes containing each allele. Had Mendel been incorrect about the heterozygous genotype of the $F_1$, or incorrect about the role of chance in producing the frequency of alleles in gametes, the result of the test cross would be different. If the round-seed plant were homozygous *RR* rather than *Rr*, all of the progeny of the cross would have the *Rr* genotype and would produce round peas. If the placement of alleles into gametes was not random, the phenotypes of test-cross progeny would not display a 1:1 ratio.

| Table 2.2 | Test-Cross Results from Mendel's Experiments | | |
|---|---|---|---|
| **Test Cross** | **Test-Cross Progeny** | | **Ratio** |
| | Dominant | Recessive | |
| Round seed (*Rr*) × wrinkled seed (*rr*) | 193 round (*Rr*) | 192 wrinkled (*rr*) | 1.01:1 |
| Yellow seed (*Gg*) × green seed (*gg*) | 196 yellow (*Gg*) | 189 green (*gg*) | 1.04:1 |
| Purple flower (*Pp*) × white flower (*pp*) | 85 purple (*Pp*) | 81 white (*pp*) | 1.05:1 |
| Tall plants (*Tt*) × short plants (*tt*) | 87 tall (*Tt*) | 79 short (*tt*) | 1.10:1 |
| TOTAL | 561 | 541 | 1.04:1 |

Genetic Insight  Test-cross analysis performed by Mendel showed that $F_1$ plants are heterozygous. This finding supports the hypothesis that alleles segregate when the $F_1$ generation form gametes.

## Hypothesis Testing by $F_2$ Self-Fertilization

A second pivotal component of Mendel's segregation hypothesis concerns the genotypes of $F_2$ progeny. Specifically, Mendel's hypothesis predicts that $F_2$ plants with the dominant phenotype can be either homozygous or heterozygous. His hypothesis further predicts that the plants are twice as likely to be heterozygous as homozygous. Look at Figure 2.6, for example, and notice that one-half of the $F_2$ progeny are heterozygous, whereas one-quarter of the $F_2$ progeny are homozygous for the dominant allele. Thus, among $F_2$ plants with the dominant phenotype (i.e., excluding $F_2$ plants with the recessive phenotype), two-thirds of the plants are heterozygous and one-third are homozygous for the dominant allele.

Mendel used a self-fertilization experiment to test the validity of his proposal that heterozygotes and homozygotes occur at a 2:1 ratio among dominant $F_2$ plants (**Figure 2.8**). He reasoned that self-fertilized $F_2$ plants could be identified as homozygous if they produced only progeny with the same phenotype. In contrast, self-fertilization of heterozygous $F_2$ plants would produce some progeny with the dominant phenotype and a smaller number with the recessive phenotype, in a 3:1 ratio.

Mendel tested his segregation hypothesis by self-fertilizing $F_2$ plants of the *dominant phenotype,* examining the progeny of each of these self-fertilizations to determine whether they exhibited the dominant phenotype

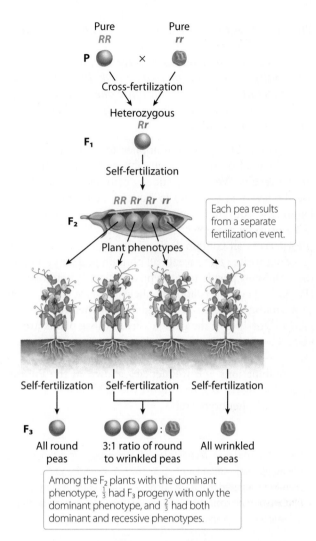

**Figure 2.8 Determination of the genotype of $F_2$ plants by the production of $F_3$ progeny.** $F_2$ plants are self-fertilized and their seeds are scored. Among the dominant (round) $F_2$, approximately one-third are expected to be homozygous for the dominant allele (*RR*). These plants produce progeny that have only round peas. The remaining two-thirds of the dominant $F_2$ are expected to be heterozygous, and produce both round and wrinkled peas in progeny. All $F_2$ wrinkled peas are homozygous recessive (*rr*) and produce only wrinkled peas as progeny.

only or both phenotypes. The results of his seven $F_2$ dominant self-fertilization experiments are shown in **Table 2.3**. Mendel's largest sample was for seed shape; he self-fertilized 565 round-seeded $F_2$ plants. In this experiment he found that 193 of the plants (34.2%) produced only round peas in progeny, demonstrating that these plants are homozygous for the dominant allele (*RR*). Self-fertilization of the other 372 round-pea–producing $F_2$ plants (65.8%) produced both round peas and wrinkled peas in progeny plants. The ratio 372:193 is very close to the 2:1 ratio of heterozygous to homozygous genotypes that Mendel predicted would constitute the dominant, round-pea–producing $F_2$ plants.

| Table 2.3 | Results of Mendel's Experiments to Identify $F_2$-Plant Genotypes by Their $F_3$ Progeny | | |
|---|---|---|---|
| Trait[a] | Heterozygous $F_2$ Plants[b] | Homozygous $F_2$ Plants[c] | Ratio[d] |
| Seed shape | 372 | 193 | 1.93:1 |
| Seed color | 353 | 166 | 2.13:1 |
| Flower color | 64 | 36 | 1.78:1 |
| Pod shape | 71 | 29 | 2.45:1 |
| Pod color | 125 | 75 | 1.67:1 |
| Flower position | 67 | 33 | 2.03:1 |
| Plant height | 72 | 28 | 2.57:1 |
| TOTAL | 1124 | 560 | 2.01:1 |

[a] Mendel self-fertilized only $F_2$ plants with the dominant phenotype in this experiment.
[b] $F_2$ plants were heterozygous if the $F_3$ progeny they produced by self-fertilization had both dominant and recessive phenotypes.
[c] $F_2$ plants were homozygous if the $F_3$ progeny they produced by self-fertilization had only the dominant phenotype.
[d] The expected ratio of heterozygous to homozygous $F_2$ plants was 2.00:1.

Mendel's self-fertilization results consistently show a 2:1 ratio among dominant $F_2$ plants for each of the seven traits examined. These results validate the proposal that gametes unite at random to produce progeny. Taken together, the test-cross experiments and the dominant $F_2$ self-fertilization experiments represent successfully designed and executed independent experiments for testing components of Mendel's segregation hypothesis. In these tests, Mendel made predictions about the experimental outcomes and then verified the results by counting the progeny produced. The resulting data supported his segregation hypothesis and illustrate how Mendel anticipated modern scientific methods, using approaches that would not be consistently applied to genetic experiments for several decades (**Genetic Analysis 2.1**).

---

Genetic Insight  Self-fertilization of $F_2$ plants having the dominant phenotype validates the proposal that their heterozygous-to-homozygous ratio is 2:1 and supports the suggestion that gametes unite at random to produce progeny.

---

## 2.3 Dihybrid and Trihybrid Crosses Reveal the Independent Assortment of Alleles

Each of the seven traits investigated by Mendel showed the same pattern of hereditary transmission that is explained by the law of segregation. The uniformity of phenotype proportions in $F_1$, $F_2$, test-cross, and self-fertilization progeny suggests that the same mechanism is responsible for allelic segregation in each one of the selected traits; but what about the inheritance of two or

more traits simultaneously? Is there a pattern or ratio of phenotypes that allowed Mendel to propose a transmission mechanism when two or more genes are examined at the same time?

### Dihybrid-Cross Analysis of Two Genes

To test the simultaneous transmission of two traits in the pea plant, Mendel performed a series of **dihybrid crosses,** crosses between organisms that differ for two traits. These tests followed an experimental strategy that paralleled his investigation of allelic segregation of single traits.

As **Figure 2.9** illustrates, Mendel began each dihybrid cross with pure-breeding lines. Having determined, for example, that round pea shape is dominant to wrinkled shape and that yellow pea color is dominant to green

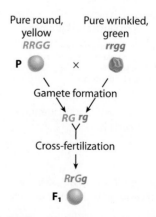

**Figure 2.9 Dihybrid-cross analysis.** Parental plants that are pure-breeding for two traits are cross-fertilized to produce $F_1$ progeny that are dihybrid and display the two dominant phenotypes round and yellow.

The presence of short hairs on the leaves of tomato plants is a dominant trait controlled by the allele $H$. The corresponding recessive trait, smooth leaf, is found in plants with the genotype $hh$. The table at right shows the progeny of three independent crosses of parental plants with genotypes and phenotypes that are unknown.

Examine the distributions of phenotypes in the progeny of each cross, and determine the parental genotypes for each cross. Use a Punnett square to diagram Cross 1.

| Cross | Number of Progeny | |
|---|---|---|
| | Hairy Leaf | Smooth Leaf |
| 1 | 32 | 11 |
| 2 | 42 | 45 |
| 3 | 0 | 24 |

| Solution Strategies | Solution Steps |
|---|---|

### Evaluate

1. Identify the topic this problem addresses and explain the nature of the requested answer.

2. Identify the critical information given in the problem.

> **TIP:** The numbers of progeny with each phenotype can be expressed as a ratio.

1. The problem presents the leaf-form phenotypes of progeny produced by three separate crosses of parental plants with unknown genotypes and phenotypes. The answer must identify parental genotypes and phenotypes for each cross and use a Punnett square to diagram Cross 1.

2. The information given for each cross is the number of progeny with hairy (dominant) and smooth (recessive) leaves. Interpretation of the phenotype ratio of progeny is required to determine parental genotypes and phenotypes.

> **PITFALL:** Genetics experiments produce finite numbers of progeny, so phenotypes may vary from expected ratios. Don't expect to see precise ratios in real data.

### Deduce

3. Examine the progeny of Cross 1, and determine the approximate ratio of progeny phenotypes.

4. Examine the progeny of Cross 2, and determine the approximate ratio of progeny phenotypes.

5. Examine the progeny of Cross 3, and determine the approximate ratio of progeny phenotypes.

3. Ratio of phenotypes in Cross 1 progeny:

$$\frac{32}{11} = 2.91:1.$$

This is an approximate 3:1 ratio. The recessive phenotype appears in about $\frac{1}{4}$ of the progeny $\left(\frac{11}{43}\right)$, and the remaining $\frac{3}{4}$ $\left(\frac{32}{43}\right)$ have the dominant phenotype.

4. Ratio of phenotypes for Cross 2:

$$\frac{42}{45} = 0.93:1$$

This is an approximate 1:1 ratio in which the dominant phenotype is seen in about one-half of the progeny $\left(\frac{42}{97}\right)$ and the recessive phenotype is seen in the other half of the progeny $\left(\frac{45}{97}\right)$.

5. Cross 3 produced only the recessive phenotype, so the ratio is 0:1.

### Solve

6. Based on the results of Cross 1, identify the genotypes and phenotypes of the parental plants in the cross. Construct a Punnett square to illustrate this cross.

> **TIP:** There are two alleles for this gene, and three genotypes are possible. The recessive phenotype is found in plants with the $hh$ genotype, whereas the dominant phenotype will be found in plants that are $Hh$ and $HH$.

6. The recessive progeny in this cross have the genotype $hh$, so each parent in Cross 1 must carry a copy of $h$. The dominant progeny are either $HH$ or $Hh$. The 3:1 progeny phenotype ratio is consistent with a parental cross $Hh \times Hh$.

The Punnett square for this cross is consistent with the observed 3:1 ratio:

| | $H$ | $h$ |
|---|---|---|
| $H$ | $HH$ | $Hh$ |
| $h$ | $Hh$ | $hh$ |

7. Based on the results of Cross 2, identify the genotypes and phenotypes of the parents.

7. Both parental plants in Cross 2 carry at least one copy of $h$. The 1:1 progeny ratio is consistent with the ratio expected for a test cross of a heterozygous organism to one that is homozygous recessive. This cross is $Hh \times hh$.

8. Based on the results of Cross 3, identify the parental genotypes and phenotypes.

8. Cross 3 produces only $hh$ progeny. This is expected for a pure-breeding cross between two homozygous organisms. This cross is $hh \times hh$.

For more practice, see Problems 10, 14, and 29.

color, Mendel proposed that pure-breeding plants producing round, yellow peas have the genotype *RRGG* and that pure-breeding plants for the recessive phenotypes wrinkled and green have the genotype *rrgg*. Gametes produced by the round, yellow plant contain one allele for each type of gene and are *RG*. In contrast, gametes from the wrinkled, green plant are *rg*. Mendel's model predicts that all of the $F_1$ progeny will therefore have the dihybrid genotype *RrGg*. These $F_1$ are heterozygous for two traits and display the dominant parental phenotypes round and yellow.

Heterozygous $F_1$ dihybrids (*RrGg*) have received alleles *R* and *G* from the round, yellow pure-breeding parent and alleles *r* and *g* from the pure-breeding wrinkled, green parent. If the assortment of alleles for each type of gene is independent, gametes produced by these $F_1$ plants are equally likely to contain *any* combination of one allele for seed shape and one allele for seed color. Probabilities of each combination of alleles for each type of gene are predicted by recognizing that four combinations of alleles will be found in the gametes—*RG*, *Rg*, *rG*, and *rg*—and that each combination is expected to occur with a frequency of $\frac{1}{4}$.

**Figure 2.10** shows a diagrammatic aid called the **forked-line diagram** that is used to determine gamete genotypes and frequencies. The forked-line diagram illustrates that one-half of all gametes produced by an *RrGg* plant will contain *R* and one-half will contain *r*. If the segregation of *G* and *g* is independent of the *R* and *r* alleles, then one-half of the gametes containing *R* will also carry *G* and the other half will carry *g*. The same is true for *r*-bearing gametes; one-half will carry *G* and the remaining half will carry *g*. The frequency of each of the four gamete genotypes is $(\frac{1}{2})(\frac{1}{2}) = \frac{1}{4}$.

A Punnett square can be used to illustrate the random union of these four different gametes to produce $F_2$ progeny (**Figure 2.11**). Each gamete has a predicted frequency of $\frac{1}{4}$, and each cell of the Punnett square has a predicted frequency of $(\frac{1}{4})(\frac{1}{4}) = \frac{1}{16}$. Among $F_2$ progeny, four phenotypes are observed, displaying either (1) both dominant phenotypes, (2) the dominant phenotype for one trait and the recessive phenotype for the other (there are

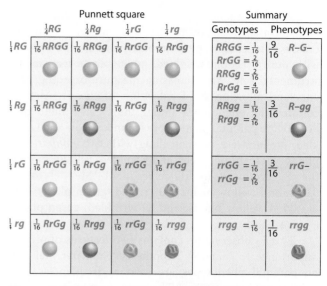

**Figure 2.11  Independent assortment of alleles at two loci.** Self-fertilization or crossing of dihybrid $F_1$ (*RrGg*) to one another produces nine genotypes distributed in a 9:3:3:1 phenotypic ratio among $F_2$ progeny.

two versions of this), or (3) both recessive phenotypes. The $F_2$ phenotypes appear in the ratio $\frac{9}{16}:\frac{3}{16}:\frac{3}{16}:\frac{1}{16}$.

By examining the $F_2$ phenotype proportions, we can see the relationship between the 3:1 ratio for each trait and the 9:3:3:1 ratio when the two traits are considered simultaneously. When pea shape and pea color are considered individually, monohybrid crosses produce $F_2$ that are $\frac{3}{4}$ dominant and $\frac{1}{4}$ recessive. The cross of two dihybrids also yields proportions of $\frac{3}{4}$ dominant to $\frac{1}{4}$ recessive for each trait, making the prediction of phenotypic ratios among the $F_2$ for both traits combined a problem of combinatorial arithmetic. Figure 3.11 reminds us that genotypes falling into the *R*– and the *G*– classes each occur in $\frac{3}{4}$ of the progeny, while *rr* and *gg* 0genotype classes each occur in $\frac{1}{4}$ of the progeny. The dash in the genotypes *R*– and *G*– is a "blank" that could be filled by either a second copy of the dominant allele or a copy of the recessive allele. In either case, the resulting genotype—for example, *RR* or *Rr*—produces the dominant phenotype. The co-occurrence of the two dominant phenotypes (round, yellow) is therefore expected to have a frequency of $(\frac{3}{4})(\frac{3}{4}) = \frac{9}{16}$, the two recessive phenotypes (wrinkled, green) will occur with a frequency of $(\frac{1}{4})(\frac{1}{4}) = \frac{1}{16}$, and the two phenotypic classes that display one dominant and one recessive trait (round, green and wrinkled, yellow) will each be found in a frequency of $(\frac{3}{4})(\frac{1}{4}) = \frac{3}{16}$.

This outcome illustrates **Mendel's law of independent assortment,** also known as **Mendel's second law.**

**The law of independent assortment**  *During gamete formation, the segregation of alleles at one locus is independent of the segregation of alleles at another locus.*

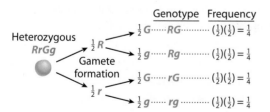

**Figure 2.10  The forked-line method for determining gamete genotype frequency.**  Chance is responsible for the independent assortment of alleles included in four genetically different gametes.

Mendel reached his conclusions regarding independent assortment on the basis of numerous dihybrid crosses. The cross of pure-breeding round, yellow plants with pure-breeding wrinkled, green plants was an instrumental one. After crossing the pure-breeding parents and allowing self-fertilization of the $F_1$, Mendel counted the phenotypes among the $F_2$ and found that both of the parental phenotypes (round, yellow and wrinkled, green) were present along with two *nonparental* phenotypes: round, green and wrinkled, yellow. Among the $F_2$ produced in his experiment, Mendel found 315 round, yellow plants; 108 round, green plants; 101 wrinkled, yellow plants; and 32 wrinkled, green plants (**Figure 2.12**).

This $F_2$ observation contains two features of pivotal importance to Mendel's hypothesis. First, parental and nonparental phenotypes are seen at frequencies that differ from one another. The most numerous class of $F_2$ progeny display the dominant parental phenotypes for each trait, round and yellow. The smallest class of $F_2$ progeny have the two recessive parental phenotypes, wrinkled and green; and the two nonparental $F_2$ classes (round, green and wrinkled, yellow) are intermediate and approximately equal in number. From these numbers, Mendel recognized that the ratios between the dominant

and recessive forms of each trait followed the familiar 3:1 pattern. In looking at pea shape, for example, Mendel found that $423\,(315 + 108)$ plants were round and that $133\,(101 + 32)$ plants were wrinkled. The ratio 423:133 reduces to 3.18:1. Similarly, for pea color he found a ratio of $416\,(315 + 101)$ yellow to $140\,(108 + 32)$ green—a ratio of 2.97:1. Considering each trait individually, the cross of heterozygous $F_1$ plants has produced an $F_2$ generation in which $\frac{3}{4}$ of the progeny have the dominant phenotype and $\frac{1}{4}$ have the recessive phenotype.

Second, Mendel predicted that if alleles at each locus unite at random to produce the $F_2$, then the expected $F_2$-plant phenotypes will occur in predictable frequencies. He hypothesized that $F_2$ progeny displaying the two dominant traits (round and yellow) will occur at a frequency of $\left(\frac{3}{4}\right)\left(\frac{3}{4}\right) = \frac{9}{16}$. Similarly, progeny carrying the two recessive traits (wrinkled and green) are expected at a frequency of $\left(\frac{1}{4}\right)\left(\frac{1}{4}\right) = \frac{1}{16}$, and each of the nonparental phenotypes are expected at a frequency of $\left(\frac{3}{4}\right)\left(\frac{1}{4}\right) = \frac{3}{16}$. Independent assortment of alleles at the two genes therefore leads to an expected distribution among the $F_2$ of

| round, yellow | $R-G-$ | $\frac{9}{16}$ |
|---|---|---|
| round, green | $R-gg$ | $\frac{3}{16}$ |
| wrinkled, yellow | $rrG-$ | $\frac{3}{16}$ |
| wrinkled, green | $rrgg$ | $\frac{1}{16}$ |

Mendel's count of 315 round, yellow; 108 round, green; 101 wrinkled, yellow; and 32 wrinkled, green (see Figure 2.12) can be converted to a ratio by dividing each number by 32, the value of the smallest class. The division by 32 reduces Mendel's observed ratio to $9.84 : 3.38 : 3.16 : 1$, which is a close fit to the 9:3:3:1 ratio predicted by his model. From this result, Mendel hypothesized that independent assortment in a dihybrid organism produces four different gamete genotypes at equal frequencies. Random union of the gametes then produces four phenotypic classes as a result of dominance relationships at each locus, and the ratio of these $F_2$ phenotypic classes is expected to be 9:3:3:1 (**Genetic Analysis 2.2**).

---

**Genetic Insight** A ratio of 9:3:3:1 is expected among the $F_2$ progeny of a dihybrid cross as a result of the independent assortment of alleles at two loci.

---

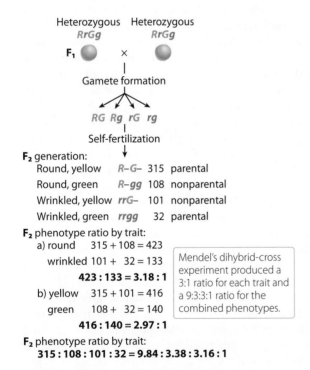

**F₂ phenotype ratio by trait:**
    $315 : 108 : 101 : 32 = 9.84 : 3.38 : 3.16 : 1$

**Figure 2.12 Phenotype proportions in the progeny of a dihybrid cross performed by Mendel.** For each trait considered individually, progeny display an approximate 3:1 ratio of the dominant to the recessive phenotype. When the two traits are considered simultaneously, a phenotypic ratio of 9:3:3:1 is expected.

## Testing Independent Assortment by Test-Cross Analysis

To test his hypothesis that combinations of pea shape and color are determined by the independent assortment of alleles, Mendel once again turned to test-cross analysis. Having proposed that the $F_1$ plants with round, yellow

In a mammalian species, long fur and the appearance of white spots are produced by dominant alleles *F* and *S*, respectively, which are found at two independently assorting loci. At these loci, the genotype *ff* produces short fur, and the genotype *ss* produces solid fur color. Given the parental genotypes for each of the following crosses, determine the expected proportions of all progeny phenotypes.

|  | Male |  | Female |
|---|---|---|---|
| Cross 1: | *FF Ss* | × | *Ff ss* |
| Cross 2: | *ff Ss* | × | *Ff Ss* |
| Cross 3: | *Ff Ss* | × | *Ff Ss* |

| Solution Strategies | Solution Steps |
|---|---|

### Evaluate

1. Identify the topic this problem addresses and explain the nature of the requested answer.

2. Identify the critical information given in the problem.

1. This is a transmission genetic problem in which parental genotypes are given. Answers must predict the phenotypes of progeny and their expected proportions. These are predicted by determining the parental gametes and their proportions.

2. Genotypes of parents are given for each cross. The genotypes are used to predict the genotypes of parental gametes and the gamete proportions.

### Deduce

3. For Cross 1, identify the genetically different gametes that can be produced by each parent and calculate the predicted proportion of each gamete.

**TIP:** A forked-line diagram is a useful tool for predicting the alleles in gametes and gamete frequencies.

3. Each of the parents can produce two genetically different gametes at predicted frequencies of $\frac{1}{2}$ each.

**Cross 1**

Male

$1F \Big\langle \begin{array}{l} \frac{1}{2}S \cdots FS \cdots (1)(\frac{1}{2}) = \frac{1}{2} \\ \frac{1}{2}s \cdots Fs \cdots (1)(\frac{1}{2}) = \frac{1}{2} \end{array}$

Female

$\frac{1}{2}F \longrightarrow 1s \cdots Fs \cdots (\frac{1}{2})(1) = \frac{1}{2}$
$\frac{1}{2}f \longrightarrow 1s \cdots fs \cdots (\frac{1}{2})(1) = \frac{1}{2}$

4. Identify the content and frequency of the genetically different gametes produced by the parents in Cross 2.

4. The male produces two types of gametes at a predicted frequency of $\frac{1}{2}$ each. The female produces four genetically different gametes at frequencies of $\frac{1}{4}$ each.

**Cross 2**

Male

$1f \Big\langle \begin{array}{l} \frac{1}{2}S \cdots fS \cdots (1)(\frac{1}{2}) = \frac{1}{2} \\ \frac{1}{2}s \cdots fs \cdots (1)(\frac{1}{2}) = \frac{1}{2} \end{array}$

Female

$\frac{1}{2}F \Big\langle \begin{array}{l} \frac{1}{2}S \cdots FS \cdots (\frac{1}{2})(\frac{1}{2}) = \frac{1}{4} \\ \frac{1}{2}s \cdots Fs \cdots (\frac{1}{2})(\frac{1}{2}) = \frac{1}{4} \end{array}$

$\frac{1}{2}f \Big\langle \begin{array}{l} \frac{1}{2}S \cdots fS \cdots (\frac{1}{2})(\frac{1}{2}) = \frac{1}{4} \\ \frac{1}{2}s \cdots fs \cdots (\frac{1}{2})(\frac{1}{2}) = \frac{1}{4} \end{array}$

5. Predict the gamete content and frequencies for the parents in Cross 3.

5. Both parents are dihybrids that produce four genetically different gametes at frequencies of $\frac{1}{4}$ each.

**Cross 3**

Male

$\frac{1}{2}F \Big\langle \begin{array}{l} \frac{1}{2}S \cdots FS \cdots (\frac{1}{2})(\frac{1}{2}) = \frac{1}{4} \\ \frac{1}{2}s \cdots Fs \cdots (\frac{1}{2})(\frac{1}{2}) = \frac{1}{4} \end{array}$

$\frac{1}{2}f \Big\langle \begin{array}{l} \frac{1}{2}S \cdots fS \cdots (\frac{1}{2})(\frac{1}{2}) = \frac{1}{4} \\ \frac{1}{2}s \cdots fs \cdots (\frac{1}{2})(\frac{1}{2}) = \frac{1}{4} \end{array}$

Female

$\frac{1}{2}F \Big\langle \begin{array}{l} \frac{1}{2}S \cdots FS \cdots (\frac{1}{2})(\frac{1}{2}) = \frac{1}{4} \\ \frac{1}{2}s \cdots Fs \cdots (\frac{1}{2})(\frac{1}{2}) = \frac{1}{4} \end{array}$

$\frac{1}{2}f \Big\langle \begin{array}{l} \frac{1}{2}S \cdots fS \cdots (\frac{1}{2})(\frac{1}{2}) = \frac{1}{4} \\ \frac{1}{2}s \cdots fs \cdots (\frac{1}{2})(\frac{1}{2}) = \frac{1}{4} \end{array}$

### Solve

6. Construct a Punnett square for Cross 1 and predict the progeny phenotypes and proportions.

6. The predicted Cross 1 progeny are $\frac{1}{2}$ long, spotted and $\frac{1}{2}$ long, solid.

| ♀ \ ♂ | FS | Fs |
|---|---|---|
| **Fs** | FFSs | FFss |
| **fs** | FfSs | Ffss |

7. Construct a Punnett square for Cross 2 and predict the progeny phenotypes and proportions.

7. The progeny predicted from Cross 2 are $\frac{3}{8}$ long, spotted; $\frac{1}{8}$ long, solid; $\frac{3}{8}$ short, spotted; and $\frac{1}{8}$ short, solid.

| ♀ \ ♂ | fS | fs |
|---|---|---|
| **FS** | FfSS | FfSs |
| **Fs** | FfSs | Ffss |
| **fS** | ffSS | ffSs |
| **fs** | ffSs | ffss |

8. Construct a Punnett square for Cross 3 and predict the progeny phenotypes and proportions.

8. The progeny produced by Cross 3 are predicted to be $\frac{9}{16}$ long, spotted; $\frac{3}{16}$ long, solid; $\frac{3}{16}$ short, spotted; and $\frac{1}{16}$ short, solid.

| ♀ \ ♂ | FS | Fs | fS | fs |
|---|---|---|---|---|
| **FS** | FFSS | FFSs | FfSS | FfSs |
| **Fs** | FFSs | FFss | FfSs | Ffss |
| **fS** | FfSS | FfSs | ffSS | ffSs |
| **fs** | FfSs | Ffss | ffSs | ffss |

For more practice, see Problems 6, 12, and 27.

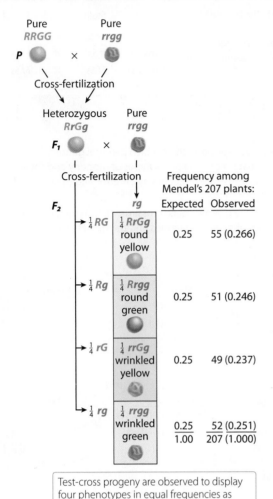

| | | Frequency among Mendel's 207 plants: | |
|---|---|---|---|
| | | Expected | Observed |
| $\frac{1}{4}$ RG | $\frac{1}{4}$ RrGg round yellow | 0.25 | 55 (0.266) |
| $\frac{1}{4}$ Rg | $\frac{1}{4}$ Rrgg round green | 0.25 | 51 (0.246) |
| $\frac{1}{4}$ rG | $\frac{1}{4}$ rrGg wrinkled yellow | 0.25 | 49 (0.237) |
| $\frac{1}{4}$ rg | $\frac{1}{4}$ rrgg wrinkled green | 0.25 / 1.00 | 52 (0.251) / 207 (1.000) |

Test-cross progeny are observed to display four phenotypes in equal frequencies as expected by application of Mendel's laws.

**Figure 2.13 Mendel's test cross to verify independent assortment.** Mendel predicted and observed an approximate 1:1:1:1 ratio among progeny, supporting his hypothesis of independent assortment.

seeds were dihybrid and had the genotype *RrGg*, he predicted that the test cross of a dihybrid (*RrGg*) to a pure-breeding wrinkled, green plant (*rrgg*) would produce four offspring phenotypes at a frequency of $\frac{1}{4}$ each. **Figure 2.13** shows that the dihybrid F$_1$ plant was expected to produce four different gamete genotypes. Recalling the logic of the forked-line diagram, remember that one-half of the gametes are expected to contain *R* and one-half to contain *r*. Gametes carry *G* and *g* independently of *R* or *r*, meaning that four different combinations of these alleles are possible in gametes: *RG*, *rG*, *Rg*, and *rg*, each occurring at an expected frequency of $(\frac{1}{2})(\frac{1}{2}) = \frac{1}{4}$. In contrast, the homozygous recessive green, wrinkled (*rrgg*) plant can produce only an *rg* gamete. In the figure, we see that the test-cross progeny are expected to have four genotypes, each corresponding to a different phenotype. The predicted progeny are expected to be $\frac{1}{4}$ *RrGg* (round, yellow), $\frac{1}{4}$ *Rrgg* (round, green), $\frac{1}{4}$ *rrGg* (wrinkled, yellow), and $\frac{1}{4}$ *rrgg* (wrinkled, green).

Mendel performed this cross, and his results almost exactly matched expectation. He found that the 207 test-cross progeny were composed of 55 round, yellow; 51 round, green; 49 wrinkled, yellow; and 52 wrinkled, green plants. This result confirmed the dihybrid genotype of the F$_1$ plant and supported the hypothesis that alleles for pea shape assort independently of those for pea color during gamete formation and that gametes unite at random to form offspring.

**Genetic Insight** A 1:1:1:1 ratio is the expected result of a test cross between a dihybrid organism and one that is homozygous recessive for alleles at two independent loci.

## Testing Independent Assortment by Trihybrid-Cross Analysis

Mendel further tested the hypothesis of independent assortment by examining the results of a **trihybrid cross**, a cross involving three traits—in this case, seed shape, seed color, and flower color. He began this experiment by crossing a pure-breeding round, yellow, purple-flowered parental plant (*RRGGWW*) to a pure-breeding wrinkled, green, white-flowered plant (*rrggww*). **Figure 2.14** illustrates the cross of pure-breeding parental strains and the resulting F$_1$ progeny, which display the dominant phenotypes round, yellow, and purple. The F$_1$ are presumed to be trihybrid (*RrGgWw*). The presumptive trihybrid F$_1$ plants were then crossed to produce F$_2$ plants, and the results were compared to expectations.

The forked-line diagram in Figure 2.14 shows the number and expected frequency of gamete genotypes. In the general case, the number of different gamete genotypes is expressed as $2^n$, where *n* = the number of genes involved. In this example, there are three genes (*n* = 3), and $2^3 = 8$ different combinations of alleles possible for the three traits in gametes from the trihybrid plant. The frequency of each gamete genotype is determined as $(\frac{1}{2})^n$, or $(\frac{1}{2})^3 = \frac{1}{8}$. To predict the number of genetically different gametes and their frequencies, the exponent 3 is used because there are three genes being examined in the experiment. In arithmetic computations like these, the exponent value usually indicates the number of genes.

Figure 2.14 illustrates a way of using the forked-line method to predict the expected frequency of the eight phenotypic classes of this trihybrid cross. For the general case where there are two phenotypes (dominant and recessive) for each trait, there are $2^n$ phenotypes in the F$_2$. Once again, *n* = the number of genes. In this example, there are $2^3 = 8$ phenotypes in the F$_2$ progeny. Computation of each expected phenotype frequency is based on the expected frequencies of $\frac{3}{4}$ dominant and $\frac{1}{4}$ recessive for each trait. The expected frequency of each trihybrid class is the product of three fractions representing the predicted probabilities of the dominant or recessive form for each trait. For the eight F$_2$ phenotypes

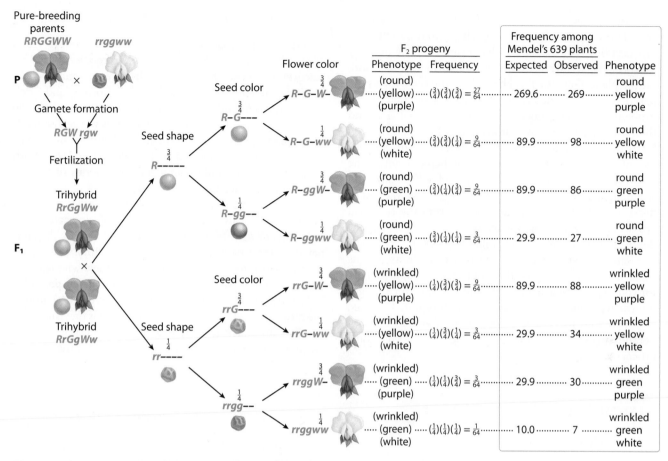

**Figure 2.14 Trihybrid cross to verify independent assortment.** The forked-line method can be used to determine the expected phenotype frequencies produced by a trihybrid cross. Expected and observed results for the F$_2$ generation of Mendel's trihybrid-cross experiment supported his hypothesis of independent assortment.

from a trihybrid cross, the expected phenotype ratio is $\frac{27}{64} : \frac{9}{64} : \frac{9}{64} : \frac{3}{64} : \frac{9}{64} : \frac{3}{64} : \frac{3}{64} : \frac{1}{64}$.

Mendel used this combinatorial thinking to predict the outcome of an experimental trihybrid cross. His experimental results for this test are given in Figure 2.14 for 639 F$_2$ progeny from the cross of round, yellow, purple-flowered F$_1$ plants. Mendel predicted the number of progeny expected in each phenotype class by multiplying the expected proportion times the sample size, 639. His results were remarkably close to expectation. The close match of these observed and expected values provides a second piece of independent evidence supporting the law of independent assortment.

Taken together, Mendel's analyses of the transmission of single traits and the joint transmission of two or three independent traits represented a major advance in the scientific understanding of hereditary transmission. The law of segregation and the law of independent assortment are the most fundamental principles of genetic transmission in diploid organisms, and they form the foundation of our understanding of transmission, molecular, and population genetics. We will see the results of these two laws again and again as we explore the details of hereditary transmission in diploid organisms—organisms that, like humans and pea plants, have two copies of each chromosome.

## Probability Calculations in Genetics Problem Solving

The predicted F$_2$-phenotype ratio from a trihybrid cross seems complicated, and at first you might not see clearly why that is the expected distribution. The key to understanding the calculation demonstrated in Figure 2.14 is to realize that each independently assorting locus truly can be treated independently of others.

Let's look at the progeny-phenotype distribution for a dihybrid cross. We expect that for each trait individually, $\frac{3}{4}$ of the progeny will display the dominant phenotype and $\frac{1}{4}$ the recessive phenotype. We could use a Punnett square to determine the phenotypic distribution of the two traits in combination as we did in Figure 2.11, but the independence of each gene gives us a quicker way to calculate the distribution of phenotypes: by their probability. In this case, the expected progeny phenotype proportions can be obtained by multiplying the two ratios—$\left(\frac{3}{4} : \frac{1}{4}\right)\left(\frac{3}{4} : \frac{1}{4}\right)$—to yield the expected ratio of $\frac{9}{16} : \frac{3}{16} : \frac{3}{16} : \frac{1}{16}$, or 9:3:3:1. We can use the same approach to predict the ratio among F$_2$ progeny of a trihybrid cross or any other kind of cross involving independently assorting genes.

Another advantage to using probability for solving genetic problems is its easy adaptability to different sorts of questions. For example, what proportion of progeny produced by self-fertilization of a trihybrid yellow, round, purple plant ($GgRrWw$) will have the same *genotype* as the parental plant? To determine the answer, we identify the probability of the genotype for each individual trait and then multiply those three probabilities together. At each locus the cross is heterozygous by heterozygous, so one-half of the progeny are expected to be heterozygous. The probability that offspring of a trihybrid self-fertilization will be trihybrid is therefore $(\frac{1}{2})(\frac{1}{2})(\frac{1}{2}) = \frac{1}{8}$. If we wanted to determine the proportion of progeny from the trihybrid cross that are $rrGGWw$, we again treat the loci independently and calculate the probability as $(\frac{1}{4})(\frac{1}{4})(\frac{1}{2}) = \frac{1}{32}$.

The problems at the end of this chapter, as well as Genetic Analysis 2.3, provide a number of opportunities for you to practice using the principles of transmission genetics. As **Experimental Insight 2.1** points out, however, opportunities to collect evidence of Mendel's laws of heredity may be as close as the produce aisle of your local grocery store ).

## The Rediscovery of Mendel's Work

In 1900, after remaining virtually unknown for 34 years, Mendel's experimental results and interpretations were rediscovered almost simultaneously by three botanists working independently of one another. Carl Correns and Erich von Tschermak both worked on *Pisum sativum*, the same plant Mendel had used, and Hugo de Vries worked on a different plant species. Each of the three identified the hereditary principles Mendel had first described in 1865. With support from the contemporaneous discoveries of the behavior of chromosomes during meiotic cell division, followed quickly by confirming evidence from other species of plants and animals, the basic principles of segregation and independent assortment were widely and rapidly disseminated in the first decade of the 20th century. Data from Mendel, Correns, Von Tschermak, and others who later investigated the $F_2$ segregation of yellow and green pea color in *Pisum* have been combined to give a ratio of 3.01:1 among the more than 200,000 yellow and green seeds they analyzed.

The approach to genetic analysis we describe in this chapter is often dubbed Mendelian genetics for the obvious

---

## Experimental Insight 2.1

### Mendelism in the Produce Aisle

Many of the appealing characteristics of fruits and vegetables available in grocery stores and at farmer's markets are the result of intensive selective breeding. For example, in recent years many new vegetable varieties have been introduced into the marketplace. Among these is a variety of corn that goes by several names, including "bicolor," "peaches and cream," and "yellow and white." Most of the kernels on a cob of bicolored corn are yellow, but a sizable number are white. With close inspection and a little quantitative analysis, you should be able to identify the genetic mechanism that produces this variation in color.

An ear of corn is a mini-genetic experiment: Each kernel on the ear, like each pea in a pod, is a separate seed, produced by a fertilization event independent of the events that produced

adjacent kernels. This means that each mature ear of corn carries hundreds of progeny for analysis.

Bicolor corn originates with the cross of two pure-breeding corn lines, one producing yellow kernels and the other producing white kernels. The yellow plant is $WW$, and the white plant is $ww$. When seed company geneticists cross these parental stocks, the kernels on the $F_1$ plants are yellow and have the heterozygous $Ww$ genotype. This $F_1$ seed is allowed to mature and is packaged for sale to farmers and home gardeners, who plant it to produce a crop. The seed is commonly labeled "hybrid," meaning "monohybrid," to reflect the heterozygosity at the kernel-color locus. Owing to segregation of alleles at the kernel-color locus, the plants that grow from this $F_1$ seed produce both yellow ($W-$) and white ($ww$) kernels on each ear.

If you saw some of this corn in your grocery store, how would you verify that the genetic basis of its yellow and white kernels is the segregation of two alleles at a single locus? The answer is that you would count the number of yellow kernels and the number of white kernels on ears of bicolor corn with the expectation of a ratio of approximately 3:1 between the yellow and white kernels.

A recent genetics class examined several dozen ears of bicolor corn and counted 7506 yellow kernels and 2376 white kernels. Among the total of 9882 kernels there are 75.96% yellow and 24.04% white, a ratio of 3.16:1. You will use these data in Problem 20 at the end of the chapter to do a statistical test to see if the observed data fit the hypothesis that this trait is the product of the segregation of alleles at a single locus.

The next time you shop for fruits and vegetables, keep in mind that you are looking at Mendelian genetics in action!

For the same mammalian species and the same traits described in Genetic Analysis 2.2 a cross between a male that has long, solid-colored fur and a female that has short, spotted fur produces eight offspring. The offspring are 2 long, spotted; 2 short, solid; 2 long, solid; and 2 short, spotted. Given the phenotypes of the parents and the distribution of offspring phenotypes, determine the genotypes of parents and offspring.

| Solution Strategies | Solution Steps |
|---|---|

### Evaluate

1. Identify the topic this problem addresses and explain the nature of the requested answer.

2. Identify the critical information given in the problem.

1. The problem requires the determination of parental genotypes and progeny genotypes based on the phenotypes of parents and the proportions of progeny with different phenotypes.

2. In this mammalian species, long fur is dominant to short fur and spotted fur color is dominant to solid fur color. Each parent is homozygous recessive for one trait and is dominant for the other trait. The progeny display a 1:1:1:1 ratio of phenotypes.

> **TIP:** Use the known and placeholder genotypes for parents and progeny phenotype ratios to completely identify parental genotypes.

### Deduce

3. Record what is known about the <u>parental genotypes</u> by writing homozygous recessive alleles for the recessive trait and writing a dominant allele and a "blank" as a placeholder for the <u>dominant trait</u>.

3. The long, solid parent is *F–ss*, carrying at least one dominant (*F–*) allele for long fur and homozygous recessive alleles (*ss*) for solid coat. The short, spotted parent is *ffS–*, carrying homozygous recessive alleles (*ff*) for fur length and at least one dominant allele (*S–*) for spotted coat.

> **PITFALL:** You cannot assume to know the genotype of an organism with the dominant phenotype without segregation information. Use general genotype forms *F–* and *S–* as placeholders for the homozygous dominant or heterozygous genotypes.

4. Infer what is known about progeny genotypes by writing homozygous recessive alleles or dominant alleles with a "blank" placeholder.

4. The inferred progeny genotypes are

    *F–S–* long, spotted
    *ffss* short, solid
    *F–ss* long, solid
    *ffS–* short, spotted

5. Determine the <u>phenotype ratio</u> of long fur to short fur among the progeny of the cross.

5. Four long fur and four short, a 1:1 ratio of dominant and recessive phenotypes.

> **TIP:** Traits assorting independently can be analyzed individually. Assess segregation based on progeny phenotype ratios for one trait at a time.

6. Determine the phenotype ratio of spotted fur to solid fur among the progeny.

6. Four progeny have spotted fur and four have solid fur, a 1:1 ratio of phenotypes.

### Solve

7. Determine the parental genotypes necessary to produce progeny with the observed ratio of long to short fur.

7. To produce the recessive short fur phenotype, each parent must contribute a recessive (*f*) allele. The female parent with short fur is *ff*, and the male parent with long fur must be heterozygous (*Ff*) for this gene. The genotype of the male parent with long, solid-colored fur is *Ffss*.

8. Determine the parental genotypes necessary to produce the observed ratio of spotted to solid coat.

8. The male parent with the recessive phenotype solid coat contributes a recessive (*s*) allele. The female parent with spotted coat must be heterozygous (*Ss*). The short, spotted female has the genotype *ffSs*.

9. Verify the parental genotypes in this cross by using a Punnett square analysis.

> **PITFALL:** To avoid errors, use a Punnett square to verify that the parental genotypes you assign will produce progeny in the observed ratio.

9. For the cross *Ffss* × *ffSs*, each parent produces two genetically different gametes at frequencies of $\frac{1}{2}$ each. The Punnett square predicts four different progeny genotypes and phenotypes in a 1:1:1:1 ratio.

| ♀ \ ♂ | *Fs* | *fs* |
|---|---|---|
| *fS* | *FfSs* Long, spotted | *ffSs* Short, spotted |
| *fs* | *Ffss* Long, solid | *ffss* Short, solid |

For more practice, see Problems 3, 16, and 40.

## Experimental Insight 2.2

### Naudinian Genetics, Anyone?

Before Mendel, many "plant hybridists" had experimented with pea plants and other plants, attempting to discern the mechanisms of plant reproduction and the process of hereditary transmission of traits. Mendel cited the work of several early hybridists in his 1866 paper.

Several of these plant hybridists came close to discovering the hereditary principles that today bear Mendel's name, none succeeded fully. For example, in 1823, Thomas Andrew Knight determined that gray seed coat is dominant to white and that self-fertilization of certain gray-seeded plants produces both gray and white seed in progeny plants. In 1822, John Goss, working with a pea variety that had blue and white seeds, reported that crossing a pure-breeding white-seeded plant with a pure-breeding blue-seeded plant produced only blue seeds in first-generation plants, and that self-fertilization then produced a second generation with a mixture of white and blue seeds in plants. Carl Friedrich Gaertner came tantalizingly close to explaining segregation in 1827 when he reported results of a cross between pure-breeding gold-kernel maize and pure-breeding red-striped maize. All the $F_1$ had gold kernels, and among the $F_2$, 328 plants had only gold kernels and 103 had red-striped kernels. If Gaertner had been able to correctly interpret his data, he would have identified a 3.18:1 ratio in the $F_2$. Alas, he never did, and missed his "golden" opportunity to explain simple heredity.

Similar fates befell other plant hybridists, but arguably the one who came closest to explaining heredity prior to Mendel was Charles Naudin, who in 1863 seemed poised to beat Mendel to the punch by 2 years. In that year, Naudin reported the following:

▪ The results of reciprocal crosses are identical. (Similar observations by Mendel were important in his identification of the particulate nature of hereditary factors.)
▪ $F_1$ progeny display a single phenotype. (As Mendel reported 2 years later.)
▪ $F_2$ progeny display two phenotypes. (These observations are the result of the segregation of alleles.)
▪ The hereditary units for traits are separated in pollen and ovum formation. (This concept was fundamental to the segregation observation of Mendel.)
▪ Nonparental combinations of phenotypes appear in the $F_2$ generation. (This is identical to Mendel's independent assortment observation.)

After making these observations, why wasn't Naudin able to propose a hereditary mechanism to explain them? The answer is that Naudin, like his predecessors and others who would follow, failed to quantify his results. Naudin did not report the number of plants falling into different phenotypic categories, and he was therefore unable to recognize the ratios between phenotypic classes that are the key to interpreting hereditary transmission. Without quantitative data, Naudin was unable to formulate a testable hypothesis.

Alas, poor Naudin! Were it not for his failure to see the necessity of quantifying experimental results, we might well be discussing Naudinian genetics in this chapter instead of Mendelian genetics!

---

reason that Gregor Mendel was the first scientist to offer a mechanism to explain the hereditary patterns he observed. However, as noted in the chapter introduction, Mendel was not the first person to make these observations. **Experimental Insight 2.2** shows why, if he had not failed to quantify the results of his crosses of pea plants, Charles Naudin, a contemporary of Mendel's, might have been the first scientist to succeed at explaining heredity.

## 2.4 Probability Theory Predicts Mendelian Ratios

Mendel recognized that chance (or random probability, the same process that determines the outcome of coin flips and rolls of the dice) is the arithmetic principle underlying the segregation of alleles for a given gene and governing the independent assortment of alleles for genes at different loci. The preceding discussions have demonstrated that the basic rules of Mendelian inheritance are actually those of random probability theory. In fact, the Mendelian probabilities we have discussed earlier are most clearly expressed by four rules of probability theory—the *product rule*, the *sum rule*, *conditional probability*, and *binomial probability*. In this section, we look more closely at these rules that describe and predict the outcome of genetic events governed by the rules of chance.

### The Product Rule

If two or more events are independent of one another, their joint probability, the likelihood of their simultaneous or consecutive occurrence, is the product of the probabilities of each one individually. The **product rule,** also called the **multiplication rule,** describes these circumstances.

You have already used the product rule several times in determining the outcomes of genetic crosses. For example, in Figures 2.6 and 2.7 the product rule is used to determine that the chance of producing an $F_2$ plant with the recessive phenotype by the cross of heterozygous $F_1$ plants that are *Gg* or *Rr* is $\left(\frac{1}{2}\right)\left(\frac{1}{2}\right) = \frac{1}{4}$. Similarly, in Figure 2.10, the probability of producing gametes with each different genotype is predicted by applying the product rule in the forked-line diagram. Likewise, in Figure 2.11 predicting the probability that $F_2$ offspring will be homozygous recessive from a cross of $F_1$ dihybrid plants with the genotype $RrGg$ is predicted by applying the product rule.

## The Sum Rule

The **sum rule,** also called the **addition rule,** defines the joint probability of occurrence of any of two or more equivalent events by summing the probabilities of each event. This rule is applied when more than one outcome satisfies the conditions of the probability question.

You applied the sum rule to several genetic calculations in the preceding section. For example, in Figure 2.6 the probability that $F_2$ progeny of the cross $Gg \times Gg$ will be heterozygous is determined by adding the chance of obtaining either of the two possible ways of obtaining heterozygous offspring: $\left(\frac{1}{4}\right) + \left(\frac{1}{4}\right) = \frac{1}{2}$. Similarly, in Figure 2.11, the probability that the $F_2$ progeny of the cross of dihybrid heterozygotes ($RrGg$) have the two dominant phenotypes is obtained by applying the sum rule. From Figure 2.11 the probability is $\left(\frac{1}{16}\right) + \left(\frac{2}{16}\right) + \left(\frac{2}{16}\right) + \left(\frac{4}{16}\right) = \left(\frac{9}{16}\right)$.

Genetic Insight  The product rule is used to determine the joint probability of two or more independent events occurring simultaneously or consecutively. The sum rule is used when two or more equivalent outcomes satisfy the conditions of the probability question.

## Conditional Probability

Probability questions in genetic experiments can be asked before a cross is made. The product rule and the sum rule might, for example, be used to predict the likelihood of obtaining a certain genotype or phenotype from a cross. Certain probability questions are asked *after* a cross has been made and might address the probability that an organism has a particular genotype given that the organism has a particular phenotype. This kind of probability is called **conditional probability,** and it is applied when specific information about the outcome modifies, or "conditions," the probability calculation.

A typical genetic example of conditional probability would be to take the $F_2$ progeny of the cross like $Gg \times Gg$ and ask, "What is the probability that yellow-seeded progeny plants are heterozygous $Gg$ like the parents?" (see Figure 2.6). Yellow seed is present in $\frac{3}{4}$ of the progeny, but this phenotypic class contains two genotypes, $GG$ and $Gg$, that are not equally frequent. In this case, the genotype $Gg$ is found in $\frac{2}{3}$ of the yellow $F_2$ progeny. The other yellow $F_2$ are $GG$. Under the conditional criterion that the only progeny phenotype considered is yellow seeds, the answer to the question posed earlier is that the yellow-seeded progeny of the cross have a $\frac{2}{3}$ probability of being $Gg$.

Another application of conditional probability is the question, "If the yellow-seeded $F_2$ are allowed to self-fertilize, what proportion of them are expected to breed true?" This question is similar to the one Mendel asked as he devised an independent test of his segregation hypothesis (see Table 2.3). True-breeding $F_2$ progeny must be homozygous, and in his seed-color experiment, only those progeny with the genotype $GG$ meet this conditional contingency. Since the genotype $GG$ is found in one-third of the yellow-seeded $F_2$, the same proportion of true-breeding plants is expected as a result of self-fertilization.

Genetic Insight  A conditional probability is the probability of an event occurring contingent upon a certain additional circumstance or set of circumstances. In genetic analysis, the additional requirement often concerns the phenotype of the progeny of a cross. The effect of the additional circumstance is to alter the number of possible outcomes or the likelihood of reproductive outcomes.

## Binomial Probability

In determining the outcomes of certain genetic events, just one event need be predicted. An example is the question, "What is the chance a couple produces a daughter?" The answer is obtained by assuming that the father has a $\frac{1}{2}$ chance of donating an X chromosome and producing a daughter and an equal $\frac{1}{2}$ chance of donating a Y chromosome to produce a son, and that male and female offspring are equally likely. Other questions concerning genetic outcomes require that we assess the probability of a combination or sequence of such events (events for which there are two possible outcomes each time). For example, determining the probabilities of different combinations of boys and girls in sets of siblings or the risk of the recessive phenotype in one or more children of a couple who are each heterozygous carriers of a recessive disease requires computation of a particular combination of events that each have two alternative outcomes. To make these determinations, we use **binomial probability** calculations, expanding the binomial expression to reflect the number of outcome combinations and the probability of each combination.

**Construction of a Binomial Expansion Formula**  A binomial expression contains two variables, each representing the frequency of one of two alternative outcomes. We can express the likelihood of one outcome as having a frequency $p$ and the alternative outcome as having a frequency $q$. Since the events $p$ and $q$ are the only outcomes possible, the sum of the two frequencies is $(p + q) = 1$. If we are examining the probabilities of the outcomes for a series of two alternative events, such as multiple flips of a coin or several successive children born to a couple, we can expand the binomial to the power of the number of successive events ($n$) to calculate the probabilities. The binomial expansion formula is written as $(p + q)^n$.

In some kinds of probability problems, the values of the binomial variables $p$ and $q$ will be equal; that is, $p = q = \frac{1}{2}$, as in the probability of producing a boy or a girl. In other cases, the two binomial values will not be equal, as in the probability that heterozygous parents will produce a child with a recessive trait. Let's use a combinatorial probability to predict the likelihood of different numbers of boys and girls produced when a couple has three children. A combinatorial approach allows us to list all the possible birth orders of boys and girls and to group them according to the total numbers of boys and girls in each set of three siblings. The following table shows that there are $2^3$ or eight different birth orders of boys and girls. This conclusion is determined based on two possible outcomes (a boy or a girl) for three successive events. Assuming the probabilities of having a boy or having a girl are $\frac{1}{2}$, each different order has a probability of $\left(\frac{1}{2}\right)^3 = \frac{1}{8}$. The outcomes can be grouped into four sets that each contain a different total number of boys and girls.

| 0 Boys 3 Girls | 1 Boy 2 Girls | 2 Boys 1 Girl | 3 Boys 0 Girls |
|---|---|---|---|
| GGG | GGB | GBB | BBB |
| | GBG | BGB | |
| | BGG | BBG | |
| Probability: $\frac{1}{8}$ | $\frac{3}{8}$ | $\frac{3}{8}$ | $\frac{1}{8}$ |

We can see that there is only one order in which to get either three boys (BBB) or three girls (GGG), and each has a probability of $\frac{1}{8}$. Notice that we use the product rule to obtain each probability. But what about the cases of 2 boys and 1 girl or 2 girls and 1 boy, where there are three different birth orders (the orders of boys and girls among the siblings) for each outcome? Here we recognize that *each* birth order has a probability of $\left(\frac{1}{2}\right)^3 = \frac{1}{8}$ and that we must sum up all similar outcomes to determine the probability of 1 or 2 boys or girls in three consecutive siblings. In each of these cases, using the sum rule, the probability is $\left(\frac{1}{8}\right) + \left(\frac{1}{8}\right) + \left(\frac{1}{8}\right) = \frac{3}{8}$.

Arithmetically, we use the binomial expansion to the third power $[(p + q)^3]$ to represent the three successive siblings. Assuming that the probability of one outcome is $p$ and the probability of the other outcome is $q$, then the general case for the binomial expands as follows:

$$(p + q)^3 = p^3 + 3p^2q + 3pq^2 + q^3$$

The values being added on the right side of the equality are the frequencies of the four sets of outcomes $p$ and $q$.

**Application of Binomial Probability to Progeny Phenotypes** Binomial probability and the binomial expansion can be used whenever a probability question addresses a repeating series of events that have two alternative outcomes. Let's look at the production of yellow and green peas in pods with six peas each. In this example, the dominant allele $G$ determines yellow color and the recessive allele $g$ determines green color. The cross producing progeny peas is a self-fertilization of a yellow-seeded heterozygous ($Gg$) plant. The probability that a seed is yellow is $\frac{3}{4}$, since the genotype would be either $GG$ or $Gg$, and the probability that the seed is green, and therefore has the $gg$ genotype, is $\frac{1}{4}$. We will use the variable $p$ to represent the probability of yellow seeds and the variable $q$ to represent the probability of green seeds.

In our example of pea pods with six seeds that are produced by crossing heterozygous ($Gg$) parental plants, there are two possible color outcomes for each pea and six peas per pod, for a total of $2^6$ or 64 combinations. Counting the total number of yellow and green peas in each pod, there are seven categories that each have a different number of yellow and green peas per pod, as we discuss momentarily.

The application of binomial expansion to complex genetic calculations requires repetition and precision in the use of the product rule and the sum rule. However, a convenient shortcut called **Pascal's triangle** eliminates the repetitive calculations required for multiple expansions of the binomial probability equation and can be used for any number of expansions between 0 and the $n$th power to yield the size of each possible class and the total number of classes possible (**Figure 2.15**). Let's return to

**Figure 2.15 Pascal's triangle of binomial coefficients ($p + q$) raised to the $n$th power.** Each line of the table shows the distribution of the total number of combinations for a given value of $n$ (number of events). For example, for $(p + q)^2$, use the $n = 2$ line, which predicts a total of four outcome combinations distributed in a 1:2:1 or $\frac{1}{4}:\frac{1}{2}:\frac{1}{4}$ ratio. Applications using the highlighted lines, $n = 4$ and $n = 6$, are discussed in the text.

| $n$ (number of events) | | | | | | | Binomial coefficients | | | | | | | Total number of combinations |
|---|---|---|---|---|---|---|---|---|---|---|---|---|---|---|
| 0 | | | | | | | 1 | | | | | | | 1 |
| 1 | | | | | | 1 | | 1 | | | | | | 2 |
| 2 | | | | | 1 | | 2 | | 1 | | | | | 4 |
| 3 | | | | 1 | | 3 | | 3 | | 1 | | | | 8 |
| 4 | | | 1 | | 4 | | 6 | | 4 | | 1 | | | 16 |
| 5 | | 1 | | 5 | | 10 | | 10 | | 5 | | 1 | | 32 |
| 6 | 1 | | 6 | | 15 | | 20 | | 15 | | 6 | | 1 | 64 |
| 7 | 1 | 7 | 21 | 35 | 35 | 21 | 7 | 1 | | | | | | 128 |
| 8 | 1 | 8 | 28 | 56 | 70 | 56 | 28 | 8 | 1 | | | | | 256 |
| 9 | 1 | 9 | 36 | 84 | 126 | 126 | 84 | 36 | 9 | 1 | | | | 512 |
| 10 | 1 | 10 | 45 | 120 | 210 | 252 | 210 | 120 | 45 | 10 | 1 | | | 1024 |
| 11 | 1 | 11 | 55 | 165 | 330 | 462 | 462 | 330 | 165 | 55 | 11 | 1 | | 2048 |
| 12 | 1 | 12 | 66 | 220 | 495 | 792 | 924 | 792 | 495 | 220 | 66 | 12 | 1 | 4096 |

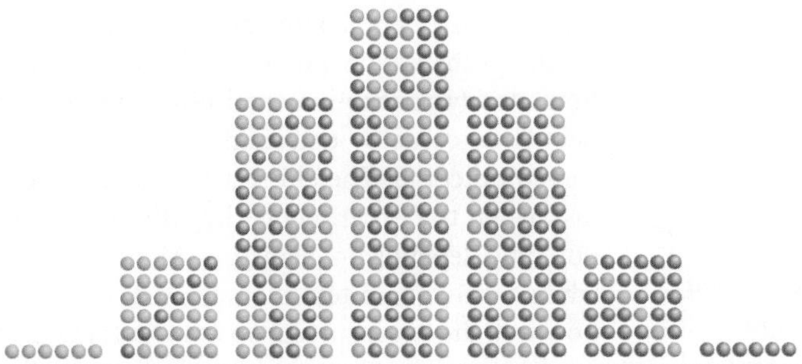

**Figure 2.16  Binomial-probability calculation of seed-color phenotype in six-seeded pods.** Pascal's triangle has been used to find the coefficients for the binomial equation expanded to $n = 6$. The 64 different outcomes are displayed in seven classes, and the equation is used to compute the expected frequency of each class.

| Seed-color outcome class | 6 yellow 0 green | 5 yellow 1 green | 4 yellow 2 green | 3 yellow 3 green | 2 yellow 4 green | 1 yellow 5 green | 0 yellow 6 green | |
|---|---|---|---|---|---|---|---|---|
| Number of combinations leading to occurrence | 1 | 6 | 15 | 20 | 15 | 6 | 1 | $= 64$ |
| Probability of occurrence for outcome class | $p^6$ | $6p^5q$ | $15p^4q^2$ | $20p^3q^3$ | $15p^2q^4$ | $6pq^5$ | $q^6$ | $= 1.00$ |
| Frequency of occurrence for outcome class $(p = \frac{3}{4}, q = \frac{1}{4})$ | 0.178 | 0.356 | 0.297 | 0.132 | 0.033 | 0.004 | 0.0002 | $= 1.00$ |

our pea pod example of binomial probability to see how Pascal's triangle is used.

**Figure 2.16** makes use of the values taken from the $n = 6$ line of Pascal's triangle (highlighted in Figure 2.15). These coefficients of the binomial expansion for $n = 6$ give the proportions of each of the seven outcome classes for this example. The coefficients are 1, 6, 15, 20, 15, 6, and 1, and they add up to a total of 64 different combinations. The coefficients are used to multiply the binomial probability of each outcome class. For this case where $p = \frac{3}{4}$ and $q = \frac{1}{4}$, the expected frequency of obtaining six yellow peas in a pod, for example, is calculated as $1(p^6)$, or $\left(\frac{3}{4}\right)^6 = 0.178$; for pods containing 3 yellow and 3 green peas, the frequency is $20\left[\left(\frac{3}{4}\right)^3\left(\frac{1}{4}\right)^3\right] = 0.132$; the proportion of pods containing 2 yellow and 4 green peas is $15\left[\left(\frac{3}{4}\right)^2\left(\frac{1}{4}\right)^4\right] = 0.033$; and so on. The complete set of expected frequencies for different combinations of seed color is shown at the bottom of Figure 2.16. Notice that the sum of category probabilities and the sum of category frequencies are each 1.00. This correspondence verifies that all possible outcomes have been taken into account.

---

**Genetic Insight**  Binomial probability predicts the likelihood of possible combinations in a sample of a specified size when each member of the sample can represent one of two outcomes. Variables $p$ and $q$ represent the overall frequencies of the two outcomes, and the binomial formula $(p + q)$ can be expanded to the $n$th power $[(p + q)^n]$ to correspond to the size of the sample. Pascal's triangle is a useful aid in determining the number and frequency of outcome classes.

---

## 2.5 Chi-Square Analysis Tests the Fit between Observed Values and Expected Outcomes

Sections 2.1 through 2.4 contain numerous examples of how the principles of probability, which make predictions about random events, can be used to predict the likelihood of different outcomes of genetic crosses. Genetic experiments like the ones described, and like the ones Mendel conducted, make predictions based on the hypothesis that chance determines the transmission of traits. To assess the validity of this hypothesis, however, geneticists must be able to compare the outcomes they obtain in their experiments to the outcomes that might be expected to occur. For example, are Mendel's $F_2$ results in Table 2.1 compatible with his segregation hypothesis predicting a 3:1 phenotype ratio?

Scientists must be able to make objective comparisons of observed and expected results to test genetic hypotheses. Qualitative statements such as "the observed results seem to be close to the results we expected" are unacceptable for scientific work. Instead, a quantitative approach, or in this case a statistical approach, is necessary to objectively compare the results obtained from a cross with the results that are predicted by probability. Mendel did not have appropriate statistical tools available to him. But in the early 1900s, the *chi-square test* was derived as a statistical test for comparison of observed experimental results with the results that may be expected when chance is generating the outcome. This section describes the chi-square test and its application to the analysis of genetic data, including some of Mendel's $F_2$ results. We begin, however, with a brief discussion of a *normal* or *Gaussian* distribution, on which chi-square analysis is based.

## The Normal Distribution

In large samples, outcomes that are predicted by chance have a **normal (Gaussian) distribution.** A normal distribution is a binomial distribution that is often called a "bell-shaped curve" because of the general shape of the curve the data form when they are graphed (**Figure 2.17**).

A normal distribution contains all the possible experimental outcomes. The **mean (μ)** is the average outcome, and other outcomes are distributed around the mean. The tall central segment of the curve nearest the mean represents the outcomes with the highest probability of occurrence. The probability of experimental outcomes gets smaller toward the farthest left and right portions of the curve. The probability of a particular experimental outcome is quantified by a measurement called the **standard deviation (σ).** In a normal distribution, approximately 68.2% of all outcome values fall within one standard deviation of the mean, 95.4% of outcomes fall within two standard deviations of the mean, and 99.8% of outcomes fall within three standard deviations of the mean (Figure 2.17). The observed result of a particular experiment can be compared to the normal distribution to determine the probability of that particular experimental observation compared to all possible outcomes in the distribution using σ, the standard deviation, as a guide.

By convention, observed experimental outcomes that have a probability of less than 5% (<0.05)—that is, a probability that is more than two standard deviations away from the mean—are often considered to show *statistically significant* difference between the observed outcome and the expected outcome. Chi-square analysis tests for statistically significant deviation in genetic experimental results.

## Chi-Square Analysis

The **chi-square ($\chi^2$) test** is the most common statistical method used in genetics experiments for comparing observed experimental outcomes to the results expected based on the probability hypothesis. Chi-square testing quantifies how closely an experimental observation matches the expected outcome by determining the probability of the observed outcome. The chi-square test is appropriate for this task when the experimental hypothesis used to predict the outcome depends on chance, as Mendelian ratios do. Thus, when a chi-square test is conducted, the test is measuring how well the experimental observations match experimental predictions. The chi-square test has proven flexible and accurate in measuring the fit between observed and expected experimental results across a wide range of experiments.

The chi-square value for the analysis of a given experiment is obtained in two steps. First, the difference between the number observed and number expected in each outcome category is squared and divided by the number expected in the category; and second, the values obtained for each outcome class are summed. The $\chi^2$ formula is

$$\chi^2 = \sum \frac{(O-E)^2}{E}$$

where $O$ is the observed number of offspring in each outcome class, $E$ is the number expected for each class, and the summation ($\Sigma$) is taken over all possible outcome classes.

The size of the chi-square value for an experiment is dependent on the three parameters of experimental sample size, number of outcome classes, and the number of observations in each outcome class, so it stands to reason that experiments with large numbers of outcome classes or more experimental observations recorded for each outcome class tend to have larger chi-square values than those found in experiments with lower numbers in each class. Simply stated, the addition of more or larger values to obtain a chi-square value leads to greater sums. Consequently, chi-square values are not directly comparable from one experiment to the next. Instead, each experimental chi-square value is interpreted in terms of the normal distribution of expected results for an experiment *of that size.*

The interpretation is done by means of a **probability value (P value),** which is a quantitative expression of the probability that the results of another experiment of the same size and structure will *deviate as much or more from expected results by chance.* P values in chi-square analysis are directly related to the probability of experimental outcomes in a normal distribution. High values for $P$ (values close to 1) are associated with low $\chi^2$ values. Low chi-square values occur when the observed and expected results are very similar. A high $P$ value indicates that chance alone is likely to explain the deviations of experimental observations from expected values. Thus, an experiment producing a $P$ value of 0.90 means that observed and expected results are close together and that 90% of all possible $\chi^2$ values are equal to or greater than the value obtained in the experiment. On the other hand, low $P$ values correspond to high chi-square values. They indicate substantial difference between observed and expected outcomes. The greater the

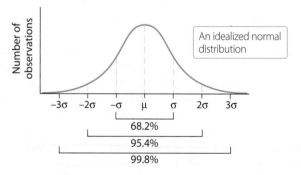

**Figure 2.17 Graphing the distribution of chance outcomes produces a normal distribution.** The standard deviation (σ) is used to characterize the scatter of possible outcomes around the mean (μ).

difference between observed and expected results of an experiment, the greater the $\chi^2$ value and the lower the $P$ value.

The statistical interpretation of a chi-square value is determined by identifying the $P$ value for each experiment, and the $P$ value is dependent on the number of **degrees of freedom ($df$)** in the experiment being examined. For each experiment, the $df$ value is equal to the number of outcome classes ($n$) minus 1, or . In a statistical sense, $df$ is equal to the number of independent variables in an experiment. For example, suppose we were conducting a chi-square test of 100 coin flips. There are two outcome classes, heads and tails, each of which we expect to see 50 times. However, once we record the number of events in one class, say 54 heads, the number of events in the second class becomes dependent on that first number. In our coin flip example, if we flip a coin 100 times and there are 54 heads recorded, the other 46 flips must be tails. Here the number of degrees of freedom is because, while there are two possible outcomes, the value of one is always dependent on the value of the other.

Table 2.4 is a table of chi-square values for different degrees of freedom, which are shown along the left-hand margin of the table. The corresponding $P$ values are listed along the top margin. To determine the $P$ value for the chi-square value from an experiment, the first step is to determine the number of degrees of freedom. The second step is to locate the chi-square value in the horizontal line to the

right of that number of degrees of freedom. The $P$ value for the result of the experiment in question is then found at the top of the column containing the chi-square value.

Interpretation of chi-square results is based on the corresponding $P$ value. A statistically significant result from chi-square analysis is defined as one for which the $P$ value is *less than 0.05*. This means that there is less than a 5% chance ($<0.05$) of obtaining the experimental observation by chance. By convention, when any experimental result has less than a 5% probability, the hypothesis of chance is *rejected*. In other words, if the $P$ value is less than 0.05, the difference between the observed and expected results is considered statistically significant, and the experimental hypothesis is rejected. Conversely, $P$ values greater than 0.05 indicate a nonsignificant deviation between observed and expected values. These values result in *failure to reject* the chance hypothesis.

## Chi-Square Analysis of Mendel's Data

Modern statistical methods allow us to do something Mendel could not do—test his experimental data for its compatibility with the predictions of the laws of segregation and independent assortment. Table 2.1 contains data from Mendel for $F_2$ segregation of the seven traits he tested. In the first row of the table, we see that Mendel

| Table 2.4 | The Chi-Square Table | | | | | | | | |
|---|---|---|---|---|---|---|---|---|---|
| **Probability ($P$) Value** | | | | | | | | | |
| *df* | 0.95 | 0.90 | 0.70 | 0.50 | 0.30 | 0.20 | 0.10 | 0.05 | 0.01 | 0.001 |
| *1* | 0.004 | 0.016 | 0.15 | 0.46 | 1.07 | 1.64 | 2.17 | 3.84 | 6.64 | 10.83 |
| *2* | 0.10 | 0.21 | 0.71 | 1.39 | 2.41 | 3.22 | 4.61 | 5.99 | 9.21 | 13.82 |
| *3* | 0.35 | 0.58 | 1.42 | 2.37 | 3.67 | 4.64 | 6.25 | 7.82 | 11.35 | 16.27 |
| *4* | 0.71 | 1.06 | 2.20 | 3.36 | 4.88 | 5.99 | 7.78 | 9.49 | 13.28 | 18.47 |
| *5* | 1.15 | 1.61 | 3.00 | 4.35 | 6.06 | 7.29 | 9.24 | 11.07 | 15.09 | 20.52 |
| *6* | 1.64 | 2.20 | 3.83 | 5.35 | 7.23 | 8.56 | 10.65 | 12.59 | 16.81 | 22.46 |
| *7* | 2.17 | 2.83 | 4.67 | 6.35 | 8.38 | 9.80 | 12.02 | 14.07 | 18.48 | 24.32 |
| *8* | 2.73 | 3.49 | 5.53 | 7.34 | 9.52 | 11.03 | 13.36 | 15.51 | 20.09 | 26.13 |
| *9* | 3.33 | 4.17 | 6.39 | 8.34 | 10.66 | 12.24 | 14.68 | 16.92 | 21.67 | 27.88 |
| *10* | 3.94 | 4.87 | 7.27 | 9.34 | 11.78 | 13.44 | 15.99 | 18.31 | 23.21 | 29.59 |
| *11* | 4.58 | 5.58 | 8.15 | 10.34 | 12.90 | 14.63 | 17.28 | 19.68 | 24.73 | 31.26 |
| *12* | 5.23 | 6.30 | 9.03 | 11.34 | 14.01 | 15.81 | 18.55 | 21.03 | 26.22 | 32.91 |
| *13* | 5.89 | 7.04 | 9.93 | 12.34 | 15.12 | 16.99 | 19.81 | 22.36 | 27.69 | 34.53 |
| *14* | 6.57 | 7.79 | 10.82 | 13.34 | 16.22 | 18.15 | 21.06 | 23.69 | 29.14 | 36.12 |
| *15* | 7.26 | 8.55 | 11.72 | 14.34 | 17.32 | 19.31 | 22.31 | 25.00 | 30.58 | 37.70 |

Fail to reject chance hypothesis                                       Reject chance hypothesis

*Note:* Chi-square values are in the body of the table, degrees of freedom are at the far-left side, and probability values are at the top of each column of chi-square values.

scored 7324 $F_2$ seeds for round or wrinkled phenotypes. Among these, he counted 5474 round and 1850 wrinkled. Based on the predictions of his segregation hypothesis, Mendel expected that 75% of the $F_2$ would be round and the remaining 25% wrinkled. That means he expected $(7324)(0.75) = 5493$ round seeds and $(7324)(0.25) = 1831$ wrinkled seeds. There is 1 degree of freedom in the experiment, and the chi-square is calculated as

$$\chi^2 = (5474 - 5493)^2/5493 + (1850 - 1831)^2/1831$$
$$= 0.066 + 0.197 = 0.263$$

For $df = 1$, the $P$ value falls between 0.50 and 0.70 (see Table 2.4). This is well above the cutoff value of 0.05 and consequently represents a nonsignificant deviation. Therefore, Mendel's $F_2$ data for seed shape are consistent with the predictions of the law of segregation.

Figure 2.12 provides data Mendel collected on seed shape and seed color that we can use to test whether his results were consistent with his predictions of independent assortment. Based on the predicted 9:3:3:1 ratio, the 556 $F_2$ produced by Mendel would be expected to have the following distribution.

| | | |
|---|---|---|
| Round, yellow | $(556)(0.5625) =$ | 312.75 |
| Round, green | $(556)(0.1875) =$ | 104.25 |
| Wrinkled, yellow | $(556)(0.1875) =$ | 104.25 |
| Wrinkled, green | $(556)(0.0625) =$ | 34.75 |
| | | 556.00 |

The chi-square value is calculated as

$$\chi^2 = (315 - 312.75)^2/312.75 + (108 - 104.25)^2/104.25$$
$$+ (101 - 104.25)^2/104.25 + (32 - 34.75)^2/34.75$$
$$= 0.016 + 0.135 + 0.101 + 0.218 = 0.470$$

In this case, $df = 3$, and the $P$ value falls between 0.90 and 0.95. This indicates a nonsignificant deviation because the $P$ value is above the 0.05 cutoff value. Mendel's $F_2$ data for seed color and seed shape are therefore also consistent with the predictions of independent assortment. A third example of chi-square analysis, using trihybrid-cross results from one of Mendel's experiments, is shown in **Table 2.5**. From statistical analysis of these data we conclude that Mendel's results are consistent with the predictions of segregation and independent assortment.

---

Genetic Insight The chi-square ($\chi^2$) test determines whether experimental data conform to the expectations of a chance hypothesis by calculating a $P$ (probability) value for the experiment. High $P$ values correspond to low $\chi^2$ values and indicate close correspondence of experimental observations and expectations. The chance hypothesis is rejected if the $P$ value for an experimental result is less than 0.05.

---

| Table 2.5 | Chi-Square Analysis of Mendel's Trihybrid-Cross Data | |
|---|---|---|
| **Mendel's Observation**[a] | | **Number Expected** |
| Phenotype | Number | |
| Round, yellow, purple | 269 | 269.58 |
| Round, yellow, white | 98 | 89.86 |
| Round, green, purple | 86 | 89.86 |
| Round, green, white | 27 | 29.95 |
| Wrinkled, yellow, purple | 88 | 89.86 |
| Wrinkled, yellow, white | 34 | 29.95 |
| Wrinkled, green, purple | 30 | 29.95 |
| Wrinkled, green, white | 7 | 9.98 |
| | 639 | 638.99 |

$$\chi^2 = (269 - 269.58)^2/269.58 + (98 - 89.86)^2/89.86$$
$$+ (86 - 89.86)^2/89.86 + (27 - 29.95)^2/29.95$$
$$+ (88 - 89.86)^2/89.86 + (34 - 29.95)^2/29.95$$
$$+ (30 - 29.95)^2/29.95 + (7 - 9.98)^2/9.98$$
$$= 2.67$$
$$df = 7$$
$$P \text{ value} > 0.90$$

[a] Data are taken from Figure 2.14.

## 2.6 Autosomal Inheritance and Molecular Genetics Parallel the Predictions of Mendel's Hereditary Principles

During the first decade of the 20th century, immediately after the rediscovery of Mendel's rules of hereditary transmission, biologists began to extend Mendel's findings to species other than pea plants. They also identified exceptions to Mendelian hereditary principles (Chapter 4). In this final section, we apply Mendelian principles to the transmission of certain traits in humans. In addition, we consider the correspondence of molecular genetics findings to Mendelian inheritance and explore the underlying causes of four of the traits that Mendel studied.

**Autosomal inheritance** refers to the transmission of genes that are carried on autosomes, the chromosomes (22 pairs in humans) that are found in both males and females. Because of the two copies of each autosome in our genome, each organism carries two copies (alleles) of each autosomal gene. The alleles on homologous chromosomes can be identical, in which case a person has a homozygous genotype; or the alleles can be different, producing a heterozygous genotype. Autosomal inheritance allows us to see Mendel's law of segregation and law of independent assortment in

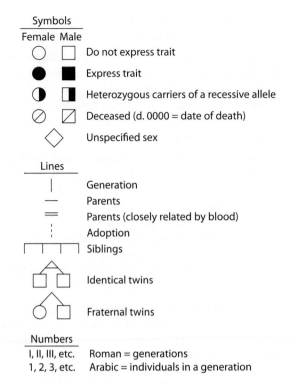

## Symbols

| Female | Male | |
|---|---|---|
| ○ | □ | Do not express trait |
| ● | ■ | Express trait |
| ◐ | ◪ | Heterozygous carriers of a recessive allele |
| ⊘ | ⧄ | Deceased (d. 0000 = date of death) |
| ◇ | | Unspecified sex |

### Lines

| | |
|---|---|
| │ | Generation |
| — | Parents |
| = | Parents (closely related by blood) |
| ⋮ | Adoption |
| ⊓⊓⊓ | Siblings |
| ⟨△⟩ | Identical twins |
| ⟨○ □⟩ | Fraternal twins |

### Numbers

| | |
|---|---|
| I, II, III, etc. | Roman = generations |
| 1, 2, 3, etc. | Arabic = individuals in a generation |

**Figure 2.18  Common pedigree symbols.**

action. Autosomes are distinct from the sex chromosomes (X and Y chromosomes), and autosomal inheritance follows different patterns than does the inheritance of genes on sex chromosomes (see Chapter 3).

Pedigrees, or family trees, are a kind of symbolic shorthand used to trace the inheritance of traits in humans and in animals such as horses, dogs, cats, cattle, and others. In standard pedigree notation, males are represented by squares and females by circles (**Figure 2.18**). A filled circle or square indicates that the phenotype of interest is present. A line through a symbol indicates the person is deceased. Parents are connected to each other by a horizontal line from which a vertical line descends to their progeny. Individuals in a pedigree are numbered by a Roman numeral (I, II, III, etc.) to indicate their generation combined with an Arabic numeral (1, 2, 3, etc.) that identifies each organism in a generation.

## Autosomal Dominant Inheritance

The pedigree in **Figure 2.19** shows characteristics commonly observed for **autosomal dominant inheritance** of a disease. Notice the following six characteristics:

1. **Each individual who has the disease has at least one affected parent.** Anyone carrying at least one copy of a dominant allele will display the dominant phenotype. Therefore, any disease or disorder caused by a dominant allele is seen in successive generations (this characteristic is described as a vertical pattern of transmission). In Figure 2.19, all 13 affected children in generations II, III, and IV have at least one affected parent. The only exceptions to this rule are (1) the occurrence of a new mutation in a child and (2) a person with the dominant mutation entering the family through marriage. The pedigree shows no evidence of a new mutation, but individual III-16 marries into the family and has the dominant mutation.

2. **Males and females are affected in equal numbers.** Mutations carried on an autosome are equally likely to occur in either sex. Among the total of 15 affected individuals in the figure, 7 are male and 8 are female.

3. **Either sex can transmit the disease allele.** Eight parents in Figure 2.19 have transmitted the disease to one or more children. Three of the transmitting parents are male and four are female.

4. **In crosses in which one parent is affected and the other is not, approximately half the offspring express the disease.** Diseases caused by dominant mutations are usually rare in populations, and most affected individuals are heterozygous. A cross between one affected parent and one unaffected parent can most often be genetically interpreted as a heterozygous-by-homozygous cross, expected to produce a 1:1 ratio between phenotypes. In this family, there are six crosses between an affected person who is heterozygous and an unaffected person who is homozygous for the recessive allele. Among the 19 children produced by these crosses, 10 of the children are affected and 9 are unaffected.

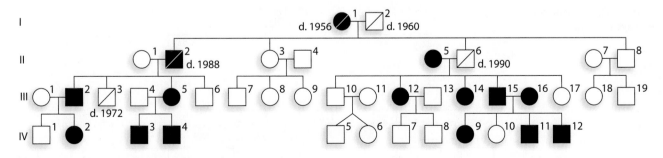

**Figure 2.19  Autosomal dominant inheritance.**

5. **Two unaffected parents will not have any children with the disease.** Dominant phenotypes require the presence of at least one copy of the dominant allele. If each parent has the recessive ("normal") phenotype, they must each be homozygous for the recessive allele, and all their offspring should also be homozygous. Three crosses of this kind are shown in the pedigree, and all seven resulting children have the normal phenotype. New mutation is an exception to this rule, but it is not seen in this family.

6. **Two affected parents may produce unaffected children.** If each parent is heterozygous, the expected ratio between affected and unaffected children is 3:1. There is one mating like this shown in the pedigree, and three of the four resulting children are affected. The mating of two heterozygous affected parents presents a one-in-four chance of producing a child homozygous for the mutant allele and a one-in-four chance of producing a child homozygous for the wild-type allele.

## Autosomal Recessive Inheritance

**Figure 2.20** shows a human pedigree displaying the characteristics commonly observed for **autosomal recessive inheritance** of a disease. There are six key features to notice:

1. **Individuals who have the disease are often born to parents who do not.** A child with the disease (the recessive phenotype) must have inherited one copy of the recessive allele from each parent. Moreover, it is common for children with the disease to have been produced by parents with the dominant (normal) phenotype who are heterozygous. Four affected family members, IV-5, IV-6, IV-10, and V-3, are the children of heterozygous carrier parents.

2. **If only one parent has the disorder, the risk that a child has the disorder depends on the genotype of the other parent.** The affected parent is homozygous recessive and must pass a copy of the recessive allele to each child. If the unaffected parent is heterozygous, the risk that a given child will be affected is $\frac{1}{2}$. If the unaffected parent is homozygous for the dominant allele, all children will be unaffected heterozygotes.

3. **If both parents have the disorder, all children will have the disorder.** If both parents are homozygous recessive, all their offspring will have the same homozygous genotype. The four affected siblings in the last generation of the idealized pedigree inherit their disorder in this way.

4. **The sex ratio of affected offspring is expected to be equal.** Males and females are equally likely to be homozygous for the recessive allele. The sex of a child is independent of the likelihood that the homozygous recessive genotype occurs at the autosomal locus. (Recall that equal sex ratios are also seen in autosomal dominant disorders.) In the example pedigree there are a total of eight affected individuals—four males and four females.

5. **The disease is usually not seen in each generation; but if an affected child is produced by unaffected parents, the risk to subsequent children of the couple is $\frac{1}{4}$.** If both parents have the dominant phenotype, they can produce a child with the recessive phenotype only if they are each heterozygous. This is usually rare in a population, so production of affected children is rare. If an affected child is born to a healthy couple, however, each parent is a heterozygous carrier of the recessive disease allele, and the disease risk to each additional child is $\frac{1}{4}$. In the example pedigree, the recessive condition is confined to the fourth and fifth generations.

6. **If the disease or disorder is rare in the population, unaffected parents of an affected child are more likely to be related to one another.** Individuals who are related to one another can carry identical alleles as a result of their shared ancestry. If the recessive allele is present in the family, the sharing of alleles through common ancestry increases the probability that related individuals might both be carriers of the recessive allele in comparison to the population at large. In Figure 3.20, the two affected parents of the four affected siblings are related to one another.

**Figure 2.20 Autosomal recessive inheritance.**

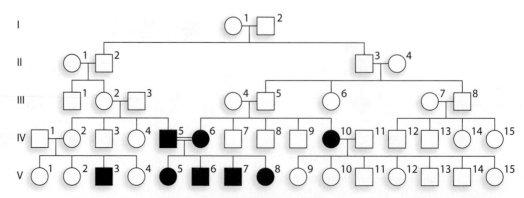

## Molecular Genetics of Mendel's Traits

The rediscovery of Mendel's traits continues to the present day using methods of molecular genetics to identify the genes responsible for the phenotypic variation Mendel studied in his traits. These molecular analyses, the first of which was published in 1990, describe the nucleic acid (DNA and RNA) variation and the polypeptide (protein and enzyme) variation responsible for Mendel's traits. A cornerstone of modern genetics is the seamless integration of the principles of transmission genetics with those of molecular genetic analysis. In the minds of geneticists today, the transmission of alleles that produce morphologic variation is equated with the transmission of variable DNA sequences that act through mRNA to produce protein variants responsible for the different morphologic forms of a trait. From this perspective, transmissio genetic analysis and molecular genetic analysis are not distinct pursuits. Rather, they are parallel scientific approaches that afford different ways of examining the same outcome—two sides of the same coin, if you will—with the pattern of transmission of morphologic variants traceable through examination of the hereditary molecules DNA, RNA, and protein.

Identifying these genes and determining how molecular variation in them produces morphologic variation in pea plants requires the demonstration that (1) allelic variation coincides with morphologic variation, (2) DNA variation in the alleles produces different protein products, (3) the protein products from each allele have different structures that lead to different functional capabilities, and (4) the functional differences between the protein products of different alleles account for the observed morphological variation in pea plants. The molecular differences between the alleles also usually clarify why the alleles are dominant or recessive relative to one another.

Mendel did not leave any neatly labeled packets of seeds for later researchers to analyze, so the process of pinpointing the exact traits he examined and the genes and proteins responsible for them has been complicated. Nevertheless, to date researchers have succeeded in identifying the genes responsible for four of Mendel's seven traits. Moreover, for each gene, the specific mutations producing the mutant alleles have been determined. In each case, the mutations significantly reduce or entirely eliminate production or function of the wild-type polypeptide. As a result of this loss of function, the mutant alleles of Mendel's traits are recessive to the dominant wild-type alleles. For each gene, a plant that inherits one wild-type allele and one mutant allele produces enough normal polypeptide to generate the wild-type phenotype. In contrast, plants that are homozygous for the mutant allele have the recessive mutant phenotype (**Table 2.6**).

**Seed Shape (Round and Wrinkled, Gene *R*)** In 1990, research published by Madan Bhattacharyya and colleagues described analysis of a gene responsible for round and wrinkled seed shape. Originally called gene *R*, and later renamed *Sbe1*, the gene produces the starch-branching enzyme that helps convert a prevalent linear form of starch called amylose into a complex branched form of starch called amylopectin. In plants that are homozygous or heterozygous for the wild-type allele, enough starch-branching enzyme is produced to convert most of the available amylose to amylopectin. In plants that are homozygous for the recessive allele, however, only a small percentage of amylose is converted to amylopectin. The recessive wrinkled seed phenotype is produced as an indirect result of the high concentration of amylose in wrinkled seeds. Amylose readily loses sugar molecules, and in developing wrinkled seeds, the high concentration of free sugar leads the seeds to import an excessive amount of water that swells the developing seeds. As seeds mature they naturally dehydrate. The maturing wrinkled seeds lose much more water than do maturing round seeds, resulting in a partial collapse of the wrinkled seed membranes that does not occur in round seeds.

**Stem Length (Tall and Short, Gene *Le*)** In 1997, two research groups, one led by David Martin and the other by Diane Lester, determined that a gene called *Le* produces the variation in stem length that Mendel saw as tall and short plants by controlling growth of the main stem of the plant. This *Le* gene produces giberellin 3-β hydroxylase, an enzyme that catalyzes one step of the multistep biochemical pathway synthesizing the plant growth hormone giberellin. Wild-type plants are able to produce giberellin and can grow tall, but a base substitution mutation in the mutant allele results in an amino acid substitution in the mutant enzyme. The amino acid change inactivates the function of the mutant enzyme and results in a blockage of the giberellin synthesis pathway. The loss of function of this enzyme produces plants with the short mutant phenotype.

**Seed Color (Yellow and Green, Gene *I*)** Two studies published in 2007, one by Ian Armstead and colleagues and the other by Sylvain Aubry and colleagues, identified the *Sgr* gene, known as "stay-green," that produces mutant green seeds rather than wild-type yellow seeds in plants that are homozygous for a mutation of the gene. In this case, the polypeptide product of *Sgr* is an enzyme that catalyzes a step in the breakdown of chlorophyll, a green-colored compound. Chlorophyll breakdown normally occurs as seeds mature, and results in the yellow color of wild-type seeds. A mutation prevents production of a functional enzyme, and the absence of its activity in the chlorophyll breakdown pathway results in the retention of green color in mutant seeds.

**Flower Color (Purple and White, Gene *A*)** Most recently, in 2010, the gene responsible for the white-flower mutation in Mendel's pea plants, a gene originally designated gene *A*, was identified. A research group led by Roger Hellens

**Table 2.6    Identification and Molecular Characterization of Mendel's Traits**

| Trait | Gene and Gene Product | Wild-Type Allele and Function | Mutant Allele and Function | Reference |
|---|---|---|---|---|
| Seed shape (round and wrinkled seeds) | The gene is *Sbe1*, producing starch-branching enzyme. | The dominant wild-type allele (*R*) produces starch-branching enzyme that converts amylase, a linear starch, into amylopectin, a complex branched starch. | The recessive mutant allele (*r*) contains an inserted segment about 800 base pairs in length. The transcript of the mutant allele does not produce an enzyme product, resulting in a loss of function. | Bhattacharyya, M. K., et al. 1990. *Cell* 60: 115–122. |
| Stem length (tall and short plants) | The gene is *Le*, producing gibberellin 3β-hydroxylase (G3βH). | G3βH produced by the dominant allele *Le* converts a precursor in the synthesis of the plant growth hormone gibberellin that causes plants to grow tall. | The recessive mutant *le* allele contains a base substitution that results in an amino acid change. The mutant G3βH has less than 5% the activity of the wild-type product and produces little gibberellin, leading to short plants. | Lester, D. R., et al. 1997. *Plant Cell* 9: 1435–1443. Martin, D. N., et al. 1997. *Proc. Natl. Acad. Sci., USA* 94: 8907–8911. |
| Seed color (yellow seed and green seed) | The gene was originally named *I* gene and was later renamed *Sgr* (called "stay green"). The gene produces an enzyme that helps break down chlorophyll. | The dominant wild-type allele (*I*) produces an enzyme that catalyzes one step in the chlorophyll breakdown pathway, which turns wild-type seeds yellow as they mature. | The recessive mutant allele (*i*) contains two base substitutions and a base pair insertion. The resulting mutant polypeptide has no function, leading to a blockage of the chlorophyll breakdown pathway and causing mutant seeds to retain their immature green color. | Armstead, I., et al. 2007. *Science* 315: 73. Aubry, S., et al. 2008. *Plant Mol. Biol.* 67: 243–256. |
| Flower color (purple flower and white flower) | Originally named gene *A* and renamed *bHLH*, the gene produces a protein that activates transcription of target genes. | The dominant wild-type allele (*A*) produces a protein that activates transcription of genes required to synthesize the purple-colored plant pigment called anthocyanin. | The recessive mutant allele (*a*) contains a base substitution that results in production of abnormal mRNA. The mutant mRNA does not produce the transcription-activating protein, thus blocking anthocyanin production and resulting in the development of white flowers. | Hellens, R. P., et al. 2010. *PLoS One* 5: 1–8. |

determined that mutation of the *bHLH* gene in pea plants produces mutant white flowers rather than wild-type purple flowers. The protein product of *bHLH* is a transcription factor protein that interacts with other proteins to activate the transcription of certain genes. In this case, the genes targeted for transcription activation are active in the pathway that normally produces the purple-colored plant pigment anthocyanin. Wild-type plants produce enough of the gene product (the transcription factor protein) to activate transcription of anthocyanin-producing genes. Plants that are homozygous for mutations of this gene, however, are unable to activate transcription of the pigment-producing genes. These plants lack the purple anthocyanin pigment, and so their flowers are white.

A common feature of each of the genes controlling Mendel's traits is that the wild-type alleles are dominant to mutant alleles that are recessive. This is a consequence of a loss of function on the part of mutant alleles. For each gene, one or two copies of the wild-type allele results in the wild-type phenotype, whereas the mutant phenotype is produced in plants that are homozygous for the mutant allele. (We develop this relationship between alleles, and explore other kinds of dominance relationships in Section 4.1.)

The central conclusions from these molecular studies identifying genes Mendel examined in his crosses are that (1) the inheritance of allelic variants precisely parallels the pattern of transmission of morphological variation and (2) morphologic variation in pea plants results from differences in the structure and function of the proteins produced by the alleles. Molecular genetic analysis has led to (3) identification of the DNA-sequence differences between alleles, determination of the impact of those differences on mRNA, and description of the alteration of protein structures resulting from each mRNA; and (4) functional analysis of the protein product of each allele to describe the role it plays in producing the phenotype.

---

**Genetic Insight** The identification of genes responsible for Mendel's traits expands the understanding of hereditary transmission by establishing a molecular basis for variation seen in plants' phenotypes.

---

## CASE STUDY

### Inheritance of Sickle Cell Disease in Humans

The Online Mendelian Index of Man (OMIM) is a continuously updated public information catalog providing up-to-date information on more than 18,000 human hereditary traits. OMIM can be accessed at www.ncbi.nlm.nih.gov/omim. Each trait listed in the OMIM catalog has a unique identifier number. One trait, named sickle cell disease (SCD), OMIM number 603903, is the subject of later discussion (see Chapter 10), which introduces several important research techniques and uses them to describe the discovery and analysis of the molecular basis of SCD and the evolution of the mutant allele. Here we examine the hereditary transmission of SCD, which is caused by a base substitution mutation in the $\beta$-globin gene. The base substitution alters the $\beta$- globin protein and results in the inheritance of SCD as an autosomal recessive condition. The inheritance of the $\beta^S$ variant and SCD can be traced by identifying the phenotypes of family members and displaying them in a pedigree, or family tree. The pedigrees shown in **Figure 2.21** identify females with circles and males with squares, and are typical of a family in which SCD is inherited. Blue circles and squares indicate family members who do not have the trait being traced; a pink circle or a square that is filled indicates a person with the trait (in this case, SCD). Mating is identified by the horizontal line connecting a male symbol and a female symbol. The offspring of a union are connected to that horizontal line by vertical lines of descent. Groups of siblings are joined to one another by bracketed lines. The individuals in a pedigree are assigned numbers that indicate their generation (Roman numerals) and their birth order in the generation (Arabic numerals). In Figure 2.4a, the father and mother are identified as I-1 and I-2. Their daughter II-4 is affected by SCD, as indicated by a filled circle. Her siblings, individuals II-1, II-2, and II-3, are healthy.

The pedigree in Figure 2.21a identifies the genotype for the $\beta$-globin gene. Each person carries two copies of the $\beta$-globin gene in each member of a certain family. Note that person II-1 is homozygous for $\beta^A$, the wild-type allele. Alternatively, siblings II-2 and II-3 and the parents in the pedigree, I-1 and I-2, are heterozygous and carry alleles $\beta^A$ and the mutant allele $\beta^S$.

The child II-4 is homozygous for $\beta^S$ and has SCD. This disorder is a recessive trait because the phenotype is displayed only in a person who is homozygous for the allele that produces it. In contrast, the dominant, wild-type phenotype is produced by the presence of either one or two copies of $\beta^A$. In this family, each parent has the dominant, wild-type phenotype, but the appearance of a child with the recessive trait means that each parent must be a heterozygous carrier of a recessive allele.

(a)

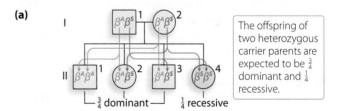

The offspring of two heterozygous carrier parents are expected to be $\frac{3}{4}$ dominant and $\frac{1}{4}$ recessive.

(b)

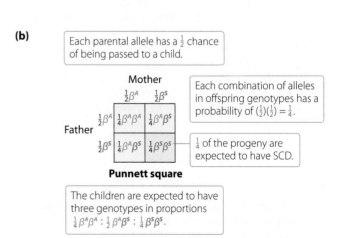

Each parental allele has a $\frac{1}{2}$ chance of being passed to a child.

Each combination of alleles in offspring genotypes has a probability of $(\frac{1}{2})(\frac{1}{2}) = \frac{1}{4}$.

$\frac{1}{4}$ of the progeny are expected to have SCD.

The children are expected to have three genotypes in proportions $\frac{1}{4}\beta^A\beta^A : \frac{1}{2}\beta^A\beta^S : \frac{1}{4}\beta^S\beta^S$.

**Figure 2.21 Hereditary transmission of sickle cell disease.** (a) Each parent passes one allele to each child. (b) Three genotypes are expected to occur among the children in the proportions shown.

Figure 2.21b illustrates the idealized transmission of alleles from heterozygous parents to offspring in generation II using a Punnett square. Each of the two alleles carried by a heterozygote has a  chance of being transmitted to an offspring. Chance dictates that four different combinations of alleles can be transmitted from these parents to their children. The arrows in the figure indicate the parental origin of alleles in the homozygous and heterozygous children of this couple. Notice that three of the four children have the dominant phenotype, being either homozygous for the dominant allele ($\beta^A\beta^A$) or heterozygous ($\beta^A\beta^S$), and that one of the four children has the homozygous $\beta^S\beta^S$ genotype and therefore suffers from SCD.

The ratio of $\frac{3}{4}$ dominant to $\frac{1}{4}$ recessive is the 3:1 ratio of phenotypes that, as we saw repeatedly in this chapter, is the expected statistical outcome of crosses between two heterozygous organisms. Each allele transmitted from a heterozygote has a $\frac{1}{2}$ chance of being passed to a child. Any one of the four combinations of alleles transmitted to a child is expected to occur with a frequency of $(\frac{1}{2})(\frac{1}{2}) = \frac{1}{4}$; thus, the frequency of children with SCD produced by heterozygous carrier parents is $\frac{1}{4}$. The three genotypes in the children are expected to occur in the ratio $\frac{1}{4}\,\beta^A\beta^A : \frac{11}{22}\beta^A\beta^S : \frac{1}{4}\,\beta^S\beta^S$. These genotypes can be distinctly identified using DNA- and protein-based analysis. (We describe these molecular techniques and explore other details of SCD in Chapter 10.)

---

## SUMMARY

### 2.1 Gregor Mendel Discovered the Basic Principles of Genetic Transmission

▌ A broad education in science and mathematics prepared Gregor Mendel to design hybridization experiments that could reveal the principles of hereditary transmission.

### 2.2 Monohybrid Crosses Reveal the Segregation of Alleles

▌ Mendel's experimental design had five important features: controlled crosses, use of pure-breeding parental strains, examination of discreet traits, quantification of results, and the use of replicate and reciprocal crosses.

▌ Crosses between pure-breeding parental plants with different phenotypes produce monohybrid $F_1$ progeny with the dominant phenotype.

▌ Monohybrid crosses produce a 3:1 ratio of the dominant to the recessive phenotype among $F_2$ progeny and demonstrate the operation of the law of segregation.

▌ The law of segregation states that two alleles at a locus will separate from one another during gamete formation, each allele has an equal probability of inclusion in a gamete, and gametes unite at random during reproduction.

▌ Mendel used test-cross analysis to demonstrate that $F_1$ plants are monohybrid, and he used the self-fertilization of $F_2$ plants with the dominant phenotype to demonstrate that the latter have a 2:1 ratio of heterozygotes to homozygotes.

### 2.3 Dihybrid and Trihybrid Crosses Reveal the Independent Assortment of Alleles

▌ The $F_2$ progeny of dihybrid $F_1$ plants display a 9:3:3:1 phenotype ratio that demonstrates the operation of the law of independent assortment.

▌ Mendel used trihybrid-cross analysis to demonstrate that alleles of multiple genes are transmitted in accordance with the predictions of the law of independent assortment.

### 2.4 Probability Theory Predicts Mendelian Ratios

▌ The product rule of probability is used to determine the likelihood of two or more independent events occurring simultaneously or consecutively. The joint probability is determined by multiplying the probabilities of the independent events.

▌ The sum rule of probability is applied when two or more outcomes are possible. The individual probabilities of the outcomes are added together to determine the joint probability.

▌ Conditional probability is the probability of outcomes that are contingent on particular conditions.

▌ Binomial probability theory describes the distribution of outcomes of an experiment in terms of the number of outcome classes and the frequency of each class. Pascal's triangle is a convenient tool for determining the distribution of binomial outcomes.

### 2.5 Chi-Square Analysis Tests the Fit between Observed Values and Expected Outcomes

▌ The chi-square test ($\chi^2$) compares observed results with the results predicted by a genetic hypothesis that is based on chance.

▌ The result of the chi-square test determines how closely predictions match results.

▌ The significance of a chi-square value is determined by the $P$ (probability) value corresponding to the number of degrees of freedom in the experiment.

### 2.6 Autosomal Inheritance and Molecular Genetics Parallel the Predictions of Mendel's Hereditary Principles

▌ Traits transmitted by autosomal inheritance are equally likely in males and females.

▌ Autosomal dominant inheritance produces a vertical pattern of transmission in which each organism with the dominant trait has at least one parent with the trait.

■ Traits transmitted in an autosomal recessive pattern are usually distributed in a horizontal pattern in which off-spring with the recessive trait frequently descend from parents that are heterozygous and have the dominant phenotype.

■ Molecular analysis of four of Mendel's traits illustrates how transmission genetic analysis and molecular genetic analysis characterize the same hereditary processes at different levels.

## KEYWORDS

artificial cross-fertilization *(p. 28)*
autosomal dominant inheritance *(p. 51)*
autosomal inheritance *(p. 50)*
autosomal recessive inheritance *(p. 52)*
binomial probability *(p. 45)*
blending theory *(p. 27)*
chi-square test ($\chi^2$ test) *(p. 47)*
conditional probability *(p. 45)*
controlled genetic cross *(p. 29)*
degrees of freedom (*df*) *(p. 49)*
dihybrid cross *(p. 35)*
dominant phenotype *(p. 30)*
$F_1$, $F_2$, $F_3$ generation *(p. 29)*
forked-line diagram *(p. 37)*
gametes *(p. 32)*

genotypic ratio (1:2:1 ratio) *(p. 32)*
heterozygous genotype
  (heterozygote) (p. 31)
homozygous genotype
  (homozygote) (p. 31)
law of independent assortment (Mendel's
  second law) *(p. 37)*
law of segregation (Mendel's first law),
  *(p. 32)*
mean ($\mu$) *(p. 48)*
monohybrid cross *(p. 32)*
normal (Gaussian) distribution *(p. 48)*
parental generation (P generation) *(p. 29)*
particulate inheritance *(p. 31)*
Pascal's triangle *(p. 46)*

pedigree *(p. 51)*
phenotypic ratio (3:1 ratio and 9:3:3:1
  ratio) *(p. 32)*
product rule (multiplication rule) *(p. 44)*
Punnett square *(p. 32)*
pure-breeding (true-breeding) *(p. 29)*
*P* value (probability value) *(p. 48)*
recessive phenotype *(p. 30)*
reciprocal cross *(p. 29)*
replicate cross *(p. 29)*
standard deviation (s) *(p. 48)*
sum rule (addition rule) *(p. 45)*
test cross (test-cross analysis) *(p. 30)*
transmission genetics *(p. 42)*
trihybrid cross *(p. 40)*

## PROBLEMS

*For instructor-assigned tutorials and problems, go to www.masteringgenetics.com.*

### Chapter Concepts

*For answers to selected even-numbered problems, see Appendix: Answers.*

1. Compare and contrast the following terms:
   a. dominant and recessive
   b. genotype and phenotype
   c. homozygous and heterozygous
   d. monohybrid cross and test cross
   e. dihybrid cross and trihybrid cross

2. For the cross $BB \times Bb$, what is the expected genotype ratio? What is the expected phenotype ratio?

3. For the cross $Aabb \times aaBb$, what is the expected genotype ratio? What is the expected phenotype ratio?

4. In mice, black coat color is dominant to white coat color. In the pedigree below, mice with a black coat are represented by darkened symbols, and those with white coats are shown as open symbols. Using allele symbols $B$ and $b$, determine the genotypes for each mouse.

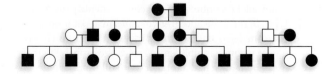

5. Two parents plan to have three children. What is the probability that the children will be two girls and one boy?

6. Consider the cross $AaBbCC \times AABbCc$.
   a. How many different gamete genotypes can each organism produce?
   b. Use a Punnett square to predict the expected ratio of offspring phenotypes.
   c. Use the forked-line method to predict the expected ratio of offspring phenotypes.

7. If a chi-square test produces a chi-square value of 7.83 with 4 degrees of freedom,
   a. in what interval range does the *P* value fall?
   b. is the result sufficient to reject the chance hypothesis?
   c. above what chi-square value would you reject the chance hypothesis for an experiment with 7 degrees of freedom?

8. Determine whether the statements below are true or false. If a statement is false, provide the correct information or revise the statement to make it correct.
   a. If a dihybrid cross is performed, the expected genotypic ratio is 9:3:3:1.
   b. A student uses the product rule to predict that the probability of flipping a coin twice and getting a head and then a tail is $\frac{1}{4}$.
   c. A test cross between a heterozygous parent and a homozygous recessive parent is expected to produce a 1:1 genotypic and phenotypic ratio.
   d. The outcome of a trihybrid cross is predicted by the law of segregation.

e. Reciprocal crosses that produce identical results demonstrate that a strain is pure-breeding.

f. If a woman is heterozygous for albinism, an autosomal recessive condition that results in the absence of skin pigment, the proportion of her gametes carrying the allele that allows pigment expression is expected to be 75%.

g. The progeny of a trihybrid cross are expected to have one of 27 different genotypes.

h. If a dihybrid $F_1$ plant is self-fertilized,
   (1) $\frac{9}{16}$ of the progeny will have the same phenotype as the $F_1$ parent.
   (2) $\frac{1}{16}$ of the progeny will be true-breeding.
   (3) $\frac{1}{2}$ of the progeny will be heterozygous at one or both loci.

9. In the datura plant, purple flower color is controlled by a dominant allele $P$. White flowers are found in plants homozygous for the recessive allele $p$. A purple-flowered datura plant is self-fertilized, and its progeny are 28 purple-flowered plants and 10 white-flowered plants.

   a. What is the genotype of the original purple-flowered plant?

   b. If the 28 purple-flowered progeny are self-fertilized, what proportion of them will breed true?

10. The dorsal pigment pattern of frogs can be either "leopard" (white pigment between dark spots) or "mottled" (pigment between spots appears mottled). The trait is controlled by an autosomal gene. Males and females are selected from pure-breeding populations, and a pair of reciprocal crosses is performed. The cross results are shown below.

   Cross 1:    P: Male leopard × female mottled
              $F_1$: All mottled
              $F_2$: 70 mottled, 22 leopard

   Cross 2:    P: Male mottled × female leopard
              $F_1$: All mottled
              $F_2$: 50 mottled, 18 leopard

   a. Which of the phenotypes is dominant? Explain your answer.

   b. Compare and contrast the results of the reciprocal crosses in the context of autosomal gene inheritance.

   c. In the $F_2$ progeny from both crosses, what proportion is expected to be homozygous? What proportion is expected to be heterozygous?

   d. Propose two different genetic crosses that would allow you to determine the genotype of one mottled frog from the $F_2$ generation.

11. Black skin color is dominant to pink skin color in pigs. Two heterozygous black pigs are crossed.

   a. What is the probability that their offspring will have pink skin?

   b. What is the probability that the first and second offspring will have black skin?

   c. If these pigs produce a total of three piglets, what is the probability that two will be pink and one will be black?

12. A male mouse with brown fur color is mated to two different female mice with black fur. Black female 1 produces a litter of 9 black and 7 brown pups. Black female 2 produces 14 black pups.

   a. What is the mode of inheritance of black and brown fur color in mice?

   b. Choose symbols for each allele, and identify the genotypes of the brown male and the two black females.

13. Figure 2.13 shows the results of Mendel's test-cross analysis of independent assortment. In this experiment, he first crossed pure-breeding round, yellow plants to pure-breeding wrinkled, green plants. The round yellow $F_1$ are crossed to pure-breeding wrinkled, green plants. Use chi-square analysis to show that Mendel's results do not differ significantly from those expected.

14. An experienced goldfish breeder receives two unusual male goldfish. One is black rather than gold, and the other has a single tail fin rather than a split tail fin. The breeder crosses the black male to a female that is gold. All the $F_1$ are gold. She also crosses the single-finned male to a female with a split tail fin. All the $F_1$ have a split tail fin. She then crosses the black male to $F_1$ gold females and, separately, crosses the single-finned male to $F_1$ split-finned females. The results of the crosses are shown below.

**Black Male × $F_1$ Gold Female**

| Gold | Black |
| --- | --- |
| 32 | 34 |

**Single-Finned Male × $F_1$ Split-Finned Female**

| Split Fin | Single Fin |
| --- | --- |
| 41 | 39 |

   a. What do the results of these crosses suggest about the inheritance of color and tail fin shape in goldfish?

   b. Is black color dominant or recessive? Explain. Is single tail dominant or recessive? Explain.

   c. Use chi-square analysis to test your hereditary hypothesis for each trait.

15. The pedigree below shows the transmission of albinism (absence of skin pigment) in a human family.

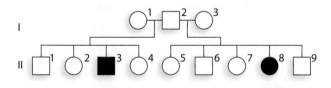

   a. What is the most likely mode of transmission of albinism in this family?

   b. Using allelic symbols of your choice, identify the genotypes of the male and his two mates in generation I.

   c. The female I-1 and her mate, male I-2, had four children, one of whom has albinism. What is the probability that they could have had a total of four children with *any other outcome* except one child with albinism and three with normal pigmentation?

   d. What is the probability that female I-3 is a heterozygous carrier of the allele for albinism?

e. One child of female I-3 has albinism. What is the probability that any of the other four children are carriers of the allele for albinism?

16. A geneticist crosses a pure-breeding strain of peas producing yellow, wrinkled seeds with one that is pure-breeding for green, smooth seeds.

   a. Use a Punnett square to predict the $F_2$ progeny that would be expected if the $F_1$ are allowed to self-fertilize.
   b. What proportion of the $F_2$ progeny are expected to have yellow seeds? Wrinkled seeds? Green seeds? Smooth seeds?
   c. What is the expected phenotype distribution among the $F_2$ progeny?

17. Suppose an $F_1$ plant from Problem 16 is crossed to the pure-breeding green, smooth parental strain. Use a forked-line diagram to predict the phenotypic distribution of the resulting progeny.

18. In pea plants, the appearance of flowers along the main stem is a dominant phenotype called "axial" and is controlled by an allele $T$. The recessive phenotype, produced by an allele $t$, has flowers only at the end of the stem and is called "terminal." Pod form displays a dominant phenotype "inflated," controlled by an allele $C$, and a recessive

"constricted" form, produced by the $c$ allele. A pure-breeding axial, constricted plant is crossed to one that is pure-breeding terminal, inflated.

   a. The $F_1$ progeny of this cross are allowed to self-fertilize. What is the expected phenotypic distribution among the $F_2$ progeny?
   b. All of the $F_2$ progeny with terminal flowers are saved and allowed to self-fertilize to produce a partial $F_3$ generation. What is the expected phenotypic distribution among these $F_3$ plants?
   c. If an $F_1$ plant from the initial cross described above is crossed to a plant that is terminal, constricted, what is the expected distribution among the resulting progeny?
   d. If the plants with terminal flowers produced by the cross in part (c) are saved and allowed to self-fertilize, what is the expected phenotypic distribution among the progeny?

19. If two six-sided dice are rolled, what is the probability that the total number of spots showing is

   a. 4?
   b. 7?
   c. greater than 5?
   d. an odd number?

## Application and Integration

*For answers to selected even-numbered problems, see Appendix: Answers.*

20. Experimental Insight 2.1 describes data on the kernel color distribution of bicolor corn, collected by a genetics class like yours. To test the hypothesis that the kernel color of bicolor corn is the result of the segregation of two alleles at a single genetic locus, the class counted 9882 kernels and found that 7506 were yellow and 2376 were white. Use chi-square analysis to evaluate the fit between the segregation hypothesis and the class results.

21. The pedigree below shows the transmission of a phenotypic character.

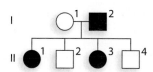

Using $B$ to represent a dominant allele and $b$ to represent a recessive allele,

   a. give the genotype(s) possible for each member of the family, assuming the trait is autosomal dominant.
   b. give the genotype(s) possible for each member of the family, assuming the trait is autosomal recessive.
   c. extend the pedigree by giving II-4 a mate and 4 children. Modify the pedigree as necessary to make it consistent with the transmission of a dominant phenotype. Do the same for the transmission of a recessive phenotype.

22. The seeds in bush bean pods are each the product of an independent fertilization event. Green seed color is dominant to white seed color in bush beans. If a heterozygous

plant with green seeds self-fertilizes, what is the probability that 6 seeds in a single pod of the progeny plant will consist of

   a. 3 green and 3 white seeds?
   b. all green seeds?
   c. at least 1 white seed?

23. List all the different gametes that are possible from the following genotypes.

   a. *AABbCcDd*
   b. *AabbCcDD*
   c. *AaBbCcDd*
   d. *AabbCCdd*

24. Organisms with the genotypes *AABbCcDd* and *AaBbCcDd* are crossed. What are the expected proportions of the following progeny?

   a. *A–B–C–D–*
   b. *AabbCcDd*
   c. a phenotype identical to either parent
   d. *A–B–ccdd*

25. In humans, the ability to bend the thumb back beyond vertical is called hitchhiker's thumb and is recessive to the inability to do so (OMIM 274200). Also, the presence of attached earlobes is recessive to unattached earlobes (OMIM 128900). In the pedigree shown, the left half of the circle or square is filled if the person has the dominant non-hitchhiker's thumb and empty if hitchhiker's thumb is present. The right half of the symbol is filled if the person has unattached earlobes and is empty if earlobes are attached. Use allelic symbols $H$ and $h$ for the thumb and $E$ and $e$ for earlobes, and identify the genotypes for each family member.

26. In the fruit fly *Drosophila*, a rudimentary wing called "vestigial" and dark body color called "ebony" are inherited at independent loci and are recessive to their dominant wild-type counterparts, full wing and gray body color. Dihybrid wild-type males and females are crossed, and 3200 progeny are produced. How many progeny flies are expected to be found in each phenotypic class?

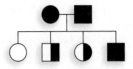

27. In pea plants, plant height, seed shape, and seed color are governed by three independently assorting loci. The three loci show dominance and recessiveness with tall ($T$) dominant to short ($t$), round ($R$) dominant to wrinkled ($r$), and yellow ($G$) dominant to green ($g$).

    a. If a true-breeding tall, wrinkled, yellow plant is crossed to a true-breeding short, round, green plant, what phenotypic ratios are expected in the $F_1$ and $F_2$?
    b. What proportion of the $F_2$ are expected to be tall, wrinkled, yellow? *ttRRGg*?
    c. What proportion of the $F_2$ that produce round, green seeds (regardless of the height of the plant) are expected to breed true?

28. A variety of pea plant called Blue Persian produces a tall plant with blue seeds. A second variety of pea plant called Spanish Dwarf produces a short plant with white seed. The two varieties are crossed, and the resulting seeds are collected. All of the seeds are white; and when planted, they produce all tall plants. These tall $F_1$ plants are allowed to self-fertilize. The results for seed color and plant stature in the $F_2$ generation are as follows:

| $F_2$ Plant Phenotype | Number |
|---|---|
| Blue seed, tall plant | 97 |
| White seed, tall plant | 270 |
| Blue seed, short plant | 33 |
| White seed, short plant | 100 |
| TOTAL | 500 |

    a. Which phenotypes are dominant, and which are recessive? Why?
    b. What is the expected distribution of phenotypes in the $F_2$ generation?
    c. State the hypothesis being tested in this experiment.
    d. Examine the data in the table by the chi-square test, and determine whether they conform to expectations of the hypothesis.

29. In tomato plants, the production of red fruit color is under the control of an allele $R$. Yellow tomatoes are *rr*. The dominant phenotype for fruit shape is under the control of an allele *T*, which produces two lobes. Multilobed fruit, the recessive phenotype, have the genotype *tt*. Two different crosses are made between parental plants of unknown genotype and phenotype. Use the progeny phenotype ratios to determine the genotypes and phenotypes of each parent.

Cross 1 Progeny:  $\frac{3}{8}$  two-lobed, red
  $\frac{3}{8}$  two-lobed, yellow
  $\frac{1}{8}$  multilobed, red
  $\frac{1}{8}$  multilobed, yellow

Cross 2 Progeny:  $\frac{1}{4}$  two-lobed, red
  $\frac{1}{4}$  two-lobed, yellow
  $\frac{1}{4}$  multilobed, red
  $\frac{1}{4}$  multilobed, yellow

30. A male and a female are each heterozygous for both cystic fibrosis (CF) and phenylketonuria (PKU). Both conditions are autosomal recessive, and they assort independently.
    a. What proportion of the children of this couple will have neither condition?
    b. What proportion of the children will have either PKU or CF?
    c. What proportion of the children will be carriers of one or both conditions?

31. In a sample of 640 families with 6 children each, the distribution of boys and girls is as shown in the following table:

| Number of families | 9 | 63 | 147 | 204 | 151 | 56 | 10 |
|---|---|---|---|---|---|---|---|
| Number of girls | 0 | 1 | 2 | 3 | 4 | 5 | 6 |
| Number of boys | 6 | 5 | 4 | 3 | 2 | 1 | 0 |

    a. Are the numbers of boys to girls in these families consistent with the expected 1:1 ratio? Support your answer by chi-square analysis.
    b. Is the distribution of the numbers of boys and girls in the families consistent with the expectations of binomial probability? Support your answer.

32. A sample of 120 families with 4 children each in which both parents are carriers of an autosomal recessive mutation for cystic fibrosis (CF) produces the following distribution of children with and without cystic fibrosis:

| Number of families | 16 | 52 | 32 | 18 | 2 |
|---|---|---|---|---|---|
| Children with CF | 0 | 1 | 2 | 3 | 4 |
| Children free of CF | 4 | 3 | 2 | 1 | 0 |

    a. Is the total number of children with CF in these families consistent with the expected ratio? Support your answer.
    b. What is the expected distribution of the number of families with 0 through 4 children with CF in this sample under the assumptions of binomial probability?
    c. Is the distribution of families with 0 through 4 children with CF consistent with the ratios expected under binomial probability? Support your answer.

33. A woman expressing a dominant phenotype is heterozygous (*Dd*) at the locus.
    a. What is the probability that the dominant allele carried by the woman will be inherited by a grandchild?
    b. What is the probability that two grandchildren of the woman who are first cousins to one another will each inherit the dominant allele?

c. Draw a pedigree that illustrates the transmission of the dominant trait from the grandmother to two of her grandchildren who are first cousins.

34. Two parents who are each known to be carriers of an autosomal recessive allele have four children. None of the children has the recessive condition. What is the probability that one or more of the children is a carrier of the recessive allele?

35. An organism having the genotype $AaBbCcDdEe$ is self-fertilized. Assuming the loci assort independently, determine the following proportions:

   a. gametes that are expected to carry only dominant alleles
   b. progeny that are expected to have a genotype identical to that of the parent
   c. progeny that are expected to have a phenotype identical to that of the parent
   d. gametes that are expected to be $ABcde$
   e. progeny that are expected to have the genotype $AabbCcDdE–$

36. A man and a woman are each heterozygous carriers of an autosomal recessive mutation of a disorder that is fatal in infancy. They both want to have multiple children, but they are concerned about the risk of the disorder appearing in one or more of their children. In separate calculations, determine the probabilities of the couple having five children with 0, 1, 2, 3, 4, and all 5 children being affected by the disorder.

37. For a single dice roll, there is a $\frac{1}{6}$ chance that any particular number will appear. For a pair of dice, each specific combination of numbers has a probability of $\frac{1}{36}$ of occurring. Most total values of two dice can occur more than one way. As a test of random probability theory, a student decides to roll a pair of six-sided dice 300 times and tabulate the results. She tabulates the number of times each different total value of the two dice occurs. Her results are the following:

| Total Value of Two Dice | Number of Times Rolled |
|---|---|
| 2 | 7 |
| 3 | 11 |
| 4 | 23 |
| 5 | 36 |
| 6 | 42 |
| 7 | 53 |
| 8 | 40 |
| 9 | 38 |
| 10 | 30 |
| 11 | 12 |
| 12 | 8 |
| TOTAL | 300 |

The student tells you that her results fail to prove that random chance is the explanation for the outcome of this experiment. Is she correct or incorrect? Support your answer.

38. You have four guinea pigs for a genetic study. One male and one female are from a strain that is pure-breeding for short brown fur. A second male and female are from a strain that is pure-breeding for long white fur. You are asked to perform two *different* experiments to test the proposal that short fur is dominant to long fur and that brown is dominant to white. You may use any of the four original pure-breeding guinea pigs or any of their offspring in experimental matings. Design two different experiments (crossing different animals and using different combinations of phenotypes) to test the dominance relationships of alleles for fur length and color, and make predictions for each cross based on the proposed relationships. Anticipate that the litter size will be 12 for each mating and that female guinea pigs can produce three litters in their lifetime.

39. Galactosemia is an autosomal recessive disorder caused by the inability to metabolize galactose, a component of the lactose found in mammalian milk. Galactosemia can be partially managed by eliminating dietary intake of lactose and galactose. Amanda is healthy, as are her parents, but her brother Alonzo has galactosemia. Brice has a similar family history. He and his parents are healthy, but his sister Brianna has galactosemia. Amanda and Brice are planning a family and seek genetic counseling. Based on the information provided, complete the following activities and answer the questions.

   a. Draw a pedigree that includes Amanda, Brice, their siblings, and parents. Identify the genotype of each person, using $G$ and $g$ to represent the dominant and recessive alleles, respectively.
   b. What is the probability that Amanda is a carrier of the allele for galactosemia? What is the probability that Brice is a carrier? Explain your reasoning for each answer.
   c. What is the probability that the first child of Amanda and Brice will have galactosemia? Show your work.
   d. If the first child has galactosemia, what is the probability that the second child will have galactosemia? Explain the reasoning for your answer.

40. Sweet yellow tomatoes with a pear shape bring a high price per basket to growers. Pear shape, yellow color, and terminal flower position are recessive traits produced by alleles $f$, $r$, and $t$, respectively. The dominant phenotypes for each trait—full shape, red color, and axial flower position—are the product of dominant alleles $F$, $R$, and $T$. A farmer has two pure-breeding tomato lines. One is full, yellow, terminal and the other is pear, red, axial. Design a breeding experiment that will produce a line of tomato that is pure-breeding for pear shape, yellow color, and axial flower position.

41. A cross between a spicy variety of *Capsicum annum* pepper and a sweet (nonspicy) variety produces $F_1$ progeny plants that all have spicy peppers. The $F_1$ are crossed, and among the $F_2$ plants are 56 that produce spicy peppers and 20 that produce sweet peppers. Dr. Ara B. Dopsis, an expert on pepper plants, discovers a gene designated *Pun1* that he believes is responsible for spicy versus sweet flavor of peppers. Dr. Dopsis proposes that a dominant allele $P$ produces spicy peppers and that a recessive mutant allele $p$ results in sweet peppers.

   a. Are the data on the parental cross and the $F_1$ and $F_2$ consistent with the proposal made by Dr. Dopsis? Explain why or why not, using $P$ and $p$ to indicate probable genotypes of pepper plants.

b. Assuming the proposal is correct, what proportion of the spicy $F_2$ pepper plants do you expect will be pure-breeding? Explain your answer.

42. Alkaptonuria is an infrequent autosomal recessive condition. It is first noticed in newborns when the urine in their diapers turns black upon exposure to air. The condition is caused by the defective transport of the amino acid phenylalanine through the intestinal walls during digestion. About 4 people per 1000 are carriers of alkaptonuria.

    Sara and James had never heard of alkaptonuria and were shocked to discover that their first child had the condition. Sara's sister Mary and her husband Frank are planning to have a family and are concerned about the possibility of alkaptonuria in one of their children.

    The four adults (Sara, James, Mary, and Frank) seek information from a neighbor who is a retired physician. After discussing their family histories, the neighbor says, "I never took genetics, but I know from my many years in practice that Sara and James are both carriers of this recessive condition. Since their first child had the condition, there is a very low chance that the next child will also have it, because the odds of having two children with a recessive condition are very low. Mary and Frank have no chance of having a child with alkaptonuria because Frank has no family history of the condition." The two couples each have babies and *both* babies have alkaptonuria.

    a. What are the genotypes of the four adults?
    b. What was incorrect about the information given to Sara and James? What is incorrect about the information given to Mary and Frank?
    c. What is the probability that the second child of Mary and Frank will have alkaptonuria?
    d. What is the chance that the third child of Sara and James will be free of the condition?
    e. The couples are worried that one of their grandchildren will inherit alkaptonuria. How would you assess the risk that one of the offspring of a child with alkaptonuria will inherit the condition?

43. Humans vary in many ways from one another. Among many minor phenotypic differences, there are five independently assorting traits that have a dominant and a recessive phenotype. The traits are (1) forearm hair (alleles $F$ and $f$)—the presence of hair on the forearm is dominant to the absence of hair on the forearm; (2) earlobe form (alleles $E$ and $e$)—unattached earlobes are dominant to attached earlobes; (3) widow's peak (alleles $W$ and $w$)—a distinct "V" shape to the hairline at the top of the forehead is dominant to a straight hairline; (4) hitchhiker's thumb (alleles $H$ and $h$)—the ability to bend the thumb back beyond vertical is dominant and the inability to do so is recessive; and (5) freckling (alleles $D$ and $d$)—the appearance of freckles is dominant to the absence of freckles.

    If a couple with the genotypes $Ff\,Ee\,Ww\,Hh\,Dd$ and $Ff\,Ee\,Ww\,Hh\,Dd$ have children, what is the chance the children will inherit the following characteristics?
    a. the same phenotype as the parents
    b. four dominant traits and one recessive trait
    c. all recessive traits
    d. the genotype $Ff\,EE\,Ww\,hh\,dd$

44. In chickens, the presence of feathers on the legs is due to a dominant allele ($F$), and the absence of leg feathers is due to a recessive allele ($f$). The comb on the top of the head can be either pea-shaped, a phenotype that is controlled by a dominant allele ($P$), or a single comb controlled by a recessive allele ($p$). The two loci assort independently. Assume that a pure-breeding rooster that has feathered legs and a single comb is crossed with a pure-breeding hen that has no leg feathers and a pea-shaped comb. The $F_1$ are crossed to produce the $F_2$. Among the resulting $F_2$, however, only birds with a single comb and feathered legs are allowed to mate. These chickens mate at random to produce $F_3$ progeny. What are the expected genotypic and phenotypic ratios among the resulting $F_3$ progeny?

45. A pure-breeding fruit fly with the recessive mutation cut wing, caused by the homozygous $cc$ genotype, is crossed to a pure-breeding fly with normal wings, genotype $CC$. Their $F_1$ progeny all have normal wings. $F_1$ flies are crossed, and the $F_2$ progeny have a 3:1 ratio of normal wing to cut wing. One male $F_2$ fly with normal wings is selected at random and mated to an $F_2$ female with normal wings. Using all possible genotypes of the $F_2$ flies selected for this cross, list all possible crosses between the two flies involved in this mating, and determine the probability of each cross.

# Cell Division and Chromosome Heredity

3

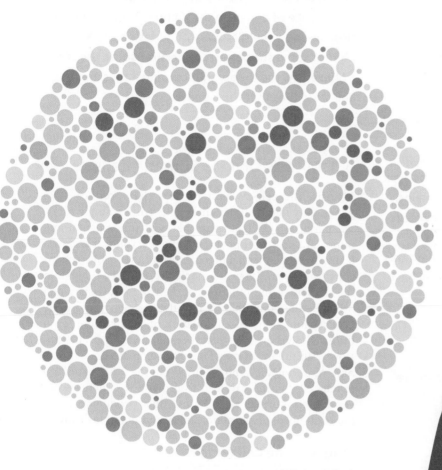

The most common form of color blindness is an X-linked recessive condition detected in individuals who are unable to see the number 22 in this image.

## ESSENTIAL IDEAS

- The cell cycle consists of interphase, during which cells carry out regular functions and replicate their DNA, and M phase, the cell-division segment of the cycle.

- Mitosis divides somatic cells and produces two genetically identical daughter cells.

- Meiosis occurs in germ-line cells and produces four genetically different haploid cells that mature to form gametes.

- The separation of chromosomes and sister chromatids during meiosis is the mechanical basis of Mendel's laws of segregation and independent assortment.

- The chromosome theory of heredity identified chromosomes as the cell structures containing genes.

- Sex determination is controlled by chromosomal and genetic factors that vary among species.

- Dosage compensation equalizes the expression of sex-linked genes of males and females of animal species.

A couple of decades or so ago, at the moment of conception that culminated in your birth, two gametes united to form the single fertilized cell—the zygote—from which you developed. Your sex was determined in that instant by the sex chromosome carried by the fertilizing sperm—an X chromosome if you are female or a Y chromosome if you are male. Shortly after fertilization, cell division began, and over the next few hours produced a tiny zygote consisting of two cells, then four cells, then eight cells, and so on, as it moved down the fallopian tube toward the uterus. In the first days of your development, these cell divisions produced hundreds of exact genetic replicas of the original cell. About 1 week after fertilization, the zygote implanted into the uterine wall; and within 2 weeks of conception, genetically controlled processes of cell differentiation and cell specialization began to form the first embryonic organs and structures. These processes eventually determined the structure and function of each cell in your body (see Chapter 20).

Since then, your body has produced thousands of generations of cells. The mechanism of cell division that produced most of them, **mitosis,** is an ongoing process that with each division creates two identical **daughter cells** that are exact genetic replicas of the parental cell they are derived from. Mitosis is responsible for the growth of your body; it repairs the damage and injury your body sustains, and it maintains your body by producing new cells to replace those that undergo programmed cell death (apoptosis). While you have been reading this passage, approximately 200 cells in your body have undergone mitotic division.

Of the trillions of cells in your body, most are **somatic cells,** the cells that form all your organs, structures, and tissues. Somatic cells of most eukaryotes contain multiple sets of chromosomes. The most common multiple of chromosome sets in animal nuclei is two, and the number of chromosomes present as homologous pairs in such nuclei is called the **diploid number.** Your somatic cell nuclei contain 46 chromosomes each, in 23 homologous pairs, so your diploid number is 46. The diploid number varies among species (each species having its characteristic number of pairs) and so is identified nonspecifically as $2n$. The value $n$ represents the **haploid number** of chromosomes, a value that is one-half the diploid number and is the number of chromosomes contained in the nuclei of gametes, the nonsomatic cells.

**Gametes,** produced from **germ-line cells,** are the germinal, or reproductive, cells: sperm and egg in animals, or pollen and egg in plants. Germ-line cells divide by the process known as **meiosis** that is different in several ways from mitosis.

In this chapter, we examine both mitosis and meiosis, and we look closely at the connection between meiotic cell division and Mendel's laws

of heredity. We also explore patterns of sex determination in eukaryotes and look at processes that equalize the expression of genes carried on **sex chromosomes,** the chromosomes that determine sex. In addition, we study the special patterns of inheritance of genes on the X chromosome, and we describe how the discovery of genes on the X chromosome supported the **chromosome theory of heredity,** the theory that chromosomes are the cell structures that carry genes.

## 3.1 Mitosis Divides Somatic Cells

Mitosis, the cell-division process that produces two genetically identical daughter cells from a single original parental cell, is among the most fundamental and important processes occurring in eukaryotes. It is genetically controlled with a precise script that enables organisms to grow and develop normally as well as maintain the structures and functions of their organs, tissues, and other bodily components. Life depends on the orderly progression and proper regulation of cell division. If too little cell division takes place, or cell division occurs too slowly, an organism may fail to develop at all, or it may have morphologic abnormalities. On the other hand, too much cell division can lead to growth of structures beyond their normal boundaries, thus producing other kinds of morphologic abnormality and possible death.

### Stages of the Cell Cycle

Cell division is regulated by genetic control of the **cell cycle,** the life cycle cells must pass through in order to replicate their DNA and divide. Since well-regulated cell division is such an integral part of life, it will not surprise you to learn that the cell cycles of all eukaryotes are similar, and that much of the molecular machinery that controls the cell cycle is evolutionarily conserved in plants and animals. The striking similarity of cell cycle control genes and processes in plants and animals, and the sharing of many of these genes with Bacteria and Archaea, is powerful evidence that all life evolved from a single common ancestor.

The eukaryotic cell cycle is divided into two principal phases—**M phase,** a short segment of the cell cycle during which cells divide, and **interphase,** the longer period between one M phase and the next (**Figure 3.1a**). Interphase consists of three successive stages, $G_1$, S, and $G_2$. During these stages, respectively, the cell expresses its genetic information, replicates its chromosomes, and prepares for entry into M phase. M phase is divided into substages that correspond to the progress of the cell during its division.

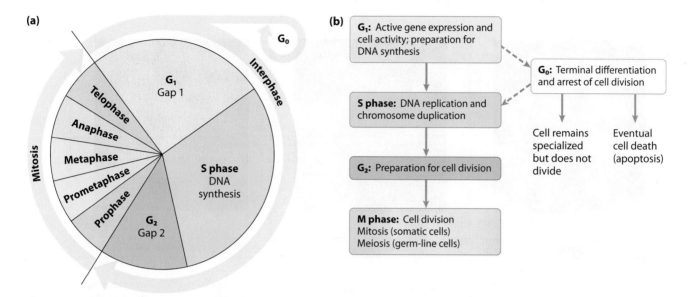

**Figure 3.1  The cell cycle.** **(a)** The cell cycle is divided into interphase and M phase, which are each further subdivided. **(b)** An overview of cell cycle activities.

When viewed under a light microscope, somatic cells in interphase may appear rather placid, but their outward appearance gives little indication of the complex activity taking place inside. During the **$G_1$** (or **Gap 1**) **phase** of interphase, for example, cells actively transcribe and translate all the protein products necessary for normal cellular structure and function (**Figure 3.1b**). Gene expression occurs throughout the rest of the cell cycle as well, but the process is most active during the $G_1$ stage. Of course, cells of different types vary in how many genes they express, in how they function in the body, and in how they interact with other cells. Consequently, the duration of $G_1$ varies. Some types of cells are rapidly dividing and spend only a short time, perhaps as little as a few hours, in $G_1$. Other cells linger in $G_1$ for periods of days, weeks, or more.

As they approach the end of $G_1$, cells follow one of two alternative paths. Most cells enter the **S phase,** or **synthesis phase,** during which DNA replication (DNA synthesis) takes place. On the other hand, a small subset of specialized cells transition from $G_1$ into a nondividing state called **$G_0$** ("G zero"), a kind of semiperpetual $G_1$-like state in which cells express their genetic information and carry out normal functions but do not progress through the cell cycle (see Figure 3.1b). Several kinds of cells in your body, including certain cells in your eyes and bones, reach a mature state of differentiation, enter $G_0$, and rarely if ever divide again. Most $G_0$ cells maintain their specialized functions until they enter programmed cell death (apoptosis) and die. Cells only rarely leave $G_0$ and resume the cell cycle.

DNA replication takes place during S phase and results in a doubling of the amount of DNA in each nucleus and the creation of two *sister chromatids* for each chromosome. Entry into the S phase almost always commits the cell to proceeding through the remainder of the cycle and then dividing. The completion of S phase brings about the transition to the **$G_2$,** or **Gap 2,** phase of the cell cycle, during which cells prepare for division. Interphase ends when cells enter M phase, from which two identical daughter cells emerge.

The successive generations of cells produced through mitosis as one cell cycle follows the next are known as cell lines or cell lineages. Each cell line or cell lineage contains identical cells (i.e., clones) that are all descended from a single founder cell. Mitosis ensures that the genetic information in cells is faithfully passed to successive generations of cell lineages. Occasional mutations occur in individual cells, however, and these are also perpetuated during the proliferation of the cell line.

## Substages of M Phase

M phase follows interphase and is divided into five substages—**prophase, prometaphase, metaphase, anaphase,** and **telophase**—whose principal features are described in **Figure 3.2**. These five substages accomplish two important functions of cell division—karyokinesis and cytokinesis. **Karyokinesis** is the equal partitioning of the chromosomal material in the nucleus of the parental cell between the nuclei of the two daughter cells. This process requires first that each of the chromosomes in the nucleus be fully and accurately duplicated and then that the duplicate copies of each chromosome be separated so that one copy goes to the nucleus of one daughter cell and the second copy goes to the other daughter nucleus. Karyokinesis is followed by **cytokinesis,** the partitioning of the cytoplasmic contents of the parental cell into the daughter cells. Cytokinesis does not demand the same degree of equivalency required in karyokinesis. The cytoplasm of the parental cells contains an abundance of the proteins and organelles that the daughter cells require in order to function, so the division of this material need not be equal.

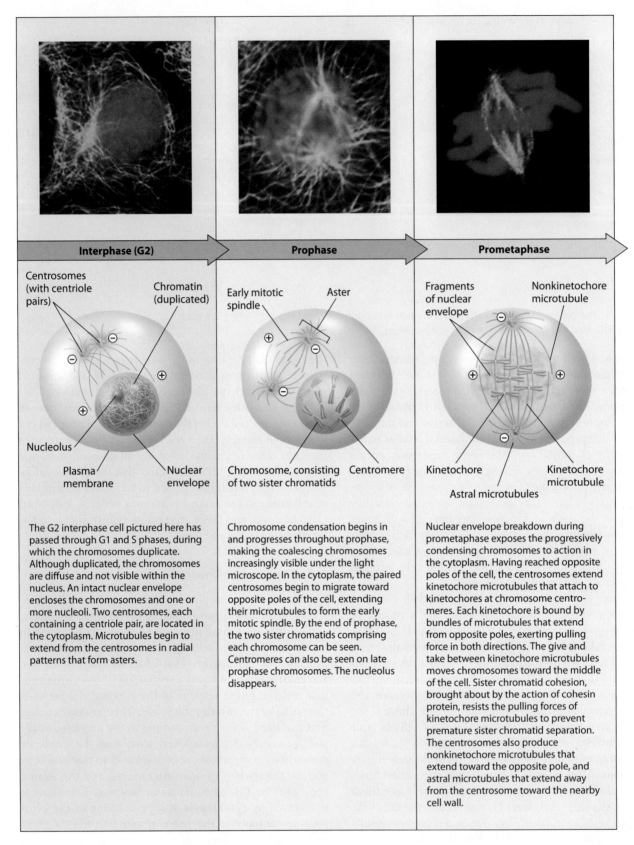

**Interphase (G2)**

Centrosomes (with centriole pairs)

Chromatin (duplicated)

Nucleolus

Plasma membrane

Nuclear envelope

The G2 interphase cell pictured here has passed through G1 and S phases, during which the chromosomes duplicate. Although duplicated, the chromosomes are diffuse and not visible within the nucleus. An intact nuclear envelope encloses the chromosomes and one or more nucleoli. Two centrosomes, each containing a centriole pair, are located in the cytoplasm. Microtubules begin to extend from the centrosomes in radial patterns that form asters.

**Prophase**

Early mitotic spindle

Aster

Chromosome, consisting of two sister chromatids

Centromere

Chromosome condensation begins in and progresses throughout prophase, making the coalescing chromosomes increasingly visible under the light microscope. In the cytoplasm, the paired centrosomes begin to migrate toward opposite poles of the cell, extending their microtubules to form the early mitotic spindle. By the end of prophase, the two sister chromatids comprising each chromosome can be seen. Centromeres can also be seen on late prophase chromosomes. The nucleolus disappears.

**Prometaphase**

Fragments of nuclear envelope

Nonkinetochore microtubule

Kinetochore

Astral microtubules

Kinetochore microtubule

Nuclear envelope breakdown during prometaphase exposes the progressively condensing chromosomes to action in the cytoplasm. Having reached opposite poles of the cell, the centrosomes extend kinetochore microtubules that attach to kinetochores at chromosome centromeres. Each kinetochore is bound by bundles of microtubules that extend from opposite poles, exerting pulling force in both directions. The give and take between kinetochore microtubules moves chromosomes toward the middle of the cell. Sister chromatid cohesion, brought about by the action of cohesin protein, resists the pulling forces of kinetochore microtubules to prevent premature sister chromatid separation. The centrosomes also produce nonkinetochore microtubules that extend toward the opposite pole, and astral microtubules that extend away from the centrosome toward the nearby cell wall.

**Figure 3.2 Interphase and the five stages of mitosis.**

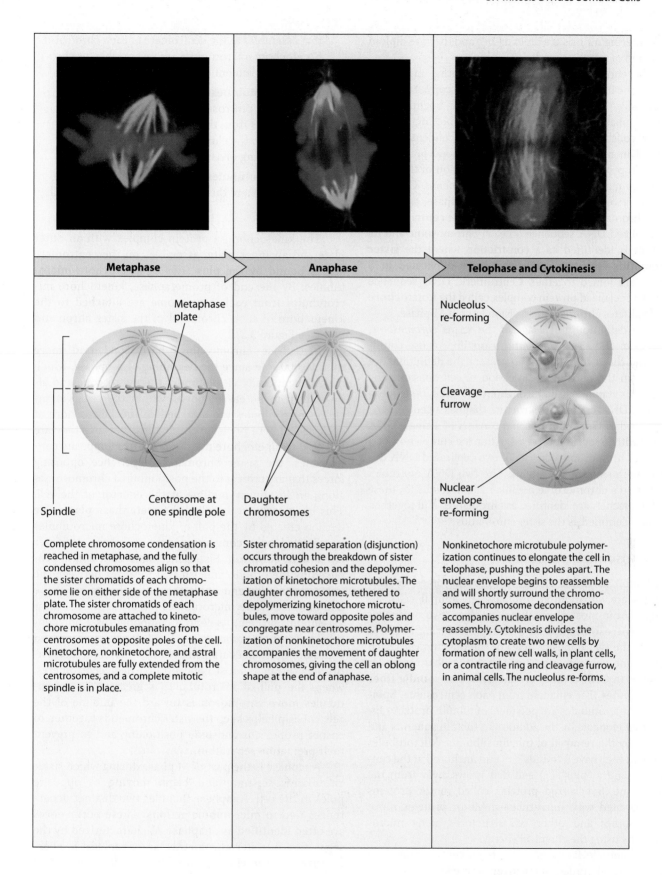

**Metaphase**

Metaphase plate

Spindle

Centrosome at one spindle pole

Complete chromosome condensation is reached in metaphase, and the fully condensed chromosomes align so that the sister chromatids of each chromosome lie on either side of the metaphase plate. The sister chromatids of each chromosome are attached to kinetochore microtubules emanating from centrosomes at opposite poles of the cell. Kinetochore, nonkinetochore, and astral microtubules are fully extended from the centrosomes, and a complete mitotic spindle is in place.

**Anaphase**

Daughter chromosomes

Sister chromatid separation (disjunction) occurs through the breakdown of sister chromatid cohesion and the depolymerization of kinetochore microtubules. The daughter chromosomes, tethered to depolymerizing kinetochore microtubules, move toward opposite poles and congregate near centrosomes. Polymerization of nonkinetochore microtubules accompanies the movement of daughter chromosomes, giving the cell an oblong shape at the end of anaphase.

**Telophase and Cytokinesis**

Nucleolus re-forming

Cleavage furrow

Nuclear envelope re-forming

Nonkinetochore microtubule polymerization continues to elongate the cell in telophase, pushing the poles apart. The nuclear envelope begins to reassemble and will shortly surround the chromosomes. Chromosome decondensation accompanies nuclear envelope reassembly. Cytokinesis divides the cytoplasm to create two new cells by formation of new cell walls, in plant cells, or a contractile ring and cleavage furrow, in animal cells. The nucleolus re-forms.

Cells entering mitosis are diploid (2*n*), and they are diploid at the end of mitosis as well.

The chromosomes are so diffuse during interphase that they cannot be seen by light microscopy. However, progressive chromosome condensation, beginning in early prophase, condenses chromosomes into increasingly tighter coils until a maximum level of condensation is reached in metaphase. The nuclear envelope breakdown occurs in prophase. Progressive condensation of the chromosomes makes them visible in mid-prophase. Chromosome centromeres become visible in prophase, as do the sister chromatids of each chromosome. The **centromere** is a specialized DNA sequence on each chromosome and its location is identified as a constriction where the **sister chromatids**—the two copies that were duplicated in S phase—are joined together. Centromeric DNA sequence binds a specialized protein complex called the **kinetochore** that facilitates chromosome division later in M phase.

The definition and usage of the terms *chromosome*, *chromatid*, and *sister chromatid* sometimes cause confusion, and this is a good time to present the definitions we will use in the remaining discussion of cell division. The term *chromosome* is used throughout the cell cycle to identify each DNA-containing structure that has a centromere. At the end of $G_1$, a chromosome consists of a single DNA duplex with associated proteins. After the completion of S phase, a chromosome consists of two replicated DNA duplexes with associated proteins. The two DNA molecules making up a chromosome are identical. Individually, these DNA molecules are identified as chromatids; and together, they are identified as the sister chromatids.

## Chromosome Distribution

In addition to visible changes to chromosomes, cellular changes are apparent in prophase. In animal cells, although not in most plants, fungi, or algae, two organelles called **centrosomes** appear that migrate during M phase to form the two opposite poles of the dividing cell. Each centrosome contains a pair of subunits called centrioles (Figure 3.3). Centrosomes are the source of **spindle fiber microtubules** that emanate from each centrosome. Spindle fiber microtubules are polymers of tubulin protein subunits that elongate by the addition of tubulin subunits and shorten by the removal of tubulin subunits. Microtubules are polar; they have a "minus" (−) end anchored at the centrosome and a "plus" (+) end that grows away from the centrosome. Specialized proteins called motor proteins are associated with microtubules. Motor proteins move chromosomes and other cell structures along microtubules by using the chemical energy.

The spindle fibers emanate from centrosomes in a 360° pattern identified as the **aster.** Three kinds of microtubules are identified in cells:

1. **Kinetochore microtubules** embed in the protein complex called the kinetochore (described shortly) that assembles at the centromere of each chromatid. Kinetochore microtubules are responsible for chromosome movement during cell division.

2. **Polar microtubules** extend toward the opposite pole of their centrosome and overlap with polar microtubules from that pole. These microtubules contribute to the elongation of the cell and to cell stability during division.

3. **Astral microtubules** grow toward the membrane of the cell, where they attach and contribute to cell stability.

The kinetochore, a protein complex with an outer plate and an inner plate, assembles on the centromere and is bound by the plus ends of kinetochore microtubules. By the end of prometaphase, kinetochore microtubules from each centrosome are attached to the kinetochore of each chromatid of the sister chromatid pair (see Figure 3.3).

Metaphase chromosomes have condensed more than 10,000-fold since the beginning of prophase, which makes them easily visible under the microscope and allows them to be easily moved within the cell. Kinetochore microtubules are attached to the kinetochore at each centromere of sister chromatids. Because they are tethered to kinetochore microtubules from opposite centrosomes, the sister chromatids experience opposing forces that are critical to the positioning of chromosomes along an imaginary midline at the equator of the cell. This imaginary line is called the **metaphase plate.** The tension created by the pull of kinetochore microtubules is balanced by a companion process known as **sister chromatid cohesion.** Sister chromatid cohesion is produced by the protein cohesin that localizes between the sister chromatids and holds them together to resist the pull of kinetochore microtubules (Figure 3.4). Cohesin is a 4-subunit protein; its central component is a polypeptide known most commonly as Scc 1, for sister chromatid cohesion. Cohesin coats sister chromatids along their entire length but is most concentrated near centromeres, where the pull of microtubules is greatest. As microtubules move chromosomes toward the midline of the cell, cohesin helps keep the sister chromatids together, to ensure proper chromosome positioning and to prevent their premature separation.

Anaphase is the part of M phase during which sister chromatids separate and begin moving to opposite poles in the cell. Anaphase includes two distinct occurrences tied to microtubule actions. These occurrences are often identified as anaphase A, characterized by the separation of sister chromatids, and anaphase B, characterized by the elongation of the cell into an oblong shape. Anaphase A begins abruptly with two simultaneous events.

In anaphase A, the enzyme separase initiates cleavage of Scc 1 polypeptide in cohesin, thus breaking down the

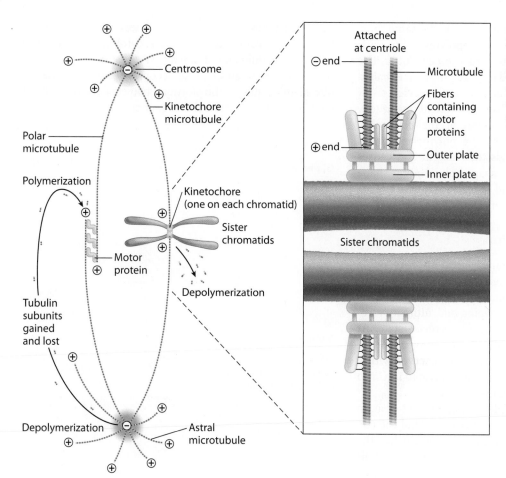

Figure 3.3 **Microtubules in dividing cells emanate from centrosomes.** Astral microtubules and polar microtubules control cell shape, and kinetochore microtubules attach to chromosome kinetochores.

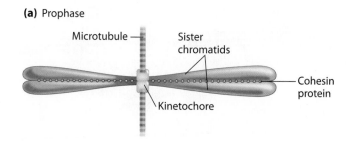

**(a)** Prophase

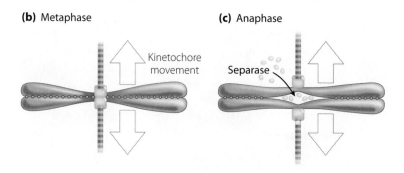

**(b)** Metaphase

**(c)** Anaphase

Figure 3.4 **Sister chromatid cohesion during mitosis.** Cohesin protein generates cohesion between sister chromatids (a) and (b). At anaphase (c), separase protein digests cohesin and allows sister chromatids to separate.

connection between sister chromatids. At the same time, kinetochore microtubules begin to depolymerize to initiate chromosome movement toward the centrioles. The separation of sister chromatids in anaphase A is called chromosome **disjunction.** As anaphase progresses, sister chromatids complete their disjunction and eventually congregate around the centrosomes at the cell poles. Anaphase B is characterized by the polymerization of polar microtubules that extends their length and causes the cell to take on an oblong shape. The oblong shape facilitates cytokinesis at the end of telophase, which leads to the formation of two daughter cells.

## Completion of Cell Division

In telophase, nuclear membranes begin to reassemble around the chromosomes gathered at each pole, eventually enclosing the chromosomes in nuclear envelopes. Chromosome decondensation begins and ultimately returns chromosomes to their diffuse interphase state. At the same time, microtubules disassemble. As telophase comes to an end, two identical nuclei are observed within a single elongated cell that is about to be divided into two daughter cells by the process of cytokinesis.

In animal cells, a contractile ring composed of actin microfilaments creates a cleavage furrow around the circumference of the cell; the contractile ring pinches the cell in two (**Figure 3.5**). In plant cells, cytokinesis entails the construction of new cell walls near the cellular midline. In both plant and animal cells, cytokinesis divides the cytoplasmic fluid and organelles.

Mitosis separates replicated copies of sister chromatids into identical nuclei, thus forming two genetically identical daughter cells. **Figure 3.6** shows four chromosomes in a cell

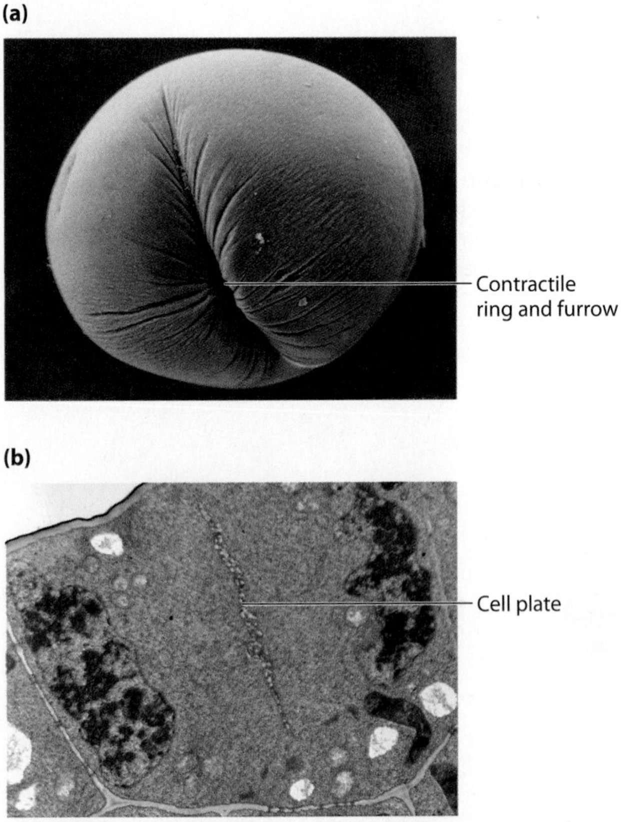

**(a)**

Contractile ring and furrow

**(b)**

Cell plate

**Figure 3.5  Cytokinesis in animal cells (a) and plant cells (b).**

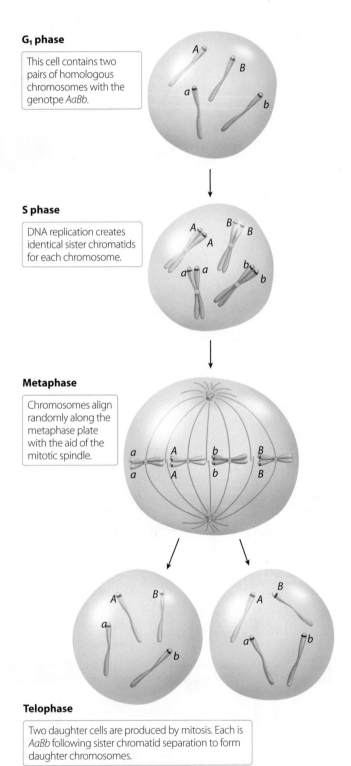

**G₁ phase**

This cell contains two pairs of homologous chromosomes with the genotpe *AaBb*.

**S phase**

DNA replication creates identical sister chromatids for each chromosome.

**Metaphase**

Chromosomes align randomly along the metaphase plate with the aid of the mitotic spindle.

**Telophase**

Two daughter cells are produced by mitosis. Each is *AaBb* following sister chromatid separation to form daughter chromosomes.

**Figure 3.6  An overview of mitosis.**

of an organism that is dihybrid (*AaBb*) for genes on the chromosomes shown. The figure follows major events of the cell cycle, showing the generation of sister chromatids in S phase, chromosome alignment on the metaphase plate in metaphase, and the production of two identical (*AaBb*) daughter cells at the end of telophase. Notice that the diploid (2*n*) number of chromosomes is maintained throughout the cell cycle.

---

**Genetic Insight**   The cell cycle is divided into interphase and M phase. Interphase contains the stages $G_1$, S, and $G_2$. M phase consists of five substages—prophase, prometaphase, metaphase, anaphase, and telophase. Mitosis separates sister chromatids to form two genetically identical daughter cells.

---

## Cell Cycle Checkpoints

Cell biologists find that no matter what the duration of the cell cycle, most cells follow the same basic program; this suggests that common, genetically controlled signals drive the cell cycle. The identity of the genes and proteins controlling the cell cycle comes not from normal cells, but from the study of cell lineages possessing mutations that affect their progression through the cell cycle. These studies have produced important insights into genetic control of the cell cycle, and in recent decades, biologists have discovered the identities and functions of many genes responsible for cell cycle control. What has been learned about genetic control of the cell cycle can be applied to the study of normal cell division as well as to the study of cell division abnormalities such as those displayed in cancer.

As cells move through the cell cycle, readiness to progress from one stage to the next is regularly assessed. Numerous **cell cycle checkpoints,** four of which are illustrated in Figure 3.7a, are monitored for cell readiness by protein interactions. A common mechanism for this monitoring is carried out by protein complexes that join a protein kinase with a second protein known as a cyclin protein. Protein kinases catalyze protein phosphorylation—the addition of a phosphate group transferred from a nucleotide triphosphate such as ATP or GTP to a target protein. Phosphorylation changes the conformation of target proteins and can either activate or inactivate the target protein. Protein kinases are usually present continuously in cells at relatively steady concentrations. **Cyclin proteins,** however, are so named because their concentrations are cyclic and linked to cell cycle stage. Cyclin protein production is stimulated by growth factor proteins that are produced by other cells. The protein kinase components of these complexes are activated only when they associate with a cyclin; thus, the protein kinases are called **cyclin-dependent kinases,** or **Cdks.** In their activated state, cyclin–Cdk complexes phosphorylate numerous

**(a)**

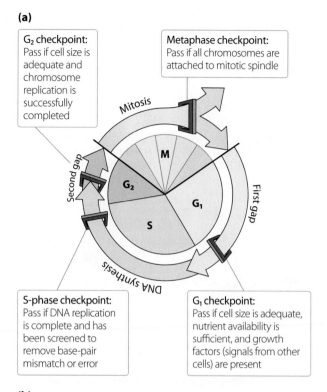

G₂ checkpoint:
Pass if cell size is adequate and chromosome replication is successfully completed

Metaphase checkpoint:
Pass if all chromosomes are attached to mitotic spindle

S-phase checkpoint:
Pass if DNA replication is complete and has been screened to remove base-pair mismatch or error

G₁ checkpoint:
Pass if cell size is adequate, nutrient availability is sufficient, and growth factors (signals from other cells) are present

**(b)**

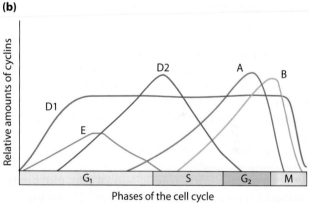

**Figure 3.7   Cell cycle checkpoints and cyclin proteins.** (a) Events at four major cell cycle checkpoints. (b) The relative amounts of cyclin proteins vary through the cell cycle.

target proteins and regulate cell cycle progression at various checkpoints.

Multiple Cdks and cyclins form a variety of cyclin–Cdk complexes (Figure 3.7b). For example, cyclin B joins together with a Cdk known as Cdk1 to form the cyclin B–Cdk1 complex that is required to phosphorylate numerous proteins required to initiate M phase of the cell cycle. In addition to stimulating cell division, the cyclin B–Cdk1 complex also activates an enzyme that degrades cyclin B, causing the rapid decrease in cyclin B level shown at the right-hand side of Figure 3.7b.

Cdk2 and Cdk4 proteins form additional cyclin–Cdk complexes that regulate other checkpoints in the cell cycle. Cdk2 joins with cyclin E to form a cyclin E–Cdk2 complex that is active at the $G_1$-S checkpoint. Separately,

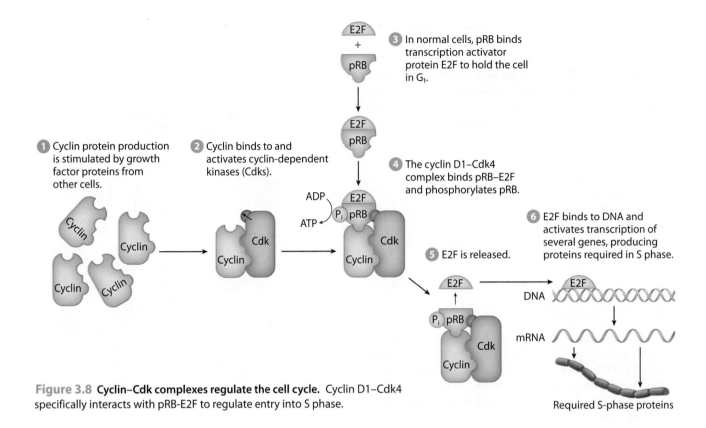

**Figure 3.8 Cyclin–Cdk complexes regulate the cell cycle.** Cyclin D1–Cdk4 specifically interacts with pRB-E2F to regulate entry into S phase.

Cdk2 joins with cyclin A to form the cyclin A–Cdk2 complex that is active at the S-G$_2$ checkpoint. Similarly, Cdk4 joins with cyclin D1, forming cyclin D1–Cdk4 that is active at the G$_1$-S checkpoint. Separately, Cdk4 pairs with cyclin D2 to form cyclin D2–Cdk4 that is active later in the cell cycle.

One prominent target of cyclin D1–Cdk4 is the retinoblastoma protein (pRB) that is produced by the retinoblastoma 1 (*RB1*) gene. In normal cells, pRB binds a transcription activator protein known as E2F, and together the pRB–E2F complex blocks cell cycle progression from G$_1$ to S phase (**Figure 3.8**). The cyclin D1–Cdk4 complex phosphorylates pRB, causing it to release E2F. Free E2F binds to DNA and activates the transcription of several genes that produce proteins essential in S phase. In other words, active cyclin D1–Cdk4 allows the cell to pass through the G$_1$ checkpoint and enter S phase by releasing E2F that is otherwise bound to unphosphorylated RB.

In the manner described, the presence of unphosphorylated pRB in a cell acts as a brake on the cell cycle, halting it at the G$_1$ checkpoint and preventing progression to S phase. pRB is one of many proteins that are known as tumor suppressors for their role in blocking the cell cycle. The *RB1* gene that produces pRB is known as a **tumor suppressor gene.** Numerous other genes are also classified as tumor suppressor genes because, in a functional sense, their protein products normally block progression of the cell cycle. In contrast, the expression of cyclin D1, from the *cyclin D1* gene, leads to formation of the cyclin D1–Cdk4 complex that *stimulates* cell cycle progression

from G$_1$ to S phase. Cyclin D1 is one of many examples of proteins produced by genes known as **proto-oncogenes.** When expressed, proto-oncogenes stimulate cell cycle progression.

## Cell Cycle Mutations and Cancer

Controlling cell division frequency is an essential activity of normal growth and development. Normal cells are infused with a blood supply that delivers oxygen and nutrients, but they proliferate only when signals from proteins called growth factors activate growth. Normal cells are also responsive to neighboring cells, and their growth is moderated so that the best interests of the whole organisms can take precedence. Cancer, on the other hand, is characterized by out-of-control cell proliferation that leads to tumor formation and the overgrowth of cancerous cells that invade and displace normal cells. Loss of cell cycle control is a fundamental mechanism leading to cancer development.

As examples of the loss of cell cycle control in cancer, let's consider two kinds of mutations that alter the normal interaction of cyclin D1–Cdk4 and pRB. The first category of mutations are those that either increase the number of copies of cyclin D1 by duplicating the *cyclin D1* gene, or significantly increase the level of transcription of *cyclin D1*. These mutations lead to higher-than-normal levels of cyclin D1. Since Cdk4 is continuously available in cells, overproduction of cyclin D1 causes uncontrolled entry into S phase by continuous phosphorylation of pRB and release of E2F to stimulate S-phase-related gene transcription. Mutations of this kind are frequently identified in parathyroid tumors, B-cell lymphomas, and other cancers in humans. Parallel observations of significantly increased production by mutation of the *cyclin E* gene are detected in breast cancer, colon cancer, and certain leukemias.

A second kind of mutation of this interaction that may lead to cancer is mutation of the *RB1* gene and the production of abnormal pRB. Mutation of *RB1* that produces pRB protein that binds weakly or not at all to E2F contributes to the development of several cancers, including those of the lung, bladder, breast, and bone. The mechanism of cancer causation is uncontrolled entry into S phase by unbound E2F. Mutation of *RB1* is also the cause of a cancer of light-sensitive cells of the retina in the eye. The cancer, called retinoblastoma, occurs in early childhood and forms tumors of rapidly proliferating cells in the retina. Retinoblastoma is rare; it occurs in 1 in 15,000 children. It occurs in two forms: a hereditary type, meaning that a child inherits a mutation of *RB1* from a parent, and a sporadic type in which *RB1* mutations are not inherited. Retinoblastoma occurs only when *both copies* of *RB1* are mutated; thus, the development of retinoblastoma is an example of a recessive cancer phenotype. In hereditary retinoblastoma, one *RB1* mutation is inherited; this means that all cells of the body, including retinal cells, carry one mutant gene. The acquisition of the second mutation of the wild-type copy of *RB1* occurs at a somatic level: The wild-type *RB1* gene undergoes mutation in any of the millions of cells in either retina. This second mutation produces the recessive genotype that leads to retinoblastoma development. Sporadic retinoblastoma also requires that both *RB1* genes undergo mutation. In sporadic retinoblastoma, however, both copies of the gene are wild type at fertilization, and mutation must alter the two copies of the gene *in the same retinal cell.* If both copies of *RB1* are hit by mutation, retinoblastoma develops.

---

**Genetic Insight**  Cell division is tightly regulated by protein-based signaling and interactions at multiple checkpoints that mark the transition from one cell cycle stage to the next. Loss of cell cycle control occurs when mutations alter normal checkpoint protein activity, and such mutations are frequently associated with cancer development.

---

# 3.2 Meiosis Produces Gametes for Sexual Reproduction

Reproduction is a basic requirement of living organisms. In more than three centuries of observation, biologists have identified a dizzying array of reproductive methods, mechanisms, and behaviors in animals, plants, and microbes. Even so, reproduction can be divided into two broad categories: (1) asexual reproduction, in which organisms reproduce without mating, giving rise to progeny that are genetically identical to their parent; and (2) sexual reproduction, in which reproductive cells called gametes are produced and unite during fertilization.

Bacteria and archaea reproduce exclusively by asexual reproduction. These organisms are haploid; they usually have just a single chromosome. Cell division follows shortly after the completion of chromosome replication, when each cell produces two genetically identical daughter cells. Single-celled eukaryotes, such as yeast, can reproduce either sexually or asexually. Asexual reproduction in yeast is similar to cell division in bacteria. A haploid cell undergoes DNA replication and distributes a copy of each chromosome to identical daughter cells. While yeast spend most of their life cycle in a haploid state and actively reproduce as haploids, it is also common for two haploid yeast cells to fuse and form a diploid cell that produces gametes (called spores) by meiosis.

In contrast to single-celled eukaryotes, multicellular eukaryotes reproduce predominantly by sexual means. In most animal species and dioecious plants, males and females each carry distinct reproductive tissues and structures. Mating requires the production of haploid gametes from both male structures and female structures. The union of haploid gametes produces diploid progeny. In monoecious plant species, including the *Pisum sativum* that Mendel worked with, male and female reproductive tissues are present in each plant, and self-fertilization is the common mode of reproduction. In sexually reproducing organisms, specialized germ-line cells undertake meiosis to produce haploid gametes, the reproductive cells. Female gametes are produced by the ovary in female animals or by the ovule in plants. Male germ-line cells are located in testes in animals, where they produce sperm, and in anthers on plants, where they produce pollen. These descriptions are broadly true for most plants and animals, but there are many exceptions including the observation of asexual reproduction in several species of fish, rotifers (small aquatic organisms), and salamanders. Male ants, bees, and wasps also have haploid somatic cells, and their processes of gamete production are distinctive.

## Meiosis versus Mitosis

Meiosis shares numerous features that are similar or identical to events in mitosis. For example, interphase of all cells is the same. Interphase of the germ-line cell cycle contains stages $G_1$, S, and $G_2$ that are indistinguishable from those in somatic cells. Similarly, the actions and functions of subcellular structures such as centrosomes and the microtubules they produce are the same in all cells. Nor is mitosis exclusive to somatic cells. Germ-line cells of plants and animals are created and maintained by

| Table 3.1 | Comparison of Mitosis and Meiosis | |
|---|---|---|
| **Characteristic** | **Mitosis** | **Meiosis** |
| Purpose | Produce genetically identical cells for growth and maintenance | Produce gametes for sexual reproduction that are genetically different |
| Location | Somatic cells | Germ-line cells |
| Mechanics | One round of division following one round of DNA replication | Two rounds of division (meiosis I and meiosis II) following a single round of DNA replication<br>The mechanical basis of Mendel's laws of heredity |
| Homologous chromosomes | Do not pair<br>Rarely undergo recombination | Synapsis during prophase I<br>Must recombine during prophase I |
| Sister chromatids | Attach to spindle fibers from opposite poles in metaphase<br>Separate and migrate to opposite poles at anaphase | Attach to spindle fibers from the same pole in metaphase I<br>Migrate to the same pole in anaphase I<br>Attach to spindle fibers from opposite poles in metaphase II<br>Separate and migrate to opposite poles in anaphase II |
| Product | Two genetically identical diploid daughter cells that continue to divide by mitosis | Four genetically different haploid cells that mature to form gametes and unite to form diploid zygotes |

mitotic division. These cells undertake meiosis solely for the purpose of producing gametes. Meiosis is distinguished from mitosis by the activities taking place during meiotic M phase and by the production of four haploid gametes. **Table 3.1** compares and contrasts numerous differences in the processes and outcomes of mitosis and meiosis that are described in the following sections.

Meiotic interphase is followed by two successive cell-division stages known as **meiosis I** and **meiosis II.** There is no DNA replication between these meiotic cell divisions, so the result of meiosis is the production of four haploid daughter cells (**Figure 3.9**). In meiosis I, homologous chromosomes separate from one another, reducing the diploid number of chromosomes ($2n$) to the haploid

**Figure 3.9  An overview of meiosis.**

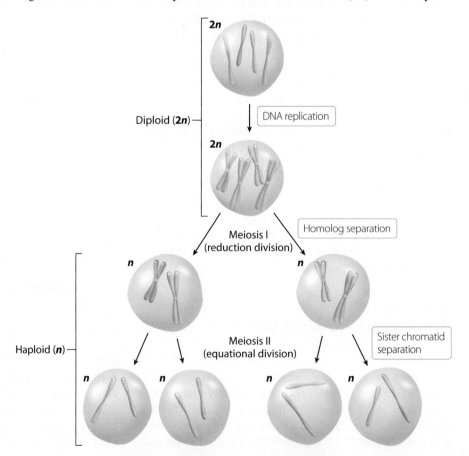

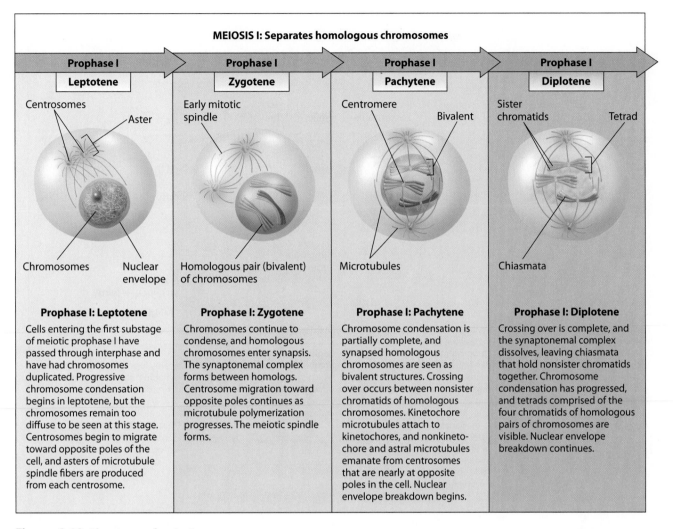

**Figure 3.10  The stages of meiosis.**

number (*n*). In meiosis II, sister chromatids separate to produce four haploid gametes, each with one chromosome of the diploid pair.

Following the completion of meiosis, each gamete contains a single nucleus holding a haploid chromosome set. The gametes of the two sexes are often dramatically different in size and morphology, however. Female gametes are generally much larger than male gametes and have a haploid nucleus, a large amount of cytoplasm, and a full array of organelles. In contrast, male gametes contain a haploid nucleus but very little cytoplasm and virtually no organelles. As the fertilized ovum begins mitotic division, the organelles and cytoplasmic structures provided by the maternal gamete support its early zygotic growth.

## Meiosis I

In addition to the two successive rounds of meiotic cell division contrasting with the single division in mitosis, meiosis features contact—and frequently recombination—between homologous chromosomes. These events occur

rarely, if at all, in mitosis. The daughter cells produced by meiosis are genetically different, in contrast to the genetically identical daughter cells produced by mitosis. These important differences between mitosis and meiosis are rooted in three hallmark events that occur in meiosis I:

1. Homologous chromosome pairing
2. Crossing over between homologous chromosomes
3. Segregation (separation) of the homologous chromosomes that reduces chromosomes to the haploid number

Meiosis I is divided into four stages: prophase I, metaphase I, anaphase I, and telophase I. Homologous chromosome pairing and recombination take place in prophase I; thus, this stage is subdivided into five substages—leptotene stage, zygotene stage, pachytene stage, diplotene stage, and diakinesis stage—to more accurately trace the interactions and recombination of homologous chromosomes. **Figure 3.10** describes these stages and substages in detail.

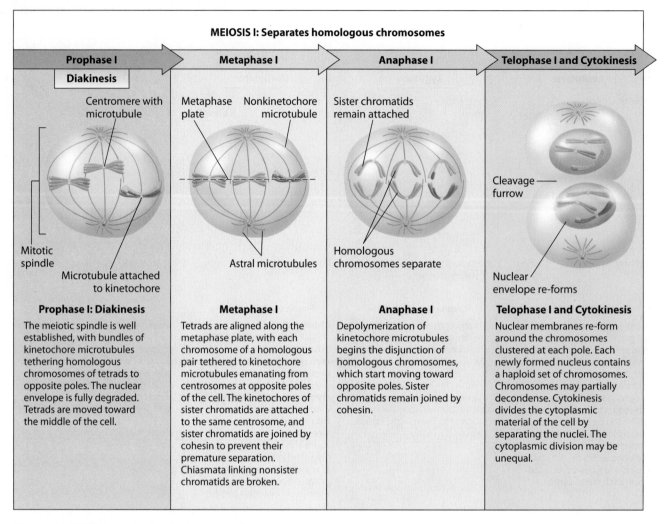

**MEIOSIS I: Separates homologous chromosomes**

| Prophase I | Metaphase I | Anaphase I | Telophase I and Cytokinesis |
|---|---|---|---|
| **Diakinesis** | | | |

Prophase I:
Diakinesis — Centromere with microtubule, Microtubule attached to kinetochore, Mitotic spindle

Metaphase I — Metaphase plate, Nonkinetochore microtubule, Astral microtubules

Anaphase I — Sister chromatids remain attached, Homologous chromosomes separate

Telophase I and Cytokinesis — Cleavage furrow, Nuclear envelope re-forms

**Prophase I: Diakinesis**

The meiotic spindle is well established, with bundles of kinetochore microtubules tethering homologous chromosomes of tetrads to opposite poles. The nuclear envelope is fully degraded. Tetrads are moved toward the middle of the cell.

**Metaphase I**

Tetrads are aligned along the metaphase plate, with each chromosome of a homologous pair tethered to kinetochore microtubules emanating from centrosomes at opposite poles of the cell. The kinetochores of sister chromatids are attached to the same centrosome, and sister chromatids are joined by cohesin to prevent their premature separation. Chiasmata linking nonsister chromatids are broken.

**Anaphase I**

Depolymerization of kinetochore microtubules begins the disjunction of homologous chromosomes, which start moving toward opposite poles. Sister chromatids remain joined by cohesin.

**Telophase I and Cytokinesis**

Nuclear membranes re-form around the chromosomes clustered at each pole. Each newly formed nucleus contains a haploid set of chromosomes. Chromosomes may partially decondense. Cytokinesis divides the cytoplasmic material of the cell by separating the nuclei. The cytoplasmic division may be unequal.

Figure 3.10  The stages of meiosis *continued*.

Chromosome condensation begins during leptotene, the first substage of prophase I, when the meiotic spindle is formed by microtubules emanating from the centrosomes, which are moving to positions at opposite ends of the cell. The nuclear membrane breaks down in zygotene, and the first hallmark feature of meiosis occurs—homologous chromosome **synapsis,** the alignment of homologous chromosome pairs. Synapsis initiates formation of a protein bridge called the **synaptonemal complex,** a tri-layer protein structure that maintains synapsis by tightly binding *nonsister chromatids* of homologous chromosomes to one another (**Figure 3.11**).

**Nonsister chromatids** are chromatids belonging to different members of a homologous pair of chromosomes. The binding of nonsister chromatids by a synaptonemal complex draws the homologs into close contact (synapsis). The synaptonemal complex contains two lateral elements, each consisting of proteins adhered to a chromatid from a different member of a pair of homologous chromosomes as well as a central element that joins

the lateral elements. The function of the synaptonemal complex is to properly align homologous chromosomes before their separation, and then to facilitate recombination between homologous chromosomes.

Chromosome condensation continues in pachytene, and sister chromatids of each chromosome can be visually distinguished by light microscopy. At this stage, the paired homologs are called a tetrad in recognition of the four chromatids that are microscopically visible in each homologous pair. Within the central element of the synaptonemal complex, new structures called **recombination nodules** appear at intervals.

Recombination nodules play a pivotal role in **crossing over** of genetic material between nonsister chromatids of homologous chromosomes. The number of recombination nodules correlates closely with the average number of crossover events along each homologous chromosome arm. Two important observations have been made about recombination nodules. First, their appearance and location within the synaptonemal

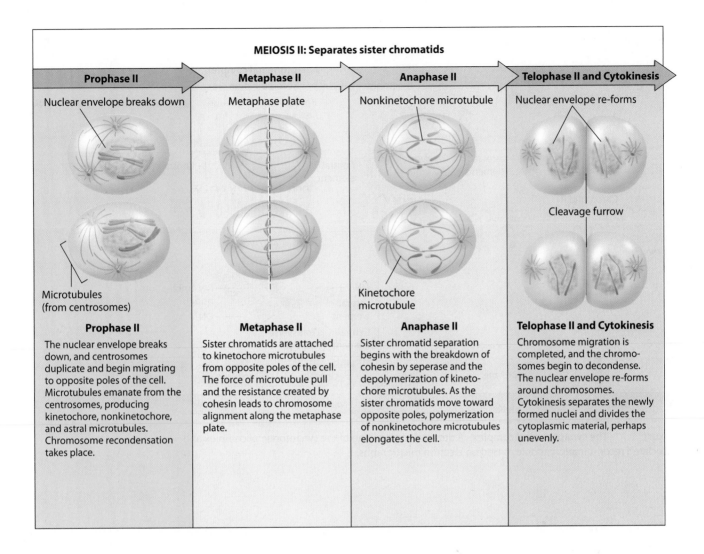

**MEIOSIS II: Separates sister chromatids**

**Prophase II**

Nuclear envelope breaks down

Microtubules
(from centrosomes)

**Prophase II**

The nuclear envelope breaks down, and centrosomes duplicate and begin migrating to opposite poles of the cell. Microtubules emanate from the centrosomes, producing kinetochore, nonkinetochore, and astral microtubules. Chromosome recondensation takes place.

**Metaphase II**

Metaphase plate

**Metaphase II**

Sister chromatids are attached to kinetochore microtubules from opposite poles of the cell. The force of microtubule pull and the resistance created by cohesin leads to chromosome alignment along the metaphase plate.

**Anaphase II**

Nonkinetochore microtubule

Kinetochore microtubule

**Anaphase II**

Sister chromatid separation begins with the breakdown of cohesin by seperase and the depolymerization of kinetochore microtubules. As the sister chromatids move toward opposite poles, polymerization of nonkinetochore microtubules elongates the cell.

**Telophase II and Cytokinesis**

Nuclear envelope re-forms

Cleavage furrow

**Telophase II and Cytokinesis**

Chromosome migration is completed, and the chromosomes begin to decondense. The nuclear envelope re-forms around chromosomes. Cytokinesis separates the newly formed nuclei and divides the cytoplasmic material, perhaps unevenly.

complex is coincident with the timing and location of crossing over; and second, recombination nodules seem to be present in organisms that undergo crossing over and absent in those that do not. Cell biologists have concluded that recombination nodules are aggregations of enzymes and proteins that are required to carry out genetic exchange between the nonsister chromatids of homologous chromosomes during pachytene. We discuss the genetic consequences of crossing over (Chapter 5) and the molecular processes of crossing over (Chapter 12).

The chromosomes continue to condense in diplotene as the synaptonemal complex begins to dissolve. The dissolution allows homologs to pull apart slightly, revealing contact points between nonsister chromatids. These contact points are called **chiasmata** (singular: **chiasma**), and they are located along chromosomes where crossing over has occurred. Chiasmata mark the locations of DNA-strand exchange between nonsister chromatids of homologous chromosomes.

Cohesin protein is present between sister chromatids to resist the pulling forces of kinetochore microtubules (**Figure 3.12**). In diakinesis, kinetochore microtubules actively move synapsed chromosome pairs toward the metaphase plate, where the homologs will align side by side.

The chiasmata between homologous chromosomes are dissolved with the onset of metaphase I. This process completes crossing over between homologous chromosomes.

Homologous chromosomes align on opposite sides of the metaphase plate in metaphase I. Kinetochore microtubules from one centrosome attach to the kinetochores of *both* sister chromatids of one chromosome. Meanwhile, kinetochore microtubules from the other centrosome attach to the kinetochores of the sister chromatids of the homolog. Karyokinesis takes place in anaphase I as homologous chromosomes separate from one another and are dragged to opposite poles of the cell (see Figure 3.10). The sister chromatids of each chromosome remain firmly joined by cohesin. Nuclear membrane reformation takes place in telophase I, when a haploid set of chromosomes

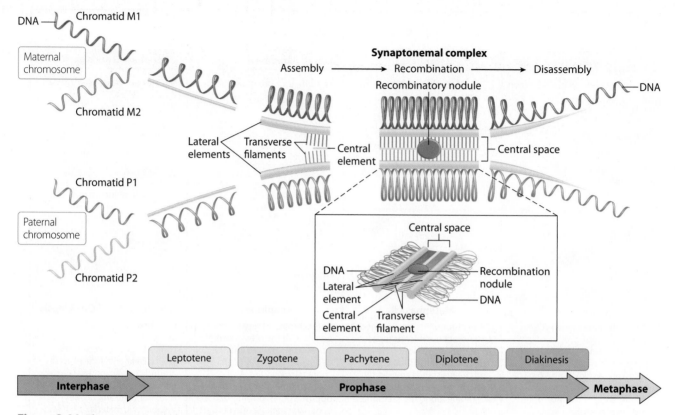

**Figure 3.11 The synaptonemal complex.** A detailed line drawing of the synaptonemal complex and associated recombination nodules based on electron micrographs.

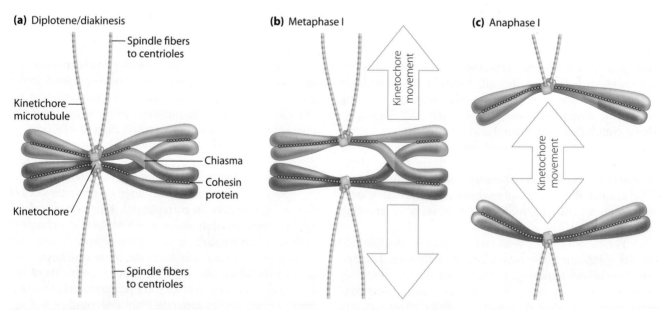

**Figure 3.12 Homolog separation in meiosis I. (a)** In diplotene and diakinesis of prophase I, crossing over between homologs is complete and contact between homologs (chiasmata) are resolved. **(b)** Spindle fibers pull chromosomes to align them on the metaphase plate. Cohesin protein adheres sister chromatids against the pull of spindle fibers. **(c)** Homologous chromosomes separate at anaphase I.

are enclosed at each pole of the cell. Cytokinesis follows the completion of telophase I.

Homologous chromosome disjunction (separation) in meiosis I reduces the number of chromosomes at each pole to the haploid number, so that one representative of each homologous pair of chromosomes is present. The first meiotic division is known as the *reduction division,* to signify the reduction of chromosome number from diploid to haploid.

## Meiosis II

The second meiotic division divides each haploid product of meiosis I by separating sister chromatids from one another in a process that is reminiscent of mitosis, except that the number of chromosomes in each cell is one-half the number observed in mitosis. The products of meiosis II mature to form the gametes that contain a haploid set of chromosomes. The four stages of meiosis II—prophase II, metaphase II, anaphase II, and telophase II—are shown and described in Figure 3.10.

Meiosis II bears a general resemblance to mitosis in that kinetochore microtubules from opposite centrosomes attach to the kinetochores of sister chromatids. Also, as in mitosis, in meiosis II the chromosomes align randomly along the metaphase plate. Furthermore, sister chromatid separation is accompanied by cohesin breakdown, the action of motor proteins, and depolymerization of microtubules. Cytokinesis takes place at the end of telophase II. Four genetically distinct haploid cells, each carrying one chromosome that represents each homologous pair, are the products of meiosis II.

**Genetic Insight**   Meiosis occurs in the M phase of the germ-line cell cycle. Homologous chromosomes synapse and may cross over in early prophase I. Homologs segregate in meiosis I, producing two haploid cells. Sister chromatids separate in meiosis II, producing four genetically distinct haploid gametes.

## The Mechanistic Basis of Mendelian Ratios

The separation of homologous chromosomes and sister chromatids in meiosis constitutes the mechanical basis of Mendel's laws of segregation and independent assortment. The connection between meiosis and Mendelian hereditary principles was first suggested, independently, by Walter Sutton and Theodor Boveri in 1903. Based on microscopic observations of chromosomes during meiosis, Sutton and Boveri proposed two important ideas. First, meiosis was the process generating Mendel's rules of heredity; and second, genes were located on chromosomes. Over the next two decades, work on numerous species proved these hypotheses to be correct.

We can understand segregation by following a pair of homologous chromosomes through meiosis in a heterozygous organism. The organism in **Figure 3.13**, for example, has the *Aa* heterozygous genotype. DNA replication in S phase creates identical sister chromatids for each chromosome. At metaphase I, the homologs align on opposite sides of the metaphase plate; and at anaphase I, the homologs separate from one another.

**Figure 3.13   Meiosis and the law of segregation.**

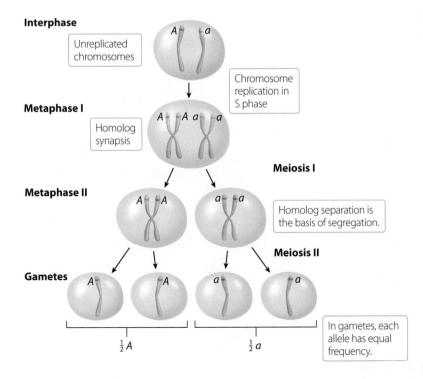

Interphase
Unreplicated chromosomes

Chromosome replication in S phase

Metaphase I
Homolog synapsis

Meiosis I

Metaphase II
Homolog separation is the basis of segregation.

Meiosis II

Gametes
$\frac{1}{2} A$     $\frac{1}{2} a$

In gametes, each allele has equal frequency.

This movement segregates the chromosome composed of two *A*-bearing chromatids from the chromosome bearing the two *a*-containing chromatids. Following these cells through to the separation of sister chromatids in meiosis II, we find that among the four gametes are two containing the *A* allele and two containing *a*. This outcome explains the 1:1 ratio of alleles that the law of segregation predicts for gametes of a heterozygous organism.

The independent assortment of alleles is illustrated by the behavior of two pairs of homologs during meiotic division in an organism, as demonstrated in **Figure 3.14** using the *AaBb* dihybrid genotype. Once again, S phase creates two identical sister chromatids for each chromosome. In metaphase I, however, two equally likely arrangements of the two homologous pairs can occur. In each arrangement, the homologous chromosomes are on opposite sides of the metaphase plate. Obviously, when a cell undergoes meiosis,

only one or the other of these alternative arrangements will occur; thus, each cell undergoing metaphase I of meiosis will have either "arrangement I" or "arrangement II." Over a large number of meiotic divisions, arrangement I and arrangement II are equally frequent. Arrangement I has chromosomes carrying dominant alleles on one side of the metaphase plate, and chromosomes carrying recessives on the opposite side. Arrangement II has a dominant-bearing and a recessive-bearing chromosome on each side of the metaphase plate. The first meiotic division segregates *A* from *a* and *B* from *b* to create the haploid products of meiosis I division.

If we now follow each haploid product of meiosis I through the meiosis II division, we see that the four gametes produced by arrangement I have the genotypes *AB* and *ab* in equal frequency. In contrast, the four gametes produced by arrangement II have the genotypes *Ab* and *aB* in equal frequency. Taking both possible arrangements

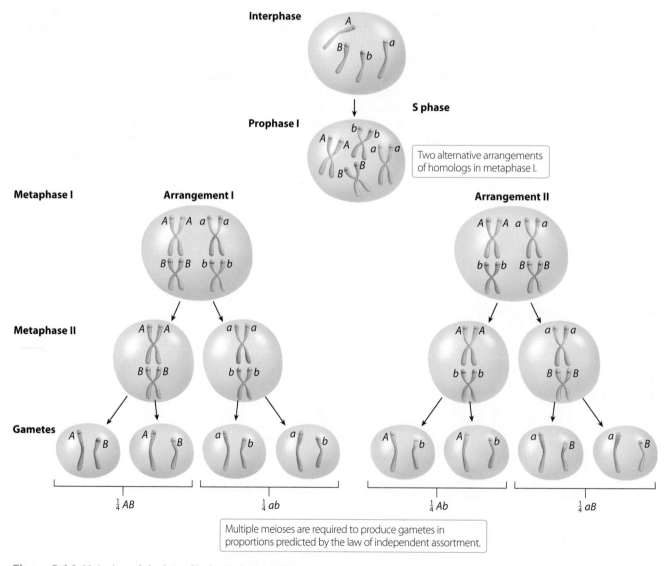

**Figure 3.14 Meiosis and the law of independent assortment.**

of homologous chromosomes at metaphase I into account, eight gametes are generated with four equally frequent genotypes. Each of the gamete genotypes—*AB*, *Ab*, *aB*, and *ab*—is produced in a frequency of 25%. The result of a large number of meiotic divisions in an *AaBb* dihybrid is a 1:1:1:1 ratio among gametes, as expected by Mendel's law of independent assortment.

---

Genetic Insight   Mendel's laws of segregation and independent assortment are products of the arrangement and separation of homologous chromosomes and sister chromatids in meiosis. The genotypes of gametes produced by meiosis match the frequencies of gamete genotypes predicted by Mendel's laws of heredity.

---

### Segregation in Single-Celled Diploids

We have seen that in sexually reproducing plants and animals, (1) the segregation of alleles can be explained by the disjunction of homologous chromosomes in meiosis I, and (2) independent assortment results from the different combinations of alleles to be found among the many gametes produced by an organism. Direct support of these conclusions is observed in the sexual reproduction of single-celled organisms such as yeast, which form diploid genomes for the purpose of sexual reproduction.

The yeast species *Saccharomyces cerevisiae* (also known as baker's yeast) is an example of an organism that can live and reproduce as a haploid but that can also form a diploid genome and produce gametes. Meiosis in *S. cerevisiae* produces four haploid gametes, called **spores,** that are temporarily contained in a sac-like structure called an **ascus.** The individual haploid gametes contained in an ascus can be separated and grown individually to reveal the alleles they contain.

*S. cerevisiae*, like all yeast, can reproduce by either sexual or asexual means (**Figure 3.15**). Asexual reproduction takes place in haploid cells by a process called budding, in which a haploid daughter cell grows out of the progenitor (parental) cell. Following DNA replication, sister chromatids separate and move into separate nuclei. One nucleus moves into a small bud that forms the daughter cell and is pinched off from the progenitor cell by cytokinesis. The newly formed bud has the same haploid genotype as its progenitor cell.

Sexual reproduction in *S. cerevisiae* is induced by starvation conditions and involves the union of two haploid yeast cells that are of different *mating types.* The mating types, called MATa and MATα, result from a difference in gene expression. Only the cross MATa × MATα produces a diploid strain, and meiosis in diploids produces the ascus containing four gametes.

To demonstrate these events, let us look at a visible marker of allelic variation in yeast. The wild-type allele ($ADE^+$) for synthesis of the amino acid adenine leads to the growth of a white yeast colony. In contrast, mutant alleles ($ade^-$) that block adenine synthesis produce the growth of red-colored colonies. The red color appears in $ade^-$ mutants due to the buildup of an intermediate product in the adenine synthesis pathway that is metabolized in $ADE^+$ cells.

When the haploid cross MATa $ADE^+$ × MATα $ade^-$ is made, the resulting diploid has the heterozygous genotype $ADE^+/ade^-$. Meiosis in this heterozygous strain produces an ascus containing four haploid spores that can be separated and grown independently to form colonies. The plate illustrated in Figure 3.14 shows two red yeast colonies and two white colonies, consistent with an equal 1:1 ratio of alleles segregated during meiosis in the heterozygous organism. This simple demonstration shows that each of the haploid spores in the ascus carries a copy of one of the daughter chromosomes produced during meiosis and thus offers direct evidence that chromosome disjunction in meiosis is the basis of Mendelian ratios.

**Genetic Analysis 3.1** on page 83 gives you practice identifying the principles of Mendelian transmission in meiotic cell division.

## 3.3 The Chromosome Theory of Heredity Proposes That Genes Are Carried on Chromosomes

The early 20th century was a time of rapid expansion of genetic knowledge, fueled by the rediscovery of Mendel's hereditary principles and by Sutton and Boveri's proposal that chromosome behavior in meiosis mirrors hereditary transmission of genes. Biologists were hard at work testing the new "gene hypotheses" of segregation and independent assortment. They examined heredity in a range of organisms, searching for evidence that confirmed earlier findings and for exceptions that would allow clarification and expansion of the newly articulated rules of heredity.

Thomas Hunt Morgan, initially skeptical of the gene hypothesis, began working on the tiny fruit fly *Drosophila melanogaster*. Morgan intended to rigorously test Mendel's rules in a natural species, not a domesticated one like *Pisum sativum*. Unlike Mendel, however, Morgan had no readily available phenotypic variants to examine. So, he and his students set out from their laboratory at Columbia University in New York City to the then rural landscape of Long Island to attract fruit flies by hanging buckets of rotting fruit on trees. Once captured and transported back to the laboratory, the flies were examined under the microscope to identify phenotypic variants. Flies captured from the wild were almost invariably of the same phenotype for each trait examined, and Morgan's group referred to these phenotypes as the "wild type." We use the term

**Haploid yeast life cycle**

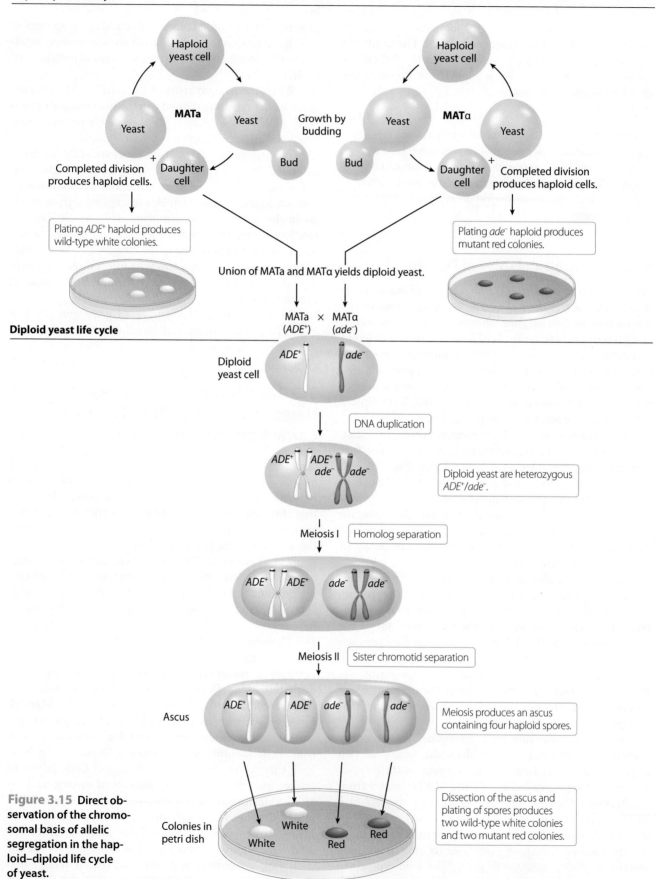

Diploid yeast cell

DNA duplication

Diploid yeast are heterozygous *ADE+*/*ade−*.

Meiosis I    Homolog separation

Meiosis II    Sister chromotid separation

Ascus

Meiosis produces an ascus containing four haploid spores.

Colonies in petri dish

Dissection of the ascus and plating of spores produces two wild-type white colonies and two mutant red colonies.

**Figure 3.15  Direct observation of the chromosomal basis of allelic segregation in the haploid–diploid life cycle of yeast.**

A diploid organism has the genotype $D_1D_2E_1E_2$. Gene D and gene E are on different chromosomes. In the diagrams requested, label each copy of each allele on chromosomes and sister chromatids.

**a.** Diagram *any correct* mitotic metaphase, illustrate these chromosomes, and label the alleles.

**b.** Diagram *any correct* meiotic metaphase I, illustrate these chromosomes, and label the alleles.

**c.** Describe the differences between the diagrams with respect to homolog and chromosome alignment.

**d.** Compare the outcome of mitosis with the outcome of meiosis in terms of the number of chromosomes and the genotype of the cells produced.

| Solution Strategies | Solution Steps |
|---|---|

### Evaluate

1. Identify the topic this problem addresses and describe the nature of the required answer.

1. This problem concerns comparisons of mitosis and meiosis. Parts (a) and (b) require illustration of chromosome alignments at metaphase. Part (c) requires an explanation of mitotic and meiotic metaphase I differences, and part (d) requires comparison of the outcomes of mitosis and meiosis.

2. Identify the critical information given in the problem.

2. The organism is identified as a dihybrid for a pair of autosomal genes on different chromosomes.

**TIP:** Heterozygous organisms carry different alleles on homologous chromosomes, but the alleles on sister chromatids are identical.

### Deduce

3. Identify the placement of alleles on homologous chromosomes and on sister chromatids in the organism described following the completion of S phase.

3. Sister chromatids carry identical alleles as a result of DNA replication in S phase. Thus, for example, sister chromatids of a single chromosome each carry a copy of *D1*. Likewise, identical alleles are carried on each set of sister chromatids.

4. Review the overall patterns of chromosome alignment along the metaphase plate during mitotic and meiotic divisions.

4. During mitotic metaphase, chromosomes align in single file and in an arbitrary order along the metaphase plate. In meiotic metaphase I, homologs align opposite one another along the metaphase plate.

### Solve

5. Diagram chromosome alignment during mitotic metaphase.

Answer a

5. Any order of the four chromosomes in single file along the metaphase plate is a correct order. One example is shown.

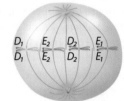

6. Diagram any correct chromosome alignment during meiotic metaphase I.

Answer b

6. Homologous chromosomes align opposite one another along the metaphase plate in meiotic metaphase I. The two correct arrangements of order of homologous chromosomes are shown.

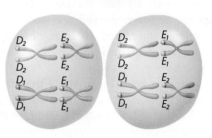

7. Describe the diagram differences with respect to homologs.

Answer c

7. Homologous chromosomes synapse in meiosis, but not in mitosis. The consequence of synapsis is that homologs align next to one another and on opposite sides of the metaphase plate in metaphase I. The absence of synapsis in mitosis leads chromosomes to align in any order along the metaphase plate in mitotic metaphase.

8. Describe the different outcomes of mitosis and meiosis.

Answer d

8. Mitosis produces two diploid daughter cells that are genetically identical to one another and to the parental cell they are derived from. Meiosis produces four haploid daughter cells that are genetically different.

For more practice, see Problems 1, 5, and 33.

*wild-type* today to signify the phenotype that is the most common in a population.

Morgan found *Drosophila* an easy organism to maintain and reproduce in the laboratory. He kept them in small glass bottles filled with a semisolid mixture of cornmeal, sugar, and water. The life cycle of *Drosophila* is between 12 and 14 days depending on growth conditions, so 25 to 30 generations could be raised in a year. Morgan took advantage of this rapid reproduction to raise large numbers of flies over many generations, searching for occasional de novo (i.e., newly occurring) mutant phenotypes in his laboratory-reared populations and also in flies captured in the wild. Over several years, he found many phenotypic variants that he used for performing and analyzing controlled genetic crosses between selected male and female fruit flies. We'll soon look at one set of crosses that led Morgan to conclude that genes are carried on chromosomes. First, however, we discuss some additional background for the theory that genes are on chromosomes.

## X-Linked Inheritance

While Sutton and Boveri were observing chromosome movements during meiosis, a researcher named Nettie Stevens was beginning a microscopic study to determine whether differences in chromosomes were evident between males and females of a species of beetles, *Tenebrio molitor*. In *T. molitor*, Stevens found that diploid cells of female beetles contained 20 large chromosomes, but diploid cells of males contained only 19 large chromosomes and 1 small chromosome. When examining the chromosomes in *T. molitor* eggs and sperm, Stevens observed that all eggs contain 10 large chromosomes. Her examination of sperm, however, showed that about half the sperm she examined contained 10 large chromosomes while the other half contained 9 large chromosomes and 1 small chromosome.

Stevens went on to study the chromosomes in somatic cells and gametes of other insects, and she concluded that sex-dependent hereditary differences are due to the presence of two large X chromosomes in females and one X chromosome and a much smaller Y chromosome in males. **Sex-linked inheritance** refers to the hereditary transmission of genes on the sex chromosomes. Stevens proposed that sex chromosomes in ova of *T. molitor* are always of the same type—each ovum contains a copy of every autosomal chromosome and one X chromosome. On the other hand, sperm can carry one copy of every autosome and either an X chromosome or a Y chromosome. Stevens suggested that the presence of either an X or a Y chromosome in sperm determines the sex of offspring, and that the equal frequency of X- and Y-bearing sperm account for the equal proportions of male and female offspring seen in crosses. Stevens was one of the first biologists to examine the

transmission of sex-linked traits, and her studies of *T. molitor* were the first to propose a chromosomal basis for sex determination.

In 1910, Thomas Hunt Morgan began a series of experiments in *Drosophila* that would validate Nettie Stevens's proposal that X and Y chromosomes help determine sex and would also provide evidence suggesting that genes are carried on chromosomes. The experiments began when Lilian Morgan, Thomas Hunt Morgan's wife and an important contributor to the laboratory group, found a mutant male *Drosophila* with white eyes in a bottle of wild-type flies that had been maintained in the lab for about a year. This white-eyed male stood out as a mutant because in *Drosophila*, wild-type flies have eyes the color of red bricks (**Figure 3.16**).

The mutant white-eyed male was crossed to a wild-type, red-eyed female. The cross produced 1237 $F_1$ flies with red eyes—a result indicating dominance of the wild type over the mutant. Subsequently, the $F_1$ were crossed to one another to produce an $F_2$ that were expected to have a 3:1 ratio of red eyes to white eyes. Among the $F_2$ were 2459 red-eyed females, 1011 red-eyed males, and 782 white-eyed males (cross A in **Figure 3.17**). No white-eyed females appeared in the $F_2$. Clearly, the $F_2$ result differed significantly from expectation, and white eyes seemed to be linked to male sex.

The unexpected result from this cross prompted a closer look at transmission of white eyes. When Morgan crossed the original white-eyed male and one of his $F_1$ red-eyed daughters, this cross produced approximately equal frequencies of red-eyed and white-eyed males and females (cross B in Figure 3.17). The final cross was a reciprocal cross between a white-eyed female and a wild-type, red-eyed male. The $F_1$ of the reciprocal cross were red-eyed females and white-eyed males (cross C in Figure 3.17). The $F_2$ contained equal proportions of red-eyed and white-eyed males and females.

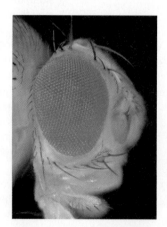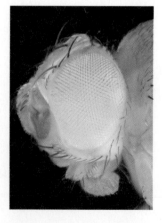

**Figure 3.16 X-linked eye-color phenotypes in *Drosophila melanogaster.*** Red eyes (left) are produced by a dominant wild-type allele. White eyes (right) are produced by a recessive mutant allele.

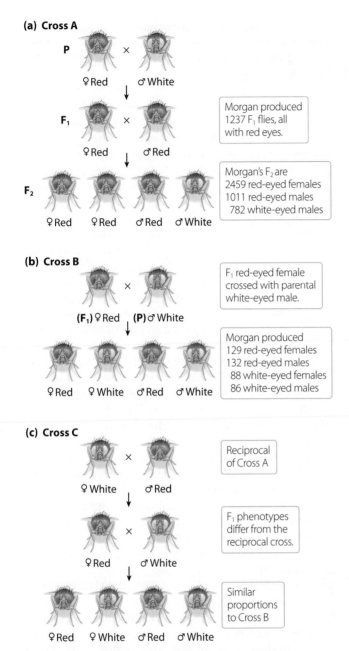

**(a) Cross A**

P    ♀ Red   ×   ♂ White

Morgan produced 1237 $F_1$ flies, all with red eyes.

$F_1$   ♀ Red   ×   ♂ Red

Morgan's $F_2$ are
2459 red-eyed females
1011 red-eyed males
782 white-eyed males

$F_2$   ♀ Red   ♀ Red   ♂ Red   ♂ White

**(b) Cross B**

$(F_1)$ ♀ Red   ×   $(P)$ ♂ White

$F_1$ red-eyed female crossed with parental white-eyed male.

Morgan produced
129 red-eyed females
132 red-eyed males
88 white-eyed females
86 white-eyed males

♀ Red   ♀ White   ♂ Red   ♂ White

**(c) Cross C**

♀ White   ×   ♂ Red

Reciprocal of Cross A

♀ Red   ×   ♂ White

$F_1$ phenotypes differ from the reciprocal cross.

♀ Red   ♀ White   ♂ Red   ♂ White

Similar proportions to Cross B

**Figure 3.17 Three *Drosophila* crosses performed by Morgan to determine X-linkage of the gene for eye color. (a)** Cross A determines that all $F_1$ flies and all female $F_2$ flies have red (wild-type) eye color. One-half of $F_2$ males have red eyes and one-half have white eyes. **(b)** Cross B verifies that $F_1$ females are heterozygous for eye color alleles by the finding of a 1:1 ratio of red eye to white eye among progeny males and females. **(c)** Cross C is the reciprocal of cross A, producing a different result in the $F_1$ and $F_2$ generations.

Diagrams of the crosses in Figure 3.17 are illustrated in **Figure 3.18**, where *w* represents the recessive allele for white eye and $w^+$ the dominant allele for red eye. The differences between reciprocal crosses observed by Morgan are not anticipated by Mendel's laws of heredity. In fact, recall that Mendel performed many reciprocal crosses and found no differences in the phenotype proportions.

Morgan realized that transmission of X chromosomes in *Drosophila* could account for the appearance of white and red eyes in his crosses if the X chromosome carried a gene for eye color. In cross A, the single X chromosome of a white-eyed male carries a recessive allele designated *w*. The X chromosome is present along with a Y chromosome in the genome of the male fruit fly. X chromosomes of females each carry a dominant allele $w^+$ that produces red eye color. The $F_1$ of this cross are red-eyed males that are $w^+$Y and red-eyed females that are $w^+w$. The $F_2$ of this cross contain equal proportions of white-eyed (*w*Y) and red-eyed ($w^+$Y) males and red-eyed females that are, in equal proportions, $w^+w^+$ and $w^+w$. By the same logic, cross B in which the original white-eyed male to his $F_1$ red-eyed daughter ($w^+w$) is expected to produce equal proportions of red and white eyes in males and females. Cross C between a white-eyed female and a red-eyed male provides a genetic test of this hypothesis. The cross produces red-eyed female and white-eyed male $F_1$ progeny as well as equal proportions of red- and white-eyed males and females in the $F_2$.

Morgan's analysis of these experiments describes **X-linked inheritance,** a term identifying the transmission of genes carried on the X chromosome. Morgan proposed X-linked inheritance as the mode of transmission of eye color in *Drosophila*. In his model of X-linked inheritance, Morgan proposes that males are **hemizygous,** a term meaning "half zygous," for X-linked genes. This term is used because males have a single X chromosome; therefore, unlike females, males cannot be homozygous or heterozygous for X-linked genes. Hemizygous males inherit their X chromosome from their mother; moreover, they express any allele on their X chromosome, since the Y chromosome does not carry genes that are homologous to those on the X chromosome. In contrast to males, females have two X chromosomes and can display heterozygous and homozygous genotypes for X-linked genes, just as they can for autosomal genes. Note also that males can transmit either the X chromosome or the Y chromosome, but that the X chromosome is passed exclusively to female progeny and the Y chromosome exclusively to male progeny. In contrast, females can transmit either X chromosome to any of their offspring.

**Genetic Insight** Inheritance of X-linked traits produces different outcomes for reciprocal crosses. Males are hemizygous for X-linked traits, whereas females are homozygous or heterozygous. Males inherit X chromosomes from their mothers and transmit X chromosomes exclusively to their female offspring.

## Analysis of Nondisjunction

Morgan's work on *Drosophila* eye color led him to propose the chromosome theory of heredity, hypothesizing that genes are carried on chromosomes. Morgan's observations

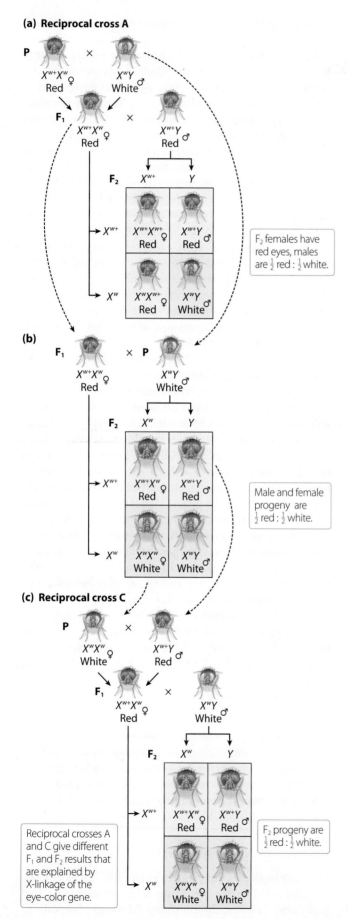

**(a) Reciprocal cross A**

**(b)**

**(c) Reciprocal cross C**

Reciprocal crosses A and C give different $F_1$ and $F_2$ results that are explained by X-linkage of the eye-color gene.

**Figure 3.18** **The X-linked genetic model of Morgan's eye-color inheritance experiments in *Drosophila*.** X and Y chromosome segregation in **(a)** cross A, **(b)** cross B, and **(c)** cross C (from Figure 3.17).

of eye-color inheritance were consistent with the chromosome theory, but the validation of the hypothesis required finding an indisputable link between a unique phenotype and the presence or absence of a particular chromosome. Calvin Bridges, a student of Morgan, studied fruit flies with unexpected eye-color phenotypes and abnormal chromosome numbers and provided proof of the chromosome theory of heredity.

Bridges focused his study on cross C (see Figures 3.17 and 3.18), between a white-eyed female (*ww*) and a red-eyed male ($w^+$Y). Nearly all the progeny from this cross had the expected phenotype and were either red-eyed females ($w^+w$) or white-eyed males (*w*Y), but about 1 in every 2000 $F_1$ flies had an "exceptional phenotype"—a term used to identify progeny with unexpected characteristics. Specifically, the exceptional flies were either white-eyed *females* or red-eyed *males*. Bridges's detection of exceptional progeny left him with two questions to answer: (1) how could the exceptional progeny be explained, and (2) did the appearance of exceptional progeny provide the information necessary to test the hypothesis that genes are on chromosomes?

The answer to the first question came when Bridges looked at chromosomes of the exceptional progeny under the microscope. He saw the exceptional females had three sex chromosomes—two X chromosomes and one Y chromosome (XXY) **(Figure 3.19)**. As we discuss in the next section, fruit flies with two X chromosomes are females, even if there happens to be a Y chromosome as well, as there is in this case. Bridges also observed an abnormal number of chromosomes in exceptional males. They carried a single X chromosome but no Y chromosome (XO). Fruit flies with one X chromosome are male, regardless of whether they carry a Y chromosome. Based on his observations, Bridges proposed that the Y chromosome carried by exceptional females came from the male parent, the only source of a Y chromosome in the cross, and that both X chromosomes in these exceptional females came from the mother, giving the exceptional females two copies of the *w* allele and white eye color. Bridges used similar logic to suggest that the single X chromosome in exceptional males came from the male parent that passed the $w^+$ allele. The exceptional males with a single X chromosome expressed the $w^+$ allele as red eyes.

According to Bridges's proposal, the exceptional phenotypes and unusual karyotypes were the result of rare mistakes in meiosis caused by the failure of X chromosomes to separate properly in either the first or second meiotic division in females. Failed chromosome separation is called **nondisjunction.** Notice in Figure 3.19 that nondisjunction also produces XXX or YO

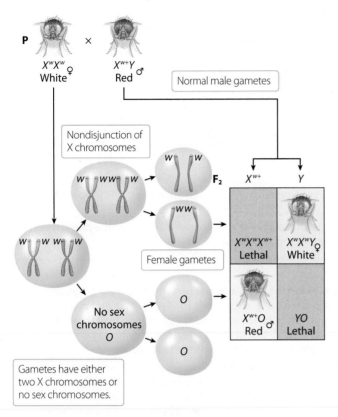

**Figure 3.19** **Exceptional progeny observed by Calvin Bridges result from X-chromosome nondisjunction.**

is controlled by the sex chromosome contributed by the heterogametic parent, whereas phenotypic sex is a matter of appropriate gene expression and the development of sex characteristics. In this section, we examine the patterns and processes of chromosomal and phenotypic sex determination in several organisms.

## Sex Determination in *Drosophila*

Bridges's study of X-chromosome nondisjunction and his proof of the chromosome theory of heredity also provided information about sex determination in *Drosophila*. Bridges consistently found that flies with one X chromosome are male and that those with two X chromosomes are female. In *Drosophila*, the number of X chromosomes appears to be a critical component in determining sex, and the number of Y chromosomes, or even the absence of a second sex chromosome, seems not to disrupt the pattern of sex determination. Thus, in *Drosophila*, flies with the sex-chromosome constitutions XY, XYY, and XO are all male, whereas flies that are XX or XXY are female.

Bridges's *Drosophila* data identified the ratio of X chromosomes to the number of haploid sets of autosomes (1X:2A in males and 2X:2A in females) as pivotal in determining sex. Bridges called this the **X/A ratio,** or the **X/autosome ratio.** In reality, the X/A ratio is too simplistic to explain *Drosophila* sex determination. *Drosophila* sex is determined by regulatory proteins that relay the number of X chromosomes present in nuclei of cells in *Drosophila* embryos. These proteins control expression of the *sex-lethal (Sxl)* gene in XX flies. As we discuss in a later chapter, Sxl protein controls the expression of additional genes that drive sex development (see Chapter 8).

## Mammalian Sex Determination

Like *Drosophila*, placental mammals have two kinds of sex chromosomes identified as X and Y. Unlike *Drosophila*, however, sex determination in placental mammals depends on the presence or absence of the Y chromosome. A single gene on the Y chromosome, abbreviated *SRY* (sex-determining region of Y), initiates a series of events that lead to male sex-phenotype development in the embryo. Consequently, mammalian embryos that have one or more Y chromosomes (XY, XXY, and XYY, for example) and therefore express *SRY* will develop as males. Conversely, embryos carrying only X chromosomes (XX, XO, and XXX, for example) and lacking *SRY* expression will develop as females.

*SRY* is a transcription factor protein that elicits a cascade of gene transcription and developmental events that ultimately produce male internal and external structures. Early mammalian embryos contain twin clusters of tissue identified as undifferentiated gonads that can develop into either ovaries or testes. Connected to the

progeny. Bridges never saw these progeny, however, because YO progeny fail to develop, and XXX is usually lethal. Bridges's observations provide conclusive proof of the chromosome theory of heredity by showing that the white (*w*) allele segregates with the X chromosome during normal meiosis and during nondisjunction. **Genetic Analysis 3.2** gives you some practice spotting X-linked inheritance.

---

**Genetic Insight** The chromosome theory of heredity states that genes are carried on chromosomes. The theory is supported by the discovery of *Drosophila* with exceptional eye-color phenotypes and abnormalities of sex chromosome number.

---

# 3.4 Sex Determination Is Chromosomal and Genetic

The term **sex determination** encompasses the genetic and biological processes that produce the male and female characteristics of a species. The sex of most organisms is identified on two levels: as chromosomal sex, the presence of sex chromosomes associated with male and female sex in a species; and as phenotypic sex, the internal and external morphology found in each sex. Chromosomal sex is determined at the moment of fertilization and

A male fruit fly from a pure-breeding stock with yellow body color and full wing size is crossed to a female from a pure-breeding stock with gray body and vestigial wings. The cross progeny consists of males with yellow body color and full-sized wings and females with gray body color and full-sized wings.

**a.** Determine the mode of inheritance of each trait.

**b.** Give genotypes for parental flies and the male and female progeny using clearly defined allele designations of your choice.

| Solution Strategies | Solution Steps |
|---|---|

**Evaluate**

1. Identify the topic this problem addresses and describe the nature of the required answer.

2. Identify the critical information given in the problem.

1. The patterns of transmission of two *Drosophila* traits and the genotypes of organisms are to be determined based on the number and proportions of male and female $F_1$ progeny with the traits.

2. Pure-breeding parental phenotypes are given along with the phenotypes of male and female progeny in the $F_1$.

**Deduce**

3. Consider the $F_1$ phenotype results in light of the parental phenotypes.

   **TIP:** Cross results that appear equally in both sexes are consistent with autosomal inheritance. Sex-dependent differences in a cross suggest sex-linked inheritance.

4. Hypothesize the modes of inheritance of body color and wing form from the $F_1$ data.

   **TIP:** Test the hypothesized mode of inheritance by comparing the predicted and observed $F_1$ progeny ratios.

3. All $F_1$ progeny have full-sized wings and none have vestigial wings, suggesting that full-sized wing is dominant. The $F_1$ males are exclusively yellow-bodied, whereas $F_1$ females are exclusively gray-bodied. The $F_1$ male body color is identical to that of the parental female, whereas the $F_1$ females' body color is identical to that in the male parent.

4. The observation of one body color in $F_1$ males and another in females suggests this is an X-linked trait. Since hemizygous males have yellow body and females have gray body, it is likely that gray body is dominant and yellow body is recessive. The $F_1$ results for wing form are the same for both sexes, suggesting that this trait is autosomal.

**Solve**

5. Test the proposed mode of transmission of wing form.

6. Test the mode of transmission of body color.

   **TIP:** Compare observed and expected $F_2$ progeny to test the hypothesized mode of inheritance.

7. Determine genotypes for parental and $F_1$ flies. Use $y^+$ for yellow body, $y$ for gray body, $v^+$ for full wing, and $v$ for vestigial wing.

5. The $F_1$ of both sexes have full-sized wings, consistent with an autosomal trait. The pure-breeding full-winged parent transmits the dominant alleles to all progeny, and the pure-breeding vestigial parent transmits the recessive allele. The $F_1$ are predicted to be heterozygous and display the dominant trait.

6. The sex-dependent difference in body color among $F_1$ males and females strongly suggests this trait is X-linked. The $F_1$ males inherit the maternal recessive allele for yellow body color and express the trait because they are hemizygous. $F_1$ females inherit a recessive allele on the maternal X chromosome and a dominant allele on the paternal X and are heterozygous, thus displaying the dominant phenotype.

7. The genotypes of pure-breeding parents are $X^y/X^y$; $+/+$ for yellow-bodied, full-winged females and $X^{y+}/Y$; $v/v$ for gray-bodied, vestigial-winged males. The $F_1$ females are $X^y/X^{y+}$; $+/v$, and $F_1$ males are $X^y/Y$; $+/v$.

For more practice, see Problems 6, 12, and 15.

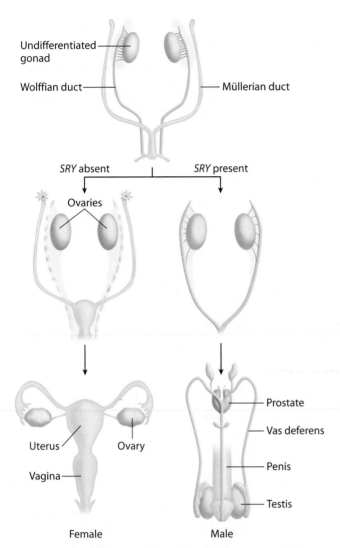

**Figure 3.20  Mammalian sex determination is initiated by the Y-linked *SRY* gene.**

undifferentiated gonads are two sets of tissues called the Wolffian ducts and the Müllerian ducts. Only one of these tissues develops. Wolffian ducts can develop to form male sexual structures. Alternatively, Müllerian ducts can develop to form female sexual structures. In male embryos, expression of *SRY* initiates testicular development of the undifferentiated gonads through MPG action. Interstitial cells in the developing testes synthesize two male androgenic hormones, testosterone and dihydrotestosterone (DHT), that help drive Wolffian duct development and ultimately lead to formation of internal and external male sexual structures. Separately, in specialized cells called sustentacular cells, *SRY* stimulates production of Müllerian-inhibitory factor (MIF) that degrades Müllerian ducts to prevent development of female sexual structures (**Figure 3.20**).

Female embryos do not carry a Y chromosome and therefore lack expression of *SRY*. The current model suggests that the absence of *SRY* suppresses the expression of genes that lead to male development and, instead, leads to

expression of genes that stimulate the undifferentiated gonad tissue to develop into ovaries and cause Müllerian ducts to develop into female sexual structures.

While *SRY* is a necessary gene in mammalian sex development, it is not sufficient by itself to direct sexual development. Mutations of X-linked and autosomal genes mentioned in **Experimental Insight 3.1** on page 91 have been identified as causes of abnormalities of human sexual development.

## Diversity of Sex Determination

You are familiar with the XX and XY chromosome designation signifying that females carry two X chromosomes and males carry an X chromosome and a Y chromosome. In many bird species, some reptiles, certain fish, and moths and butterflies, however, females carry two different sex chromosomes, and males carry two sex chromosomes that are the same. To avoid confusion with the XX/XY system, a different lettering system called the **Z/W system** is used. In the Z/W system, males are identified as having two Z sex chromosomes, or a sex chromosome composition of ZZ. In contrast, females have two different sex chromosomes and are identified as ZW. The letters Z and W are used to highlight the different sex-chromosome compositions associated with each sex in the species affected.

The sex-chromosome differences in the Z/W system produce different results from reciprocal crosses involving Z-linked genes, just as there are reciprocal cross differences for X-linked genes. **Figure 3.21** shows reciprocal crosses between pure-breeding hens (female) and roosters (male) involving a Z-linked dominant allele for barred feathers ($Z^B$) and its recessive counterpart, nonbarred feathers ($Z^b$). The $F_1$ results of the reciprocal crosses reveal differences. Cross A produces barred hens ($Z^B W$) and barred roosters ($Z^B Z^b$) in the $F_1$, whereas cross B produces nonbarred hens ($Z^b W$) and barred roosters ($Z^B Z^b$). The $F_2$ results of these crosses also yield differences. Among the $F_2$ of cross A, all roosters are barred, whereas hens display a 1:1 ratio of barred:nonbarred. Among the $F_2$ of cross B, there is a 1:1 ratio of barred to nonbarred in both sexes. We can conclude that the mechanism of transmission of Z-linked genes in the Z/W system is the same as that in the XX/XY system except that the patterns are the reverse of those in placental mammals.

Sex chromosome content is even more unusual in monotremes like the platypus, an egg-laying mammal that is native to Australia. Because monotremes are mammals, it was thought at one time that they had a mechanism of sex determination similar to the XX/XY system in placental mammals. When platypus chromosomes were examined, however, biologists discovered that males have five sets of X and Y sex chromosomes, whereas females have five sets of X chromosomes. Male platypus sex chromosomes are represented as $X_1Y_1X_2Y_2X_3Y_3X_4Y_4X_5Y_5$ and female

**(a) Cross A**

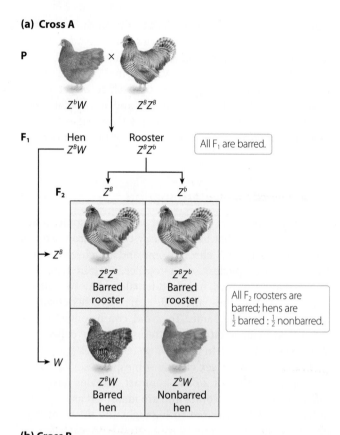

All F₁ are barred.

All F₂ roosters are barred; hens are ½ barred : ½ nonbarred.

**(b) Cross B**

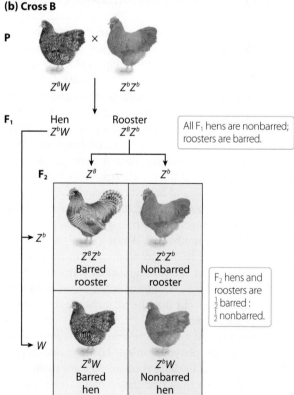

All F₁ hens are nonbarred; roosters are barred.

F₂ hens and roosters are ½ barred : ½ nonbarred.

Reciprocal crosses A and B yield different F₁ and F₂ results, indicating that feather form is sex linked.

**Figure 3.21 ZW inheritance of feather form in poultry is revealed by analysis of reciprocal crosses. (a)** A hemizygous female (hen) with recessive nonbarred feathers crossed to a pure-breeding male (rooster) with dominant barred feathers produces F₁ progeny and male F₂ progeny with barred feathers. F₂ females display a 1:1 ratio of barred to nonbarred. **(b)** The reciprocal cross produces different results in the F₁ and F₂ generations, consistent with sex-linked ZW transmission of feather form.

platypus sex chromosomes as $X_1X_1X_2X_2X_3X_3X_4X_4X_5X_5$. Multiple sets of sex chromosomes have also been documented in some plant species, termites, and spiders.

---

**Genetic Insight** Sex determination in *Drosophila* is based on the X/A ratio. Male sex in placental mammals is determined by the presence or absence of a Y chromosome and the *SRY* gene. The Z/W system in birds, fishes, and some insects, and multiple sex chromosomes in monotremes and other organisms, are examples of other sex-determination mechanisms.

---

## 3.5 Human Sex-Linked Transmission Follows Distinct Patterns

The inheritance of mutant alleles on the human X chromosome produces mutant phenotypes in two simple patterns. **X-linked recessive** inheritance is the hereditary pattern that determines white eye color in *Drosophila*. With this mode of inheritance, females homozygous for the recessive allele, and hemizygous males whose X chromosome carries the recessive allele, display the recessive phenotype. The alternative mode of X-linked transmission is **X-linked dominant** inheritance, in which heterozygous females and males hemizygous for the dominant allele express the dominant phenotype.

Keep in mind that the terms *recessive* and *dominant* refer specifically to the expression of the traits in females that are either homozygous or heterozygous for each X-linked gene. Hemizygous males express any allele on their X chromosome, regardless of the hereditary pattern in females. In addition, the probability of chromosome transmission to offspring is not the same for the two sexes. Female X-linked transmission is identical to autosomal transmission, whereas hemizygous males always transmit their X chromosome to female offspring and their Y chromosome to male offspring.

### Expression of X-Linked Recessive Traits

X-linked recessive traits are expressed in hemizygous males who carry the recessive allele and in females who are homozygous for the recessive allele. Because hemizygous males express the single copy of a recessive X-linked allele in their phenotype, one of the hallmarks of X-linked recessive

## Experimental Insight **3.1**

### Mutations Altering Human Sex Development

Many genes in addition to *SRY* direct human sexual development. Here we identify three other genes whose mutation affects the production or cell-signaling capacity of the male androgenic hormones testosterone and DHT (dihydrotestosterone) and results in abnormal sexual development. These conditions have different causes and distinctive consequences. From a medical perspective, ambiguous gender identification is a consequence of the conditions. In personal terms, significant psychosocial issues of self and of gender identity confront individuals with each of these conditions.

Androgen insensitivity syndrome (AIS) (OMIM 300068) is caused by mutations of the X-linked *AR* (androgen receptor) gene. *AR* is pivotal in producing androgen receptors on androgen-sensitive cells. AIS individuals are XY, have a fully functional *SRY* gene, and produce normal amounts of testosterone and DHT. In the absence of androgen receptors, however, testosterone and DHT cannot bind to cells, which therefore do not initiate the gene expression that accompanies male sexual development. Due to this deficit, individuals with AIS have an external phenotype that appears to be female (i.e., sex reversal); but internal reproductive structures do not develop as either male or female, thus rendering AIS individuals sterile. Androgen insensitivity prevents development of male sexual structures, whereas *SRY*-initiated MIF production degrades the Müllerian ducts and blocks the development of female sexual structures.

When genes operating in the biochemical pathway controlling testosterone and DHT are mutated, improper androgen levels occur, and individuals can exhibit *pseudohermaphroditism*—a term referring to the appearance of nonfunctional forms of both male and female structures in a single person. Pseudohermaphrodites are sterile. The autosomal recessive disorder 5-alpha-reductase deficiency (OMIM 607306) produces a form of pseudohermaphroditism due to mutation of the steroid 5-alpha-reductase-2 (*SRD5A2*) gene. *SRD5A2* produces 5-alpha-reductase enzyme that helps convert testosterone to DHT. Individuals with 5-alpha-reductase deficiency are XY, have a wild-type *SRY* gene, undergo Wolffian duct development, and express MIF. Wolffian duct development produces male internal structures, but the inability to convert testosterone to DHT results in the absence of external male structures. At birth, individuals with 5-alpha-reductase deficiency appear to be female. At puberty, however, the adrenal glands begin testosterone production that leads to secondary male sexual characteristics such as deepening of the voice, facial hair growth, and development of a masculine physique.

Lastly, mutation of *CYP21*, a gene producing the enzyme 21-hydroxylase, causes the most common form of autosomal recessive congenital adrenal hyperplasia (CAH) (OMIM 201910). Functional 21-hydroxylase participates in depletion of testosterone and DHT; thus, its mutation leads to accumulation of testosterone and DHT. *CYP21* mutation produces pseudohermaphroditism in males and females due to high androgen levels. Boys with CAH enter puberty as early as 3 years of age and display male musculature, enlarged penis, and testes growth. Girls with CAH are born with an enlarged clitoris that can be mistaken for a small penis. While normal internal female reproductive anatomy is present, CAH females experience male-like facial hair growth and deepening voice at puberty. Menstruation does not occur, due to excessive androgen levels.

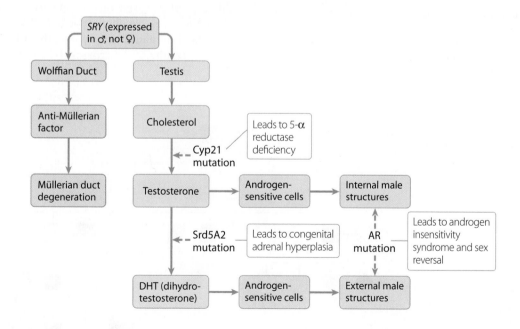

| Table 3.2 | A Short List of Human X-Linked Dominant and X-Linked Recessive Traits[a] |
|---|---|
| **Disease** | **Symptom** |
| *X-Linked Dominant Disorders* | |
| Amelogenesis imperfecta (OMIM 301200) | Abnormal tooth-enamel development and distribution |
| Congenital generalized hypertrichosis (OMIM 307150) | Extensive hair distribution on the face and body |
| Hypophosphatemia (OMIM 307800) | Phosphate deficiency causing rickets (bowleggedness) |
| Rett syndrome (OMIM 312750) | Mental retardation and neurodevelopmental defects |
| *X-Linked Recessive Disorders* | |
| Anhidrotic ectodermal dysplasia (OMIM 305100) | Absence of teeth, hair, and sweat glands |
| Color blindness (red-green) (OMIM 303800) | Color-perception deficiency |
| Fragile X syndrome (OMIM 300624) | Mental retardation and neurodevelopmental defects |
| Hemophilia A (OMIM 306700) | Blood-clotting abnormality |
| Lesch-Nyhan syndrome (OMIM 300322) | Mental retardation with self-mutilation and spastic cerebral palsy |
| Muscular dystrophy (Becker type, OMIM 300376) and Duchenne type (OMIM 310200) | Progressive muscle weakness |
| Ornithine transcarbamylase deficiency (OMIM 311250) | Mental deterioration due to ammonia accumulation with protein ingestion |
| Retinitis pigmentosa (OMIM 300029) | Night blindness, constricted visual field |

[a] OMIM = Online Mendelian Inheritance of Man (see Chapter 1 for discussion).

inheritance is the observation that many more males than females express the traits. **Table 3.2** lists several X-linked disorders, including color blindness that affects perception of red and green color (see the chapter opener photo) and is the subject of the Case Study at the end of the chapter, and hemophilia A, a blood-clotting disorder that we discuss in more detail later. Four features characterizing X-linked recessive inheritance are illustrated in **Figure 3.22**.

1. Many more males than females have the recessive phenotype, as a result of male hemizygosity. There are 10 recessive males and 2 recessive females.

2. If a recessive male mates with a homozygous dominant female, all progeny have the dominant phenotype. All female offspring are heterozygous carriers, and all male offspring are hemizygous for the dominant allele. See the progeny resulting from the cross I-1 × I-2.

3. Matings of recessive males and carrier females produce the recessive phenotype in half the offspring and the dominant phenotype in the other half. See the results of the crosses II-13 × II-14 and III-1 × III-2.

**Figure 3.22  X-linked recessive inheritance produces four characteristic features.**

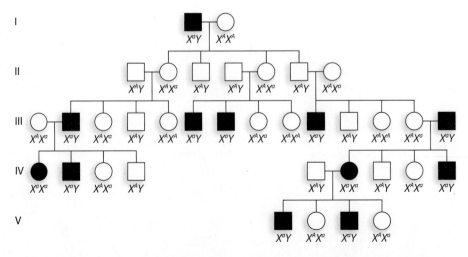

4. Mating of a homozygous recessive female and a hemizygous dominant male produces male progeny with the recessive phenotype, and female offspring who have the dominant phenotype and are carriers of the recessive allele. See the results of the cross IV-5 × IV-6.

Hemophilia A, a serious blood-clotting disorder, is caused by mutation of an X-linked gene called *factor VIII (F8)* that produces a blood-clotting protein called factor VIII protein. Hemophilia A is transmitted in an X-linked recessive manner, most often by a carrier mother who passes the mutant allele to an affected son. In typical X-linked recessive fashion, approximately half the sons of carrier mothers have the disease. In these families, the disease often appears to "skip" a generation because the mutant allele is passed from affected father to carrier daughter and on to an affected grandson.

In some families, a de novo (newly occurring) mutation of the *F8* gene is responsible for the appearance of hemophilia. An example occurred in the royal families of England and Europe: An apparent de novo mutation of the *F8* gene affected Queen Victoria of England (**Figure 3.23**). Victoria had four sons, one of whom had hemophilia, along with five daughters, two of whom were

known carriers. Victoria's carrier daughters had normal blood clotting but introduced the mutation to the royal families of Russia, Germany, and Spain through intermarriage. These daughters passed the mutation to their sons who had hemophilia and to their daughters who were carriers like their mothers.

## X-Linked Dominant Trait Transmission

Transmission of traits controlled by X-linked dominant alleles has three distinctive characteristics:

1. Heterozygous females mated to wild-type males transmit the dominant allele to half their progeny of each sex.

2. Dominant hemizygous males mated to homozygous recessive females transmit the dominant trait to *all* their daughters, but to *none* of their sons.

3. The trait appears about equally frequently in males and females, since just a single copy of the allele is necessary to produce the dominant phenotype.

Congenital generalized hypertrichosis (CGH) is a rare and dramatic X-linked dominant disorder in humans that displays each of these characteristics. The

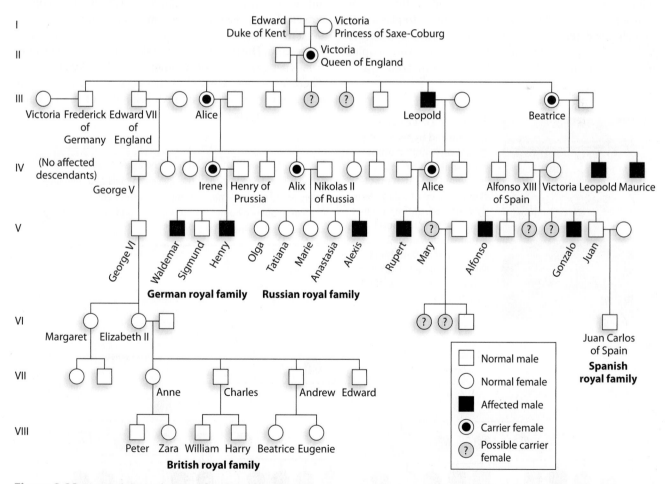

**Figure 3.23** **Hemophilia in the royal families of Europe.**

condition substantially increases the number of hair follicles on the body and produces much more body hair than normal, both in males and females (Figure 3.24a). Females with CGH have a recognizable phenotype, but face and body hair is less extensive and tends to be present in patches, for reasons we discuss at the end of the chapter. A partial pedigree of a family with CGH illustrates the transmission of the dominant alleles by the woman III-1 to about half her children and transmission of the allele by the man II-2 to all his daughters but none of his sons (Figure 3.24b). Genetic Analysis 3.3 examines the segregation of X-linked alleles in *Drosophila*.

---

**Genetic Insight**   Transmission of X-linked recessive inheritance leads to the recessive phenotype in many more hemizygous males than females. X-linked dominant alleles are transmitted from females to offspring in a pattern identical to autosomal dominant transmission. Males transmit X-linked dominant alleles to all daughters, but to no sons.

---

## Y-Linked Inheritance

The Y chromosome is found only in males, and **Y-linked** genes are transmitted in a male-to-male pattern. In mammals, fewer than 50 genes are found on the Y chromosome; and like *SRY*, those genes are likely to play a role in male sex determination or development. In addition to *SRY*, the human Y chromosome contains several copies of genes in the *YRRM* family (Y chromosome RNA recognition motif), and a gene dubbed *DAZ* (deleted in azoospermia, the inability to produce sperm). A segment of the Y chromosome containing *DAZ* is deleted in some men who are unable to produce sperm.

Females never carry a Y chromosome; so from an evolutionary perspective, it makes sense that the genes carried on a Y chromosome should be male-specific, having either to do with male sex determination or reproduction. Since males carry only a single copy of the Y chromosome, it was thought that males are hemizygous for these genes, as they are for genes on their X chromosome. However, recent sequencing of the human Y chromosome revealed that most of the genes on the Y chromosome are duplicated there, because the Y chromosome was formed from two segments containing the same series of genes. Since these two regions are on the same chromosome, they do not recombine during meiosis as do regions on homologous chromosomes.

The human X and Y chromosomes have a few genes in common located in two regions of homology called the **pseudoautosomal region (PAR)** that allow the chromosomes to pair as homologs during meiosis (Figure 3.25). These regions, designated PAR1 and PAR2, are located at the ends of X and Y chromosomes. They are "pseudoautosomal" because their pattern of inheritance resembles autosomal transmission—there are two homologous copies of each region per genome, even though the chromosomal regions are on sex chromosomes rather than autosomes. During meiosis in males, the X and Y chromosomes synapse with one another using the PAR segments. There is evidence that, like other regions of chromosome homology in synapsis, the PAR segments of X and Y undergo crossing over.

---

**Genetic Insight**   Y-chromosome transmission is male to male. Few genes except those specific to male determination are found on the Y chromosome. The X and Y chromosomes share sequence homology only in the pseudoautosomal regions (PARs) that permit synapsis.

---

**(a)**

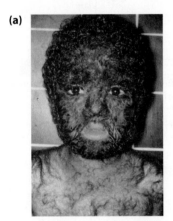

**Figure 3.24  Congenital generalized hypertrichosis (CGH), an X-linked dominant trait in humans.  (a)** A boy with CGH. **(b)** A modified pedigree of a large family with CGH. In the single instance of transmission from an affected male (II-2), notice that all daughters (III-5 to III-8) have CGH. The 6-year-old boy in panel (a) is IV-5.

**(b)**

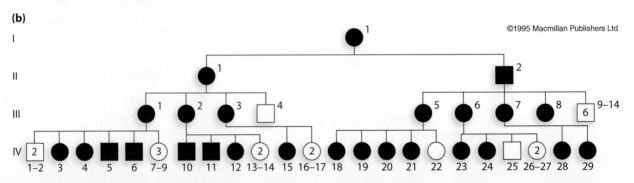

©1995 Macmillan Publishers Ltd

In fruit flies, yellow body ($y$) is recessive to gray body ($y^+$), and the trait of body color is inherited on the X chromosome. Vestigial wing ($v$) is recessive to full-sized wing ($v^+$), and the trait has autosomal inheritance. A cross of a male with yellow body and full wings to a female with gray body and full wings is made. Based on analysis of the progeny of the cross shown below, determine the genotypes of parental and progeny flies.

| Phenotype | Number of Males | Number of Females |
|---|---|---|
| Yellow body, full wing | 296 | 301 |
| Yellow body, vestigial wing | 101 | 98 |
| Gray body, full wing | 302 | 298 |
| Gray body, vestigial wing | 101 | 103 |
| | 800 | 800 |

| Solution Strategies | Solution Steps |
|---|---|

**Evaluate**

1. Identify the topic this problem addresses and describe the nature of the required answer.

2. Identify the critical information given in the problem.

1. The parental and progeny genotypes for an autosomal and an X-linked trait are to be determined based on the number and proportions of male and female progeny with the traits.

2. Parental phenotypes are given along with the number of male and female progeny with each phenotype.

**Deduce**

3. Identify the phenotype ratios for body color and for wing form in progeny.

3. Among the progeny, the ratio of full to vestigial wing is approximately 3:1 in each sex—598:202 in males and 599:201 in females. In contrast, the ratio of yellow body to gray body is approximately 1:1 in each sex—397:403 in males and 399:401 in females.

4. Determine the phenotype proportions in the progeny.

4. Among the 1600 progeny, the phenotype frequencies are the same for males and females:

| Phenotype | Number Proportion |
|---|---|
| Yellow, full | 296/800 and 301/800 $= \frac{3}{8}$ |
| Yellow, vestigial | 101/800 and 98/800 $= \frac{1}{8}$ |
| Gray, full | 302/800 and 298/800 $= \frac{3}{8}$ |
| Gray, vestigial | 101/800 and 103/800 $= \frac{1}{8}$ |

> **TIP:** Certain phenotype ratios of progeny are characteristic of crosses between specific genotypes.

5. Deduce the origins of the progeny ratios.

5. A 3:1 ratio for an autosomal trait is the predicted result of the cross of two heterozygotes. A 1:1 ratio for X-linked phenotypes involving a male parent with the recessive trait is the predicted result if the dominant female is heterozygous. For the autosomal and X-linked traits in this cross, a phenotype proportion of $\frac{3}{8}$ is predicted for the dominant autosomal trait ($\frac{3}{4}$) and either of the X-linked phenotypes ($\frac{1}{2}$). The $\frac{1}{8}$ proportion is predicted for the recessive autosomal trait ($\frac{1}{4}$) and either X-linked phenotype ($\frac{1}{2}$).

**Solve**

6. Assign genotypes to $F_2$ flies.

6. Based on the information above, the genotypes are as follows:

| Phenotype | Male Genotype | Female Genotype |
|---|---|---|
| Yellow, full | $X^y/Y; +/\_$ | $X^y/X^y; +/\_$ |
| Yellow, vestigial | $X^y/Y; v/v$ | $X^y/X^y; v/v$ |
| Gray, full | $X^{y+}/Y; +/\_$ | $X^{y+}/X^y; +/\_$ |
| Gray, vestigial | $X^{y+}/Y; v/v$ | $X^{y+}/X^y; v/v$ |

For more practice, see Problems 20, 22, and 28.

**(a)**

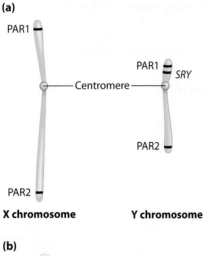

**X chromosome**        **Y chromosome**

**(b)**

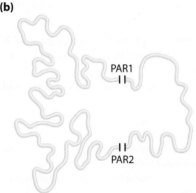

**Figure 3.25 The pseudoautosomal regions of the X and Y chromosomes.**

## 3.6 Dosage Compensation Equalizes Dosage of Sex-Linked Genes

In organisms with sex chromosomes, there is an imbalance between the sexes in the copy number of genes on the sex chromosomes. In *Drosophila* and placental mammals,

females have two copies of each X-linked gene, one on each X chromosome, whereas males have just a single copy of each X-linked gene. In animals, gene–dosage balance is essential for normal embryonic development and normal biological processes. Any mechanism that compensates for differences in the number of copies of genes due to the different chromosome constitutions of males and females is called **dosage compensation.** There are at least three dosage compensation mechanisms that equalize X-linked gene expression between male and female animals. Table 3.3 shows dosage compensation mechanisms in animals. In this section, we focus attention on dosage compensation in placental mammals.

### Random X-Chromosome Inactivation in Placental Mammals

Placental mammals use *random X inactivation* as their dosage compensation mechanism. Early in mammalian gestational development, about 2 weeks after fertilization in humans, when the female zygote consists of a few hundred cells, one of the two X chromosomes in each somatic cell of a female is randomly inactivated. This idea was first proposed in 1962 by Mary Lyon in her **random X inactivation hypothesis,** also known as the **Lyon hypothesis.** In approximately half the somatic cells in a female embryo, the maternally derived X chromosome is inactivated; and in the other half of somatic cells, inactivation silences the paternally derived X chromosome. At the end of this process, each somatic cell of a female has one active X chromosome that is equally likely to be the maternal X or the paternal X.

Random X inactivation takes place in every cell with two or more X chromosomes. Thereafter, the inactive chromosome can be seen as a tightly condensed mass adhering to the nuclear wall. The inactive X chromosome is known as a **Barr body,** having first been visualized by Murray Barr in 1949.

| Table 3.3 | Mechanisms of Dosage Compensation in Animals | | |
|---|---|---|---|
| **Animal** | **Sex Chromosomes** | | **Dosage Compensation Mechanism** |
| | Males | Females | |
| Fruit fly | XY | XX | Expression of X-linked genes in males is doubled relative to female X-linked gene expression. |
| Roundworm | XO | XX[a] | Gene expression of each X chromosome in the hermaphrodite ("female") is decreased to one-half that of the X chromosome in the male. |
| Marsupial mammals | XY | XX | The paternally derived X chromosome is inactivated in all female somatic cells. |
| Placental mammals | XY | XX | One X chromosome is randomly inactivated in each female somatic cell. |

[a] XX worms are hermaphrodites.

Once it occurs in an ancestral cell, X inactivation is permanent in the somatic cell descendants. As a consequence of random X inactivation, the normal female placental mammal is, in terms of X chromosomes, a mosaic of two kinds of cells. One cell type expresses the maternally derived X chromosome, and the other expresses the paternally derived X chromosome (Figure 3.26). Each individual cell expresses the allelic information of only one of those chromosomes. In the organism as a whole, however, the alleles of both chromosomes are expressed in approximately equal concentration.

In most cases, the silencing of one X chromosome in each cell of a female has no detectable effect on the function of a tissue or on the phenotype. Occasionally, however, female carriers of X-linked recessive traits display a phenotypic manifestation of the recessive allele. Calico and tortoiseshell coat-color patterning in female cats is a product of mosaicism created by random X inactivation (Figure 3.27). Females with an allele for black coat color on one X chromosome and yellow coat color on the homologous X chromosome have black and yellow patches of fur corresponding to portions of skin where each X chromosome is active. The sizes and the distribution of the orange and black sectors of these cats reflect the locations of the clonal descendants of the cells in which each X chromosome was originally inactivated. The specific pattern of X inactivation is unique to each female cat

**Figure 3.27  Calico coat, produced by X inactivation in female cats.**

embryo, and the patterns of cellular migration are variable as well. As a result, each adult female calico or tortoiseshell cat has a unique pattern of black and orange sectors marking its coat.

Random X inactivation requires a gene on the X chromosome called the *X-inactivation–specific transcript (XIST)* that encodes a large RNA molecule. *XIST* RNA spreads out from the gene, "painting" the X chromosome as it accumulates. X chromosomes that are painted with *XIST* RNA have all, or nearly all, of their genes silenced. The *XIST* RNA accumulates only on the one chromosome transcribing the gene and does not spread to the homologous X chromosome. In other words, *XIST* acts only in cis (on the same chromosome) but not in trans (on the homologous chromosome). Examination of inactivated chromosomes in the nucleus detects *XIST* RNA coating the Barr body in a nucleus.

**Genetic Insight**  Compensation for differences in dosage of X-linked genes resulting from different male and female sex-chromosome constitutions is routine in animals. Different mechanisms of compensation are seen in different species. Humans and other placental mammals undergo random X inactivation, which silences one X chromosome at random in each female cell.

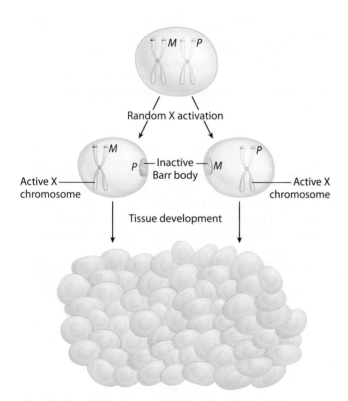

**Figure 3.26  Random X inactivation in female placental mammals.**

## CASE STUDY

### John Dalton's Eyes Help Solve the Mystery of Color Blindness

John Dalton (1766–1844) was a chemist, best known for his development of the atomic theory of matter. In recognition of his seminal achievements in chemistry, a unit for measuring atomic mass, the dalton, was named in his honor. John Dalton was also color-blind. He wrote extensively about his lack of complete color perception, but knew nothing of the hereditary or molecular basis of his color blindness. Dalton was so interested in his condition that he directed that after his death, his eyes should be preserved in case they might one day prove useful in investigating color blindness.

Dalton's form of color blindness—the most common form affecting the ability to perceive color in the mid-wavelength portion of the visible spectrum—is an X-linked recessive condition. The chapter opener photo is a color chart of the kind used to detect this type of color blindness. If you are among the approximately 8% of the population that has a red-green color perception deficiency, you are unable to see the number 22.

Two closely related X-linked genes producing photosensitive proteins called opsins are the source of red-green color blindness. One gene produces red opsin protein, the other green opsin protein. Opsins embed in the membrane of the cone cells, where their sensitivity to light in different parts of the visible spectrum is responsible for our ability to see color. The green opsin protein is sensitive to light in the medium-wavelength portion of the visible spectrum, and red opsin absorbs light at long wavelengths. Blue opsin protein, encoded by an autosomal gene on chromosome 7, is sensitive to short-wavelength light. Mutations of this gene are much rarer than those of red and green opsin genes, and they produce a distinct type of color blindness.

Red and green opsin genes are located near one another on the X chromosome. The genes are 98% identical in their DNA sequences and produce proteins that differ at just 15 of their 329 amino acids. Occasional mutations of the red or green opsin genes produce abnormal opsin proteins that alter color perception. But the most common mutations causing red and green color blindness are due to the misalignment and unequal crossover of red and green opsin genes during prophase I of female meiosis (**Figure 3.28a**). The misalignment results from the high degree of DNA sequence similarity between the genes. The outcomes of these unequal crossover events are a maternal X chromosome with extra genetic material that carries a red opsin gene and two copies of the green opsin gene, and an X chromosome that is missing genetic material and has a red opsin gene but not a green opsin gene. Transmission of the chromosome lacking a green opsin gene can produce the form of color blindness experienced by Dalton.

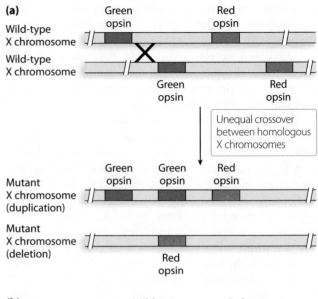

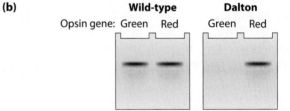

**Figure 3.28 The molecular genetics of John Dalton's color blindness.** (a) Wild-type green opsin and red opsin genes on the X chromosome undergo unequal crossover to produce mutant chromosomes with a duplicated green opsin gene or a missing green opsin gene. (b) Analysis of DNA isolated from Dalton's preserved eyes reveals the missing green opsin gene.

John Dalton's desire to contribute to the understanding of color blindness came to fruition in 1995—150 years after his death—through a report by David Hunt and his colleagues on their analysis of DNA extracted from one of Dalton's long-preserved eyes. Hunt and his colleagues examined Dalton's DNA for the presence of red opsin and green opsin genes. They demonstrated that Dalton possessed the red opsin gene but not green opsin gene, confirming Dalton's self-diagnosis of color blindness (**Figure 3.28b**). Dalton's X chromosome was able to use the intact red opsin gene to produce red opsin protein, but the lack of a green opsin gene left him without green opsin protein.

## 3.1 Mitosis Divides Somatic Cells

▌ The cell cycle has two principal phases: interphase, whose stages are $G_1$, S, and $G_2$; and M phase, during which cell division occurs.

▌ Mitosis is the process of division for somatic cells. Mitosis contains five substages: prophase, prometaphase, metaphase, anaphase, and telophase.

▌ Mitosis contains a single cell division and separates sister chromatids into diploid daughter cells that are genetically identical to one another and to the parental cell they are derived from.

▌ The cell cycle is under tight genetic control. Regulatory molecules control the transition from one stage of the cycle to the next by acting at genetically controlled checkpoints to monitor cell cycle transitions.

▌ Mutation of cell cycle control genes is associated with cancer development.

## 3.2 Meiosis Produces Gametes for Sexual Reproduction

▌ Meiosis contains two cell divisions, designated meiosis I and meiosis II.

▌ During meiosis I (the "reduction division"), homologous chromosomes are separated to produce haploid daughter cells that carry one chromosome from each homologous pair of chromosomes.

▌ The meiosis II division separates sister chromatids and produces four genetically different haploid daughter cells that form gametes.

▌ During prophase I, homologous chromosomes synapse with the aid of the synaptonemal complex. Homologous chromosomes can cross over to exchange genetic material during this substage.

▌ Mendel's laws of segregation and independent assortment find their mechanical basis in the patterns of separation of chromosomes and sister chromatids during meiosis.

## 3.3 The Chromosome Theory of Heredity Proposes That Genes Are Carried on Chromosomes

▌ The chromosome theory of heredity proposes that genes are carried on chromosomes and are faithfully transmitted through gametes to successive generations.

▌ Thomas Hunt Morgan's identification of X-linked transmission of white eye color in *Drosophila* and Calvin Bridges's analysis of exceptional phenotypes produced by X-chromosome nondisjunction demonstrated the validity of the chromosome theory of heredity.

## 3.4 Sex Determination Is Chromosomal and Genetic

▌ Mechanisms of sex determination take many forms in animals. *Drosophila* sex is determined by the ratio of expression of X-linked and autosomal genes, whereas human sex is determined by the presence of *SRY* on the Y chromosome.

▌ Sex-chromosome patterns are diverse among organisms. Birds, fishes, and some insects have Z and W sex chromosomes, and monotremes have multiple sex chromosomes.

## 3.5 Human Sex-Linked Transmission Follows Distinct Patterns

▌ Human X-linked dominant inheritance and X-linked recessive inheritance are identifiable, respectively, by the pattern of male transmission and the pattern of male expression of traits.

▌ Genes on the Y chromosome are transmitted exclusively from male to male.

## 3.6 Dosage Compensation Equalizes Dosage of Sex-Linked Genes

▌ Dosage compensation balances the level of expression of sex-linked genes and is critical for normal animal development. Mechanisms for achieving dosage compensation vary among species.

▌ Random inactivation of one X chromosome in each cell of placental mammalian females is controlled by an X-inactivation center on the X chromosome.

## KEYWORDS

ascus (spores) *(p. 81)*
aster *(p. 68)*
Barr body *(p. 96)*
Cdks (cyclin-dependent kinases) *(p. 71)*
cell cycle (cell cycle checkpoints)
  *(pp. 64, 71)*
centromere *(p. 68)*
centrosome *(p. 68)*

chiasma (chiasmata) *(p. 77)*
chromosome theory of heredity *(p. 64)*
crossing over *(p. 76)*
cyclin protein *(p. 71)*
cytokinesis *(p. 65)*
daughter cell *(p. 64)*
diploid number of chromosomes (2*n*)
  *(p. 64)*

disjunction *(p. 70)*
dosage compensation *(p. 96)*
gamete (germ-line cell) *(p. 64)*
haploid number of chromosomes (*n*) *(p. 64)*
hemizygous *(p. 85)*
interphase ($G_1$ phase, S phase, $G_2$ phase)
  *(p. 65)*
karyokinesis *(p. 65)*

## PROBLEMS

*For instructor-assigned tutorials and problems, go to www.masteringgenetics.com.*

### Chapter Concepts

*For answers to selected even-numbered problems, see Appendix: Answers.*

1. Examine the following diagrams of cells from an organism with diploid number $2n = 6$, and identify what stage of M phase is represented.

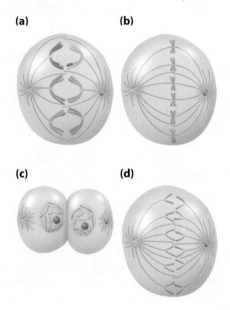

**(a)**    **(b)**

**(c)**    **(d)**

2. Our closest primate relative, the chimpanzee, has a diploid number of $2n = 48$. For each of the following stages of M phase, identify the number of chromosomes present in each cell.
   a. end of mitotic telophase
   b. meiotic metaphase I
   c. end of meiotic anaphase II
   d. early mitotic prophase
   e. mitotic metaphase
   f. early prophase I

3. In a test of his chromosome theory of heredity, Morgan crossed an $F_1$ female *Drosophila* with red eyes to a male with white eyes (see Figure 3.16, cross B). Morgan presumed the female was heterozygous for the X-linked recessive white-eye allele and that the male was hemizygous for white eye. The results of the cross were 129 red-eyed females, 132 red-eyed males, 88 white-eyed females, and 86 white-eyed males. Test these data by chi-square analysis, and determine if the results are consistent with Morgan's hypothesis that *Drosophila* eye color is an X-linked trait.

4. Tension between sister chromatids is essential to ensure their efficient separation at mitotic anaphase or in meiotic anaphase II. Explain why sister chromatid cohesion is important, and discuss the role of the proteins cohesin and separase in sister chromatid separation.

5. The diploid number of the hypothetical animal *Geneticus introductus* is $2n = 36$. Each diploid nucleus contains 3 ng of DNA in $G_1$.
   a. What amount of DNA is contained in each nucleus at the end of S phase?
   b. Explain why a somatic cell of *Geneticus introductus* has the same number of chromosomes and the same amount of DNA at the beginning of mitotic prophase as one of these cells does at the beginning of prophase I of meiosis.
   c. Complete the following table by entering the number of chromosomes and amount of DNA present per cell at the end of each stage listed.

| End of Cell Cycle Stage | Number of Chromosomes | Amount of DNA |
|---|---|---|
| Telophase I | | |
| Mitotic anaphase | | |
| Telophase II | | |

6. An organism has alleles $R_1$ and $R_2$ on one pair of homologous chromosomes, and it has alleles $T_1$ and $T_2$ on another pair. Diagram these pairs of homologs at the end of metaphase I, the end of telophase I, and the end of telophase II, and show how meiosis in this organism produces gametes in expected Mendelian proportions. Assume no crossover between homologous chromosomes.

7. Explain how the behavior of homologous chromosomes in meiosis parallels Mendel's law of segregation for autosomal alleles $D$ and $d$. During which stage of M phase do these two alleles segregate from one another?

8. Suppose crossover occurs between the homologous chromosomes in the previous problem. At what stage of M phase do alleles $D$ and $d$ segregate?

9. Alleles $A$ and $a$ are on one pair of autosomes, and alleles $B$ and $b$ are on a separate pair of autosomes. Does crossover between one pair of homologs affect the expected proportions of gamete genotypes? Why or why not? Does

crossover between both pairs of chromosomes affect the expected gamete proportions? Why or why not?

10. How many Barr bodies are found in a normal human female nucleus? In a normal male nucleus?

11. Describe the role of the following structures or proteins in cell division:

a. microtubules
b. cyclin-dependent kinases
c. kinetochores
d. synaptonemal complex

## Application and Integration

*For answers to selected even-numbered problems, see Appendix: Answers.*

12. A woman's father has ornithine transcarbamylase deficiency (OTD), an X-linked recessive disorder producing mental deterioration if not properly treated. The woman's mother is homozygous for the wild-type allele.

a. What is the woman's genotype? (Use *D* to represent the dominant allele and *d* to represent the recessive allele.)
b. If the woman has a son with a normal man, what is the chance the son will have OTD?
c. If the woman has a daughter with a man who does not have OTD, what is the chance the daughter will be a heterozygous carrier of OTD? What is the chance the daughter will have OTD?
d. Identify a male with whom the woman could produce a daughter with OTD.
e. For the instance you identified in part (d), what proportion of daughters produced by the woman and the man are expected to have OTD? What proportion of sons of the woman and the man are expected to have OTD?

13. In humans, hemophilia (OMIM 306700) is an X-linked recessive disorder that affects the gene for factor VIII protein, which is essential for blood clotting. The dominant and recessive alleles for the *factor VIII* gene are represented by *H* and *h*. Albinism is an autosomal recessive condition that results from mutation of the gene producing tyrosinase, an enzyme in the melanin synthesis pathway. *A* and *a* represent the tyrosinase alleles. A healthy woman named Clara (II-2), whose father (I-1) has hemophilia and whose brother (II-1) has albinism, is married to a healthy man named Charles (II-3), whose parents are healthy. Charles's brother (II-5) has hemophilia, and his sister (II-4) has albinism. The pedigree is shown below.

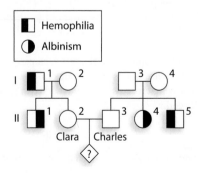

a. What are the genotypes of the four parents (I-1 to I-4) in this pedigree?
b. Determine the probability that the first child of Clara and Charles will be a
   i. boy with hemophilia
   ii. girl with albinism

iii. healthy girl
iv. boy with both albinism and hemophilia
v. boy with albinism
vi. girl with hemophilia
c. If Clara and Charles's first child has albinism, what is the chance the second child has albinism? Explain why this probability is higher than the probability you calculated in part (b).

14. A wild-type male and a wild-type female *Drosophila* with red eyes and full wings are crossed. Their progeny are shown below.

| Males | Females |
|---|---|
| $\frac{3}{8}$ full wing, red eye | $\frac{3}{4}$ full wing, red eye |
| $\frac{3}{8}$ miniature wing, red eye | $\frac{1}{4}$ purple eye, full wing |
| $\frac{1}{8}$ purple eye, full wing | |
| $\frac{1}{8}$ miniature wing, purple eye | |

a. Using clearly defined allele symbols of your choice, give the genotype of each parent.
b. What is/are the genotype(s) of females with purple eye? Of males with purple eye and miniature wing?

15. A woman with severe discoloration of her tooth enamel has four children with a man who has normal tooth enamel. Two of the children, a boy (B) and a girl (G), have discolored enamel. Each has a mate with normal tooth enamel and produces several children. G has six children, four boys and two girls. Two of her boys and one of her girls have discolored enamel. B has seven children, four girls and three boys. All four of his daughters have discolored enamel, but all his boys have normal enamel. Explain the inheritance of this condition.

16. In a large metropolitan hospital, cells from newborn babies are collected and examined microscopically over a 5-year period. Among approximately 7500 newborn males, one Barr body is seen in the nuclei of six babies. All other newborn males have none. Among 7500 female infants, four have two Barr bodies in each nucleus, two have no Barr bodies, and the rest have one. What is the cause of the unusual number of Barr bodies in a small number of male and female infants?

17. In cats, tortoiseshell coat color appears in females. A tortoiseshell coat has patches of dark brown fur and patches of orange fur that each in total cover about half the body but have a unique pattern in each female. Male cats can be either dark brown or orange, but a male cat with tortoiseshell coat is rarely produced. Two sample crosses between

males and females from pure-breeding lines produced the tortoiseshell females shown below.

Cross I     P: dark brown male × orange female

            F1: orange males and tortoiseshell females

Cross II    P: orange male × dark brown female

            F1: dark brown males and tortoiseshell females

a. Explain the inheritance of dark brown, orange, and tortoiseshell coat colors in cats.
b. Why are tortoiseshell cats female?
c. The genetics service of a large veterinary hospital gets referrals for three or four male tortoiseshell cats every year. These cats are invariably sterile and have underdeveloped testes. How are these tortoiseshell male cats produced? Why do you think they are sterile?

18. The gene causing Coffin-Lowry syndrome (OMIM 303600) was recently identified and mapped on the human X chromosome. Coffin-Lowry syndrome is a rare disorder affecting brain morphology and development. It also produces skeletal and growth abnormalities, as well as abnormalities of motor control. Coffin-Lowry syndrome affects males who inherit a mutation of the X-linked gene. Most carrier females show no symptoms of the disease but a few carriers do. These carrier females are always less severely affected than males. Offer an explanation for this finding.

19. Four eye-color mutants in *Drosophila*—apricot, brown, carnation, and purple—are inherited as recessive traits. Red is the dominant wild-type color of fruit-fly eyes. Eight crosses (A to H) are made between parents from pure-breeding lines.

| Cross | Parents | | $F_1$ Progeny | |
|-------|---------|------|---------------|------|
|       | Female  | Male | Female | Male |
| A | Apricot | Red | Red | Apricot |
| B | Brown | Red | Red | Red |
| C | Red | Purple | Red | Red |
| D | Red | Apricot | Red | Red |
| E | Carnation | Red | Red | Carnation |
| F | Purple | Red | Red | Red |
| G | Red | Brown | Red | Red |
| H | Red | Carnation | Red | Red |

a. Which of these eye-color mutants are X-linked recessive and which are autosomal recessive? Explain how you distinguish X-linked from autosomal heredity.
b. Predict $F_2$ phenotype ratios of crosses A, B, D, and G.

20. For each pedigree below,
a. Identify which simple pattern of hereditary transmission (autosomal dominant, autosomal recessive, X-linked dominant, or X-linked recessive) is most likely to have occurred. Give genotypes for individuals involved in transmitting the trait.
b. Determine which other pattern(s) of transmission is/are possible. For each possible mode of transmission, specify the genotypes necessary for transmission to occur.

c. Identify which pattern(s) of transmission is/are impossible. Specify why transmission is impossible.

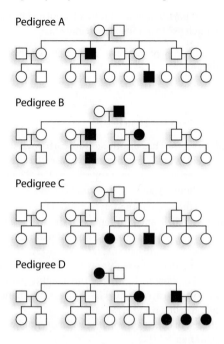

Pedigree A

Pedigree B

Pedigree C

Pedigree D

21. Use the pedigrees below to depict transmission of (a) an X-linked recessive trait and (b) an X-linked dominant trait. Give genotypes for each person in each pedigree. Carefully design each transmission pattern so that pedigree (a) cannot be confused with autosomal recessive transmission and pedigree (b) cannot be confused with autosomal dominant transmission. Identify the transmission events that eliminate the possibility of autosomal transmission for each pedigree.

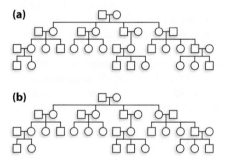

(a)

(b)

22. In a breed of domestic cattle, horns can appear on males and on females. Males and females can also be hornless. The following crosses are performed with parents from pure-breeding lines.

| Cross I | Cross II |
|---------|----------|
| Parents: horned male × hornless female | Parents: hornless male × horned female |
| $F_1$: males horned, females hornless | $F_1$: males horned, females hornless |
| $F_2$: males are $\frac{3}{4}$ horned, $\frac{1}{4}$ hornless | $F_2$: males are $\frac{3}{4}$ horned, $\frac{1}{4}$ hornless |
| females are $\frac{1}{4}$ horned, $\frac{3}{4}$ hornless | females are $\frac{1}{4}$ horned, $\frac{3}{4}$ hornless |

Explain the inheritance of this phenotype in cattle, and assign genotypes to all cattle in cross I.

23. In domestic chickens, sex is determined by the ZW system. A Z-linked gene determining feather formation has two alleles: *F* produces fast feathering, and *f* produces slow feathering. Give the genotypes of parental and progeny male (rooster) and female (hen) chickens for the following crosses:

    a. Parents: fast-feathering rooster × slow-feathering hen
       Progeny: Roosters and hens are each $\frac{1}{2}$ fast-feathering, $\frac{1}{2}$ slow-feathering.
    b. Parents: slow-feathering rooster × fast-feathering hen
       Progeny: All roosters are fast-feathering; all hens are slow-feathering.

24. In a species of fish, a black spot on the dorsal fin is observed in males and females. A fish breeder carries out a pair of reciprocal crosses and observes the following results.

    | Cross I | Parents: | black-spot male × nonspotted female |
    |---|---|---|
    | | Progeny: | 22 black-spot males |
    | | | 24 black-spot females |
    | | | 25 nonspotted males |
    | | | 21 nonspotted females |
    | Cross II | Parents: | nonspotted male × black-spot female |
    | | Progeny: | 45 spotted males |
    | | | 53 nonspotted females |

    a. Why does this evidence support the hypothesis that a black spot is sex linked?
    b. Identify which sex is homogametic and which is heterogametic. Give genotypes for the parents in each cross, and explain the progeny proportions in each cross.

25. Lesch-Nyhan syndrome (OMIM 300322) is a rare X-linked recessive disorder that produces severe mental retardation, spastic cerebral palsy, and self-mutilation.

    a. What is the probability that the first son of a woman whose brother has Lesch-Nyhan syndrome will be affected?

    b. If the first son of the woman described in (a) is affected, what is the probability that her second son is affected?
    c. What is the probability that the first son of a man whose brother has Lesch-Nyhan syndrome will be affected?

26. In humans, *SRY* is located near a pseudoautosomal region (PAR) of the Y chromosome, a region of homology between the X and Y chromosomes that allows them to synapse during meiosis in males and is a region of crossover between the chromosomes. The diagram below shows *SRY* in relation to the pseudoautosomal region.

    About 1 in every 25,000 newborn infants is born with sex reversal; the infant is either an apparent male, but with two X chromosomes, or an apparent female, but with an X and a Y chromosome. Explain the origin of sex reversal in human males and females involving the *SRY* gene. (*Hint*: See Experimental Insight 3.1 for a clue about the mutational mechanism.)

27. In an 1889 book titled *Natural Inheritance* (Macmillan, New York), Francis Galton, who investigated the inheritance of measurable (quantitative) traits, formulated a law of "ancestral inheritance." The law stated that each person inherits approximately one-half of his or her genetic traits from each parent, about one-quarter of the traits from each grandparent, one-eighth from each great grandparent, and so on. In light of the chromosome theory of heredity, argue either in favor of Galton's law or against it.

28. In *Drosophila*, the X-linked echinus eye phenotype disrupts formation of facets and is recessive to wild-type eye. Autosomal recessive traits vestigial wing and ebony body assort independently of one another. Examine the progeny from the three crosses shown below, and identify the genotype of parents in each cross.

| Parental Phenotype | | Progeny Phenotype | Proportion | |
|---|---|---|---|---|
| Female | Male | | Female | Male |
| a. Wild type | Echinus | Wild type | $\frac{3}{8}$ | $\frac{3}{8}$ |
| | | Echinus | $\frac{3}{8}$ | $\frac{3}{8}$ |
| | | Vestigial | $\frac{1}{8}$ | $\frac{1}{8}$ |
| | | Echinus, vestigial | $\frac{1}{8}$ | $\frac{1}{8}$ |
| b. Wild type | Wild type | Vestigial, ebony | $\frac{2}{32}$ | $\frac{1}{32}$ |
| | | Vestigial | $\frac{6}{32}$ | $\frac{3}{32}$ |
| | | Ebony | $\frac{6}{32}$ | $\frac{3}{32}$ |
| | | Wild type | $\frac{18}{32}$ | $\frac{9}{32}$ |
| | | Echinus, vestigial, ebony | 0 | $\frac{1}{32}$ |
| | | Echinus, vestigial | 0 | $\frac{3}{32}$ |
| | | Echinus, ebony | 0 | $\frac{3}{32}$ |
| | | Echinus | 0 | $\frac{9}{32}$ |

| Parental Phenotype | | Progeny Phenotype | Proportion | |
| --- | --- | --- | --- | --- |
| Female | Male | | Female | Male |
| c. Ebony | Echinus | Echinus, vestigial, ebony | $\frac{1}{32}$ | $\frac{1}{32}$ |
| | | Echinus, vestigial | $\frac{3}{32}$ | $\frac{3}{32}$ |
| | | Echinus, ebony | $\frac{3}{32}$ | $\frac{3}{32}$ |
| | | Echinus | $\frac{9}{32}$ | $\frac{9}{32}$ |
| | | Vestigial, ebony | $\frac{1}{32}$ | $\frac{1}{32}$ |
| | | Vestigial | $\frac{3}{32}$ | $\frac{3}{32}$ |
| | | Ebony | $\frac{3}{32}$ | $\frac{3}{32}$ |
| | | Wild type | $\frac{9}{32}$ | $\frac{9}{32}$ |

29. A wild-type *Drosophila* male and a female with wild-type phenotype are crossed, producing 324 female progeny and 161 male progeny. All their progeny are wild type.

    a. Propose a genetic hypothesis to explain these data.
    b. Design an experiment that will test your hypothesis, using the wild-type progeny identified above. Describe the results you expect if your hypothesis is true.

30. Red-green color blindness in humans is inherited as an X-linked recessive condition. Consider reciprocal crosses between a color-blind parent and a parent with normal color vision in which the dominant allele is identified as $C$ and the recessive allele as $c$. Cross I is $Cc \times cY$, and cross II is $cc \times CY$. Determine the phenotypes and their proportions in progeny produced by each cross. Explain why the reciprocal cross results are consistent with an X-linked recessive inheritance but not with an autosomal recessive inheritance of color blindness.

31. While examining a young tortoiseshell cat, you and the veterinarian you are interning with get a surprise—the cat is male, not female! From your undergraduate genetics course, you recall that tortoiseshell coats are produced by the random X inactivation that takes place in mammalian females. The veterinarian orders a chromosome analysis of the cat and finds that he is XXY: He has two X chromosomes and one Y chromosome. Help the veterinarian figure out how a tortoiseshell cat could be male., (*Hint:* Think about X-inactivation in mammals with two X chromosomes.)

32. *Drosophila* has a diploid chromosome number of $2n = 8$, which includes one pair of sex chromosomes (XX in females and XY in males) and three pairs of autosomes. Consider a *Drosophila* male that has a copy of the $A_1$ allele on its X chromosome (the Y chromosome is the homolog) and is heterozygous for alleles $B_1$ and $B_2$, $C_1$ and $C_2$, and $D_1$ and $D_2$ of genes that are each on a different autosomal pair. In the diagrams requested below, indicate the alleles carried on each chromosome and sister chromatid. Assume that no crossover occurs between homologous chromosomes.

    a. What is the genotype of cells produced by mitotic division in this male?
    b. Diagram *any correct* alignment of chromosomes at mitotic metaphase.
    c. Diagram *any correct* alignment of chromosomes at metaphase I of meiosis.
    d. For the metaphase I alignment shown in (c), what gamete genotypes are produced at the end of meiosis?
    e. How many different metaphase I chromosome alignments are possible in this male? How many genetically different gametes can this male produce? Explain your reasoning for each answer.

# Gene Interaction

$4$

Several genes interact to produce readily observable traits of tomatoes such as color, size, and shape. Multiple genes and environmental factors contribute to tomato characteristics such as taste.

## ESSENTIAL IDEAS

▪ Dominance relationships between alleles have a molecular basis. The biological effect of gene products determines what type of dominance is observed.

▪ Gene expression can be affected by nongenetic (environmental) factors and also as a consequence of factors related to sex.

▪ Gene expression can be affected by interactions with other genes, causing characteristic changes in Mendelian ratios.

▪ Mutation of different genes can produce the same effect on phenotype. The number of genes causing mutation of a phenotype is discovered by genetic complementation analysis.

Mendel's laws of segregation and independent assortment encapsulate the basic rules of genetic transmission in diploid organisms. We see the results of these rules through the analysis of the relative proportions of progeny with different phenotypes from crosses. We can also glimpse the hereditary transmission of DNA, RNA, and protein variation through analyses of Southern, northern, and western blots. On a mechanical level, we portray the physical basis of these rules through the movement and segregation of homologous chromosomes and sister chromatids during meiosis.

Mendel's success in identifying and describing these two hereditary laws was partly due to his use of traits whose phenotypic characteristics are determined exclusively by inheritance of alleles for single genes. In interpreting the inheritance of these traits, he did not have to contend with phenotypic variation introduced by other genes or by environmental (nongenetic) factors. In Mendel's experiments, each trait was decided by a single pair of alleles, one fully dominant and one fully recessive, at each of seven genes. The simple case in which just two alleles influence a trait is, however, quite rare in nature. Although a diploid organism can have no more than two alleles at a locus (because such individuals have just two copies of each chromosome), there may be many alleles for a single locus within a population. Rarely do geneticists observe just two alleles segregating at a locus, with one allele completely dominant to the other, with only the alleles at a single locus controlling the phenotype, and with environmental factors playing a minimal role in determining the phenotypic character. In most cases, phenotype determination is more complicated because one or more additional circumstances affect the outcome. Collectively, these circumstances are identified as *gene interactions;* this phrase refers to any of several ways different genes can collaborate or interact with one another or with nongenetic (environmental) factors to influence the expression of a phenotypic character. Among the most important of these interactions are the following:

▌ There may be more than two alleles for a given locus within the population.

▌ Dominance of one allele over another may not be complete.

▌ Two or more genes may affect a single trait.

▌ The expression of a trait may be dependent on the interaction of two or more genes, on the interaction of genes with nongenetic factors, or both.

In this chapter, we examine patterns of phenotypic variation that result from the occurrence of each of these circumstances. Our discussions demonstrate that while traits arising through gene interactions do not always exhibit the classic Mendelian ratios

(described in Chapter 2), the observed ratios can nevertheless be explained by Mendelian principles.

## 4.1 Interactions between Alleles Produce Dominance Relationships

Mendel wisely chose to examine traits presenting in one of two alternative forms. One form of each trait he studied displays complete dominance over the other form. Complete dominance makes the phenotype of a heterozygous organism indistinguishable from that of an organism homozygous for the dominant allele; thus, only organisms homozygous for the recessive allele display the recessive phenotype. The complete dominance of one allele also results in the exclusive expression of the dominant phenotype among the heterozygous $F_1$ progeny of a cross between pure-breeding homozygous parents, while the $F_2$ progeny display a 3:1 ratio of dominant to recessive phenotypes. We now know that the phenotypes of the seven traits that Mendel studied are controlled by two alternative alleles at seven different genes. In the cases that have been examined at the molecular level, the dominant alleles reflect the wild-type function of the gene, while the recessive alleles encode gene products with reduced or no functional activity.

Questions concerning the molecular basis of dominant and recessive alleles drove genetic research in the early and mid-20th century, including questions of how dominance of an allele could be ascertained, why certain mutations are recessive whereas others are dominant, and whether mutations always cause genes to lose function or whether mutations can impart new or additional functions to alleles.

### The Molecular Basis of Dominance

A character is called dominant if it is seen in the homozygous and heterozygous genotypes, and it is called recessive if it is observed only in a single homozygous genotype. In this sense, dominance and recessiveness have a phenotypic basis. The phenotypes are, however, a consequence of the activities of proteins produced by the alleles of a gene. In this sense, dominance and recessiveness have a molecular basis. The dominance of one allele over another is determined by the activity of the protein products of the allele—by the manner in which the protein products of alleles work to produce the phenotype.

Let's compare two examples to illustrate the molecular basis of dominance and recessiveness. In both examples, a wild-type allele produces an active enzyme and a mutant allele produces either very little enzyme or none at all. In the first example the mutant allele is recessive, but in the second example the mutant allele is dominant. In the first example, gene $R$ has a dominant wild-type allele $R^+$ and a recessive mutant allele $r$. Gene $R$ produces an enzyme that must generate 40 or more units of catalytic activity to drive a critical reaction step. Successful completion of this step produces the wild-type phenotype, whereas failure to complete the step generates a mutant phenotype. Each copy of allele $R^+$ produces 50 units of enzyme activity. The mutant allele $r$ produces no functional enzyme and has 0 units of activity. Homozygous $R^+R^+$ organisms produce 100 units of enzyme activity (50 units from each copy of $R^+$), far exceeding the minimum required to achieve the wild-type phenotype. Heterozygous organisms ($R^+r$) produce a total of 50 units of enzyme activity, which is sufficient to produce the wild-type phenotype. Homozygous $rr$ organisms produce no enzymatic action, however, and display the mutant phenotype. Based on its ability to catalyze the critical reaction step and produce the wild-type phenotype in either a homozygous ($R^+R^+$) and heterozygous ($R^+r$) genotype, $R^+$ is dominant over $r$. Dominant wild-type alleles of this kind are identified as **haplosufficient** since one (haplo) copy is sufficient to produce the wild-type phenotype.

The second example involves gene $T$, for which the wild-type allele is recessive to a mutant allele. Gene $T$ produces an enzyme required to catalyze a critical reaction step that produces a wild-type phenotype if it is completed. The inability to complete the reaction step results in a mutant phenotype. For the reaction step in question, 18 units of enzyme activity are required. The wild-type allele $T_1$ produces 10 units of activity. A mutant allele, $T_2$, generates 5 units of enzyme activity. Homozygous $T_1T_1$ organisms generate 20 units of catalytic enzyme activity, enough to catalyze the critical reaction step and produce the wild-type phenotype. Heterozygous organisms, on the other hand, produce only 15 units of enzymatic activity and have the mutant phenotype because they fall short of the 18 units required to catalyze the reaction step. Similarly, homozygous $T_2T_2$ organisms, which produce 10 units of enzyme activity, also have a mutant phenotype. In this case, the mutant allele $T_2$ is dominant over the wild-type allele $T_1$ since both the heterozygous ($T_1T_2$) and homozygous ($T_2T_2$) organisms have a mutant phenotype. In cases like this, the wild-type allele is identified as **haploinsufficient** because a single copy is not sufficient to produce the wild-type phenotype.

## Effects of Mutation

Genetic analysis often focuses on rare mutations and other infrequent phenomena. In many instances, the study of these rare events provides clues to the underlying causes of commonly occurring events that are not yet understood. In the case of any genetic mutation, a central question concerns the precise mechanism through which the mutation disrupts normal gene function.

From a functional perspective, the wild-type phenotype is produced in organisms with two copies of the wild-type allele (**Figure 4.1a**). In comparison to the level of activity of the protein products of the wild-type allele, mutant alleles can often be placed into either a *loss-of-function* or a *gain-of-function* category. A **loss-of-function mutation** results in a significant decrease or in the complete loss of the functional activity of a gene product. This common mutational category contains mutations like those described in the $R$ gene example above. Loss-of-function mutant alleles are usually recessive, but under certain circumstances, they may be dominant.

**Gain-of-function mutations** identify alleles that have acquired a new function or are altered to express substantially more activity than the wild-type allele. Gain-of-function mutations are almost always dominant and usually produce dominant mutant phenotypes in heterozygous organisms. As a consequence of their newly acquired functions, certain gain-of-function mutations are lethal in a homozygous state.

**Loss-of-Function Mutations**    As the previous discussion suggests, mutations resulting in a loss of function vary in the extent of loss of normal activity of the gene product. A loss-of-function mutation that results in a complete loss of gene function in comparison to the wild-type gene product is identified as a **null mutation,** also known as an **amorphic mutation** (**Figure 4.1b**). The word *null* means "zero" or "nothing," and the word *amorphic* means "without form." These mutant alleles produce no functional gene product and are often lethal in a homozygous genotype. The elimination of functional gene products can result from various types of mutational events, including those that block transcription, produce a gene product that lacks activity, or result in deletion of all or part of the gene.

Alternatively, a mutation resulting in partial loss of gene function may be identified as a **leaky mutation,** also known as a **hypomorphic mutation** (**Figure 4.1c**). *Hypomorphic* means "reduced form"; like the term *leaky*, it implies that a small percentage of normal functional capability is retained by the mutant allele but at a lower level than is found for the wild-type allele. The severity of the phenotypic abnormality depends on the residual level of activity from the leaky mutant allele. A greater percentage of activity from a leaky allele results in a less severely affected phenotype than when the mutation incurs a more substantial loss of function. Both null and hypomorphic loss-of-function mutations are often recessive and homozygous lethal. Dominant loss-of-function mutations are also known to occur.

Certain loss-of-function mutations produce dominant mutant phenotypes through alterations in the function of a multimeric protein of which the mutant polypeptide forms a part (**Figure 4.1d**). Multimeric proteins, composed of two or more polypeptides that join together to form a

**(a) Wild type**

Homozygous

Alleles

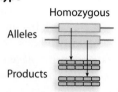

Products

The expression of the products of wild-type alleles produces wild-type phenotype.

**(b) Loss of function:** Null/amorphic mutation

Homozygous        Heterozygous

Alleles

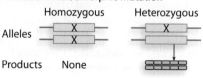

Products    None

Null alleles produce no functional product. Homozygous null organisms have mutant (amorphic) phenotype due to absence of the gene product. Heterozygous organisms produce less functional gene product than homozygous wild-type organisms and may have mutant phenotype. See text for discussion of dominant versus recessive mutations.

**(c) Loss of function:** Leaky/hypomorphic mutation

Homozygous        Heterozygous

Alleles

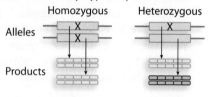

Products

Leaky mutant alleles produce a small amount of wild-type gene product. Homozygous organisms have a mutant (hypomorphic) phenotype. Heterozygous organisms may also be mutant.

**(d) Loss of function:** Dominant negative mutation

Homozygous        Heterozygous

Alleles

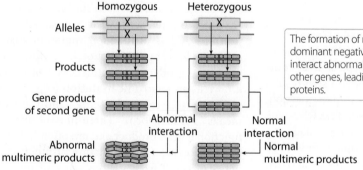

Products

Gene product of second gene

Abnormal interaction          Normal interaction

Abnormal multimeric products          Normal multimeric products

The formation of mulitmeric proteins is altered by dominant negative mutants whose products interact abnormally with the protein products of other genes, leading to malformed multimeric proteins.

**(e) Gain of function:** Hypermorphic mutation

Homozygous        Heterozygous

Alleles

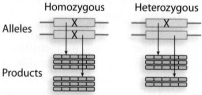

Products

Excessive expression of the gene product leads to excessive gene action. The mutant phenotype may be more severe or lethal in the homozygous genotype than in the heterozygous genotype.

**(f) Gain of function:** Neomorphic mutation

Homozygous        Heterozygous

Alleles

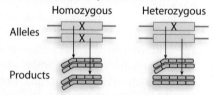

Products

The mutant allele has novel function that produces a mutant phenotype in homozygous and heterozygous organisms, and may be more severe in homozygous organisms.

**Figure 4.1  The functional consequences of mutation.** (a) Wild-type. (b), (c), and (d) Loss-of-function mutations. (e) and (f) Gain-of-function mutations.

functional protein, are particularly subject to **dominant negative mutations** as a consequence of some change that prevents the polypeptides from interacting normally to produce a functional protein. A multimeric protein that contains an abnormal polypeptide may suffer a reduction or total loss of functional capacity. Mutations of this kind are dominant due to the substantial loss of function of the multimeric protein. These mutations are characterized as "negative" due to the spoiler effect of the abnormal polypeptide on the multimeric protein.

An example of dominant negative mutation is seen in the human hereditary disorder osteogenesis imperfecta (OMIM 116200, 116210, and 116220), which is caused by defects in the bone protein collagen and has multiple forms with different severity. Collagen protein is composed of three interwoven polypeptide strands—two polypeptides from the *COL1A1* gene and one polypeptide from the *COL1A2* gene. The trimeric collagen protein is subject to dominant negative mutation as a consequence of *COL1A1* mutations that produce a defective polypeptide. The trimeric structure of collagen and the 2:1 ratio of incorporation of COL1A1 polypeptide over COL1A2 polypeptide means that in individuals who are homozygous wild type for *COL1A2* and heterozygous for *COL1A1* mutation, most collagen protein contains one or two mutant COL1A1 proteins. As a result, most collagen protein is defective, and osteogenesis imperfecta develops.

**Gain-of-Function Mutations**    Mutations resulting in a gain of function fall into two categories that depend on the functional behavior of the new mutation. **Hypermorphic** ("greater than wild-type form") **mutations** produce more gene activity per allele than the wild type (**Figure 4.1e**) and are usually dominant. The gene product of a hypermorphic allele is indistinguishable from that of the wild-type allele, but it is present in a greater amount and thus induces a higher level of activity. The excess concentration is the functional equivalent of overdrive, pushing processes forward more rapidly, at the wrong time, in the wrong place, or for a longer time than normal. Hypermorphic mutants often result from regulatory mutations that increase gene transcription, block the normal response to regulatory signals that silence transcription, or increase the number of gene copies by gene duplication. The severity of phenotypic effect may coincide with the genotype such that mutation homozygotes display a more severely affected phenotype than is observed in heterozygotes.

Gain-of-function mutations resulting from **neomorphic** ("new form") **mutations** acquire novel gene activities not found in the wild type (**Figure 4.1f**) and are usually dominant. The gene products of neomorphic mutants are functional, but have structures that differ from the wild-type gene product. The altered structures lead the mutant protein to function differently than the wild-type protein. Homozygotes for a neomorphic allele

may exhibit a more severely affected phenotype than do heterozygotes.

Our description of the molecular basis of dominance and of loss-of-function and gain-of-function mutations provides a conceptual basis for understanding how different patterns of dominance relationships can develop among alleles of a gene. These concepts apply to all diploid organisms, but the notational systems that identify genes and alleles are variable among species. These different notational systems developed in the early years of genetics research when genetic experiments were carried out on distinct taxonomic groups. Geneticists studying fruit flies developed one notation system for identifying wild-type and mutant alleles, geneticists studying yeast developed another, and geneticists studying plants developed another. As the table inside the front cover illustrates, each model organism has its own unique style of gene description and nomenclature. The different notation systems cause confusion for students of genetics because they follow different rules for naming and identifying genes and alleles. The table inside the front cover contains the rule systems we follow throughout this book.

---

**Genetic Insight**    Dominance relationships between alleles are determined by actions or effectiveness of the gene products in producing phenotypes. Activity levels of loss-of-function mutations and gain-of-function mutations are determined by comparisons to the activity level of the products of the corresponding wild-type allele.

---

## Incomplete Dominance

Mendel's description of inheritance of traits controlled by a dominant and a recessive allele of single genes is a simple hereditary process that is relatively rare in nature. Commonly, however, the dominance of one allele over another is not complete. **Incomplete dominance,** also known as **partial dominance,** identifies such circumstances. When incomplete dominance exists among alleles, the phenotype of the heterozygous organism is distinctive; it falls between the phenotypes of the homozygotes on a continuum of some kind and is typically more similar to one homozygous phenotype than the other. When traits display incomplete dominance, two pure-breeding parents with different phenotypes produce $F_1$ heterozygotes having a phenotype different from that of either parent. The $F_1$ phenotype is intermediate between the parental forms, although it may more closely resemble one parental phenotype than the other.

In previous discussions we used a notational system in which an uppercase letter—for example, *A* indicates a dominant allele, and the same letter in lowercase—*a*—designates a recessive allele. In incomplete dominance systems, the relationship between alleles is different, so a different notational system—one that avoids implying

**Figure 4.2  Incomplete dominance in flowering time of pea plants. (a)** Allele $T_2$ is incompletely dominant over allele $T_1$ as indicated by the late flowering time of $T_1T_2$ plants. **(b)** Segregation of alleles $T_1$ and $T_2$.

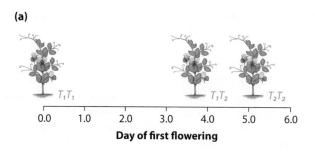

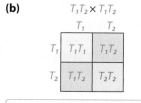

$\frac{1}{4}T_1T_1$  Early flowering (Day 0.0)
$\frac{1}{2}T_1T_2$  Intermediate flowering (Day 3.7)
$\frac{1}{4}T_2T_2$  Late flowering (Day 5.2)

dominance or recessiveness—is used. In the nomenclature system for incomplete dominance, alleles are symbolized with either upper- or lowercase letters plus a suffix that may be a number or a letter. For example, pairs of alleles with incomplete dominance can be designated $A1$ and $A2$, $B^1$ and $B^2$, $d_1$ and $d_2$, and $w^a$ and $w^b$.

Genetic research has identified innumerable examples of incomplete dominance in animals and plants; one example is the trait described as flowering time in Mendel's pea plants (*Pisum sativum*). In peas, the first appearance of flowers is under the genetic control of a locus that we will call $T$ (for flowering *time*). The earliest-flowering strain of pea plants has the homozygous genotype $T_1T_1$; the flowering time of this strain is described as day 0.0. The latest-flowering strain is homozygous $T_2T_2$, and it flowers 5.2 days later on average than $T_1T_1$ plants. A cross of pure-breeding, early-flowering and late-flowering strains produces $T_1T_2$ heterozygous progeny that begin to flower 3.7 days later on average than the earliest-flowering strain (**Figure 4.2a**).

Geneticists can tell that, in this case, flowering time is controlled by a single locus because self-fertilization of $T_1T_2$ plants produces a 1:2:1 ratio of early-, intermediate-, and late-flowering progeny (**Figure 4.2b**). We say the $T_2$ allele is partially dominant, but not completely dominant, to $T_1$ because the heterozygous phenotype is distinct from either homozygous phenotype but more closely resembles the late-flowering strain.

---

Genetic Insight  Incomplete dominance between alleles produces an F₁ hybrid phenotype that is intermediate between the parent phenotypes but is typically more similar to one parent than to the other.

---

## Codominance

**Codominance,** like incomplete dominance, leads to a heterozygous phenotype different from the phenotype of either homozygous parent. Unlike incomplete dominance, however, codominance is characterized by the detectable expression of both alleles in heterozygotes. Codominance is most clearly identified when the protein products of both alleles are detectable in heterozygous organisms, typically by means of some sort of molecular

analysis such as gel electrophoresis or a biochemical assay that can distinguish between the different proteins. We explore the details of these types of molecular analysis in a later discussion (see Chapter 10).

More than one pattern of dominance between the alleles of a gene can occur under certain circumstances. In the following section, we examine the codominance of two alleles and the recessiveness of a third allele of the gene determining human blood type.

**Dominance Relationships of ABO Alleles**  One physiological attribute many of us know about ourselves is our blood type, which is type A, type B, type AB, or type O. All of us have one of these four common blood types that result from alleles at the ABO blood group gene located on chromosome 9 (OMIM 110300). There are three alleles in all human populations, and combinations of the alleles can occur. Most combinations of different ABO alleles result in complete dominance of one allele, but one combination results in codominance.

The three alleles of the ABO gene are identified as $I^A$, $I^B$, and $i$, and the four blood groups are phenotypes produced by different combinations of these alleles. On the basis of genotype–phenotype (i.e., blood type) correlation, geneticists have concluded that $I^A$ and $I^B$ have complete dominance over $i$, and that $I^A$ and $I^B$ are codominant to one another. The complete dominance of $I^A$ and $I^B$ to $i$ is indicated by the identification of blood type A in individuals whose genotype is $I^AI^A$ or $I^Ai$, and of blood type B in individuals whose genotype is $I^BI^B$ or $I^Bi$. The completely recessive nature of the $i$ allele is confirmed by the observation that only $ii$ homozygotes have blood type O. Lastly, codominance of $I^A$ and $I^B$ to one another is confirmed by the observation that blood type AB occurs only in individuals who have the heterozygous genotype $I^AI^B$.

ABO blood type is identified by a reaction between an ABO antigen—a sugar moiety embedded on the surface of red blood cells—and an antibody—a molecule produced by the immune system that binds to a specific antigen protein. Blood typing makes use of an antigen–antibody reaction to determine if a specific antigen is present on red blood cells. A positive reaction occurs when the antibody detects its antigen target. The antibody binds the antigen and also attaches to other antigen-bound antibodies, causing red blood cells to form visible clumps. Clumping indicates that the antibody has detected its antigen target, whereas an absence of clumping indicates that blood does not contain the antigen target of the antibody.

To test for ABO blood type, two antisera—one called "anti-A antiserum" and containing purified anti-A antibody, the other called "anti-B antiserum" and containing purified anti-B antibody—are placed in separate depressions on a microscope slide, and a drop of the blood to be typed is added to each depression. A person with blood type A shows clumping with anti-A antiserum but not with anti-B (**Figure 4.3**). Conversely, blood type B is identified when clumping occurs with anti-B but not with anti-A. If clumping occurs with both antisera, the blood type is AB. Clumping with neither antiserum identifies blood type O.

Proper cross-matching of blood type is essential for safe blood transfusion. In reality, several antigens produced by different genes determine the suitability of donor and recipient blood for transfusion, and hospitals and clinics must carefully compare donor and recipient blood to identify the possibility of adverse reactions before transfusion takes place. The general rule for safe blood transfusion is that the recipient blood must not contain an antibody that reacts with an antigen in the donated blood. When such a reaction occurs, hemolysis can occur and blood clots produced by clumping blood cells form at the site of transfusion. These adverse reactions can potentially cause life-threatening complications.

The antibodies anti-A and anti-B develop in humans from birth, but people do not carry an antibody if they also carry the corresponding antigen. Thus people with blood type A, who have the A antigen, also carry the anti-B antibody. People with blood type B have the B antigen and the anti-A antibody. Those with blood type AB have both antigens and neither anti-A nor anti-B antibody. Finally, people with blood type O have neither A nor B antigen and have both anti-A and anti-B antibody.

**The Molecular Basis of Dominance and Codominance of ABO Alleles**    The two ABO blood group antigens on the surfaces of red blood cells each have a slightly different molecular structure. The antigens are glycolipids that contain a lipid component and an oligosaccharide component. The lipid portion of the antigen is anchored in the red blood cell membrane, and the segment protruding outside the cell contains the oligosaccharide. Initially, the oligosaccharide is composed of five sugar molecules and is called the H antigen. It results from the activity of an enzyme produced by the *H* gene (**Figure 4.4**). The H antigen is present on the surfaces of all red blood cells, and it can be further modified, in two alternative ways, by the addition of a sixth sugar; or it can be left unmodified. The final modification of the H antigen depends on the enzymatic activity of the protein product of the ABO blood group locus.

Two alternative sugars can be added to the H antigen by the gene products of the $I^A$ or $I^B$ alleles, respectively. If the $I^A$ allele is present in the genotype, it produces the gene product α-3-N-acetyl-D-galactosaminyltransferase, or simply, "A-transferase." A-transferase catalyzes the addition of the sugar N-acetylgalactosamine to the H antigen, producing a six-sugar oligosaccharide known as the A antigen. The $I^B$ allele, on the other hand, produces α-3-D-galactosyltransferase, commonly called "B-transferase," which catalyzes the addition of a different sugar, galactose, and produces a six-sugar oligosaccharide known as the B antigen. The molecular basis of the differences between the A and B alleles are several nucleotide differences that change four amino acids of the resulting transferase enzymes and alter enzymatic activity. In contrast, the *i* allele is due to a single base-pair deletion and is a null allele that does not produce a functional gene product capable of adding a sixth sugar to the H antigen.

At the cellular level, anti-A antibody recognizes the N-acetylgalactosamine addition mediated by $I^A$, and anti-B antibody identifies the galactose addition produced by the action of $I^B$. Neither of these antibodies has any reactivity with the unmodified H antigen; so unmodified H antigen, present in individuals with blood type O, is not recognized by either antibody. Either one or two copies of the $I^A$ or the $I^B$ allele in a genotype is sufficient to produce an ABO antigen detectable by anti-A or anti-B antibodies. Both $I^A$ and $I^B$ are dominant to *i*, since $I^A$ and $I^B$ produce enzymes that modify the H antigen but *i* does not. On the other hand, the $I^AI^B$ genotype leads to production of both A-transferase and B-transferase, resulting in the addition of N-acetylgalactosamine to some H antigens and the addition of galactose to other H antigens. In the $I^AI^B$

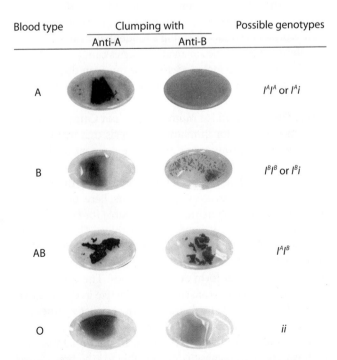

| Blood type | Clumping with | | Possible genotypes |
|---|---|---|---|
| | Anti-A | Anti-B | |
| A | | | $I^AI^A$ or $I^Ai$ |
| B | | | $I^BI^B$ or $I^Bi$ |
| AB | | | $I^AI^B$ |
| O | | | *ii* |

**Figure 4.3  ABO blood type.** Blood type is determined by mixing a drop of blood with a drop of anti-A or anti-B antiserum.

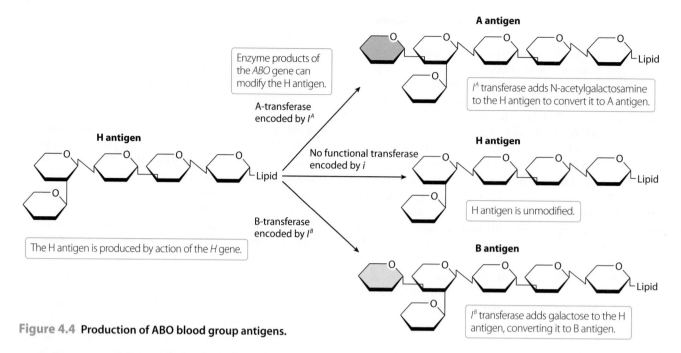

**Figure 4.4  Production of ABO blood group antigens.**

genotype, all red blood cells carry both types of H-antigen modifications; about half of the red cell-surface antigens are A antigens, and the rest are B antigens. In the heterozygous $I^A I^B$ genotype, therefore, the action of both alleles is detected in the phenotype, leading to the conclusion that $I^A$ and $I^B$ are codominant to one another.

Many nonhuman primates have a blood group system that is essentially identical to the human ABO blood group system. ABO blood groups have been identified in the great apes (chimpanzee, gorilla, and orangutan) as well as in numerous Old World monkey species, including macaques (genus *Macaca*) and baboons (genus *Papio*). Two important evolutionary observations derive from this finding. First, the ABO blood group is a long-standing feature of the immune system genetics in primates, one that evolved early in the ancestral history of primates and was retained over tens of millions of years as primates diversified. Second, the retention of the ABO blood group system in primates demonstrates the importance of this immune system response in protecting primates from infectious and foreign antigens. Natural selection has played a preeminent role in maintaining this system. The ABO blood group genes are one example of the shared evolutionary history that can be identified through the examination of the taxonomic distribution of genes in lineages. **Genetic Analysis 4.1** examines the inheritance of blood group phenotypes where alleles have a variety of dominance relationships.

**Genetic Insight**  Codominance between alleles occurs when the expression of both alleles is detected in heterozygous organisms that have a phenotype distinguishable from either homozygote.

## Allelic Series

Diploid genomes contain pairs of homologous chromosomes; thus, each individual organism can possess at most two alleles at a locus. In populations, however, the number of alleles is theoretically unlimited, and some genes have scores of alleles. At the population level, a locus possessing three or more alleles is said to have multiple alleles. The ABO blood group locus, with its three alleles, is one example of multiple alleles. Like the *ABO* gene, other multiple-allelic loci display a variety of dominance relationships among the alleles. Commonly, an order of dominance emerges among the alleles, based on the activity of each allele's protein product, forming a sequential series known as an **allelic series.** Alleles in an allelic series can be completely dominant or completely recessive, or they can display various forms of incomplete dominance or codominance.

**The C-Gene System for Mammalian Coat Color**  Genetic analysis of coat color in mammals reveals that many genes are required to produce and distribute pigment to the hair follicles or skin cells, where they are displayed as coat color or skin color. While various interactions among these genes can modify color expression, we focus here on just one gene, the *C* (color) gene that is responsible for coat color in mammals such as cats, rabbits, and mice. This gene has dozens of alleles that have been identified in more than 80 years of genetic analysis, but we limit our discussion to just four alleles that form an allelic series. The *C* gene produces the enzyme tyrosinase, which is active in the first two steps of a multistep biochemical pathway that synthesizes the pigment melanin, which imparts coat color in furred mammals and skin color in humans. In the initial melanin pathway steps, tyrosinase is responsible for the breakdown (catabolism) of the amino acid tyrosine.

The MN blood group in humans is an autosomal codominant system with two alleles, *M* and *N*. Its three blood group phenotypes, M, MN, and N, correspond to the genotypes *MM, MN,* and *NN*. The ABO blood group assorts independently of the MN blood group.

A male with blood type O and blood type MN has a female partner with blood type AB and blood type N. Identify the blood types that might be found in their children, and state the proportion for each type.

| Solution Strategies | Solution Steps |
|---|---|
| **Evaluate** | |
| 1. Identify the topic this problem addresses and describe the nature of the required answer. | 1. The problem concerns the inheritance of two blood types. The gene determining ABO blood type carries three alleles: $I^A$ and $I^B$ are codominant to one another and dominant to *i*. The MN blood group gene carries two alleles that are codominant. The answer requires finding the possible blood types, and their expected proportions, of the children of parents whose blood types are given. |
| 2. Identify the critical information given in the problem. | 2. The blood types of the parents are given. |
| **Deduce** | |
| 3. Deduce the blood group genotypes of the male parent. | 3. The male has blood types O and MN. Type O results from homozygosity for the recessive *i* allele, whereas MN is produced in heterozygotes carrying both alleles. The male genotype is *ii MN*. |
| 4. Deduce the blood group genotypes of the female parent. | 4. The female has blood groups AB and N. The AB blood type is found in heterozygotes, and blood type N in homozygotes. The female blood group genotype is $I^A I^B$ *NN*. |
| **Solve** | |
| 5. Identify the gamete genotypes and their frequencies for the male. | 5. Independent assortment predicts two gamete genotypes for the male: All gametes contain *i*, half carry *M*, and half carry *N*. |
| 6. Identify the female gamete genotypes and their frequencies. | 6. Independent assortment predicts two gamete genotypes for the female: All gametes contain *N*, half contain $I^A$, and half contain $I^B$. |
| 7. Predict the progeny genotypes and phenotypes. **TIP:** Use a Punnett square to evaluate this cross. | 7. |

| ♀ \ ♂ | *Mi* | *Ni* |
|---|---|---|
| $NI^A$ | $MNI^Ai$ Blood types: MN and *A* | $NNI^Ai$ Blood types: N and *A* |
| $NI^B$ | $MNI^Bi$ Blood types: MN and *B* | $NNI^Bi$ Blood types: N and *B* |

For more practice, see Problems 6, 9, and 21.

---

The *C*-gene alleles form an allelic series that is determined by the phenotypes of offspring of various matings. Allele *C* is dominant to all other alleles of the gene, and any genotype with at least one copy of *C* produces wild-type coat color. These genotypes are written as *C*– to indicate that regardless of the second allele in the genotype, the phenotype is dominant. Three other alleles producing tyrosinase enzymes with reduced or no tyrosinase activity form an allelic series with *C* (Figure 4.5). The allele $c^{ch}$ produces a phenotype called chinchilla, a diluted coat color. The $c^h$ allele produces the Himalayan phenotype, characterized by fully pigmented extremities (paws, tail, nose, and ears) but virtually absent pigmentation on other parts of the body. The Himalayan phenotype is the "Siamese" coat-color pattern often seen in cats, rabbits, and mice.

Finally, the *c* allele produces a protein product with no enzymatic activity. This is a fully recessive null allele, and homozygosity for it produces an albino phenotype.

Crosses between animals with different genotypes at the *C* gene reveal the dominance relations of the alleles. For example, in crosses A, B, and C in Figure 4.6, complete dominance of *C* over other alleles in the series is demonstrated by the finding that all of the progeny of an animal with the genotype *CC* have full color, regardless of the genotype of the mate. The dominance order of alleles in the series is revealed by the pattern of 3:1 ratios obtained from crosses of various heterozygous genotypes shown in Figure 4.6. In cross D, chinchilla is shown to be partially dominant over Himalayan. Most of the coat of these animals has diluted (chinchilla) color, and the Himalayan

**Figure 4.5** Allelic series for coat-color determination in mammals.

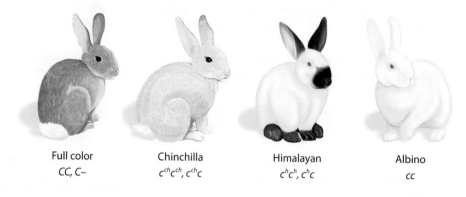

| Full color | Chinchilla | Himalayan | Albino |
|:---:|:---:|:---:|:---:|
| $CC, C-$ | $c^{ch}c^{ch}, c^{ch}c$ | $c^h c^h, c^h c$ | $cc$ |

pattern has darker color of paws, face, and tail. Cross E shows that chinchilla is completely dominant over albino. Himalayan is completely dominant over albino (cross F). The dominance relationships within this allelic series locus can be expressed as $C > c^{ch} > c^h > c$.

**The Molecular Basis of the *C*-Gene Allelic Series**
Tyrosinase enzymes produced by different *C*-gene alleles have distinctive levels of catabolic activity that are the basis for the dominance relationships between the alleles. The allele *C* is a dominant wild-type allele producing fully active tyrosinase that is defined as 100% activity. The percentage of wild-type tyrosinase activity produced by each allele explains the order observed for the allelic series. Biochemical examination reveals that the enzyme produced by the $c^{ch}$ hypomorphic allele has less than 20% of the active than the wild-type enzyme. In the homozygous $c^{ch}c^{ch}$ genotype or heterozygous genotypes $c^{ch}c^h$ or $c^{ch}c$, only a small amount of melanin is synthesized. This leads to a decreased amount of pigment, and it has the effect of muting the coat color.

The tyrosinase enzyme produced by the hypomorphic $c^h$ (Himalayan) allele is unstable and is inactivated at a temperature very near the normal body temperature of most mammals. This type of gene product is an example of a **temperature-sensitive allele.** Cats with the Siamese coat-color pattern are familiar examples of the action of this temperature-sensitive allele. The parts of cats that are farthest away from the core of the body (the paws, ears, tail, and tip of the nose) at most times tend to be slightly cooler than the trunk. At these cooler extremities, the temperature-sensitive tyrosinase produced by the $c^h$ allele remains active, producing pigment in the hairs there. However, in the warmer central portion of the body, the slightly higher temperature is enough to cause the tyrosinase produced by the $c^h$ allele to denature, or unravel. This inactivates the enzyme and leads to an absence of pigment in the central portion of the body. Animals that are $c^h c^h$ or $c^h c$ have the Himalayan phenotype. The final allele in the series, *c*, is a null allele that does not produce functional tyrosinase. Homozygotes for this allele are unable to initiate the catabolism of tyrosine. This leads to an absence of melanin and produces the condition known as albinism.

**Genetic Insight** Phenotypes produced by the interactions of multiple alleles in an allelic series reveal hierarchical relationships that depend on the biological action of allelic products. Dominance is associated with the allele producing the most active gene product, and the placement of other alleles in the series is dependent on the relative activity of their respective gene products.

## Lethal Mutations

Certain single-gene mutations are so detrimental that they cause death early in life or terminate gestational development. These life-ending mutations are categorized as lethal and are caused by a **lethal mutation.** Lethal alleles are often inherited as recessive mutants, recessive lethal alleles that kill only homozygotes. As a rule, recessive lethal alleles have variable frequencies in populations, and they may persist in some populations over a long period of time. Natural selection can eliminate copies of the allele when they occur in homozygous genotypes; however, recessive lethal alleles are "hidden" by dominant wild-type alleles in heterozygous genotypes, thus evading natural selection. Under certain circumstances, heterozygous carriers of a recessive lethal allele have a natural selection advantage (see Chapter 10).

Lethal alleles are often detected as distortions in segregation ratios, where one or more classes of expected progeny are missing. For example, in plant and animal crosses between two organisms heterozygous for a recessive lethal allele, the phenotype of the progeny is 3:1 (viable : dead). The dead offspring are homozygous for a recessive lethal mutation. These progeny might not be seen at all, due to embryonic lethality, or they may be stillborn or die very young. Of the viable offspring, two-thirds are expected to be heterozygous for the lethal allele and one-third are expected to be homozygous for the dominant wild-type allele (**Figure 4.7**).

In flowering plants, the effects of lethal alleles can be observed directly. For example, mutation of the *RPN1a* gene that encodes a subunit of the 26S proteosome, a multi-protein complex involved in protein degradation, is an example of a loss-of-function null mutation (*rpn1a*) that

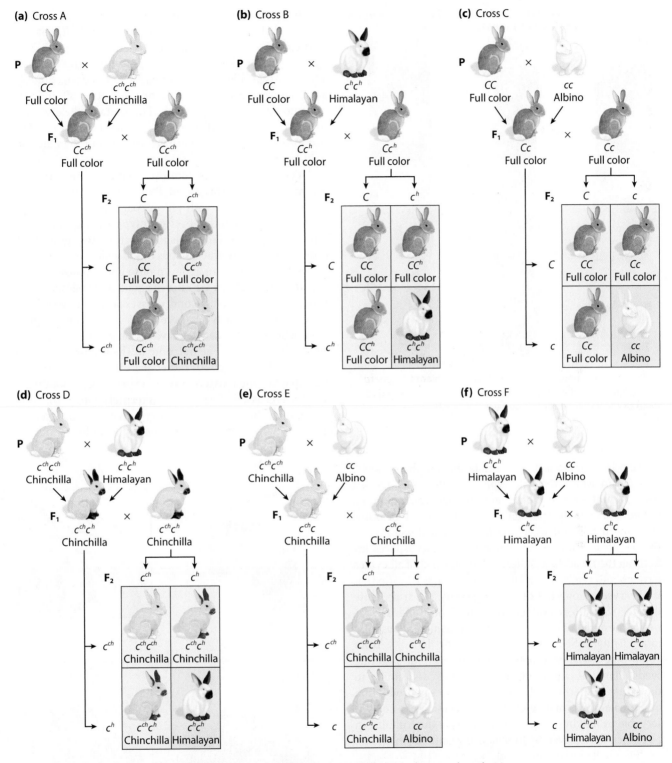

**Figure 4.6 The genetics of *C*-gene dominance.** (a)–(f) Crosses A to F illustrate the complete dominance of *C* and the complete recessiveness of *c*, and establish the allelic series as $C > c^{ch} > c^h > c$.

results in embryonic lethality in *Arabidopsis thaliana* and other plant species. In an *RPN1a/rpn1a* × *RPN1a/rpn1a* cross, a 3:1 segregation ratio of living seeds (*RPN1a*) to dead seeds (*rpn1a/rpn1a*) can be observed in the fruit. When the living seeds are planted, approximately two-

thirds are heterozygous for the lethal allele (*RPN1a/rpn1a*) and one-third are homozygous for the wild-type allele (*RPN1a/RPN1a*).

Lethal mutations that result in female gametophytic lethality are also detectable in flowering plants. Consider

| Wild type | Embryo lethal (*RPN1a/ rpn1a* × *RPN1a/rpn1a*) 3:1 | Gametophyte lethal (*FER/fer* × male) 1:1 |
|---|---|---|

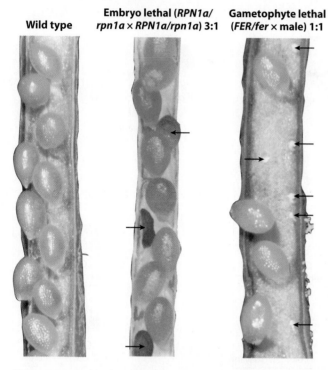

**Figure 4.7  Evidence of lethal mutations in plants.** Gametophytic lethality is detected by observing a 1:1 ratio of living to dead seeds. Arrows indicate undeveloped seeds.

a plant heterozygous for a female gametophytic allele, *FER/fer*, in which the wild-type *FER* allele was derived from its mother, and the mutant *fer* allele came from its father. During megasporogenesis, half of all megaspores will inherit the *FER* allele and the other half will inherit the *fer* allele. Embryo sacs derived from megaspores inheriting the *fer* allele will die, so that only half of all ovules develop into seeds. The alleles segregate in a 1:1 ratio that is observed among the developing seeds in a fruit. Note that the 1:1 ratio is a direct observation of Mendelian ratios in the gametes of a heterozygous organism. Thus a 1:1 ratio distinguishes female gametophytic lethality from embryonic lethality, which results in a 3:1 ratio among seeds. Plants usually produce pollen in excess, similar to the excess of sperm production relative to egg production in animals; thus, male gametophytic lethality is not observable by looking at developing seeds in the fruit. It can be detected, however, by looking for plants in which half of all the pollen grains are dead.

In contrast, lethal alleles in animals are usually detected by a distortion in segregation ratios. The first case of a lethal allele was identified in 1905 by Lucien Cuenot, who studied a lethal mutation in mice carrying a dominant mutation for yellow coat color. In mice, wild-type coat color is a brown color, called "agouti" (*a-GOO-tee*), produced by the presence of yellow and black pigments in each hair shaft (**Figure 4.8a**). Agouti hairs are black at the base and tip, with yellow pigment in the central portion of

the shaft. Yellow coat color is seen when yellow pigment is deposited along the entire length of the hair shaft, not just in the middle portion as it is in agouti (**Figure 4.8b**). The *Agouti* gene produces the yellow pigment found in hairs. The wild-type allele for agouti coat color is designated *A*, and its normal activity leads to the production of a moderate amount of yellow pigment. The mutant allele, designated $A^Y$, is a gain-of-function dominant mutation that produced substantially more yellow pigment than did the wild-type allele.

The $A^Y$ mutation is dominant, but true-breeding yellow mice cannot be produced. From a genetic perspective, this means that mice with yellow coat color are heterozygous ($AA^Y$) and that the $A^YA^Y$ genotype is lethal in embryonic development. From this information, two important observations about the genetics of the yellow allele can be made. First, mating an agouti mouse and a yellow mouse will *always* result in a 1:1 ratio of agouti and yellow among progeny (**Figure 4.9a**). Second, crosses between two yellow mice (both of which are necessarily heterozygous) produce evidence of the recessive lethal nature of the $A^Y$ allele (**Figure 4.9b**). The outcome of these crosses is a 2:1 ratio of yellow to agouti, rather than the 3:1 ratio that is anticipated when heterozygotes expressing a dominant allele are crossed. The genetic interpretation of this observation is that alleles of heterozygous yellow mice segregate normally in gamete formation and unite at random

**(a)** Agouti hair color

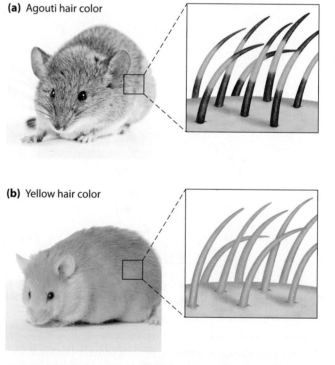

**(b)** Yellow hair color

**Figure 4.8  Coat color in mice.** (a) Wild-type agouti coat color is a mixture of black and yellow pigment in hair shafts. (b) Yellow coat occurs when yellow pigment produced by the overly active mutant allele $A^Y$ displaces black pigment.

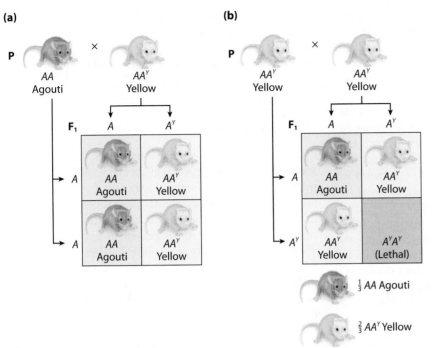

**Figure 4.9 Dominance and lethality of $A^Y$. (a)** A 1:1 ratio identified $A^Y$ as a dominant mutant allele. **(b)** The lethality of $A^Y$ in the homozygous genotype results in a 2:1 ratio of yellow to agouti in the cross of yellow-coated heterozygous mice.

to produce a 1:2:1 ratio at conception, but that $A^YA^Y$ zygotes do not survive gestation. Recessive lethality of $A^Y$ prevents embryonic development of homozygotes, eliminating that class among progeny and resulting in the 2:1 ratio seen among progeny of heterozygous parents.

Nearly a century after Cuenot first identified homozygous lethality of the mutant $A^Y$ allele, the molecular basis of the lethality was identified. Much to the surprise of geneticists, the lethality had little to do with yellow coat color itself; instead, yellow coat was an almost inadvertent consequence of a mutation that deleted part of a gene near the coat-color gene. The mutation producing the $A^Y$ allele results from a deletion that affects two genes, the *Agouti* gene and a neighboring gene identified as *Raly*. *Raly* produces a protein that is essential for mouse embryo development. Each gene has its own promoter. The wild-type *Raly* promoter drives a high level of transcription, whereas the *Agouti* gene promoter is considerably less actively transcribed (**Figure 4.10**). The dominant mutation producing yellow coat color comes about by a deletion of approximately 120,000 bp that deletes the entire *Raly* gene and the *Agouti* gene promoter, thus bringing the *Agouti* gene under the control of the *Raly* promoter. The *Raly* promoter drives a high level of *Agouti* gene transcription that results in excess yellow pigment that displaces black pigment in hair shafts and leads to the mutant yellow phenotype. Heterozygotes with the $AA^Y$ genotype have yellow coats and survive due to haplosufficiency of the single copy of *Raly*. Homozygous $A^YA^Y$ mice are unable to produce the essential protein product from the *Raly* gene and fail to develop, resulting in the skewed 2:1 Mendelian ratio that characterizes the progeny of two heterozygous yellow-coated mice.

## Sex-Limited Traits

The sex of an organism can exert an influence on its gene expression. One consequence of such influence is the potential limitation of gene expression to one sex but not the other in a pattern called **sex-limited gene expression.** Differences in gene expression between the sexes can result in the appearance of these **sex-limited traits.** Both sexes typically carry the genes for sex-limited traits, but the genes are expressed in just one sex.

In mammals, for example, the development of breasts and the ability to produce milk are traits limited to females. Horn development is a trait limited to males in some species of sheep, cows, and other hoofed animals.

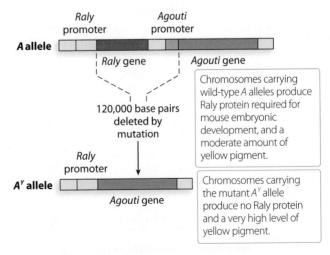

**Figure 4.10 Mutation of *Raly* and *Agouti* producing yellow coat.**

Behavioral traits in some species, particularly traits related to mating, are also strongly influenced by sex. For example, the courtship behavior of crowned cranes includes an elaborate display of body positioning, neck intertwining, and vocalization that is performed differently by males and females of the species.

The mechanism that limits the expression of a trait to just one sex is most often the differential influence of hormones acting as intercellular regulators of gene expression. In the case of male canary vocalization, for example, changes in male singing patterns are initiated in late winter by an increase in male hormones released by the brain in response to increased day length and warmer temperatures. These hormones stimulate enlargement of the testes and increased production of testosterone, which in turn stimulates the development of neurons in the brain that elaborate the song center, induce the development of muscles in the vocalization area of the throat, and allow males to produce sex-limited vocalization to attract mates.

## Sex-Influenced Traits

**Sex-influenced traits** are those in which the phenotype corresponding to a particular genotype differs depending on the sex of the organism carrying the genotype. Hormones are thought to influence the differential expression of genotypes in the sexes.

The appearance of a chin beard versus the absence of a beard, the beardless phenotype, in certain goat breeds is an example of a sex-influenced trait. Bearding is inherited as an autosomal trait determined by two alleles, $B$ and $b$, which are present in three genotypes in each sex. In both sexes, $BB$ homozygotes are beardless, and homozygotes of either sex with the $bb$ genotype are bearded. It is thought that androgenic hormones are a principal factor influencing the bearded phenotype. The effect of different levels of androgenic hormones on bearding in the sexes is seen by comparing females and males with the heterozygous genotype ($Bb$). Heterozygous males have a beard, whereas heterozygous females are beardless. **Figure 4.11** illustrates the results of a cross between two heterozygotes that produces different ratios of bearded to beardless males and females. Mendelian inheritance occurs, but as a consequence of sex-influenced expression, the cross yields a 3:1 ratio of bearded to beardless males and a 3:1 ratio of beardless to bearded females.

## Delayed Age of Onset

From an evolutionary perspective, it is easy to understand that a dominant lethal allele can be efficiently eliminated by the action of natural selection. Even so, there are numerous examples of dominant lethal hereditary conditions, and a pertinent evolutionary genetic question concerns how these mutations persist in populations. One answer is that some dominant lethal alleles sidestep natural selection by having a **delayed age of onset;** the abnormalities they produce do not appear until after affected organisms have had an opportunity to reproduce and transmit the mutation to the next generation.

One prominent example of delayed age of onset of a dominant lethal allele in humans is the condition called Huntington disease (HD). This progressive neuromuscular disorder, usually fatal within 10 to 15 years of diagnosis, is caused by mutation of a gene near one end of chromosome 4. (We have much more to say about the symptoms and progression of HD in Chapter 5, where we also discuss the mapping of the HD gene, and in Chapter 16, where we discuss the cloning of the HD gene). The HD mutant allele persists in the population because symptoms do not begin in about half of all cases until the person's late thirties or early forties, well after most people have begun having children (**Figure 4.12**).

---

Genetic Insight  Lethal alleles cause the death of organisms and may be either dominant or recessive. Death may occur early or late in life. When death occurs during gestation, the presence of a lethal allele is often detected by the absence of an expected class of progeny or by the alteration of seed or seedling ratios.

---

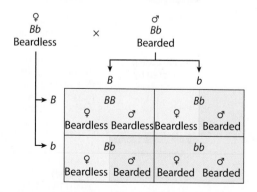

**Figure 4.11  Sex-influenced inheritance of beard appearance in goats.** The bearding allele (*b*) is expressed differently in males and females.

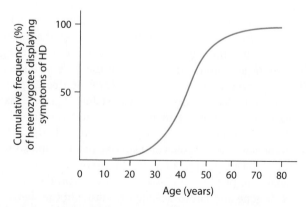

**Figure 4.12  The age-of-onset curve for Huntington disease (HD).**

# 4.2 Some Genes Produce Variable Phenotypes

To interpret phenotype ratios and identify the distribution of genotypes among phenotypic classes, geneticists make the assumption that phenotypes differ because their underlying genotypes differ. This assumption is valid only to the extent that a particular genotype always produces the same phenotype. If the strict correlation between genotype and phenotype does not consistently hold true—if instead the same genotype can produce different phenotypes—the usual reasons are gene–environment interaction or interactions with alleles of other genes in the genome.

In this section, we describe two phenomena, referred to as *incomplete penetrance* and *variable expressivity*, in which phenotypic variation occurs among organisms with the same genotype. In addition, we look at specific instances of environmental influence on gene expression that is often associated with incomplete penetrance or variable expressivity.

## Incomplete Penetrance

When the phenotype of an organism is consistent with the organism's genotype, the organism is said to be **penetrant** for the trait. In such a case, if the organism carries a dominant allele for the trait in question, the dominant phenotype is displayed. Sometimes an organism with a particular genotype fails to produce the corresponding phenotype, in which case the organism is **nonpenetrant** for the trait.

Traits for which nonpenetrant individuals occasionally or routinely occur are identified as displaying **incomplete penetrance.** The human condition known as polydactyly ("many digits") is an autosomal dominant condition that displays incomplete penetrance. Individuals with polydactyly have more than five fingers and toes—the most common alternative number is six (**Figure 4.13**). Polydactyly occurs in hundreds of families around the world, and in these families the dominant allele is nonpenetrant in about 25–30% of individuals who carry it. Most people who carry the dominant mutant polydactyly allele have extra digits; but at least one in

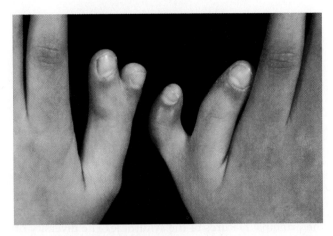

**Figure 4.13** Polydactyly, an autosomal dominant trait with incomplete penetrance.

four people with the mutant allele do not have extra digits and instead express the normal five digits. The gene mutated to produce polydactyly was recently identified (see Chapter 20).

**Figure 4.14** shows a family in which polydactyly segregates as a dominant mutation. Nine individuals in the family carry a copy of the polydactyly allele. Six of them are penetrant for the phenotype (meaning that they express the phenotype), but at least three family members— II-6, II-10, and III-10—are nonpenetrant. Each of these individuals has a child or grandchild with polydactyly; thus, each carries the dominant allele for polydactyly but is nonpenetrant for the condition. When nonpenetrant individuals are relatively common, the magnitude of frequency of penetrance can be quantified. Penetrance values vary among different families; but for the family shown in Figure 4.14, the penetrance of polydactyly is $\frac{6}{9}$, or 66.7%, which is about the average seen worldwide among hundreds of families with polydactyly.

## Variable Expressivity

Sometimes the discrepancy between genotype and phenotype is a matter of degree rather than presence or absence.

**Figure 4.14** Incomplete penetrance for polydactyly. Three nonpenetrant individuals (II-6, II-10, and III-10) are seen in this family.

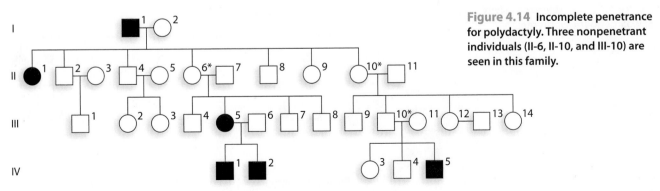

\* Nonpenetrant individual

In the phenomenon of **variable expressivity,** the same genotype produces phenotypes that vary in the degree or magnitude of expression of the allele of interest.

Waardenburg syndrome is a human autosomal dominant disorder displaying variable expressivity. Individuals with Waardenburg syndrome may have any or all of four principal features of the syndrome: (1) hearing loss, (2) differently colored eyes, (3) a white forelock of hair, and (4) premature graying of hair. In the Waardenburg pedigree shown in **Figure 4.15**, notice that the circles and squares representing family members with Waardenburg syndrome may be entirely or only partly colored. Each quadrant of the symbols represents one of the principal features of the syndrome. The diversity of symbol darkening demonstrates the variation in expressivity of Waardenburg syndrome in this family. Molecular genetic analysis tells us that each family member with Waardenburg syndrome carries exactly the same dominant allele, yet among the eight affected members of the family, there are six different patterns of phenotypic expression.

It is often difficult to pinpoint the cause of incomplete penetrance or variable expressivity. Three kinds of interactions may be responsible: (1) other genes that act in ways that modify the expression of the mutant allele, (2) environmental or developmental (i.e., nongenetic) factors that interact with the mutant allele to modify its expression, and (3) some combination of other genes and environmental factors interacting to modify expression of the mutation. In inbred laboratory strains of model genetic organisms, variation in genetic factors can be eliminated experimentally to allow separation of gene–gene and gene–environment variability, something that cannot be done in organisms such as humans.

## Gene–Environment Interactions

Genes control virtually all of the differences observed between species. The genome of an organism lays out the body plan and biochemical pathways of the organism, and it controls the progress of development from conception to death. But genes alone are not responsible for

**Genetic Insight** In incomplete penetrance, a genotype is not expressed by every organism in which it is present. In variable expressivity, the organisms that share a genotype express the corresponding phenotype to different degrees. Both are explained by genetic and/or nongenetic interactions that modify or prevent the consistent expression of a genotype.

all the variation seen between organisms. The environment, the myriad of physical substances and conditions an organism encounters at different stages of life, is the other essential contributor to observable variation between organisms. **Gene–environment interaction** is the result of the influence of environmental factors (i.e., nongenetic factors) on the expression of genes and on the phenotypes of organisms.

As an example, consider the tall and short pure-breeding lines of pea plants studied by Mendel. Inherited genetic variation dictates that one line will produce tall plants and the other line will produce short plants, but the environment in which the individual plants are grown also has a significant influence on plant height. Environmental factors such as variations in water, light, soil nutrients, and temperature each influence plant growth. It is not hard to imagine that genetically identical plants of a type adapted to temperate zones might grow to different heights if one plant has an ideal growth environment while the other faces a hot, arid environment with poor soil.

Phenotypic expression of genotypes can also depend on the interaction of genetically controlled developmental programs and external factors operating on organisms. For example, the seasonal change in coat color observed in arctic mammals that are nearly white in winter but have darker coats in spring and summer results from an interaction between numerous genes and external environmental cues such as day length and temperature. Similarly, environmental cues that induce plants to bloom in the spring trigger changes in gene expression that stimulate the growth and development of multiple plant structures, including flowers and reproductive structures.

**Figure 4.15 Variable expressivity of Waardenburg syndrome.**

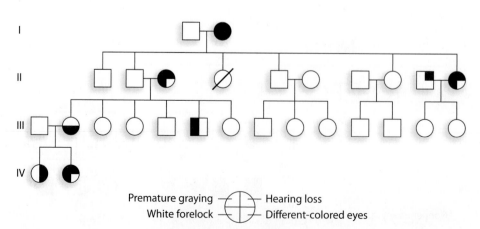

Premature graying ——— Hearing loss
White forelock ——— Different-colored eyes

Such capacities to make seasonal changes evolved by aiding the survival of these organisms, and they suggest that gene–environment interaction is pivotal in understanding and interpreting phenotypic variation.

**Environmental Modification to Prevent Hereditary Disease**   A prime example of gene–environment interaction in humans is actually a case of environmental intervention that is commonly practiced to prevent the development of the human autosomal recessive condition known as phenylketonuria (PKU). This case illustrates that the same alleles may produce different phenotypes in different environments. PKU is caused by the absence of the enzyme phenylalanine hydroxylase, which catalyzes the first step of the pathway that breaks down the amino acid phenylalanine, a common component of dietary protein.

At one time, PKU accounted for thousands of cases of severe mental retardation every year. PKU occurred in 1 out of 10,000 to 1 out of 20,000 newborns in most populations around the world. Infants with PKU are normal at birth, but over the first several months of life the body's inability to break down phenylalanine becomes toxic to developing neurons. As neurons die, mental and motor capacities are irretrievably lost, making full manifestation of PKU inevitable. In the 1960s, a simple blood test became available to detect PKU in the first days of life. The test identifies the disease before the disease has had a chance to manifest itself and begin to damage the body. PKU is one of dozens of rare hereditary disorders that newborn infants are routinely screened for in U.S. hospitals.

The key to preventing PKU, following early detection of the disease in newborns, is the severe restriction of phenylalanine in the diet. Because phenylalanine is an amino acid and is a component of many proteins, babies with PKU are given a diet consisting of specially selected and processed proteins that have had phenylalanine removed. An infant who is started on the phenylalanine-free diet soon after birth and kept on it through adolescence avoids the complications of PKU and will develop and function normally despite having PKU. Thousands of people with PKU are living fully normal and productive lives today, thanks to this simple environmental modification that prevents the expression of the devastating PKU phenotype. In this case, people who are homozygous recessive for the mutant PKU allele do not express the trait if they are raised in a largely phenylalanine-free environment.

Dietary hazards abound for children and young adults with PKU, particularly in the form of the artificial sweetener known as aspartame. This sweetener is made by a chemical reaction that fuses the amino acids phenylalanine and aspartic acid to form a compound we perceive to taste sweet. Once consumed, aspartame is quickly broken down into its two constituent amino acids, and phenylalanine is released. Regular intake of aspartame is dangerous for those with PKU; for this reason, a dietary caution reading "Phenylketonurics: Contains phenylalanine" appears on the packaging of food products containing aspartame. Look for it on the next artificially sweetened product you pick up!

## Pleiotropic Genes

**Pleiotropy** is the alteration of multiple, distinct traits of an organism by a mutation in a single gene. Most mutations displaying pleiotropy do so either by altering the development of phenotypic features through the direct action of the mutant protein or as a secondary result of a cascade of problems stemming from the mutation. Mendel unknowingly encountered a case of pleiotropy. Two of the traits he considered for his studies were the inheritance of purple versus white flower color (see Figure 2.1) and the inheritance of a gray versus a white seed coat. Upon noticing that plants with white flowers invariably also have white seed coats, whereas purple-flowered plants always have gray seed coats, he correctly surmised that the inheritance of these traits had the same genetic basis. Today we know that flower color, seed-coat color, and the appearance of color at leaf axils (where the leaf attaches to the stem) result from the production of the purple pigment anthocyanin. Mutations that block anthocyanin production are pleiotropic because they leave several plant structures without color and produce mutant white phenotypes for multiple traits.

Pleiotropy through the direct action of a mutant protein product is frequently encountered in studies of development. One example is the activity of the *Drosophila* hormone called juvenile hormone (JH), which is active throughout the *Drosophila* life cycle and influences numerous attributes of development and reproduction. Increased production or increased activity of JH has been shown to prolong developmental time, decrease adult body size, promote early sexual maturity, raise fecundity (the ability to produce offspring), and decrease life span. An evolutionary tradeoff is associated with changes in JH level or activity. On the one hand, producing more JH can lead to production of more offspring through earlier sexual maturity and higher fecundity. On the other hand, body size decreases and life span is shortened by increased JH activity.

Pleiotropy in sickle cell disease (SCD) is an example of the phenotypically diverse secondary effects that can occur due to a mutant allele. SCD (OMIM 603903) is an autosomal recessive condition caused by mutation of the β-globin gene that, in turn, affects the structure and function of hemoglobin, the main oxygen-carrying molecule in red blood cells (see Chapter 10). Many of the red blood cells of people with SCD take on a sickle shape and cause numerous physical problems and complications (**Figure 4.16**).

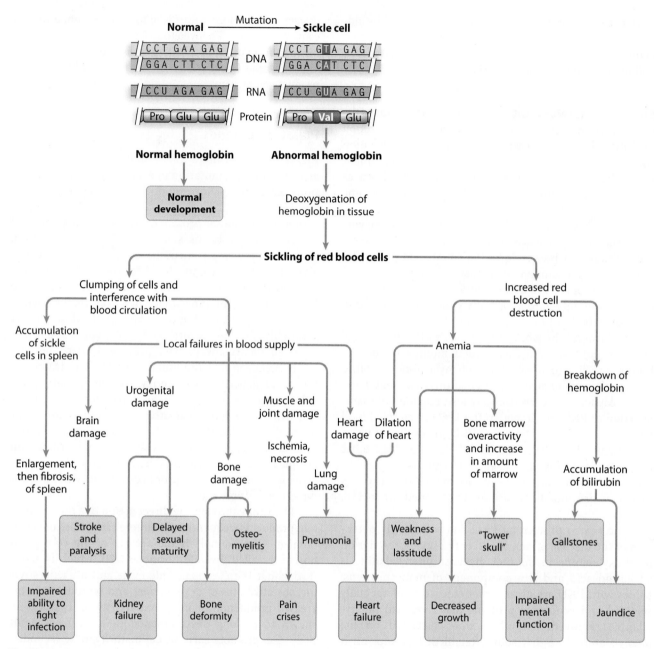

**Figure 4.16** **Pleiotropy in sickle cell disease.**  The sickling of red blood cells has a range of phenotypic consequences.

Genetic Insight  A pleiotropic allele is one that affects many phenotypic traits. Pleiotropic alleles can work either directly, by influencing the development of multiple processes, or indirectly, by damaging organs and body structures.

## 4.3 Gene Interaction Modifies Mendelian Ratios

No gene operates alone to produce a phenotypic trait. Rather, genes work together to build the complex structures and organ systems of plants and animals. What we see as a phenotype is the physical manifestation of the action of many genes that have each played a role and have worked in complex but coordinated ways to produce a trait or structure. At the cellular and molecular levels, the mutual reliance of genes on one another requires each gene to carry out its activity in the right place, at the right time, and at the appropriate level. For example, the products of several genes interact to produce flower color. Similarly, a complex phenotypic attribute like the ability to hear requires many genes to produce the various structures that convert acoustical vibrations into the electrical impulses we perceive as sound. In this section, we look in detail at **gene interaction,** the collaboration of multiple genes in the production of a single phenotypic character or a group of related characteristics. First, however, let's examine the genetic control of phenotypes from a perspective we have not yet explored.

## Gene Interaction in Pathways

It is common for biologists to describe phenotypic characters as single-gene traits. This designation means that different forms of a trait can be transmitted to offspring by the segregation of alleles of a single gene. Phenotypic characteristics such as pea flower color and pea shape are examples of single-gene traits inherited as the result of allelic variation at single loci.

The term *single-gene trait* conveniently summarizes the observation that inherited variation for one gene can produce a mutant phenotype rather than a wild-type phenotype. The term is not an accurate depiction of genetic reality, however, as the following example of *Drosophila* eye color makes clear. Numerous genes contribute to production of the wild-type red eye color of *Drosophila*. Geneticists know this to be the case because many distinct mutant eye-color phenotypes have been mapped to different genes. We will consider just three of these genes. Two of the genes produce different eye-color pigments, and the third gene transports pigments to eye cells. The *brown* gene produces an enzyme that operates in a pathway synthesizing a vermilion-colored (bright red) pigment. The gene carries a dominant wild-type allele $b^+$ and a recessive null mutant allele $b$, and flies that are $bb$ have brown-colored eyes. The gene is named after the mutant phenotype it is associated with. The *vermilion* gene produces an enzyme that is active in a pathway synthesizing a brown pigment. The wild-type allele $v^+$ is dominant over the null mutant allele $v$. Flies that are $vv$ have vermilion-colored eyes. The *white* gene produces a pigment-transporting protein from the dominant allele $w^+$ that carries pigments to the eye. A mutant protein from the $w$ allele is incapable of pigment transportation, and flies that do not produce the protein have white eyes.

Production of wild-type proteins from all three genes is necessary to produce wild-type eye color, and hereditary eye color mutations result from the mutation of one or more of the genes (**Figure 4.17**). Wild-type eye color is the result of synthesis of brown and vermilion pigments and the transportation of both pigments to eye cells, where they are blended. Mutation of any one or more of these genes results in a mutant phenotype. This example demonstrates that multiple genes are active in pathways determining different biological properties. Inherited variation of one gene can block a segment of a pathway and produce a mutation attributable to a single gene, but such a finding does not negate the importance of the action of multiple genes affecting each trait.

Three distinct types of genetic pathways can be identified. The eye-color example just described illustrates a **biosynthetic pathway.** Biosynthetic pathways are networks of interacting genes that produce a molecule or compound as their end product. Compounds such as pigments, amino acids, nucleotides, hormones, and so on are examples of the products of biosynthetic pathways.

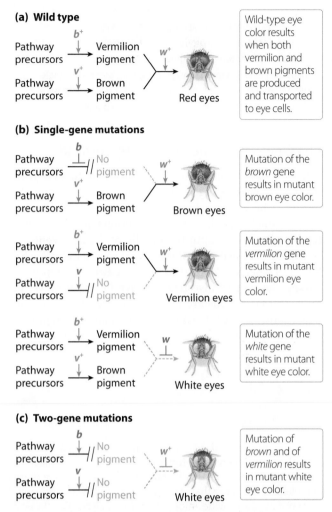

**Figure 4.17** **Interacting genes control eye color in**
*Drosophila.* (a) Wild-type (red) eye color requires activity of
three genes. (b) Mutation of any gene produces a distinctive
mutant phenotype. (c) Double mutation of *brown* and *vermilion*
produces white eyes.

*Signal transduction pathways* are a second type of multiple-gene pathway. These pathways are responsible for reception of chemical signals, such as hormones, that are generated outside a cell and initiate a response inside a cell. Signal transduction operates through the release of a signaling molecule that is part of a sequence of steps culminating in the activation or repression of gene expression in response to an intracellular or extracellular signal.

Finally, *developmental pathways* consist of genes that direct the growth, development, and differentiation of body parts and structures. Numerous developmental pathways have been identified in organisms, and the functions of their genes have been determined by experimental analyses of mutant phenotypes. Geneticists use this analytic approach, known as *genetic dissection*, to identify the step-by-step events making up a genetic pathway. The use of genetic dissection to analyze a biosynthetic pathway is explored in the next section. Examples of signal transduction and developmental pathways are presented in later discussions (see Chapter 20).

> Genetic Insight   Gene interaction is the rule, not the exception, for phenotype production: No gene operates alone to produce a phenotype, even though hereditary variation at one locus may lead to segregation patterns consistent with single-gene traits. Networks of collaborating genes make up biosynthetic, signal transduction, or developmental pathways.

## The One Gene–One Enzyme Hypothesis

The concept of biosynthetic pathways originated with Archibald Garrod's suggestion in 1908 that the inability to produce the enzyme homogentisic acid oxidase is the cause of the human hereditary condition known as alkaptonuria. It was not until the middle of the 20th century, however, that the details of specific biosynthetic pathways began to emerge. George Beadle and Edward Tatum were among the first to investigate biosynthetic pathways, in research that laid the groundwork for the later definition and examination of signal transduction and developmental pathways.

Beadle and Tatum's experiment studied growth variants of the fungus *Neurospora crassa*, and its details are described in **Experimental Insight 4.1**. The idea behind their experiments was simple—to generate single-gene mutations in *Neurospora* and interpret the normal function of genes by observing the phenotypic consequences of their mutation. The famous hereditary proposal known as the **one gene–one enzyme hypothesis** came out of these experiments. It says that each gene produces an enzyme, and each enzyme has a specific functional role in a biosynthetic pathway that produces a phenotype. Beadle and Tatum observed that single-gene mutations block the completion of biosynthetic pathways and lead to the production of mutant phenotypes. Their hypothesis proposed that each mutant phenotype was attributable to the loss or defective function of a specific enzyme. Since each enzyme defect was inherited as a single-gene defect, the one gene–one enzyme hypothesis identifies the direct connection between genes, proteins, and phenotypes.

The one gene–one enzyme concept has undergone adjustments since its proposal to account for three

---

## Experimental Insight 4.1

### The One Gene–One Enzyme Hypothesis

George Beadle and Edward Tatum's experiments had the goal of describing gene function. Their work took place at about the time DNA was being identified as the hereditary molecule, and more than a decade before DNA structure was identified. To provide information for analysis, Beadle and Tatum devised an experiment that would induce single-gene mutations in the filamentous fungus *Neurospora crassa* and then studied the mutants to determine how mutations altered *Neurospora* growth. Recall that *Neurospora* can grow as a haploid, or two haploid cells can fuse to form and grow as diploids that undergo meiosis (see Chapter 2).

In the first step of their experiments, Beadle and Tatum grew numerous genetically identical cultures of haploid wild-type fungi that were irradiated to induce random mutations ❶. The irradiated conidia (asexually produced fungal spores) were mated with wild-type haploids. The resulting diploids underwent meiosis to produce haploid spores that were grown in a two-step process to identify mutants. The diploids could also be tested to confirm the presence of a single-gene mutation by observation of a 3:1 ratio in their progeny. Irradiated haploid spores were grown first on a *complete growth medium* that contains a rich mixture of nutrients and supplements and is capable of supporting the growth of wild-type and mutant fungi ❷. In the second step, growing fungi were picked from colonies on the complete medium and transferred to a *minimal growth medium* that supplies only the minimal constituents needed to support the growth of wild-type fungi ❸. Mutant fungi are identified because they grow on complete medium, but they are unable to grow on a minimal growth medium.

With numerous mutants in hand, Beadle and Tatum were able to address questions of which genes were mutated by

another two-part process. They identified the chemical category of the compound that cannot be produced and then determined the specific missing compound. An example of this analysis is illustrated in steps ❹ and ❺, where growth analysis tests a mutant for its ability to grow on various kinds of *supplemented minimal media*. These are growth media that have had one or more compounds added to them to support the growth of specific kinds of mutants. Step 4 shows one mutant that grows only on medium that has been supplemented with all 20 of the common amino acids; this result indicates that the strain lacks the ability to synthesize one or more amino acids. The specific defect in this mutant strain is tested in step ❺ using 20 different supplemented minimal media, each supplemented with one amino acid. One mutant grows on minimal medium supplemented with methionine (met), thus identifying the strain as one that is unable to synthesize methionine. This strain is described as being *Met−* ("Met minus" or "methionine minus"), to identify the defective pathway. The wild type is able to synthesize methionine and is identified as *Met+* ("Met plus" or "methionine plus").

By testing hundreds of independent mutants in this way, Beadle and Tatum discovered that most mutants carried single mutations that could be overcome by supplementing minimal growth media with one particular compound. This finding led them to propose that single mutations prevented mutants from completing a specific step of a biochemical pathway. Based on this outcome, they proposed that single-gene mutations altered the ability of mutants to produce one enzyme critical in a particular biosynthetic pathway. The correlation between single-gene mutations and single defects in biosynthetic pathways is the basis of the one gene–one enzyme hypothesis.

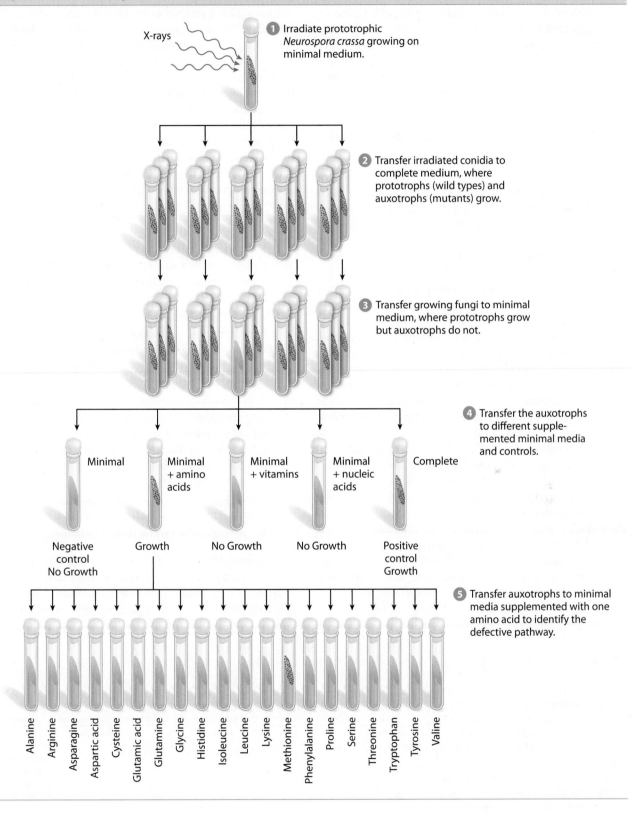

X-rays

**1** Irradiate prototrophic *Neurospora crassa* growing on minimal medium.

**2** Transfer irradiated conidia to complete medium, where prototrophs (wild types) and auxotrophs (mutants) grow.

**3** Transfer growing fungi to minimal medium, where prototrophs grow but auxotrophs do not.

**4** Transfer the auxotrophs to different supplemented minimal media and controls.

Minimal

Minimal + amino acids

Minimal + vitamins

Minimal + nucleic acids

Complete

Negative control No Growth

Growth

No Growth

No Growth

Positive control Growth

**5** Transfer auxotrophs to minimal media supplemented with one amino acid to identify the defective pathway.

Alanine

Arginine

Asparagine

Aspartic acid

Cysteine

Glutamic acid

Glutamine

Glycine

Histidine

Isoleucine

Leucine

Lysine

Methionine

Phenylalanine

Proline

Serine

Threonine

Tryptophan

Tyrosine

Valine

observations: (1) Some protein-producing genes do not produce enzymes, but produce transport proteins, structural proteins, and regulatory proteins; (2) some genes produce RNAs rather than proteins; and (3) some proteins (e.g., β-globin) must join with other proteins to acquire a function. Despite these modifications, Beadle and Tatum's fundamental conclusion linking each gene to a particular product is valid and forms the basis for understanding of gene function.

## Genetic Dissection to Investigate Gene Action

Beadle and Tatum's experiments opened the way to investigation of the roles of individual-gene mutations in biosynthetic pathways. These investigations began with three assumptions about biosynthetic pathways that have proven to be correct: (1) Biosynthetic pathways consist of sequential steps, (2) completion of one step generates the substrate for the next step in the pathway, and (3) completion of every step is necessary for production of the end product of the pathway. These assumptions support the conclusion that wild-type strains are able to complete each pathway step, and that mutant strains are unable to complete a pathway because one or more pathway steps are blocked by mutation.

**Genetic dissection** in this context is an experimental approach that tests the ability of a mutant to execute each step of a biosynthetic pathway and assembles the steps of a pathway by determining the point at which the pathway is blocked in each mutant. The strategy of genetic dissection is illustrated for a *met–* strain in **Figure 4.18** using experimental data collected in 1947 by Norman Horowitz on four independently isolated *Neurospora crassa met–* mutants.

The goals of Horowitz's genetic dissection analysis were to (1) determine the number of intermediate steps within the methionine biosynthetic pathway, (2) determine the order of steps in the pathway, and (3) identify the step affected by each mutation. In designing his experiment, Horowitz relied on previous biochemical work identifying homoserine as the first compound in the methionine biosynthetic pathway and identifying cysteine, homocysteine, and cystathionine as later intermediates in the pathway. Horowitz tested the control prototroph (met+) and

four methionine-requiring auxotrophs (Met 1 to Met 4) for their ability to grow on (1) minimal medium, (2) minimal medium plus cysteine only, (3) minimal medium plus cystathionine only, (4) minimal medium plus homocysteine only, and (5) minimal medium plus methionine only. Figure 4.18a shows growth (+) or no growth (−) of the four *met–* mutants and the wild-type strain (*met+*) on each of the experimental media. The wild-type strain grows on all media, since supplementation of minimal medium with any of the intermediates has no effect on its growth. Each methionine mutant grows on minimal medium plus methionine, the end product of the biosynthetic pathway, but they show different growth patterns with other supplemented media. The following is an analysis of each mutant:

1. Met 1 grows only on minimal medium plus methionine, thus indicating that a mutation in the last step of the pathway prevents conversion of the final intermediate product to methionine. Only the addition of methionine to minimal medium bypasses the pathway block.

2. Met 2 exhibits growth with supplementation by either methionine or homocysteine, thus indicating a block at the step that produces homocysteine. This result also tells us that homocysteine is the substrate converted to methionine in the biosynthetic pathway.

3. Met 3 grows on minimal medium supplemented with either methionine, homocysteine, or cystathionine, but not on minimal medium plus cysteine. This tells us that *Met 3* is blocked at the step that produces cystathionine and that cystathionine precedes homocysteine in the pathway.

4. Met 4 grows with any supplementation of minimal medium. This tells us that *Met 4* is defective at a step that precedes the production of cysteine.

Figure 4.18b shows the steps of the biosynthetic pathway for methionine as determined by analysis of these mutants. The pathway step that is blocked in the mutant is identified based on the logic that supplementation by a compound needed *after* the blockage will permit growth,

**Figure 4.18 Genetic dissection of methionine biosynthesis pathway.**
**(a)** Growth of a wild-type strain and four independent *met–* mutant strains on minimal medium and various supplemented minimal media. For each mutant, the compound that accumulates is the one that immediately precedes the point of blockage. **(b)** The order of intermediate compounds in the methionine biosynthesis pathway and the step blocked in each *met–* mutant strain.

**(a)** Experimental data

| Mutant strain | Growth Medium | | | | | Compound accumulating in mutant |
|---|---|---|---|---|---|---|
| | Minimal medium | Minimal + cysteine | Minimal + cystathionine | Minimal + homocysteine | Minimal + methionine | |
| Control prototroph | + | + | + | + | + | None |
| Met 1 | − | − | − | − | + | Homocysteine |
| Met 2 | − | − | − | + | + | Cystathionine |
| Met 3 | − | − | + | + | + | Cysteine |
| Met 4 | − | + | + | + | + | Homoserine |

**(b)** Order of intermediates in pathway

```
        Met 4         Met 3            Met 2             Met 1
          ↓             ↓                ↓                 ↓
Homoserine → Cysteine → Cystathionine → Homocysteine → Methionine
```

whereas adding a compound used *before* the blockage will not aid growth. The blocked step is also identified by the substance that accumulates in the auxotroph: In each mutant, a different intermediate substance builds up because the step that would convert it to the next intermediate in the pathway is defective. Accumulation of cysteine by *met 3*, cystathionine by *met 2*, and homocysteine by *met 1* supports the assignment of these mutants to specific steps in the pathway. **Genetic Analysis 4.2** illustrates genetic dissection of a biosynthetic pathway by assessment of the growth habits of auxotrophs.

---

Genetic Insight   Mutant genes encode defective proteins that block individual steps of a biosynthetic pathway. Genetic dissection analyzes nutrient requirements of mutants to determine the order of steps in a biosynthetic pathway and identify the step blocked in each mutant.

---

## Epistasis and Its Results

Genes contributing to different steps of a multistep pathway work together to produce the end product of the pathway. Because of this interaction, mutation of one gene may prevent completion of the pathway and production of the end product. In other words, gene interaction can result in one gene influencing whether and how other pathway genes are expressed or how they function.

Gene interactions occur in various forms, and most produce distinctive progeny phenotype ratios as a result of the specific interaction mechanisms. These altered ratios of wild-type and mutant phenotypes are caused by **epistasis,** the name given to gene interactions—called **epistatic interactions**—in which an allele of one gene modifies or prevents the expression of alleles at another gene. A minimum of two genes are required for epistasis. The genes that interact by epistasis are involved in producing a particular phenotypic characteristic, and they usually participate in the same pathway. Epistasis is most readily detected among progeny of dihybrid crosses where both genes carry dominant and recessive alleles. In these cases, independent assortment predicts a 9:3:3:1 ratio of four phenotypes in the $F_2$ progeny, but epistasis results in fewer than four phenotypes. This reduction in the number of $F_2$ phenotype classes occurs because different genotype classes have the same phenotype. In other words, the hallmark of epistatic interaction in a dihybrid cross is modification of the 9:3:3:1 ratio due to the combining of two or more genotype classes into a single phenotypic class.

Epistasis results from mutation in pathways that require a specific activity from every gene in the pathway for the wild-type phenotype to be produced. Given the possible outcomes of dihybrid crosses, there are six ways the $F_2$ phenotype proportions can be rearranged by epistasis. All six altered ratios have been seen in plants or animals. **Figure 4.19** gives an overview of these patterns, showing the modification of dihybrid ratios that characterizes each form of epistasis. The remainder of this discussion provides a brief description and example of each of the epistatic patterns. First, however, we describe a dihybrid cross involving genes contributing to eye color in the fruit fly *Drosophila melanogaster* in which there is *no interaction* between the genes to alter the resulting 9:3:3:1 phenotypic ratio.

**No Interaction (9:3:3:1 ratio)**   Epistasis is most easily identified through specific deviations from the expected 9:3:3:1 ratio among the $F_2$ progeny of a dihybrid cross involving dominant and recessive alleles. This expected $F_2$ ratio results from the action of two independently assorting genes *in the absence of epistasis*—that is, when the genes do not interact to change the expression of one or the other. The contributions of the *brown, vermilion,* and *white* genes to wild-type (red) *Drosophila* eye color are illustrated in Figure 4.17. In this example, we limit our consideration to variation of the *brown* gene and the *vermilion* gene to show the outcome expected from independent assortment of two genes contributing to a specific phenotypic character.

**Figure 4.19** **Patterns resulting from epistatic gene interaction.**

Four $zmt^-$ bacterial mutants ($zmt$-1 to $zmt$-4), each with a single-gene mutation, are available for study. Five intermediates in the zmt synthesis pathway have been identified (D, F, M, R, and S), but their order in the pathway is not known. Each mutant is tested for its ability to grow on minimal medium supplemented with one of the intermediate compounds. All mutants grow when zmt is added to minimal medium, and the wild-type strain grows under all growth conditions tested. Find the order of intermediates in the zmt-synthesis pathway, and identify the step that is blocked in each mutant strain. In the growth table at right, "+" indicates growth and "−" indicates no growth.

| Mutant Strain | Added to Minimal Medium | | | | | | |
|---|---|---|---|---|---|---|---|
| | D | F | M | R | S | Nothing | zmt |
| Wild type | + | + | + | + | + | + | + |
| $zmt$-1 | − | − | − | − | + | − | + |
| $zmt$-2 | − | + | + | + | + | − | + |
| $zmt$-3 | − | + | − | − | + | − | + |
| $zmt$-4 | − | + | + | − | + | − | + |

| Solution Strategies | Solution Steps |
|---|---|

### Evaluate

1. Identify the topic this problem addresses and describe the nature of the required answer.

2. Identify the critical information given in the problem.

1. This problem deals with mutants of the zmt-synthesis pathway and requires an analysis of the defect in each mutant as well as ordering of the intermediates in the zmt-synthesis pathway.

2. The problem provides growth information for wild-type $zmt^+$ bacteria as well as four $zmt^-$ mutant strains when plated on minimal medium and media individually supplemented with zmt or one of five intermediates in the zmt-synthesis pathway.

### Deduce

3. Compare and evaluate the patterns of growth supported by the supplements.

TIP: zmt is the final product of the pathway, and it supports growth of all $zmt^-$ mutants.

4. Identify the final product of the pathway and next-latest pathway intermediate compound.

3. All mutants grow with zmt supplementation and with supplementation by compound S. None grows without any supplementation, and none obtains growth support from compound D. Compounds F, M, and R each support growth of one or more mutants.

4. zmt is the last compound synthesized. Compound S also supports the growth of all mutants and is likely the immediate precursor of zmt.

### Solve

5. Identify the first compound synthesized in the pathway.

6. Identify the second, third, and fourth compounds synthesized in the pathway.

TIP: Medium supplemented with an intermediate compound that occurs after the pathway step that is blocked by a mutant will support growth.

5. Compound D does not support growth of any of the $zmt^-$ mutants and likely occurs before any of the synthesis steps affected by mutations. Compound D is the first compound in the pathway.

6. Compound R supports the growth of mutant $zmt$-2, indicating the compound bypasses the step blocked in $zmt$-2. Compound R likely follows compound D in the pathway, and $zmt$-2 is defective in its ability to convert D to R. $zmt$-2 grows on intermediate compounds that occur after its point of pathway blockage, but not on compound D that comes before the $zmt$-2 blockage.

Compound M supports growth of $zmt$-2 and $zmt$-4, and bypasses the blockage in both mutants. Growth of $zmt$-4 is not supported by compounds D or R that occur before the conversion step blocked in $zmt$-4. The conclusion is that compound M follows R and that $zmt$-4 is unable to convert R to M. Compounds F, M, and S each support growth of $zmt$-4, and each bypasses the blockage.

Compound F supports growth of $zmt$-3 and follows compound M in the pathway. $zmt$-3 is unable to convert M to F. Compound S supports new growth of $zmt$-1, indicating that it follows compound F in the pathway and that $zmt$-1 fails to convert compound F to S.

TIP: To confirm this solution, verify that growth of each mutant is supported by supplementation with compounds that follow the blockage but not by supplementation with compounds that precede the blockage.

7. Assemble the zmt-synthesis pathway, and identify the mutants at each pathway step.

7.

$zmt$-2　　$zmt$-4　　$zmt$-3　　$zmt$-1

D ⟶ R ⟶ M ⟶ F ⟶ S ⟶ zmt

For more practice, see Problems 4, 16 , 17, and 27.

In other words, there is no interaction between the genes and the outcome is predicted by chance.

The analysis begins with the mating of a pure-breeding brown-eyed fly ($b/b$; $v^+/v^+$) to a pure-breeding vermilion-eyed fly ($b^+/b^+$; $v/v$). Writing the genotypes in the forms shown separates the alleles on homologous chromosomes by a slash (/) and separates the genes on different chromosomes by the semicolon. In this example, parental flies are assumed to have wild-type function of the $w^+$ allele, although the genotype for this and other wild-type genes is not shown. The $F_1$ progeny have wild-type eye color and are dihybrid ($b^+/b$; $v^+/v$) (**Figure 4.20**). Progeny in the $F_2$ generation have four eye-color phenotypes, as predicted by independent assortment. Red eye color (wild type) is observed in $\frac{9}{16}$ of the progeny, brown and vermilion eyes are each seen in $\frac{3}{16}$ of the $F_2$, and the white-eyed phenotype appears in $\frac{1}{16}$ of the $F_2$ progeny. These latter progeny are double mutants that are unable to synthesize any colored pigment. The eye color phenotype is white, but the mutational basis of white eye in this case is different from that of flies with mutation of the white gene that transports pigment to the eye (see Figure 4.17). The 9:3:3:1 phenotypic ratio provides evidence that two independently assorting genes contribute to the eye-color phenotype. This ratio indicates that the genes *are not* undergoing epistatic interaction with one another.

The simplest examples of epistasis are interactions between two genes, each with a dominant and a recessive allele. **Foundation Figure 4.21** illustrates six patterns of epistasis for the interaction of two genes. As we describe these patterns here, and as you examine Figure 4.21, notice that the phenotypic ratios observed for each trait result from the combining of the 9:3:3:1 genotype categories. (Refer to Figure 4.19 for an overview of these epistatic patterns.)

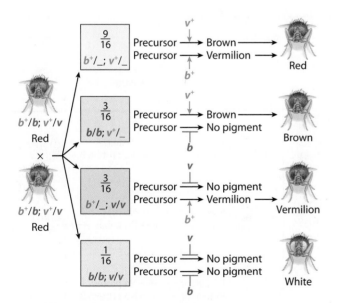

**Figure 4.20  No gene interaction in the production of eye color in *Drosophila*.** A 9:3:3:1 ratio results from the independent assortment of alleles in a dihybrid cross of red-eyed flies with the genotype $b^+/b$; $v^+/v$.

**Complementary Gene Interaction (9:7 ratio)**  William Bateson and Reginald Punnett (of Punnett square fame) were the first biologists to document a deviation from the expected 9:3:3:1 $F_2$ progeny ratio of a dihybrid cross resulting from the epistatic interaction of two genes. In experiments conducted on sweet peas (*Lathyrus odoratus*), an ornamental plant different from Mendel's edible pea (*Pisum sativum*), Bateson and Punnett began by crossing two pure-breeding white-flowered lines. The $F_1$ generation yielded a surprise—all of the progeny plants had purple flowers. When Bateson and Punnett crossed $F_1$ plants, the $F_2$ produced a ratio of $\frac{9}{16}$ purple-flowered plants to $\frac{7}{16}$ white-flowered plants.

Bateson and Punnett recognized that their results could be explained if two genes interacted with one another to produce sweet pea flower color. Assuming two genes are responsible for a single pigment that gives the sweet pea flower its purple color, each parental line—represented by the genotypes *ccPP* and *CCpp*—is pure-breeding for white flowers as a result of homozygosity for recessive alleles at one of the genes. The cross of these two lines of pure-breeding white parents produces dihybrid purple-flowered $F_1$ plants—genotype *CcPp*—because the dominant allele at each locus enables completion of each step of the pathway leading to the synthesis of purple pigment. Independent assortment of alleles results in four genotypic classes, *C–P–*, *ccP–*, *C–pp*, and *ccpp*, produced in the 9:3:3:1 ratio that is expected from a dihybrid cross. Among the $F_2$, however, only the $\frac{9}{16}$ carry the *C–P–* genotype that confers the ability to produce purple pigment. The remaining $\frac{7}{16}$ of the $F_2$ are homozygous either for one of the recessive alleles *c* and *p* or for both sets of alleles. None of these plants are able to synthesize pigment, due to the absence of functional gene products from one or both loci, and they all have the same mutant phenotype.

A 9:7 phenotypic ratio results from **complementary gene interaction** that requires genes to work in tandem to produce a single product. Figure 4.21 ❶ shows that at the molecular level, purple flower color in sweet peas is produced when the pigment anthocyanin is deposited in petals. The production of the purple-flowered $F_1$ progeny and the 9:7 $F_2$ ratio is explained by the independent assortment of two genes, *C* and *P*, that produce gene products controlling different steps of the anthocyanin-synthesis pathway. Since anthocyanin production requires the action of the product of *C* as well as the product of *P*, both steps must be successfully completed for anthocyanin production and deposition in flower petals. On the other hand, any recessive homozygous genotype at the *C* locus, the *P* locus, or both loci results in blockage of the pathway and production of white flowers containing no pigment.

The ability of two mutants with the same mutant phenotype to produce progeny with the wild-type phenotype is called *genetic complementation*, and it indicates that more than one gene is involved in determining the phenotype. We discuss the details of genetic complementation in the last section of this chapter.

## Epistatic Ratios

### 1 Complementary gene interaction

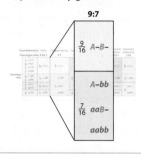

**9:7**

$\frac{9}{16}$ A–B–

A–bb

$\frac{7}{16}$ aaB–

aabb

Complementary gene interaction occurs when genes must act in tandem to produce a phenotype. The wild-type action from both genes is required to produce the wild-type phenotype. Mutation of one or both genes produce a mutant phenotype.

Example: sweet pea flower color

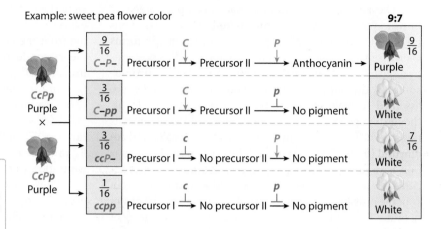

**9:7**

$\frac{9}{16}$ C–P– : Precursor I $\xrightarrow{C}$ Precursor II $\xrightarrow{P}$ Anthocyanin → Purple $\frac{9}{16}$

$\frac{3}{16}$ C–pp : Precursor I $\xrightarrow{C}$ Precursor II $\xrightarrow{p}$ No pigment → White

$\frac{3}{16}$ ccP– : Precursor I $\xrightarrow{c}$ No precursor II $\xrightarrow{P}$ No pigment → White $\frac{7}{16}$

$\frac{1}{16}$ ccpp : Precursor I $\xrightarrow{c}$ No precursor II $\xrightarrow{p}$ No pigment → White

CcPp Purple × CcPp Purple

### 2 Duplicate gene interaction

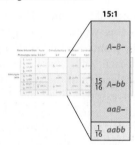

**15:1**

A–B–

$\frac{15}{16}$ A–bb

aaB–

$\frac{1}{16}$ aabb

Duplicate gene interaction allows dominant alleles of either duplicate gene to produce the wild-type phenotype. Only organisms with homozygous mutations of both genes have a mutant phenotype.

Example: bean flower color

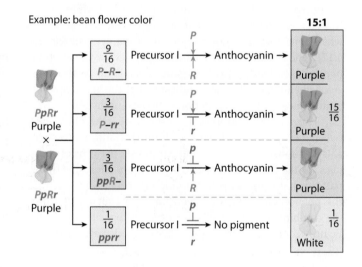

**15:1**

$\frac{9}{16}$ P–R– : Precursor I $\xrightarrow{P}{R}$ Anthocyanin → Purple

$\frac{3}{16}$ P–rr : Precursor I $\xrightarrow{P}{r}$ Anthocyanin → Purple $\frac{15}{16}$

$\frac{3}{16}$ ppR– : Precursor I $\xrightarrow{p}{R}$ Anthocyanin → Purple

$\frac{1}{16}$ pprr : Precursor I $\xrightarrow{p}{r}$ No pigment → White $\frac{1}{16}$

PpRr Purple × PpRr Purple

### 3 Dominant gene interaction

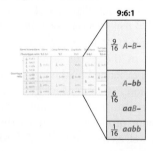

**9:6:1**

$\frac{9}{16}$ A–B–

$\frac{6}{16}$ A–bb, aaB–

$\frac{1}{16}$ aabb

Dominant gene interaction occurs between genes that each contribute to a phenotype producing one. One phenotype if dominant alleles are present at each gene, a second phenotype if recessive alleles are homozygous for either gene, and a third phenotype if recessive homozygosity occurs at both genes.

Example: squash fruit shape

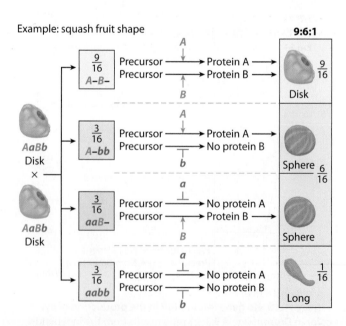

**9:6:1**

$\frac{9}{16}$ A–B– : Precursor $\xrightarrow{A}$ Protein A → / Precursor $\xrightarrow{B}$ Protein B → Disk $\frac{9}{16}$

$\frac{3}{16}$ A–bb : Precursor $\xrightarrow{A}$ Protein A → / Precursor $\xrightarrow{b}$ No protein B → Sphere

$\frac{3}{16}$ aaB– : Precursor $\xrightarrow{a}$ No protein A → / Precursor $\xrightarrow{B}$ Protein B → Sphere $\frac{6}{16}$

$\frac{3}{16}$ aabb : Precursor $\xrightarrow{a}$ No protein A / Precursor $\xrightarrow{b}$ No protein B → Long $\frac{1}{16}$

AaBb Disk × AaBb Disk

## 4 Recessive epistasis

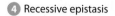

**9:3:4**

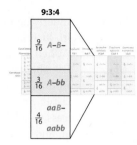

Recessive epistasis occurs when recessive alleles at one gene mask or reduce the expression of alleles at the interacting locus.

Example: labrador retriever coat color

**9:3:4**

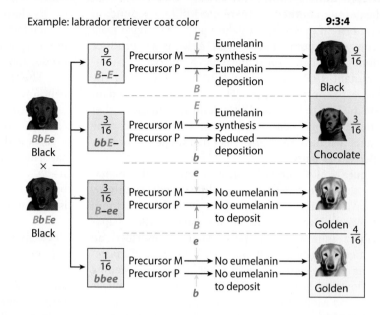

| Genotype | Pathway | Phenotype |
|---|---|---|
| $\frac{9}{16}$ B–E– | Precursor M → Eumelanin synthesis; Precursor P → Eumelanin deposition | Black $\frac{9}{16}$ |
| $\frac{3}{16}$ bbE– | Precursor M → Eumelanin synthesis; Precursor P → Reduced deposition | Chocolate $\frac{3}{16}$ |
| $\frac{3}{16}$ B–ee | Precursor M → No eumelanin; Precursor P → No eumelanin to deposit | Golden |
| $\frac{1}{16}$ bbee | Precursor M → No eumelanin; Precursor P → No eumelanin to deposit | Golden $\frac{4}{16}$ |

BbEe Black × BbEe Black

## 5 Dominant epistasis

**12:3:1**

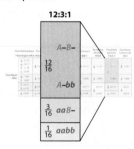

In dominant epistasis, a dominant allele of one gene masks or reduces the expression of alleles of a second gene.

Example: foxglove flower color

**12:3:1**

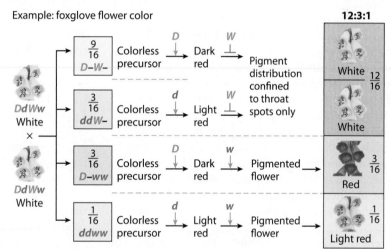

| Genotype | Pathway | Phenotype |
|---|---|---|
| $\frac{9}{16}$ D–W– | Colorless precursor → Dark red → Pigment distribution confined to throat spots only | White $\frac{12}{16}$ |
| $\frac{3}{16}$ ddW– | Colorless precursor → Light red → Pigment distribution confined to throat spots only | White |
| $\frac{3}{16}$ D–ww | Colorless precursor → Dark red → Pigmented flower | Red $\frac{3}{16}$ |
| $\frac{1}{16}$ ddww | Colorless precursor → Light red → Pigmented flower | Light red $\frac{1}{16}$ |

DdWw White × DdWw White

## 6 Dominant suppression

**13:3**

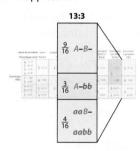

Dominant suppression occurs when the dominant allele of one gene suppresses the expression of a dominant allele of a second gene.

Example: chicken feather color

**13:3**

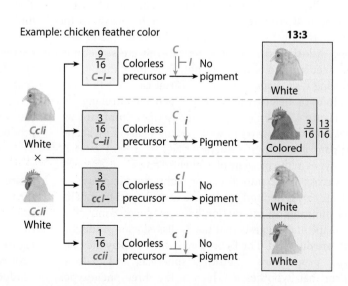

| Genotype | Pathway | Phenotype |
|---|---|---|
| $\frac{9}{16}$ C–I– | Colorless precursor → No pigment | White |
| $\frac{3}{16}$ C–ii | Colorless precursor → Pigment | Colored $\frac{3}{16}$ $\frac{13}{16}$ |
| $\frac{3}{16}$ ccI– | Colorless precursor → No pigment | White |
| $\frac{1}{16}$ ccii | Colorless precursor → No pigment | White |

CcIi White × CcIi White

**Duplicate Gene Action (15:1 ratio)** Two genes that duplicate one another's activity constitute a redundant genetic system in which any genotype possessing at least one copy of a dominant allele at *either* locus will produce the dominant phenotype. Only when homozygous recessive mutant alleles are present at both loci does the recessive phenotype appear. The genes in a redundant system are said to have **duplicate gene action;** they either encode the same gene product, or they encode gene products that have the same effect in a single pathway or compensatory pathways.

Figure 4.21 ❷ provides an illustration and explanation of duplicate gene action identified inadvertently by Gregor Mendel in an experiment involving flower color in bean plants. Near the end of his famous 1866 paper describing inheritance in peas, Mendel described an experiment with beans that began with the cross of a pure-breeding purple-flowered bean plant to a pure-breeding white-flowered bean plant. The $F_1$ plants all had purple flowers, and Mendel probably assumed that flower color determination in beans would follow the same pattern as in peas. Among the 32 $F_2$ plants Mendel produced, however, 31 had purple flowers and only 1 had white flowers. Among the $F_2$ plants, $\frac{15}{16}$ have a genotype containing at least one copy of either $P$ or $R$, and only $\frac{1}{16}$ have the genotype *pprr* and the white-flowered phenotype.

Figure 4.21 ❷ shows that a dominant allele at either locus is capable of catalyzing the conversion of a precursor to anthocyanin and producing the dominant phenotype. Conversely, if homozygous recessive alleles are present at both loci, no functional gene product is produced, and the synthesis pathway is not completed. White flowers result from the absence of pigment in the $\frac{1}{16}$ of the $F_2$ progeny that are homozygous recessive for alleles of both genes.

**Dominant Gene Interaction (9:6:1 ratio)** Fruit shape in summer squash is classified as either long, spherical, or disk shaped. Plants that bear long fruit are consistently pure-breeding, indicating that these plants are homozygous for genes controlling fruit shape. On the other hand, plants producing disk-shaped fruit or spherical fruit are sometimes pure-breeding and sometimes not, indicating that plants producing disk-shaped or spherical fruit can be either homozygous or heterozygous for the genes controlling the trait. Figure 4.21 ❸ illustrates and describes **dominant interaction** between two genes controlling squash fruit shape. Dominant interaction is characterized by a 9:6:1 ratio of phenotypes in the progeny of a dihybrid cross.

A cross of two pure-breeding plants producing spherical fruit can generate $F_1$ that have disk-shaped fruit. This result indicates an interaction between genes controlling fruit shape and suggests that the $F_1$ disk-shape–producing plants are dihybrid. The $F_2$ progeny, which display the phenotypic proportions $\frac{9}{16}$ disk, $\frac{6}{16}$ spherical, and $\frac{1}{16}$ long, confirm that hypothesis. Which of the three phenotypes occurs depends on whether a dominant allele is present for both genes, one gene, or neither gene. In the $F_2$ generation,

plants with at least one dominant allele at each locus ($A–B–$) have disk-shaped fruit, plants with recessive alleles at each locus ($aabb$) produce long fruit, and plants that are homozygous recessive at either of the loci ($A–bb$ or $aaB–$) produce spherical fruit.

The molecular model of the events underlying dominant interaction assumes that each gene produces a different protein that contributes to fruit shape. When dominant allelic action produces both proteins, disk-shaped fruit is generated. If only one of the proteins is produced, spherical fruit results, as for the genotypic classes $aaB–$ and $A–bb$. Plants that are homozygous for recessive alleles of both genes ($aabb$) produce neither protein, and long fruit is the result.

**Recessive Epistasis (9:3:4 ratio)** Black, chocolate, and yellow coat colors in Labrador retrievers result from the interaction of two genes, one that produces pigment and another that distributes the pigment to hair follicles. This form of gene interaction, in which homozygosity for a recessive allele at one locus can mask the phenotypic expression of a second gene, is called **recessive epistasis** and has the characteristic 9:3:4 ratio of phenotypes illustrated by Figure 4.21 ❹.

Crossing pure-breeding chocolate parents to pure-breeding yellow ones produces $F_1$ progeny with black coats. That the $F_1$ progeny are dihybrid is revealed by the $F_2$ generation, in which $\frac{9}{16}$ of the progeny carry the genotypes in the $B–E–$ class and have black coats, $\frac{3}{16}$ have a genotype that is $bbE–$, resulting in chocolate-colored coats, and $\frac{4}{16}$ carry genotypes that are either $B–ee$ or $bbee$ and have yellow coats.

The molecular explanation for this genetic system is tied to production of the hair pigment melanin. Dogs can produce eumelanin that gives hair a black or brown color and pheomelanin that gives hair a reddish or yellowish tone. The $B$ gene is *TYRP1* that controls melanin distribution. The wild-type allele $B$ produces full melanin distribution, the mutant allele $b$ has reduced distribution. Gene $E$ is *MC1R* that controls eumelanin synthesis. Allele $E$ permits synthesis; the mutant allele $e$ does not. Dogs that are $B\_E\_$ produce and deposit large amounts of eumelanin and have black coats. Dogs that are $bbE\_$ produce eumelanin but deposit less due to their $bb$ genotype. These dogs have chocolate (brown) coat. Dogs that are homozygous ee are unable to produce eumelanin and instead produce only pheomelanin. These dogs have yellow coat color.

**Dominant Epistasis (12:3:1 ratio)** Determination of flower color in foxgloves provides an example of **dominant epistasis,** where a dominant allele at one locus masks the expression of alleles at a second locus described in Figure 4.21 ❺. In foxgloves, a wild-type allele $d$ produces a light red pigment seen in flowers, and a mutant allele $D$ produces a dark red flower pigment. At another gene, allele $w$ is a wild-type allele that distributes pigment throughout the flower. A mutant allele $W$ restricts pigment to the throat of the flower. Dihybrid $F_1$ plants (*DdWw*) have white flowers with dark red spots in the flower throat. Dominant epistasis is revealed in the $F_2$ by a 12:3:1 ratio of white-flowered plants with throat spots (*D–W–* and *ddW–*), dark-red–flowered plants (*D–ww*), and light-red–flowered plants (*ddww*). In foxgloves, allele *W* exerts an epistatic effect by preventing the distribution of flower color outside the throat.

**Dominant Suppression (13:3 ratio)**   Our final example of epistatic gene interaction is **dominant suppression,** illustrated in Figure 4.21 ⑥. Dominant suppression is similar to dominant epistasis but occurs when a dominant allele of one gene completely suppresses the phenotypic expression of alleles of another gene. In chickens, for example, feather color requires a dominant allele *C*. Chickens that are homozygous for a recessive allele *c* have white feathers. The *C* allele can have its color-producing action suppressed by a dominant suppressor allele, *I*. The recessive allele *i* does not exert suppression. Crosses between pure-breeding colored chickens (*CCii*) and pure-breeding white chickens (*ccII*) produce white-feathered $F_1$ that are dihybrid (*CcIi*). Production of the $F_2$ results in a 13:3 ratio that is characteristic of dominant suppression. Chickens carrying a *cc* genotype are unable to produce feather color, and those carrying *C–* along with *I–* have feather color production suppressed. Only chickens with the *C–ii* genotype are able to produce colored feathers.

Figure 4.21 ⑥ shows that the product of allele *C* converts a colorless precursor into pigment, whereas the allele *c* product is inactive and fails to convert the precursor, resulting in white feather color for *cc* genotypes. Dominant suppression of *C* by the product of *I* prevents pigment production in chickens with the *C–I–* genotype. The homozygous genotype *ii* is unable to suppress color in the *C–*. **Genetic Analysis 4.3** tests your ability to analyze crosses involving epistatic gene interaction.

## 4.4 Complementation Analysis Distinguishes Mutations in the Same Gene from Mutations in Different Genes

Suppose you are a geneticist working in California, and you have identified a recessive mutation causing petunia flowers to be white rather than the wild-type purple color. A friend of yours, also a geneticist, is working on petunias in the Netherlands and contacts you because she has also identified a recessive mutation resulting in white-flowered petunias. Since there has been no contact between California petunias and Netherland petunias, the mutations have arisen independently. When geneticists encounter organisms with the same mutant phenotype, two initial questions are (1) do these organisms have mutations of the same gene or of different genes, and (2) how many genes are responsible for the mutations observed?

We have already seen that mutations of different genes can produce the same, or very similar, abnormal phenotypes. This phenomenon is known as **genetic heterogeneity.** We have also seen that a mating of two organisms with the same or a similar abnormal phenotype can sometimes produce offspring with the wild-type phenotype. This phenomenon is called **genetic complementation,** and it occurs when mutant organisms carry mutations of different genes that produce the same abnormal phenotype. In contrast, if the two mutations are in the same gene, offspring of a cross between the two mutants will have a mutant phenotype; this situation is known as a failure of genetic complementation. In this section, we describe how to distinguish whether two independent mutations are in the same gene or in different genes.

An analytic approach called genetic complementation testing examines the relation between two or more recessive mutations affecting one phenotypic attribute. Researchers use it to determine whether two recessive mutations are in the same gene or in different genes. It also provides information on the number of different genes that can produce the mutant phenotype. Here we limit our discussion to testing eukaryotic genomes, using eye color in *Drosophila* as an example. Strategies for complementation testing in bacteria and bacterial viruses (bacteriophage) differ somewhat from those used in plants and animals (Chapter 7).

Genetic complementation testing crosses pure-breeding mutants for a recessive mutation and examines the phenotype of cross progeny. The heterozygous $F_1$ progeny of these crosses are then examined for the wild-type or mutant phenotypes. If wild-type progeny are produced, genetic complementation has occurred, and the conclusion is that the mutant alleles are of different genes. On the other hand, if the mutant alleles are of the same gene, the progeny of two pure-breeding mutants will have a mutant phenotype. This result indicates that no genetic complementation has taken place.

As an example, we examine genetic complementation testing using two genes affecting *Drosophila* eye color, both of which we have discussed previously: the *vermilion* gene, whose product produces eye-color pigment, and the *white* gene, whose product produces the eye-color pigment transport protein. Both genes are located on the X chromosome in *Drosophila*. The sequential action of the gene products in eye-color production is illustrated in **Figure 4.22a**. Genetic complementation is illustrated by the production of wild-type (red) female progeny from the cross of a pure-breeding female with vermilion eyes to a pure-breeding male with white eyes. No genetic complementation occurs when a pure-breeding apricot female and a pure-breeding buff male are crossed. All progeny have mutant eye colors.

Genetic complementation analysis utilizes numerous crosses of different pure-breeding mutants to one another to determine if the progeny are mutant (no genetic complementation) or wild type (genetic complementation). A

Dr. Ara B. Dopsis, a famous plant geneticist, decides to try his hand at iris propagation. He selects two pure-breeding irises, one red and the other blue, and crosses them. To his surprise, all $F_1$ plants have purple flowers. He decides to create more purple irises by self-fertilizing the $F_1$ irises. Dr. Dopsis produces 320 $F_2$ plants consisting of 182 with purple flowers, 59 with blue flowers, and 79 with red flowers.

**a.** From the information available, describe the genetic phenomenon that produces the phenotypic ratio observed in the $F_2$ plants. Identify the number of genes that are involved in this trait.

**b.** Using clearly defined symbols of your own choosing, identify the genotypes of parental and $F_1$ plants.

| Solution Strategies | Solution Steps |
|---|---|
| **Evaluate** | |
| 1. Identify the topic this problem addresses and state what is required in the answer. | 1. This problem concerns an interpretation of $F_1$ and $F_2$ results, the identification of the genetic mechanism responsible for the observation, and the assignment of genotypes to parental and $F_1$ plants in a manner consistent with the genetic mechanism. |
| 2. Identify the critical information given in the problem. | 2. The problem states that the blue- and red-flowered parents are pure-breeding and that their $F_1$ are exclusively purple flowered. Among the $F_2$ purple is predominant, but red and, to a lesser extent, blue are also observed. |
| **Deduce** | |
| 3. Deduce the potential genetic mechanisms that could account for producing purple-flowered $F_1$ plants from the pure-breeding red and blue parental plants. | 3. Two potential mechanisms are suggested by these data. First, a single gene with incomplete dominance might generate a phenotype in $F_1$ heterozygous plants that is different from that of either homozygous parent. Second, two genes displaying an epistatic interaction might account for a phenotype in an $F_1$ dihybrid that is distinct from either pure-breeding parent. |
| 4. Determine the <u>relative</u> phenotype proportions predicted by the possible genetic mechanisms and evaluate the observed phenotype ratio. <br><br> **TIP:** Compare the relative percentages of each phenotype to see which genetic model most closely predicts the observed percentages. | 4. A single-gene model predicts that the self-fertilization of an $F_1$ heterozygote will result in a 1:2:1 (25%:50%:25%) in the $F_2$. A two-gene epistasis model producing three $F_2$ phenotypes could be dominant gene interaction (9:6:1 ratio), dominant epistasis (12:3:1 ratio), or recessive epistasis (9:4:3 ratio). Recessive epistasis predictions are a closer match to observations than dominant epistasis predictions. Recessive epistasis predicts phenotype percentages of approximately 56%:25%: 19%. The observed ratio of $F_2$ phenotypes is $\frac{182}{320} = 56.8\%$ purple, $\frac{79}{320} = 24.7\%$ red, and $\frac{59}{320} = 18.4\%$ blue. |
| **Solve** | **Answer a** |
| 5. Identify the genetic mechanism most likely to account for the outcomes of these crosses. | 5. Comparison of the $F_2$ predictions of the single-gene incomplete dominance model and the two-gene recessive epistasis model determines that recessive epistasis is a better predictor of relative progeny proportions. The likely genetic model explaining these data is recessive epistasis. (Note that the number of $F_2$ observed in each category can be compared to the number expected by chi-square analysis.) |
| 6. Assign genotypes to parental and $F_1$ plants. | 6. Using symbols $A$ and $a$ for one gene and $B$ and $b$ for the second gene, the genotypes of plants are <br><br> Parents: $aaBB$ (red) and $AAbb$ (blue) <br> $F_1$: $AaBb$ (purple) |

For more practice, see Problems 5, 10, 22, and 31.

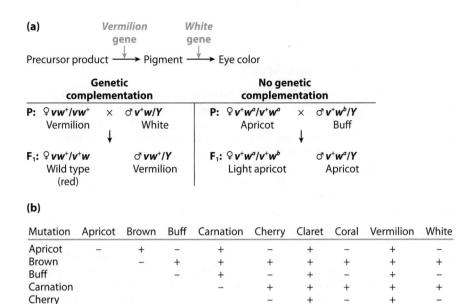

**(b)**

| Mutation | Apricot | Brown | Buff | Carnation | Cherry | Claret | Coral | Vermilion | White |
|---|---|---|---|---|---|---|---|---|---|
| Apricot | − | + | − | + | − | + | − | + | − |
| Brown |  | − | + | + | + | + | + | + | + |
| Buff |  |  | − | + | − | + | − | + | − |
| Carnation |  |  |  | − | + | + | + | + | + |
| Cherry |  |  |  |  | − | + | − | + | − |
| Claret |  |  |  |  |  | − | + | + | + |
| Coral |  |  |  |  |  |  | − | + | − |
| Vermilion |  |  |  |  |  |  |  | − | + |
| White |  |  |  |  |  |  |  |  | − |

**Complementation**

| Group | Mutant (allele) |
|---|---|
| I | Apricot (***w^a***), buff (***w^b***), cherry (***w^{ch}***), coral (***w^{co}***), white (***w***) |
| II | Carnation (***c***) |
| III | Claret (***cl***) |
| IV | Brown (***b***) |
| V | Vermilion (***v***) |

Figure 4.22 **Genetic complementation and no genetic complementation involving the *Drosophila* eye color genes *vermilion* and *white*. (a)** The cross of pure-breeding vermilion to pure-breeding white shows genetic complementation by production of wild-type eye color in the $F_1$. The cross between pure-breeding apricot and pure-breeding buff produces no genetic complementation in the $F_1$ that have mutant eye color. **(b)** Genetic complementation testing among nine distinct *Drosophila* eye color mutants reveals five complementation groups corresponding to five genes. Five mutant alleles of *white* mutually fail to complement and are assigned to the same gene. The other four mutants each complement one another, and the *white* gene mutants and are assigned to their own gene.

table of genetic complementation testing data shown in Figure 4.22b indicates whether the cross of parental mutant phenotypes produces wild-type progeny (indicated in the table by plus symbols: +), or mutant progeny (indicated in the table by minus symbols: −). Any given pair of mutants that *complement* one another by producing wild-type progeny are mutations of *different genes*. (Recall the results of complementary gene action illustrated in Figure 4.21 ❶.) In contrast, the cross of mutant parents produces only the mutant phenotype in progeny when the mutations *fail to complement* one another and are mutations of the *same gene*.

Complementation analysis of the *Drosophila* eye-color mutation results displayed in Figure 4.22b focuses on crosses that *fail to complement* as these are the result of mutations that are in the same gene. Mutations that mutually fail to complement one another are identified as a **complementation group,** consisting of one or more mutant alleles of a single gene. A complementation group consists of mutants whose phenotypes consistently fail to complement one another and that complement mutants in other complementation groups. In the genetic context, a "complementation group" is synonymous with a "gene," because the mutant alleles of each complementation group all affect the same phenotypic characteristic. Thus, in genetic complementation analysis, the number of complementation groups equals the number of genes.

In the complementation testing data in Figure 4.22b, apricot, buff, cherry, coral, and white all exhibit a mutual failure to complement. This result identifies the five mutants as occurring in the same gene. (Historically, white was the first mutation identified and is the name the gene has become known by.) Geneticists conclude that apricot, buff, cherry, coral, and white are mutant alleles of the *white* (*w*) gene in *Drosophila*. These mutations form complementation group I. In contrast, the mutations brown, carnation, claret, and vermilion each complement all other mutants. This observation tells investigators that they are not alleles of another mutant, but that instead each mutant represents a separate gene. Each of these mutants forms its own complementation group (i.e., complementation groups II through V). Among the nine *Drosophila* eye-color mutants examined, five genes (five complementation groups) are identified. One gene is represented by five mutants, and the other four genes are represented by one mutation each.

---

**Genetic Insight** Complementation occurs when two mutants produce wild-type progeny, indicating that parental organisms carry mutations of different genes. By contrast, failure to complement occurs when two organisms carry mutations in the same gene.

---

## CASE STUDY

### Identification of Xeroderma Pigmentosum Complementation Groups

In this case study, we examine the use of genetic complementation analysis to identify the number of genes involved in a rare but genetically heterogeneous human condition called xeroderma pigmentosum (XP). XP is characterized by severe sensitivity to ultraviolet (UV) irradiation from sunlight and by up to a thousandfold increase in the rate of sun-induced skin cancer. While the experimental approaches to complementation testing in humans are necessarily different from those employed for laboratory organisms, the interpretations of "crosses" follow the same processes.

People with XP are deficient in a type of DNA repair called nucleotide excision repair (NER) that would otherwise protect their skin from the UV-induced damage that leads to cancer. In NER, a short section of DNA containing a UV-induced lesion is removed, and the gap is filled by new DNA (see Chapter 12).

Research work that began in the late 1970s identified seven complementation groups representing seven different genes that are mutated in different forms of XP. Two approaches identify the existence of the seven complementation groups. Anthony Andrews and his colleagues obtained cultured skin cells from XP patients and from normal controls and tested the ability of the cells to grow after exposure to measured doses of UV irradiation (**Figure 4.23**). The cells were exposed to UV light at a wavelength of 254 nm for different amounts of time, and their growth was measured as the percentage of original cells able to form colonies after UV exposure. These researchers identified five distinct patterns of response to UV exposure that are designated as complementation groups A to E.

Other researchers measured the response of cultured XP cells to UV exposure by determining the level of NER taking place in XP cell cultures taken from different XP individuals in comparison to normal cells. The results showed that XP cell lines vary in their levels of NER from less than 5% of normal to about 50% of normal. These results could be due to the mutations being in different genes or, alternatively, to different hypomorphic alleles of the same gene.

Genetic complementation analysis was used in the study of XP cell cultures with low NER to identify cell lineages carrying different XP gene mutations. To do this, two cells from lineages with low NER were fused to form a heterokaryon, a hybrid cell with two nuclei. A heterokaryon contains all the genetic information from both contributing cells. The experimental rationale is that if the two cells contain mutations of *different* genes, the heterokaryon will experience genetic complementation that would be detected as normal or near normal levels of NER; but if the mutations are in the same gene, NER will be about the same in the heterokaryon as in the individual cell lines. This analysis of NER levels in XP heterokaryons ultimately indicated seven complementation groups of XP genes.

Each of the seven XP-associated genes has had its function identified and its position mapped in the human genome in the last decade or so. Four of the genes produce proteins that are required to remove a segment of the strand of DNA damaged by UV irradiation as part of the DNA repair process. Proteins from two other XP-associated genes are required to recognize UV-induced DNA damage and the seventh gene produces a protein that binds to the DNA lesion once it is located. The knowledge of the identity of the seven XP-associated genes has led to the finding that other cancer-associated hereditary diseases also involve mutations of one or another of the XP-associated genes. (see Chapter 12).

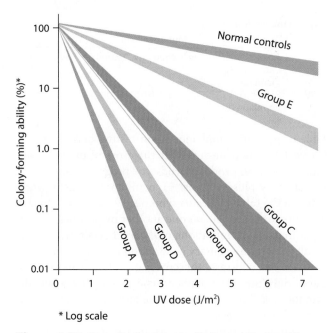

\* Log scale

**Figure 4.23** **Growth of cultured cells from patients with xeroderma pigmentosum (XP).** Five XP complementation groups are identified based on growth ability.

---

SUMMARY    Mastering GENETICS™    *For activities, animations, and review quizzes, go to the study area at www.masteringgenetics.com.*

### 4.1 Interactions between Alleles Produce Dominance Relationships

▪ Loss-of-function mutations decrease or eliminate gene activity. Gain-of-function mutations can cause over-expression or result in new functions.

▪ Incomplete dominance produces heterozygotes with phenotypes that differ from those of either homozygote but are closer to one homozygous phenotype than the other.

▪ Codominant alleles are both equally detected in the heterozygous phenotype.

- The interaction of allelic products determines the dominance relationship between alleles.
- ABO blood types are produced by alleles whose protein products produce dominance or codominance depending on the genotype.
- Multiple alleles of a single gene can display a variety of dominance relationships that establish an allelic series.
- Lethal alleles can kill gametes, can prevent the gestational development of certain classes of progeny, or can have their lethal effect later in life.
- In sex-limited and sex-influenced traits, alleles are manifested differently in each sex.

## 4.2 Some Genes Produce Variable Phenotypes

- In incomplete penetrance, an allele does not always have the expected effect on the phenotype.
- In variable expressivity, organisms with the same genotype have different degrees of phenotypic expression.
- Pleiotropic mutations affect two or more distinct and seemingly independent attributes of the phenotype.

## 4.3 Gene Interaction Modifies Mendelian Ratios

- Epistasis is revealed by six alternative ratios that modify the 9:3:3:1 ratio expected among the progeny of a dihybrid cross.
- Epistasis types and ratios are complementary gene interaction (9:7), duplicate gene interaction (15:1), dominant gene interaction (9:6:1), recessive epistasis (9:3:4), dominant epistasis (12:3:1), and dominant suppression (13:3).

## 4.4 Complementation Analysis Distinguishes Mutations in the Same Gene from Mutations in Different Genes

- In genetic heterogeneity, mutations in different genes can produce the same phenotype.
- Genetic complementation produces progeny with the wild-type phenotype from parents that are pure-breeding for similar mutant phenotypes. The detection of genetic complementation means the mutations occur in different genes.
- The failure to detect genetic complementation from the cross of two similar mutant organisms identifies the mutant alleles as being carried by the same gene.

## KEYWORDS

allelic series *(p. 112)*
biosynthetic pathway *(p. 123)*
codominance *(p. 110)*
complementation group *(p. 135)*
delayed age of onset *(p. 118)*
dominant negative mutation *(p. 109)*
epistatic interaction (epistasis) *(p. 127)*
    complementary gene interaction (9:7 ratio) *(p. 129)*
    dominant interaction (9:6:1 ratio) *(p. 132)*
    dominant epistasis (12:3:1 ratio) *(p. 132)*
    dominant suppression (13:3 ratio) *(p. 134)*
    duplicate gene action (15:1 ratio) *(p. 132)*

recessive epistasis (9:3:4 ratio) *(p. 132)*
gain-of-function mutation *(p. 107)*
gene–environment interaction *(p. 120)*
gene interaction *(p. 122)*
genetic complementation *(p. 134)*
genetic dissection *(p. 126)*
genetic heterogeneity *(p. 134)*
haploinsufficient *(p. 107)*
haplosufficient *(p. 107)*
hypermorphic mutation *(p. 109)*
incomplete dominance (partial dominance) *(p. 109)*
incomplete penetrance (nonpenetrant, penetrant) *(p. 119)*

leaky mutation (hypomorphic mutation) *(p. 107)*
lethal mutation (lethal allele) *(p. 114)*
loss-of-function mutation *(p. 107)*
neomorphic mutation *(p. 109)*
null mutation (amorphic mutation) *(p. 107)*
one gene–one enzyme hypothesis *(p. 124)*
pleiotropy *(p. 121)*
sex-influenced trait (sex-influenced expression) *(p. 118)*
sex-limited trait (sex-limited gene expression) *(p. 117)*
temperature-sensitive allele *(p. 114)*
variable expressivity *(p. 120)*

## PROBLEMS

*For instructor-assigned tutorials and problems, go to www.masteringgenetics.com.*

### Chapter Concepts

*For answers to selected even-numbered problems, see Appendix: Answers.*

1. Define and distinguish *incomplete penetrance* and *variable expressivity*.

2. Define and distinguish *epistasis* and *pleiotropy*.

3. When working on barley plants, two researchers independently identify a short-plant mutation and develop homozygous recessive lines of short plants. Careful measurements of the height of mutant short plants versus normal tall plants indicate that the two mutant lines have the same height. How would you determine if these two mutant lines carry mutation of the same gene or of different genes?

4. Fifteen bacterial colonies growing on a complete medium are replica-plated to a minimal medium. Twelve of the colonies grow on minimal medium.

   a. Using terminology from the chapter, characterize the 12 colonies that grow on minimal medium and the 3 colonies that do not.

   b. The three colonies that do not grow on minimal medium are replica-plated to minimal medium plus the amino acid serine (min + Ser), and all three colonies grow. Characterize these three colonies.

c. The serine biosynthetic pathway is a three-step pathway in which each step is catalyzed by the enzyme product of a different gene, identified as enzymes A, B, and C in the diagram below.

$$\text{3-Phosphoglycerate} \xrightarrow{\text{Enzyme A}} \underset{\text{(3-PHP)}}{\text{3-Phospho-hydroxypyruvate}} \xrightarrow{\text{Enzyme B}}$$

$$\underset{\text{(3-PS)}}{\text{3-Phosphoserine}} \xrightarrow{\text{Enzyme C}} \underset{\text{(Ser)}}{\text{Serine}}$$

Mutant 1 grows only on min + Ser. In addition to growth on min + Ser, mutant 2 also grows on min + 3-PHP and min + 3-PS. Mutant 3 grows on min + 3-PS and min + Ser. Identify the step of the serine biosynthesis pathway at which each mutant is defective.

5. In a type of parakeet known as a "budgie," feather color is controlled by two genes. A yellow pigment is synthesized under the control of a dominant allele $Y$. Budgies that are homozygous for the recessive $y$ allele do not synthesize yellow pigment. At an independently assorting gene, the dominant allele $B$ directs synthesis of a blue pigment. Recessive homozygotes with the $bb$ genotype do not produce blue pigment. Budgies that produce both yellow and blue pigments have green feathers; those that produce only yellow pigment or only blue pigment have yellow or blue feathers, respectively; and budgies that produce neither pigment are white (albino).

a. List the genotypes for green, yellow, blue, and albino budgies.
b. A cross is made between a pure-breeding green budgie and a pure-breeding albino budgie. What are the genotypes of the parent birds?
c. What are the genotype(s) and phenotype(s) of the $F_1$ progeny of the cross described in part (b)?
d. If $F_1$ males and females are mated, what phenotypes are expected in the $F_2$, and in what proportions?
e. The cross of a green budgie and a yellow budgie produces offspring that are 12 green, 4 blue, 13 yellow, and 3 albino. What are the genotypes of the parents?

6. The ABO and MN blood groups are given below for four sets of parents (1 to 4) and four children (a to d). Recall that the ABO blood group has three alleles, $I^A$, $I^B$, and $i$. The MN blood group has two codominant alleles, $M$ and $N$. Using your knowledge of these genetic systems, match each child with every set of parents who might have conceived the child, and exclude any parental set that could not have conceived the child.

| | Mother | | Father | |
|---|---|---|---|---|
| | ABO | MN | ABO | MN |
| 1 | O | M | B | M |
| 2 | B | N | B | N |
| 3 | AB | MN | B | MN |
| 4 | A | N | B | MN |

| | Children | |
|---|---|---|
| | ABO | MN |
| a | B | M |
| b | O | M |
| c | AB | MN |
| d | B | N |

7. The wild-type color of horned beetles is black, although other colors are known. A black horned beetle from a pure-breeding strain is crossed to a pure-breeding green female beetle. All of their $F_1$ progeny are black. These $F_1$ are allowed to mate at random with one another, and 320 $F_2$ beetles are produced. The $F_2$ consists of 179 black, 81 green, and 60 brown. Use these data to explain the genetics of horned beetle color.

8. Two genes interact to produce various phenotypic ratios among $F_2$ progeny of a dihybrid cross. Design a different pathway explaining each of the $F_2$ ratios below, using hypothetical genes $R$ and $T$ and assuming that the dominant allele at each locus catalyzes a different reaction or performs an action leading to pigment production. The recessive allele at each locus is null (loss-of-function). Begin each pathway with a colorless precursor that produces a white or albino phenotype if it is unmodified. The ratios are for $F_2$ progeny produced by crossing wild-type $F_1$ organisms with the genotype $RrTt$.

a. $\frac{9}{16}$ dark blue : $\frac{6}{16}$ light blue : $\frac{1}{16}$ white
b. $\frac{12}{16}$ white : $\frac{3}{16}$ green : $\frac{1}{16}$ yellow
c. $\frac{9}{16}$ green : $\frac{3}{16}$ yellow : $\frac{3}{16}$ blue : $\frac{1}{16}$ white
d. $\frac{9}{16}$ red : $\frac{7}{16}$ white
e. $\frac{15}{16}$ black : $\frac{1}{16}$ white
f. $\frac{9}{16}$ black : $\frac{3}{16}$ gray : $\frac{4}{16}$ albino
g. $\frac{13}{16}$ white : $\frac{3}{16}$ green

9. The ABO blood group assorts independently of the Rhesus (Rh) blood group and the MN blood group. Three alleles, $I^A$, $I^B$, and $i$, occur at the $ABO$ locus. Two alleles, $R$, a dominant allele producing Rh+, and $r$, a recessive allele for Rh−, are found at the $Rh$ locus, and codominant alleles $M$ and $N$ occur at the $MN$ locus. Each gene is autosomal.

a. A child with blood types A, Rh−, and M is born to a woman who has blood types O, Rh−, and MN and a man who has blood types A, Rh+, and M. Determine the genotypes of each parent.
b. What proportion of children born to a man with genotype $I^A I^B$ $Rr$ $MN$ and a woman who is $I^A i$ $Rr$ $NN$ will have blood types B, Rh−, and MN? Show your work.
c. A man with blood types B, Rh+, and N says he could not be the father of a child with blood types O, Rh−, and MN. The mother of the child has blood types A, Rh+, and MN. Is the man correct? Explain.

10. In rats, gene B produces black coat color if the genotype is $B-$, but black pigment is not produced if the genotype is $bb$. At an independent locus, gene D produces yellow pigment if the genotype is $D-$, but no pigment is produced when the genotype is $dd$. Production of both pigments results in brown coat color. If neither pigment is produced, coat color is cream. Determine the genotypes of parents of litters with the following phenotype distributions.

a. 4 brown, 4 black, 4 yellow, 4 cream
b. 3 brown, 3 yellow, 1 black, 1 cream
c. 9 black, 7 brown

11. In the rats identified in problem 10, a third independently assorting gene involved in determination of coat color in rats is the $C$ gene. At this locus, the genotype $C-$ permits expression of pigment from genes $B$ and $D$. The $cc$ genotype, however, prevents expression of coat color and

results in albino rats. For each of the following crosses, determine the expected phenotype ratio of progeny.

a. $BbDDCc \times BbDdCc$
b. $BBDdcc \times BbddCc$
c. $bbDDCc \times BBddCc$
d. $BbDdCC \times BbDdCC$

12. Using the information provided in Problems 10 and 11, determine the genotype and phenotype of parents that produce the following progeny:

a. $\frac{9}{16}$ brown : $\frac{3}{16}$ black : $\frac{4}{16}$ albino
b. $\frac{3}{8}$ black : $\frac{3}{8}$ cream : $\frac{2}{8}$ albino
c. $\frac{27}{64}$ brown : $\frac{16}{64}$ albino : $\frac{9}{64}$ yellow : $\frac{9}{64}$ black : $\frac{3}{64}$ cream
d. $\frac{3}{4}$ brown : $\frac{1}{4}$ yellow

13. Total cholesterol in blood is reported as the number of milligrams (mg) of cholesterol per 100 milliliters (mL) of blood. The normal range is 180–220 mg/100 mL. A gene mutation altering the function of cell-surface cholesterol receptors restricts the ability of cells to collect cholesterol from blood and draw it into cells. This defect results in elevated blood cholesterol levels. Individuals who are heterozygous for a mutant allele and a wild-type allele have levels of 300–600 mg/100 mL, and those who are homozygous for the mutation have levels of 800–1000 mg/100 mL. Identify the genetic term that best describes the inheritance of this form of elevated cholesterol level, and justify your choice.

14. Flower color in snapdragons results from the amount of the pigment anthocyanin in the petals. Red flowers are produced by plants that have full anthocyanin produc-

tion, and ivory-colored flowers are produced by plants that lack the ability to produce anthocyanin. The allele *An1* has full activity in anthocyanin production, and the allele *An2* is a null allele. Dr. Ara B. Dopsis, a famous genetic researcher, crosses pure-breeding red snapdragons to pure-breeding ivory snapdragons and produces $F_1$ progeny plants that have pink flowers. He proposes that this outcome is the result of incomplete dominance, and he crosses the $F_1$ to test his hypothesis. What phenotypes does Dr. Dopsis predict will be found in the $F_2$, and in what proportions?

15. A plant line with reduced fertility comes to the attention of a plant breeder who observes that seed pods often contain a mixture of viable seeds that can be planted to produce new plants, and withered seeds that cannot be sprouted. The breeder examines numerous seed pods in the reduced fertility line and counts 622 viable seeds and 204 nonviable seeds.

a. What single-gene mechanism best explains the breeder's observation?
b. Propose an additional experiment to test the genetic mechanism you propose. If your hypothesis is correct, what experimental outcome do you predict?

16. In cattle, an autosomal mutation called *Dexter* produces calves with short stature and short limbs. Embryos that are homozygous for the *Dexter* mutation have severely stunted development and either spontaneously abort or are stillborn. What progeny phenotypes do you expect from the cross of two *Dexter* cows? What are the expected proportions of the expected phenotypes?

## Application and Integration

*For answers to selected even-numbered problems, see Appendix: Answers.*

17. The coat color in mink is controlled by two codominant alleles at a single locus. Red coat color is produced by the genotype $R_1R_1$, silver coat by the genotype $R_1R_2$, and platinum color by $R_2R_2$. White spotting of the coat is a recessive trait found with the genotype $ss$. Solid coat color is found with the $S-$ genotype.

a. What are the expected progeny phenotypes and proportions for the cross $SsR_1R_2 \times ssR_2R_2$?
b. If the cross $SsR_1R_2 \times SsR_1R_1$ is made, what are the progeny phenotypes, and in what proportions are they expected to occur?
c. Two crosses are made between mink. Cross 1 is the cross of a solid, silver mink to one that is solid, platinum. Cross 2 is between a spotted, silver mink and one that is solid, silver. The progeny are described in the table below. Use these data to determine the genotypes of the parents in each cross.

18. Strains of petunias come in four pure-breeding colors: white, blue, red, and purple. White petunias are produced when plants synthesize no flower pigment. Blue petunias and red petunias are produced when plants synthesize blue or red pigment only. Purple petunias are produced in plants that synthesize both red *and* blue pigment. The mixture of red and blue makes purple. Flower-color pigments are synthesized by gene action in two separate pigment-producing biochemical pathways. Pathway I contains gene *A* that produces an enzyme to catalyze conversion of a colorless pigment designated white$_1$ to blue pigment. In Pathway II, the enzymatic product of gene *B* converts the colorless pigment designated white$_2$ to red pigment. The two genes assort independently.

<pre>
                       gene A
Pathway I:    White₁ ──────→ Blue
                                      +      =      Purple
Pathway II:   White₂ ──────→ Red
                       gene B
</pre>

a. What are the possible genotype(s) for pure-breeding red petunias?
b. What are the possible genotype(s) for true-breeding blue petunias?

| Cross | Offspring | | | | | |
| | Spotted, platinum | Spotted, silver | Spotted, red | Solid, platinum | Solid, silver | Solid, red |
| --- | --- | --- | --- | --- | --- | --- |
| 1 | 2 | 3 | 0 | 6 | 5 | 0 |
| 2 | 3 | 7 | 2 | 4 | 5 | 3 |

c. True-breeding red petunias are crossed to pure-breeding blue petunias, and all the $F_1$ progeny have purple flowers. If the $F_1$ are allowed to self-fertilize and produce the $F_2$, what is the expected phenotypic distribution of the $F_2$ progeny? Show your work.

19. Feather color in parakeets is produced by the blending of pigments produced from two biosynthetic pathways shown below. Four independently assorting genes (*A, B, C,* and *D*) produce enzymes that catalyze separate steps of the pathways. For the questions below, use an uppercase letter to indicate a dominant allele producing full enzymatic activity and a lowercase letter to indicate a recessive allele producing no functional enzyme. Feather colors produced by mixing pigments are green (yellow + blue) and purple (red + blue). Red, yellow, and blue feathers result from production of one colored pigment, and white results from absence of pigment production.

$$
\begin{array}{lccc}
 & \text{Enzyme A} & \text{Enzyme B} & \\
\text{Pathway I:} & \text{Compound I} \longrightarrow & \text{Compound II} \longrightarrow & \text{Compound III} \\
 & \text{(colorless)} & \text{(red)} & \text{(yellow)}
\end{array}
$$

$$
\begin{array}{lccc}
 & \text{Enzyme C} & \text{Enzyme D} & \\
\text{Pathway II:} & \text{Compound X} \longrightarrow & \text{Compound Y} \longrightarrow & \text{Compound Z} \\
 & \text{(colorless)} & \text{(colorless)} & \text{(blue)}
\end{array}
$$

a. What is the genotype of a pure-breeding purple parakeet strain?

b. What is the genotype of a pure-breeding yellow strain of parakeet?

c. If a pure-breeding blue strain of parakeet (*aa BB CC DD*) is crossed to one that is pure-breeding purple, predict the genotype(s) and phenotype(s) of the $F_1$. Show your work.

d. If $F_1$ birds identified in part c are mated at random, what phenotypes do you expect in the $F_2$ generation? What are the ratios among phenotypes? Show your work.

20. Brachydactyly type D is a human autosomal dominant condition in which the thumbs are abnormally short and broad. In most cases, both thumbs are affected, but occasionally just one thumb is involved. The accompanying pedigree shows a family in which brachydactyly type D is segregating. Filled circles and squares represent females and males who have involvement of both thumbs. Half-filled symbols represent family members with just one thumb affected.

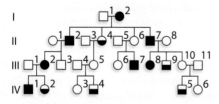

a. Is there any evidence of variable expressivity in this family? Explain.

b. Is there evidence of incomplete penetrance in this family? Explain.

21. A male and a female mouse are each from pure-breeding albino strains. They have a litter of 10 pups, all of which have normal pigmentation. The $F_1$ pups are crossed to one another to produce 56 $F_2$ mice, of which 31 are normally pigmented and 25 are albino.

a. Using clearly defined allele symbols of your own choosing, give the genotypes of parental and $F_1$ mice. What genetic phenomenon explains these parental and $F_1$ phenotypes?

b. What genetic phenomenon explains the $F_2$ results? Use your allelic symbols to explain the $F_2$ results.

22. Xeroderma pigmentosum (XP) is an autosomal recessive condition characterized by moderate to severe sensitivity to ultraviolet (UV) light. Patients develop multiple skin lesions on UV-exposed skin, and skin cancers often develop as a result. XP is caused by deficient repair of DNA damage from UV exposure.

a. Many genes are known to be involved in repair of UV-induced DNA damage, and several of these genes are implicated in XP. What genetic phenomenon is illustrated by XP?

b. A series of 10 skin-cell lines was grown from different XP patients. Cells from these lines were fused, and the heterokaryons were tested for genetic complementation by assaying their ability to repair DNA damage caused by a moderate amount of UV exposure. In the table below, + indicates that the fusion cell line performs normal DNA damage mutation repair, and 0 indicates defective DNA repair. Use this information to determine how many DNA-repair genes are mutated in the 10 cell lines, and identify which cell lines share the same mutated genes.

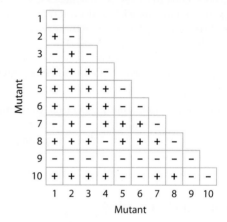

23. Three strains of green-seeded lentil plants appear to have the same phenotype. The strains are designated $G_1$, $G_2$, and $G_3$. Each green-seeded strain is crossed to a pure-breeding yellow-seeded strain designated Y. The $F_1$ of each cross are yellow; however, self-fertilization of $F_1$ plants produces $F_2$ with different proportions of yellow- and green-seeded plants as shown below.

| Parental Strain | | $F_1$ Phenotype | $F_2$ Phenotype | |
|---|---|---|---|---|
| Green | Yellow | | Green | Yellow |
| $G_1$ | Y | All yellow | $\frac{1}{4}$ | $\frac{3}{4}$ |
| $G_2$ | Y | All yellow | $\frac{7}{16}$ | $\frac{9}{16}$ |
| $G_3$ | Y | All yellow | $\frac{37}{64}$ | $\frac{27}{64}$ |

a. For what number of genes are variable alleles segregating in the $G_1 \times Y$ cross? The $G_2 \times Y$ cross? In the $G_3 \times Y$ cross? Explain your rationale for each answer.

b. Using the allele symbols *A* and *a*, *B* and *b*, and *D* and *d* to represent alleles at segregating genes, give the genotypes of parental and $F_1$ plants in each cross.

c. For each set of $F_2$ progeny, provide a genetic explanation for the yellow:green ratio. What are the genotypes of yellow and green $F_2$ lentil plants in the $G_2 \times Y$ cross?

d. If green-seeded strains $G_1$ and $G_3$ are crossed, what is the phenotype and the genotype of $F_1$ progeny?

e. What proportion of the $F_2$ are expected to be green? Show your work.

f. If strains $G_2$ and $G_3$ are crossed, what will be the phenotype of the $F_1$?

g. What proportion of the $F_2$ will have yellow seeds? Show your work.

24. Blue flower color is produced in a species of morning glories when dominant alleles are present at two gene loci, $A$ and $B$. (Plants with the genotype $A-B-$ have blue flowers.) Purple flowers result when a dominant allele is present at only one of the two gene loci, $A$ or $B$. (Plants with the genotypes $A-bb$ and $aaB-$ are purple.) Flowers are red when the plant is homozygous recessive for each gene (i.e., $aabb$).

a. Two pure-breeding purple strains are crossed, and all the $F_1$ plants have blue flowers. What are the genotypes of the parental plants?

b. If two $F_1$ plants are crossed, what are the expected phenotypes and frequencies in the $F_2$?

c. If an $F_1$ plant is backcrossed to one of the pure-breeding parental plants, what is the expected ratio of phenotypes among progeny? Why is the phenotype ratio the same regardless of which parental strain is selected for the backcross?

25. The following crosses are performed between morning glories whose flower color is determined as described in Problem 23. Use the segregation data to determine the genotype of each parental plant.

| Parental Phenotypes | Offspring Phenotypes |
|---|---|
| a. blue × blue | $\frac{3}{4}$ blue : $\frac{1}{4}$ purple |
| b. purple × purple | $\frac{1}{4}$ blue : $\frac{1}{2}$ purple : $\frac{1}{4}$ red |
| c. blue × red | $\frac{1}{4}$ blue : $\frac{1}{2}$ purple : $\frac{1}{4}$ red |
| d. purple × red | $\frac{1}{2}$ purple : $\frac{1}{2}$ red |
| e. blue × purple | $\frac{3}{8}$ blue : $\frac{1}{2}$ purple : $\frac{1}{8}$ red |

26. Two pure-breeding strains of summer squash producing yellow fruit, $Y_1$ and $Y_2$, are each crossed to a pure-breeding strain of summer squash producing green fruit, $G_1$, and to one another. The following results are obtained:

| Cross | P | $F_1$ | $F_2$ |
|---|---|---|---|
| I | $Y_1$ (yellow) × $G_1$ (green) | All yellow | $\frac{3}{4}$ yellow : $\frac{1}{4}$ green |
| II | $Y_2$ (yellow) × $G_1$ (green) | All green | $\frac{3}{4}$ green : $\frac{1}{4}$ yellow |
| III | $Y_1$ (yellow) × $Y_2$ (yellow) | All yellow | $\frac{13}{16}$ yellow : $\frac{3}{16}$ green |

a. Examine the results of each cross and predict how many genes are responsible for fruit-color determination in summer squash. Justify your answer.

b. Using clearly defined symbols of your choice, give the genotypes of parental, $F_1$, and $F_2$ plants in each cross.

c. If the $F_1$ of crosses I and II are mated, predict the phenotype ratio of the progeny.

27. Marfan syndrome is an autosomal dominant disorder in humans. It results from mutation of the gene on chromosome 15, which produces the connective tissue protein fibrillin. In its wild-type form, fibrillin gives connective tissues, such as cartilage, elasticity. When mutated, however, fibrillin is rigid and produces a range of phenotypic complications, including excessive growth of the long bones of the leg and arm, sunken chest, dislocation of the lens of the eye, and susceptibility to aortic aneurysm, which can lead to sudden death in some cases.

Different sets of symptoms are seen among various family members, as shown in the pedigree below. Each quadrant of the circles and squares represents a different symptom, as the key indicates.

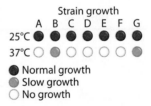

Long bones — Sunken chest
Lens dislocation — Aortic aneurysm

Since all cases of Marfan syndrome are caused by mutation of the fibrillin gene, and all family members with Marfan syndrome carry the same mutant allele, how do you explain the differences shown in the pedigree?

28. Yeast are single-celled eukaryotic organisms that grow in culture as either haploids or diploids. Diploid yeast are generated when two haploid strains fuse together. Seven haploid strains of yeast exhibit similar growth habit: at 25°C, each strain grows normally; but at 37°C, they show different growth capabilities. The table below displays the growth pattern.

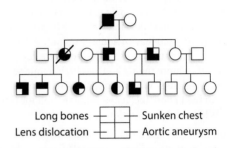

Strain growth
A B C D E F G
25°C
37°C
● Normal growth
● Slow growth
○ No growth

a. Describe the nature of the mutation affecting each of these mutant yeast strains. Explain why strains B and G display different growth habit at 37°C than the other strains.

b. Each of the mutant pairs of haploid yeast is fused, and the resulting diploids are tested for their ability to grow at 37°C. The results of the growth experiment are shown below.

**37°C growth data**

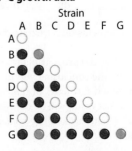

Strain
A B C D E F G
A
B
C
D
E
F
G

How many different genes are mutated among these seven yeast strains? Identify the strains that represent each gene mutation.

29. During your work as a laboratory assistant in the research facilities of Dr. O. Sophila, a world-famous geneticist, you come across an unusual bottle of fruit flies. All the flies in the bottle appear normal when they are in an incubator set at 22°C. When they are moved to a 30°C incubator, however, a few of the flies slowly become paralyzed; and after about 20 to 30 minutes, they are unable to move. Returning the flies to 22°C restores their ability to move after about 30 to 45 minutes.

    With Dr. Sophila's encouragement, you set up 10 individual crosses between single male and female flies that exhibit the unusual behavior. Among 812 progeny, 598 exhibit the unusual behavior and 214 do not. When you leave one of the test bottles in the 30°C incubator too long, you discover that more than 2 hours at high temperature kills the paralyzed flies. When you tell this to Dr. Sophila, he says, "Ah ha! I know the genetic explanation for this condition." What is his explanation?

30. Dr. Ara B. Dopsis and Dr. C. Ellie Gans are performing genetic crosses on daisy plants. They self-fertilize a blue-flowered daisy and grow 100 progeny plants that consist of 55 blue-flowered plants, 22 purple-flowered plants, and 23 white-flowered plants. Dr. Dopsis believes this is the result of segregation of two alleles at one locus and that the progeny ratio is 1:2:1. Dr. Gans thinks the progeny phenotypes are the result of two epistatic genes and that the ratio is 9:3:4.

    The two scientists ask you to resolve their conflict by performing chi-square analysis on the data for *both* proposed genetic mechanisms. For each proposed mechanism, fill in the values requested on the form the researchers have provided for your analysis.

    a. Use the form below to calculate chi square for the 1:2:1 hypothesis of Dr. Sophila.

| Phenotype | Observed | Expected |
|---|---|---|
| Blue | 55 | _____ |
| Purple | 22 | _____ |
| White | 23 | _____ |
| Chi-square value: _____ df: _____ $p$ value > _____ | | |

    b. Use the form below to calculate chi square for the 9:3:4 hypothesis of Dr. Gans.

| Phenotype | Observed | Expected |
|---|---|---|
| Blue | 55 | _____ |
| Purple | 22 | _____ |
| White | 23 | _____ |
| Chi-square value: _____ df: _____ $p$ value > _____ | | |

    c. What is your conclusion regarding these two genetic hypotheses?

    d. Using any of the 100 progeny plants, propose a cross that will verify the conclusion you proposed in part (c). Plants may be self-fertilized, or one plant can be crossed to another. What result will be consistent with the 1:2:1 hypothesis? What result will be consistent with the 9:3:4 hypothesis?

31. Human ABO blood type is determined by three alleles ($I^A$, $I^B$, and $i$) whose gene products modify the H antigen produced by protein activity of an independently assorting *H* gene. A rare abnormality known as the "Bombay phenotype" is the result of epistatic interaction between the gene for the ABO blood group and the *H* gene. Individuals with the Bombay phenotype appear to have blood type O based on the inability of both anti-A antibody and anti-B antibody to detect an antigen. The apparent blood type O in Bombay phenotype is due to the absence of H antigen as a result of homozygous recessive mutations of the *H* gene. Individuals with the Bombay phenotype have the *hh* genotype. Use the information above to make predictions about the outcome of the cross shown below.

$$I^A I^B \, Hh \times I^A I^B \, Hh$$

32. In rabbits, albinism is an autosomal recessive condition caused by the absence of the pigment melanin from skin and fur. Pigmentation is a dominant wild-type trait. Three pure-breeding strains of albino rabbits, identified as strains 1, 2, and 3, are crossed to one another. In the table below, $F_1$ and $F_2$ progeny are shown for each cross. Based on the available data, propose a genetic explanation for the results. As part of your answer, create genotypes for each albino strain using clearly defined symbols of your own choosing. Use your symbols to diagram each cross, giving the $F_1$ and $F_2$ genotypes.

| | Cross | $F_1$ Progeny | $F_2$ Progeny |
|---|---|---|---|
| Cross A | strain 1 × strain 2 | 56 albino | 192 albino |
| Cross B | strain 1 × strain 3 | 72 pigmented | 181 pigmented, 139 albino |
| Cross C | strain 2 × strain 3 | 34 pigmented | 89 pigmented, 72 albino |

33. Dr. O. Sophila, a close friend of Dr. Ara B. Dopsis, reviews the $F_2$ results Dr. Dopsis obtained in his experiment with iris plants described in Genetic Analysis 4.3. Dr. Sophila thinks the $F_2$ progeny demonstrate that a single gene with incomplete dominance has produced a 1:2:1 ratio. Dr. Dopsis insists his proposal of recessive epistasis producing a 9:4:3 ratio in the $F_2$ is correct. To test his proposal, Dr. Dopsis examines the $F_2$ data under the assumptions of the single-gene incomplete dominance model using chi-square analysis. Calculate and interpret this chi-square value. Can Dr. Dopsis reject the single-gene incomplete dominance model on the basis of this analysis? Explain why or why not.

# Genetic Linkage and Mapping in Eukaryotes

# 5

Thomas Hunt Morgan, Nobel laureate (1933), discovered sex-linked inheritance, identified genetic linkage, proposed crossing over between homologous chromosomes, and developed the concept of gene mapping by recombination analysis.

## ESSENTIAL IDEAS

- Genetic linkage occurs between genes that lie so close to one another on a chromosome that alleles are unable to assort independently.

- Genetic linkage produces significantly more progeny with parental phenotypes and significantly fewer progeny with nonparental phenotypes than are expected by chance.

- Crossing over between homologous chromosomes results in recombination of alleles on chromosomes in gametes.

- Geneticists use the frequency of recombination between genes to construct gene maps identifying the relative order of and distance between genes on chromosomes.

- Cytological evidence demonstrates that recombination results from crossing over between homologous chromosomes.

- Specialized statistical methods aid in mapping human genes.

- Genome evolution can be studied by examining genetic linkage.

- Mitotic crossover is rare and can result in the localized appearance of distinctive phenotypes.

I n 1933, Thomas Hunt Morgan won the Nobel Prize for Physiology or Medicine—partly for his work establishing the chromosome theory of heredity (Chapter 3), but also for his role in identifying and explaining *genetic linkage and recombination*, and their application to *genetic linkage mapping*, which we discuss in this chapter. Morgan, like all successful scientists, was assisted by dedicated colleagues who included many exceptional students and other scientists. Among them were Calvin Bridges, whose work we discussed in connection with the chromosome theory of heredity (Chapter 3), and Alfred Sturtevant, who as an undergraduate researcher in Morgan's laboratory became the first person to use genetic linkage data to assemble a genetic map. A number of less well-remembered researchers, including Morgan's wife Lilian, were also important members of the research enterprise.

The work of Morgan, his colleagues, and numerous others led to the validation of three foundational theories in genetics. First, the work validated the chromosome theory of heredity (see Chapter 3), and it expanded the theory by showing that each chromosome carries multiple genes in a specific order. Second, the research validated the concept of the gene as a physical entity that is an integral part of a chromosome, and it led to work that expanded understanding of gene structure and demonstrated that genes are composed of nucleotides between which recombination may occur. Third, the work validated evolutionary theory by confirming that closely related species have a similar number of chromosomes and a similar arrangement of genes on chromosomes. The work led to an expansion of evolutionary theory that showed that recombination provides a mechanism by which variation in chromosome number and the arrangement of genes on chromosomes can accrue as species diverge from a common ancestor.

The observations and analysis of genetic linkage, recombination, and genetic linkage mapping are the focus of this chapter. We touch on the connection between the investigation of gene mapping and chromosome evolution in this chapter as well, but postpone most of the discussion.

## 5.1 Linked Genes Do Not Assort Independently

Genes that are located on the same chromosome are called **syntenic genes.** When two syntenic genes are so close to one another that their alleles are unable to assort independently, the genes are said to be linked to one another. This **genetic linkage** produces a distinctive pattern of gamete genotypes that can be quantified and analyzed to map the locations of genes on chromosomes. The alleles of syntenic genes can be reshuffled by crossing over between homologous chromosomes to produce **recombinant chromosomes.** In studies of

linked genes, chromosomes that do not undergo crossing over to reshuffle the alleles under study are identified as **parental chromosomes,** or **nonrecombinant chromosomes.** This discovery, made more than a century ago, opened the door to the development of **genetic linkage mapping,** which plots the positions of genes on chromosomes. Over the last century, new methods for identifying and mapping genes have been added to the analytical arsenal of genetics, but the importance of genetic linkage and its mapping applications remains undiminished.

Mendelian genetic ratios such as 3:1 and 9:3:3:1 are the products of segregation and independent assortment of alleles of genes for which chance determines the probabilities of gamete genotypes and the results of gamete union. Even when these independently assorting genes are subject to epistatic interactions, the rules of probability describe the distribution of the contributing alleles and can be used to interpret the resulting ratios (see Section 4.3). Oftentimes, two genes that assort independently do so because they are located on different chromosomes.

Syntenic genes can also assort independently, if they are far apart on a chromosome. In this situation, crossing over occurs frequently enough between the genes to randomize the combinations of alleles produced during meiosis. Syntenic genes that are in close proximity to one another, however, do not assort independently but continue to reside on the same chromosome as it segregates from its homolog during cell division. The alleles of linked genes tend to segregate together rather than assorting independently.

The connection that causes alleles of linked genes to segregate together during meiosis can be broken by crossing over. Recall that homologous chromosomes synapse and form the synaptonemal complex in prophase I (see Figure 3.11). The recombination nodules, consisting of proteins and enzymes, that form part of this complex can generate crossing over by facilitating the breakage, exchange, and reunion of segments of homologous chromosomes. This recombination of chromosome segments reshuffles the alleles carried at linked genes, resulting in haploid gametes that contain different combinations of alleles of syntenic genes than were present in the diploid cell that began meiosis.

The following observations and conclusions about genetic linkage are essential to understanding the phenomenon. We discuss them in the following paragraphs and then expand on the same fundamental ideas throughout the remainder of the chapter.

1. Linked genes are always syntenic, and they are always located near one another on a chromosome. When syntenic genes are so far apart on the chromosome that crossing over between them generates independent assortment of the alleles, the genes are not linked.

2. Genetic linkage leads to the production of a significantly greater number of gametes containing chromosomes with parental combinations of alleles than would be expected under assumptions of independent assortment, and to a significantly smaller number of gametes containing chromosomes with alleles that are different from the parental combinations.

3. Crossing over is less likely to occur between linked genes that are close to one another than between genes that are farther apart on a chromosome. The frequency of crossing over is roughly proportionate to the distance between genes, a relationship that allows genes to be mapped.

## Indications of Genetic Linkage

Genetic linkage can be recognized by comparing the observed frequencies of gamete genotypes, or progeny phenotypes, with the frequencies expected under the assumptions of independent assortment. If genes are linked, parental gametes—also known as nonrecombinant gametes—that contain parental combinations of the alleles will be produced significantly more often than predicted by chance. The

excess parental gametes may also result in progeny in which *parental* phenotypes for the genes occur significantly more often than predicted by chance.

**Figure 5.1** illustrates the identification of genetic linkage by comparing the frequencies of gamete genotypes for two crosses, one illustrating independent assortment and the other illustrating genetic linkage. In Figure 5.1a, gene *A* and gene *B* are on different chromosomes, and alleles of the genes assort independently. The parental organisms are *AABB* and *aabb*, and their gametes *AB* and *ab* are the parental gametes. The $F_1$ progeny are dihybrid (*AaBb*), and independent assortment predicts these dihybrids will produce four genetically different gametes in a ratio of 1:1:1:1. Notice that the frequency of parental gametes (*AB* and *ab*) is 50%, and that the frequency of nonparental gametes (*Ab* and *aB*) is also 50%.

Figure 5.1b illustrates gamete–genotype production for syntenic genes *D* and *E* that are linked. The *DDEE* parent produces parental gametes that are *DE*, and the *ddee* parent produces *de* gametes. The dihybrid $F_1$ progeny are *DdEe*, carrying alleles *D* and *E* on one chromosome and *d* and *e* on the homolog. This arrangement of alleles can be written *DE/de*, with the slash ("/") separating the alleles carried on

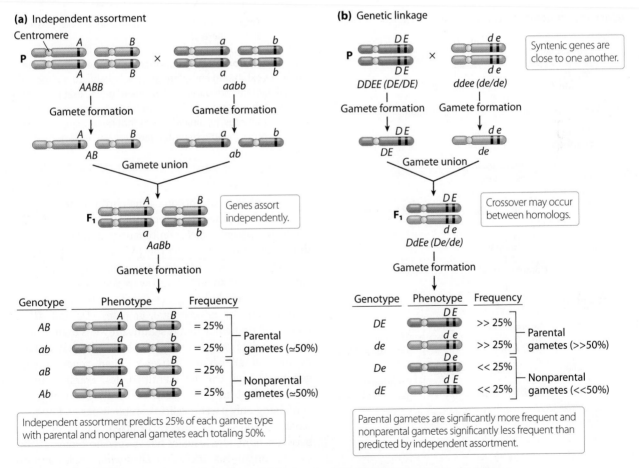

**Figure 5.1** **Independent assortment versus genetic linkage.** **(a)** For this dihybrid, four genetically different gametes are expected at 25% each when the genes assort independently. **(b)** When genes are linked, parental gametes are much more frequent than expected by chance and are more frequent than nonparental gametes.

**(a)** Complete genetic linkage (no crossover)

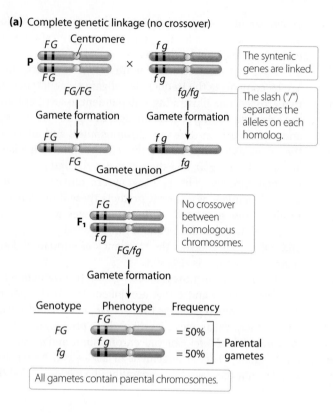

The syntenic genes are linked.

The slash ("/") separates the alleles on each homolog.

No crossover between homologous chromosomes.

| Genotype | Phenotype | Frequency | |
|---|---|---|---|
| FG | FG | = 50% | Parental gametes |
| fg | fg | = 50% | |

All gametes contain parental chromosomes.

**(c)** Incomplete genetic linkage (crossover in 40% of gametes)

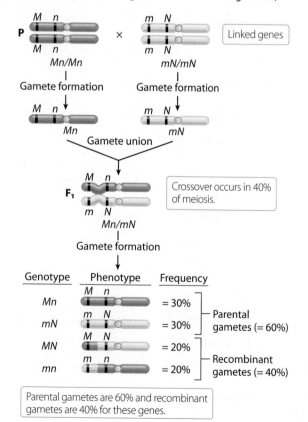

Linked genes

Crossover occurs in 40% of meiosis.

| Genotype | Phenotype | Frequency | |
|---|---|---|---|
| Mn | M n | = 30% | Parental gametes (= 60%) |
| mN | m N | = 30% | |
| MN | M N | = 20% | Recombinant gametes (= 40%) |
| mn | m n | = 20% | |

Parental gametes are 60% and recombinant gametes are 40% for these genes.

**(b)** Incomplete genetic linkage (crossover in 20% of gametes)

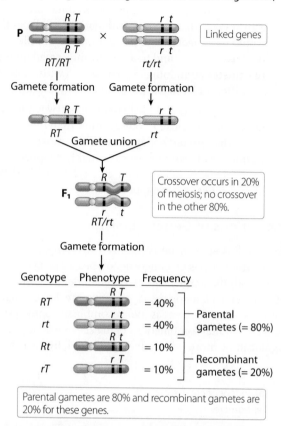

Linked genes

Crossover occurs in 20% of meiosis; no crossover in the other 80%.

| Genotype | Phenotype | Frequency | |
|---|---|---|---|
| RT | R T | = 40% | Parental gametes (= 80%) |
| rt | r t | = 40% | |
| Rt | R t | = 10% | Recombinant gametes (= 20%) |
| rT | r T | = 10% | |

Parental gametes are 80% and recombinant gametes are 20% for these genes.

**Figure 5.2 Complete versus incomplete genetic linkage.**
**(a)** Genes exhibiting complete genetic linkage do not recombine and all gametes are parental. **(b)** Linked genes that recombine in 20% of meioses produce 20% nonparental gametes and 80% parental gametes. **(c)** Linked genes that recombine in 40% of meioses produce 60% parental gametes and 40% nonparental gametes.

each member of the homologous chromosome pair. With genetic linkage, the rate of recombination among the alleles is low, and parental allele combinations usually stay together during meiosis, leading to the production of parental gametes (*DE* and *de*) at a combined frequency that is significantly greater than 50%. The low frequency of crossing over between closely linked genes results in the production of recombinant, or nonparental, gametes (*De* and *dE*) at a combined frequency that is significantly less than 50%.

Complete genetic linkage is observed when no recombination at all occurs between linked genes. Complete genetic linkage can be identified, for example, in cases where a dihybrid produces two equally frequent gametes containing only parental allele combinations and no recombinant gametes (**Figure 5.2a**). The absence of recombination between homologs usually has a specific biological basis. Certain organisms, including *Drosophila* males and other males in the insect order *Diptera* (of which *Drosophila* is a member), exhibit complete genetic linkage. There is no recombination between homologous chromosomes in these male flies. The biological basis of the absence of recombination in these organisms remains unknown.

Incomplete genetic linkage is far more common for linked genes. The resulting recombination between the homologs produces a mixture of parental and nonparental gametes. In the $F_1$ dihybrid shown in **Figure 5.2b**, recombination produces four genetically different gametes, of which two are parental and two are nonparental (recombinant). The two parental gametes each have approximately the same frequency, and their total is significantly greater than 50% of all gametes. In this example, the frequency of each parental gamete ($RT$ and $rt$) is 40%, and the total frequency of parental gametes is 80%. Recombinant gametes, which have nonparental combinations of alleles, are approximately equal in frequency to one another and constitute significantly less than 50% of all gametes. In this case, a total of 20% of gametes are recombinant: 10% of the gametes are $Rt$ and 10% are $rT$. Since the relative proportions of parental and recombinant gametes depend on the frequency of crossing over between linked genes, the proportions differ among pairs of linked genes. Note that the percentages of different gametes obtained for the cross in Figure 5.2c are different from those in Figure 5.2b.

The **recombination frequency**, expressed as the variable $r$, identifies the rate of recombination for a given pair of linked genes. The value of $r$ is expressed as

$$r = \frac{\text{number of recombinant gametes}}{\text{total number of meioses}}$$

Recombination frequency varies between different pairs of syntenic genes, depending roughly on the distance separating the genes on the chromosome. Comparing Figure 5.2b and Figure 5.2c, for example, we see that recombination frequency is 20% ($r = 0.20$) in Figure 5.2b and 40% ($r = 0.40$) in Figure 5.2c. The greater recombination frequency in Figure 5.2c compared to Figure 5.2b is most likely the consequence of a greater distance between genes $N$ and $M$ than between genes $T$ and $R$. The correlation between recombination frequency and gene distance can be expressed in two equivalent ways: (1) crossing over occurs at a higher rate between genes that are separated by a greater distance, and at a lower rate for genes that are closer together; and (2) linked genes with higher recombination frequencies are more distant from one another than linked genes with lower recombination frequencies. There are some caveats to this generalization, however, as we discuss in later sections.

---

**Genetic Insight**  When genes are linked, gametes carry parental combinations of the alleles significantly more often than predicted by chance and carry nonparental combinations significantly less often than expected. The recombination frequency ($r$) between linked genes expresses the proportion of recombinant (nonparental) gametes, and recombination frequencies increase along with the distance between genes.

## The Discovery of Genetic Linkage

William Bateson, an early champion of Mendelian genetics and Reginald Punnett, after whom the Punnett square is named, reported a series of experiments on sweet peas in 1905, 1906, and 1908. Those experiments opened a new chapter in genetics by drawing attention to genetic linkage. Bateson and Punnett studied the traits of flower color and the shape of pollen grains in sweet peas, first as independent traits and then together in the same plants.

When the traits were studied separately, the genes for flower color and pollen shape obeyed the rules of segregation—generating 3:1 phenotypic ratios among the $F_2$, for example. But Bateson and Punnett went on to study both traits in the same plants, intending to test the law of independent assortment. They crossed pure-breeding purple-flowered, long-pollen plants ($PPLL$) to pure-breeding red-flowered, round-pollen plants ($ppll$). As expected, the $F_1$ consisted exclusively of red-flowered, long-pollen plants, and these plants were crossed to obtain the $F_2$. But then, instead of the 9:3:3:1 ratio predicted by the independent assortment hypothesis, a far larger than expected portion of $F_2$ progeny showed parental combinations of phenotypes, and many fewer showed nonparental combinations (**Table 5.1**).

In the $F_2$, Bateson and Punnett observed that the two parental phenotypes—purple, long and red, round—were substantially in excess of expected frequencies, and that the two nonparental phenotypes—purple, round and red, long—were substantially less frequent than expected. This observation led Bateson and Punnett to suggest that the two combinations of alleles carried in the parents—$PL$ and $pl$—remained together very frequently when they were passed through gametes to subsequent generations by an unknown mechanism. Bateson and Punnett described these alleles as exhibiting "coupling." They described the appearance of new, nonparental phenotypes in the $F_2$ as indicating "repulsion" of the parental alleles, to produce nonparental phenotypes in progeny.

| Table 5.1 | Bateson and Punnett's Observed and Expected Phenotypes in $F_2$ Sweet Peas | | |
|---|---|---|---|
| **Phenotype** | **Genotype** | **Number of Progeny** | |
| | | Observed | Expected (9:3:3:1 ratio) |
| Purple, long | $P\text{–}L\text{–}$ | 4831 | $(6952)(9/16) = 3910.5$ |
| Purple, round | $P\text{–}ll$ | 390 | $(6952)(3/16) = 1303.5$ |
| Red, long | $ppL\text{–}$ | 393 | $(6952)(3/16) = 1303.5$ |
| Red, round | $ppll$ | 1338 | $(6952)(1/16) = 434.5$ |
| | | 6952 | 6952.0 |

**Figure 5.3  Morgan's analysis of genetic linkage of X-linked genes for eye color (*w*) and wing form (*m*).**  The number of test-cross progeny with each phenotype are compared to expected values that are determined assuming independent assortment of the genes.

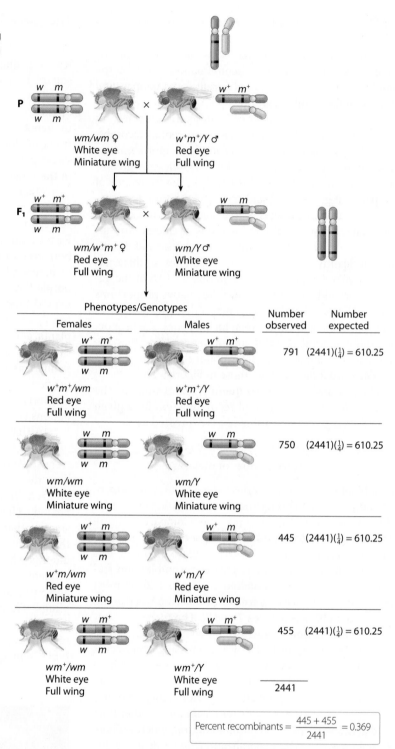

$$\text{Percent recombinants} = \frac{445 + 455}{2441} = 0.369$$

In 1911, Morgan performed the first of many crosses that confirmed and explained the observation of coupling and repulsion identified by Bateson and Punnett. Morgan had by this time identified several genes on the X chromosome of the fruit fly, including *w* (white eye) and *m* (miniature wing). **Figure 5.3** illustrates that Morgan crossed a female pure-breeding for white eyes and miniature wings (*wm/wm*) with hemizygous wild-type males displaying red eye and full wing (*w⁺m⁺/Y*). The slash, /, used in these genotypes separates alleles carried on ho-

mologous X chromosomes in the female or separates the X chromosome and Y chromosome in the male. The $F_1$ progeny were dihybrid wild-type females (*w⁺m⁺/wm*) and white, miniature (*wm/Y*) hemizygous males.

Morgan then produced an $F_2$ generation, predicting a 1:1:1:1 ratio based on the assumption of independent assortment of the genes. Instead, Morgan found substantial deviation from expectations. As in the Bateson and Punnett experiment, Morgan observed that parental phenotypes predominated (791 + 750 = 1541, or 63.1%) and

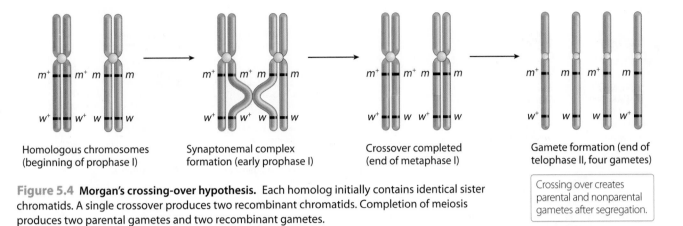

**Figure 5.4 Morgan's crossing-over hypothesis.** Each homolog initially contains identical sister chromatids. A single crossover produces two recombinant chromatids. Completion of meiosis produces two parental gametes and two recombinant gametes.

> Crossing over creates parental and nonparental gametes after segregation.

that fewer than the expected number of nonparental phenotypes were produced. The recombination frequency for this experiment is $r = 445 + 455/2441 = 0.0369$, or 36.9%. Notice that the two parental phenotypes are observed in an approximate 1:1 ratio (791:750), as are the nonparental phenotypes (455:445), as expected from segregation.

Based on this result, Morgan proposed that parental phenotypes are produced when the gametes of the $F_1$ female carry chromosomes with the same alleles as in the parents, in this case $w^+m^+$ and $wm$. Eggs containing parental alleles unite with sperm carrying $w$ and $m$ on the X chromosome or carrying the Y chromosome, and parental phenotypes (the same phenotypes as in the P generation flies) are produced. Conversely, nonparental phenotypes are the result of recombination between homologous X chromosomes during $F_1$ female meiosis (**Figure 5.4**). The production of recombinant chromosomes carrying either $w^+m$ or $wm^+$ required the physical rearrangement (recombination) of homologous X chromosomes. The union of eggs containing recombinant X chromosomes with sperm produced $F_2$ with nonparental phenotypes. Morgan confirmed this explanation through the examination of many other pairs of linked genes on the fruit fly X chromosome.

## Detecting Autosomal Genetic Linkage through Test-Cross Analysis

Turning his attention to autosomal genes, and employing 20/20 hindsight, Morgan realized that Bateson and Punnett had detected genetic linkage but were unable to explain it because, with respect to experimental design, *they had performed the wrong cross!* The $F_2$ progeny in the Bateson and Punnett experiment fell into four phenotypic classes, but three of those classes contained multiple genotypes, owing to the dominance relationships among the alleles (see Figure 2.11). Bateson and Punnett were unable to determine which alleles in the progeny derived from each $F_1$ parent because they had no way of ascertaining the high frequency of parental combinations of alleles and the low frequency of recombinants in $F_1$ gametes.

Morgan realized that the linkage of autosomal genes in *Drosophila* could be fully interpreted through the use of **two-point test-cross analysis** in which a dihybrid $F_1$ fly is crossed to a pure-breeding mate with the recessive phenotypes. The "two points" in these analyses are the two genes being tested. In two-point test-cross analysis, the homozygous recessive fly contributes only recessive alleles to test-cross progeny. In contrast, the dihybrid fly can contribute either a dominant allele of a gene, in which case the progeny display the dominant phenotype, or the recessive allele, thus producing the recessive form of the trait.

In one experiment, Morgan used test-cross analysis to examine genetic linkage of autosomal genes affecting eye color and wing shape. *Drosophila* eye color is red if an autosomal dominant allele $pr^+$ is present, whereas the recessive purple eye color is produced when the only allele present is $pr$. Full-sized wing is the product of an autosomal dominant allele $vg^+$, and its recessive counterpart, vestigial wing, is determined by the allele $vg$. Morgan crossed fruit flies that are pure-breeding for red eyes and full wing with pure-breeding purple-eyed, vestigial-winged flies (**Figure 5.5a**). The $F_1$ were uniformly red eyed and full winged ($pr^+ vg^+/pr vg$). Morgan then test-crossed dihybrid $F_1$ females to purple-eyed, vestigial-winged males ($pr vg/pr vg$). In this cross, males contributed only recessive alleles ($pr$ and $vg$), but females could produce any one of four gamete genotypes. The alleles of the female gamete thus controlled the phenotype of test-cross progeny. If the female contributed a dominant allele to progeny, the phenotype for that trait was dominant; and conversely, if the donated female allele was recessive, the phenotype was recessive. Test-cross progeny phenotypes corresponded directly to the alleles contributed by $F_1$ females, thus making it possible to unambiguously identify the allelic content of chromosomes in female gametes.

Under the assumption of independent assortment, dihybrid females should produce four equally frequent gametes, and test-cross progeny are expected to have four phenotypes distributed in a 1:1:1:1 ratio (see Figure 2.13). With genetic linkage however, parental combinations of alleles occur preferentially in gametes, producing

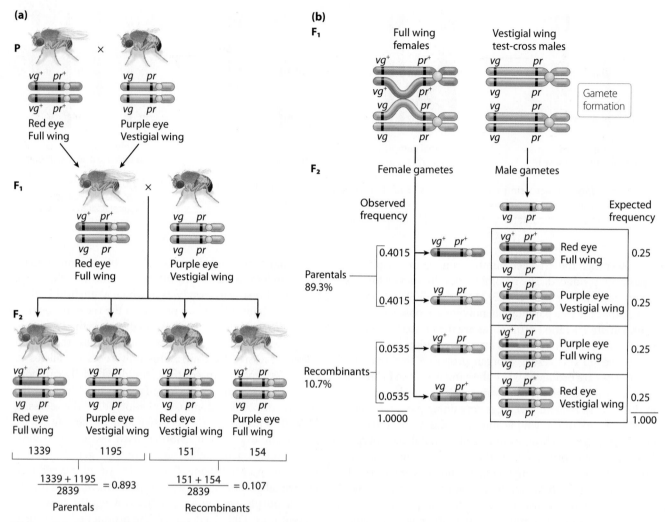

**Figure 5.5  Morgan's test-cross analysis of genetic linkage between autosomal genes.** (a) Dihybrid $F_1$ females ($pr^+vg^+/pr\,vg$) are test-crossed to males homozygous for recessive mutant purple eye color and vestigial wing ($pr\,vg/pr\,vg$), permitting identification of progeny as carrying either a parental or a recombinant chromosome. (b) Single crossover during female meiosis leads to parental and recombinant gametes at frequencies specified by recombination or by chance, and gamete union produces test-cross progeny.

test-cross progeny with a significant excess of parental phenotypes and a significant deficit of nonparental phenotypes.

Morgan's test-cross progeny displayed the four expected phenotypes, but in numbers that deviated dramatically from expected Mendelian proportions. Among test-cross progeny, 89.3% were parental, and just 10.7% were recombinant. The nonrecombinant progeny classes were found in approximately a 1:1 ratio (1339:1195), as were the recombinant classes (154:151); thus, the two parental chromosomes were transmitted equally frequently, as were the two recombinant chromosomes. **Figure 5.5b** shows that among the 89.3% of parental female gametes, one-half, or 44.65%, are predicted to be of each parental type. Similarly, among the 10.7% of gametes that are recombinant, each recombinant type is predicted with a frequency of 5.35%.

In the years immediately following Morgan's explanation of genetic linkage, other biologists, working on plant species and animal species, used test-cross analysis to verify Morgan's hypothesis. The collective results of these experimental observations can be summarized as follows:

1. Genetic linkage is a physical relationship between genes that are located near one another on a chromosome.

2. Recombination occurs between linked genes on homologous chromosomes in significantly less than 50% of meiotic divisions. Significantly more than 50% of gametes contain parental combinations of alleles.

3. The recombination frequency varies among linked genes and is roughly proportionate to the distance between genes on a chromosome.

**Genetic Analysis 5.1** takes you through the identification of parental and recombinant progeny and the determination of recombination frequency.

In tomato plants (*Lycopersicon esculentum*), red fruit color (*T–*) is dominant to tangerine color (*tt*), and smooth leaf (*H–*) is dominant to hairy leaf (*hh*). Both genes are located on chromosome 7, and they have a recombination frequency of 20%. A pure-breeding plant producing tangerine-colored fruit and smooth leaves is crossed to a pure-breeding red-fruited, hairy-leaved plant. The $F_1$ are test crossed to a pure-breeding tangerine-fruited, hairy plant. What are the expected genotypes, phenotypes, and phenotype proportions among test-cross progeny?

| Solution Strategies | Solution Steps |
|---|---|
| **Evaluate** | |
| 1. Identify the topic of this problem and state the nature of the required answer. | 1. This problem concerns the prediction of test-cross progeny for linked genes. The answer requires that the expected frequency of each possible category of test-cross progeny be predicted from the information given about recombination frequency between the genes. |
| 2. Identify the critical information given in the problem. | 2. Dominant and recessive phenotypes, the genotypes of two pure-breeding parental plants, and the recombination frequency between genes controlling two traits are given in the problem. |
| **Deduce** | |
| 3. Identify the genotypes of parental plants and their gametes. | 3. Each parent is pure-breeding for a dominant and a recessive trait: <br> Tangerine, smooth = *ttHH* <br> Red, hairy = *TThh* <br> Parental gametes = all *tH* from one parent and all *Th* from the other |
| 4. Identify the genotype and phenotype of $F_1$ plants, and determine the parental arrangements of alleles. | 4. $F_1$ are dihybrid (*tH/Th*) and have the two dominant phenotypes (red and smooth). The pure-breeding parents have contributed chromosomes carrying *tH* and *Th*. |
| **Solve** | |
| 5. Determine the number and frequency of $F_1$ gametes, given the recombination frequency of 20%. | 5. Four genetically different gametes are possible: *tH, Th, TH,* and *th*. Among these gametes, 20% will be recombinants and 80% parentals (100% − 20% = 80%). Chance predicts that the two parental gametes (*tH* and *Th*) are produced at equal frequency. Likewise, the two recombinant gametes (*TH* and *th*) are produced at equal frequency. The expected gamete frequencies are <br><br> Parentals: *tH* = (0.80)(1/2) = 0.40 <br> *Th* = (0.80)(1/2) = 0.40 <br> Recombinants: *TH* = (0.20)(1/2) = 0.10 <br> *th* = (0.20)(1/2) = 0.10 |

**TIP:** With genetic linkage, parental combinations of alleles are significantly greater than 50% of the gametes.

| | |
|---|---|
| 6. Determine the expected outcome of the test cross. | 6. Test-cross progeny are expected to be 40% each tangerine, smooth and red, hairy; and 20% each red, smooth and tangerine, hairy. |

|  |  |  | *th* | (1.0) | Test-cross progeny |
|---|---|---|---|---|---|
| Parental | 0.40 | *tH* | *tH/th* | 0.40 | Tangerine, smooth — 40% |
|  | 0.40 | *Th* | *Th/th* | 0.40 | Red, hairy — 40% |
| Recombinant | 0.10 | *TH* | *TH/th* | 0.10 | Red, smooth — 10% |
|  | 0.10 | *th* | *th/th* | 0.10 | Tangerine, hairy — 10% |

For more practice, see Problems 5, 6, and 12.

## 5.2 Genetic Linkage Mapping Is Based on Recombination Frequency between Genes

An important outcome of Morgan's studies of linked genes in *Drosophila* was his recognition that more parental than recombinant progeny occurred, and that the proportion of recombinants varied considerably from one pair of linked genes to another. Morgan summarized this idea in 1911, stating, "The proportions that result are not so much the expression of a numerical system as of the relative location of the factors (genes) in the chromosome." Morgan was saying that chance (independent assortment) was not the sole factor determining the relative proportions of gametes produced by an organism, because the close proximity of linked genes on a chromosome led to a much higher proportion of parental gametes and a much lower proportion of nonparental gametes than were expected by chance. His intuition was correct. In this section, we examine methods for constructing genetic maps from recombination data for two linked genes, and in the next section, we'll move on to consider the mapping of three linked genes.

### The First Genetic Linkage Map

In the context of early 20th-century biology, Morgan's idea that genes were on chromosomes was not novel. For example, Sutton, Boveri, and others had noted the parallel between hereditary transmission and chromosome division. But biologists at the time did not know either the structure of genes or how they were encoded on chromosomes (see Section 3.4). Morgan was the first to demonstrate that genes are on chromosomes, however, and his proposal that the recombination frequency for a linked pair of genes might correspond to the *distance* between those genes on a chromosome was a novel idea.

Morgan viewed genes as inhabiting fixed locations on chromosomes. Like cities along a road, the order of genes could be determined and their distances could be quantified. If his hypothesis were correct, he reasoned, then recombination frequencies could be used to produce a genetic linkage map depicting gene order along a chromosome and to calculate a quantitative index of linear distances between genes. As Morgan discussed his ideas about recombination frequency and gene distances, Alfred Sturtevant, then an undergraduate student working in Morgan's laboratory, had an epiphany. In a 1965 book, Sturtevant recalled the moment:

> In the latter part of 1911, in a conversation with Morgan, I suddenly realized that the variations in strength of linkage, already attributed by Morgan to differences in the spatial separation of genes, offered the possibility of determining sequences in the linear dimension of a chromosome. I went home and spent most of the night (to the neglect of my

| Gene Pairs | Recombination Frequency |
|---|---|
| Yellow (*y*) and white (*w*) | 214/21,736 = 0.010 |
| Yellow (*y*) and vermilion (*v*) | 1464/4551 = 0.322 |
| Vermilion (*v*) and white (*w*) | 471/1584 = 0.297 |
| Vermilion (*v*) and miniature (*m*) | 17/573 = 0.030 |
| Miniature (*m*) and white (*w*) | 2062/6116 = 0.337 |
| White (*w*) and rudimentary (*r*) | 406/898 = 0.452 |
| Rudimentary (*r*) and vermilion (*v*) | 109/405 = 0.269 |

**Table 5.2  Sturtevant's Recombination Data for Five X-Linked Genes in *Drosophila***

other undergraduate homework) in producing the first chromosome map.

Sturtevant used the results of numerous two-point test-cross experiments on five X-linked genes in *Drosophila* to create the first genetic linkage map. He based his map-building approach on the idea that smaller recombination frequencies indicated genes residing closer to each other on the chromosome, and larger recombination frequencies indicated greater distances between genes on the chromosome. To construct his genetic map, Sturtevant used the data in Table 5.2. His finished recombination map is illustrated in Figure 5.6. In the century since Sturtevant first compiled his map, millions of progeny fruit flies have been analyzed for X-chromosome recombination. The accumulated data have led to slight modifications in Sturtevant's estimated recombination frequencies but have not necessitated any changes in gene order. Sturtevant assembled his map using logic of the kind demonstrated in the following four steps:

1. Of the genes tested, the pair with the smallest recombination frequency, and therefore in closest proximity, are the gene producing white eye (*w*) and the gene carrying yellow (*y*) body. With their recombination frequency of just 1%, they must be at almost the same spot on the chromosome.

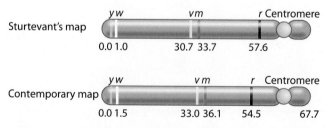

**Figure 5.6  The first linkage map.** The original *Drosophila* X chromosome map of five genes assembled by Alfred Sturtevant (top) and the contemporary X-chromosome map for *Drosophila* based on current data (bottom). Sturtevant's map is based in part on the recombination frequencies given in Table 5.2.

2. Vermilion (*v*) is more distant from yellow (32.2% recombination) than it is from white (29.7%), suggesting the order *y–w–v*.

3. Miniature (*m*) is close to vermilion (3% recombination) but is more distant from white (33.7% recombination) than is vermilion. Adding vermilion to the gene map produces the order *y–w–v–m*.

4. Rudimentary (*r*) is very distant from white (45.2% recombination) and also fairly distant from vermilion (26.9% recombination). This information places rudimentary on the opposite side of the map from white, yielding the final map *y–w–v–m–r*.

## Map Units

As we examine our map of the *Drosophila* X chromosome (Figure 5.6), the correlation between recombination frequency and physical distance on chromosomes becomes easier to understand. The recombination frequencies between genes on a chromosome can even be converted into units of physical distance, using the concept of a **map unit (m.u.)**. A map unit is also known as a **centiMorgan (cM)** in honor of Thomas Hunt Morgan's contribution to recombination mapping. It is common (at least in introductory genetics courses) to use the equivalency

1% recombination = 1 m.u. or 1 cM of distance between linked genes

This is an approximation, and not a very good one for certain regions of particular genomes, as we discuss in a later section. Despite its shortcomings, however, it is accurate enough for our instructional purposes in this textbook.

> **Genetic Insight** Genetic linkage maps give the relative positions of genes along a chromosome. Assuming recombination frequency is proportionate to the distance between genes, one map unit (or 1 cM) equals approximately 1% recombination.

## Chi-Square Analysis of Genetic Linkage Data

In our discussion of genetic linkage data, we have noted that when genes are linked, *significantly* more parental phenotypes than recombinant phenotypes are found among progeny. But how can we tell whether the observed data constitute evidence of genetic linkage rather than a simple case of chance variation from expected values? The question is settled by the use of chi-square analysis of observed and expected values to identify statistically significant differences. (Section 2.6 describes the chi-square test and demonstrates the calculation and interpretation of chi-square *p*, or probability, values.)

As an example, let's revisit the data obtained by Morgan on the *w* gene affecting eye color and the *m* gene controlling wing form in *Drosophila*, presented in Figure 5.3. The cross of $F_1$ dihybrid females ($wm/w^+m^+$) to white-eyed, miniature-winged males ($wm/Y$) produces an $F_2$ generation that would have been expected to display a 1:1:1:1 phenotypic ratio. This ratio is based on the assumption that independent assortment determines the alleles contained in female gametes. Using the observed and expected values, we calculate the chi-square value as follows:

$$\chi^2 = \frac{(791 - 610.25)^2}{610.25} + \frac{(750 - 610.25)^2}{610.25} + \frac{(445 - 610.25)^2}{610.25} + \frac{(455 - 610.25)^2}{610.25} = 169.79$$

There are 3 degrees of freedom (df = 3) in this problem, and the corresponding *p* value is $p < 0.005$ (see Table 2.4, page 49). This result indicates a significant deviation of observed from expected results, suggesting that chance is not responsible for the observed distribution. Combined with the observation that the two phenotypes that exceed the expected number are parental, these data are consistent with the presence of genetic linkage between the genes.

# 5.3 Three-Point Test-Cross Analysis Maps Genes

Two-point test-cross analysis is an effective way to calculate the distance between two linked genes, but it is not the most effective way to build genetic maps containing multiple genes. However, by expanding the idea of test-cross analysis to **three-point test-cross analysis,** geneticists can efficiently map three linked genes simultaneously.

## Finding the Relative Order of Genes by Three-Point Mapping

Let's consider a three-point test cross between a trihybrid organism ($a^+ab^+bc^+c$) and an organism that is homozygous recessive for the three traits (*aabbcc*). The configuration of alleles in the trihybrid does not have to be known at the start, since the three-point analysis will deduce the configuration of alleles on parental chromosomes as part of the process.

Incomplete genetic linkage of three genes in a trihybrid produces eight genetically different gamete genotypes. This is the same number of genetically different gametes expected if we assume independent assortment; but, unlike the expectations for independent assortment, the gamete frequencies are unequal if the genes are linked. Among the eight gamete genotypes are two parental genotypes that are significantly more frequent than expected by chance as well as six recombinant genotypes, each detected less often than expected. Assuming, for the purposes of this example, that the three linked genes are in the order *a–b–c*, we can identify parental and recombinant gametes by the relative frequencies of the corresponding test-cross progeny classes.

**(a) Test cross 1**

**(b) Test cross 2**

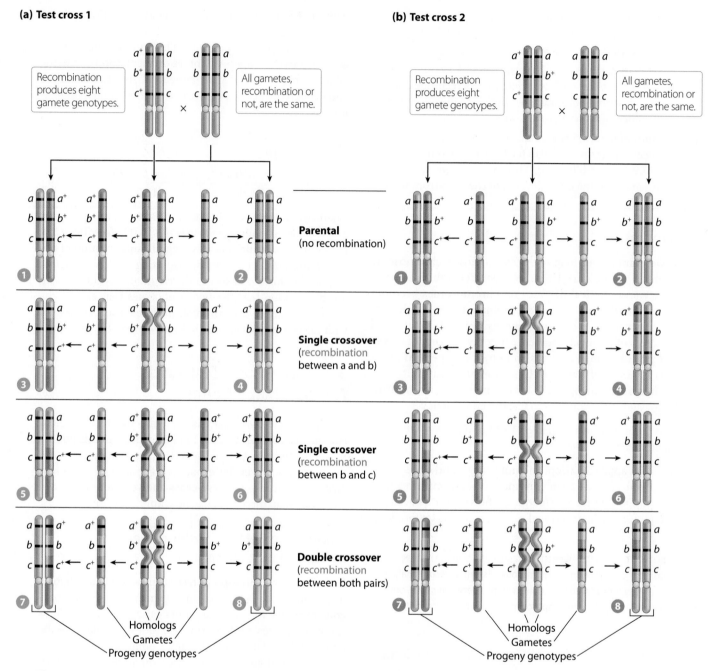

**Figure 5.7   Three-point test crosses for different allele configurations in a trihybrid parent crossed to a triple recessive parent.   (a)** In test cross 1, parental chromosomes carry the three wild-type and the three recessive alleles. Gametes with these alleles are parental and produce progeny with parental phenotypes. Single and double recombinant gametes lead to test-cross progeny displaying recombination. **(b)** A different configuration of alleles on parental chromosomes of the trihybrid in test cross 2 produces parental and recombinant progeny that are different from those in test cross 1.

Imagine that Test cross 1 mates a trihybrid organism with the genotype $a^+b^+c^+/abc$ to one that is $abc/abc$ (**Figure 5.7a**). Test cross 2 shows an alternative arrangement of alleles on parental chromosomes, mating the trihybrid $a^+bc^+/ab^+c$ to an organism with genotype $abc/abc$ (**Figure 5.7b**). In Test cross 1, parental gametes ($a^+b^+c^+$ and $abc$) are produced when no crossovers occur between the genes, and the resulting progeny have either the three wild-type or three

recessive phenotypes. A single crossover occurring between genes $a$ and $b$ produces two recombinant gametes, $a^+bc$ and $ab^+c^+$, and progeny with the corresponding patterns of phenotypes. Likewise, single crossover between genes $b$ and $c$ also produces two recombinant gametes, $a^+b^+c$ and $abc^+$, and corresponding progeny. A double-crossover event that causes crossing over both between $a$ and $b$ and between $b$ and $c$ will produce a pair of double-crossover gametes, $a^+bc^+$

and $ab^+c$, and progeny with the corresponding mixtures of wild-type and recessive traits.

Test cross 2 produces the same eight gamete genotypes obtained from test cross 1, but the alleles start out arranged differently on the parental chromosomes. Thus, the parental and recombinant gamete genotypes in this test cross are different from those in the first test cross. In this test cross, the parental gametes are $a^+bc^+$ and $ab^+c$. The single-crossover gametes are $a^+b^+c$ and $abc^+$ for crossover between genes $a$ and $b$. Single crossover between genes $b$ and $c$ produces gametes $a^+bc$ and $ab^+c^+$. A double-crossover causing recombination between each pair of genes produces double-crossover gametes $a^+b^+c^+$ and $abc$.

As expected when genes are linked, each of the six recombinant gametes is observed at a frequency that is significantly less than predicted by chance. Single-crossover gametes form at frequencies determined by the relative distances between gene pairs. Within each single-crossover class, the two gametes will be equally frequent. Double-crossover gametes will be the least frequent class because *both* crossover events must occur. As within each single-crossover class, the two kinds of double-crossover gametes are produced at equal frequency.

---

Genetic Insight    Progeny of a test cross of three linked genes fall into eight phenotypic classes. Parental classes are most common and more frequent than would be predicted by chance. All recombinant classes are less frequent than would be predicted by chance. Double recombinants are the least frequent classes, and numbers of single recombinants occur proportionately to the distance between genes.

---

## Constructing a Three-Point Recombination Map

To illustrate the use of three-point test-cross data for constructing a genetic map, we will now analyze the data from a 1935 study by Rollins Emerson of genetic linkage in maize (*Zea mays*). Emerson tested three genes: the gene producing the phenotypes green seedling ($V-$) and yellow seedling ($vv$), the gene producing rough leaf ($Gl-$) and glossy leaf ($gl\ gl$), and the gene for normal fertility ($Va-$) and variable fertility ($va\ va$).

Maize was an important genetic experimental organism in the first half of the 20th century because of the large number of variable genetic traits it possesses, the ease with which large numbers of plants can be grown in a single season, the ability of researchers to control matings in a manner similar to Mendel's, and the production of large numbers of seeds from each cross. On an ear of corn, each kernel is a seed produced by the union of gametes; thus, a single ear can carry hundreds of progeny seeds, each the product of independent fertilization, and a small number of plants can yield tens of thousands of progeny seeds for analysis.

Emerson crossed pure-breeding wild-type plants with the dominant phenotypes green seedling, rough leaves, and normal fertility (*V Gl Va/V Gl Va*) to pure-breeding plants with the recessive phenotypes yellow seedling, glossy leaves, and variable fertility (*v gl va/v gl va*). The cross produced $F_1$ trihybrid plants with the dominant phenotypes and the genotype *V Gl Va/v gl va* that carries three dominant alleles on one chromosome and three recessive alleles on the homolog. The $F_1$ were then test-crossed to pure-breeding yellow, glossy, variable plants (*v gl va/v gl va*). The test-cross progeny are shown in Table 5.3. To create a genetic map that

| Table 5.3 | Emerson's Three-Point Test-Cross Analysis | | |
|---|---|---|---|
| **Parental cross:** | *V Gl Va/V Gl Va*<br>Green, rough, normal | × | *v gl va/v gl va*<br>yellow, glossy variable |
| **Test cross:** | *V Gl Va/v gl va*<br>Green, rough, normal | × | *v gl va/v gl va*<br>yellow, glossy, variable |

**Test-cross progeny:**

| Phenotype | Number Observed | Number Expected | Genotype (♀ gamete/♂ gamete) |
|---|---|---|---|
| 1. Yellow, rough, normal | 60 | 90.75 | *v Gl Va/v gl va* |
| 2. Yellow, glossy, normal | 48 | 90.75 | *v gl Va/v gl va* |
| 3. Yellow, rough, variable | 4 | 90.75 | *v Gl va/v gl va* |
| 4. Yellow, glossy, variable | 270 | 90.75 | *v gl va/v gl va* |
| 5. Green, rough, normal | 235 | 90.75 | *V Gl Va/v gl va* |
| 6. Green glossy, normal | 7 | 90.75 | *V gl Va/v gl va* |
| 7. Green, rough, variable | 40 | 90.75 | *V Gl va/v gl va* |
| 8. Green, glossy, variable | 62 | 90.75 | *V gl va/v gl va* |
| | 726 | 726 | |

places the three genes in correct relative order and to calculate recombination frequencies between gene pairs, we ask and answer five questions about these data:

1. Are the data consistent with the proposal of genetic linkage?
2. What are the alleles on parental chromosomes?
3. What is the gene order on the chromosome?
4. What are the recombination frequencies of the gene pairs?
5. Is the frequency of double crossovers consistent with independence of the single crossovers?

### Question 1: Are the Data Consistent with the Proposal of Genetic Linkage?

Under the assumptions of independent assortment, trihybrid plants produce eight genetically different gametes at a frequency of 0.125, or 1/8, each, and test-cross progeny are expected in eight equally frequent phenotypic classes. In this experiment, with 726 test-cross progeny, the expected number of progeny in each class would be $(726)(0.125) = 90.75$. Chi-square analysis comparing observed and expected numbers of progeny in each class yields a chi-square value in excess of 800. There are $(8 - 1) = 7$ degrees of freedom, and the corresponding $p$ value is $p < 0.005$. From this result, we conclude that the observed distribution of test-cross progeny deviates significantly from expectation, and we reject the independent assortment hypothesis as the explanation of these data.

If the deviation in this experiment is due to genetic linkage, then we would expect the numbers of progeny having parental phenotypes to be excessively high. Comparing the observed and expected values in each test-cross class shows that only two phenotype classes exceed expected numbers: the green, rough, normal class and the yellow, glossy, variable class. These are the two parental phenotypes. From this analysis, we conclude that the data are consistent with genetic linkage: the distribution of test-cross progeny deviates significantly from expectation, and only parental phenotypes are seen more often than expected by chance.

### Question 2: What Are the Alleles on Parental Chromosomes?

We can answer this question in two ways. The simpler approach is to use the phenotype information available about pure-breeding parental plants in the cross. The parent plants were pure-breeding dominant and pure-breeding recessive. From this information, we know that trihybrid $F_1$ plants have the dominant alleles on one chromosome and the recessive alleles on the homologous chromosome. The genetic structure of the test cross is *V Gl Va/v gl va × v gl va/v gl va*, and so the alleles on parental chromosomes must be *V Gl Va* and *v gl va*. Test-cross progeny Classes 4 and 5 in Table 5.3 are parentals.

The second approach is necessary when we do not know the phenotypes of parents or when the alleles on each chromosome are not known. In this approach, test-cross data are used to determine parental chromosomes.

The data in Table 5.3 indicate that the test-cross progeny in Class 5—green, rough, normal (*V Gl Va/v gl va*)—and in Class 4—yellow, glossy, variable (*v gl va/v gl va*)—exceed expected frequency and are therefore the parental classes. Both approaches tell us the same story: The parental chromosomes carry alleles *V Gl Va* and *v gl va*.

### Question 3: What Is the Gene Order on the Chromosome?

With parental chromosomes identified, the six remaining classes must be recombinants: four are single-crossover classes, and two are double crossovers. Double-crossover progeny will be the least frequent of all classes, because *both* crossover events must occur simultaneously to produce **double recombinants,** or **double crossovers.** From progeny numbers, we may presume that the smallest classes, Class 3—yellow, rough, variable—and Class 6—green, glossy, normal—are the probable double recombinants. We can use these predictions to test possible gene orders on parental chromosomes.

For these three genes there are only three possible gene orders: (1) *va–v–gl*, (2) *v–va–gl*, or (3) *va–gl–v*. There are no data to assist us in determining the left-to-right orientation of the chromosome, so the difference between these gene orders is defined entirely by which gene is in the *middle—v, va,* or *gl*—and which two genes flank the middle gene. Each gene order could be written in the opposite direction, since each is a *relative* order of the three genes. For example, *va–v–gl* and *gl–v–va* are equivalent gene orders because each has *v* as the middle gene.

There are two ways to determine the gene order. One procedure is to list each gene order possible for the parental chromosomes, draw the corresponding double crossover chromosomes, and then determine whether the double crossover gametes produced by this activity match the predicted double crossover progeny. If a match is not seen, the gene order is incorrect, but if a match is found, the correct gene order has been identified.

1. Possible gene order *va–v–gl*

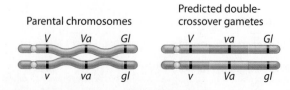

Result: Double-crossover gametes obtained from this gene order are not those predicted from the data.

Conclusion: The proposed gene order is incorrect; *v* is not the middle gene.

2. Possible gene order *v–va–gl*

Result: Double-crossover gametes obtained from this gene order are not those predicted from the data.

Conclusion: The proposed gene order is incorrect; *va* is not the middle gene.

3. Possible gene order *va–gl–v*

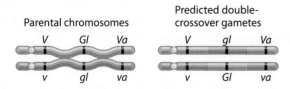

Result: Double-crossover gametes obtained from this gene order match those predicted from the data.

Conclusion: This proposed gene order is correct: *gl* is the middle gene, and the gene order may be written as either *v–g–va* or *va–gl–v*. This analysis confirms that test-cross progeny Classes 3 and 6 are double-crossover progeny.

The second method for determining gene order is a shortcut approach that requires some familiarity with recombination. Looking back at Figure 5.7, note that if we compare parental and double-crossover chromosomes, the alleles of the outside genes appear to remain the same while the middle allele appears to switch. In other words, when we compare one parental chromosome with one double-recombinant chromosome, two alleles match and one does not. The odd one out is the allele in the middle. If a trihybrid parent has alleles arranged as $a^+b^+c^+/abc$, then double crossover produces gametes that are $a^+bc^+/ab^+c$. Parental alleles $a^+$ and $c^+$ match one double recombinant, and alleles $b$ and $b^+$ are switched. Similarly, the second parental gamete has alleles $a$ and $c$ that match the other double recombinant. Alleles of the middle gene, $b$ and $b^+$, have switched in the double recombinant compared to the parental chromosome.

Remember, we have already identified the parental and double-crossover phenotypic groups by their numbers. We now look at the double crossovers to see which two alleles match parental phenotypes and to see which allele changes and is therefore the middle gene. In our data set, double-recombinant chromosomes are *V gl Va* and *v Gl va*. In this case, alleles of the *gl* gene have switched, indicating that *gl* is the middle gene. Based on this approach, parental chromosomes are *V Gl Va* and *v gl va*.

**Question 4: What Are the Recombination Frequencies of the Gene Pairs?**  Taking the gene pairs one at a time, we calculate the recombination frequencies by counting the total number of crossovers that occur between the genes of that pair. Every crossover event between the two genes is counted, whether the event occurs by itself (a single crossover) or simultaneously with another event (a double crossover). In this case, there are 11 double recombinants, each with one crossover between *v* and *gl* and one crossover between *gl* and *va*, for a total of 22 crossover events. Single-crossover progeny are predicted on the basis of parental

chromosomes having the gene order *v–gl–va* Between *v* and *gl*, a single crossover produces the following.

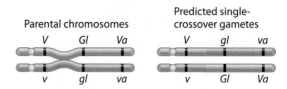

Test-cross progeny carrying these recombinant chromosomes have the phenotypes yellow, rough, normal (Class 1) and green, glossy, variable (Class 8). The recombination frequency is calculated as the sum of all single and double recombinants for this gene pair divided by the total number of progeny: 60 + 62 + 4 + 7/726 = 0.183, or 18.3%. Therefore, the distance between these genes is approximately 18.3 cM.

Single crossover between *gl* and *va* produces the following.

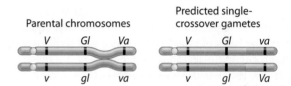

Test-cross progeny carrying these chromosomes are found in Class 2 (yellow, glossy, normal) and Class 7 (green, rough, variable). Recombination frequency $r$ = 48 + 40 + 4 + 7/726 = 0.136, or 13.6%. The intergenic distance is approximately 13.6 cM.

Recombination between the flanking markers, *va* and *v*, is calculated by counting all crossovers between the genes. For these genes, recombination between *v* and *va* is $r$ = 60 + 62 + 48 + 40 + 22/726 = 0.320, or 32%.

**Question 5: Is the Frequency of Double Crossovers Consistent with Independence of the Single Crossovers?**  In most tests of genetic linkage, the number of double crossovers is *less* than the number expected. The reduction is caused by an effect called **interference (I),** which limits the number of crossovers that can occur in a short length of chromosome. Interference, which we discuss further in Section 5.4, is quantified by comparing the number or frequency of observed double-crossover events to the number or frequency expected, assuming each crossover event occurs independently. In Emerson's data set, there are 11 double crossovers among test-cross progeny, or (11/726) = 0.015 (1.5%). If each crossover were independent, expected double-crossover frequency would be the product of the two single-crossover frequencies, (0.183)(0.136) = 0.025 (2.5%). The expected number of double-crossover progeny would therefore be (0.025)(726) = 18.2. Observed double recombinants are divided by expected double recombinants, producing a value known as the **coefficient of coincidence (c).** Either the numbers or the frequencies of observed and expected double recombinants can be used to determine $c$:

$$c = \frac{\text{observed double recombinants}}{\text{expected double recombinants}}$$

$$= 11/18.2 = 0.60 \text{ (using numbers)}$$

or

$$= 0.015/0.025 = 0.60 \text{ (using frequencies)}$$

Interference is defined as $I = 1 - c$, so for this data set $I = 1 - 0.60 = 0.40$. Interference identifies the proportion of double recombinants that are expected but *are not produced* in the experiment (the difference between expectation and actuality). In this case, the number of double recombinants was 40% lower than expected. Interference is a very common observation in most regions of most genomes. On occasion, however, certain regions of some genomes generate *more* double recombinants than expected. In these cases $I < 0$, a situation called **negative interference.** Interference will be $I = 0$ when the observed and expected double crossovers are equal. The molecular basis of interference is not well understood, although current research shows that there is a mechanical limit that restricts the number of recombination events in a particular region of a chromosome.

---

Genetic Insight  Genetic linkage maps are constructed in a five-step process that determines (1) significant deviation of observed results compared with expected results, (2) parental chromosome genotypes, (3) double-crossover chromosome genotypes and gene order, (4) recombination frequencies between gene pairs, and (5) interference in formation of double crossovers.

---

## Determining Gamete Frequencies from Genetic Maps

The same principle used to construct genetic linkage maps—the relation between relative distances and recombination frequency—can be used to make predictions in the opposite direction, that is, to determine the expected frequencies of recombinant gametes on the basis of completed genetic linkage maps.

In **Figure 5.8a**, two linked genes have a recombination frequency of 10%. For the dihybrid organism $AB/ab$, two gametes ($AB$ and $ab$) are parental, and two ($Ab$ and $aB$) are recombinant. Recombinant gametes equal 10% of total gametes, and each recombinant is expected to occur with the same frequency. The probability is calculated as $(1/2)(0.010) = 0.05$ for each recombinant gamete. In this calculation, 1/2 is the probability that each recombinant chromosome has a one-half chance of appearing in a gamete, and 0.010 is the probability of recombination between the genes. Conversely, parental gametes $AB$ and $ab$ are formed at a frequency equal to 100% minus 10%, or 90% of total gametes. Parental gametes are also expected at equal frequency—in this case $(1/2)(0.90)$, or 45% each.

Gamete frequencies for three linked genes are predicted in a similar manner. In **Figure 5.8b**, genes $a$ and $b$ are shown along with a third gene, $c$, located 20 cM from gene $b$. To predict gamete frequencies, we make the assumption that interference is $I = 0$ to simplify the calculation of the number of recombinants. For the trihybrid $ABC/abc$, parental gametes are produced when crossover does not occur in either gene interval. The probability of *no crossovers* between genes $a$ and $b$ is 90% (0.9), and between $b$ and $c$ it is 80% (0.8). Considering both gene pairs, the proportion of nonrecombinant gametes is $(0.9)(0.8) = 0.72$.

**(a)**

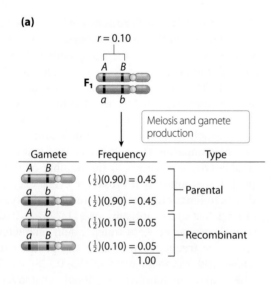

**(b)**

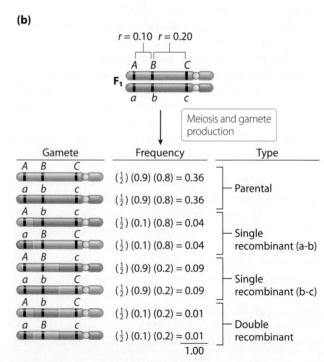

**Figure 5.8  Gamete genotype frequencies calculated from genetic linkage data.** (a) Gamete frequencies predicted from a map of two linked genes. (b) Gametes predicted from a map of three linked genes assuming interference is zero ($I = 0$).

There are two equally frequent parental gametes, each with an expected frequency of $(0.72)(0.5) = 0.36$. Recombination frequency is 10% (0.1) between $a$ and $b$. Two single recombinants between genes $a$ and $b$ have an expected frequency of $(0.1)(0.8)(0.5) = 0.04$ each. Similarly, single recombinants between genes $b$ and $c$ have expected frequencies of $(0.9)(0.2)(0.5) = 0.09$ each. Each of the double-recombinant gametes, $AbC$ and $aBc$, are expected with a frequency of $(0.1)(0.2)(0.5) = 0.01$. The sum of frequencies of the eight predicted gamete genotypes is 1.0, indicating that all gametes have been counted.

## 5.4 Recombination Results from Crossing Over

Morgan's hypothesis of recombination by crossing over between homologous chromosomes has stood the test of time and is now universally accepted. When he proposed it, Morgan's model fit nicely with a 1909 observation by F. A. Janssens, who captured a view of meiotic chromosomes under the microscope and suggested that the chiasmata seen between homologous chromosomes might be points of recombination. Clear proof of the hypothesis of gene recombination by chromosome exchange was not obtained, however, until 20 years after Morgan proposed it. In 1931, research published by Harriet Creighton and Barbara McClintock on crossing over in corn (*Zea mays*), and a nearly simultaneous report by Curt Stern on crossing over in *Drosophila*, provided direct evidence that gene recombination and physical exchange between homologous chromosomes went hand-in-hand.

### Cytological Evidence of Recombination

Creighton and McClintock studied recombination between homologous copies of chromosome 9 in corn that were distinguished by two genetic markers—the genes controlling kernel color (*c1*) and starch type (*wx*) in *Zea mays*—and by two cytological markers—structural differences in the homologous copies of chromosome 9 that were observed under the microscope. One copy of chromosome 9 had the normal microscopic appearance and carried alleles *c1* and *Wx*. The homologous copy of chromosome 9 carried alleles *C1* and *wx* and was cytologically altered in two ways. On the end nearer *C1*, the chromosome had a darkly staining region called a "knob"; on the other end, near *wx*, the chromosome carried a fragment of chromosome 8 that had been transferred by a chromosome-rearrangement event called *translocation* (we explore this event in Chapter 13). Creighton and McClintock obtained cytological evidence that recombination involved the physical exchange between homologous chromosomes by detecting genetic recombinants (chromosomes carrying the alleles *C1* and *Wx* or carrying the alleles *c1* and *wx*) that were also cytologically rearranged chromosomes (**Figure 5.9**).

Just a few weeks after Creighton and McClintock reported their evidence of a link between chromosome

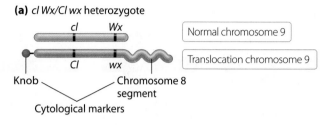

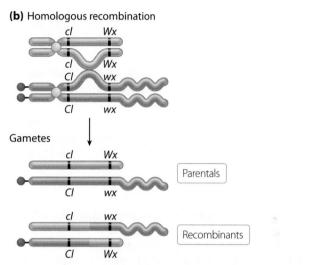

**Figure 5.9  Cytological proof from *Zea mays* that recombination results from crossing over.**  Progeny displaying recombinant phenotypes are also seen to carry physically rearranged chromosomes.

rearrangement and genetic recombination, Curt Stern reported similar findings in *Drosophila*. The combined genetic and chromosomal recombination analyses in corn and fruit fly provided convincing evidence that genetic recombination between homologous chromosomes is accompanied by physical exchange between the chromosomes in plants and in animals.

### Limits of Recombination along Chromosomes

Creighton, McClintock, and Stern showed convincingly that crossover is accompanied by chromosome breakage and rejoining. Morgan and Sturtevant's work, supported by data from several of their contemporaries, established that the relative distance between two linked genes on a chromosome influences the frequency of recombination between them. Two important questions about the likelihood and frequency of crossing over derive from these observations. First, why does distance between genes influence recombination frequency? And second, is there an upper limit to the frequency of recombinant gametes for a pair of linked genes?

The answer to the first question is that in early prophase I, points of crossing over are established at recombination nodules that occur along the synaptonemal complex. Two genes that are close to one another are less likely to have a recombination nodule between them and are less likely to recombine than are a pair of genes separated by a greater distance on a chromosome.

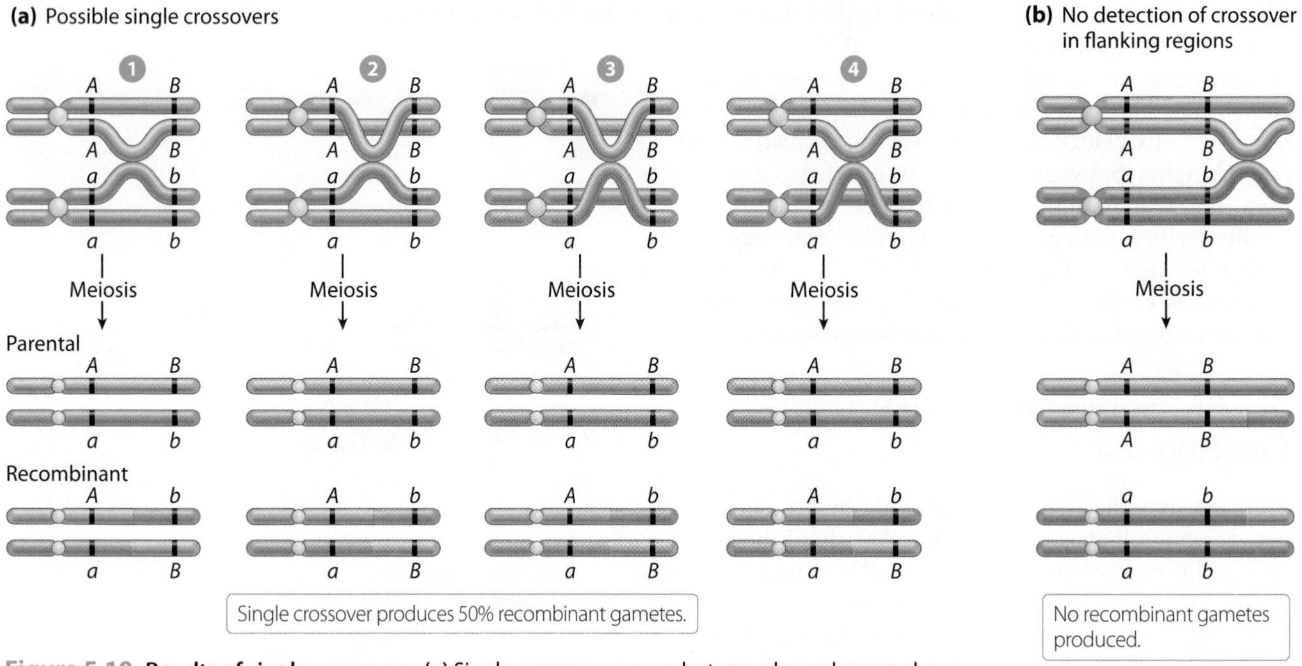

**(a)** Possible single crossovers

**(b)** No detection of crossover
in flanking regions

Single crossover produces 50% recombinant gametes.

No recombinant gametes produced.

**Figure 5.10  Results of single crossover.  (a)** Single crossovers occur between homologous chromosomes in multiple ways. Each meiosis produces two parental chromosomes and two recombinant chromosomes, thus 50% of gametes can carry recombinant chromosomes. **(b)** Single crossover taking place outside the chromosome region being tested does not reveal recombinant chromosomes.

Recombination occurs after DNA replication has been completed, when each member of a homologous chromosome pair is composed of two sister chromatids. This is the four-strand stage. Single crossovers involve one chromatid from each homolog. There are four equivalent ways this process can occur, and all four events produce the same outcome—two parental gametes and two recombinant gametes (**Figure 5.10a**). Crossovers that occur between nonsister chromatids but not between the loci tested will not leave genetic evidence of recombination (**Figure 5.10b**).

There are three patterns of double crossover between two genes. The outcomes of each pattern are unique with respect to the number of recombinant gametes produced. **Two-strand double crossover** produces no recombinants, because two recombination events between a pair of genes do not produce genetic evidence of recombination in the form of a recombinant gamete (**Figure 5.11a**). A **three-strand double crossover,** involving three of the sister chromatids, can happen in two ways that each produce the same genetic outcome—two parental and two recombinant chromosomes in gametes (**Figure 5.11b**). When a **four-strand double crossover** occurs, all four chromosomes in gametes are recombinant (**Figure 5.11c**).

In answer to the second question we posed earlier, recombination between a pair of linked genes is limited to 50% of the gametes. As we have seen, of the four gametes produced by single crossover, two are recombinant gametes (have the nonparental genotype) and so result in a total of 50% recombinant. Likewise, summing the outcomes of the example two-, three-, and four-strand double crossovers

shown in Figure 6.11 gives a total of 8/16 (50%) recombinant gametes. This establishes an upper limit of 50% as the frequency of both parental and nonparental genotypes in gametes. Most instances of genetic linkage produce substantially more than 50% parental chromosomes and substantially less than 50% nonparental. The smallest proportions of recombinant chromosomes are associated with the most tightly linked genes (i.e., the genes that are closest together), and the recombinant proportions increase as the distance between genes becomes greater. Recombination frequencies between *linked* genes can increase *up to* 50% as the distance between genes gets larger, and the corresponding frequency of parental chromosomes *decreases* to 50%. Thus, frequencies of recombination between linked genes are always less than 50%. Once there is sufficient distance between syntenic genes, however, crossover randomizes the combinations of alleles on chromosomes, and the pattern becomes that of independent assortment. In other words, syntenic genes that are far apart assort independently.

**Genetic Analysis 5.2** on page 162 presents the results of test crosses involving three linked genes and takes you through the determination of recombination frequencies between the genes.

---

Genetic Insight  Recombination frequency between two genes has an upper limit of 50%, no matter how far apart the genes are on a chromosome. Recombination frequency between *unlinked* genes equals 50%, but it is *less than* 50% for linked genes.

**(a)** Two-strand double crossover
(three equivalent ways, one position held constant)

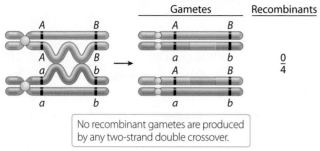

No recombinant gametes are produced
by any two-strand double crossover.

**(b)** Three-strand double crossover (one position held constant)

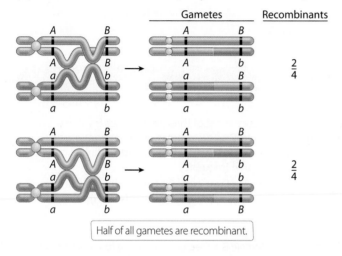

Half of all gametes are recombinant.

**(c)** Four-strand double crossover (one position held constant)

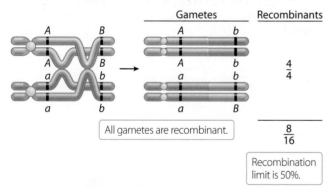

All gametes are recombinant.

$$\frac{8}{16}$$

Recombination
limit is 50%.

**Figure 5.11** **Results of double crossover.** Double crossovers between two genes involving two, three, or all four chromatids result collectively in a maximum of 50% recombinant gametes.

## Recombination within Genes

Our discussion thus far describes how the linear order of genes along chromosomes can be determined based on crossover *between* genes. Does crossover ever occur *within* genes? The answer is yes.

Crossing over within genes, called **intragenic recombination,** is an infrequent event that is detected through the examination of large numbers of progeny, usually for evidence of recombination between homologs carrying different mutant alleles of the same gene. Since the site of mutation within the gene is different for each allele, intragenic recombination produces one wild-type recombinant chromosome and one double-mutant chromosome.

Melvin Green and Kathleen Green were the first to report intragenic recombination in a 1949 study of the *Drosophila* gene for an X-linked recessive mutant eye phenotype called "lozenge," which disrupts the number and pattern of facets on the eye of the fly. Several different mutations of the lozenge gene each produce a distinctive lozenge phenotype. The Green's (a husband and wife team), following up on work begun a few years earlier by Clarence Oliver, used lozenge-eyed females, each carrying two different lozenge-producing alleles, $lz^{BS}$ and $lz^g$, on the homologous copies of their X chromosomes (**Figure 5.12**). The lozenge mutations are located at different positions within the lozenge gene; thus, each mutant allele has mutant DNA sequence at the site of mutation but has wild-type DNA sequence in the rest of the gene. Rare intragenic recombination leads to one double-mutant X chromosome carrying both lozenge mutations in a single gene, and a wild-type X chromosome with a lozenge gene that contains neither mutation. The double-mutant chromosomes produce a phenotype that is distinct from either of the mutations alone. Green's detected fewer than 20 double-mutant X chromosomes and the wild-type X chromosome in more than 16,000 progeny of the lozenge-eyed females, but the result was sufficient to verify intragenic recombination.

## Biological Factors Affecting Accuracy of Genetic Maps

Inherent in the use of recombination frequency as a measure of approximate distance between genes along a chromosome is the assumption that genetic distance and physical distance are proportional throughout the genome

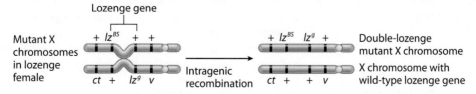

**Figure 5.12** **Intragenic recombination in the lozenge eye gene of *Drosophila*.** Progeny resulting from intragenic recombination can be detected by a distinct lozenge phenotype produced by the double-mutant chromosome or by having wild-type eyes. The genes *ct* and *v* are used to verify intragenic recombination.

Dr. O. Sophila, a famous geneticist, is evaluating genetic linkage among three X-linked genes in *Drosophila*. At these genes, red eye ($v^+$) is dominant to vermilion eye ($v$); full wing ($r^+$) is dominant to rudimentary wing ($r$); and gray body color ($y^+$) is dominant to yellow ($y$). Dr. Sophila has the results of three test crosses. Help Dr. Sophila identify which pairs of genes are linked, and calculate the recombination fraction between linked genes.

**Test Cross I:**

♀ *yv/++* (gray body, red eye) ×
♂ *yv/Y* (yellow body, vermilion eye)

| Progeny | Number |
|---|---|
| Yellow, vermilion | 338 |
| Gray, red | 332 |
| Yellow, red | 160 |
| Gray, vermilion | 170 |
| | 1000 |

**Test Cross II:**

♀ *vr/++* (red eye, full wing) × ♂ *vr/Y*
(vermilion eye, rudimentary wing)

| Progeny | Number |
|---|---|
| Vermilion, rudimentary | 396 |
| Red, full | 389 |
| Vermilion, full | 110 |
| Red, rudimentary | 105 |
| | 1000 |

**Test Cross III:**

♀ *yr/++* (gray body, full wing) × ♂ *yr/Y*
(yellow body, rudimentary wing)

| Progeny | Number |
|---|---|
| Yellow, rudimentary | 246 |
| Gray, full | 252 |
| Yellow, full | 259 |
| Gray, rudimentary | 243 |
| | 1000 |

| Solution Strategies | Solution Steps |
|---|---|

**Evaluate**

1. Identify the topic of this problem and state the nature of the required answer.

1. This problem involves the assessment of three test crosses involving X-linked genes. The answer requires determination of genetic linkage versus independent assortment for each gene pair and, for linked genes, the calculation of recombination frequency.

2. Identify the critical information given in the problem.

2. The genotypes and phenotypes of test-cross flies are given, and the number of test-cross progeny in each phenotypic category is also given.

**Deduce**

3. Determine the test-cross results expected under the assumption of independent assortment.

3. In each cross, the dihybrid female is expected to produce four genetically different gametes at frequencies of 25% each and the progeny are expected to display four phenotypes in a 1:1:1:1 ratio (250 each). In Test Cross I, for example, the following results are expected and expected results are similar for each test cross.

| Phenotype | Female | Male | Number |
|---|---|---|---|
| Yellow, vermilion | *yv/yv* | *yv/Y* | 250 |
| Gray, red | *yv/y⁺v⁺* | *y⁺v⁺/Y* | 250 |
| Yellow, red | *yv/yv⁺* | *yv⁺/Y* | 250 |
| Gray, vermilion | *yv/y⁺v* | *y⁺v/Y* | 250 |

**TIP:** Chi-square analysis could be used to test the statistical significance of deviations between observed and expected outcomes.

**Solve**

4. Examine each cross and determine if there is evidence of genetic linkage between the gene pairs.

4. Test Cross I and Test Cross II show clear deviation from the predicted ratio, with parental categories substantially greater than 250 each and nonparental categories substantially less than 250 each. The progeny of Test Cross III are distributed in numbers consistent with the independent assortment prediction. These statements are based on chi-squared analysis that is not shown.

5. Calculate the recombination fractions between linked pairs of genes.

5. In Test Cross I, the recombinant progeny are yellow, red and gray, vermilion. $r = 160 + 170/1000 = 0.330$, indicating that these genes are linked and are separated by 33 mu.

In Test-Cross II, the recombinant phenotypes are vermilion, full and red, rudimentary. The recombination frequency is $r = 110 + 105/1000 = 0.215$, or approximately 21.5 mu.

For more practice, see Problems 2, 4, and 28.

and that recombination frequencies for given genes are constant among all members of a species. However, studies in numerous species indicate that age, environment, and sex may affect recombination frequency. For example, advancing age of female fruit flies decreases the frequency of crossover between gene pairs; more crossovers between a specific pair of genes are seen in younger females than in older. Female *Drosophila* crossover frequency is also affected by temperature. Growth of a fruit-fly colony at 22°C is optimal for recombination, and increases or decreases of temperature from optimum can change crossover frequency. Restricting dietary levels of calcium and magnesium, important cofactors for enzymes that interact with DNA, also decreases crossover frequency in fruit flies

The most dramatic impact on recombination frequency in animals, however, is connected to sex. Recombination frequency differs for males and females of most animal species and follows a general pattern in which the heterogametic sex, the sex with two different sex chromosomes (most often males), has a lower rate of recombination than the homogametic sex, the sex with two fully homologous sex chromosomes (most often females). The higher recombination frequency in the homogametic sex is a genome-wide phenomenon and *is not* limited to the sex chromosomes. Fruit flies display an extreme version of this phenomenon—female fruit flies undergo homologous recombination while male fruit flies undergo no recombination at all!

These observations are seen across the taxonomic spectrum, including in humans. Human females experience more crossing over than human males, resulting in a larger recombination map in females. A detailed recombination and genome sequencing analysis of human chromosome 19 exemplifies this phenomenon. Chromosome

19 is composed of about 65 megabases (Mb), or 65 million base pairs, in both male and female genomes (**Figure 5.13**). However, the length of the chromosome as determined by adding the estimated recombination distances along the entire length of the chromosome is a larger number of map units in females than in males. Also notice that recombination frequencies are greater in regions at the ends of the chromosome in males but are greater in females in central chromosome regions. For the human genome as a whole, the female genetic map contains about 4400 cM, and the male map about 2700 cM. Geneticists studying the human genome usually produce a "sex averaged" human genetic map that is slightly larger than 3500 cM.

Among different species, the number of nucleotide base pairs per map unit varies. For example, the human genome consists of a little less than 3 billion base pairs of DNA and the sex-averaged genome contains about 830,000 bp/cM. In contrast, the *Arabidopsis* genome contains about 200,000 bp/cM; thus, recombination is about four times as frequent in *Arabidopsis* as it is in humans.

---

**Genetic Insight**  Crossover frequency in most organisms is affected by factors of age, sex, and environment. The heterogametic sex generally experiences less homologous recombination than the homogametic sex.

---

## Correction of Genetic Map Distances

Because so many factors affect crossing over and recombination in eukaryotic genomes, it is reasonable to ask whether recombination frequencies and map distances calculated on the basis of observed recombination between gene pairs are in fact fully accurate representations of the actual numbers of recombination events. The answer is no. Experimental evidence indicates that the map distances calculated between two randomly selected genes usually *underestimate* the physical distance between genes, largely because of undetected crossovers between the genes. The farther apart two syntenic genes are, the greater the inaccuracy, because double crossovers between a pair of genes are not detected as recombinant for flanking markers.

A single crossover between genes $A$ and $B$ in a dihybrid ($AB/ab$) produces two parental gametes ($AB$ and $ab$) and two recombinant gametes ($Ab$ and $aB$). As illustrated in Figure 5.11, however, a double crossover between the same genes produces crossover gametes that are not recombinant for flanking markers and are indistinguishable from parentals. These crossover-nonrecombinant gametes are not counted when recombination frequency between genes is calculated, because they are not observed. Larger distances between genes provide greater opportunity for double crossover and thus greater likelihood of crossover-nonrecombinant gametes.

In theory, the relationship between recombination frequency and map distance is linear, but this is not the

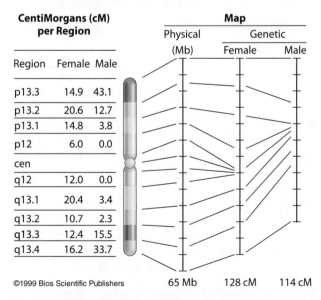

| CentiMorgans (cM) per Region | | | Map | | |
|---|---|---|---|---|---|
| | | | Physical | Genetic | |
| | | | (Mb) | Female | Male |
| Region | Female | Male | | | |
| p13.3 | 14.9 | 43.1 | | | |
| p13.2 | 20.6 | 12.7 | | | |
| p13.1 | 14.8 | 3.8 | | | |
| p12 | 6.0 | 0.0 | | | |
| cen | | | | | |
| q12 | 12.0 | 0.0 | | | |
| q13.1 | 20.4 | 3.4 | | | |
| q13.2 | 10.7 | 2.3 | | | |
| q13.3 | 12.4 | 15.5 | | | |
| q13.4 | 16.2 | 33.7 | | | |

©1999 Bios Scientific Publishers     65 Mb     128 cM     114 cM

**Figure 5.13  Physical distance versus recombination distance on human male and female chromosome 19.** In most sexually reproducing organisms, the heterogametic sex has fewer recombination events and a shorter recombination map than does the homogametic sex. Data adapted from J.L. Weber et al (1993).

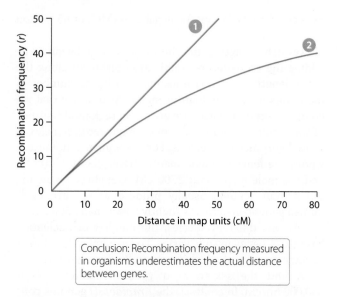

Conclusion: Recombination frequency measured in organisms underestimates the actual distance between genes.

**Figure 5.14 The relationship between recombination frequency and physical distance between genes.** Line 1 traces a linear relationship between recombination frequency and the physical distance between linked genes. Line 2 traces the observed correspondence between recombination frequency and physical distance.

case in reality. Line 1 in **Figure 5.14** depicts a linear relationship between recombination frequency and the distance in map units (cM). In contrast, line 2 illustrates that relationship as measured in organisms. The lines diverge at about 8 cM, indicating that the relationship between recombination frequency and map distance is linear only for linked genes that are separated by less than 8 cM, and that observed recombination frequencies usually underestimate the physical distance between genes.

The central problem in correlating recombination frequency with the number of recombination events is the difficulty of identifying the number of meioses that produce each possible number of crossovers—zero, one, two, three, four, and so on. In an attempt to correctly model different recombination classes and to accurately assess the correlation between recombination frequency and crossover, J. B. S. Haldane developed a **mapping function** in 1919 that correlates map distance and recombination frequency between gene pairs. The Haldane mapping function is based on the Poisson distribution, which is a formula expressing the probability of a rare event, assuming that successive events are independent. The Poisson distribution predicts the probability of zero, one, two, three, four, or more recombination events in a chromosome region, where each crossover during meiosis is counted. The Poisson formula is

$$f_n = (\ln^{-m} m^n)/n!$$

where

$f_n$ = the frequency of samples with $n$ number of events
$n$ = the actual number of events (i.e., crossovers) in the sample

$\ln$ = the base of natural logarithms (approximately 2.72)
$m$ = the average number of events in the sample
 ! = the factorial symbol (i.e., 4! = 4 × 3 × 2 × 1)

Using the formula for the Poisson distribution, we can express the frequency of the zero-crossover category ($n = 0$) as

$$\ln^{-m}(m^0/0!)$$

In this case, $m^0$ and 0! equal 1, so the frequency of the zero-recombination category is $\ln^{-m}$. This function can be used to correlate the recombination frequency with the actual number of crossover events. If we identify $1 - \ln^{-m}$ as the frequency of meioses with more than zero (one, two, three, four, etc.) crossovers and recognize that in those meioses experiencing crossover, 50% (1/2) of all products will be recombinant, we can express the function as

$$r = \frac{1}{2}(1 - \ln^{-m})$$

With this formula, recombination frequencies are converted to a corrected number of recombination events expressed in map units. For example, if the distance between two genes is measured as $r = 0.20$, application of Haldane's correction produces a corrected distance of 26 m.u. between the genes.

One concern raised about Haldane's mapping function is that it may overestimate the actual recombination frequency when interference occurs. Damodar Kosambi developed a modified mapping function to correct map distance in species with interference, and it has become one of the most widely applied improvements. The Kosambi formulation assumes that interference decreases as a linear function of distance between genes. **Figure 5.15**

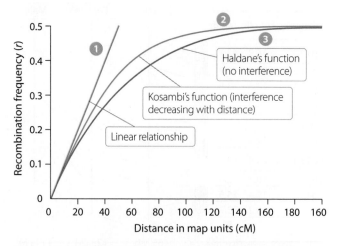

**Figure 5.15 Mapping functions correct estimates of distance between linked genes.** The linear relationship ❶ and mapping corrections by Kosambi ❷ and Haldane ❸ are based on different assumptions about the relationship between recombination frequency and the physical distance between linked genes.

compares the uncorrected linear relationship between crossover and recombination frequency ($r = cM$) with Kosambi's mapping function and with the Haldane mapping function.

Different mapping functions have been specifically designed to deal with particular kinds of recombination data in various species. For example, mapping functions that efficiently correct mapping data for organisms like *Drosophila*, which produces large numbers of progeny, may not be fully appropriate for correcting human mapping data, where the number of progeny is much smaller.

---

Genetic Insight Mapping functions correct the underestimate of physical distance between genes that is generated by analysis of genetic mapping data. Genetic distance and physical distance are similar over short intervals but become more dissimilar as distance between genes increases.

---

## 5.5 Linked Human Genes Are Mapped Using Lod Score Analysis

Until relatively recently, the human genetic map was rather sparse. Humans cannot be studied through controlled matings and in any case produce much smaller numbers of offspring than do organisms like *Drosophila* and *Zea mays*. Consequently, gene-mapping methods developed and used successfully to map genes in model organisms are difficult to apply to human gene mapping. Historically, X-linked genes, by virtue of their unique patterns of transmission, were the first and easiest human genes to map, whereas progress in mapping human autosomal genes was hampered by a scarcity of known polymorphic genetic markers, such as blood group antigens and blood proteins.

Human genome mapping changed significantly in the mid-1980s, facilitated both by the emergence of molecular genetic methods to identify polymorphic DNA markers and by advances in gene-mapping software. Different types of polymorphic DNA markers, including restriction fragment length variants (RFLPs) and single nucleotide polymorphisms (SNPs) (described in Chapter 10), ultimately made thousands of new human genetic markers available for study in linkage analysis. Combined with sophisticated statistical techniques and modern computer power, the use of polymorphic DNA markers has given geneticists the ability to effectively map human genes by genetic linkage analysis.

The availability of large numbers of DNA markers on each chromosome led first to the identification of **linkage groups,** clusters of syntenic genes that are linked to one another, and then to assignment of chromosomal locations to linkage groups. The discovery of genetic linkage between a genetic marker with a known chromosome location and any member of a linkage group assigns the linkage group to a chromosome location near the genetic marker. Different linkage groups on the same chromosome can then be organized into maps of chromosome segments and whole chromosomes.

### Allelic Phase

Efforts to map human genes often focus on finding the chromosomal locations of disease-causing genes. This is a common first step toward the eventual cloning and sequencing of a gene that may be the cause of hereditary disease. A strategy known as *functional cloning,* or *reverse genetics* (discussed in Chapter 17), can be used to map a gene whose function is not known. Once the location of the gene is identified, the gene can be cloned and sequenced, and the sequence can be examined for clues to the normal function of the gene and to the mechanisms by which gene mutation produces inherited abnormalities.

To map genes, parental and recombinant chromosomes must be identified, and one of the first obstacles researchers encounter in the effort to map human genes is the difficulty of determining **allelic phase,** the arrangement of alleles of linked genes on homologous parental chromosomes. **Figure 5.16** illustrates the problem of determining allelic phase in families and the importance of key individuals in this process. The two pedigrees in the figure are identical in structure and in the distribution of

**(a)**

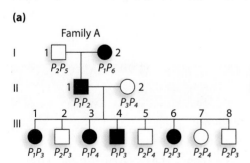

Allelic phase is known in family A by tracing the transmission of the disease allele (*D*) and the *A₁* genetic marker allele from I-2 to II-1 and to III-1, III-3 and III-4; III-6 is a probable recombinant.

**(b)**

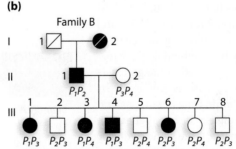

Allelic phase is not known in family B because the disease allele carried by II-1 could be on either the chromosome carrying genetic marker allele *A₁* or the chromosome carrying *A₂*.

**Figure 5.16 Allelic phase analysis in human families A and B.**

an autosomal dominant hereditary disease indicated by shaded symbols. The alleles of the gene determining the disease phenotype are $D$ and $d$. In addition to allelic information for the disease locus, the pedigrees show allelic information for a closely linked polymorphic DNA marker that has six alleles identified as $P_1$ to $P_6$.

Allelic phase is known to be $P_1 D$ in Family A because the affected woman in generation I (I-2) transmits marker allele $P_1$ along with the dominant disease allele ($D$) to her daughter, II-1. The unaffected father in generation I (I-1) is homozygous for the recessive wild-type allele ($dd$) at the disease locus and heterozygous for DNA marker alleles $P_2$ and $P_5$. Allelic phase in II-1 is $P_1 D/P_2 d$; the chromosome on the left of the solidus (/) is maternal, the chromosome on the right paternal. Considering that her mate (II-2) is $P_3 d/P_4 d$, we can identify the transmission of parental and recombinant gametes from II-1 to her children in generation III. Children III-1, III-3, and III-4 inherited a maternal chromosome carrying $P_1 D$ to produce their disease and either the $P_3$ or $P_4$ allele along with $d$ on their paternal chromosome. On the other hand, III-2, III-5, III-7, and III-8 inherited alleles $P_2$ and $d$ on their maternal chromosome and either $P_3$ or $P_4$ along with $d$ on their paternal chromosome. Child III-6 has apparently inherited a recombinant chromosome carrying alleles $P_2$ and $D$ from her mother along with a paternal chromosome carrying $P_3$ and $d$ on the paternal chromosome.

The pedigree for Family B does not allow identification of allelic phase. In this family, there is no marker information for generation I, and thus allelic phase for II-1 is unknown. She could either be $P_1 D/P_2 d$ or $P_1 d/P_2 D$. For the purposes of genetic linkage analysis, each possible phase must be treated as equally likely. With allelic phase in II-1 unknown, we cannot be certain which of her children have inherited parental chromosomes and which carry recombinants. If II-1 is $P_1 D/P_2 d$, her children III-1 to III-5, and III-7 and III-8 are parental, and III-6 is recombinant. Alternatively, if she is $P_1 d/P_2 D$, then III-1 to III-5 and III-7 and III-8 are recombinant and III-6 is parental.

## Lod Score Analysis

Although it is not possible to unambiguously identify and count recombinants in pedigrees like Family B, a statistical method developed by Newton Morton in 1955, and refined and expanded since then, allows geneticists to calculate the overall probability of genetic linkage. Morton's method determines whether genetic linkage exists between genes for which allelic phase is unknown by comparing the likelihood of obtaining the genotypes and phenotypes observed in a pedigree if two genes are linked versus the likelihood of getting the same pedigree outcomes if the genes assort independently. The ratio of these two likelihoods gives the "odds" of genetic linkage, and the **log**arithm of the **od**ds ratio generates the **lod score,** a statistical value representing the probability of genetic linkage between the genes.

The numerator of the odds ratio that yields the lod score is the likelihood that the distribution of phenotypes and genotypes in the pedigree is produced by genetic linkage between the genes. The denominator is the likelihood of the same pedigree outcomes assuming independent assortment between the genes (i.e., no genetic linkage). Lod score analysis evaluates each pedigree and determines the likelihood of genetic linkage for many different recombination frequencies, each expressed as a variable called the **θ value** ("theta value"). Using input data on each family member that identifies presence or absence of the disease and the genotype at a potentially linked marker gene, software programs calculate the likelihoods of genetic linkage versus no linkage between the genes and compute lod scores for each θ value specified by the investigator. The θ values are any recombination frequency between θ = 0 (complete genetic linkage) and θ = 0.50 (independent assortment). The programs determine lod scores, and because they are log values, the lod scores for a given θ value in different families can be added together. After analyzing all available family data, the lod scores for each θ value are summed and the highest lod score value obtained in a study is designated $Z_{max}$. The $Z_{max}$ corresponds to the θ value that is the most likely recombination frequency between the genes tested.

For each θ value tested, the lod score will be positive if the likelihood of genetic linkage is greater than the likelihood of independent assortment because, in that case, the numerator value (likelihood assuming genetic linkage) is greater than the denominator value (likelihood assuming independent assortment). Conversely, if the pedigree is more likely to be produced by independent assortment than by genetic linkage, the independent assortment likelihood will be larger than the genetic linkage likelihood, and the lod score will be negative.

Lod scores are calculated using the assumption that if two genes have a recombination frequency equal to θ, the probability that a particular gamete is recombinant is also equal to θ, and the probability that a gamete is nonrecombinant is $1 - θ$. Table 5.4 shows calculated lod score values for the two families shown in Figure 5.16. For each child

| Table 5.4 | Lod Score Values for the Families in Figure 5.16 | | | | |
|---|---|---|---|---|---|

**Family A (Phase Known)**

| θ value | 0 | 0.1 | 0.2 | 0.3 | 0.4 | 0.5 |
|---|---|---|---|---|---|---|
| Lod score | $-\infty$ | 1.09 | 1.03 | 0.80 | 0.46 | 0.0 |

**Family B (Phase Unknown)**

| θ value | 0 | 0.1 | 0.2 | 0.3 | 0.4 | 0.5 |
|---|---|---|---|---|---|---|
| Lod score | $-\infty$ | 0.79 | 0.73 | 0.50 | 0.19 | 0.0 |

in generation III, the probability that the gamete from the mother is parental is $1 - \theta$, and the probability that a recombinant gamete is transmitted from mother to child is $\theta$. Since allelic phase is known for Family A, only the known phase is tested. In contrast, Family B does not have a known allelic phase; thus, each possible phase is assumed to be equally likely. In the Family B lod score computation, each phase is tested and is part of the numerator. Because a known allelic phase produces more genetic linkage information, the lod scores for Family A are greater than the lod scores for Family B. In the context of lod score analysis, Family A is identified as the more informative of the two pedigrees.

A lod score is a statistic that can argue in favor of genetic linkage, if the probability of genetic linkage is sufficiently greater than the probability of independent assortment, or it can argue against genetic linkage, if the probability of independent assortment is sufficiently greater than the linkage probability. Lod scores can be interpreted for individual families, or they can be added together for as many families as are analyzed. In either case, lod score significance is interpreted by the following parameters:

1. A lod score of 3.0 or greater is considered significant evidence *in favor* of genetic linkage. A lod score of 3.0 or more indicates significant odds of genetic linkage at each $\theta$ value at which it occurs. The $\theta$ values identified as significant indicate the most likely number of centiMorgans between linked genes.

2. Lod score values of less than $-2.0$ represent significant evidence *against* genetic linkage. Any lod score values for single or multiple families less than $-2.0$ reject genetic linkage at each $\theta$ value with that result.

3. Lod score values between 3.0 and $-2.0$ are inconclusive, neither affirming nor rejecting genetic linkage between the genes examined. Inconclusive results can be revised as additional data are collected.

The three lod score curves shown in **Figure 5.17** illustrate that lod score results may produce different patterns depending on the level of information available for the pedigree and on the actual relationship between the genes tested. Curve ❶ displays data with a maximum lod score value $(Z_{max})$ of about 4.0 at $\theta = 0.23$, suggesting the two genes are separated by 23 cM. The lod scores are significantly positive in the range of 18 to 30 map units. The curve provides significant evidence against genetic linkage at $\theta < 0.5$. Curve ❷ results from there being very little genetic linkage information, and its lod scores are inconclusive at all distances. Curve ❸ rejects genetic linkage at $\theta$ values less than 0.12 but is inconclusive through the rest of the linkage range.

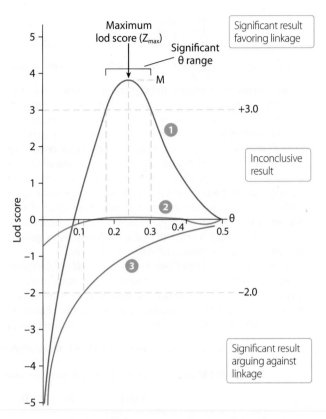

**Figure 5.17  Sample lod score curves.** Lod score values (vertical axis) are plotted against recombination fractions ($\theta$ values; horizontal axis) for three hypothetical lod score analyses.

A number of more comprehensive software programs permitting multipoint linkage analysis have been developed to simultaneously analyze genetic linkage data for multiple genes and genetic markers. Multipoint linkage analysis tests all possible gene orders to identify the most likely order of linked genes. **Experimental Insight 5.1** discusses the application of lod score analysis in the mapping of *BRCA1*, a gene whose mutation can increase susceptibility to breast and ovarian cancer in women. **Genetic Analysis 5.3** on page 169 guides you through the interpretation of lod score values for linkage between a disease-causing gene and a linked DNA genetic marker.

---

**Genetic Insight**  Lod score analysis identifies the probability of genetic linkage between genes having a specified recombination frequency versus the probability that the genes assort independently.

---

## 5.6 Genetic Linkage Analysis Traces Genome Evolution

Evolution changes genomes as populations evolve and as species diverge from common ancestors. These changes include alterations of allele frequencies (Chapter 22), changes in the number and structure of chromosomes (Chapter 13), and the acquisition of new genes or gene

## Experimental Insight 5.1

### Mapping a Gene for Breast and Ovarian Cancer Susceptibility

Most cases of cancer develop through the acquisition of multiple mutations in somatic cells, meaning that there is no inherited mutation that increases the likelihood of cancer development. In some families, however, the frequent occurrence of a particular kind of cancer in a pattern consistent with single-gene inheritance can suggest the hereditary transmission of a mutant allele that increases the susceptibility of individuals to the cancer. The identity, indeed the very existence of these genes, is not known until they are conclusively shown to contribute to cancer development. One research strategy to identify cancer-susceptibility genes seeks genetic linkage of susceptibility genes to genetic markers that have a known chromosome location.

In the late 1970s, Mary Claire King and several collaborators devised a strategy in a search for a gene whose mutation could increase susceptibility to breast and ovarian cancer in families. King and her colleagues sought to maximize the chance of finding such a cancer-susceptibility gene by carefully selecting

families in which multiple cases of breast and ovarian cancers appeared at young ages, and in which occasional cases of bilateral cancer occurred (affecting both breasts or both ovaries in a single patient) in patterns consistent with an autosomal dominant inheritance of disease susceptibility.

King initially looked for genetic linkage between inherited cancer susceptibility and biochemical markers such as polymorphic blood proteins and enzymes. None of the dozens of biochemical markers screened produced significant evidence of genetic linkage to a breast and ovarian cancer susceptibility gene. In the early 1990s, however, King and her colleagues turned to the use of DNA genetic markers. Then, in 1994, they identified genetic linkage between a group of tightly clustered DNA markers on human chromosome 17 and a gene named *Breast Cancer 1* (*BRCA1*). Lod score analysis of chromosome 17, as summarized in the following table, revealed that the candidate gene has a $Z_{max}$ value of 21.68 at $\theta = 0.13$.

**Lod Score Data for Linkage of *BRCA1* to Chromosome 17q in Humans**

| Genetic Marker | Lod Scores at Recombination ($\theta$) Values | | | | | | | |
|---|---|---|---|---|---|---|---|---|
| | 0.001 | 0.01 | 0.05 | 0.10 | 0.20 | 0.30 | $Z_{max}$ | $\theta_{max}$ |
| D17S250 | −11.98 | −8.96 | −1.20 | 3.81 | 7.30 | 6.65 | 7.42 | 0.23 |
| D17S579 | −1.43 | 1.62 | 8.55 | 12.08 | 12.55 | 9.17 | 13.02 | 0.16 |
| D17S588 | 8.23 | 11.39 | 18.35 | 21.33 | 20.15 | 14.79 | 21.68 | 0.13 |
| NME1 | −1.41 | 0.75 | 6.01 | 8.70 | 9.13 | 6.76 | 9.45 | 0.16 |
| D17S74 | −39.15 | −31.73 | −13.34 | −2.73 | 6.32 | 7.50 | 7.67 | 0.27 |

*Source:* Data from J. Hall et al. (1994).

Five genetic markers that are part of a multipoint linkage analysis are shown. *BRCA1* is most likely close to the middle of this linkage group, near the DNA marker gene D17S588.

Subsequent studies have identified and cloned the *BRCA1* gene and determined that it participates with a second gene called *BRCA2* in DNA mutation repair. A large number of mutations of *BRCA1* have been identified, and some of them dramatically increase the likelihood that a woman will

develop breast or ovarian cancer. Other mutations of *BRCA1* do not appear to significantly increase breast or ovarian cancer risk. A good deal of work remains to be done to clarify the role of this gene in breast and ovarian cancer development, but the research strategy designed by King demonstrates the power of genetic linkage analysis for locating genes of interest. (We discuss more about *BRCA1* and *BRCA2* in Chapter 12).

---

functions (Chapters 12 and 21). In this section, we discuss one application of genetic linkage to the study of evolution.

As populations age, one would expect recombination to randomize the combinations of alleles on chromosomes. When this expected randomization does not occur, evolution is frequently the cause. The specific array of alleles in a set of linked genes on a single chromosome is called a **haplotype** (a contraction of "**haplo**id geno**type**"). Because the alleles in a haplotype belong to linked genes, they tend to be passed together during meiosis. Homologous chromosomes carried by an organism can contain different haplotypes. Haplotypes can consist of any combination of linked genes producing

molecular genetic variation—SNPs, for example—or morphological variation. Haplotypes that are defined by SNP loci usually span regions of 10,000 to 100,000 base pairs, whereas haplotypes for genes producing morphological variation tend to be much larger, spanning up to several million base pairs. Using letters A through F to specify linked SNP loci, and primed (′) and unprimed letters to distinguish the alleles of these sequences, we can specify two sample haplotypes for the same region on homologous chromosomes as

$$...A'\,B\,C'\,D\,E\,F'...$$
$$...A\,B'\,C\,D'\,E'\,F...$$

In a study of human families with an autosomal dominant disease caused by a gene whose location is unknown, geneticists use lod score analysis to test linkage between the disease gene and a variable DNA genetic marker. Provide a complete interpretation of the lod score data displayed in the following table, and identify the most likely distance between the marker gene and the disease gene.

| | | | | | | | $\theta$ Value | | | | | | |
|------|-------|-------|------|------|------|------|------|------|------|-------|-------|-------|------|
| 0.0 | 0.01 | 0.02 | 0.03 | 0.04 | 0.05 | 0.06 | 0.08 | 0.10 | 0.15 | 0.20 | 0.30 | 0.40 | 0.50 |
| $-\infty$ | −6.95 | −1.10 | 0.20 | 1.22 | 2.25 | 7.23 | 7.02 | 5.11 | 4.23 | −2.01 | −6.84 | −9.91 | 0.0 |

| Solution Strategies | Solution Steps |
|---|---|
| **Evaluate** | |
| 1. Identify the topic of this problem and state the nature of the required answer. | 1. This problem concerns lod score analysis assessing genetic linkage between a variable DNA genetic marker and a gene carrying a dominant mutation producing a disease. The answer requires interpretation of the lod score values, identification of potential genetic linkage, and determination of the most likely distance between the DNA marker gene and the disease gene. |
| 2. Identify the critical information given in the problem. **TIP:** A lod score greater than +3.0 provides statistically significant evidence in favor of genetic linkage of the genes, whereas a lod score less than −2.0 provides evidence against linkage of the genes. Lod scores between +3.0 and −2.0 are not significant. | 2. Lod score values are given for 14 $\theta$ values (map units between genes). |
| **Deduce** | |
| 3. Identify significant lod score values in the lod score table and locate $Z_{max}$. | 3. Significant evidence against genetic linkage occurs at $\theta \leq 0.01$ and at $\theta \geq 0.20$. Conversely, significant results in favor of genetic linkage are seen at $\theta = 0.06$ to $\theta = 0.15$. The $Z_{max}$ value is 7.23 and corresponds to $\theta = 0.06$ (6 m.u.). |
| **Solve** | |
| 4. Interpret the meaning of the lod scores for genetic linkage. | 4. The data support genetic linkage between the marker gene and the disease gene at recombination distances of between 6 m.u. and 15 m.u. Linkage between the genes is rejected at less than 2 m.u. and at more than 20 m.u. The lod score results between 2 m.u. and 5 m.u. are inconclusive. |
| 5. Identify the most likely distance between the DNA marker gene and the disease gene. **TIP:** The maximum lod score value corresponds to a specific distance between genes that is identified by its $\theta$ value. | 5. The $Z_{max}$ value is 7.23 at $\theta = 0.06$, thus identifying the most likely distance between the disease gene and gene A as 6 m.u. |

For more practice, see Problems 18, 22, 29, and 30.

Over multiple generations, crossing over is expected to occur among the original haplotypes to produce new haplotypes that occur at frequencies determined by chance. In other words, for genes in a population, the genotype for a chromosome at one gene is expected to be independent of its genotypes for other genes. When this occurs, the chromosome region is said to be in **linkage equilibrium.** This means that knowing the alleles at one gene does not help predict the alleles present at other genes on the chromosome.

As an example, let's consider two SNP genes A and B in the haplotypes above. Assuming that the frequencies of alleles at SNP A are $A = 0.70$ and $A' = 0.30$ and at SNP B are $B = 0.20$ and $B' = 0.80$, we can use chance to predict haplotypes. For the A SNP and the B SNP, the predicted haplotypes and frequencies are

$$A'B' = (0.30)(0.80) = 0.24$$
$$A'B = (0.30)(0.20) = 0.06$$
$$A\,B' = (0.70)(0.80) = 0.56$$
$$A\,B = (0.70)(0.20) = \underline{0.14}$$
$$= 1.00$$

When linkage equilibrium is not observed, the frequencies of certain haplotypes in a population deviate significantly from the frequencies expected. This situation is identified as **linkage disequilibrium,** and it frequently occurs as a consequence of evolutionary processes operating on a population. Two different evolutionary processes are common causes of linkage disequilibrium. Migration can produce linkage disequilibrium if haplotypes have been recently introduced into a population and there has not been a sufficient number of generations for crossing over to randomize alleles. Natural selection is the second common cause of linkage disequilibrium. If an allele in a haplotype is favored by natural selection, the allele will increase in frequency in the population. The other alleles in the haplotype will also be favored because of their close proximity to the favored allele. Recombination may be slow to randomize alleles in haplotypes containing an allele favored by natural selection, but linkage disequilibrium will persist over an extended number of generations.

---

Genetic Insight   Haplotypes are combinations of alleles of linked genes or other genetic markers occurring in a defined segment of a chromosome. Haplotypes are expected to occur in combinations and frequencies predicted by linkage equilibrium, but they may occur in linkage disequilibrium when too little time has passed for crossing over to randomize them or when natural selection operates in favor of certain haplotypes.

---

## 5.7 Genetic Linkage in Haploid Eukaryotes Is Identified by Tetrad Analysis

The genetic mapping experiments conducted in maize, *Drosophila*, humans, and other diploid organisms have allowed biologists to develop extensive genetic maps for many species. They are a triumph of scientific reasoning and the careful execution of experimental design. As successful as these experiments have been, however, certain other organisms have life cycles that allow the genotypes of individual gametes to be studied more directly, without requiring interpretation of the expression of traits among the progeny of controlled crosses. For this research, geneticists depend on eukaryotic microorganisms such as the class *Ascomycetes* that includes bread mold (*Neurospora crassa*) and yeast (*Saccharomyces cerevisiae*).

*Ascomycetes* species spend most of their life cycle in a haploid state, dividing by mitosis to produce new cells. For example, haploid yeast cells of *Saccharomyces cerevisiae* undergo mitotic division during the vegetative portion of the life cycle, reproducing new haploid cells that bud off from parental cells (**Figure 5.18**). Diploid yeast form by the union of two genetically different haploid *mating types*. (We discuss the genetics of this process in Chapter 15). The diploid cells undergo meiosis, producing four haploid **ascospores** contained within a saclike structure called an **ascus**. The four ascospores in an ascus are called a **tetrad**. In yeast, the ascospores are not arranged in any particular order, so the structure is called an **unordered tetrad.** Within each tetrad, two of the ascospores are **a** mating type and two are **a** mating type. At maturity, the ascus ruptures in an event known as sporulation, and spores are released to grow as haploids. In laboratory studies, mature ascospores can be removed from

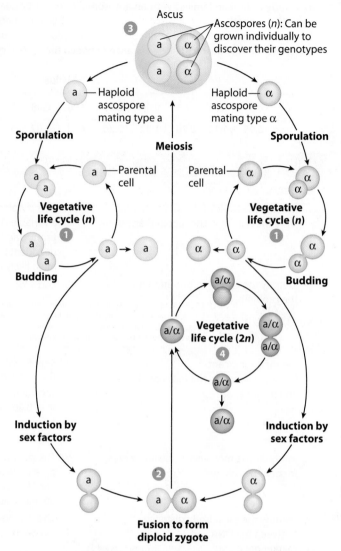

**Figure 5.18   The life cycle of yeast *Saccharomyces cerevisiae*.** ❶ Haploid yeast grow by vegetative propagation. ❷ Yeast of different mating types can fuse to produce diploids. ❸ Haploid ascospores are produced by meiosis in diploid yeast. ❹ Diploid strains propagate by vegetative growth.

their ascus and grown as haploids in culture to discover their genotypes. This process is called **tetrad analysis.**

### Analysis of Unordered Tetrads

Suppose a dihybrid yeast cell with the genotype $a^+ab^+b$ is produced by fusing two haploid cells with genotypes $a^+b^+$ and $ab$. If the genes are on different chromosomes, two equally likely arrangements of chromosomes occur in metaphase I, labeled "Alternative I" and "Alternative II" in **Figure 5.19a**. If no crossover occurs between homologs, each tetrad contains ascospores with two genotypes. Ascospores produced by the alternative I arrangement of metaphase chromosomes contain the same alleles as were found in the parental haploids ($a^+b^+$ and $ab$, in this case). Tetrads with these two ascospore genotypes are known as **parental ditypes (PD).** Tetrads that undergo

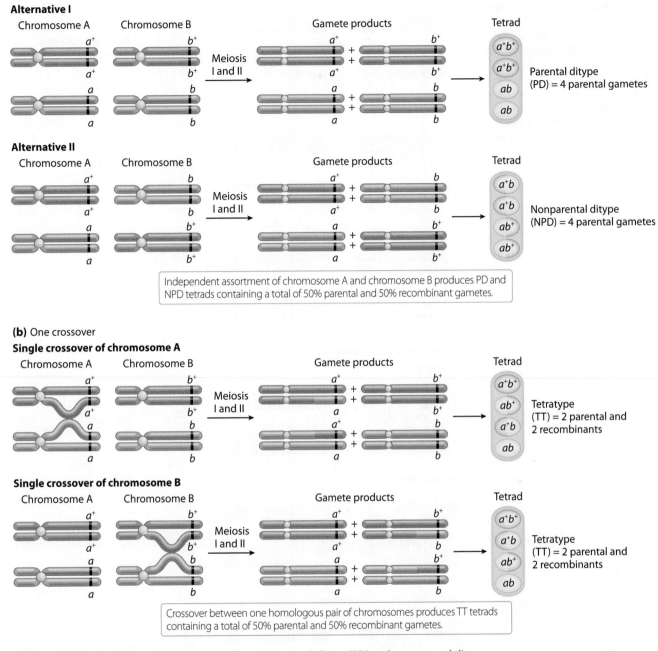

**Figure 5.19  Tetrad results for unlinked genes.** (a) Parental ditype (PD) and nonparental ditype (NPD) tetrads are the products of segregation and independent assortment. Each ascus contains two genetically different types of ascospore. (b) Single crossovers between either homologous pair of chromosomes produce tetratype (TT) tetrads that contain four genetically different ascospores.

the alternative II metaphase chromosome arrangement produce ascospores that have different genotypes than the parents. These tetrads are called **nonparental ditypes (NPD).** If crossing over occurs between either of the homologous chromosome pairs, the tetrad contains ascospores with four different genotypes and is known as a **tetratype (TT)** (Figure 5.19b).

Now let's consider what is observed when the genes are linked. In Figures 5.10 and 5.11, we saw that several

types of single and double crossover can occur between homologous chromosomes in diploids; Figure 5.20 illustrates the tetrad combinations that result from no crossover and from single and various double crossovers between a pair of homologous chromosomes carrying alleles $a^+b^+$/$ab$ at linked loci. The figure illustrates that for these linked genes, all three tetrad types form, but PD and TT tetrads are each more frequent than NPD. PD tetrads are most common, being produced when no crossover

**Figure 5.20 Tetrad formation with linked genes is determined by the occurrence or type of crossover.** (a) No crossing over produces the parental ditype. (b) Single crossover produces the tetratype. (c) Two-strand double crossover produces the parental ditype. (d) Three-strand double crossover produces the tetratype. (e) Four-strand double crossover produces the nonparental ditype.

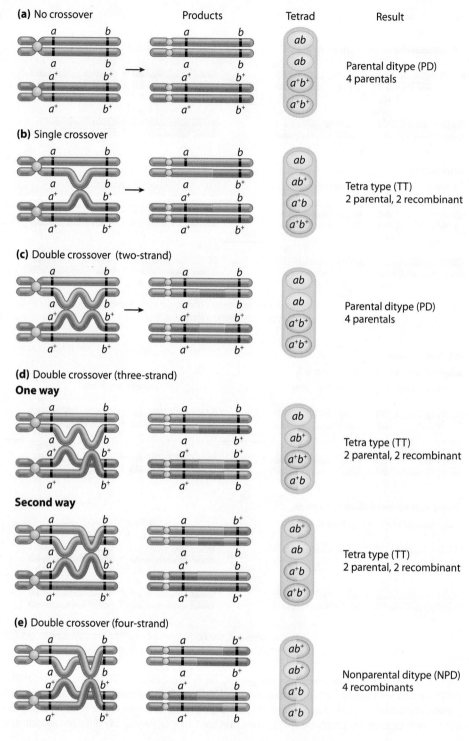

occurs between genes and when two-strand double crossover takes place. TT tetrads are less frequent than PD, occurring when single crossovers or three-strand double crossovers take place. NPD tetrads are least frequent, forming only when four-strand double crossover occurs. Genetic linkage produces the tetrad expectation PD > TT > NPD.

Genetic linkage analysis in tetrads is based on the relative frequencies of different tetrad types rather than an assessment of individual progeny. The formula used to determine recombination frequency in tetrad analysis (familiar from our previous assessments of genetic linkage) is

$$r = \frac{\text{number of recombinants (100)}}{\text{total number of progeny}}$$

An example of this analysis comes from a study that examined tetrads produced by fusion of haploid strains

| Table 5.5 | Recombination Calculation in Tetrads | | |

**Genotype:** $pdx\ pan^+/pdx^+\ pan$

| | **Tetrad Types** | | |
| | PD | TT | NPD |
| --- | --- | --- | --- |
| | $pdx\ pan^+$ | $pdx\ pan^+$ | $pdx\ pan$ |
| Ascospore | $pdx\ pan^+$ | $pdx\ pan$ | $pdx\ pan$ |
| Genotypes | $pdx^+\ pan$ | $pdx^+\ pan^+$ | $pdx^+\ pan^+$ |
| | $pdx^+pan$ | $pdx^+\ pan$ | $pdx^+\ pan^+$ |
| Number | 28 | 20 | 1    = 49 |

$pdx\ pan^+ \times pdx^+\ pan$. The data in Table 5.5 show that among 49 tetrads analyzed, 28 are PD, 20 are TT, and 1 is NPD. A close examination of Figure 5.20 reveals that in tetrads, recombinant chromosomes are found in one-half the ascospores of TT tetrads and all the ascospores of NPD tetrads. On this basis, tetrad recombination frequency is determined using

$$r = \frac{(\frac{1}{2}\,TT) + NPD}{\text{total tetrads}}$$

Recombination frequency for this example is therefore

$$r = \frac{[(\frac{1}{2})(20) + 1]}{49} = 0.224\ (22.4\%)$$

## Ordered Ascus Analysis

Fungi such as *Neurospora crassa* follow the same basic haploid–diploid life cycle as yeast but produce an ascus with eight haploid ascospores rather than four. In *Neurospora*, the fusion of two haploid fungi forms a diploid meiocyte that undergoes meiotic divisions to generate four haploid products aligned in a tetrad ascus. Mitotic division of the ascospores immediately follows completion of meiosis, forming an eight-member ascus (Figure 5.21). The two members of each mitotically produced pair of daughter spores are adjacent to one another in the *Neurospora* octad, and the octad is called an **ordered ascus.** Consequently, the arrangement of daughter spores reflects the identity and orientation of the alleles carried by each chromatid in metaphase I. An ordered ascus can be dissected before sporulation, and haploid spores can be removed one by one to determine their genotype. In this way, each product of meiosis is identified, and its spatial relationship to other meiotic products is determined.

Ordered ascus analysis can be used to map the distance between linked genes and the position of a gene relative to the centromere of its chromosome. Gene-to-centromere distance is calculated based on the segregation of homologous chromosomes in meiosis I and of sister chromatids in meiosis II. In an $a^+a$ meiocyte in which no crossover occurs between the gene and the centromere, alleles segregate in meiosis I. Completion of meiosis and the mitotic division produces an ordered ascus with four spores of one type grouped in the top half of the ascus and spores of the other type filling the bottom half (Figure 5.22). This pattern of segregation is called **first-division segregation,** to signify the separation of alleles $a^+$ and $a$ in the first meiotic division. In the absence of crossover, none of the spores in first-division segregation asci are recombinant.

If crossover takes place, alleles $a^+$ and $a$ are not separated until the second meiotic division, a pattern called **second-division segregation.** If crossover occurs between the gene and centromere, a single crossover produces one of four different octad patterns, depending on the orientation of chromatids during meiosis. One example is illustrated in Figure 5.23a where the ordered ascus has a 2:2:2:2 ratio. Alternative chromosome orientations

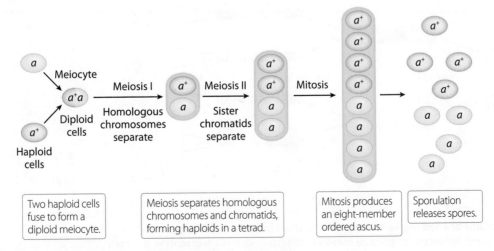

Figure 5.21 **Ordered ascus production in the fungus *Neurospora crassa*.**

**Figure 5.22 First-division segregation in ordered ascus formation.**

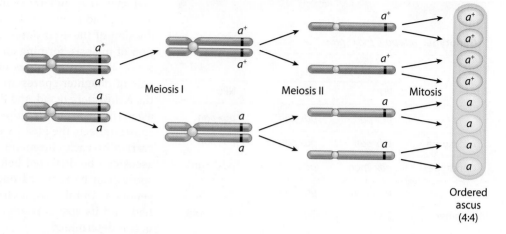

accompanied by single crossover produce three additional ordered ascus patterns that group identical mitotic products next to one another (**Figure 5.23b**). In each case, the overall 1:1 ratio of the two alleles is seen among the eight ascospores—only the order of spores differs. The relative proportion of second-division segregation asci is used to calculate the map distance (in centiMorgans) between a gene and the centromere via the formula

$$x\,\text{cM} = \frac{\frac{1}{2}(\text{number of second-division segregation asci})}{\text{total number of asci}} \times 100$$

This calculation is equivalent to counting the number of recombinant spores and dividing by the total number of progeny, because one-half the spores in second-division segregation asci are recombinant. **Figure 5.24** provides an example using *Neurospora crassa*. Wild-type fungi that grow as buff-colored colonies with normal growth habit are mated to mutants that grow as orange colonies with fluffy growth habit. As computed in the figure, the distance from the centromere to the color gene is 16.5 cM, and the distance from the centromere to the growth-habit gene is 30.7 cM.

**Genetic Insight** In eukaryotes whose gametes are contained in a single ascus, gamete proportions are used to calculate recombination frequencies.

## 5.8 Mitotic Crossover Produces Distinctive Phenotypes

Our discussion of crossing over and recombination has been limited to events that occur during meiosis. You may have wondered whether crossing over occurs during mitosis, and if so, what its consequences are. Synapsis of homologous chromosomes during mitosis occurs only occasionally in animals; thus, there is little opportunity for recombination to occur. In certain cases, however, homologous recombination does occur during mitosis. The rate of **mitotic crossover** varies considerably among organisms, but its consequences have been revealed through some fascinating examples.

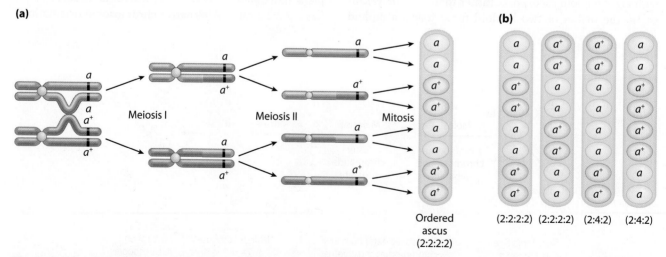

**Figure 5.23 Second-division segregation in ordered ascus formation.** (a) Single crossover produces a 2:2:2:2 ordered ascus. (b) Multiple outcomes of second-division segregation can occur, depending on the chromatids involved in crossing over.

| P (genotype) | F1 (genotype) | Trait | First Division (D1) | Second Division (D2) | Combined (D1 + D2) | Distance from Centromere to Trait $\frac{[\frac{D2}{2}]}{[D1+D2]} \times 100 = cM$ | Gene Map |
|---|---|---|---|---|---|---|---|
| $C^+g^+$ | $C^+g^+/cg$ | Color (c) | 73 | 36 | 109 | $\frac{[\frac{36}{2}]}{[109]} \times 100 = 16.5$ | 30.7 cM |
| $cg$ | | Growth (g) | 42 | 67 | 109 | $\frac{[\frac{67}{2}]}{[109]} \times 100 = 30.7$ | 16.5 cM, c g |

**Figure 5.24 Calculation of centromere-to-gene distance in *Neurospora crassa*.**

The first well-documented example of mitotic crossover came in 1936, when Curt Stern studied *Drosophila* crosses of two X-linked recessive traits, yellow body color (*y*) and short, twisted bristles called singed (*sn*). Stern crossed females homozygous for wild-type (gray) body color and singed bristles ($y^+ sn/y^+ sn$) with yellow-bodied, normal-bristled males ($y sn^+/Y$) and obtained dihybrid F₁ females that had wild-type body color and bristle form ($y^+ sn/y sn^+$). Close examination of a small number of F₁ females revealed an unexpected phenotype. These females had wild-type body color and wild-type bristles over most of the body but had small patches of either yellow body color or singed bristles. Even more surprising, some females had a patch of yellow body *and* a patch of singed bristles, and when they did, the patches were always *adjacent* to one another in a pattern called a twin spot (**Figure 5.25**). Among

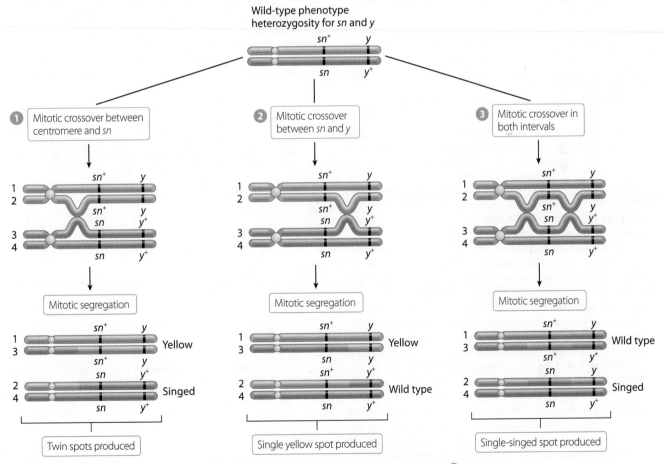

**Figure 5.25 Mitotic crossover.** In *Drosophila* crosses analyzed by Curt Stern, twin spots ❶, single yellow spot ❷, and single singed spot ❸ were produced by mitotic crossing over followed by a particular segregation pattern during mitotic cell division. In each case, the chromatids and their centromeres are numbered prior to crossing over. The numbers used after crossing over show the segregation patterns that produce the identified mitotic crossover phenotypes.

these three unusual spotting patterns, twin spot was about twice as common as single yellow spotting and single yellow spotting was much more common than single-singed spotting.

In formulating an explanation for the odd patches and their different frequencies, Stern reasoned that since the twin spots were always side by side, they must result from reciprocal events. He realized that rare crossover between homologous chromosomes during mitosis could explain twin spots, and it could also be a source of both kinds of single spots as well. Stern proposed that mitotic crossover events like those illustrated in Figure 5.25 were responsible for single and twin spots in *Drosophila*. Twin spotting is explained by mitotic

crossover between *sn* and the centromere if the particular pattern of chromosome segregation illustrated in Figure 5.25 takes place. Mitotic crossover between *y* and *sn* followed by the chromosome segregation shown produces single yellow spots. The double crossover and chromosome segregation pattern shown are required to produce single-singed spot. Twin spotting is the most common observation because the map distance between *sn* and the centromere is 45 cM. In contrast, the distance between *y* and *sn* is 21 cM, so twin spotting is about twice as common as single yellow spot. The double crossover producing single-singed spot is less frequent than either single-crossover, thus single-singed spot is the least frequent phenotype.

## CASE STUDY

### Mapping the Huntington Disease Gene

In the fall of 1934, a young graduate student named Lenore Sabin entered Thomas Hunt Morgan's genetics laboratory to begin a 2-year project studying the genetics of reproduction in fruit flies. Shortly after graduating, Sabin married Milton Wexler, and together they had two daughters, Alice and Nancy. Unknown to her at the time, Lenore Sabin Wexler carried a genetic time bomb—a mutation passed from her father to her and her three brothers that would produce the fatal autosomal dominant neuromuscular disorder Huntington disease (HD) in each of them.

Wexler's diagnosis, her struggle with the disease, and, ultimately, her death were the impetus for her husband Milton Wexler to form the Hereditary Disease Foundation that played a prominent role in funding and promoting the genetic research that mapped the *HD* gene. Wexler's condition inspired her daughter Nancy to become a genetics researcher and an advocate for HD research. The family's experience with HD led Alice to write a book chronicling her family's struggle with the disease and the Wexler family's involvement in the search for the disease gene (Wexler, 1995).

In the first attempts to map the *HD* gene by genetic linkage analysis, researchers searched for genetic linkage between the gene and biochemical markers (blood protein and enzyme genes). The results of these investigations were consistently negative, producing lod scores in excess of −2.0. However, these negative results eliminated about 20% of the human genome as the possible location of the *HD* gene. Negative lod score results tell researchers where the gene *is not* located, but if the gene is to be identified and its abnormalities characterized, its correct location must be found.

In the early 1980s, partially under the auspices of the Hereditary Disease Foundation, Nancy Wexler and a large group of colleagues embarked on a quest to map the *HD* gene using lod scores to test genetic linkage between the *HD* gene and a large number of newly discovered DNA genetic markers that had known chromosome locations. The research subjects were members of a very large extended Venezuelan family containing hundreds of cases of HD. The researchers collected and analyzed DNA genetic marker variation from 3000 individuals and assembled a pedigree containing more

than 12,000 members. Two members of Wexler's collaborative group, James Gusella and Michael Connealy, separately used lod score analysis to test genetic linkage between DNA genetic markers and the *HD* gene in two large American families that each contained many cases of HD. In 1983, a combined lod score analysis of the Venezuelan and American families identified genetic linkage between the *HD* gene and a DNA genetic marker known as D4S10, located at one end of human chromosome (Figure 5.26). The linkage of the *HD*

**(a)**

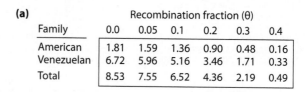

| Family | Recombination fraction ($\theta$) | | | | | |
|---|---|---|---|---|---|---|
| | 0.0 | 0.05 | 0.1 | 0.2 | 0.3 | 0.4 |
| American | 1.81 | 1.59 | 1.36 | 0.90 | 0.48 | 0.16 |
| Venezuelan | 6.72 | 5.96 | 5.16 | 3.46 | 1.71 | 0.33 |
| Total | 8.53 | 7.55 | 6.52 | 4.36 | 2.19 | 0.49 |

**(b)**

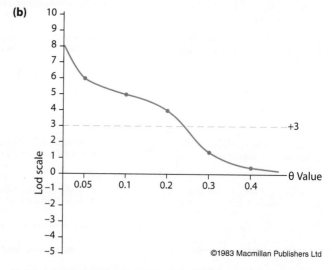

©1983 Macmillan Publishers Ltd

**Figure 5.26  Lod score data for the D4S10 marker and the Huntington disease gene.  (a)** The lod score table indicates significant evidence of genetic linkage from $\theta = 0.00$ to at least $\theta = 0.20$. **(b)** A curve of lod score values on $\theta$ values shows $Z_{max}$ at $\theta = 0.00$ and significantly positive lod scores up to $\theta = 0.25$.

gene to D4S10 is supported by calculating a $Z_{max}$ value of 8.53 at $\theta = 0.0$ for the combined Venezuelan and American families.

In the decade that followed, additional genetic linkage analysis in more families refined the location of the *HD* gene relative to D4S10. Based on the expanded analysis, a $Z_{max}$ value at $\theta = 0.03$ determined that the map distance between D4S10 and the *HD* gene is about 3 cM. Numerous additional DNA genetic markers linked to the *HD* gene were also identified and used to construct a complex genetic map of this region of chromosome 4. Armed with a genetic map pinpointing the location of the *HD* gene, researchers were able to isolate and clone the gene in the mid-1990s. This led to the eventual identification of the *huntingtin* (*HTT*) gene, the identification of the huntingtin protein produced by *HTT*, and determination of the mutational process—a process called *DNA triplet expansion* (Chapter 12).

Lenore Sabin Wexler, who worked closely with Thomas Hunt Morgan, in whose laboratory genetic linkage was first described and explained, played an important role in the genetic linkage study that mapped the disease gene she carried. Through her family's foundation and the exhaustive efforts of her husband and daughters, Lenore Wexler was at the center of the genetic linkage analysis that mapped *HTT*.

## SUMMARY

MasteringGENETICS

*For activities, animations, and review quizzes, go to the study area at www.masteringgenetics.com.*

### 5.1 Linked Genes Do Not Assort Independently

- Genetic linkage identifies genes that are so close to one another on a chromosome that their alleles do not assort independently.

- With genetic linkage, parental combinations at frequencies that are significantly greater than those predicted by chance and nonparental combinations are much less frequent than expected.

- William Bateson and Reginald Punnett first observed genetic linkage when they noticed high numbers of parental phenotypes in $F_2$ progeny.

- Thomas Hunt Morgan performed test-cross analysis of linked genes to demonstrate that linkage violates independent assortment and that crossover between homologous chromosomes is responsible for the production of recombinant gametes.

- Crossover frequency between linked genes is correlated with the distance between genes on a chromosome. Crossover occurs less often between genes that are close together than between genes that are farther apart.

- In crosses involving linked genes, the two parental phenotypes are observed in progeny in approximately equal frequencies. The two recombinant phenotypes also occur at approximately equal frequency.

### 5.2 Genetic Linkage Mapping Is Based on Recombination Frequency between Genes

- The correlation between physical map distance and recombination frequency permits gene mapping based on recombination frequency.

### 5.3 Three-Point Test-Cross Analysis Maps Genes

- Three or more genes can be mapped by test-cross analysis. In a three-point cross, parental phenotypes are most frequent, double recombinants are least frequent, and the four phenotypes resulting from two single-recombination events are of intermediate frequency that depends on the actual distance between genes.

- Genetic linkage maps are constructed in five steps:
  1. Find significantly higher proportions of parental phenotypes than predicted by chance.
  2. Identify the alleles on parental chromosomes (the most common classes).
  3. Identify double recombinants (the least frequent classes), comparing them to parental chromosomes to determine gene order.
  4. Calculate recombination frequencies between genes.
  5. Calculate interference with the occurrence of double crossovers.

- Recombination frequency usually underestimates the physical distance between genes. Mapping functions are used to correct these estimates.

### 5.4 Recombination Results from Crossing Over

- Studies correlating genetic recombination with the visible recombination of distinctive physical structures on chromosomes support the idea that crossing over causes recombination.

- Crossing over occurs at the four-strand stage in prophase I of meiosis, after completion of DNA replication. Two nonsister chromatids of homologous chromosomes exchange parts in two-strand single crossovers. Two, three, or all four chromatids can be involved in double crossovers.

- Recombination occurs within genes as well as between genes. Several biological properties of organisms affect recombination. In animals, the heterogametic sex experiences less recombination genome-wide than the homogametic sex.

### 5.5 Linked Human Genes Are Mapped Using Lod Score Analysis

- Statistical approaches such as lod score analysis detect evidence of linkage in small families.

- Lod score analysis determines the likelihood of genetic linkage between genes at specified recombination values ($\theta$ values). A cumulative lod score of +3.0 or more is statistically significant evidence in favor of genetic linkage between two genes. Lod scores of −2.0 or less represent significant evidence against genetic linkage.

### 5.6 Genetic Linkage Analysis Traces Genome Evolution

- Linkage disequilibrium between alleles of linked genes may indicate that evolution is acting on the genome region.

## 5.7 Genetic Linkage in Haploid Eukaryotes Is Identified by Tetrad Analysis

▌ In certain eukaryotic microorganisms, the products of individual meiotic cell divisions are contained within an ascus. Parental and recombinant gametes contained in an ascus can be analyzed to map genes.

## 5.8 Mitotic Crossover Produces Distinctive Phenotypes

▌ Mitotic crossing over is a rare event that produces patches of tissue with unusual phenotype.

## KEYWORDS

allelic phase *(p. 165)*
ascus (ascospore) *(p. 170)*
coefficient of coincidence (c) *(p. 157)*
double recombinant (double crossover), two-, three-, or four- stranded *(pp. 156, 160)*
first-division segregation *(p. 173)*
genetic linkage (genetic linkage mapping), complete and incomplete *(p. 144)*
haplotype *(p. 168)*
interference (*I*), *(p. 157)*
intragenic recombination *(p. 161)*

linkage equilibrium and disequilibrium *(p. 169)*
linkage group *(p. 165)*
lod score (log of the odds ratio) *(p. 166)*
mapping function *(p. 164)*
map unit (m.u.), centiMorgan (cM) *(p. 153)*
mitotic crossover *(p. 174)*
negative interference *(p. 158)*
nonparental ditype (NPD) *(p. 171)*
ordered ascus *(p. 173)*
parental (nonrecombinant) chromosome or gamete *(p. 144)*
parental ditype (PD) *(p. 170)*

recombinant (nonparental) chromosome or gamete *(p. 144)*
recombination frequency (r) *(p. 147)*
second-division segregation *(p. 173)*
syntenic genes *(p. 144)*
tetrad *(p. 170)*
tetrad analysis *(p. 170)*
tetratype (TT) *(p. 171)*
theta value (θ value) *(p. 166)*
three-point test-cross analysis *(p. 153)*
two-point test-cross analysis *(p. 149)*
unordered tetrad *(p. 170)*
$Z_{max}$ *(p. 166)*

## PROBLEMS        *For instructor-assigned tutorials and problems, go to www.masteringgenetics.com.*

### Chapter Concepts

*For answers to selected even-numbered problems, see Appendix: Answers.*

1. Make a diagram illustrating the alleles on homologous chromosomes for the following genotypes, assuming in each case that the genes reside on the same chromosome in the order written:
   a. *AB/ab*
   b. *aBc/abC*
   c. *DFg/DFG*
   d. the gametes produced by an organism with the genotype *Rt/rT*
   e. progeny of the cross *Rt/rT* × *rt/rt*

2. In a diploid species of plant, the genes for plant height and fruit shape are syntenic and separated by 18 m.u. Allele *D* produces tall plants and is dominant to *d* for short plants, and allele *R* produces round fruit and is dominant to *r* for oval fruit.
   a. A plant with the genotype *DR/dr* produces gametes. Identify gamete genotypes, label parental and recombinant gametes, and give the frequency of each gamete genotype.
   b. Give the same information for a plant with the genotype *Dr/dR*.

3. A pure-breeding tall plant producing oval fruit as described in Problem 2 is crossed to a pure-breeding short plant producing round fruit.
   a. The F₁ are crossed to short plants producing oval fruit. What are the expected proportions of progeny phenotypes?
   b. If the F₁ identified in part (a) are crossed to one another, what proportion of the F₂ are expected to be short and produce round fruit? What proportion are expected to be tall and produce round fruit?

4. Genes *E* and *H* are syntenic in an experimental organism with the genotype *EH/eh*. Assume that during each meiosis, one crossover occurs between these genes. No homologous chromosomes escape crossover, and none undergo double crossover. Are genes *E* and *H* genetically linked? Why or why not? What is the proportion of parental gametes produced by meiosis?

5. In tomato plants, purple leaf color is controlled by a dominant allele *A*, and green leaf by a recessive allele *a*. At another locus, hairy leaf *H* is dominant to hairless leaf *h*. The genes for leaf color and leaf texture are separated by 16 m.u. on chromosome 5. On chromosome 4, a gene controlling leaf shape has two alleles: a dominant allele *C* that produces cut-leaf shape and a recessive allele *c* that produces potato-shaped leaf.
   a. The cross of a purple, hairy, cut plant heterozygous at each gene to a green, hairless, potato plant produces the following progeny:

| Phenotype | Frequency, % |
|---|---|
| Purple, hairy, cut | 21 |
| Purple, hairy, potato | 21 |
| Green, hairless, cut | 21 |
| Green, hairless, potato | 21 |
| Purple, hairless, cut | 4 |
| Purple, hairless, potato | 4 |
| Green, hairy, cut | 4 |
| Green, hairy, potato | 4 |
|  | 100 |

Give the genotypes of parental and progeny plants in this experiment.

b. Fully explain the number and frequency of each phenotype class.

6. In *Drosophila*, the map positions of genes are given in map units numbering from one end of a chromosome to the other. The X chromosome of *Drosophila* is 66 m.u. long. The X-linked gene for body color—with two alleles, $y^+$ for gray body and $y$ for yellow body—resides at one end of the chromosome at map position 0.0. A nearby locus for eye color, with alleles $w^+$ for red eye and $w$ for white eye, is located at map position 1.5. A third X-linked gene, controlling bristle form, with $f^+$ for normal bristles and $f$ for forked bristles, is located at map position 56.7. Each gene resides on the X chromosome, and at each locus the wild-type allele is dominant over the mutant allele.

a. In a cross involving these three X-linked genes, do you expect any gene pair(s) to show genetic linkage? Explain your reasoning.

b. Do you expect any of these gene pair(s) to assort independently? Explain your reasoning.

c. A wild-type female fruit fly with the genotype $y^+w^+f/ywf^+$ is crossed to a male fruit fly that has yellow body, white eye, and forked bristles. Predict the frequency of each progeny phenotype class produced by this mating.

d. Explain how each of the predicted progeny classes is produced.

7. Genes *A*, *B*, and *C* are linked on a chromosome and found in the order *A-B-C*. Genes *A* and *B* recombine with a frequency of 8%, and genes *B* and *C* recombine at a frequency of 24%. For the cross $a^+b^+c/abc^+ \times abc/abc$, predict the frequency of progeny. Assume interference is zero.

8. Gene *G* recombines with gene *T* at a frequency of 7%, and gene *G* recombines with gene *R* at a frequency of 4%.

a. Draw two possible genetic maps for these three genes, and identify the recombination frequencies predicted for each map.

b. Assuming any desired genotype is available, propose a genetic cross whose result could be used to determine which of the proposed genetic maps is correct.

9. Genes *A*, *B*, *C*, *D*, and *E* are linked on a chromosome and occur in the order given. The test cross $Ae/aE \times ae/ae$ indicates the genes recombine with a frequency of 28%.

a. If 1000 progeny are produced by the test cross, determine the number of progeny in each outcome class.

b. Previous genetic linkage crosses have determined that recombination frequencies for these genes are 6% for genes *A* and *B*, 4% for genes *B* and *C*, 10% for genes *C* and *D*, and 11% for genes *D* and *E*. The sum of these frequencies between genes *A* and *E* is 31%. Why does the recombination distance between these genes, determined by adding the intervals between adjacent linked genes, differ from the distance determined by the test cross?

10. Syntenic genes can assort independently. Explain this observation.

11. The recombination frequency between linked genes is less than 50%. Why is 50% recombination the maximum value?

12. On the Drosophila X chromosome, the dominant allele $y^+$ produces gray body color and the recessive allele $y$ produces yellow body. This gene is linked to one controlling full eye shape by a dominant allele $lz^+$ and lozenge eye shape with a recessive allele $lz$. These genes recombine with a frequency of approximately 28%. The *L* gene is linked to gene *F* controlling bristle form, where the dominant is long bristles and the recessive is forked bristles. The *L* and *F* genes recombine with a frequency of approximately 32%.

a. Using any genotypes you choose, design two separate crosses, one to test recombination between genes *Y* and *L* and the second between genes *L* and *F*. Assume 1000 progeny are produced by each cross, and give the number of progeny in each outcome category. (In setting up your crosses, remember that *Drosophila* males do not undergo recombination.)

b. Can any cross reveal genetic linkage between gene *Y* and gene *F*? Why or why not?

c. Why is "independent assortment" the genetic term that best describes the observations of a genetic cross between gene *L* and gene *F*?

## Application and Integration

*For answers to selected even-numbered problems, see Appendix: Answers.*

13. Researchers cross a corn plant that is pure-breeding for the dominant traits colored aleurone (*C1*), full kernel (*Sh*), and waxy endosperm (*Wx*) to a pure-breeding plant with the recessive traits colorless aleurone (*c1*), shrunken kernel (*sh*), and starchy (*wx*). The resulting $F_1$ plants were crossed to pure-breeding colorless, shrunken, starchy plants. Counting kernels from about 30 ears of corn, The following results are reported.

| Kernel Phenotype | Number |
| --- | --- |
| Colored, shrunken, starchy | 116 |
| Colored, full, starchy | 601 |
| Colored, full, waxy | 2538 |
| Colored, shrunken, waxy | 4 |
| Colorless, shrunken, starchy | 2708 |
| Colorless, full, starchy | 2 |
| Colorless, full, waxy | 113 |
| Colorless, shrunken, waxy | 626 |
| | 6708 |

a. Why are these data consistent with genetic linkage among the three genes?

b. Perform a chi-square test to determine if these data show significant deviation from the expected phenotype distribution.

c. What is the order of these genes in corn?

d. Calculate the recombination fraction between the gene pairs.

e. What is the interference value for this data set?

14. Nail–patella syndrome is an autosomal disorder affecting the shape of nails on fingers and toes, as well as the structure of kneecaps. The pedigree below shows the transmission of nail–patella syndrome in a family along with ABO blood type.

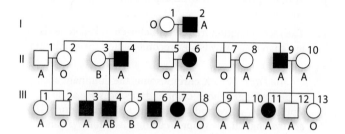

a. Is nail–patella syndrome a dominant or a recessive condition? Explain your reasoning.

b. Does this family give evidence of genetic linkage between nail–patella syndrome and ABO blood group? Why or why not?

c. Using $N$ and $n$ to represent alleles at the nail–patella locus and $I^A$, $I^B$, and $I$ to represent ABO alleles, write the genotypes of I-1 and I-2 as well as their five children in generation II.

d. Why does III-6 have nail–patella syndrome and III-8 not have the disorder? Give genotypes for these two individuals.

e. Why does III-11 have nail–patella syndrome and III-12 not have nail–patella syndrome? Give genotypes for these two individuals.

15. Three dominant traits of corn seedlings, tunicate seed ($T–$), glossy appearance ($G–$), and liguled stem ($L–$), are studied along with their recessive counterparts, nontunicate ($tt$), nonglossy ($gg$), and liguleless ($ll$). A trihybrid plant with the three dominant traits is crossed to a nontunicate, nonglossy, liguleless plant. Kernels on ears of progeny plants are scored for the traits, with the following results:

| Phenotype | Number |
| --- | --- |
| Tunicate, glossy, liguled | 102 |
| Tunicate, glossy, liguleless | 106 |
| Tunicate, nonglossy, liguled | 18 |
| Tunicate, nonglossy, liguleless | 20 |
| Nontunicate, glossy, liguled | 22 |
| Nontunicate, glossy, liguleless | 23 |
| Nontunicate, nonglossy, liguled | 99 |
| Nontunicate, nonglossy, liguleless | 110 |
|  | 500 |

a. Is there evidence of genetic linkage among any of these gene pairs? If so, identify the evidence.

b. Is there evidence of independent assortment among any of these gene pairs? If so, identify the evidence.

c. Using the gene symbols given above, write the genotypes of $F_1$ and $F_2$ plants.

d. If evidence of linkage is present, calculate the recombination fraction(s) from the data presented.

e. Could all three genes be carried on the same chromosome? Discuss why or why not.

16. In a diploid plant species, an $F_1$ with the genotype $Gg\ Ll\ Tt$ is test-crossed to a pure-breeding recessive plant with the genotype $gg\ ll\ tt$. The offspring genotypes are as follows:

| Genotype | Number |
| --- | --- |
| Gg Ll Tt | 621 |
| Gg Ll tt | 3 |
| Gg ll Tt | 64 |
| Gg ll tt | 109 |
| gg Ll Tt | 103 |
| gg Ll tt | 67 |
| gg ll Tt | 7 |
| gg ll tt | 626 |
|  | 1600 |

a. What is the order of these three linked genes?

b. Calculate the recombination fractions between each pair of genes.

c. Why is the recombination fraction for the outside pair of genes not equal to the sum of recombination fractions between the adjacent gene pairs?

d. What is the interference value for this data set?

e. Explain the meaning of this $I$ value.

17. The map below indicates the order of genes on a chromosome and the number of map units between adjacent gene pairs. Give the expected frequency of gametes produced by organisms with the following genotypes. Assume one map unit equals 1% recombination and that interference is zero.

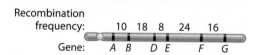

a. $De/dE$

b. $Ad/aD$

c. $DeF/dEf$

d. $BdE/bde$

e. $AEG/aeg$

f. $ABDE/abde$

18. The Rh blood group in humans is determined by a gene on chromosome 1. A dominant allele produces Rh+ blood type, and a recessive allele generates Rh−. Elliptocytosis is an autosomal dominant disorder that produces abnormally shaped red blood cells that have a short life span resulting in hereditary anemia. A large family with elliptocytosis is tested for genetic linkage of Rh blood

group and the disease. The lod score data below are obtained for the family.

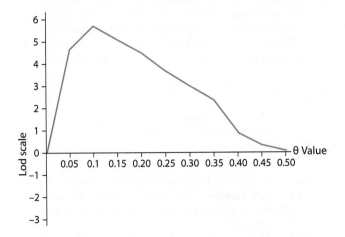

a. From these data, can you conclude that Rh and elliptocytosis loci are genetically linked in this family? Why or why not?

b. What is $Z_{max}$ for this family?

c. Over what range of $\theta$ do lod scores indicate significant evidence in favor of genetic linkage?

19. Genetic linkage mapping for a large number of families identifies 4% recombination between the genes for Rh blood type and elliptocytosis. At the Rh locus, alleles $R$ and $r$ control Rh+ and Rh− blood types. Allele $E$ producing elliptocytosis is dominant to the wild-type recessive allele $e$. Tom and Terri each have elliptocytosis, and each is Rh+. Tom's mother has elliptocytosis and is Rh− while his father is healthy and has Rh+. Terri's father is Rh+ and has elliptocytosis; Terri's mother is Rh− and is healthy.

a. What is the probability that the first child of Tom and Terri will be Rh− and have elliptocytosis?

b. What is the probability that a child of Tom and Terri who is Rh+ will have elliptocytosis?

20. *Neurospora* with the genotype $a^+ a$ form tetrads in the following frequencies:

| Tetrad | Number |
| --- | --- |
| $a^+a^+a\,a$ | 192 |
| $a\,a\,a^+a^+$ | 208 |
| $a\,a^+a\,a^+$ | 23 |
| $a\,a^+a^+a$ | 27 |
| $a^+a\,a^+a$ | 29 |
| $a^+a\,a\,a^+$ | 21 |
| | 500 |

a. What is the distance between the gene and the centromere?

b. Diagram the meiosis producing the tetrad class $a\,a\,a^+a^+$.

c. Diagram the meiosis producing the tetrad class $a^+a\,a\,a^+$.

21. Gene $R$ and gene $T$ are genetically linked. Answer the following questions concerning a dihybrid organism with the genotype $Rt/rT$:

a. If $r = 0.20$, give the expected frequencies of gametes produced by the dihybrid.

b. Determine the gamete frequencies if a two-strand double crossover occurs between the genes.

c. Determine the genotypes of gametes produced by a three-strand double crossover in this dihybrid organism.

d. Determine the genotypes of gametes produced by a four-strand double crossover in this dihybrid.

22. T. H. Morgan's data on eye color and wing form, shown in Figures 5.3 and 5.6, reveal genetic linkage between the two genes. Test this genetic linkage data with chi-square analysis, and show that the results are significantly different from the expectation under the assumption of independent assortment.

23. A wild-type trihybrid soybean plant is crossed to a pure-breeding soybean plant with the recessive phenotypes pale leaf (l), oval seed (r), and short height (t). The results of the three-point test cross are shown below. Traits not listed are wild type.

| Phenotype | Number |
| --- | --- |
| Pale | 648 |
| Pale, oval | 64 |
| Pale, short | 10 |
| Pale, oval, short | 102 |
| Oval | 6 |
| Oval, short | 618 |
| Short | 84 |
| Wild type | 98 |
| | 1630 |

a. What are the alleles on each homologous chromosome of the parental wild-type trihybrid soybean plant? Place the alleles in their correct gene order. Use $L$, $R$, and $T$ to represent dominant alleles and $l$, $r$, and $t$ for recessive alleles.

b. Calculate the recombination fraction between the adjacent pairs of genes.

c. Calculate the interference value for these data.

24. The boss in your laboratory has just heard of a proposal by another laboratory that genes for eye color and the length of body bristles may be linked in *Drosophila*. Your lab has numerous pure-breeding stocks of *Drosophila* that could be used to verify or refute genetic linkage. In *Drosophila*, red eyes ($c+$) are dominant to brown eyes ($c$), and long bristles ($d+$) are dominant to short bristles ($d$). Your lab boss asks you to design an experiment to test the genetic linkage of eye color and bristle-length genes, and to begin by crossing a pure-breeding line homozygous for red eyes and short bristles to a pure-breeding line that has brown eyes and long bristles.

a. Give the genotypes of the pure-breeding parental flies, and the genotype(s) and phenotype(s) of the $F_1$ progeny they produce.

b. In your experimental design, what is the genotype and phenotype of the line you propose to cross to the $F_1$ to obtain the most useful information about genetic linkage between the eye color and bristle-length genes? Explain why you make this choice.

c. Assume the eye color and bristle-length genes are separated by 28 m.u. What are the approximate frequencies of phenotypes expected from the cross you proposed in part (b)?

d. How would the results of the cross differ if the genes are not linked?

25. In rabbits, chocolate-colored fur (*w+*) is dominant to white fur, straight fur (*c+*) is dominant to curly fur (*c*), and long ear (*s+*) is dominant to short ear (*s*). The cross of a trihybrid rabbit with straight, chocolate-colored fur and long ears to a rabbit that has white, curly fur and short ears produces the following results:

| Phenotype | Number |
|---|---|
| White, short, straight | 13 |
| Chocolate, long, straight | 165 |
| Chocolate, long, curly | 13 |
| White, long, straight | 82 |
| Chocolate, short, straight | 436 |
| Chocolate, short, curly | 79 |
| White, short, curly | 162 |
| White, long, curly | 450 |
| | 1400 |

a. Determine the order of the genes on the chromosome, and identify the alleles that are present on each of the homologous chromosomes in the trihybrid rabbits.

b. Calculate the recombination frequencies between each of the adjacent pairs of genes.

c. Determine the interference value for this cross.

26. The following progeny are obtained from a test cross of a trihybrid wild-type plant to a plant with the recessive phenotypes compound leaves (*c*), intercalary leaflets (*i*), and green fruits (*g*). (Traits not listed are wild type.) The test-cross progeny are as follows:

| Phenotype | Number |
|---|---|
| Compound leaves | 324 |
| Compound leaves, intercalary leaflets | 32 |
| Compound leaves, green fruits | 5 |
| Compound leaves, intercalary leaflets, green fruits | 51 |
| Intercalary leaflets | 3 |
| Intercalary leaflets, green fruits | 309 |
| Green fruits | 42 |
| Wild type | 49 |
| | 815 |

a. Determine the order of the three genes, and construct a genetic map that identifies the correct order and the alleles carried on each chromosome in the trihybrid parental plant.

b. Calculate the frequency of recombination between the adjacent pairs of genes in the map.

c. How many double-crossover progeny are expected among the test-cross progeny? Calculate the interference for this cross.

27. In tomatoes, the allele *T* for tall plant height is dominant to dwarf allele *t*, the *P* allele for smooth skin is dominant to the *p* allele for peach fuzz skin, and the allele *R* for round fruit is dominant to the recessive *r* allele for oblong fruit. The genes controlling these traits are linked on chromosome 1 in the tomato genome, and the genes are arranged in the order and with the recombination frequencies shown.

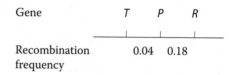

| Gene | *T* | *P* | *R* |
|---|---|---|---|
| Recombination frequency | | 0.04 | 0.18 |

a. A pure-breeding tall, peach fuzz, round plant is crossed to a pure-breeding plant that is dwarf, smooth, oblong. What are the gamete genotypes produced by each of these plants?

b. What is the genotype and phenotype of the $F_1$ progeny of this cross?

c. What are the genotypes of gametes produced by the $F_1$, and what is the predicted frequency of each gamete?

d. The $F_1$ are test-crossed to dwarf, peach fuzz, oblong plants, and 1000 test-cross progeny are produced. What are the phenotypes of test-cross progeny, and what number of progeny is expected in each class?

28. Neurofibromatosis 1 (NF1) is an autosomal dominant disorder inherited on human chromosome 17. Part of the analysis mapping the *NF1* gene to chromosome 17 came from genetic linkage studies testing segregation of *NF1* and DNA genetic markers on various chromosomes. A DNA marker with two alleles, designated *1* and *2*, is linked to *NF1*. The pedigree below shows segregation of *NF1* (darkened symbols) and gives genotypes for the DNA marker for each family member.

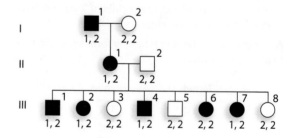

a. Determine the alleles for the *NF1* gene and the DNA marker gene on each chromosome carried by the four family members in generation I and generation II. Use *N* for the dominant *NF1* allele and *n* for the recessive allele and assume I-1 is heterozygous for the disease allele (*Nn*).

b. Based on the phase of alleles on chromosomes in generation II, is there any evidence of recombination among the eight offspring in generation III? Explain.

c. What is the estimated recombination frequency between the *NF1* gene and the DNA marker?

29. A 2006 genetic study of a large American family (Ikeda et al., 2006) identified genetic linkage between DNA markers on chromosome 11 and the gene producing the autosomal dominant neuromuscular disorder spinocerebellar ataxia

type 5 (*SCA5*). The following lod score data are taken from the 2006 study:

| | \multicolumn{6}{c}{Theta (θ) Value} | | | | | |
| --- | --- | --- | --- | --- | --- | --- |
| | 0.01 | 0.05 | 0.10 | 0.20 | 0.30 | 0.40 |
| *SCA5* and DNA marker *A* | 11.02 | 12.26 | 11.94 | 10.04 | 7.26 | 3.77 |
| *SCA5* and DNA marker *B* | 0.35 | 0.94 | 1.07 | 0.99 | 0.75 | 0.43 |

a. Does either group of lod scores indicate statistically significant odds in favor of genetic linkage? Explain your answer.
b. What is the maximum value for each set of lod scores?
c. Based on the available information, is DNA marker *A* linked to the gene producing SCA5? Explain your answer.
d. Based on available information, is DNA marker *B* linked to the gene for SCA5? Explain your answer.

30. A *Drosophila* experiment examining potential genetic linkage of X-linked genes studies a recessive eye mutant (echinus), a recessive wing-vein mutation (crossveinless), and a recessive bristle mutation (scute). The wild-type phenotypes are dominant. Trihybrid wild-type females (all have the same genotype) are crossed to hemizygous males displaying the three recessive phenotypes. Among the 20,785 progeny produced from these crosses are the phenotypes and numbers listed in the table. Any phenotype not given is wild type.

| Phenotype | Number |
| --- | --- |
| 1.  Echinus | 8576 |
| 2.  Scute | 977 |
| 3.  Crossveinless | 716 |
| 4.  Echinus, scute | 681 |
| 5.  Scute, crossveinless | 8808 |
| 6.  Scute, crossveinless, echinus | 4 |
| 7.  Echinus, crossveinless | 1002 |
| 8.  Wild type | 1 |
| | 20,764 |

a. Determine the gene order and identify the alleles on the homologous X chromosomes in the trihybrid females.
b. Calculate the recombination frequencies between each of the gene pairs.
c. Compare the recombination frequencies and speculate about the source of any apparent discrepancies in the recombination data.
d. Use chi-square analysis to demonstrate that the data in this experiment are not the result of independent assortment.

31. As part of their analysis of intragenic recombination, Melvin Green and Kathleen Green studied lozenge-eyed females with the mutation $lz^{46}$ on one X chromosome and the mutation $lz^g$ on the homologous X chromosome. The $lz^g$-bearing X chromosome also carried recessive mutations for cut wing (*ct*) and vermilion-colored eye (*v*). These females were mated to cut wing males that had vermilion-colored, lozenge-shaped eyes. The chromosomes of these flies are depicted below.

a. Diagram the recombination event within the *lz* gene and draw the resulting recombinant X chromosomes, illustrating the *lz* alleles and the flanking markers on each chromosome.
b. What are the phenotypes of progeny male flies carrying *lz* intragenic recombinants? (A double-mutant X chromosome carrying both $lz^g$ and $lz^{46}$ produces a compound lozenge eye that has a different appearance than either the $lz^g$- or $lz^{46}$-derived eye.)

32. In experiments published in 1918 that sought to verify and expand the genetic linkage and recombination theory proposed by Morgan, Thomas Bregger studied potential genetic linkage in corn (*Zea mays*) for genes controlling kernel color (colored is dominant to colorless) and starch content (starchy is dominant to waxy). Bregger performed two crosses. In Cross 1, pure-breeding colored, starchy-kernel plants (*C1 Wx/C1 Wx*) were crossed to plants pure-breeding for colorless, waxy kernels (*c1 wx/c1 wx*). The F$_1$ of this cross were test-crossed to colorless, waxy plants. The test-cross progeny are as follows:

| Phenotype | Number |
| --- | --- |
| Colored, waxy | 310 |
| Colored, starchy | 858 |
| Colorless, waxy | 781 |
| Colorless, starchy | 311 |
| | 2260 |

In Cross 2, plants pure-breeding for colored, waxy kernels (*C1 wx/C1 wx*) and colorless, starchy kernels (*c1 Wx/c1 Wx*) were mated, and their F$_1$ were test-crossed to colorless, waxy plants. The test-cross progeny are as follows:

| Phenotype | Number |
| --- | --- |
| Colored, waxy | 340 |
| Colored, starchy | 115 |
| Colorless, waxy | 92 |
| Colorless, starchy | 298 |
| | 845 |

a. For each set of test-cross progeny, determine whether genetic linkage or independent assortment is more strongly supported by the data. Explain the rationale for your answer.
b. Calculate the recombination frequency for each of the progeny groups.
c. Are the results of these two experiments mutually compatible with the hypothesis of genetic linkage? Explain why or why not.
d. Merge the two sets of progeny data and determine the combined recombination frequency.

# 6

# Genetic Analysis and Mapping in Bacteria and Bacteriophage

## ESSENTIAL IDEAS

▪ Bacterial conjugation is a one-way transfer of genetic material from a donor cell to a recipient cell. Three types of donor cells can conjugate with recipient cells to transfer donor DNA.

▪ Donor bacterial genetic maps are derived from conjugation analysis.

▪ A particular type of bacteria conjugation can produce bacteria with genomes that are partially diploid.

▪ Transformation is the absorption of extracellular DNA across the cell wall and membrane of a recipient bacterial cell, and its analysis leads to mapping of donor bacterial genes.

▪ Transduction, mediated by bacteriophages, is the transfer of DNA from a donor bacterial cell to a recipient cell, and its analysis leads to mapping of donor bacterial genes.

▪ Fine-structure genetic analysis of a bacteriophage genome demonstrated that DNA nucleotide base pairs are the fundamental units of mutation and recombination.

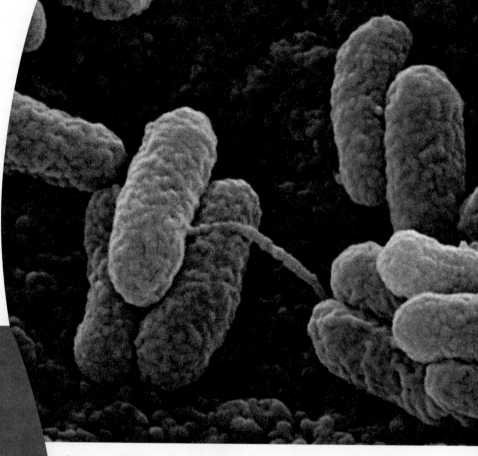

Bacteria transfer DNA to one another by multiple mechanisms, including the process of gene transfer called conjugation shown here. The bacterial "donor" (center left) transfers DNA through a tube that connects it to a bacterial "recipient" (lower right).

Here's a disturbing little secret of human life: Your body contains approximately 100 trillion cells, but only about 10 trillion of them are yours! The other 90% of the cells you carry around are bacteria, fungi, and other forms of microscopic life. Many of these biological hitchhikers perform useful, even essential, functions. For example, you carry hundreds of species of bacteria in your gut that collectively have a mass of more than 3 pounds. Without these intestinal bacteria, your digestion of carbohydrates would be impaired, and your ability to manufacture essential nutrients such as vitamin $B_{12}$ and vitamin K would be disabled. The bacteria teeming in your digestive tract also help keep potentially harmful bacteria at bay by vigorously competing for available nutrients. Similarly, the millions of bacteria that currently reside on your skin (yes, even though you showered recently!) help keep your skin healthy by competing with infectious bacteria. Despite this normal and healthy competition, harmful bacteria can gain access to our bodies. Occasionally even our normally helpful microbial passengers turn against us and cause illness, infection, or in extreme cases, death.

Given the biological, medical, and technological importance of bacteria and other microorganisms, it is no wonder they are studied intensively in modern genetics, using the bacterium *Escherichia coli* and yeast *Saccharomyces cerevisiae* as model genetic organisms. The relative ease of studying microorganisms fueled revolutionary change in genetics in the latter half of the 20th century. Much of the initial information in molecular genetics and many of the methods of genetic analysis pioneered in the study of bacteria have proven valuable in the study of more complex organisms.

In this chapter, we examine the genetics of bacteria and learn how maps of their genomes are constructed. We take a genetic approach in our discussion, focusing on genetic techniques that are used to map genes in bacterial genomes. Genome sequencing has become available on more than 1000 bacterial genomes in recent years, and these sequencing databases have enormous utility (Chapter 18). Our focus here, however, is on investigating and understanding how genetic analysis is applied to the study of gene transfer and mapping in bacterial and bacteriophage genomes.

We begin by looking at three mechanisms by which DNA can be transferred from one bacterium to another. After showing how analysis of these processes helps microbial geneticists locate the positions of genes on the bacterial chromosome, the chapter turns to a discussion of bacteriophages, the viruses that infect bacterial cells. It describes experiments that led to a fine-structure map of a bacteriophage genome and provided an essential bridge between transmission genetics and modern molecular genetics.

## 6.1 Bacteria Transfer Genes by Conjugation

Bacteria propagate by binary fission, a process in which the bacterial chromosome replicates, and a copy is distributed to each of the progeny cells along with a share of the contents of the dividing cell. In a matter of hours, this form of clonal propagation can generate a "colony" containing thousands of genetically identical bacteria cells. The ability of bacteria to produce colonies of clones, however, does not mean that bacteria never recombine genetically. A series of studies in the 1940s and 1950s identified and described the three mechanisms of gene transfer and recombination between bacteria that are a focus of this chapter.

Bacteria are a highly diverse taxonomic group, and they are essential for genetic study. Among the features that make bacteria so useful to geneticists are the following:

- **Genomic simplicity.** Most bacterial genomes contain fewer genes and fewer base pairs in their haploid genomes than do other organisms.

- **Uncomplicated genotypes.** The haploid genome of most bacteria allows all mutations to be observed directly, without interference from dominance interactions between alleles.

- **Rapid generation times.** Bacteria reproduce prolifically; their generation times can be measured in minutes.

- **Large numbers of progeny.** Enormous numbers of clonal progeny can be examined, increasing the likelihood that statistically rare events will be observed.

- **Ease of propagation.** Microbes may be grown either in liquid culture or on culture plates. The cultures are easy and inexpensive to maintain, and they require little laboratory space.

- **Numerous heritable differences.** Mutants are easily created, identified, isolated, and manipulated for examination.

A central activity of interest in this chapter is the propensity of bacteria to transfer genetic material from one to another. Transfer occurs by three processes: *conjugation*, the transfer of replicated DNA from a donor bacterium to a recipient bacterium; *transformation*, the uptake of DNA from the environment by a recipient bacterium; and *transduction*, the transfer of DNA from a donor bacterium to a recipient bacterium by way of a viral vector. Each of these mechanisms involves a *one-way transfer* of genetic material from a bacterial *donor* cell to a *recipient* cell. The transferred DNA is either an extrachromosomal *plasmid* or a portion of the donor bacterial chromosome. Often, the plasmids transferred into recipient cells bring new genes that change the growth behavior of recipient cells. Alternatively, plasmids may carry a *second copy* of genes already on the bacterial chromosome.

When bacterial chromosome DNA from the donor cell is transferred to a recipient bacterium, the homologous parts of the donor and recipient DNA molecules can undergo recombination that leads to a change in the genotype of the recipient cell. Regardless of the nature of the DNA transferred from donor cells to recipient cells, a key to understanding the process is to remember that it is a one-way street: Genetic material moves from donor to recipient.

## Characteristics of Bacterial Genomes

Bacterial genomes are usually composed of a *single* chromosome that carries primarily essential genes—those necessary for the species' metabolic and growth activities. The **bacterial chromosome** is usually a covalently closed, *circular* molecule of double-stranded DNA. In keeping with the small size of the genome—from a few hundred thousand to several million base pairs—the bacterial chromosome, too, is usually quite *small,* likewise varying from a few hundred thousand to several million base pairs. Occasional exceptions are found. For example, the bacterial genus *Vibrio* has a genome containing two bacterial chromosomes, designated chromosome I and chromosome II, and chromosome I is considerably larger than chromosome II (see Chapter 11).

In addition to the main bacterial chromosome, most bacteria also carry multiple copies of **plasmids,** small double-stranded circular DNA molecules containing nonessential genes that are used infrequently or under specialized conditions not ordinarily encountered by the species (**Figure 6.1**). Plasmids vary widely in their number of genes and their total number of base pairs, but they are always considerably smaller than bacterial chromosomes. Plasmids are described as extrachromosomal DNA, meaning they are generally separate from the bacterial chromosome, although we will encounter some exceptions as the chapter proceeds.

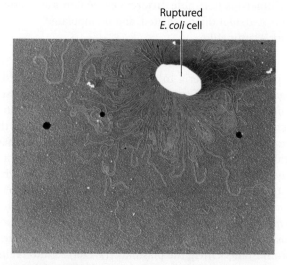

Ruptured
*E. coli* cell

**Figure 6.1  Bacterial chromosome and plasmids.**  A ruptured *E. coli* cell has released its chromosomal DNA along with multiple plasmids (red).

Many different kinds of naturally occurring plasmids are found in bacteria, and each contains several genes. One plasmid we are about to discuss, called an **F (fertility) plasmid,** contains genes that promote its own transfer from a donor bacterium to a recipient. Another type of plasmid we discuss, known as an **R (resistance) plasmid,** carries antibiotic resistance genes that can be transferred from donors to recipients. Plasmids are easily modified in the laboratory to produce specific characteristics or to carry particular genes that are useful in a wide range of recombinant DNA applications (see Chapters 16 and 17).

Many plasmids are able to replicate on their own, independent of the replication of the bacterial chromosome. Consequently, the number of plasmids in a single bacterial cell can increase rapidly and these plasmids are often identified as "high copy number" plasmids. The actual number is variable and is largely dependent on the type of plasmid in question. Low-copy-number plasmids are frequently unable to replicate on their own because their replication is tied to that of the bacterial chromosome. These so-called low-copy-number plasmids are present in 1 or 2 copies per bacterial cell, whereas 50 or more high-copy-number plasmids may be found in each cell. As you will soon see, high-copy-number plasmids play a pivotal role in conjugation and in the analysis of bacterial gene transfer and gene mapping.

## Conjugation Identified

Bacterial DNA transfer was first identified by Joshua Lederberg and Edward Tatum in 1946. They used two triple-auxotrophic strains of *E. coli* that had different nutritional requirements for growth (see Section 4.3 for a review of prototrophy and auxotrophy). The researchers first established three separate bacterial cultures growing, initially, in a complete medium (**Figure 6.2**). In culture ❶, they grew an auxotrophic strain called Y-24, which has the genotype *bio⁻ leu⁺ cys⁻ phe⁻ thr⁺ thi⁺*. Because of its genotype, the Y-24 strain requires addition of the vitamin biotin (*bio*) and the amino acids cysteine (*cys*) and phenylalanine (*phe*) to a minimal medium for growth. In culture ❷, they placed an auxotrophic strain called Y-10, which has the genotype *bio⁺ leu⁻ cys⁺ phe⁺ thr⁻ thi⁻*. The Y-10 strain requires addition of the vitamin thiamine (*thi*) and the amino acids leucine (*leu*) and threonine (*thr*) for growth. Culture ❸ contained an equal mixture of both Y-10 and Y-24.

Each culture was allowed to grow. Then approximately 10⁹ cells from each culture were plated onto dishes of minimal medium, where a prototrophic (wild-type) genotype is required for growth. Lederberg and Tatum saw no growth on Plates 1 and 2, which contained cells transferred from culture ❶ and culture ❷, respectively. These results were consistent with the nutritional requirements of Y-24 and Y-10, and indicated that all the cells transferred to those plates were auxotrophs. Plate 3, however, developed about 100 growing colonies! These colonies

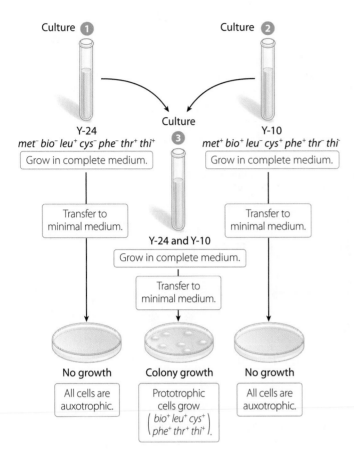

**Figure 6.2 Lederberg and Tatum's detection of recombination between auxotrophic *E. coli* cells.** Auxotrophic bacterial strains ❶ Y-24 and ❷ Y-10 each contain multiple mutations and grow on complete medium, but not on minimal medium. ❸ Mixing the strains leads to the formation of prototrophic bacteria that grow on minimal medium.

grew from bacterial cells that had somehow acquired the prototrophic genotype ($bio^+ leu^+ cys^+ phe^+ thr^+ thi^+$).

Lederberg and Tatum were certain that this outcome did not result from the reversion (reverse mutation) of auxotrophs to prototrophs (reversion is mutation that produces a wild-type allele from a mutant allele). First, the odds of that many genes reverting at once are prohibitively small. Second, plates 1 and 2 served as "negative control" plates. If reversion were responsible, these plates would show colony growth. Instead of reversion, the researchers claimed there had been a transfer of genetic information. More specifically, they proposed that one auxotrophic strain was transferring some of its prototrophic alleles to the other auxotrophic strain when the two strains were mixed, and that the second strain was replacing its auxotrophic alleles by incorporating the prototrophic information from the first strain.

Lederberg and Tatum hypothesized that physical contact between bacteria was necessary for gene transfer, but their original experiment did not provide direct evidence that this might be so. Four years later, Bernard Davis replicated the work in a way that showed the necessity of contact between bacterial cells for gene transfer to

take place. For his experiment, Davis constructed a U-tube with a fine glass filter separating one arm from the other (**Figure 6.3**). The filter was a glass disk with very small pores that allowed passage of small molecules such as nutrients but not bacterial cells. A cotton ball plugging one end of the U-tube and a rubber stopper connected to an air line at the other allowed Davis to move the material in the tube by alternating suction and pressure. The tube contained a culture of *E. coli* strain Y-10 on one side of the glass disk and a culture of strain 58–161, auxotrophic for methionine synthesis ($met^-$), on the other side of the disk.

From previous experiments, Davis knew that without a barrier, prototrophic bacteria are produced by gene transfer between these two bacterial strains. After alternating suction and pressure for several hours, Davis plated bacterial samples from each side of the U-tube onto minimal medium and found no growth from either side of the U-tube. This lack of growth was an indication that cells on either side of the disk retained their auxotrophy. Davis concluded that physical contact between bacterial cells is required for gene transfer to take place.

Microscopic studies have confirmed the physical union between bacteria hypothesized by Lederberg and Tatum and supported by Davis. This process of gene transfer is called **conjugation** (**Figure 6.4**). One of the participating bacteria, known as a **donor cell,** transfers some of its genetic information to the other cell, known as a **recipient cell.** The genetic information is conveyed by way of a hollow tube known as a **conjugation pilus** or **conjugation tube** that physically connects donor and recipient.

## Transfer of the F Factor

In 1953, William Hayes discovered that the bacteria interacting in Lederberg and Tatum's and in Davis's experiments did not contribute equally to the genetic outcome, as they do in a genetic cross between eukaryotes. Instead, the process was unequal: One of the bacteria, the donor, appeared to transfer some of its genetic information to the other bacterium, the recipient. Hayes identified this difference when he came across a bacterial strain that had formerly acted as a donor but had lost its donor ability for an unknown reason. In subsequent crosses of his "sterilized" former donor strain, Hayes saw that the strain was able to reacquire its donor state if it conjugated with other donors. This finding led Hayes to conclude that a one-way transfer of genetic information takes place between donors and recipients.

Hayes further proposed that the ability to act as a donor was hereditary and was determined by a "fertility factor" (F factor) that was transferable from donors to recipients. Donors are designated as **F⁺ (F⁺ cells)** to indicate their possession of an F factor, and recipients are identified as **F⁻ (F⁻ cells)** and lack the F factor. In the years after Hayes proposed the existence of the F factor, microbiologists identified the F factor as the F plasmid (fertility plasmid).

**Figure 6.3 Davis's U-tube experiment, showing that genetic recombination requires cell-to-cell contact.** Auxotrophic bacterial strains Y-10 and 58-161 are unable to grow on minimal medium, but produce some prototrophs that grow on minimal medium when they make contact following mixing. Prototrophs are not produced when the auxotrophs are placed in a U-tube, indicating that direct contact is required to generate prototrophic bacteria.

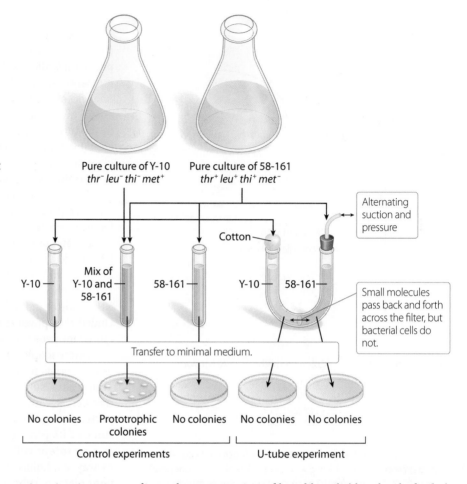

Pure culture of Y-10
$thr^- leu^- thi^- met^+$

Pure culture of 58-161
$thr^+ leu^+ thi^+ met^-$

Alternating suction and pressure

Cotton

Y-10   Mix of Y-10 and 58-161   58-161   Y-10   58-161

Small molecules pass back and forth across the filter, but bacterial cells do not.

Transfer to minimal medium.

No colonies   Prototrophic colonies   No colonies   No colonies   No colonies

Control experiments   U-tube experiment

Microbiologists today know that conjugation is controlled by genes carried on the F plasmid. As a consequence, only donor cells initiate conjugation. Recipient cells (F⁻ cells) are unable to initiate conjugation. Consequently, conjugation occurs between a donor cell and a recipient, but not between two donor cells. F factor genes direct the construction of hair-like pili (the plural of *pilus*) that have sensory functions. One pilus becomes specialized to serve as the conjugation pilus that connects donor and recipient, forming the conduit across which DNA from the donor cell is transferred (see Figure 6.4). Ultimately, three kinds of cells are seen in conjugation: a donor cell that contains an F plasmid and donates genetic information, a recipient cell that receives DNA from a donor cell but does not contain a functional F factor, and the **exconjugant cell** that is produced by conjugation. An exconjugant cell is essentially a recipient cell that has had its genetic content modified by receiving DNA from a donor cell.

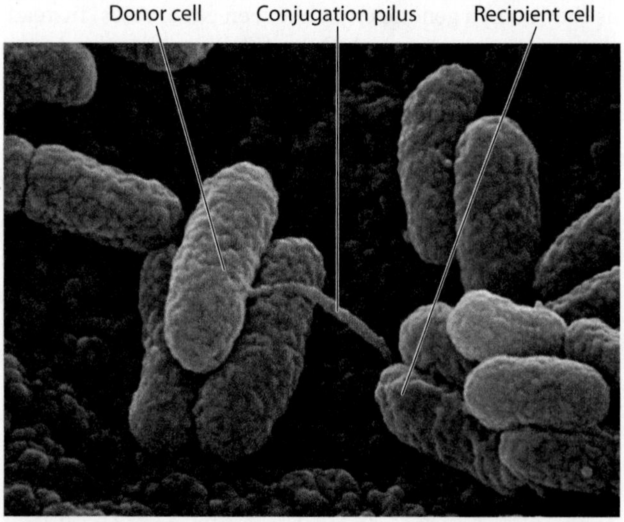

The F factor is some 100 kb in length, and about 35% of its sequence is devoted to about 40 genes that control conjugation (**Figure 6.5**). The F plasmid genes that play a role in *E. coli* conjugation are given four-letter designations consisting of the prefix *tra* or *trb* followed by a capital letter. Much of the remainder of the F factor consists of four insertion sequence (IS) elements: one copy of IS2, two copies of IS3, and one copy of the very large IS1000. **Insertion sequence (IS) elements** are mobile segments of bacterial DNA that are capable of transposing themselves throughout the bacterial genome and have an important functional role in bacterial gene transfer (see Chapter 12).

Conjugation between an F⁺ donor and an F⁻ recipient transfers a copy of the F factor and produces exconjugants

Donor cell   Conjugation pilus   Recipient cell

**Figure 6.4 Bacterial conjugation.** The donor bacterium makes contact with the recipient and transfers DNA through the conjugation pilus.

**(a)** Genes important in F factor transfer

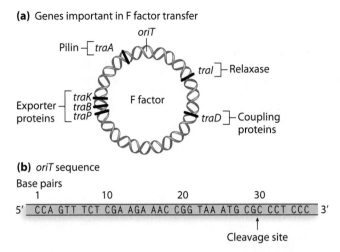

**(b)** *oriT* sequence

Base pairs

```
     1          10          20          30
5'  CCA GTT TCT CGA AGA AAC CGG TAA ATG CGC CCT CCC  3'
                                    ↑
                              Cleavage site
```

**Figure 6.5  F plasmid structure.** **(a)** Several genes important in F factor transfer are shown, as well as the origin of transfer (*oriT*). **(b)** The 38-bp sequence of *oriT*, including the cleavage site.

that are F⁺ donors, as illustrated in **Figure 6.6.** Conjugation begins with contact between the F⁺ and the F⁻ cell, initiated by the formation of a conjugation pilus. Conjugation pili are composed of pilin protein, produced by the *traA* gene on the F factor (see Figure 6.5). Circular DNA elements like the F factor that can replicate independently of the bacterial chromosome or, as we discuss in the following section, can integrate into the bacterial chromosome and replicate as part of the chromosome, are also termed an **episome.**

Shortly after contact is established by the conjugation pilus, gene expression from the F factor produces a protein complex called the relaxosome. This protein complex binds to a specialized F factor sequence called the **origin of transfer (*oriT*).** At *oriT,* the relaxosome catalyzes cleavage of one phosphodiester bond on one DNA strand, called the **T strand,** to signify that this is the strand transferred to the recipient cell. DNA cleavage at *oriT* defines a 3′ end and a 5′ end on the T strand and initiates some unwinding of the DNA duplex in the vicinity of *oriT.*

T strand unwinding releases most of the components of the relaxosome, but one protein, called relaxase, the product of the *traI* gene, binds to the free 5′ end of the T strand DNA to form a nucleoprotein complex. The nucleoprotein complex at the 5′ end of the T strand provides a critical recognition signal for a protein called the coupling protein, the product of the *traD* gene, which takes a position near the entry of the conjugation pilus. The nucleoprotein complex binds briefly to the coupling protein and then affiliates with several proteins of the exporter complex that move the nucleoprotein complex and the T strand across the conjugation pilus and into the recipient cell.

T strand transfer across the conjugation pilus is accompanied by a specialized process of DNA replication, known as **rolling circle replication,** inside the donor cell. In this specialized unidirectional replication process, one strand of DNA is spooled off across the conjugation pilus

while, within the donor, the remaining DNA strand serves as the template for unidirectional synthesis of a replacement DNA strand. In the recipient, the spooled-off DNA strand also acts as a template for DNA synthesis. We discuss the molecular details of DNA replication in Chapter 7.

Rolling circle replication begins at *oriT,* where the single-stranded break in DNA exposed the 3′ hydroxyl end of the T strand. At this exposed 3′ hydroxyl end, DNA polymerase adds new nucleotides, utilizing the complementary, intact (unbroken) DNA strand as a template. The new DNA replication taking place during rolling circle replication eventually displaces the 5′ end of the T strand, freeing it to be transferred across the conjugation pilus into the recipient cell.

Completion of rolling circle replication in the donor cell restores the donor's double-stranded F factor, leaving that cell's F⁺ donor state intact. Meanwhile, inside the recipient cell, the imported T strand acts as a template directing the synthesis of a complementary DNA strand. At the conclusion of this process, the two ends of *oriT* join to circularize the molecule, completing the creation of an F factor in the recipient. With the presence of an F factor, the formerly F⁻ recipient cell is converted to an F⁺ donor cell.

**Table 6.1** emphasizes two noteworthy features of F⁺ × F⁻ conjugation. First, complete transfer of the F factor converts the F⁻ recipient cell to an F⁺ donor cell. Second, no donor bacterial chromosomal genes are transferred during this conjugation process. Only the F factor DNA is transferred to an F⁻ recipient cell by an F⁺ donor cell. You may recall that Lederberg and Tatum provided clear evidence of chromosomal gene transfer from one bacterial strain to another, and Davis showed that conjugation was required for the transfer to occur. However, F⁺ × F⁻ conjugation *is not responsible* for the observations of Lederberg and Tatum; the logical conclusion is that there must be some other type of conjugation, perhaps involving different kinds of bacterial donor cells, to transfer bacterial chromosomal genes from a donor cell to a recipient cell.

---

Genetic Insight  Bacterial conjugation is controlled by F factor genes that direct formation of a conjugation tube between donor cells and recipient cells. Donor–recipient conjugation leads to DNA transfer by rolling circle replication. F⁺ donor cells transfer only F factor DNA to F⁻ recipient cells.

---

## Formation of an Hfr Chromosome

The experimental evidence described to this point demonstrates that *E. coli* possess the ability to transfer genes on the chromosome of the donor bacterium to the chromosome of the recipient bacterium. Contact between the donor and the recipient bacteria is required for gene transfer. However, this ability cannot be explained by conjugation involving an F⁺ donor because in F⁺ × F⁻ conjugation, only genes on the F plasmid are transferred,

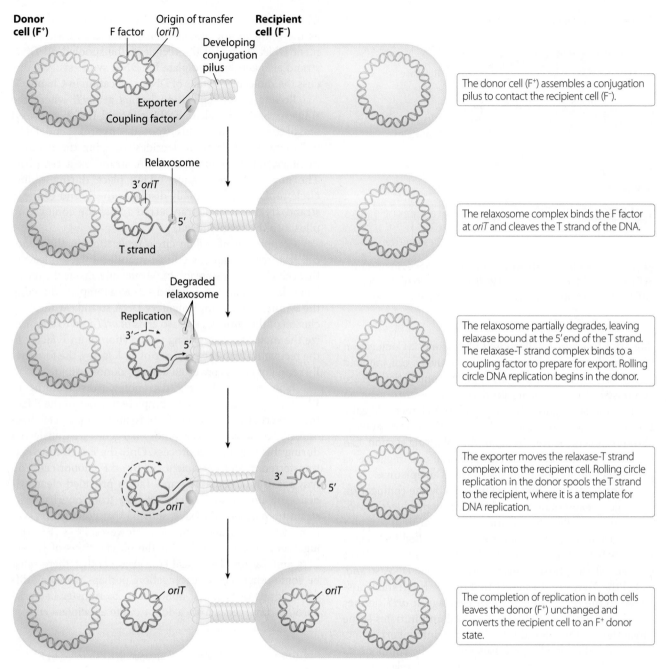

**Figure 6.6 Conjugation of F⁺ and F⁻ cells.** Rolling circle replication transfers a single strand of the F factor, beginning at *oriT*, from a donor cell to a recipient cell, where it is replicated to convert the recipient cell (F⁻) to an F⁺ donor.

| Table 6.1 | Outcomes of Bacterial Conjugation | |
|---|---|---|
| **Conjugation** | **Outcome** | |
| | Exconjugant Converted to Donor State? | Donor Bacterial Genes Transferred to Exconjugant? |
| F⁺ × F⁻ | Yes, F⁻ → F⁺ | No |
| Hfr × F⁻ | No | Yes |
| F′ × F⁻ | Yes, F⁻ → F′ | Yes |

not genes on the donor chromosome. The dilemma facing microbiologists was how to reconcile the results of the Lederberg and Tatum experiment and the Davis experiment with what was then known about F⁺ donors.

The resolution came from an experiment in 1953 by Luigi Luca Cavalli-Sforza, who found that a previously unknown form of donor bacteria was responsible for the gene transfers observed by Lederberg, Tatum, and Davis. Working with mutagenized donor *E. coli*, Cavalli-Sforza identified donor strains that transferred donor bacterial genes to recipient bacteria at an extraordinarily high rate.

Cavalli-Sforza labeled these bacterial strains **high-frequency recombination,** or **Hfr, strains** to indicate the high rate at which Hfr donor genes recombined with the chromosome of $F^-$ recipients. Cavalli-Sforza also determined that conjugation involving Hfr donors and $F^-$ recipients virtually never converted the recipients to $F^+$ or Hfr donors.

Microscopic examination of Cavalli-Sforza's Hfr strain, and of other Hfr strains isolated from nature, revealed an important difference in the configuration of the F factor. Instead of being an extrachromosomal plasmid, the F factor in Hfr strains is integrated into the bacterial chromosome, forming an **Hfr chromosome** (Figure 6.7). The formation of Hfr chromosomes is rare: Only about 1 in every 100,000 $F^+$ cells converts to an Hfr cell. The integration event forming the Hfr chromosome takes place at IS elements. F plasmids can have multiple IS elements, and bacterial chromosomes have several IS elements. On this basis, 20 to 30 distinct Hfr chromosomes can be formed in some bacterial species. The actual number varies among bacterial strains.

Two attributes of their integrated F factors in Hfr chromosomes distinguish one Hfr from another. First, the *location* of F factor integration varies between Hfr strains: It can occur at any of the IS sites present on the bacterial chromosome. Second, the integrated F factor can have one of two different *orientations* at each integration

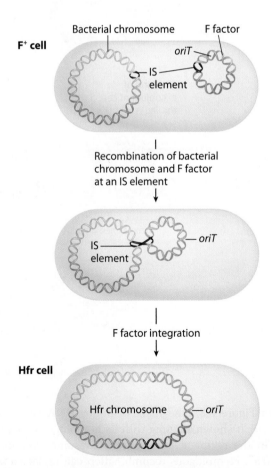

**F⁺ cell**

Bacterial chromosome     F factor

*oriT*

IS element

Recombination of bacterial chromosome and F factor at an IS element

IS element     *oriT*

F factor integration

**Hfr cell**

Hfr chromosome     *oriT*

**Figure 6.7  Hfr chromosomes.** Hfr cells carry an Hfr chromosome that is created when an F factor integrates into an insertion sequence (IS) in the bacterial chromosome.

location. The integration of an F factor to form a new Hfr chromosome occurs just once, establishing an Hfr strain with a site of F factor insertion and an orientation of the F factor that are fixed characteristics of all bacteria of the resulting Hfr lineage. Both location and orientation of the F factor are important to consider in mapping bacterial genes in Hfr chromosomes, as we discuss in Section 6.2.

## Hfr Gene Transfer

Hfr bacteria transfer genetic material to recipient cells by the same rolling circle replication process seen in $F^+ \times F^-$ conjugation. As in $F^+ \times F^-$ conjugation, the relaxosome binds to *oriT* and cuts the T strand to initiate unwinding and transfer of the T strand to the recipient. A portion of the integrated F factor is transferred first, followed by the bacterial chromosomes and finally by the remainder of the integrated F factor. In theory the entire Hfr chromosome could be transferred during Hfr $\times$ $F^-$ conjugation, but in reality this is impossible. The normal movement of bacteria will break the conjugation pilus long before Hfr transfer is completed. Complete Hfr transfer would take approximately 2 hours to accomplish—a period of time that far exceeds the duration of conjugation. In conjugation experiments, the duration of conjugation is stochastic. Some conjugation events are very short, others quite long, and others of intermediate duration.

The segment of T strand DNA that is successfully transferred into the recipient cell is used as template DNA to generate a double-stranded linear fragment. At whatever point the conjugation pilus ruptures, conjugation is interrupted, and T strand transfer and replication cease. Figure 6.8 illustrates conjugation between an Hfr with the genotype $thr^+ leu^- str^S$ and an $F^-$ with the genotype $thr^- leu^+ str^R$ (the function of $str^R$ and $str^S$ is explained momentarily). Within the recipient cell, the donor DNA is a linear double-stranded DNA fragment containing a portion of the F factor and a segment of donor bacterial DNA that was adjacent to *oriT*. Without the complete *oriT* sequence, the linear DNA cannot circularize; and since only a portion of the F factor is transferred, Hfr donors cannot convert $F^-$ recipient cells to a donor state (see Table 6.1). However, before the linear segment of donated donor DNA undergoes enzymatic degradation in the recipient cell, it can undergo homologous recombination with the recipient chromosome. The new exconjugant cell, formerly the recipient cell, may thus acquire one or more genes from the donor bacterial chromosome.

Conjugation experiments mix one strain of donor bacteria in a culture vessel with a different strain of recipient bacteria. Exconjugants produced within the vessel can be identified by their acquisition of donor genes. Exconjugants are identified by their genotypes that are distinct from those of either the donor strain or the recipient strain. Exconjugants are identified by their growth on a **selective growth medium,** a medium containing compounds that permit only exconjugants with specific genotypes to grow and that also prevent the growth of donor cells and recipient cells.

**Figure 6.8 Hfr conjugation and exconjugant detection.** An Hfr chromosome fragment transferred during interrupted mating between an Hfr donor cell to an F⁻ recipient cell can undergo homologous recombination with the recipient chromosome. Exconjugants are detected on selective growth media, such as the minimal medium shown here.

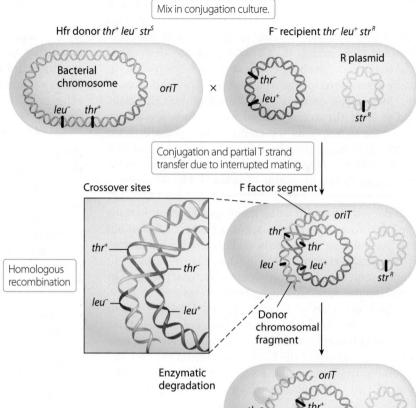

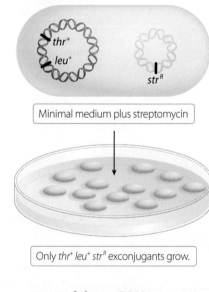

In experiments of this kind, antibiotic sensitivity and resistance is used as a tool to control growth of bacteria. In the recipient cells, resistance to the antibiotic streptomycin ($str^R$) comes from a gene carried on an extrachromosomal R plasmid (see Figure 6.8). The donor cell is streptomycin sensitive ($str^S$), but this is due to the *absence of an R plasmid*, not to the presence of an allele for streptomycin sensitivity. Streptomycin resistance is therefore a genotypic attribute of recipient and exconjugant cells but not of donor cells, and its presence in the selective growth medium will kill donors cells so they do not grow and potentially confuse the analysis. In Cavalli-Sforza's research, antibiotics were used in the selective medium to permit growth of bacterial cells carrying resistance genes and to kill bacteria that did not carry resistance. Likewise, nutritional compounds such as amino acids were used to support the growth of certain auxotrophic strains.

As an example, consider a conjugation experiment involving an Hfr strain that is susceptible to streptomycin ($str^S$) and carries the alleles $thr^+$ and $leu^-$ (for biosynthesis of the amino acid threonine and the inability to synthesize leucine). Imagine that the F⁻ strain is unable to synthesize threonine ($thr^-$) but capable of leucine synthesis ($leu^+$) and resistant to streptomycin ($str^R$). The selective medium necessary to grow and isolate exconjugants in this case is a minimal medium plate with added streptomycin. The streptomycin in the selective medium kills $str^S$ donor cells, and the absence of threonine prevents growth of nonrecombinant recipient cells. All growing cells on the selection plate are $thr^+$ $leu^+$ $str^R$, a genotype that could occur only in exconjugants.

In Figure 6.8, a segment of donor DNA containing $thr^+$ $leu^-$ is shown aligning with its homologous counterpart in the recipient bacterial chromosome, containing $thr^-$ $leu^+$. Homologous recombination can replace a segment of the recipient chromosome with a homologous segment of DNA from the donor chromosome. In the case shown here, two crossovers transfer $thr^+$ from the donor DNA into the recipient chromosome, so that exconjugants have the genotype $thr^+ leu^+ str^R$.

With or without homologous recombination to form an exconjugant, the ultimate fate of linear DNA in bacteria cells is enzymatic degradation through the action of nuclease enzymes. If nucleases reach the donated DNA before it can pair and recombine with the recipient chromosome, exconjugant formation is blocked. If recombination does take place, an exconjugant chromosome forms, and the segment of the recipient chromosome that was spliced out during recombination is degraded along with the remainder of the donated DNA.

For our purposes, conjugation between an Hfr donor cell and an F⁻ recipient cell has two key outcomes. First, the transfer of one or more donor alleles into the recipient chromosome by homologous recombination forms an exconjugant chromosome. Second, the F factor is not transferred in full during conjugation, and therefore the F⁻ recipient cell is not converted to a donor state (see Table 6.1).

---

Genetic Insight In Hfr donors, the F factor has been integrated into an insertion sequence. These donors are able to transfer donor genes but are unable to convert the recipient to a donor state.

---

# 6.2 Interrupted Mating Analysis Produces Time-of-Entry Maps

We have noted that Hfr chromosomes are too long to be fully transferred from a donor cell to a recipient cell. As a consequence, **interrupted mating,** the cessation of conjugation caused by breakage of the conjugation tube, takes place during naturally occurring conjugation. Interrupted matings stop conjugation before the Hfr chromosome can be completely transferred from the donor to the recipient. Several decades ago, researchers realized that if experimental conjugation was tested for gene transfer at timed intervals, it would be possible to map the order of donor genes, and to determine the distances between genes. This experimental strategy is called **time-of-entry mapping.** Each Hfr strain used in time-of-entry mapping experiments will transfer genes in a specific order that is a characteristic of the strain. The *order* of gene transfer and the *time* of the first appearance of recombinants for each gene are functions of the gene's proximity to the origin of transfer (*oriT*). As a result, genes that are closest to the 5′ end of the T strand cross the conjugation pilus shortly after conjugation begins, while genes that are more distant from the 5′ end of the T strand will cross the conjugation pilus later in time. Genes closest to *oriT* are also more frequently transferred than are genes that are more distant from *oriT*. The result is that genes that are closest to *oriT* recombine into exconjugant chromosomes at earlier times and in greater numbers than genes that are

distant from *oriT*. The number of minutes between the beginning of conjugation and the appearance of a particular recombinant is identified as the "time of entry" of the gene of interest. This measure, reported as minutes of conjugation, can be used to determine the order of genes on the Hfr chromosome in a time-of-entry map.

## Time-of-Entry Mapping Experiments

In 1956, Ellie Wollman, Francois Jacob, and William Hayes used conjugation data from the F⁻ strain P678 and the Hfr strain HfrH to demonstrate the utility of interrupted mating for time-of-entry mapping. In this experiment, P678 is $str^R$, resistant to the antibiotic streptomycin and HfrH is $str^S$, streptomycin sensitive. The donor and recipient genotypes for six genes studied are given in **Table 6.2**. Two of these genes had known locations: the genes for threonine and leucine synthesis (*thr* and *leu*), which are closer to the origin of transfer in HfrH than any of the genes tested. The goal of this experiment was to map the positions of *azi, tonA, lac,* and *galB* relative to *thr, leu,* and to determine the distance between genes in minutes of conjugation.

The experiment begins by mixing of donor and recipient bacterial strains to initiate conjugation. Every few minutes, a small sample of the culture is removed and agitated to break any conjugation pili, interrupt the mating, and stop the process of DNA transfer. The sample bacteria are plated on growth plates containing different supplemental compounds in the medium to determine if exconjugants have formed by recombination between the recipient chromosome and homologous donated DNA. The first recombinant alleles in exconjugants are, as expected, $thr^+$ and $leu^+$. The researchers select for these exconjugants by plating cells on a medium that lacks leucine and threonine but contains streptomycin, and therefore will permit the growth of only $leu^+ thr^+ str^R$ exconjugants. The order of the other four genes is determined using these $leu^+ thr^+ str^R$ exconjugants.

| Table 6.2 | Genotypes of *E. coli* Strains F⁻ P678 and HfrH |
|---|---|
| **HfrH** | **F⁻ P678** |
| $thr^+$ (prototrophic for threonine) | $thr^-$ (auxotrophic for threonine) |
| $leu^+$ (prototrophic for leucine) | $leu^-$ (auxotrophic for leucine) |
| $azi^R$ (resistant to sodium azide) | $azi^S$ (susceptible to sodium azide) |
| $tonA^R$ (resistant to phage T1 infection) | $tonA^S$ (sensitive to phage T1 infection) |
| $lac^+$ (able to utilize lactose) | $lac^-$ (unable to utilize lactose) |
| $galB^+$ (able to utilize galactose) | $galB^-$ (unable to utilize galactose) |

**Figure 6.9 Time-of entry mapping. (a)** Recombinants are identified by screening exconjugants for donor allele acquisition at regular intervals and plotting their time of entry into the exconjugant chromosome. **(b)** Donor alleles *leu*⁺ and *thr*⁺ appear in exconjugants within 3 minutes of conjugation initiation. Other donor alleles follow according to their order on the chromosome. **(c)** The Hfr chromosome time-of-entry map is assembled from the recombinant data.

Samples from the conjugation mixture are taken every few minutes and plated on the selective medium that identifies those with the *leu*⁺ *thr*⁺ *str*^R genotype. Exconjugants with this genotype are then placed on a second plate to determine which other donor alleles have undergone recombination.

Figure 6.9a shows the results of this experiment, which are interpreted in Figure 6.9b: Exconjugants carrying the donor *azi* allele appear 8 minutes after conjugation begins, *tonA* recombinants appear at 10 minutes, *lac* recombinants appear at 16 minutes, and *galB* recombinants are the last to appear, at 25 minutes. The order of these four genes and the distances in minutes between them are combined to produce the time-of-entry genetic map for HfrH (Figure 6.9c).

Time-of-entry mapping is an effective approach for mapping genes near the 5′ end of the T strand. However, the genetic mapping information obtainable from a single Hfr strain is limited. First, because the conjugation pilus is broken and mating is interrupted, the likelihood of gene transfer drops off quickly with distance from *oriT*. Second, an Hfr strain can transfer genes in just one direction.

To obtain experimental information about gene order and distances between genes on the bacterial chromosome of a given species, multiple Hfr strains with different sites of episome insertion and different orientations of the episome are examined. Each IS element on the bacterial chromosome constitutes a different *location* of F factor integration, and each integration location transfers a different gene first. The donor chromosome shown in Figure 6.10a illustrates six genes and six IS elements. Each IS element is a potential site for F factor integration, and the first gene to transfer will be different for each integration site. In addition, at each IS element, the episome can be oriented in either of two directions (Figure 6.10b). Thus, F factor *orientation* is a second factor determining the order of gene transfer for an Hfr strain. Once F factor insertion location and orientation occur, they are fixed characteristics of the Hfr strain that do not change. This gives each Hfr strain a consistent and determinable order of gene transfer.

Figure 6.10b illustrates F factor integration and gene transfer from IS1 in orientation 1. In this orientation, the gene transfer order will be *leu-thr-gal-phe-cys-val*. In the figure, the four ends of the double-stranded episomes are labeled I, II, III, and IV; the 5′-to-3′ polarity of strands is also indicated. Recall that relaxosome binding to *oriT* leads to cleavage of the T strand, which in Figure 6.10b is illustrated with ends I and II. The 5′ end of the T strand

**(a)** Donor allele appearance

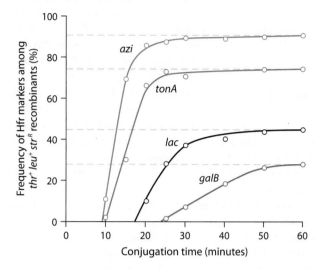

**(b)** Conjugation progression

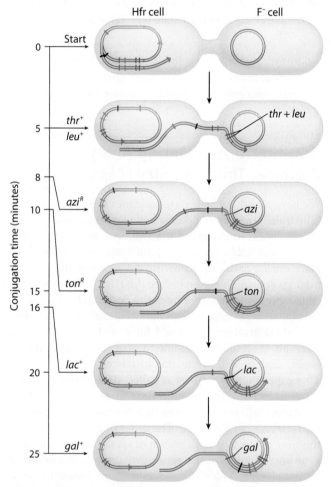

**(c)** Hfr chromosome map

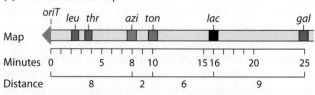

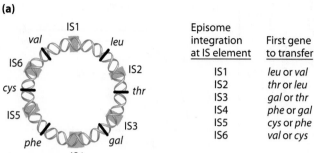

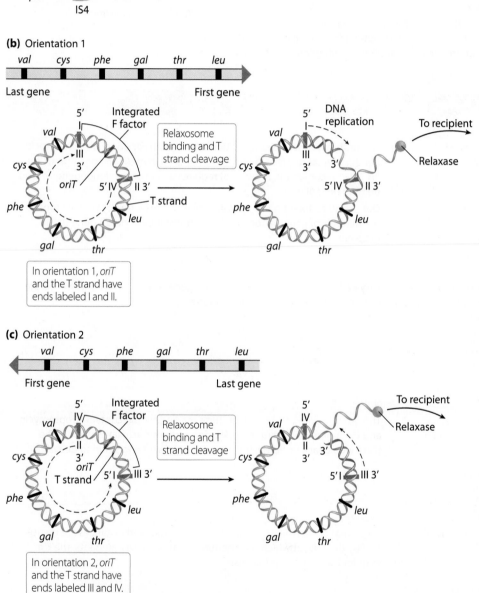

**(a)**

| Episome integration at IS element | First gene to transfer |
|---|---|
| IS1 | *leu* or *val* |
| IS2 | *thr* or *leu* |
| IS3 | *gal* or *thr* |
| IS4 | *phe* or *gal* |
| IS5 | *cys* or *phe* |
| IS6 | *val* or *cys* |

**Figure 6.10  F factors integrate at IS sites in one of two orientations.  (a)** A model bacterial chromosome with six insertion sequences (IS1 to IS6) and six nearby marker genes. **(b)** One F factor orientation into IS1 transfers the *leu* gene first. **(c)** The alternative F factor orientation at IS1 transfers the *val* gene first. Relaxase attaches to the free end of OriT at the beginning of transfer.

**(b)** Orientation 1

In orientation 1, *oriT* and the T strand have ends labeled I and II.

**(c)** Orientation 2

In orientation 2, *oriT* and the T strand have ends labeled III and IV.

moves across the conjugation pilus with *leu* as the first gene following the episome fragment. **Figure 6.10c** shows the same simplified bacterial chromosome with insertion at IS1 in orientation 2, where the T strand is the inner strand (still with ends I and II). Orientation 2 is the opposite of orientation 1, and it transfers genes in the opposite order. When the T strand is cleaved and its 5′ end moves across the conjugation pilus, the first marker gene to transfer will be *val*, followed by *cys-phe-gal-thr-leu.*

**Genetic Analysis 6.1** guides you through time-of-entry mapping for an Hfr conjugation experiment.

## Consolidation of Hfr Maps

In Hfr maps, an arrowhead is used to indicate the orientation of the integrated F factor. You can think of the arrowhead as indicating the tip of a DNA strand that is the first part to enter and move across the conjugation pilus. The first gene

An interrupted mating experiment is carried out in *E. coli* to map genes for biosynthesis of the amino acids threonine (*thr*), leucine (*leu*), glutamic acid (*glu*), and alanine (*ala*). An Hfr strain that is *his⁺ thr⁺ leu⁺ glu⁺ ala⁺ str^S* transfers *his* very early and is sensitive to the antibiotic streptomycin. It is mated to an F⁻ strain with the genotype *his⁻ thr⁻ leu⁻ glu⁻ ala⁻ str^R*. A time-of-entry profile for *thr, leu, glu,* and *ala* is shown at right.

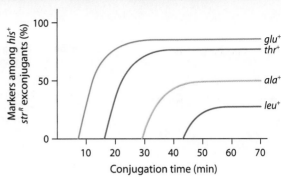

a. Exconjugants that are *his⁺* and *str^R* are initially selected for additional experimental analysis. What compounds must be present or absent in growth plates to allow exconjugants containing these selected markers to grow?

b. Use the data provided to deduce the order of genes transferred in this Hfr strain and to identify the distances in minutes. Identify the order of genes on the donor chromosome and indicate the approximate location of the *his* gene.

| Solution Strategies | Solution Steps |
|---|---|
| **Evaluate** | |
| 1. Determine the topic this problem addresses and describe the nature of the required answer. | 1. The problem concerns conjugation between an Hfr donor and an F⁻ recipient. Answer (a) requires identification of growth medium constituents for a *his⁺, str^R* exconjugant, answer (b) requires a map of the donor genes based on their time of entry. |
| 2. Identify the critical information given in the problem. | 2. Donor and recipient genotypes are given. A time-of-entry profile identifies the minutes of conjugation needed to transfer each donor gene to the recipient. |
| **Deduce** | |
| 3. Determine the significance of the very early transfer of *his⁺* in the context of developing a time-of-entry map.<br><br>TIP: Genes that are closer to *oriT* have earlier and more frequent opportunities to transfer to the recipient and to appear as recombinants in exconjugants than do genes that are distant from *oriT*. | 3. Very early transfer of *his⁺* indicates the gene is close to *oriT* and will be the first gene to cross the conjugation tube. |
| **Solve** | **Answer a** |
| 4. Identify the compounds needed to allow growth of exconjugants with the selected markers *his⁺* and *str^R*.<br><br>TIP: To select exconjugants that are *his⁺* and *str^R*, growth plates must provide conditions in which only the exconjugants that are resistant to streptomycin and able to synthesize histidine can grow. | 4. The growth plate used to select these markers would contain streptomycin and the amino acids threonine, leucine, glutamic acid, and alanine. The plate would lack histidine, thus requiring the growing strain to be *his⁺*. |
| | **Answer b** |
| 5. Construct a time-of-entry map based on the conjugation data. | 5. Given that *his* transfers first, and that gene order and distances are identified by the time at which recombinants appear in exconjugants, the Hfr map for this strain is as follows: |

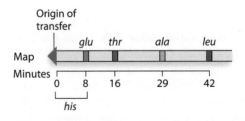

For more practice, see Problems 15, 16, and 26.

to follow the arrowhead into the recipient is closest to *oriT* and crosses the conjugation pilus first and most frequently among all donor genes. This leads it to be the first gene to recombine and the gene that recombines in the highest frequency. **Figure 6.11a** shows the locations and orientations of F factors in six *E. coli* Hfr strains. Each strain was used to construct a time-of-entry map, and the maps are consolidated by identifying overlapping regions to form a single circular map of the donor chromosome. Note that as a result of the two orientations possible at IS locations, some Hfrs transfer genes in one direction whereas others transfer genes in the opposite direction. For example, HfrH transfers genes in the direction opposite to that of Hfr2 and Hfr3.

Time-of-entry mapping was practiced for decades before genomic sequencing became a reality. During that period, many bacterial chromosomes were mapped by time of entry, including the chromosome of the model genetic organism *E. coli*. With the advent of genomic sequencing, however, it became possible to identify every nucleotide base pair, and every gene, in a genome. The accuracy and validity of Hfr mapping can be demonstrated by comparing a small segment of *E. coli* genomic sequence with the corresponding segment of the *E. coli* time-of-entry map. **Figure 6.11b**

compares a segment of the *E. coli* time-of-entry map with the corresponding segment of the chromosome produced by genomic sequencing. The comparison reveals exact correlation of gene placement and gene order.

Let's practice consolidating time-of-entry maps into a larger map of a circular chromosome using the following data on gene transfer from four different Hfr strains. For each strain, the genes are listed in order of transfer. The first gene transferred is at the top and the last gene transferred is at the bottom, and the minutes of conjugation are given in parentheses for each gene. The genes mentioned in the following discussion are presented in color.

| Hfr Strain | | | |
| --- | --- | --- | --- |
| Hfr1 | Hfr2 | Hfr3 | Hfr4 |
| *serR* (2) | *nadB* (8) | *tyrT* (4) | *serR* (4) |
| *leuY* (10) | *proL* (17) | *fumC* (12) | *pheR* (12) |
| *asnB* (15) | *fumC* (29) | *proL* (24) | *cysE* (25) |
| *serC* (20) | *tyrT* (37) | *nadB* (33) | *leuU* (37) |
| *tyrT* (27) | *serC* (44) | *leuU* (46) | *nadB* (50) |
| *fumC* (35) | *asnB* (49) | *cysE* (58) | *proL* (59) |

**(a)** Data collected from Hfr strains for construction of time-of-entry map

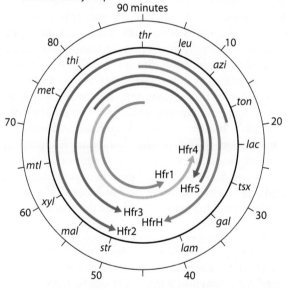

**Figure 6.11  Consolidated Hfr map of *E. coli*. (a)** Data for six Hfr strains identify F factor orientations and gene transfer order. Overlapping Hfrs produces a consolidated map of the chromosome. **(b)** Comparison of a segment of an Hfr time-of-entry map with a genomic sequence map. A 5-minute segment (minutes 40–45) of the *E. coli* time-of-entry map is shown in comparison to a segment of approximately 500,000 base pairs of the *E. coli* genome derived from *E. coli* genomic sequencing. Selected genes between 42.5 minutes and 44.5 minutes on the time-of-entry map (upper) are aligned with their positions in the genome sequence map (lower) to illustrate the compatibility of the two mapping approaches.

**(b)** Comparing segments of Hfr time-of-entry maps and sequenced genome

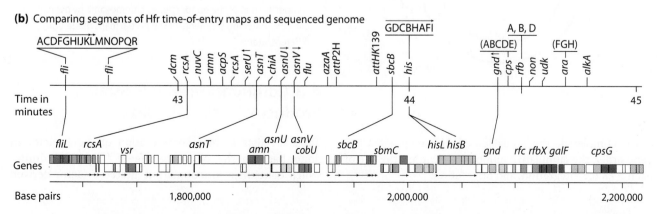

The data set from each Hfr strain is used to generate a partial map showing gene order, the distance in minutes between genes, and the orientation of the integrated F factor. The individual Hfr maps are then consolidated to show each F factor integration site, its orientation, and the gene order and distances in minutes. We anticipate that the minutes of conjugation between a given pair of genes will be the same in each Hfr strain transferring the gene pair. For example, Hfr strains 1, 2, and 3 each transfer the gene pair *tyrT-fumC*, and in each strain the genes are 8 minutes apart, no matter the orientation of the episome.

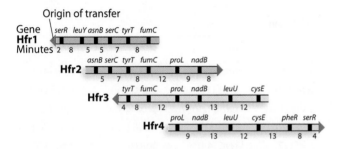

Continuation of the overlap process leads eventually to closure of the circle and completion of the chromosome map (as did the compilation of overlaps presented in Figure 6.11a). In the above table, for example, notice that Hfr1 and Hfr4 share *serR* as the gene nearest the site of insertion. This is the connection that allows us to close the circular map. To begin construction of the circular map, we will assume that Hfr1 transfers genes in a clockwise direction.

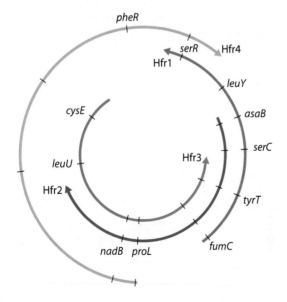

Once completed, the consolidated Hfr map identifies gene order, the cumulative number of minutes, the site of each F factor integration, and orientation:

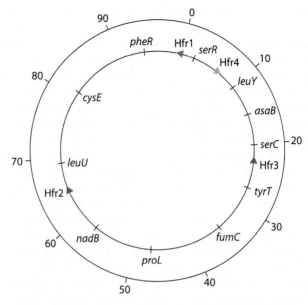

While conjugation mapping is an accurate way to determine gene order and to approximate the distance between genes, it is not precise enough to accurately map closely linked genes. The rapid rate at which Hfr strains transfer DNA makes precise measurement between linked genes difficult. For example, in *E. coli* conjugation, the chromosome transfers tens of thousands of nucleotides a minute. *Escherichia coli* genome analysis tells biologists that the average gene is about 1000 base pairs in length, which means that many genes may be transferred during each minute of conjugation. Since the differences in the time of entry of closely linked genes may be only a few seconds, the order of those genes may not be resolvable by conjugation mapping experiments alone. Two other mechanisms of DNA transfer between bacteria, transformation and transduction, facilitate high-resolution mapping of closely linked genes. Section 6.4 discusses gene mapping by transformation, and Section 6.5 describes gene mapping by transduction. First, however, we finish our explorations of the various configurations possible for the F factor.

---

Genetic Insight  Time-of-entry mapping identifies the order of and distance between genes transferred from each Hfr donor strain. Individual Hfr maps are consolidated into a composite map of the donor species chromosome by overlapping the matching segments of the partial maps made from different strains.

---

## 6.3 Conjugation with F′ Strains Produces Partial Diploids

Table 6.1 lists a third configuration of the F factor in donor bacteria, that of the so-called **F′** ("F prime") **donor,** which contains a functional but altered F factor derived from imperfect excision of the F factor out of the Hfr chromosome. The integration event that creates an Hfr chromosome depends on interactions between matching IS elements of the F factor and of the bacterial chromosome, and when this process is reversed, the F factor can once again become an extrachromosomal F⁺ factor. Occasionally, however, the excision event is imprecise, and the excised F factor—in this case called an **F′ factor**—contains all of its own DNA plus a segment of bacterial chromosomal DNA from the region adjacent to the integration site (**Figure 6.12a**). An F′ factor can carry a variable length of bacterial DNA. Donor cells carrying an F′ factor are called **F′ cells.**

Like the other forms of conjugation described above, conjugation between an F′ donor and an F⁻ recipient follows the by-now-familiar process of relaxosome complex binding to *oriT*, cleavage of the T strand, and movement of the T strand across the conjugation pilus with its 5′ end leading the way. Cells with small F′ factors are more likely to transfer the entire F′ factor than are cells with large bacterial chromosome inclusions. Consequently, small inclusions are usually transferred in their entirety.

If the entire F′ chromosome is transferred, both parts of *oriT* are transferred, allowing the F′ factor to circularize in the recipient cell. At the completion of F′ factor transfer in such cases, the recipient cell, now containing a complete F′ factor, is converted to an F′ donor (see Table 6.1). It has acquired copies of all the donor chromosomal genes carried on the F′ factor. Because the newly received chromosomal genes are homologs of genes already present on the recipient bacterial chromosome, the resulting exconjugants are **partial diploids.** The diploid portion of the genome is limited to the genes present in two copies, one on the exconjugant chromosome and the second on the F′ factor. No homologous recombination is necessary to produce these partially diploid genotypes, and partial diploidy is retained as a characteristic of these exconjugants and their descendants.

**Figure 6.12b** illustrates the creation of a partial diploid exconjugant carrying two alleles of the *lac* gene. The *lac*⁺ allele on the F′ factor enables the cell to use lactose for growth, whereas the mutant *lac*⁻ allele on the exconjugant chromosome is unable to function in lactose utilization. In this partial diploid, the *lac*⁺ allele is dominant over the *lac*⁻ allele. Partial diploids of this type have been used in genetic studies to examine the mode of action of genes in bacteria and to dissect the regulation of coordinated gene action in bacterial metabolism and growth (see Chapter 14).

**Genetic Analysis 6.2** on page 200 guides you through an analysis of donor and recipient bacterial strains and the identification of donor types through the analysis of three conjugation experiments.

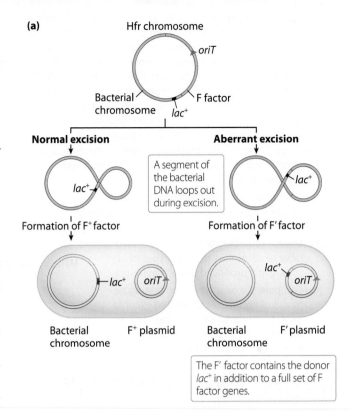

(a)

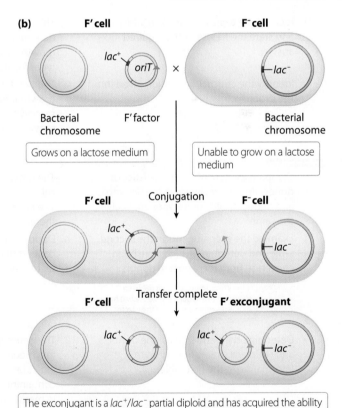

(b)

**Figure 6.12  F factor excision from Hfr integration.  (a)** Normal excision (left) restores an Hfr to an F⁺, whereas aberrant excision (right) forms an F′ plasmid in an F′ donor cell. **(b)** F′ × F⁻ conjugation produces an exconjugant that is a partial diploid *lac*⁺/*lac*⁻.

In *E. coli*, the abilities to utilize the sugar lactose, synthesize the amino acid methionine, and resist the antibiotic streptomycin are conferred by alleles *lac⁺*, *met⁺*, and *str^R*. Mutant alleles produce bacteria that are susceptible to streptomycin (*str^S*), are unable to grow on media containing lactose (*lac⁻*), and require methionine supplementation for growth (*met⁻*). *E. coli* strains are identified as donors or recipients in the first table, which also contains information on their ability to grow under various conditions. The second table contains growth information for the exconjugants of mating between donor and recipient strains. In each table, "+" indicates growth and "−" indicates no growth. "Min" signifies a minimal medium, and supplemented minimal medium plates are indicated by, for example, "Min+*met*" (minimal medium plus methionine). "Lac" indicates a plate containing only lactose as the sugar.

a. Use the growth information in the first table to determine the genotype of each strain at the *lac*, *met*, and *str* loci.

b. Use the growth information in the second table to determine the genotypes of exconjugants produced by each mating.

c. Compare the genotypes and mating behavior of donors, recipient, and exconjugants to determine whether each donor is F⁺, Hfr, or F′. Explain your rationale for each donor identification.

| Strain | Type | Strain Growth | | | | |
|--------|------|-----|-----|---------|-------------|-------------|
| | | Min | *lac* | Min+*met* | Min+*met*+*str* | *lac*+*met*+*str* |
| A | Donor | + | + | + | − | − |
| B | Donor | + | + | + | − | − |
| C | Donor | + | + | + | − | − |
| D | Recipient | − | − | + | + | − |

| Mating | Exconjugant Growth | | | | Are the Exconjugants Donors? |
|--------|---------|-------------|---------|-------------|------|
| | Min+*str* | Min+*met*+*str* | *lac*+*str* | *lac*+*met*+*str* | |
| A × D | + | + | − | − | Yes |
| B × D | − | + | − | − | Yes |
| C × D | + | + | − | + | No |

## Solution Strategies

## Solution Steps

### Evaluate

1. Identify the topic this problem addresses and state the nature of the required answer.

1. This is a conjugation problem in which genotypes of donors and a recipient are determined by growth characteristics. Donor types (F⁺, Hfr, F′) are identified by growth characteristics of exconjugants. The answers require identifying genotypes for *lac, met,* and *str* for each donor, recipient, and exconjugant.

2. Identify the critical information given in the problem.

2. The two tables identify growth characteristics. The first table contains growth information on three donors (A, B, and C) and a recipient (D). The second table contains growth information on the exconjugants of mating between each donor and the recipient.

### Deduce

3. Compare the growth characteristics of donors and the recipient in the first table, and deduce which genotypes are likely the same.

3. The growth characteristics of the three donor strains (A, B, and C) are identical on each kind of medium. These three strains have the same genotype. The recipient, strain D, has a different set of growth characteristics and therefore a different genotype.

4. Examine the exconjugants in the second table and determine which have been converted from recipients to donors.

4. Donor A and donor B transfer a complete F sequence to the recipient and convert the exconjugant to a donor. Donor C does not transfer the complete F sequence, so the C × D exconjugant is not converted to a donor.

> TIP: When an exconjugant has been converted to a donor state, we know it has received a complete copy of the F factor.

### Solve

5. Determine the genotypes of the donor and recipient strains from growth information in the first table.

Answer a

5. The genotype shared by donor strains A, B, and C is *met⁺ lac⁺ str^S*. The minimal medium contains glucose. Growth of donor strains in this medium indicates their prototrophy for methionine (*met⁺*). Growth in the lactose-containing medium indicates they are *lac⁺*. The inability of donors to grow in media containing streptomycin indicates they are *str^S*.

The recipient genotype is *met⁻ lac⁻ str^R*. It is unable to grow on the minimal (glucose-containing) medium, but it can grow on glucose plus methionine, indicating it is *met⁻*. It also grows on the minimal medium plus methionine and streptomycin, indicating that it is *str^R*. Lactose utilization is tested on the medium containing lactose plus methionine and streptomycin. Here it fails to grow, indicating it is *lac⁻*.

6. Determine the genotypes of exconjugants from growth information in the second table.

6. Using analysis similar to that employed above, we conclude that the exconjugant genotypes are

A × D $met^+ lac^- str^R$, conversion to donor

B × D $met^- lac^- str^R$, conversion to donor

C × D $met^- lac^+ str^R$, no conversion

7. Identify each donor by donor type and explain the rationale for each identification.

**TIP:** Compare the genotypes of exconjugants to the recipient genotype to determine if one or more donor alleles have been transferred during conjugation. Use Table 6.1 for help in categorizing each donor.

7. A × D exconjugants have acquired $met^+$ and have undergone conversion to a donor state. F′ donors can transfer an allele and convert the recipient, so we conclude that strain A is an F′ donor. Exconjugants of the B × D mating retain the recipient genotype, but they are converted to a donor state. $F^+$ donors produce this result, so strain B is an $F^+$ donor. The C × D conjugation produces exconjugants that have acquired $lac^+$ but have not undergone conversion. This is a characteristic of Hfr donors, so we conclude that strain C is Hfr.

For more practice, see Problems 17 and 21.

---

Genetic Insight  F′ donor cells are created by the aberrant excision of an integrated F factor from an Hfr chromosome. The F′ chromosome contains the complete F factor sequence plus one or more bacterial genes that were adjacent to the F factor. Conjugation of F′ and F⁻ cells produces exconjugants that have the F′ donor state and that are partially diploid from the transfer of donor genes.

---

## 6.4 Bacterial Transformation Produces Genetic Recombination

**Transformation** occurs when a recipient cell takes up a fragment of donor cell DNA from the surrounding growth medium. The DNA fragment passes through the wall and membrane of the recipient cell and is incorporated into the recipient cell chromosome by homologous recombination. Transformation is a naturally occurring mechanism that can be used to produce accurate maps of bacterial genes, including those that are closely linked and not readily mapped by conjugation experiments. The recipient cell taking up transforming DNA is identified as competent, meaning able to internalize exogenous (donor) DNA. Transformation is also used as a laboratory technique by molecular biologists seeking to introduce DNA into bacteria. In these cases, transformation is used to introduce plasmids into bacteria. A common application of plasmid transformation is DNA cloning (see Chapter 16).

### Steps in Transformation

Transformation is a four-step process, as illustrated in **Figure 6.13**. It is preceded by the **lysis,** or breakage, of a donor cell and the release of fragmented DNA from the donor chromosome. The transforming DNA is double stranded and can be taken up by a recipient bacterial cell.

The passage of double-stranded transforming DNA across the recipient cell wall and cell membrane is accompanied by degradation of one of the strands (step ❶ of Figure 6.13). The remaining strand of transforming DNA aligns with, or "invades," a complementary region of the recipient chromosome ❷. The alignment triggers the action of several enzymes that excise one strand of the recipient chromosome and replaces it with the transforming strand. This recombination event forms heteroduplex DNA: One strand is derived from the recipient cell, and the approximately complementary transforming strand is derived from the bacterial donor ❸. After the subsequent DNA replication and cell-division cycle ❹, one daughter cell is a transformed cell, also called the **transformant.** It contains a chromosome carrying the transforming strand and its newly synthesized complementary strand. The other daughter cell retains the recipient chromosome and is not genetically altered.

### Mapping by Transformation

Transforming DNA is usually shorter than about 100,000 bp (100 kb) in length. For a bacterial species like *E. coli*, which has a genome of $4 \times 10^6$ bp of DNA and approximately 5000 genes, the transforming DNA may have 1, 2, or as many as 50 genes. Even at maximum lengths, transforming DNA from the donor cell represents only 1 to 2% of the total genome of the recipient cell. Consequently, transformation is useful for mapping genes that are closely linked. To be mapped by transformation, two or more genes must be transferred into the recipient on the same fragment of transforming DNA. Thus, genetic analysis focuses on **cotransformation,** the

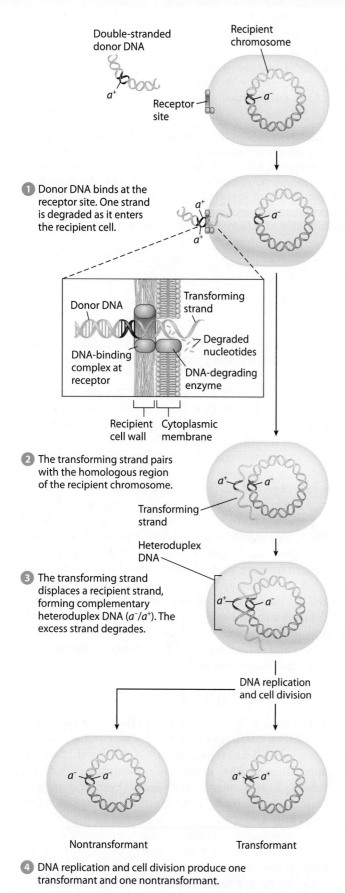

**① Donor DNA binds at the receptor site. One strand is degraded as it enters the recipient cell.**

**② The transforming strand pairs with the homologous region of the recipient chromosome.**

**③ The transforming strand displaces a recipient strand, forming complementary heteroduplex DNA ($a^-/a^+$). The excess strand degrades.**

**④** DNA replication and cell division produce one transformant and one nontransformant.

**Figure 6.13 Transformation of a competent bacterium ($a^-$) by donor DNA ($a^+$).**

simultaneous transformation of two or more genes. For cotransformation to occur, the crossover events must incorporate closely linked genes on a single fragment of transforming DNA.

To explore how cotransformation mapping works, let's first consider the *E. coli* genes *tyrP* and *purA*, separated by nearly 3 million base pairs on the chromosome and therefore too far apart to be cotransformed on the same transforming DNA fragment. Suppose each gene individually is transformed at a frequency of 0.002 (i.e., 2 cells per 1000). Since these genes are unlinked, the only way for them to be cotransformed is by two random, independent transformation events, each brought about by a separate piece of transforming DNA. We therefore expect the rate of cotransformation of *tyrP* and *purA* to be *the product of the two transformation frequencies,* or $(0.002)(0.002) = 4 \times 10^{-6}$. These two independent transformations require a total of four crossover events, two for each transforming DNA fragment.

In contrast, two linked genes that are close neighbors on the donor chromosome can easily end up on the same fragment of transforming DNA. A single pair of crossover events can then integrate both genes into the recipient chromosome with a cotransformation frequency that is about the same order of magnitude as transformation of a single gene. Moreover, the cotransformation frequency will always be substantially *greater than the product of the individual transformation frequencies.* **Figure 6.14** gives cotransformation data for three linked genes in *E. coli*: *purR*, *val*, and *pdxH*. These genes are found in a region of about 25,000 bp pairs in length. Suppose once again that the individual transformation frequency for each gene is 0.002. If the genes were not linked, an independent cotransformation frequency of about $10^{-6}$ would be expected for any pair of these genes. As the figure shows, however, the observed cotransformation frequencies are significantly greater, supporting the hypothesis of close linkage for these genes on the *E. coli* chromosome.

The data in Figure 6.14 also provide clues to the order of these three genes. We expect that the flanking genes (the genes on each end of the map) will have the lowest cotransformation frequency, and that the middle gene will have a higher cotransformation frequency when paired with either of the flanking genes. Our data show *pdxH* and *val* to have the lowest cotransformation frequency; thus, we know that these are the flanking genes. *purR* is the middle gene, and its cotransformation frequency with either *pdxH* or with *val* is substantially greater than the cotransformation frequency for the two flanking genes. In addition, because of the higher cotransformation frequency for *purR* and *val* than for *purR* and *pdxH*, we assume that the distance between *purR* and *val* is smaller than the distance between *purR* and *pdxH*.

| Experiment | Cotransformation genes | Cotransformation frequency |
|---|---|---|
| 1 | *pdxH* and *val* | 0.08 |
| 2 | *purR* and *val* | 0.81 |
| 3 | *purR* and *pdxH* | 0.68 |

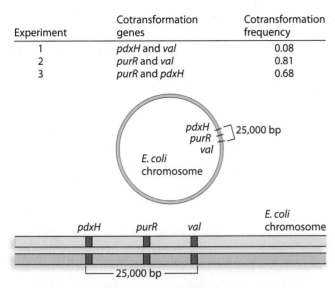

**Figure 6.14  Cotransformation mapping among genes *pdxH*, *purR*, and *val*.** These three genes are encoded in a segment of approximately 25,000 base pairs. Cotransformation frequency is lowest for the most distant pair of genes (*pdxH* and *val*) and highest for the closest gene pair (*purR* and *val*).

Genetic Insight  Recipient cells are transformed by uptake of donor DNA from the medium. The order and linkage of genes on a donor chromosome are mapped on the basis of cotransformation frequencies. Cotransformation frequencies that are significantly greater than those expected from independent transformation imply genetic linkage.

## 6.5 Bacterial Transduction Is Mediated by Bacteriophages

**Transduction** is the transfer of genetic material from a donor bacterial cell and the integration of that material into a recipient bacterial cell by way of a bacteriophage acting as a vector. To accomplish this transfer, a bacteriophage must infect the donor cell, and a few of the progeny phages must errantly package a fragment of the donor bacterial chromosome rather than a complete copy of the phage chromosome.

Following lysis of the original bacterial host cell, phages carrying the mispackaged bacterial DNA attach to a new host cell (the recipient cell) and inject the donor chromosome fragment. Inside the recipient, homologous recombination can take place between the donated fragment and the recipient chromosome. In this section, we review the life cycles of bacteriophages (*phages*, for short) that infect *E. coli*. Then we consider *cotransduction mapping*, a powerful technique for mapping bacterial genomes.

## Bacteriophage Life Cycles

Bacteriophage particles are generally less than 1% the size of the bacterial cells they attack. Their outer structure is a protein coat composed of an icosahedral head, a hollow protein sheath, and in some phages, a set of appendages called tail fibers (Figure 6.15). The phage's head houses its rudimentary genome, composed of a single chromosome ranging in size from about 5000 to 100,000 base pairs. The replication of phage DNA, the transcription of phage genes, and the translation that produces phage proteins are dependent on numerous proteins and enzymes found in the host bacterial cells, which the phages must infect in order to reproduce.

Bacteriophages employ a variety of mechanisms to attack bacteria. All of the mechanisms make use of bacterial proteins that evolved in the bacteria for other purposes than as a means of phage entry. For example, λ phage uses the maltose-binding protein of *E. coli* as a site of attachment. Maltose-binding protein studs the surface of *E. coli* cells, which use it to sense the presence of the sugar maltose in the growth medium. Thus, when studying the infection of *E. coli* by λ phage, microbiologists add maltose to the growth medium as a means of enhancing the phage infection rate.

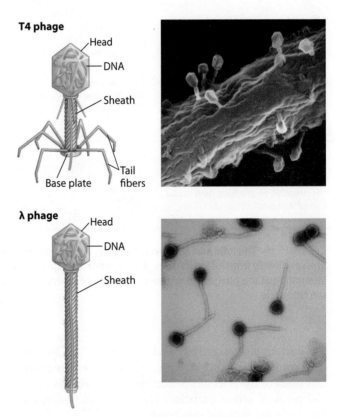

**Figure 6.15  T4 bacteriophage and λ phage structures.** Bacteriophage consist of proteinaceous head filled with DNA, a sheath, and tail fibers in some phage.

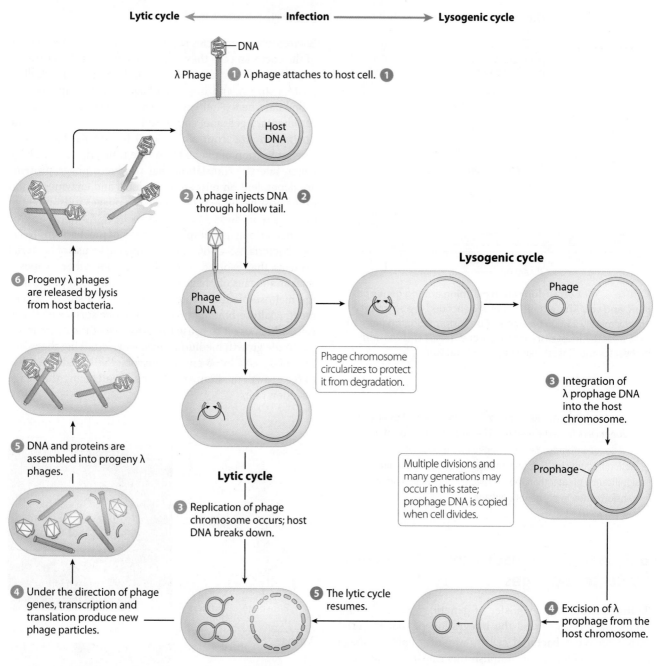

**Figure 6.16  The lytic and lysogenic life cycles of a temperate bacteriophage.** The lytic cycle progresses directly from infection through phage reproduction to lysis. The lysogenic cycle features the integration of the phage into the host chromosome where it resides until excision and resumption of the lytic cycle.

Bacteriophages actively seek out and attach to host cells, commencing a six-step process called the **lytic cycle,** that leads to the lysis of the host cell. Lysis releases up to 200 progeny phage particles. The steps composing the lytic cycle are depicted in **Figure 6.16.**

❶ **Attachment of the phage particle to the host cell.**

❷ **Injection of the phage chromosome into the host cell.** Injection is quickly followed by circularization

of the phage chromosome, to protect it from enzymatic degradation.

❸ **Replication of phage DNA,** using numerous host proteins and enzymes. A copy of the phage chromosome is required for each of the eventual progeny phage particles, which generally number between 50 and 200.

❹ **Transcription and translation of phage genes,** using numerous host proteins, enzymes, and ribosomes. Heads, sheaths, and tail fibers for all progeny particles must be synthesized and assembled.

⑤ **Packaging of phage chromosomes into phage heads.** This step is commonly accompanied by fragmentation of the host chromosome. Occasional mispackaging of a fragment of the host chromosome into a phage head can follow chromosome fragmentation.

⑥ **Lysis of the host cell,** resulting in the death of the host and the release of progeny phage particles.

Certain bacteriophages—classified as **temperate phages,** of which $\lambda$ phage is the best-known example—are capable of a temporary, alternative life cycle that leads to the integration of the phage chromosome into the bacterial host chromosome. The integration process is termed **lysogeny.** Environmental and growth conditions are largely what initiate a **lysogenic cycle.** Lysogeny can persist for many bacterial replication and division cycles, but it eventually comes to an end, and the lytic cycle resumes. (We discuss the details and genetic regulation of this alternation between life cycles in Chapter 14.) Five steps characterizing the lysogenic cycle are shown in Figure 6.16.

❶ **Attachment of the phage particle to the host cell.**

❷ **Injection of the phage chromosome into the host cell,** followed by phage-chromosome circularization.

❸ **Integration of the phage chromosome into the host chromosome.** This process is site specific, meaning that it occurs at a specific DNA sequence found in both the phage and bacterial chromosomes. Once integrated into the host chromosome, the phage DNA is termed the **prophage.** The prophage remains stably integrated at the same location for multiple cycles of bacterial chromosome replication and cell division.

❹ **Excision of the prophage.** In response to an environmental signal, such as a high dose of ultraviolet irradiation, the prophage reverses its integration and is excised intact. This event is usually an exact reversal of the site-specific integration, but rare mistakes in prophage excision lead to a specific kind of abnormal phage that may contain host genetic material.

❺ **Resumption of the lytic cycle,** beginning with phage-chromosome replication.

---

**Genetic Insight** Bacteriophages infect and cause lysis of the host bacterial cell after invading the cell and using bacterial proteins, enzymes, and structures for phage-DNA replication, transcription, and translation. Temperate phages are capable of an alternative life cycle called the lysogenic cycle, in which prophages integrate into the host chromosome.

---

## Discovery of Transduction

In transduction, the transfer of genetic material from a donor cell to a recipient cell is followed by homologous recombination of the donor fragment and the recipient chromosome. The completion of this process creates a **transductant** strain that is the product of transduction. The conditions leading to transduction are rare and require an error in the packaging of a phage chromosome during lytic infection. Two types of transduction occur. **Generalized transduction** is the transfer of *any* bacterial gene, and it can be accomplished by bacteriophages that do not discriminate between phage and host DNA. On the other hand, **specialized transduction** is accomplished only by temperate phages, and the genes transferred are limited to those very close to the site of prophage integration.

Transduction was first identified in a 1952 experiment by Norman Zinder and Joshua Lederberg. The experiment was designed to examine conjugation between two strains of the bacterium *Salmonella typhimurium*. In their experiment, Zinder and Lederberg mixed the double-auxotrophic strain $his^-$ $met^-$ $phe^+$ $trp^+$ $tyr^+$, which requires histidine and methionine for growth, with the triple auxotroph $his^+$ $met^+$ $phe^-$ $trp^-$ $tyr^-$, which grows only with supplementation of phenylalanine, tryptophan, and tyrosine. As expected, Zinder and Lederberg found a low frequency of prototrophic recombinants, consistent with their hypothesis that conjugation was the process transferring genes from donors to recipients.

Zinder and Lederberg then performed a U-tube experiment with the same auxotrophic strains that had a similar structure to the Davis U-tube experiment described in Section 6.1 (see Figure 6.3). They did not expect to see any prototrophs result, because the glass disk in the U-tube would prevent the movement of bacterial cells back and forth across the disk. To their surprise, however, Zinder and Lederberg found that their U-tube experiment did produce prototrophs. This outcome indicated that genes were being transferred by a mechanism that *did not* require the cell-to-cell contact of conjugation. Apparently, the agent responsible for the transfer was small enough to pass through the glass filter and therefore had to have been just a small fraction of the size of the *Salmonella* cells in the experiment.

Zinder and Lederberg knew that one of the *Salmonella* cultures was contaminated with a bacteriophage identified as P22, and they speculated that the phage might have played a role in the unusual gene transfer in the U-tube experiment. The researchers conducted additional U-tube experiments with glass filters having various pore sizes and found that when the pore size was larger than the size of P22, prototrophs were produced, but not when the pore size was smaller than P22. They were also able to show that the material passing through the glass filter was larger than a DNA molecule, thus ruling out transformation as the process creating prototrophs in the U-tube experiment. Zinder and Lederberg concluded that the presence of P22 and the exclusion of transformation as the mechanism of gene transfer meant that the P22 phage was carrying bacterial DNA. They named their newly discovered process of genetic transfer *transduction*.

## Generalized Transduction

In the six decades since Zinder and Lederberg discovered generalized transduction, numerous kinds of generalized transducing phages have been identified. **Generalized transducing phages** are formed when a random piece of donor bacterial DNA of the appropriate length is mistakenly packed into the phage head instead of a similarly sized length of phage DNA. This occasional error in DNA packaging occurs because the packing mechanism that inserts DNA into the phage head discriminates DNA by its length (in base pairs) rather than by sequence. Generalized transducing phages can carry any segment of donor DNA, since the process of mistaken packaging is random.

The phage P1 is a well-studied bacteriophage that infects *E. coli* and is a prolific producer of generalized transducing phage. This phage was initially chosen for intensive study of its transduction ability because it has a large genome of nearly 100,000 bp (100 kb). To produce progeny generalized transducing phage, P1 must capture segments of donor bacterial DNA that are almost exactly 100 kb, a length that is nearly 2% of the *E. coli* chromosome. Analysis of P1 infections tells us that about 1 in 50 progeny of a P1 infection are generalized transducing phages.

**Figure 6.17** illustrates generalized transduction in seven steps (combining attachment and injection into a single first step):

❶ A normal P1 phage attaches to a donor bacterial cell and injects its chromosome into the cell.

❷ Replication of the phage chromosome is followed by transcription and translation to produce phage proteins. Fragmentation of the bacterial chromosome precedes the packaging of phage chromosomes into phage heads.

❸ Assembly of progeny phage, including packing of phage heads, is largely normal, but a few progeny phages receive a random fragment of the donor bacterial chromosome that is approximately the same length as the phage chromosome. These abnormal progeny phage are *generalized transducing phage*.

❹ Host-cell lysis releases normal and generalized transducing phage.

❺ Generalized transducing phage attach to new recipient cells and inject the fragment of donor DNA.

❻ In each recipient cell, homologous recombination occurs between the fragment of donor DNA and the

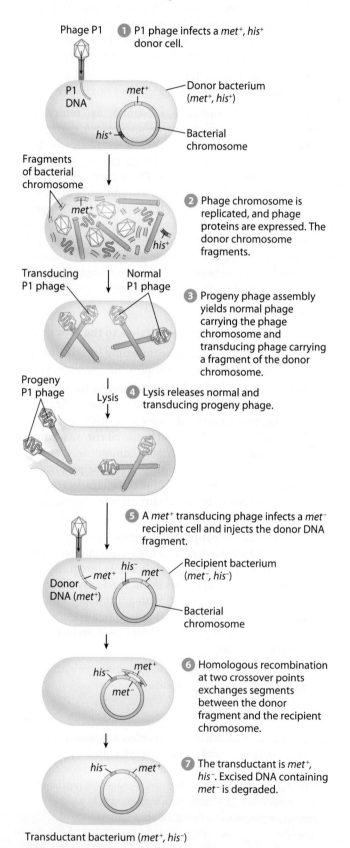

❶ P1 phage infects a *met⁺*, *his⁺* donor cell.

❷ Phage chromosome is replicated, and phage proteins are expressed. The donor chromosome fragments.

❸ Progeny phage assembly yields normal phage carrying the phage chromosome and transducing phage carrying a fragment of the donor chromosome.

❹ Lysis releases normal and transducing progeny phage.

❺ A *met⁺* transducing phage infects a *met⁻* recipient cell and injects the donor DNA fragment.

❻ Homologous recombination at two crossover points exchanges segments between the donor fragment and the recipient chromosome.

❼ The transductant is *met⁺*, *his⁻*. Excised DNA containing *met⁻* is degraded.

Transductant bacterium (*met⁺*, *his⁻*)

**Figure 6.17 Transduction by P1 phage.** Transducing phage are generated by the mistaken packaging of a fragment of the donor bacterium's DNA into a phage head (❸). Transductant bacteria are produced by homologous recombination between the introduced fragment of donor DNA and the recipient bacterial chromosome (❻ and ❼).

recipient chromosome. Pairs of crossover events are required to splice the donor fragment into the recipient chromosome and excise a homologous segment of the chromosome. The excised chromosome fragment is degraded by enzymes.

**7** A stable transductant strain results.

## Cotransduction

The donor cell in the transduction experiment shown in Figure 6.17 has the genotype $met^+$ $his^+$, and the recipient is $met^-$ $his^-$. The bacterial culture in which this experiment takes place will contain millions of bacteria, most of which are not transduced. In addition, many cells may be transduced with donor alleles that are not tested for in the experiment. The transductants detected in this particular experiment are those in which either one or both of the $met^+$ or $his^+$ alleles are transduced.

Transductants having either the genotype $met^+$ $his^-$ or the genotype $met^-$ $his^+$ offer evidence that each allele can be individually transduced. In addition, a certain number of transductants will undergo simultaneous transduction of both genes to produce $met^+$ $his^+$ transductants. These cells have undergone **cotransduction** of both donor alleles. The frequency of cotransduction, called **cotransduction frequency,** depends on how close the two genes are to one another on the donor chromosome. The closer the genes are, the higher the probability of cotransduction (thus, the higher the cotransduction frequency), and the farther apart the genes are, the lower the cotransduction probability. If, for example, an experimenter carried out the transduction cross in Figure 6.17 and identified 200 transductants for $met^+$, the experimenter could determine the frequency of cotransduction by then identifying how many of those $met^+$ transductants were also transduced (i.e., were cotransduced) for $his^+$. If the analysis determined that 28 of the 200 $met^+$ transductants were also transduced for $his^+$, the cotransduction frequency for those genes is 14% $(\frac{28}{200})$.

To succeed in finding cotransductants in an experiment, researchers may have to genotype large numbers of colonies. To reduce the number of colonies that must be genotyped in such experiments, a two-step strategy is used that first identifies cells transduced with one donor allele and then screens those transductants for the acquisition of additional donor alleles. The first step employs a **selected marker screen** to identify transductants for one of the donor alleles of interest. Transductants for the selected marker are then screened a second time, for a second donor allele, in an **unselected marker screen.** The goal is to determine the percentage of transductants for the selected marker that are also transduced for the unselected marker, while reducing unnecessary colony genotyping.

## Cotransduction Mapping

Genetic map construction in bacteria uses cotransduction frequencies to determine the relative order of three or

**(a)** Contransduction frequencies

| Donor genotype | Recipient genotype | Selected marker | Unselected marker | Percent cotransduction of unselected marker with $cys^+$ |
|---|---|---|---|---|
| $cys^+ trpE^+$ | $cys^- trpE^-$ | $cys^+$ | $trpE^+$ | 63 |
| $cys^+ trpC^+$ | $cys^- trpC^-$ | $cys^+$ | $trpC^+$ | 53 |
| $cys^+ trpB^+$ | $cys^- trpB^-$ | $cys^+$ | $trpB^+$ | 47 |
| $cys^+ trpA^+$ | $cys^- trpA^-$ | $cys^+$ | $trpA^+$ | 46 |

**(b)** *trp* operon map

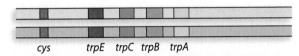

cys       trpE  trpC  trpB  trpA

**Figure 6.18** **Yanofsky's cotransduction frequency analysis and mapping of *trp* operon genes in *E. coli*.** **(a)** Cotransduction frequencies of $cys^+$ and a gene of the *trp* operon are determined in separate selected marker–unselected marker experiments. **(b)** Yanofsky's proposed map of the *trp* operon.

more genes. In **cotransduction mapping,** the frequency of cotransduction is greater for genes that are close together and is lower for genes that are farther apart. The reason is that any two genes on the donor chromosome have two chances to be separated by a chromosomal event. The first separation chance comes when the donor chromosome fragments. Genes that are close together are more likely to be on the same donor chromosome fragment than genes that are far apart. The second chance for separation comes during homologous recombination. Once again, genes that are close together on the donor fragment are less likely to be separated by a crossover event than genes that are far apart on the fragment.

Let's look at two studies that test the order of the same four genes in *E. coli.* **Figure 6.18** provides cotransduction data for experiments performed in 1959 by Charles Yanofsky on genes that are part of the *tryptophan operon,* a cluster of genes involved in the synthesis of the amino acid tryptophan that share a single promoter. (We discuss operons in detail in Chapter 14). For the current discussion, you only need to know that genes in an operon are transcribed under the control of a single promoter and are much closer to one another than genes that have their own promoters.

Yanofsky used the selected–unselected marker approach to determine cotransduction frequencies for each of four genes in the tryptophan operon (*trpA, trpB, trpC,* and *trpE*) and a gene outside the operon, *cys.* Yanofsky performed four crosses, each with a donor strain that was $cys^+$ and prototrophic for one *trp* gene. His recipient strains were each $cys^-$ and auxotrophic for the *trp* gene being tested. At the time he began his experiments, Yanofsky knew that *cys* lies outside the tryptophan operon, and he constructed his experiments to measure the cotransduction frequency between *cys* and the *trp* gene of interest. In each experiment, $cys^+$ was the selected marker used to identify informative transductants. The unselected marker was the *trp* allele from the donor. Yanofsky acquired data to determine the cotransduction frequency of $cys^+$ and the unselected *trp* marker.

In his first experiment, he determined that among $cys^+$ transductants, 63% are cotransduced for $TrpE^+$. In his second experiment, he found 53% cotransduction between $cys^+$ and $trpC^+$. Yanofsky concluded that *trpE* is closer to *cys* than is *trpC* based on the higher cotransduction frequencies for *cys* and *trpE* than for *cys* and *trpC*. Cotransduction frequencies for *cys* and *trpB* and for *cys* and *trpA* are not sufficiently different to determine gene order, but based on cotransduction frequencies, *trpA* and *trpB* are each more distant from *cys* than are *trpE* and *trpC*. Yanofsky proposed a genetic map of the tryptophan operon with the order *cys-trpE-trpC-trpB-trpA*.

The second study was conducted to test the order of these genes and either corroborate or refute Yanofsky's proposed gene map. In this study the donor bacterial genotype is $cys^+ \ trpC^- \ trpB^-$ and the recipient genotype is $cys^- \ trpC^+ \ trpB^+$. Transductants are selected for $cys^+$ transduction, and the transductants are then screened to determine their genotypes for *trpC* and *trpB*. The genotypes of 302 $cys^+$ transductants are shown in Table 6.3. Cotransductants for the donor *cys* and *trpC* alleles have the genotype $cys^+ \ trpC^-$ and are found in class 1, which has 139 cotransductants, and class 2, which has 18. The *cys–trpC* cotransduction frequency is therefore $\frac{139}{302} + \frac{18}{302} = 0.52$, or 52%. Similarly, cotransduction of *cys* and *trpB* is identified by the genotype $cys^+ \ trpB^-$. Transductant classes 1 and 4 have this cotransductant genotype, and the cotransduction frequency is $\frac{139}{302} + \frac{4}{302} = 0.47$, or 47%.

To test Yanofsky's proposed *Trp* operon map, the crossover events required to produce each cotransductant class are identified. **Figure 6.19** illustrates the locations of four crossover points used in different combinations for each cotransductant class. Transductants acquiring $cys^+$ must undergo crossover at point 1 plus at least one additional point. The precise location of crossover point 1 can vary over a large expanse of the chromosome to the left of *cys*. The second crossover

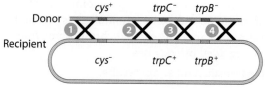

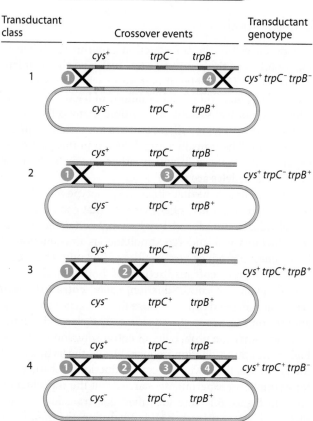

**Figure 6.19  A test of Yanofsky's proposed *Trp* operon gene map.** The approximate locations of possible crossovers are numbered 1 through 4. For each cotransductant genotype, the required crossover sites are identified.

point must occur to the right of *cys* in any of three locations: at location 2, within a relatively large distance between *cys*, which is outside the operon, and *trpC* within the operon; at point 3, a very small space in the operon between *trpC* and *trpB*; or at point 4, a large region to the right of *trpB*. Three different double-crossover combinations generate transductant classes 1, 2, and 3, and transductant Class 4 is produced by a quadruple recombination requiring crossover at all four points. The quadruple crossover is expected to be the least frequent of the combinations producing cotransductants. This study verifies Yanofsky's proposed *Trp* operon map for two reasons. First, cotransduction frequencies for *cys–trpC* and for *cys–trpB* are almost identical in the two studies (53% versus 52% for *cys–trpC*, and 46 versus 47% for *cys–trpB*), placing *trpC* closest to *cys* in both. Second, the quadruple

| Table 6.3 | Test of Yanofsky's Proposed *trp* Operon Gene Order | |
|---|---|---|
| **Transductant Class** | **Transductant Genotype** | **Number** |
| 1 | $cys^+ \ trpC^- \ trpB^-$ | 139 |
| 2 | $cys^+ \ trpC^- \ trpB^+$ | 18 |
| 3 | $cys^+ \ trpC^+ \ trpB^+$ | 141 |
| 4 | $cys^+ \ trpC^+ \ trpB^-$ | 4 |
| TOTAL | | 302 |

recombination event is expected to occur less frequently than any of the double crossover events.

---

Genetic Insight  Cotransduction occurs when two or more genes are transduced simultaneously from a donor cell. The frequency of cotransduction is used to construct genetic maps, on the principle that cotransduction frequencies are higher for genes that are closer together and are lower for genes farther apart.

---

## Specialized Transduction

As described above, temperate bacteriophages, such as lambda (λ) phage, have the ability to lysogenize their host by integrating into the host chromosome to create a prophage. The site of integration is a DNA sequence called the **att** site (for "attachment") that is identical in the bacterial chromosome and the phage chromosome. The 15-bp sequence is called *attP* in lambda phage (the *P* stands for phage) and *attB* (*B* for bacteria) in its host *E. coli* bacterium (**Figure 6.20**). A specialized phage enzyme recognizes the *att* sites and makes a staggered cut there. The complementary single-stranded ends of cleaved *att* DNA reanneal as the prophage integrates, to create an *att* sequence at each end of the integrated prophage.

Because the *attB* and *attP* sequences are identical, the excision of a prophage is almost always the exact reversal of prophage integration. Occasionally, however, excision is inaccurate: It removes only a portion of the prophage and, along with it, a portion of the adjacent bacterial chromosome. Aberrant excision of a prophage forms a **specialized transducing phage** (**Figure 6.21**). In *E. coli*, *attB* is located between the genes *galK* and *bioA*; thus, aberrant prophage excision occurring in one direction will capture the bacterial *gal*⁺ gene to form the λ*dgal*⁺ specialized transducing phage (d is for defective), and in the other direction will capture the bacterial *bioA* gene, to form the λ*dbio*⁺ specialized transducing phage.

Both kinds of specialized transducing phage are defective for certain attributes of phage growth and behavior. The λ*dgal*⁺ phage is missing several essential genes, so while it can infect host cells, it cannot complete either the lytic or lysogenic cycle. On the other hand, λ*dbio*⁺ phages are not missing any essential genes, but they lack genes necessary for lysogeny. Thus, λ*dbio*⁺ phages are exclusively lytic.

The study of specialized transducing phage has led to the discovery of a significant source of gene transfer between bacteria. In addition to the mechanisms we discuss in this chapter, microbiologists now know that specialized transduction is a source of gene transfer between bacteria of the same species and even between bacteria of different species.

**Genetic Analysis 6.3** on page guides you through an analysis of a transduction to determine gene order in a donor strain.

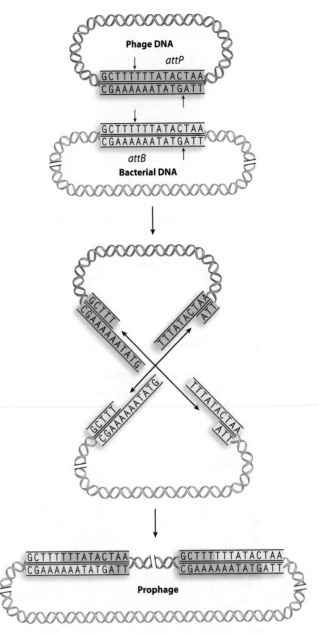

**Figure 6.20  Creation of λ prophage by recombination between identical *attB* and *attP* DNA sequences.**

# 6.6 Bacteriophage Chromosomes Are Mapped by Fine-Structure Analysis

Before DNA was identified as the hereditary material, many biologists regarded genes as indivisible units of heredity that could not be subdivided by recombination. This idea derives from Mendel's original description of "particulate inheritance" of traits. Before knowing the molecular structure of DNA, biologists had difficulty describing how recombination within a gene could occur. Geneticists knew that different mutations could affect a single gene, and had data from the 1949 study of intragenic recombination of the *Drosophila* lozenge eye mutation by Melvin and Kathleen Green showing that

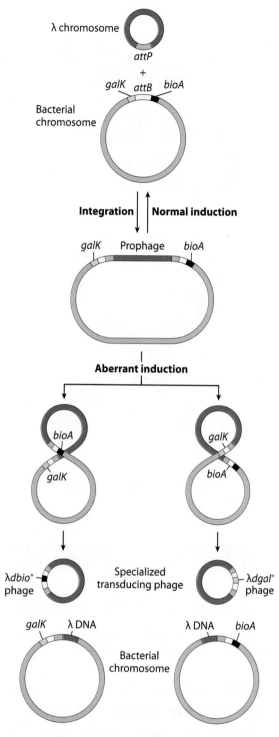

**Figure 6.21** **Patterns of λ prophage induction.** The λ phage integrates into the host bacterial to form the prophage by site-specific recombination between the *attP* and *attB* sites (upper). Normal prophage induction precisely reverses integration and restores *attB* and *attP* sequences (middle). Aberrant induction (lower) produces specialized transducing phage λ*dbio*⁺ or λ*dgal*⁺, depending on the direction of aberrant induction.

different mutations can occupy unique locations within a gene (see Figure 5.12). But what remained lacking was a refined understanding of the internal structure, or fine structure, of genes.

Beginning in the early 1950s, Seymour Benzer changed scientists' understanding of genes with a series of experiments that revealed the existence of a **genetic fine structure,** a phrase referring to the composition of genes at the level of their molecular building blocks. Benzer demonstrated that the building blocks of genes were responsible for both mutation and recombination. The publication of his principal conclusions coincided with the identification of the molecular structure of DNA. When the functional subunits of DNA were revealed to be nucleotides, it was impossible to miss the connection between them and Benzer's fine structure.

Benzer focused on two questions. First, was the gene the fundamental unit of mutation, or could components of genes be mutated? Second, was recombination a process occurring only between genes, or did recombination also occur between the components of genes? Benzer studied these questions using the *rII* region of the T4 bacteriophage. Genes in the *rII* region determine whether and how the phage will lyse its *E. coli* host. Lysis is examined using a bacterial lawn, a solid coating of bacteria on the surface of a growth medium. If the growing bacteria are exposed to a bacteriophage, infected cells lyse and progeny phages are released. Progeny phages infect new host cells, and as the infection–lysis–infection cycle continues, a bacteria-free spot called a plaque—a hole in the bacterial lawn—appears on the growth medium.

Benzer showed that two genes, *rIIA* and *rIIB*, control the ability of T4 phages to lyse *E. coli* host cells. Those T4 phages carrying wild-type copies of *rIIA* and *rIIB* lyse multiple strains of *E. coli,* leading to the production of small plaques (**Figure 6.22**). On the other hand, phages with mutation of either *rIIA* or *rIIB* form large, irregularly shaped plaques on *E. coli* strain B, but they are unable to form any plaques on *E. coli* K12 (λ).

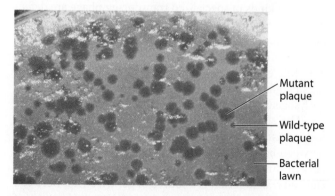

**Figure 6.22** **Plaque formation by *rII* wild types and mutants.** On a bacterial lawn of *E. coli* B strain, small, circular wild-type plaques are formed by T4 phages with a wild-type *rII* region. Large, irregular mutant plaques are formed by T4 phages with *rII* mutations.

In *E. coli*, *thr*⁺ and *leu*⁺ are prototrophic alleles that control synthesis of the amino acids threonine and leucine. The auxotrophic alleles are defective in their ability to synthesize these amino acids. Bacteria carrying the *azi*$^R$ allele are resistant to the antibiotic compound azide, and those carrying *azi*$^S$ are susceptible. *E. coli* with the genotype *thr*⁺ *leu*⁺ *azi*$^R$ are infected with the P1 phage. Progeny phages are collected and used to infect bacteria with the genotype *thr*⁻ *leu*⁻ *azi*$^S$, and the cells are then placed on media selective for one or two of the donor markers in a transduction experiment. The table at right identifies the selected markers and gives the frequency of cotransduction of unselected markers for each experiment. From the information provided, determine the order of the three genes on the donor chromosome.

| Experiment | Selected Marker(s) | Unselected Marker(s) |
|---|---|---|
| 1 | *leu*⁺ | *azi*$^R$ = 50%, *thr*⁺ = 4% |
| 2 | *thr*⁺ | *azi*$^R$ = 0%, *leu*⁺ = 4% |
| 3 | *leu*⁺ and *thr*⁺ | *azi*$^R$ = 2% |

| Solution Strategies | Solution Steps |
|---|---|
| **Evaluate** | |
| 1. Identify the topic this problem addresses and state the nature of the required answer. | 1. This is a cotransduction problem with cotransduction frequencies that are used to determine the order of three genes in the donor . |
| 2. Identify the critical information given in the problem. | 2. The results of three transduction experiments are given. Each experiment has a different gene as the selected marker. |
| **Deduce** | |
| 3. Describe the advantage of using the selected–unselected marker experimental approach. | 3. Selecting for transduction of one of the genes of interest and then evaluating transductants for the other gene(s) reduces the number of plates that must be evaluated and simplifies the experimental analysis. |
| 4. Interpret the <u>results</u> of each experiment.<br><br>**TIP:** Cotransduction frequencies are highest for genes that are closest together on the bacterial chromosome. | 4. Experiment 1 indicates close proximity of *leu* and *azi*, and a greater distance between *leu* and *thr*. Experiment 2 suggests the same more distant relationship between *thr* and *leu*, but also shows no cotransduction between *thr* and *azi*. Experiment 3 informs us that cotransduction of all three donor alleles occurs at a low frequency. We can interpret this to mean that the segment of chromosome containing these genes is small enough to form a single fragment for transduction. |
| **Solve** | |
| 5. Combine your observations to identify the <u>order</u> of these three genes.<br><br>**TIP:** Crossovers occur in pairs during the homologous recombination that accompanies transduction. When three genes are involved, a quadruple crossover is less frequent than any of the double crossovers. | 5. Putting the results of these experiments together, we can identify cotransduction of *thr* and *azi* (shown at 0% in experiment 2) as the quadruple crossover cotransductant. All other events are a result of double crossover. The quadruple crossover event is expected to be least frequent among the cotransductants. On this basis, *leu* can be identified as the middle gene of the three tested. The gene map is shown below, and the four crossover intervals are identified. |

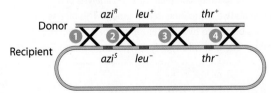

The crossover events accounting for each cotransduction detected in the experiments are shown below.

| Cotransduction | Crossovers |
|---|---|
| *azi*$^R$ and *leu*⁺ | 1 and 3 |
| *leu*⁺ and *thr*⁺ | 2 and 4 |
| *azi*$^R$, *leu*⁺, and *thr*⁺ | 1 and 4 |

For more practice, see Problems 9, 18, and 22.

Benzer used several different mutagens to produce almost 20,000 *rII* mutants that he studied in three ways. First, he used *genetic complementation analysis*, which showed that there are two genes in the *rII* region. Second, he mapped different mutations of *rIIA* and different mutations of *rIIB*, thus showing that *intragenic recombination* was possible and could be used to establish the locations of different mutations in each gene. Finally, Benzer developed *deletion mapping* to refine the genetic map. The following discussions examine each of these achievements individually.

## Genetic Complementation Analysis

To identify the number of genes in the *rII* region, Benzer performed genetic complementation analysis, coinfecting K12 (λ) bacteria with different pairs of *rII* mutants. When two *rII* mutants exhibiting genetic complementation coinfect K12 (λ) bacteria, plaques form on the bacterial lawn, indicating that wild-type lysis has been restored. This result identifies the mutants as mutations of different genes. Coinfections by *rII* mutants that did not lead to plaque formation on K12 (λ) represented a failure to complement, and these pairs were identified as mutations of the same gene. These mutants of a single gene are alleles of one another. Benzer identified two genetic complementation groups, which he designated A and B, and these led him to identify two genes in *rII—rIIA* and *rIIB*.

Subsequent analysis revealed that each gene produces a protein and that both proteins are required for lysis. Figure 6.23a illustrates genetic complementation for one pair of *rII* mutants. One mutant produces functional A protein and the other produces functional B protein, thus providing all the protein components necessary to carry out lysis. Genetic complementation produces a large number of plaques in infected bacterial lawns, but the individual progeny phages released following lysis remain mutant. Figure 6.23b illustrates a failure of mutants to complement. In this example, both mutants carry a mutation of *rIIB*.

---

**Genetic Insight** Genetic complementation analysis of *rII* lysis mutants reveals two complementation groups. This means the mutant alleles belong to two different genes, designated *rIIA* and *rIIB*, that are needed for wild-type lysis by bacteriophage T4.

---

## Intragenic Recombination Analysis

On rare occasions, Benzer observed that two lysis mutants that fail to complement (i.e., mutants of the same gene) nonetheless produce a few plaques of K12 (λ). He proposed that these plaques were produced by wild-type phage that resulted from rare intragenic recombination between two mutants whose chromosomes carry mutations in different locations in a single gene (Figure 6.24). One of the resulting recombinant chromosomes carries a double mutation, and the other is wild type. Wild-type

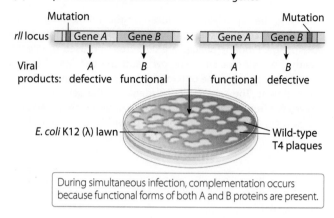

**(a)** Complementation of mutations in different genes

During simultaneous infection, complementation occurs because functional forms of both A and B proteins are present.

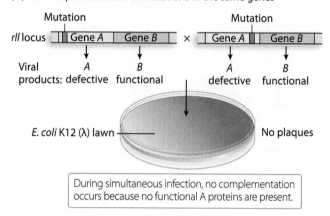

**(b)** No complementation of mutations in the same genes

During simultaneous infection, no complementation occurs because no functional A proteins are present.

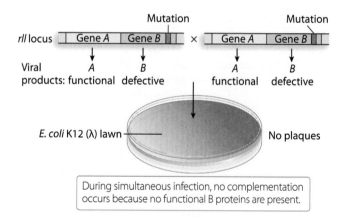

During simultaneous infection, no complementation occurs because no functional B proteins are present.

**Figure 6.23 Genetic complementation analysis for *rII* lysis.** **(a)** Genetic complementation of two lysis-defective phage mutants occurs when the mutants carry mutations of different genes. Genetic complementation is revealed by the formation of wild-type plaques on K12 (λ) bacteria. **(b)** No complementation occurs in lysis-defective mutants that carry mutations of the same gene.

chromosomes are found in progeny phages, that carry out wild-type lysis.

Based on a determination of the number of cells in an experimental flask and counting the number of K12 (λ) plaques subsequently produced, Benzer was able to calculate the intragenic recombination frequency within

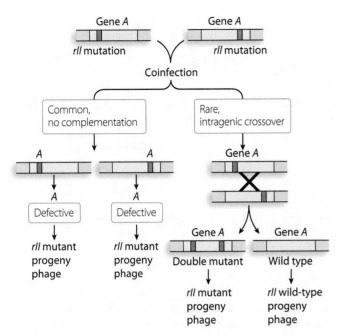

**Figure 6.24** **Simultaneous coinfection of a host cell by two noncomplementing *rIIA* mutants.** No complementation (left) is the common and expected outcome. Rarely, however, intragenic recombination (left) produces wild-type and double-mutant progeny phage.

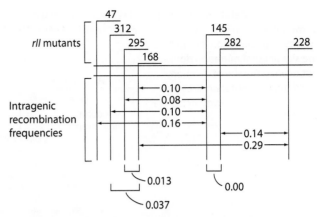

**Figure 6.25** **Intragenic recombination frequencies between mutations occurring in a segment of the *rIIA* gene in T4 bacteriophage.**

the *rII* gene for a given pair of mutations. Reasoning that reciprocal recombination was more likely to occur between two mutations that are distant within a gene, and less likely between mutations that are closer within a gene, Benzer was able to convert the observed number of plaques into a frequency of recombination with which he mapped *rII* mutations (**Figure 6.25**). Because of the large number of observations he made, Benzer was able to conclude that if no wild-type recombinants were obtained, the mutations occurred in the same nucleotide.

---

Genetic Insight  Intragenic recombination takes place between noncomplementing *rII* lysis mutants. The calculated recombination frequencies are used to create detailed recombination maps of phage lysis genes.

---

## Deletion Mapping Analysis

Benzer's mutagenesis of *rII* generated two types of mutants: **revertible mutants,** which could undergo spontaneous reversion back to wild type, and **nonrevertible mutants,** which *never* reverted. Revertible mutations are caused by DNA base-sequence substitutions (point mutations), which can be changed back to wild-type sequence by reversion. On the other hand, nonrevertible mutations are partial deletion mutations, in which part of the gene sequence is lost. A deleted DNA sequence cannot be restored by reversion.

Using a technique called **deletion mapping,** Benzer took advantage of this difference between revertible and nonrevertible mutants to map the position of individual *rII*

mutations. Deletion mapping relies on the production of wild-type phage by intragenic recombination between a revertible mutant and nonrevertible mutant (see Figure 6.24). When one mutant is revertible and the other is nonrevertible, the ability to form wild-type intragenic recombinants depends on the locations of the mutations. **Figure 6.26a** illustrates reversion to wild type through intragenic recombination between a point mutation and a deletion mutation whose locations *do not overlap*. In contrast, **Figure 6.26b** shows that if the locations of the point mutation and the deletion mutation *overlap* one another, the production of wild-type intragenic recombinants is impossible. Wild-type recombinants are not formed in this case, because the deletion mutant cannot provide the wild-type sequence to replace the mutated sequence in the point mutant.

In research published between 1955 and 1962, Benzer conducted deletion mapping of almost 20,000 *rII* mutants. He infected bacteria with phage carrying individual revertible mutations (point mutations), paired one at a time with phage carrying different nonrevertible mutations (deletion mutations). After analyzing the patterns of formation of wild-type recombinants, he produced a fine-structure genetic map depicting more than 300 discrete sites within the *rIIA* and *rIIB* genes.

In 1961, Benzer published a fine-structure map containing 1612 point mutations of *rIIA* and *rIIB* (**Figure 6.27**). Two features of this map are of interest. First, the mutations are scattered throughout *rIIA* and *rIIB*, suggesting the genes are composed of subunits that are individually mutable. Second, the distribution of the mutations is nonrandom. More than 100 point mutations aggregate in region A6c, and region B4 is the site of more than 500 independent point mutations. These sites are *mutational hotspots* that can be brought about by several circumstances (Chapter 12).

Several of Benzer's deletions are shown, and his mapping strategy is outlined, in **Figure 6.28**. Thirty-two deletion mutants in two groups called Series I and Series II are shown in Figure 6.27a. In Figure 6.27b, an *rIIA* point mutant is tested for its ability to form wild-type

**Figure 6.26 Deletion mapping of mutants in the *rII* region.** Wild-type recombinants form if the site of point mutation does not overlap the site of deletion, but if the two mutation sites overlap no wild-type recombinants are possible.

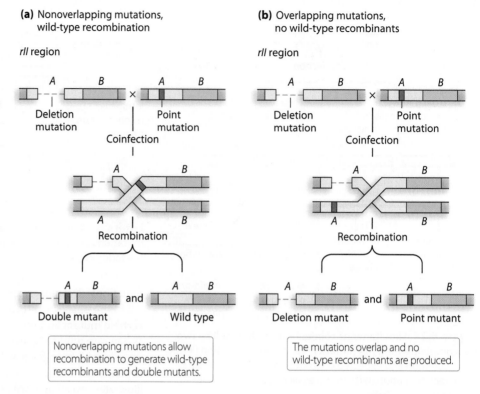

recombinants with the seven Series I deletion mutants and a subset of three Series II deletion mutants. Series I mutants are used first, to determine which of the six segments of *rIIA* (A1 to A6) contains the point mutant. The point mutant in this example forms wild-type recombinants with deletion mutant *638* but not with any of the six other mutants tested. The only *rIIA* region present in *638* that is absent in the other mutants is segment A6, leading to the conclusion that the point mutation occurs

in the A6 segment of *rIIA*. The A6 region is subdivided into four segments (A6a to A6d). The three partial deletion mutants of Series II are then selected for the final step in the mapping. In the Series II analysis, we see that the point mutant does not form wild-type recombinants with *PB230* and *P18* but is able to do so with *164*. The smallest interval that is missing from *PB230* and *P18* but present in *164* is the a2 region of *rIIA6*. This point mutation therefore maps to *rIIA6a2*.

**Figure 6.27 A genetic map showing the location of revertible (point) mutants of the *rII* region.** This mutational map assembled by Benzer places more than 1600 mutants in the *rII* region and identifies hotspots where mutations are particularly common.

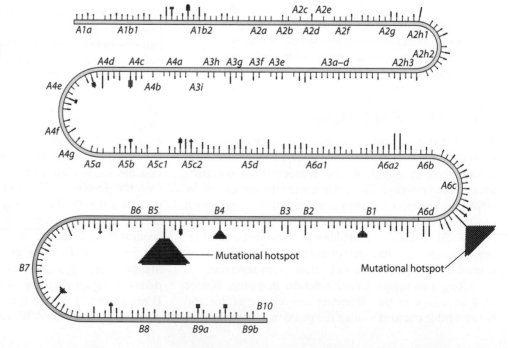

(a)

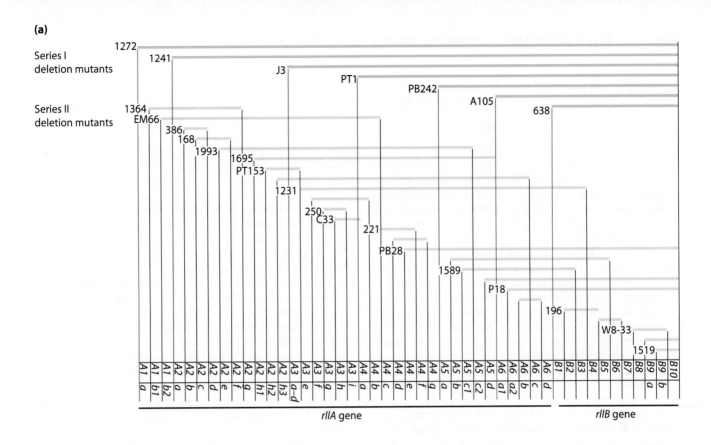

(b)

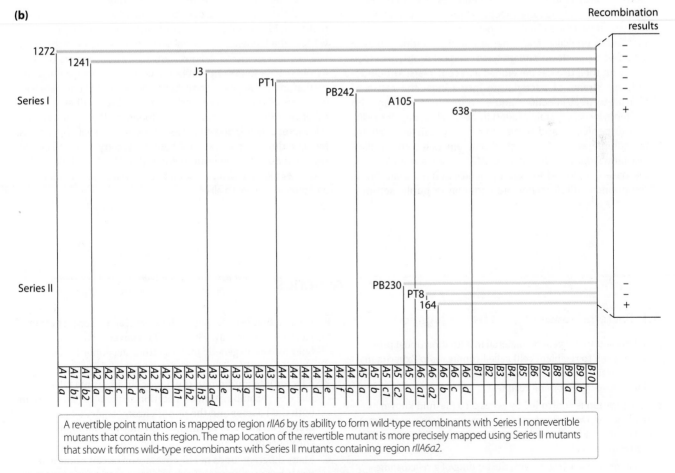

A revertible point mutation is mapped to region *rIIA6* by its ability to form wild-type recombinants with Series I nonrevertible mutants that contain this region. The map location of the revertible mutant is more precisely mapped using Series II mutants that show it forms wild-type recombinants with Series II mutants containing region *rIIA6a2*.

**Figure 6.28 Deletion mapping in the *rII* region.** **(a)** Seven Series I partial deletion mutants of the *rII* region and 25 Series II partial deletion mutants subdivide the *rII* region into 47 segments. **(b)** Deletion mapping analysis of an *rIIA* point (revertible) mutant to region *rIIA6a2* by its ability to form wild-type recombinants (+) and its inability to form wild-type recombinants (−) with partial deletion of Series I and Series II mutants.

## CASE STUDY

### The Evolution of Antibiotic Resistance

Alexander Fleming got a little sloppy with his sterile technique one day in 1929 and made a mistake that has since saved millions of lives. Fleming was working with *Staphylococcus*, a common bacterial strain that causes a serious and potentially fatal "staph" infection when it enters the body through a cut or abrasion. On the fateful day, Fleming unknowingly contaminated his *Staphylococcus* culture with a fungus.

Normally, fungal cells reproduce in culture along with bacterial cells and are noticed when the culture is spread on plates. Fleming's contaminating fungus was different, however, because when Fleming spread his contaminated culture on plates, only fungal colonies grew—there were no bacterial colonies! The fungus had killed the bacterial cells in the culture. Recognizing this as an important, if inadvertent, discovery, Fleming quickly identified the fungus as *Penicillium* and gave the compound that killed *Staphylococcus* the name penicillin.

In the 1930s, Howard Florey showed that penicillin was an effective antibiotic against a broad spectrum of infectious bacteria. At the beginning of World War II, Florey directed a major "scale-up" project to put penicillin into mass production. Penicillin proved tremendously effective at preventing what otherwise might have been fatal bacterial infections.

Today, penicillin and other antibiotics continue to save lives, but a significant and growing medical problem associated with their use has developed in recent years: Antibiotic-resistant strains of bacteria are increasingly the cause of difficult-to-treat infections and even death. More than 95% of *Staphylococcus* strains found in hospitals today are resistant to penicillin, and some strains carry resistance alleles to multiple antibiotics. Such strains are called methicillin-resistant *Staphylococcus aureus* (MRSA), since methicillin is the base compound for several penicillin-related antibiotic compounds. MRSA strains are common in public settings.

They infect people in hospitals, medical clinics, and locker rooms, killing thousands around the world each year. The strains are so common that you may have MRSA on your skin at this moment!

What happened to bring about this shift? Where once medical science had the ability to kill most infectious bacteria, our current situation is that any of us might incur a potentially life-threatening MRSA infection. The answer has two parts: the evolution of antibiotic resistance, and the rapid transfer of antibiotic resistance genes within and among bacterial species by conjugation, transduction, and transformation.

The evolution of antibiotic resistance in bacteria is a real-world example of evolution by selection. In this case, selection pressure is applied by the use (and sometimes misuse and overuse) of antibiotics used to treat humans and animals. For example, treatment with antibiotics can eliminate antibiotic-sensitive bacteria but may leave a residual pool of antibiotic-resistant bacteria in the body after treatment. The routine use of antibiotics in animal feed can also encourage the proliferation of antibiotic-resistant bacteria.

To make matters worse, the propensity of bacteria to exchange DNA by conjugation, transformation, and transduction makes rapid dissemination of antibiotic resistance possible even without the tendency of natural selection to increase resistant-allele frequencies in populations. Particularly troubling are the ability of bacteria to take up antibiotic resistance genes by transformation and the ability of some bacterial species to conjugate with members of other species and even with bacteria belonging to other genera. Drug-resistant genes found on R (resistance) plasmids spread rapidly and effectively through bacterial populations. Multiple-drug-resistant strains frequently carry more than one type of R plasmid. The bottom line is that MRSA bacteria strains, and other multiple-drug resistant bacterial strains are a significant and growing threat to all of us.

## SUMMARY

*For activities, animations, and review quizzes, go to the study area at www.masteringgenetics.com.*

### 6.1 Bacteria Transfer Genes by Conjugation

▪ Bacteria transfer genetic material in a unidirectional process (donor cell to recipient cell) called conjugation. Experimental analysis determined that conjugation requires direct contact between donor and recipient.

▪ Conjugation is controlled by genes on a plasmid known as an F factor. Donor bacteria that carry an extrachromosomal F factor are $F^+$ cells, and bacteria without an F factor are $F^-$, or recipient, cells.

▪ F factor transfer begins with the binding of a relaxosome protein complex at the transfer origin (*oriT*) and cleavage of one strand of F factor DNA, the T strand. Rolling circle DNA replication transfers the F factor from the donor cell to the recipient cell across a conjugation pilus.

▪ Conjugation between an $F^+$ donor and an $F^-$ recipient transfers the F factor only. The $F^-$ cell is converted to an $F^+$ cell but receives no genetic material from the donor bacterial chromosome.

▪ F factor integration into the donor chromosome takes place by recombination at insertion sequences (IS) found in both the F factor and the donor chromosome. F factor integration creates an Hfr (high-frequency recombination) chromosome.

▪ Many different kinds of Hfr chromosomes can occur in a single bacterial species. Each Hfr has a particular orientation and site of integration.

▪ Conjugation between an Hfr donor and an $F^-$ recipient transfers a portion of the F factor and a segment of donor

DNA. The donor segment undergoes homologous recombination with the recipient chromosome. Exconjugants receive donor bacterial genes but are not converted to a donor state.

## 6.2 Interrupted Mating Analysis Produces Time-of-Entry Maps

▌ Time-of-entry maps are created for each Hfr strain by interrupted mating studies that identify the order of entry of donor genes and determine the distance (in minutes) between transferred genes.

▌ Hfr maps for a given bacterium are consolidated to form a genetic map of the donor chromosome as a whole.

## 6.3 Conjugation with F′ Strains Produces Partial Diploids

▌ F′ donor strains are created when excision of an F factor from Hfr integration removes F factor DNA along with adjacent donor chromosome DNA.

▌ Conjugation between an F′ donor and an F⁻ recipient generates partial diploidy in exconjugants.

## 6.4 Bacterial Transformation Produces Genetic Recombination

▌ Extracellular fragments of DNA released when a donor bacterial cell lyses can be absorbed across the cell membrane of a competent recipient cell as transforming DNA.

▌ Transforming DNA undergoes homologous recombination with the recipient chromosome to produce transformants that have acquired donor DNA.

▌ Closely linked genes could be cotransformed, and genetic maps of these genes are generated by analysis of transformation.

## 6.5 Bacterial Transduction Is Mediated by Bacteriophages

▌ Bacteriophage infection of a host bacterial cell can lead to lysis of the host cell.

▌ Temperate bacteriophages can undergo site-specific integration into the host chromosome by lysogeny.

▌ Generalized transducing phages are created when a phage particle mistakenly packages a segment of a bacterial chromosome during lysis of the host cell.

▌ Recipient cells undergo generalized transduction when donor DNA introduced by a generalized transducing phage recombines with the recipient chromosome. Any donor genes can be transduced during generalized transduction.

▌ Cotransduction mapping determines the order of genes on the donor chromosome.

▌ Specialized transducing phages are produced by the aberrant excision of a lysogenic prophage that removes a portion of the prophage and an adjacent segment of host DNA. Specialized transduction is limited to transduction of genes adjacent to the site of prophage integration.

## 6.6 Bacteriophage Chromosomes Are Mapped by Fine-Structure Analysis

▌ Seymour Benzer used genetic complementation analysis to determine that two genes make up the *rII* region controlling T4 bacteriophage lysis of *E. coli.*

▌ Analysis of intragenic recombination, and deletion mapping of more than 1600 *rIIA* and *rIIB* mutants, led to the conclusion that DNA nucleotides are the fundamental unit of recombination.

## KEYWORDS

attachment site (*att* site) (*p. 209*)
bacterial chromosome (*p. 186*)
conjugation (*p. 187*)
conjugation pilus (conjugation tube) (*p. 187*)
cotransduction (cotransduction frequency, cotransduction mapping) (*p. 207*)
cotransformation (*p. 201*)
deletion mapping (*p. 213*)
donor cell (bacterial donor) (*p. 187*)
episome (*p. 189*)
exconjugant cell (*p. 188*)
F (fertility) factor (F plasmid) (*p. 186*)
F⁺ cell (F⁺ donor) (*p. 187*)
F⁻ (F⁻ cells) (*p. 187*)
F′ cell (F′ donor) (*pp. 198, 199*)
F′ factor (*p. 199*)

generalized transduction (generalized transducing phage) (*pp. 205, 206*)
genetic fine structure (*p. 210*)
Hfr (high-frequency recombination) cell (Hfr donor) (Hfr chromosome) (*p. 191*)
interrupted mating (*p. 193*)
IS (insertion sequence) element (*p. 188*)
lysogenic cycle (lysogeny) (*p. 205*)
lytic cycle (lysis) (*pp. 201, 204*)
nonrevertible mutants (*p. 213*)
origin of transfer (*oriT*) (*p. 189*)
partial diploid (*p. 199*)
plasmid (*p. 186*)
prophage (*p. 205*)
R (resistance) plasmid (*p. 186*)

recipient cell (F⁻ cell) (*p. 187*)
revertible mutant (*p. 213*)
rolling circle replication (*p. 189*)
selected marker screen (*p. 207*)
selective growth medium (*p. 191*)
specialized transduction (specialized transducing phage) (*pp. 205, 209*)
T strand (*p. 189*)
temperate phage (*p. 205*)
time-of-entry mapping (*p. 193*)
transductant (*p. 205*)
transduction (*p. 203*)
transformant (*p. 201*)
transformation (*p. 201*)
unselected marker screen (*p. 207*)

## Chapter Concepts

*For answers to selected even-numbered problems, see Appendix: Answers.*

1. For bacteria that are F$^+$, Hfr, F′, and F$^-$, answer the following.
   a. Describe the state of the F factor.
   b. Which of these cells are donors? Which is the recipient?
   c. Which of these donors can convert exconjugants to a donor state?
   d. Which of these donors can transfer a donor gene to exconjugants?
   e. Describe the results of conjugation (i.e., changes in the recipient and the exconjugant) that allow detection of the state of the F factor in a donor strain.
   f. Describe a "partial diploid" and how it originates.

2. The flow diagram shown below identifies possible relationships between bacterial strains in various F factor states. For each of the four links in the diagram, provide a description of the events involved in the transition.

$$\begin{array}{ccccc} & 1 & & 2 & \quad 4 \\ F^- & \to & F^+ & \to & Hfr \to F' \\ & & & \leftarrow & \\ & & & 3 & \end{array}$$

3. Conjugation between an Hfr cell and an F$^-$ cell does not usually result in conversion of exconjugants to the donor state. Occasionally however, the result of this conjugation is two Hfr cells. Explain how this occurs.

4. Bacteria transfer genes by conjugation, transduction, and transformation. Compare and contrast these mechanisms. In your answer, identify which if any processes involve homologous recombination and which if any do not.

5. Explain the importance of the following features in conjugating donor bacteria:

   a. the origin of transfer
   b. the conjugation pilus
   c. homologous recombination
   d. the relaxosome
   e. relaxase
   f. T strand DNA
   g. pilin protein

6. Describe the difference between the bacteriophage lytic cycle and lysogenic cycle.

7. Describe what is meant by the term *site-specific recombination* as used in identifying the processes that lead to the integration of temperate bacteriophages into host bacterial chromosomes during lysogeny or to the formation of specialized transducing phage.

8. What is a prophage, and how is a prophage formed?

9. How is the frequency of cotransduction related to the relative positions of genes on a bacterial chromosome? Draw a map of three genes and describe the expected relationship of cotransduction frequencies to the map.

10. Describe the differences between genetic complementation and recombination as they relate to the detection of wild-type lysis by a mutant bacteriophage.

11. Among the mechanisms of gene transfer in bacteria, which one is capable of transferring the largest chromosome segment from donor to recipient? Which process generally transfers the smallest donor segments to the recipient? Explain your reasoning for both answers.

## Application and Integration

*For answers to selected even-numbered problems, see Appendix: Answers.*

12. Seven deletion mutations (1 to 7 in the table below) are tested for their ability to form wild-type recombinants with five point mutations (a to e). The symbol + indicates that wild-type recombination occurs, and – indicates that wild types are not formed. Use the data to construct a genetic map of the order of point mutations, and indicate the segment deleted by each deletion mutation.

**Deletion Mutation**

| Point Mutation | 1 | 2 | 3 | 4 | 5 | 6 | 7 |
|---|---|---|---|---|---|---|---|
| a | – | + | – | – | + | + | – |
| b | + | + | + | – | + | – | – |
| c | + | + | + | + | – | – | – |
| d | – | + | + | – | + | – | – |
| e | + | – | – | – | + | + | – |

13. An *rII* lysis mutation caused by a point mutation is tested against several deletion mutations shown in Figure 6.28 for its ability to form wild-type recombinants. The deletion mutants are divided into two groups, Series I and Series II. In the "result" column of the table below, + indicates the formation of wild-type recombinants and – indicates that wild types do not form. In the first part of your answer, use the Series I data exclusively to identify the segment of the

| Series I | | Series II | |
|---|---|---|---|
| Deletion Mutation | Result | Deletion Mutation | Result |
| 1272 | – | 1364 | + |
| 1241 | – | EM66 | – |
| J3 | – | 386 | + |
| PT1 | + | 168 | + |
| PB242 | + | 1993 | – |
| A105 | + | 1695 | – |
| 638 | + | PT153 | + |
| | | 1231 | – |
| | | C33 | + |

*rII* region containing the lysis mutant tested. In the second part of your answer, use the Series II data to refine the point mutation location. Explain your rationale for mutation location assignments for both the Series I and the Series II data.

14. Suppose you have an *rII* lysis mutant that maps to segment *A2h2*. Use the Series I and Series II deletion mutants identified in the problem above, and fill out the "results" columns with the + and − designations expected for the *A2h2* mutant.

15. Five Hfr strains from the same bacterial species are analyzed for their ability to transfer genes to F⁻ recipient bacteria. The data shown below list the origin of transfer (*oriT*) for each strain and give the order of genes, with the first gene on the left and the last gene on the right. Use the data to construct a circular map of the bacterium.

| Hfr Strain | Genes Transferred |
|---|---|
| Hfr 1 | *oriT met ala lac gal* |
| Hfr 2 | *oriT met leu thr azi* |
| Hfr 3 | *oriT gal pro trp azi* |
| Hfr 4 | *oriT leu met ala lac* |
| Hfr 5 | *oriT trp thr leu met* |

16. An interrupted mating study is carried out on Hfr strains 1, 2, and 3 identified in the problem above. After conjugation is established, a small sample of the mixture is collected every minute for 20 minutes to determine the distance between genes on the chromosome. Results for each of the three Hfr strains are shown below. The total duration of conjugation (in minutes) is given for each transferred gene.

| Hfr strain 1 | *oriT* | *met* | *ala* | *lac* | *gal* |
|---|---|---|---|---|---|
| Duration (min) | 2 | 8 | 13 | 17 | |
| Hfr strain 2 | *oriT* | *met* | *leu* | *thr* | *azi* |
| Duration (min) | 2 | 7 | 10 | 17 | |
| Hfr strain 3 | *oriT* | *gal* | *pro* | *trp* | *azi* |
| Duration (min) | 3 | 8 | 14 | 19 | |

a. For each Hfr strain, draw a time-of-entry profile like the one in Figure 6.9a.
b. Using the chromosome map you prepared in answer to Problem 15, determine the distance in minutes between each gene on the map.
c. Explain why *azi* is the last gene of strain 2 to transfer in the 20 minutes of conjugation time. How many minutes of conjugation time would be needed to allow the next gene on the map to transfer from Hfr strain 2?
d. Write out the interrupted mating results you would expect after 20 minutes of conjugation for Hfr strains 4 and 5. Use the format shown at the beginning of this problem.
e. In minutes, what is the total length of the chromosome in the donor species?

17. An Hfr strain with the genotype *cys⁺ leu⁺ met⁺ str^S* is mated with an F⁻ strain carrying the genotype *cys⁻ leu⁻ met⁻ str^R*. In an interrupted mating experiment, small samples of the conjugating bacteria are withdrawn every 3 minutes for 30 minutes. The withdrawn cells are shaken vigorously to stop conjugation and then placed on three different selection media, composed as follows:

Medium 1: Minimal medium plus leucine, methionine, and streptomycin

Medium 2: Minimal medium plus cysteine, methionine, and streptomycin

Medium 3: Minimal medium plus cysteine, leucine, and streptomycin

a. What donor gene is the selected marker in each medium?
b. List all possible bacterial genotypes growing on each medium.
c. What is the purpose of adding streptomycin to each selection medium?

The table below shows the number of colonies growing on each selection medium. The sampling time indicates how many minutes have passed since conjugation began.

| Sampling Time (minutes) | Number of Colonies | | |
|---|---|---|---|
| | Plate 1 | Plate 2 | Plate 3 |
| 3 | 0 | 0 | 0 |
| 6 | 0 | 0 | 0 |
| 9 | 0 | 62 | 0 |
| 12 | 0 | 87 | 0 |
| 15 | 51 | 124 | 0 |
| 18 | 79 | 210 | 62 |
| 21 | 109 | 250 | 85 |
| 24 | 144 | 250 | 111 |
| 27 | 152 | 250 | 122 |
| 30 | 152 | 250 | 122 |

d. Determine the order of donor genes *cys, leu,* and *met* from the interrupted mating data.
e. Suppose a fourth selection medium containing leucine and streptomycin is prepared. At what sampling time do you expect the first-growing colonies to appear? Explain your reasoning.

18. A triple-auxotrophic strain of *E. coli* having the genotype *phe⁻ met⁻ ara⁻* is used as a recipient strain in a transduction experiment. The strain is unable to synthesize its own phenylalanine or methionine, and it carries a mutation that leaves it unable to utilize the sugar arabinose for growth. The recipient is crossed to a prototrophic strain with the genotype *phe⁺ met⁺ ara⁺*. The table below shows the selected marker and gives cotransduction frequencies for the unselected markers.

| Selected Marker | Selected Colonies Containing the Unselected Marker (%) | | |
|---|---|---|---|
| | *phe⁺* | *met⁺* | *ara⁺* |
| *met⁺* | 4 | − | 7 |
| *phe⁺* | − | 2 | 51 |
| *met⁺, phe⁺* | − | − | 79 |
| *ara⁺* | 68 | 5 | − |

a. Identify the compounds present in each of the selective media.

b. Use the cotransduction data to determine the order of these genes.

19. Arabinose is a sugar used by $ara^+$ bacteria as the sole carbon source in a growth medium. Two different mutations, $ara1^-$ and $ara2^-$, each prevent the utilization of arabinose for growth in auxotrophic bacteria. Conjugation experiments are conducted using donor and recipient bacteria auxotrophic for arabinose utilization but are otherwise prototrophic. Following conjugation, $10^6$ exconjugants are placed on growth media containing only arabinose as a carbon source. The results are shown below.

| Experiment | Donor | Recipient | Number of $ara^+$ Colonies |
|---|---|---|---|
| 1 | $ara1-$ | $ara1-$ | 0 |
| 2 | $ara2^-$ | $ara2-$ | 0 |
| 3 | $ara1^-$ | $ara2^-$ | 7 |

a. What is the most likely biological explanation for the presence of prototrophic colonies in experiment 3 but not in the other experiments?

b. Suppose that in a follow-up experiment using $ara1$, $ara2$, and a third arabinose mutation, $ara3$, cotransformation frequencies are determined for each pair of mutants. For $ara1$ and $ara2$, the cotransformation frequency is 0.00008; for $ara1$ and $ara3$, cotransformation takes place at a frequency of 0.00018; and cotransformation of $ara2$ and $ara3$ occurs at a frequency of 0.00014. Place the mutations in order on the arabinose gene.

20. An attribute of growth behavior of eight bacteriophage mutants (1 to 8) is investigated in experiments that establish coinfection by pairs of mutants. The experiments determine whether the mutants complement one another (+) or fail to complement (−). These eight mutants are known to result from point mutation. The results of the complementation tests are shown below.

**Mutations**

|  | 1 | 2 | 3 | 4 | 5 | 6 | 7 | 8 |
|---|---|---|---|---|---|---|---|---|
| 1 | − | + | + | + | − | + | + | − |
| 2 |  | − | + | + | + | + | + | + |
| 3 |  |  | − | + | + | + | − | + |
| 4 |  |  |  | − | + | − | + | + |
| 5 |  |  |  |  | − | + | + | − |
| 6 |  |  |  |  |  | − | + | + |
| 7 |  |  |  |  |  |  | − | + |
| 8 |  |  |  |  |  |  |  | − |

a. How many genes are represented by these mutations?

b. Identify the mutants of each gene.

c. In each coinfection above that is identified as a failure to complement (−), researchers see evidence of recombination producing wild-type growth. How do the researchers distinguish between wild-type growth resulting from complementation and wild-type growth that is due to recombination?

d. A new mutation, designated 9, fails to complement mutants 1, 3, 5, 7, and 8. Wild-type recombinants form between mutant 9 and mutations 3, 5, and 8; however, no wild-type recombinants form between mutant 9 and mutations 1 and 7. What kind of mutation is mutant 9? Explain your reasoning.

e. New mutation 10 fails to complement mutants 1, 4, 5, 6, 8, and 9. Mutant 10 forms wild-type recombinants with mutants 1, 5, and 6, but not with mutants 4 and 8. Mutant 9 and mutant 10 form wild-type recombinants. What kind of mutation is mutant 10? Explain your reasoning.

f. Gene mapping information identifies mutants 2 and 3 as the flanking markers in this group of genes. Assuming these mutations are on opposite ends of the gene map, determine the order of mutations in the region of the chromosome.

21. Synthesis of the amino acid histidine is a multistep anabolic pathway that uses the products of 13 genes ($hisA$ to $hisM$) in E. coli. Two independently isolated $his^-$ E. coli mutants, designated $his1^-$ and $his2^-$, are studied in a conjugation experiment. A $his^+$ F′ donor strain that carries a copy of the $hisJ$ gene on the plasmid is mated with a $his1^-$ recipient strain in experiment 1 and with a $his2^-$ recipient in experiment 2. The exconjugants are grown on plates lacking histidine. Growth is observed among the exconjugants of experiment 2 but not among those of experiment 1.

a. Why is growth observed in experiment 2 but not in experiment 1?

b. What is the genotype of exconjugants in experiment 2?

22. The phage P1 is used as a generalized transducing phage in an experiment combining a donor strain of E. coli of genotype $leu^+ phe^+ ala^+$ and a recipient strain that is $leu^- phe^- ala^-$. In separate experiments, transductants are selected for $leu^+$ (experiment A), for $phe^+$ (experiment B), and for $ala^+$ (experiment C). Following selection, transductant genotypes for the unselected markers are identified.

a. What compound or compounds are added to the minimal medium to select for transductants in experiments A, B, and C?

Selection experiment results below show the frequency of each genotype.

| Experiment A | | Experiment B | | Experiment C | |
|---|---|---|---|---|---|
| $phe^- ala^-$ | 26% | $leu^- ala^-$ | 65% | $leu^- phe^-$ | 71% |
| $phe^+ ala^-$ | 50% | $leu^+ ala^-$ | 48% | $leu^+ phe^-$ | 21% |
| $phe^- ala^+$ | 19% | $leu^- ala^+$ | 0% | $leu^- phe^+$ | 0% |
| $phe^+ ala^+$ | 3% | $leu^+ ala^+$ | 4% | $leu^+ phe^+$ | 3% |

b. Determine the order of genes on the donor chromosome.

c. Diagram the crossover events that form each of the transductants in experiment A.

d. In experiment B, why are there no transductants with the genotype $leu^- ala^+$?

23. A series of seven point mutations are mapped along the $rIIA$ gene and then tested for their ability to form wild-type recombinants with $rII$ partial-deletion mutants. In the table, "+" indicates the formation of wild-type recombinants, and "−" indicates that wild types do not form. Use the data to show the length and endpoints of each deletion as accurately as you can.

| *rIIA* point mutants | 37 | 46 | 21 | 19 | 34 | 27 | 12 |
|---|---|---|---|---|---|---|---|
| Mutant map: | | | | | | | |

**Deletion Mutants** — **Point Mutants**

| Deletion Mutants | 12 | 19 | 21 | 27 | 34 | 37 | 46 |
|---|---|---|---|---|---|---|---|
| B622 | + | + | − | + | + | + | − |
| CT48 | − | + | + | − | − | + | + |
| MB101 | + | + | + | + | + | − | − |
| VG14 | + | − | + | + | + | + | + |
| N220 | + | − | − | + | − | + | + |

24. Five *rII* partial-deletion mutants are mapped and then tested for their ability to form wild-type recombinants with six point mutants. The extent and endpoints of deletion mutants are shown below the *rII* region of the chromosome.

 a. Use the data in Table A to place each point mutation as precisely as you can along the chromosome.

 b. Use the complementation data in Table B to determine where the division between *rIIA* and *rIIB* is located on the *rII* region.

rII region _____

Deletion mutations

M12    ⌊_____⌋

C19         ⌊_____⌋

W42                  ⌊_____⌋

L36              ⌊_____⌋

R22                  ⌊_____⌋

**Table A**

| Point Mutants | Deletion Mutants | | | | |
|---|---|---|---|---|---|
| | C19 | L36 | M12 | R22 | W42 |
| 55 | + | + | − | + | + |
| 67 | + | − | + | − | − |
| 74 | + | + | + | − | − |
| 82 | − | + | − | + | + |
| 85 | + | + | + | − | + |
| 91 | − | − | + | + | + |

**Table B**

| Deletion Mutant | Complemented by | |
|---|---|---|
| | *rIIA* | *rIIB* |
| C19 | + | − |
| L36 | − | − |
| M12 | + | − |
| R22 | − | + |
| W42 | − | + |

c. Based on the data and on your analysis, draw a complementation table for the five point mutants 55, 67, 74, 82, and 85. (Skip mutant 91 for this problem.)

d. Add mutant 91 to your complementation table (assume it maps to *rIIA*).

25. Three genes, *pip*, *zap*, and *rag*, are located near minute 32 on the chromosome of a bacterial species. To determine the order of these genes, you perform a cotransformation experiment and calculate the cotransformation frequencies of the three possible combinations of two of these genes. The results are shown below.

| Experiment | Cotransformed Genes | Cotransformation Frequency |
|---|---|---|
| 1 | *zap* and *rag* | 0.84% |
| 2 | *zap* and *pip* | 0.14% |
| 3 | *rag* and *pip* | 0.62% |

a. What is the gene order?

b. Which two genes are most closely linked?

c. A newly discovered gene *zig* is tested for linkage to *rag*. A double auxotrophic bacterial strain with the genotype *rig⁻ rag⁻* is transformed by a prototrophic bacterial strain (*zig⁺ rag⁺*). The individual transformation frequencies for the genes are 0.018% for *zig* and 0.012% for *rag*. In a screen of $10^8$ cells, there are two *zig⁺ rag⁺* cotransformants. Do the results indicate that *zig* and *rag* are linked? Why or why not?

26. Hfr strains that differ in integrated F factor orientation and site of integration are used to construct consolidated bacterial chromosome maps. The data below show the order of gene transfer for five strains.

| Hfr Strain | Order of Gene Transfer (first → last) |
|---|---|
| Hfr A | *oriT – thr – leu – azi – ton – pro – lac – ade* |
| Hfr B | *oriT – mtl – xyl – mal – str – his* |
| Hfr C | *oriT – ile – met – thi – thr – leu – azi – ton* |
| Hfr D | *oriT – his – trp – gal – ade – lac – pro – ton* |
| Hfr E | *oriT – thi – met – ile – mtl – xyl – mal – str* |

a. Identify the overlaps between Hfr strains. Identify the orientations of F factors relative to one another.

b. Draw a consolidated map of the bacterial chromosome. (*Hint:* Begin by placing the insertion site for Hfr A at the 2 o'clock position and arranging the genes *thr-leu-azi-* ... in clockwise order.)

# 7 DNA Structure and Replication

## ESSENTIAL IDEAS

▪ Seventy-five years of observations and analysis culminated in the identification of DNA as the hereditary molecule.

▪ DNA is a double-stranded molecule consisting of four kinds of nucleotides, abbreviated A, T, C, and G, that are held together by a mechanism of complementary base pairing.

▪ DNA replication faithfully duplicates the genome by a semiconservative process that progresses bidirectionally from each origin of replication.

▪ Origins of replication are defined by their nucleotide sequence. Numerous proteins and enzymes act in concert to produce two identical DNA duplexes.

▪ Laboratory techniques based on a molecular understanding of DNA replication perform targeted replication of short DNA sequences and sequence DNA.

The laboratory method known as polymerase chain reaction (PCR) is made possible by *Taq* polymerase that was first isolated from *Thermus aquaticus* bacteria living in near-boiling conditions in Yellowstone National Park. The inset photo (upper left) shows growing *T. aquaticus*.

The central dogma of biology identifies DNA as the repository of genomic information for organisms and its central role in the production of RNA transcripts of genes and of polypeptides produced by translation of mRNA. DNA's role in these processes requires its faithful replication, and that is the subject of this chapter.

In Chapter 1, we reviewed the primary and secondary structures of DNA and RNA and the fundamentals of DNA replication. In this chapter, we discuss the structure of DNA in greater detail and extend the earlier description to include the molecular processes occurring in DNA replication. We also examine two analytical methods—polymerase chain reaction (PCR) and dideoxy DNA sequencing—that were developed as an outcome of the understanding of replication. The Case Study at the end of the chapter describes the use of PCR and dideoxy sequencing to identify and analyze the mutation associated with Huntington disease (OMIM 143100), an autosomal dominant disorder in humans.

## 7.1 DNA Is the Hereditary Molecule of Life

When scientists speak of the "hereditary molecule" of a species, they mean the molecular substance that carries and conveys the species' genetic information. Our contemporary understanding of hereditary transmission and the evolution of species is rooted in the knowledge that DNA is the hereditary molecule of all organisms. Long before the hereditary role of DNA was established, however, research had identified five essential characteristics of hereditary material. The hereditary material must be

1. Localized to the nucleus and a component of chromosomes
2. Present in a stable form in cells
3. Sufficiently complex to contain the genetic information that directs the structure, function, development, and reproduction of organisms
4. Able to accurately replicate itself so that daughter cells contain the same information as parental cells
5. Mutable, undergoing mutation at a low rate that introduces genetic variation and serves as a foundation for evolutionary change

### Chromosomes Contain DNA

The weakly acidic substance known today as DNA was first noticed in 1869, when Friedrich Miescher isolated it from the nuclei of white blood cells in a mixture of nucleic acids and proteins he called "nuclein." Miescher made little progress in determining the composition of nuclein, however, and the substance was little studied over the next several decades.

In the 1870s, microscopic studies identified the fusion of male and female nuclei during reproduction. Shortly thereafter, chromosomes were first observed in cell nuclei. Advances in microscopy led to the observation that the nuclei of different species contain different numbers of chromosomes, and to accurate descriptions of the equal chromosome contributions of males and females to reproduction. The earliest suggestion that DNA was the hereditary material came from one such study by Edmund Wilson in 1895. After accurately documenting that sperm and egg cells contribute the same number of chromosomes during reproduction, Wilson speculated, "The precise equivalence of the chromosomes contributed by the sexes is a physical correlative of the fact that the two sexes play, on the whole, equal parts in hereditary transmission, and it seems to show that the chromosomal substance, the chromatin, is to be regarded as the physical basis of inheritance. Now, chromatin is known to be closely similar to, if not identical with[,] a substance known as nuclein ($C_{29} H_{49} N_9 P_3 O_{22}$, according to Miescher), which analysis shows to be a tolerably definite chemical composed of nucleic acid (a complex organic acid rich in phosphorus) and albumin. And thus we reach the remarkable conclusion that inheritance may, perhaps, be effected by the physical transmission of a particular chemical compound from parent to offspring."

In 1900, Mendel's hereditary principles were rediscovered (see Chapter 1). Shortly thereafter, in 1903, Wilson's student Walter Sutton and, independently, Theodor Boveri accurately described the parallels between, on the one hand, homologous-chromosome and sister-chromatid separation and, on the other hand, the inheritance of genes.

Over the next 20 years, the nucleus and chromosomes were a focus of biological investigations of heredity. By 1920, the principal constituent of nuclein was identified as DNA, and the basic chemistry of DNA was deciphered. The molecule was determined to be a polynucleotide consisting of four repeating subunits—the four DNA nucleotides—held together by covalent bonds. The four DNA nucleotides are adenine (A), thymine (T), cytosine (C), and guanine (G).

In 1923, DNA was localized to chromosomes. This discovery made DNA a candidate for the hereditary material, but DNA is not the sole constituent of chromosomes. Proteins are in high concentration in chromosomes; RNA is present in the nucleus and around chromosomes; and other compounds, including lipids and carbohydrates, were also considered as potential candidates for the hereditary material at one time or another. In fact, some early researchers, including, eventually, Edmund Wilson, thought protein was potentially a better candidate for the hereditary material than DNA. They noted that protein is composed of 20 different amino acids, whereas DNA has only 4 kinds of nucleotides. The protein proponents suggested that the "20-letter alphabet" of protein could contain more information than the "4-letter alphabet" of DNA. It was against this backdrop that the results of three experiments conducted between 1928 and 1952 combined to identify DNA—not RNA, protein, or another chemical constituent of cells—as the hereditary material of organisms.

### The Transformation Factor

Frederick Griffith, a British physician with an interest in epidemiology, studied pneumonia infection in mice and published a lengthy research report in 1928 describing his findings. Modern biology focuses on just the few pages of Griffith's long report that provided indirect evidence that DNA is the molecule responsible for conveying hereditary characteristics in bacteria.

Griffith studied strains of the bacterium *Pneumococcus*, which causes fatal pneumonia in mice, and found the strains to have a smooth (S) appearance under a microscope owing to the presence of a protective capsule (**Figure 7.1**). On the other hand, *Pneumococcus* strains that do not cause disease are identifiable by their rough (R) appearance. Rough bacterial strains have a mutant allele of the polysaccharide gene, which results in a weakened and easily broken capsule. This single gene mutation thus leaves R bacteria vulnerable to attack by mouse immune system antibodies.

The S and R forms of *Pneumococcus* occur in four antigenic types of the bacteria, identified as I, II, III, and IV. Each antigenic type elicits a different immune response from the mouse immune system as a result of the presence of several genetic differences. A single mutation of the polysaccharide gene can convert an S strain to an R strain *of the same antigenic type*—for example, converting an SII strain to an RII strain—but the antigenic type cannot be changed by a single mutation. In other words, mutation alone cannot change RII bacteria into SIII.

Griffith's most important observations are derived from four infection tests he performed using S and R

**Figure 7.1  Appearance of smooth versus rough colonies of *Pneumococcus*.**

bacterial strains of different antigenic types (**Figure 7.2**). Griffith showed that injecting mice with strain SIII, for example, quickly produces illness and death (❶). Griffith cultured blood taken from dead mice and identified the presence of SIII bacteria. Griffith also determined that injection of "heat-killed" SIII bacteria that are killed using high heat and pressure does not induce illness ❷, nor does using any R strain ❸. Griffith's most significant result ❹ came when he injected a mixture of heat-killed

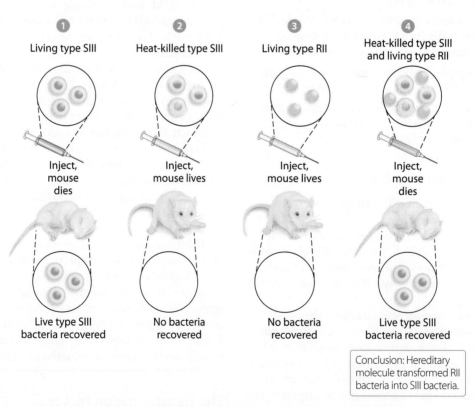

Conclusion: Hereditary molecule transformed RII bacteria into SIII bacteria.

**Figure 7.2  Frederick Griffith's experiment identifying a "transformation factor" responsible for heredity.** ❶ Injection of living SIII bacteria kills mice. ❷ Heat-killed SIII do not kill mice, nor do living RII bacteria ❸. ❹ Coinjection of a mixture of heat-killed SIII and living RII bacteria results in mouse death by SIII infection.

SIII strain and living RII strain. He found that most of the mice died from pneumonia, and when he tested blood cultures from the dead mice, he detected living SIII bacteria. Knowing that this outcome could not have been the result of a simple mutational event, Griffith proposed that a molecular component he called the transformation factor was responsible for transforming RII into SIII.

Griffith concluded that the transforming factor was a molecule that carried hereditary information, but he was unable to suggest the class of molecule. Today biologists know that the process identified by Griffith is a naturally occurring process called *transformation,* which is used by bacteria to transfer DNA (see Section 6.4).

## DNA Is the Transformation Factor

Shortly after Griffith published his report on the transformation factor, Martin Dawson, working with Oswald Avery, developed an in vitro transformation procedure to mix living R cells with a purified extract of cellular material derived from heat-killed SIII cells containing the transformation factor. Biochemical assays indicated that the SIII extract consisted mostly of DNA, along with a small amount of RNA and trace amounts of proteins, lipids, and polysaccharides.

The most direct evidence that DNA was the transformation factor came from an experiment performed by Avery and his colleagues MacLeod and McCarty (Figure 7.3).

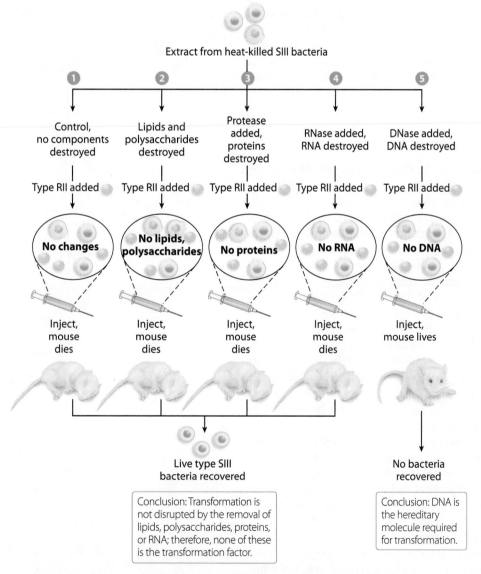

**Figure 7.3 Avery, MacLeod, and McCarty's use of in vitro transformation to identify DNA as the most likely hereditary molecule.** A purified extract from heat-killed SIII bacteria successfully transforms RII cells in the control experiment ❶. Destruction of lipids and polysaccharides ❷, proteins ❸, or RNA ❹ does not affect transformation; however, destruction of DNA ❺ prevents transformation.

This experiment identified the role of DNA in transformation by eliminating lipids, polysaccharides, protein, RNA, and DNA one at a time from the SIII extract. In each experimental trial, the SIII extract was treated to remove one component at a time, and the treated extract was mixed with RII cells. The in vitro transformation reaction was allowed to take place, and the resulting cells were injected into mice. Blood from the injected mice was then cultured to determine if SIII cells were present—an indication that transformation had occured.

Figure 7.3 shows that in vitro transformation takes place and mice are killed by SIII infection in the control experiment ❶, and when lipids and polysaccharides ❷, proteins ❸, or RNA ❹ are removed from the extract. In contrast to the other results, experiment ❺, which uses DNase to specifically degrade DNA, does not result in transformation. Blood drawn from mice injected with DNase-treated extract reveals no living SIII bacteria—a clear indication that transformation is blocked by the destruction of DNA. Based on these observations Avery, MacLeod, and McCarty correctly concluded that DNA is the transformation factor and the probable hereditary material.

## DNA Is the Hereditary Molecule

Avery, MacLeod, and McCarty's work convinced many biologists that DNA was the long-sought hereditary material, and a great deal of research in the late 1940s and early 1950s was devoted to deducing the physical structure of DNA. Biologists realized that once the structure of DNA was known, the chemical nature of genes would be identified, and biological research would move into the realm of genetic molecular biology. As clear and convincing as the work of Avery and his colleagues seems in retrospect, however, there were several unanswered questions about the role of DNA in heredity. There was also a need to demonstrate directly that the presence of a specific DNA molecule induces the appearance of a particular phenotype. That evidence came in a 1952 report by Alfred Hershey and Martha Chase, who showed that DNA, but not protein, is responsible for bacteriophage infection of bacterial cells.

**Bacteriophages,** also known as **phages,** are viruses that infect bacteria. Phages such as T2, for example, consist of a protein shell with a tail segment that attaches to a host bacterial cell and a head segment that contains DNA. T2 phages are among the many bacteriophage that do not carry any RNA. Like other viruses, T2 must infect host cells, bacterial cells in this case, in order to reproduce. Infection by a phage begins with the injection of phage DNA into a bacterial cell; the emptied phage shell remains attached to the outside of the bacterial cell. Inside the infected bacterial cell, phage DNA then replicates, and phage proteins are produced

(Figure 7.4). These proteins are assembled into progeny phages that are each filled with a copy of the phage chromosome. Progeny phages are released by lysis of the host bacterial cell.

In their experiment, Hershey and Chase took advantage of an essential difference between the chemical composition of DNA and protein to confirm the hereditary role of DNA (Figure 7.5). Proteins contain large amounts of sulfur but almost no phosphorus; conversely, DNA

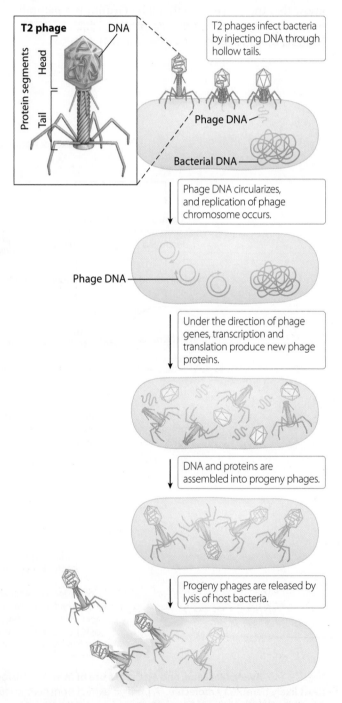

**Figure 7.4  The infection of bacteria by bacteriophage T2.**

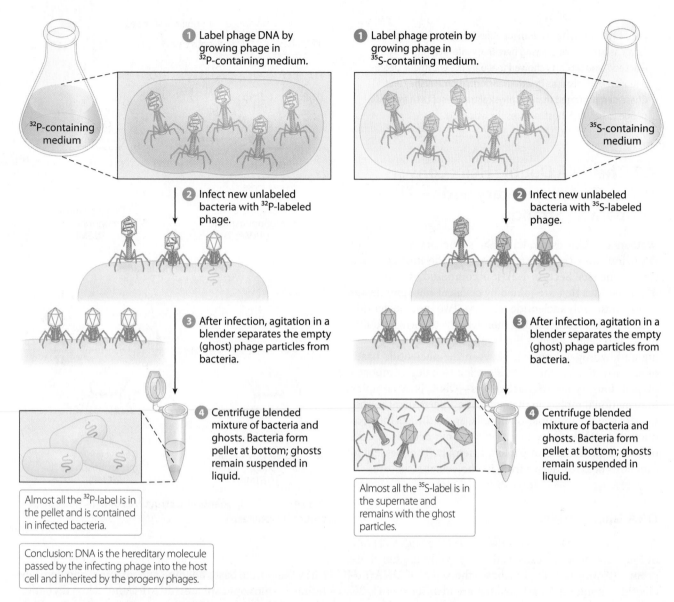

**Figure 7.5   Hershey-Chase experiment showing DNA to be the molecule in bacteriophages that causes lysis of infected bacterial cells.**

contains a large amount of phosphorus but no sulfur. To prepare phages for the experiment, Hershey and Chase initially grew phage cultures in different growth media. One growth medium contained $^{35}$S, the radioactive form of sulfur, to label protein ❶; the other environment contained radioactive phosphorus, $^{32}$P, to label DNA ❶. The researchers used radioactively labeled phages from each medium to infect unlabeled host bacterial cells in parallel experiments ❷ ❷.

After a short time, each mixture was agitated in a blender to separate bacterial cells from the now empty phage shells. Such empty phage shells are called "ghosts" ❸ ❸. The relatively large bacterial cells were easily separated from the ghosts by centrifugation. The heavier bacteria collect in a pellet at the bottom of the centrifuge

tube, while the lighter ghosts remain suspended in the supernatant. Testing each fraction for radioactivity revealed that virtually all the $^{32}$P label was associated with newly infected bacterial cells and almost none with ghost particles ❹. On the other hand, the $^{35}$S label was found in the ghost-particle fraction, and only trace amounts were found associated with the bacterial pellet ❹. This result demonstrates that phage DNA, but not phage protein, is transferred to host bacterial cells and directs the synthesis of phage DNA and proteins, the assembly of progeny phage particles, and ultimately the lysis of infected cells. The experiment demonstrated that the transformation factor identified previously by Griffith was DNA; it also showed that Avery, MacLeod, and McCarty were correct in concluding that DNA is the hereditary material.

## 7.2 The DNA Double Helix Consists of Two Complementary and Antiparallel Strands

Watson and Crick's model of the secondary structure of DNA indicates that in some respects, the molecule is a simple one (see Section 1.2). It is composed of four kinds of nucleotides that are joined by covalent phosphodiester bonds into polynucleotide chains. Two polynucleotide chains come together along their lengths to form a double helix, also called a DNA duplex, by complementary pairing and hydrogen bonding between the nucleotide bases of each strand. Yet for all its simplicity—being composed of just four types of nucleotides—DNA is a complex informational molecule that serves as a permanent repository of genetic information in cells, and it directs the production of RNA molecules that carry out actions in cells or carry information for protein assembly. These essential functions of DNA derive from its molecular structure.

### DNA Nucleotides

A nucleotide in DNA has three parts: (1) a sugar, (2) one of four nitrogenous bases, and (3) up to three phosphate groups (**Figure 7.6**). Deoxyribose, the sugar of DNA nucleotides, contains 5 carbons that are identified as 1', 2', 3', 4', and 5'. An oxygen atom connects the 1' carbon to the 4' to form a five-sided (pentose) ring, and the 5' carbon projects outward from the 4' carbon (and from the ring). A nucleotide base is attached to the 1' carbon; a hydroxyl group (OH) is attached to the 3' carbon; and a single phosphate molecule, or a chain of phosphates up to three molecules long, is attached at the 5' carbon. Deoxyribose carries a hydrogen molecule at the 2' carbon instead of a hydroxyl (OH) group. This is the basis for naming the sugar *deoxy*ribose.

The nitrogenous bases in DNA are of two structural types—a single-ringed form called a pyrimidine, and a double-ringed form called a purine. Cytosine (C) and thymine (T) are pyrimidines, and adenine (A) and guanine (G) are purines. DNA nucleotides that are part of a polynucleotide chain have one phosphate group that forms the covalent phosphodiester bond with the adjacent nucleotide in the strand. Deoxyadenosine 5'-monophosphate (dAMP) and deoxyguanosine 5'-monophosphate (dGMP)

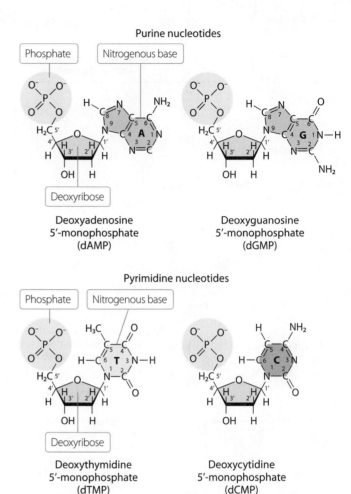

**Figure 7.6  Components and structures of DNA nucleotide monophosphates.**

carry the purine bases adenine and guanine, and deoxycytidine 5'-monophosphate (dCMP) and deoxythymidine 5'-monophosphate (dTMP) carry the pyrimidine bases cytosine and thymine. Collectively, these are identified as the **deoxynucleotide monophosphates (dNMPs)**, where *N* can refer to any of the four nucleotide bases. In contrast, free (reactive) DNA nucleotides that are not part of a polynucleotide chain carry a string of three phosphate groups at the 5' carbon and are identified as dATP, dGTP, dCTP, and dTTP. Collectively, these are the **deoxynucleotide triphosphates (dNTPs)**.

Individual nucleotides are assembled into a polynucleotide chain by the enzyme DNA polymerase, which catalyzes the formation of a phosphodiester bond between the 3' hydroxyl group of one nucleotide and the 5' phosphate group of an adjacent nucleotide (**Figure 7.7**). Two of the three phosphates of a dNTP are removed (as a pyrophosphate group) during phosphodiester bond formation, leaving the nucleotides of a polynucleotide chain in their monophosphate form. Each polynucleotide chain has a **sugar-phosphate**

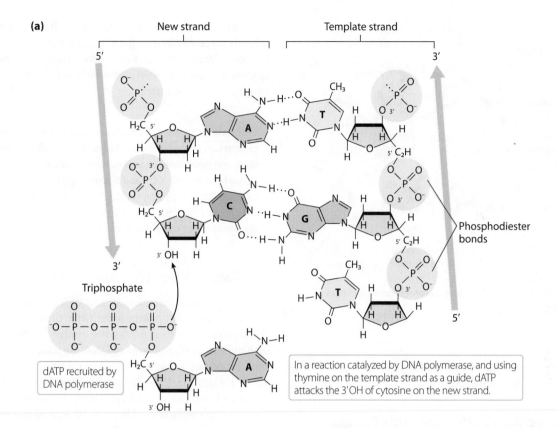

**(a)**

dATP recruited by DNA polymerase

In a reaction catalyzed by DNA polymerase, and using thymine on the template strand as a guide, dATP attacks the 3′OH of cytosine on the new strand.

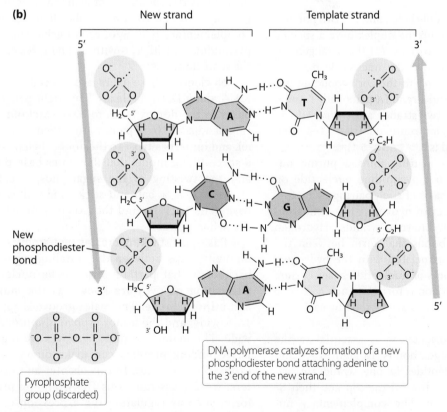

**(b)**

New phosphodiester bond

Pyrophosphate group (discarded)

DNA polymerase catalyzes formation of a new phosphodiester bond attaching adenine to the 3′ end of the new strand.

**Figure 7.7  DNA strand elongation.  (a)** Nucleotides complementary to the template strand are added to the 3′ end of the new strand by DNA polymerase. **(b)** DNA nucleotide triphosphates are recruited by DNA polymerase, which uses catalytic action to remove two phosphates (the pyrophosphate group) and form a new phosphodiester bond.

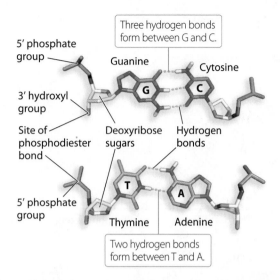

5′ phosphate group

Three hydrogen bonds form between G and C.

Guanine

Cytosine

3′ hydroxyl group

Site of phosphodiester bond

Deoxyribose sugars

Hydrogen bonds

5′ phosphate group

Thymine

Adenine

Two hydrogen bonds form between T and A.

**Figure 7.8** **Hydrogen bonding between complementary nucleotides.**

**backbone** consisting of alternating sugar and phosphate groups throughout its length.

## Complementary DNA Nucleotide Pairing

DNA is most stable as a double helix, and the two polynucleotide strands that make up the duplex have a specific relationship that follows two rules: (1) the arrangement of the nucleotides is such that the nucleotide bases of one strand are *complementary* to the corresponding nucleotide bases on the second strand (A pairs with T and G pairs with C), and (2) the two strands are *antiparallel* in orientation (one strand is, for example 5′-ATCG-3′ and the complementary strand is 3′-TAGC-5′).

Complementary base pairing joins a purine nucleotide on one strand to a pyrimidine nucleotide on the other. The chemical basis of such pairing is the formation of a stable number of hydrogen (H) bonds between the bases of the different strands. Hydrogen bonds are noncovalent bonds that form between the partial charges that are associated with the hydrogen, oxygen, and nitrogen atoms of nucleotide bases (**Figure 7.8**). Two stable hydrogen bonds form for each A-T base pair, and three hydrogen bonds are formed by each G-C base pair.

Antiparallel strand orientation is essential to the formation of stable hydrogen bonds. In Figures 7.7 and 7.8, notice that the nucleotides in one strand are oriented with their 5′ carbon toward the top and their 3′ carbon toward the bottom. The complementary nucleotides in the other strand are antiparallel; that is, their 5′-to-3′ orientations run in the *opposite* direction. Antiparallel orientation of complementary strands brings the partial charges of complementary nucleotides into

alignment to form hydrogen bonds. If complementary strands were to align in parallel (i.e., with their 5′ and 3′ carbons facing in the same direction), the charges of complementary nucleotides would repel, and no hydrogen bonds would form.

---

Genetic Insight  DNA is composed of four nucleotides containing bases of two structural types: purines (adenine and guanine) and pyrimidines (cytosine and thymine). Phosphodiester bonds between the 3′ OH and 5′ phosphate of adjacent nucleotides form single polynucleotide strands. Complementary base pairing (A with T, C with G) in antiparallel strands forms double-stranded nucleic acid molecules.

---

## The Twisting Double Helix

The DNA double helix has an axis of helical symmetry, an imaginary line that passes lengthwise through the core of the double helix and marks the center of the molecule. The molecular dimensions of DNA are measured using the unit called an angstrom (Å). One angstrom is equal to $10^{-10}$ meters, or 1 ten-billionth of a meter. In DNA, the distance from the axis of symmetry to the outer edge of the sugar-phosphate backbone is 10 Å, and the molecular diameter is 20 Å at any point along the length of the helix (**Figure 7.9a**). The 20-Å molecular diameter results from complementary pairing of each purine with the correct pyrimidine (A with T, G with C), and gives each base pair the same dimension.

Nucleotide base pairs are spaced at intervals of 3.4 Å along DNA duplexes. This tight packing of DNA bases in the duplex leads to **base stacking,** the offsetting of adjacent base pairs so that their planes are parallel, and imparts a twist to the double helix. **Figure 7.9b** is a space-filling model that illustrates base-pair stacking and the twisting of the sugar-phosphate backbones. **Figure 7.9c** is a ball-and-stick model illustrating how base pairs twist around the axis of symmetry to create the helical spiral.

Base-pair stacking creates two grooves in the double helix, gaps between the spiraling sugar-phosphate backbones that partially expose the nucleotides. The alternating grooves are known as the **major groove** and **minor groove.** The major groove is approximately 12 Å wide, and the minor groove is approximately 6 Å wide. The major and minor grooves are regions where DNA-binding proteins can make direct contact with exposed nucleotides. In this chapter and in later chapters, we discuss functions DNA-binding proteins perform, such as regulating the expression of genes and controlling the onset and progression of DNA replication. Most of these functions depend on the presence of characteristic sequences of DNA nucleotides. DNA-binding proteins gain access to DNA nucleotides in

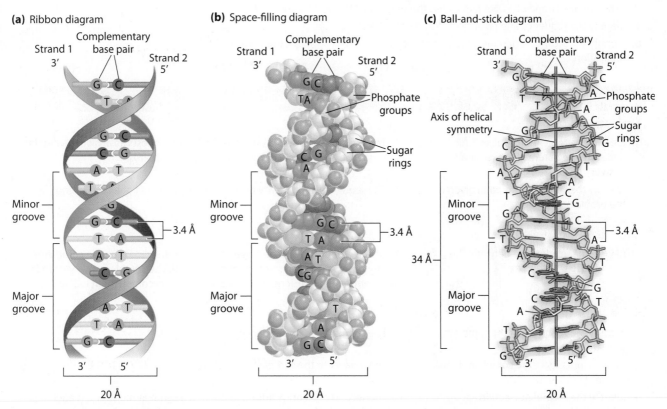

**(a)** Ribbon diagram    **(b)** Space-filling diagram    **(c)** Ball-and-stick diagram

**Figure 7.9   The DNA double helix.  (a)** Ribbon diagram, **(b)** space-filling diagram, and **(c)** ball-and-stick diagram show the sugar-phosphate backbones, base pairs, major and minor grooves, and dimensions of the DNA duplex.

major and minor grooves of the molecule. **Genetic Analysis 7.1** explores relationships between complementary DNA strands.

## 7.3 DNA Replication Is Semiconservative and Bidirectional

Given the role of DNA as an information repository and an information transmitter, the integrity of the nucleotide sequence of DNA is of paramount importance. Each time DNA is copied, the new version must be a precise duplicate of the original version. The high fidelity of DNA replication is essential to reproduction and to the normal development of biological structures and functions. Without faithful DNA replication, the information of life would become hopelessly garbled by rapidly accumulating mutations that would threaten survival.

Considering the importance of DNA throughout the biological world, it was no surprise to discover that the general mechanism of DNA replication is the same in all organisms. This universal process evolved in the earliest life-forms and has been retained for billions of years. As organisms diverged and became more complex, however, an array of differences did develop among DNA

replication proteins and enzymes. Despite the diversification of these specific components of DNA replication, three attributes of DNA replication are shared by all organisms:

1.  Each strand of the parental DNA molecule remains intact during replication.

2.  Each parental strand serves as a template directing the synthesis of a complementary, antiparallel daughter strand.

3.  Completion of DNA replication results in the formation of two identical daughter duplexes, each composed of one parental strand and one daughter strand.

### Three Competing Models of Replication

In their famous 1953 paper describing the structure of DNA, Watson and Crick concluded with the observation, "It has not escaped our notice that the specific base-pairing we have proposed immediately suggests a possible copying mechanism for the genetic material." Specifically, Watson and Crick recognized that a consequence of complementary base pairing was that nucleotides on one strand of the duplex could be used to identify the nucleotides of the

A portion of one strand of a DNA duplex has the sequence 5'-ACGACGCTA-3'.

**a.** Identify the sequence and polarity of the other DNA strand.

**b.** Identify the *second* nucleotide added if the sequence given is used as a template for DNA replication.

| Solution Strategies | Solution Steps |
|---|---|
| **Evaluate** | |
| 1. Identify the topic this problem addresses, and state the nature of the required answer. | 1. The question concerns a DNA sequence and requests an answer giving the sequence and polarity of the complementary strand. |
| 2. Identify the critical information given in the problem. | 2. The sequence and polarity are given for a portion of one DNA strand. |
| **Deduce** | |
| 3. Review the relationship between the strands of a DNA duplex. | 3. DNA is a double helix composed of single strands that contain complementary base pairs (A pairs with T, and G with C). The complementary strands are antiparallel (i.e., one strand is 5′ to 3′, and its complement is 3′ to 5′). |
| **Solve** | |
| 4. Identify the sequence of the complementary strand. | 4. The complementary sequence is TGCTGCGAT. |
| 5. Give the polarity of the complementary strand. | 5. The polarity of the complementary strand is 3'-TGCTGCGAT-5'. |
| 6. Identify the second nucleotide added during DNA replication of the given sequence.  <br> TIP: DNA polymerase catalyzes the addition of a new nucleotide to the 3′ end of a growing strand. | 6. The second nucleotide added to the newly synthesized strand is adenine, which is complementary to thymine on the template strand. |

For more practice, see Problems 5, 9, 17, and 19.

---

other strand. Watson and Crick presumed that DNA replication used the nucleotide sequence of each strand to form a new duplex, and they envisioned that each DNA strand could act as a template. Watson and Crick did not know the precise mechanism by which template-based replication took place, but their recognition of the likelihood that it does take place raised the crucial question of what the exact mechanism of replication might be.

Almost immediately after the DNA structure was identified, three competing models of DNA replication emerged (**Figure 7.10**). The models shared the idea that the two original strands (the parental strands) of the duplex act as templates to direct the assembly of newly synthesized DNA by complementary base pairing. The models also predicted that the completion of DNA replication produced two identical DNA duplexes (daughter duplexes). The models differed, however, in describing the makeup of the daughter duplexes. The ❶ **semiconservative DNA replication** model—which proved to be correct—proposed that each daughter duplex contains one original parental strand of DNA and one complementary, newly synthesized daughter strand. The ❷ **conservative DNA replication**

model predicts that one daughter duplex contains the two strands of the parental molecule and the other contains two newly synthesized daughter strands. Lastly, the ❸ **dispersive DNA replication** model predicts that each daughter duplex is a composite of interspersed parental duplex segments and daughter duplex segments.

## The Meselson-Stahl Experiment

In 1958, Matthew Meselson and Franklin Stahl took advantage of the newly developed method of high-speed cesium chloride (CsCl) density gradient ultracentrifugation to decipher the mechanism of DNA replication in an experiment of beautiful simplicity. In this analytical method, a tube filled with a CsCl mixture is subjected to high ultracentrifuge speeds that exert thousands of gravities of separating force, creating a graded variation in density—a density gradient—throughout the CsCl mixture. When substances are placed in the CsCl gradient and ultracentrifugation takes place, the substances migrate at a rate determined by their molecular density. This technique is capable of separating molecules that have only slightly different molecular weights.

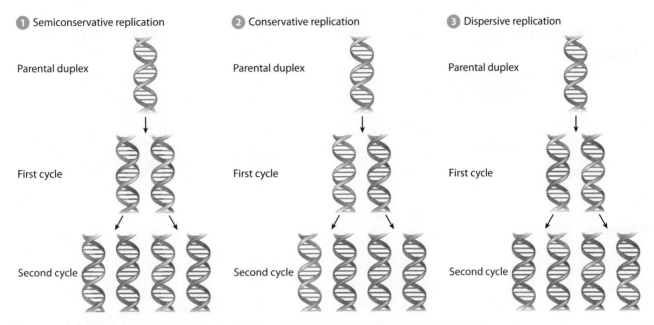

**Figure 7.10** **Three proposed mechanisms of DNA replication tested by Meselson and Stahl.** The results expected for two cycles of DNA replication are shown for each model.

The Meselson and Stahl experiment took advantage of the exquisite sensitivity of CsCl ultracentrifugation to compare parental and daughter DNA molecules. They began by growing *Escherichia coli* in a growth medium containing the heavy isotope of nitrogen, $^{15}$N, for many generations to fully saturate parental DNA with heavy-isotope-containing nitrogen. Eventually, all DNA duplexes in the bacteria growing in that medium contained only the heavy nitrogen isotope. These duplexes were designated $^{15}$N/$^{15}$N to signify the incorporation of $^{15}$N in both strands of the duplex. (By the same token, a DNA duplex composed of two strands containing only $^{14}$N, the normal isotope of nitrogen, is designated $^{14}$N/$^{14}$N, and a duplex with one strand containing each isotope is designated $^{15}$N/$^{14}$N.) DNA collected for CsCl gradient analysis from this starting generation, designated generation 0, was exclusively $^{15}$N/$^{15}$N. Next, some of these $^{15}$N-labeled *E. coli* were transferred to a new growth medium containing only $^{14}$N. At various intervals through successive DNA replication cycles, DNA was collected from a few cells on the $^{14}$N medium for analysis.

The lower portion of Figure 7.11 shows the results of CsCl gradient analysis of DNA collected from generation 0 and after each of three cycles of DNA replication. As shown by the molecular depictions in the upper portion of the figure, the results are consistent with the semiconservative model only. The conservative model predicted DNA molecules with two distinct densities after generation 1 ($^{15}$N/$^{15}$N and $^{14}$N/$^{14}$N). The results reject this model. Similarly, the dispersive model predicted a single DNA density in all generations. The generation 2 results reject this replication model. Within a few years of Meselson and Stahl's

identification of semiconservative replication in bacteria, the mechanism was identified experimentally in eukaryotes as well, solidifying the idea that all life shares the same general process of DNA replication, as a consequence of life's single origin and the evolutionary connections among living things. **Genetic Analysis 7.2** explores the replication analysis of "Martian microbe" DNA with an experimental approach similar to Meselson and Stahl's.

## Origin of Replication in Bacterial DNA

Solving the riddle of the basic mechanism of DNA replication introduced new questions about how replication is initiated and how it progresses. Does replication commence at specific points on each chromosome? If so, how many such points does a chromosome have? Does DNA replication progress in one direction or in both directions from a replication origin? Experimental evidence clearly demonstrates that DNA replication is most often *bidirectional*, progressing in both directions from a single **origin of replication** in bacterial chromosomes and from multiple origins of replication in eukaryotic chromosomes.

In 1963, John Cairns reported the first evidence of a single origin of DNA replication in *E. coli*. Cairns grew bacteria in a medium containing a radioactive isotope of thymine ($^3$H-thymine) and then placed their chromosomes, extracted during replication, on X-ray film to produce an autoradiographic image (Figure 7.12). The radioactive decay of $^3$H is slow and took several months to produce an image. Once the exposure was complete, Cairns examined his autoradiographs by microscopy and found dark lines that revealed the pattern of replicating DNA molecules. He concluded that loop A

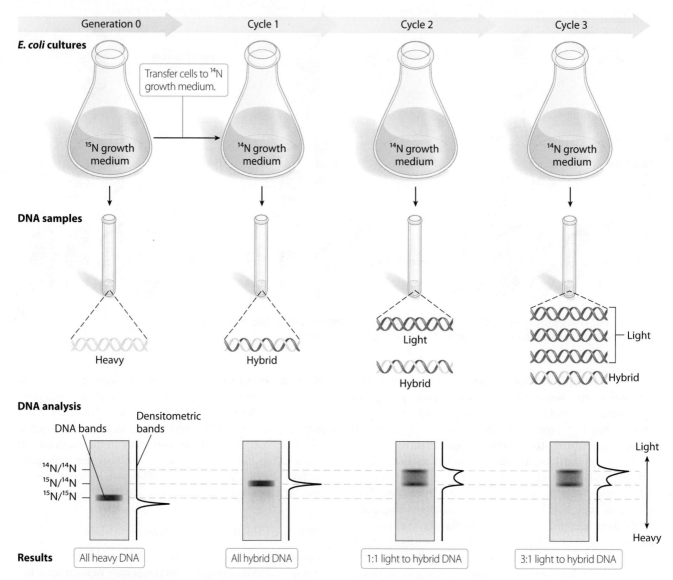

**Figure 7.11** **The Meselson-Stahl experimental results.** Photographs of DNA bands in centrifuge tubes and densitometry scans (lower) identify the duplex DNA composition at each stage and are consistent only with semiconservative DNA replication. The semiconservative replication process is interpreted for each replication cycle.

and loop B in the photograph are replicated DNA, that loop C is unreplicated DNA, and that replication proceeded from a single origin of replication. In Cairns's images, the shape of the replicating chromosome was reminiscent of the Greek letter theta; hence, the name applied to these replicating chromosomes is θ **structure** ("theta structure").

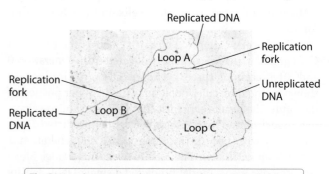

The DNA is elongating at the replication forks, with both strands of unreplicated DNA (blue) being used as a template. This structure is consistent with DNA replication from a single origin.

**Figure 7.12** **Bacterial DNA replication.** Cairns's autoradiographic image captures replicated DNA (red) and unreplicated DNA (blue). The replicating chromosome structure indicates that replication initiates from a single origin.

Suppose future exploration identifies a microbe living in the polar ice of Mars. Analysis reveals that the microbe contains double-stranded DNA. An experiment like Meselson and Stahl's is done to determine how the DNA is replicated. Below are a control densitometry scan and the densitometry scans of DNA collected after each of three replication cycles. The positions and heights of peaks reflect the relative amounts of DNA of different densities in each sample collected. What is the mechanism of DNA replication in this hypothetical Martian microbe?

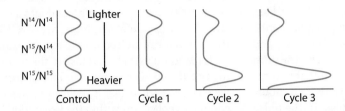

| Solution Strategies | Solution Steps |
|---|---|
| **Evaluate** | |
| 1. Identify the topic this problem addresses, and state the nature of the required answer. | 1. This question concerns the mechanism of DNA replication of a hypothetical extraterrestrial organism. The answer will identify how the DNA replicates. |
| 2. Identify the critical information given in the problem. | 2. An experiment with a design like that of Meselson and Stahl's generates information on the composition of replicated DNA. A control graph showing densitometry results for $N^{15}/N^{15}$, $N^{15}/N^{14}$, and $N^{14}/N^{14}$ DNA is provided, along with the results of three DNA replication cycles in the organism. |

**TIP:** Review the predictions of the conservative, semiconservative, and dispersive models of DNA replication.

| | |
|---|---|
| **Deduce** | |
| 3. Inspect and <u>evaluate</u> the results of the first DNA replication cycle. | 3. Two peaks of equal height are seen, indicating that DNA of two different densities is present in equal amounts after one replication cycle. The densities indicate DNA compositions of $^{15}N/^{15}N$ and $^{14}N/^{14}N$. No $N^{15}N^{14}$ DNA is detected. |
| **Solve** | |
| 4. Deduce possible mechanism(s) of DNA replication based on the first replication cycle result. | 4. This result is consistent with the predictions of conservative replication and inconsistent with the predictions of semiconservative and dispersive replication. |
| 5. Inspect and evaluate the result of the second replication cycle in light of the conclusion from the previous step. | 5. The same two forms of DNA are present, but there is now approximately three times as much $^{15}N/^{15}N$ DNA as $^{14}N/^{14}N$ DNA. This is consistent with the prediction of the conservative model of DNA replication. |
| 6. Evaluate the result of DNA replication cycle 3, and confirm that your interpretation is correct. | 6. After the completion of cycle 3, the ratio of DNA molecules is skewed substantially in favor of $^{15}N/^{15}N$. After three replication cycles, the conservative model predicts a molecular ratio of 7:1 of $N^{15}/N^{15}$ to $N^{14}/N^{14}$. We therefore conclude that the microbe has a conservative mode of replication. |

For more practice, see Problems 20 and 22.

## Evidence of Bidirectional DNA Replication

The θ structures in Cairns's autoradiographs are consistent with either unidirectional or bidirectional DNA replication, although within a few years the replication mechanism was shown to be bidirectional. In **bidirectional DNA replication** of circular bacterial chromosomes, new DNA is synthesized in both directions from a single origin of replication, creating an expanding **replication bubble,** as shown in **Figure 7.13a**. At each end of the replication bubble is a **replication fork** where new DNA nucleotides are added to elongating daughter strands. The growth of the replication bubble forms the theta structure captured in Cairns's autoradiographic images. Replication is completed when the two replication forks encounter one another on the opposite side of the circular chromosome from the origin of replication. With completion of replication, the two daughter duplexes are separate chromosomes.

The localization of origin of replication and replication terminus sequences on opposite sides of the chromosome, provided by Raymond Rodriguez and colleagues in 1973, support the bidirectional replication model (Figure 7.13b).

In 1968, Joel Huberman and Arthur Riggs used a technique called pulse-chase labeling to produce the first experimental evidence of bidirectional replication in bacteria (Figure 7.14). In pulse-chase labeling experiments, cells are exposed alternately to high levels (the pulse) and low levels (the chase) of a radioactive tracer. The tracer in this case is $^3$H-thymine. The tracer is used so that autoradiographic images can detect where radioactivity is incorporated into the replicating DNA molecules. In the pulse periods of the experiment, when the concentration of $^3$H-thymine is high, a large amount of radioactivity is incorporated into replicating DNA; conversely, in the chase periods, when the $^3$H-thymine level is low, little radioactivity is incorporated into replicating DNA. Autoradiography shows dark tracks where high levels are present and light tracks where levels are low. The bidirectional replication model predicts alternating dark and light tracks in *both directions* from replication origins during a pulse-chase labeling experiment that will be symmetrical around an origin of replication. This is the result obtained by Huberman and Riggs. Subsequent experiments in various eukaryotes and bacteria provided

unambiguous evidence that DNA replication progresses bidirectionally from replication origins.

Additional support for the bidirectional orientation of DNA replication comes from biochemical studies of the DNA polymerase responsible for most *E. coli* DNA replication. This DNA polymerase is capable of incorporating about 1000 nucleotides per second into a newly synthesized strand. At this rate of synthesis, the $4 \times 10^6$ nucleotides of the genome can be replicated in approximately 2000 seconds (33 minutes). This is close to the minimum generation time of *E. coli*. The enzymatic rate of the molecule would have to be twice as fast if replication was unidirectional to complete replication within the generation time. In contrast to bacteria, the rate of catalytic activity of eukaryotic DNA polymerase is approximately 2000 to 4000 nucleotides per minute, less than a tenth the rate in *E. coli*. Eukaryotes have genomes many times larger than *E. coli*, and multiple chromosomes to replicate, so one can logically conclude they replicate their genomes from multiple origins of replication on each chromosome.

## Multiple Replication Origins in Eukaryotes

Autoradiograph analysis reveals multiple origins of replication on eukaryotic chromosomes, and direct observation by electron microscopy confirms it (Figure 7.15). Most

**(a)** Bidirectional replication

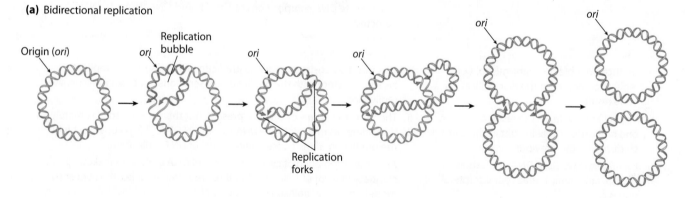

**(b)** Unidirectional replication

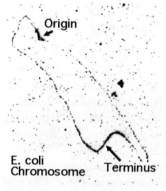

**Figure 7.13 Bidirectional DNA replication. (a)** Cairns' autoradiographic images are consistent with a bidirectional DNA replication model. **(b)** The locations of origin of replication and replication terminus regions on opposite sides of the *E. coli* chromosome.

**(a)** Result of pulse-labeling experiment

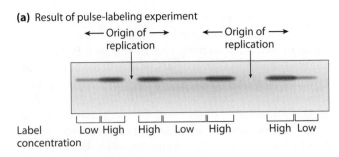

**(b)** Interpretation according to bidirectional model

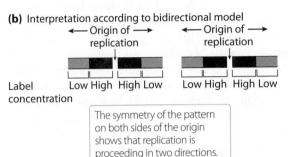

The symmetry of the pattern on both sides of the origin shows that replication is proceeding in two directions.

**Figure 7.14 Pulse-chase labeling evidence of bidirectional DNA replication. (a)** The results of pulse-chase labeling. **(b)** Interpretation of results according to the bidirectional model.

large eukaryotic genomes contain thousands of origins of replication, separated on average, on each chromosome, by 40,000 to 50,000 base pairs (bp). Current estimates indicate that the human genome contains more than 10,000 origins of replication that are spaced 30 to 300 kilobases (kb) apart. Eukaryotic replication origins are not all initiated at the same moment, and among different types of cells, the length of S phase is variable, meaning that the rate of progression of DNA replication varies among cells of different types. Rapidly dividing cells replicate their

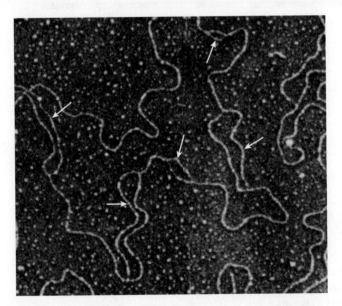

**Figure 7.15 Multiple origins of replication on a single chromosome from *Drosophila melanogaster*.** The arrows point to replication bubbles, which are expanding bidirectionally.

DNA more quickly (i.e., have shorter S phase) than do slowly dividing cells. In addition, experimental evidence identifies "early-replicating" (i.e., early in S phase) and "late-replicating" (late in S phase) segments of large eukaryotic genomes. Early-replicating genome segments appear to contain many expressed genes, whereas late-replicating regions contain many fewer expressed genes. In *Drosophila*, for example, late-replicating regions include chromosome segments immediately surrounding centromeres, where few expressed genes are located.

Regardless of its initiation, timing, and pace, all DNA in each eukaryotic cell is replicated at the end of S phase. The end products of replication of each eukaryotic chromosome are a pair of identical DNA duplexes that are sister chromatids. The sister chromatids will remain joined through $G_2$ and will be separated at anaphase of the upcoming M phase.

Genetic Insight  DNA replication in bacterial chromosomes has a single origin of replication and progresses bidirectionally around the circular DNA molecule. Eukaryotic genomes undergo bidirectional replication from multiple origins that may initiate replication at different times during S phase. Eukaryotic DNA replication produces sister chromatids.

# 7.4 DNA Replication Precisely Duplicates the Genetic Material

A great deal of what molecular biologists know about DNA replication comes from the study of bacteria, particularly *E. coli*. We reviewed some of the basic steps of DNA replication (Chapter 1), and in this section we provide additional details of the process. Much of this information is applicable to eukaryotes as well, owing to the shared evolutionary history of the Bacteria, Archaea, and Eukarya. Despite the strong similarities across taxonomic groups, however, the process of DNA replication in bacteria is not identical to that in eukaryotes. As we discuss the molecular mechanics of DNA replication in this section, we will note the important features that differ substantially between the members of the two domains.

We also wish to offer a cautionary note about discussions of DNA replication. Although the initial parts of our replication discussion identify individual enzymes and proteins, do not be misled into thinking of these proteins as solo actors that enter and leave the replication fork at will. Instead, they are part of large, complex aggregations of proteins and enzymes called *replisomes* that assemble at each replication fork. In *E. coli*, for example, the replisomes active in DNA replication contain approximately 30 distinct proteins and enzymes. Later in the section, we describe how one replisome at each replication fork carries out the nearly simultaneous replication of both template strands.

## DNA Sequences at Replication Origins

Origins of DNA replication contain sequences that attract replication enzymes. The best-characterized origin-of-replication sequence is from *E. coli* and is designated *oriC*. This sequence contains approximately 245 bp of DNA, is AT-rich (i.e., has a preponderance of adenine and thymine base pairs, presumably because less energy is required for their denaturation), and is subdivided by three 13-bp sequences, so-called 13-mers, followed by four 9-bp sequences, called 9-mers (**Figure 7.16a**).

In comparison to *oriC*, other bacterial species have origin-of-replication sequences containing similar, but not identical DNA sequences. This similarity of DNA sequences at replication origins is a product of evolutionary conservation of DNA sequences. Natural selection acts to maintain sequence similarity because the function of the conserved sequence region is essential to the survival of the organism. In other words, natural selection maintains sequences of DNA within a region that performs an essential function. Comparisons of such sequences within and among related species usually leads to the identification of **consensus sequences,** which consist of the nucleotides found *most often* at each position of DNA in the conserved region.

A consensus sequence is the portion of a conserved region that carries out the essential function that is the focus of natural selection. The 13-mer and 9-mer consensus sequences that are part of *oriC* have been maintained by natural selection because they have essential functional roles in replication initiation, as we explain in the following section. Beyond the presence of the consensus sequences themselves, natural selection also acts to maintain specific spacing between the consensus sequences. Spacing is important because DNA-binding proteins must assemble at sites where consensus sequences are located. Different proteins are attracted to different consensus sequences, and each protein must have the physical space to bind to DNA and to interact with the other proteins bound to the consensus sequence region. In summary, DNA sequences that play essential functional roles have their overall sequence composition conserved by natural selection. Short segments of these regions of conserved sequences are consensus sequences where DNA proteins bind to initiate their action. The spacing between consensus sequence elements is also important to allow room on the DNA molecule for protein binding and interaction between bound proteins.

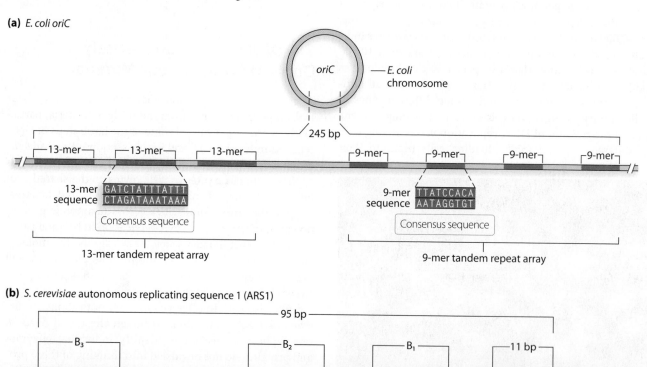

**(a)** *E. coli oriC*

**(b)** *S. cerevisiae* autonomous replicating sequence 1 (ARS1)

**Figure 7.16  Origin of replication sequences in *E. coli* and yeast.**  (a) *OriC* in *E. coli* contains three 13-mer and four 9-mer consensus sequences in a region of 245 base pairs of conserved sequence. (b) The yeast ARS1 origin of replication contains a consensus 11-bp segment and regions B₁, B₂, and B₃ spanning 95 base pairs of conserved sequence. A solidus (/) between nucleotides of consensus sequences (e.g., A / T) indicates that the two nucleotides are equally common at this position.

Among eukaryotic organisms, the yeast *Saccharomyces cerevisiae* has the most fully characterized origin-of-replication sequences. In yeast, the multiple origins of replication are identified as autonomously replicating sequence (ARS). There is overall conservation of DNA sequence, and the organization of ARSs is similar throughout the yeast genome. ARS1 in yeast has been fully sequenced (**Figure 7.16b**). Within the 95 bp of ARS1 is an 11-bp consensus sequence and three other regions ($B_1$, $B_2$, and $B_3$) of conserved DNA sequences that differ somewhat from one another and from the 11-bp consensus sequence region. Much less is known about the DNA sequences at replication origins in other eukaryotic species, but it is presumed that they contain consensus sequences at the sites of action of the DNA-binding proteins that initiate DNA replication.

## Replication Initiation in Bacteria

DNA replication in *E. coli* requires that replication-initiating enzymes locate and bind to the consensus sequences in *oriC*. In *E. coli*, three enzymes, DnaA, DnaB, and DnaC, bind at *oriC* and initiate DNA replication (**Figure 7.17**). The first to bind is DnaA, attaching to the 9-mer components of *oriC*. The DnaA bends DNA and breaks (hydrolyzes) hydrogen bonds in the A-T rich 13-mer region of *oriC*, creating an open complex, a short region where strands of the duplex are separated. Then DnaB, carried to *oriC* by DnaC, attaches to both strands in the open complex. The DnaB is a **helicase** protein that uses ATP energy to hydrolyze hydrogen bonds joining complementary nucleotides. This hydrolysis separates the DNA strands and unwinds the double helix. The unwound strands of DNA would seek maximum stability by reannealing, re-forming complementary double-stranded DNA, except for the presence of **single-stranded binding protein (SSB)**. Single-stranded binding protein prevents reannealing of the separated strands, keeping them available to serve as templates for new DNA synthesis.

For the DNA strands of a circular chromosome, unwinding creates torsional stress that accumulates as the unwound region gets larger and as DNA replication progresses. Without free ends, covalently closed circular DNA readily accumulates superhelical twists, called **supercoiled DNA**, that resemble an over-twisted rubber band (**Figure 7.18a**). The accumulating stress could break the molecule at random locations, potentially leading to a breakdown of DNA replication. This potentially lethal event is prevented by enzymes known as **topoisomerases** that catalyze a controlled cleavage and rejoining of DNA, thus enabling over-twisted strands to unwind (**Figure 7.18b**).

The DNA polymerase enzyme that is primarily responsible for synthesizing new DNA strands adds nucleotides to daughter strands that are complementary

and antiparallel to template DNA strands. These new nucleotides are added to the 3' end of the growing daughter strand, and the overall direction of daughter strand elongation is 5' to 3'. Curiously, however, DNA polymerases are unable to *initiate* DNA strand synthesis on their own. To perform its catalytic activity, a DNA polymerase requires the presence of a primer sequence, a short single-stranded segment that begins a daughter strand and provides a 3'-OH end to which a new DNA nucleotide can be added by DNA polymerase. To satisfy the requirement for a primer, DNA replication is initiated by a specialized RNA polymerase, called **primase,** that synthesizes a short RNA primer, described in the following paragraph.

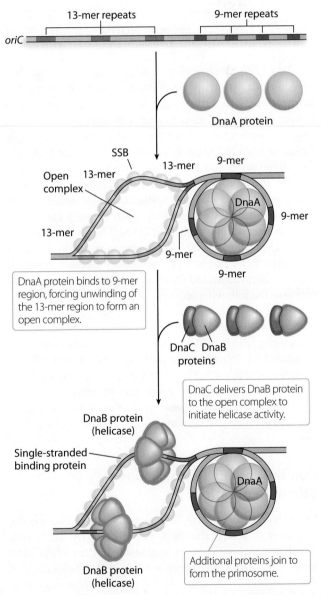

**Figure 7.17  Replication initiation at *oriC*, requiring DnaA, DnaB, and DnaC proteins.**

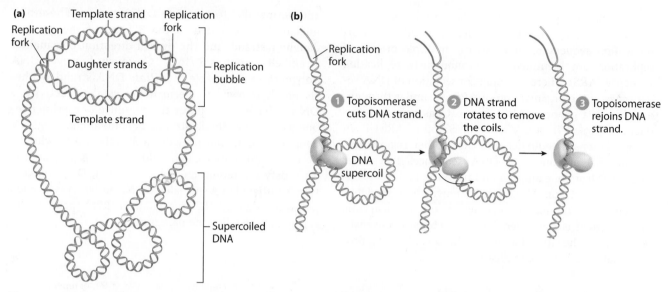

**Figure 7.18** DNA supercoiling in bacteria (a) and its cutting and release by topoisomerase (b).

In *E. coli* DNA replication, primase and a number of additional proteins join DnaA at *oriC* to form the **primosome.** The functional role of the primosome is to deliver primase and its necessary accessory proteins to an origin of replication, where they synthesize a short single strand of RNA called the **RNA primer.** Measuring just one dozen to two dozen nucleotides in length, RNA primers provide the 3′ OH needed for DNA polymerase activity. RNA primers contain the nucleotide base uracil (U), in place of thymine. Consequently, RNA primers cannot remain as part of fully replicated DNA. Thus, while they are essential in allowing DNA polymerase to begin its DNA synthesis, RNA primers are temporary and are removed from newly synthesized DNA strands by a process we describe in the following section.

---

Genetic Insight  DNA replication in *E. coli* is initiated at specific DNA sequences, designated *oriC*. The DnaA and DnaB proteins bind to *oriC* and initiate helicase activity that unwinds DNA strands in preparation for replication. The primosome delivers primase to *oriC* for synthesis of a short RNA primer that is used to initiate DNA replication.

---

## Continuous and Discontinuous Strand Replication

Each strand of parental DNA acts as a template for the synthesis of a new daughter strand of DNA. In *E. coli,* daughter DNA strands are synthesized at the replication fork by the **DNA polymerase III (pol III) holoenzyme,** the principal DNA-synthesizing enzyme. *Holoenzyme* is the general term used for multiprotein complexes in which a core enzyme is associated with additional protein components that complete its structure and lead to its

function. The pol III holoenzyme begins its work at the 3′-OH end of an RNA primer and rapidly synthesizes new DNA with a sequence complementary to the template-strand nucleotides. Pol III adds new nucleotides to a daughter strand as long as there are complementary nucleotides on the template strand to direct nucleotide addition to the daughter strand.

Experimental evidence indicates that most of the enzymes we are describing as participating in DNA replication are part of a single large protein complex at each replication fork called the **replisome.** There is one replisome at each replication fork, and each contains, among other components, two copies of pol III. In each replisome, one pol III carries out the 5′-to-3′ synthesis of one daughter strand *continuously,* in the *same* direction in which the replication fork progresses. The second pol III enzyme in a replisome carries out synthesis of the other daughter strand. The continuously elongated daughter strand is called the **leading strand** (Figure 7.19). Notice that Figure 7.19 divides the replication bubble into four quadrants. The upper right and lower left quadrants contain leading strands.

The daughter strands in the upper left and lower right quadrants shown in Figure 7.19 have a 5′-to-3′ direction of elongation that runs *opposite* to the direction of movement of the replication fork. These daughter strands are elongated *discontinuously,* in short segments, each of which is initiated by an RNA primer. The discontinuously synthesized daughter strand is called the **lagging strand.** Thus in Figure 7.19, the lower right and upper left quadrants of the replication bubble contain lagging strands.

Reiji Okazaki detected the synthesis of short fragments of DNA in the replication of the lagging strand. He observed that early in bacterial replication, newly synthesized DNA segments on one strand are 1000 to 2000 nucleotides long, while later in replication the

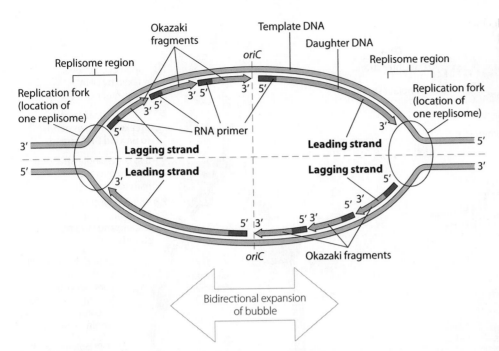

**Figure 7.19 The replication bubble.** Bidirectional expansion is driven by DNA synthesis at each replication fork. One replisome containing two DNA pol III enzymes operates at each fork to replicate each daughter strand.

newly synthesized segments are much longer. Okazaki's discovery suggested that short segments of DNA are synthesized and that these short segments are joined together as replication progresses. The short segments of newly replicated DNA are called **Okazaki fragments,** and they are the result of discontinuous synthesis of DNA on the lagging strand.

In Figure 7.19, notice that each daughter strand contains a segment characterized as leading strand that adjoins a segment characterized as lagging strand. All daughter strands are composed of adjoining leading and lagging segments, and they will ultimately be structurally identical.

## RNA Primer Removal and Okazaki Fragment Ligation

To complete DNA replication, RNA primers must be removed and replaced with DNA, and Okazaki fragments must be joined together to form complete DNA strands. These tasks are accomplished by the enzymes *DNA polymerase I* and *DNA ligase* that are each part of the replisome complex at each replication fork.

When DNA pol III on the lagging strand reaches an RNA primer, thus running out of template, it leaves a single-stranded gap between the last DNA nucleotide of the newly synthesized daughter strand and the first nucleotide of the RNA primer (**Figure 7.20**). The pol III, having very low affinity for these DNA–RNA single-stranded gaps, is then replaced by **DNA polymerase I (pol I),** which has high affinity for such gaps (Figure 7.20, ❶). The DNA pol I removes nucleotides of the RNA primer one by one and replaces them with DNA nucleotides, beginning with the 5′ nucleotide of the RNA primer and progressing in the 3′ direction until all the RNA

nucleotides in the primer have been replaced by DNA nucleotides complementary to the template strand.

The pol I enzyme possesses two activities that accomplish the removal of RNA nucleotides and their replacement by DNA nucleotides. DNA pol I first uses its **5′-to-3′ exonuclease activity** to remove the 5′-most nucleotide from the RNA primer. This creates one open space opposite the template, which is then filled with the correct DNA nucleotide by the **5′-to-3′ polymerase activity** of DNA pol I. The pol I removes each RNA primer nucleotide and replaces each with a DNA nucleotide. In so doing, pol I continually pushes the single-stranded gap in the 3′ direction, eventually replacing all of the RNA primer nucleotides with DNA nucleotides.

Once the entire RNA primer is replaced, a remaining single-stranded gap sits between two DNA nucleotides. At this point, **DNA ligase,** having exclusive and very high affinity for DNA–DNA single-stranded gaps, is attracted to the gap and there performs its single task of forming a phosphodiester bond between the two DNA nucleotides that joins two Okazaki fragments. Both pol I and DNA ligase are active on leading *and* lagging strands. The level of activity is greater on lagging strands, however, where every 1000 to 2000 nucleotides, they are needed to join Okazaki fragments during replication of *E. coli* DNA. **Figure 7.21** incorporates all the features just described into an overview of DNA replication.

---

**Genetic Insight**  DNA polymerase III synthesizes *E. coli* DNA beginning at 3′-OH ends of RNA primers and carries out the bulk of DNA replication. DNA polymerase I replaces pol III to remove RNA primer nucleotides and replace them with DNA nucleotides. DNA ligase catalyzes formation of phosphodiester bonds that join DNA segments and completes replication.

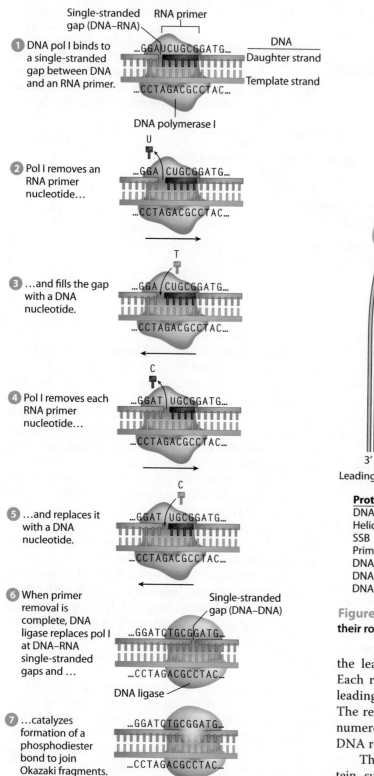

Figure 7.20 **Removal and replacement of RNA primer nucleotides and ligation of Okazaki fragments.**

## Simultaneous Synthesis of Leading and Lagging Strands

As we have seen, the replisome components include two DNA pol III holoenzymes, one of which synthesizes

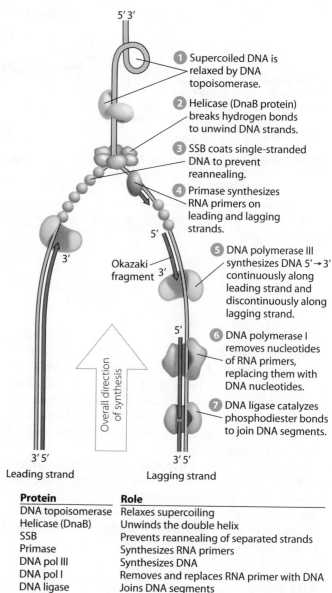

| Protein | Role |
|---|---|
| DNA topoisomerase | Relaxes supercoiling |
| Helicase (DnaB) | Unwinds the double helix |
| SSB | Prevents reannealing of separated strands |
| Primase | Synthesizes RNA primers |
| DNA pol III | Synthesizes DNA |
| DNA pol I | Removes and replaces RNA primer with DNA |
| DNA ligase | Joins DNA segments |

Figure 7.21 **The principal proteins of DNA replication and their roles.**

the leading strand and the other the lagging strand. Each replisome complex carries out replication of the leading strand and the lagging strand simultaneously. The replisome also includes pol I and ligase, as well as numerous other components that collectively carry out DNA replication.

The DNA pol III holoenzyme itself contains 11 protein subunits. The two pol III core polymerases are each tethered to a different copy of the τ (tau) protein (**Figure 7.22**). The τ proteins are joined to a five-protein complex known as the **clamp loader.** Two additional proteins form the **sliding clamp,** a protein structure that can close around double-stranded DNA during replication. The sliding clamp, with its diameter of approximately 50 Å, has a "doughnut hole" of about 35 Å that encircles the DNA (**Figure 7.23a**).

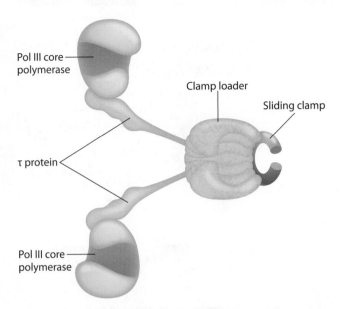

**Figure 7.22  DNA polymerase III holoenzyme.** The complex contains two DNA polymerase core enzymes attached to τ (tau) arms, and the clamp loader, shown holding a sliding clamp.

Each sliding clamp locks onto a DNA template strand and there affiliates with DNA pol III core enzyme, firmly anchoring the enzyme to the template to carry out the bulk of replication (**Figure 7.23b**). The clamp is the key to the enzyme's high level of activity. Pol III on DNA without a sliding clamp has very low processivity. When no more template is available, the DNA pol III is dropped by the sliding clamp and replaced by DNA pol I, which as we have seen removes RNA primers and replaces them with DNA.

**(a)** Two views of the sliding clamp

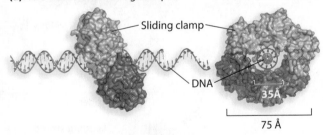

**(b)** Sliding clamp operation

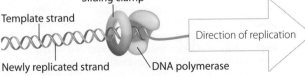

**Figure 7.23  The DNA sliding clamp. (a)** Two views of the sliding clamp, one showing the clamp and DNA polymerase on DNA in profile (left) and the other showing DNA through the "doughnut hole" of the sliding clamp (right). **(b)** The sliding clamp–DNA polymerase complex has high processivity during replication.

**Foundation Figure 7.24** presents a model of how the DNA pol III holoenzyme coordinates the simultaneous synthesis of leading and lagging strands at a replication fork. The outline of this model was proposed in the early 1960s by Arthur Kornberg to explain the experimental observation that a single large protein complex at each replication fork carries out replication of both strands of DNA. Known both as the Kornberg model and as the "trombone" model, it has been revised and updated in the decades since it was first proposed. The trombone model depicts the activity of the clamp loader in providing a mechanism for the continuous synthesis of leading strand regions and for the grasping, synthesis, and release of lagging strand regions by DNA pol III–sliding clamp complexes affiliated with each arm of the clamp loader. This model provides a mechanism by which a single replisome can advance with the replication fork and synthesize both daughter strands as it proceeds. In summary, replisomes contain multiple DNA polymerase enzymes and a large number of accessory proteins that operate in a rapid and highly coordinated manner to carry out DNA synthesis.

Bacteria and eukaryotes each possess multiple polymerases for DNA replication (**Table 7.1**). Other polymerases are primarily involved in repair of DNA mutations or in recombination. For example, *E. coli* has three DNA polymerases that are principally involved in mutation repair, and eukaryotes have eight DNA polymerases that engage in mutation repair of various kinds.

## DNA Proofreading

Accurate replication of DNA is essential for the survival of organisms. The introduction of errors into a DNA sequence during replication could create potentially lethal

| Table 7.1 | Properties of Bacterial and Eukaryotic DNA Replication Polymerases |
|---|---|
| **Polymerase** | **Functions** |
| *Bacterial polymerases* | |
| Primase | RNA primer synthesis |
| I | RNA primer removal, proofreading, mutation repair |
| III | DNA replication, proofreading |
| *Eukaryotic polymerases* | |
| α | Primer synthesis |
| δ | Lagging strand synthesis, proofreading, DNA mutation repair |
| ε | Leading strand synthesis, proofreading, DNA mutation repair |
| γ | Mitochondrial DNA replication and mutation repair |

## The Trombone Model of DNA Replication

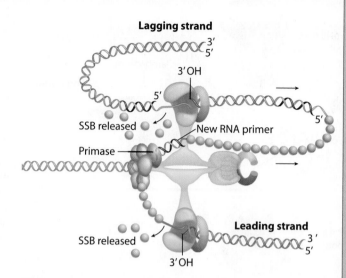

**1** DNA helicase denatures the parental duplex, and SSB coats leading strand and lagging strand templates. The leading strand DNA pol III–sliding clamp complex synthesizes the leading strand continuously. The lagging strand pol III–sliding clamp complex synthesizes an Okazaki fragment.

**2** Primase binds the lagging strand template and synthesizes a new RNA primer. SSB is released ahead of leading strand and lagging strand synthesis, and ahead of RNA primer synthesis.

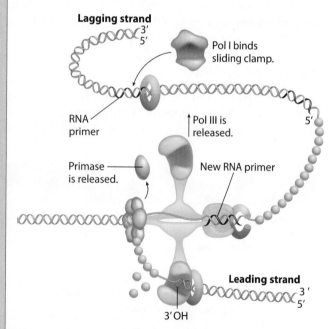

**3** Lagging strand DNA pol III completes synthesis of an Okazaki fragment and is released by the sliding clamp. A DNA pol I replaces pol III to begin removal of the RNA primer and replacement of RNA nucleotides by DNA nucleotides.

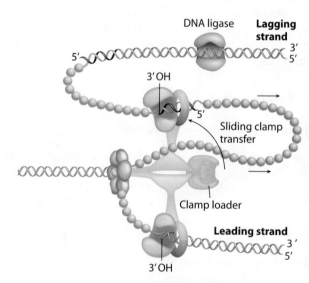

**4** DNA ligase joins Okazaki fragments. The clamp loader places a new sliding clamp near the 3' end of the RNA primer on the newly primed lagging strand. Lagging strand DNA pol III binds the sliding clamp and initiates synthesis of a new Okazaki fragment.

mutations. While this occasionally happens, DNA replication is remarkably accurate and is not a major source of mutation, largely because DNA polymerases are generally able to undertake **DNA proofreading** to be sure replication is accurate. As a result of DNA proofreading, mutations due to DNA replication errors occur about once every billion ($10^9$) nucleotides in wild-type *E. coli*. To put this number into perspective, consider this textbook as an analogy. It contains about 800 pages, each holding about 5000 "bits" of information (letters, punctuation marks, spaces, etc.) for a total of $4 \times 10^6$ bits per book. It would take 250 books, each the size of this one, to equal $10^9$ bits of information. If each bit were equal to a DNA nucleotide, the error rate for DNA replication would be like having *one* typographical error in all 250 books!

This extraordinary accuracy is the work of the multifunctional DNA polymerases that have the ability not only to synthesize DNA (5′-to-3′ polymerase activity) but also to "proofread" newly synthesized DNA for accuracy and remove erroneous nucleotides (see Table 7.1). This proofreading ability resides in the **3′-to-5′ exonuclease activity** of DNA polymerases capable of removing some of the newly laid daughter strand sequence.

Polymerases like pol III and pol I have a structure somewhat like an open hand: A "thumb" and "fingers" hold the template and daughter strands in the "palm," where 5′-to-3′ polymerase activity is centered (**Figure 7.25**). When a replication error occurs, the mismatched DNA bases of the template and daughter strands are unable to hydrogen bond properly. As a result, the 3′-OH end of the daughter strand becomes displaced, blocking the further addition of nucleotides and inducing rotation of the daughter strand into the 3′-to-5′ exonuclease site at the "heel" of the hand. Several nucleotides, including the mismatched one, are then removed from the 3′ end of the daughter strand, after which the daughter strand rotates back to the polymerase site in the palm and replication resumes. Like their counterparts in bacteria, the principal DNA replication polymerases in eukaryotes also have proofreading ability to help ensure the accuracy of DNA replication.

---

Genetic Insight  DNA replication is carried out by replisome complexes located at each replication fork. Replisomes carry numerous proteins and enzymes that act quickly and efficiently to synthesize DNA. DNA polymerase activity is driven by sliding clamp proteins that lock polymerase to template strands, and polymerases are capable of proofreading their work to ensure high fidelity of DNA replication.

---

## Eukaryotic DNA Polymerases

The shared evolutionary history of bacteria and eukaryotes is reflected in the general similarities of their mechanisms of DNA replication. Even so, as the genomes have

**(a)** DNA polymerase error

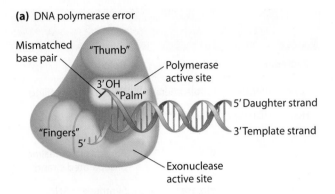

**(b)** Exonuclease removal of mismatched base pair

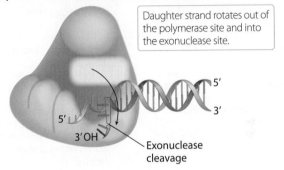

**(c)** Daughter strand resumes DNA synthesis

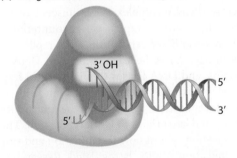

**Figure 7.25  DNA polymerase proofreading activity. (a)** A replication error by polymerase. **(b)** Polymerase shifts on newly synthesized DNA to utilize its 3′-to-5′ exonuclease activity. **(c)** The polymerase resumes 5′-to-3′ synthesis.

differentiated over time, specific enzymes have evolved specialized functions that are unshared by the other domain. For example, eukaryotes have acquired more than a dozen kinds of DNA polymerase in the nucleus. Three of these polymerases, α (alpha), δ (delta), and ε (epsilon), play pivotal roles in eukaryotic DNA replication. A fourth DNA polymerase, γ (gamma), is limited to mitochondria, where it replicates the mitochondrial chromosome (see Table 7.1).

Synthesis of RNA primers is carried out by DNA polymerase α, which contains four subunits whose sizes range from 48 kilodaltons (kD) to 180 kD. Polymerase δ carries out lagging strand DNA synthesis, and polymerase ε carries out leading strand DNA synthesis. Each of these polymerases interacts with a protein known as **proliferating cell nuclear antigen (PCNA)** that functions

| Table 7.2 | Bacterial and Eukaryotic Replication Proteins and Enzymes | |
|---|---|---|
| **Bacterial** | **Eukaryotic** | **Role** |
| Topoisomerase | Topoisomerase | Relaxes supercoiling |
| Helicase (DnaB) | Helicase (Mcm) | Unwinds the double helix |
| SSB | RPA | Prevents reannealing of separated strands |
| Primase | Primase | Synthesizes RNA primers |
| DNA pol III | DNA pol α, pol δ, pol ε | Synthesizes DNA |
| DNA pol I | RNase H | Removes and replaces RNA primer |
| DNA ligase | DNA ligase I | Joins DNA segments |
| Sliding clamp | PCNA | Binds polymerase |
| γ complex | RFC | Connects polymerases, coordinates clamp loader |

as the sliding clamp in eukaryotic DNA replication. The DNA polymerase–PCNA complex is reminiscent of the DNA pol III–holoenzyme complex we encountered in *E. coli* replication. As in *E. coli* replication, DNA replication in eukaryotes utilizes DNA helicase to unwind DNA at replication origins, and DNA ligases to join Okazaki fragments. However, unlike the single chromosome with its single origin of replication in *E. coli*, the eukaryotic genome has early and late origins of DNA replication. This finding raises questions about how the initiation and timing at those replication origins is regulated. Researchers have identified several equivalent DNA replication proteins and enzymes through comparison of bacteria and eukaryotes, and they continue to address unresolved questions about eukaryotic DNA replication (Table 7.2).

## Telomeres

Linear chromosomes, such as those in the nuclei of your cells, present a peculiar problem with regard to DNA replication—they cannot be replicated all the way to their ends! Instead, eukaryotic chromosomes get progressively shorter with each replication cycle.

This apparent defect in the replication process is a consequence of an RNA primer being located at one end of the lagging strand and thus not able to be replaced by DNA. This results in the lagging strand being shorter than its template strand, causing the chromosome to become shorter with each replication cycle (Figure 7.26). Because the daughter strand is missing nucleotides, the complementary parent (template) strand has a corresponding single-stranded overhang that is easily broken off. The same fate does not befall the leading strand, which can be fully replicated to its end.

The loss of DNA with each replication cycle sounds ominous, but the problem is solved by the presence at chromosome ends of repetitive DNA sequences called **telomeres.** Telomeres do not contain protein-coding genes, but instead are made up of repeats that are most often 6-bp sequences repeated hundreds or thousands of times to give the telomere a length of 2 to 20 kb, depending on the species. Since its sequences are repetitive and contain no genetic information, portions of the telomere can safely be lost in each replication cycle, without consequence to the organism. Gel electrophoresis of telomeric DNA has documented the progressive shortening of telomere length during cell culture.

Telomeres are synthesized by the ribonucleoprotein **telomerase,** a combination of several proteins and a molecule of RNA. The telomerase RNA molecule is encoded by a distinct gene and acts as the template for the telomeric DNA repeat sequence. Elizabeth Blackburn and Carol Greider discovered both telomeres and telomerase in 1987 and along with Jack Szostak were awarded the 2009 Nobel Prize in Physiology or Medicine for their work.

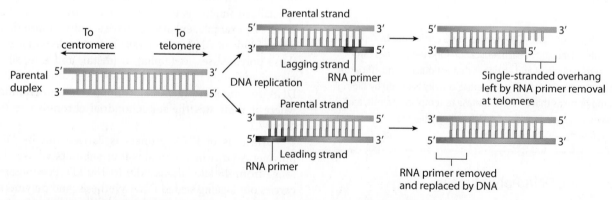

**Figure 7.26  Loss of DNA at telomeres.** Leading strands are synthesized to the ends of linear chromosomes, but lagging strands are shortened each replication cycle, when RNA primer sequence at the telomere end of the template strand is removed.

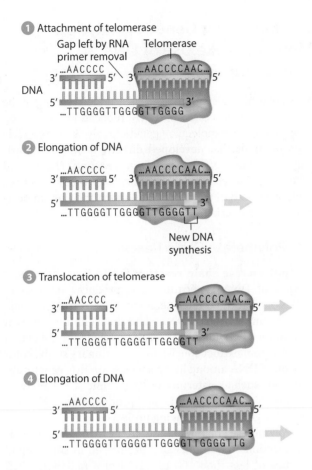

**1** Attachment of telomerase

**2** Elongation of DNA

New DNA synthesis

**3** Translocation of telomerase

**4** Elongation of DNA

**Figure 7.27  Telomerase synthesizes repeating telomeric sequence.**

**Figure 7.27** depicts the mechanism of telomerase action deduced from the study of the ciliated protozoan *Tetrahymena*. The repetitive sequence 5'-TTGGGG-3' is the characteristic telomeric repeat sequence of *Tetrahymena*. The template RNA in the *Tetrahymena* telomerase contains the repeat AACCCC that is used to elongate the telomere of one strand enough for the eukaryotic α polymerase to synthesize additional RNA primers (see Table 7.1), allowing new DNA replication that can fill out the chromosome ends.

In the decades since Blackburn and Greider identified telomere structure and this mechanism for their maintenance, similar repeating telomeric sequences have been detected in all eukaryotes. For example, the human telomeric repeat sequence is 5'-TTAGGG-3', and it is encoded by a telomeric RNA molecule with the complementary repetitive sequence 3'-AAUCCC-5'. In humans, telomeric sequence is repeated 250 to 1500 times at chromosome ends. The same telomeric sequence and template DNA sequence are found in vertebrates, protozoans (*Trypanosoma*), yeast (*Saccharomyces*), fungus (*Neurospora*), and plants (*Arabidopsis*). This represents an example of convergent evolution of DNA sequences. Convergent evolution is a mechanism producing similar traits or, in this case, DNA sequences among distantly related organisms due to similar adaptation or natural selection pressure.

The importance of telomerase activity in germ-line cells has been demonstrated in experimental mouse lines that are mutated to be homozygous for loss-of-function mutations of the *TERT* (*telomerase reverse transcriptase*) gene, the gene that encodes telomerase. These homozygous mutant mice are relatively normal when interbred for up to three generations, but severe developmental and fertility defects are detected in the fourth and fifth inbred generations. *TERT* loss-of-function homozygosity is lethal by the seventh generation, meaning that no inbred *TERT*-deficient mice can be maintained by inbreeding for more than six generations.

The molecular explanation for the delayed phenotypic effect of *TERT* inactivation is that each successive generation of inbreeding in the homozygous mutant line leads to the loss of telomeric DNA. It is now evident that genetic mechanisms monitor telomere length, and that telomere length is a kind of chronometer that keeps track of the age of a cell. Once the shortening reaches a critical point, the cell is directed into the apoptotic pathway, the mechanism of programmed cell death that removes old or damaged cells from an organism. This phenomenon is thought to be the explanation for a long-standing observation in cell biology that most normal cells survive in culture for between about 30 to 50 cell divisions before entering a crisis phase, where their division first slows and then stops altogether, and the cells die.

## Telomeres, Aging, and Cancer

Considering the importance of telomere length to chromosome stability, cell longevity, and reproductive success, it may surprise you to learn that telomerase activity is limited to only a few kinds of cells in eukaryotes. Telomerase is active in germ-line cells, where it functions to ensure that gametes pass on full-length chromosomes. Telomerase activity is also detected in some stem cells, thus enabling the cells that differentiate from those stem cells to have full-length chromosomes. In contrast, telomerase activity is virtually nonexistent in differentiated somatic cells, the kinds of cells that have finite life spans and make up nearly all the cells of most body organs and tissues. In somatic cells, genes responsible for producing telomerase are turned off, and almost no telomerase activity is detectable. This accounts for the finite life span of somatic cells in cell culture first observed in 1965 by Leonard Hayflick, who found that the number of cell divisions of cultured cells is dependent on the source of the cells. This limitation on the growth of most cells in culture is known as the Hayflick limit.

The connection between telomerase inactivity and normal aging of cells prompted geneticists to look at human premature aging conditions for evidence of mutations affecting telomere formation or telomerase activity. People with a condition known as Werner syndrome

(OMIM 277700) experience early onset graying of the hair, skin wrinkling, osteoporosis, and several other changes associated with individuals several decades older. Molecular geneticists identified a mutation of *RECQL2*, a gene producing a helicase protein required for telomerase activity, as the gene responsible for Werner syndrome. Patients with Werner syndrome experience chromosome instability and premature aging as a result of abnormally short telomeres. In the human condition called dyskeratosis congenita (OMIM 305000), patients have abnormalities of skin and nails, occasional loss of vision and hearing, and abnormalities of blood cell production that are a frequent cause of death. The *DKC1* gene responsible for dyskeratosis congenita affects the activity of genes responsible for normal telomerase function. Defective telomerase activity and shortened telomeres are thought to be at the root of dyskeratosis congenita.

In contrast to the importance of telomerase activity to maintaining normal telomere length as chromosomes are passed through the germ line, what is the consequence of abnormal *reactivation* of telomerase activity in somatic cells? Such an event can lead aging cells to continue to proliferate, allowing them to escape programmed cell death by apoptosis. This is exactly what seems to happen in many kinds of cancer, where mutations reactivate the expression of *TERT* and reintroduce telomerase activity into cells where *TERT* is normally silent.

Recent studies of gene expression in human cancer cells find that mutations reactivating *TERT* are among the most frequent mutations in cancers of all types. In cancers of the internal organs, including lung, breast, stomach, ovary, kidney, bladder, uterus, testis, and prostate, 78% to 100% of advanced-cancer cells show evidence of reactivation of telomerase activity. This is a highly significant increase over the 0% to 3% rate of telomerase reactivation in normal somatic cells. In the cancer cells, the reactivation of telomerase activity appears to stabilize telomere length, disrupting the normal program of progressive telomere shortening that would lead to apoptosis. This extended life span may allow affected cells to acquire additional mutations associated with cancer development and cancer advancement.

---

**Genetic Insight**  Telomeres are repetitive DNA sequences added to the ends of linear chromosomes through the activity of telomerase, a ribonucleoprotein that synthesizes telomeres. Telomerase is normally active in germ-line cells and inactive in somatic cells. Telomeres shorten each replication cycle, and shortening leads to cell death by apoptosis. Telomerase mutations cause premature aging syndromes, and aberrant reactivation of telomerase activity is commonly found in cancer cells.

---

# 7.5 Molecular Genetic Analytical Methods Make Use of DNA in Replication Processes

Molecular biologists have used their understanding of the enzymes and processes of DNA replication to develop new methods of molecular genetic analysis. Two widely used methods that developed directly from this knowledge are the *polymerase chain reaction* (PCR) and *dideoxyribonucleotide DNA sequencing*. In this section, we look at both of these methods and at their use in deciphering DNA variation.

## The Polymerase Chain Reaction

The **polymerase chain reaction (PCR)** is an automated version of DNA replication that produces millions of copies of a short, targeted segment of DNA from a miniscule amount of original DNA. The almost limitless uses of PCR in modern biological research include the collection of DNA from extinct species for evolutionary study; comparison of DNA among living species; forensic genetic applications such as paternity testing, crime scene analysis, and individual identification; and production of DNA segments for genome sequencing projects.

Polymerase chain reactions are in vitro DNA-replication reactions that are usually performed in small liquid volumes of less than 100 μL (1/10 of a milliliter). PCR reactions require double-stranded DNA containing the target sequence that is to be copied, a supply of the four DNA nucleotides, a heat-stable DNA polymerase, and two different single-stranded DNA primers (described below). These PCR components are mixed with a buffer solution at the beginning of the reaction.

The DNA polymerase most often used in PCR is called *Taq* polymerase, and it is named after the thermophilic bacterial species *Thermus aquaticus* that was first collected in Yellowstone National Park. This bacterium lives in hot springs at near-boiling conditions, having evolved heat-stable proteins that remain active at these temperatures. The heat stability of *Taq* DNA polymerase is important to the efficiency of PCR.

The PCR reaction itself closely resembles DNA replication, and like the replication reaction in cells, PCR uses two different, short, single-stranded DNA sequences called **PCR primers** to provide a start point for *Taq* polymerase synthesis. PCR primers, like RNA primers in cellular replication, are generally 12 to 24 nucleotides in length. They base-pair to different DNA strands, binding on *opposite sides* of the region of DNA to be copied in PCR. The primer binding sites are at the 5′ and 3′ boundaries of most of the replication products.

Polymerase chain reactions occur as a series of three-step DNA replication cycles that result in "amplification," meaning the replication in large numbers, of the target DNA sequence (**Figure 7.28**). Each step of a typical PCR cycle lasts

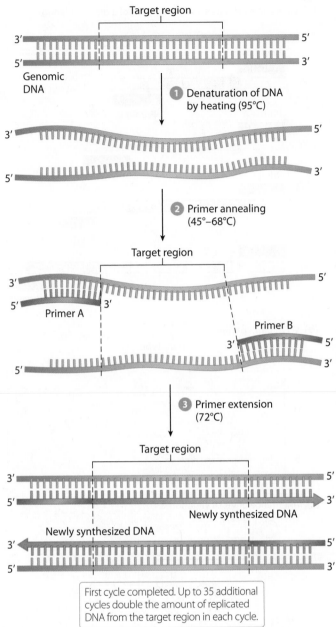

**Figure 7.28 The three-step cycle of the polymerase chain reaction (PCR).**

from 30 seconds to several minutes, and 30 to 36 is a typical number of cycles. Each complete PCR cycle doubles the number of copies of the target DNA sequence, so beginning with a single copy of double-stranded target sequence, 30 PCR cycles can yield $2^{30}$, or more than 10 billion, copies of the target sequence! The steps of each PCR cycle are as follows:

❶ *Denaturation.* The reaction mixture is heated to approximately 95°C, causing double-stranded DNA to *denature* into single strands as the hydrogen bonds between complementary strands break down.

❷ *Primer annealing.* The reaction temperature is reduced to between about 45°C and 68°C to allow

*primer annealing,* the hybridization of the two primers to complementary sequences that bracket the target sequence.

❸ *Primer extension.* Raising the temperature of the reaction to 72°C allows *primer extension,* during which *Taq* DNA polymerase synthesizes DNA, beginning at the 3′ end of each primer and taking approximately 1 minute for every 1000 bp synthesized.

PCR has an enormous variety of applications, but it also has limitations, the most important of which are (1) the requirement of some knowledge of the sequences needed for primers and (2) that amplification products longer than 10 to 15 kb are difficult to produce. In most cases, the length limitations on PCR restrict its use to the study of selected DNA segments or individual genes. The requirement for primer sequence information can be satisfied by informed guesses about the sequences likely to occur at primer binding sites or by using primers from one species to amplify similar sequences in another species. For example, a biologist wanting to study DNA-sequence similarity between species could use a pair of primers that amplify a *Drosophila* gene to examine the human genome for a related gene. There may be one or more base-pair mismatches between the *Drosophila* primers and the human DNA sequences they bind to, but the mismatches need not prevent primer annealing if the temperature of the PCR reaction is lowered during step 2 of the reaction. The lower temperature can increase the stability of hybridization of the primers and their target sequences enough to allow the former to prime the PCR amplification.

The polymerase chain reaction makes it practical to obtain large quantities of DNA from a particular gene for molecular analysis. The technique PCR usually takes place in small plastic tubes that are specially designed for this purpose, and has revolutionized many aspects of biology, including molecular genetics, recombinant DNA analysis, evolutionary genetics, and forensic biology.

## Separation of PCR Products

The PCR process selectively amplifies only the fragment of DNA bounded by the two primers. Those amplified fragments are then separated from the rest of the reaction mixture by gel electrophoresis (see Chapter 10) and visualized by staining with EtBr (ethidium bromide). The size of PCR products is measured in base pairs, and any variability in their length results from differences in the number of nucleotides between the two primer binding sites. These differences can be exploited in genetic analysis. As an example, let's look at an analysis of short repeating sequences of DNA that are frequently used as one kind of genetic marker. Known as a variable number tandem repeat (VNTR), this type of marker contains end-to-end repeating DNA sequences that are each up to 20 bp in length. These types of genetic markers are also known as short tandem repeat polymorphisms (STRPs).

Figure 7.29a shows four hypothetical VNTR alleles ($V_1$ to $V_4$) that might be found in a population. The alleles differ in the number of repeats of the DNA sequence they carry. The repeats are consecutively numbered in the figure. The PCR primers bind to the same sequences for each allele. The primers bind outside the repeat region, so amplification of each allele produces a DNA fragment of a characteristic length that is determined by the number of DNA repeats the allele contains.

There are four alleles for this VNTR gene, and there are 10 possible genotypes. In Figure 7.29b, gel electrophoresis of PCR-amplified DNA fragment bands shows that each genotype has a distinctive band number and composition. Each homozygous genotype has a single band and each heterozygous genotype has two bands. The bands are identified by their repeat number.

The inheritance of the VNTR alleles follows a codominant pattern in which both alleles are detected in heterozygous genotypes (Figure 7.29c). In this family, each parent transmits one allele to each child and as a consequence of the different heterozygous genotypes of the parents, each allele in a child can be traced to one of the parents.

Genetic Insight  The polymerase chain reaction (PCR) produces large quantities of sequence copied from particular genes in a small in vitro DNA replication reaction. PCR-generated DNA is used in a variety of molecular genetic analyses.

## Dideoxynucleotide DNA Sequencing

The ultimate description of any DNA molecule consists of its precise sequence of bases. Such a sequence contains not only the codons of genes that can be transcribed and translated into proteins but also all the information required for expressing and regulating genes and for the replication of the DNA itself. DNA-sequencing technologies have revolutionized biology by making DNA sequence information readily available. While the implications for molecular biology and genetics are obvious, DNA-sequencing technology has also found broad applications in agriculture, medicine, and evolutionary biology. DNA-sequencing technologies have changed rapidly as laboratory and computer technology have combined to make sequencing faster and cheaper by orders of magnitude. A long-term goal of genome science is the "$1000 genome"—a term that means the development of molecular and computer technologies capable of producing complete individual genome sequences for each of us for roughly $1000 each. In 2010, two independent research groups took a big step toward

(a) Each allele produces a PCR fragment of a different length.

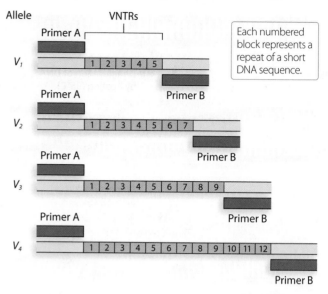

(b) VNTR band patterns

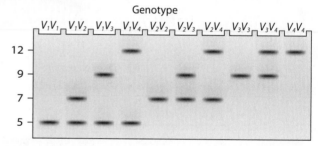

(c) Inheritance of VNTR variation

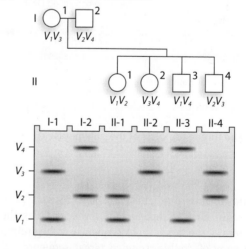

Figure 7.29  PCR amplification of variable number tandem repeat (VNTR) alleles.  (a) Four VNTR alleles ($V_1$ to $V_4$) are characterized by different numbers of identical DNA repeat sequences. (b) Ten genotypes are possible at the VNTR gene, each having a unique pattern of PCR-fragment sizes. One band is seen for each homozygous genotype and two bands for each heterozygous genotype. (c) Hereditary transmission of VNTR alleles follows a codominant pattern.

that goal by completing the first projects designed to completely sequence the genome of a single person. The cost was approximately $50,000, orders of magnitude smaller than the millions of dollars required to produce the draft of the human genome in 2000. Advances in genome sequencing are expected to continue unabated, and the cost will continue to drop. It is likely that within your lifetime you will have the opportunity to have your own personal genome sequenced. In fact, you could even be one of those doing the sequencing!

The first DNA-sequencing protocols were developed in 1977, one by Allan Maxam and Walter Gilbert and another by Fred Sanger. Of the two methods, Sanger's was more amenable to automation and it is the basis for the development of high-throughput approaches to genome sequencing that is the method of choice today. These types of genetic markers are also known as short tandem repeat polymorphisms (STRPs). Here we describe Sanger's dideoxynucleotide DNA-sequencing technique. Chapters 16 and 18 discuss more recent advances in DNA sequencing.

**Dideoxynucleotide DNA sequencing—dideoxy sequencing,** for short—is Sanger's DNA replication method. It is based on cellular DNA replication reactions and uses DNA polymerase to replicate new DNA from a single-stranded template (the strand to be sequenced) beginning at a primer sequence attached to the template strand. In dideoxy sequencing reactions, the four standard deoxynucleotide (dNTP) components of DNA, in large amounts, are mixed with smaller amounts of a **dideoxynucleotide triphosphate (ddNTP).** Dideoxynucleotides differ from deoxynucleotides in lacking two oxygen atoms (*dideoxy* means "two deoxygenated sites") rather than the usual one deoxygenated site. Whereas dNTPs are deoxygenated at the 2′ carbon and have a hydroxyl group (OH) at the 3′ carbon, ddNTPs have hydrogen (H) atoms rather than hydroxyl groups at the 2′ *and* 3′ carbons (**Figure 7.30a**). The absence of a hydroxyl group at the 3′ carbon in ddNTP prevents the ddNTP from forming a phosphodiester bond to elongate a DNA strand. Incorporation of a ddNTP by DNA polymerase into a growing strand is a chain-terminating event that blocks further strand elongation (**Figure 7.30b**). Dideoxy sequencing therefore produces a large number of partial replication products, each terminated by incorporation of a ddNTP at a different site in the sequence.

In preparation for dideoxy sequencing, many copies of the DNA fragment to be sequenced are obtained in single-stranded form, usually by denaturing double-stranded DNA. Samples of the fragment are then placed in four parallel replication reactions. Each reaction mixture contains a single-stranded DNA primer, DNA polymerase, large amounts of each of the four standard nucleotides (dATP, dGTP, dCTP, and dTTP), and a small amount of *one* dideoxynucleotide, either that of adenine (ddATP), thymine (ddTTP), cytosine (ddCTP), or guanine (ddGTP).

**(a)**

Chemical structure

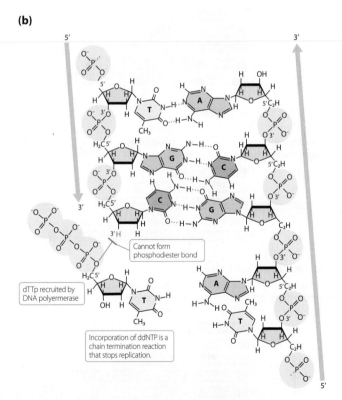

Deoxynucleotide triphosphate (dNTP)

Dideoxynucleotide triphosphate (ddNTP)

**(b)**

**Figure 7.30 Nucleotides used in DNA sequencing reactions. (a)** Dideoxynucleotides (ddNTPs) are deoxygenated at both the 2′ and 3′ carbons and cannot be used to elongate DNA. **(b)** The incorporation of a dideoxynucleotide of cytosine (ddCTP) terminates the replication reaction.

The four parallel DNA-sequencing reactions shown in **Figure 7.31** are used to sequence the DNA fragment shown at the top of the figure. As each reaction begins, a

**(a)** ddCTP Reaction ("C" lane)

> Incorporation of dCTP allows the chain to continue growing, but incorporation of ddCTP terminates chain elongation.

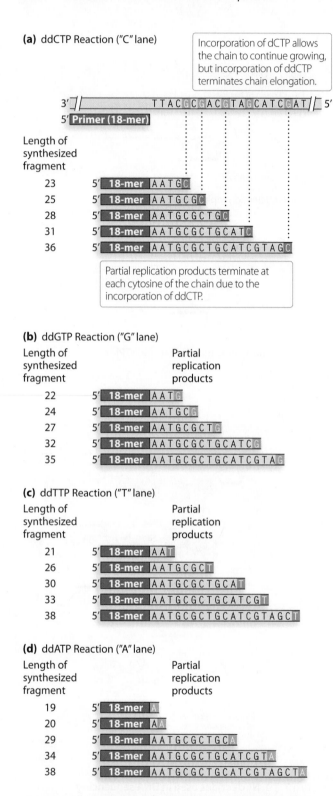

Length of synthesized fragment

| | |
|---|---|
| 23 | 5′ 18-mer A A T G C |
| 25 | 5′ 18-mer A A T G C G C |
| 28 | 5′ 18-mer A A T G C G C T G C |
| 31 | 5′ 18-mer A A T G C G C T G C A T C |
| 36 | 5′ 18-mer A A T G C G C T G C A T C G T A G C |

> Partial replication products terminate at each cytosine of the chain due to the incorporation of ddCTP.

**(b)** ddGTP Reaction ("G" lane)

| Length of synthesized fragment | Partial replication products |
|---|---|
| 22 | 5′ 18-mer A A T G |
| 24 | 5′ 18-mer A A T G C G |
| 27 | 5′ 18-mer A A T G C G C T G |
| 32 | 5′ 18-mer A A T G C G C T G C A T C G |
| 35 | 5′ 18-mer A A T G C G C T G C A T C G T A G |

**(c)** ddTTP Reaction ("T" lane)

| Length of synthesized fragment | Partial replication products |
|---|---|
| 21 | 5′ 18-mer A A T |
| 26 | 5′ 18-mer A A T G C G C T |
| 30 | 5′ 18-mer A A T G C G C T G C A T |
| 33 | 5′ 18-mer A A T G C G C T G C A T C G T |
| 38 | 5′ 18-mer A A T G C G C T G C A T C G T A G C T |

**(d)** ddATP Reaction ("A" lane)

| Length of synthesized fragment | Partial replication products |
|---|---|
| 19 | 5′ 18-mer A |
| 20 | 5′ 18-mer A A |
| 29 | 5′ 18-mer A A T G C G C T G C A |
| 34 | 5′ 18-mer A A T G C G C T G C A T C G T A |
| 38 | 5′ 18-mer A A T G C G C T G C A T C G T A G C T A |

**Figure 7.31 DNA sequencing reactions. (a)** A target region of DNA is located by binding a single-stranded primer of 18 nucleotides (an "18-mer") that carries a 5′ label. Replication products terminated by ddCTP each have a different length. **(b)** Replication products terminated by ddGTP. **(c)** Termination products generated by ddTTP. **(d)** Termination products generated by ddATP.

single-stranded 18-mer primer binds to template DNA. Using the five nucleotides available in each reaction, DNA polymerase replicates the DNA fragment by adding nucleotides beginning at the 3′-OH end of the primer. The primers used in dideoxy sequencing are labeled with either radioactive phosphorus ($^{32}$P) or with a fluorescent label on their 5′ ends to facilitate detection of the DNA fragments produced in the sequencing reaction. In Figure 7.31a, DNA synthesis from a template strand progresses until incorporation of the reaction mixture's ddCTP terminates chain elongation. A few replicating fragments incorporate ddCTP at the first opportunity (the C nearest the 3′ end of the primer), but most incorporate dCTP at this site and continue replication. Some of these fragments incorporate ddCTP at the next opportunity and stop replicating, while most others incorporate dCTP and continue replication. Replication proceeds this way, halting for a few fragments each time a G appears on the template strand and a C is incorporated into the newly synthesized fragment. The result from this reaction is a series of partially replicated fragments whose replication is halted at each site of C incorporation.

The three other reaction mixtures containing ddGTP, ddTTP, and ddATP likewise produce a series of partial replication products that all end with their particular ddNTP (Figure 7.31b–d). Upon the completion of the four parallel sequencing reactions, partial replication DNA products will occur for every nucleotide in the template.

After the replication reactions are complete, the contents of each reaction are loaded into separate lanes of a DNA electrophoresis gel. Following completion of gel electrophoresis, the DNA sequence can be determined by examining the different-sized replication products spread across the four gel lanes. The bands shown in **Figure 7.32a** are visible in an autoradiograph because the primers that begin each fragment are end-labeled. The shortest fragment seen is in the A lane at the bottom, indicating that the first ddNTP nucleotide added to the 3′ end of the primer was ddATP. The second-shortest fragment is also in the A lane, indicating that chains to which ddATP was added in the second position terminated elongation there. The third-shortest fragment in the gel is in the T lane, and the fourth-shortest in this example is in the G lane. So far, the sequence of nucleotides in the synthesized DNA is AATG.

By continuation of this analytical process, the DNA sequence of the synthesized strand is "read" from the gel in the 5′-to-3′ direction (the direction in which a replicating strand elongates), as demonstrated in Figure 7.32a. The "inferred strand" is the template strand, which is complementary and antiparallel to the sequenced strand. **Figure 7.32b** shows an autoradiograph of a dideoxysequencing gel and shows a portion of the sequence read near the middle of the gel at the left.

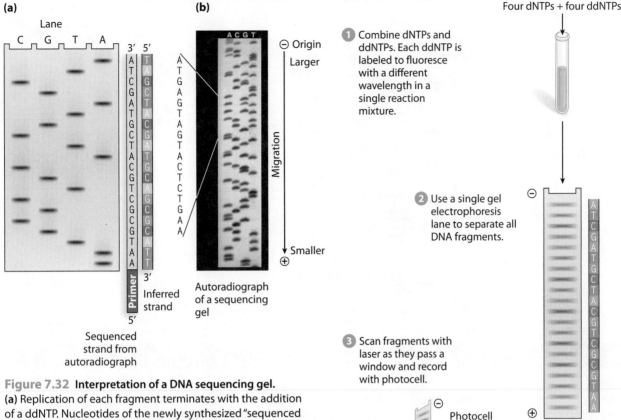

**Figure 7.32  Interpretation of a DNA sequencing gel.**
**(a)** Replication of each fragment terminates with the addition of a ddNTP. Nucleotides of the newly synthesized "sequenced strand" are read off the autoradiograph, and the 5′-to-3′ polarity of the strand corresponds to the smaller-to-larger fragment-length direction. The "inferred strand" is the template strand and it is complementary and antiparallel to the "sequenced strand." **(b)** A photograph of a dideoxy sequencing gel.

Manual dideoxy sequencing, as described above, is a labor-intensive process that today has been largely supplanted by high throughput, automated DNA sequencing and powerful computational software and hardware that can run 24 hours a day, 365 days a year, and assemble genomic sequence at the rate of 10,000 to 20,000 bp per hour! Automated DNA sequencers use a single reaction mixture containing the four dNTPs and all four dideoxynucleotides, each type labeled with a fluorescent compound of a different wavelength. A laser light is used to excite the fluorescent tag on each replication fragment as it migrates past a particular point on the electrophoresis gel, and the wavelengths are recorded by a photocell and transmitted to computer memory. Successive scans of each lane of the sequencing gel are run to build up an accumulated fluorescence pattern that identifies the DNA sequence of the fragment (**Figure 7.33**). We describe recent advances in DNA sequencing technology in Chapter 18. **Genetic Analysis 7.3** tests your skills at interpreting dideoxy sequencing results.

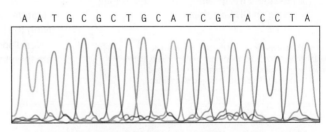

**Figure 7.33  Automated DNA sequencing.** ❶ Sequencing reactions contain the four dNTPs and the four ddNTPs. Each ddNTP is labeled with a distinctive fluorescent tag. ❷ Replication products are separated by gel electrophoresis, and as each band passes a window a laser excites the fluorescent label (❸). Excitation is captured by a photocell and software interprets the corresponding nucleotide and assembles sequence information (❹).

From the dideoxy DNA sequencing gel shown below, deduce the sequence and strand polarities of the DNA duplex fragment.

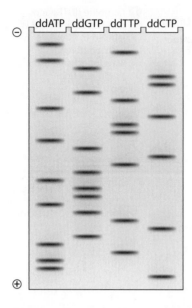

## Solution Strategies

## Solution Steps

### Evaluate

1. Identify the topic this problem addresses, and state the nature of the required answer.

2. Identify the critical information given in the problem.

1. This question concerns dideoxynucleotide DNA sequencing. The answer requires interpretation of a DNA sequencing gel to determine the double-stranded sequence of a fragment of DNA, including strand polarities.

2. A dideoxynucleotide DNA sequencing gel is shown.

### Deduce

3. Review the essential steps of dideoxynucleotide DNA sequencing.

4. Examine the gel and identify the "beginning" of DNA synthesis.

**TIP:** DNA fragments toward the bottom of the gel (nearer the positive pole) are shorter than fragments higher up in the gel. The sequence of the synthesized strand shown in the gel is 5′ at the bottom and 3′ at the top.

3. DNA polymerase incorporates nucleotides in four parallel reactions. Each reaction includes the four normal DNA nucleotides (dNTPs) and one labeled dideoxynucleotide (ddNTP). Incorporation of a dNTP allows continued strand synthesis, but incorporation of a ddNTP terminates synthesis.

4. The 3′ end of the primer is used to initiate DNA synthesis. The first nucleotide incorporated during synthesis is cytosine, as determined by identifying the location of the smallest synthesized fragment: the "C" lane. The second and third nucleotides are both adenine. The first three nucleotides are therefore 5′-CAA-3′.

### Solve

5. Write the rest of the sequence (along with the polarity) of the synthesized strand shown in the gel.

6. Determine the sequence and polarity of the template strand used for DNA synthesis.

5. The synthesized strand is

   5′-[primer]-CAATAGCTGAGGAGTCGATTCATGCCGATA-3′.

6. The template DNA strand is

   3′-GTTATCGACTCCTCAGCTAAGTACGGCTAT-5′.

For more practice, see Problems 28, 29, 30, and 34.

# CASE STUDY

## Use of PCR and DNA Sequencing to Analyze Huntington Disease Mutations

Both PCR and DNA sequencing analysis have been used to study the gene identified as *HD* that is mutated in Huntington disease (OMIM 143100). *HD* encodes the huntingtin protein that is expressed in brain cells and in other cells of the body. The normal function of wild-type huntingtin is not known, but it interacts with dozens of other proteins. In mutant form, huntingtin appears to aggregate with itself and other proteins, hastening the death of neurons in the brain that lead to the motor abnormalities—progressive loss of motor control by unintentional and uncontrollable movement—that are characteristic of the disease.

Huntington disease is one of several human trinucleotide repeat disorders that are caused by increases in the length of gene sections containing end-to-end repeats of three nucleotides. A CAG trinucleotide region of *HD* that encodes the amino acid glutamine produces a polyglutamine tract in the wild-type allele. The length of the polyglutamine tract is increased in mutant huntingtin protein as a result of an increased number of CAG repeats in mutant alleles.

Wild-type *HD* genes vary in the number of CAG repeats, ranging from 6 to 28 repeats in the general population. *HD* alleles with 28 to 35 CAG repeats do not cause disease, but as a consequence of the increased CAG number, the alleles are unstable and prone to further expansion. Alleles that have 36 to 40 CAG repeats have expanded beyond the normal range and the huntingtin protein produced by these alleles can behave abnormally and can result in disease symptoms that show reduced penetrance. Individuals who carry 36 to 40 CAG repeats might or might not develop HD. If they do, disease symptoms have a late age of onset and progress slowly. Individuals with *HD* alleles containing more than 40 CAG repeats have HD that can develop at any time from the late teens onward. **Figure 7.34** shows dideoxy DNA sequencing analysis of the CAG repeat segment of the *HD* gene for a wild-type allele with 21 CAG repeats and for a mutant allele with 48 CAG repeats.

The polymerase chain reaction provides a way of visualizing the CAG triplet repeat expansion and of following the transmission of alleles in the families of people with HD. Employing primers that bind on opposite sides of the CAG repeat region, researchers amplify fragments of DNA by PCR and separate them by gel electrophoresis. The binding sites of the PCR primers are identical for all alleles, but differences are seen in the lengths of amplified PCR products because of different numbers of CAG repeats between the primer binding sites. Amplified DNA fragments containing the primers are shorter if they are generated from wild-type DNA sequences than from mutant alleles, because wild-type alleles have a smaller number of repeats than do mutant alleles. In the Huntington disease family shown in **Figure 7.35**, each person with HD is heterozygous and carries one wild-type allele with fewer than 36 repeats of the CAG sequence and one expanded allele with more than 36 repeats. In contrast, family members who do not have HD are homozygous or heterozygous for alleles with fewer than 36 CAG repeats. These and similar molecular methods are used to assess the number of CAG repeats in *HD* for presymptomatic genetic testing of people at risk for inheriting Huntington disease.

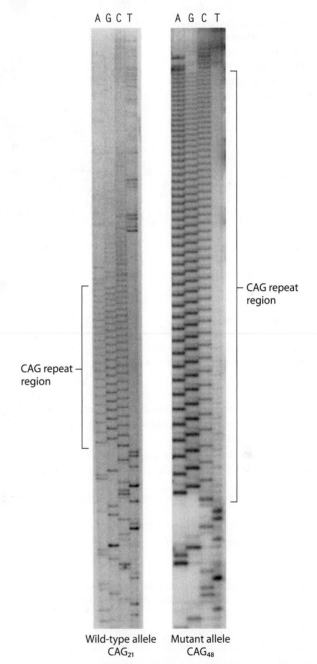

**Figure 7.34 Dideoxy DNA sequencing of the *HD* gene.** Gel electrophoresis results of dideoxy sequencing of a wild-type *HD* allele with 21 CAG repeats is compared to the results for an *HD* allele with 48 CAG repeats. Data from the Huntington Disease Collaborative Research Group (1993).

At-risk individuals can be tested before disease symptoms appear and can be told whether they carry an expanded *HD* allele. These methods can also be used to identify the presence of a CAG expansion of *HD* in individuals diagnosed with Huntington disease by clinicians.

**Figure 7.35 CAG expansion of the *HD* gene detected by Southern blot analysis of PCR-amplified DNA.** Each family member represented by a filled circle or square has Huntington disease. PCR analysis of the *HD* gene reveals each such person to be heterozygous for HD and to carry one disease-producing allele with more than 36 CAG repeats.

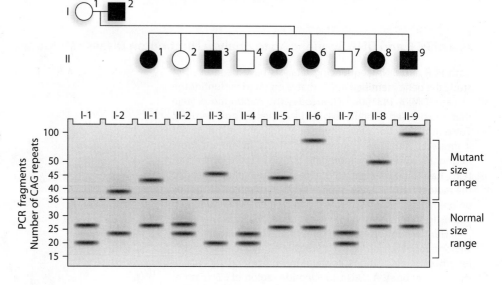

---

SUMMARY

MasteringGENETICS™    *For activities, animations, and review quizzes, go to the study area at www.masteringgenetics.com.*

## 7.1 DNA Is the Hereditary Molecule of Life

▌ F. Griffith determined in 1928 that a molecular transformation factor was responsible for transformation of living R bacteria into an S form.

▌ In 1944, O. Avery, C. MacLeod, and M. McCarty's study of in vitro transformation caused by an S cell extract identified DNA as the transformation factor and strongly suggested it is the hereditary material.

▌ A. Hershey and M. Chase determined in 1952 that bacteriophage T2 uses DNA, not protein, to reproduce within host E. coli cells.

## 7.2 The DNA Double Helix Consists of Two Complementary and Antiparallel Strands

▌ The DNA nucleotides consist of the five-carbon sugar deoxyribose, a phosphate group, and one of four nitrogen-containing nucleotide bases.

▌ The DNA nucleotide bases are the purines adenine and guanine, and the pyrimidines cytosine and thymine.

▌ Phosphodiester bonds form between 5′ phosphate and 3′OH groups to join nucleotides into polynucleotide chains.

▌ Complementary base pairs consist of a purine and a pyrimidine. In DNA, A and T form two stable hydrogen bonds, whereas G and C form three stable hydrogen bonds.

▌ Complementary nucleic acid strands are antiparallel.

▌ The stacking of base pairs in DNA imparts helical counting the number of triplet repeats characterizing each allele. This also makes twisting that creates major grooves and minor grooves in the duplex.

## 7.3 DNA Replication Is Semiconservative and Bidirectional

▌ Experimental evidence demonstrates that DNA replication is semiconservative, meaning each daughter molecule receives one parental strand and one newly synthesized strand that was produced using the parental strand as a template.

▌ Most DNA replication is bidirectional. A replication bubble with replication forks at each end expands as replication progresses.

▌ Bacterial genomes have a single replication origin, whereas eukaryotic genomes have many origins of replication.

▌ Eukaryotic replication origins initiate asynchronously during 0S phase.

▌ Eukaryotic DNA replication produces sister chromatids.

## 7.4 DNA Replication Precisely Duplicates the Genetic Material

▌ Replication begins at specific locations that are defined by conserved DNA sequences and consensus DNA sequences.

▌ DNA replication begins with the synthesis of an RNA primer by primase, followed by synthesis of leading and lagging DNA strands by DNA polymerase.

▌ To complete replication, RNA primers are removed by DNA polymerase, and DNA segments are joined by DNA ligase.

▌ DNA polymerases not only replicate DNA but also proof-read newly synthesized DNA for accuracy.

▌ Eukaryotic chromosomes have repetitive sequences called telomeres at their ends that shorten with each replication in somatic cell cycles.

▌ Telomerase is a ribonucleoprotein that synthesizes telomeric repeat sequences to maintain telomere length in germ-line and stem cells.

## 7.5 Molecular Genetic Analytical Methods Make Use of DNA in Replication Processes

▌ The polymerase chain reaction (PCR) is used to produce large numbers of copies of target DNA sequences.

▌ Dideoxynucleotide DNA sequencing is used to determine the sequence of DNA fragments.

## KEYWORDS

bacteriophages (phages) *(p. 226)*
base stacking *(p. 230)*
bidirectional DNA replication *(p. 235)*
clamp loader *(p. 242)*
consensus sequences *(p. 238)*
conservative DNA replication *(p. 232)*
deoxynucleotide 5′-monophosphates (dNMPs) *(p. 228)*
deoxynucleotide 5′-triphosphates (dNTPs) *(p. 228)*
dideoxy DNA sequencing *(p. 251)*
dideoxynucleotide triphosphates (ddNTPs) *(p. 251)*
dispersive DNA replication *(p. 232)*
DNA ligase *(p. 241)*
DNA polymerase (pol I, pol III, 5′-to-3′ polymerase activity) *(pp. 240, 241)*

DNA proofreading (3′-to-5′ exonuclease activity) *(pp. 241, 245)*
DNA replication (semiconservative, conservative, dispersive) *(p. 232)*
helicase *(p. 239)*
lagging strand *(p. 240)*
leading strand *(p. 240)*
major groove *(p. 230)*
minor groove *(p. 230)*
Okazaki fragment *(p. 241)*
origin of replication *(p. 233)*
PCR primers *(p. 248)*
polymerase chain reaction (PCR) *(p. 248)*
primase *(p. 239)*
primosome *(p. 240)*
proliferating cell nuclear antigen (PCNA) *(p. 245)*

replication bubble *(p. 235)*
replication fork *(p. 235)*
replisome *(pp. 237, 240)*
RNA primer *(p. 240)*
semiconservative DNA replication *(p. 232)*
single-stranded binding (SSB) protein *(p. 239)*
sliding clamp (proliferating cell nuclear antigen [PCNA]) *(p. 242)*
sugar-phosphate backbone *(p. 228)*
supercoiled DNA *(p. 239)*
telomerase *(p. 246)*
telomere *(p. 246)*
θ (theta) structure *(p. 234)*
topoisomerases *(p. 239)*

## PROBLEMS

Mastering**GENETICS**™

*For instructor-assigned tutorials and problems, go to www.masteringgenetics.com.*

### Chapter Concepts

*For answers to selected even-numbered problems, see Appendix: Answers.*

1. What results from the experiments of Frederick Griffith provided the strongest support for his conclusion that a transformation factor is responsible for heredity?

2. Explain why Avery, MacLeod, and McCarty's in vitro transformation experiment showed that DNA, but not RNA or protein, is the hereditary molecule.

3. Hershey and Chase selected the bacteriophage T2 for their experiment assessing the role of DNA in heredity because T2 contains protein and DNA, but not RNA. Explain why T2 was a good choice for this experiment.

4. Explain how the Hershey and Chase experiment identified DNA as the hereditary molecule.

5. One strand of a fragment of duplex DNA has the sequence 5′-ATCGACCTGATC-3′.
   a. What is the sequence of the other strand in the duplex?
   b. Identify the bond that joins one nucleotide to another in the DNA strand.
   c. Is the bond in part (b) a covalent or a noncovalent bond?
   d. Which chemical groups of nucleotides react to form the bond in part (b)?
   e. What enzymes catalyze the reaction in part (d)?
   f. Identify the bond that joins one strand of a DNA duplex to the other strand.
   g. Is the bond in part (f) a covalent or a noncovalent bond?
   h. What term is used to describe the pattern of base pairing between one DNA strand and its partner in a duplex?

   i. What term is used to describe the polarity of two DNA strands in a duplex?

6. The principles of complementary base pairing and antiparallel polarity of nucleic acid strands in a duplex are universal for the formation of nucleic acid duplexes. What is the chemical basis for this universality?

7. For the following fragment of DNA, determine the number of hydrogen bonds and the number of phosphodiester bonds present:

   5′-ACGTAGAGTGCTC-3′
   3′-TGCATCTCACGAG-5′

8. Figures 1.6 and 1.7 present simplified depictions of nucleotides containing deoxyribose, a nucleotide base, and a phosphate group (see pages 7 and 10). Use this simplified method of representation to illustrate the sequence 3′-AGTCGAT-5′ and its complementary partner in a DNA duplex.
   a. What kind of bond joins the C to the G within a single strand?
   b. What kind of bonds join the C in one strand to the G in the complementary strand?
   c. How many phosphodiester bonds are present in this DNA duplex?
   d. How many hydrogen bonds are present in this DNA duplex?

9. Consider the sequence 3'-ACGCTACGTC-5'.

   a. What is the double-stranded sequence?
   b. What is the total number of covalent bonds joining the nucleotides in each strand?
   c. What is the total number of noncovalent bonds joining the nucleotides of the complementary strands?

10. DNA polymerase III is the main DNA-synthesizing enzyme in bacteria. Describe how it carries out its role of elongating a strand of DNA.

11. You are participating in a study group preparing for an upcoming genetics exam, and one member of the group proposes that each of you draw the structure of two DNA nucleotides joined in a single strand. The figures are drawn and exchanged for correction. You receive the drawing below to correct.

   a. Identify and correct at least five things that are wrong in the depiction of each nucleotide.
   b. What is wrong with the way the nucleotides are joined?
   c. How would you draw this single-stranded segment?

12. Explain how RNA participates in DNA replication.

13. A sample of double-stranded DNA is found to contain 20% cytosine. Determine the percentage of the three other DNA nucleotides in the sample.

14. Bacterial DNA polymerase I and DNA polymerase III perform different functions during DNA replication.

   a. Identify the principal functions of each molecule.
   b. If mutation inactivated DNA polymerase I in a strain of E. coli, would the cell be able to replicate its DNA? If so, what kind of abnormalities would you expect to find in the cell?
   c. If a strain of E. coli acquired a mutation that inactivated DNA polymerase III function, would the cell be able to replicate its DNA? Why or why not?

15. Diagram a replication fork in bacterial DNA and label the following structures or molecules.

   a. DNA pol III
   b. helicase
   c. RNA primer
   d. origin of replication
   e. leading strand (label its polarity)
   f. DNA pol I
   g. topoisomerase
   h. SSB protein
   i. lagging strand (label its polarity)
   j. primase
   k. Okazaki fragment

16. Which of the following equations are true for the percentages of nucleotides in double-stranded DNA?

   a. $(A + G)/(C + T) = 1.0$
   b. $(A + T)/(G + C) = 1.0$
   c. $(A)/(T) = (G)/(C)$
   d. $(A)/(C) = (G)/(T)$
   e. $(A)/(T) = (G)/(C)$

17. Which of the following equalities is not true for double-stranded DNA?

   a. $(G + T) = (A + C)$
   b. $(G + C) = (A + T)$
   c. $(G + A) = (C + T)$

18. List the order in which the following proteins and enzymes are active in E. coli DNA replication: DNA pol I, SSB, ligase, helicase, DNA pol III, and primase.

19. Two viral genomes are sequenced, and the following percentages of nucleotides are identified:

   Genome 1: A = 28%, C = 22%, G = 28%, T = 22%
   Genome 2: A = 22%, C = 28%, G = 28%, T = 22%

   What is the structure of DNA in each genome?

## Application and Integration

*For answers to selected even-numbered problems, see Appendix: Answers.*

20. Matthew Meselson and Franklin Stahl demonstrated that DNA replication is semiconservative in bacteria. Briefly outline their experiment and its results for two DNA replication cycles, and identify how the alternative models of DNA replication were excluded by the data.

21. What is the significance of θ structures observed by John Cairns in his experiments on bacterial DNA replication? What do θ structures tell us about the process of DNA replication in bacteria?

22. Joel Huberman and Arthur Riggs used pulse labeling to examine the replication of DNA in mammalian cells. Briefly describe the Huberman-Riggs experiment, and identify how the results exclude a unidirectional model of DNA replication.

23. Why do the genomes of eukaryotes, such as *Drosophila*, need to have multiple origins of replication, whereas bacterial genomes, such as that of *E. coli*, have only a single origin?

24. Bloom syndrome (OMIM 210900) is an autosomal recessive disorder caused by mutation of a DNA helicase. Among the principal symptoms of the disease are chromosome instability and a propensity to develop cancer. Explain these symptoms on the basis of the helicase mutation.

25. How does rolling circle replication (see Chapter 6, page 189) differ from bidirectional replication?

26. Telomeres are found at the ends of eukaryotic chromosomes.

a. What is the sequence composition of telomeres?

b. How does telomerase assemble telomeres?

c. What is the functional role of telomeres?

d. Why is telomerase usually active in germ-line cells but not in somatic cells?

27. A family consisting of a mother (I-1), a father (I-2), and three children (II-1, II-2, and II-3) are genotyped by PCR for a region of an autosome containing repeats of a 10-bp sequence. The mother carries 16 repeats on one chromosome and 21 on the homologous chromosome. The father carries repeat numbers of 18 and 26.

a. Following the illustration style of Figure 7.29c, which aligns members of a pedigree with their DNA fragments in a gel, draw a DNA gel containing the PCR fragments generated by amplification of DNA from the parents (I-1 and I-2). Label the size of each fragment.

b. Identify all the possible genotypes of children of this couple by specifying PCR fragment lengths in each genotype.

c. What genetic term best describes the pattern of inheritance of this DNA marker? Explain your choice.

28. In a dideoxy DNA sequencing experiment, four separate reactions are carried out to provide the replicated material for DNA-sequencing gels. Reaction products are usually run in gel lanes labeled A, T, C, and G.

a. Identify the nucleotides used in the dideoxy DNA-sequencing reaction that produces molecules for the A lane of the sequencing gel.

b. How does PCR play a role in dideoxy DNA sequencing?

c. Why is incorporation of a dideoxynucleotide during DNA sequencing identified as a "replication-terminating" event?

29. The following dideoxy DNA-sequencing gel is produced in a laboratory.

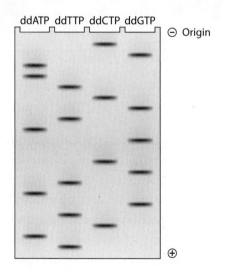

What is the double-stranded DNA sequence of this molecule? Label the polarity of each strand.

30. Using an illustration style and labeling similar to that in Problem 29, draw the electrophoresis gel containing dideoxy sequencing fragments for the DNA template strand 3'-AGACGATAGCAT-5'.

31. A PCR reaction begins with one double-stranded segment of DNA. How many double-stranded copies of DNA are present after the completion of 10 amplification cycles? After 20 cycles? After 30 cycles?

32. DNA replication in early *Drosophila* embryos occurs about every 5 minutes. The *Drosophila* genome contains approximately $1.8 \times 10^8$ base pairs. Eukaryotic DNA polymerases synthesize DNA at a rate of approximately 40 nucleotides per second. Approximately how many origins of replication are required for this rate of replication?

33. Three independently assorting autosomal minisatellite markers are used to assess the paternity of a colt (C) recently born to a quarter horse mare (M). Blood samples are drawn from the mare, her colt, and three possible male sires ($S_1$, $S_2$, and $S_3$). DNA at each marker locus is amplified by PCR, and a DNA electrophoresis gel is run for each marker. Amplified DNA bands are visualized in each gel by ethidium bromide staining. Gel results are shown below for each marker.

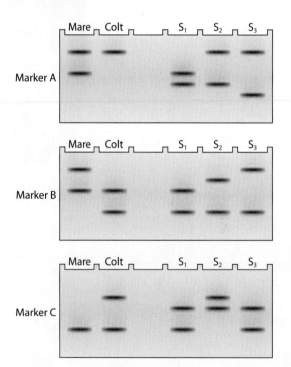

Evaluate the data and determine if any of the potential sires can be excluded. Explain the basis of exclusion, if any, in each case.

34. A sufficient amount of a small DNA fragment is available for dideoxy sequencing. The fragment to be sequenced contains 20 nucleotides following the site of primer binding, as shown in the next column:

5'-ATCGCTCGACAGTGACTAGC-[primer site]-3'

Dideoxy sequencing is carried out, and the products of the four sequencing reactions are separated by gel electrophoresis. Draw the bands you expect will appear on the gel from each of the sequencing reactions.

# 8

# Molecular Biology of Transcription and RNA Processing

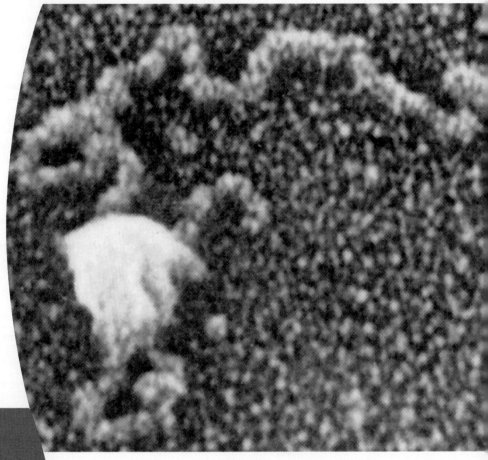

An electron micrograph of spliceosomes engaged in intron splicing.

## ESSENTIAL IDEAS

■ Ribonucleic acid (RNA) molecules are transcribed from genes and are classified either as messenger RNA or as one of several types of functional RNA.

■ Bacterial transcription is a four-step process that begins with promoter recognition by RNA polymerase and ends with the completion of transcript synthesis.

■ Eukaryotes use different RNA polymerases to transcribe different kinds of RNA. Each type of polymerase initiates transcription at a different type of promoter.

■ Eukaryotic RNAs undergo three processing steps after transcription. Alternative events during and after transcription allow different transcripts and proteins to be produced from the same DNA sequence.

Wilhelm Johansson introduced the term *gene* in 1909 to describe "the fundamental unit of inheritance." Johansson's definition encompasses the understanding that genes contain genetic information and are passed from one generation to the next, and that genes are the basis of the fundamental structural, functional, developmental, reproductive, and evolutionary properties of organisms. This basic definition of the gene remains valid today, more than a century after being coined; but our knowledge of molecular genetics has expanded enormously, refining our understanding of the structure and function of genes and clarifying the roles genes play in producing traits.

We previously introduced the central dogma of biology, which describes the flow of genetic information as going from DNA to RNA to protein (see Figure 1.8). This formula conveys that DNA is the repository of genetic information, which is converted through *transcription* into RNA and then *translated* into protein. Transcription is the process by which RNA polymerase enzymes and other transcriptional proteins and enzymes use the template strand of DNA to synthesize a complementary RNA strand. Translation is the process by which *messenger RNA* is used to direct protein synthesis.

The subject of this chapter is transcription and the production of RNA. This topic is closely tied to the process of translation that is the subject of the following chapter. As a consequence of shared evolutionary history and common ancestry of bacteria, archaea, and eukaryotes, and the fundamentality of transcription and translation to gene expression, these processes in bacteria have a number of general features in common with eukaryotes. On the other hand, differences among the members of these kingdoms, including differences in cell structure, gene structure, and genome organization, lead to significant differences in how their genes are transcribed and translated.

Multiple types of RNA are introduced and described here, but the principal focus of discussion is messenger RNA, the transitory molecule that carries genetic information in the form of the code that becomes translated into proteins. The discovery of mRNA and of its role in the central dogma raised numerous questions: How is a gene recognized by the transcription machinery? Where does transcription begin? Which strand of DNA is transcribed? Where does transcription end? How much transcript is made? How is RNA modified after transcription? We answer these questions in the chapter and set the answers in a context that compares and contrasts the process of transcription in bacterial and eukaryotic genomes.

## 8.1 RNA Transcripts Carry the Messages of Genes

In the late 1950s, with the structure of DNA in hand, molecular biology researchers focused on identifying and describing the molecules and mechanisms responsible for conveying the genetic message of DNA. RNA was known to be chemically similar to DNA and present in abundance in all cells, but its diversity and biological roles remained to be discovered. Some roles were strongly suggested by cell structure. For example, in eukaryotic cells, DNA is located in the nucleus, whereas protein synthesis takes place in the cytoplasm, suggesting that DNA could not code directly for proteins but RNA perhaps could. Bacteria, however, lack a nucleus, so an open research question was whether bacteria and eukaryotes used similar mechanisms and similar molecules to convey the genetic message for protein synthesis. The search was on to identify the types of RNA in cells and to identify the mechanisms by which the genetic message of DNA is conveyed for protein synthesis.

### RNA Nucleotides and Structure

Both DNA and RNA are polynucleotide molecules composed of nucleotide building blocks. One principal difference between the molecules is the single-stranded structure of RNA versus the double-stranded structure of DNA. The single-stranded structure of RNA molecules can lead them to adopt folded secondary structures by complementary base pairing of segments of the molecule. In certain instances, folded secondary structures are essential to RNA function, as discussed below.

The RNA nucleotides, like those of DNA, are composed of a sugar, a nucleotide base, and one or more phosphate groups. Each RNA nucleotide carries one of four possible nucleotide bases, but RNA nucleotides have two critical chemical differences in comparison to DNA nucleotides. The first difference concerns the identity of the RNA nucleotide bases. The purines adenine and guanine in RNA are identical to the purines in DNA. Likewise, the pyrimidine cytosine is identical in RNA and DNA. In RNA, however, the second pyrimidine is **uracil** (U) rather than the thymine carried by DNA. The four RNA **ribonucleotides** are shown in Figure 8.1. The structure of uracil is similar to that of thymine; but notice, by comparing the structure of uracil in Figure 8.1 with that of thymine in Figure 7.6, that thymine has a methyl group ($CH_3$) at the 5 position of the pyrimidine ring, whereas uracil does not. In all other respects, uracil matches thymine, and when uracil undergoes base pairing, its complementary partner is adenine.

The second chemical difference between RNA and DNA is the presence of the sugar **ribose** in RNA rather than the deoxyribose occurring in DNA. The ribose gives

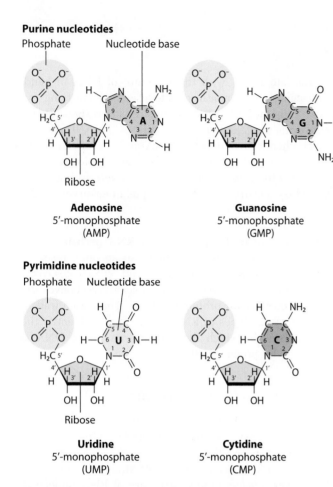

**Figure 8.1 The four RNA ribonucleotides.** Shown in their monophosphate forms, each ribonucleotide consists of the sugar ribose, one phosphate group, and one of the nucleotide bases adenine, guanine, cytosine, and uracil.

RNA its name (ribonucleic acid). Compare the ribose molecules shown in Figure 8.1 to deoxyribose in Figure 7.6, and notice that ribose carries a hydroxyl group (OH) not found in deoxyribose at the 2′ carbon of the ring. Except for this difference, ribose and deoxyribose are identical, having a nucleotide base attached to the 1′ carbon and a hydroxyl group attached to the 3′ carbon.

The similarity of the sugars of RNA and DNA leads to the formation of essentially identical sugar-phosphate backbones in the molecules. RNA strands are assembled by formation of phosphodiester bonds, between the 5′ phosphate of one nucleotide and the 3′ hydroxyl of the adjacent nucleotide, that are identical to those found in DNA (**Figure 8.2**). RNA is synthesized from a DNA template strand using the same purine-pyrimidine complementary base pairing described for DNA except for the pairing between adenine of DNA with uracil of RNA. **RNA polymerase** enzymes catalyze the addition of each ribonucleotide to the 3′ end of the nascent strand and form phosphodiester bonds between a triphosphate group at the 5′carbon of one nucleotide and the hydroxyl group at the 3′ carbon of the adjacent nucleotide, eliminating two

phosphates (the pyrophosphate group), just as in DNA synthesis. Compare Figure 8.2 to Figure 7.7 to see the similarity of these nucleic acid synthesis processes.

## Identification of Messenger RNA

In the late 1950s, researchers used the pulse-chase technique (see Section 7.3) to follow the trail of newly synthesized RNA in eukaryotic cells. The "pulse" step of this technique exposes cells to radioactive nucleotides that become incorporated into newly synthesized nucleic acids (see Chapter 7). After a short incubation period to incorporate the labeled nucleotides, a "chase" step removes unincorporated radioactive nucleotides by replacing them with unlabeled nucleotides. An experimenter can then observe the location and movement of the labeled nucleic acid to determine the pattern of its movement and its ultimate destination and fate.

In 1957, microbiologist Elliot Volkin and geneticist Lazarus Astrachan used the pulse-chase method to examine transcription in bacteria immediately following infection by a bacteriophage. Exposing newly infected bacteria to radioactive uracil, they observed rapid incorporation of the label, indicating a burst of transcriptional activity. In the chase phase of the experiment, when radioactive uracil was removed, Volkin and Astrachan found that the radioactivity quickly dissipated, indicating that the newly synthesized RNA broke down rapidly. They concluded that the synthesis of a type of RNA with a very short life span is responsible for the production of phage proteins that drive progression of the infection.

Similar pulse-chase experiments were soon conducted with eukaryotic cells. In these experiments, cells were pulsed with radioactive uracil that was then chased with nonradioactive uracil. Immediately after the pulse, radioactivity was concentrated in the nucleus, indicating that newly synthesized RNA has a nuclear location. Over a short period, radioactivity migrated to the cytoplasm, where translation takes place. The radioactivity dissipated after lingering in the cytoplasm for a period of time. These experiments led researchers to conclude that the RNA synthesized in the nucleus was likely to act as an intermediary carrying the genetic message of DNA to the cytoplasm for translation into proteins.

The discovery of mRNA was capped in 1961 when an experiment by the biologists Sydney Brenner, Francois Jacob, and Matthew Meselson identified an unstable form of RNA as the genetic messenger. Brenner and his colleagues designed an experiment using the bacteriophage T2 and *Escherichia coli* to investigate whether phage protein synthesis requires newly constructed ribosomes, or whether phage proteins could be produced using existing bacterial ribosomes and a messenger molecule to encode the proteins. The experiment found that newly synthesized phage RNA associates with bacterial ribosomes to produce phage proteins. The RNA that directed the protein synthesis

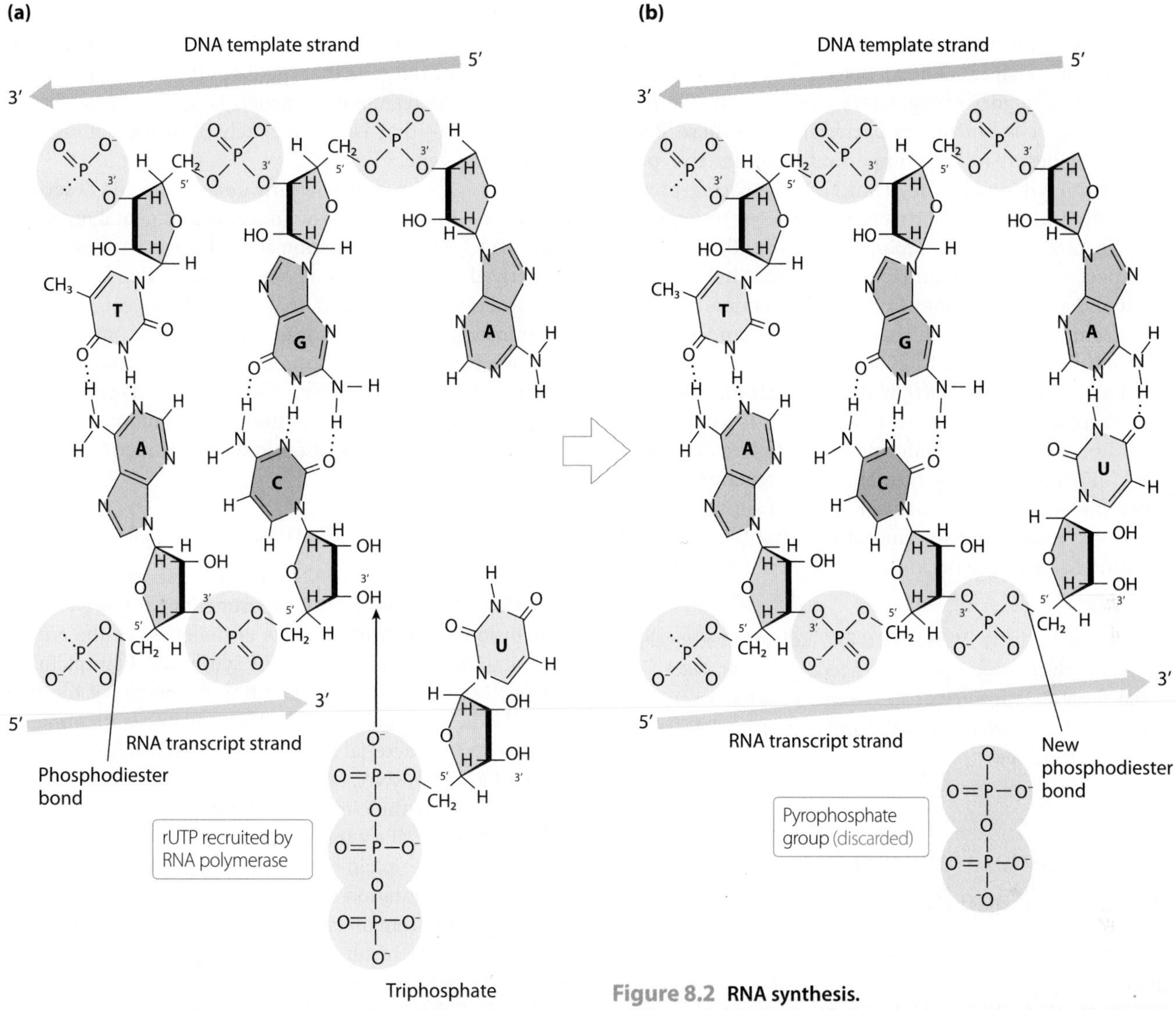

**(a)**

**(b)**

Figure 8.2  **RNA synthesis.**

formed and degraded quickly, leading the experimenters to conclude that a phage "messenger" RNA with a short half-life is responsible for protein synthesis during infection.

## RNA Classification

All RNAs are transcribed from RNA-encoding genes. The different types of RNA are constructed from the same building blocks but perform a variety of roles in cells. In light of these different roles, RNAs are divided into two general categories—*messenger RNA* and *functional RNA*. Genes transcribing **messenger RNA (mRNA)** are protein-producing genes, and their transcripts direct protein synthesis by the process of translation. Messenger RNA is the short-lived intermediary form of RNA that conveys the genetic message of DNA to ribosomes for translation. Messenger RNA is the only form of RNA that undergoes translation. Transcription of mRNA and post-transcriptional processing of mRNA are principal areas of focus in this chapter.

**Functional RNAs** perform a variety of specialized roles in the cell. The functional RNAs carry out their activities in nucleic acid form and are not translated. Two major categories of functional RNA are active in bacterial and eukaryotic translation. **Transfer RNA (tRNA)** is encoded in dozens of different forms in all genomes. Each tRNA is responsible for binding a particular amino acid that it carries to the ribosome. There the tRNA interacts with mRNA and deposits its amino acid for inclusion in the growing protein chain. **Ribosomal RNA (rRNA)** combines with numerous proteins to form the ribosome, the molecular machine responsible for translation. Certain bacterial rRNA molecules interact with mRNA to initiate translation. (The roles of tRNA and rRNA in translation are addressed in Chapter 9.)

Three additional types of functional RNA perform specialized functions in eukaryotic cells only. **Small nuclear RNA (snRNA)** of various types is found in the nucleus of eukaryotic cells, where it participates in mRNA processing. Certain snRNAs unite with nuclear proteins to form ribonucleoprotein complexes that are responsible for intron removal. We discuss these activities in later sections of this chapter. **Micro RNA (miRNA),** a recently recognized category of regulatory RNA, is active in plant and animal cells but not in bacterial cells. Micro RNAs have a widespread and important role in the post-transcriptional regulation of mRNA, regulating protein production through a process called *RNA interference* (see Chapter 15). Finally, **small interfering RNA (siRNA)** is a specialized type of functional RNA that helps protect plant and animal genomes from the production of viruses and from the spread of transposable genetic elements within the genome (see Chapter 12.)

Lastly, certain RNAs in eukaryotic cells have catalytic activity. In contrast to DNA, which is exclusively a repository of genetic information, catalytically active RNA molecules can catalyze biological reactions. Called **ribozymes,** catalytically active RNAs can activate cellular reactions, including the removal of introns in a process identified as *self-splicing*, described later in the chapter.

---

Genetic Insight   Many kinds of RNA are transcribed in cells. The two major categories are messenger and functional RNA. Functional RNAs perform specific roles related to translation, pre-mRNA processing, or regulating gene expression. Only mRNA is translated.

---

## 8.2 Bacterial Transcription Is a Four-Stage Process

Transcription is the synthesis of a single-stranded RNA molecule by RNA polymerase. The polymerase uses one strand of DNA, the **template strand,** to assemble a complementary, antiparallel strand of ribonucleotides (see Figure 1.9). The **coding strand** of DNA, also known as the **nontemplate strand,** is complementary to the template strand. The gene—that is, the stretch of DNA regions that produces an RNA transcript—contains several segments with distinct functions (**Figure 8.3**). The **promoter** of the gene is immediately **upstream**—that is, immediately 5′ to the start of transcription, which is identified as corresponding to the +1 nucleotide. The promoter sequence is a transcription-regulating sequence that controls the access of RNA polymerase to the gene. The coding region is the portion of the gene that is transcribed into mRNA

and contains the information needed to synthesize the protein product of the gene. The **termination region** is the portion of the gene that regulates the cessation of transcription. The termination region is located immediately **downstream**—that is, immediately 3′ to the coding segment of the gene.

Transcription is most clearly understood and described in bacteria, and *E. coli* is the model experimental organism from which most information about bacterial transcription is derived. In this section, we examine the four stages of transcription in *E. coli* that serve as a general model for transcription in all bacteria. We also compare and contrast bacterial transcription with transcriptional activities in eukaryotic genomes. The essential stages of transcription in *E. coli* are (1) promoter recognition, (2) chain initiation, (3) chain elongation, and (4) chain termination.

### Bacterial RNA Polymerase

A single type of *E. coli* RNA polymerase catalyzes transcription of all RNAs. The experimental support for this conclusion stems from analysis of the effect of the antibiotic rifampicin on bacterial RNA synthesis. Rifampicin inhibits RNA synthesis by preventing formation of the first phosphodiester bond in the RNA chain. In rifampicin-sensitive ($rif^S$) bacterial strains, synthesis of all three major types of RNA (mRNA, tRNA, and rRNA) is inhibited in the presence of rifampicin. In contrast, rifampicin-resistant ($rif^R$) bacteria actively transcribe DNA into the three major RNAs when rifampicin is present. Molecular analysis identifies a single mutation of RNA polymerase in $rif^R$ strains that allows it to remain catalytically active when exposed to rifampicin. This result is consistent with the conclusion that a single RNA polymerase is active in bacterial transcription. Subsequent molecular studies have confirmed the presence of a single bacterial RNA polymerase.

Bacterial RNA polymerase is composed of a pentameric (five-polypeptide) **core enzyme** that binds to a sixth polypeptide, called the **sigma subunit (σ),** when the polymerase becomes active. In its active form, the RNA polymerase is described as a holoenzyme, meaning an intact complex with full enzymatic capacity. **Figure 8.4** shows a common type of sigma subunit known as σ$^{70}$, but there are other sigma subunits in *E. coli*.

The core enzyme is large, weighing approximately 390,000 daltons (390 kD) and consisting of two identical α subunits; one subunit each of the β and β′ subunits; and

**Figure 8.3  A general diagram of gene structure and associated nomenclature.**

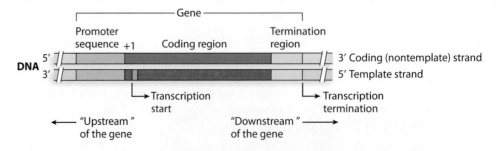

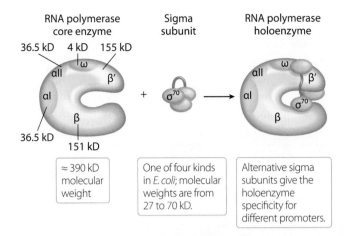

**Figure 8.4  Bacterial RNA polymerase core enzyme plus a sigma (σ) subunit form the fully active holoenzyme.**

an ω (omega) subunit. It is the RNA-synthesizing portion of the RNA polymerase holoenzyme. By itself, the core enzyme can transcribe DNA template-strand sequence into RNA sequence, but the core enzyme is unable to bind to a promoter or initiate RNA synthesis without a sigma subunit. The joining of the sigma subunit to the core enzyme to form a holoenzyme induces a conformational shift in the core segment that enables it to bind specifically to particular promoter consensus sequences (defined below). Several different types of sigma subunits, so-called **alternative sigma subunits,** have been identified in bacteria. Alternative sigma subunits impart different types of conformational changes to the core enzyme that in each case enable it to bind to a different promoter consensus sequence, as we discuss below.

## Bacterial Promoters

A promoter is a double-stranded DNA sequence that is the binding site for RNA polymerase. Promoters regulate transcription since RNA polymerase cannot initiate transcription without binding to promoter sequence. Bacterial promoters are located a short distance upstream of the coding sequence, typically within a few nucleotides of the start of transcription (the +1 nucleotide). RNA

polymerase is attracted to promoters by the presence of **consensus sequences,** short regions of highly similar DNA sequences located in the same position relative to the start of transcription in different gene promoters.

Although promoters are double stranded, promoter consensus sequences are written in a single-stranded shorthand form that gives the 5′-to-3′ sequence of the coding (non-template) strand of DNA (**Figure 8.5**). At the −10 position of the *E. coli* promoter is the **Pribnow box sequence,** or the **−10 consensus sequence,** consisting of 6 bp having the consensus sequence 5′-TATAAT-3′. The Pribnow box is separated by about 25 bp from another 6-bp region, the **−35 consensus sequence,** identified by the nucleotides 5′-TTGACA-3′. The nucleotide sequences that occur upstream, downstream, and between these consensus sequences are highly variable and contain no other consensus sequences. Thus, in a functional sense, the −10 (Pribnow) and −35 consensus sequences are important because of their nucleotide content, their location relative to one another, and their location relative to the start of transcription. In contrast to the consensus sequences themselves, the nucleotides between −10 and −35 are important as spacers between the consensus elements, but their specific sequences are not critical.

Natural selection has operated to retain strong sequence similarity in consensus regions and to retain the position of the consensus regions relative to the start of transcription. The effectiveness of evolution in maintaining promoter consensus sequences is illustrated by comparison with the sequences between and around −10 and −35, which are not conserved and which exhibit considerable variation. In addition, the spacing between the sequences and their placement relative to the +1 nucleotide is stable. RNA polymerase is a large molecule that binds to −10 and −35 consensus sequences and occupies the space between and immediately around the sites. Crystal structure models show that the enzyme spans enough DNA to allow it to contact promoter consensus regions and reach the +1 nucleotide. Once bound at a promoter in this fashion, RNA polymerase can initiate transcription. **Genetic Analysis 8.1** guides you through the identification of promoter consensus regions.

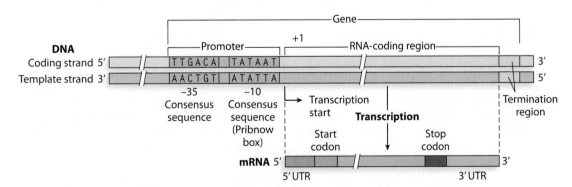

**Figure 8.5  Bacterial promoter structure.** Two promoter consensus sequences, the Pribnow box at −10, and the −35 sequence are essential promoter regulatory elements.

DNA sequences in the promoter region of 10 *E. coli* genes are shown. Sequences at the −35 and −10 sites are boxed.

**a.** Use the sequence information provided to deduce the −35 and −10 consensus sequences.

**b.** Speculate on the relative effects on transcription of a mutation in a promoter consensus region versus a mutation in the sequence between consensus regions.

| Gene | −35 region | | −10 region | +1 |
|------|------------|--|------------|-----|
| A2   | AATGC | TTGACT | CTGTAGCGGGAAGGCG-- | TATTAT | GCACACC-C | CGC |
| bio  | AAAAC | GTGTTT | TTTGTTGTTAATTCGGTG | TAGACT | TGT---AA | ACCT |
| his  | AGTTC | TTGCTT | TCTAACGTGAAAGTGGTT | TAGGTT | AAAAGAC-A | TCA |
| lac  | CAGGC | TTTACA | CTTTATGCTTCCGGCTCG | TATGTT | GTG-TGG-A | ATT |
| lacI | GAATG | GCGCAA | AACTTTTCGCGGTATGG- | CATGAT | AGCGCCC-G | GAA |
| leu  | AAAAG | TTGACA | TCCGTTTTTGTATCCAG- | TAACTC | TAAAAGC-A | TAT |
| recA | AACAC | TTGATA | CTGTATGAGCATACAG-- | TATAAT | TGCTTC--A | ACA |
| trp  | AGCTG | TTGACA | ATTAATCATCGAACTAG- | TTAACT | AGTACGC-A | AGT |
| tRNA | AACAC | TTTACA | GCGGGCCGTCATTTGA-- | TATGAT | GCGCCCC-G | CTT |
| X1   | TCCGC | TTGTCT | TCCTAGGCCGACTCCC-- | TATAAT | GCGCCTCCA | TCG |

## Solution Strategies

## Solution Steps

### Evaluate

1. Identify the topic this problem addresses, and describe the nature of the required answer.

2. Identify the critical information provided in the problem.

1. This question concerns bacterial promoters. The answer requires identification of consensus sequences for −35 and −10 regions of promoters, and speculation about the consequences of promoter mutations.

2. The problem provides promoter sequence information for 10 *E. coli* genes and identifies the segment of each promoter containing the −10 and −35 regions.

### Deduce

3. Examine the −10 and −35 sequences of these promoters, and look for common patterns.

   **TIP:** A consensus sequence identifies the most common nucleotide at each position in a DNA segment.

3. The −10 and −35 sites are the location of RNA polymerase binding during transcription initiation. Count the numbers of A, T, C, and G in each position in the boxed regions.

### Solve

4. Determine the consensus sequence at the −10 and −35 regions.

   **TIP:** Identify the most commonly occurring nucleotide in each position of the 6-nucleotide consensus region of these genes.

**Answer a**

4. At the −10 site, and moving left to right (toward +1), the most common nucleotides in each position in the consensus region, and the number of times they occur in that position, are

$$\text{T A T A A T}$$
$$(9)\ (9)\ (6)\ (5)\ (5)\ (9)$$

At the −35 site, also moving left to right (toward the +1), the most common nucleotides in each position, and the number of times they occur in that position, are

$$\text{T T G A C A}$$
$$(8)\ (9)\ (8)\ (6)\ (6)\ (6)$$

**Answer b**

5. Compare and contrast the likely effects of consensus sequence mutations with those of mutations occurring between consensus regions.

5. Mutation in a consensus sequence is likely to alter the efficiency with which a protein binds to the promoter, and to decrease the amount of gene transcription. In contrast, mutations between consensus sequences are unlikely to alter gene transcription, because the sequences in these intervening regions do not bind tightly to RNA polymerase.

For more practice, see Problems 4, 7, and 18.

## Transcription Initiation

RNA polymerase holoenzyme initiates transcription through a process involving two steps. In the first step, the holoenzyme makes an initial loose attachment to the double-stranded promoter sequence and then binds tightly to it to form the **closed promoter complex** (❶ in **Figure 8.6**). In the second step, the bound holoenzyme unwinds approximately 18 bp of DNA around the −10 consensus sequence to form the **open promoter complex** (❷ in Figure 8.6). Following formation of the open

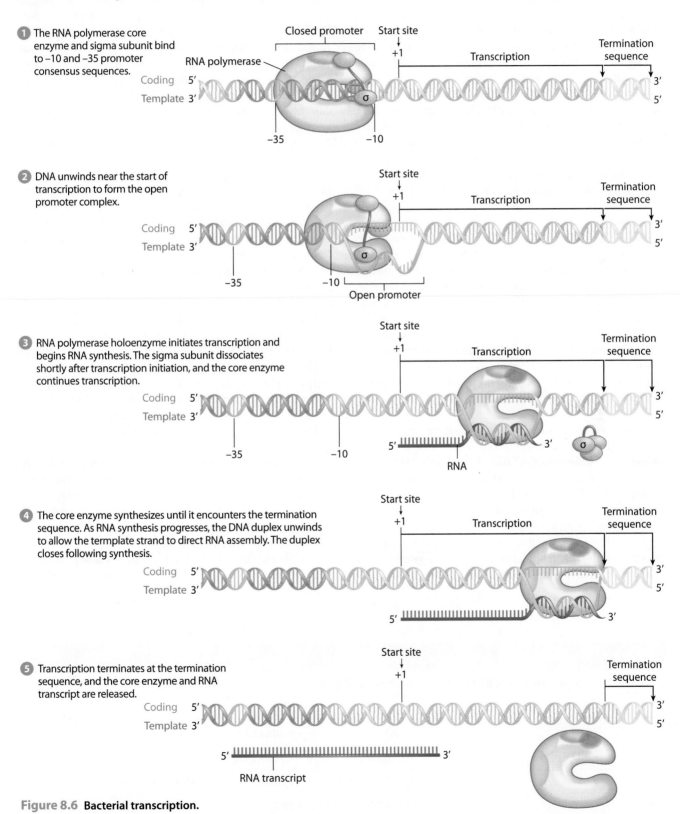

**Figure 8.6  Bacterial transcription.**

promoter complex, the holoenzyme progresses downstream to initiate RNA synthesis at the +1 nucleotide on the template strand of DNA (❸ in Figure 8.6).

Bacterial promoters often differ from the consensus sequence by one or more nucleotides, and some are different at several nucleotides. Since considerable DNA-sequence variation occurs among promoters, it is reasonable to ask how RNA polymerase is able to recognize promoters and reliably initiate RNA synthesis. For an answer, we turn to the sigma subunits that confer promoter recognition and chain-initiation ability on RNA polymerase.

Four alternative sigma subunits identified in *E. coli* are named according to their molecular weight (Table 8.1). Each alternative sigma subunit leads to recognition of a different set of −10 and −35 consensus sequences by the holoenzyme. These different consensus sequence elements are found in promoters of different types of genes; thus, the sigma subunit that it becomes attached to determines the specific gene promoters a holoenzyme will recognize.

The sigma subunit $\sigma^{70}$ is the most common in bacteria. It recognizes promoters of "housekeeping genes," the genes whose protein products are continuously needed by cells. Because of the constant need for their products, housekeeping genes are continuously expressed. Subunits $\sigma^{54}$ and $\sigma^{32}$ recognize promoters of genes involved in nitrogen metabolism and genes expressed in response to environmental stress such as heat shock and are utilized when the action of these genes is required. The fourth sigma subunit, $\sigma^{28}$, recognizes promoters for genes required for bacterial chemotaxis (chemical sensing and motility).

The specificity of each type of sigma subunit for different promoter consensus sequences produces RNA polymerase holoenzymes that have different DNA-binding specificities. Microbial geneticists estimate that each *E. coli* cell contains about 3000 RNA polymerase holoenzymes at any given time, and that each of the four kinds of sigma subunits are represented to differing degrees among them. Because sigma subunits readily attach and detach from core enzymes in response to changes in environmental conditions, the organism is able to change its transcription patterns to adjust to different conditions.

## Transcription Elongation and Termination

Upon reaching the +1 nucleotide, the holoenzyme begins RNA synthesis by using the template strand to direct RNA assembly. The holoenzyme remains intact until the first 8 to 10 RNA nucleotides have been joined. At that point, the sigma subunit dissociates from the core enzyme, which continues its downstream progression (❸ in Figure 8.6). The sigma subunit itself remains intact and can associate with another core enzyme to transcribe another gene.

Downstream progression of the RNA polymerase core is accompanied by DNA unwinding ahead of the enzyme to maintain approximately 18 bp of unwound DNA (❹ in Figure 8.6). As the RNA polymerase passes, progressing at a rate of approximately 40 nucleotides per second, the DNA double helix reforms in its wake. When transcription of the gene is completed, the 5′ end of the RNA trails off the core enzyme (❺ in Figure 8.6).

The end product of transcription is a single-stranded RNA that is complementary and antiparallel to the template DNA strand. The transcript has the same 5′-to-3′ polarity as the coding strand of DNA, the strand complementary to the template strand. The coding strand and the newly formed transcript also have identical nucleotide sequences, except for the presence of uracil in the transcript in place of thymine in the coding strand. For this reason, gene sequences are written in 5′-to-3′ orientation as single-stranded sequences based on the coding strand of DNA. This allows easy identification of the mRNA sequence of a gene by simply substituting U for T.

Gene transcription is not a one-time event, and shortly after one round of transcription is initiated, a second round begins with new RNA polymerase–promoter interaction. Following sigma subunit dissociation and core enzyme synthesis of 50 to 60 RNA nucleotides, a new holoenzyme can bind to the promoter and initiate a new round of transcription while the first core enzyme continues along the gene. In addition, if the transcript under construction is mRNA,

| Table 8.1 | *Escherichia coli* RNA Polymerase Sigma Subunits | | | |
|---|---|---|---|---|
| **Subunit** | **Molecular Weight (Daltons)** | **Consensus Sequence** | | **Function** |
| | | −35 | −10 | |
| $\sigma^{28}$ | 28 | TAAA | GCCGATAA | Flagellar synthesis and chemotaxis |
| $\sigma^{32}$ | 32 | CTTGAA | CCCCATTA | Heat shock genes |
| $\sigma^{54}$ | 54 | CTGGPyАPyPu | TTGCA | Nitrogen metabolism |
| $\sigma^{70}$ | 70 | TTGACA | TATAAT | Housekeeping genes |

the 5′ end is immediately available to begin translation. In contrast, transcripts that are functional RNAs, such as transfer and ribosomal RNA, must await the completion of transcription before undergoing the folding into secondary structures that readies them for cellular action.

---

Genetic Insight *Escherichia coli* RNA polymerase holoenzyme is a five-subunit core enzyme paired with one of four types of sigma subunits that give each polymerase specificity for certain consensus DNA sequences located at positions −10 and −35 relative to the +1 position where transcription starts. The binding of RNA polymerase to double-stranded DNA leads to duplex unwinding, formation of an open promoter complex, and initiation of transcription.

---

## Transcription Termination Mechanisms

Termination of transcription in bacterial cells is signaled by a DNA termination sequence that usually contains a repeating sequence producing distinctive 3′ RNA sequences. Termination sequences are downstream of the stop codon; thus, they are transcribed after the coding region of the mRNA and so are not translated. Two transcription termination mechanisms occur in bacteria. The most common is **intrinsic termination,** a mechanism dependent only on the occurrence of specialized repeat sequences in DNA that induce the formation in RNA of a secondary structure leading to transcription termination. Less frequently, bacterial gene transcription terminates by **rho-dependent termination,** a mechanism characterized by a different terminator sequence and requiring the action of a specialized protein called the **rho protein.**

**Intrinsic Termination** Most bacterial transcription termination occurs exclusively as a consequence of termination sequences encoded in DNA—that is, by intrinsic termination. Intrinsic termination sequences have two features. First, they are encoded by a DNA sequence containing an **inverted repeat,** a DNA sequence repeated in opposite directions but with the same 5′-to-3′ polarity. **Figure 8.7** shows the inverted repeats ("repeat 1" and "repeat 2") in a termination sequence, separated by a short spacer sequence that is not part of either repeat. The second feature of intrinsic termination sequences is a string of adenines on the template DNA strand that begins at the 5′ end of the repeat 2 region. Transcription of inverted repeats produces mRNA with complementary segments that are able to fold into a short double-stranded stem ending with a single-stranded loop. This secondary structure is a **stem-loop structure,** also known as a **hairpin.** A string of uracils complementary to the adenines on the template strand immediately follows the stem-loop structure at the 3′ end of the RNA.

The formation of a stem-loop structure followed immediately by a poly-U sequence near the 3′ end of RNA

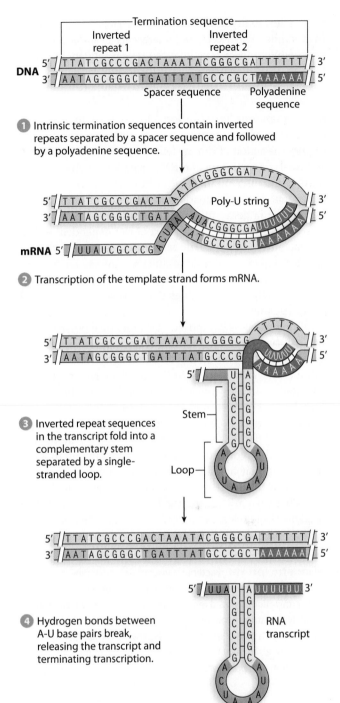

**Figure 8.7 Intrinsic termination of transcription is driven by the presence of inverted repeat DNA sequences.**

causes the RNA polymerase to slow down and destabilize. In addition, the 3′ U-A region of the RNA–DNA duplex contains the least stable of the complementary base pairs. Together, the instability created by RNA polymerase slowing and the U-A base pairs induces RNA polymerase to release the transcript and separate from the DNA. To draw a simple analogy, the behavior of RNA polymerase during intrinsic termination of transcription is like that of a bicycle rider at slow speed. Slow forward momentum creates instability and eventually the rider loses balance. In a similar way, RNA polymerase is destabilized as it slows while transcribing inverted repeat sequences, and it falls off DNA when the transcript is released where A-U base pairs form.

**Rho-Dependent Termination**   In contrast to the more common intrinsic termination, certain bacterial genes require the action of rho protein to bind to nascent mRNA and catalyze separation of mRNA from RNA polymerase to terminate transcription. Genes whose transcription is rho-dependent have termination sequences that are distinct from those in genes utilizing intrinsic termination. Stem-loop structures often form as part of rho-dependent termination, but rho-dependent terminator sequences do not have a string of uracil residues. Instead, the sequences contain a **rho utilization site,** or **rut site,** which is a stretch of approximately 50 nucleotides that is rich in cytosine and poor in guanine.

Rho protein is composed of six identical polypeptides and has two functional domains, both of which are utilized during the two-step process of transcription termination. The first step is initiated when rho protein is activated by an ATP molecule that binds to one functional domain of rho. Activated rho protein utilizes its second domain to bind to the rut site of the RNA transcript. Using ATP-derived energy, rho then moves along the mRNA in the 3′ direction, eventually catching up to RNA polymerase that has slowed near a terminator sequence. As the rho travels, it catalyzes the breakage of hydrogen bonds between mRNA and the DNA template strand. The bond breakage releases the transcript from the RNA polymerase and induces the polymerase to release the DNA.

---

Genetic Insight   Bacterial transcription termination is dependent on the presence of certain DNA sequences. In intrinsic termination, inverted repeat DNA sequences followed immediately by a string of adenines produce an mRNA stem-loop followed by a poly-U string. In rho-dependent termination, a rho protein binds to mRNA and catalyzes the release of the transcript from RNA polymerase.

---

# 8.3 Eukaryotic Transcription Uses Multiple RNA Polymerases

Eukaryotic transcription progresses through the same stages as bacterial transcription, but several factors make the process more complex. First, eukaryotic promoters and consensus sequences are considerably more diverse than in *E. coli,* and eukaryotes have three different RNA polymerases that recognize different promoters, transcribe different genes, and produce different RNAs. Second, the molecular apparatus assembled at promoters to initiate and elongate transcription is more complex in

eukaryotes. Third, eukaryotic genes contain introns and exons, requiring extensive post-transcriptional processing of mRNA. Finally, eukaryotic DNA is permanently associated with a large amount of protein to form a compound known as *chromatin*. The chromatin composition of eukaryotic genomes directly affects transcription. Chromatin plays a central role in regulating eukaryotic transcription by way of changing the strength of protein association with DNA that can permit or block RNA polymerase and transcription factor access to promoters. In later chapters, we discuss chromatin structure and introduce the role chromatin plays in permitting or blocking transcription (Chapter 11) and explore the functional role of chromatin in the regulation of gene expression in eukaryotes (Chapter 15). In this section, however, we do not consider the role of chromatin in transcription; instead, our discussion is limited to the molecular processes of transcription itself.

## Eukaryotic Polymerases

Three different RNA polymerases transcribe distinct classes of RNA coded by eukaryotic genomes: **RNA polymerase I (RNA pol I)** transcribes three ribosomal RNA genes, **RNA polymerase II (RNA pol II)** is responsible for transcribing messenger RNAs that encode polypeptides as well as for transcribing most small nuclear RNA genes, and **RNA polymerase III (RNA pol III)** transcribes all transfer RNA genes as well as one small nuclear RNA gene and one ribosomal RNA gene. RNA pol II and RNA pol III are responsible for miRNA and siRNA synthesis.

Eukaryotic RNA polymerases share similarities of sequence and function with *E. coli* RNA polymerase and with the RNA polymerase of archaea as well. Each of the eukaryotic RNA polymerases contains large subunits that are homologous to the five subunits of the *E. coli* RNA polymerase core enzyme (Table 8.2). The archaeal RNA polymerase also contains five subunits with homology to bacterial RNA polymerase subunits. Eukaryotic and archaeal RNA polymerases are more complex than bacterial RNA polymerase in that they contain 6 to 11 additional protein subunits. Overall, however, all RNA polymerases have a similar shape: they are reminiscent of a "hand" with protein "fingers" to help grasp DNA, and with a "palm" in which polymerization takes place (see Figure 7.21). These similarities of RNA polymerase structure and function are a direct result of the shared evolutionary history of bacteria, archaea, and eukaryotes.

## Consensus Sequences for RNA Polymerase II Transcription

RNA polymerase II transcribes polypeptide-coding genes into mRNA. The promoters for these genes are numerous and highly diverse, with different overall lengths and differences in the number and type of consensus sequences

**Table 8.2     RNA Polymerase Protein Subunits**

| Bacterial Core | Archaeal Core | Eukaryotic Cores | | |
|---|---|---|---|---|
| | | RNA pol I | RNA pol II | RNA pol III |
| β′ | A′/A″ | RPA1 | RPB1 | RPC1 |
| β | B | RPA2 | RPB2 | RPC2 |
| α$^I$ | D | RPC5 | RPB3 | RPC5 |
| α$^{II}$ | L | RPC9 | RPB11 | RPC9 |
| ω | K [+6 other subunits] | RPB6 [+9 other subunits] | RPB6 [+7 other subunits] | RPB6 [+11 other subunits] |

prominent among the sources of promoter variation. Given these characteristics, it is reasonable to ask how RNA polymerases locate promoter DNA for different genes.

Three lines of investigation help researchers to identify and characterize promoters of different polypeptide-coding genes: (1) promoters are identified by determining which DNA sequences are bound by proteins associated with RNA pol II during transcription, (2) putative promoter sequences from different genes are compared to evaluate their similarities, and (3) mutations that alter gene transcription are examined to identify how DNA base-pair changes affect transcription. **Research Technique 8.1**

## Research Technique 8.1

### Band Shift Assay to Identify Promoters

PURPOSE  The functional action of promoters in transcription depends on consensus DNA sequences that bind RNA polymerase and transcription factor proteins. To locate promoters, molecular biologists first scan DNA for potential promoter consensus sequences and then determine that the sequence binds transcriptionally active proteins.

MATERIALS AND PROCEDURES  Fragments of DNA containing suspected promoter consensus sequence are examined by two experimental methods. The first, called *band shift assay,* verifies that the sequence of interest binds proteins. The second, called *DNA footprint protection assay,* identifies the exact location of the protein-binding sequence.

In band shift assay, two identical samples of DNA fragments that contain suspect consensus sequence are analyzed. One DNA sample is a control to which no transcriptional proteins are added. The experimental DNA sample, on the other hand, has transcriptional proteins added. Both the control and the experimental DNA samples are subjected to electrophoresis.

DNA footprint protection also begins with two identical samples of DNA fragments containing suspected consensus sequences. All fragments are end-labeled with $^{32}$P. The experimental DNA is mixed with transcriptional proteins, but the control sample is not. Both samples are exposed to DNase I that randomly cuts DNA that is not protected by protein. The samples are run in separate lanes of an electrophoresis gel, and each end-labeled fragment produced is identified by autoradiography.

DESCRIPTION  In band shift assay result, notice that the electrophoretic mobility of experimental DNA is slower than that of control DNA. This is the anticipated result if the

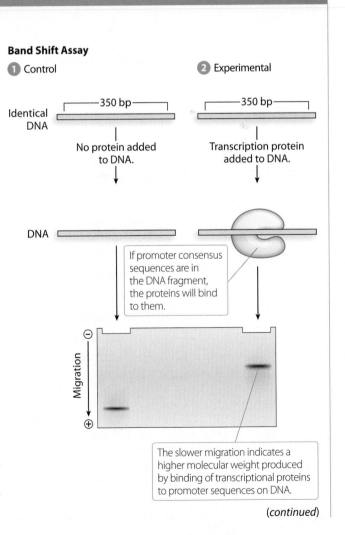

**Band Shift Assay**

① Control          ② Experimental

Identical DNA — 350 bp —          — 350 bp —

No protein added to DNA.          Transcription protein added to DNA.

DNA

If promoter consensus sequences are in the DNA fragment, the proteins will bind to them.

Migration ⊖ / ⊕

The slower migration indicates a higher molecular weight produced by binding of transcriptional proteins to promoter sequences on DNA.

*(continued)*

experimental sample contains consensus sequence that is bound by transcriptional proteins. The bound protein increases the molecular weight of the experimental sample and slows its migration relative to the same DNA without bound protein. In the DNA footprint protection assay, notice that the experimental DNA lane contains a gap in which no DNA fragments appear. The gap represents "footprint protection" for the portion of the fragment that is protected from DNase I digestion by bound transcriptional proteins. No such

protection occurs for the control fragment that is randomly cleaved.

CONCLUSION Evidence from these two methods constitutes necessary but not sufficient evidence that the DNA fragment contains a promoter. The final piece of evidence that a DNA fragment contains a promoter rests on mutational analysis that identifies functional changes caused by mutations of specific nucleotides of promoter consensus sequences (see Figure 8.11).

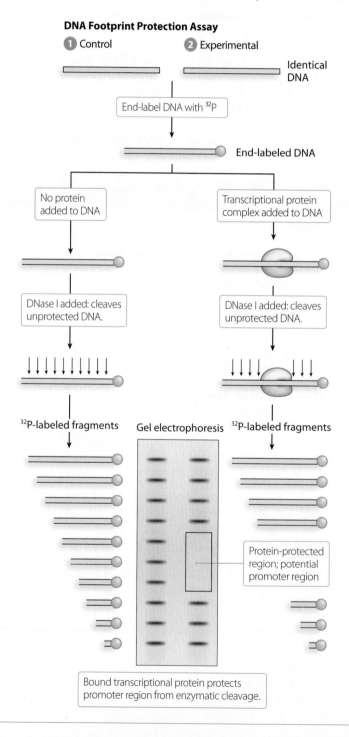

**DNA Footprint Protection Assay**

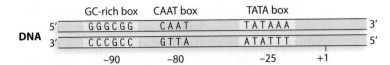

discusses the experimental identification and analysis of promoters.

The most common eukaryotic promoter consensus sequence, the *TATA box*, is shown in **Figure 8.8** as part of a set of three consensus segments that were the first eukaryotic promoter elements to be identified. A **TATA box,** also known as a **Goldberg-Hogness box,** is located approximately at position $-25$ relative to the beginning of the transcriptional start site. Consisting of 6 bp with the consensus sequence TATAAA, it is the most strongly conserved promoter element in eukaryotes. The figure shows two additional consensus sequence elements that are more variable in their frequency in promoters. A 4-bp consensus sequence identified as the **CAAT box** is most commonly located near $-80$ when it is present in the promoter. An upstream GC-rich region called the **GC-rich box,** with a consensus sequence GGGCGG located $-90$ or more upstream of the transcription start, has a frequency that is less than that of CAAT box sequences.

Comparison of eukaryotic promoters reveals a high degree of variability in the type, number, and location of consensus sequence elements (**Figure 8.9**). Some promoters contain all three of the consensus sequences identified above, others contain one or two of these consensus elements, some contain none at all, and many contain other types of consensus sequence elements altogether. For example, the thymidine kinase gene contains TATA, CAAT, and GC-rich boxes along with an octamer (OCT) sequence, called an OCT box. The histone *H2B* gene contains two OCT boxes in addition to a TATA box and a pair of GC-rich boxes. Each of these consensus sequence elements play important roles in the binding of *transcription factors,* a group of transcriptional proteins described below.

## Promoter Recognition

RNA polymerase II recognizes and binds to promoter consensus sequences in eukaryotes with the aid of proteins called **transcription factors (TF).** The TF proteins bind to promoter regulatory sequences and influence transcription initiation by interacting, directly or indirectly, with RNA polymerase. Transcription factors that influence mRNA transcription, and therefore interact with RNA pol II, are given the designation TFII. Individual TFII proteins also carry a letter designation, such as A, B, or C.

In most eukaryotic promoters, the TATA box is the principal binding site for transcription factors during promoter recognition. At the TATA box, a protein called TFIID, a multisubunit protein containing **TATA-binding protein (TBP),** binds the TATA box sequence, and subunits of a protein called **TBP-associated factor (TAF).** The assembled TFIID binds to the TATA box region to form the **initial committed complex** (**Figure 8.10**). Next, TFIIB, TFIIF, and RNA polymerase II join the initial committed complex to form the **minimal initiation complex,** which in turn is joined by TFIIE and TFIIH to form the **complete initiation complex.** The complete initiation complex contains multiple proteins that are commonly identified as "general transcription factors." Once assembled, the complete initiation complex directs RNA polymerase II to the $+1$ nucleotide on the template strand, where it begins the assembly of messenger RNA.

While most of the eukaryotic genes that have been examined have a TATA box and undergo TBP binding, there is evidence that some metazoan genes may use a related factor called TLF (*TBP-like factor*). The complexity of TBP, TLF, and associated proteins is analogous to the different sigma factors in prokaryotic systems, thus allowing differential recognition of promoters in eukaryotes.

## Detecting Promoter Consensus Elements

The diversity of eukaryotic promoters begs an important question: How do researchers verify that a segment of DNA is a functionally important component of a promoter? The research has two components; the first, outlined in Research Technique 8.1, is discovering the presence and location of DNA sequences that transcription factor proteins will bind to. The second component

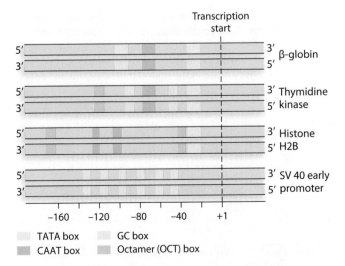

**Figure 8.9  Examples of eukaryotic promoter variability.**

**1** TAF and TBP form TFIID and bind the TATA box.

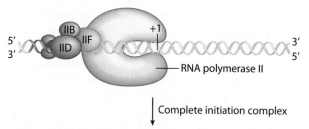

**2** The addition of TFIIBB, RNA polymerase II, and TFIIF forms the minimal initiation complex.

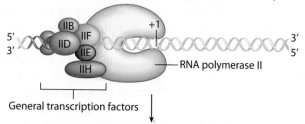

**3** TFIIE and TFIIH join to form the complete initiation complex.

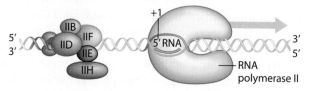

General transcription factors

**4** RNA polymerase II is released from the complete initiation complex to begin transcription.

**Figure 8.10  Eukaryotic transcription by RNA polymerase II.**

involves mutational analysis to confirm the functionality of the sequence. Researchers produce many different point mutations in the DNA sequence under study and then compare the level of transcription generated by each mutant promoter sequence with transcription generated by the wild-type sequence.

Figure 8.11 shows a synopsis of promoter mutation analysis from an experiment performed by the molecular biologist Richard Myers and colleagues on a mammalian β-globin gene promoter. These researchers produced mutations of individual base pairs in TATA box, CAAT box, and GC-rich sequences, and of nucleotides between the consensus sequences, to identify the effect of each individual mutation on the relative transcription level of the gene. They found that base-pair mutations in each of the three consensus regions significantly decreased the transcription level of the gene. In contrast, mutations outside the consensus regions had nonsignificant effects on transcription level. Such results show the functional importance of specific DNA sequences in promoting transcription and confirm a functional role in transcription for TATA box, CAAT box, and GC-rich sequences.

---

**Genetic Insight** Eukaryotic RNA polymerase II transcribes protein-coding genes into messenger RNA. Transcription factors aggregate at the TATA box to prepare a binding site for RNA polymerase II. Upstream promoter elements play a crucial role in elevating the level of transcription.

---

## Enhancers and Silencers

Promoters alone are often not sufficient to initiate transcription of eukaryotic genes, and other regulatory sequences are needed to drive transcription. This is particularly the case for multicellular eukaryotes that have different numbers and patterns of expressed genes in different cells and tissues, and that change their patterns

**Figure 8.11  Mutation analysis of the β-globin gene promoter.** Functional consensus sequences are identified by detecting mutations in three regions containing TATA box, CAAT box, and upstream GC-rich box sequences (shaded in the figure) that substantially reduce the relative transcription level.

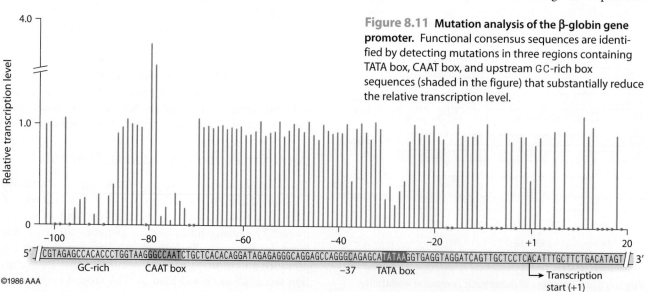

of gene expression as the organisms grow and develop. These *tissue-specific* or *developmental* types of transcriptional regulation are fully discussed in later chapters (Chapters 15 and 20), but here we highlight two categories of DNA transcription-regulating sequences that lead to differential expression of genes.

**Enhancer sequences** are one important group of DNA regulatory sequences that increase the level of transcription of specific genes. Enhancer sequences bind specific proteins that interact with the proteins bound at gene promoters, and together promoters and enhancers drive transcription of certain genes. In many situations, enhancers are located upstream of the genes they regulate; but enhancers can be located downstream as well. Some enhancers are relatively close to the genes they regulate, but others are thousands to tens of thousands of base pairs away from their target genes. Thus, important questions for molecular biologists are what proteins are bound to enhancers, and how enhancer sequences regulate transcription of the gene given their different distances from the start of transcription?

The answers are that enhancers bind activator proteins and associated coactivator proteins to form a protein "bridge" that bends the DNA and links the complete initiation complex at the promoter to the activator–coactivator complex at the enhancer (**Figure 8.12**). The bend produced in the DNA may contain dozens to thousands of base pairs. The action of enhancers and the proteins they bind dramatically increases the efficiency of RNA pol II in initiating transcription, and as a result increases the level of transcription of genes regulated by enhancers.

On the other side of the transcription-regulating spectrum are **silencer sequences,** DNA elements that can act at a great distance to repress transcription of their target genes. Silencers bind transcription factors called repressor proteins, inducing bends in DNA that are similar to what is seen when activators and coactivators bind to enhancers—except with the consequence of reducing the transcription of targeted genes. Like enhancers, silencers can be located upstream or downstream of a target gene and can reside up to several thousand base pairs away from it. Thus enhancers and silencers may operate by similar general mechanisms but with opposite effects on transcription.

## Transcription Factors and Signal Transduction

Two additional factors affect eukaryotic tissue-specific and developmental gene transcription: the presence of specific transcription factor proteins, and *signal-transduction pathways* that communicate the need for specific regulatory molecules such as transcription factors. If a particular transcription factor necessary for transcription of a gene is available, the gene can be transcribed, assuming that the other protein components of the transcriptional apparatus are present. In contrast, if the specific transcription factor is missing, gene transcription will not take place even if all the other required components of the transcription apparatus are assembled. Distinct transcription factor proteins thus play a pivotal role in regulating eukaryotic gene transcription.

Transcription factor availability is in turn dependent on the synthesis of the transcription factors themselves and on their availability in the cell. Transcription factor synthesis can be tightly regulated, and different types of cells may contain distinct arrays of transcription factors. This leads to different patterns of gene expression in cells. For example, long muscle cells have a characteristic pattern of gene transcription that produces specific cellular structure and function. In contrast, the neurons that excite long muscle cells have a distinct structure and function, and they transcribe a different pattern of genes. The presence of different transcription factors in these cells partially accounts for their different patterns of gene transcription.

In addition to regulated synthesis of transcription factors, the availability of activated transcription factors is controlled through **signal-transduction pathways,** sequential events that release a regulatory molecule inside cells in response to signals received outside cells. Signal-transduction pathways utilize transmembrane proteins that have an interaction domain outside the cell, a binding domain inside the cell, and a transmembrane region that traverses the cell membrane (**Figure 8.13**). The interaction domain is the receptor that receives a stimulating message. Stimuli can be molecules such as hormones or growth factors, or they can be physical agents such as light or temperature. Stimulation of the interaction domain signals that transcription of a particular gene or genes is needed. On stimulation, the binding domain that holds a transcription factor or other regulatory molecule releases the molecule

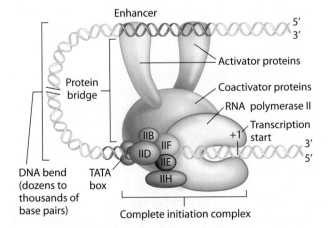

**Figure 8.12 Enhancers activate transcription in cooperation with promoters.** A protein bridge composed of transcriptional proteins forms between enhancer and promoter sequences that may be separated by thousands of nucleotides.

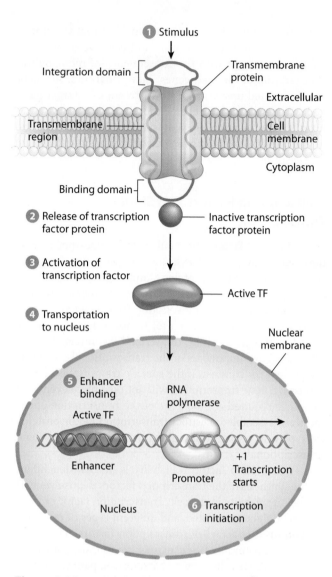

**Figure 8.13 Signal transduction.** Stimulation ❶ received by a transmembrane receptor protein induces the release of a transcription factor ❷ that is activated ❸ and transported into the nucleus ❹, where it binds an enhancer sequence ❺ and participates in the initiation of transcription of a targeted gene ❻.

into the cytoplasm of the cell. Once activated, the transcription factor is transported to the nucleus, where it binds to a regulatory DNA sequence such as an enhancer to help initiate transcription of the targeted gene. Transcription factors and signal-transduction pathways are quite diverse; we discuss more about their roles in transcription in later chapters (Chapters 15 and 20).

---

**Genetic Insight** Eukaryotic gene transcriptional control often requires the binding of regulatory proteins to DNA sequences such as enhancers and silencers. Transcription factor proteins are essential to transcription, and their synthesis differs among cell types. Signal-transduction pathways are an important avenue for release of specific transcription factors.

---

## RNA Polymerase I Promoters

The genes for rRNA are transcribed by RNA polymerase I, utilizing a transcription initiation mechanism similar to that used by RNA pol II. RNA polymerase I is the most specialized eukaryotic RNA polymerase as it transcribes a limited number of genes. It is recruited to upstream promoter elements following the initial binding of transcription factors, and it transcribes ribosomal RNA genes found in the **nucleolus,** a nuclear organelle containing rRNA and multiple copies of the genes encoding rRNAs. Nucleoli (the plural of *nucleolus*) contain many tandem (meaning "end to end") repeat copies of rRNA genes. In *Arabidopsis*, for example, each nucleolus contains about 700 copies of rRNA genes. Nucleoli play a key role in the manufacture of ribosomes. At nucleoli, transcribed ribosomal RNA genes are packaged with proteins to form the large and small ribosomal subunits.

Promoters recognized by RNA pol I contain two similar functional sequences near the start of transcription. The first is the **core element,** stretching from $-45$ to $+20$ and bridging the start of transcription, and the second is the **upstream control element,** spanning nucleotides $-100$ to $-150$ (**Figure 8.14**). The core element is essential for transcription initiation, and the upstream control element increases the level of gene transcription. Both of these elements are rich in guanine and cytosine; DNA sequence comparisons show that all upstream control elements have the same base pairs at approximately 85 percent of nucleotide positions, and the same is true of all core elements. Two upstream binding factor 1 (UBF1) proteins bind the upstream control element. A second protein complex, known as sigma-like factor 1 (SL1) protein, binds the core element. This complex recruits RNA pol I to the core element, to initiate transcription of rRNA genes.

## RNA Polymerase III Promoters

The remaining eukaryotic RNA polymerase, RNA polymerase III, is primarily responsible for transcription of tRNA genes. However, it also transcribes one rRNA and other RNA-encoding genes. Each of these genes has a promoter structure that differs significantly from the structure of promoters recognized by RNA pol I or RNA pol II. Small nuclear RNA genes have three upstream elements, whereas the genes for 5S ribosomal RNA and transfer RNA each contain two **internal promoter elements** that are *downstream* of the start of transcription.

The upstream elements of small nuclear RNA genes are a TATA box, a **promoter-specific element (PSE),** and an octamer (OCT) (**Figure 8.15a**). A small number of transcription factors—TFIIIs, in this case—bind to these elements and recruit RNA polymerase III, which initiates transcription in a manner similar to that of the other polymerases.

**①** The core element initiates transcription, and the upstream control element increases transcription efficiency.

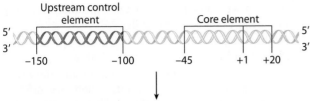

**②** UBF1 and SL1 bind to upstream control and core elements.

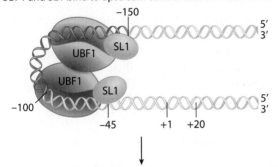

**③** RNA pol I is recruited to the core element to initiate transcription.

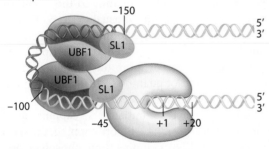

**Figure 8.14** Promoter consensus sequences for transcription initiation by RNA polymerase I.

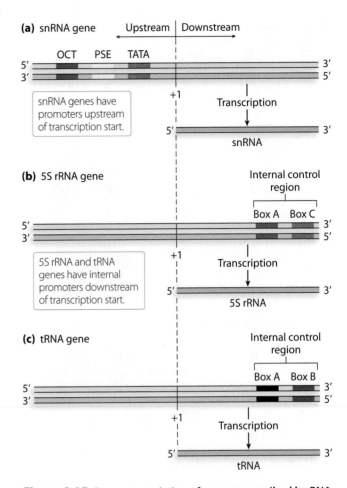

**Figure 8.15** Promoter variation of genes transcribed by RNA polymerase III.

The genes for 5S ribosomal RNA and transfer RNA have internal promoter elements called **internal control regions (ICRs)**; see **Figure 8.15b and c**. The ICRs are two short DNA sequences—designated box A and box B in some genes and box A and box C in other genes—located downstream of the start of transcription, between nucleotides +55 and +80 (**Figure 8.16**). To initiate transcription, box B or box C is bound by TFIIIA, which facilitates the subsequent binding of TFIIIC to box A. TFIIIB then binds to the other transcription factors. In the final initiation step, RNA polymerase III binds to the transcription factor complex and overlaps the +1 nucleotide. With RNA polymerase correctly positioned, transcription begins approximately 55 bp upstream of the beginning of box A, at the +1 nucleotide.

## Termination in RNA Polymerase I or III Transcription

Each of the eukaryotic RNA polymerases utilizes a different mechanism to terminate transcription. Here we

briefly describe termination in transcription by RNA pol I or RNA pol III, leaving termination of RNA pol II transcription for more extensive discussion in a later section. Transcription by RNA polymerase III is terminated in a manner reminiscent of *E. coli* transcription termination. The RNA pol III transcribes a terminator sequence that creates a string of uracils in the transcript. The poly-U string is similar to the string that occurs in bacterial intrinsic termination (see page 269). The RNA pol III terminator sequence does not contain an inverted repeat, however, so no stem-loop structure forms near the 3' end of RNA.

Transcription by RNA pol I is terminated at a 17-bp consensus sequence that binds **transcription-terminating factor I (TTFI)**. The binding site for TTFI is the DNA consensus sequence

$$\text{AGGTCGACCAG}^A/_T{}^A/_T\text{NTCG}$$

In this sequence, adenine and thymine are equally likely to appear at two adjacent sites, as indicated by the diagonal lines; N signifies a location at which all four nucleotides are more or less equally frequent. A large rRNA precursor transcript is cleaved about 18 nucleotides upstream of the

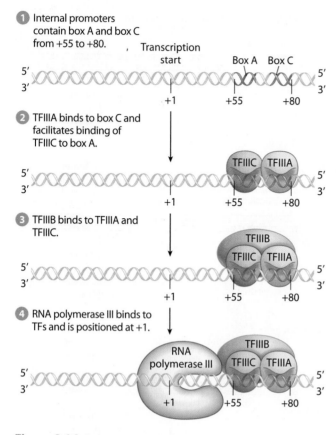

**1** Internal promoters contain box A and box C from +55 to +80.

**2** TFIIIA binds to box C and facilitates binding of TFIIIC to box A.

**3** TFIIIB binds to TFIIIA and TFIIIC.

**4** RNA polymerase III binds to TFs and is positioned at +1.

**Figure 8.16  Promoter internal control regions for transcription by RNA polymerase III.**

TTFI binding site, so the consensus sequence does not appear in the mature transcript.

---

**Genetic Insight**  RNA pol I and RNA pol III initiate transcription from promoters that differ from one another and from those utilized by RNA pol II. The mechanisms of transcriptional termination also differ for each type of RNA polymerase.

---

## 8.4 Post-Transcriptional Processing Modifies RNA Molecules

Bacterial and eukaryotic transcripts differ in several ways. For example, eukaryotic transcripts are more stable than bacterial transcripts. The half-life of a typical eukaryotic mRNA is measured in hours to days, whereas bacterial mRNAs have an average half-life measured in seconds to minutes. A second difference is the separation, in time and in location, between transcription and translation. Recall that in bacteria the lack of a nucleus leads to coupling of transcription and translation. In eukaryotic cells, on the other hand, transcription takes place in the nucleus, and translation occurs later at free ribosomes or at those attached to the rough endoplasmic reticulum in the cytoplasm. A third difference is the

presence of introns in eukaryotic genes that are absent from bacterial genes. Each of these differences comes into play as we consider post-transcriptional modifications of mRNA in eukaryotic cells.

In this section, we discuss post-transcriptional processing of multiple types of RNA, but we emphasize three sequential processing steps that modify the initial eukaryotic gene mRNA transcript, called **pre-mRNA,** into **mature mRNA,** the fully processed mRNA that migrates out of the nucleus to the cytoplasm for translation. These modification steps are (1) **5′ capping,** the addition of a modified nucleotide to the 5′ end of mRNA; (2) **3′ polyadenylation,** cleavage at the 3′ end of mRNA and addition of a tail of multiple adenines to form the **poly-A tail**; and (3) **intron splicing,** RNA splicing to remove introns and ligate exons.

### Capping 5′ mRNA

After RNA pol II has synthesized the first 20 to 30 nucleotides of the mRNA transcript, a specialized enzyme, guanylyl transferase, adds a guanine to the 5′ end of the pre-mRNA, producing an unusual 5′-to-5′ bond that forms a triphosphate linkage. Additional enzymatic action then methylates the newly added guanine and may also methylate the next one or more nucleotides of the transcript. This addition of guanine to the transcript and the subsequent methylation is known as 5′ capping.

Guanylyl transferase initiates 5′ capping in three steps depicted in **Figure 8.17**. Before capping, the terminal 5′ nucleotide of mRNA contains three phosphate groups, labeled $\alpha$, $\beta$, and $\gamma$ in Figure 8.17. Guanylyl transferase first removes the $\gamma$ phosphate, leaving two phosphates on the 5′ terminal nucleotide **1**. The guanine triphosphate containing the guanine that is to be added loses two phosphates ($\gamma$ and $\beta$) to form a guanine monophosphate **2**. Then, guanylyl transferase joins the guanine monophosphate to the mRNA terminal nucleotide to form the 5′-to-5′ triphosphate linkage **3**. Methyl transferase enzyme then adds a methyl ($CH_3$) group to the 7-nitrogen of the new guanine, forming 7-methylguanosine ($m^7G$). Methyl transferase may also add methyl groups to 2′−OH nearby nucleotides of mRNA.

The 5′ cap has several functions, including (1) protecting mRNA from rapid degradation, (2) facilitating mRNA transport across the nuclear membrane, (3) facilitating subsequent intron splicing, and (4) enhancing translation efficiency by orienting the ribosome on mRNA.

### Polyadenylation of 3′ Pre-mRNA

Termination of transcription by RNA pol II is not fully understood. What is known is that the 3′ end of mRNA is not generated by transcriptional termination. The 3′ end of the pre-mRNA is created by enzymatic action that removes a segment from the 3′ end of the transcript and

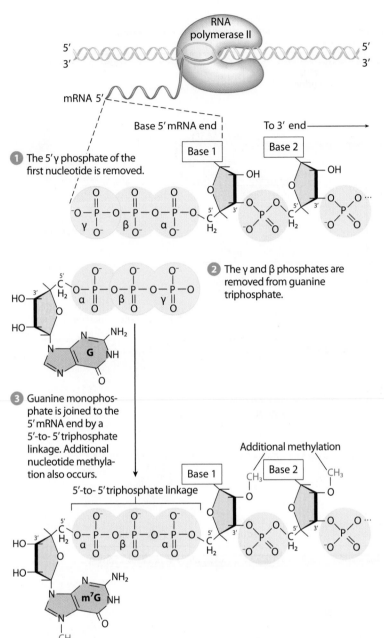

**Figure 8.17** **Capping the 5′ end of eukaryotic pre-mRNA.**

**1** The 5′ γ phosphate of the first nucleotide is removed.

**2** The γ and β phosphates are removed from guanine triphosphate.

**3** Guanine monophosphate is joined to the 5′ mRNA end by a 5′-to- 5′ triphosphate linkage. Additional nucleotide methylation also occurs.

replaces it with a string of adenine nucleotides, the poly-A tail. This step of pre-mRNA processing is thought to be associated with subsequent termination of transcription.

**Figure 8.18** illustrates these steps. Polyadenylation begins with the binding of a factor called cleavage and polyadenylation specificity factor (CPSF) near a six-nucleotide mRNA sequence, AAUAAA, that is downstream of the stop codon and thus not part of the coding sequence of the gene. This six-nucleotide sequence is known as the **polyadenylation signal sequence.** The binding of cleavage-stimulating factor (CstF) to a uracil-rich sequence several dozen nucleotides downstream of the polyadenylation signal sequence quickly follows, and the binding of two other cleavage factors, CFI and CFII, and polyadenylate polymerase (PAP) enlarges the complex **1**. The pre-mRNA is then cleaved 15 to 30 nucleotides downstream of the

polyadenylation signal sequence **2**. The cleavage releases a transcript fragment bound by CFI, CFII, and CstF, which is later degraded **3**. The 3′ end of the cut pre-mRNA then undergoes the enzymatic addition of 20 to 200 adenine nucleotides that form the 3′ poly-A tail through the action of CPSF and PAP **4**. After addition of the first 10 adenines, molecules of poly-A-binding protein II (PABII) join the elongating poly-A tail and increase the rate of adenine addition **5**. The 3′ poly-A tail has several functions, including (1) facilitating transport of mature mRNA across the nuclear membrane, (2) protecting mRNA from degradation, and (3) enhancing translation by enabling ribosomal recognition of messenger RNA.

Certain eukaryotic mRNA transcripts do not undergo polyadenylation. The most prominent of these are transcripts of genes producing *histone proteins*, which are

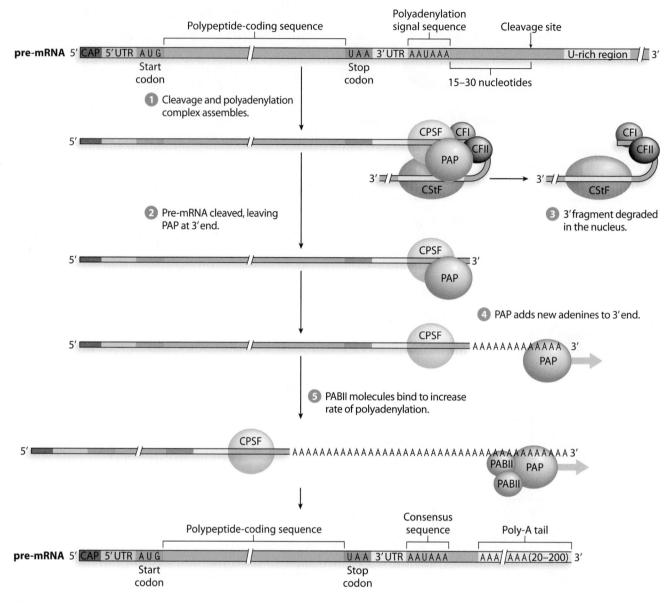

**Figure 8.18** Polyadenylation of the 3′ end of eukaryotic pre-mRNA.

key components of *chromatin,* the DNA–protein complex that makes up eukaryotic chromosomes (see Chapter 11). On these and other "tailless" mRNAs, the 3′ end contains a short stem-loop structure reminiscent of the ones seen in the intrinsic transcription termination mechanism of bacteria. There may be an evolutionary connection between bacterial transcription termination and stem-loop formation on "tailless" eukaryotic mRNAs.

## Pre-mRNA Intron Splicing

The third step of pre-mRNA processing is intron splicing, which consists of removing intron segments from pre-mRNA and ligating the exons. Intron splicing requires

exquisite precision to remove all intron nucleotides accurately without intruding on the exons, and without leaving behind additional nucleotides, so that the mRNA sequence encoded by the ligated exons will completely and faithfully direct synthesis of the correct polypeptide. As an example of the need for precision in intron removal, consider the following "precursor string" made up of exon-like blocks of letters forming three-letter words interrupted by unintelligible intron-like blocks of letters. If editing removes the "introns" accurately, the "edited string" can be divided into its three-letter words to form a "sentence." If an error in editing were to remove too many or too few letters, a nonsense sentence would result.

| | intron | intron |
| --- | --- | --- |
| Precursor string: | youmaynox p g h r c y e o m t p wtipthepp f x w u b i j r d l z m c o l z otandsipthetea | |
| Edited string: | youmaynowtipthepotandsipthetea | |
| Sentence: | you may now tip the pot and sip the tea | |

The finding that introns interrupt the genetically informative segments of eukaryotic genes was a stunning discovery reported independently by the molecular biologists Richard Roberts and Phillip Sharp in 1977. Nothing known about eukaryotic gene structure at the time suggested that most eukaryotic genes are subdivided into intron and exon elements. Roberts and Sharp shared the 1993 Nobel Prize in Physiology or Medicine for their codiscovery of "split genes" in the eukaryotic genome.

Sharp's research group discovered the split nature of eukaryotic genes by using a technique known as R-looping. In this method, DNA encoding a gene is isolated, denatured to single-stranded form, and then mixed with the mature mRNA transcript from the gene. Regions of the gene that encode sequences in mature mRNA will be complementary to those sequences in the mRNA and will hybridize with them to form a DNA–mRNA duplex. However, DNA segments encoding introns will not find complementary sequences in mature mRNA and will remain single-stranded, looping out from between the hybridized sequences.

**Figure 8.19** shows a map of the *hexon* gene studied in R-looping experiments by Sharp and colleagues. The experimental results, photographed by electron microscopy, reveal four DNA–mRNA hybrid regions where exon DNA sequence pairs with mature mRNA sequence. Three single-stranded R-loop sequences are introns which do not pair with mRNA.

## Splicing Signal Sequences

Eukaryotic pre-mRNA contains specific short sequences that define the 5′ and 3′ junctions between introns and their neighboring exons. In addition, there is a consensus sequence near each intron end to assist in its accurate identification. The **5′ splice site** is located at the 5′ intron end, where it abuts an exon (**Figure 8.20**). This site contains a consensus sequence with a nearly invariant GU dinucleotide forming the 5′-most end of the intron. The consensus sequence includes the last three nucleotides of the adjoining exon, as well as the four or five nucleotides that follow the GU in the intron. At the **3′ splice site** on the opposite end of the intron, a consensus sequence of 11 nucleotides contains a pyrimidine-rich region and a nearly invariant AG dinucleotide at the 3′-most end of the intron. The third consensus sequence, called the branch site, is located 20 to 40 nucleotides upstream of the 3′ splice site. This consensus sequence is pyrimidine-rich

and contains an invariant adenine, called the **branch point adenine,** near the 3′ end.

Mutation analysis shows that these consensus sequences are critical for accurate intron removal. Mutations altering nucleotides in any of the three consensus regions can produce abnormally spliced mature mRNA. The abnormal mRNAs—too short if exon sequence is mistakenly removed, too long if intron sequence is left behind, or altered in other ways that result in improper reading of mRNA sequence—produce proteins with incorrect sequences of amino acids (see Chapter 12).

Introns are removed from pre-mRNA by an snRNA–protein complex called the **spliceosome.** The spliceosome is something like a molecular workbench to which pre-mRNA is attached while spliceosome subunit components cut and splice it in a four-step process that, first, cleaves the 5′ splice site; second, forms a **lariat intron structure** that binds the 5′ intron end to the

**(a)**

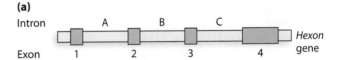

**(b)**

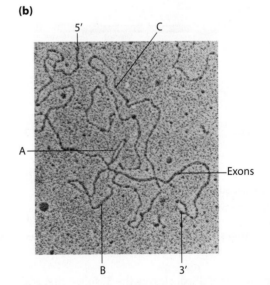

**Figure 8.19  R-loop experimental analysis. (a)** The *hexon* gene contains four exons (1 to 4) and three introns (A to C). **(b)** Electron micrographs show hybridization of mature mRNA with exon sequences of denatured *hexon* DNA. Intron sequences are not hybridized and remain single stranded.

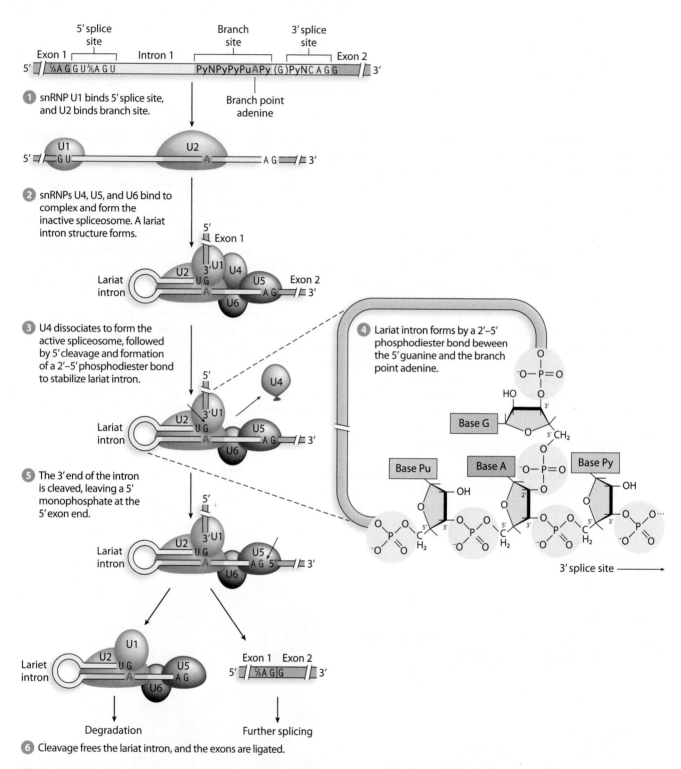

**Figure 8.20  Intron splicing in eukaryotic pre-mRNA.**  Spliceosome assembly and intron removal.

branch point adenine; third, cleaves the 3′ splice site; and finally, ligates exons and releases the lariat intron to be degraded to its nucleotide components. An electron micrograph of spliceosomes in action is seen in the opener photo for this chapter.

Figure 8.20 illustrates the steps of nuclear pre-mRNA splicing, beginning with the aggregation of five small

nuclear ribonucleoproteins (snRNPs; pronounced "snurps") to form a spliceosome. The snRNPs are snRNA–protein subunits designated U1 to U6. The spliceosome is a large complex made up of multiple snRNPs, but its composition is dynamic; it changes throughout the different stages of splicing when individual snRNPs come and go as particular reaction steps are carried out.

Spliceosome components are recruited to 3′ and 5′ splice sites by specialized proteins called **SR proteins.** These proteins are rich in serine (single-letter abbreviation S) and arginine (single-letter abbreviation R), and they bind to sequences in exons known as **exonic splicing enhancers (ESEs)** (Figure 8.21). The binding of SR protein to ESEs helps ensure that spliceosome components bind efficiently to authentic 3′ and 5′ splice sites that are nearby, rather than binding to incorrect splice sites away from exons. The SR proteins are essential to splicing and are produced in a variety of forms, some being present in all cells and others only in certain cells. These proteins may play a role in *alternative intron splicing* that we discuss in a later section.

## Coupling of Pre-mRNA Processing Steps

Each intron–exon junction is subjected to the same spliceosome reactions, raising the question of whether there is a particular order in which introns are removed from pre-mRNA—or whether U1 and U2 search more or less randomly for 5′ splice-site and branch-site consensus sequences, inducing spliceosome formation when they happen to encounter an intron. The answer is that introns appear to be removed one by one, but not necessarily in order along the pre-mRNA. For example, a study of intron splicing of the mammalian *ovomucoid* gene demonstrates the successive steps of intron removal. The *ovomucoid* gene contains eight exons and seven introns. The pre-mRNA transcript is approximately 5.6 kb, and the mature mRNA is reduced to 1.1 kb. Northern blot analysis of *ovomucoid* pre-mRNAs at various stages of intron removal illustrates that each intron is removed separately, rather than all introns being removed at once. The order of intron removal does not precisely match their 5′-to-3′ order in pre-mRNA.

The three steps of pre-mRNA processing are tightly coupled. In comprehensive models developed over the last decade or so, the carboxyl terminal domain (CTD) of RNA polymerase II plays an important role in this coupling by functioning as an assembly platform and regulator of pre-mRNA processing machinery. The CTD is located at the site of emergence of mRNA from the polymerase and contains multiple heptad (seven-member) repeats of amino acids that can be phosphorylated.

Binding of processing proteins to the CTD allows the mRNA to be modified as it is transcribed.

Current models propose that "gene expression machines" consisting of RNA polymerase II and an array of pre-mRNA-processing proteins are responsible for the coupling of transcription and pre-mRNA processing. Foundation Figure 8.22 illustrates this gene expression machine model. The CTD of RNA polymerase II associates with multiple proteins that carry out capping (CAP), intron splicing (SF), and polyadenylation (pA) so that the processes of transcription and pre-mRNA processing occur simultaneously. At the initiation of transcription, phosphorylation (P) along the CTD assists the binding of 5′-capping enzymes, which carry out their capping function and then dissociate. During transcription elongation, specific transcription elongation factors bind the CTD and facilitate splicing-factor binding. This coupling of transcription and splicing helps explain the efficiency with which introns are identified.

---

Genetic Insight  Eukaryotic pre-mRNA processing takes place in the nucleus and is necessary for producing mature mRNA for translation. Specialized enzymes add a 5′ cap, produce a 3′ poly-A tail, and remove introns from pre-mRNA. These processes are controlled by proteins and are closely coupled owing to the formation of large protein complexes that carry out the processing of pre-mRNA.

---

## Alternative Transcripts of Single Genes

Before the complete sequencing of the human genome in the early 2000s, estimates of the number of human genes varied, having been as high as 80,000 to 100,000 genes 20 years or so earlier. A principal reason for this prediction was that human cells produce well over 100,000 distinct polypeptides. It came as something of a surprise, then, when gene annotation of the human genome revealed just over 22,000 genes. The difference between the number of genes and the number of polypeptides is mirrored by similar findings in other eukaryotic genomes, especially those of mammals. It is common for large eukaryotic genomes to express more proteins than there are genes in the genomes. Three transcription-associated mechanisms can account for the ability of single DNA sequences to produce more than one polypeptide: (1) pre-mRNA can be spliced in alternative patterns in different types of cells; (2) alternative promoters can initiate transcription at distinct +1 start points in different cell types; and (3) alternative locations of polyadenylation can produce different mature mRNAs. Collectively, these varied processes are identified as **alternative pre-mRNA processing.**

**Alternative intron splicing** is the mechanism by which post-transcriptional processing of identical pre-mRNAs in different cells can lead to mature mRNAs with different combinations of exons. These alternative mature

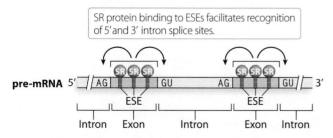

**Figure 8.21  SR-protein recruitment of spliceosome components to 5′ and 3′ splice sites.**

## The Gene Expression Machine Model for Coupling Transcription with pre-mRNA Processing

**1** A preinitiation complex (PIC) with affiliated capping (CAP), polyadenylation (pA), splicing factor (SF), and transcription factor proteins (TF) bind the carboxyl terminal domain (CTD) of RNA pol II.

**2** RNA pol II initiates transcription. PIC dissociates, leaving the pre-mRNA processing proteins on the CTD. CAP proteins carry out 5′ capping.

**3** Capping proteins dissociate and pre-mRNA elongates.

**4** Spliceosome complexes affiliate with pre-mRNA with the aid of SF proteins. Intron splicing takes place as RNA pol continues elongation of mRNA.

**5** Polyadenylation proteins identify the pA signal sequence and carry out polyadenylation. Transcription terminates. Splicing continues to completion.

**6** Fully processed mature mRNA dissociates from RNA pol II and is transported to cytoplasm for translation.

mRNAs produce different polypeptides. In other words, alternative splicing is a mechanism by which a single DNA sequence can produce more than one protein. Alternative splicing is common in mammals—approximately 70 percent of human genes are thought to undergo alternative splicing—but it is less common in other animals, and it is rare in plants.

The products of the human *calcitonin/calcitonin gene-related peptide* (*CT/CGRP*) gene exemplify the process of alternative splicing (**Figure 8.23a**). The *CT/CGRP* gene produces the same pre-mRNA transcript in many cells, including thyroid cells and neuronal cells.

The transcript contains six exons and five introns and includes two alternative polyadenylation sites, one in exon 4 and the other following exon 6. In thyroid cells, *CT/CGRP* pre-mRNA is spliced to form mature mRNA containing exons 1 through 4, using the first poly-A site for polyadenylation. Translation produces calcitonin, a hormone that helps regulate calcium. In neuronal cells, the same pre-mRNA is spliced to form mature mRNA containing exons 1, 2, 3, 5, and 6. Polyadenylation takes place at the site that follows exon 6, since exon 4 is spliced out as though it were an intron. Translation in neuronal cells produces the hormone CGRP.

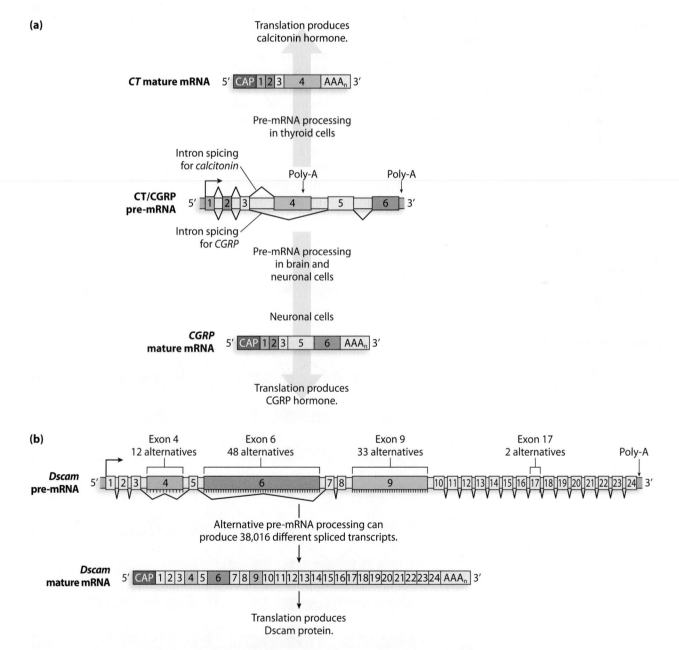

**Figure 8.23  Alternative splicing.  (a)** The calcitonin/calcitonin gene-related protein (*CT/CGRP*) gene is transcribed into either calcitonin or CGRP. **(b)** *Dscam* pre-mRNA contains numerous alternatives for exons 4, 6, 9, and 17. Combinatorial splicing could generate as many as 38,016 different mature mRNAs.

One of the most complex patterns of alternative splicing occurs in the *Drosophila Dscam* gene, which produces a protein directing axon growth in *Drosophila* larvae. Mature mRNA from *Dscam* contains 24 exons, but as shown in **Figure 8.23b**, numerous alternative sequences can be used as exons 4, 6, 9, and 17. In total, more than 38,000 different alternative splicing arrangements of *Dscam* are possible, although not all are observed in the organism.

Alternative splicing is largely controlled by variation of SR proteins in different types of cells. Cell-type-specific SR proteins can direct patterns of intron splicing that are unique to certain cells by directing the assembly of spliceosome components at the 3′ and 5′ splice sites required for production of a certain mature mRNA.

The use of **alternative promoters** occurs when more than one sequence upstream of a gene can bind transcription factors and initiate transcription. Similarly, **alternative polyadenylation** is possible when genes contain more than one polyadenylation signal sequence that can activate 3′ pre-mRNA cleavage and polyadenylation. Alternative promoters and alternative polyadenylation are driven by the variable expression of transcriptional or polyadenylation proteins in a cell-type-specific manner. Like the variable expression of SR proteins that leads to alternative splicing, the variable

expression of transcriptional and polyadenylation proteins generates characteristic mature mRNAs from specific genes in particular cells. The result is that transcription of a single gene may lead to the production of several different mature mRNAs in different types of cells, and to their translation into distinct proteins in each of those cell types.

A comprehensive example of a single gene for which all three alternative mechanisms operate to produce distinct polypeptides in different cells is that of the rat α-tropomyosin (α-*Tm*) gene that produces nine different mature mRNAs and, correspondingly, nine different tropomyosin proteins from a single gene. **Figure 8.24a** shows a map of α-*Tm*. The gene contains 14 exons, including alternatives for exons 1, 2, 6, and 9. The gene has two promoters (identified as $P_1$ and $P_2$) as well as five alternative polyadenylation sites (identified as $A_1$ to $A_5$). The nine distinct mature mRNAs from α-*Tm* are produced in muscle cells (two forms), brain cells (three forms), and fibroblast cells (four forms); see **Figure 8.24b**. Each different mature mRNA illustrates a unique pattern of promoter selection, intron splicing, and choice of polyadenylation site. All mature mRNAs, and their corresponding tropomyosin proteins, contain the genetic information of exons 3, 4, 5, 7, and 8; however, they may contain distinct information in the alternative exons that

**Figure 8.24 Alternative pre-mRNA processing of the rat α-tropomyosin gene.** Alternative splicing patterns are indicated by the arched lines connecting exons. Nine distinct mature mRNAs produced by different types of muscle, brain, and fibroblast cells each produce a different tropomyosin protein.

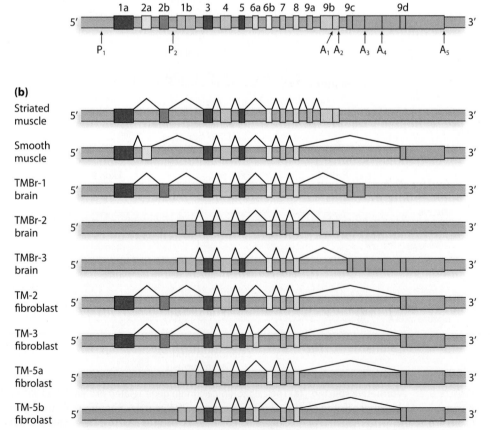

depends largely on the cell-type-specific selection of promoter and polyadenylation site.

In striated muscle cells, for example, promoter $P_1$ and polyadenylation site $A_2$ are used. The mature mRNA includes the alternative exons 1a, 2b, 6b, 9a, and 9b. In contrast, tropomyosin in smooth muscle cells utilizes promoter $P_1$ and polyadenylation site $A_5$, and its mature mRNA contains exons 1a, 2a, 6b, and 9d. Brain cells produce three different tropomyosin proteins, each of which are translated from differentially spliced pre-mRNAs that also utilize different polyadenylation sites. In addition, two forms of the brain cell tropomyosin proteins are translated from mRNAs that utilize promoter $P_2$, and one from an mRNA utilizing $P_1$. Among the four different tropomyosin proteins produced in fibroblasts, the mRNAs all use polyadenylation site $A_5$, but they differ in selection of $P_1$ versus $P_2$, and alternative splicing occurs as well. **Genetic Analysis 8.2** guides you through analysis of the results of alternative mRNA processing.

---

**Genetic Insight**  Alternative splicing, alternative promoters, and alternative polyadenylation sites can produce different mature mRNA transcripts, and distinct proteins, from single genes. These mechanisms allow certain eukaryotes to produce many more proteins than they have genes in their genomes.

---

## Intron Self-Splicing

In addition to introns that are spliced by spliceosomes, RNAs can contain introns that self-catalyze their own removal. Three categories of self-splicing introns, designated group I, group II, and group III introns, have been identified. The molecular biologist Thomas Cech and his colleagues discovered group I introns in 1981, when they observed that a 413-nucleotide precursor mRNA of an rRNA gene from the protozoan *Tetrahymena* could splice itself without the presence of any protein. Following up on this initial observation, Cech and others have shown that group I introns are large, self-splicing ribozymes (catalytically active RNAs) that catalyze their own excision from certain mRNAs and also from tRNA and rRNA precursors in bacteria, simple eukaryotes, and plants. **Intron self-splicing** takes place by way of two transesterification reactions (**Figure 8.25** ❶, ❷) that excise the intron and allow exons to ligate ❸. Cech and Sidney Altman shared the 1989 Nobel Prize in Physiology or Medicine for their contributions to the discovery and description of the catalytic properties of RNA.

Group II introns, which are also self-splicing ribozymes, are found in mRNA, tRNA, and rRNA of fungi, plants, protists, and bacteria. Group II introns form highly complex secondary structures containing many stem-loop arrangements. Their self-splicing takes place in a lariat-like manner utilizing a branch point nucleotide that in many cases is adenine. It is thought that nuclear

❶ Exon-intron base pairing. The G-binding site nucleotide attacks the UpA bond, bonding to the adenine and cleaving exon A.

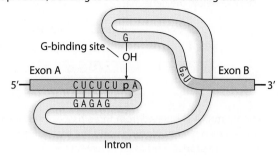

❷ The 3′ end of exon A attacks the GpU bond at the intron–exon junction.

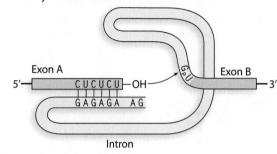

❸ The intron is released, and exons ligate.

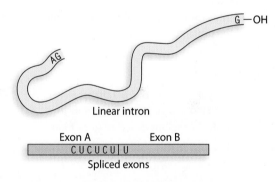

**Figure 8.25  Self-splicing of group I introns.**

pre-mRNA splicing may have evolved from group II self-splicing introns.

Group III introns and group II introns are similar in having elaborate secondary structures and lariat-like splicing structures that utilize a branch point nucleotide. Group III introns are much shorter than group II introns, however, and their secondary structures are different from those of group II introns.

## Ribosomal RNA Processing

In bacteria and eukaryotes, rRNAs are transcribed as large precursor molecules that are cleaved into smaller RNA molecules by removal and discarding of spacer sequences intervening between the sequences of the different RNAs. The *E. coli* genome contains seven copies of an rRNA gene. Each gene copy is transcribed into a single 30S precursor RNA that is processed by the removal of

The *JLB-1* gene, expressed in several human organs, contains seven exons (1 to 7) and six introns (A to F). Three oligonucleotide probes (I to III), hybridizing to exons 2, 4, and 7, respectively, are indicated by asterisks below the gene map:

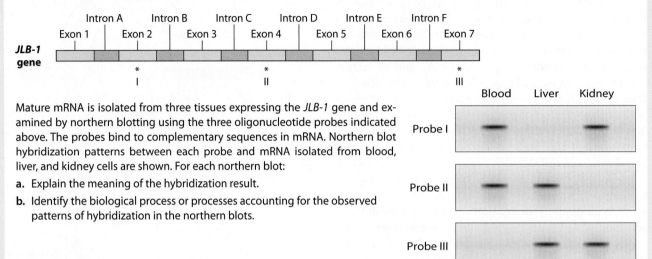

Mature mRNA is isolated from three tissues expressing the *JLB-1* gene and examined by northern blotting using the three oligonucleotide probes indicated above. The probes bind to complementary sequences in mRNA. Northern blot hybridization patterns between each probe and mRNA isolated from blood, liver, and kidney cells are shown. For each northern blot:

a. Explain the meaning of the hybridization result.

b. Identify the biological process or processes accounting for the observed patterns of hybridization in the northern blots.

| Solution Strategies | Solution Steps |
|---|---|

**Evaluate**

1. Identify the topic this problem addresses, and state the nature of the required answer.

2. Identify the critical information provided in the problem.

1. This problem concerns the production of mature mRNAs from a single human gene expressed in different organs. The answer requires identification of the specific mechanisms responsible for the data obtained from each organ.

2. The problem gives gene structure, the binding location of each of three molecular probes hybridizing the gene, and the results of three northern blot analyses of mature mRNA from different organs.

**Deduce**

3. Identify the regions of *JLB-1* that are anticipated to be part of the pre-mRNA.

4. Identify the regions expected to be found in mature mRNA.

3. Pre-mRNA from this gene is anticipated to include all intron and exon sequences.

4. Exon segments are expected in mature mRNA, along with modification at the 5′ mRNA end (capping) and the 3′ end (poly-A tailing).

**Solve**

5. Determine the hybridization pattern of molecular probes in each organ.

TIP: Hybridization of a probe occurs when the probe finds its target sequence. The absence of hybridization indicates that the target sequence for a probe is not present.

Answer a

5. Blood: Probes I and II hybridize, but probe III does not. This result indicates that exons 2 and 4 are present in the mature mRNA of blood, but exon 7 is not.

Liver: Probe I fails to hybridize to mRNA from liver, indicating that exon 1 is missing from liver mRNA. Probes II and III hybridize liver mRNA, indicating that exons 4 and 7 are included in the mature transcript.

Kidney: Probe II does not hybridize kidney mRNA, indicating that exon 4 is missing from it. Probes I and III find hybridization targets, indicating that exons 2 and 7 are present in the transcript.

Answer b

6. Interpret the hybridization patterns in each tissue and identify the process or processes that reasonably account for the observed patterns.

TIP: Alternative promoters, alternative polyadenylation sites, and alternative splicing are three mechanisms that lead eukaryotic genomes to generate distinct proteins from the same gene.

6. Blood: The absence of exon 7 is most likely due to either the use of an alternative polyadenylation site that generates 3′ cleavage of pre-mRNA ahead of exon 7 or to differential splicing that removes exon 7 from pre-mRNA during intron splicing.

Liver: The absence of exon I is most likely due either to use of an alternative promoter that initiates transcription at a point past exon 1 or to differential splicing of liver pre-mRNA.

Kidney: The absence of exon 4 is most likely the result of differential splicing of pre-mRNA.

For more practice, see Problems 2, 3, and 8.

intervening sequences to yield 5S, 16S, and 23S rRNAs, along with several tRNA molecules (**Figure 8.26a**). All seven gene copies produce the same three rRNAs, but each gene generates a different set of tRNAs.

Eukaryotic genomes have hundreds of rRNA genes clustered in regions of repeated genes on various chromosomes. Each gene produces a 45S precursor

rRNA that contains an external transcription sequence (ETS) and two internal transcription sequences (ITS1 and ITS2) that are removed by processing. The transcript is processed in multiple steps to yield three rRNA molecules weighing 5.8S, 18S, and 28S (**Figure 8.26b**). Eukaryotic genomes differ somewhat in the steps that process the 45S pre-rRNA transcript. In general,

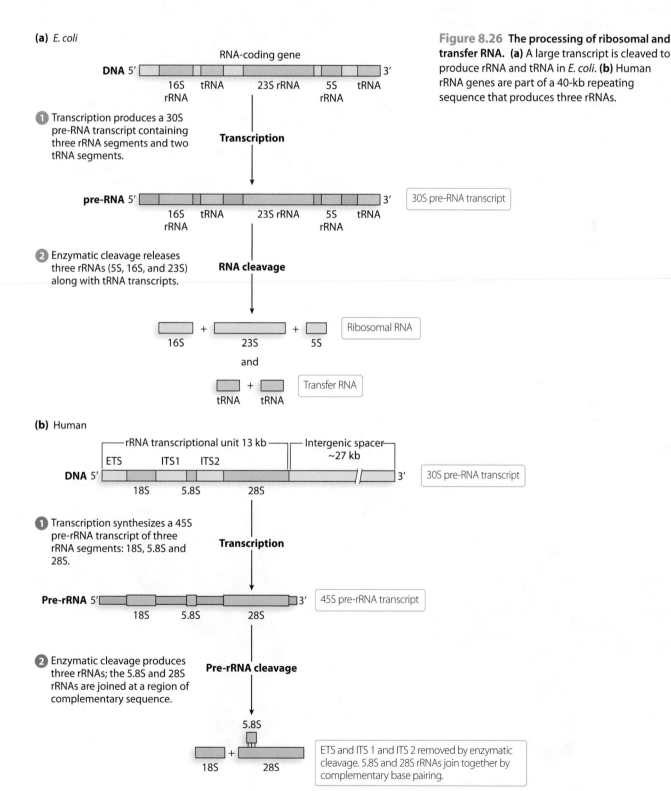

**Figure 8.26** **The processing of ribosomal and transfer RNA.** **(a)** A large transcript is cleaved to produce rRNA and tRNA in *E. coli.* **(b)** Human rRNA genes are part of a 40-kb repeating sequence that produces three rRNAs.

however, the 45S transcript is cleaved to a 41S intermediate from which the 18S transcript is then removed, followed by cleavage that produces the 28S and 5.8S transcripts. The 5.8S and 28S products pair with one another and become part of the same ribosomal subunit. After processing, the resulting rRNAs fold into complex secondary structures and are joined by proteins to form ribosomal subunits. Some chemical modifications of rRNA, particularly methylation of selected nucleotide bases, occur after completion of transcription.

## Transfer RNA Processing

Processing of tRNA is different in bacteria and eukaryotes. Each type of tRNA has distinctive nucleotides and a specific pattern of folding, but all tRNAs have similar structures and functions (**Figure 8.27**). Some bacterial transfer RNA molecules are produced simultaneously with rRNAs, as described above (see Figure 8.26a). Other

tRNAs are transcribed as part of a large pre-tRNA transcript that is then cleaved to yield multiple tRNA molecules. In eukaryotes, tRNA genes occur in clusters on specific chromosomes. Each eukaryotic tRNA gene is individually transcribed by RNA polymerase III, and a single pre-tRNA is produced from each gene.

The number of different tRNAs produced depends on the type of organism. In bacteria, the exact number of different tRNAs varies, but it is usually substantially *less than* 61, the number of codons found in mRNA. At a minimum, each species must have at least 20 different tRNAs, one for each amino acid, but most produce at least 30 to 40 different tRNAs. The low number of different tRNAs (compared to number of codons) results from a phenomenon called *third base wobble*, a relaxation of the "rules" of complementary base pairing at the third base of codons (see Chapter 9). Although third base wobble plays a role in reducing the number of distinct tRNA genes needed in eukaryotic genomes,

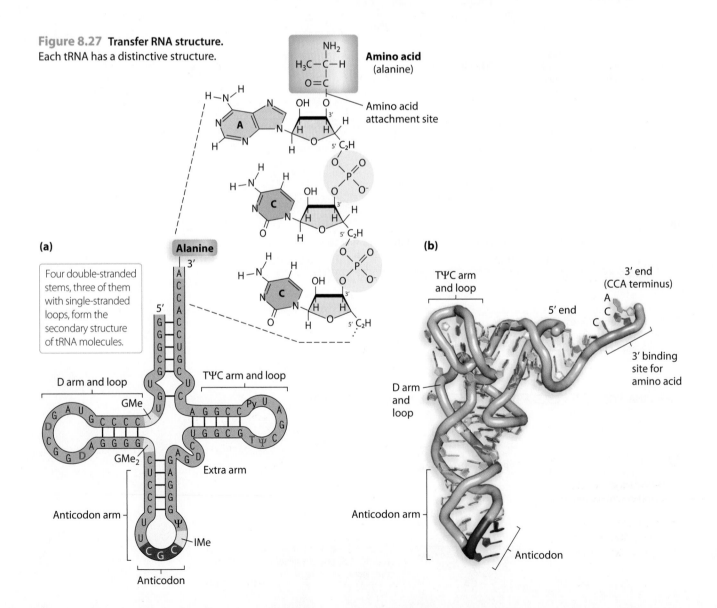

**Figure 8.27 Transfer RNA structure.**
Each tRNA has a distinctive structure.

eukaryotes nevertheless produce a larger number of different tRNAs than bacteria do. Some eukaryotic genomes contain a full complement of 61 different tRNA genes, one corresponding to each codon of the genetic code.

Bacterial tRNAs require processing before they are ready to assume their functional role of transporting amino acids to the ribosome. The precise processing events differ somewhat among tRNAs, but several features are common. First, many tRNAs are cleaved from large precursor tRNA transcripts to produce several individual tRNA molecules. Second, nucleotides are trimmed off the 5′ and 3′ ends of tRNA transcripts to prepare the mature molecule. Third, certain individual nucleotides in different tRNAs are chemically modified to produce a distinctive molecule. Fourth, tRNAs fold into a precise three-dimensional structure that includes four double-stranded stems, three of which are capped by single-stranded loops; each stem and loop constitutes an "arm" of the tRNA molecule. Fifth, tRNAs undergo post-transcriptional addition of bases. The most common addition is three nucleotides, CCA, at the 3′ end of the molecule. This region is the binding site for the amino acid the tRNA molecule transports to the ribosome. Figure 8.27 shows tRNA$_{Ala}$, which carries alanine. The CCA terminus is indicated, along with chemically modified nucleotides in each arm that are characteristic of this tRNA. Both a two-dimensional and a three-dimensional representation are shown.

Eukaryotic tRNAs undergo processing modifications similar to those of bacterial tRNAs. In addition, however, eukaryotic pre-tRNAs may contain small introns that are removed during processing. For example, an intron 14 nucleotides in length is removed from the precursor molecule by a specialized nuclease enzyme that cleaves the 5′ and 3′ splice sites of tRNA introns. The cleaved tRNA then refolds to form the anticodon stem, and the enzyme RNA ligase joins the 5′ and 3′ ends of the tRNA.

---

Genetic Insight   Ribosomal and transfer RNA transcripts are processed by enzymatic action to form functional rRNAs and tRNAs. Certain mRNAs are capable of self-splicing to yield mature mRNA.

---

## Post-Transcriptional RNA Editing

A firmly established tenet in the central dogma of biology is the role of DNA as the repository and purveyor of genetic information. Notwithstanding the modifications made to precursor RNA transcripts after transcription, a fundamental principle of biology is that DNA dictates the sequence of mRNA nucleotides and controls the order of amino acids in proteins. And yet, in the mid-1980s, a phenomenon called **RNA editing** was uncovered that is

responsible for post-transcriptional modifications that change the genetic information carried by mRNA.

Two kinds of RNA editing occur. In one kind of RNA editing, uracils are inserted into edited mRNA with the assistance of a specialized RNA called **guide RNA (gRNA)**. A guide RNA, transcribed from a separate RNA-encoding gene, contains a sequence complementary to the region of mRNA that it edits. With the aid of a protein complex, a portion of guide RNA pairs with complementary nucleotides of pre-edited mRNA and acts as a template to direct the insertion (and occasionally the deletion) of uracil (**Figure 8.28**). Guide RNA releases edited mRNA after editing is complete. The protein translated from edited mRNA may differ from the protein produced from unedited transcript.

The second kind of RNA editing is by base substitution, and frequently consists of the replacement of cytosine with uracil (C-to-U editing) in mRNA. This type of RNA editing has been identified in mammals, most land plants, and several single-celled eukaryotes. The consequence of C-to-U RNA editing is demonstrated by the protein product of the mammalian *apolipoprotein B* gene

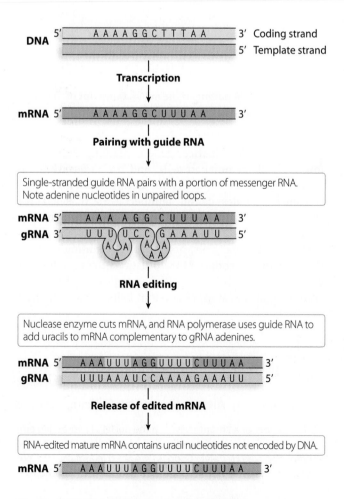

**Figure 8.28  Guide RNA directs RNA editing.**

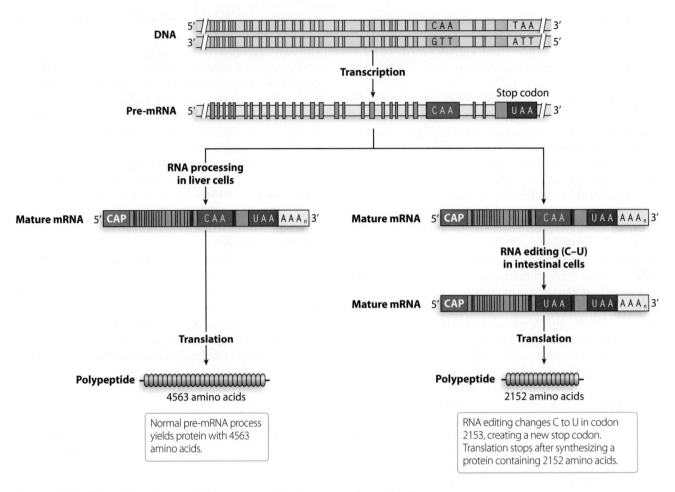

**Figure 8.29   RNA editing of the mRNA transcript of the human *apolipoprotein B* gene.**

(**Figure 8.29**). An identical gene containing 29 exons is found in all mammalian cells, and the same mRNA is transcribed in all tissues. Part of this messenger RNA sequence includes codon number 2153 that has the sequence CAA and is translated as glutamine in liver apolipoprotein B, a protein consisting of 4563 amino acids. In intestinal cells, however, RNA editing changes the cytosine in codon 2153 to a uracil, converting the codon to UAA. This C-to-U change produced by RNA editing creates a "stop" codon that halts translation after

the assembly of the first 2152 amino acids of intestinal apolipoprotein B.

---

**Genetic Insight**   RNA editing makes post-transcriptional changes in mRNA sequence by base substitution (usually C to U) or by insertion, most often of uracil, into the transcript. Both types of RNA editing can alter the protein product translated from edited mRNA.

---

## CASE STUDY

### Sexy Splicing: Alternative mRNA Splicing and Sex Determination in *Drosophila*

The number of X chromosomes in the nuclei of *Drosophila* embryos is critical in sex determination, but the X/autosome (X/A) ratio proposed by Calvin Bridges (X/A = 1.0 in females and X/A = 0.5 in males) is not accurate (see Section 3.4). Instead, the

process involves differential gene expression and pre-mRNA splicing. The molecular basis of *Drosophila* sex determination depends on a series of steps that begins with the transcription activation of the *sex-lethal* (*Sxl*) gene, includes alternative

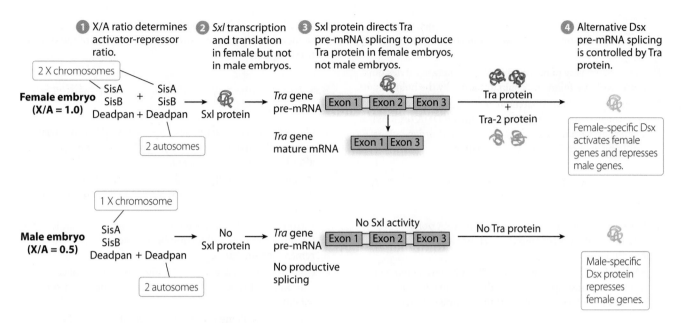

**Figure 8.30** The X/A ratio determines gene transcription and transcript splicing pattern to determine sex in fruit flies.

splicing of the pre-mRNA transcript of the *transformer* (*Tra*) gene, and culminates with one of two alternative splicing variants of the pre-mRNA transcripts of the *double-sex* (*Dsx*) gene. The Dsx protein directs further transcription activation and repression, leading to female or to male development.

The X/A ratio in fly embryos initially influences the transcription and translation of two X-linked activator proteins called SisA and SisB, and an autosomal gene producing a transcription repressor protein called Deadpan (**Figure 8.30**). Since the genes producing SisA and SisB are X-linked, early female embryos produce twice as much of each activator as do early male embryos, and the ratio of SisA + SisB to Deadpan differs between female and male embryos. In early female embryos, the ratio of SisA + SisB protein to Deadpan protein leads to transcription of the *Sxl* gene and to the production of Sxl protein. *Sxl* transcription is repressed in male embryos and no Sxl protein is produced.

Sxl protein is a splicing regulator that operates on the pre-mRNA transcript of the *Tra* gene. In female embryos, *Tra* pre-mRNA is spliced to produce a functional Tra protein. In male embryos, the absence of Sxl protein leads to alternative *Tra* pre-mRNA splicing that does not produce functional Tra protein. The Tra protein is also a splicing regulator that operates on the pre-mRNA of *Dsx* along with a second protein known as Tra-2. In female embryos, Tra protein and Tra-2 protein splices *Dsx* pre-mRNA in one alternative variant, which when translated produces female-specific Dsx protein. Female-specific Dsx activates transcription female-specific genes and represses transcription of male-specific genes to produce female flies. Tra protein is absent in male embryos, and *Dsx* pre-mRNA is spliced in an alternative variant. Dsx protein in male embryos represses female-specific genes and allows transcription of unrepressed male-specific genes, leading to male sex development.

SUMMARY

*For activities, animations, and review quizzes, go to the study area at www.masteringgenetics.com.*

## 8.1 RNA Transcripts Carry the Messages of Genes

- RNA molecules are synthesized by RNA polymerases using as building blocks the RNA nucleotides A, G, C, and U to form single-stranded sequences complementary to DNA template strands.

- Messenger RNA is the transcript that undergoes translation to produce proteins. Five other major forms of functional RNA are transcribed, and may undergo modification, but are not translated.

## 8.2 Bacterial Transcription Is a Four-Stage Process

- Transcription has four stages: promoter recognition, chain initiation, chain elongation, and chain termination.

- A single RNA polymerase transcribes all bacterial genes. This polymerase is a holoenzyme composed of a five-subunit core enzyme and a sigma subunit that aids the recognition of different forms of bacterial promoters.

- Bacterial promoters have two consensus sequence regions located upstream of the transcription start at approximately −10 and −35.
- The core enzyme of bacterial RNA polymerase carries out RNA synthesis following chain initiation by the holoenzyme.
- Transcription of most bacterial genes terminates by an intrinsic mechanism that depends only on DNA terminator sequences. Certain bacterial genes have a rho-dependent mechanism of transcription termination.

## 8.3  Eukaryotic Transcription Uses Multiple RNA Polymerases

- Eukaryotic cells contain three types of RNA polymerases that transcribe mRNA and the various classes of functional RNA.
- RNA polymerase II transcribes mRNA by interaction with numerous transcription factors that lead the enzyme to recognize promoters controlling transcription of polypeptide-coding genes.
- Promoters recognized by RNA polymerase II have a TATA box and additional regulatory elements that bind transcription factors and RNA pol II during transcription initiation.
- Eukaryotic promoter regulatory elements are recognized by their consensus sequences.
- Tissue-specific and developmental modifications in transcription are regulated by enhancer and silencer sequences.
- RNA polymerase I uses exclusive transcription factors to recognize upstream consensus sequences of ribosomal RNA genes.

- RNA polymerase III recognizes promoter consensus sequences that are upstream and downstream of the start of transcription.

## 8.4  Post-Transcriptional Processing Modifies RNA Molecules

- 5′ capping of eukaryotic messenger RNA adds a methylated guanine through the action of guanylyl transferase shortly after transcription is initiated.
- Polyadenylation at the 3′ end of eukaryotic messenger RNA is signaled by an AAUAAA sequence and is accomplished by a complex of enzymes.
- Intron splicing is controlled by cellular proteins that identify introns and exons and form spliceosome complexes that remove introns and ligate exons.
- Consensus sequences at the 5′ splice site, the 3′ splice site, and the branch point serve as guides during intron splicing.
- Alternative splicing is regulated by cell-type-specific variation of proteins that identify introns and exons.
- Some RNA molecules have catalytic activity and are able to self-splice introns without the aid of proteins.
- Ribosomal and transfer RNA molecules are generated by cleavage of large precursor molecules transcribed in bacterial and eukaryotic genomes.
- RNA editing is a post-transcriptional altering of nucleotide sequence, causing the transcripts to differ from the corresponding template DNA sequence.

## KEYWORDS

3′ polyadenylation (3′ poly-A tailing) (p. 278)
3′ splice site (p. 281)
5′ capping (p. 278)
5′ splice site (p. 281)
−35 consensus sequence (p. 265)
alternative mRNA processing (alternative intron splicing, promoters, polyadenylation) (pp. 283, 286)
branch point adenine (p. 281)
CAAT box (p. 273)
closed promoter complex (p. 267)
coding strand (p. 264)
complete initiation complex (p. 273)
consensus sequence (p. 265)
core element (p. 276)
core enzyme (p. 264)
downstream (p. 264)
enhancer sequence (p. 275)
exonic splicing enhancers (ESEs) (p. 283)

functional RNAs (tRNA, rRNA, snRNA, miRNA, siRNA, ribozymes) (pp. 263, 264)
GC-rich box (p. 273)
guide RNA (gRNA) (p. 291)
initial committed complex (p. 273)
internal control region (ICR) (p. 277)
internal promoter element (p. 276)
intrinsic termination (p. 269)
intron self-splicing (p. 287)
intron splicing (p. 278)
inverted repeat (p. 269)
lariat intron structure (p. 281)
mature mRNA (p. 278)
messenger RNA (mRNA) (p. 263)
micro RNA (miRNA) (p. 264)
minimal initiation complex (p. 273)
nucleolus (nucleoli) (p. 276)
nontemplate strand (p. 264)
open promoter complex (p. 267)

polyadenylation signal sequence (p. 279)
precursor mRNA (pre-mRNA) (p. 278)
Pribnow box (−10 consensus sequence) (p. 265)
promoter (p. 264)
promoter-specific element (PSE) (p. 276)
rho-dependent termination (rho protein) (p. 269)
rho utilization site (rut site) (p. 270)
ribonucleotides (A, U, G, and C) (p. 261)
ribose (p. 261)
ribosomal RNA (rRNA) (p. 263)
RNA editing (p. 291)
RNA polymerase (RNA pol, RNA pol I, RNA pol II, RNA pol III) (pp. 262, 270)
sigma (σ) subunit (alternative sigma subunits) (pp. 264, 265)
signal-transduction pathway (p. 275)

## PROBLEMS

*For instructor-assigned tutorials and problems,
go to www.masteringgenetics.com.*

### Chapter Concepts

*For answers to selected even-numbered problems, see Appendix: Answers.*

1. In terms of the central dogma of molecular biology,
   a. What is a gene?
   b. Why are genes for rRNA and tRNA considered to be genes even though they do not produce polypeptides?

2. In one to two sentences each, describe the three processes that commonly modify eukaryotic pre-mRNA.

3. Answer these questions concerning promoters.
   a. What role do promoters play in transcription?
   b. What is the common structure of a bacterial promoter with respect to consensus sequences?
   c. What consensus sequences are detected in the mammalian β-globin gene promoter?
   d. Eukaryotic promoters are more variable than bacterial promoters. Explain why.
   e. What is the meaning of the term *alternative promoter*? How does the use of alternative promoters affect transcription?

4. The diagram below shows a DNA duplex. The template strand is identified, as is the location of the +1 nucleotide.

   a. Assume this region contains a gene transcribed in a bacterium. Identify the location of promoter consensus sequences and of the transcription termination sequence.
   b. Assume this region contains a gene transcribed to form mRNA in a eukaryote. Identify the location of the most common promoter consensus sequences.
   c. If this region is a eukaryotic gene transcribed by DNA polymerase III, where are the promoter consensus sequences located?

5. The following is a portion of an mRNA sequence:

   3'-AUCGUCAUGCAGA-5'

   a. During transcription, was the adenine at the left-hand side of the sequence the first or the last nucleotide added to the portion of mRNA shown? Explain how you know.
   b. Write out the sequence and polarity of the DNA duplex that encodes this mRNA segment. Label the template and coding DNA strands.
   c. Identify the direction in which the promoter region for this gene will be located.

6. Compare and contrast the properties of DNA polymerase and RNA polymerase, listing at least three similarities and at least three differences between the molecules.

7. The DNA sequences shown below are from the promoter regions of six bacterial genes. In each case, the last nucleotide in the sequence (highlighted) is the +1 nucleotide that initiates transcription.
   a. Examine these sequences and identify the Pribnow box sequence at approximately −10 for each promoter.
   b. Determine the consensus sequence for the Pribnow box from these sequences.

| | |
|---|---|
| Gene 1 ... | TTCCGGCTCGTATGTTGTGTGGA ... |
| Gene 2 ... | CGTCATTTGATATGATGCGCCCG ... |
| Gene 3 ... | CCACTGGCGGTGATACTGAGCACA ... |
| Gene 4 ... | TTTATTGCAGCTTATAATGGTTACA ... |
| Gene 5 ... | TGCTTCTGACTATAATAGACAGGG ... |
| Gene 6 ... | AAGTAAACACGGTACGATGTACCACA ... |

8. Between the transcripts of bacterial and eukaryotic genes, there are differences in the transcripts themselves, in whether the transcripts are modified before translation, and in how the transcripts are modified. For each of these three areas of contrast, describe what the differences are and why the differences exist.

9. Describe the two types of transcription termination found in bacterial genes. How does transcription termination differ for eukaryotic genes?

10. What is the role of enhancer sequences in transcription of eukaryotic genes? Speculate about why enhancers are not part of transcription of bacterial genes.

11. Describe the difference between intron sequences and spacer sequences, such as the spacer sequence depicted in Figure 8.26b.

12. Draw a bacterial promoter and label its consensus sequences. How does this promoter differ from a eukaryotic promoter transcribed by RNA polymerase II? By RNA polymerase I? By RNA polymerase III?

13. How do SR proteins help guide pre-mRNA intron splicing? What is meant by the term *alternative splicing*, and how does variation in SR protein production play a role?

14. Three genes identified in the diagram as *A*, *B*, and *C* are transcribed from a region of DNA. The 5'-to-3' transcription of genes *A* and *C* elongates mRNA in the right-to-left direction, and transcription of gene *B* elongates mRNA in the left-to-right direction. For each gene, identify the coding strand for each gene by designating it as an "upper strand" or "lower strand in the diagram."

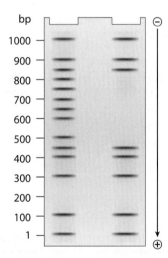

## Application and Integration

*For answers to selected even-numbered problems, see Appendix: Answers.*

15. The eukaryotic gene *Gen-100* contains four introns labeled A to D. Imagine that *Gen-100* has been isolated and its DNA has been denatured and mixed with polyadenylated mRNA from the gene.
    a. Illustrate the R-loop structure that would be seen with electron microscopy.
    b. Label the introns.
    c. Are intron regions single stranded or double stranded? Why?

16. The segment of the bacterial *TrpA* gene involved in intrinsic termination of transcription is shown below.

    ```
    3'-TGGGTCGGGGCGGATTACTGCCCCGAAAAAAAACTTG-5'
    5'-ACCCAGCCCCGCCTAATGACGGGGCTTTTTTTTGAAC-3'
    ```

    a. Draw the mRNA structure that forms during transcription of this segment of the *TrpA* gene.
    b. Label the template and coding DNA strands.
    c. Explain how a sequence of this type leads to intrinsic termination of transcription.

17. A 2-kb fragment of *E. coli* DNA contains the complete sequence of a gene for which transcription is terminated by the rho protein. The fragment contains the complete promoter sequence as well as the terminator region of the gene. The cloned fragment is examined by band shift assay (see Research Technique 8.1). Each lane of a single electrophoresis gel contains the 2-kb cloned fragment under the following conditions:

    Lane 1: 2-kb fragment alone

    Lane 2: 2-kb fragment plus the core enzyme

    Lane 3: 2-kb fragment plus the RNA polymerase holoenzyme

    Lane 4: 2-kb fragment plus rho protein

    a. Diagram the relative positions expected for the DNA fragments in this gel retardation analysis.
    b. Explain the relative positions of bands in lanes 1 and 3.
    c. Explain the relative positions of bands in lanes 1 and 4.

18. A 3.5-kb segment of DNA containing the complete sequence of a mouse gene is available. The DNA segment contains the promoter sequence and extends beyond the polyadenylation site of the gene. The DNA is studied by band shift assay (see Research Technique 8.1), and the following gel bands are observed.

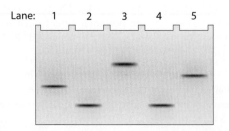

Match these conditions to a specific lane of the gel.
    a. 3.5-kb fragment plus TFIIB and TFIID
    b. 3.5-kb fragment plus TFIIB, TFIID, TFIIF, and RNA polymerase II
    c. 3.5-kb fragment alone
    d. 3.5-kb fragment plus RNA polymerase II
    e. 3.5-kb fragment plus TFIIB

19. A 1.0-kb DNA fragment from the 5' end of the mouse gene described in the previous problem is examined by DNA footprint protection analysis (see Research Technique 8.1). Two samples are end-labeled with $^{32}$P, and one of the two is mixed with TFIIB, TFIID, and RNA polymerase II. Both samples are exposed to DNase I. The results are displayed below.

a. What length of DNA is bound by the transcriptional proteins? Explain how the gel results support this interpretation.
b. Draw a diagram of this DNA fragment, bound by the transcriptional proteins showing the approximate position of proteins along the fragment. Use the illustration style seen in Research Technique 8.1 as a model.
c. Explain the role of DNase I. The DNA exposed to proteins is run in the right-hand lane of the gel. Control DNA is in the left-hand lane.

20. Wild-type *E. coli* grow best at 37°C but can grow efficiently up to 42°C. An *E. coli* strain has a mutation of the sigma subunit that results in an RNA polymerase holoenzyme that is stable and transcribes at wild-type levels at 37°C. The mutant holoenzyme is progressively destabilized as the temperature is raised, and it completely denatures and ceases to carry out transcription at 42°C. Relative to

wild-type growth, characterize the ability of the mutant strain to carry out transcription at

a. 37°C

b. 40°C

c. 42°C

d. What term best characterizes the type of mutation exhibited by the mutant bacterial strain? (*Hint*: The term was used in Chapter 4 to describe the Himalayan allele of the mammalian *C* gene.)

21. A mutant strain of *Salmonella* bacteria carries a mutation of the rho protein that has full activity at 37°C but is completely inactivated when the mutant strain is grown at 40°C.

a. Speculate about the kind of differences you would expect to see if you compared a broad spectrum of mRNAs from the mutant strain grown at 37°C and the same spectrum of mRNAs from the strain when grown at 40°C.

b. Are all mRNAs affected by the rho protein mutation in the same way? Why or why not?

22. The human β-globin wild-type allele and a certain mutant allele are identical in sequence except for a single base-pair substitution that changes one nucleotide at the end of intron 2. The wild-type and mutant sequences of the affected portion of pre-mRNA are

|  | Intron 2 | Exon 3 |
|---|---|---|
| wild type | 5'-CCUCCCACAG | CUCCUG-3' |
| mutant | 5'-CCUCCCACUG | CUCCUG-3' |

a. Speculate about the way in which this base substitution causes mutation of β-globin protein.

b. This is one example of how DNA sequence change occurring somewhere other than in an exon can produce mutation. List other kinds of DNA sequence changes occurring outside exons that can produce mutation. In each case, characterize the kind of change you would expect to see in mutant mRNA or mutant protein.

23. Microbiologists describe the processes of transcription and translation as "coupled" in bacteria. This term indicates that a bacterial mRNA can be undergoing transcription at the same moment it is also undergoing translation.

a. How is coupling of transcription and translation possible in bacteria?

b. Is coupling of transcription and translation possible in single-celled eukaryotes such as yeast? Why or why not?

24. A full-length eukaryotic gene is inserted into a bacterial chromosome. The gene contains a complete promoter sequence and a functional polyadenylation sequence, and it has wild-type nucleotides throughout the transcribed region. However, the gene fails to produce a functional protein.

a. List at least three possible reasons why this eukaryotic gene is not expressed in bacteria.

b. What changes would you recommend to permit expression of this eukaryotic gene in a bacterial cell?

25. The illustration below shows a portion of a gene undergoing transcription. The template and coding strands for the

gene are labeled, and a segment of DNA sequence is given. For this gene segment:

a. Superimpose a drawing of RNA polymerase as it nears the end of transcription of the DNA sequence.

b. Indicate the direction in which RNA polymerase moves as it transcribes this gene.

c. Write the polarity and sequence of the RNA transcript from the DNA sequence given.

d. Identify the direction in which the promoter for this gene is located.

Coding strand    5'
Template strand    3'

26. DNA footprint protection (described in Research Technique 8.1) is a method that determines whether proteins bind to a specific sample of DNA and thus protect part of the DNA from random enzymatic cleavage by DNase I. A 400-bp segment of cloned DNA is thought to contain a promoter. The cloned DNA is analyzed by DNA footprinting to determine if it has the capacity to act as a promoter sequence. The gel shown below has two lanes, each containing the cloned 400-bp DNA fragment treated with DNase I to randomly cleave unprotected DNA. Lane 1 is cloned DNA that was mixed with RNA polymerase II and several TFII transcription factors before exposure to DNase I. Lane 2 contains cloned DNA that was exposed only to DNase I. RNA pol II and TFIIs were not mixed with DNA before adding DNase I.

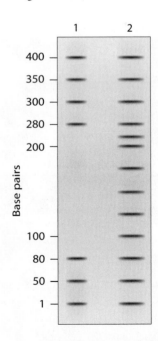

a. Explain why this gel provides evidence that the cloned DNA may act as a promoter sequence.

b. Approximately what length is the DNA region protected by RNA pol II and TFIIs?

c. What additional genetic experiments would you suggest to verify that this region of cloned DNA contains a functional promoter?

# The Molecular Biology of Translation

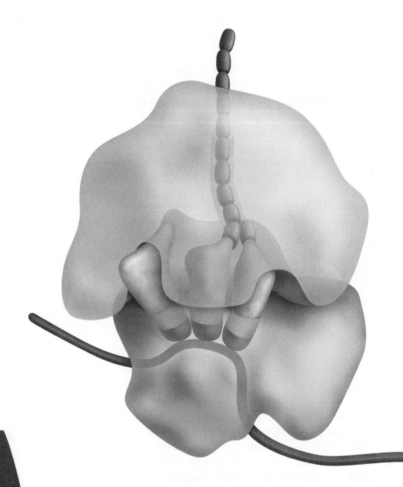

Ribosomes use codon sequences of messenger RNA to direct the assembly of polypeptides during translation. This rendering of a ribosome engaged in translation is based on recent crystal structure analysis and accurately shows the large subunit (top) and small subunit (bottom), the track of mRNA through the small subunit, the spaces for E, P, and A sites into which tRNAs fit, and the egress of the polypeptide through the large subunit.

## ESSENTIAL IDEAS

■ Translation is the cellular process of polypeptide production carried out by ribosomes under the direction of mRNA.

■ Ribosomes assemble on mRNA and initiate translation at the start codon.

■ Polypeptide elongation and termination at the stop codon are similar in bacteria and eukaryotes.

■ Transfer RNA molecules carry amino acids to ribosomes, which assemble polypeptides with the aid of ribosomal proteins.

■ A virtually universal genetic code comprising 64 mRNA codons directs polypeptide assembly.

■ In eukaryotes, polypeptides undergo posttranslational processing and are sorted into vesicles for transport to cellular destinations or for secretion.

Long before the discovery that DNA is the hereditary molecule, biologists had established the relationship between genes and proteins. In 1902, Archibald Garrod was the first to explicitly draw this connection when he proposed that the human hereditary disorder alkaptonuria was caused by an inherited defect in the enzyme homogentisic acid oxidase. As Garrod and other biologists expanded their exploration of the gene–protein connection, they found evidence that hereditary variation was closely tied to variations in proteins. Principal among the biologists who developed this connection were George Beadle and Edward

Tatum, whose research established the "one gene–one enzyme" hypothesis (Chapter 5).

In this chapter, we turn our attention to translation, the mechanism by which the messenger RNA (mRNA) transcripts of genes are used to assemble amino acids into polypeptide strings that form proteins. Translation is carried out by ribosomes that bring together mRNA transcripts and transfer RNA (tRNA) molecules that carry amino acids and facilitate the assembly of polypeptides, strings of amino acids.

The story of how polypeptides are produced by translation, and the story of how scientists came to understand the process, offers intriguing insight into the design of molecular genetic experiments. In this chapter, we describe some of these experiments and examine the molecular biology of translation. We also look at posttranslational processes that are instrumental in producing functional proteins and guiding them to their appropriate destinations in cells. The chapter concludes with a case study describing the action of commonly used antibiotics that interfere with bacterial translation.

## 9.1 Ribosomes Are Translation Machines

Polypeptides are strings of amino acids that are joined together by covalent peptide bonds. The order of amino acids in a polypeptide is directed by the nucleotide sequences of mRNA. Polypeptide assembly is orchestrated by ribosomes, which are ribonucleoprotein "machines" containing multiple molecules of ribosomal RNA (rRNA) and dozens of proteins. Bacterial and eukaryotic ribosomes are composed of two subunits that assemble into a ribosome as translation begins. Ribosomes bind mRNA and provide an environment for complementary base pairing between mRNA codon sequences and the anticodon sequences of tRNA. (In Chapter 1 and Figure 1.11, we review these basic mechanical features of translation). **Figure 9.1** encapsulates the essential elements of translation. Ribosomes translate mRNA in the 5′ → to 3′ direction, beginning with the start codon and ending with a stop codon. At each triplet codon, complementary base pairing between mRNA and tRNA

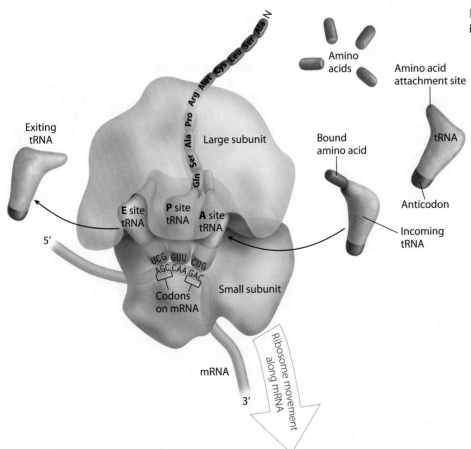

**Figure 9.1** **Translation overview in bacteria and eukaryotes.**

**Figure 9.2 Alignment of DNA, mRNA, and polypeptide.**

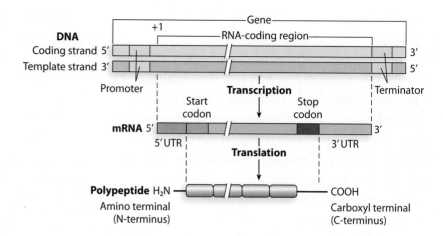

determines which amino acid is added to the nascent (growing) polypeptide. The start codon and stop codon define the boundaries of the translated segment of mRNA. The resulting polypeptides have an N-terminal (amino-terminal) end corresponding to the 5′ end of mRNA and a C-terminal (carboxyl-terminal) end that corresponds to the 3′ end of mRNA (Figure 9.2).

Figure 9.2 identifies two segments of the mRNA transcript that do not undergo translation. Between the 5′ end of mRNA and the start codon is a segment known as the **5′ untranslated region,** abbreviated **5′ UTR.** The region between the stop codon and the 3′ end of the molecule is the **3′ untranslated region,** or **3′ UTR.** As we describe in later sections, the 5′ UTR contains sequences that help initiate translation, and the 3′ UTR contains sequences associated with transcription termination and, in eukaryotes, with modifying the 3′ end of mRNA.

## Bacterial and Eukaryotic Ribosome Structures

The specific molecules composing bacterial and eukaryotic ribosomes differ, but the overall structures and functions of the ribosomes are similar. Ribosomes in bacteria and eukaryotes perform three essential tasks:

1. Bind messenger RNA and identify the start codon where translation begins.

2. Facilitate the complementary base pairing of mRNA codons and tRNA anticodons that determines amino acid order in the polypeptide.

3. Catalyze peptide bond formation between amino acids during polypeptide formation.

Differences in ribosomal composition between bacteria and eukaryotes include the number and sequence of rRNA molecules and the number and type of ribosomal proteins. The ribosomes of both kinds of organisms have important structural similarities, however; they are each composed of two subunits, called the **large ribosomal subunit** and the **small ribosomal subunit.** Each subunit size is measured in Svedberg units (S), which describe the speed of sedimentation of a substance during centrifugation. Keep in mind that

these Svedberg units are not additive when the subunits are combined, because sedimentation is a composite property that is affected by size, shape, and hydration state.

The ribosomes of *E. coli* are the most thoroughly studied bacterial ribosomes and serve as a model for general ribosome structure (Figure 9.3a). The small subunit of bacterial ribosomes has a Svedberg value of 30S. It contains 21 proteins and a single 16S rRNA composed of 1541 nucleotides. The large subunit of the bacterial ribosome is a 50S particle composed of 31 proteins, a small 5S rRNA containing 120 nucleotides, and a large 23S rRNA containing 2904 nucleotides. When fully assembled, the intact bacterial ribosome has a Svedberg value of 70S.

Both the large and small subunits contribute to the formation of three regions that play important functional roles during translation: the **peptidyl site,** or **P site,** the **aminoacyl site,** or **A site** and, the **exit site,** or **E site.** The P site holds a tRNA to which the nascent polypeptide is attached. The A site binds a new tRNA molecule carrying the next amino acid to be added to the polypeptide. The E site provides an avenue of egress for tRNAs as they leave the ribosome after their amino acid has been added to the polypeptide chain. Ribosomes also form a channel through which the polypeptide emerges. In addition, there is a channel through the large subunit through which the nascent polypeptide is extruded from the ribosome (see Figure 9.1).

Among eukaryotes, mammalian ribosomes are the most fully characterized (Figure 9.3b). The small 40S ribosomal subunit contains approximately 35 proteins and a single 18S rRNA composed of 1874 nucleotides. The large mammalian ribosomal subunit has a Svedberg value of 60S and contains 45 to 50 proteins, along with three molecules of rRNA. The rRNA molecules have values of 5S (120 nucleotides), 5.8S (160 nucleotides), and 28S (4718 nucleotides). The intact mammalian ribosome has a Svedberg value of 80S. Like the bacterial ribosome, the intact mammalian ribosome possesses a P site, an A site, an E site, and a channel for polypeptide egress.

The proteins contained in ribosomal subunits can be separated from one another by a specialized type of

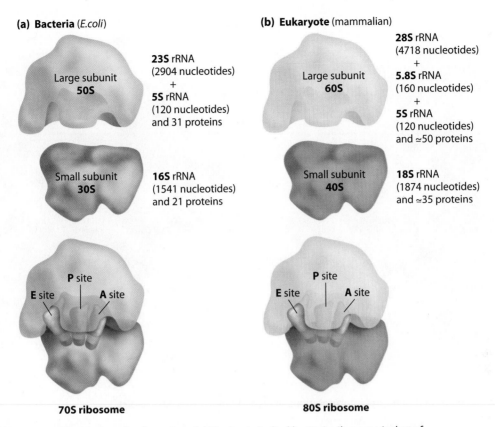

**(a) Bacteria** (*E.coli*)

Large subunit
**50S**

**23S** rRNA
(2904 nucleotides)
+
**5S** rRNA
(120 nucleotides)
and 31 proteins

Small subunit
**30S**

**16S** rRNA
(1541 nucleotides)
and 21 proteins

**P** site
**E** site    **A** site

**70S ribosome**

**(b) Eukaryote** (mammalian)

Large subunit
**60S**

**28S** rRNA
(4718 nucleotides)
+
**5.8S** rRNA
(160 nucleotides)
+
**5S** rRNA
(120 nucleotides)
and ≈50 proteins

Small subunit
**40S**

**18S** rRNA
(1874 nucleotides)
and ≈35 proteins

**P** site
**E** site    **A** site

**80S ribosome**

**Figure 9.3  Ribosomes of bacteria and eukaryotes.  (a)** The best-studied bacteria ribosome is that of *E. coli*, and **(b)** the best-studied eukaryotic ribosomes are mammalian.

electrophoresis called two-dimensional gel electrophoresis. The 21 proteins that are part of the small ribosomal subunit in *E. coli* and the 31 proteins found in the large ribosomal subunit are efficiently separated by this method. **Research Technique 9.1** (see page 304) describes how two-dimensional gel electrophoresis is used to characterize the proteins found in *E. coli* ribosomal subunits.

### The Three-Dimensional Structure of the Ribosome

Ribosomes are small—a mere 25 nanometers (nm) in diameter—so small that almost 10,000 of them can fit in the space occupied by the period at the end of this sentence. No one has ever "seen" a ribosome, but in recent years structural biologists have used powerful molecular imaging techniques to explore the three-dimensional configuration of many biological molecules and structures, including ribosomes and ribosomal subunits, at levels of resolution that are measured in ångströms (Å). These structural analyses have clarified how ribosomal subunits fit together, and they have produced a detailed understanding of ribosomal interactions with mRNA and tRNA.

Structural analysis of ribosomes and other molecular complexes in cells is made possible by a technique known as cryo-electron microscopy (cryo-EM), pioneered by Robert Glaeser in the 1970s and perfected by Jacques Dubochet in the 1980s. Cryo-EM uses liquid nitrogen or liquid ethane,

with temperatures nearly −200°C, to instantaneously freeze macromolecules and thus preserve them in their native state. A frozen macromolecule is then placed on a microcaliper and scanned from various angles by electron beams that use specialized software to create a three-dimensional picture of molecular structure. Cryo-EM creates exquisitely precise three-dimensional images of ribosome structure—much like CAT-scan imaging of the human body—revealing details down to a few angstroms in size (**Figure 9.4**). These images have identified the location and dimensions of the E, A, and P sites, for example, and have clarified the mechanical activities of ribosomes during translation.

## 9.2 Translation Occurs in Three Phases

Translation occurs in three phases: initiation, elongation, and termination. The three phases are generally similar in bacteria and eukaryotes, and yet they differ in several ways, particularly during translation initiation, where distinct mechanisms are used to identify the start codon.

### Translational Initiation

Translation initiation in bacteria and eukaryotes begins when the small ribosomal subunit binds near the 5′ end of mRNA and identifies the start codon sequence. In the next stage, the **initiator tRNA,** the tRNA carrying the

**(a)**

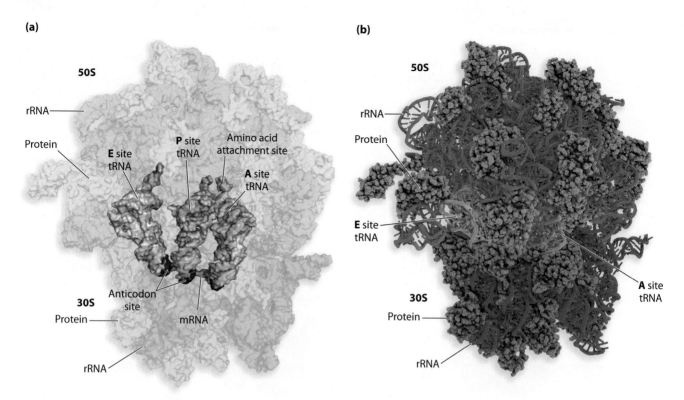

**(b)**

**Figure 9.4  Three-dimensional computer interpretations of cryo-EM-generated data depict ribosome structure.**

first amino acid of the polypeptide, binds to the start codon. In the final stage of initiation, the large subunit joins the small subunit to form an intact ribosome, and translation begins. During these stages, *initiation factor proteins* help control ribosome formation and binding of the initiator tRNA, and guanosine triphosphate (GTP) provides energy. The tRNAs used during translation each carry a specific amino acid and are identified as **charged tRNAs.** In contrast, a tRNA without an amino acid is **uncharged.** Specialized enzymes discussed in a later section are responsible for recognizing different tRNAs and charging each one with the correct amino acid.

Starting translation at the authentic (correct) start codon is essential for translation of the correct polypeptide. Errant translation starting at the wrong codon, or even at the wrong nucleotide of the start codon, may produce an abnormal polypeptide and result in a nonfunctional protein. Thus, critical questions for biologists studying translation initiation were these: How does the ribosome locate the authentic start codon? And if more than one AUG (start codon) sequence occurs near the 5′ end of the mRNA, how is the authentic start codon identified? Bacteria and eukaryotes use different mechanisms to identify the authentic start codon.

## Bacterial Translational Initiation

In *E. coli,* six critical molecular components come together to initiate the translation process: (1) mRNA, (2) the small

ribosomal subunit, (3) the large ribosomal subunit, (4) the initiator tRNA, (5) three essential initiation factor proteins, and (6) GTP.

For most of translational initiation in bacteria, the 30S ribosomal subunit is affiliated with an **initiation factor (IF)** protein called IF3, which prevents the 30S subunit from binding to the 50S subunit (**Figure 9.5**). The small subunit–IF3 complex binds near the 5′ end of mRNA, searching for the AUG sequence that serves as the start codon. The **preinitiation complex** forms when the authentic start codon sequence is identified by base pairing that occurs between the 16S rRNA in the 30S ribosome and a short mRNA sequence located a few nucleotides upstream of the start codon in the 5′ UTR of mRNA (Figure 9.5, ❶). John Shine and Lynn Dalgarno identified the location and sequence of this region in 1974, and it is named the **Shine–Dalgarno sequence** in recognition of their work.

The Shine–Dalgarno sequence is a purine-rich sequence of about six nucleotides located three to nine nucleotides upstream of the start codon. A complementary pyrimidine-rich segment containing the sequence UCCUCC is found near the 3′ end of 16S rRNA, and it pairs with the Shine–Dalgarno sequence to position the mRNA on the 30S subunit (see Figure 9.5). The Shine–Dalgarno sequence is another example of a consensus sequence. Like the consensus sequences we describe for promoters (Chapter 8) the Shine–Dalgarno sequence has a characteristic nucleotide composition and a

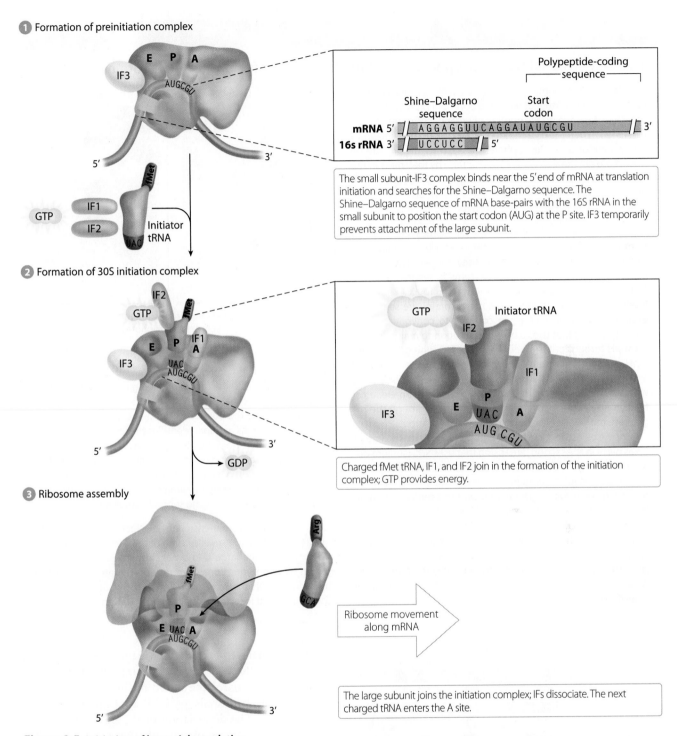

**1** Formation of preinitiation complex

The small subunit-IF3 complex binds near the 5′ end of mRNA at translation initiation and searches for the Shine–Dalgarno sequence. The Shine–Dalgarno sequence of mRNA base-pairs with the 16S rRNA in the small subunit to position the start codon (AUG) at the P site. IF3 temporarily prevents attachment of the large subunit.

**2** Formation of 30S initiation complex

Charged fMet tRNA, IF1, and IF2 join in the formation of the initiation complex; GTP provides energy.

**3** Ribosome assembly

Ribosome movement along mRNA

The large subunit joins the initiation complex; IFs dissociate. The next charged tRNA enters the A site.

**Figure 9.5  Initiation of bacterial translation.**

precise position relative to the start codon, but its exact nucleotide sequence varies slightly from one mRNA to another (**Figure 9.6**).

In the next step of translation initiation (Figure 9.5, **2**), the initiator tRNA binds to the start codon at what will be part of the P site after ribosome assembly. The amino acid on the initiator tRNA is a modified methionine called **N-formylmethionine (fMet)**; thus, the charged initiator tRNA is abbreviated **tRNA^fMet**. This tRNA has a 3′-UAC-5′ anticodon sequence that is a complementary mate to the start codon sequence. An initiation factor (IF) protein designated IF-2 and a molecule of GTP are bound to tRNA^fMet. Initiation factor 1 (IF-1) also joins the complex to forestall attachment of the 50S subunit. At this point, the **30S initiation complex,** consisting of mRNA bound to the 30S subunit, tRNA^fMet located at the start codon, three initiation factors, and a molecule of GTP, has been formed.

## Research Technique **9.1**

### Two-Dimensional Gel Electrophoresis and the Identification of Ribosomal Proteins

**PURPOSE** All ribosomes are composed of two subunits that are each a complex mixture of rRNA and dozens of proteins. One approach to determining the number of proteins contained in each ribosomal subunit uses a method of electrophoresis known as two-dimensional gel electrophoresis to separate the proteins by their charge in the first dimension and then by their mass in the second dimension. Two-dimensional gel electrophoresis produces a distinctive "protein fingerprint" that distributes each ribosomal protein to a different location in the two-dimensional gel.

**MATERIALS AND PROCEDURES** Ribosomes are isolated from cells, the subunits are separated, and the subunits are treated to dissociate the proteins they contain. The mixture containing liberated ribosomal proteins is separated in the first dimension by a version of gel electrophoresis known as isoelectric focusing. In this procedure, proteins are separated exclusively by their charge. In contrast to conventional gel electrophoresis that uses a buffered solution to maintain constant pH throughout the gel, isoelectric focusing gels contain a pH gradient. A protein's pH environment affects its charge, and every protein has a pH—called the isoelectric point—at which it has neutral charge and cannot move in an electrical field. In isoelectric focusing, proteins migrate through the pH gradient to their isoelectric point, where they stop.

Once isoelectric focusing is complete, protein separation takes place in the second dimension that uses SDS (sodium dodecyl sulfate) gel electrophoresis. SDS is a strong anionic detergent that denatures proteins by disrupting the interactions that keep them folded. Denatured proteins migrate through the gel at a rate determined by their mass, that is, by the number of amino acids they contain. In the SDS gel dimension of two-dimensional gel electrophoresis, each protein has a unique starting point corresponding to its isoelectric point. Proteins with large mass (more amino acids) migrate a short distance in the second dimension, whereas proteins with small mass (fewer amino acids) migrate a greater distance.

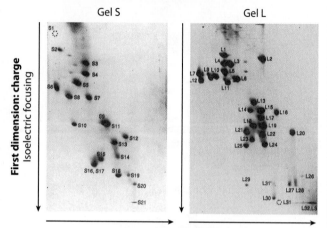

**DESCRIPTION** A pair of two-dimensional electrophoresis gels, one containing proteins of the small subunit of the *E. coli* ribosome (gel S) and the other containing proteins of the large subunit (gel L), reveal protein spots (the protein fingerprint) corresponding to the positions of proteins that make up each ribosomal subunit. Each spot identifies the location of a unique protein that differs from the other proteins in the gel by a combination of charge and mass. The proteins in gel S are identified as S1 to S21, and in gel L as L1 to L32. These experimental results indicate that the small subunit of bacterial ribosomes contains 21 proteins, and the large subunit contains 32 proteins.

**CONCLUSION** Two-dimensional gel electrophoresis identifies 21 proteins in the small subunit of the *E. coli* ribosome and 32 proteins in the large ribosomal subunit. Each protein obtained by two-dimensional electrophoresis can be subjected to additional biochemical examination to specifically identify the protein and investigate its role in translation.

|  | Shine–Dalgarno sequence | Start codon |
|---|---|---|
| *E. coli araB* | UUUGGAUUGGAGUGAAACGAUGGCGAUUGCA | 3' |
| *E. coli lacI* | CAAUUCAGGGGUGGUGAAUAUGAAACCAGUA | |
| *E. coli lacZ* | UUCACACAGGAAACAGCUAUGACCAUGAUU | |
| *E. coli thrA* | GGUAACCAGGUAACAAGGAUGCGAGUGUUG | |
| *E. coli trpA* | AGCACGAGGGGAAAUCUGAUGGAACGCUAC | |
| *E. coli trpB* | AUAUGAAGGAAAGGAACAAUGACAACAUUA | |
| λ phage *cro* | AUGUACUAAGGAGGUUGUAUGGAACAACGC | |
| R17 phage A protein | UCCUAGGAGGUUUGACCUAUGCGAGCUUUU | |
| Oβ phage A replicase | UAACUAAGGAUGAAAUGCAUGUCUAAGACA | |
| φX174 phage A protein | AAUCUUGGAGGCUUUUUUAUGGUUCGUUCU | |
| *E. coli* RNA polymerase B | AGCGAGCUGAGGAACCCUAUGGUUUACUCC | |

**Figure 9.6 The Shine–Dalgarno consensus binding sequence.** The AUG start codon sequence (orange) is near the Shine–Dalgarno region (gold), which binds to the 3' end of 16S rRNA.

In the final step of initiation (Figure 9.5, ❸), the 50S subunit joins the 30S subunit to form the intact ribosome. The energy for the union of the two subunits is derived from hydrolysis of GTP to GDP (guanosine diphosphate). The dissociation of IF1, IF2, and IF3 accompanies the joining of subunits that creates the **70S initiation complex.** This complex is a fully active ribosome with a P site, an A site, an E site, and a channel for exit of the polypeptide. The first tRNA (tRNA$^{fMet}$) is already paired with mRNA at the P site, and the open A site contains the second codon and is awaiting the next charged tRNA.

**Eukaryotic Translational Initiation** The eukaryotic 40S ribosomal subunit complexes with the **eukaryotic initiation factor (eIF)** proteins eIF1A and eIF3 and with a charged tRNA$^{Met}$ to form the preinitiation complex

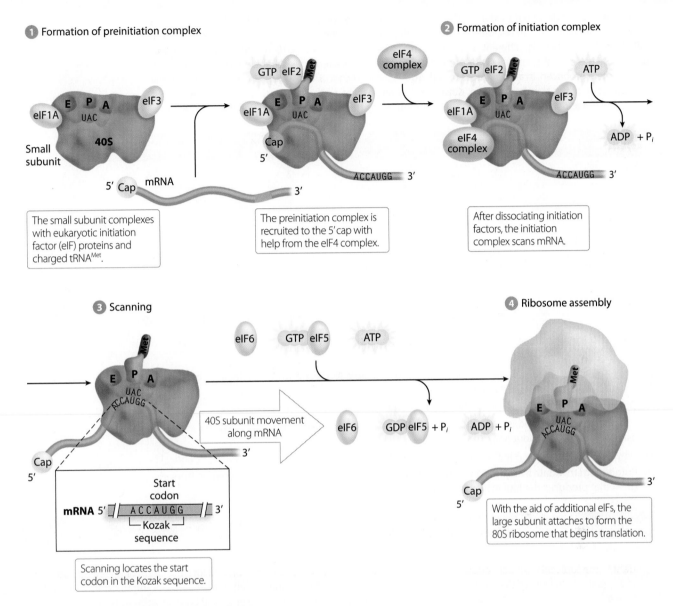

**① Formation of preinitiation complex**

The small subunit complexes with eukaryotic initiation factor (eIF) proteins and charged tRNA^Met.

The preinitiation complex is recruited to the 5′ cap with help from the eIF4 complex.

**② Formation of initiation complex**

After dissociating initiation factors, the initiation complex scans mRNA.

**③ Scanning**

40S subunit movement along mRNA

Start codon

mRNA 5′ ACCAUGG 3′
Kozak sequence

Scanning locates the start codon in the Kozak sequence.

**④ Ribosome assembly**

With the aid of additional eIFs, the large subunit attaches to form the 80S ribosome that begins translation.

**Figure 9.7  Initiation of eukaryotic translation.**

(Figure 9.7, ①). The preinitiation complex is then recruited to the 5′-cap region of mRNA, located at the end of the 5′ UTR. In step ②, the preinitiation complex joins the eIF4 complex, a group of at least four eIF4 proteins that assembles at the 5′ cap independently of translational initiation. Together, these components comprise the **initiation complex.**

Once the initiation complex is formed, it embarks on a process called **scanning** (Figure 9.7, ③), in which it uses ATP hydrolysis to move the small ribosomal subunit through the 5′ UTR in search of the start codon. About 90% of eukaryotic mRNAs use the first AUG encountered by the initiation complex as the start codon, but the remaining 10% use the second or, in some cases, the third AUG as the start codon. The initiation

complex is able to accurately locate the authentic start codon because the codon is embedded in a consensus sequence that reads

5′-ACC**AUG**G-3′

(the start codon itself is shown in bold). This consensus sequence is called the **Kozak sequence** after Marilyn Kozak, who discovered it in 1978.

Locating the start codon leads to recruitment of the 60S subunit to the complex, using energy derived from GTP hydrolysis. This final step ④ in the formation of the 80S ribosome is accompanied by dissociation of the eIF proteins. In the 80S ribosome, the initiator tRNA^Met is located at the P site; the A site is vacant, awaiting arrival of the second tRNA (**Genetic Analysis 9.1**).

In an investigation designed to identify the consensus sequence containing the AUG codon that initiates translation of eukaryotic mRNA, Marilyn Kozak (1986) compared the amounts of protein produced from mRNA molecules having different single-base mutations flanking the AUG. Protein production was gauged by the optical density (OD) of protein bands in electrophoretic gels. Higher OD values indicated more protein produced. In all sequences tested, AUG is the start codon, and its adenine (A) was labeled the +1 nucleotide of the translated region. Kozak created single-base mutants at nucleotide positions −4, −3, −2, −1, and +4 and measured the OD of protein produced by translation of the resulting mutant mRNAs.

| Table A | Position −3 and +4 Mutants | | | | | |
|---|---|---|---|---|---|---|
| −3 | G | A | T | C | G | A |
| −2 | C | C | C | C | C | C |
| −1 | C | C | C | C | C | C |
| +1 | A | A | A | A | A | A |
| +2 | U | U | U | U | U | U |
| +3 | G | G | G | G | G | G |
| +4 | T | T | G | G | G | G |
| OD | 0.7 | 2.6 | 0.9 | 0.9 | 3.1 | 5.0 |

| Table B | Position −2 and −1 Mutants | | | |
|---|---|---|---|---|
| −3 | A | A | A | A |
| −2 | C | C | G | G |
| −1 | A | A | A | A |
| +1 | A | A | A | A |
| +2 | U | U | U | U |
| +3 | G | G | G | G |
| OD | 3.3 | 1.8 | 1.9 | 2.0 |

a. Looking just at the nucleotides in positions −3 and +4 (Table A), decide which nucleotides give the highest level of protein production.

b. Describe the impact of each nucleotide (A, T, C, and G) in the −3 position.

c. Looking just at nucleotides at position −2 and −1 (Table B), decide which nucleotides give the highest level of protein production.

d. Why did Kozak use only A in the −3 position to test the effects of nucleotides at positions −2 and −1?

e. Putting together data from both Table A and Table B, give the sequence of the mRNA region from −3 to +4 that produces the highest level of translation.

| Solution Strategies | Solution Steps |
|---|---|
| **Evaluate** | |
| 1. Identify the topic this problem addresses and explain the nature of the requested answer. | 1. This problem involves examination and interpretation of the effects on translation of mRNA sequence differences surrounding the start codon. The answer requires identifying the effects of base substitutions on translation and identifying the mRNA sequence corresponding to the highest translation level. |
| 2. Identify the critical information given in the problem. <br><br> **TIP:** Notice that AUG is the start codon sequence in all mutants tested. As a consequence, differences in OD result from differences among the surrounding nucleotides. | 2. Two tables provide mRNA sequence for different sequence variants. For each variant, an OD value describes the approximate level of protein produced by translation of the sequence. Higher OD values correspond to more protein production. |
| **Deduce** | |
| 3. Identify the <u>constant</u> and variable nucleotides displayed in Table A. | 3. In Table A, the nucleotide C is constant at positions −1 and −2, and position +3 is always G. Nucleotide variability is limited to positions −3 and +4. |
| 4. Identify the constant and variable nucleotides shown in Table B. | 4. In Table B, only the nucleotide at the −2 position varies; all other nucleotides are constant. |
| **Solve** | Answer a |
| 5. Specify the nucleotides in the −3 and +4 positions (Table A) that give the highest OD. | 5. In Table A, the presence of A in position −3 and G in position +4 produces the highest OD value. At the +4 position, G produces two high OD values and two low ODs, and T produces one high and one low OD. |

For more practice, see Problems 32, 33, and 34.

6. Assess how each nucleotide in the −3 position affects OD.

Answer b

6. At position −3, A produces the highest and the third-highest OD values; G produces the second-highest and the lowest OD; T and C produce the same low OD value.

Answer c

7. Evaluate how nucleotide differences at the −1 and −2 positions (Table B) affect OD.

7. In Table B, a C in position −2 and an A in position -1 produce the highest OD. Considering only the variable position −2, C produces higher OD values than does G.

Answer d

8. Explain the decision to base Table B evaluations only on sequences with A in the −3 position.

8. Adenine is selected as the nucleotide in position −3 for Table B evaluations based on the high average OD value for this nucleotide in comparison to other nucleotides. The average OD for A in the −3 position is $\frac{(5.0 + 2.6)}{2} = 3.8$ versus the next-highest average of $\frac{(3.1 + 0.7)}{2} = 1.9$ for G in the −3 position.

Answer e

9. Identify the start codon consensus sequence that results in the highest level of translation.

TIP: Compare OD values and nucleotide differences from both tables to determine the most efficient consensus sequence.

9. Data from the two tables combined identify the sequence ACCAUGG (start codon in bold) as the most efficient consensus sequence for the start codon. For the nucleotide positions immediately surrounding the start codon, A is most efficient at −3, C is more efficient than G at −2, C is more efficient than A at −1, and G is more efficient than T at +4.

---

Genetic Insight Translation initiation is a multistep process that assembles the ribosome from its subunits. In bacterial mRNA, the Shine–Dalgarno sequence near the start codon is initially bound by the small ribosomal subunit, which is later joined by the large ribosomal subunit. In eukaryotic mRNA, the small ribosomal subunit attaches near the 5′ cap and "scans" the mRNA, searching for the authentic start codon located in the Kozak sequence.

## Polypeptide Elongation

Elongation, the second phase of translation, begins with the recruitment of **elongation factor (EF)** proteins into the initiation complex. Elongation factors facilitate three steps of polypeptide synthesis:

1. Recruitment of charged tRNAs to the A site
2. Formation of a peptide bond between sequential amino acids
3. Translocation of the ribosome in the 3′ direction along mRNA

GTP hydrolysis provides the energy for each step of elongation in bacteria (**Figure 9.8**) and eukaryotes (**Figure 9.9**). Moreover, the steps in the elongation process are the same in both types of organisms: although the elongation factors differ, the ribosomal P, A, and E sites of both organisms serve nearly identical functions. The rates of elongation are also similar; bacteria add about 20 new amino acids per second to a nascent polypeptide chain, and eukaryotes elongate the polypeptide at a rate of 15 amino acids per second. Lastly, numerous studies indicate high fidelity of translation in prokaryotes and eukaryotes. An error rate of approximately one amino acid in each 10,000 added to polypeptides is estimated for bacteria.

**Polypeptide Elongation in Bacteria** In bacteria, several different elongation factor proteins (EFs) and other ribosomal proteins carry out elongation in a series of steps depicted in Figure 9.8. The energy required for these steps is generated by hydrolysis, the cleavage of phosphate molecules from nucleotide triphosphate compounds. Hydrolysis releases energy and converts nucleotide triphosphates to nucleotide diphosphates (i.e., GTP → GDP). In step ❶ (tRNA recruitment), charged tRNAs affiliated with an EF inspect the open A site. The tRNA with the correct anticodon sequence enters the A site. In step ❷, the enzyme peptidyl transferase catalyzes peptide bond formation between the amino acid at the P site and the newly recruited amino acid at the A site. This elongates the polypeptide and transfers the polypeptide to the tRNA at the A site. The tRNA at the P site departs the ribosome through the E site. In step ❸, EFs translocate the ribosome by moving it in the 3′ direction on mRNA. This translocation step is exactly one codon in length, that is, three nucleotides. Translocation moves the tRNA formerly at the A site to the P site, and opens the A site for binding by a charged tRNA with the correct anticodon sequence.

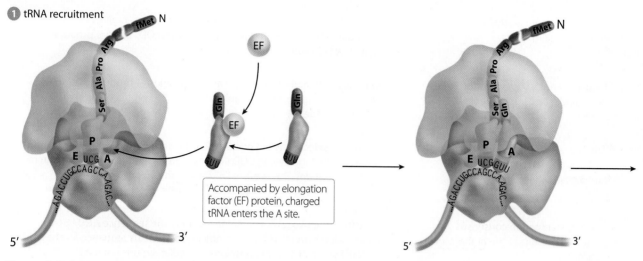

**Figure 9.8** **Bacterial translation elongation.**

**Elongation of Eukaryotic Polypeptides** Distinct elongation factors carry out polypeptide elongation in eukaryotes. But as in bacterial translation, hydrolysis provides the energy for polypeptide elongation, and the steps of polypeptide elongation are nearly identical. Step ❶ in Figure 9.9 illustrates the recruitment of the appropriate charged tRNA to the eukaryotic ribosome A site. Many different charged tRNAs, each associated with EF protein, may inspect the A site. However, only the tRNA with the correct anticodon sequence binds the codon. In step ❷, peptidyl transferase catalyzes peptide bond formation. This switches the polypeptide to the tRNA at the A site and releases the tRNA from the P site. Translocation steps the ribosome along mRNA in step ❸ of polypeptide elongation. The newly opened A site is ready to host base pairing for the next codon–anticodon combination.

## Translation Termination

The elongation cycle continues until one of the three stop codons, UAG, UGA, or UAA, enters the A site of the ribosome. There are no tRNAs with anticodons complementary to stop codons, so the entry of a stop codon into the A site is a translation-terminating event. Bacteria and eukaryotes both use **release factors (RF)** to bind a stop codon in the A site (**Figure 9.10**). The polypeptide bound to tRNA at the P site is released by hydrolysis of GTP, which is complexed with the RF. Polypeptide release causes ejection of the RF from the P site and leads to the separation of the ribosomal subunits.

In bacteria, two release factors, RF1 and RF2, recognize stop codons. RF1 recognizes UAG and UAA, and RF2 recognizes UAA and UGA. Eukaryotic translation is terminated by the action of a single release factor called eRF

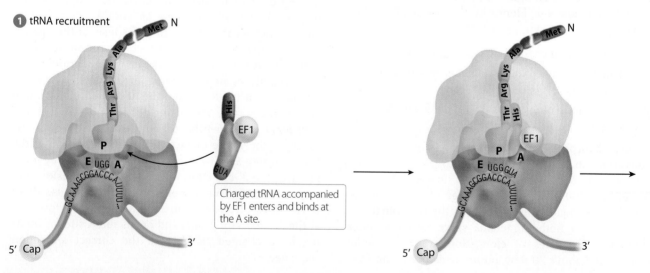

**Figure 9.9** **Eukaryotic translation elongation.**

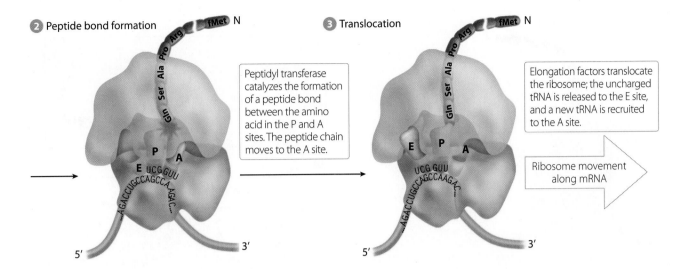

② Peptide bond formation

Peptidyl transferase catalyzes the formation of a peptide bond between the amino acid in the P and A sites. The peptide chain moves to the A site.

③ Translocation

Elongation factors translocate the ribosome; the uncharged tRNA is released to the E site, and a new tRNA is recruited to the A site.

Ribosome movement along mRNA

(eukaryotic release factor) that recognizes all three stop codons (**Genetic Analysis 9.2**).

Genetic Insight    Polypeptides elongate one amino acid at a time. Charged tRNAs enter the A site and use anticodon sequences to base pair with mRNA codons. Peptide bonds form between amino acids at the A site and P site. Ribosomes move 5' to 3' along mRNA. Translation is terminated by release factor binding to a stop codon.

## 9.3 Translation Is Fast and Efficient

With mRNA transcripts of hundreds to thousands of genes in cells, translation is an active and ongoing process that must efficiently initiate, elongate, and terminate polypeptide synthesis. In recent decades, research has uncovered several aspects of the translation machinery that help explain the speed, accuracy, and efficiency of polypeptide production.

### The Translational Complex

Cell biologists estimate that each bacterial cell contains about 20,000 ribosomes, collectively constituting nearly one-quarter of the mass of the cell. The number of ribosomes per eukaryotic cell is variable, but it too is in the tens of thousands. Given these numbers, it is not surprising that translation is almost never a matter of a solitary ribosome translating a single mRNA. Rather, electron micrographs reveal structures called **polyribosomes,** containing groups of ribosomes that are all actively translating the same mRNA (**Figure 9.11**). Each ribosome in the polyribosome structure independently synthesizes a polypeptide, markedly increasing the efficiency of utilization of an mRNA.

Bacterial polyribosomes have a maximum density of about one ribosome per 80 nucleotides, which results in the even spacing of ribosomes seen in Figure 9.11. The ribosomes closest to the 5' end of the mRNA can be seen to have synthesized the shortest length of polypeptide, while ribosomes closer to the 3' mRNA end are farther

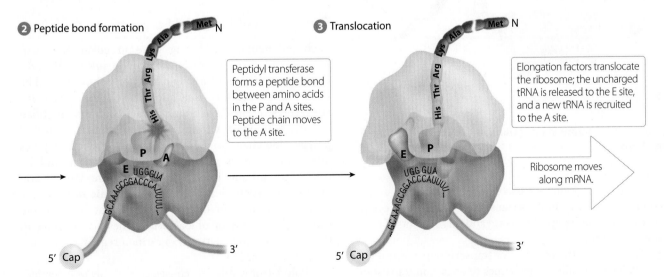

② Peptide bond formation

Peptidyl transferase forms a peptide bond between amino acids in the P and A sites. Peptide chain moves to the A site.

③ Translocation

Elongation factors translocate the ribosome; the uncharged tRNA is released to the E site, and a new tRNA is recruited to the A site.

Ribosome moves along mRNA.

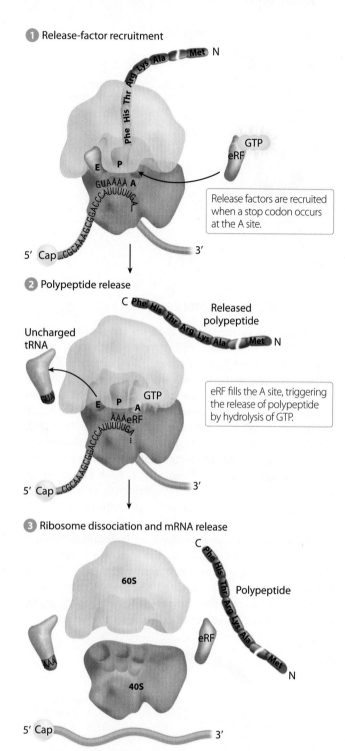

1 Release-factor recruitment

Release factors are recruited when a stop codon occurs at the A site.

2 Polypeptide release

Released polypeptide

Uncharged tRNA

eRF fills the A site, triggering the release of polypeptide by hydrolysis of GTP.

3 Ribosome dissociation and mRNA release

60S

Polypeptide

eRF

40S

5′ Cap

3′

**Figure 9.10   Termination of translation by release factor (RF) proteins.**

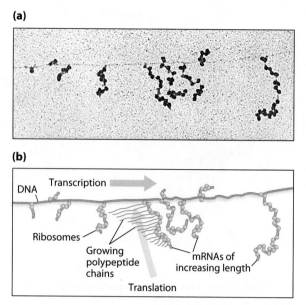

**(a)**

**(b)**

DNA   Transcription

Ribosomes

Growing polypeptide chains

mRNAs of increasing length

Translation

**Figure 9.11   Polyribosomes.** **(a)** Electron micrograph of multiple ribosomes simultaneously translating mRNA molecules. Ribosomes that are closest to the stop codon have the longest polypeptides. **(b)** Artist rendition of the polyribosome electron micrograph. Transcription and translation are coupled in bacteria. Even as DNA is being transcribed into mRNA by RNA polymerase, polyribosomes attach to the still-growing strands of mRNA and begin translating them into polypeptides.

along in translation. As each ribosome reaches a stop codon, RF binding initiates translation-termination events that result in polypeptide release and dissociation of ribosomal subunits.

In bacteria, the coupling of transcription and translation (Chapter 8) allows ribosomes to engage in translation of the 5′ region of mRNAs whose 3′ end is still under construction by RNA polymerase. This coupling is observed in Figure 9.11. Transcription occurs along DNA in the left-hand to right-hand direction. Translation of the mRNA transcripts begins before transcription is complete. In eukaryotes, however, transcription and translation are uncoupled. Transcription takes place in the nucleus, where pre-mRNA is processed to form mature mRNA. Translation occurs in the cytoplasm after release of mature mRNA.

## Translation of Bacterial Polycistronic mRNA

Each polypeptide-producing gene in eukaryotes produces monocistronic mRNA, meaning mRNA that directs the synthesis of a single kind of polypeptide. The scanning model for translation described earlier for eukaryotes implies that a single start codon is identified in eukaryotic mRNA to initiate synthesis of one kind of polypeptide chain. In contrast, groups of bacterial genes often share a single promoter, and the resulting mRNA transcript contains information that synthesizes several different polypeptides. These **polycistronic mRNAs** are produced as part of operon systems that regulate the transcription of sets of bacterial genes functioning in the same metabolic pathway (a form of regulation we discuss in Chapter 15).

Polycistronic mRNAs consist of multiple polypeptide-producing segments—multiple cistrons—that each contain

A portion of an mRNA encoding C-terminal amino acids and the stop codon of a wild-type polypeptide is

5'-…CAACUGCCUGACCCACACUUAUCACUAAGUAGCCUAGCAGUCUGA…-3'

The wild-type amino acid sequence encoded by this portion of mRNA uses the codon 5'-CAA-3' to encode the amino acid Asn. The remainder of the amino acids follow the same reading frame.

N…Asn-Cys-Leu-Thr-His-Thr-Tyr-His-C

The C-terminal ends of three mutant proteins produced by alleles of this gene are as follows.

Mutant 1:   N…Asn-Cys-Leu-Thr-His-Thr-C
Mutant 2:   N…Asn-Cys-Leu-Thr-His-Thr-Tyr-His-Lys-C
Mutant 3:   N…Asn-Cys-Leu-Thr-His-Thr-Tyr-His-Tyr-Ser-Ser-Leu-Ala-Val-C

Identify the mutational events that produce each of the mutant proteins.

| Solution Strategies | Solution Steps |
|---|---|
| **Evaluate** | |
| 1. Identify the topic this problem addresses and explain the nature of the requested answer. | 1. This problem concerns evaluation of the C-terminal end of a wild-type protein sequence and the mRNA segment that encodes it and comparison of the wild-type protein to three mutant proteins to determine the alteration producing each mutant. The answers require the identification of specific mRNA sequence changes leading to each mutant protein. |
| 2. Identify the critical information given in the problem. | 2. In this problem the C-terminal end of a wild-type protein and the mRNA sequence that encodes it are given. Also given are the C-terminal sequences of three mutant proteins encoded by mutant mRNA sequences derived by alteration of the wild-type sequence. |

TIP: Any of three stop codons (UAG, UGA, or UAA) terminates translation immediately after the codon specifying the amino acid at the C terminus of a polypeptide.

| **Deduce** | |
|---|---|
| 3. Use the genetic code to identify the codons corresponding to wild-type amino acids and to identify the stop codon. | 3. Two codons, AAC and AAU, encode asparagine (Asn). If we skip the 5'-most nucleotide of the mRNA sequence and begin reading at the A in the second position, the first codon is AAC followed by UGC-CUG-ACC-CAC-ACU-UAU-CAC-UAA. These codons encode the wild-type amino acids, and UAA is the stop codon. |
| 4. Compare each mutant polypeptide to the wild type and determine which codon contains the mutation. | 4. Mutant 1—The polypeptide sequence is truncated two amino acids short of the normal stop codon. The Tyr codon (UAU) appears to have changed to a stop codon.

Mutant 2—The wild-type sequence is extended by the addition of lysine (Lys), indicating that mutation changed the stop codon to a codon specifying Lys and is now followed immediately by a new stop codon.

Mutant 3—The wild-type sequence is extended by six amino acids. This suggests another mutation affected the stop codon. |

TIP: Examine the wild-type nucleotide sequence at the place where mutation is expected to have occurred, and identify ways in which base substitution, insertion, or deletion could have had the observed effect on the amino acid sequence.

| **Solve** | |
|---|---|
| 5. Identify the mutation and its consequence for translation in Mutant 1. | 5. Two different base substitutions altering the tyrosine (Tyr) codon UAU to a stop codon could cause Mutant 1. The wild-type UAU codon was most likely altered by base substitution to form either a UAA or a UAG stop codon. |
| 6. Identify the mutation and its consequence in Mutant 2. | 6. Lysine (Lys), which was added to the mutant polypeptide, is encoded by AAA or AAG. Deletion of the U from the wild-type stop codon would produce an AAG codon followed by UAG, a stop codon. |
| 7. Identify the mutation and its consequence in Mutant 3. | 7. Tyrosine, specified by codons UAU and UAC, is found in place of the normal stop codon. This is followed by a serine codon (UCN or AGU/C), rather than the GUA (Val) that follows the "in-frame" stop codon in the wild type. A base-pair insertion that adds a U or a C into the third position of the normal UAA stop codon forms a UAU or a UAC tyrosine (Tyr) codon. The altered reading frame from that point would then read AGU (Ser), followed by AGC (Ser), CUA (Leu), GCA (Ala), GUC (Val), and UGA (stop). |

For more practice, see Problems 5, 11, 16, and 29.

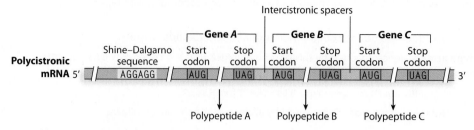

**Figure 9.12 Polycistronic mRNA.** A polycistronic mRNA is a transcript of multiple genes and will produce a polypeptide from each gene. Whether a single ribosome or multiple ribosomes are necessary to produce the polypeptides depends on the length of intercistronic spacers between the polypeptide-coding sequences.

a Shine–Dalgarno site (for ribosome binding and translation initiation) and start and stop codons. An intercistronic spacer sequence that is not translated separates the cistrons of polycistronic mRNA (Figure 9.12).

Intercistronic spacers are variable in length: Some are just a few nucleotides long, and others are 30 to 40 nucleotides long. If the intercistronic spacer is three or four nucleotides, short enough to be spanned by a ribosome, the ribosome may remain intact after completing synthesis of one polypeptide. Thus ribosomes may not necessarily dissociate from the mRNA after encountering a stop codon. Instead, the ribosome can proceed immediately to the next start codon and begin translation of a new polypeptide. On the other hand, if the intercistronic spacer is more than six nucleotides or so in length, the ribosome dissociates and new translation initiation must occur to translate the next polypeptide encoded by the polycistronic mRNA.

## 9.4 The Genetic Code Translates Messenger RNA into Polypeptide

Nucleic acids and amino acids are chemically very different compounds, and there is no *direct* mechanism by which mRNA could synthesize a polypeptide. Nevertheless, the genetic information carried in the nucleotide sequences of mRNA does provide a means by which the amino acid sequences of polypeptides can be specified. The "genetic code" is the name used to describe the correspondence between mRNA codon sequences and individual amino acids.

As mentioned earlier, translation of the code into specific polypeptides depends on transfer RNA (tRNA) to carry amino acids to the ribosome. There, tRNA pairs with mRNA by complementary base pairing between mRNA codon nucleotides and tRNA anticodon nucleotides. Once the correct tRNA is bound by a codon, it transfers its amino acid to the end of a growing polypeptide chain. Transfer RNA molecules facilitate the translation of genetic information from one chemical language (nucleic acid) to another (amino acid). That is, tRNA is an adaptor molecule that interprets and then acts on the information carried in mRNA.

Our basic review of translation and the genetic code depicted a triplet code: Groups of three consecutive mRNA nucleotides form codons that each correspond to one amino acid. The genetic code contains 64 different codons, more than enough to encode the 20 common amino acids used to construct polypeptides. The greater number of codons than amino acids leads to *redundancy* of the genetic code, as evidenced by the observation that single amino acids are specified by from one to as many as six different codons. This redundancy is explained by aspects of how tRNA molecules interact with mRNA codons, the subject of this section.

### The Genetic Code Displays Third-Base Wobble

The triplet genetic code is a biological example of the principle that the simplest hypothesis is the most likely to be correct: During the late 1950s, arithmetic logic led many researchers to conclude that the genetic code was most likely triplet. This simple solution to the question of how amino acid sequences could be coded by nucleic acid sequences posits that a doublet genetic code (two nucleotides per codon) could produce just 16 ($4^2$) combinations of codons, which is not enough different combinations to specify 20 amino acids. On the other hand, a quadruplet genetic code would generate $4^4$, or 256, different combinations of codons—far too many for the needs of genomes. In contrast, a triplet genetic code, yielding $4^3$, or 64 different codons, provides enough variety to encode 20 amino acids with some, but not excessive, redundancy (Figure 9.13 and Table 9.1; also see Table 1.2). Among the 64 codons, 61 specify amino acids, and the remaining 3 are the stop codons that terminate translation. Only two amino acids, methionine (Met), with the codon AUG; and tryptophan (Trp), with the codon UGG; are encoded by single codons. The other 18 amino acids are specified by two to six codons. Codons that specify the same amino acid are called **synonymous codons.**

Each transfer RNA molecule carries a particular amino acid to the ribosome, where complementary base pairing between each mRNA codon sequence and the corresponding anticodon sequence of a correct tRNA takes place. Note that this complementary base pairing

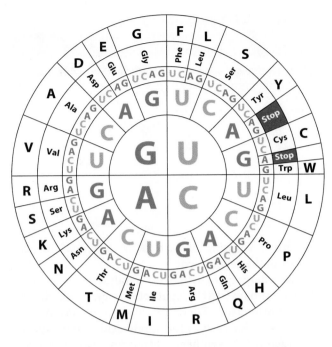

**Figure 9.13 The genetic code.** To read this circular table of the genetic code, start with the inner ring, which contains the nucleotide in the first position (5' nucleotide) of a codon. The second-position nucleotide is in the second ring, and the third-position nucleotide is in the third ring. Three-letter and one-letter abbreviations for the corresponding amino acids occupy the outermost rings.

| Table 9.1 | Redundancy of the Genetic Code | | |
|---|---|---|---|
| **Amino Acid** | **Abbreviation** | | **Codons** |
| | 3-letter | 1-letter | |
| Alanine | Ala | A | GCA, GCC, GCG, GCU |
| Arginine | Arg | R | AGA, AGG, CGA, CGC, CGG, CGU |
| Asparagine | Asn | N | AAC, AAU |
| Aspartic acid | Asp | D | GAC, GAU |
| Cysteine | Cys | C | UGC, UGU |
| Glutamic acid | Glu | E | GAA, GAG |
| Glutamine | Gln | Q | CAA, CAG |
| Glycine | Gly | G | GGA, GGC, GGG, GGU |
| Histidine | His | H | CAC, CAU |
| Isoleucine | Ile | I | AUA, AUC, AUU |
| Leucine | Leu | L | UUA, UUG, CUA, CUC, CUG, CUU |
| Lysine | Lys | K | AAA, AAG |
| Methionine | Met | M | AUG |
| Phenylalanine | Phe | F | UUC, UUU |
| Proline | Pro | P | CCA, CCC, CCG, CCU |
| Serine | Ser | S | AGC, AGU, UCA, UCC, UCG, UCU |
| Threonine | Thr | T | ACA, ACC, ACG, ACU |
| Tryptophan | Trp | W | UGG |
| Tyrosine | Tyr | Y | UAC, UAU |
| Valine | Val | V | GUA, GUC, GUG, GUU |

requires antiparallel alignment of the mRNA and tRNA strands. Consider the codon sequence for aspartic acid (Asp), 5'-GAC-3'. Base-pairing rules predict that the tRNA anticodon sequence is 3'-CUG-5' (**Figure 9.14**). Asp is also specified by a synonymous codon, 5'-GAU-3', that pairs with tRNA carrying the anticodon sequence 5'-CUA-3'. Transfer RNA molecules with different anticodon sequences for the same amino acid are called **isoaccepting tRNAs.**

Does the presence of synonymous codons and isoaccepting tRNAs mean that a genome must provide 61 different tRNA genes and transcribe a tRNA molecule to match each codon? The answer is no. In fact, most genomes have 30 to 50 different tRNA genes. How does a genome that encodes fewer than 61 different tRNA molecules recognize all 61 functional codons? The answer lies in relaxation of the strict complementary base-pairing rules at the third base of the codon. The mechanics of translation provide for flexibility in the pairing of the third base, the 3'-most nucleotide, of the codon. **Third-base wobble** is the name given to the mechanism that relaxes the requirement for complementary base pairing between the third base of a codon and the corresponding nucleotide of its anticodon.

How does third-base wobble work? The answer is found in the chemical structures of nucleotides that hydrogen bond in base-pairing reactions. A careful look at

synonymous codons reveals a pattern to the chemical structure of the third bases in cases of wobble. With the exception of the AUA codon for isoleucine (Ile) and the UGG codon for tryptophan (Trp), synonymous codons can

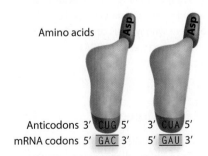

**Figure 9.14 Codon–anticodon pairing.** A pair of isoaccepting aspartic acid tRNAs illustrates complementary antiparallel base-pairing of codon and anticodon sequences.

be grouped into pairs that have the same two nucleotides in the first and second positions and differ only at the third base, where the synonymous codons either both carry a purine (A or G) or both carry a pyrimidine (C or U). For example, consider the synonymous pairs of codons for histidine (His) and glutamine (Gln; see Figure 9.13 and Table 9.1). The first two bases of each of these codons are C and A. Both His codons have a pyrimidine at the third position, whereas the Gln codons have a purine in the third position. As you look at other pairs of synonymous codons in Table 9.1, notice that they also differ only by carrying the alternative purine or pyrimidine nucleotide at the third position.

Amino acids specified by four synonymous codons, such as alanine (Ala), valine (Val), and glycine (Gly), display an analogous pattern: Each amino acid is represented by two pairs of synonymous codons, and the members of each pair differ in the third position only, by carrying the alternate purine or pyrimidine. The pattern continues in arginine (Arg), serine (Ser), and leucine (Leu), each of which is specified by six synonymous codons. These sets of codons each consist of three pairs, each pair having the same nucleotides in the first two positions and differing by having the alternate purine or pyrimidine in the third position.

Third-base wobble occurs through flexible base pairing between the wobble nucleotide—that is, the 3′ nucleotide of a codon—and the 5′ nucleotide of an anticodon. At the wobble position, base pairing between the nucleotides of the codon and the anticodon need not be complementary. They must, however, involve a purine and a pyrimidine. Third-base wobble pairings are summarized in Table 9.2. The wobble nucleotides in anticodons can be any of the RNA nucleotides or may be the modified nucleotide **inosine (I).** Inosine is structurally similar to G but lacks the amino group attached to guanine's 2 carbon. Because of this difference, inosine base-pairs with either purines or pyrimidines. Figure 9.15 shows three examples of third-base wobble, in which three tRNA molecules collectively recognize seven different codons.

---

Genetic Insight  The triplet genetic code consists of 61 codons specifying 20 common amino acids and 3 stop codons. The code is redundant because 18 of the amino acids have two or more synonymous codons. Synonymous codons are recognized by isoaccepting tRNAs that may utilize third-base wobble.

---

## Charging tRNA Molecules

Transfer RNA molecules are transcribed from tRNA genes. Recall the three-dimensional structure of tRNAs (see Figure 8.35) and the CCA terminus at the 3′ end of tRNA molecules as the site of attachment of an amino acid. Each tRNA carries only one of the 20 amino acids,

| Table 9.2 | Third-Base Wobble Pairing between Codon and Anticodon Nucleotides |
| --- | --- |
| **3′ Nucleotide *of Codon*** | **5′ Nucleotide *of Anticodon*** |
| A or G | U |
| G | C |
| U | A |
| U or C | G |
| U, C, or A | I |

and correct charging of each tRNA is crucial for the integrity of the genetic code.

The charging of tRNAs is catalyzed by enzymes called **aminoacyl-tRNA synthetases** or, more simply, **tRNA synthetases.** There are 20 different tRNA synthetases, one for each of the amino acids. To charge an uncharged tRNA, synthetase catalyzes a two-step reaction that forms a bond between the carboxyl group of the amino acid and the 3′ hydroxyl group of adenine in the CCA terminus. Experimental analysis reveals that the recognition of isoaccepting tRNAs by tRNA synthetase is a complex process that does not follow a single set of rules. Mutations in any of the four arms of tRNA, or in the anticodon sequence itself, render a tRNA unrecognizable to its tRNA synthetase.

Studies of structural interactions between tRNA synthetases and their tRNAs show tRNA synthetase to be a large molecule that contacts several parts of a tRNA as part of the recognition process. These contact points can include the anticodon sequence and the other arms and loops of the tRNA (Figure 9.16). Once in

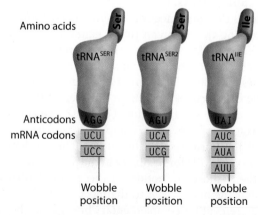

**Figure 9.15   Effect of wobble.** Wobble base pairing reduces the number of different tRNAs required during translation. In this example, two different tRNAs, each carrying serine, use wobble to recognize two distinct codons. A single isoleucine-carrying tRNA uses wobble to recognize three isoleucine codons.

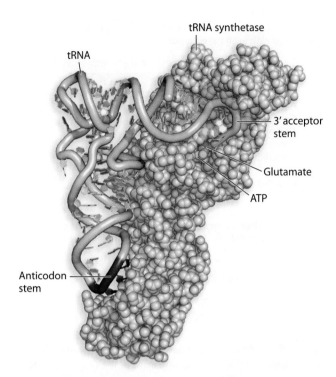

**Figure 9.16 Interaction of aminoacyl-tRNA synthetase with tRNA.** Aminoacyl-tRNA synthetase contacts multiple points on tRNA.  ATP and the 3′ acceptor stem of tRNA fit in a cleft that also accommodates the amino acid.

contact with tRNA synthetase, the tRNA acceptor stem fits into an active site of tRNA synthetase. The active site contains the amino acid that will be added to the tRNA acceptor stem and ATP that provides energy for amino acid attachment.

The specific relationships between tRNAs and tRNA synthetases is a crucial aspect of translation that is essentially invariably accurate and underlies the function of the genetic code. Errors in charging tRNAs are rare. Such mistakes, called mischarging, occur in 1 in 50,000 to 1 in 100,000 charged tRNAs. The level of mischarging is so low because tRNA synthetase operates a "proofreading" system that usually discards the incorrect amino acid before it is attached to tRNA.

---

Genetic Insight  Twenty different tRNA synthetases, one for each amino acid, are responsible for recognizing the correct isoaccepting tRNAs to which each synthetase attaches its amino acid.

---

## 9.5 Experiments Deciphered the Genetic Code

A remarkable set of experiments performed in the early 1960s deciphered the genetic code and opened the way for biologists to understand the molecular processes that convert a messenger RNA nucleotide sequence into a polypeptide. At the time, biologists knew *what* the hereditary material was (DNA), and they knew *what* molecule

conveyed the genetic message to ribosomes for translation (mRNA), but they did not know *how* the protein-coding information carried by messenger RNA was deciphered during the assembly of polypeptides. Several questions had to be answered about the structural nature of the genetic code before the code itself could be deciphered. The three most important questions, listed here, are examined in the sections below:

1. Do neighboring codons overlap one another, or is each codon a separate sequence?
2. How many nucleotides make up a messenger RNA codon?
3. Is the polypeptide-coding information of messenger RNA continuous, or is coding information interrupted by gaps?

### No Overlap in the Genetic Code

To address the question of whether mRNA carried overlapping codons, experimenters reasoned that if the genetic code were nonoverlapping, each messenger RNA nucleotide would be part of one and only one codon, whereas in an overlapping genetic code, a nucleotide could be part of multiple codons. An overlapping genetic code would provide an efficient way to package a large amount of information using a small number of nucleotides, but it would also restrict the flexibility of the genetic code and could potentially increase the impact of single-nucleotide mutations.

Consider the partial messenger RNA sequence

…ACUAAG…

If the genetic code is triplet and nonoverlapping (recall that a doublet code does not provide enough codons to specify 20 amino acids, and a quadruplet code provides far too many), this partial sequence produces two codons, each specifying an amino acid:

| codon | 1 | 2 |
|---|---|---|
| | …ACU | AAG… |
| amino acid | 1 | 2 |

In an overlapping triplet genetic code, on the other hand, these six nucleotides would spell out four complete codons and two partial codons. The sequence would fully encode four amino acids and contribute to the coding of two others:

…ACUAAG…

| amino acid | 1 | ACU |
|---|---|---|
| | 2 | CUA |
| | 3 | UAA |
| | 4 | AAG |
| | 5 | AG… |
| | 6 | G… |

In 1957, based on his analysis of the available information on amino acid sequences of proteins, Sidney Brenner became convinced that an overlapping triplet genetic code was impossible because it was too restrictive. We can use the codon AAG to illustrate Brenner's reasoning. If the genetic code were overlapping, the codon immediately upstream of an AAG codon would always have to carry two adenine nucleotides in the middle and 3′ nucleotide positions to correspond to the first two positions of AAG. The triplet preceding an AAG codon could therefore be AAA, CAA, UAA, or GAA; but even if all four of those codons specified a different amino acid, an overlapping code would restrict the amino acid upstream of the one specified by AAG to being one of only four other amino acids. In one test, Brenner examined the upstream neighbor of each AAG lysine in a polypeptide and found 17 different amino acids in that position. Brenner's conclusion was that an overlapping genetic code restricted evolutionary flexibility and was unsupported by biochemical observations.

Conclusive evidence of a nonoverlapping genetic code came from a 1960 study of single-nucleotide substitutions induced by the mutation-producing compound nitrous oxide. Heinz Fraenkel-Conrat and his colleagues studied the effect of nitrous oxide on the coat protein of tobacco mosaic virus (TMV). Nitrous oxide causes mutations by inducing single base-pair substitutions in DNA that lead to mutant mRNA molecules with one nucleotide base change compared to wild-type mRNA. A single base change in mRNA would alter *three consecutive codons* if the genetic code were overlapping, but just a *single codon* if the genetic code were nonoverlapping (**Figure 9.17a**). A single base substitution in an overlapping genetic code could therefore change three consecutive amino acids in a protein, whereas in a nonoverlapping genetic code, a single base substitution could change just one amino acid. Fraenkel-Conrat's analysis revealed that only single amino acid changes occurred as a result of mutation by nitrous oxide (**Figure 9.17b**). This result is consistent with that predicted for a nonoverlapping genetic code, and it is inconsistent with the prediction for an overlapping genetic code.

## A Triplet Genetic Code

Proof of a triplet genetic code came in 1961 when Francis Crick, Leslie Barnett, Sidney Brenner, and R. J. Watts-Tobin used the compound proflavin to create mutations in a gene called *rII* in T4 bacteriophage. Proflavin causes mutations by inserting or deleting single base pairs from DNA. This deletion leads to the absence of single nucleotides from mRNA, thus changing the reading frame of the mRNA. **Reading frame** refers to the specific codon sequence as determined by the point at which the grouping of nucleotides into triplets begins.

To more fully grasp the importance of the experiment by Crick and his colleagues, we offer an analogy to guide

**(a)** An overlapping genetic code would change three consecutive codons with each base mutation.

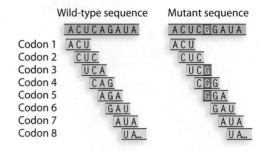

**(b)** A nonoverlapping genetic code would change one codon with each base mutation.

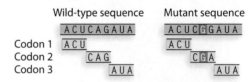

**Figure 9.17** **Proof that the genetic code is nonoverlapping.** The sequence of the last 10 amino acids at the C-terminal end of a TMV protein contained a single amino acid change following the induction of base-substitution mutation. This result conforms to the prediction of the nonoverlapping model of the genetic code.

understanding of the experimental results. Just as the string of three-letter words below becomes unintelligible upon the addition or deletion of a single letter, a single-nucleotide change shifts by one nucleotide the triplet reading frame used to decipher the information in a gene. The addition or deletion of nucleotides changes the reading frame and produces a mutation called a **frameshift mutation.** Single-letter additions or deletions garble the translated message by changing the reading frame:

wild-type: YOUMAYNOWSIPTHETEA ("you may now sip the tea")
mutant (addition): YOUMA c YNOWSIPTHETEA ("you ma c yno wsi pth ete a")
mutant (deletion): YOUMAYNO | | SIPTHETEA ("you may nos ipt het ea")

Frameshift mutations can be reverted—that is, the correct reading frame can be restored—if a second mutation in a different location within the same gene restores the reading frame. This second mutation, called a **reversion mutation,** counteracts ("reverses") the reading frame disruption by inserting a nucleotide, if the initial mutation was a deletion, or by deleting a nucleotide, if the initial mutation was an insertion. Reversion mutations can restore most of the wild-type reading frame so that a portion of the amino acid sequence of the polypeptide encoded by the gene will be wild type. The reading frame, and thus the amino acid sequence, between the two mutation sites will, however, be altered. For example, here is how the two frameshift mutations shown above might be reverted:

mutant (addition): YOUMA C YNOWSIPTHETEA (you mac yno wsi pth ete a)

reversion mutant (deletion): YOUMA C YNO | | SIPTHETEA ("you mac yno sip the tea")

mutant (deletion): YOUMAYNO | | SIPTHETEA ("you may nos ipt het ea")

reversion (addition): YOUMAYNOSIP R THE TEA ("you may nos ipr the tea")

Crick and his colleagues analyzed numerous bacteriophage proflavin-induced *rII*-gene mutants, designating each addition mutant as a (+) and each deletion mutation as a (−). Their approach to data interpretation in this analysis was unusual in two respects. First, they *assumed* that the genetic code had a nonoverlapping triplet structure and interpreted their data from that perspective. In other words, they had already reasoned their way to a conclusion about the structure of the genetic code, and they sought data that could confirm their reasoning. Second, because the available tools of molecular analysis were not sophisticated enough to tell Crick and his colleagues whether a particular mutant resulted from an insertion or a deletion of nucleotide base pairs, they *guessed* that the first *rII*-gene mutant they examined, a mutation designated FC 0, resulted from insertion ("FC" stands for Francis Crick). Designating FC 0 as a (+) mutation turned out to be a correct guess. Based on their assumptions that (1) the genetic code is a nonoverlapping triplet and (2) FC 0 is an insertion (+) mutation, the data reported by Crick and colleagues supported the notion that the genetic code is based on nucleotide triplets.

Data on six of these mutants is displayed in **Table 9.3**. Each mutant is designated either (+) or (−). Any combination of a (+) mutant and a (−) mutant generates a wild-type revertant. In each case, the initial mutation causes a frameshift mutation, and the reversion mutation restores the reading frame. On the other hand, any two (+) mutants or any two (−) mutants together leave the *rII* gene in a mutant state because the second mutation does not restore the reading frame. The triplet structure of the genetic code is further demonstrated by the observation that the reading frame is restored by the presence of *three* (+) mutations or *three* (−) mutations. For example, the total of three insertions restores the reading frame in the following sentence:

triple mutant (addition):
YOUMA C YNOW T S L IPTHETEA ("you ma c yno w t s l ip the tea")

## No Gaps in the Genetic Code

In their 1961 research, Crick and colleagues also suggested that the genetic code is read as a continuous string of mRNA nucleotides uninterrupted by any kind of gap, space, or pause. If a gap or spacer were present

| Table 9.3 | Phenotype Resulting from Various Combinations of Proflavin-Induced Base-Pair Insertion (+) and Deletion (−) Mutations at the *rII* Locus of Bacteriophage T4 |
|---|---|

| Combined Mutations | +/− Designations | Result |
|---|---|---|
| FC 0, FC 1 | + − | Wild-type revertant |
| FC 0, FC 21 | + − | Wild-type revertant |
| FC 40, FC 1 | + − | Wild-type revertant |
| FC 58, FC 1 | + − | Wild-type revertant |
| FC 0, FC 40, FC 58 | + + + | Wild-type revertant |
| FC 1, FC 21, FC 23 | − − − | Wild-type revertant |
| FC 0, FC 40 | + + | *rII* mutant |
| FC 0, FC 58 | + + | *rII* mutant |
| FC 1, FC 21 | − − | *rII* mutant |
| FC 1, FC 23 | − − | *rII* mutant |

between mRNA codons, the mRNA transcript might be represented as follows (*x* indicates the gap between codons):

YOUxMAYxNOWxSIPxTHExTEAx ("you may now sip the tea")

If the genetic code were structured in some such way, with each codon set off from its neighbors, insertion or deletion of a nucleotide would not cause the kind of frameshift mutation that Crick and colleagues had observed. Instead, insertion or deletion of nucleotides could be expected to alter the affected codon but not the identity of adjoining codons. For example, consider the following insertion mutation, where the separation between codons confines the alteration to a single word:

YOUx,MA T Yx,NOWx,SIPx,THEx,TEAx, ("you ma t y now sip the tea")

---

Genetic Insight  The genetic code is a triplet code composed of three nucleotides per codon. Codons do not overlap one another, and there are no gaps or markers between codons.

---

## Deciphering the Genetic Code

The genetic code was deciphered in a series of experiments performed between 1961 and 1965. This remarkable 4-year period in biology was highlighted by extensive collaborative and competitive international research that culminated in the assembly of a simple table containing the instructions shared by all organisms for translating mRNA nucleotide sequences into polypeptide sequences. Deciphering the genetic code was a milestone in establishing the mechanism of the central dogma of biology (DNA → RNA → protein) and laying the molecular foundation for modern genetic research. This triumph of deductive reasoning was instantaneously recognized for its profound significance, and it resulted in the awarding of a Nobel Prize in Physiology or Medicine to Har Gobind Khorana and Marshall Nirenberg in 1968.

Once it had been established that the genetic code consists of triplets, researchers sprang to the task of establishing which triplets are associated with each amino acid in the process of translation. Nirenberg and Johann Heinrich Matthaei performed a simple experiment in 1961 that laid the groundwork for later experiments in deciphering the genetic code. Their experimental design was straightforward: Construct synthetic strings of repeating nucleotides, and use an in vitro translation system to translate the sequence into a polypeptide. For example, Nirenberg and Matthaei synthesized an artificial mRNA containing only uracils, known as a poly(U). They devised an in vitro translation system composed of the known cellular components of bacterial translation—ribosomes, charged transfer RNA molecules, and essential translational proteins. Regardless of where translation might begin along the poly(U) mRNA, the only possible codon it contained was UUU. The researchers were therefore hoping to determine which amino acid corresponds to the UUU codon.

Twenty separate in vitro translations of poly(U) mRNA were carried out, each time using a pool of 19 unlabeled amino acids and one amino acid labeled with radioactive carbon ($C^{14}$). To determine which amino acid is encoded by poly(U) mRNA, Nirenberg and Matthaei used a different radioactive amino acid in each translation. They detected production of a highly radioactive polypeptide after conducting translation in a system containing radioactively labeled phenylalanine (**Figure 9.18**). The radioactive polypeptide was poly-phenylalanine (poly-Phe). Since the only possible triplet codon in mRNA is UUU, Nirenberg and Matthaei reasoned that 5'-UUU-3' codes for phenylalanine. They went on to construct poly(A), poly(C), and poly(G) synthetic mRNAs and identified 5'-AAA-3' as a codon for lysine (Lys), 5'-CCC-3' as a proline (Pro) codon, and 5'-GGG-3' as a codon for glycine (Gly) (**Table 9.4**).

Khorana adapted the experimental strategy of Nirenberg and Matthaei to synthesize mRNA molecules that contained di-, tri-, and tetranucleotide repeats. His construction of repeat-sequence mRNAs allowed him to define many additional codons (see Table 10.4). For example, Khorana used the dinucleotide repeat UC to form a synthetic mRNA with the sequence

$$5'-UCUCUCUCUCUCUCUC-3'$$

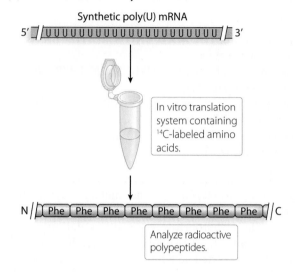

**(a)** In vitro translation of synthetic mRNA

Synthetic poly(U) mRNA

5' UUUUUUUUUUUUUUUUUUUUUUU 3'

In vitro translation system containing ¹⁴C-labeled amino acids.

N [ Phe | Phe | Phe | Phe | Phe | Phe | Phe | Phe ] C

Analyze radioactive polypeptides.

**(b)**  Incorporation of ¹⁴C-labeled phenylalanine into polypeptides

| Synthetic mRNA | Radioactivity (counts/min) |
|---|---|
| None | 44 |
| Poly(U) | 39,800 |
| Poly(A) | 50 |
| Poly(C) | 38 |

**Figure 9.18  Use of synthetic mRNAs to determine genetic code possibilities. (a)** Synthetic poly(U) mRNA is translated in vitro in the presence of individual ¹⁴C-labeled amino acids. A polypeptide consisting of phenylalanine is formed. **(b)** These radioactivity counts demonstrate that only poly(U) synthetic mRNA incorporates radioactive phenylalanine into a polypeptide.

This mRNA can be translated in either a reading frame that begins with uracil or a reading frame that begins with cytosine. In both cases, the reading frame produces alternating UCU-CUC codons. Khorana identified the amino acids of the resulting polypeptide and found it contained alternating serine (Ser) and leucine (Leu), but he could not be sure which codon specified Ser and which encoded Leu. Following a similar strategy, Khorana also examined the polypeptides produced by mRNAs composed of repeating AG dinucleotides (5'-...AGAGAGAG...-3'), as well as dinucleotides of UG and AC. He found that two alternating amino acids made up the resulting polypeptides, thus narrowing the potential correspondences between codons and amino acids.

When Khorana used mRNA containing trinucleotide repeats, most of these mRNAs produced three different polypeptides that each consisted of only one kind of amino acid. For example, the reading frame for poly-UUC can begin with either of the uracils or with cytosine. Messenger RNA is read as consecutive UUC codons if the first uracil initiates the reading frame, as UCU if the second uracil begins the reading frame, or as CUU if cytosine is at the start of the reading frame. Although the different reading frames each produced a polypeptide containing one amino acid, Khorana was again unsure which codon specified which amino acid.

| Table 9.4 | Polypeptide Production from Synthetic mRNAs | | |
|---|---|---|---|
| **Synthetic mRNA** | **mRNA Sequence** | **Polypeptides Synthesized** | **Observation** |
| Repeating nucleotides | Poly-U  UUUU... | Phe- Phe- Phe... | Polypeptides have one amino acid. |
| | Poly-C  CCCC... | Pro- Pro- Pro | |
| | Poly-A  AAAA... | Lys-Lys-Lys | |
| | Poly-G  GGGG... | Gly-Gly-Gly | |
| Repeating dinucleotides | Poly-UC  UCUC... | Ser-Leu-Ser-Leu | Polypeptides have two alternating amino acids. |
| | Poly-AG  AGAG... | Arg-Glu-Arg-Glu | |
| | Poly-UG  UGUG... | Cys-Val-Cys-Val | |
| | Poly-AC  ACAC... | Thr-His-Thr-His | |
| Repeating trinucleotides | Poly-UUC  UUCUUCUUC... | Phe-Phe...and Ser-Ser...and Leu-Leu... | Three polypeptides have one amino acid each. |
| | Poly-AAG  AAGAAGAAG... | Lys-Lys...and Arg-Arg...and Glu-Glu | |
| | Poly-UUG  UUGUUGUUG... | Leu-Leu...and Cys-Cys...and Val-Val | |
| | Poly-UAC  UACUACUAC... | Tyr-Tyr...and Thr-Thr...and Leu-Leu | |
| Repeating tetranucleotides | Poly-UAUC  UAUCUAUC... | Tyr-Leu-Ser-Ile-Tyr-Leu-Ser-Ile | Some polypeptides have four repeating amino acids. |
| | Poly-UUAC  UUACUUAC... | Leu-Leu-Thr-Tyr-Leu-Leu-Thr-Tyr | |
| | Poly-GUAA  GUAAGUAA... | None (UAA stop codon) | Must include one or more stop codons. |
| | Poly-GAUA  GAUAGAUA... | None (UAG stop codon) | |

*Note:* Data adapted from Khorana (1967).

Khorana's analysis of repeating tetranucleotides confirmed the redundancy of the genetic code and the existence of stop codons. Poly-UUAC contains four repeating codons regardless of which nucleotide initiates the reading frame. The codons UUA - UAC - ACU - CUU repeat in an ongoing pattern. The polypeptide product contains four repeating amino acids, Leu-Leu-Thr-Tyr. It was clear that *two* of the four codons specify leucine, although Khorana could not determine which two codons are redundant. The tetranucleotide repeat GUAA yields an mRNA with four repeating codons, GUA - UAA - AAG - AGU, but generates *no* polypeptide. From this the researchers deduced that one of these codons is a stop codon.

Nirenberg and Philip Leder contributed the final piece of the genetic code puzzle in 1964 when they devised an experiment to resolve the ambiguities of codon identity remaining from Khorana's experiments. Nirenberg and Leder synthesized many different mini-mRNAs that were each just three nucleotides in length (**Figure 9.19**). The tiny mRNAs were added individually to in vitro translation systems containing ribosomes, along with 19 unlabeled amino acids and one [14]C-labeled amino acid, all attached to different transfer RNA molecules. The mRNA formed a complex with the ribosome, and the tRNA charged with the corresponding amino acid. Each in vitro mixture was

then poured through a filter that captured the large ribosome–mRNA–tRNA complexes but permitted noncomplexed molecules of mRNA or tRNA to pass through. The filter was subsequently tested to determine if the three-nucleotide mRNA sequence bound a transfer RNA with the radioactive amino acid. Nirenberg and Leder tested all 64 combinations of nucleotides with their tiny mRNA system and were able to identify codon–amino acid correspondences for the entire genetic code. In addition, they identified the nucleotide composition of the three stop codons, UAA, UAG, and UGA (**Genetic Analysis 9.3**).

**Genetic Insight** The genetic code was deciphered through a series of experiments that utilized synthetic mRNAs and an in vitro translation system. Of 64 possible codons, 61 correspond to one of the 20 common amino acids; the remaining 3 are stop codons that signal a halt to translation. The code is redundant, meaning that most amino acids are encoded by more than one codon.

## The (Almost) Universal Genetic Code

In astonishing testimony to a single origin of life on Earth and to the power of evolution to maintain virtually complete uniformity over hundreds of millions of years, every

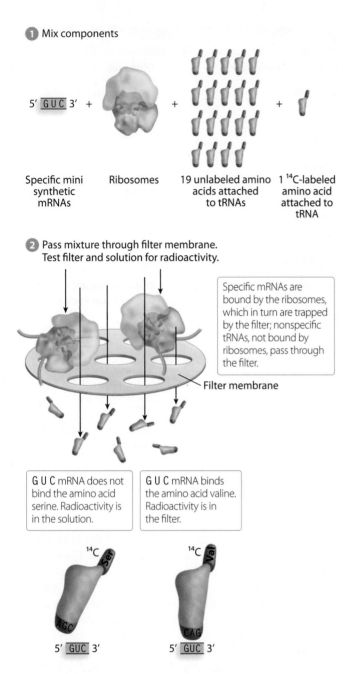

**1** Mix components

5′ GUC 3′  +  Ribosomes  +  19 unlabeled amino acids attached to tRNAs  +  1 ¹⁴C-labeled amino acid attached to tRNA

Specific mini synthetic mRNAs

**2** Pass mixture through filter membrane. Test filter and solution for radioactivity.

Specific mRNAs are bound by the ribosomes, which in turn are trapped by the filter; nonspecific tRNAs, not bound by ribosomes, pass through the filter.

Filter membrane

GUC mRNA does not bind the amino acid serine. Radioactivity is in the solution.

GUC mRNA binds the amino acid valine. Radioactivity is in the filter.

5′ GUC 3′

5′ GUC 3′

**Figure 9.19  Deciphering the genetic code with synthetic mini mRNAs.** For the synthetic mini mRNA GUC, a ¹⁴C-labeled serine tRNA does not hybridize within the ribosome to form a complex, and radioactivity is located in the pass-through solution. ¹⁴C-labeled valine tRNA does hybridize to the GUC mini mRNA within the ribosome. The mRNA-ribosome-tRNA complex is caught by the filter membrane, where radioactivity is detected.

living organism uses the same genetic code to synthesize polypeptides. In all living things, from bacteria to humans, the hereditary script carried by any given mRNA is translated by a similar mechanism and produces the same polypeptide. The universality of the genetic code makes it possible to use bacterial systems to express biologically important protein products found in plants or animals.

The production of human insulin to treat diabetes and of factor VIII protein to treat hemophilia are two of numerous examples of recombinant human gene cloning that are possible in part because bacteria and humans use the same genetic code for translation.

As with most general rules, however, there are a few exceptions to the universality of the genetic code; thus, biologists characterize the genetic code as *almost* universal. The exceptions are found principally in mitochondria, which are specially adapted to life within plant and animal cells, but two exceptions occur in free-living organisms as well (**Table 9.5**). The near universality of the genetic code presents two important evolutionary questions. First, why has the genetic code remained essentially unchanged in living organisms; and second, why have changes evolved in mitochondria? The answer to the first question is that natural selection pressure against codon change is intense. A single codon change would dramatically alter the composition of almost every polypeptide an organism produces. Countless evolutionary examples tell us that nearly all of the changes that occur would be deleterious, and many would be lethal. Simply stated, a change in the genetic code would alter the rules of the game of life, and natural selection prevents such changes.

Do the exceptions to the universality of the genetic code in mitochondria suggest that natural selection occasionally operates less intensively? The answer appears to be yes. The genomes of mitochondria found in plant and animal cells are small compared to nuclear genomes. They carry up to about 25 genes and produce about 10 to 15 polypeptides. Any disruption caused by a change in the mitochondrial genetic code is likely to be limited, since

| Table 9.5 | Genomes Using Modifications of the Universal Genetic Code | | |
|---|---|---|---|
| Codon | Universal Code | Unusual Code | Genome |
| AGA, AGG | Arg | Stop | Mitochondria in plants, animals, and yeast |
| AUA, AUU | Ile | Met | Mitochondria in plants, animals, and yeast |
| UGA | Stop | Trp | Mitochondria in plants, animals, and yeast |
| CUN[a] | Leu | Thr | Mitochondria in yeast |
| UAA, UAG | Stop | Gln | Green algae, protozoa |
| UGA | Stop | Cys | Protozoa |

N[a] = any third-position nucleotide.

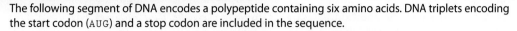

The following segment of DNA encodes a polypeptide containing six amino acids. DNA triplets encoding the start codon (AUG) and a stop codon are included in the sequence.

```
5'-…CCCAGCCTAGCCTTTGCAAGAGGCCATATCGAC…-3'
3'-…GGGTCGGATCGGAAACGTTCTCCGGTATAGCTG…-5'
```

**a.** Identify the sequence and polarity of the mRNA encoded by this gene.

**b.** Determine the amino acid sequence of the polypeptide, and identify the N- and C-terminal ends of the polypeptide.

**c.** Base-substitution mutation changes the first transcribed G of the template strand to an A. How does this alter the polypeptide?

| Solution Strategies | Solution Steps |
|---|---|
| **Evaluate** | |
| 1. Identify the topic this problem addresses and explain the nature of the requested answer. | 1. This problem concerns the identification of DNA coding and template DNA strands, the protein encoded by DNA, and an evaluation of a mutation of the DNA sequence. The answer requires identification of the DNA strands, identification of start and stop codons, and determination of the amino acid sequence of wild-type and mutant proteins. |
| 2. Identify the critical information given in the problem. | 2. DNA sequence that includes a start (AUG) codon and a stop codon is given. |
| **Deduce** | |
| 3. Identify the start codon by inspecting both DNA strands for 3'-TAC-5' that potentially encodes a start (AUG) codon on the template strand. <br><br> TIP: The AUG start codon is the most common codon for translation initiation and is encoded by the DNA triplet 3'-TAC-5'. | 3. Scanning both DNA strands in their 3' to 5' direction identifies a single 3'-TAC-5' sequence. The sequence is on the upper strand of the sequence beginning with the seventh nucleotide from the right. |
| 4. Survey the putative template strand identified in the previous step and determine if DNA triplets 3'-ATC-5', 3'-ACT-5', and 3'-ATT-5' encoding possible stop codons occur as the seventh codon of an mRNA sequence. <br><br> TIP: The stop codons UAG, UGA, and UAA are encoded by DNA triplets 3'-ATC-5', 3'-ACT-5', and 3'-ATT-5'. | 4. Since just one DNA triplet encoding a start codon is present, a scan of the strand finds a 3'-ATC-5' triplet sequence encoding a UAG stop codon. The start and stop triplets are highlighted on the template strand: <br><br> Stop Start 5'-CCCAGC CTA GCCTTTGCAAGAGGC CAT ATCGAC-3' |
| **Solve** | |
| 5. Identify the mRNA sequence encoding the six amino acids of the polypeptide. <br><br> TIP: The mRNA sequence can be determined from either the coding strand or the template strand of DNA. | TIP: Substituting U for T on the coding strand produces mRNA sequence. Alternatively, arranging RNA nucleotides complementary to the template strand and assigning antiparallel polarity produces mRNA. <br><br> Answer a <br> 5. The mRNA sequence is <br><br> 5'-AUG GCC UCU UGC AAA GGC UAG-3' |
| 6. List the amino acid sequence of the polypeptide. | Answer b <br> 6. The polypeptide sequence is <br><br> N-Met-Ala-Ser-Cys-Lys-Gly-C |
| 7. Identify the effect of the G → A base substitution on the polypeptide. | Answer c <br> 7. Substituting the first transcribed G → A alters the second codon of mRNA by changing GCC → GUC and substitutes valine (Val) for alanine (Ala) in the second position of the polypeptide sequence. |

For more practice, see Problems 1, 5, 29, 30, and 31.

the number of genes affected is so small. In addition, there are many mitochondria per cell, providing "backup copies" of the mitochondrial genome. If a change in the genetic code severely disrupts the function of one mitochondrion, others are present in the cell to carry out normal activities. Thus, changes were able to accrue in the mitochondrial genetic code, most likely in codons that are not extensively used.

## Transfer RNAs and Genetic Code Specificity

In our discussion of the genetic code and polypeptide assembly at the ribosome, we describe the specific base-pair interaction between the anticodon sequence of charged tRNA and the codon sequence of mRNA as the key to incorporating the correct amino acid into the polypeptide. But how did biologists determine that the specificity of the genetic code resides in the tRNA–mRNA interaction and not in the recognition of the amino acid carried by tRNA?

The answer came from a simple and clever experiment by Francois Chapeville and several colleagues in 1962. The researchers began by preparing normal cysteine-charged tRNAs. This complex is designated Cys-tRNA$^{Cys}$. The researchers then treated Cys-tRNA$^{Cys}$ with the compound Raney nickel that removes an SH group from cysteine and converts it to alanine. This treatment produces Ala-tRNA$^{Cys}$ in which alanine rather than cysteine is attached to tRNA$^{Cys}$. When Chapeville and colleagues used Ala-tRNA$^{Cys}$ in an in vitro translation reaction, the polypeptide contained alanine rather than cysteine in amino acid positions that would normally carry cysteine. In other words, Ala-tRNA$^{Cys}$ efficiently paired with mRNA codons specifying cysteine and deposited alanine in the nascent polypeptide, even though the mRNA sequence specified cysteine.

Two important conclusions come from this experiment. First, the genetic code derives its specificity through the complementary base-pair interaction of tRNA and mRNA. The amino acid carried by charged tRNA does not play a role in determining which amino acids are incorporated into polypeptides. Rather, tRNA alone—acting through the base-pairing interaction of its anticodon with the codon of mRNA—gives specificity to the genetic code. The second conclusion is to demonstrate the importance of the fidelity of aminoacyl-tRNA synthetases in correctly recognizing their cognate tRNAs and charging them with the proper amino acid.

---

**Genetic Insight** The genetic code is near universal and relies on properly charged tRNAs to carry the correct amino acids and to bind codons to deliver amino acids to the nascent polypeptide.

## 9.6 Translation Is Followed by Polypeptide Processing and Protein Sorting

Translation produces polypeptides, but the production of functional proteins from these polypeptides is often not completed until the polypeptides are chemically modified and transported to the cellular or extracellular locations where they are active. Two categories of posttranslational events modify and transport polypeptides. One is **posttranslational polypeptide processing,** which modifies polypeptides into functional proteins through the removal or chemical alteration of amino acids after translation is completed. The other, **protein sorting,** is a separate, genetically controlled process that uses **signal sequences,** also called **leader sequences,** short sequences of amino acids at the N-terminal end of a polypeptide, to sort proteins and direct them to their cellular destinations. Both categories of processing are common in both bacteria and eukaryotes. Protein sorting is needed in bacterial cells because of the many proteins specifically destined for the cell membrane. In eukaryotes, however, protein sorting is far more complex than in bacteria, dispatching proteins to particular cellular organelles, such as the chloroplast, mitochondrion, lysosome, and nucleus as well as causing certain proteins to be secreted from the cells.

### Posttranslational Processing

The removal of one or more amino acids from a polypeptide is a common form of posttranslational polypeptide processing. Earlier in the chapter, we identified AUG as the usual start codon and noted that it encodes the modified amino acid N-formylmethionine (fMet) in bacterial cells and methionine in eukaryotes. Yet fMet is never found in functional bacterial proteins, and amino acids other than methionine are frequently the first amino acid of polypeptides in eukaryotes. The absence of fMet from functional bacterial proteins is the result of posttranslational cleavage of fMet from each bacterial polypeptide (**Figure 9.20a**). Similarly, methionine is usually removed as part of posttranslational processing in eukaryotes, and the new N-terminal amino acid is acetylated as part of the process.

In addition to N-terminal amino acids, other amino acid residues can be modified as well. One of the most common modifications of individual amino acids is performed by enzymes known as kinases that carry out phosphorylation of proteins by adding a phosphate group to individual amino acids (**Figure 9.20b**). This is an important regulatory process that can switch a protein from an inactive to an active form, or vice versa. Other enzymes may add methyl groups, hydroxyl groups, or acetyl groups to individual amino acids of polypeptides. The addition of carbohydrate side chains to polypeptides to form a glycoprotein is another important kind of posttranslational modification. For example, in one kind of posttranslational modification,

**(a)** Cleavage of N-terminal amino acids

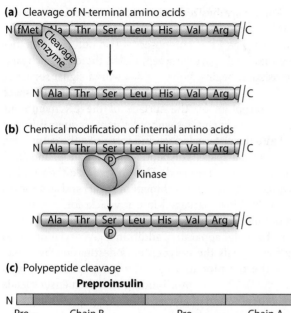

**(b)** Chemical modification of internal amino acids

**(c)** Polypeptide cleavage

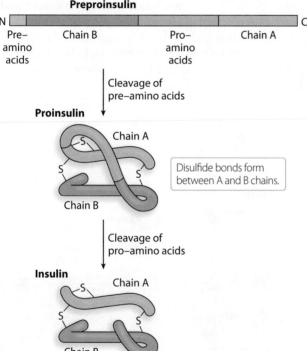

**Figure 9.20  Examples of posttranslational processing.**

the H substance is altered by the protein products of the $I^A$ and $I^B$ alleles of the ABO blood group gene (Chapter 4).

Posttranslational processing may include the cleavage of a polypeptide into multiple segments that each form functional proteins or that aggregate after elimination of one or more segments to form a functional protein. Production of the hormone insulin, which facilitates transport of glucose into cells, includes two posttranslational modification stepsthat remove segments of the original polypeptide (Figure 9.20c). The polypeptide product translated from the insulin gene is called preproinsulin. It is an inactive protein that contains a leader segment, called the pre–amino acid segment, at the N-terminal end and a connecting segment, called the pro–amino acid segment, that separates the

A-chain segment and the B-chain segment, the two functional pieces of the polypeptide. During posttranslational processing of preproinsulin, the pre–amino acids of the signal sequence are removed, after the polypeptide is transported through the cell membrane, to form proinsulin. Three disulfide bonds form within and between the A-chain and B-chain segments, followed by polypeptide cleavage that removes the pro–amino acid segment. What results is a functional insulin molecule consisting of 20 amino acids in the A-chain segment and 31 amino acids in the B-chain segment.

## Protein Sorting

Like the passengers in a busy airline terminal, proteins in a cell have different destinations, to which they travel with the aid of a "ticket" that tells the cell where to transport them. The destination is often an organelle or the cell membrane; in certain cases, the polypeptide is destined for transport out of the cell. The ticket that communicates the destination of a polypeptide is a signal sequence of 15 to 20 or so amino acids at the N-terminal end.

First articulated in the early 1970s by Gunther Blobel, the **signal hypothesis** proposes that the first 15 to 20 amino acids of many polypeptides contain an "address label" in the form of a signal sequence that designates the protein's destination in the cell. Blobel's hypothesis proposed that the signal sequence directs proteins to the endoplasmic reticulum (ER), where they are sorted for their cellular destinations (Figure 9.21). In a manner somewhat

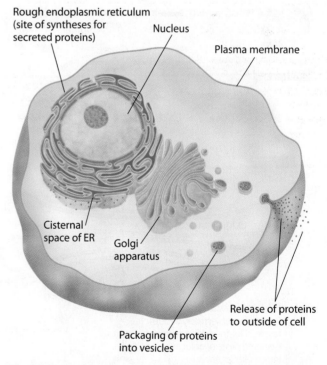

**Figure 9.21  Sites for polypeptide packaging and processing.** Polypeptides move from ribosomes into the rough endoplasmic reticulum, where they are packaged into vesicles that transport them to the Golgi apparatus for additional processing.

analogous to the way mail is sorted for postal delivery by its postal code or zip code, signal sequences at the N-terminal end of newly synthesized proteins facilitate their uptake by the ER and identify their ultimate destination.

Blobel's signal hypothesis has received broad experimental support. For example, shortly after Blobel proposed his hypothesis, Cesar Milstein and colleagues concocted an in vitro translation system that lacked ER, and they used it to produce proteins that are normally secreted by cells. Milstein found that the proteins in his in vitro system were 18 to 20 amino acids longer than the same proteins produced in vivo and then secreted by cells. The extra amino acids were at the N-terminal end of the proteins. Follow-up in vitro experiments used the same system but with ER included and, in contrast, produced proteins that did not carry the extra amino acids but were identical to the secreted versions of proteins normally found in cells. Experimental work since the 1970s has documented the composition and processing of signal sequences of polypeptides destined for secretion.

The ER and the Golgi apparatus are each a kind of sprawling industrial complex for the manufacture and packaging of proteins (Figure 9.21). Polypeptides destined for secretion are synthesized at the rough endoplasmic reticulum, the name given to parts of the ER where ribosomes dot the surface. The ER is composed of two outer membranes with an internal space between, called the cisternal space. As translation begins, polypeptides with a signal sequence have their N-terminal end pushed into the cisternal space through receptors on the surface of the ER membrane (**Figure 9.22**). Polypeptides without a signal sequence are not forced into the cisternal space and remain in the cytoplasm. In the cisternal space, polypeptides form a loop, and their signal sequence is removed. The polypeptide then undergoes glycosylation (the attachment of sugar) and as it moves through the ER is packaged into a vesicle for transport a short distance to the Golgi apparatus.

In the Golgi apparatus, additional glycosylation takes place that signals the polypeptide's destination by determining the receptor to which the polypeptide will bind (**Figure 9.23**). The receptor binding that then ensues leads directly to inclusion of glycosylated polypeptides in transport vesicles that bud off the Golgi apparatus. The vesicles travel to their cellular destinations, including organelles and the cell membrane (the latter is where protein secretion occurs).

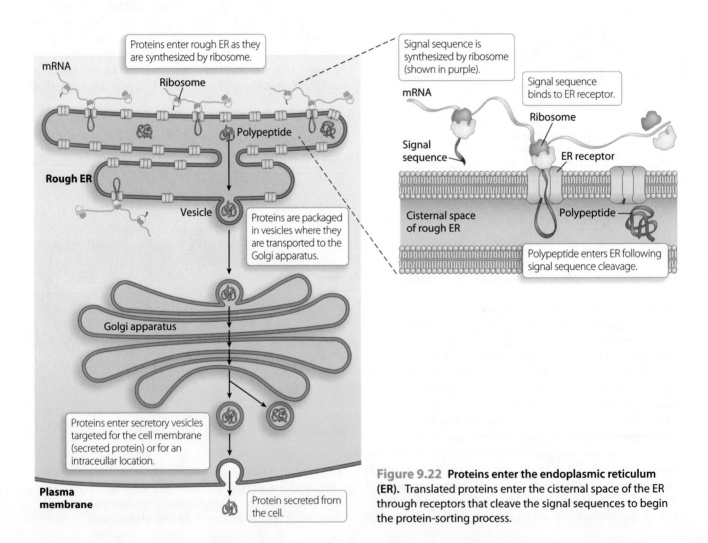

**Figure 9.22 Proteins enter the endoplasmic reticulum (ER).** Translated proteins enter the cisternal space of the ER through receptors that cleave the signal sequences to begin the protein-sorting process.

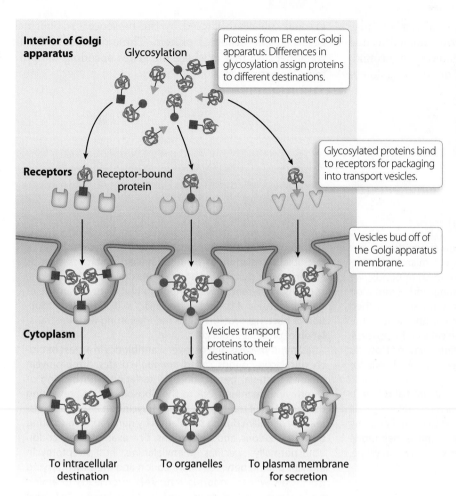

**Interior of Golgi apparatus**

Glycosylation

Proteins from ER enter Golgi apparatus. Differences in glycosylation assign proteins to different destinations.

**Receptors**    Receptor-bound protein

Glycosylated proteins bind to receptors for packaging into transport vesicles.

Vesicles bud off of the Golgi apparatus membrane.

**Cytoplasm**

Vesicles transport proteins to their destination.

To intracellular destination

To organelles

To plasma membrane for secretion

**Figure 9.23  Protein sorting in the Golgi apparatus.** Proteins are targeted for different destinations in the cell by different carbohydrate "tags" attached during processing in the Golgi apparatus. The tags bind to specific receptors, causing proteins to be packaged into appropriate vesicles in the Golgi. Vesicle proteins interact with receptors at the protein destination.

---

Genetic Insight  Polypeptides undergo posttranslational cleavage and chemical modification of amino acids. Eukaryotic proteins are labeled for transport to various cellular locations by the presence of a signal sequence. The signal sequence allows nascent polypeptides entry into the cisternal space of the ER, where they are packaged into vesicles for transport to the Golgi apparatus. In the Golgi, polypeptides undergo glycosylation and are packaged into transport vesicles for delivery to cellular destinations.

## Destruction of Incorrectly Folded Proteins

Mutations cause protein defects by changing the amino acid sequence of a polypeptide. The defective proteins fail to function normally, most often because they fold improperly, forming an unstable structure or one lacking functional sites of enzymatic activity, binding for transport, or other tasks performed by proteins. Within the ER, misfolded proteins are identified and bound by molecules called **chaperones** that perform an important

role in protein folding in the ER. Chaperones affiliate with proteins during the folding process and dissociate when the correctly folded structure is attained. If correct folding does not occur, the chaperones remain irreversibly bound to the misfolded proteins, resulting in the sequestration of such proteins in the ER. The sequestered chaperone–misfolded protein complexes are usually destroyed.

There are some exceptions to this general rule, however, that result in the destruction of particular types of cells that accumulate large amounts of mutant proteins. Diseases resulting from the death of cells due to accumulation of misfolded proteins are classified as conformational diseases. The name of this broad category of genetic disease comes from the process of "protein conformation" through which proteins attain their folded form.

Several human protein conformational diseases are known, and they are often neurodegenerative disorders and dementias. For example, Alzheimer disease is produced by the accumulation of misfolded β-amyloid protein that leads to the destruction of certain brain cells. A familial form of Parkinson disease produced by the accumulation of misfolded α-synuclein protein also results in the destruction of certain brain cells. Similarly, Huntington disease is produced by the destruction of neurons due to the accumulation of misfolded huntingtin protein. Each of these diseases has a late age of onset that is due to gradual cell death caused by the presence of abnormal aggregates of misfolded protein. These diseases are neuronal because that is where the genes encoding the proteins are expressed.

## CASE STUDY

### Antibiotics and Translation Interference

We have all taken antibiotics at various times during our lives to counteract a painful or persistent microbial infection. As a result of the efficiency of these compounds, we have experienced rapid relief of symptoms and elimination of the infection. These beneficial effects are accomplished by selective cell death. Specifically, the antibiotic kills microorganisms without harming our own cells in the process. What is the biochemical basis of antibiotic action? How do antibiotic compounds specifically target microbial cells for destruction?

You will probably not be surprised to learn that different antibiotics target different aspects of microbe biology to kill microorganisms. But you may be surprised to learn that many different antibiotics target microbial translation as their mode of action (Table 9.6). Familiar antibiotics such as tetracycline, streptomycin, and chloramphenicol target different stages of microbial translation, as do less familiar antibiotics such as erythromycin, puromycin, and cycloheximide. Each antibiotic contains a different active compound that takes advantage of unique features of bacterial translation to disrupt the production of bacterial proteins while not interfering with the translation of proteins in our cells.

*Streptomycin* is one of several antibiotics in a class of biochemical compounds called *aminoglycosides*. Streptomycin inhibits bacterial translation by interfering with binding of N-formylmethionine tRNA to the ribosome, thus preventing the initiation of translation. Streptomycin can also cause misreading of mRNA during translation by generating mispairing between codons and anticodons. For example, the codon UUU normally specifies phenylalanine, but streptomycin induces pairing between a UUU codon and the tRNA carrying isoleucine, whose codon is AUU. This error leads to amino acid changes in proteins and potentially to defective protein activity. Other aminoglycosides, such as neomycin, kanamycin, and gentamycin, also cause mispairing between codons and anticodons and can generate defective proteins. *Erythromycin* also impairs bacterial translation, but it does so in a very different way. It binds to the 50S (large) subunit in the tunnel from which the newly synthesized polypeptide emerges. In this manner, erythromycin blocks the passage of the polypeptide out of the ribosome. This causes the ribosome to stall on mRNA and bring translation to a halt. Table 10.6 provides details about these and other actions of antibacterial agents.

Single-celled eukaryotic microorganisms, such as fungi, can also cause human infections. To fight these infections, antibiotics such as puromycin and cycloheximide that target translational activities of eukaryotic cells are used. *Puromycin* has a three-dimensional structure similar to that of the 3′ end of a charged tRNA. It stops translation of bacterial and eukaryotic mRNAs by binding at the ribosomal A site and acting as an analog of charged tRNA. When puromycin is bound at the A site, its amino group forms a peptide bond with the carboxyl group of the P-site amino acid. However, puromycin does not contain a carboxyl group. This difference prevents formation of any additional peptide bonds and brings translation to a halt. *Cycloheximide* exclusively blocks eukaryotic translation by binding to the 60S subunit and inhibiting peptidyl transferase activity, much like chloramphenicol does to bacterial peptidyl transferase.

| Table 9.6 | Antibiotic Inhibitors of Protein Synthesis |
|---|---|
| **Antibiotic** | **Inhibitory Action** |
| Chloramphenicol | Blocks polypeptide formation by inhibiting peptidyltransferase in the 70S ribosome (antibacterial action) |
| Erythromycin | Blocks translation by binding to 50S subunit and inhibiting polypeptide release (antibacterial action) |
| Streptomycin | Inhibits translation initiation and causes misreading of mRNA by binding to the 30S subunit (antibacterial action) |
| Tetracycline | Binds to the 30S subunit and inhibits binding of charged tRNAs (antibacterial action) |
| Cycloheximide | Blocks polypeptide formation by inhibiting peptidyltransferase activity in the 80S ribosome (antieukaryote action) |
| Puromycin | Causes premature termination of translation by acting as an analog of charged tRNA (antibacterial and antieukaryote action) |

## 9.1 Ribosomes Are Translation Machines

▎ Translation takes place at the ribosome, where mRNA codons are coupled to transfer RNA anticodons by complementary base pairing.

▎ Polypeptides are linear strings of amino acids that are joined to one another by peptide bonds that are assembled at ribosomes.

▎ Polypeptides have an N-terminal (amino) end and a C-terminal (carboxyl) end.

▎ Ribosomes are composed of two subunits that each consist of ribosomal RNA and numerous proteins.

▎ Ribosomes have three functional sites of action: the P site, where the polypeptide is held; the A site, where tRNA molecules bind to add their amino acid to the end of the polypeptide; and the E site, which provides an exit point for uncharged tRNAs.

## 9.2 Translation Occurs in Three Phases

▎ Bacterial translation is initiated with the binding of the Shine–Dalgarno sequence on the 5′ mRNA end to a complementary sequence of nucleotides on the 3′ end of the 16S rRNA in the small ribosomal subunit. The nearby start codon is the site where translation commences.

▎ In eukaryotic mRNA, the 5′ cap is the binding site for eukaryotic initiation factors that cause the small ribosomal subunit to begin scanning in search of the start codon, which is part of the Kozak sequence.

▎ Peptidyl transferase joins amino acids to one another in the growing polypeptide chain, and elongation factor proteins help translocate the ribosome along mRNA.

▎ During polypeptide synthesis, charged tRNAs enter the A site, and peptidyl transferase catalyzes peptide bond formation, transferring the polypeptide from the P-site tRNA to the A-site tRNA. Elongation factor proteins translocate the ribosome, shifting the tRNA–polypeptide complex from the A site to the P site and opening the A site for the next charged tRNA.

▎ Translation terminates when a stop codon enters the A site. Release factor proteins, rather than tRNA, bind to stop codons. Release factors cause release of the polypeptide and lead to the dissociation of the ribosome from mRNA.

## 9.3 Translation Is Fast and Efficient

▎ An mRNA undergoes simultaneous translation by several ribosomes that attach to it sequentially to form a polyribosome.

▎ Usually, a ribosome will dissociate from mRNA upon encountering a stop codon, but small intercistronic spacers in some bacterial polycistronic mRNAs permit a ribosome to translate two or more polypeptides sequentially from the mRNA before dissociating.

## 9.4 The Genetic Code Translates Messenger RNA into Polypeptide

▎ The genetic code is redundant, meaning that most amino acids are specified by more than one codon. Redundancy of the genetic code is made possible by third-base wobble that relaxes the strict complementary base-pairing requirements at the third base of the codon.

▎ Specialized enzymes called aminoacyl-tRNA synthetases catalyze the addition of a specific amino acid to each tRNA.

## 9.5 Experiments Deciphered the Genetic Code

▎ In vitro experimental analysis demonstrates that the genetic code is triplet and does not contain gaps or overlaps.

▎ Each mRNA codon is composed of three consecutive nucleotides. Of the 64 codons contained in the genetic code, 61 specify amino acids and 3 correspond to stop codons.

▎ The genetic code was deciphered by analysis of in vitro translation of synthetic messenger RNA.

▎ The genetic code is essentially universal among living organisms. The few exceptions to the genetic code are found mainly in mitochondria.

▎ Properly charged tRNAs play the central role in converting mRNA sequence into polypeptide sequence.

## 9.6 Translation Is Followed by Polypeptide Processing and Protein Sorting

▎ Formation of functional proteins occurs after translation is completed. Amino acids may be chemically modified, or amino acids may be removed from polypeptides during posttranscriptional polypeptide processing.

▎ Proteins in eukaryotic cells are sorted to their cellular destinations by signal sequences at their N-terminal ends. Signal sequences are removed from polypeptides in the ER, and polypeptides destined for different sites in the cell are differentially glycosylated before being packaged for transport to the Golgi apparatus.

▎ In the Golgi apparatus, polypeptides are packaged into transport vesicles for shipment to their cellular destinations.

## KEYWORDS

3′ untranslated region (3′ UTR) *(p. 300)*

5′ untranslated region (5′ UTR) *(p. 300)*

30S initiation complex *(p. 303)*

70S initiation complex *(p. 304)*

aminoacyl site (A site) *(p. 300)*

aminoacyl-tRNA synthetase (tRNA synthetase) *(p. 314)*

chaperone *(p. 325)*

charged tRNA *(p. 302)*
elongation factor (EF) *(p. 307)*
eukaryotic initiation factor (eIF) *(p. 304)*
exit site (E site) *(p. 300)*
frameshift mutation *(p. 316)*
initiation complex *(p. 305)*
initiation factor (IF) *(p. 302)*
initiator tRNA *(p. 301)*
inosine (I) *(p. 314)*
isoaccepting tRNA *(p. 313)*
Kozak sequence *(p. 305)*

large ribosomal subunit *(p. 300)*
N-formylmethionine (fMet; tRNA^fMet)
    *(p. 303)*
peptidyl site (P site) *(p. 300)*
polycistronic mRNA *(p. 310)*
polyribosome *(p. 309)*
posttranslational polypeptide processing
    *(p. 322)*
preinitiation complex *(p. 302)*
protein sorting *(p. 322)*
release factor (RF) *(p. 308)*

reading frame *(p. 316)*
reversion mutation *(p. 316)*
scanning *(p. 305)*
Shine–Dalgarno sequence *(p. 302)*
signal hypothesis *(p. 323)*
signal sequence (leader sequence)
    *(p. 322)*
small ribosomal subunit *(p. 300)*
synonymous codon *(p. 312)*
third-base wobble *(p. 313)*
uncharged tRNA *(p. 302)*

---

## PROBLEMS

*For instructor-assigned tutorials and problems,
go to www.masteringgenetics.com.*

### Chapter Concepts

*For answers to selected even-numbered problems, see Appendix: Answers.*

1. Some proteins are composed of two or more polypeptides. Suppose the DNA template strand sequence
   `3'-TACGTAGGCTAACGGAGTAAGCTAACT-5'` produces a polypeptide that joins in pairs to form a functional protein.
   a. What is the amino acid sequence of the polypeptide produced from this sequence?
   b. What term is used to identify a functional protein like this one formed when two identical polypeptides join together?

2. In the experiments that deciphered the genetic code, many different synthetic mRNA sequences were tested.
   a. Describe how the codon for phenylalanine was identified.
   b. What was the result of studies of synthetic mRNAs composed exclusively of cytosine?
   c. What result was obtained for synthetic mRNAs containing AG repeats, that is,

      `AGAGAGAG...`?

   d. Predict the results of experiments examining `GCUA` repeats.

3. Several lines of experimental evidence pointed to a triplet genetic code. Identify three pieces of information that supported the triplet hypothesis of genetic code structure.

4. Outline the events that occur during initiation of translation in *E. coli*.

5. A portion of a DNA template strand has the base sequence

   `5'-...ACGCGATGCGTGATGTATAGAGCT...-3'`

   a. Identify the sequence and polarity of the mRNA transcribed from this fragmentary template strand sequence.
   b. Determine the amino acid sequence encoded by this fragment. Identify the N- and C-terminal directions of the polypeptide.
   c. Which is the third amino acid added to the polypeptide chain?

6. Describe three features of tRNA molecules that lead to their correct charging by tRNA synthetase enzymes.

7. Identify the amino acid carried by tRNAs with the following anticodon sequences.

   a. `5'-UAG-3'`
   b. `5'-AAA-3'`
   c. `5'-CUC-3'`
   d. `5'-AUG-3'`
   e. `5'-GAU-3'`

8. For each of the anticodon sequences given in the previous problem, identify the other codon sequence to which it could potentially pair using third base wobble.

9. What is the role of codons `UAA`, `UGA`, and `UAG` in translation? What events occur when one of these codons appears at the A site of the ribosome?

10. Compare and contrast the composition and structure of bacterial and eukaryotic ribosomes, identifying at least three features that are the same and three features that are unique to each type of ribosome.

11. Consider translation of the following mRNA sequence:

    `5'-...AUGCAGAUCCAUGCCUAUUGA...-3'`

    a. Diagram translation at the moment the fourth amino acid is added to the polypeptide chain. Show the ribosome; label its A, P, and E sites; show its direction of movement; and indicate the position and anticodon triplet sequence of tRNAs that are currently interacting with mRNA codons.
    b. What is the anticodon triplet sequence of the next tRNA to interact with mRNA?
    c. What events occur to permit the next tRNA to interact with mRNA?

12. The diagram of an eukaryotic ribosome shown below contains several errors.

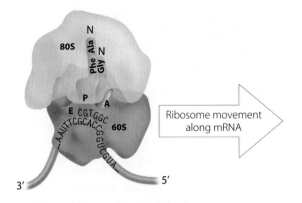

a. Examine the diagram carefully, and identify each error.

b. Redraw the diagram, and correct each error using the mRNA sequence shown.

13. Third-base wobble allows some tRNAs to recognize more than one mRNA codon. Based on this chapter's discussion of wobble, what is the *minimal* number of tRNA molecules necessary to recognize the following amino acids?

a. leucine

b. arginine

c. isoleucine

d. lysine

14. The genetic code contains 61 codons to specify the 20 common amino acids. Many organisms carry fewer than 61 different tRNA genes in their genomes. These genomes take advantage of isoaccepting tRNAs and the rules governing third-base wobble to encode fewer than 61 tRNA genes. Use these rules to calculate the *minimal* number of tRNA genes required to specify all 20 of the common amino acids.

15. The three major forms of RNA (mRNA, tRNA, and rRNA) interact during translation.

a. Describe the role each form of RNA performs during translation.

b. Which of the three types of RNA might you expect to be the least stable? Why?

c. Which form of RNA is least stable in eukaryotes? Why is this form least stable?

d. Compared to the average stability of mRNA in *E. coli*, is mRNA in a typical human cell more stable or less stable? Why?

## Application and Integration

19. In an experiment to decipher the genetic code, a poly-AC mRNA (ACACACAC…) is synthesized. What pattern of amino acids would appear if this sequence were to be translated by a mechanism that reads the genetic code as

a. a doublet without overlaps?

b. a doublet with overlaps?

c. a triplet without overlaps?

d. a triplet with overlaps?

e. a quadruplet without overlaps?

f. a quadruplet with overlaps?

20. Identify and describe the steps that lead to the secretion of proteins from eukaryotic cells.

16. The figure below contains sufficient information to fill in every row. Use the information provided to complete the figure.

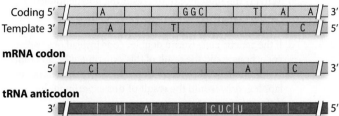

17. The line below represents a mature eukaryotic mRNA. The accompanying list contains many sequences or structures that are part of eukaryotic mRNA. A few of the items in the list, however, are not found in eukaryotic mRNA. As accurately as you can, show the location, on the line, of the sequences or structures that belong on eukaryotic mRNA; then, separately, list the items that are not part of eukaryotic mRNA.

5′ _____ 3′

a. stop codon

b. poly-A tail

c. intron

d. 3′ UTR

e. promoter

f. start codon

g. AAUAAA

h. 5′ UTR

i. 5′ cap

j. termination sequence

18. After completing Problem 17, carefully draw a line below the mRNA to represent its polypeptide product in accurate alignment with the mRNA. Label the N-terminal and C-terminal ends of the polypeptide. Carefully draw two lines above and parallel to the mRNA, and label them "coding strand" and "template strand." Locate the DNA promoter sequence. Identify the locations of the +1 nucleotide and of a transcription termination sequence.

*For answers to selected even-numbered problems, see Appendix: Answers.*

21. The amino acid sequence of a portion of a polypeptide is

N…Cys-Pro-Ala-Met-Gly-His-Lys…C.

a. What is the mRNA sequence encoding this polypeptide fragment? Use N to represent any nucleotide, Pu to represent a purine, and Py to represent a pyrimidine. Label the 5′ and 3′ ends of the mRNA.

b. Give the DNA template and coding strand sequences corresponding to the mRNA. Use the N, Pu, and Py symbols as placeholders.

22. Har Gobind Khorana and his colleagues performed numerous experiments translating synthetic mRNAs. In one

experiment, an mRNA molecule with a repeating UG dinu-cleotide sequence was assembled and translated.

a. Write the sequence of this mRNA and give its polarity.
b. What is the sequence of the resulting polypeptide?
c. How did the polypeptide composition help confirm the triplet nature of the genetic code?
d. If the genetic code were a doublet code instead of a triplet code, how would the result of this experiment be different?
e. If the genetic code was overlapping rather than nonover-lapping, how would the result of this experiment be different?

23. An experiment by Khorana and his colleagues translated a synthetic mRNA containing repeats of the trinucelotide UUG.

a. How many reading frames are possible in this mRNA?
b. What is the result obtained from each reading frame?
c. How does the result of this experiment help confirm the triplet nature of the genetic code?

24. The human β-globin polypeptide contains 146 amino acids. How many mRNA nucleotides are required to en-code this polypeptide?

25. The mature mRNA transcribed from the human β-globin gene is considerably longer than the sequence needed to encode the 146–amino acid polypeptide. Give the names of three sequences located on the mature β-globin mRNA but not translated.

26. Figure 9.6 contains several examples of the Shine–Dalgarno sequence. Using the seven Shine–Dalgarno sequences from *E. coli*, determine the consensus sequence and identify its location relative to the start codon.

27. Figure 9.20 shows three posttranslational steps required to produce the sugar-regulating hormone insulin from the starting polypeptide product preproinsulin.

a. A research scientist is interested in producing human in-sulin in the bacterial species *E. coli*. Will the genetic code allow the production of human proteins from bac-terial cells? Explain why or why not.
b. For human insulin, explain why it is not feasible to insert the entire human insulin gene into *E. coli* and anticipate the production of insulin.
c. Recombinant human insulin (made by inserting human DNA–encoding insulin into *E. coli*) is one of the most widely used recombinant pharmaceutical products in the world. What segments of the human insulin gene are used to create recombinant bacteria that produce human insulin?

28. A DNA sequence encoding a five–amino acid polypeptide is given below.

…ACGGCAAGATCCCACCCTAATCAGACCGTACCATTCACCTCCT…
…TGCCGTTCTAGGGTGGGATTAGTCTGGCATGGTAAGTGGAGGA…

a. Locate the sequence encoding the five amino acids of the polypeptide, and identify the template and coding strands of DNA.
b. Give the sequence and polarity of the mRNA encoding the polypeptide.

c. Give the polypeptide sequence, and identify the N-terminus and C-terminus.
d. Assuming the sequence above is a bacterial gene, identify the region encoding the Shine–Dalgarno sequence.
e. What is the function of the Shine–Dalgarno sequence?

29. A portion of the coding strand of DNA for a gene has the sequence

5'-…GGAGAGAATGAATCT…-3'

a. Write out the template DNA strand sequence and polar-ity as well as the mRNA sequence and polarity for this gene segment.
b. Assuming the mRNA is in the correct reading frame, write the amino acid sequence of the polypeptide using three-letter abbreviations and, separately, the amino acid sequence using one-letter abbreviations.

30. A eukaryotic mRNA has the following sequence. The 5' cap is indicated in italics (*CAP*), and the 3' poly(A) tail is indicated by italicized adenines.

5'-*CAP*CCAAGCGUUACAUGUAUGGAGAGAAUGAAACUG-AGGCUUGCCACGUUUGUUAAGCACCUAUGCUACCG*AAAAAAAA AAAAAAAAAAAAAAAAA*-3'

a. Locate the start codon and stop codon in this sequence.
b. Determine the amino acid sequence of the polypeptide produced from this mRNA. Write the sequence using the three-letter and one-letter abbreviations for amino acids.

31. Diagram a eukaryotic gene containing three exons and two introns, the pre-mRNA and mature mRNA transcript of the gene, and a partial polypeptide that contains the fol-lowing sequences and features. Carefully align the nucleic acids, and locate each sequence or feature on the appropri-ate molecule.

a. the AG and GU dinucleotides corresponding to intron–exon junctions
b. the +1 nucleotide
c. the 5' UTR and the 3' UTR
d. the start codon sequence
e. a stop codon sequence
f. a codon sequence for the amino acids Gly-His-Arg at the end of exon 1 and a codon sequence for the amino acids Leu-Trp-Ala at the beginning of exon 2

32. The following table contains DNA-sequence information compiled by Marilyn Kozak (1987). The data consist of the percentage of A, C, G, and T at each position among the 12 nucleotides preceding the start codon in 699 genes from various vertebrate species, and as the first nucleotide after the start codon. The start codon occupies positions +1 to +3, and the +4 nucleotide occurs immediately after the start codon. Use the data to determine the consensus sequence for the 13 nucleotides (−12 to −1 and +4) sur-rounding the start codon in vertebrate genes.

| Position | −12 | −11 | −10 | −9 | −8 | −7 | −6 | −5 | −4 | −3 | −2 | −1 | [start] | +4 |
|---|---|---|---|---|---|---|---|---|---|---|---|---|---|---|
| Percent A | 23 | 26 | 25 | 23 | 19 | 23 | 17 | 18 | 25 | 61 | 27 | 15 | [AUG] | 23 |
| Percent C | 35 | 35 | 35 | 26 | 39 | 37 | 19 | 39 | 53 | 2 | 49 | 55 | [AUG] | 16 |
| Percent G | 23 | 21 | 22 | 33 | 23 | 20 | 44 | 23 | 15 | 36 | 13 | 21 | [AUG] | 46 |
| Percent T | 19 | 18 | 18 | 18 | 19 | 20 | 20 | 20 | 7 | 1 | 11 | 9 | [AUG] | 15 |

33. The following table lists α-globin and β-globin gene sequences for the 12 nucleotides preceding the start codon and the first nucleotide following the start codon. The data are for 16 vertebrate globin genes reported by Kozak (1987). The sequences are written from −12 to +4 with the start codon sequence in capital letters.

| | Gene Sequence | |
|---|---|---|
| | −12 | start +4 |
| **α-Globin Family** | | |
| Human adult | agagaacccaccATGg | |
| Human embryonic | caccctgccgccATGt | |
| Baboon | ccagcgcgggcATGg | |
| Mouse adult | caggaagaaaccATGg | |
| Rabbit adult | gaaggaaccaccATGg | |
| Goat embryonic | tcagctgccaccATGt | |
| Duck adult | ggagctgcaaccATGg | |
| Chicken embryonic | ctctcctgcacaATGg | |
| **β-Globin Family** | | |
| Human fetal | agtccagacgccATGg | |
| Human embryonic | aggcctggcatcATGg | |
| Rabbit adult | aaacagacagaATGg | |
| Rabbit embryonic | agaccagacatcATGg | |
| Chicken adult | ccaaccgccgccATGg | |
| Chicken embryonic | cccgctgccaccATGg | |
| *Xenopus* adult | tcaactttggccATGg | |
| *Xenopus* larval | tctacagccaccATGg | |

Use the data in this table to

a. Determine the consensus sequence for the 16 selected α-globin and β-globin genes.
b. Compare the consensus sequence for these globin genes to the consensus sequence derived from the larger study of 699 vertebrate genes in Problem 32.

34. The six nucleotides preceding the start codon and the first nucleotide after the start codon in eukaryotes exhibit strong sequence preference as determined by the percentages of nucleotides in the −6 to −1 positions and the +4 position. Use the data given in the table for Problem 33 to determine the seven nucleotides that most commonly surround the start in vertebrates.

# 10 The Integration of Genetic Approaches: Understanding Sickle Cell Disease

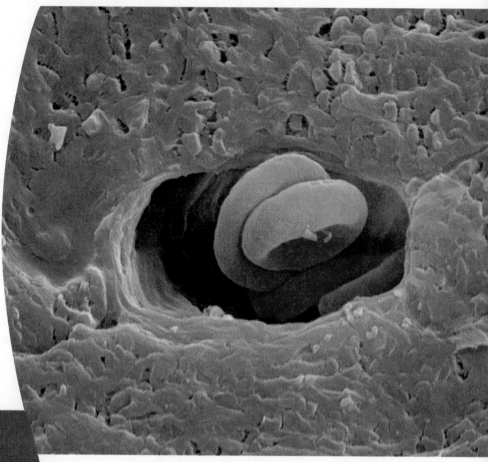

Normal red blood cells barely squeeze through narrow capillaries, but sickle-shaped red blood cells can block blood flow in capillaries.

## ESSENTIAL IDEAS

■ Progress in understanding the human hereditary anemia called sickle cell disease shows the power of combining analytical approaches from transmission genetics, molecular genetics, and evolutionary genetics.

■ A mutant allele of one of the two genes forming the red blood cell protein hemoglobin causes abnormalities that lead to sickle cell disease.

■ The transmission of sickle cell disease in families parallels molecular genetic analysis of globin gene and protein variation.

■ The geographic distribution of the mutation producing sickle cell disease is attributable to natural selection pressure exerted in malaria-rich environments.

In previous chapters, we describe and analyze gene transmission and function, the structure and function of DNA, the processes of gene expression, and the role of evolution in genetics. Each of these aspects of modern genetics contributes to the broad explanatory power of the science, a power achieved specifically through the integration of these principles and approaches. This chapter is designed to bring the integration of these genetic analysis approaches into focus using the human hereditary disorder sickle cell disease as an example.

The chapter has a second purpose as well. In the course of illustrating how analyses of hereditary transmission, molecular genetic variation, and evolution contribute to a comprehensive understanding of sickle cell disease, it also describes gel electrophoresis and related experimental methods that are commonly applied to the analysis of DNA, RNA, and protein variation. These methods are part of the basic "toolkit" of genetic analysis and can be used to obtain substantial information about nucleic acid and protein variation.

## 10.1 An Inherited Hemoglobin Variant Causes Sickle Cell Disease

**Sickle cell disease (SCD),** also known as sickle cell anemia, has been intensively investigated for more than a century, and its study has generated a revolution in genetics. Not only was SCD among the first genetic disorders shown to be caused by an inherited defect in a protein molecule, but the discovery of its cause—several years before DNA was identified as the hereditary molecule—helped pave the way for the molecular era in genetics. The identification of sickle cell disease as a "molecular disease" promoted the idea that inherited diseases have a molecular basis and played a key role in establishing the molecular nature of mutations. Investigation of SCD led to the description of the molecular basis of the disease and ultimately to an explanation of the role natural selection plays in the evolution and maintenance of the disease-causing allele in populations.

SCD is a potentially fatal autosomal recessive disorder caused by an abnormality in the structure and function of **hemoglobin (Hb),** the main oxygen-carrying protein in red blood cells. The hemoglobin defect producing SCD shortens the life span of red blood cells, producing severe anemia (an abnormally low number of red blood cells) that reduces the ability of blood to deliver oxygen to tissues. Oxygen deprivation causes tissue damage and tissue death throughout the body, accompanied by significant muscle pain and accumulated damage to organs.

The hemoglobin variant causing SCD is one of hundreds of different variant hemoglobin alleles occurring in people around the world, and inherited variations in hemoglobin are the most common type of hereditary abnormality found in humans. Hundreds of millions of people carry mutant alleles that alter the structure or function of hemoglobin molecules. Most of these alleles are rare. But a few, such as the mutant allele causing SCD, are common in certain populations. The SCD allele is common in multiple populations around the Mediterranean region, in the Middle East, and in Africa, and the mutant allele has formed and evolved independently in each of these regions.

### The First Patient with Sickle Cell Disease

Several principles of molecular genetics have their origin in the study of hemoglobin and the genes that produce it, including the concept of a molecular disease—a designation bestowed on SCD by Linus Pauling in 1954 (and explained below). A good place to begin our discussion, however, is with an event that occurred more than a century ago—December 1904, to be precise—when Walter Noel, a 20-year-old man of African origin, was admitted to Presbyterian Hospital in New York City suffering from severe anemia and debilitating muscle pain. Noel had arrived in New York City a year or so earlier from the island of Grenada in the West Indies, and he had just begun the first year of a dentistry training program when he was admitted to the hospital.

The physician in charge of Noel's case was an intern named Ernest Irons, who was supervised by a more experienced physician named James Herrick. Irons drew blood from Noel, examined it under a microscope, and was shocked to see that many of Noel's red blood cells had a peculiar elongated and sickled shape that contrasted starkly with the circular, biconcave shape of normal red blood cells (**Figure 10.1**).

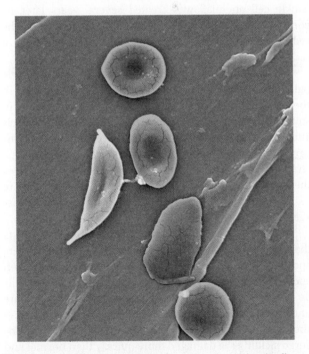

**Figure 10.1  Red blood cell shape.** Normal red blood cells have a biconcave shape, whereas sickle-shaped red blood cells are elongated.

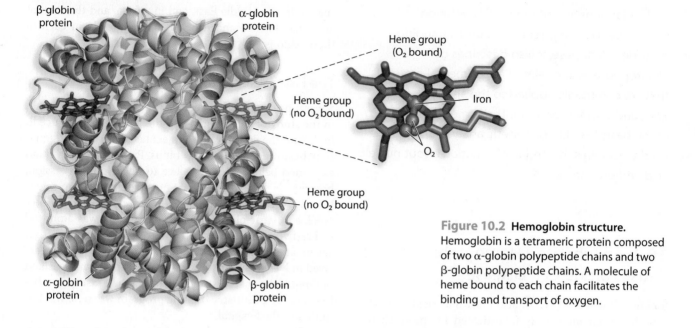

**Figure 10.2 Hemoglobin structure.** Hemoglobin is a tetrameric protein composed of two α-globin polypeptide chains and two β-globin polypeptide chains. A molecule of heme bound to each chain facilitates the binding and transport of oxygen.

Noel recovered from this initial bout with the illness. Over the next two and a half years, he was to be readmitted several times and treated for the same symptoms. After completing his dentistry training, he returned to Grenada, where he practiced dentistry until he died 9 years later at the age of 32. In 1910, Herrick published a paper describing Walter Noel's case. The paper was the first clinical description of SCD, although the disorder had no name at the time Herrick described it. Its original name, "sickle cell anemia," was created several years later by combining *sickle,* for the characteristic deformity of the red blood cells, and *anemia,* for the chronic shortage of red blood cells in most patients.

Sickle cell disease patients experience severe muscle pain when the number of sickle-shaped red blood cells in their circulation is large enough to impede blood flow in small blood vessels and capillaries. These blood vessels and capillaries are barely wide enough for normal biconcave red blood cells to move through in single file (see the chapter opener photo). The reduced blood flow deprives the surrounding tissues of oxygen, causing immediate pain and potential long-term damage to organs and tissues due to oxygen deprivation.

Red blood cells are oxygen transportation specialists that are pumped to the lungs where they pick up oxygen, and then through the circulatory system to carry oxygen and other molecules throughout the body. Red blood cells do not contain nuclei and cannot divide. They circulate until they are damaged and removed from circulation—about 100 to 120 days on average. Red blood cells that undergo sickling are damaged more quickly than normal and have a shorter life span. Unfortunately, the body's red blood cell production capacity is limited. The accelerated rate of loss of red blood cells in SCD results in chronic anemia as one of the symptoms of the disorder.

## Hemoglobin Structure

Hemoglobin molecules are tetramers, protein structures consisting of four proteins joined together (**Figure 10.2**). The hemoglobin tetramer contains two protein chains from each of two different **globin genes** that are encoded on separate chromosomes in the human genome. Each molecule of the most common form of hemoglobin consists of two **α-globin** (pronounced *AL-fa GLOBE-in*) proteins, produced by the α-globin gene, and two **β-globin** (*BAY-ta GLOBE-in*) proteins, produced by the β-globin gene. This particular composition, denoted $\alpha_2\beta_2$, is identified as hemoglobin A, or HbA, where *Hb* is an abbreviation for *hemoglobin* and *A* designates the most common form. Each of the four proteins in hemoglobin carries one iron-containing molecule of heme, a compound that undergoes reversible binding with a molecule of oxygen. Thus each hemoglobin molecule can bind and transport four oxygen molecules.

The α-globin and β-globin genes are members of a group of genes that share strong structural and functional similarities and produce similar globin proteins. The organization of the two genes is also very similar (**Figure 10.3**). Both genes contain three exons and two introns. The **α-globin** gene encodes a polypeptide containing 141 amino acids, and the polypeptide encoded by the **β-globin** gene contains 146 amino acids.

---

**Genetic Insight** Hemoglobin consists of four polypeptides, two from the α-globin gene and two from the β-globin gene. These genes are members of gene clusters that share strong similarities of sequence, structure, and function.

---

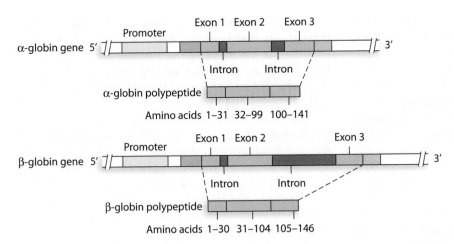

**Figure 10.3 Globin proteins and their genes.** The α-globin and β-globin genes each contain three exons and two introns. The amino acids encoded by each exon are indicated by the numbers describing their places in the final polypeptide chain.

## Globin Gene Mutations

The globin genes may be the most intensively studied genes in the human genome, and the existence and distribution of α-globin and β-globin gene variants are well documented in most human populations. At present, nearly 500 different allelic variants of the α-globin and β-globin genes are known. Nearly all of these globin gene variants are rare. Some are so rare that they exist only in a single family. There are a few notable exceptions, however, and these more common variants provide well-researched examples of some hereditary processes that you are likely to have studied in previous biology courses. They also give us the chance to explore how globin gene variants affect hemoglobin structure and function.

In a century of research since Herrick's description of Walter Noel's SCD, physicians and human biologists have fully explored the heredity, molecular basis, and evolution of the disorder. Today biologists know that SCD is a common hereditary anemia caused by a single base-pair substitution in the β-globin gene sequence. The mutant allele, designated $\beta^S$, produces a β-globin protein that contains a single amino acid difference from the normal **$\beta^A$ allele.** Individuals with SCD carry two **$\beta^S$ alleles** and do not have the $\beta^A$ allele. They have the genotype $\beta^S\beta^S$ and produce only mutant β-globin chains. When two mutant β-globin proteins join two normal α-globin proteins, the hemoglobin molecules formed are abnormal. The loss of normal hemoglobin function in homozygotes for the mutant allele results in the inheritance of SCD as an autosomal recessive condition. (See Case Study at the end of Chapter 3.)

The hemoglobin contained in red blood cells of people with SCD is less stable than wild-type (normal) hemoglobin when oxygen concentration is low. This instability can cause hemoglobin to collapse into linear, crystal-like molecules at low oxygen levels. Due to its high concentration in red blood cells, the collapse of hemoglobin into linear structures deforms red cells into the sickle shapes first seen by Ernest Irons.

Individuals who are heterozygous carriers of SCD have the genotype $\beta^A\beta^S$. All their hemoglobin tetramers contain two normal α-globin proteins, but some contain two $\beta^A$ proteins, some contain two $\beta^S$ proteins, and others contain one of each type of β-globin protein. Consequently, a small percentage of the red blood cells of heterozygous individuals can acquire a sickle-shaped form when oxygen level is low, as it is when red blood cells are returning to the heart. This condition shortens the average life span of red blood cells in heterozygotes; but since it affects only a small percentage of red blood cells, it does not cause the anemia seen in those with SCD. Heterozygous carriers are sometimes identified as having "sickle cell trait", and, while symptoms are generally mild, severe complications can occur under circumstances in which the availability of oxygen is reduced or the need for oxygen is high. Concerns about potential health consequences for athletes who are heterozygous carriers of sickle cell trait have been expressed in several ways. For example, in 2010, following the deaths of ten student athletes with sickle cell trait over the previous decade, the National Collegiate Athletic Association (NCAA) implemented a policy requiring student athletes to be tested for sickle cell trait or to sign a waiver of the test.

## 10.2 Genetic Variation Can Be Detected by Examining DNA, RNA, and Proteins

Having examined the physiology and inheritance patterns of SCD, we can now explore some molecular genetics techniques for analyzing the $\beta^S$ and $\beta^A$ alleles as well as the mRNA and proteins that are produced by the alleles. We also consider the analysis of specific types of DNA sequence variation. The molecular methods discussed in this chapter are useful in a wide range of genetic analyses.

### Gel Electrophoresis

In 1949, James Neel used transmission genetic analysis to demonstrate that SCD is a recessive disorder. That same year, Linus Pauling and his colleagues published the first description of the molecular basis of SCD and coined the term *molecular disease* to describe it. They used the term to denote a disease caused by a variation in the molecular structure of a protein.

Pauling isolated hemoglobin from people having the various genotypes and used the analytical technique of **gel electrophoresis** to separate the hemoglobin molecules of each type. Gel electrophoresis is a technique for separating different protein or nucleic acid molecules from one another in an electrical field on the basis of their charge, size, and shape (**Figure 10.4**). A gel support matrix is created by molding a liquid inside a form, typically a plastic casting tray. A "comb" is placed in the liquid as it is poured into the form, to produce "wells," or depressions, in the gel. In the form, the liquid solidifies into a flexible semisolid.

The sample wells are small reservoirs into which biological samples, such as proteins or nucleic acid (DNA or RNA), are loaded. Usually, multiple wells are employed, each marking the **origin of migration** for one of the samples and thus serving as the starting point for one of the "lanes" of the gel. After biological samples are loaded into the wells, an electrical current is applied to the gel by connecting a positive electrode to one end and a negative electrode to the other. The samples migrate through the matrix of tiny pores and passageways created by the solidification of the gel. Molecules make their way from the origin of migration near the negatively charged end of the gel toward the positive charge at the opposite end.

The materials most commonly used to form electrophoresis gels are **agarose,** a form of cellulose, and **polyacrylamide,** a synthetic material made by a polymerization reaction between chemical compounds. Neither of these substances interacts with proteins or nucleic acids as they move through the gel, so the rates of migration are determined entirely by the characteristics of the molecules in each sample.

In gel electrophoresis, biological molecules that have electric charge migrate toward the end having the opposite charge. Most biological molecules, including DNA, RNA, and most proteins, have negative charge and migrate toward the positive end. Therefore, the origin of migration is usually placed near the negative end. Proteins with positive charge migrate toward the negative end, so when they are being studied, the origin of migration will be placed near the positive end.

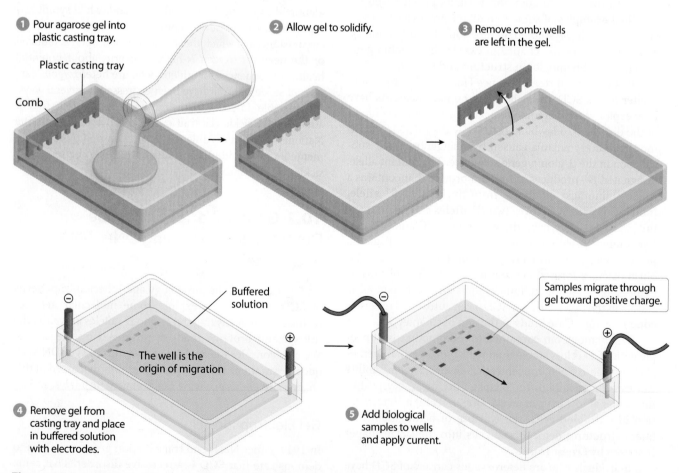

**Figure 10.4  Apparatus and procedure for gel electrophoresis.**

Molecular movement through the electrophoresis gel is driven by the flow of electricity. Molecules stop moving when current flow is turned off. In electrophoretic gels, molecule movement occurs at a rate that depends on three parameters of molecular structure. Each of these parameters individually is important in determining how a particular molecule migrates, but they can also interact with one another to produce a characteristic migration rate for each molecule. The parameters are as follows.

■ **Molecular weight**—Smaller molecules (i.e., proteins with fewer amino acids or nucleic acids with fewer nucleotides) migrate more quickly than larger molecules. This characteristic is an important determinant of electrophoretic migration of all biological molecules, and it is the main parameter in DNA and RNA migration.

■ **Molecular charge**—Molecules with greater negative charge migrate toward the positive pole more rapidly than molecules with less negative charge. Variation in molecular charge of proteins is imparted by amino acid composition and is an important characteristic influencing protein migration. In contrast, nucleic acids have negative charge that derives from the sugar-phosphate backbone. This negative charge is proportionate to mass and thus does not contribute to differences in migration rate among nucleic acid molecules of different lengths.

■ **Molecular shape (molecular conformation)**—Tightly condensed, globular molecules migrate more quickly than linear molecules. Protein migration can be strongly influenced by conformation; however, when nucleic acids are being compared, the only migration differences caused by molecular shape occur in comparisons of linear and circular DNA.

Pauling's electrophoretic analysis of hemoglobin proteins purified from red blood cells showed that proteins produced by individuals with different β-globin genotypes have different **electrophoretic mobility,** a term that describes either the rate of a molecule's electrophoretic migration or its final position of the protein in the gel. We saw above that a protein molecule's size (molecular weight), charge, and shape (conformation) determine its electrophoretic mobility. In Pauling's analysis of hemoglobin protein, each allele was seen to produce a different protein with a characteristic electrophoretic mobility; in other words, as each type of protein migrated through the gel, it formed a separate **band,** an aggregation of protein with a single electrophoretic mobility, that could be visualized by staining the gel with protein stain (**Figure 10.5a**).

The protein band seen in the $\beta^S\beta^S$ lane had lower electrophoretic mobility (smaller distance migrated from the origin) than the protein band detected in the $\beta^A\beta^A$ lane. Only a single band is detected in each of these lanes, suggesting that all the protein in the lane is identical. In contrast, when an electrophoresis lane contains protein from a heterozygous ($\beta^A\beta^S$) individual, the protein in that lane separates into two bands, each corresponding to the electrophoretic mobility of the protein band in one of the lanes containing protein from a homozygote.

Pauling then used a technique called densitometry to show that a single kind of β-globin protein is present in lanes containing protein from a homozygous individual, and that two kinds of protein are present in lanes containing protein taken from heterozygotes (**Figure 10.5b**). Densitometry quantifies the amount of protein present in a gel lane by measuring how much light is blocked from passing through the gel by the presence of a band of protein. Comparison of the positions of the protein bands and densitometry peaks in each homozygous lane, and the detection of two bands and two peaks at those positions in the heterozygous lane confirm that homozygous individuals produce a single form of β-globin protein that differs depending on the homozygous genotype, and that heterozygotes produce proteins of both types in approximately equal concentration.

The importance of Pauling's work is twofold. First, it introduced laboratory methods for the detection of distinct forms of globin protein; and second, it demonstrated that

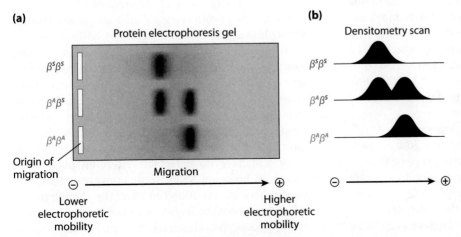

**(a)**

Protein electrophoresis gel

$\beta^S\beta^S$

$\beta^A\beta^S$

$\beta^A\beta^A$

Origin of migration

Migration

⊖ ⟶ ⊕

Lower electrophoretic mobility

Higher electrophoretic mobility

**(b)**

Densitometry scan

$\beta^S\beta^S$

$\beta^A\beta^S$

$\beta^A\beta^A$

⊖ ⟶ ⊕

**Figure 10.5 Gel electrophoresis of hemoglobin proteins. (a)** Individuals with the three genotypes $\beta^S\beta^S$, $\beta^A\beta^S$, and $\beta^A\beta^A$ are analyzed. The single bands in the $\beta^S\beta^S$ and $\beta^A\beta^A$ lanes indicate that each homozygous individual produces a single type of protein. The detection of two protein bands in the $\beta^A\beta^S$ lane indicates that both alleles are expressed in heterozygotes. **(b)** Each genotype produces a unique pattern of protein electrophoretic mobility that is also reflected by densitometry results.

hemoglobin variation explains the inheritance of SCD as a molecular disease. Pauling's study was the first to show that the inheritance patterns of disorders in pedigrees parallel those of the transmission of molecular variation. His work also illustrates that among heterozygous carriers, molecular evidence often supports the expression of both alleles, even if the abnormal morphology characteristic of a disorder is present only in individuals who are homozygous for a recessive allele. In short, Pauling was the first to draw attention to a fundamental principle of genetics: Hereditary morphologic variation has a molecular basis.

---

Genetic Insight  Gel electrophoresis separates biological molecules (nucleic acids and proteins) using an electrical field. In protein electrophoresis, electrophoretic mobility is a function of the size, charge, and shape of each protein molecule.

---

## Hemoglobin Peptide Fingerprint Analysis

In 1957, Vernon Ingram published a description of the molecular basis of SCD based on analysis of hemoglobin protein. At that time, the notion that SCD is a molecular disease caused by an inherited alteration of hemoglobin structure was already firmly established. What Ingram had set out to clarify was the molecular difference in hemoglobin that caused the difference in hemoglobin behavior. Hemoglobin was then considered to be composed of two identical "molecular segments," each containing almost 300 amino acids. Today we know that each "molecular segment" actually has two parts—an α-globin chain, with 141 amino acids, and a β-globin chain, with 146 amino acids. We also know that the complete hemoglobin molecule is a tetramer, composed of two α-globin polypeptide chains and two β-globin polypeptide chains.

Ingram examined hemoglobin structural variation with a two-step approach called **peptide fingerprint analysis.** To prepare it for fingerprint analysis, the hemoglobin protein is first broken into many fragments by chemical treatment. The protein fragments are then subjected to electrophoresis to separate the fragments in one direction, or dimension, on a gel. Next the hemoglobin fragments are separated in a second dimension, perpendicular to the first, by **chromatography,** which uses a solvent to carry fragments with different amino acid composition to different final positions. At the end of these two separations, the locations of the protein "spots" serve as a kind of "fingerprint" of the protein.

When Ingram compared the fingerprint spots of different hemoglobin samples, he found that just a single peptide fragment from the hemoglobin of people with SCD (genotype $\beta^S\beta^S$) was different in the hemoglobin of people who were homozygous for the wild-type allele (genotype $\beta^A\beta^A$). Further analysis showed that the difference was a change in a single amino acid residue: In those with SCD,

the amino acid valine (Val) substitutes for glutamic acid (Glu) in amino acid position number 6 of the 146 amino acids in the β-globin protein chain. As confirmation of his conclusion, Ingram found that the hemoglobin fingerprints for heterozygous carriers of SCD (genotype $\beta^A\beta^S$) had spots corresponding to both the glutamic-acid-containing portion of wild-type hemoglobin (the product of the $\beta^A$ allele) and the valine-containing portion of mutant hemoglobin (the product of the $\beta^S$ allele). **Genetic Analysis 10.1** guides you through genotype identification by protein gel electrophoresis.

---

Genetic Insight  The red blood cell defect responsible for sickle cell disease results from a single amino acid substitution in the β-globin protein. This mutation alters the electrophoretic mobility of hemoglobin protein and is detectable by electrophoretic analysis and peptide fingerprint analysis.

---

## Identification of DNA Sequence Variation

With the identification of hemoglobin protein structure and the amino acid sequences of the α-globin and β-globin chains, scientists were ready to combine the analysis of hemoglobin variation with analysis of DNA and mRNA sequences to explain how nucleic acid variation produces SCD. Before we can examine this research, however, some additional description of nucleic acid and of protein electrophoretic analysis is required. This subsection and the next present some background information on the identification of DNA sequence variability using DNA-digesting enzymes and gel electrophoresis. These techniques are tools with many applications in DNA analysis. After the discussion, we return to the analysis of SCD.

DNA sequences are linear strings of the nucleotides adenine (A), guanine (G), cytosine (C), and thymine (T). Scientists compare sequences from different organisms by aligning them side by side and noting the number, location, and type of nucleotide sequence differences. Genomic analysis has determined that the most common kind of DNA sequence difference between organisms of the same species consists of differences of single nucleotides, a type of difference called a **single nucleotide polymorphism** (**SNP;** pronounced *snip*). SNPs are prevalent in the genomes of all organisms. The human genome, for example, contains millions of SNPs scattered among the approximately 3 billion base pairs (bp) that constitute our genome. By their prevalence, SNPs have become an important category of genetic marker (genetic markers are important tools in genetic mapping, described in Chapter 5). SNPs usually occur in unexpressed regions of genomes and have no detectable effect on phenotype. Occasionally, however, SNPs occur in expressed regions of genes, where the variation can affect the phenotype.

Whether or not the sequence variation at a SNP locus affects a phenotypic character, the allelic sequence is

Individuals homozygous for the $\beta^A$ hemoglobin allele are assigned the genotype AA. Those with SCD are homozygous for the mutant $\beta^S$ allele and are identified as SS. An individual designated AS is $\beta^A\beta^S$. A second β-globin gene mutation designated $\beta^C$ has a DNA base-pair substitution that leads to a single amino acid substitution in the β-globin polypeptide chain. Individuals who are $\beta^C\beta^C$ are designated CC. The $\beta^C$ and $\beta^A$ DNA sequences differ by a single base pair, and the polypeptides differ by a single amino acid, just as $\beta^S$ and $\beta^A$ polypeptides differ by a single amino acid.

The diagram at right illustrates the electrophoretic mobility of hemoglobin protein from individuals with the AA, SS, and CC genotypes. The diagram also illustrates bands for two individuals with unknown genotypes and has space to fill in the bands for the SC genotype.

**a.** Interpret the hemoglobin protein band patterns for Unknown 1 and Unknown 2, and identify the genotype of each person.

**b.** Draw the hemoglobin protein band pattern expected for an individual who is SC.

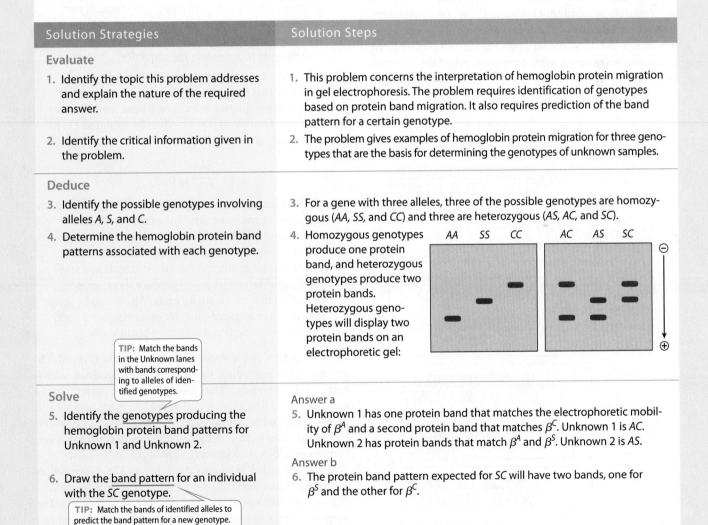

| Solution Strategies | Solution Steps |
|---|---|
| **Evaluate** | |
| 1. Identify the topic this problem addresses and explain the nature of the required answer. | 1. This problem concerns the interpretation of hemoglobin protein migration in gel electrophoresis. The problem requires identification of genotypes based on protein band migration. It also requires prediction of the band pattern for a certain genotype. |
| 2. Identify the critical information given in the problem. | 2. The problem gives examples of hemoglobin protein migration for three genotypes that are the basis for determining the genotypes of unknown samples. |
| **Deduce** | |
| 3. Identify the possible genotypes involving alleles A, S, and C. | 3. For a gene with three alleles, three of the possible genotypes are homozygous (AA, SS, and CC) and three are heterozygous (AS, AC, and SC). |
| 4. Determine the hemoglobin protein band patterns associated with each genotype. | 4. Homozygous genotypes produce one protein band, and heterozygous genotypes produce two protein bands. Heterozygous genotypes will display two protein bands on an electrophoretic gel: |

TIP: Match the bands in the Unknown lanes with bands corresponding to alleles of identified genotypes.

| | |
|---|---|
| **Solve** | **Answer a** |
| 5. Identify the genotypes producing the hemoglobin protein band patterns for Unknown 1 and Unknown 2. | 5. Unknown 1 has one protein band that matches the electrophoretic mobility of $\beta^A$ and a second protein band that matches $\beta^C$. Unknown 1 is AC. Unknown 2 has protein bands that match $\beta^A$ and $\beta^S$. Unknown 2 is AS. |
| | **Answer b** |
| 6. Draw the band pattern for an individual with the SC genotype. | 6. The protein band pattern expected for SC will have two bands, one for $\beta^S$ and the other for $\beta^C$. |

TIP: Match the bands of identified alleles to predict the band pattern for a new genotype.

For more practice, see Problems 4, 5, and 31.

transmitted from one generation to the next. **Figure 10.6** shows two DNA sequences representing two SNP alleles that are identical except for the highlighted base pairs. An A‑T base pair is found in allele $S_1$, and a G‑C pair specifies allele $S_2$. Individual organisms in a population can be homozygous ($S_1S_1$ or $S_2S_2$) or heterozygous ($S_1S_2$) for these SNP alleles. The pattern of hereditary transmission of SNP alleles follows the same pattern as alleles of expressed genes, with each parent contributing one allele to offspring.

The complete sequencing and surveying of a genome in search of SNP variation is accomplished by genome sequencing (Chapter 18). For certain genetic analyses involving SNPs, however, it is not necessary to examine complete genome sequences. For these analyses, SNP variation can be detected using a special class of DNA-digesting enzymes that act only on specific DNA sequences. Known as **restriction endonucleases**—or, more commonly, **restriction enzymes**—these enzymes act like precise molecular scissors. First, the enzymes recognize a precise DNA nucleotide sequence of a few base pairs, called the enzyme's **restriction sequence,** and then they cut each strand of

DNA at the restriction sequence, in a specific manner. When long DNA molecules containing multiple restriction sequences are treated with a restriction enzyme, many fragments of DNA are produced. The number of restriction fragments produced by a given restriction enzyme is characteristic for a given sequence of DNA. Inherited variability in the number or the length (in base pairs) of restriction fragments is called **restriction fragments length polymorphism (RFLP).**

Hundreds of different restriction enzymes have been identified since they were first discovered in the 1960s, and each restriction enzyme shares three general properties:

1. Each enzyme exclusively recognizes its own restriction sequence, consisting of a precise 5′-to-3′ nucleotide order on each DNA strand. For example, the restriction enzyme *Eco*RI exclusively recognizes the restriction sequence 5′-GAATTC-3′. Because the restriction sequence for each restriction endonuclease is precise, any variation blocks the ability of the restriction enzyme to recognize the sequence.

2. Restriction sequences are usually palindromes, meaning that each strand of the double-stranded restriction sequence has the same nucleotide order (running from 5′ to 3′). The double-stranded *Eco*RI restriction sequence is

    5′-GAATTC-3′
    3′-CTTAAG-5′

3. A restriction enzyme cuts each strand of its restriction sequence in the same way. For example, *Eco*RI cuts each strand of DNA between the G and the A of the restriction sequence (**Figure 10.7**). Some restriction enzymes, like *Eco*RI, cut the DNA strands in a staggered, or offset, manner and produce short, single-stranded ends called **sticky ends.** Alternatively, some restriction enzymes do not generate staggered cuts, and the resulting fragments have **blunt ends.**

Restriction sequences are listed in **Table 10.1**, which groups them according to whether they produce sticky ends or blunt ends. The sticky ends are short single strands protruding at each end of the fragment, capable of base-pairing with complementary sticky ends on other fragments. Restriction enzymes are also used to form recombinant DNA molecules (Chapter 16).

SNP variation is one of the kinds of DNA-sequence change that can destroy or create a restriction sequence. When this occurs—that is, when SNP variation creates or eliminates a restriction sequence—the number of restriction fragments produced from the DNA may change, and the length of DNA restriction fragments (as measured by the number of base pairs) may also change. Such a SNP-based difference affecting any restriction site is an inherited genetic variant that can be identified by observing variation in the number or length of DNA restriction fragments that are identified following gel

**(a)**

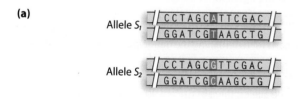

Allele $S_1$
CCTAGCATTCGAC
GGATCGTAAGCTG

Allele $S_2$
CCTAGCGTTCGAC
GGATCGCAAGCTG

**(b)**

Genotype:          Sequence:

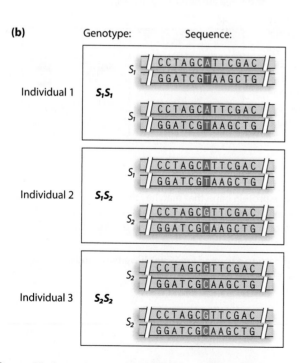

Individual 1    $S_1S_1$

$S_1$
CCTAGCATTCGAC
GGATCGTAAGCTG

$S_1$
CCTAGCATTCGAC
GGATCGTAAGCTG

Individual 2    $S_1S_2$

$S_1$
CCTAGCATTCGAC
GGATCGTAAGCTG

$S_2$
CCTAGCGTTCGAC
GGATCGCAAGCTG

Individual 3    $S_2S_2$

$S_2$
CCTAGCGTTCGAC
GGATCGCAAGCTG

$S_2$
CCTAGCGTTCGAC
GGATCGCAAGCTG

**Figure 10.6 Single nucleotide polymorphism (SNP). (a)** At a SNP locus, two alleles differ by one base pair. Allele $S_1$ contains an A‑T base pair (green), and allele $S_2$ contains a G‑C base pair (purple). **(b)** Three genotypes result from these two alleles.

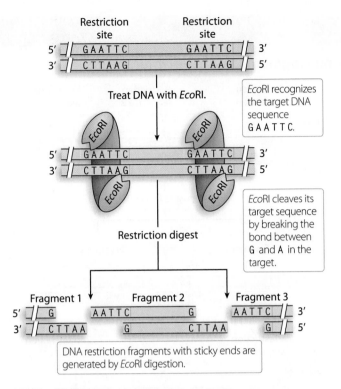

**Figure 10.7** Restriction digestion by *Eco*RI.

electrophoresis. Restriction fragment length is typically recorded in **kilobases (kb)**. One kilobase is equal to 1000 nucleotide bases. **Research Technique 10.1** discusses the origin and inheritance of DNA fragment length variation.

| Table 10.1 | Example Restriction Enzymes | |
|---|---|---|
| **Restriction Endonuclease** | **Source Organism** | **Restriction Sequence** |
| *Producers of sticky ends* | | |
| *Eco*RI | *Escherichia coli* | 5'-GᴬATTC-3' |
| | | 3'-CTTAAG-5' |
| *Bam*HI | *Bacillus amyloliquifaciens* | 5'-GᴳGATCC-3' |
| | | 3'-CCTAGG-5' |
| *Hind*III | *Haemophilus influenzae* | 5'-AᴬAGCTT-3' |
| | | 3'-TTCGAA-5' |
| *Dde*I | *Desulfovibrio desulfuricans* | 5'-CᶜTNAG-3' |
| | | 3'-GANTC-5' |
| *Producers of blunt ends* | | |
| *Pvu*II | *Proteus vulgaris* | 5'-CAGᶜCTG-3' |
| | | 3'-GTCGAC-5' |
| *Sma*I | *Serratia marcescens* | 5'-CCCᶜGGG-3' |
| | | 3'-GGGCCC-5' |

*Note:* N = any nucleotide (A, T, C, or G); ∧ and ∨ indicate cleavage locations.

**Genetic Insight** Single nucleotide polymorphisms (SNPs) are single-nucleotide, base-pair differences that can occur anywhere in a genome. SNP variation can affect the number and length of restriction fragments.

## Molecular Probes

The use of electrophoretic analysis for detecting DNA restriction fragment variation, variation in mRNA transcripts from expressed genes, or variation in the polypeptide products of genes is complicated by the sheer number of restriction fragments, mRNA molecules, or protein molecules in a sample under analysis. Treating human genomic DNA with a restriction enzyme like *Eco*RI, whose restriction sequence is common in the genome, can produce hundreds of thousands of restriction fragments. Similarly, isolating mRNA molecules or protein molecules from cells yields a large number of different products. Without methods for identifying specific substances—whether specific DNA sequences, mRNA transcripts, or protein products—electrophoretic analysis would be hopelessly complex.

A compound called **ethidium bromide (EtBr)** allows researchers to identify the location of DNA fragments or RNA molecules in electrophoresis gels. Because EtBr stains all DNA or RNA in a gel, however, it is not specific to the sequence of any particular gene. When EtBr is added to electrophoresis gels containing DNA or RNA samples, it intercalates into the molecules. Exposure to ultraviolet light causes the EtBr to emit fluorescent light, so that the EtBr-stained DNA or RNA can be visualized (**Figure 10.8a**). Similarly, general protein stains—stains that bind to any protein—can be used to discover the location of proteins in an electrophoresis gel (**Figure 10.8b**).

Two especially fortunate innovations in gel electrophoresis have made the identification of individual proteins, mRNAs, and DNA fragments possible. The first is the development of methods for blotting, a general name for the transfer of nucleic acids or proteins from an electrophoresis gel to a membrane that can withstand rigorous treatment and analysis. **Southern blotting** (named after its inventor, Edwin Southern) is the term applied to DNA transfer; **northern blotting** (named by tongue-in-cheek analogy with Southern blotting) identifies the transfer of mRNA; and **western blotting** is the term identifying the transfer of proteins.

The second innovation is the development of **molecular probes.** These are antibodies or single-stranded nucleic acids that specifically bind to target molecules, thus making possible the identification of a given protein or specific DNA or RNA sequence. Locating a particular nucleic acid sequence or a specific protein from a heterogeneous pool of molecules in an electrophoresis gel is similar to trying to find the unique string of letters that

## Research Technique 10.1

### The Production and Detection of DNA Restriction Fragment Length Polymorphisms

**PURPOSE** Restriction digestion followed by DNA gel electrophoresis is one method for detecting variation of DNA sequence that alters the number or the relative positions of RFLP sequences. Variation in the number or length of restriction fragments can result from DNA sequence changes that alter a restriction sequence, making it unrecognizable, or that create a new restriction sequence. RFLP changes can also result from the insertion or deletion of DNA between restriction sequences that increase or decrease the length of restriction fragments.

**MATERIALS AND PROCEDURES** DNA is isolated from cells and treated with one or more restriction enzymes to produce DNA restriction fragments. The restriction fragments are then separated by DNA gel electrophoresis, causing the fragments to be visualized as "bands" on the gel. Laboratory methods described in Research Technique 10.2 can also aid in the identification of specific restriction fragments.

**DESCRIPTION** DNA sequence variation altering the number or length of restriction fragments (RFLPs) produces distinctive restriction fragments for each allele. Organisms that are homozygous for DNA sequence at a restriction site produce the same restriction fragment from homologous chromosomes and produce one DNA band on a gel. Heterozygous organisms have different DNA sequences, produce a different restriction fragment from each chromosome, and have two DNA bands on a gel. Some examples of these processes are shown to the right.

**CONCLUSION** DNA base substitution changes that alter a restriction sequence and the insertion or deletion of DNA between two restriction sequences are the principal ways DNA sequence alterations can produce RFLPs. RFLP alleles form genotypes whose DNA restriction fragments produce distinctive patterns in gel electrophoresis. Each genotype has a distinctive combination of band number and band size on the gel.

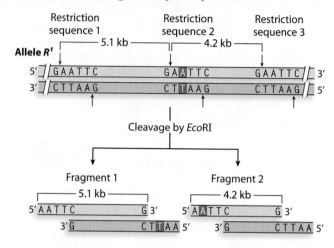

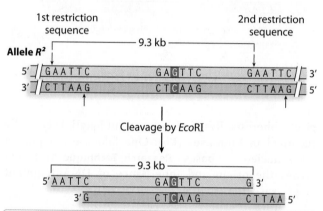

**RFLP Variation.** Two homologous regions of DNA are identical except for a SNP that produces a base-pair substitution (highlighted) in restriction site 2 of one chromosome. DNA treated with *Eco*RI cuts allele $R^1$ at three restriction sites (1, 2, and 3) and forms two small DNA restriction fragments of 5.1 and 4.2 kb. The base substitution in the DNA sequence of the $R^2$ allele eliminates restriction site 2, and the DNA is cut only at sites 1 and 3, resulting in a single DNA restriction fragment of 9.3 kb.

**Figure 10.8 Visualization of nucleic acids and proteins in gels. (a)** Nucleic acid molecules (DNA and RNA) are visualized by binding ethidium bromide (EtBr) to them. EtBr fluoresces when excited by ultraviolet light, revealing bands of DNA in the gel. **(b)** General protein stains (such as coomassie blue, shown here) bind to proteins in electrophoretic gels to reveal the locations of protein bands.

**(a)**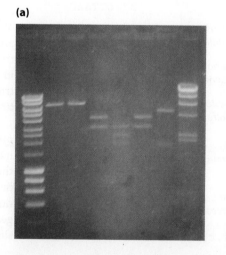

**(b)**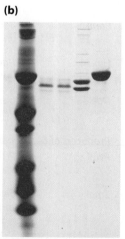

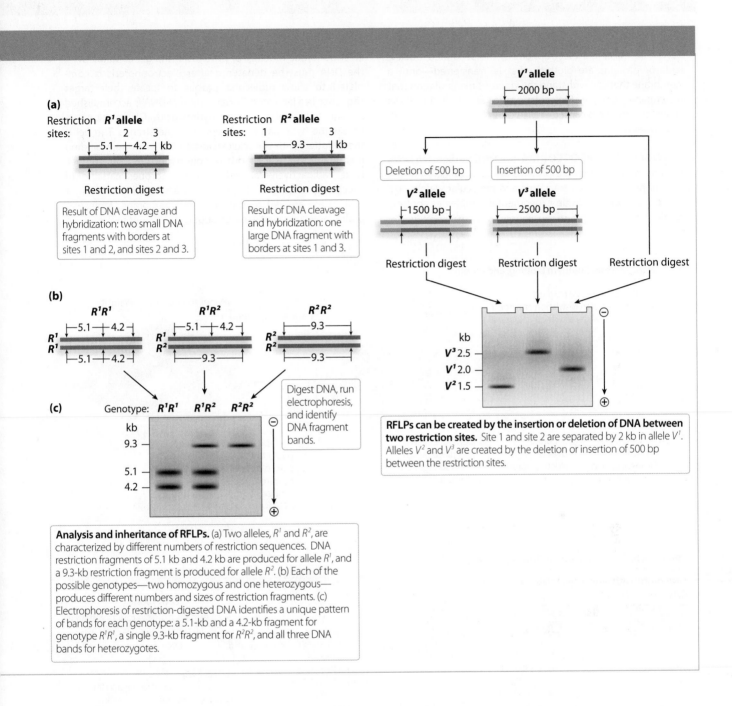

**Analysis and inheritance of RFLPs.** (a) Two alleles, $R^1$ and $R^2$, are characterized by different numbers of restriction sequences. DNA restriction fragments of 5.1 kb and 4.2 kb are produced for allele $R^1$, and a 9.3-kb restriction fragment is produced for allele $R^2$. (b) Each of the possible genotypes—two homozygous and one heterozygous—produces different numbers and sizes of restriction fragments. (c) Electrophoresis of restriction-digested DNA identifies a unique pattern of bands for each genotype: a 5.1-kb and a 4.2-kb fragment for genotype $R^1R^1$, a single 9.3-kb fragment for $R^2R^2$, and all three DNA bands for heterozygotes.

**RFLPs can be created by the insertion or deletion of DNA between two restriction sites.** Site 1 and site 2 are separated by 2 kb in allele $V^1$. Alleles $V^2$ and $V^3$ are created by the deletion or insertion of 500 bp between the restriction sites.

comprise a short series of words in a voluminous text document. Scanning each block of letters for the correct string is almost impossible without a tool for targeting the desired sequence. Just as word processing programs locate a desired string of letters by using a "find" command that searches all strings for a match, biologists use molecular probes to identify target nucleic acid sequences or target proteins following electrophoresis.

In the search for a target DNA molecule in a Southern blot, the molecular probe is a short, single-stranded DNA fragment, and the target molecule is a region of DNA that contains a sequence complementary to the probe sequence. Similarly, single-stranded molecular probes detect target mRNAs in northern blots by the complementary base pairing of probe and a segment of the target nucleotide sequence. The pairing of complementary nucleic acid strands of the probe and the target sequence is called **hybridization**. In contrast to the nucleic acid probes used to detect DNA or RNA target sequences, molecular probes used to detect target proteins in western blots are antibodies—immune system proteins that bind only to specific target proteins. Descriptions of Southern, northern, and western blotting, and the use of different kinds of molecular probes to identify specific nucleic acids or proteins on the blots, are provided in **Research Technique 10.2**.

# Blotting and Probing Nucleic Acid and Protein Molecules

**PURPOSE** After gel electrophoresis, the separated nucleic acids or proteins are blotted—that is, transferred—onto a membrane that can withstand the vigorous manipulation that accompanies analysis. Molecular probes are applied to blots to detect sequences carried in DNA or RNA, and to detect specific proteins.

**MATERIALS AND PROCEDURES** Restriction-digested DNA, isolated mRNA, or isolated proteins are first subjected to gel electrophoresis. Known standards and molecular weight size markers are run alongside experimental samples as controls to identify the length of nucleic acids or to identify the

electrophoretic mobility of proteins. If the gel contains DNA, the DNA must be denatured after electrophoresis is completed to allow molecular probes to locate their target sequence in a later step. Denaturation of DNA is accomplished by bathing the gel in a sodium hydroxide (NaOH) solution that breaks the hydrogen bonds between the strands. The gel is then blotted with a nucleic-acid-binding or protein-binding membrane that will absorb sample molecules from the gel. Next, radioactively labeled molecular probes capable of specifically binding the target nucleic acid sequence or target protein are applied to the prepared membrane. Molecular probe molecules that are not bound to a target molecule on

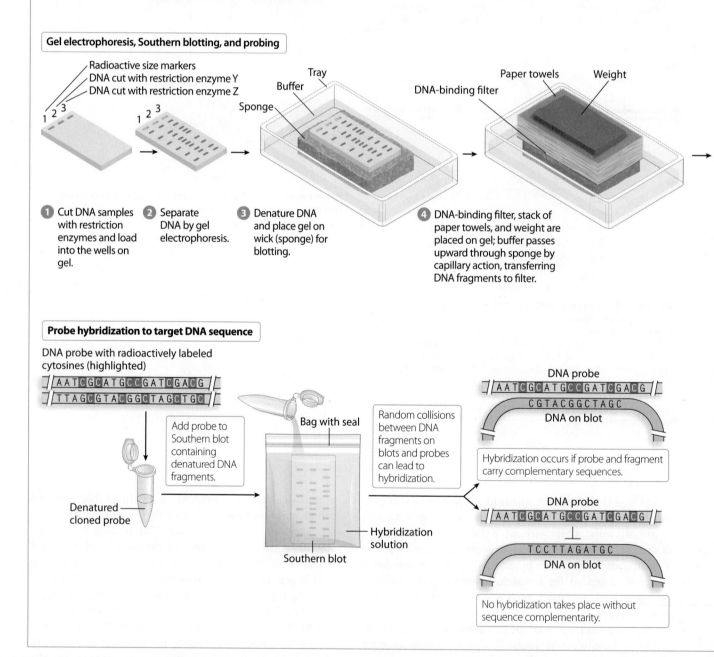

**Gel electrophoresis, Southern blotting, and probing**

1. Cut DNA samples with restriction enzymes and load into the wells on gel.
2. Separate DNA by gel electrophoresis.
3. Denature DNA and place gel on wick (sponge) for blotting.
4. DNA-binding filter, stack of paper towels, and weight are placed on gel; buffer passes upward through sponge by capillary action, transferring DNA fragments to filter.

**Probe hybridization to target DNA sequence**

DNA probe with radioactively labeled cytosines (highlighted)

Add probe to Southern blot containing denatured DNA fragments.

Random collisions between DNA fragments on blots and probes can lead to hybridization.

Hybridization occurs if probe and fragment carry complementary sequences.

No hybridization takes place without sequence complementarity.

the blot are washed away. Subsequently, autoradiography using X-ray film captures the location of any bound molecular probe by detecting the radiation. Alternatively, nonradioactive or fluorescent labels can be linked to molecular probes for subsequent detection. Similar methods are used to prepare Southern blots of restriction-digested DNA, northern blots of mRNA, and western blots of protein, except that neither RNA nor protein is denatured before blotting.

DESCRIPTION  Southern blotting is named after its developer, Edwin Southern, and uses single-stranded molecular probes to detect denatured DNA on the blot by complementary base pairing. Northern blotting detects membrane-bound mRNAs using single-stranded molecular probes in a manner similar to that of Southern blotting. Western blotting detects proteins with the use of antibodies that specifically bind to target proteins.

CONCLUSION  Southern, northern, and western blots are produced by similar methods and use molecular probes to detect sample molecules or sequences of interest. Labeled molecular probes bind to specific target sequences or molecules and are detected in autoradiographs or other analyses of blots that serve as a permanent record of the results of gel electrophoresis.

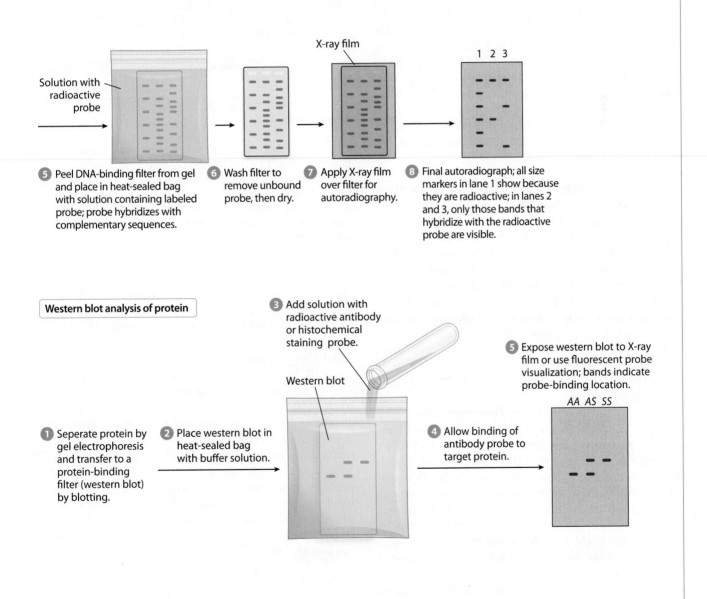

Solution with radioactive probe

X-ray film

1 2 3

5 Peel DNA-binding filter from gel and place in heat-sealed bag with solution containing labeled probe; probe hybridizes with complementary sequences.

6 Wash filter to remove unbound probe, then dry.

7 Apply X-ray film over filter for autoradiography.

8 Final autoradiograph; all size markers in lane 1 show because they are radioactive; in lanes 2 and 3, only those bands that hybridize with the radioactive probe are visible.

**Western blot analysis of protein**

3 Add solution with radioactive antibody or histochemical staining probe.

Western blot

5 Expose western blot to X-ray film or use fluorescent probe visualization; bands indicate probe-binding location.

AA  AS  SS

1 Seperate protein by gel electrophoresis and transfer to a protein-binding filter (western blot) by blotting.

2 Place western blot in heat-sealed bag with buffer solution.

4 Allow binding of antibody probe to target protein.

## Electrophoretic Analysis of Sickle Cell Disease

Like the hundreds of other mutations of the α-globin and β-globin genes that affect humans, the mutation producing SCD is a DNA sequence change that leads to an mRNA transcript differing from the wild type and, ultimately, to the production of a mutant form of β-globin protein. Specifically, through genetic studies spanning a period of 50 years, scientists discovered that a change in a single DNA base leads to a single-base difference in mRNA transcripts and to β-globin proteins that differ at just one of the 146 amino acids that comprise them.

**Figure 10.9** shows this key portion of the DNA, mRNA, and amino acid sequences of the wild-type ($\beta^A$) and mutant ($\beta^S$) alleles. The single-nucleotide difference between the alleles is the result of a SNP of the type we described above. In comparison to the wild-type allele, the mutant $\beta^S$ allele contains a single DNA base-pair substitution in the sixth DNA triplet of the coding sequence. This substitution leads to a single-nucleotide change in codon 6 of mRNA and to a protein with valine (Val) rather than glutamic acid (Glu) as the sixth amino acid in the β-globin polypeptide chain.

### Southern Blot Analysis of β-Globin Gene Variation

The $\beta^S$ SNP is unusual in that it occurs in the coding sequence of the gene, whereas most SNPs occur in noncoding segments of the genome. We can detect the SNP in the $\beta^S$ allele because it destroys a restriction sequence, leading to an RFLP that is revealed by Southern blot analysis.

Either two or three restriction sequences for the restriction endonuclease *Dde*I can occur near the β-globin gene, depending on the allele. *Dde*I recognizes the double-stranded restriction sequence 5'-CTNAG-3', where N indicates that any of the four nucleotides (A, T, C, or G) can occur in the middle of the 5-bp sequence as long as the variable nucleotide is flanked by C-T and A-G dinucleotide combinations.

**Figure 10.10** shows three *Dde*I restriction sites, labeled 1, 2, and 3, in the $\beta^A$ allele. All three *Dde*I restriction sequences are cleaved, producing two DNA fragments of 1150 bp and 200 bp for the DNA region shown. Southern blotting of DNA from the $\beta^A$ allele produces two DNA bands corresponding to fragment lengths of 1150 bp and 200 bp. The target sequence for the molecular probe is split between two restriction fragments by DNA cleavage at *Dde*I site 2, and the probe hybridizes to both the 1150-bp and the 200-bp restriction fragments from $\beta^A$ alleles.

In contrast, in **Figure 10.11**, the $\beta^S$ allele is shown to contain two *Dde*I restriction sequences, labeled 1 and 3. The middle restriction sequence, labeled 2 in the illustration, is missing from the $\beta^S$ allele as a result of the base-pair substitution that produces the SNP. Only *Dde*I restriction sites 1 and 3 are cleaved in DNA carrying the $\beta^S$ allele;

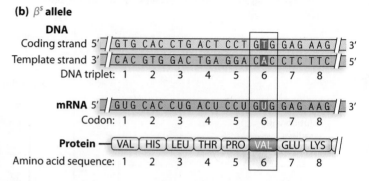

**(a)** $\beta^A$ **allele**

**(b)** $\beta^S$ **allele**

**Figure 10.9 SCD mutation in the DNA sequence of the β-globin gene.** DNA, mRNA, and amino acid sequences spanning the first eight amino acids of **(a)** the wild-type $\beta^A$ allele and **(b)** the $\beta^S$ allele are shown. A single nucleotide polymorphism occurs in DNA triplet 6 (boxed), causing a change in the sixth codon of mRNA and a change in the sixth amino acid of the polypeptide from Glu to Val.

site 2 is not recognized by *Dde*I because of the SNP variation. This cleavage produces a single restriction fragment of 1350 bp in DNA carrying the $\beta^S$ sequence. The length of this fragment is the sum of the lengths of the two restriction fragments detected from the $\beta^A$ allele (i.e., 1150 bp + 200 bp). Southern blot analysis of $\beta^S$-allele DNA produces a single DNA restriction fragment, measuring 1350 bp (1.35 kb) in length. Because *Dde*I site 2 is altered by SNP variation, the entire molecular probe target sequence is contained on a single 1350-bp (1.35-kb) restriction fragment (**Figure 10.12**). People who are $\beta^A\beta^A$ have bands of 1150 bp (1.15 kb) and 200 bp (0.20 kb) detected by the probe. Those who are $\beta^S\beta^S$ have a single band of 1350 bp (1.35 kb), and those who have $\beta^A\beta^S$ produce all three bands because they carry both alleles. **Genetic Analysis 10.2** guides your interpretation of Southern blot analysis.

---

**Genetic Insight** A single base change (a SNP) in the DNA sequence mutates the $\beta^A$ allele to the $\beta^S$ allele and also removes a *Dde*I restriction site. The base substitution leads to RFLP variation that is detected in Southern blot analysis.

---

**βᴬ-allele sequence**

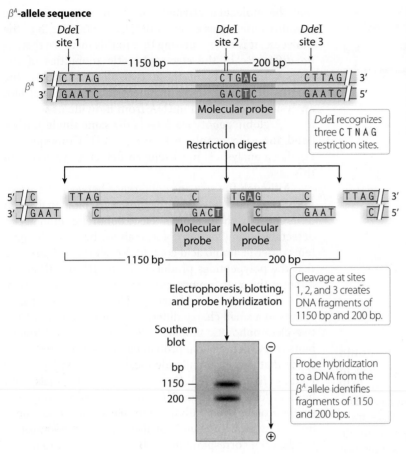

**Figure 10.10** ***Dde*I restriction digestion and Southern blotting of wild-type β-globin gene.** Restriction digestion and Southern blot analysis of βᴬ-allele DNA sequence identifies two DNA fragments that are hybridized by the molecular probe. A restriction fragment of 1150 bp (1.15 kb) is produced by cleavage at sites 1 and 2, and the 200-bp fragment is produced by cleavage at sites 2 and 3.

**βˢ-allele sequence**

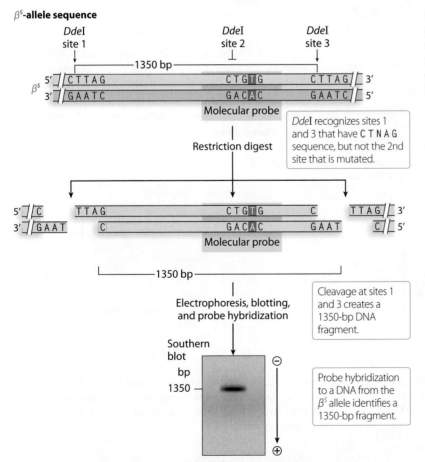

**Figure 10.11** **The single-nucleotide polymorphism in the βˢ allele.** Base-pair substitution inactivates *Dde*I site 2, and only sites 1 and 3 are cleaved. The molecular probe detects a single 1350-bp (1.35-kb) fragment in Southern blot analysis.

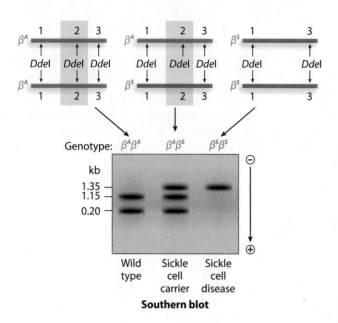

**Figure 10.12  RFLP results for β-globin genotypes.**

### Northern and Western Blot Analysis of the β-Globin Gene Transcript and Protein

The DNA sequences of the $\beta^A$ and $\beta^S$ alleles are identical except for the SNP that distinguishes the sequences of one from the other. Upon transcription, each allele produces an mRNA molecule containing 664 nucleotides. The single-nucleotide substitution that differentiates the two alleles does not alter the *length* of the mRNA transcript. Considering that the molecular attribute producing electrophoretic mobility differences among mRNAs is total length of the molecule, it is not surprising that in this instance there is *no difference* in the electrophoretic mobilities of the mRNA transcripts of these two alleles, because the lengths of their mRNAs are identical. A northern blot analysis performed on mRNA from individuals with the three β-globin genotypes detects the same single-mRNA band for each genotype (**Figure 10.13**). Consequently, northern analysis is not useful in detecting variation in this case.

Although the sequence difference between these two mRNAs is not detectable by northern blot analysis, a difference in their protein electrophoretic mobility is detectable using western blot analysis, because the proteins differ in amino acid content. Recall from Figure 10.9 that the polypeptides produced by the $\beta^A$ and $\beta^S$ alleles differ at the sixth amino acid position of their respective 146-member amino acid strings. The amino acid change results in a small charge difference that produces distinctive electrophoretic mobilities for the proteins. Western blots reveal hemoglobin protein bands for the three genotypes that are essentially identical to the band patterns Pauling first detected (**Figure 10.14**). Individuals with homozygous genotypes $\beta^A\beta^A$ and $\beta^S\beta^S$ each produce a single protein band with different electrophoretic mobility, and heterozygous individuals ($\beta^A\beta^S$) have two protein bands, each corresponding to the polypeptide product of a different allele.

> **Genetic Insight**  Northern blot analysis produces identical results for genotypes $\beta^A\beta^A$, $\beta^S\beta^S$, and $\beta^A\beta^S$, because the lengths of mRNA transcripts from the two alleles ($\beta^A$ and $\beta^S$) are identical. In contrast, western blot analysis yields detectable electrophoretic mobility differences between the protein products of the different alleles.

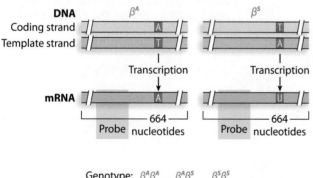

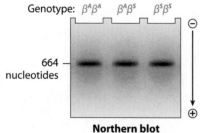

**Figure 10.13  Northern blot analysis of human β-globin mRNA.**  Transcription produces an mRNA that is 664 nucleotides in length for both alleles. The results of northern blot analysis are therefore identical for the three genotypes.

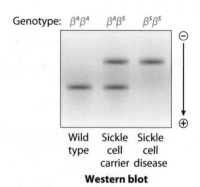

**Figure 10.14  Western blot analysis of human β-globin protein.**  Single protein bands are seen in western blot analysis of $\beta^A\beta^A$ and $\beta^S\beta^S$ homozygotes; two protein bands are detected for heterozygotes.

The 6-kb segment of DNA shown contains the *Bca* gene. The hybridization location for a molecular probe complementary to a portion of the gene is indicated. The locations of five *Eco*RI restriction sequences are also indicated, and the distances (in kilobases) between restriction sites are given.

a. If this 6-kb region is digested with *Eco*RI, how many DNA fragments are generated? How many nucleotide base pairs are expected in each of the resulting restriction fragments?

b. Which restriction fragment(s) will contain all or part of the *Bca* gene?

c. DNA from the 6-kb segment is digested with *Eco*RI, and the resulting fragments are separated by DNA gel electrophoresis. Which of the restriction fragments will be bound by the molecular probe and seen as bands in the Southern blot? Which fragments will not be detected by Southern blotting? Explain your answer.

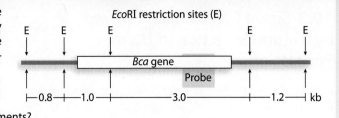

*Eco*RI restriction sites (E)

| Solution Strategies | Solution Steps |
|---|---|
| **Evaluate** | |
| 1. Identify the topic this problem addresses and explain the nature of the required answer. | 1. This problem concerns restriction digestion of a fragment of a gene and detection of restriction fragments with a molecular probe for a portion of the gene of interest. The answer requires identification of the length of restriction fragments that will and will not be detected by the probe. |
| 2. Identify the critical information given in the problem. | 2. The locations of five *Eco*RI restriction sites and the distances between the sites are given. The segment of the gene bound by the molecular probe is identified. |
| **Deduce** | |
| 3. Assess the relationship of the molecular probe to the gene, and assess the kilobase scale in relation to the *Eco*RI restriction sites and restriction fragments. | 3. The molecular probe binds to the longer of the two restriction fragments that contain part of the *Bca* gene. The sum of kilobase pairs in all the *Eco*RI restriction fragments equals 6.0 kb. |
| **Solve** | Answer a |
| 4. Determine the number and length (in base pairs) of restriction fragments. | 4. Digestion with *Eco*RI produces four restriction fragments with lengths that are 0.8 kb (800 bp), 1.0 kb (1000 bp), 3.0 kb (3000 bp), and 1.2 kb (1200 bp). |
| | Answer b |
| 5. Identify the DNA fragments that contain portions of the *Bca* gene. | 5. The 1.0- and the 3.0-kb restriction fragments contain segments of the *Bca* gene. |
| | Answer c |
| 6. Identify the DNA fragment that will hybridize with the molecular probe. | 6. Only the 3.0-kb restriction fragment contains the sequence hybridized by the molecular probe. This fragment will be seen on the Southern blot. |
| **TIP:** Molecular probes hybridize to target regions that contain complementary base sequences. | |
| 7. Explain why one fragment is hybridized by the probe and why other fragments are not. | 7. The DNA sequence complementary to the molecular probe sequence is completely contained on the 3.0-kb restriction fragment, so this fragment binds the probe. None of the other three restriction fragments contains a sequence complementary to the molecular probe, so although they are separated from one another by DNA gel electrophoresis, they are not hybridized by the probe and are not seen on the Southern blot. |
| **PITFALL:** Avoid confusion by remembering that DNA fragments that do not contain sequences complementary to a molecular probe cannot hybridize with the probe. | |

For more practice, see Problems 15, 23, and 25.

## 10.3 Sickle Cell Disease Evolved by Natural Selection in Human Populations

Dozens of variant alleles of hemoglobin genes produce one form or another of hereditary anemia. According to the World Health Organization, hereditary anemias are the most common of all human genetic diseases; they occur in an estimated 250 to 300 million people around the world. Most of the globin-gene mutations causing hereditary anemia are rare, but a few are found in high frequency in certain populations. The $\beta^S$ allele occurs in frequencies as high as 15% in several indigenous populations of Africa, the Middle East, and the Indian subcontinent. Population and evolutionary genetic analysis verifies that the allele arose independently in each region and has evolved by the same evolutionary process in each locality. Examples of other β-globin alleles found in high frequency are $\beta^C$, primarily in populations from West Africa, and $\beta^E$, in populations from Southeast Asia and the Pacific Islands.

The high frequencies of $\beta^S$, $\beta^C$, and $\beta^E$ are consistent with the conclusion that natural selection is working to increase the occurrence of these alleles. Population studies over the last 50 years have firmly established malaria as the agent of natural selection leading to a high frequency of these β-globin gene alleles in certain populations. An environment where malaria is endemic favors the survival and reproduction of individuals who are heterozygous for $\beta^A$ and one of the mutant alleles over the other genotypes. In other words, individuals who are $\beta^A\beta^S$, $\beta^A\beta^C$, or $\beta^A\beta^E$ have a survival and reproductive advantage over individuals who are homozygous $\beta^A\beta^A$ (and therefore succumb more easily to malaria) and over those who are homozygous for the mutant alleles (and therefore suffer from hereditary anemia).

### Malaria Infection

Malaria is a potentially fatal infectious disease caused by protozoans. One of the most common and most serious forms of malaria is caused by *Plasmodium falciparum*. This protozoan is carried by the mosquito vector *Anopheles gambeii*, which transfers the protozoan to animals, including humans, when it bites them. The symptoms of malaria include high fever and other problems that can cause death if not effectively treated. Once infected with *P. falciparum*, a person can suffer recurrences of malaria throughout life. As a consequence, victims of the disease are less healthy than their uninfected counterparts and are susceptible to other diseases as well. Overall, malaria victims experience higher morbidity (illness) and mortality (death) and produce fewer children than does the rest of the population.

*Plasmodium falciparum* and the mosquito that carries it flourish in tropical environments, and therefore malaria is endemic to the tropics. *P. falciparum* embryos live in their mosquito hosts, but they do not begin larval development until they are transferred to a mammalian host. Once inside a mammalian host, the plasmodium begins to mature, first in the liver of the host animal and later in the red blood cells.

### Heterozygous Advantage

One of the best-documented examples of natural selection in the evolution of human populations has been the relationship observed between malaria and the $\beta^S$ allele. Numerous anthropological and epidemiological studies have recorded the effects of malaria on the evolution of $\beta^S$ and SCD in African populations, and in other populations in Southern Europe and Asia (**Figure 10.15**). The central finding of these studies is that heterozygotes with the genotype $\beta^A\beta^S$ survive and reproduce more dependably than other genotypes in environments where malaria is common.

The improved survival and reproduction of heterozygotes can be explained at a cellular level by the selective advantage that heterozygotes derive from the shortened average life span of their red blood cells. The average red cell life span in these individuals is shortened due to the presence of a certain amount of mutant β-globin protein and the consequent formation of a small number of sickle-shaped red blood cells. The shorter red cell life spans interrupt the developmental cycle of *Plasmodium* larvae by preventing many of the immature parasites from reaching maturity. As a result, heterozygotes suffer fewer cases of malaria than are experienced by $\beta^A\beta^A$ homozygotes, and when they do get malaria, their disease is less severe.

On a population level, individuals with SCD (homozygous for the mutant gene) survive and reproduce very poorly due to their hemoglobin disorder. Those who are $\beta^A\beta^A$ also have lower reproductive fitness than do heterozygous carriers, because of the ravages of malaria. The result is that natural selection favors heterozygous carriers and causes populations to evolve a gene pool that includes large proportions of both alleles. This **heterozygous advantage** seen for $\beta^A\beta^S$ individuals is balanced by the disadvantage to those with SCD. The action of these conflicting trends produces an overall increase in the $\beta^S$ allele until it reaches a stable **equilibrium frequency,** where the gain and loss of $\beta^S$ alleles is equal. The term **balanced polymorphism** is used to describe situations like this, in which the loss of an allele because of selection against one of its phenotypes is balanced by natural selection in favor of the allele for another phenotype.

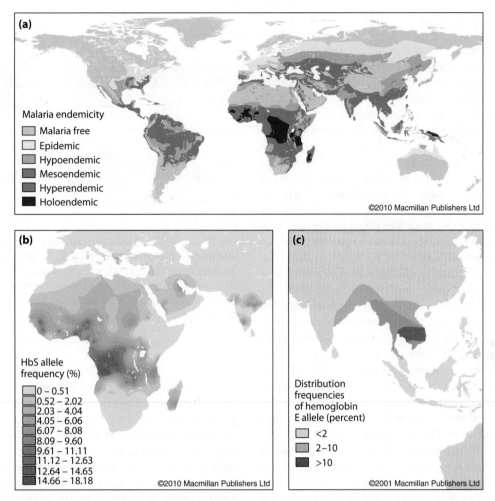

**Figure 10.15 The distribution of malaria and sickle cell disease.** (a) Colored areas indicate the regions of the world where malaria is an endemic disease. (b) Frequency distribution of the $\beta^S$ allele in some of the human populations occupying the malarial belt. (c) The distribution of the $\beta^E$ allele in Southeast Asia.

In research concerning heterozygous advantage in the evolution of $\beta^S$, three findings are particularly important:

1.  The frequency of SCD carriers rises with increasing age in the population. Studies of genotype frequencies in malaria-afflicted populations find the frequency of $\beta^A\beta^S$ heterozygotes to be lower in children than in adolescents, and to be lower in adolescents than in adults. In other words, individuals with $\beta^S\beta^S$ and $\beta^A\beta^A$ genotypes are being lost from these populations at younger ages than are heterozygotes.

2.  Heterozygous women produce a greater average number of children than do women who are $\beta^A\beta^A$. This is an indication that the overall health of heterozygotes is better, leading them to reproduce more efficiently.

3.  Across the "malaria belt," the portion of the tropics where malaria is common, the $\beta^S$ allele has developed

and evolved at least three times independently in different populations. Some human biologists believe the genetic evidence supports four separate mutation and evolution events. These independent evolutionary events account for the presence of $\beta^S$ in high frequency in populations in the Middle East, the region surrounding the Mediterranean Ocean, parts of the Indian subcontinent, and parts of Africa.

Genetic Insight  The $\beta^S$ allele has evolved by natural selection and has attained a high frequency in several populations because heterozygous carriers survive and reproduce more efficiently than do others in the populations. In these populations, the selective advantage of the $\beta^S$ allele in the heterozygous genotype is counteracted by the loss of alleles due to SCD to produce a balanced polymorphism.

## Evolution of $\beta^C$ and $\beta^E$

Additional support for the role of natural selection in the evolution of globin genes comes from the study of two other β-globin gene alleles that are present at high frequencies in other populations in the malaria belt. Mutant β-globin alleles $\beta^C$ and $\beta^E$ have evolved due to the natural selection pressure of malaria in much the same way $\beta^S$ has evolved.

The mutation known as $\beta^C$ likely occurred thousands of years ago on the west coast of Africa. This mutation is a base substitution of a nucleotide immediately adjacent to the site of the $\beta^S$ mutation in the sixth DNA triplet of the β-globin gene (Figure 10.16a). The effect of the mutation is to change the sixth amino acid of the β-globin protein from glutamic acid to lysine.

Although the mutation affects the same amino acid position altered in $\beta^S$, the complications of the mutation are not as severe as those seen in SCD. Homozygosity for the $\beta^C$ mutation does not produce severe anemia and is rarely fatal. Like SCD carriers, however, heterozygotes with the genotype $\beta^A\beta^C$ are more resistant to malaria than are $\beta^A\beta^A$ homozygotes. This situation leads to the spread of the $\beta^C$ mutation by a process of natural selection parallel to that seen for $\beta^S$.

On the other side of the malaria belt, in Southeast Asia and the adjacent Pacific Islands, another β-globin gene mutation, $\beta^E$, is prevalent. $\beta^E$ is a base substitution mutation that alters amino acid 26 of the β-globin protein, changing it from glutamic acid to lysine (Figure 10.16b). The anemia seen in $\beta^E\beta^E$ homozygotes is severe, but the selection it exerts against $\beta^E$ is balanced by the greater resistance of heterozygous carriers of the allele to malaria.

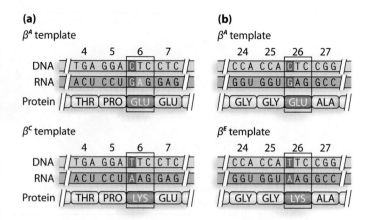

**Figure 10.16  Sequence comparisons of $\beta^A$ (the wild type) and mutant β-globin alleles $\beta^C$ and $\beta^E$.** (a) $\beta^S$ and $\beta^C$ are base-substitution mutants that alter amino acid position 6, changing glutamic acid (Glu) to lysine (Lys) in $\beta^C$. (b) The base substitution mutant $\beta^E$ changes amino acid 26 from glutamic acid (Glu) to lysine (Lys).

Like the $\beta^S$ variant that has been our focus throughout this chapter, the $\beta^C$ and $\beta^E$ variants are distributed across the malarial belt that spans much of the tropical regions surrounding the equator. Clinical and epidemiological studies confirm that all three β-globin gene variants are advantageous in the heterozygous state because they reduce the incidence and intensity of malarial disease in carriers. The incidence of hereditary disease produced by homozygosity is balanced by the improved odds of survival and reproduction for carriers of these globin gene variants.

## CASE STUDY

### Transmission and Molecular Genetic Analysis of Thalassemia

Autosomal recessive forms of a hereditary anemia called thalassemia result from mutations of globin genes that create an imbalance in the ratio of the **α-globin** to **β-globin** polypeptides. The imbalance reduces the amount of hemoglobin that can form and generates anemia. Owing to differences between mutant alleles, **thalassemias** exhibit varying levels of severity, from mild to fatal. One particular form of thalassemia is common on the Mediterranean island of Sardinia.

The Sardinian thalassemia mutation (OMIM 141900) is a DNA nucleotide base substitution $(G\text{-}C \rightarrow A\text{-}T)$ in the 39th codon (corresponding to the 39th amino acid) of the β-globin gene (Figure 10.17a). The mutation changes the 39th codon of the transcript from 5'-CAG-3', coding for the amino acid glutamine (Gln), to the sequence 5'-UAG-3', which is a stop codon. This change results in the premature termination of translation of β-globin protein after the first 38 amino acids. The truncated protein is not functional. Consequently, individuals who are homozygous for the mutant allele have no β-globin

protein. Their ability to form hemoglobin is greatly diminished, causing severe anemia. Heterozygotes also have diminished capacity to produce hemoglobin, and they suffer from chronic anemia. Since heterozygotes have one wild-type β-globin allele, however, their anemia is less severe than in homozygotes.

Wild-type β-globin alleles have two recognition sites for restriction endonuclease *MaeI* (restriction sequence 5'-CTAG-3') in the vicinity of the gene (Figure 10.17b). The base-substitution mutation in DNA triplet 39 creates a new *MaeI* restriction site that is not found in the wild-type sequence. Whereas the wild-type allele contains two *MaeI* restriction sites separated by approximately 1500 base pairs (1.5 kb), the mutant allele sequence contains a third *MaeI* restriction site that cleaves the 1.5-kb region into two DNA fragments of 0.5 kb and 1.0 kb. Southern blot analysis of *MaeI*–digested β-globin DNA utilizes a molecular probe that binds near one end of the β-globin gene. The probe binds a 1.5-kb DNA fragment produced by *MaeI* treatment of the

**(a) DNA sequence variation of β-globin alleles**

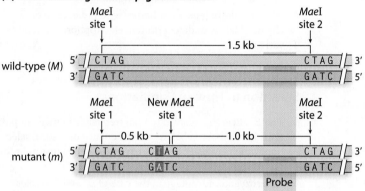

**(b) Restriction digestion of β-globin alleles**

**(c) Southern blot analysis of β-globin allele variation**

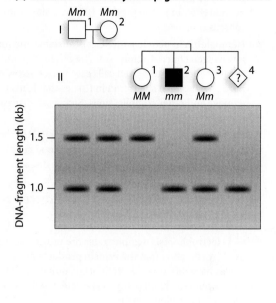

**Figure 10.17   Molecular genetic analysis of variation at DNA triplet 39 of the β-globin gene in Sardinian β-thalassemia. (a)** DNA sequences of the wild-type (*M*) and mutant (*m*) β-globin alleles from triplet 36 through 41. **(b)** Restriction maps of wild-type and triplet-39 mutant alleles show a new *Mae*I restriction site in the mutant allele. The location of molecular probe binding identifies a 1.5-kb DNA fragment for the wild-type allele (*M*) and a 1.0-kb fragment for the mutant allele (*m*). **(c)** Southern-blot analysis of a family showing segregation of wild-type and triplet-39 mutant alleles. Heterozygous (*Mm*) parents produce children with all three genotypes. The genotype of II-4 is discussed in the text.

wild-type allele and a 1.0-kb DNA fragment produced by *Mae*I treatment of the mutant allele. The 0.5-kb DNA fragment is also produced by *Mae*I digestion of the mutant allele, but that fragment is not detected in Southern blot analysis because it is not bound by the molecular probe.

In terms of the presence or absence of the Sardinian mutant allele, three genotypes are possible at this locus, each having a unique restriction fragment banding pattern detectable by Southern blotting. Homozygotes for the wild-type allele (*MM*) produce a single Southern blot band of 1.5 kb. Homozygotes with severe anemia have the mmgenotype and produce a single DNA band of 1.0 kb. Heterozygotes (*Mm*) produce both bands, since they carry both alleles.

Figure 10.17c shows a nuclear family pedigree that is consistent with an autosomal pattern of inheritance of thalassemia. The pedigree symbols for each family member are located directly above the Southern blot lane containing that person's DNA. The Southern blot detects a different pattern of DNA bands for each genotype. The figure also illustrates Southern blot results for DNA obtained from a fetus (the diamond-shaped symbol identified as II-4) being carried by I-2. This analysis is a prenatal molecular diagnostic test for Sardinian thalassemia that is based on the Southern blot band differences among the three possible genotypes.

---

SUMMARY          Mastering**GENETICS**™     *For activities, animations, and review quizzes, go to the study area at www.masteringgenetics.com.*

## 10.1 An Inherited Hemoglobin Variant Causes Sickle Cell Disease

▌ Hemoglobin is an abundant protein in red blood cells, transporting oxygen throughout the body. Its structure is tetrameric, composed of two polypeptides encoded by the α-globin gene cluster and two polypeptides encoded by the β-globin gene cluster.

▌ Mutation of genes frequently leads to abnormal structure and function of proteins. Mutations of α-globin or β-globin genes often produce hereditary anemia, the most common category of hereditary disease known in humans.

▌ Sickle cell disease (SCD) is a common hereditary anemia in humans caused by homozygosity for the $\beta^S$ allele of the β-globin gene. Individuals with SCD have the genotype

$\beta^S\beta^S$. The globin protein produced by $\beta^S$ differs from the normal β-globin gene product ($\beta^A$) by a single amino acid substitution.

▌ Hemoglobin in people with SCD is unstable and linearizes at low oxygen concentration, distorting the red blood cell into a sickle shape. The distorted cells can block narrow capillaries, producing oxygen starvation in tissues that leads to tissue damage and other complications. Sickle cell disease leads to premature death of red blood cells.

▌ Heterozygous carriers of the $\beta^S$ mutation ($\beta^A\beta^S$) have a small percentage of sickle-shaped red blood cells but do not suffer symptoms or complications of the disease.

## 10.2  Genetic Variation Can Be Detected by Examining DNA, RNA, and Proteins

▌ Gel electrophoresis demonstrates the molecular basis of SCD by revealing that the protein products of the $\beta^A$ and $\beta^S$ alleles have different electrophoretic mobilities. Distinctive electrophoretic band patterns are detected for many genotypes of the β-globin locus.

▌ The single amino acid substitution caused by the $\beta^S$ allele is a valine in place of a glutamine in the β-globin protein.

▌ DNA analysis identifies the $\beta^S$ mutation as a base-pair substitution in the β-globin gene that produces a single nucleotide polymorphism (SNP) and eliminates a *Dde*I restriction site.

▌ RFLP variation distinguishes the $\beta^A$ and $\beta^S$ alleles.

▌ The pattern of inheritance of RFLPs parallels that of alleles at the β-globin gene.

▌ DNA restriction fragments are detected by transferring denatured DNA fragments from electrophoresis gels to a permanent membrane in the Southern blotting process.

▌ In Southern blots, single-stranded nucleic acid probes labeled with radioactive or chemical markers hybridize with complementary target sequences in DNA fragments.

▌ The presence and position of a DNA fragment hybridized by a molecular probe is revealed by the appearance of bands in Southern blot analysis.

▌ Northern blotting is similar to Southern blotting but examines mRNA for differences in length.

▌ Western blotting uses antibodies with radioactive or chemical labels to detect protein electrophoretic mobility variation.

## 10.3  Sickle Cell Disease Evolved by Natural Selection in Human Populations

▌ The $\beta^S$ allele has evolved to high frequency in many populations in the malaria belt as a consequence of natural selection, which favors $\beta^A\beta^S$ heterozygotes as the most fit in the malarial environment.

▌ Heterozygous advantage in the case of $\beta^A$ and $\beta^S$ alleles stems from disruption of the malarial parasite life cycle, a result of the somewhat shorter average life span of red blood cells in heterozygotes.

▌ Mutations of the β-globin gene, including $\beta^C$ and $\beta^E$, appear to have evolved in distinct populations by processes similar to those that established $\beta^S$ in human populations.

## KEYWORDS

α-globin gene (α-globin gene cluster, α-globin chain) *(p. 334)*

β-globin gene (β-globin gene cluster, β-globin chain) *(p. 334)*

$\beta^A$ allele *(p. 335)*

$\beta^S$ allele *(p. 335)*

agarose (agarose gel) *(p. 336)*

balanced polymorphism *(p. 350)*

band (in electrophoresis gel) *(p. 337)*

blunt end *(p. 340)*

chromatography *(p. 338)*

electrophoretic mobility *(p. 337)*

equilibrium frequency *(p. 350)*

ethidium bromide (EtBr) *(p. 341)*

gel electrophoresis *(p. 336)*

hemoglobin (Hb) *(p. 333)*

heterozygous advantage *(p. 350)*

hybridization (of molecular probe) *(p. 343)*

kilobase (kb) *(p. 341)*

molecular probe (probe) *(p. 341)*

northern blotting *(p. 341)*

origin of migration *(p. 336)*

peptide fingerprint analysis *(p. 338)*

polyacrylamide (polyacrylamide gel) *(p. 336)*

restriction endonuclease (restriction enzyme) *(p. 340)*

restriction fragment length polymorphism (RFLP) *(p. 340)*

restriction sequence *(p. 340)*

sickle cell disease (SCD) *(p. 333)*

single nucleotide polymorphism (SNP) *(p. 338)*

Southern blotting *(p. 341)*

sticky end *(p. 340)*

thalassemias *(p. 352)*

western blotting *(p. 341)*

## PROBLEMS

*For instructor-assigned tutorials and problems, go to www.masteringgenetics.com.*

### Chapter Concepts

*For answers to selected even-numbered problems, see Appendix: Answers.*

1. Define the following terms as described in this chapter:
   a. gene cluster
   b. heterozygous advantage
   c. dominant
   d. intron
   e. hemoglobin tetramer
   f. hereditary anemia
   g. exon

h. heterozygous
i. recessive
j. molecular disease
k. restriction endonuclease
l. homozygous
m. gel electrophoresis
n. restriction fragment length polymorphism
o. SNP
p. electrophoretic mobility
q. Southern blot
r. molecular probe
s. northern blot
t. antibody probe
u. western blot

2. Using sickle cell disease as an example, describe the similarities and differences between the terms *genetic disease* and *molecular disease*. How are molecular or genetic diseases different from diseases that are caused by an infectious organism such as a bacterium?

3. Compare and contrast the contributions of Neel, Pauling, and Ingram to our understanding of the genetic and molecular bases of sickle cell disease.

4. Why do differences in protein electrophoretic mobility often result from changes to protein amino acid sequences? How can electrophoretic mobility differences arise between the protein products of different alleles?

5. Electrophoretic analysis of hemoglobin from a person with normal HbA and a person with hereditary anemia reveals no difference in the electrophoretic mobility. How can this occur?

6. Many types of hereditary anemia result from single amino acid substitutions affecting one of the hemoglobin protein chains. For example, the wild-type β-globin allele has the template–DNA sequence CTC at triplet 6, which encodes the amino acid glutamic acid (Glu) at position 6 of the β-globin protein (see Figure 10.9). The mutant allele producing $\beta^S$ contains the DNA sequence CAC and encodes valine (Val) at β-globin position 6, and the $\beta^C$ mutation contains TTC in DNA and encodes lysine (Lys) at position 6. The table below shows several other β-globin gene mutants that are the result of single amino acid substitutions. Use the information provided and Table 1.2 to determine the wild-type template– DNA sequence and the template sequence for each mutant.

| β-Globin Form | Position | Amino Acid |
|---|---|---|
| $\beta^A$ (wild type) | 7 | Glu |
| Siriraj | 7 | Lys |
| San Jose | 7 | Gly |
| $\beta^A$ (wild type) | 58 | Pro |
| Ziguinchor | 58 | Arg |
| $\beta^A$ (wild type) | 145 | Tyr |
| Bethesda | 145 | His |
| Fort Gordon | 145 | Asp |

7. A single base substitution creates the α-globin gene mutant Hb Constant Spring (Hb$^{CS}$), whose product contains 172 amino acids. Wild-type α-globin protein contains 141 amino acids. The wild-type mRNA carries the codon CGU to encode arginine (Arg) as the final amino acid of the chain, followed by the stop codon UAA. The Hb$^{CS}$ mutant produces mRNA that has the sequence CGUCAA in this region. Explain how the single DNA base substitution in Hb$^{CS}$ can lead to production of a protein that contains 31 more amino acids than the wild type has.

8. Wild-type β-globin protein is composed of 146 amino acids. A β-globin gene mutant known as Hb Cranston contains 157 amino acids. Partial mRNA sequences of $\beta^A$ and Hb Cranston ($\beta^{Cr}$) are shown. The numbers indicate amino acid positions. Identify the mutation that causes $\beta^{Cr}$, and describe how the mutation leads to a longer than normal β-globin protein chain.

```
       144 145 146 Stop
βA  AAG UAU CAC UAA GCU CGC UUU CUU GCU GUC
    CAA UUU CUA UUA A

       144 145 146 147         150
βCr AAG AGU AUC ACU AAG CUC GCU UUC UUG
             155 156 157 Stop
      CUG UCC AAU UUC UAU UAA
```

9. Describe why sickle cell disease is considered to be a recessive genetic disorder.

10. What molecular parameter causes DNA fragments to have different electrophoretic mobility? What parameter causes different mobilities in mRNA? What parameters cause different mobilities in proteins?

11. How is an autoradiograph produced from a Southern blot?

12. Both Southern blotting and northern blotting can reveal information about the DNA fragments or RNA molecules being examined, but the positions of nucleic acid bands in one kind of blot cannot be directly compared with those in the other. Why?

13. The target sequence on a fragment of DNA is 3'-ATATCGCACGGACT-5'. What is the sequence and polarity of an equivalent-length molecular probe used to detect this target sequence? Explain why the molecular probe you have proposed will detect the targeted sequence.

14. The $\beta^S$ allele occurs in a central West African population at a frequency of 15%. The same allele occurs in a population from the southern tip of Africa at a frequency of less than 1%. Speculate about the reason for the different frequencies of the allele in these two African populations.

## Application and Integration

*For answers to selected even-numbered problems, see Appendix: Answers.*

15. The family represented in the pedigree and Southern blot below has been evaluated for the presence and distribution of the $\beta^S$ allele. Use the information in the Southern blot and the explanation provided in the chapter to identify the phenotype and determine the genotype of each person tested.

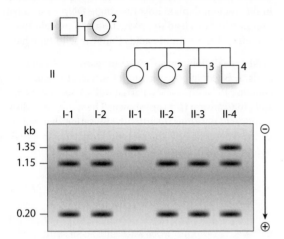

16. Suppose the mating couple (I-1 and I-2) shown in Problem 15 are expecting a fifth child.
    a. Is it possible that their fetus could have sickle cell disease? If so, what is the probability? If not, explain why not.
    b. Fetal DNA is collected and analyzed by Southern blotting. The fetus has a single DNA band that is 1.35 kb in length. What is your interpretation of this result? Explain your answer.

17. What are restriction endonucleases, and why are they useful in identifying DNA sequence variation?

18. Following restriction digestion, DNA fragments produced by digestion with certain enzymes have "sticky ends," while fragments produced by digestion using other enzymes have "blunt ends." Distinguish the meaning of these two terms.

19. The double-stranded DNA sequence below is part of a restriction fragment you wish to detect by autoradiography.

    5'-…ATTCATGACGGACTATTCGAGAGCTGATGCAT…-3'
    3'-…TAAGTACTGCCTGATAAGATCTCGACTACGTA…-5'

    Identify which of the following molecular probes is the best choice for achieving the desired hybridization reaction. Indicate where on the upper or lower strand the probe will hybridize.
    a. 3'-TGATATCGTACCGAA-5'
    b. 5'-TGCCTGATAAGATCT-3'
    c. 3'-ACAGCCTAGTAAGAT-5'
    d. 3'-ACTGCCTGATAAGCT-5'

20. Restriction enzymes recognize specific double-stranded DNA sequences that have the same sequence on both strands. For example, the restriction sequence for *Bam*H I is 5' GGATCC 3' on each DNA strand and for *Sma*I is 5' CCCGGG 3' on each strand. A single phosphodiester bonds on each strand is cut at the same place in the sequence on each strand. Explain how restriction enzymes are able to recognize the same sequence and cut the sequence in the same place on each DNA strand.

21. Four alleles of a variable DNA marker gene produce different-sized DNA fragments as follows: $R^1 = 4$ kb, $R^2 = 13$ kb, $R^3 = 10$ kb, $R^4 = 7$ kb.
    a. Identify the genotypes of individuals depicted in lanes 1, 2, and 3 in the gel shown.

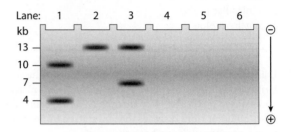

    b. In lanes 4, 5, and 6, draw the band patterns expected for individuals with, respectively, genotypes $R^1R^3$, $R^3R^4$, and $R^1R^1$.

22. Consider this DNA sequence:

    5'-TTCGAATTCGACTCAGGATCCTACAAGTTTCAT-3'
    3'-AAGCTTAAGCTGAGTCCTAGGATGTTCAAACTA-5'

    Which of the following restriction sites are present in this sequence? Draw a box around each restriction sequence.
    a. *Eco*RI (5'-GAATTC-3')
    b. *Bam*HI (5'-GGATCC-3')
    c. *Hind*III (5'-AAGCTT-3')

23. Two probes designated probe A and probe B hybridize very near one another in a region of DNA that contains DNA fragment length variation when digested with the restriction enzyme *Hind*III. Four maps show the location and intervening distances in kilobases of *Hind*III restriction sites and the binding locations of probes A and B. The maps correspond to alleles $H^1$ to $H^4$.

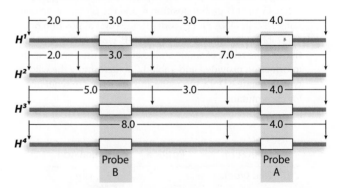

    a. For the genotype $H^1H^3$, what DNA bands are expected in an autoradiogram using probe A? Using probe B? Using both probes together?
    b. For the genotype $H^2H^4$, what band pattern is expected using probe A? Probe B? Using both probes together?
    c. Suppose a woman with the genotype $H^1H^3$ has a child with a man whose genotype is $H^2H^4$. What are the four possible genotypes for a child of this couple?

d. What are the sizes of bands produced by each possible child of this couple using probe A? Using probe B?

24. Plants of a particular species can either have the dominant wild-type phenotype, tall, or the recessive phenotype called dwarf. Genetic analysis has identified the stature gene, and DNA analysis of tall plants yields a DNA fragment of 7.5 kb corresponding to a portion of the gene. The wild-type gene map is illustrated, along with an autoradiograph showing DNA restriction fragments from a normal parental plant (T; lane 1), a tall plant that carries a copy of the mutant allele (T; lane 2), the two DNA fragment patterns observed in tall progeny plants (T; lanes 3 and 4), and the DNA fragment pattern seen in dwarf plants (D; lane 5).

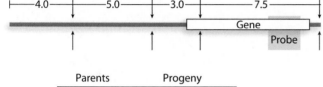

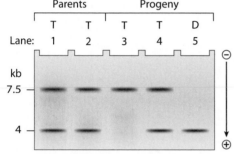

a. Propose two mutational events that could cause the small DNA fragment that represents the mutant allele.
b. In northern blot analysis of mRNA, what mRNA differences would you anticipate for your two proposed mutational mechanisms?

25. A second strain of dwarf plants has a different mutation of the same gene identified in Problem 24. In the second strain, plants carrying a copy of the mutant allele produce a DNA restriction fragment of 10.5 kb, rather than the 7.5-kb fragment. DNA fragments produced by digestion of DNA from tall carrier plants are shown below in lanes 1 and 2, fragments from tall progeny of carriers are shown in lanes 3 and 4, and DNA from dwarf plants is shown in lane 5.

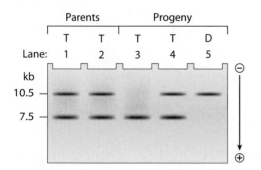

a. What mutational mechanism is most likely responsible for the production of abnormal DNA fragment length corresponding to this mutant allele? Explain your reasoning.
b. In comparison to the length of mRNA from the normal allele, will the mRNA from this mutant allele most likely be longer, shorter, or about the same length? Explain your answer.

26. During gel electrophoresis of linear DNA molecules, why do longer molecules move more slowly than shorter molecules? What determines the difference in electrophoretic mobility of mRNA molecules?

27. What three features of proteins are most important in determining their electrophoretic mobility? Based on your answer, describe how single amino acid substitutions can change the electrophoretic mobility of a protein.

28. In molecular biology, restriction endonucleases isolated from bacteria are used to cleave DNA into fragments. What functional role do you think restriction endonucleases serve in the bacteria from which they are derived? (*Hint:* The restriction endonuclease produced by a bacterium cannot cut the bacterium's DNA.)

29. A complete plant gene containing four introns and five exons is carried on a 6.0-kb DNA fragment. DNA sequencing analysis finds that this fragment contains 1000 base pairs that flank the transcribed region of the gene and 5000 base pairs that are transcribed. Four introns contain 3500 base pairs, and five exons contain 1500 base pairs. Northern blot analysis is performed on mRNA of this gene using a probe that binds to a portion of one of the exons. mRNA isolated from the cytoplasm of cells is compared to mRNA isolated from cell nuclei on the northern blot. Do you expect that all the mRNAs will be a uniform length, or will mRNA molecules of multiple lengths be detected on the northern blot?

30. Two male hounds, identified in the figure as ♂1 and ♂2, got loose one night at Wet Noses Puppy Farm. A female (♀A in the figure) got pregnant and had a litter of three puppies (P1, P2, and P3). The owner of Wet Noses is desperate to know which male is the father and has electrophoretic analysis of a variable DNA genetic marker (shown in the figure) to guide in the identification. The owner thinks ♂1 is the father of the three puppies. Is the owner correct? Explain your answer.

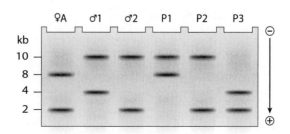

31. The map below illustrates three alleles in a genome segment. The alleles differ in the number and location of restriction sequences. Restriction digestion and Southern blotting with the molecular probe whose hybridization location is indicated results in detection of a single DNA band for each allele.

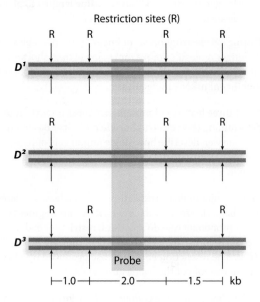

a. List the size (in kilobases) of DNA bands detected by Southern blotting of restriction-digested DNA from organisms with the genotypes $D^2D^2$, $D^2D^3$, and $D^3D^3$.

b. Restriction-digested DNA from two organisms is analyzed by Southern blotting. Restriction fragments of 2.0 and 3.5 kb are observed on the Southern blot of one organism, and bands of 2.0 and 3.0 kb are observed for the other. What are the genotypes of these organisms?

c. Organisms with the genotype $D^1D^1$ are identified by the detection of a 2.0-kb DNA band on a Southern blot. Why does this genotype produce a single detectable band, and why are the 1.0- and 1.5-kb restriction fragments not detected in Southern blotting?

# Chromosome Structure

# 11

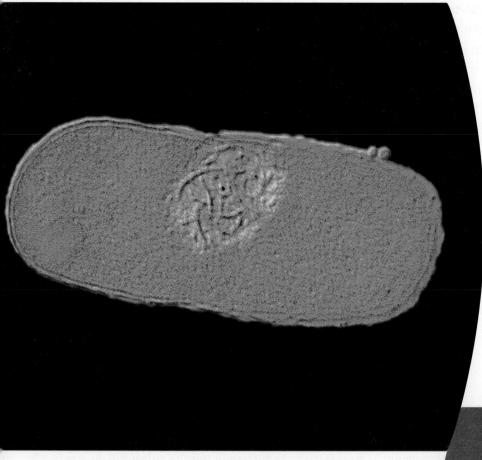

The *E. coli* chromosome is found in the nucleoid of the bacteria cell.

## ESSENTIAL IDEAS

- Associated proteins and supercoiling compress the bacterial chromosome into a nucleoid region within the cell.

- Eukaryotic chromosomes contain large amounts of protein that organize and condense chromosomes.

- Chromosome bands uniquely identify each chromosome but vary with chromosome condensation level.

- Variations in organization and condensation within eukaryotic chromosomes are associated with transcription.

The genome of a species is the total amount of hereditary information in an entire set of its chromosomes. Chromosomes consist largely of DNA, and we described the molecular structure of DNA and the importance of its nucleotide sequence in Chapter 7. But the structure and sequence of DNA are only a partial description of the genome. Arguably even more important to the genome story is the way the DNA is organized in chromosomes.

Every chromosome carries a single, long DNA molecule. The chromosome may be singular, as in haploid bacterial species; it may be a member of a homologous pair of chromosomes, as observed in diploid species; or there may be multiple sets of chromosomes, as are found in polyploid plant species. But whatever their number and whether they belong to archaea, bacteria, or eukarya, all chromosomes contain a mixture of DNA and proteins.

The combining of protein with DNA to produce chromosomes performs four essential functions. First, it helps compact the DNA so that it will fit into the cell or nucleus. Second, it stabilizes DNA and protects it from damage. Third, it promotes chromosome condensation required for cell division. Finally, the packaging of chromosomes with proteins helps regulate DNA replication and gene transcription.

The subject of this chapter is the DNA–protein association in chromosomes and the ways in which the four essential functions identified above are accomplished. Most of the chapter focuses on eukaryotic chromosomes, which have a much more complex array of DNA-associated proteins than do bacterial chromosomes. We begin, however, with a discussion of the latter.

## 11.1 Bacterial Chromosomes Are Simple in Organization

Bacteria are haploid organisms that have single chromosomes, not homologous pairs. The customary description of the bacterial genome as consisting of a single

chromosome that is almost always circular is true for many of the most widely studied bacterial species, such as *Escherichia coli* and *Bacillus subtilis*. The *E. coli* genome consists of a single, circular chromosome containing approximately 4.6 million base pairs (Mb) that encode about 4200 genes. Nonetheless, certain bacterial species have linear chromosomes, and a few species carry more than one chromosome.

### Bacterial and Archaeal Chromosomes

The genomes of most bacterial and archaeal species consist almost exclusively of a single, covalently closed circular chromosome. However, the total length of the chromosome and the number of genes it contains vary substantially among species (Table 11.1). The single or, in those few species with more than one chromosome, largest chromosome carries essential genes that are required for reproduction, gene expression, and normal metabolic activities carried out by the organism. Bacteria also routinely carry multiple copies of one or more types of plasmid, which are extrachromosomal DNA molecules that are not part of the bacterial genome (see Section 6.1). Plasmids carry nonessential genes—genes not needed by bacteria in the completion of their normal processes and tasks.

Most bacterial and archaeal genomes contain a single chromosome that carries all of the genes that are essential to the organism. In some genomes, however, the essential genes are located on one main chromosome and one or two additional chromosomes. For example, the genome of *Borrelia burgdorferi*, the bacterial organism that causes Lyme disease, is composed of a single large linear chromosome of 910,000 bp (0.91 Mb) and a smaller circular chromosome of 530,000 bp (0.53 Mb). *Agrobacterium tumefaciens*, the bacterium causing crown gall disease in many plant species, has a 3.0-Mb circular chromosome, two much smaller circular chromosomes, and a 2.1-Mb

| Table 11.1 | Chromosome Diversity among Bacteria and Archaea | | |
|---|---|---|---|
| **Species** | **Genome Size (in Mb)** | **Number of Chromosomes** | **Chromosome Form(s)** |
| *Mycoplasma genitalium* | 0.58 | 1 | Circular |
| *Borrelia burgdorferi* | 1.4 | 2 | One circular, one linear |
| *Methanococcus jannaschii* | 1.6 | 3 | All circular |
| *Haemophilus influenzae* | 1.83 | 1 | Circular |
| *Thermotoga maritime* | 1.89 | 1 | Circular |
| *Vibrio cholerae* | 4.0 | 2 | Both circular |
| *Escherichia coli* | 4.2 | 1 | Circular |
| *Agrobacterium tumefaciens* | 5.7 | 4 | Three circular and one linear |
| *Sinorhizobium meliloti* | 6.7 | 3 | All circular |

linear chromosome. No linear chromosomes have been found to date in archaeal species, but the methane-loving archaebacterium *Methanococcus jannaschii* has three circular chromosomes in its genome—one large chromosome of 1.66 Mb and two smaller chromosomes of 58 kb and 16 kb.

The chromosomes of bacteria and archaea are densely packed to form a small region called the **nucleoid** (see chapter opener image) within the cell. Efficient organization of the chromosome into a series of tight loops makes this region remarkably small. For example, if the 4.6 Mb of the *E. coli* nucleoid were unpacked and laid out along a ruler, it would measure about 1200 µm, nearly 1000 times longer than an *E. coli* cell itself. To get a sense of this size difference, imagine trying to stuff a 62-foot-long thread into the kind of gelatin-based capsule you might take for allergy or a headache!

## Bacterial Chromosome Compaction

How does *E. coli* package a chromosome 1000 times longer than the cell itself and leave room for molecular activities such as replication, transcription, and translation? The answer is twofold. First, proteins help organize the chromosome into the loops that efficiently pack the nucleoid, and second, the circular DNA of the chromosome undergoes *supercoiling*.

Bacterial DNA is associated with two major groups of proteins, **small nucleoid–associated proteins** and **structural maintenance of chromosomes (SMC) proteins.** Several different proteins belong to the small nucleoid–associated group of proteins, and all appear to participate in DNA bending that contributes to folding and condensation of the chromosome. The small nucleoid–associated proteins whose functions are best characterized are H-NS protein and HU protein. **Figure 11.1** illustrates a possible general arrangement for H-NS and HU in securing loops of chromosomal DNA within the nucleoid. It also shows the role of the SMC proteins, which

appear to attach directly to DNA itself, holding the DNA in coils (or perhaps V-shaped configurations) that can form large nucleoprotein (i.e., nucleic acid and protein) complexes. In addition to HU, H-NS, and SMC proteins, other proteins interact in the nucleoid to compact DNA. The precise identity and individual roles of these proteins are still a subject of active investigation.

The second mechanism facilitating chromosome compaction in the nucleoid is DNA supercoiling, which occurs at the chromosomal level. Covalently closed circular chromosomes exist in various coiled forms; the least twisted is the relaxed-circle form, and the most tightly twisted is the highly supercoiled form. Relaxed-circle DNA lies in an *O* shape, resembling a large, undistorted rubber band. In contrast, DNA in the highly supercoiled state is more like a rubber band that as a result of extensive twisting has become convoluted, overlaps itself, and will not lie flat in a single plane. Supercoiling compacts the circular chromosome so that it occupies less space and thus fits more readily into the nucleoid.

**Figure 11.2** illustrates the visualization of supercoiling by electromicroscopy and by gel electrophoresis. The electron micrograph in Figure 11.2a shows two circular chromosomes from the same bacterial species, one highly supercoiled and the other in a relaxed-circle structure. Figure 11.2b shows gel electrophoresis results for relaxed-circle DNA, highly supercoiled DNA, and DNA in various states of supercoiling. The more tightly compacted highly supercoiled DNA has much greater electrophoretic mobility (i.e., it migrates farther from the origin) than relaxed-circle DNA.

Supercoiled DNA exists naturally in circular chromosomes of bacteria, but it can become "overwound"—as it does during DNA replication, for example. Overwinding puts unsustainable torsional stress on DNA, and if the stress is not relieved, DNA will break at random. Supercoiled DNA is partially unwound by bacterial gyrase enzymes, also called topoisomerases, that relax supercoiled DNA in a controlled manner (See Chapter 7, for a discussion of overwound supercoiled DNA created during replication and the role of topoisomerase in unwinding the molecule).

Two forms of DNA supercoiling are detected in organisms. **Negative supercoiling** occurs by twisting DNA around its axis in the *opposite* direction from that of the double helical twist. The DNA duplex has a right-handed twist, so negative supercoiling twists the chromosome to the left. Negatively supercoiled DNA is the predominant form in most organisms. A few species of archaea, however, have a chromosome with **positive supercoiling** that is twisted in the *same* direction as the double helical twist. These species are primarily hyperthermophiles, a term used to identify forms that live in very high temperatures. Hyperthermophiles live in very hot liquid environments such as hot springs. A specialized topoisomerase in these species induces positive supercoiling. It is thought that

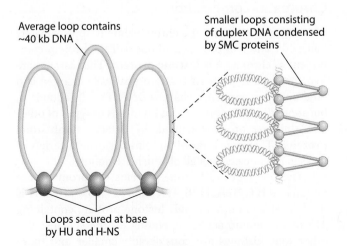

Average loop contains ~40 kb DNA

Smaller loops consisting of duplex DNA condensed by SMC proteins

Loops secured at base by HU and H-NS

**Figure 11.1 Bacterial chromosome compaction.**

**Figure 11.2 Circular DNA of bacteria in multiple forms. (a)** An electron micrograph shows supercoiled and relaxed-circle chromosomes. **(b)** The coiling of circular bacterial DNA determines its electrophoretic mobility. In lane 1, highly supercoiled DNA has a much higher electrophoretic mobility than the same relaxed-circle DNA. In lane 2, 5 minutes of treatment with topoisomerase to relax supercoiling produces many different coiled forms of the chromosome. In lane 3, 30 minutes of topoisomerase treatment converts much of the DNA to relaxed circle.

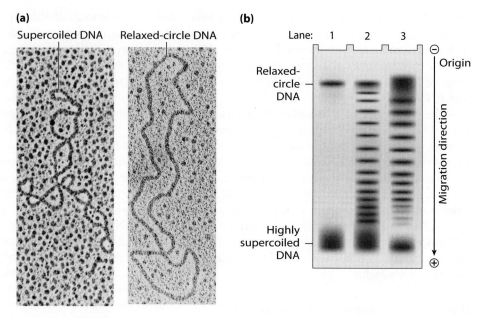

positive supercoiling of the chromosome provides additional stability against the degradation effects of high heat that might otherwise denature the duplex by destabilizing the hydrogen bonds.

---

Genetic Insight    Circular bacterial chromosomes are compacted to form the nucleoid by proteins that produce loops in DNA and by negative supercoiling of the chromosome.

---

## 11.2 Eukaryotic Chromosomes Are Organized as Chromatin

We have noted repeatedly that eukaryotic cells differ from bacterial cells in numerous ways. With regard to chromosomes, the most significant differences are the presence of multiple pairs of chromosomes in the eukaryotic nucleus and the organization of eukaryotic DNA into *chromatin*. Like its bacterial counterpart, each eukaryotic chromosome contains one long DNA double helix; but a different and more diverse array of proteins are associated with it, either complexing directly with the DNA or forming a structural support for the chromosome. The nuclear proteins that organize eukaryotic DNA and produce the chromatin structure are essential components of the genome structure and function. They provide a mechanism for efficient chromosome condensation and segregation during cell division and for chromosome organization that is integral to the regulation of gene transcription.

Every nucleated cell in your body is the beneficiary of a remarkable feat of biomolecular engineering attributable to the proteins bound to your DNA. Each nucleus contains approximately 6 billion base pairs of DNA, divided into 46 chromosomes. If all the chromosomes were

taken from the nucleus of a single one of your cells, and their DNA was stripped of proteins and unwound to a relaxed state, the DNA molecules laid end to end would span 1.8 meters—nearly 6 feet. This is more than 260,000 times larger than the diameter of the nucleus itself. The DNA from your shortest chromosome alone would be almost 15,000 times longer than the nuclear diameter! Returning to the analogy of the medicinal capsule mentioned above in connection with the *E. coli* chromosome, a capsule representing a human nucleus would contain 46 pieces of thread, representing the 46 human chromosomes, with a combined length of 625 feet! It seems remarkable that this level of compaction can occur at all and still allow the DNA to be replicated and transcribed and the chromosomes to be efficiently separated during cell division. But occur it does, thanks to **chromatin,** the complex containing DNA and proteins that combine to form eukaryotic chromosomes.

### Chromatin Composition

By weight, each eukaryotic chromosome is approximately half DNA and half proteins. About half the protein component of chromatin is **histone protein.** The histones are five small, basic proteins that are positively charged and bind tightly to negatively charged DNA. Equally abundant, but more diverse, is an array of hundreds of types of other DNA-binding proteins named, by default, **nonhistone proteins.** This large array of proteins performs a variety of tasks in the nucleus, not all of which are defined.

The five types of histone proteins in chromatin are designated **H1, H2A, H2B, H3,** and **H4** (Table 11.2). H1 is the largest and most variable histone protein, containing 215 to 244 amino acids, depending on the species. The other four histones are considerably smaller and more uniform in size, containing between 102 and 129 amino

| Table 11.2 | Histone Protein Characteristics | | | |
|---|---|---|---|---|
| Histone [a] | Basic/Acidic Amino Acids | Molecular Weight (D) | Number of Amino Acids | Location |
| H1 | 5.4 | 23,000 | 224 | Linker DNA |
| H2A | 1.4 | 13,960 | 129 | Nucleosome |
| H2B | 1.7 | 13,774 | 125 | Nucleosome |
| H3 | 1.8 | 15,273 | 135 | Nucleosome |
| H4 | 2.5 | 11,236 | 102 | Nucleosome |

[a] Histone proteins from calf thymus gland.

acids. Among eukaryotes, there is very strong conservation of the amino acid sequence of histone proteins, suggesting significant evolutionary pressure to retain the structure and function of each. A comparison of H4 amino acid sequences in cows and pea plants, which diverged from the common ancestor of land plants and animals more than 500 million years ago, identifies just two amino acid differences among the 102 amino acids in the protein. The comparison tells us that since the time when plants and animals last shared a common ancestor, extraordinarily strong evolutionary pressure has maintained H4 DNA and amino acid sequence identity among organisms. This extreme example of evolutionary conservation speaks to the importance of histones in eukaryotic chromosome organization.

Histones are the principal agents in chromatin packaging. Two molecules each of four histones—H2A, H2B, H3, and H4—join together to form an octameric (eight-member) **nucleosome** (Foundation Figure 11.3). Nucleosomes self-assemble following the transcription and translation of histone genes that are each present in multiple copies in eukaryotic genomes. The translated proteins first assemble into dimers containing two different histones each: H2A-H2B histone dimers contain one molecule each of histone 2A and histone 2B, and H3-H4 dimers contain one molecule each of histone 3 and histone 4. Current evidence indicates that nucleosomes are formed in steps that begin with two H3-H4 dimers assembling to form a histone tetramer. Once the tetramer is formed, two H2A-H2B dimers associate to form the final octameric nucleosome. The fully assembled nucleosome is a flat-ended structure approximately 11 nm in diameter by 5.7 nm thick (see Figure 11.3a). A span of DNA approximately 146 bp long wraps one and three-quarter turns around each nucleosome core particle. This wrapping of DNA around the nucleosome is the first level of DNA condensation, and it condenses the DNA approximately sevenfold.

The 146 bp of DNA wrapped around a core particle is called **core DNA.** Electron micrographs of chromatin fibers in their most decondensed state show a regular series of circular structures strung together by connecting filaments (see Figure 11.3b). This form of chromatin is identified as the "beads on a string" morphology of chromatin. The "beads" are nucleosomes that are a little more than 11 nm in diameter, and the "string" is called **linker DNA.** The length of linker DNA segments varies somewhat among organisms, although in each species, the nucleosomes occur at regular intervals. In the yeast *Saccharomyces cerevisiae*, linker DNA is 13 to 18 bp in length. Linker DNA is about 35 bp long in the fruit fly *Drosophila*. In humans and other mammals, linker DNA spans about 40 to 50 bp; and in sea urchins, linker DNA is very long—approximately 110 bp. We have seen that core DNA is 146 bp in length, and when we add the length of linker DNA to that, the number of base pairs in each nucleosome repeat is approximately 160 to 260. The length of the nucleosome repeat is consistent within the genome of a given species. This beads-on-a-string form of chromatin is identified as the **10-nm fiber,** since the diameter of nucleosomes is approximately 10 nm.

Cryogenic electron microscopy (cryo-EM) imaging has produced detailed images of nucleosome structure and revealed the likely points of interaction between the octameric nucleosome core particle and core DNA. Timothy Richmond and his colleagues have described the crystal structure of the nucleosome observed at 2.8-Å resolution (Figure 11.4). The 146 bp of DNA wrap around the core particle in 1.65 turns. Specific molecular interactions take place between the N-terminal tails of histone proteins forming the nucleosome and the core DNA. In a later chapter, we discuss these interactions in detail and consider the role they play in chromatin structure and the regulation of gene expression (see Chapter 15).

The 10-nm fiber is an unnatural state for chromatin. To achieve it, chromatin must be chemically treated and held in conditions that are not found in cells. Under normal cellular conditions, chromatin forms the **30-nm fiber,** which is 6 times more condensed than the 10-nm fiber (see Figure 11.3c). Electron micrographs and molecular modeling help us visualize how the 30-nm fiber is assembled. If we consider the 10-nm fiber to be a kind of primary structure for chromatin, then the 30-nm fiber is a secondary structure. It is produced by coalescence of the 10-nm fiber into a cylindrical filament of coiled nucleosomes that is often called the **solenoid structure.** Each turn of the solenoid contains six to eight nucleosomes. The diameter of the solenoid is approximately 34 nm.

The histone protein H1 plays a key role in stabilizing the 30-nm fiber. The long N-terminal and C-terminal ends of the H1 protein attach to adjacent nucleosome core particles. H1 protein pulls the nucleosomes into

FOUNDATION FIGURE 11.3

## Condensing the Nuclear Material

The hierarchy of chromatin organization and chromosome condensation.

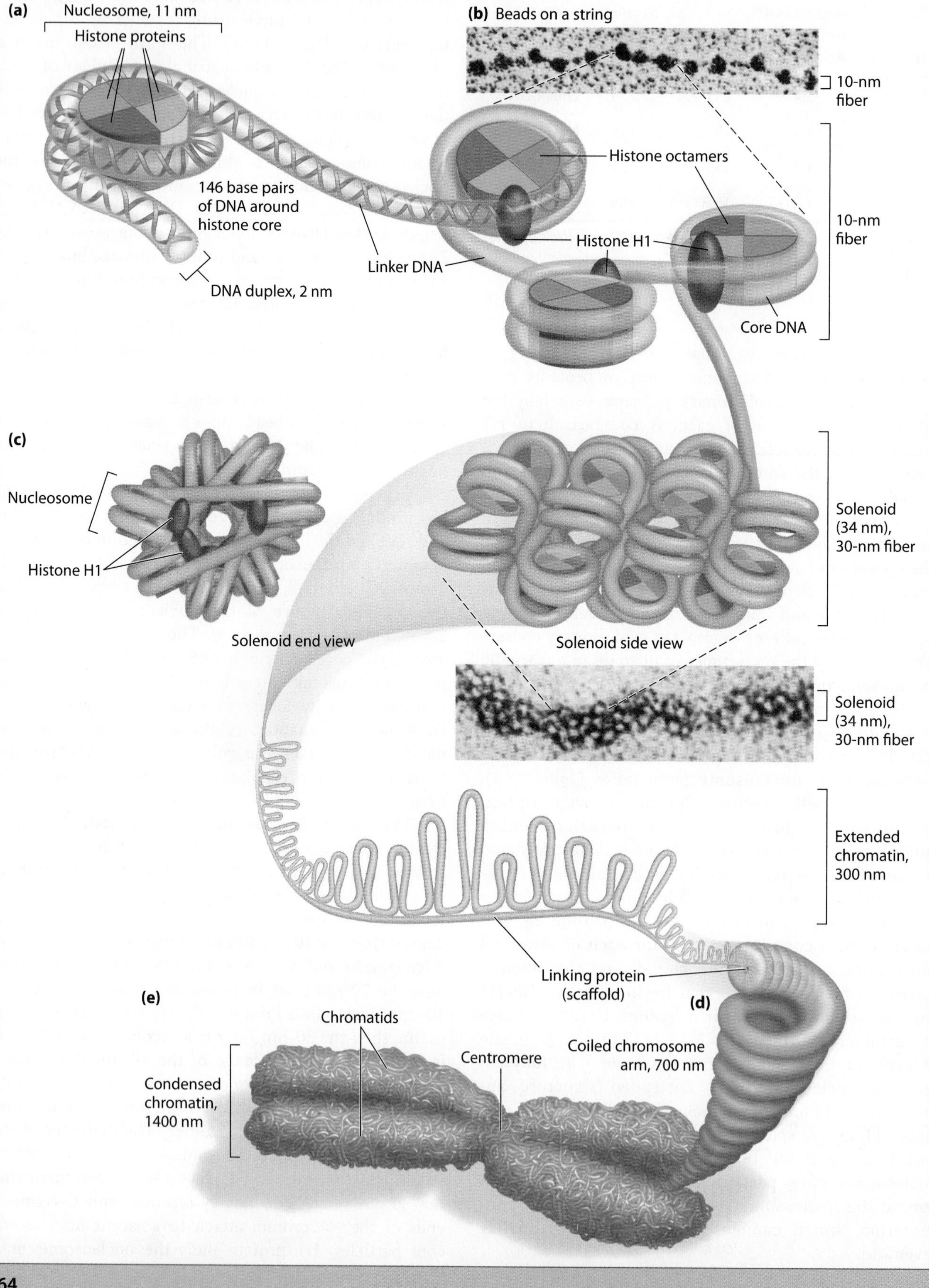

(a)
Nucleosome, 11 nm
Histone proteins
146 base pairs of DNA around histone core
DNA duplex, 2 nm
Linker DNA
(b) Beads on a string
10-nm fiber
Histone octamers
Histone H1
10-nm fiber
Core DNA
(c)
Nucleosome
Histone H1
Solenoid end view
Solenoid side view
Solenoid (34 nm), 30-nm fiber
Solenoid (34 nm), 30-nm fiber
Extended chromatin, 300 nm
Linking protein (scaffold)
(d)
(e)
Chromatids
Centromere
Coiled chromosome arm, 700 nm
Condensed chromatin, 1400 nm

The plant species *Arabidopsis thaliana* has a genome containing approximately 100 million bp of DNA. For this problem, assume *Arabidopsis* has a core-DNA length of 145 bp and a linker-DNA length of 55 bp.

**a.** Determine the approximate number of nucleosomes in each nucleus.

**b.** Determine approximately how many molecules of histone protein H4 are found in each nucleus.

| Solution Strategies | Solution Steps |
|---|---|
| **Evaluate** | |
| 1. Identify the topic this problem addresses, and explain the nature of the required answer. | 1. This problem asks about the number of nucleosomes per nucleus and about the histone composition of nucleosomes. The answer requires approximate numbers of nucleosomes and of histone H4 molecules per nucleus. |
| 2. Identify the critical information given in the problem. | 2. The approximate genome size of *A. thaliana* is given in base pairs, as are the lengths of its core and linker DNA. |
| **Deduce** | |
| 3. Describe the number of DNA base pairs that wrap around each <u>nucleosome</u>, and state the approximate number in the span between nucleosomes. | 3. The core DNA wrapping a nucleosome is 145 bp in length. Linker DNA between nucleosomes is approximately 55 bp in length. In total, there is one nucleosome for about every 200 bp of DNA. |
| TIP: The nucleosome is wrapped by core DNA, and the spans between nucleosomes consist of linker DNA. | |
| 4. Describe nucleosome composition. | 4. Nucleosomes are octamers of histone protein consisting of two molecules each of H2A, H2B, H3, and H4. |
| **Solve** | Answer a |
| 5. Calculate the number of nucleosomes in each *A. thaliana* nucleus. | 5. If we estimate that a new nucleosome associates with DNA about every 200 bp, the approximate number of nucleosomes per nucleus is $$\frac{1 \times 10^7 \text{ nucleotides/nucleus}}{2 \times 10^2 \text{ nucleotides/nucleosome}} = 5 \times 10^5 \text{ nucleosomes/nucleus}$$ |
| | Answer b |
| 6. Calculate the number of molecules of H4 in the nucleosomes of an *Arabidopsis* nucleus. | 6. There are 2 H4 molecules per nucleosome, thus $(2)(5 \times 10^5) = 10^6$, or 1 million H4 molecules, per nucleus. |

For more practice, see Problems 4, 7, and 11.

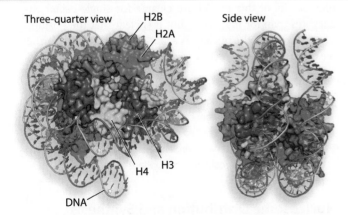

**Three-quarter view** · H2B · H2A · **Side view** · H4 · H3 · DNA

**Figure 11.4 Nucleosome structure.** Computer-generated rendering of the X-ray crystal structure of the nucleosome at 2.8-Å resolution by cryo-electron microscopy. The eight histone molecules are visible in a color-coded array.

an orderly solenoid array and lines the inside of the structure. Experimental analysis shows that chromatin from which H1 has been removed can form 10-nm fibers but not 30-nm fibers. Chromatin exists in a 30-nm-fiber state or a more condensed state during interphase. **Genetic Analysis 11.1** guides you through an interpretation of chromatin organization.

**Genetic Insight** The basic unit of eukaryotic chromatin consists of core DNA wrapped around histone proteins H2A, H2B, H3, and H4 that form an eight-member nucleosome core particle that is wrapped by 146 bp of DNA to form nucleosomes. The histone protein H1 associates with linker DNA and participates in higher order chromatin condensation.

## Higher Order Chromatin Structure

Beyond the 30-nm stage of condensation, chromatin compaction explains the structure of interphase chromosomes and the process of chromosome condensation in M phase. Nonhistone proteins perform several functions in both phases. Interphase chromosome structure results from the formation of looped domains of chromatin similar to supercoiled bacterial DNA (see Figure 11.3d). The loops are variable in size, containing from tens to hundreds of kilobase pairs and consisting of 30-nm-fiber DNA looped on a category of nonhistone proteins that are the foundation of chromosome shape. The diameter of looped chromatin is approximately 300 nm, so looped chromatin is called the **300-nm fiber.** With continued condensation, the chromatin loops form the sister chromatids. In metaphase, chromosome condensation reaches its zenith, resulting in chromosomes that are easily visualized by microscopy (see Figure 11.3e).

The **chromosome scaffold** is a filamentous nonhistone protein framework that gives chromosomes their shape. This scaffold is in some ways like the steel superstructure that provides the shape, strength, and support for a building. **Figure 11.5** shows the protein scaffold of a metaphase chromosome after being stripped of DNA. The shape of the scaffold is clearly reminiscent of the metaphase chromosome structure, consisting of sister chromatids joined at the centromere, which is visible as a constriction near the midpoint of the scaffold.

Chromatin loops containing 20,000 to 100,000 bp are anchored to the chromosome scaffold by other nonhistone proteins at sites called **matrix attachment regions (MARs);** see **Figure 11.6**. The **radial loop–scaffold model** predicts that the chromatin loops gather into rosette-like structures and are further compressed by nonhistone proteins. The total compaction of chromatin achieved by metaphase is approximately a 250-fold compaction of the already condensed 300-nm fiber.

Higher order chromosome condensation plays a critical role in two distinctive features of eukaryotic genetics. First, the general process of chromosome condensation compacts chromosomes to a degree that allows them to be efficiently separated at anaphase. Second, the chromatin loops formed during condensation play a role in regulating gene expression. Recent analysis of DNA binding to the chromosome scaffold indicates that certain repetitive DNA sequences are common at MARs. These sequences, called ATC sequences, are rich in A-T base pairs and have a high concentration of C in one strand. ATC sequences are found throughout the genome. Consequently, they can attach to the MARs in different patterns in different tissues. Experimental evidence indicates that active transcription takes place in chromatin loops, particularly in segments of loops that are distant from MARs. Thus, larger loops tend to have more active transcription than small loops. The positioning of ATC sequences throughout the genome appears to play a role in cell-type-dependent patterns of chromatin looping of a

|———— 2 µm ————|

**Figure 11.5** **The chromosome scaffold of a metaphase chromosome.** Stripped of chromatin, the remaining chromosome scaffold is composed of nonhistone proteins whose overall shape is reminiscent of a metaphase chromosome.

given chromosome that can lead to expression of certain genes in one type of cell but not in another. For example, if gene *A* is designated for expression in a certain type of cell but gene *B* is not, gene *A* will be found far away from an MAR, whereas gene *B* will be close to an MAR. The molecular details of this model are clearer for single-celled eukaryotes than for mammals, but it is clear that the position of a gene within the nucleus is a factor in its transcription, as we discuss in a later section (see also Chapter 15).

Genetic Insight Progressive chromosome condensation during prophase results from compaction of the 30-nm fiber into higher order structures. Scaffold proteins form the chromosome superstructure for chromatin condensation that ultimately produces the characteristic shape of metaphase chromosomes.

## Nucleosome Distribution and Synthesis during Replication

Our discussion of DNA replication described the enzymatic processes necessary for synthesizing a new daughter DNA strand on a template DNA strand (see Chapter 7).

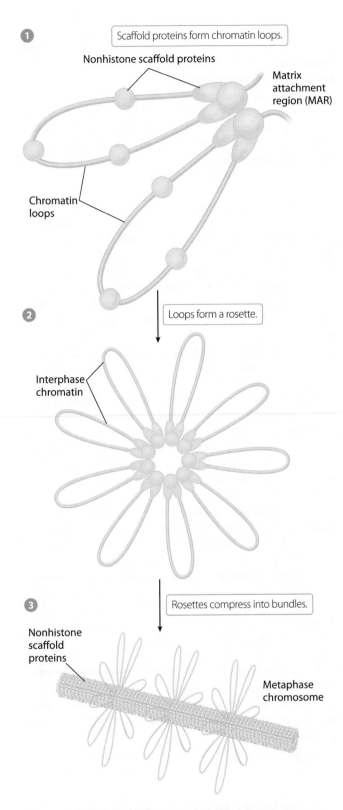

**1** Scaffold proteins form chromatin loops.

Nonhistone scaffold proteins

Matrix attachment region (MAR)

Chromatin loops

**2** Loops form a rosette.

Interphase chromatin

**3** Rosettes compress into bundles.

Nonhistone scaffold proteins

Metaphase chromosome

**Figure 11.6  The radial loop–scaffold model of chromatin condensation.** **1** Chromatin is anchored at matrix attachment regions (MARs). **2** Nonhistone proteins organize chromatin loops into rosettes. **3** Rosettes are compressed in metaphase chromosomes.

It did not, however, mention the presence of nucleosomes or the necessity of moving them temporarily to allow replication of eukaryotic genomes. Nor did it mention the equally essential process of adding nucleosomes to the newly synthesized DNA after the replication fork passes, or the assembly of new histone proteins that are needed because of the doubling of DNA as a result of replication.

The presence of nucleosomes raised several questions for researchers. Are old nucleosomes lost during replication? Are new nucleosomes synthesized during replication? Do old nucleosomes remain intact, so that nucleosomes are composed of either old histone proteins entirely or new histone proteins entirely? Whether or not nucleosomes stay intact, how are nucleosomes distributed to the sister chromatids? Experimental evidence collected by numerous investigators finds that old histones are retained either as dimers, tetramers, or individual molecules, that most nucleosomes present after replication are at least partially assembled from old nucleosome components, and that most nucleosomes contain some newly synthesized histone protein dimers.

The current model proposes that as the replication fork passes, nucleosomes break down into protein sub-assemblies—specifically, H3-H4 tetramers and H2A-H2B dimers. The H3-H4 tetramers immediately reaffiliate, more or less at random, with one of the sister chromatid products of replication. In contrast, many H2A-H2B dimers apparently become disassembled into individual histone proteins and then quickly reform into dimers with either old or newly synthesized protein partners.

Enough new synthesis of all four proteins takes place to double the number of nucleosomes. In this process, new H2A-H2B dimers and H3-H4 tetramers assemble. Some new H2A-H2B dimers join old H3-H4 tetramers already on DNA, while other new H2A-H2B dimers join new H3-H4 tetramers to form nucleosomes. Thus, about half of the nucleosomes assembled during replication are composed of old H3-H4 tetramers that are randomly distributed to the sister chromatids and combined with either new or old H2A-H2B dimers. The remaining nucleosomes contain new H3-H4 and either new or old H2A-H2B components (**Figure 11.7**).

## Chromatin Remodeling

The ubiquitous presence of nucleosomes along eukaryotic DNA presents a particular challenge to the initiation of transcription. Specifically, when DNA-binding proteins search for the appropriate consensus DNA sequences at which to initiate transcription, their binding to those sequences can be hindered by the presence of nucleosomes. To allow the binding of transcription-initiating proteins, nucleosomes must be displaced to expose promoter and

**(a)**   Nucleosome

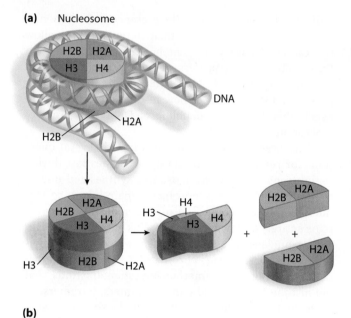

**(b)**

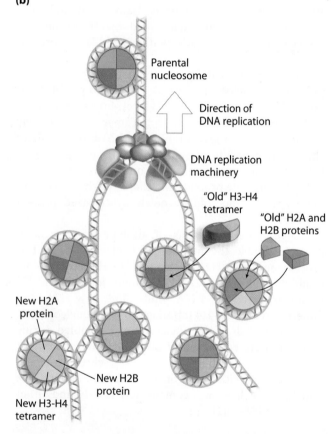

**Figure 11.7  Nucleosome inheritance after DNA replication.** Following the passage of the replication fork, "old" H3-H4 tetramers are randomly assigned to daughter strands, and newly synthesized H3-H4 tetramers inhabit strands not bound by old tetramers. Old and new H2A-H2B dimers join the tetramers to form complete nucleosomes.

other consensus DNA sequences. This nucleosome displacement process is called **chromatin remodeling** because the degree of chromatin compaction is altered by the changes to nucleosomes.

Chromatin remodeling is accomplished by chemical modifications to the individual histone proteins that make up nucleosomes. The chemical modifications, sometimes called **epigenetic marks** or **epigenetic modifications,** consist of the addition and removal of chemical groups such as methyl groups (methylation and demethylation), acetyl groups (acetylation and deacetylation), and phosphate groups (phosphorylation and dephosphorylation) near the amino-terminal (N-terminal) ends of histone proteins. The effect of these epigenetic marks is to alter how tightly DNA and nucleosomes bind together. The tighter the association, the less likely it is that transcription will occur; and, conversely, looser association between nucleosomes and DNA increases transcription. While there are exceptions, it is often the case that methylation of nucleosome N-terminal tails leads to tighter packaging of chromatin that reduces transcription, whereas acetylation of histone tails loosens chromatin and leads to more transcription. Phosphorylation is detected in both more compact and less compact chromatin. Epigenetic modifications alter the strength of association between nucleosomes and DNA and can lead to the displacement of nucleosomes (i.e., to remodeling of chromatin) to make regulatory DNA sequences more accessible to DNA-binding proteins.

The epigenetic modifications that remodel chromatin are transmitted during cell division or reproduction. Note that although they alter the ability of cells to transcribe genes and regions of the genome, they do not alter DNA sequence. The retention of old histones during DNA replication provides a mechanism for the retention of epigenetic "memory" that can be passed to daughter cells. At the same time, the epigenetic modifications of gene expression are reversible since they consist of an array of chemical groups bound to histone proteins. The number and pattern of epigenetic modifications to N-terminal histone tails is a complex but increasingly well-understood component of the regulation of gene transcription in eukaryotes (for more detail, see Chapter 15).

Genetic Insight  Nucleosomes loosen their grip on DNA during replication and are partially disassembled, and then later reconstituted. The N-terminal tails of individual histone proteins of nucleosomes are acetylated to loosen chromatin and generate more transcription, whereas histone tails are methylated to compact chromatin and reduce transcription.

# 11.3 Chromosome Regions Are Differentiated by Banding

Chromosome condensation, driven by chromatin packaging, reaches its maximum at the end of metaphase. Using microscopy, cytogeneticists can distinguish each chromosome by its overall size and shape and by the patterns of light and dark *chromosome bands* along the length of chromosomes that are produced by treating chromosomes with specific dyes and stains.

## Chromosome Structures and Banding Patterns

The centromere divides each chromosome into segments known as **chromosome arms** that are of unequal lengths. Almost invariably, one chromosome arm, called the **short arm** and also known as the **p arm,** is shorter than the other arm that is known as the **long arm** or the **q arm** (Figure 11.8). The position of the centromere determines the relative lengths of the short and long arms, leading to descriptive terms for the shapes of metaphase chromosomes. A **metacentric chromosome** has a more or less centrally located centromere and chromosome arms of similar lengths. **Submetacentric chromosomes** have a centromere nearer one end, producing one arm that is distinctly shorter than the other. The centromere of **acrocentric chromosomes** is nearly at the end of the chromosome. The "short arm" of acrocentric chromosomes is often composed of highly repetitive (so-called satellite) DNA. **Telocentric chromosomes** have a terminal centromere and no short arm.

An organized photographic display of a complete set of chromosomes for a species is called a **karyotype** and features the chromosomes grouped into homologous pairs in descending order of overall size of the autosomes. The sex chromosomes are identified separately. The chromosomes in a karyotype may be stained with various dyes to produce the distinctive banding pattern for each type of chromosome in the set. A human karyotype containing 22 pairs of autosomes (numbered 1 through 22) and one pair of sex chromosomes is presented in Figure 11.9.

During the late 1960s and early 1970s, several techniques for **chromosome banding** were developed, primarily by experimentation with human and other mammalian chromosomes. Chromosome banding allows cytogeneticists to accurately identify each chromosome and chromosome segment in a karyotype. Generating a karyotype and banding the chromosomes is a multistep process. Cells are first grown in culture and then arrested in metaphase or slightly earlier in M phase (prometaphase or late prophase) by the addition of certain chemicals. Arresting the cell cycle maximizes the number of cells in the culture that contain well-condensed chromosomes. Next, individual cells from the arrested cell culture are dropped onto a microscope slide. This bursts the cells and

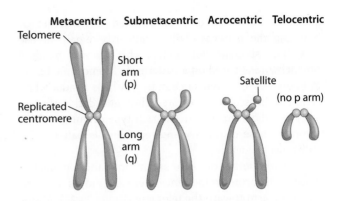

**Figure 11.8 Chromosome shape.** The position of the centromere and the ratio of the lengths of the long arm (q arm) and short arm (p arm) at metaphase determine chromosome shape.

ruptures the nuclear membrane, allowing the chromosomes to spill out.

Different stains and dyes can be used on chromosomes to make bands appear. Certain stains and dyes require the chromosomes to be treated first with the proteinase enzyme trypsin to remove some of the proteins that make up chromatin. After the microscope slide containing the banded chromosomes has been photographed, the digital micrograph is manipulated to group the homologs. The digital images of any chromosomes that are bent or twisted can be manipulated to straighten the images for karyotype assembly. In the years before digital photography was possible, micrographs were captured as analog images. Karyotypes made from these older images were created by cutting out the pictures of individual chromosomes and pasting homologs next to one another.

The chromosome banding patterns produced by different stains and dyes correlate with one another. An international symposium in Paris, France, was convened in 1971 to agree on the standard banding pattern for each human

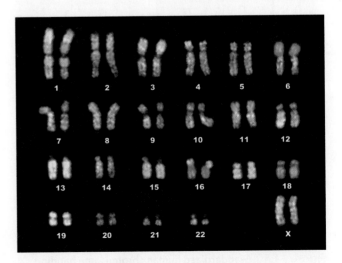

**Figure 11.9 A normal human karyotype.** Fluorescent compounds and computer-enhanced colors make each chromosome distinctive.

chromosome as well as on a standardized nomenclature for discussing the patterns. These standards remain in use today. The standard for human chromosome banding nomenclature is based on a pattern of chromosome banding known as **G (Giemsa) banding,** which results when Giemsa stain is used to create lightly and darkly stained bands in a different pattern on each type of human chromosome. These distinctive and reproducible patterns are readily recognizable.

The numbering of major and minor band regions begins at each chromosome centromere and moves outward along each arm toward the telomere (**Figure 11.10**). Major regions are subdivided to permit a designation for each light and dark band region of a chromosome. Each band is given a designation that specifies the chromosome number, chromosome arm, and band location. An example is 5q2.3.1, which is the dark band on the long arm of chromosome 5 indicated in Figure 11.10a. A karyotype of

Standard banding patterns and landmark designations for human chromosomes 1 through 5

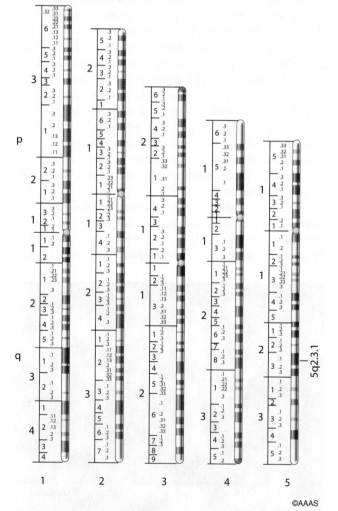

**Figure 11.10 Standardized human chromosome banding patterns.** Human chromosomes 1 to 5 in late prophase. Heterochromatic regions are shown as gray and black bands, euchromatic regions as white bands.

human metaphase chromosomes contains about 300 light and dark G bands among the 23 chromosomes (22 autosomes plus the X chromosome).

Chromosome banding by G banding and other techniques was, at one time, limited to chromosomes in metaphase. Recently, however, advanced techniques have allowed cytogeneticists to stain chromosomes earlier in the cell cycle. Chromosome banding in late prophase and in prometaphase results in as many as 2000 highly reproducible chromosome bands. Each chromosome band, whatever its size, contains many chromatin loops and between 1 million and 10 million base pairs of DNA.

When a karyotype has an abnormal number of chromosomes, the easily recognized patterns make identification of missing or extra chromosomes relatively simple. Similarly, rearrangements caused by insertion, deletion, or other changes to the structure of a chromosome are detected by alterations of the standard chromosome banding pattern. In addition, the similarities in chromosome band patterns seen in comparisons of the chromosomes of related species make it possible to trace the evolution of chromosome number and chromosome structure by reconstructing the likely karyotype of common ancestors (see Chapter 13).

## Heterochromatin and Euchromatin

Chromatin condensation varies throughout the cell cycle but also varies from one part of a chromosome to another. There is clear evidence that the latter variation is directly related to gene transcription. During interphase, chromosome regions containing genes that are actively expressed generally have a lesser degree of chromatin condensation than chromosome regions that do not contain expressed genes. These regions of active expression are identified as **euchromatin,** or as **euchromatic regions.** Most expressed genes are located in euchromatic regions, where condensation is variable during the cell cycle. Euchromatic chromosome regions are lightly staining regions of G-banded chromosomes. Conversely, chromosome regions in which chromatin is tightly condensed contain **heterochromatin** and are called **heterochromatic regions.** Heterochromatic regions contain many fewer expressed genes than do euchromatic regions. Heterochromatin is identified as darkly staining chromosome regions in G-banded chromosomes.

Two distinct classes of heterochromatin are detected. **Facultative heterochromatin** exhibits variable levels of condensation. At times, facultative heterochromatin is highly condensed, while at other times it is less so. The transcription of genes in regions of facultative heterochromatin usually correlates with periods of less compaction. **Constitutive heterochromatin,** on the other hand, is in a permanent heterochromatic state and contains very few expressed genes. Constitutive heterochromatin is predominantly composed of repetitive DNA sequences. It is particularly prominent in chromosome telomeres and in the centromeric regions of chromosomes.

Neither telomeric nor centromeric constitutive heterochromatin contains expressed genes. (See Chapter 7 for a description of how the telomeric repeats are synthesized by telomerase.)

## Centromere Structure

The repetitive DNA sequences in centromeres facilitate the binding there of the specialized kinetochore proteins and spindle fiber microtubules that function to divide homologous chromosomes and sister chromatids during cell division (see Figures 3.4 and 3.6). These repetitive sequences are organized in a form not found in any other part of chromosomes. In the early 1980s, John Carbon and Louis Clarke described centromeric DNA, or CEN sequences, in the yeast *Saccharomyces cerevisiae*.

Carbon and Clarke analyzed the DNA sequences of 16 yeast centromeres and discovered that each had a slightly different CEN sequence, even though each centromere performs the same function. Yeast CEN sequences span 112 to 120 bp and are divided into three domains, designated centromeric DNA elements (CDE) I, II, and III. **Figure 11.11a** shows four examples of yeast CEN sequences that illustrate the overall similarity but subtle variation in centromere sequences. The centromeric consensus sequences revealed in Carbon and Clarke's analysis are shown in **Figure 11.11b**. That of CDE I is an 8-bp sequence RTCACRTG, where R is either of the purines adenine or guanine. That of CDE III contains 26 bp rich in A-T. Between these elements is CDE II, varying in length from 78 to 86 bp and having more than

90% of its sequence composed of A-T base pairs. Microtubules attach to kinetochores to carry out their roles in cell division (see Chapter 3) (**Figure 11.11c**).

Following the work of Carbon and Clarke, CEN sequences were characterized in *Schizosaccharomyces pombe* and other species of yeast, and for numerous other eukaryotes as well. The CEN sequences of *S. pombe* and other yeast species are similar in length but differ somewhat in sequence from those of *S. cerevisiae*. Later research showed the CEN sequences of multicellular eukaryotes to be much larger than those found in yeast and to occur within a wide expanse of constitutive heterochromatin that characteristically clusters around the centromeres of animal and plant chromosomes. These expanded CEN sequences bind several microtubules, rather than just the one bound in yeast.

## Centromeric Heterochromatin

The centromeric DNA sequences of eukaryotes are highly repetitive and are constitutively heterochromatic. A specialized form of the histone H3 protein known as centromere protein A (CENP-A) binds centromeric DNA. CENP-A is similar to H3 from its C-terminal end through much of its length but has a very different N-terminal tail that is much longer than the one found in other H3 molecules. The CENP-A acts like H3, forming chromatin, but the extended N-terminal tails of CENP-A proteins lead to the binding of unique kinetochore proteins that bind only to centromeres. Recall that these kinetochore proteins and the structures they form are the attachment locations for microtubules that move chromosomes and chromatids during cell division. The importance of CENP-A

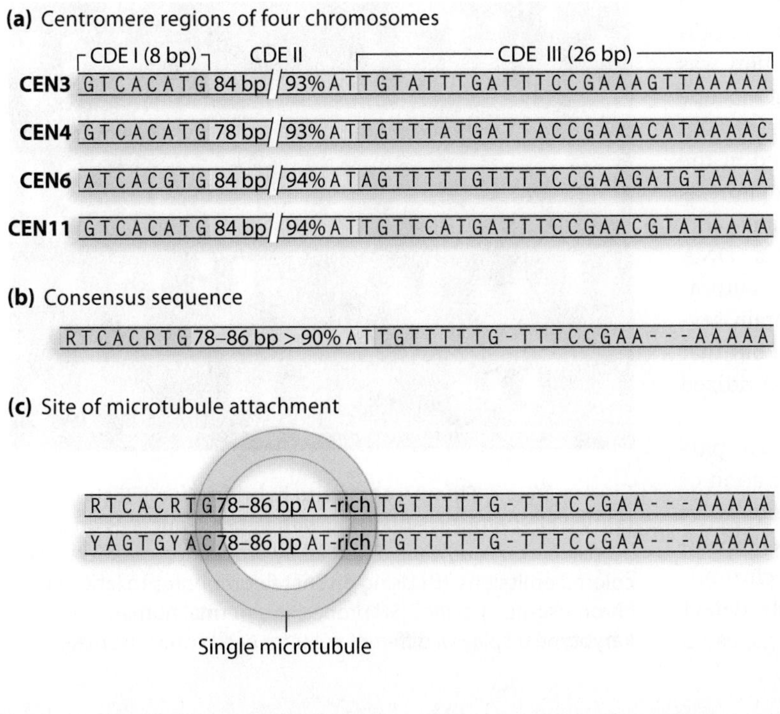

**Figure 11.11 Conserved nucleotide sequence at the yeast centromere.** (Abbreviations: R = purine, Y = pyrimidine.)

in kinetochore formation is demonstrated by mutants in which abnormalities or loss of CENP-A interferes with the creation of kinetochores.

---

Genetic Insight The centromere is a specialized region of constitutive heterochromatin that is bound by kinetochore proteins to facilitate microtubule attachment and chromosome and sister chromatid movement to poles of the cell during cell division.

---

## In Situ Hybridization

We previously discussed laboratory techniques in which a labeled molecular probe hybridizes to a target sequence of DNA or RNA (see Chapter 10). These procedures are used to aid researchers in detecting fragments of DNA containing a sequence of interest or in detecting a specific mRNA transcript. Chromosomes themselves—or, more specifically, the DNA in chromosomes—can also be targeted for hybridization by molecular probes.

Chromosomal DNA need not be fragmented in order for a gene-specific or a chromosome-specific target sequence to be detected by a molecular probe. The technique known as **in situ hybridization** uses molecular probes labeled either with radioactivity—or, more recently, with fluorescent compounds—to detect their target sequences. Many methods of in situ hybridization are used, but common early versions worked by denaturing the DNA of intact chromosomes and then bathing the chromosomes in a solution containing labeled molecular probe. Hybridization occurs between the single-stranded chromosomal DNA and the single-stranded probe.

First-generation in situ hybridization methods used radioactive nucleic acid probes that produced autoradiographs when a small piece of photographic film was placed on top of a chromosome spread on a microscope slide. Decay of $^{32}$P radioactive label in the probe exposed the photographic film, which was then developed in the same way as an autoradiograph of an electrophoresis gel. In chromosome autoradiographs, dark regions corresponded to the chromosome locations of a DNA sequence hybridized by the probe. Unfortunately, autoradiographs produced by this method usually contain several dark spots that are not sites of hybridization but that occur because of difficulty washing away unhybridized amounts of molecular probe.

Many advances have occurred over the decades, particularly through the development of a new generation of fluorescent labels, or fluorophores, for use as molecular probes. The development of fluorophores to label molecular probes greatly improved the resolution of chromosome labeling, making it possible to accurately detect each labeled chromosome and to locate specific genes on

chromosomes. Today, **fluorescent in situ hybridization (FISH)** uses molecular probes labeled with compounds that emit fluorescent light when excited by light in the visible or UV spectrum. In FISH, various compounds emitting fluorescent light in different wavelengths can be used simultaneously to label probes. An advantage of this approach is that each emission wavelength produces a different color when the chromosomes are viewed under a microscope.

Figure 11.12a illustrates the use of a FISH probes, one producing red color and the other producing green color, to identify two genes on the same human chromosome. Figure 11.12b utilizes multiple FISH probes to individually label each human chromosome. In this instance, each

**(a)**

**(b)**

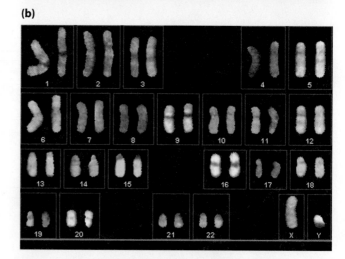

**Figure 11.12 Fluorescent in situ hybridization (FISH).**
**(a)** Two FISH probes hybridizing with target sequences on a human chromosome are detected by production of differently colored emissions. **(b)** Using distinct fluorophores to label 24 chromosome-specific FISH probes, this normal human male karyotype displays a different color for each chromosome.

chromosome-specific probe is labeled by a distinct fluorophore that emits a different color of fluorescent light. There are 24 different fluorophores: one for each autosome, one for the X chromosome, and one for the Y chromosome. The emission data from each probe is collected by a photoreceptor and processed by specialized software to produce the karyotype displaying each chromosome in a different color.

FISH techniques have numerous applications for chromosomal analyses in humans and other species. One important and practical use of these methods is the identification of the complex chromosome rearrangements often found in cancer cells, which we discuss in the Case Study that ends the chapter.

## 11.4 Chromatin Structure Influences Gene Transcription

The presence of chromatin in eukaryotes constitutes a prominent difference between eukaryotic genomes and those of bacteria. The structure of chromatin is instrumental in controlling eukaryotic gene expression. Differences in chromatin structure lead to different levels of gene transcription in different chromosome locations. In this section, we explore these variations in chromatin structure and describe how the location of a gene—for example, its presence in heterochromatin versus euchromatin—can influence whether it is transcribed. We begin with a description of chromosomes during interphase, a period of the cell cycle during which chromosomes are not well condensed and are therefore challenging to observe.

### Chromosome Territory during Interphase

An early observer of chromosomes in the nucleus was Theodore Boveri, who was one of the first to propose the connection between genes and chromosomes. Among his observations, Boveri noticed that interphase chromosomes are not uniformly arrayed within a nucleus. He suggested the variation might be related to their activity in the cell. Recent research using in situ hybridization techniques to study chromosome positioning in the interphase nucleus indicates that Boveri's suggestion was correct.

Cell biologists Thomas Cremer and Christoph Cremer used chromosome painting to investigate the arrangement of chromosomes in the nucleus during interphase and found that chromosomes are partitioned into their own chromosome territories (**Figure 11.13**). A **chromosome territory** is a small region of the nucleus that is the domain of a single chromosome. It is not bounded by any sort of membrane, nor is it demarcated in any distinctive manner. Chromosomes do not occupy exactly the same territory in each nucleus (the nucleus does not have reserved seating for each chromosome), but once confined to a territory, a chromosome does not

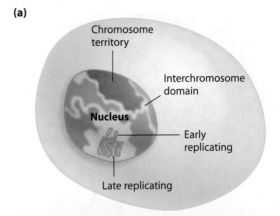

(a)

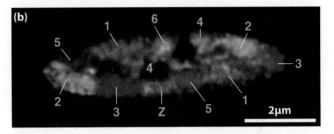

(b)

**Figure 11.13** **Chromosome territories in the eukaryotic nucleus.** **(a)** Chromosomes occupy discrete territories separated by interchromosome domains during interphase of the cell cycle. **(b)** FISH labeling of chromosomes in the nucleus of a chicken embryo reveals occupancy of discrete territories.

stray from it until the initiation of M phase of the cell cycle. Chromosomes are, however, dynamically active within their territories during interphase and can be seen to move, twist, and turn during transcription and DNA replication. The chromosomes appear to be anchored by their centromeres and perhaps to take positions that allow, for each chromosome, characteristic patterns of gene expression and other activities during interphase.

Adjacent chromosome territories are separated by an **interchromosomal domain** that contains no chromatin. These domains are channels for the movement of proteins, enzymes, and RNA molecules within the nucleus and among chromosome territories. The distribution of chromosome territories places the largest and most gene-rich chromosomes toward the center of the nucleus, while the territories of smaller chromosomes containing fewer genes are located toward the outer edges of the nucleus.

The positioning of a chromosome within its territory corresponds to the activities in which the parts of the chromosome are engaged at particular stages of interphase. For example, chromosome regions that replicate early in S phase are generally found further away from the nuclear membrane. The regions closer to the center of the nucleus are the locales of so-called early-replicating chromosome segments. In contrast, late-replicating chromosome segments, portions of chromosomes that replicate late in S phase, are found nearer to the nuclear membrane. Also, the most transcriptionally active

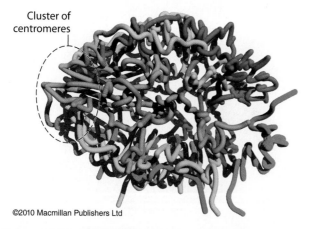

Cluster of
centromeres

©2010 Macmillan Publishers Ltd

**Figure 11.14** **A three-dimensional model of chromosomes in the yeast nucleus.** Yeast-chromosome centromeres are clustered toward one end of the nucleus; chromosome arms radiate from the centromere cluster.

chromosome regions are found closest to the border between a chromosome territory and an interchromosomal domain, presumably because of (1) greater access to proteins and enzymes needed for transcription and (2) faster dispersal of RNA transcripts after transcription is completed. While transcription occurs throughout each chromosome territory, experimental evidence suggests that transcription is most intense where they border on interchromosomal domains.

Recently, C. Anthony Blau and several colleagues have extended the Cremers' findings by developing a three-dimensional model of the 16 chromosomes in yeast haploid nuclei (**Figure 11.14**). Employing a method that differentially identifies each chromosome, the researchers were able to precisely map the location of each chromosome within the nucleus. The resulting three-dimensional map of chromosome positioning reveals that chromosome centromeres are clustered together and that the chromosome arms project away from the centromere cluster. Knowledge of the positioning of chromosomes within the nucleus will make it possible to determine how DNA sequences influence chromosome positioning and, in turn, how chromosome positioning influences the transcription and replication of sequences.

---

**Genetic Insight** Chromosomes occupy discrete chromosome territories during interphase. Gene transcription is dependent in part on the positions of chromosomes within their territories.

---

## Dynamic Chromatin Structure

Chromatin structure is dynamic; chromosome segments potentially exist in different states of compaction and relaxation during the cell cycle. In broad terms, these changes

regulate access to DNA by the proteins that carry out replication, transcription, recombination, and DNA repair. At one end of the spectrum of chromatin condensation are the heterochromatic regions of chromosomes, which are relatively inaccessible to transcriptional proteins either transiently (in the case of facultative heterochromatin) or nearly always (in the case of constitutive heterochromatin). In contrast, in euchromatic chromosome regions, transcriptional proteins and enzymes are more easily able to gain access to DNA.

Several lines of experimental evidence relate chromatin structure to gene expression. The first identification of this connection came from the observation of **position effect variegation (PEV)** affecting eye color in *Drosophila*. During the 1920s and 1930s, in tests of the effect of X-rays on *Drosophila* development, Hermann Muller identified X-ray–exposed fruit flies with a variegated pattern of eye color in which red and white patches of colored eye tissue are seen. Red patches result from expression of the wild-type $w^+$ allele for red color. White patches of fly eyes have no color, as the result of absence of $w^+$ expression.

In Muller's most important variegation experiments, he noticed that the X chromosomes of flies with variegated eye color had an abnormal structure. These X chromosomes had been broken by the damaging effects of X-rays, and the chromosome pieces had then rejoined, but with one piece inverted 180 degrees relative to its normal position. Due to the X-chromosome inversion, the *w* gene moved from its normal location near the telomere of the X chromosome, where it is actively expressed, to a new position nearer the centromere of the chromosome. The centromeric region contains constitutive heterochromatin that only briefly relaxes to allow DNA-replication enzymes and structures access to the sequences during DNA replication. Heterochromatin quickly reforms at the centromere, and the heterochromatic condensation spreads outward from there.

The extent of heterochromatin spread varies from one chromosome to the next; there are essentially no expressed genes near the centromere, so a little more or a little less heterochromatic spread is normally of no genetic consequence. In fruit flies with the X-chromosome inversion, however, the stopping point of heterochromatin spread determines whether $w^+$ is expressed or not (**Figure 11.15**). If heterochromatin spreading does not reach the new location of $w^+$, the gene will be in a euchromatic region and can be actively transcribed, producing red eye color. On the other hand, cells in which centromeric heterochromatin spreading encompasses the new location of $w^+$ do not express the gene and lack the pigment color. In this example, the *w* gene is not mutated by inversion. Instead, the opportunity to express the gene is altered by relocation of the gene to a new position that falls within a heterochromatic chromosome region.

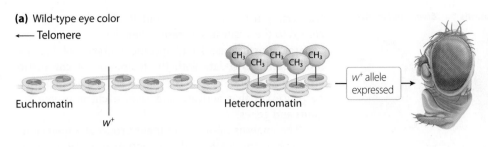

**(a)** Wild-type eye color

Euchromatin

Heterochromatin

$w^+$

$w^+$ allele expressed

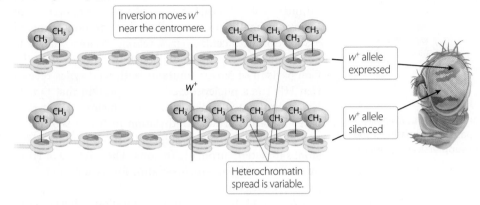

**(b)** Variegated eye color

Inversion moves $w^+$ near the centromere.

$w^+$

$w^+$ allele expressed

$w^+$ allele silenced

Heterochromatin spread is variable.

**Figure 11.15  Position effect variegation of eye color in *Drosophila*.** The $w^+$ allele is expressed in wild-type X chromosomes and in X-chromosome inversions in which heterochromatin does not spread to cover the gene. If heterochromatin covers the new gene location, $w^+$ is silenced.

Hence, in the "position effect," the expression of the gene is dependent on its position with respect to centromeric heterochromatin.

Since Muller first described position effect variegation, cell biologists have come to understand that this kind of position effect is an epigenetic phenomenon in which gene expression is determined by chromatin structure. The epigenetic state of position effect genes is transmissible during mitotic cell division. As a result, cells in which $w^+$ is positioned within centromeric heterochromatin produce descendant cells that are unable to produce pigment because the gene is silenced. Consequently, white patches appear where the descendant cells with inactivated $w^+$ are located, producing a variegated pattern of eye-color pigmentation in the eye tissue (see Figure 11.15b). No two variegated *Drosophila* eyes have the same pattern of variegation. Even the two eyes of the same variegated fly have different patterns! The occurrence of PEV indicates that (1) gene expression can be silenced by the gene's position on a chromosome, and (2) silencing is a feature of chromatin structure that is transmissible from one cell generation to the next.

---

**Genetic Insight**  Chromatin structure regulates gene expression. In position effect variegation, actively transcribed genes that are transposed into heterochromatic chromatin near the centromere undergo a form of epigenetic silencing.

---

## PEV Mutations

Genetic analysis of eukaryotic genomes reveals PEV to be a widespread phenomenon, suggesting that mechanisms controlling chromatin structure are important in the control of gene expression. Mutations modifying PEV have led to the identification of several genes and proteins that play a direct role in establishing and maintaining chromatin structures associated with gene expression and gene silencing. For example, mutations known as ***E(var)* mutations,** where *E(var)* is short for *e*nhancers of position effect *var*iegation, increase or enhance the appearance of the mutant white-eye phenotype by encouraging the spread of heterochromatin beyond its normal boundaries. The effect of *E(var)* mutation is to produce a greater number of eye cells lacking pigment (Figure 11.16). In contrast, ***Su(var)* mutations,** where *Su(var)* is short for *su*ppressors of position effect *var*iegation, restrict the spread of heterochromatin or interfere with its formation. *Su(var)* mutations increase the extent of normally pigmented regions of the eye by suppressing the emergence of white patches.

Several dozen *E(var)* and *Su(var)* mutations are known in the *Drosophila* genome, and *Su(var)* mutations have proven especially valuable in the identification of genes and proteins that modulate chromatin structure. Genetic analysis of *E(var)* and *Su(var)* mutations supports the hypothesis that chromatin structure is dynamic and is associated with gene expression. In fact,

**Variegated eye**        ***Su(var)* mutations**        ***E(var)* mutations**

| | | |
|---|---|---|
| Red patches are produced by cells in which $w^+$ is transcribed, and white patches in which $w^+$ is inactivated by heterochromatin spread. | Mutations block efficient formation of heterochromatin and leave most cells with active $w^+$ transcription. | Mutations enhance heterochromatin formation and restrict $w^+$ expression to small patches. |

**Figure 11.16** *E(var)* **and** *Su(var)* **mutations.** Mutations of genes whose protein products participate in chromatin remodeling are detected by enhancement or suppression of position effect variegation.

chromatin structure appears to oscillate: Sometimes it is in a highly condensed state in which gene transcription is silenced (i.e., heterochromatic) and sometimes it is in a more loosely condensed state that allows transcription (i.e., euchromatic), but it often exists in an intermediate state of condensation.

Collectively, the experimental analyses of suppressors and enhancers of PEV identify genes that make epigenetic "marks" on histone proteins, causing attachment and detachment of methyl, acetyl, and phosphoryl groups to amino acids of the histones. These epigenetic marks are associated with chromatin remodeling that leads to gene transcription or gene silencing. The patterns of methylation and demethylation, acetylation and deacetylation, and phosphorylation and dephosphorylation are maintained on histones and may be passed through successive generations of cells. Five important features of epigenetic modification have been identified by researchers: (1) epigenetic modifications alter chromatin structure, (2) they are transmissible during cell division, (3) they are reversible, (4) they are directly associated with gene transcription, and (5) they *do not* alter DNA sequence. In a later chapter, we return to a discussion of epigenetic modifications of chromatin and its important role in regulating gene expression (Chapter 15).

## Chromatin-Remodeling Proteins

Among the questions cell biologists have studied through *Su(var)* and *E(var)* mutational analysis are, "Which genes and proteins control chromatin condensation?" and "How does chromatin remodeling alter access to DNA by transcription factor proteins and RNA polymerase?" In regard to these questions, researchers have arrived at two general conclusions. First, specific patterns of histone modification correlate with the formation of chromatin structure. Second, chromatin state is regularly remodeled by processes that activate and silence chromosome regions and genes.

The analysis of one prominent group of *Su(var)* mutations exemplifies how the detection of defective proteins can elucidate normal functions. Some *Su(var)* mutations are caused by defective expression of heterochromatin protein-1 (HP-1), a protein found in association with centromeres, telomeres, and other heterochromatic chromosome locations in *Drosophila*. Comparison of *Su(var)* mutants with wild types reveals that HP-1 is a nucleosome-binding protein that targets lysine amino acids in position 9 of histone H3 if they carry a methyl group. Methylation of lysine 9 of H3 is one of the most common epigenetic modifications of histones in heterochromatic regions. The absence of HP-1 interferes with heterochromatin formation and suppresses variegation.

A second group of *Su(var)* mutations affects genes encoding histone methyltransferases (HMTs), enzymes responsible for catalyzing the addition of methyl groups to amino acids of histone proteins. Histone methyltransferases appear to target methylation-specific basic amino acids (e.g., arginine and lysine) in nucleosomes, attaching methyl groups to these amino acids as part of epigenetic marking of histones. As noted above, the lysine residue in position 9 of histone protein H3 is a frequent target for methylation. Upon methylation, this location is described as H3K9me, which is short for *h*istone *3*, lysine (one-letter abbreviation *K*), position 9, and *m*ethylation. If HMTs are not functioning properly, epigenetic methylation is not established, and heterochromatin formation is inhibited.

The identification of the functions of these two groups of *Su(var)* mutations led to a simple model of HP-1 and HMT function predicting that specific methylated histone locations in nucleosomes (e.g., H3K9me) are methylated by HMTs and act as sites of HP-1 binding that helps condense chromatin structure to silence gene expression (**Figure 11.17**). According to this model, *Su(var)* mutants that are defective in their silencing of $w^+$ could carry an HMT gene mutation that leads to the failure to properly methylate nucleosomes, or they could carry a mutation of the *HP-1* gene and be rendered unable to remodel chromatin to a tightly condensed form. We discuss chromatin remodeling and epigenetic modifications of chromatin more fully in connection with mechanisms that regulate transcription of eukaryotic genes (see Chapter 15). **Genetic Analysis 11.2** guides you through an analysis of chromatin structure surrounding a gene.

**Heterochromatin**

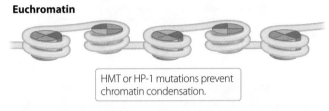

HMT and HP-1 combine to condense chromatin and block transcription.

**Euchromatin**

HMT or HP-1 mutations prevent chromatin condensation.

**Figure 11.17 HMT and HP-1.** Mutation analysis identifies the proteins HMT and HP-1 as drivers of heterochromatin formation.

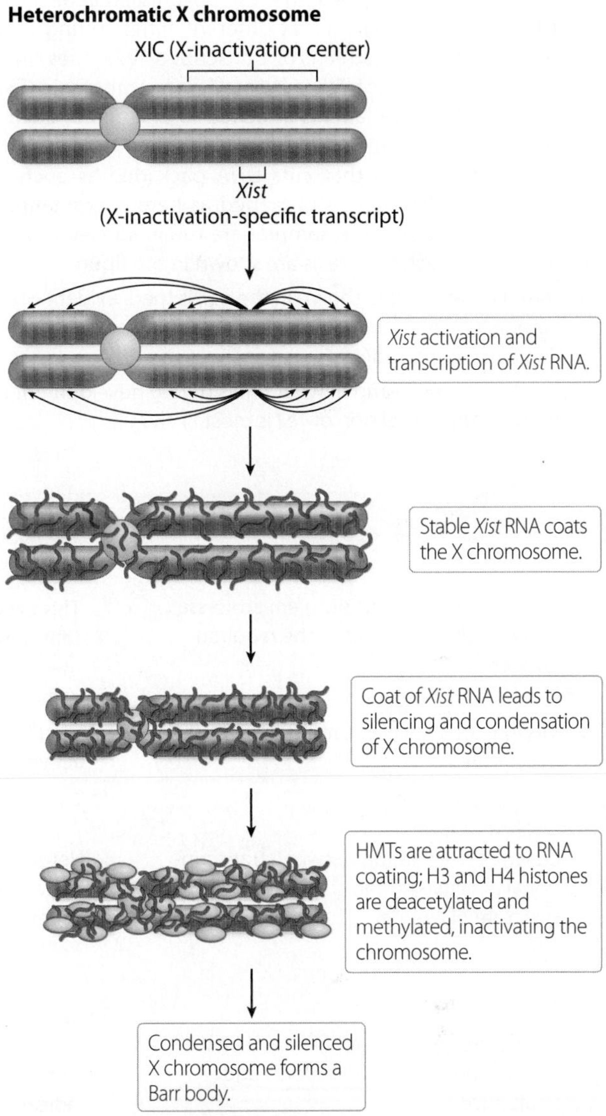

## Inactivation of Mammalian Female X Chromosomes

Sexually reproducing animals face a peculiar problem with respect to the different numbers of each sex chromosome carried by males and females of the species. We previously discussed this problem and explained that mammalian females undergo random X inactivation in each nucleus early in gestational development (see Chapter 3). Recall that random X inactivation leaves one active X chromosome that is largely euchromatic and one inactive X chromosome that is almost entirely heterochromatic in each nucleus. The heterochromatic X chromosome is almost completely silent with respect to gene expression. This highly heterochromatic X chromosome forms a Barr body in the nucleus. All descendant cells maintain the same active (euchromatic) and inactive (heterochromatic) X chromosomes, leading to the mosaic pattern of cells characteristic of mammalian females (see Figure 3.26).

Extensive studies of X inactivation in mice and humans have detected about a dozen genes on the heterochromatic (inactive) X chromosome that escape silencing. One of these genes is critically important to the establishment and maintenance of X-inactivation. The gene, called *X-inactivation-specific transcript* (*Xist*), is *active* on the heterochromatic X chromosome and is *inactive* on the euchromatic chromosome. It is located in the X-inactivation center, or XIC, of the X chromosome (**Figure 11.18**). The *Xist* gene is transcribed only on the heterochromatic chromosome, where it is active; it is not transcribed on the euchromatic X chromosome,

**Heterochromatic X chromosome**

XIC (X-inactivation center)

*Xist*
(X-inactivation-specific transcript)

*Xist* activation and transcription of *Xist* RNA.

Stable *Xist* RNA coats the X chromosome.

Coat of *Xist* RNA leads to silencing and condensation of X chromosome.

HMTs are attracted to RNA coating; H3 and H4 histones are deacetylated and methylated, inactivating the chromosome.

Condensed and silenced X chromosome forms a Barr body.

**Figure 11.18 The X-inactivation center (XIC).** The XIC contains *Xist*, which is transcribed to produce a specialized RNA that coats the X chromosome. This mechanism is responsible for random X-inactivation in mammals.

where it is inactive. The gene transcript is a specialized RNA transcript called *Xist* RNA that never leaves the nucleus and is never translated. Instead, *Xist* RNA exclusively coats the X chromosome that produces it. The *Xist* RNA coating attracts HMTs and histone deacetylases that methylate and deacetylate histone proteins H3 and H4. These epigenetic modifications are linked directly to transcriptional silencing of genes. The *Xist* RNA coating, subsequent methylation and deacetylation, and other protein-driven modifications inactivate one X chromosome and condense it into a heterochromatic state in each mammalian female nucleus.

The mouse-tissue enzyme gene *Te2* is expressed in heart (H), kidney (K), and thymus gland (T) cells at different times during mouse development. DNA that is upstream of transcribed *Te2* genes and of transcriptionally silent copies of *Te2* is isolated from embryonic (E) and adult (A) tissues. A radioactive label is attached to one end of each isolated DNA fragment. The end-labeled fragments are then treated with the enzyme micrococcal nuclease that cuts DNA packaged as euchromatin but is unable to cut DNA that is packaged as heterochromatin. The enzyme-treated, end-labeled DNA samples are run in an electrophoresis gel. The results of the electrophoresis are shown in the figure.

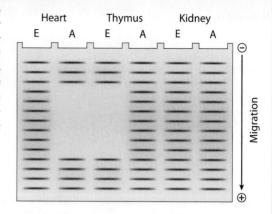

a. Based on the results shown in the gel, is there evidence that chromatin compaction and remodeling play a role in the expression of *Te2*? Explain your reasoning.

b. In which tissue(s) and at what times during development do the results indicate the expression of *Te2* is most likely?

| Solution Strategies | Solution Steps |
|---|---|
| **Evaluate** | |
| 1. Identify the topic this problem addresses, and explain the nature of the required answer. | 1. This problem deals with analysis of chromatin structure near a mouse gene that is expressed in different tissues and at different times during development. The answer requires interpreting the results of gel electrophoresis of enzyme-exposed DNA upstream of the gene. |
| 2. Identify the critical information given in the problem. | 2. Results are given for electrophoresis of DNA upstream of transcriptionally active and transcriptionally inactive copies of *Te2*. |
| **Deduce** | |
| 3. Describe the relation between transcriptional activity and the chromatin structure of a gene. | 3. Heterochromatin is densely packaged chromatin that is largely inaccessible to transcriptional proteins and contains few, if any, transcribed genes. Euchromatin, on the other hand, is more relaxed chromatin that contains transcribed genes. |
| **Solve** | **Answer a** |
| 4. Assess the gel electrophoresis results for each tissue from embryonic and adult mice.<br><br>**TIP:** The enzyme cleaves euchromatic, but not heterochromatic, DNA. | 4. Two patterns of DNA bands are observed. One pattern contains bands whose lengths increase in an unbroken continuum. This band pattern indicates that fragments of all lengths are created by enzymatic digestion. The second pattern contains smaller fragments and larger fragments, but no intermediate-sized fragments. This pattern indicates that intermediate fragment lengths are not created by enzyme treatment. |
| | **Answer b** |
| 5. Use the gel results to interpret the embryonic and adult mouse tissue expression patterns of *Te2*.<br><br>**TIP:** Gaps in the pattern of digested DNA fragments indicate that some of the chromatin is tightly bound and resistant to enzymatic digestion. They are thus a sign of heterochromatic chromatin structure that does not contain transcribed genes. | 5. A continuous pattern of DNA fragments is observed when DNA is euchromatic and likely to contain transcribed genes. The gel results are consistent with *Te2* gene expression in embryonic heart and adult thymus, and in both embryonic and adult kidney. The results suggest that *Te2* is not transcribed in adult heart or in embryonic thymus. Compact chromatin structure around the gene in adult heart and embryonic thymus gland cells could account for the observed silencing of gene expression in these tissues. |

For more practice, see Problems 17 and 25.

## CASE STUDY

### Fishing for Chromosome Abnormalities in Cancer Cells

Cancer cells are highly abnormal and typically contain numerous mutations that disrupt many fundamental cell activities, including normal control of the cell cycle, cell-to-cell interactions and communication, rate of cell division, and mutation repair processes. In addition, the genomes and the chromosomes of cancer cells are unstable and prone to further mutational changes. Chromosomal mutations in cancer cells are usually multiple and most often include deletion or duplication of all or parts of chromosomes, as well as translocations in which part of one chromosome is transferred and attached to a nonhomologous chromosome.

At one time, the chromosome abnormalities in cancer cells were identified by inspecting the banding patterns of chromosomes for changes. This process has been improved in recent years by the development of multicolor FISH techniques and the use of distinct probes and fluorophores for each chromosome. The new methodology permits more accurate detection and identification of chromosome abnormalities. Figure 11.19a shows the karyotype of a cancer cell in which FISH has revealed that multiple chromosomes are abnormal. Notice particularly that some chromosomes contain more than one color, indicating that they are actually composed of pieces from two or more nonhomologous chromosomes.

While cancer cells typically have highly unstable chromosomes that tend to produce the kind of grossly abnormal karyotype seen in Figure 11.19a, certain cancer cells are unusual in that they contain specific chromosome rearrangements that are almost diagnostic for a particular type of cancer. This is the case in chronic myelogenous leukemia (CML), where a specific reciprocal translocation between chromosomes 9 and 22 is seen in almost every case. Similarly, in a cancer called Burkitt's lymphoma, a different reciprocal translocation between chromosomes 8 and 14 is very frequently observed. Figure 11.19b shows the reciprocal translocation between chromosomes 9 and 22 found in cancer cells of people with CML. Likewise, the translocation between chromosomes 8 and 14 in people with Burkitt's lymphoma is also revealed by FISH (Figure 11.19c). The use of fluorophores emitting at different wavelengths makes it easy to identify the chromosomes involved in the translocation mutation and to pinpoint the locations of chromosome translocations. In a later chapter, we discuss the biological disturbances created by these translocations and how they contribute to the development of cancer (Chapter 13).

**(a)**

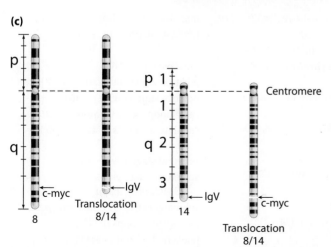

**(b)**

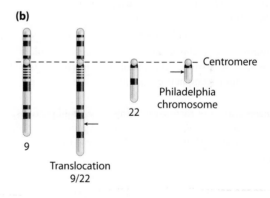

**(c)**

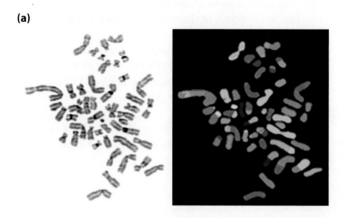

**Figure 11.19 FISH detection of chromosome rearrangements in human cancer cells. (a)** General chromosome instability leads to the frequent observation of multiple chromosome abnormalities in cancer cells that are readily observed using FISH methods (right). **(b)** A reciprocal translocation between chromosome 9 and chromosome 22 is very common in chronic myelogenous leukemia (CML). **(c)** Translocation between chromosome 8 and chromosome 14 is frequently detected in Burkitt's lymphoma cells.

## 11.1 Bacterial Chromosomes Are Simple in Organization

▌ Bacterial genomes are haploid and usually contain a single, circular chromosome. The genomes of certain bacterial species contain more than one chromosome.

▌ Bacterial chromosomes are 1000 or more times longer than the cells they reside in and are located in the nucleoid.

▌ Proteins associate with bacterial chromosomes to aid compaction.

▌ Supercoiling of circular bacterial chromosomes is the principal mechanism for compaction of the chromosome into bacterial cells.

## 11.2 Eukaryotic Chromosomes Are Organized as Chromatin

▌ Eukaryotic nuclei contain multiple chromosomes that are highly compacted.

▌ Eukaryotic chromosomes are composed of chromatin—a mixture of DNA, histone proteins, scaffold proteins, and other nonhistone proteins.

▌ Eight histone protein molecules form nucleosomes around which 146 bp of DNA wraps to form the 10-nm fiber.

▌ The 10-nm fiber condenses to form the 30-nm fiber.

▌ Nonhistone proteins form the chromosome scaffold that gives structure to chromatids and aids in additional chromosome compaction during prophase of the cell cycle.

▌ Chromatin loops form with the aid of scaffold proteins. In each different type of cell, expressed genes are more distant from anchor points on the scaffold than unexpressed genes.

## 11.3 Chromosome Regions Are Differentiated by Banding

▌ Chromosomes are categorized by structure on the basis of the centromere position and the ratio of long arm (q arm) length to short arm (p arm) length.

▌ Each chromosome has a distinctive pattern of bands, created when stains or dyes are applied.

▌ Heterochromatic DNA forms darkly staining bands that contain relatively few expressed genes.

▌ Euchromatic DNA forms lightly staining bands that contain the majority of expressed genes.

▌ The centromere consists of specialized DNA sequences that bind kinetochore proteins.

▌ Specialized molecular probes are used for in situ hybridization to locate specific genes or chromosome-specific DNA sequences. These probes often utilize fluorescent labels for detection.

## 11.4 Chromatin Structure Influences Gene Transcription

▌ During interphase, each chromosome inhabits a territory of its own in the nucleus. Chromosome positioning within the territory is tied to replication and transcription.

▌ Studies of position effect variegation (PEV) have determined that the structure of chromatin surrounding a gene directly influences transcription.

▌ Chromatin remodeling consists of specific epigenetic modifications of the N-terminal tails of histone proteins.

▌ Mutations that enhance [*E(var)*] or suppress [*Su(var)*] position effect variegation alter genes that are responsible for chromatin remodeling.

▌ Random X inactivation in mammalian females is caused by a specialized RNA transcribed from an active gene on the inactivated X chromosome.

## KEYWORDS

10-nm fiber *(p. 363)*

30-nm fiber (solenoid) *(p. 363)*

300-nm fiber *(p. 366)*

acrocentric chromosome *(p. 369)*

chromatin *(p. 362)*

chromatin remodeling *(p. 368)*

chromosome arms [long arm (q arm), short arm (p arm)] *(p. 369)*

chromosome banding (Giemsa banding, G banding) *(pp. 369, 370)*

chromosome scaffold *(p. 366)*

chromosome territory *(p. 373)*

constitutive heterochromatin *(p. 370)*

core DNA *(p. 363)*

epigenetic marks (epigenetic modifications) *(p. 368)*

euchromatin (euchromatic region) *(p. 370)*

*E(var)* mutations *(p. 375)*

facultative heterochromatin *(p. 370)*

fluorescent in situ hybridization (FISH) *(p. 372)*

heterochromatin (heterochromatic region) *(p. 370)*

histone proteins (H1, H2A, H2B, H3, H4) *(p. 362)*

in situ hybridization *(p. 372)*

interchromosomal domain *(p. 373)*

karyotype *(p. 369)*

linker DNA *(p. 363)*

matrix attachment region (MAR) *(p. 366)*

metacentric chromosome *(p. 369)*

negative supercoiling *(p. 361)*

nonhistone proteins *(p. 362)*

nucleoid *(p. 361)*

nucleosome *(p. 363)*

position effect variegation (PEV) *(p. 374)*

positive supercoiling *(p. 361)*

radial loop–scaffold model *(p. 366)*

small nucleoid–associated proteins *(p. 361)*

solenoid structure *(p. 363)*

structural maintenance of chromosomes (SMC) proteins *(p. 361)*

submetacentric chromosome *(p. 369)*

*Su(var)* mutations *(p. 375)*

telocentric chromosome *(p. 369)*

For instructor-assigned tutorials and problems, go to www.masteringgenetics.com.

## PROBLEMS

## Chapter Concepts

For answers to selected even-numbered problems, see Appendix: Answers.

1. What is a bacterial nucleoid? What is a bacterial plasmid? How are these objects similar, and how do they differ?

2. Biologists typically define bacterial genomes as "haploid," but some bacterial genomes contain more than one chromosome carrying genomic DNA. Does the description of "haploid" bacterial genomes conflict with the occurrence of bacterial genomes with more than one chromosome? Why or why not?

3. Bacterial DNA is compacted by two principal mechanisms. Identify and briefly describe each mechanism.

4. The human genome contains $2.9 \times 10^9$ base pairs. Approximately how many nucleosomes are required to organize the 10-nm-fiber structure of the human genome? Show the calculation you use to determine the answer.

5. Give descriptions for the following terms:
   a. histone proteins
   b. nucleosome
   c. CEN sequences
   d. G bands
   e. euchromatin
   f. heterochromatin
   g. epigenetic modification
   h. chromosome territory
   i. nucleoid

6. Describe the importance of light and dark G bands that appear along chromosomes.

7. In eukaryotic DNA,
   a. Where are you most likely to find histone protein H4?
   b. Where are you most likely to find histone protein H1?
   c. Along a 6000-bp segment of DNA, approximately how many molecules of each kind of histone protein do you expect to find? Explain your answer.

d. How does the role of H1 differ from the role of H3 in chromatin formation?

8. Describe the relative differences you expect between the levels of chromosome condensation in interphase and in metaphase.

9. Human late prophase karyotypes have about 2000 visible G bands. The human genome contains approximately 22,000 genes. Consider the region 5p15.1 through the end of the short arm of chromosome 5 that is identified on the late prophase chromosome in Figure 11.10, and assume the entire region is deleted. Approximately how many genes will be lost as a result of the deletion?

10. What are the two or three most essential components of a bacterial chromosome sequence? Of a eukaryotic chromosome sequence? Thinking in an evolutionary context, devise an argument to explain why these components are present.

11. What are the most essential components of yeast centromeres?

12. Do bacterial chromosomes have centromeres? Do they have telomeres? Devise an argument for each answer to explain why or why not from an evolutionary perspective.

13. Identify two ways in which telomeres and centromeres are similar.

14. Thinking about DNA sequence, describe how the sequence will change with distance from the telomere.

15. In what way does position effect variegation (PEV) of *Drosophila* eye color indicate that gene expression takes place in euchromatic regions of chromosomes but not in heterochromatic chromosome regions?

## Application and Integration

For answers to selected even-numbered problems, see Appendix: Answers.

16. As a follow-up to Genetic Analysis 11.1, in which you determined the approximate number of nucleosomes per nucleus in *Arabidopsis thaliana*, answer these questions:
    a. If the number of nucleosomes given in answer to part a of the Genetic Analysis is for the nucleus of a cell in $G_1$, how many nucleosomes do you expect in the nucleus after completion of S phase? Explain your answer.
    b. Are all of the additional nucleosomes that are present after completion of S phase composed of newly synthesized histone proteins? Explain your answer.

17. A survey of organisms living deep in the ocean reveals two new species whose DNA is isolated for analysis. DNA samples from both species are treated to remove nonhistone proteins. Each DNA sample is then treated with micrococcal nuclease that cuts DNA not protected by proteins but is unable to cut DNA bound by histone proteins. Following micrococcal nuclease treatment, DNA samples are subjected to gel electrophoresis, and the gels are stained with ethidium

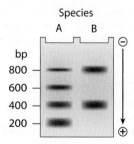

bromide to stain all DNA bands in the gel. The ethidium bromide staining patterns of DNA from each species are shown in the figure. The number of base pairs (bp) in small DNA fragments is shown at the left of the gel. Interpret the gel results in terms of chromatin organization and the spacing of nucleosomes in the chromatin of each species.

18. A eukaryote with a diploid number of $2n = 6$ carries the chromosomes shown below and numbered 1 to 6.

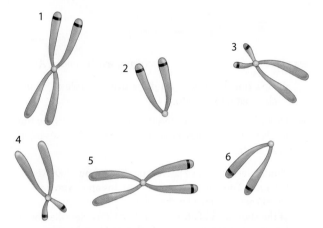

a. Carefully examine and redraw these chromosomes in any valid metaphase I alignment. Draw and label the metaphase plate, and label each chromosome by its assigned number.
b. Explain how you determined the correct alignment of homologous chromosomes on opposite sides of the metaphase plate.

19. The chromosome diagram shown below represents a eukaryotic chromosome stained by G (Giemsa) banding. Indicate the heterochromatic and euchromatic regions of the chromosome, and label the chromosome's centromeric and telomeric regions.

Centromere

a. What term best describes the shape of this chromosome?
b. Do you expect the centromeric region to contain facultative heterochromatin? Why or why not?
c. Describe the features of general sequence composition and protein binding that differentiate the centromeric region from other regions of the chromosome.
d. Why are expressed genes not found in the telomeric region of chromosomes?
e. Are you more likely to find the DNA sequence encoding the digestive enzyme amylase in a heterochromatic, euchromatic, centromeric, or telomeric region? Explain your reasoning.

20. The genome of the bacterium *Methanococcus jannaschii* contains a principal circular double-stranded chromosome composed of $1.6 \times 10^6$ bp. A geometric calculation tells us that the diameter of the circular chromosome is about 10 times the diameter of the *M. jannaschii* cell.

a. How is this principal chromosome packaged inside the cell?
b. Describe how this chromosome is packaged in the bacterial nucleoid.
c. Why is this chromosome supercoiled?

21. Micrococcal nuclease cuts DNA that is not directly associated with nucleosomes. Treatment of human DNA with micrococcal nuclease produces nucleosome repeats that are consistently 185 to 200 bp in length. Why does this result indicate that nucleosomes are evenly spaced on human DNA? What result would be obtained if nucleosomes were randomly spaced along DNA?

22. Histone protein H4 isolated from pea plants and cow thymus glands contains 102 amino acids in both cases. A total

of 100 of the amino acids are identical between the two species. Give an evolutionary explanation for this strong amino acid sequence identity based on what you know about the functions of histones and nucleosomes.

23. Studies of epigenetic modifications of histone proteins in yeast find an approximately 15–20% increase in the transcription of genes if there is acetylation of H3 and H4 in nearby nucleosomes. The studies also detect a reduction in the level of methylation of nucleosomes (hypomethylation) in the vicinity of actively transcribed genes. Speculate about the significance of the epigenetic observations for gene transcription.

24. Experimental evidence demonstrates that the nucleosomes present in a cell after the completion of S phase are composed of some "old" histone dimers and some newly synthesized histone dimers. Describe the general design for an experiment that uses a protein label such as $^{35}$S to show that nucleosomes are often a mixture of old and new histone dimers following DNA replication.

25. Micrococcal nuclease cuts DNA that is not protected by bound proteins, but is unable to cut DNA that is complexed with proteins. Human DNA is isolated, stripped of its non-histone proteins, and mixed with micrococcal nuclease. Samples are removed after 30 minutes, 1 hour, and 4 hours and run separately in gel electrophoresis. The resulting gel is stained with ethidium bromide, and the results are shown in the figure. DNA fragment sizes in base pairs (bp) are estimated by the scale to the left of the gel.

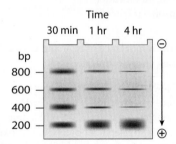

a. Examine the gel results and speculate why longer micrococcal nuclease treatment produces different results.
b. Draw a conclusion about the organization of chromatin in the human genome from this gel.

26. Genomic DNA from the nematode worm *Caenorhabditis elegans* is organized by nucleosomes in the manner typical of eukaryotic genomes, with 145 bp encircling each nucleosome and approximately 55 bp in linker DNA. When *C. elegans* chromatin is carefully isolated, stripped of nonhistone proteins, and placed in an appropriate buffer, the chromatin decondenses to the 10-nm fiber structure. Suppose researchers mix a sample of 10-nm-fiber chromatin with a large amount of the enzyme DNase I that randomly cleaves DNA in regions that are not protected by bound protein. Next, they remove the nucleosomes, separate the DNA fragments by gel electrophoresis, and stain the fragments by ethidium bromide.

a. Approximately what range of DNA fragment sizes do you expect to see in the stained electrophoresis gel? How many bands will be visible on the gel?
b. Explain the origin of DNA fragments seen in the gel.
c. How do the expected results support the 10-nm-fiber model of chromatin?

# Gene Mutation, DNA Repair, and Homologous Recombination

Peacocks normally have brightly colored feathers, but this peacock has no feather color as the result of a gene mutation.

## ESSENTIAL IDEAS

▌ Gene mutations are rare and random.

▌ Mutations change DNA sequence, alter polypeptide composition and function, and cause phenotypic variation.

▌ Spontaneous nucleotide changes can lead to mutation.

▌ Chemical mutagens and radiation damage DNA and produce mutations.

▌ DNA repair systems can directly repair DNA damage or can remove and replace damaged segments.

▌ Specialized enzymes can bypass the blockage of DNA replication due to unrepaired damage.

▌ Meiotic recombination is a programmed event accomplished by processes related to DNA damage repair.

▌ Gene conversion is a directed DNA sequence change associated with homologous recombination.

Mutation can be defined most simply as a heritable change in DNA sequence. An enormous range of DNA changes are included in this definition. At one end of the spectrum are single DNA base-pair substitutions that might or might not lead to a change in the amino acid composition of a polypeptide. At the opposite end of the spectrum is the addition or subtraction of one or more chromosome sets from the genome.

Mutations derive from damage done to DNA by chemical, physical, or biological agents, or from spontaneous changes of DNA nucleotide bases. Damage can occur in any cell at any time, and the resulting mutations can be passed along by cell division. In single-celled organisms, mutations are transmitted when the cells divide, whereas in plants and animals, mutations in somatic cells are transmitted by mitosis to daughter cells, and mutations in germ-line cells are transmitted to the next generation through gametes.

The study of DNA damage, DNA repair, and mutations is a cornerstone of genetic analysis, for numerous reasons. For example, (1) mutations are the ultimate source of new genetic variety in species and populations, and evolution would be impossible without the inherited variation they generate; (2) the inheritance patterns of allelic variants are the foundation on which the principles of heredity are based; (3) the mapping of mutations is a first step toward the construction of genetic maps of chromosomes and genomes; (4) examination of the abnormalities occurring in mutant organisms leads to discovery of the wild-type function of genes; (5) the study of DNA damage repair mechanisms is integral to understanding the biological and molecular processes that contribute to hereditary variation; (6) the repair of double-stranded DNA breaks underlies the process of crossing over between homologous chromosomes in meiosis; and (7) defective DNA repair underlies several diseases, most notably cancer. This chapter explores these topics in its discussion of gene mutation, repair of DNA damage, and the mechanism of homologous recombination.

## 12.1 Mutations Are Rare and Occur at Random

Mutations have been a focus of attention since the science of genetics first began. Recall, for example, that the inheritance of wrinkled seeds in the pea plants studied by Mendel results from homozygosity for a mutant allele that does not produce a starch-branching enzyme (see Chapter 2). In the absence of this enzyme, too many free sugar molecules are present in developing embryos that eventually form wrinkled seeds (see Table 2.6). Similarly, the short pea plant height Mendel observed occurs in plants homozygous for a mutant allele that is unable to produce a growth hormone. Likewise, Morgan's discovery of the chromosomal basis of heredity grew out of his study of the mutant white eye-color phenotype in *Drosophila* (see Chapter 3).

In each of these examples, a mutant phenotype results from the mutation of a single gene transmitted during reproduction. But in reality, every organism carries mutant alleles, whether they manifest in distinct abnormal phenotypes or not. The presence of mutations is ubiquitous in genomes and they occur in every generation. Current estimates for the human genome based on genome sequencing data indicate that as individuals we acquire one to four new mutations each generation.

Recognizing that inherited variation was the hereditary source of variation and evolution in populations, biologists had to answer two important questions. First, What is the frequency of mutations of different genes? and second, Is variation among organisms a result of randomly occurring mutations, some of which are adaptive, or does a change in the environment induce genetic change? The answer to the first question came from numerous studies of a wide range of organisms, and the answer to the second question came most prominently from a 1943 experiment by Salvador Luria and Max Delbrück that demonstrated the random nature of mutations.

### Mutation Frequency

How often do mutations occur? Are the numbers of mutations the same for each gene, or do they vary depending on the gene or on the organism? These are some of the questions geneticists asked about mutations as they sought to describe the frequency and distribution of mutations in genomes.

In bacteria and other haploid microorganisms, **mutation frequency** can be measured by the number of times mutation alters a particular gene. Mutations in these organisms creating auxotrophic nutritional deficiencies that impair the organisms' ability to grow on a minimal medium are most readily identified. Auxotrophic mutants require supplementation of a minimal medium for growth.

Mutation frequency is defined and investigated differently in sexually reproducing diploids. In these organisms, the mutation frequency is the number of mutational events in a given gene over a given unit of time; and while recessive mutations can be identified, it is more common to identify dominant mutations as they are easier to detect. Mutation frequencies are most often measured per DNA replication cycle, or per cell generation. For example, if 100,000 pollen grains are examined for variation in color, the discovery of two abnormally colored grains would be reported as a mutation rate of $2 \times 10^{-5}$. Alternatively, longitudinal studies of reproduction in animals and plants can identify the rates of

| Table 12.1 | Spontaneous Mutation Frequencies for Selected Genes | | |
|---|---|---|---|
| Organism | Character Affected | Gene | Frequency |
| **Bacteria** | | | |
| *Escherichia coli* | Lactose fermentation | $lac^+ \rightarrow lac^-$ | $2 \times 10^{-6}$ |
| | Phage T1 resistance | $T1\text{-}s \rightarrow T1\text{-}r$ | $2 \times 10^{-8}$ |
| | Histidine dependence | $his^+ \rightarrow his^-$ | $2 \times 10^{-6}$ |
| *Salmonella typhimurium* | Tryptophan independence | $trp^- \rightarrow trp^+$ | $5 \times 10^{-8}$ |
| | Histidine dependence | $his^- \rightarrow his^-$ | $2 \times 10^{-6}$ |
| | Threonine dependence | $thr^+ \rightarrow thr^-$ | $4 \times 10^{-6}$ |
| **Algae** | | | |
| *Chlamydomonas reinhardii* | Penicillin resistance | $pen\text{-}s \rightarrow pen\text{-}r$ | $1 \times 10^{-7}$ |
| | Streptomycin sensitivity | $str\text{-}r \rightarrow str\text{-}s$ | $1 \times 10^{-6}$ |
| **Fungi** | | | |
| *Neurospora crassa* | Adenine independence | $ade^- \rightarrow ade^+$ | $4 \times 10^{-8}$ |
| | Inositol independence | $inos^- \rightarrow inos^+$ | $8 \times 10^{-8}$ |
| **Plant** | | | |
| *Zea mays* (corn) | Kernel color | $C^+ \rightarrow c^-$ | $2 \times 10^{-6}$ |
| | Endosperm composition | $Su^+ \rightarrow su^-$ | $2 \times 10^{-6}$ |
| | Kernel shape | $Sh^+ \rightarrow sh$ | $1 \times 10^{-6}$ |
| **Insect** | | | |
| *Drosophila melanogaster* (fruit fly) | Eye color | $w^+ \rightarrow w$ | $4 \times 10^{-5}$ |
| | Body color | $y^+ \rightarrow y$ | $1.2 \times 10^{-6}$ |
| | Eye color | $bw^+ \rightarrow bw$ | $3 \times 10^{-5}$ |
| | Body color | $e^+ \rightarrow e$ | $2 \times 10^{-5}$ |
| **Mammal** | | | |
| *Mus musculus* (mouse) | Coat color | $b^+ \rightarrow b^-$ | $8.5 \times 10^{-4}$ |
| | Pigment pattern | $s^+ \rightarrow s^-$ | $3 \times 10^{-5}$ |
| | Coat color | $a^+ \rightarrow a^-$ | $4 \times 10^{-5}$ |
| | Eye color | $p^+ \rightarrow p^-$ | $8.5 \times 10^{-4}$ |
| *Homo sapiens* (human) | Stature (achondroplasia) | $A^+ \rightarrow A^-$ | $5 \times 10^{-5}$ |
| | Eye form (aniridia) | $AN^+ \rightarrow AN^-$ | $5 \times 10^{-6}$ |
| | Clotting (hemophilia) | $H^+ \rightarrow H^-$ | $2 \times 10^{-5}$ |
| | Huntington disease | $HUT^+ \rightarrow HUT^-$ | $2 \times 10^{-6}$ |
| | Nail-patella syndrome | $NP^+ \rightarrow NP^-$ | $5 \times 10^{-6}$ |
| | Duchenne muscular dystrophy | $DY^+ \rightarrow DY^-$ | $1 \times 10^{-4}$ |
| | Neurofibromatosis | $NF1^+ \rightarrow NF1^-$ | $1 \times 10^{-4}$ |

mutations in offspring. Dominant mutations are most easily identified by this approach for the simple reason that a single copy of a dominant mutant allele will be observable in the phenotype. **Table 12.1** lists mutation rates for selected genes in a variety of microbial, plant, and animal species.

From decades of study of frequencies of gene mutations, three essential conclusions can be drawn. First, mutation frequencies are low in all genomes, meaning that genome stability is paramount and mutations contribute slowly to inherited diversity. Second, gene mutation frequencies differ considerably among organisms, suggesting that genomes may have variable levels of tolerance for mutations and that mutation repair efficiency may vary among organisms. Lastly, mutation frequencies among different genes of a single species show variation, suggesting that there are intrinsic DNA sequence variables that lead to different mutation rates among the genes in a genome.

Each species has an average gene mutation frequency, and research indicates that the frequencies correlate with genome size. In other words, organisms with larger genomes generally have a higher mutation frequency than organisms with smaller genomes. Current estimates find that vertebrates have mutation frequencies at least 100-fold higher per generation than do bacteria. In addition to differences in mutation frequencies among organisms, mutation frequencies for certain genes within a given genome may be substantially higher than the species average. In the human genome, for example, the average mutation frequency for the average gene is on the order of 1 to 10 per million gametes. But in specific genes, such as *DYS*, which produces the human X-linked recessive disorder Duchenne muscular dystrophy, and *NF1*, which produces autosomal dominant neurofibromatosis, we see elevated mutation frequencies. Genes like these are identified as **hotspots of mutation,** individual genes or regions of genomes where mutations occur much more often than average.

Hotspots of mutation typically derive from specific characteristics of the genes or DNA sequences involved. One reason for mutational hotspots is large gene size. This is the case for *DYS* and *NF1*, which each experience mutation at a rate that is about one order of magnitude higher than the average human mutation rate ($10^{-5}$ for *DYS* and *NF1* versus the aforementioned $10^{-6}$ for the average human gene). These genes are also the two largest genes known in the human genome. *DYS*, for example, spans approximately 2.5 million bp on the X chromosome. We discuss other circumstances creating mutational hotspots in later sections.

## The Fluctuation Test

The nature of mutation was the subject of considerable speculation until nearly the mid-point of the 20th century. The speculation largely ended in 1943, when Luria and Delbrück undertook an experiment that tested the nature of a mutation that protected *E. coli* from lysis after exposure to bacteriophage. Although in the Luria and Delbrück experiment the frequency of phage-resistant bacteria was low, their number could be counted due to the large number of bacterial cells on each growth plate and the ability of phage-resistant bacteria to grow in the presence of bacteriophage.

Luria and Delbrück tested two competing theories of the nature of mutation. One, the random mutation hypothesis (that turned out to be the correct hypothesis), proposes that mutations occur at random in organisms, and predicts that separate bacterial cultures may contain different frequencies of a particular gene mutation. The alternative hypothesis, known as the adaptation mutation hypothesis, proposes that environmental change triggers an adaptive mutational response and predicts that different bacterial cultures exposed to the same environmental stress will have similar frequencies of a particular gene mutation.

Luria and Delbrück designed the "fluctuation test" for discovering whether mutations occur at random or are induced by environmental change. The mutation they examined was the acquisition of resistance to infection by bacteriophage. The presence of bacteriophage on experimental plates was the environmental factor that might have triggered an adaptive response of bacteria. Luria and Delbrück inoculated 20 small-volume bacterial cultures with bacteria from a larger culture. The 20 small cultures and the original large culture were allowed to expand to the same final concentration of bacteria and they never encountered bacteriophage during this growth phase. Once all cultures were grown, the researchers took equal volumes from each of the 20 small cultures and collected 20 samples from the large culture, spreading each sample on experimental plates containing bacteriophage. After allowing bacteria to grow, they counted the number of phage-resistant colonies on each experimental plate.

The adaptation mutation hypothesis predicts that since the bacteria on each plate were exposed at the same time to bacteriophage, each plate should produce about the same number of phage-resistant colonies if phage in the growth environment stimulated adaptive genetic change. The random mutation hypothesis, on the other hand, predicts that phage-resistant cells develop spontaneously in each culture; thus their total numbers will vary considerably among different cultures. **Figure 12.1** illustrates that more fluctuation around the average number of resistant colonies is predicted by the random mutation hypothesis, and this is precisely what Luria and Delbrück detected. From this work, they concluded that mutations occur at random and innumerable experiments on various organisms in subsequent years support this conclusion.

---

**Genetic Insight** Mutations occur at random. They are rare but vary in rate among different species, from gene to gene within a species, and among different segments of genomes.

---

## 12.2 Gene Mutations Modify DNA Sequence

Gene mutations most often change DNA sequence by substituting, adding, or deleting one or more DNA base pairs. These kinds of localized mutations occur at a specific or identifiable location in a gene and are called **point mutations.** In this section, we describe several types of point mutations that have characteristic consequences depending on the type of sequence change and the location of sequence change in a gene. These point mutations are summarized in **Table 12.2.**

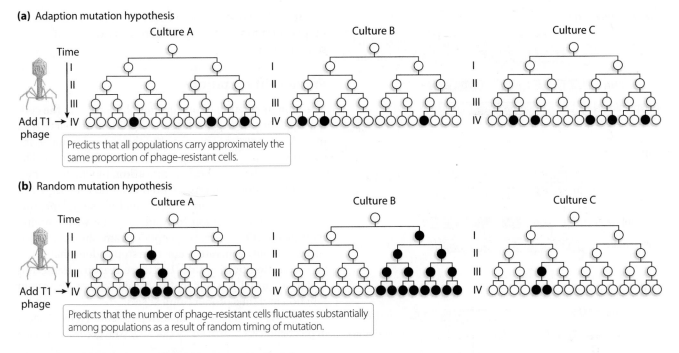

**Figure 12.1** **The fluctuation test of Luria and Delbrück.**

## Base-Pair Substitution Mutations

The replacement of one nucleotide base pair by another is a **base-pair substitution mutation.** Two types of base-pair substitutions occur: **transition mutations,** in which one purine replaces the other (i.e., A replaces G, or vice versa) or one pyrimidine replaces the other (i.e., C replaces T, or vice versa); and **transversion mutations,** in which a purine is replaced by a pyrimidine or vice versa.

Base-pair substitution mutations are further categorized at the molecular level by the manner in which they alter the informational content of the gene. Three types of base-pair mutations are commonly recognized: silent mutations, missense mutations, and nonsense mutations.

**Silent Mutation**    A base-pair substitution producing a protein with the same amino acid sequence as the wild-type protein is known as a **silent mutation,** also called a synonymous mutation. Figures 12.2a and 12.2b illustrate a silent mutation in which an A-T to G-C transition mutation changes the wild-type leucine codon (5′-UUA-3′) to a mutant codon (5′-UUG-3′) that also encodes leucine. Silent mutations are possible because the genetic code is redundant, having 2 to 6 codons for most amino acids (see Table 9.1 for lists of redundant codons).

**Missense Mutation**    A base-pair substitution that results in an amino acid change to the protein is a **missense mutation.** Figure 12.2c shows a T-A to A-T transversion

| Table 12.2 | Point Mutations |
| --- | --- |
| **Type** | **Consequence** |
| ***Coding-Sequence Mutations*** | |
| Silent | No change to amino acid sequence |
| Missense | Changes one amino acid of the polypeptide |
| Nonsense | Creates stop codon and prematurely terminates translation |
| Frameshift | Results in wrong sequence of amino acids |
| ***Regulatory Mutations*** | |
| Promoter | Changes timing or amount of gene transcription |
| Polyadenylation | Alters sequence of mRNA; may affect mRNA stability |
| Splice site | May retain intron sequence in or exclude exon sequence from mature mRNA |
| DNA replication mutation: Triplet-repeat expansion | Increases (or less often, decreases) number of repeats of DNA triplets, causing instability of mRNA or incorrect number of repeating amino acids in a protein |

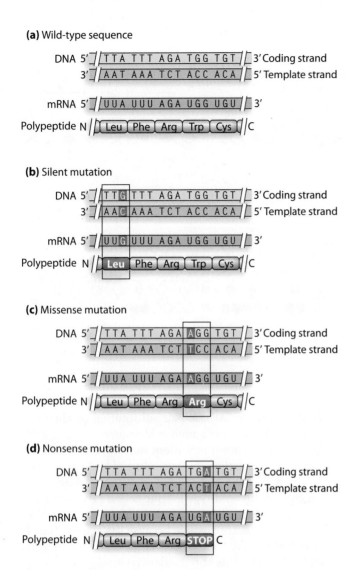

**Figure 12.2** **The consequences of base-pair substitutions.**

mutation that alters the wild-type 5'-UGG-3' codon to 5'-AGG-3', changing the amino acid from tryptophan to arginine. Protein function may be altered by a missense mutation. The specific nature of the protein change (i.e., whether it results in complete or only partial loss of protein function) depends on what kind of amino acid change takes place, and where in the polypeptide chain the change occurs.

The tall versus short trait studied in pea plants by Gregor Mendel uses a dominant allele (*Le*) of a gene required to produce the plant growth hormone gibberellin to produce tall plants. A recessive allele (*le*) of this gene produces an enzyme with less than 5 percent of the catalytic activity of the wild-type enzyme and produces short plants (see Table 2.6). The mutant allele contains a G-C to A-G transition that produces a missense mutation changing an alanine to a threonine in the gene product.

**Nonsense Mutation**    A base-pair substitution that creates a stop codon in place of a codon specifying an amino acid is a **nonsense mutation.** The GC-to-AT base-pair

substitution shown in Figure 12.2d that changes the UGG (Trp) codon to a UGA (stop) codon is an example of a nonsense mutation.

## Frameshift Mutations

Insertion or deletion of one or more base pairs in the coding sequence of a gene leads to addition or deletion of mRNA nucleotides. This can alter the reading frame of the codon sequence, beginning at the point of mutation. The result would be a **frameshift mutation,** in which the mutant polypeptide contains an altered amino acid sequence from the point of mutation to the end of the polypeptide (Figure 12.3). In addition to producing the wrong amino acids in a portion of the polypeptide, frameshift mutations also commonly generate premature stop codons that result in a truncated polypeptide. For these reasons, frameshift mutations usually result in the complete loss of protein function and thus produce null alleles.

The yellow versus green seed pods studied by Mendel in pea plants is controlled by the *SGR* (staygreen) gene (see

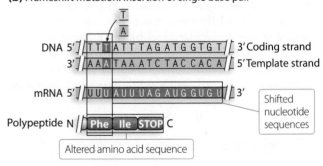

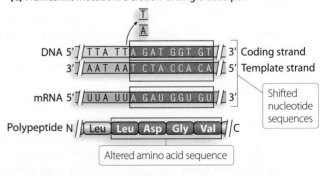

**Figure 12.3** **Frameshift mutation.**

Table 2.6). The protein product of the dominant allele participates in the chlorophyll degradation pathway, but the mutant protein has very poor function due to a mutation that inserts 6 bp of DNA and adds two amino acids to the protein.

## Regulatory Mutations

Some point mutations have the effect of reducing or increasing the amount of wild-type gene transcript and the amount of wild-type polypeptide without affecting the transcript and polypeptide sequences. These mutations, classified as **regulatory mutations**, occur in noncoding regions of genes, such as promoters, introns, and regions coding 5′-UTR and 3′-UTR segments of mRNA. None of these regions directly encodes proteins, but mutations in these regions can nonetheless produce abnormal mRNAs. Three types of regulatory mutations are commonly recognized: promoter mutations, splicing mutations, and cryptic splice sites.

**Promoter Mutations**  Promoter consensus sequences recognized by RNA polymerase II and its associated transcription factors direct the efficient initiation of transcription. Mutations that alter consensus sequence nucleotides and interfere with efficient transcription initiation are **promoter mutations**. The human β-globin gene offers multiple examples of promoter mutations, with various consequences for transcription. **Figure 12.4a** lists mutations at six positions of the human β-globin gene promoter that result in mild to moderate reduction in the amount of β-globin gene transcript and in a reduced amount of β-globin protein. Each of these promoter mutations reduces transcription, but none eliminates transcription entirely. Some promoter mutations of other genes result in the complete elimination of transcription.

**Splicing Mutations**  The DNA dinucleotide GT, on the coding strand, occurs invariably at the 5′ splice site of the intron to demarcate the boundary between the 5′ intron end and the 3′ end of an exon (the GT of coding strand DNA corresponds to the GU dinucleotide of mRNA; see Figure 8.21). In the human β-globin gene, an AG dinucleotide occurs at the 3′ end of exon 1. Each of these dinucleotides is part of the consensus sequence at which the spliceosome forms. Mutations of either of these dinucleotide sequences or of nearby nucleotides in the consensus sequence within the intron can result in splicing errors that inaccurately remove intron sequences from pre-mRNA.

The wild-type 5′ splice site of the human β-globin gene contains a GT that is essential for normal splicing, as are the next several nucleotides in the intron (**Figure 12.4b**). In intron 1 of the β-globin gene, two separate mutations that substitute the guanine of the GT dinucleotide abolish normal splicing entirely in mutations that are known as **splicing mutations**. Additionally,

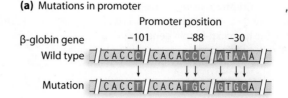

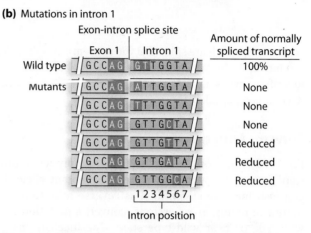

**Figure 12.4  Regulatory mutations of the human β-globin gene. (a)** These base-pair substitution mutations in the promoter reduce transcription of the gene. **(b)** These base-pair substitutions in intron 1 reduce or eliminate normal pre-mRNA splicing.

one base-pair substitution mutation of position 5 of intron 1 also prevents the production of normally spliced mRNA. The translation of abnormally spliced transcripts does not produce wild-type β-globin protein. Other base-pair substitution mutations in intron 1 result in production of a mixture of normally and abnormally spliced transcript and produce some wild-type β-globin protein.

A mutation resulting from a G-C to A-T transition mutation alters the guanine and the 5′ splice site of one intron in the *A* gene that produces purple versus white flowers studied by Mendel (see Table 2.6). This dominant allele produces a transcription activating protein that leads to production of the flower pigment anthocyanin. The splice site mutation in the mutant allele leads to abnormal splicing that retains eight additional nucleotides in mature mRNA. The mutant protein does not function in the anthocyanin pathway.

**Cryptic Splice Sites**  Certain base-pair substitution mutations produce new splice sites that replace or compete with authentic splice sites during pre-mRNA processing. These newly formed splice sites are known as **cryptic splice sites.** Intron 1 of the human β-globin gene is 130 nucleotides in length. A base-pair substitution mutation that changes G to A at position 110 of intron 1 creates an AG dinucleotide that is a cryptic splice site (**Figure 12.5**). The cryptic splice site is spliced in about 90% of the intron 1 3′ splicing events. This aberrant splicing leaves 19 additional nucleotides in the mature mRNA; these nucleotides have been removed in the other 10% of mature

**Figure 12.5  Cryptic splicing.** Base-pair substitution of G - C to A - T at position 110 of intron 1 of the human β-globin gene creates a cryptic 3′ splice site.

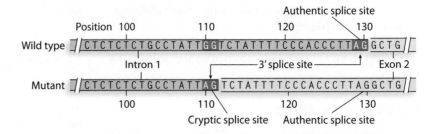

transcripts, which are spliced at the authentic 3′ splice site for intron 1. In Genetic Analysis 12.1, you can practice identifying types of mutations by the alterations they produce in polypeptides.

## Forward Mutation and Reversion

**Forward mutation,** often identified simply as "mutation," converts a wild-type allele to a mutant allele. In contrast, mutations identified as **reverse mutations** or, more commonly, as **reversions,** convert a mutation to a wild-type or near wild-type state. The mechanisms of base-pair substitution described earlier are examples of

processes that create mutation. Reversions can be caused by similar mechanisms. In one type of reversion, called a **true reversion,** the wild-type DNA sequence is restored to encode its original message by a second mutation at the same site or within the same codon (**Figure 12.6a**). Alternatively, reversion can occur by a second mutation elsewhere in the gene. **Figure 12.6b** illustrates an example of one such reversion—an **intragenic reversion,** which is a reversion that occurs through mutation elsewhere in the same gene. Here the initial mutation was caused by deletion of two base pairs, and the intragenic reversion is a compensatory insertion of two base pairs near the site of the initial mutation, restoring the allele to a near

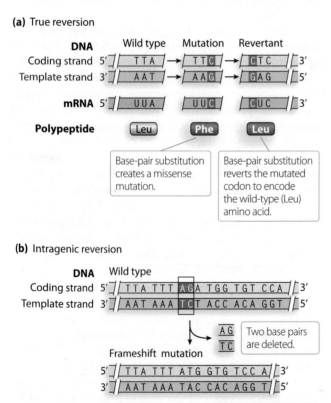

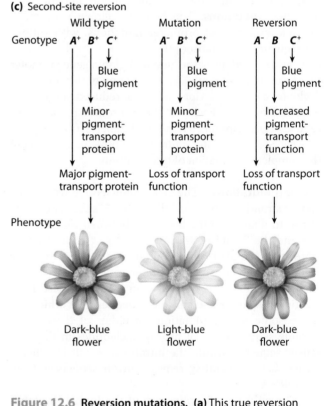

**Figure 12.6  Reversion mutations. (a)** This true reversion restores the wild-type amino acid sequence to the polypeptide. **(b)** This intragenic reversion reverts a frameshift mutation caused by a 2-bp deletion by insertion of 2 bp at a nearby site in the gene. **(c)** Second-site reversion restores a near wild-type phenotype through a compensatory mutation of a second gene.

A fragment of a polypeptide has the following wild-type amino acid sequence:

Met–His–Ala–Trp–Asn–Gly–Glu–His–Arg

The amino acid sequences of three mutants are shown below. For each mutant, identify the type of mutation that has occurred and specify how the mRNA sequence has been changed.

Mutation 1: Met–His–Ala–Trp–Lys–Gly–Glu–His–Arg

Mutation 2: Met–His–Ala

Mutation 3: Met–Met–Leu–Gly–Met–Ala–Glu–His–Arg

| Solution Strategies | Solution Steps |
| --- | --- |
| **Evaluate** | |
| 1. Identify the topic this problem addresses, and explain the nature of the required answer. | 1. This problem concerns mutations affecting the amino acid sequence of a gene. The type of change causing each mutation must be identified, and the effect of the mutation on mRNA must be described. |
| 2. Identify the critical information given in the problem. | 2. The wild-type amino acid sequence and the corresponding portions of the mutant polypeptides are given. |

**Deduce**

3. Determine the sequence of wild-type mRNA.

TIP: Use $N$ if the position could be occupied by any nucleotide, $^A/_G$ for the alternative purines, and $^U/_C$ for alternative pyrimidines.

3. The sequence of wild-type mRNA is,

$5'$-AUG CA$^U/_C$ GCN UGG AA$^U/_C$ GGN GA$^A/_G$ CA$^U/_C$$^A/_C$GN-$3'$

TIP: Use the genetic code in Table 1.2.

**Solve**

4. Compare each mutant sequence to the wild-type polypeptide, and identify the probable types of mutations.

4. Comparisons are as follows.

Mutant 1: This is a missense mutation in which the mutant polypeptide has one amino acid changed from Asn to Lys.

Mutant 2: This is a nonsense mutation in which a Trp codon is changed to a stop codon.

Mutant 3: This mutant contains alterations of five consecutive amino acids, beginning with the second amino acid (His to Met). The wild-type sequence is restored beginning with the seventh amino acid (Glu). This mutant results from two compensatory frameshift mutations. The first alters the reading frame, and the second restores it.

5. Determine the mRNA change producing the missense mutant.

5. The wild-type (Asn) codon is AA$^U/_C$, and the mutant (Lys) codon is AA$^A/_G$. This change results from a transversion mutation.

6. Determine the mRNA change producing the nonsense mutant.

6. The wild-type Trp (UGG) codon is changed to a stop codon. The change is either UGG to UGA or UGG to UAG. In either case, this is a transition mutation.

7. Determine the mRNA change producing the frameshift mutant.

7. The appearance of Met in position 2 means the second codon of the frameshift mutant is AUG. This change requires deletion of the first C of the wild-type sequence and means that U, not C, is present as the sixth nucleotide of the wild type. Beginning with Glu, the wild-type amino acid sequence is restored. This requires insertion of G immediately after the Ala codon.

For more practice, see Problems 4, 9, and 28.

wild-type form. **Figure 12.6c** illustrates an example of a **second-site reversion,** produced by mutation in a different gene. In this case, the original mutation inactivates gene *A* and results in the loss of function of the major pigment-transporting protein in a flower. A minor pigment-transporting gene, *B,* remains active, transporting a small amount of blue pigment from gene *C*. The initial mutation produces a light-blue flower. The second-site reversion is a mutation of gene *B* that increases gene transcription and thus increases production of the pigment-transporting protein. The mutation of gene *B* compensates for the mutation of gene *A* and restores the wild-type dark-blue flower phenotype. A second-site mutation of this type can also be identified as a **suppressor mutation.**

---

Genetic Insight  Point mutations are caused by substitutions, insertions, or deletions of single bases in coding segments of genes and can generate abnormal proteins. Regulatory mutations alter the amount of wild-type protein produced. Reversions counteract the phenotypic consequence of forward mutations and restore wild-type phenotype fully or substantially.

---

## 12.3 Gene Mutations May Arise from Spontaneous Events

In **spontaneous mutation,** naturally occurring mutations arise in cells without being induced by exposure of DNA to a physical, chemical, or biological agent capable of creating DNA damage. Spontaneous mutations arise primarily through errors during DNA replication and through spontaneous changes in the chemical structure of nucleotide bases.

### DNA Replication Errors

DNA replication has extraordinarily high fidelity. Replication errors resulting in base-pair mismatches between a template strand and a newly synthesized strand of DNA occur at an approximate rate of $1 \times 10^{-9}$ in wild-type *Escherichia coli*, and a similar accuracy rate is found in eukaryotic DNA replication. The overall efficiency of DNA replication is attributable to the proofreading capabilities of DNA polymerases and to the operation of DNA base-pair mismatch repair systems (Section 12.5).

An exception to the general accuracy of replication, however, is observed in genomic regions containing repetitive sequence. Replication errors in such regions change the number of base-pairs in repeating sequences and they are another source of hotspots of mutation. The repeating DNA sequences are commonly short, end-to-end repeats.

For example, stretches consisting of many end-to-end repeats of the same two nucleotides (dinucleotide repeats), of the same three nucleotides (trinucleotide repeats), or of longer repeat units are susceptible to replication errors that have the effect of either increasing or decreasing the number of repeats.

Mutations altering the number of DNA repeats occur by a process called **strand slippage.** In the mid-1960s, George Streisinger and his colleagues described the first known example of strand slippage, which generated frameshift mutations in a gene of the bacteriophage T4. Streisinger proposed that strand slippage occurs when the DNA polymerase of the replisome temporarily dissociates from the template strand as it moves across a region of repeating DNA sequence (**Figure 12.7**). During dissociation, a portion of newly replicated DNA forms a temporary double-stranded hairpin structure induced by the complementary base pairing of nucleotides in the loop. Reassociation of DNA polymerase and resumption of replication leads to re-replication of a portion of the repeat region, increasing the length of the repeat region in the daughter strand. Repeating DNA sequences are often hotspots of mutation due to strand slippage.

In the last two decades, a special category of mutations has been identified as a cause of hereditary disease in humans and other organisms. The diseases are now classed as **trinucleotide repeat disorders** (**Table 12.3**). The wild-type alleles of the genes in question normally contain a variable number of DNA trinucleotide repeats. On rare occasions, these genes undergo mutation through strand slippage events that cause the number of trinucleotide repeats to increase. For each of these examples, expansion of the number of trinucleotide repeats beyond the wild-type range results in a hereditary disorder by blocking the production of wild-type mRNA and reducing or eliminating the production of wild-type protein.

### Spontaneous Nucleotide Base Changes

DNA nucleotide bases are organic chemical structures that can occasionally convert, in what are called tautomeric shifts, to alternative structures known as tautomers. Tautomers are structures that have the same composition and general arrangement but a slight difference in bonding and placement of a hydrogen. The generation of a tautomer changes the three-dimensional structure of the nucleotide base from a more stable common form to a rare, less stable form. Tautomeric shifts affecting nitrogenous bases can lead to base-pair mismatch between a rare tautomer on one DNA strand and the common form on the complementary strand.

Mispairing of DNA nucleotides due to the presence of tautomers is the most common form of DNA replication error. **Figure 12.8a** shows standard base pairing involving the common tautomeric forms of the nucleotide

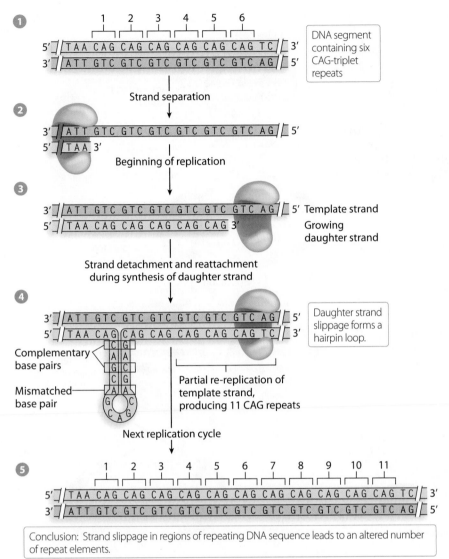

**Figure 12.7** **Strand slippage during DNA replication.**

bases. In comparison, base-pair mismatches that occur between a rare tautomeric nucleotide and a normal nucleotide are shown in Figure 12.8b. Notice in each case of mismatch that purine-pyrimidine pairing is maintained, but that the rare tautomer forms the wrong number of hydrogen bonds, leading to the mispairing.

Tautomeric shifts can lead to base-pair substitution mutations. In Figure 12.9, the rare enol form of thymine

| Table 12.3 | Human Trinucleotide Repeat Disorders | | | | |
|---|---|---|---|---|---|
| **Disease** | **OMIM Number** | **Repeat Sequence** | **Repeat Range** | | **Principal Disease Phenotype** |
| | | | Normal | Disease | |
| Fragile X syndrome | 309550 | CGG | 6–50 | 200–2000 | Mental retardation |
| Friedreich ataxia | 229300 | GAA | 6–29 | 200–900 | Loss of coordination |
| Huntington disease | 143100 | CAG | 10–34 | 40–200 | Uncontrolled movement |
| Jacobsen syndrome | 147791 | CGG | 11 | 100–1000 | Growth retardation |
| Myotonic dystrophy (type I) | 160900 | CTG | 5–37 | 80–1000 | Muscle weakness |
| Spinal and bulbar muscular atrophy | 313200 | CAG | 14–32 | 40–55 | Muscle wasting |
| Spinocerebellar ataxia (multiple forms) | 271245 | CAG | 4–44 | 45–140 | Loss of coordination |

**(a)** Standard base pairing

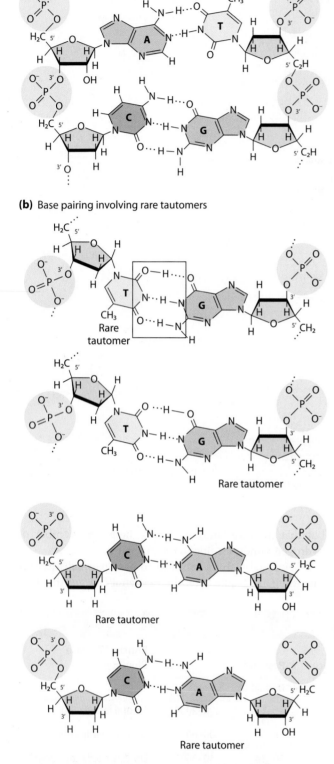

**(b)** Base pairing involving rare tautomers

**Figure 12.8 Base pairing of nucleotide tautomers.**
**(a)** Standard base pairing of common nucleotide tautomers.
**(b)** Base-pair mismatches resulting from pairing of a common and an uncommon nucleotide tautomer.

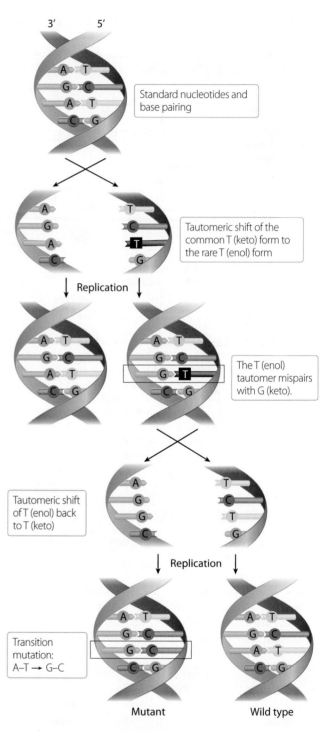

**Figure 12.9 Transition mutation arising from a tautomeric shift of the common keto form of thymine to the rare enol form.**

mispairs with guanine during DNA replication. Three stable hydrogen bonds form between these mispaired nucleotides. A tautomeric shift of thymine back to its common form followed by replication of the unrepaired base-pair mismatch produces one chromatid with a wild-type T-A base pair and the sister chromatid with a C-G base-pair substitution mutation.

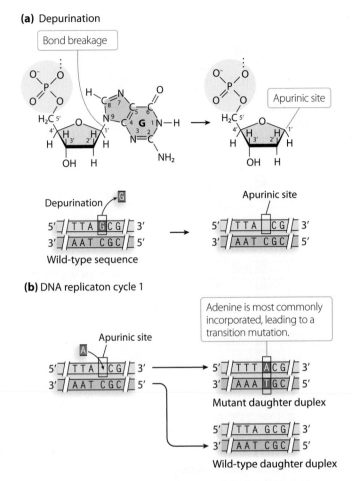

**(a)** Depurination

**(b)** DNA replicaton cycle 1

Adenine is most commonly incorporated, leading to a transition mutation.

**Figure 12.10  Depurination. (a)** Breakage of the 1′ carbon bond releases a purine and creates an apurinic site. **(b)** Adenine is most commonly used to fill apurinic sites during DNA replication, causing a G‑C to A‑T mutation in this case.

## DNA Nucleotide Lesions

Two types of spontaneous damage to individual nucleotides are associated with subsequent mutation. **Depurination** is the loss of one of the purines, adenine or guanine, from a nucleotide by breakage of the covalent bond at the 1′ carbon of deoxyribose that links the sugar to the nucleotide base (**Figure 12.10**). Living cells lose thousands of purines a day, making depurination one of the most frequent spontaneous chemical changes affecting DNA. Fortunately, nearly all of the lost purines are replaced before the next DNA replication cycle (the repair mechanism is discussed later in the chapter), but a few are not. An unrepaired lesion of this type is called an **apurinic site,** meaning a nucleotide that lacks a purine base. During replication, the apurinic site does not contain a template base, and DNA polymerase usually compensates by placing an adenine (but sometimes guanine) opposite the site. During the next DNA replication cycle, the apurinic site once again attracts an adenine to the newly synthesized strand. Meanwhile, the complementary strand

with the incorporated adenine directs synthesis of a daughter duplex that contains a base-pair substitution mutation if the purine lost was G.

**Deamination,** the loss of an amino ($NH_2$) group from a nucleotide base, is a second form of spontaneous chemical modification that can lead to mutation. Each of the DNA nucleotide bases contains an amino group. Deamination of cytosine is frequently a source of mutation. When cytosine is deaminated, the amino group is usually replaced by an oxygen atom, forming the nucleotide base uracil (**Figure 12.11a**). DNA mismatch repair readily recognizes uracil as an RNA nucleotide base and removes it from DNA. The excised uracil is replaced by cytosine, and wild-type sequence is restored.

A different scenario occurs, however, when deamination takes place on a cytosine that has been methylated—in other words, a cytosine in which a hydrogen atom at the number 5 carbon has been replaced with a $CH_3$ (methyl) group. Deamination of 5-methylcytosine (5meC) creates thymine and generates a base-pair mismatch between the newly formed thymine on one strand and the previously complementary guanine on the other strand (**Figure 12.11b**). If mismatch repair enzymes correct the mismatch before the next DNA replication cycle, two outcomes are possible: Either (1) the repair will restore the wild-type G‑C base pair, or (2) the repair will generate an A‑T base-pair transition substitution (**Figure 12.11c**). Alternatively, if mismatch repair does not occur prior to replication, the guanine-containing strand will be used to produce a daughter duplex with the wild type (G‑C) sequence, while the thymine-containing strand will be used to synthesize a daughter duplex containing a G‑C to A‑T base-pair substitution (**Figure 12.11d**).

Cytosines that are side by side with guanines in a DNA strand are joined by a phosphodiester bond. These dinucleotides are identified as CpG dinucleotides (the *p* signifies the single phosphate bond of the phosphodiester bond). Cytosines of CpG dinucleotides are frequent targets for methylation in mammalian promoters where methylation helps regulate transcription. Experimental evidence shows that CpG dinucleotides are hotspots of mutation as a result of deamination of methylated cytosine and the production of thymine.

---

**Genetic Insight**  Nucleotides undergo spontaneous chemical changes such as tautomeric shifts, depurination, and deamination that can lead to mutation.

---

## 12.4 Mutations May Be Induced by Chemicals or Ionizing Radiation

**Induced mutations** are mutations produced by interaction between DNA and a physical, chemical, or biological agent that generates damage resulting in mutation. The agents generating mutation-inducing DNA damage are

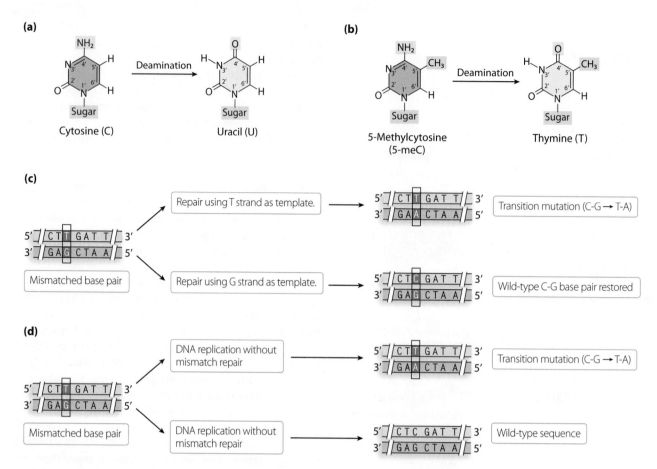

**Figure 12.11 Deamination. (a)** Unmethylated cytosine is deaminated to form uracil. **(b)** Deamination of 5-methylcytosine forms thymine that is mismatched to guanine. **(c)** Mismatch repair can create a G - C to A - T transition mutation or can remove the thymine to restore wild-type sequence. **(d)** Unrepaired mismatches are replicated to produce one wild-type chromatid and one transition mutation chromatid.

called **mutagens.** As we discuss in this section, chemical and physical mutagens interact with DNA in specific ways to create particular types of changes in sequence. Biological mutagens are primarily transposable genetic elements (see discussion in Chapter 13). Mutagens are sometimes exotic or rare, but often they are routinely present in the everyday life of an organism. For this reason, the study of mutagenesis is as important for public health and safety as it is for advancing our understanding of the biological basis of mutation and repair. We begin the discussion of mutagens by examining chemical reactions that cause mutation; move on to look in detail at the action of ultraviolet (UV) irradiation, a physical mutagen to which humans and other organisms are exposed daily; and conclude the section with a discussion of the *Ames test,* a biological procedure for assessing the mutagenic potential of chemical compounds.

## Chemical Mutagens

Chemical compounds that induce mutations do so by specific and characteristic interactions with DNA nucleotide bases. As a result, they can be classified by their mode of action on DNA. They may act as (1) nucleotide base analogs, (2) deaminating agents, (3) alkylating agents, (4) oxidizing agents, (5) hydroxylating agents, or (6) intercalating agents. Compounds in each of these categories and the types of mutations they cause are listed in Table 12.4. In sum, each mutagen reacts in a specific way with DNA and produces a consistent kind of mutation as a result.

**Nucleotide Base Analogs**   A **base analog** is a chemical compound that has a structure similar to one of the DNA nucleotides and therefore can work its way into DNA, where it pairs with a nucleotide. DNA polymerases are unable to distinguish base analogs from normal nucleotides due to their similarity in molecular size and shape. Consequently, base analogs are incorporated into DNA strands during replication. Because base analogs resemble endogenous nucleotides (i.e., the nucleotides produced by the body), the mutations they induce are transitions. For example, the compound 5-bromodeoxyuridine (BrdU) is a derivative of uracil and is very similar to thymine in size

| Table 12.4 | Examples of Mutagenic Agents and Their Consequences | |
|---|---|---|
| **Mutagen** | **Type of Agent** | **Mutagenic Event (with example)** |
| 2-Aminopurine | Base analog | Transition mutation (C - G to T - A) |
| 5-Bromodeoxyuridine | Base analog | Transition mutation (G - C to A - T) |
| Ethyl methanesulfonate | Alkylating agent | Transition mutation (G - C to A - T) |
| Hydroxylamine | Hydroxylating agent | Transition mutation (C - G to T - A) |
| Nitrous oxide | Deaminating agent | Transition mutation (C - G to T - A) |
| Oxygen radicals | Oxidative agent | Transversion mutation (G - C to T - A) |
| Acridine orange | Intercalating agent | Frameshift mutation |
| Proflavin | Intercalating agent | Frameshift mutation |

and shape. In cells, it acts as an analog of thymine. After BrdU becomes incorporated during DNA replication, it pairs with adenine (**Figure 12.12**). Replication of the template strand containing the mispaired adenine results in a C - G to A - T transversion mutation in the example shown in the figure. BrdU undergoes tautomerization much more readily than its analog, thymine. In its tautomeric form, BrdU can lead to mutations in later replication cycles as well. Other base analogs produce different base-pair substitution mutations. For example, the compound 2-aminopurine (2-AP) is a base analog of adenine that pairs with thymine and can lead to either kind of transition mutation.

**Deaminating Agents** Nitrous acid ($HNO_2$) is a deaminating agent, meaning an agent that removes an amino group ($NH_2$). It is capable of removing amino groups from any nucleotide. In most instances, deamination produces no mutagenic effect; but deamination of 5-methylcytosine (shown in Figure 12.11) is an exception, leading to G - C to A - T base-pair substitution. In addition, when nitrous

acid deaminates adenine, the product is hypoxanthine, a modified nucleotide that can mispair with cytosine and lead to a C - G to T - A base-pair substitution mutation (**Figure 12.13a**).

**Alkylating Agents** Alkylating agents add bulky side groups such as methyl ($CH_3$) and ethyl ($CH_3$–$CH_2$) groups to nucleotide bases. These added groups are known as **bulky adducts.** Ethyl methanesulfonate (EMS) is a powerful alkylating agent that adds an ethyl group to thymine, producing 4-ethylthymine, or an ethyl group to guanine, creating $O^6$-ethylguanine (**Figure 12.13b**). This and other bulky adducts interfere with normal DNA base pairing and may distort the DNA double helix.

**Hydroxylating Agents** Hydroxylation is the addition of a hydroxyl (OH) group to a recipient compound by a donor called a hydroxylating agent. Hydroxylamine is a hydroxylating agent that adds a hydroxyl group to cytosine by displacing an $H_2$, thus creating hydroxylaminocytosine (**Figure 12.13c**). Hydroxylaminocytosine

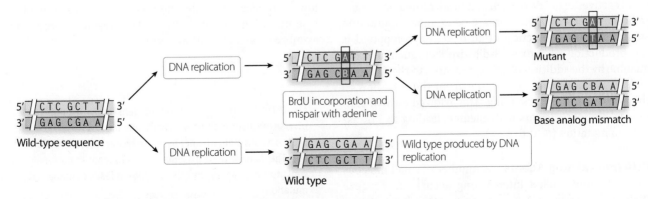

**Figure 12.12  Mutation by incorporation of the nucleotide base analog 5-bromouridine (BrdU).**

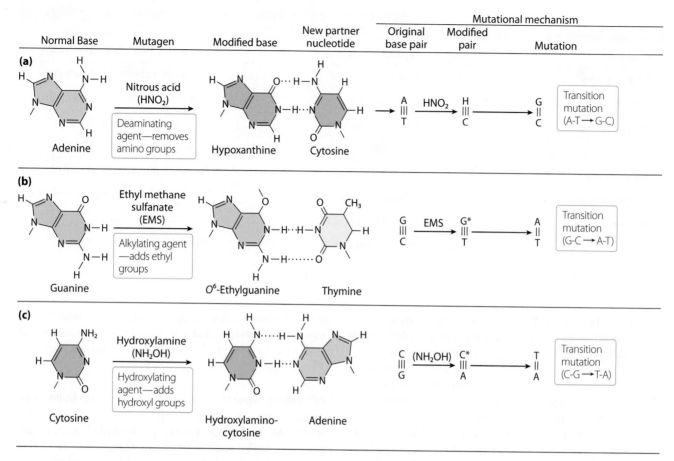

**Figure 12.13 Examples of the action of chemical mutagens.** In **(a)**, H is hypoxanthine. In **(b)** and **(c)**, the asterisks (*) denote modified nucleotides.

often pairs with guanine but frequently mispairs with adenine, leading to C-G to T-A base-pair transition mutations.

**Oxidative Reactions** Oxidation is a chemical process of electron transfer by addition of an oxygen atom or removal of an atom of hydrogen. The oxidative reactions that lead to mutation result when compounds that contain reactive forms of oxygen (often identified as oxygen radicals) react with DNA bases. One example of mutation stemming from the action of an oxidative compound is the production of 8-oxy-7,8-dihydrodeoxyguanine from guanine by the addition of an oxygen atom to the 7 carbon and transfer of a hydrogen atom from the 8 carbon to the 7 carbon of the purine ring. The oxidized form of guanine frequently mispairs with adenine, leading to a transversion mutation (G-C to T-A).

**DNA Intercalating Agents** A number of small molecular compounds called intercalating agents can squeeze their way between DNA base pairs. DNA-intercalating compounds, such as proflavin, benzo(a)pyrene (a component of cigarette smoke), and aflatoxin (a toxin found in mold-contaminated peanuts) are large compounds that can find their way between base pairs and distort the duplex (**Figure 12.14**). These compounds can also attach to nucleotide bases to form bulky adducts that contribute to DNA distortion. These helical distortions lead to DNA strand nicking that is not efficiently repaired. In the following replication cycle, the nicked strands can gain or lose one or more nucleotides. As a result, intercalating agents cause frameshift mutations.

Genetic Insight Chemical mutagens interact with DNA by diverse mechanisms, but each particular mutagen modifies DNA in a consistent manner to create characteristic DNA damage. The damage done by some mutagens leads to base-pair substitutions; damage by other mutagens leads to frameshift mutations.

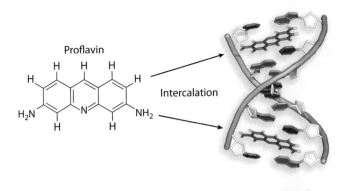

Proflavin

Intercalation

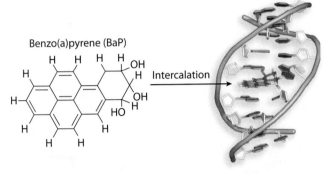

Benzo(a)pyrene (BaP)

Intercalation

**Figure 12.14  DNA intercalating agents.** Proflavin and benzo(a)pyrene intercalate into the double helix and distort its shape, leading to mutations.

## Radiation-Induced DNA Damage

Electromagnetic energy is conveyed in waves, or rays, and categorized by its wavelength, the distance between the equivalent points in its repeating troughs or peaks. Higher-energy radiation has shorter wavelengths than lower-energy radiation. One category of electromagnetic energy is visible light that has wavelengths between approximately 750 nm and 380 nm. All forms of radiant energy with wavelengths less than 380 nm, in other words all electromagnetic energy above the visible spectrum, can cause DNA damage and are mutagenic. These DNA-damaging forms of radiation are ultraviolet radiation, X-rays, gamma rays, and cosmic rays. Ultraviolet (UV) radiation, produced as a component of sunlight, is the mutagen that we and other organisms are most often exposed to.

Like the effects of chemical mutagens, the mutagenicity of UV radiation derives from specific lesions it creates in DNA. UV irradiation alters DNA nucleotides by inciting the formation of additional bonds that form aberrant structures called **photoproducts.** One type of photoproduct, called a **pyrimidine dimer,** is produced by the formation of one or two additional covalent bonds between adjacent pyrimidine dinucleotides in a strand of DNA. Two prominent kinds of pyrimidine dimers occur (**Figure 12.15**). One is a **thymine dimer,** which has two covalent bonds joining the 5 and 6 carbons of thymines

that are adjacent in the same DNA strand. The second, called a **6-4 photoproduct,** also joins adjacent thymines by formation of a bond between the 6 carbon of one thymine and the 4 carbon of the other thymine. Either of these dimeric complexes can also involve the other pyrimidine, cytosine, although cytosine is less commonly included than thymine. The formation of these dimers distorts DNA by pulling the dimerized nucleotide bases closer together and disrupting hydrogen bond formation with their complementary nucleotides on the opposite strand.

Organisms from bacteria to humans have DNA repair systems that identify and correct most pyrimidine dimers, but a few may escape repair; and when they do, DNA replication can be disrupted. Consider the problem posed during replication when pyrimidine dimers are encountered by DNA polymerase performing its synthesis and proofreading activities (see Section 7.4). DNA polymerase adds new nucleotides to nascent DNA strands by identifying the nucleotide complementary to the template strand and then catalyzing phosphodiester bond formation between the 5′ triphosphate of the new nucleotide and the 3′ OH of the previous nucleotide on the new strand. Hydrogen bond formation between complementary nucleotides on the old and newly synthesizing strand orients the 3′ OH of the previously added nucleotide so that it is in the correct position to react with the 5′ triphosphate of the incoming nucleotide. A mismatched base on a nascent DNA strand will fail to form normal hydrogen bonds with the nucleotide on the template strand, leaving the 3′ OH out of position as DNA polymerase attempts to catalyze the next phosphodiester bond. This occurrence activates the proofreading function of the DNA polymerase.

More specifically, when it encounters thymines in a dimer on the template strand, DNA polymerase attempts to add complementary adenines to the nascent DNA strand. But the first adenine fails to form the necessary hydrogen bonds, because the placement of its complementary partner is distorted. In attempting to add the second adenine, DNA polymerase identifies the mispositioned 3′ OH of the first adenine, initiates 5′-to-3′ proofreading activity, and then attempts to resume synthesis in the thymine dimer region—but with the same negative result. Continued repetition of these unsuccessful attempts to replicate across the thymine dimer causes replication to stall at this point.

How does the replication process overcome the blockage caused by the presence of pyrimidine dimers? It circumvents the problem in two ways. First, replication blockage by pyrimidine dimers induces reinitiation of DNA synthesis at an adjacent RNA primer site. This reinitiation of replication potentially leaves gaps spanning dozens to hundreds of nucleotides in newly synthesized DNA strands, but the gaps are subsequently filled by **translesion DNA synthesis,** which is carried out by specialized bypass DNA polymerases (one in bacteria and

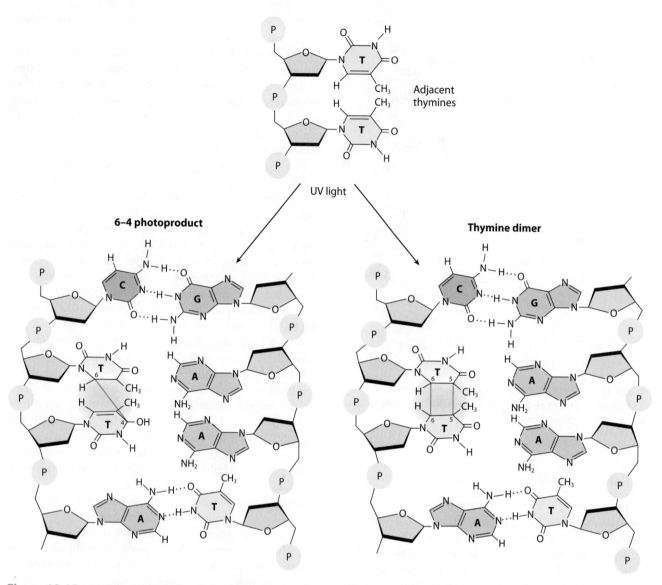

**Figure 12.15 UV photoproducts.** UV irradiation forms photoproducts from adjacent pyrimidines, distorting the double helix and potentially blocking replication.

several in eukaryotes) that can replicate across the gaps. These specialized DNA polymerases are more prone to replication error, however, because they lack proofreading ability. In fact, it is the absence of proofreading activity that allows these polymerases to carry out replication across pyrimidine dimers. Replication can thus proceed, but at the risk of introducing mutations. We discuss the process further in Section 12.6.

Irradiation that has higher energy than that of UV waves, X-rays, and radioactive materials, for example, can cause DNA damage in multiple ways, the most serious being the induction of DNA single-strand or double-strand breaks. These breaks potentially block DNA replication and thus pose a significant threat to the integrity and survival of affected cells. DNA damage of this type is dealt with by specialized strand-break repair mechanisms that we discuss later in the chapter.

**Genetic Insight** Radiation energy in wavelengths shorter than 380 nm is mutagenic. Ultraviolet radiation is a commonly encountered mutagen that creates pyrimidine dimers in DNA, and these can lead to base-pair substitution mutations. Higher-energy radiation can induce DNA strand breaks.

## The Ames Test

In our day-to-day lives, we encounter scores of naturally occurring and synthetic chemicals in the food we eat, the air we breathe, the cars we drive, and even the books we read. Each year hundreds of new chemical compounds are introduced as part of various commercial and industrial processes. How do we determine which of these chemicals pose a hazard to our health by increasing the mutation frequencies of genes? Occasionally, the mutagenic or carcinogenic potential of a compound is so great that evidence of its danger is relatively easy to identify. Much more often, however, the mutagenicity of a compound is more subtle, and careful analysis of experimental data is required to ascertain it.

For nearly 40 years, thousands of natural and synthetic compounds have been assayed for mutagenic potential by a simple biological test developed by Bruce Ames. This procedure, called the **Ames test,** exposes bacteria to experimental compounds in the presence of a mixture of purified enzymes produced by the mammalian liver. In animals, ingested chemicals are routed to the liver, where they are broken down by detoxifying enzymes. Using a critical subset of detoxifying liver enzymes called the S9 extract, the Ames test mimics the biological defense processes that take place in the liver of animals exposed to chemical compounds. During enzymatic breakdown in the liver, numerous intermediate products can be produced, some of which may be mutagenic, even if the original compound was not. The purpose of the Ames test is to detect whether the original compound or any of its normal breakdown products are mutagens.

The Ames test most commonly uses strains of the bacterium *Salmonella typhimurium* that carry mutations affecting their ability to synthesize the amino acid histidine. These bacteria are designated *his⁻* to indicate their mutation prevents histidine synthesis. They will not grow unless they are provided with a medium that is supplemented with histidine. The Ames test is designed to identify the rate of *reversion mutations* (*his⁻* to *his⁺*) that restore the ability of bacteria to synthesize their own histidine, thus eliminating the need for histidine supplementation of the growth medium.

The Ames test uses multiple *his⁻* strains of *S. typhimurium,* each carrying different kinds of mutations of histidine-synthesizing genes. Some test strains carry base-pair substitution (transition and transversion) mutations; others carry frameshift mutations. The use of these different mutant strains allows detection of compounds that induce base-pair substitution mutations as well as of those that induce frameshift mutations, by comparing reversion rates in experimental bacterial cultures exposed to potential mutagens with spontaneous reversion rates in control bacterial cultures.

In each experimental culture, the S9 extract is added to mutant strains of *S. typhimurium* with base-pair substitution or frameshift mutations and each mixture is separately plated onto a medium lacking histidine (**Figure 12.16**). The test compound, in different concentrations, is added to a filter paper disk in the center of each test plate, and the plates are incubated. To determine the frequency of spontaneous reversion from *his⁻* to *his⁺*, control cultures, containing *his⁻* bacteria and S9 are plated onto medium lacking histidine and the plates are incubated. Bacteria on the control plates are not exposed to the test compound.

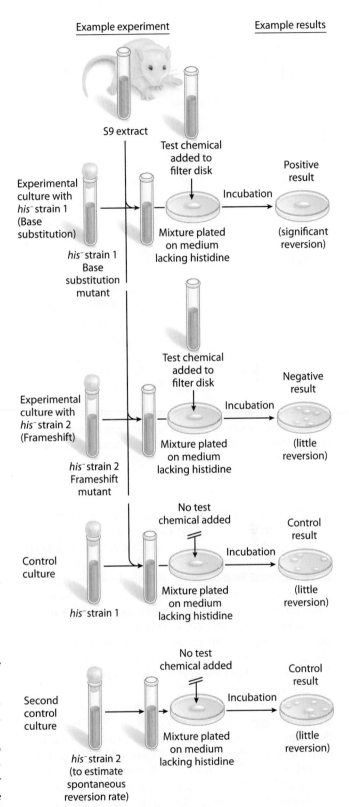

**Figure 12.16  The Ames test for potential mutagenicity of chemical compounds.**

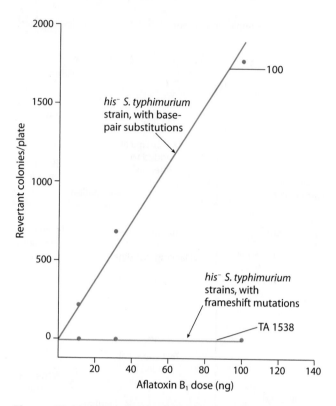

**Figure 12.17  Mutagenicity of aflatoxin B₁ determined by the Ames test.**  Aflatoxin B₁ induces a high rate of reversions in *his⁻* bacteria with base-pair substitution mutations (strain TA 100), but not in frameshift mutants (strain TA 1538).

## 12.5 Repair Systems Correct Some DNA Damage

The structural and informational integrity of DNA is under continuous assault from spontaneous chemical change and from various mutagens. Despite this ongoing challenge, organisms preserve the fidelity of their DNA by repairing most lesions that occur and leaving very few mutations to accrue. However, a species' survival depends on maintaining a delicate balance between mutation and repair. Too many mutations may doom an organism and ultimately affect survival of the species. On the other hand, too few mutations will limit the range of genetic variability and may hamper the species' ability to evolve.

Organisms have evolved multiple repair mechanisms, and often these are partially redundant with regard to the lesions they identify and repair. In broad terms, these damage repair processes fall into two categories: (1) those that directly repair DNA damage and restore it to its wild-type state; and (2) those that allow the organism to circumvent problems such as blocked DNA replication, which can occur when damage is not repaired.

### Direct Repair of DNA Damage

The most direct way to repair DNA lesions is to identify and then reverse DNA damage. Recall one such direct mechanism, proofreading by DNA polymerase (see Chapter 7), that identifies a base-pair mismatch, removes the erroneous DNA segment, and resynthesizes the sequence. Two additional repair systems carrying out direct repair of DNA damage are identified here.

**Repair of UV-Induced Photoproducts**  Pyrimidine dimers, the common photoproducts of UV-induced DNA damage, can be directly repaired by **photoreactive repair,** a DNA repair mechanism found in bacteria, single-celled eukaryotes, plants, and some animals (e.g., *Drosophila*), but not in humans. Photoreactivation utilizes the enzyme photolyase to break the bonds formed during pyrimidine dimerization. In Figure 12.18, a thymine dimer produced by UV irradiation is bound by photolyase. Visible light energy is absorbed by photolyase and is redirected to break the bonds forming the dimer. Photolyase is the product of the *phr* (photoreactive repair) gene of *E. coli. phr* gene mutations result in a substantial increase in UV-induced mutations in bacteria.

The results of an Ames test are interpreted by counting the number of growing colonies on each plate and comparing the numbers to one another and to the control plates. These data are used to determine whether a test compound is mutagenic and to develop a growth-response curve describing how the concentration of the test compound affects its mutagenicity. A positive result, indicating that the test compound is mutagenic, is indicated by a significant increase in the reversion rate in bacterial cells that have one kind of mutation over cells that have the other kind of mutation and over the spontaneous reversion rate.

In Figure 12.17, the dose-response curve reveals the strong mutagenic potential of aflatoxin B₁, a toxin produced by the fungus *Aspergillus* that grows on nuts and maize (corn). This powerful mutagen induces large numbers of base-pair substitution mutations in the *his⁻* strain designated TA 100 that contains a base-pair substitution and is highly sensitive to aflatoxin B₁. At all doses tested, the reversion rate for aflatoxin B₁ in base-pair substitution *his⁻* strains is elevated above that of the *his⁻* strain TA 1538, which contains a frameshift mutation. This result indicates that aflatoxin B₁ actively induces reversion of base-pair substitution mutants, but not of frameshift mutants. Genetic Analysis 12.2 guides you through an analysis of an Ames test of potential mutagens.

Three potentially hazardous compounds, A, B, and C, are assayed by the Ames test. Two strains of $his^-$ bacteria (1 and 2) are used. Auxotrophy in strain 1 is caused by a frameshift mutation and in strain 2 by the substitution of one base pair, resulting in a nonsense mutation. Each strain is treated with the different compounds. An S9 fraction (supernatant of solubilized rat liver enzymes) was added to each mixture of auxotrophic bacteria plus one of the compounds.

After treatment, the cells were plated on minimal medium. Control plates contain each of the two strains treated with S9 alone, without A, B, or C present. The table below shows the number of prototrophic colonies observed on the plates:

| Compound Tested | Strain 1 | Strain 2 |
|---|---|---|
| A | 904 | 6 |
| B | 5 | 4 |
| C | 3 | 680 |
| Control (no compound) | 6 | 3 |

a. Assess the growth results, and determine whether each compound is mutagenic or not.

b. Determine the type of mutation most likely induced by any mutagens.

| Solution Strategies | Solution Steps |
|---|---|
| **Evaluate** | |
| 1. Identify the topic this problem addresses, and explain the nature of the required answer. | 1. This problem concerns interpretation of the results of an Ames test of three compounds. Identify which if any of the compounds are mutagenic, and describe the nature of that mutagenicity. |
| 2. Identify the critical information given in the problem. | 2. The number of revertant colonies is given for each compound. A control result is also given. The cause of auxotrophy in each mutant strain is identified. |
| **Deduce** | |
| 3. Describe the meaning of growth results on the control plate. | 3. The control plates have had no test compound added. The growing colonies on these plates are spontaneous revertants for each of the auxotrophic tester strains. |
| 4. Deduce the meaning of growth results on each of the experimental plates. | 4. Compound A produces many revertants of strain 1, but no reversion over spontaneous levels in strain 2. Compound C generates many revertants of strain 2, but does not produce revertants at a rate greater than the control in strain 1. Compound B does not increase the reversion rate above the spontaneous level in either strain. |
| **Solve** | |
| 5. Identify the mutagenic compounds and justify your answer. | 5. Compounds A and C are mutagenic, but compound B is not. The large numbers of revertant colonies on the strain 1 test of compound A and the number of revertants on the strain 2 test of compound C identify these compounds as mutagens. Compound B does not show an increased rate of reversion relative to the background numbers on the control plates. |
| 6. Describe the nature of mutagenicity for each compound.<br><br>TIP: Base-pair substitution mutagens generally revert base-pair substitution auxotrophs, and frameshift mutagens revert frameshift auxotrophs. | 6. Compound A causes frameshift mutations by inducing a high rate of reversion of $His^-$ strain 1 auxotrophs. Compound C causes a high rate of reversions of strain 2 auxotrophs by inducing base-pair substitution reversions. |

For more practice, see Problems 26, 28, and 33.

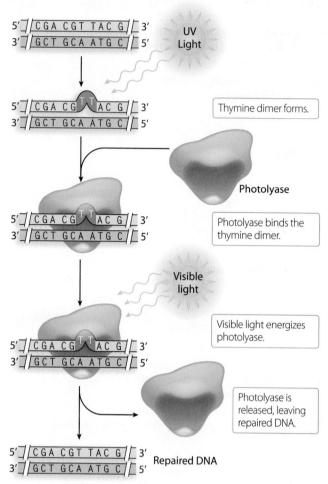

**Figure 12.18 Photoreactive repair.**

**Repair of Damage by Alkylating Agents** DNA damage by alkylating agents, such as EMS, which adds ethyl groups, or nitrosoguanidine, which adds methyl groups to nucleotide bases, is repaired by enzymes that remove the added chemical group and restore nucleotides to their normal form. For example, $O^6$-methylguanine, created when nitrosoguanidine methylates guanine, is converted back to guanine by the repair enzyme $O^6$-methylguanine methyltransferase. The enzyme has a single active site that pulls the methyl group off the nucleotide, converting the enzyme to a permanently inactive form. Each repair enzyme removes just a single methyl group. Consequently, exposure to a high dose of nitrosoguanidine can result in a methylation level that may overwhelm a cell's ability to correct the damage. Limited cell repair resources offer one explanation for the observation that some compounds produce mutations only when exposure is high.

## Nucleotide Excision and Replacement

If a modified nucleotide cannot be directly repaired, DNA's double-stranded structure can be exploited by use of the undamaged strand as the template, in a different approach to DNA repair. Multiple repair systems in bacteria and eukaryotes remove one or more nucleotides from a damaged strand and use complementary base pairing to replace excised nucleotides.

**UV Repair** UV-induced damage to DNA can be mended by **UV repair** (Figure 12.19)—excision of a short region of the strand containing the photoproduct, followed by synthesis of new DNA to replace the excised region. In *E. coli*, UV repair is carried out by the protein products of four UV repair genes called *uvr-A*, *uvr-B*, *uvr-C*, and *uvr-D* (Figure 12.20). Two molecules of UVR-A protein and one molecule of UVR-B protein bind on one strand of DNA

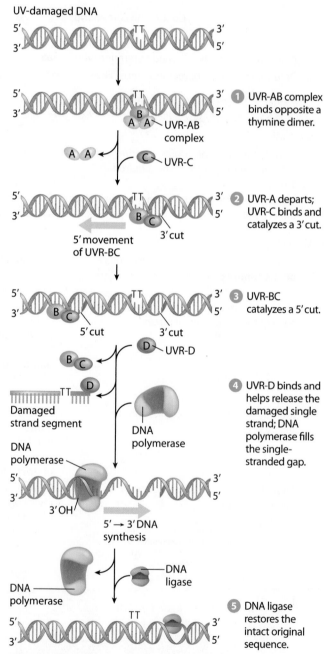

**Figure 12.19 Nucleotide excision repair.** UVR proteins remove UV-induced damage from DNA strands.

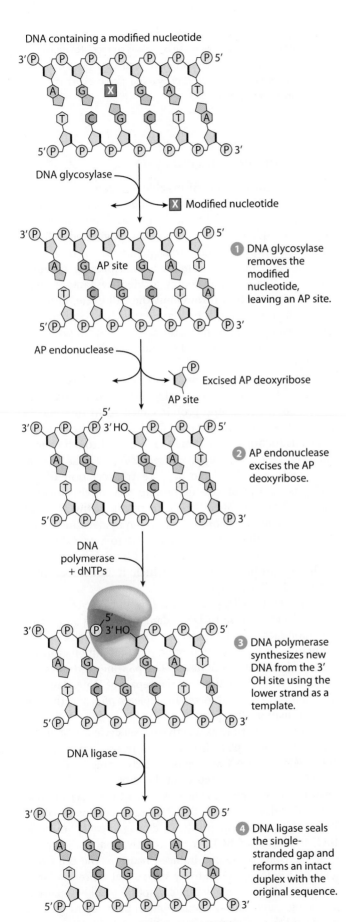

DNA containing a modified nucleotide

DNA glycosylase

X Modified nucleotide

❶ DNA glycosylase removes the modified nucleotide, leaving an AP site.

AP site

AP endonuclease

Excised AP deoxyribose

AP site

❷ AP endonuclease excises the AP deoxyribose.

DNA polymerase + dNTPs

❸ DNA polymerase synthesizes new DNA from the 3′ OH site using the lower strand as a template.

DNA ligase

❹ DNA ligase seals the single-stranded gap and reforms an intact duplex with the original sequence.

**Figure 12.20** **Nucleotide base excision repair.** DNA glycosylase and AP (apurinic) endonuclease remove chemically modified nucleotides from DNA.

opposite the site of the photoproduct. The two molecules of UVR-A dissociate from the strand, and a molecule of UVR-C joins UVR-B to form a UVR-BC complex. Each molecule of the UVR-BC complex cleaves a pair of phosphodiester bonds on the strand containing the photoproduct. UVR-B cleaves a bond about four or five nucleotides to the 3′ side of the photoproduct, and UVR-C cuts a bond about seven or eight nucleotides to the 5′ side of the lesion. The single-stranded fragment of approximately 12 nucleotides containing the photoproduct is released. UVR-D, which is a helicase, binds along with DNA pol I to the exposed 3′ OH. UVR-D unwinds DNA, and pol I uses the complementary strand to synthesize a replacement for the missing segment. In the final step of the excision repair process, DNA ligase binds to the region to seal the sugar-phosphate backbone.

**Nucleotide Base Excision Repair** A multistep repair process called **nucleotide excision repair** is initiated when enzymes known as DNA glycosylases recognize and remove a modified purine nucleotide base to produce apurinic (AP) sites (Figure 12.20). AP nuclease then removes the remainder of the apurinic nucleotide, after which the activity of DNA polymerase and DNA ligase fills the nucleotide gap and seals the sugar-phosphate backbone of the repaired strand.

## DNA Recombination Repair

DNA damage generated on a large scale by a high level of exposure to a mutagen can overwhelm the ability of repair mechanisms to correct all the resulting lesions. For example, *E. coli* exposed to a high dose of UV radiation may acquire so many photoproducts that many remain as DNA replication begins. Replication may then bypass DNA segments containing photoproducts on the template strand, because hydrogen bonds cannot form between a template nucleotide and one newly added to the synthesized strand if the template nucleotide is part of a photoproduct. Replication will reinitiate at the location of an adjacent RNA primer, but single-stranded gaps stretching over dozens of nucleotides will remain. **Recombination repair** can fill these single-stranded gaps by directing recombination of the single-stranded segment with the complementary strand using the sister chromatid as the repair template (**Figure 12.21**).

In *E. coli*, the recombinational enzyme RecA is the principal agent of recombination repair. RecA binds to the region of the single-stranded gap to form an active nucleoprotein filament. The unreplicated template strand next invades the newly synthesized duplex on the other side of the replication fork. The unreplicated template strand next invades is fully complementary to the opposite template strand, and RecA induces cleavage and transfer of the opposite template strand to fill the single-stranded gap on the unreplicated template strand. This action fills the single-stranded gap left by interrupted DNA replication.

Recombination repair does not repair the original DNA lesion. Rather, recombination repair overrides the interruption of DNA replication caused by the presence of DNA photoproducts on template strands and allows the completion of replication. Similar recombination repair mechanisms operate in eukaryotic genomes.

**Genetic Insight**  DNA repair processes correct DNA damage by directly repairing the lesion, by removing and replacing a segment of a damaged strand, or by filling gaps left over after incomplete replication.

## DNA Damage Signaling Systems

The biochemical mechanisms able to recognize the presence of DNA damage and mount a damage repair response are crucial to the health and survival of an organism. They consist of tightly regulated genetic processes involving numerous genes and proteins. In humans and other mammals, a multiprotein complex that acts as a genomic sentry to identify damage. This process is in action throughout the cell cycle, and it is especially important in regulating the $G_1$-to-S transition of the cell cycle, preventing the cell cycle from progressing to S phase until the cell has adequately repaired any mutations.

One important protein in this process is BRCA1, the first gene implicated in familial breast and ovarian cancer susceptibility. A second protein called ATM that plays a pivotal role in communicating DNA damage through signal transduction. DNA damage acquired through chemical or radiation exposure is sensed using ATM as a signal transduction molecule to activate transcription of the *p53* gene that produces the protein p53. By this mechanism, ATM activates the "p53 repair pathway" that controls cellular response to mutation by deciding either (1) to pause the cell cycle at the $G_1$-to-S transition to allow time for mutation repair or (2) to direct the cell to the apoptotic pathway, in which it undergoes programmed cell death (**Figure 12.22**).

In healthy cells, p53 level is low, but ATM increases its level in response to DNA damage. Acting as a transcription factor, p53 initiates $G_1$ arrest of the cell cycle by inducing synthesis of the protein p21 that inhibits formation of Cdk–cyclin complexes. The p53-induced pause in the cell cycle allows time for the repair of DNA damage. The completion of DNA repair depletes p53, and the cell cycle transitions to S phase.

At the same time as it activates cell cycle arrest at the $G_1$–S transition, p53 also activates transcription of the *BAX* gene. The BAX protein is a slowly acting inhibitor of the BCL2 protein that represses the apoptotic pathway. In healthy cells in which p53 level is low, *BAX* transcription is not initiated. As a result, BCL2 protein remains uninhibited and represses apoptosis. In damaged cells, however, BAX inhibition of BCL2 leads to activation of the apoptotic pathway. If the p53-induced pause in the cell cycle goes on too long, the pathway senses that there is a large amount of DNA damage that cannot be quickly repaired. The long pause allows the apoptotic pathway to go forward, and the cell undergoes programmed cell death.

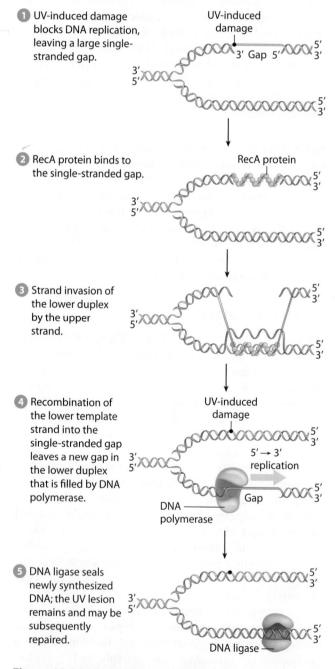

1. UV-induced damage blocks DNA replication, leaving a large single-stranded gap.

2. RecA protein binds to the single-stranded gap.

3. Strand invasion of the lower duplex by the upper strand.

4. Recombination of the lower template strand into the single-stranded gap leaves a new gap in the lower duplex that is filled by DNA polymerase.

5. DNA ligase seals newly synthesized DNA; the UV lesion remains and may be subsequently repaired.

**Figure 12.21  Recombination repair.**

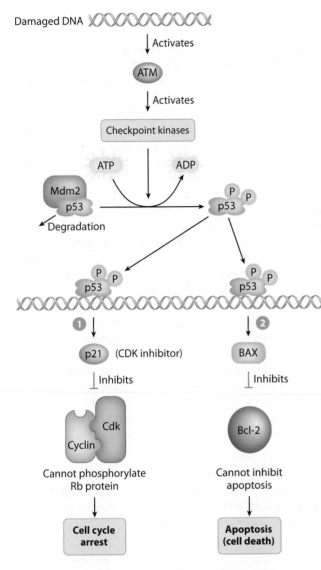

**Figure 12.22** The *p53* DNA damage repair pathway.

## DNA Damage Repair Disorders

DNA damage repair disorders, resulting from mutations in genes that participate in the repair of DNA damage or in the signaling or initiating of damage repair, cause an organism to be highly sensitive to chemical mutagens and to radiation. Such disorders greatly increase the organism's susceptibility to cancers caused by mutagen exposure. We return to this theme in the Case Study that concludes this chapter, where we discuss a connection between mutations of *p53* and the occurrence of cancer, and the role of transmission of *p53* mutation in the human familial cancer syndrome known as Li-Fraumeni syndrome (OMIM 151623). **Table 12.5** lists some human mutation repair disorders that are associated with significantly elevated risks of specific types of cancer.

---

Genetic Insight  Protein complexes that scan the genome for DNA damage, and the accompanying processes that repair or circumvent damage to DNA, are integral to cell survival and to the well-being of organisms. Defects in the damage survey or damage repair systems can be acquired through mutation or can occasionally be transmitted as mutations. Specific disorders can result from mutations of the genes that contribute to damage survey or repair.

---

## 12.6 Proteins Control Translesion DNA Synthesis and the Repair of Double-Strand Breaks

The repair mechanisms described to this point are able to repair DNA damage or overcome its effects without introducing additional errors. Not all damage repair mechanisms are error free, however, and a few are a significant

| Table 12.5 | Selected Human Mutation Repair Disorders |
|---|---|
| **Disorder and OMIM Number** | **Description** |
| Ataxia telangiectasia (208900) | Mutation of the *ATM* gene and absence of ATM protein. Poor coordination (ataxia), red marks on the face (telangiectasia), increased sensitivity to X-rays and other radiation, high cancer risk. |
| Breast–ovarian cancer (604370) | Mutation of *BRCA1*. Defective DNA repair and increased susceptibility to breast and ovarian cancer. |
| Li-Fraumeni syndrome (151623) | Mutation of *p53* and defective p53 pathway. High cancer risk. |
| Nonpolyposis colon cancer (120435) | Defective base-pair mismatch repair caused by mutation of any one of seven different genes. High risk of colon cancer. |
| Trichothiodystrophy (601675) | Mutations of any one of five gene mutations causing increased sensitivity to oxidative damage. Mental retardation, dwarfism, skin and hair abnormalities, and increased cancer risk. |
| Xeroderma pigmentosum (278700) | Defective excision repair resulting from the mutation of any one of seven UV damage repair genes. Extreme sensitivity to UV-induced damage and high skin cancer risk. |

source of DNA changes that can lead to mutation. It may not immediately be obvious why repair mechanisms that are prone to introducing additional errors were able to evolve. After all, the point of DNA repair systems is just that—to repair damage so as to maintain the integrity of the genome. This conundrum is resolved by the fact that error-prone repair mechanisms are activated only in instances of widespread DNA damage that would otherwise prevent the completion of DNA replication and might cause cell death.

## Translesion DNA Synthesis

The previous section describes recombination repair as one response of *E. coli* to widespread DNA damage. There is a second repair mechanism that can be activated in *E. coli* in response to massive DNA damage. This repair system, called SOS repair, has been known for decades but has only recently been understood at the molecular level. The system takes its name from the maritime phrase "save our ship," used when sinking was imminent. In the past, SOS repair was described as a last-ditch effort on the part of a heavily damaged bacterial cell to replicate its DNA and divide before succumbing to DNA damage. Recent research demonstrates that SOS repair is accomplished by activating specialized DNA polymerases in a process known as translesion DNA synthesis. This short-lived process allows DNA replication by alternative polymerases across lesions that block the action of DNA polymerase III (pol III), the main DNA-replicating polymerase in *E. coli*.

Translesion DNA synthesis is performed by **translesion DNA polymerases,** also called **bypass polymerases.** Bypass polymerases operate differently from pol III in several respects. First, bypass polymerases are able to replicate across DNA lesions that stall pol III. This ability is accounted for by the second difference distinguishing bypass polymerases, the absence of proofreading. In other words, bypass polymerases do not have 3'-to-5' exonuclease capacity and are unable to remove newly added nucleotides that fail to hydrogen bond with the template strand nucleotide. Due to their lack of proofreading capability, the third distinguishing feature of bypass polymerases is that they are prone to making replication errors. Finally, bypass polymerases synthesize only short segments of DNA; they fall off the template strand after synthesizing a small number of nucleotides. From these distinguishing features, molecular biologists conclude that bypass polymerases are used to complete replication that would otherwise be blocked. This comes at the price, however, of potentially introducing new mutations.

The SOS system in *E. coli* operates through a specialized bypass polymerase identified as polymerase V ("polymerase five"), or pol V. When pol III stalls at damaged DNA, RecA protein coats the template strand ahead of the lesion that is already bound by single-stranded binding protein (SSB). Recall that SSB coats the single DNA strands separated ahead of the replication fork (see Figure 7.21). The RecA protein in the DNA–RecA–SSB complex is an active form that also activates transcription of several genes, including pol V. Pol V displaces polymerase III, synthesizes a short portion of the daughter strand across the DNA lesion, and is then replaced by pol III, which resumes its normal replication activity.

Eukaryotic genomes utilize a similar mechanism for translesion DNA synthesis. In eukaryotes, however, bypass polymerases are always present in cells, so the system of regulating their access to DNA is quite different. The regulatory mechanism guiding the choice of polymerase decides which polymerase binds to PCNA, the eukaryotic sliding clamp. When eukaryotic replication stalls at a DNA lesion, a protein called Rad6 that is always present at the replication fork adds a ubiquitin (Ub) group to PCNA. This process, called ubiquitination, normally targets a protein for destruction. On PCNA, however, ubiquitination merely causes an alteration of conformation, giving the bypass polymerase a strong affinity for ubiquinated PCNA. In this process, bypass polymerase displaces normal DNA polymerase and carries out translesion synthesis of DNA. As in the SOS system, the use of bypass polymerases in eukaryotic cells is error prone because the enzyme lacks proofreading capability.

---

**Genetic Insight** Translesion DNA synthesis is an error-prone system that utilizes bypass polymerases to replicate short segments of daughter strands whose normal replication is blocked by lesions in template strand DNA.

---

## Double-Strand Break Repair

A common feature of the DNA repair mechanisms we have examined is the use of DNA polymerase and a template strand of DNA to guide the repair, replacement, or synthesis of DNA. These repair systems are effective as long as one strand of DNA is intact and can serve as a template. But what happens if *both strands* of DNA are damaged in a manner that does not provide a template strand for strand repair? Such damage is a frequent consequence of exposure to X-rays and certain types of oxygen radicals. The damage caused by these agents breaks both strands of DNA, leaving lesions that are known as double-strand breaks. Because they can cause chromosome instability and incomplete replication of the genome, double-strand breaks are potentially lethal to cells and elevate the risk of cancer and the chance of chromosome structural mutations.

To protect organisms from the unpleasant consequences of double-strand breaks, two mechanisms have evolved to carry out **double-strand break repair.** The first is an error-prone repair process known as *nonhomologous* *end joining* that repairs double-strand breaks occurring before DNA replication. The second is an error-free process called *synthesis-dependent strand annealing* that repairs double-strand breaks occurring after the completion of DNA replication.

**Nonhomologous End Joining**  If a double-stranded break damages a eukaryotic chromosome during $G_1$ of the cell cycle, replication of the damaged chromosome is blocked. Considering that DNA polymerases, even bypass polymerases, require a template strand to direct synthesis of a daughter strand, it is clear that a double-strand break is incompatible with the completion of replication. One repair alternative that allows cells to reacquire their capacity to fully replicate their genome is **nonhomologous end joining (NHEJ),** although its four-step process for repairing double-strand breaks inevitably leads to mutation (**Figure 12.23**).

In the first step, double-strand breaks are recognized by the proteins Ku70 and Ku80 that form heterodimers and attach to the broken ends of the DNA duplex. The heterodimers stabilize the broken ends against degradation by nucleases. Next, nucleases trim back (resect) the free ends of each broken strand. Resection leaves blunt ends on each side of the break. Finally, the blunt ends are ligated by a specialized ligase called ligase IV ("ligase four").

Completion of NHEJ produces an intact DNA duplex and allows replication across the repaired region in the upcoming replication cycle, but the repair is often imperfect because resection removes nucleotides that

cannot be replaced. For this reason, NHEJ is error prone. Yet, as potentially damaging as this process is, its outcome is superior to the alternatives suffered by cells that are unable to repair double-strand breaks, and it prevents more extensive loss from degradation of unprotected ends. Mutations can be generated, however, when nucleotides are lost by transcribed genes.

**Synthesis-Dependent Strand Annealing**  In eukaryotes, once DNA replication is complete, each chromosome is composed of two identical sister chromatids. We saw earlier that recombination repair takes advantage of the presence of sister chromatids to fill single-stranded gaps left when replication is blocked by UV-induced damage to the template strand. Double-stranded breaks present a different type of challenge to post-replication cells, but the presence of identical sister chromatids can be exploited in an error-free repair process known as **synthesis-dependent strand annealing (SDSA).**

As shown in **Figure 12.24**, a double-stranded break (DSB) affects one sister chromatid; the other chromatid

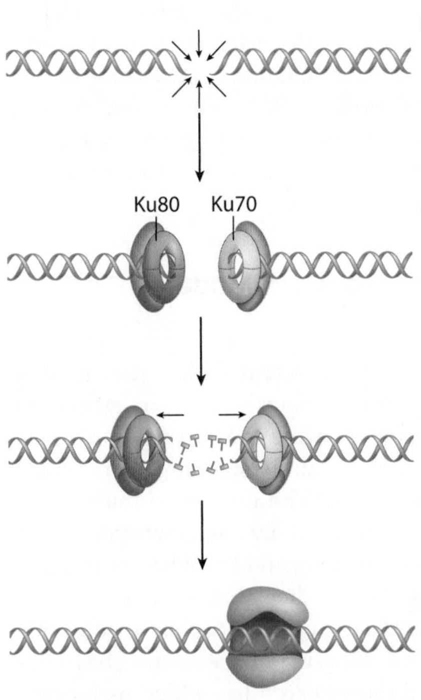

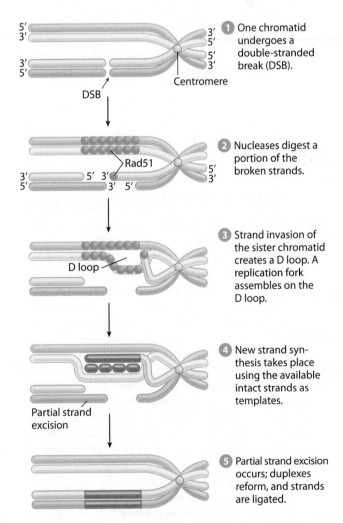

**1** X-ray or oxidative damage produces double-strand break in DNA.

**2** Ku80–Ku70 protein complex binds DNA ends.

Ku80  Ku70

**3** Ends are trimmed, resulting in a loss of nucleotides.

**4** DNA ligase IV ligates blunt ends to reform an intact duplex.

**Figure 12.23  Nonhomologous end joining.** NHEJ is an error-prone system that rejoins DNA strands following a double-stranded break.

**1** One chromatid undergoes a double-stranded break (DSB).

Centromere

DSB

**2** Nucleases digest a portion of the broken strands.

Rad51

**3** Strand invasion of the sister chromatid creates a D loop. A replication fork assembles on the D loop.

D loop

**4** New strand synthesis takes place using the available intact strands as templates.

Partial strand excision

**5** Partial strand excision occurs; duplexes reform, and strands are ligated.

**Figure 12.24  Synthesis-dependent strand annealing (SDSA).**

is intact. SDSA begins with trimming of one of the broken strands. This is followed by attachment of the protein Rad51, a homolog of the bacterial protein RecA, to the broken region to form a nucleoprotein filament. In a process reminiscent of the action of RecA in facilitating recombination repair, Rad51 binds to the strands and facilitates the invasion of the intact chromatid by the resected end of a strand from the sister chromatid. This **strand invasion** process displaces one strand of the duplex and creates a **displacement (D) loop.** DNA replication within the D loop synthesizes new DNA strands from intact template strands. The sister chromatids are reformed by dissociation and annealing of the nascent strand to the other side of the break. By accomplishing the removal of DNA in the immediate vicinity of a double-stranded break and the replacement of the excised DNA with a duplex identical to that in the sister chromatid, SDSA carries out error-free repair of double-stranded breaks.

---

**Genetic Insight**  Nonhomologous end joining is an error-prone system for the repair of double-strand DNA breaks that occur before S phase. Synthesis-dependent strand annealing takes place after completion of the S phase and utilizes the identical sister chromatid as a source of template strand sequence to synthesize a replacement duplex in an error-free manner.

---

## 12.7 Meiotic Recombination Is Controlled by Programmed Double-Strand Breaks

Mutation is the ultimate source of new inherited variation in species, but by necessity it is a process that only slowly accumulates variation. In sexually reproducing organisms, there is another much more potent source of inherited variation—meiotic recombination generates new combinations of the alleles that already exist in an organism (see Figure 6.5). This additional diversity increases the opportunities for survival of offspring.

Meiotic recombination is a homologous process that takes place early in prophase I of meiosis. Biologists and geneticists have interpreted and understood the genetic consequences of recombination for more than a century, but an understanding of meiotic recombination at the molecular level has been more elusive. Though initially discovered in the early 20th century through the work of Thomas Hunt Morgan and his colleagues, who detected recombinant chromosomes in gametes, homologous recombination could not be studied on a molecular level until the 1950s. In the decade following the determination of the double helical structure of DNA, numerous researchers attempted to construct likely models of homologous recombination. In more than 60 years since work began in earnest to describe the molecular mechanism of homologous recombination, many models have been proposed, and modification of models has been continuous. Molecular biologists continue to adjust models of recombination to match observations, but two salient points are now clear. First, meiotic recombination is a genetically controlled process initiated by enzymes that produce double-stranded DNA breaks; and second, the molecular mechanism of homologous recombination is closely related to the processes that repair double-stranded DNA breaks.

### The Holliday Model

The first viable molecular model of meiotic recombination was proposed by Robin Holliday in 1964. Known as the **Holliday model,** it offered a plausible scheme for meiotic recombination by hypothesizing that spontaneously generated single-stranded breaks in one chromatid led to invasion of a nonsister chromatid of the homologous chromosome. Holliday's scheme for breaking and rejoining DNA strands suggested that some encounters between homologous chromosomes would produce crossover chromosomes whereas others would not.

The original Holliday model ultimately proved to be too simplistic and has been superseded by more accurate models of meiotic recombination. The more recent models rely on some of the features of the Holliday model but incorporate new knowledge and steps. Perhaps the most important features distinguishing the current model of meiotic recombination from the original Holliday model are, first, that meiotic recombination is now known to be initiated by *double-stranded DNA breaks* and, second, that the double-stranded breaks initiating meiotic recombination are generated in a programmed manner by the activity of a specialized enzyme.

### The Double-Stranded Break Model of Meiotic Recombination

Experimental evidence inconsistent with earlier models of homologous recombination, or identifying events not predicted by those models, were the driving forces that shaped the evolution of understanding concerning homologous recombination. The outline of the current model was proposed in 1983 by Jack Szostak, Terry Orr-Weaver, Rodney Rothstein, and Franklin Stahl. Their model was the first to predict that double-stranded breaks were the foundation of homologous recombination. The accumulated experimental evidence has confirmed this view, and the research has added major new details to the original proposal by Szostak and his colleagues.

In the current model, homologous recombination is initiated by the protein Spo11 ("Spo eleven") that was first discovered in yeast and is now known to exist in homologous form in all eukaryotes (**Foundation Figure 12.25**, ❶). Spo11 is a dimeric protein that generates slightly asymmetric double-strand cuts in one chromatid. The protein Mrx and Exo1 associate with Spo11, and after Spo11 degrades, Mrx resects the single strands ❷. Two RecA homolog proteins, Rad5 and Dmc1 join at the trimmed region ❸. This protein complex helps form a strand-exchange assemblage, facilitating strand invasion and formation of a D loop ❹, ❺. The invading strand pairs with the complementary strand in the D loop. Outside the D loop, the two strands that appear to cross over one another form a **Holliday junction,** an interim structure proposed in the original Holliday model. Notice that there is also a **heteroduplex region,** containing two complementary strands of DNA that originated in different homologs. Also identified as **heteroduplex DNA,** these regions are a molecular signature of homologous recombination. Because the two strands of the heteroduplex DNA originate in different homologs, there may be mismatched base pairs between them. In other words, if heterozygosity is present in the DNA sequences forming a heteroduplex region, one or more base pairs will be mismatched in the heteroduplex DNA. We discuss the implications of this situation in the following section.

Extension of the invading strand and DNA synthesis within the broken strand are guided by intact template strands ❻, ❼. At this point, a second heteroduplex region has formed. The 3′ end of the invading strand next connects with the 5′ end of a strand segment that was initially part of the invading strand ❽, to form a second Holliday junction. Now the nonsister chromatids of the recombining chromosomes are interconnected to one another by the presence of **double Holliday junctions (DHJs):** The recombining chromosomes contain DHJs and two heteroduplex regions.

### Holliday Junction Resolution

The recombinational steps just described take place in prophase I of meiosis, and any connections between homologous chromosomes must be resolved in prophase, long before the homologs attach to spindle fibers in metaphase. Cutting and reconnecting single strands of interconnected homologous chromosomes resolves crossing over.

Two resolution patterns, same sense resolution and opposite sense resolution (❶ and ❷ in Figure 12.25), complete crossing over and disengage homologs so they can be separated anaphase. Same sense resolution involves either two **north-south (NS) resolution** cuts or two **east-west (EW) resolution** cuts of DNA strands to separate the homologs (see Foundation Figure 12.25). When the connection between homologs is resolved by

two NS or EW cuts, the flanking markers ($A_1$ and $B_1$ and $A_2$ and $B_2$) *do not recombine.* As a consequence, recombination of those genes is not produced, although heteroduplex regions are present. This resolution occurs only rarely. Far more common is resolution in which one Holliday junction region is resolved by a NS cut and the other by an EW cut. The resulting chromosomes are recombinant and carry $A_1$ and $B_2$ or $A_2$ and $B_1$. These recombinations are detectable among progeny where they are counted as recombinants between the genes.

---

**Genetic Insight** Homologous recombination is a genetically programmed process initiated by the protein Spo11 generating double-strand DNA breaks. Recombining homologs form DHJs and acquire heteroduplex DNA through their interaction. Resolution of DHJ connections may or may not result in genetic evidence of recombination.

---

## 12.8 Gene Conversion Is Directed Mismatch Repair in Heteroduplex DNA

Our final topic in this chapter is **gene conversion,** a process of so-called directed DNA sequence change that occurs by base-pair mismatch repair within heteroduplex DNA. These base-pair mismatches can occur when DNA sequence is heterozygous in a heteroduplex region created during meiotic recombination. In gene conversion, the "directed" change is base-pair mismatch repair that switches the nucleotide sequence of one allele to that of another allele that is already present because the organism is heterozygous in the portion of the genome where heteroduplex DNA forms. In contrast to mutation that can change one allele into any other allele, gene conversion can only switch one allele to another allele already present in a heterozygous genotype of the other allele in the heterozygous region.

Gene conversion is most readily detected in fungi that form an ascus, a sack of haploid spores that are the products of meiotic division. For example, we identified that for fungi with the genotype $a^+a$, the ratio of these alleles in spores in an eight-cell ascus is expected to be equal (4:4) (see Figure 5.23). Gene conversion changes that ratio by switching one or more alleles from one form to another: either $a^+$ to $a$, or $a$ to $a^+$. The result is an **aberrant ratio** of spores in an eight-cell ascus, commonly 5:3 or 6:2 instead of 4:4. Since gene conversion is strictly limited to conversions from one allele to the alternative form in a heterozygous genotype, it is distinct from mutation, in which an allele can be altered to almost an infinite variety of forms. Similarly, in organisms producing a four-cell ascus, a 2:2 ratio is expected for a heterozygote, and any other ratio is an aberrant ratio.

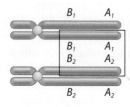

Meiotic recombination diagrammed between these nonsister chromatids of homologous chromosomes

**1** Spo11 creates double-strand break in one DNA duplex.

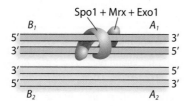

Spo1 + Mrx + Exo1

**2** Enzymatic digestion 5′→ 3′ by Mrx creates single-stranded segments.

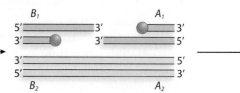

**5** Strand invasion creates one D loop and the first heteroduplex region.

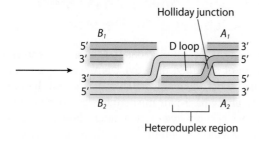

Holliday junction

D loop

Heteroduplex region

**6** Strand extension by DNA polymerase displaces D loop DNA, which pairs with complementary single-stranded DNA to form the second heteroduplex region.

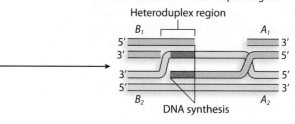

Heteroduplex region

DNA synthesis

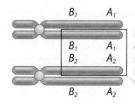

Meiotic recombination diagrammed between these nonsister chromatids of homologous chromosomes

**1** Same Sense Resolution

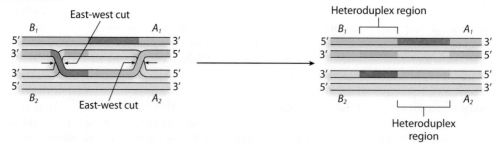

East-west cut

East-west cut

Heteroduplex region

Heteroduplex region

Same sense resolution produces offset heteroduplex regions but no recombination of flanking genes. This form of resolution occurs infrequently.

**3** Dmc1 and Rad51 assemble strand-exchange nucleoprotein filaments.

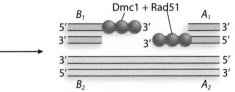

**4** The strand-exchange filaments promote strand invasion.

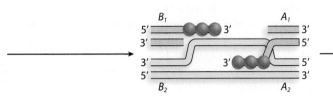

**7** Strand extension and ligation fills the single-stranded gap in the strand paired with D loop DNA.

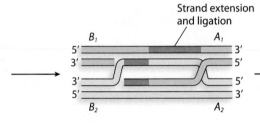

**8** Double Holliday junctions form after the nick is sealed; chromatids contain offset heteroduplexes.

Heteroduplex region

Holliday junction

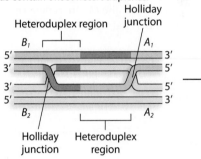

Holliday junction    Heteroduplex region

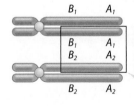

Meiotic recombination diagrammed between these nonsister chromatids of homologous chromosomes

**2** Opposite Sense Resolution

North-south cut

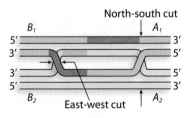

East-west cut

Heteroduplex region

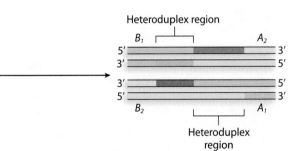

Heteroduplex region

Opposite sense resolution is very common and generates recombination of flanking genes and creates offset heteroduplex regions.

Early in the process of constructing gene conversion models and meiotic recombination models, biologists realized that gene conversion could occur only if heteroduplex DNA formed during crossing over. Thus, one goal in the development of contemporary models of meiotic recombination was to account for the formation of heteroduplex DNA followed by gene conversion in four-cell and in eight-cell asci. **Figure 12.26** illustrates the formation of heteroduplex DNA for alleles $A_1$ and $A_2$ that differ by substitution of one base pair. Allele $A_1$ carries a C-G base pair at the differing position, and $A_2$ carries an A-T base pair. Mismatches between G and A, and between C and T, are highlighted in the heteroduplex regions.

In a four-celled ascus, the repair of base-pair mismatches in heteroduplex DNA results in three aberrant ratios or patterns of spores (**Figure 12.27**). In repair option 1, both mismatches repair by converting the sequence to that of $A_2$. Conversely, in repair option 2, both mismatches repair to produce $A_1$. In each case, gene conversion has taken place, and the resulting asci contain an aberrant 3:1 ratio of alleles. In repair option 3, the pattern of mismatch repair produces an ascus with an

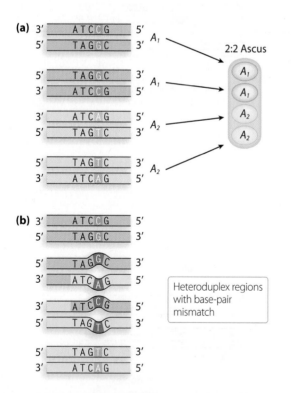

**(a)** A segment of allele $A_1$ contains a C-G base pair, whereas allele $A_2$ contains an A-T base pair at the same location. Segregation produces a 2:2 ascus. **(b)** Crossover between homologous chromosomes generates heteroduplex DNA containing G-A and C-T base-pair mismatches (in red) between the otherwise complementary strands.

**Figure 12.26 Heteroduplex DNA.**

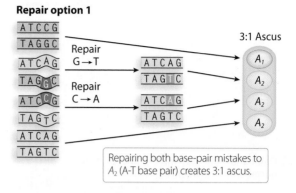

**Repair option 1**

Repairing both base-pair mistakes to $A_2$ (A-T base pair) creates 3:1 ascus.

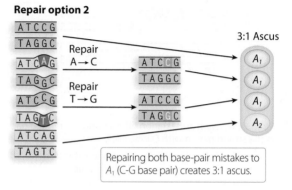

**Repair option 2**

Repairing both base-pair mistakes to $A_1$ (C-G base pair) creates 3:1 ascus.

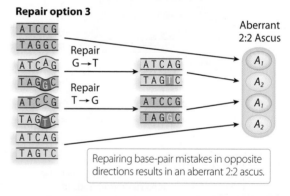

**Repair option 3**

Repairing base-pair mistakes in opposite directions results in an aberrant 2:2 ascus.

**Figure 12.27 Example mismatch repair and gene conversion patterns in a four-celled ascus.**

aberrant 2:2 ratio in which $A_1$ and $A_2$ are in alternating order instead of the like alleles being side by side, as expected normally.

The pattern of mismatch repair also determines the aberrant ratios in the eight-celled ascus by gene conversion. **Figure 12.28** shows three options for the repair of base-pair mismatches. In option 1, both mismatch repairs favor a single allele ($A_1$ in this case) and produce an ascus containing an aberrant 6:2 ratio. A similar aberrant 6:2 ratio producing an ascus containing 6 $A_2$ and 2 $A_1$ gametes occurs if both mismatches are repaired in favor of $A_2$ rather than $A_1$. In repair option 2, just one rather than both base-pair mismatches are repaired before the DNA replication cycle, resulting in an ascus containing

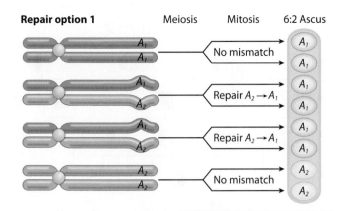

**Repair option 1** | Meiosis | Mitosis | 6:2 Ascus

Repairing both base-pair mistakes from $A_2 \to A_1$ creates a 6:2 ascus.

**Figure 12.28  Mismatch repair and gene conversion in an eight-celled ascus.**

**Repair option 2** | Meiosis | Mitosis | 5:3 Ascus

Repairing of one mistake but no repair of the other creates a 5:3 ratio.

**Repair option 3** | Meiosis | Mitosis | Aberrant 4:4 Ascus

No repair of base-pair mistakes in the heteroduplex region creates an aberrant 4:4 ratio.

an aberrant 5:3 ratio. Two different aberrant 5:3 ratios, 5 $A_1$:3 $A_2$ and 5 $A_2$:3 $A_1$, are possible, depending on the favored allele in the single mismatch repair. In repair option 3, no mismatch repair takes place. The spores are arrayed in a 3:1:1:3 pattern, a distribution called an aberrant 4:4 ratio.

**Genetic Insight**  Gene conversion is a directed process of base-pair mismatch repair in heteroduplex DNA that converts one allele to the alternative form in a heterozygous organism. Aberrant ratios of haploid spores in four-cell or eight-cell asci are the result of gene conversion.

## CASE STUDY

### Li-Fraumeni Syndrome Is Caused by Inheritance of Mutations of *p53*

Numerous studies of human cancers identify *p53* as the most commonly mutated gene in cancer cells. From the pivotal role *p53* and its protein product play in cells, it is easy to see why cells lacking *p53* function are abnormal. In the absence of

functional p53 protein, DNA damage is undetected and cells progress through $G_1$ of the cell cycle to S phase with the DNA damage present. Similarly, homozygous inactivation of *p53* interferes with the initiation of apoptosis in cases where cells have high levels of damage. By itself, homozygous mutation of *p53* does not cause cancer. Other mutations must be present to cause the rapid cell proliferation and other abnormalities that characterize cancer. Still, homozygous *p53* mutation can play a pivotal role in the accumulation of additional mutations that lead to cancer development.

In 1969, Frederick Li and Joseph Fraumeni encountered a family in which cancer ravaged each generation (**Figure 12.29**). This family stood out for its pattern of cancer that was consistent with an autosomal dominant mode of transmission and because many of the cancer cases occurred decades earlier than are typical for these types of cancer in the general population (i.e., breast cancers appeared in the 30s in affected family members as opposed to the 60s in members of the general population). Both of these features are hallmarks of cancer-prone families in which an inherited germ-line mutation increases individual susceptibility to cancer. Interestingly, however, unlike most cancer-prone families, in which one or two types of cancers predominate, this family studied by Li and Fraumeni had many types of cancer, including soft tissue sarcomas, breast cancers, brain cancer, osteosarcoma, adrenocortical cancer, and leukemia. After Li and Fraumeni's description, other families with similar patterns of mixed cancers were identified. This inherited cancer-prone condition was designated Li-Fraumeni syndrome 1 (LFS1; OMIM 151623). Study of LFS1 sparked revolutionary investigations of cancer biology and genetics that identified several genes that are frequently mutated in cancer cells, as well as investigations of the inheritance of mutations that increase cancer susceptibility.

In 1997, the first evidence of the molecular defect in LFS1 emerged with the identification of abnormalities of the *p53* gene in approximately 70% of FRS1 family members with cancer. The mutations were discovered in germ-line cells, meaning that one parent passed a mutant copy of *p53* in sperm or egg. At conception, the fertilized eggs were heterozygous, and as they developed, all cells carried one mutated and one wild-type copy of *p53*. Individual cells of mutation carriers become homozygous for *p53* mutation by the occurrence of a

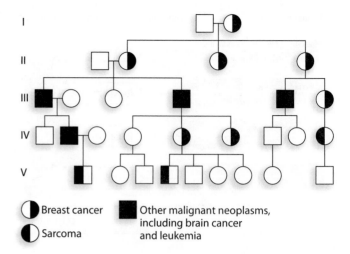

Breast cancer

Sarcoma

Other malignant neoplasms, including brain cancer and leukemia

**Figure 12.29 Li-Fraumeni syndrome.** Inherited mutations of *p53* greatly increase susceptibility to sarcoma, breast cancer, brain cancer, leukemia, and other cancers.

somatic mutation that alters the wild-type allele. The resulting homozygous mutant cells do not produce normal p53 protein, and they are unable to properly regulate the cell cycle or entry into apoptosis. Cancers develop in individuals without functioning *p53* through the accumulation of somatic mutations of other genes. The specific types of cancer that develop depend on which genes are mutated and on the tissues or cell types in which the mutations occur.

Following identification of the role of germ-line *p53* mutations in LFS1, inherited mutations of other DNA repair genes have been identified as increasing the susceptibility to certain cancers in families. For example, mutations of *BRCA1* and *BRCA2* that interact with *p53* can increase susceptibility to breast and ovarian cancer. Other inherited mutations of DNA repair genes that increase susceptibility to cancers include the disorders listed in Table 12.5.

---

SUMMARY  *For activities, animations, and review quizzes, go to the study area at www.masteringgenetics.com.*

## 12.1 Mutations Are Rare and Occur at Random

▌ Mutation is a heritable alteration in DNA sequence.

▌ Mutations result from damage done to DNA.

▌ Mutation frequencies are low in all organisms.

▌ Mutational hotspots are genes or regions where mutations occur much more often than average.

▌ The fluctuation test by Luria and Delbrück demonstrated that mutations occur at random.

## 12.2 Gene Mutations Modify DNA Sequence

▌ Base-pair substitution mutations can be either transitions or transversions.

▌ Base-pair substitutions can change one amino acid of the polypeptide, can create a new stop codon, or can leave the polypeptide unchanged.

▌ Frameshift mutations result from the insertion or deletion of one or more base pairs that shift the mRNA reading frame during translation.

▌ Regulatory mutations alter gene transcription or pre-mRNA splicing.

▌ Forward mutation alters a wild-type allele to mutant form, and reversion changes a mutant back to wild-type or near wild-type form.

## 12.3 Gene Mutations May Arise from Spontaneous Events

▌ DNA replication errors can substitute base pairs, and strand slippage can modify the number of repeats of a DNA sequence.

▌ Tautomeric shifts of nucleotide base structure can induce spontaneous base-pair substitution mutations.

■ Multiple spontaneous changes in nucleotide structure can result in mutation of DNA sequence by base-pair mismatching.

## 12.4 Mutations May Be Induced by Chemicals or Ionizing Radiation

■ Mutagenic chemicals interact in characteristic reactions with DNA nucleotides and generate specific mutations.

■ Chemical compounds may create mutations by acting as nucleotide base analogs, adding or removing side groups from nucleotides, or intercalating into DNA.

■ Energy in the ultraviolet range and higher (shorter in wavelength) is mutagenic. Ultraviolet radiation induces the formation of photoproducts that lead to base-pair substitution mutations.

■ The Ames test identifies mutagenic chemical compounds by testing for increased reversion rates in auxotrophic bacteria exposed to a test compound in the presence of detoxifying enzymes from the eukaryotic liver.

## 12.5 Repair Systems Correct Some DNA Damage

■ Direct repair of DNA lesions removes damaged nucleotides and prevents mutation.

■ Mismatched DNA nucleotides, photoproducts induced by UV radiation, and modified nucleotide side chains are removed by direct repair.

■ Excision repair, UV repair, and methyl-directed repair mechanisms remove segments of DNA single strands containing damaged nucleotides and direct new synthesis to fill the resulting single-stranded gap.

■ Genetically controlled systems monitor the genome and regulate DNA repair.

## 12.6 Proteins Control Translesion DNA Synthesis and the Repair of Double-Strand Breaks

■ SOS repair, controlled by the RecA protein, is a specialized process activated during replication in bacteria in response to widespread DNA damage.

■ Translesion DNA synthesis uses bypass polymerases to complete replication when damage is present.

■ Nonhomologous end joining repairs double-strand DNA breaks occurring before DNA replication.

■ Synthesis-dependent strand annealing repairs double-strand breaks occurring after the completion of replication.

## 12.7 Meiotic Recombination Is Controlled by Programmed Double-Strand Breaks

■ Meiotic recombination is a programmed process initiated by an enzyme that generates double-strand breaks.

■ Strand invasion and new DNA synthesis form heteroduplex DNA in both recombination participants.

■ Heteroduplex DNA contains base-pair mismatches if DNA sequences are heterozygous.

■ DNA strands forming double Holliday junctions are cut and rejoined to separate homologs before their separation in meiosis.

■ Resolution of double Holliday junctions generates heteroduplex DNA and can produce recombinant or nonrecombinant chromosomes.

## 12.8 Gene Conversion Is Directed Mismatch Repair in Heteroduplex DNA

■ Gene conversion occurs by the repair of base-pair mismatches in heteroduplex DNA.

■ Gene conversion in a four-celled or eight-celled ascus generates aberrant ratios of spores that differ from the expected 2:2 or 4:4 ratios.

## KEYWORDS

6-4 photoproduct *(p. 399)*
aberrant ratio *(p. 411)*
Ames test *(p. 401)*
apurinic site *(p. 395)*
base analog *(p. 396)*
base-pair substitution mutation *(p. 387)*
bulky adduct *(p. 397)*
bypass polymerase (translesion DNA polymerase) *(p. 408)*
cryptic splice site *(p. 389)*
deamination *(p. 395)*
depurination *(p. 395)*
displacement loop (D loop) *(p. 410)*
double Holliday junction (DHJ) *(p. 411)*
double-strand break repair *(p. 408)*
east-west (EW) resolution *(p. 411)*

forward mutation (mutation) *(p. 390)*
frameshift mutation *(p. 388)*
gene conversion *(p. 411)*
heteroduplex DNA (heteroduplex region) *(p. 411)*
Holliday junction *(p. 411)*
Holliday model *(p. 410)*
hotspot of mutation *(p. 386)*
induced mutations *(p. 395)*
intragenic reversion *(p. 390)*
missense mutation *(p. 387)*
mutagen *(p. 396)*
mutation frequency *(p. 384)*
nonhomologous end joining (NHEJ) *(p. 409)*
nonsense mutation *(p. 388)*

north-south (NS) resolution *(p. 411)*
nucleotide excision repair *(p. 405)*
photoproduct *(p. 399)*
photoreactive repair *(p. 402)*
point mutation *(p. 386)*
promoter mutation *(p. 389)*
pyrimidine dimer (thymine dimer) *(p. 399)*
recombination repair *(p. 405)*
regulatory mutation *(p. 389)*
reversion (reverse mutation) *(p. 390)*
second-site reversion *(p. 392)*
silent mutation *(p. 387)*
splicing mutation *(p. 389)*
spontaneous mutation *(p. 392)*
strand invasion *(p. 410)*

strand slippage *(p. 392)*

suppressor mutation *(p. 392)*

synthesis-dependent strand annealing (SDSA) *(p. 409)*

transition mutation *(p. 387)*

translesion DNA synthesis *(p. 399)*

transversion mutation *(p. 387)*

trinucleotide repeat disorder *(p. 392)*

true reversion *(p. 390)*

ultraviolet (UV) repair *(p. 404)*

---

PROBLEMS  *For instructor-assigned tutorials and problems, go to www.masteringgenetics.com.*

## Chapter Concepts

*For answers to selected even-numbered problems, see Appendix: Answers.*

1. Identify two general ways chemical mutagens can alter DNA. Give examples of these two mechanisms.

2. Nitrous acid and 5-BrdU alter DNA by different mechanisms. Identify each mechanism and describe how each compound creates mutation.

3. Using the adenine-thymine base pair in this DNA sequence

   ...GCTC...

   ...CGAG...

   a. Give the sequence after a transition mutation.
   b. Give the sequence after a transversion mutation.

4. The partial amino acid sequence of a wild-type protein is

   ... Arg–Met–Tyr–Thr–Leu–Cys–Ser...

   The same portion of the protein from a mutant has the sequence

   ... Arg–Met–Leu–Tyr–Ala–Leu–Phe...

   a. Identify the type of mutation.
   b. Give the sequence of the wild-type DNA template strand. Use $^A/_G$ if the nucleotide could be either purine, $^T/_C$ if it could be either pyrimidine, *N* if any nucleotide could occur at a site, or the alternative nucleotides if a purine and a pyrimidine are possible.

5. Thymine is usually in its normal keto form, but if a tautomeric shift occurs just before DNA replication, diagram the base pair that would result.

6. Ultraviolet (UV) radiation is mutagenic.
   a. What kind of DNA lesion does UV energy cause?
   b. How do UV-induced DNA lesions lead to mutation?
   c. Identify and describe two DNA repair mechanisms that remove UV-induced DNA lesions.

7. Researchers interested in studying mutation and mutation repair often induce mutations with various agents. What kinds of gene mutations are induced by
   a. Chemical mutagens? Give two examples.
   b. Radiation energy? Give two examples.

8. The effect of base-pair substitution mutations on protein function varies widely from no detectable effect to the complete loss of protein function (null allele). Why do the functional consequences of base-pair substitution vary so widely?

9. The two DNA and polypeptide sequences shown are for alleles at a hypothetical locus that produce different polypeptides, both five amino acids long. In each case, the lower DNA strand is the template strand:

   |             | 5′...ATGCATGTAAGTGCATGA...3′ |
   |-------------|-------------------------------|
   | allele $A_1$: | 3′...TACGTACATTCACGTACT...5′ |
   | $A_1$ polypeptide | N–Met–His–Val–Ser–Ala–C |

   |             | 5′...ATGCAAGTAAGTGCATGA...3′ |
   |-------------|-------------------------------|
   | allele $A_2$: | 3′...TACGTTCATTCACGTACT...5′ |
   | $A_2$ polypeptide | N–Met–Gln–Val–Ser–Ala–C |

   Based on DNA and polypeptide sequences alone, is there any way to determine which allele is dominant and which is recessive? Why or why not?

10. In numerous population studies of spontaneous mutation, two observations are made consistently: (1) most mutations are recessive, and (2) forward mutation is more frequent than reversion. What do you think are the likely explanations for these two observations?

11. Two different mutations are identified in a haploid strain of yeast. The first prevents the synthesis of adenine by a nonsense mutation of the *ade-1* gene. In this mutation, a base-pair substitution changes a tryptophan codon (UGG) to a stop codon (UGA). The second affects one of several duplicate tRNA genes. This base-pair substitution mutation changes the anticodon sequence of a $tRNA_{Trp}$ from 3′-ACC-5′ to 3′-ACU-5′.
    a. Do you consider the first mutation to be a forward mutation or a reversion? Why?
    b. Do you consider the second mutation to be a forward mutation or a reversion? Why?
    c. Assuming there are no other mutations in the genome, will this double-mutant yeast strain be able to grow on minimal medium? If growth will occur, characterize the nature of growth relative to wild type.

12. Many human genes are known to have homologs in the mouse genome. One approach to investigating human hereditary disease is to produce mutations of the mouse homologs of human genes by methods that can precisely target specific nucleotides for mutation.
    a. Numerous studies of mutations of the mouse homologs of human genes have yielded valuable information about how gene mutations influence the human disease process. In general terms, describe how and why creating mutations of the mouse homologs can give information about human hereditary disease processes.
    b. Despite the homologies that exist between human and mouse genes, some attempts to study human hereditary

disease processes by inducing mutations in mouse genes indicate there is little to be learned about human disease in this way. In general terms, describe how and why the study of mouse gene mutations might fail to produce useful information about human disease processes.

13. Answer the following questions concerning the accuracy of DNA polymerase during replication.

    a. What general mechanism do DNA polymerases use to check the accuracy of DNA replication and identify errors during replication?

    b. If a DNA replication error is detected by DNA polymerase, how is it corrected?

    c. If a replication error escapes detection and correction, what kind of abnormality is most likely to exist at the site of replication error?

    d. Identify two mechanisms that can correct the kind of abnormality resulting from the circumstances identified in part (c).

    e. If the kind of abnormality identified in part (c) is not corrected before the next DNA replication cycle, what kind of mutation occurs?

    f. DNA mismatch repair can accurately distinguish between the template strand and the newly replicated strand of a DNA duplex. What characteristic of DNA strands is used to make this distinction?

14. Apert syndrome is a human autosomal dominant condition that affects development of the head, hands, and feet. In a survey of 322,182 consecutive births in Ireland, two new cases of Apert syndrome were identified. What is the mutation rate of this gene per gamete?

15. Polydactyly is a human autosomal dominant condition that produces extra fingers and toes. Studies of hundreds of families with polydactyly have determined that penetrance for the dominant allele is 70%. Hospital-based surveys of live births find that 1 in 40,000 infants has a new case of polydactyly. Use this information to estimate the mutation rate of the gene.

16. The table shown lists the approximate new mutation rates for three autosomal dominant human diseases.

| Trait | Mutations per $10^6$ Gametes |
|---|---|
| Retinoblastoma (tumor of the retina) | 20 |
| Achondroplasia (statural dwarfism) | 80 |
| Neurofibromatosis (tumor of nervous tissue) | 220 |

    a. In a series of 50,000 consecutive live births recorded in a large metropolitan area, how many new cases of each disease are expected?

    b. Identify two molecular reasons why the rate of new mutations causing neurofibromatosis is more than 10 times greater than the mutation rate causing retinoblastoma.

17. A 1-mL sample of the bacterium *E. coli* is exposed to ultraviolet light. The sample is used to inoculate a 500-mL flask of complete medium that allows growth of all bacterial cells. The 500-mL culture is grown on the benchtop, and two equal-size samples are removed and plated on identical complete-medium growth plates. Plate 1 is immediately wrapped in a dark cloth, but plate 2 is not covered. Both plates are left at room temperature for 36 hours and then examined. Plate 2 is seen to contain many more growing colonies than plate 1. Thinking about DNA repair processes, how do you explain this observation?

18. A strain of *E. coli* is identified as having a null mutation of the *RecA* gene. What biological property do you expect to be absent in the mutant strain? What is the molecular basis for the missing property?

19. Define *gene conversion* and contrast it with *gene mutation*.

20. Some homologous recombination events produce gene conversion. Is homologous recombination a mutational event? Explain why or why not.

21. What is heteroduplex DNA, and why does it form? What is the relationship between heteroduplex DNA and gene conversion?

22. Is heteroduplex DNA always an outcome of homologous recombination? Why or why not?

23. A strain of yeast producing a four-celled ascus is heterozygous for the wild-type allele *Ala-B* and the mutant allele *ala-b*. The wild-type allele carries an A - T base pair and the mutant allele a G - C base pair at a site that is part of a heteroduplex region. Identify the events that produce the following kinds of asci.

    a. 3 *Ala-B*: 1 *ala-b*

    b. 3 *ala-b*: 1 *Ala-B*

    c. 2 *ala-b*: 2 *Ala-B*

24. Gene conversion is relatively easy to detect in four-cell and eight-cell asci of fungi, where ratios such as 3A:1a or 5a:3A indicate that gene conversion has taken place. Why is gene conversion much more difficult to detect in multicellular eukaryotes?

## Application and Integration

*For answers to selected even-numbered problems, see Appendix: Answers.*

25. Following the spill of a mixture of chemicals into a small pond, bacteria from the pond are tested and show an unusually high rate of mutation. A number of mutant cultures are grown from mutant colonies and treated with known mutagens to study the rate of reversion. Most of the mutant cultures show a significantly higher reversion rate when exposed to base analogs such as proflavin and 2-aminopurine. What does this suggest about the nature of the chemicals in the spill?

26. A geneticist searching for mutations uses the restriction endonucleases *Sma*I and *Pvu*II to search for mutations that eliminate restriction sites. *Sma*I will not cleave DNA with CpG methylation. It cleaves DNA at the restriction digestion sequence

$$\downarrow$$
$$5'\text{-CCC GGG-}3'$$
$$3'\text{-GGG CCC-}5'$$
$$\uparrow$$

*Pvu*II is not sensitive to CpG methylation. It cleaves DNA at the restriction sequence

```
5'-CAG CTG-3'

3'-GTC GAC-5'
```

a. What common feature do *Sma*I and *Sac*II share that would be useful to a researcher searching for mutations that disrupt restriction digestion?
b. What process is the researcher intending to detect with the use of these restriction enzymes?

27. A wild-type culture of haploid yeast is exposed to ethyl methanesulfonate (EMS). Yeast cells are plated on a complete medium, and 6 colonies (colonies numbered 1 to 6) are transferred to a new complete medium plate for further study. Four replica plates are made from the complete medium plate to plates containing minimal medium or minimal medium plus one amino acid (replica plates numbered 1 to 4) with the following results:

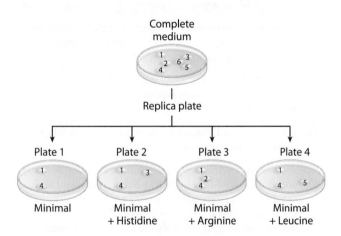

a. Identify the colonies that are prototrophic (wild type). What growth information leads to your answer?
b. Identify the colonies that are auxotrophic (mutant). What growth information leads to your answer?
c. Identify any colonies that are His⁻, Arg⁻, Leu⁻
d. For the following colonies, fill in "+" for the wild-type synthesis and "−" for the mutant synthesis of arginine, histidine, and leucine for colonies 1, 3, and 5.
e. Are there any colonies for which genotype information cannot be determined? If so, which colony or colonies?

28. A fragment of a wild-type polypeptide is sequenced for seven amino acids. The same polypeptide region is sequenced in four mutants.

| | |
|---|---|
| Wild-type polypeptide | N...Thr–His–Ser–Gly–Leu–Lys–Ala...C |
| Mutant 1 | N...Thr–His–Ser–Val–Leu–Lys–Ala...C |
| Mutant 2 | N...Thr–His–Ser–C |
| Mutant 3 | N...Thr–Thr–Leu–Asp–C |
| Mutant 4 | N...Thr–Gln–Leu–Trp–Ile–Glu–Gly...C |

a. Use the available information to characterize each mutant.
b. Determine the wild-type mRNA sequence.
c. Identify the mutation that produces each mutant polypeptide.

29. Experiments by Charles Yanofsky in the 1950s and 1960s helped characterize the nature of tryptophan synthesis in *E. coli*. In one of Yanofsky's experiments, he identified glycine (gly) as the wild-type amino acid in position 211 of tryptophan synthetase, the product of the *trpA* gene. He identified two independent missense mutants with defective tryptophan synthetase at these positions that resulted from base-pair substitutions. One mutant encoded arginine (arg) and another encoded glutamic acid (glu). At position 235, wild-type tryptophan synthetase contains serine (ser), but a base-pair substitution mutant encodes leucine (leu). At position 243, the wild-type polypeptide contains glutamine, and a base-pair substitution mutant encodes a stop codon. Identify the most likely wild-type codons for positions 211, 235, and 243. Justify your answer in each case.

30. Common baker's yeast (*Saccharomyces cerevisiae*) is normally grown at 37°C, but it will grow actively at temperatures down to approximately 20°C. A haploid culture of wild-type yeast is mutagenized with EMS. Cells from the mutagenized culture are spread on a complete-medium plate and grown at 25°C. Six colonies (1 to 6) are selected from the original complete-medium plate and transferred to two fresh complete-medium plates. The new complete plates (shown below) are grown at 25°C and 37°C. Four replica plates are made onto minimal medium or minimal plus adenine from the 25°C complete-medium plate. The new plates are grown at either 25°C or 37°C, as indicated below.

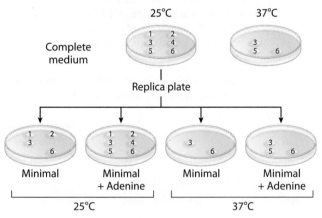

a. Which colonies are prototrophic and which are auxotrophic? What growth information is used to make these determinations?
b. Classify the nature of the mutations in colonies 1, 2, and 5.
c. What can you say about colony 4?

31. The two gels illustrated below contain dideoxynucleotide DNA-sequencing (see Chapter 7, pages 250–254) information for a segment of wild-type and mutant DNA corresponding to the N-terminal end of the protein. The start codon and the next five codons are sequenced.

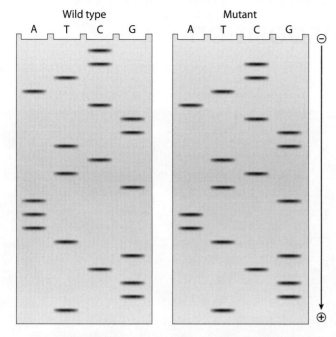

Wild type                    Mutant

a. Write the DNA sequence of both alleles, including strand polarity.
b. Identify the template and nontemplate strands of DNA.
c. Write out the mRNA sequences encoded by each template strand, and underline the start codons.
d. Determine the amino acid sequences translated from these mRNAs.
e. What is the cause of the mutation?

32. Alkaptonuria is a human autosomal recessive disorder caused by mutation of the *HAO* gene that encodes the enzyme homogentisic acid oxidase. Restriction mapping of the *HAO* gene region reveals four *Bam*HI restriction sites (B1 to B4) in the wild-type allele and three *Bam*HI restriction sites in the mutant allele. *Bam*HI utilizes the restriction sequence 5'-GGATCC-3'. The *Bam*HI restriction sequence identified as B3 is altered to 5'-GGAACC-3' in the mutant allele. The mutation results in a Ser-to-Thr missense mutation. Restriction maps of the two alleles are shown below, and the binding sites of two molecular probes (probe A and probe B) are identified.

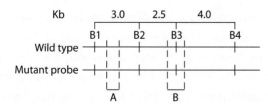

Kb          3.0      2.5      4.0
            B1    B2    B3    B4
Wild type
Mutant probe
              A            B

DNA samples taken from a mother (M), father (F), and two children (C1 and C2) are analyzed by Southern blotting of *Bam*HI-digested DNA. The resulting autoradiograph is illustrated below.

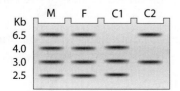

Kb        M    F    C1    C2
6.5
4.0
3.0
2.5

a. Using *A* to represent the wild-type allele and *a* for the mutant allele, identify the genotype of each family member. Identify any family member who is alkaptonuric.
b. In a separate figure, draw the autoradiograph patterns for all the genotypes that could be found in children of this couple.
c. Explain how the DNA sequence change results in a Ser-to-Thr missense mutation.

33. Two haploid strains of fungus are fused to form a diploid that produces eight-celled asci. Fungus strain A has the genotype +*ade1 his2*, and strain B is *a*++. The three genes are linked and occur in the order given.

   a. The alleles at the *A* gene locus are determined in an ascus, and the order is *aaaa++++*. Write the genotype for all three genes that you expect to find most commonly.
   b. One ascus from the diploid is of the following type:

$$+\ ade1\ his2$$
$$+\ ade1\ his2$$
$$+\ ade1\ his2$$
$$+\ ade1\ his2$$
$$a\ ade1\ +$$
$$a+\quad +$$
$$a+\quad +$$
$$a+\quad +$$

Explain the events that produced this ascus.

   c. One ascus from the diploid is of the following type:

$$a+\quad +$$
$$a+\quad +$$
$$a+\quad his2$$
$$a+\quad his2$$
$$+\ ade1\ his2$$
$$+\ ade1\ his2$$
$$+\ ade1\ his2$$
$$+\ ade1\ his2$$

Explain the events that produced this ascus.

34. Two haploid fungi are crossed and examined for evidence of recombination and gene conversion in a region of three linked genes. Strain A has the genotype $alm^+\ brp^+\ cty^+$, and strain B has the genotype $alm^-\ brp^-\ cty^-$. Most of the spores in the eight-celled asci of this species are found in a 4:4 ratio of $ala^+\ brp^+\ cty^+ : ala^-\ brp^-\ cty^-$. However, one ascus contains the following spores:

$$ala^-\ brp^-\ cty^-$$
$$ala^-\ brp^-\ cty^-$$
$$ala^-\ brp^+\ cty^-$$
$$ala^-\ brp^+\ cty^-$$
$$ala^+\ brp^+\ cty^+$$
$$ala^+\ brp^+\ cty^+$$
$$ala^+\ brp^+\ cty^+$$
$$ala^+\ brp^+\ cty^+$$

a. What is unusual about the spores in this ascus?
b. What do the spores suggest about the location of recombination and the process of gene conversion?
c. Explain how the spores produce a 6:2 ratio for alleles of the *brp* gene and 4:4 ratios for *ala* and *cty*.

# 13 Chromosome Aberrations and Transposition

## ESSENTIAL IDEAS

▌ Nondisjunction causes changes in the number of chromosomes and may result in gametes containing the wrong chromosome number.

▌ Changes in the number of sets of chromosomes alter phenotypes, and can confer evolutionary advantages.

▌ Chromosome breakage can change chromosome structure and may lead to the loss or duplication of genes.

▌ Chromosome breakage can lead to chromosome inversions and translocations.

▌ Transposable genetic elements move throughout the genome and modify genes, chromosomes, and genomes.

▌ Bacterial transposable genetic elements facilitate DNA transfer.

▌ Transposition is a source of mutation and expansion of eukaryotic genomes.

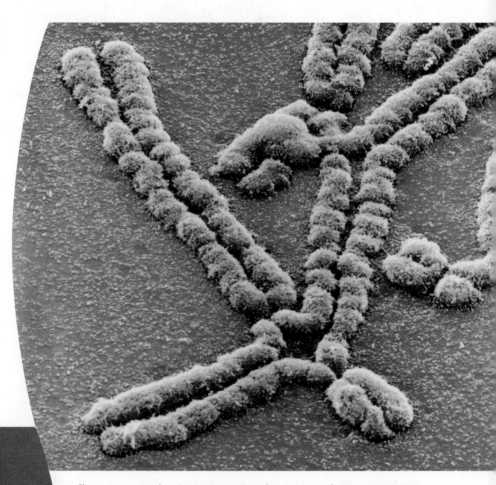

Chromosome translocations are mutations that rearrange chromosome structure. This electronmicrograph shows two pairs of homologous chromosomes that have exchanged segments and must form a tetravalent structure involving the four chromosomes in order to synapse their homologous regions during prophase I.

Something interesting is happening to the mice on Madeira, a tiny island off the western coast of Portugal: They are in the process of differentiating into two species! Madeira, about 20 miles long and 8 miles wide, has steep volcanic mountains running down the middle that form a barrier to easy mouse migration. The common house mouse (*Mus musculus*) was introduced to Madeira by sailors in the 1400s. Today, Madeira has two distinct populations of mice, one on either side of the central mountain range.

There is more than a mountain range separating these two populations, however. Both populations have undergone multiple chromosome fusions that have reduced their diploid number. The usual chromosome number for *Mus musculus* is 20 pairs ($2n = 40$). On Madeira, however, one population has $2n = 22$, and the other has $2n = 24$. Because each population has a different chromosome number, interpopulation hybrids are sterile. Such hybrids carry 23 chromosomes (11 from one parent and 12 from the other) and therefore cannot form viable gametes. This example of speciation, primarily based on differences in chromosome structure and chromosome number, has taken place over less than 600 years.

Variation and evolution at the chromosome level are genomic in scope—that is, they potentially alter the content of the genome, changing interactions between homologous chromosomes in meiosis and limiting the possibility of reproduction between organisms with chromosomal differences. This chapter addresses two distinct categories of chromosome change. The first comprises a group of alterations of chromosome number and chromosome structure known collectively as **chromosome aberrations.** On one hand, chromosome aberrations are mutational events that may seriously alter normal phenotypes or be fatal to their bearers. On the other hand, chromosome aberrations are part of the mutational process that creates genetic diversity within species, providing a mechanism for diversification and speciation, as displayed by the mice of Madeira. The second category of chromosome change is *transposition*, the movement of DNA elements within the genomes of organisms. Transposition is a source of mutation as well as a source of additional DNA sequence that can increase the size of genomes. Both chromosome aberrations and transposition contribute to evolution and speciation, by reorganizing and reshaping the content of genomes.

# 13.1 Nondisjunction Leads to Changes in Chromosome Number

Previously, we discussed the connection between Mendel's two laws of heredity and the disjunction of homologous chromosomes and sister chromatids during meiosis (see Chapter 3). In the discussion that follows, we focus on *nondisjunction* as a process of failed chromosome and sister chromatid disjunction that can result in abnormalities of chromosome number.

## Euploidy and Aneuploidy

The number of chromosomes contained in a nucleus and the relative size and shape of each chromosome are species-specific characteristics, but neither parameter is directly associated with the complexity of the organism (Table 13.1). Chromosome number varies widely among species, although closely related species tend to have similar numbers.

With a few unusual exceptions, the number of chromosomes is the same for males and females of a species and the number of chromosomes in nuclei of normal cells is a multiple of the haploid number ($n$), the number in a single set of chromosomes. Regardless of whether the total chromosome number is $2n$ (diploid), $3n$ (triploid), or a higher multiple of $n$, it is described as a **euploid** number of chromosomes if it is a whole-number multiple of the haploid number. If cells contain a number of chromosomes that is

| Table 13.1 | Chromosome Number in Selected Animal Species |
|---|---|
| **Species** | **Diploid Chromosome Number ($2n$)** |
| Carp (*Cyprinus carpio*) | 104 |
| Cat (*Felis catus*) | 38 |
| Chicken (*Gallus domesticus*) | 78 |
| Chimpanzee (*Pan troglodytes*) | 48 |
| Cow (*Bos taurus*) | 60 |
| Dog (*Canis familiarus*) | 78 |
| Frog (*Rana pipiens*) | 26 |
| Fruit fly (*Drosophila melanogaster*) | 8 |
| Horse (*Equus caballus*) | 64 |
| Human (*Homo sapiens*) | 46 |
| Mouse (*Mus musculus*) | 40 |
| Rat (*Rattus norvegicus*) | 42 |
| Rhesus monkey (*Macaca mulatta*) | 42 |

not euploid, the chromosome number is **aneuploid.** Aneuploidy occurs when one or more chromosomes are lost or gained relative to the normal euploid number. Chromosome nondisjunction is a principal cause of aneuploidy.

## Chromosome Nondisjunction

The term chromosome nondisjunction, or simply nondisjunction, applies to the failure of homologous chromosomes or sister chromatids to separate as they normally do during cell division. Nondisjunction can occur in somatic cells or in germ-line cells with similar results. If a single pair of homologous chromosomes fails to properly disjoin in a somatic cell during mitotic cell division, one of the resulting daughter cells carries an extra chromosome ($2n + 1$), and the other is missing a chromosome ($2n - 1$).

Mitotic cells of animals that contain the wrong number of chromosomes may suffer reduced viability in comparison to their counterparts with a diploid number of chromosomes. The poor survival of these cells usually limits their number in organisms, although cells with abnormal numbers of chromosomes are common in cancer where other genetic changes in addition to alterations of chromosome number play a central role in cancer cell survival and proliferation.

In contrast to the limited circumstances for changes to chromosome number in animal cells, plants frequently have substantially more tolerance for changes in chromosome number, and it is not unusual to find plant strains with more than two copies of each chromosome. We describe this situation, called *polyploidy,* in more detail in a later section.

Nondisjunction in germ-line cells produces aneuploid gametes and can lead to the production of aneuploid zygotes. Meiotic nondisjunction can occur in either meiosis I or II and most often affects just a single homologous pair or a single pair of sister chromatids. Meiosis I nondisjunction is the failure of homlogous chromosomes to separate. It results in both homologs moving to a single pole. One secondary gametophyte contains both chromosomes and the other contains neither chromosome. These gametophytes contain aneuploid chromosome numbers of $n + 1$ and $n - 1$ (this assumes only one chromosome pair is affected) (**Figure 13.1**). Meiosis II usually proceeds normally even when meiosis I is aberrant and its completion sends the sister chromatids to different gametes. The four resulting gametes each contain an aneuploid number of chromosomes. Union of an aneuploid gamete with a normal haploid gamete shown in the figure results in a fertilized egg with an aneuploid number of chromosomes that will be either **trisomic** ($2n + 1$), having three of one of the chromosomes rather than a homologous pair, or **monosomic** ($2n - 1$), having just a single copy of one of the chromosomes rather than a homologous pair.

If nondisjunction occurs in meiosis II, it typically follows a normal meiosis I. As a result, both secondary gametes contain the haploid number of chromosomes (**Figure 13.2**). Since these are separate cells, they independently divide during meiosis II; thus, if nondisjunction occurs, only one of the secondary gametes will be affected. Among the four resulting gametes, two are normal, because normal disjunction took place during each meiotic division. The other two gametes are aneuploid, containing either $n + 1$ or $n - 1$ chromosomes. Trisomic or monosomic zygotes are produced when one of these aneuploid gametes unites with a normal gamete at fertilization. Normal diploid zygotes result from fertilization of either of two haploid gametes.

**Figure 13.1  Meiosis I nondisjunction.** Homologous chromosomes fail to disjoin in meiosis I and all resulting gametes are aneuploid. Fertilization by a normal haploid gamete produces zygotes that are trisomic ($2n + 1$) or monosomic ($2n - 1$).

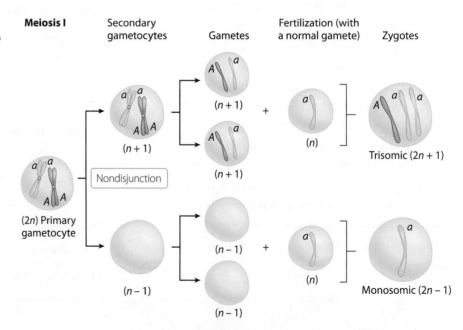

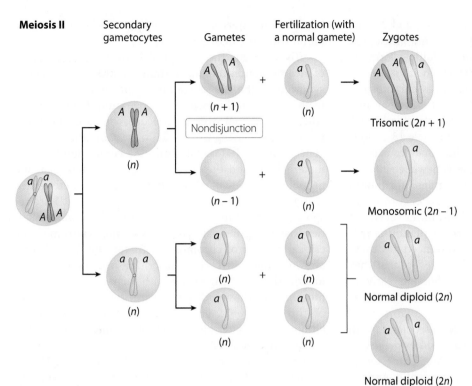

**Figure 13.2 Meiosis II nondisjunction.** Sister chromatid disjunction fails in meiosis II. Normal fertilization of the resulting gametes generates trisomic, monosomic, or normal diploid zygotes.

## Gene Dosage Alteration

In 1913, at about the same time Calvin Bridges was publishing his demonstration of the chromosome theory of heredity, Albert Francis Blakeslee and John Belling studied the phenotypic consequences of aneuploidy in the diploid ($2n = 24$) jimson weed (*Datura stramonium*), in which 12 chromosome pairs are identified as A to L. Blakeslee and Belling identified 12 phenotypically distinct lines of trisomic *Datura*, one for each of the chromosome pairs (**Figure 13.3**).

This result suggests that chromosome number is a factor in phenotype, and in the years that followed Blakeslee and Belling's report, other studies demonstrated that aneuploidy causes severe phenotypic consequences in nearly all animal species and that it affects the phenotype of many plant species. The abnormalities result from changes in **gene dosage,** the number of copies of a gene. Aneuploidy changes the dosage of *all the genes* on the affected chromosome. In a diploid where two copies of a

gene, on a homologous pair of chromosomes, generate 100% of gene dosage, a monosomic mutant has just one gene copy and just 50% of normal gene dosage for each gene on the chromosome. In contrast, a trisomic mutant has three copies and 150% of normal gene dosage for each of the genes on the chromosome.

Changes in gene dosage lead to an imbalance of gene products from the affected chromosome relative to unaffected chromosomes, and this imbalance is at the heart of alterations of normal development and the production of abnormal phenotypes. Most animals are highly sensitive to changes in gene dosage, and their developmental biology, especially within the nervous system, does not proceed normally in the presence of gene dosage imbalance. In contrast to the potential for developmental disruptions due to aneuploidy in animals, gene dosage changes are more readily tolerated in many species of plants owing to their distinct developmental programs.

## Aneuploidy in Humans

Medical practitioners and other specialists in human biology have provided a wealth of information about human aneuploidy, most of it originating with germ-line nondisjunction and the production of aneuploid gametes. Humans are enormously sensitive to the changes in gene dosage brought about by aneuploidy, and most human aneuploidies are incompatible with life. Theoretically, there are potentially 24 different kinds of trisomy in humans—one for each autosome, and one each for the X and Y chromosomes—and an equal number of potential monosomies. Yet only

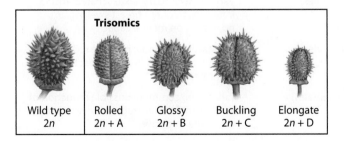

**Figure 13.3 The appearance of selected trisomics in jimson weed (*Datura stramonium*).**

autosomal trisomies of chromosomes 13, 18, and 21, and no autosomal monosomies, are seen with any measurable frequency in newborn human infants. Multiple forms of sex-chromosome trisomy are also detected with some frequency at birth along with one type of sex-chromosome monosomy (Table 13.2). A wide variety of other chromosome abnormalities occur in newborn infants as well. Each of the aneuploidy conditions identified in Table 13.2, along with the other chromosome abnormalities that occur, result in significant phenotypic abnormalities in newborn infants.

Human biologists know that trisomies and monosomies other than those listed in the table occur at conception, but the resulting zygotes almost never survive to be born alive. The explanation for this situation is that the abnormalities of development produced by these other trisomies and monosomies are so severe that they almost always lead to spontaneous abortion early in pregnancy. The best available data on human aneuploidy occurrence and survival come from studies that monitor women for minute hormone changes associated with conception and early pregnancy. These studies make two surprising observations. First, in the first trimester of pregnancy, about half of these human conceptions spontaneously abort, and second, more than half of these spontaneously terminated human pregnancies carry abnormalities of chromosome number or chromosome structure. These observations point to a surprisingly high 15% to 25% frequency of meiotic nondisjunction in

humans. Other errors producing gametes with abnormal chromosomes can occur as well.

To ascertain the biological basis for the high rate of meiotic nondisjunction in humans, trisomy 21 (Down syndrome)—the most common autosomal trisomy at birth—has been the focus of intense study. Epidemiologic studies conducted over several decades have linked the risk of a child having trisomy 21 to the age of the mother at conception. Table 13.3 illustrates the connection between maternal age and the risk of trisomy 21. Molecular and genomic analyses of Down syndrome have determined that a small number of genes on chromosome 21 are responsible for the mental retardation and heart abnormalities that are principal symptoms of the syndrome. The critical portion of chromosome 21 for Down syndrome, known as the Down syndrome critical region (DSCR), was identified by studying people with partial trisomy of chromosome 21. These individuals carry two complete copies of chromosome 21 and a small additional segment of chromosome 21 on another chromosome. These studies identify region 21q22.2 as the DSCR. In other words, Down syndrome individuals invariably carry 21q22.2 in three copies. People who carry three copies of a segment other than 21q22.2 do not display the principal Down syndrome symptoms. Among a handful of candidate genes, *DYRK*, a homolog of a gene in mice and *Drosophila* that produces dosage-sensitive learning defects, makes a major contribution to Down syndrome. In mice, increased

| Table 13.2 | Human Aneuploidies and Frequencies at Birth | | |
|---|---|---|---|
| **Aneuploidy** | **Syndrome** | **Frequency at Birth** | **Syndrome Characteristics** |
| *Autosomal Aneuploidy* | | | |
| Trisomy 13 | Patau syndrome | 1 in 15,000 | Mental retardation and developmental delay, possible deafness, major organ abnormalities, early death |
| Trisomy 18 | Edward syndrome | 1 in 8000 | Mental retardation and developmental delay, skull and facial abnormalities, early death |
| Trisomy 21 | Down syndrome | 1 in 1500 | Mental retardation and developmental delay, characteristic facial abnormalities, short stature, variable life span |
| *Sex-Chromosome Aneuploidy* | | | |
| 47, XXY | Klinefelter syndrome | 1 in 1000 (males) | Variable secondary sexual characteristics, infertility, frequent breast swelling; no impact on mental capacity |
| 47, XYY | Jacob syndrome | 1 in 1000 (males) | Tall stature common; possible reduction but not loss of fertility; no impact on mental capacity |
| 47, XXX | Triple X syndrome | 1 in 1000 (females) | Tall stature common; possible reduction of fertility; menstrual irregularity; no impact on mental capacity |
| 45, XO | Turner syndrome | 1 in 5000 (females) | No secondary sexual characteristics; infertility, short stature; webbed neck common; no impact on mental capacity |

| Table 13.3 | Risk of Down Syndrome (Trisomy 21) by Maternal Age[a] | | |
| --- | --- | --- | --- |
| Maternal Age Range | Total Live Births Studied | Trisomy 21 Births | Rate per 1000 Births |
| 15–19 | 30,272 | 18 | 0.49 |
| 20–24 | 117,593 | 87 | 0.73 |
| 25–29 | 108,746 | 96 | 0.90 |
| 30–34 | 49,487 | 72 | 1.56 |
| 35–39 | 19,522 | 73 | 4.19 |
| 40–44 | 4880 | 73 | 18.02 |
| 45–49 | 304 | 19 | 55.02 |

[a] Data adapted from E. B. Hook and A. Lindsjo, Down syndrome in live births by single year maternal age interval in a Swedish study: Comparison with results from a New York State study. *Am. J. Hum. Genet.* 30 (1978): 19–27.

dosage of the *DYRK* homolog reduces brain size. *DSCAM* is a second gene whose increased dosage is linked to Down syndrome. This gene also has homologs in mouse and *Drosophila,* where its protein product participates in the formation of the heart and components of the developing nervous system.

A different kind of change in gene dosage is seen in humans with Turner syndrome, a monosomy of the X chromosome in which there is one X chromosome but no second sex chromosome (see Table 13.2). Despite the occurrence of X-inactivation in human female embryos, two sex chromosomes are necessary for normal early development. In female embryos that are XO (Turner syndrome), the single copy of the gene *SHOX*, located in pseudoautosomal region 2 on the short arm of the X chromosome and the Y chromosome, is insufficient to direct certain aspects of normal development. The haploinsufficiency of *SHOX* appears to play a central role in producing Turner syndrome.

---

Genetic Insight  Aneuploidy usually results from chromosome nondisjunction and produces abnormalities through changes in the dosage of critical genes.

---

## Reduced Fertility in Aneuploidy

The type and extent of developmental abnormalities in an aneuploid organism are a consequence of changes in the dosage of the genes affected, but disturbed patterns of chromosome segregation during meiosis can produce another problem as well: Reduction in the number of normal haploid gametes can also impair fertility. In the case of a trisomy, chromosome segregation in meiosis is disturbed because three homologous chromosomes rather than two are aligning at synapsis and segregating during anaphase I.

In a trisomy, two patterns of homologous chromosome synapsis are possible among the three chromosomes at metaphase I (Figure 13.4)—either a **trivalent synaptic**

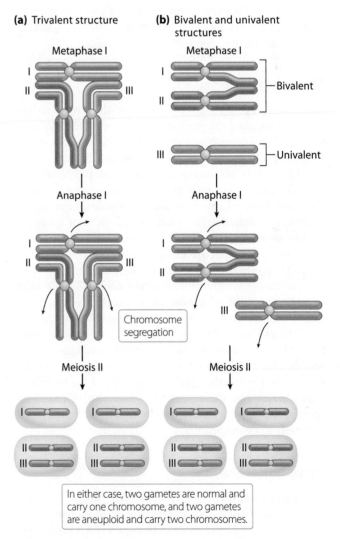

**(a)** Trivalent structure

**(b)** Bivalent and univalent structures

In either case, two gametes are normal and carry one chromosome, and two gametes are aneuploid and carry two chromosomes.

**Figure 13.4  Two meiotic patterns of segregation in trisomics.  (a)** Three chromosomes form a trivalent structure at synapsis and produce only two normal haploid gametes among the four gametes. **(b)** A bivalent and a univalent arrangement of three chromosomes also leads to just two normal haploid gametes.

**structure** or a bivalent and univalent arrangement—but there is no mechanism to divide three chromosomes equally at anaphase I. Thus two chromosomes move to one pole and one chromosome moves to the opposite pole during anaphase. On completion of meiosis, half of the gametes are haploid for the chromosome; but the remaining gametes contain two copies. This effectively reduces the number of viable gametes by approximately one-half, because the gametes with an extra chromosome will produce trisomic progeny that are unlikely to survive. A 1:1 ratio of haploid (normal) to diploid (abnormal) gametes has been observed in numerous experimental organisms. This circumstance results in a form of **semisterility,** a reduction—but not complete elimination—of fertility. Studies of occasional cases of reproduction by people with Down syndrome identify both reduced fertility and the frequent production of children with Down syndrome as results of the high rate of production of gametes with an extra copy of chromosome 21 due to the high proportion of gametes with two copies of chromosome 21.

## Mosaicism

Our discussion of random X-inactivation of mammalian females identified the phenomenon as an example of naturally occurring mosaicism, in which different cells of the organism contain differently functioning X chromosomes (see Chapter 3). Mosaicism can also develop through mitotic nondisjunction early in embryogenesis and is one of the many kinds of chromosome abnormalities that occur in newborn infants. For example, 25–30% of cases of Turner syndrome, the X-chromosome monosomy (XO), occur in females that are mosaic with some cells that are 45, XO and others that are 46, XX. Some individuals with mosaic Turner syndrome individuals carry 47, XXX cells as well. This kind of mosaicism is usually derived from mitotic nondisjunction in a 46, XX zygote (**Figure 13.5**).

In fruit flies, butterflies, and moths, sex-chromosome mosaicism produces a particular sexually ambiguous phenotype called a **gynandromorphy.** Gynandromorph sex morphology is female ("gyn") on one half of the body and male ("andro") on the other half (**Figure 13.6**). Gynandromorphy develops as a consequence of mitotic nondisjunction early in development.

## Uniparental Disomy

A rare abnormality of chromosome content called **uniparental disomy** has been identified in humans. Uniparental disomy occurs when both copies of a homologous chromosome pair originate from a single parent. It was first identified in connection with two chromosomal conditions, Angelman syndrome (AS; OMIM 105830) and Prader-Willi syndrome (PWS; OMIM 176270), that are usually the result of a partial deletion of the 15q11.12 portion of chromosome 15.

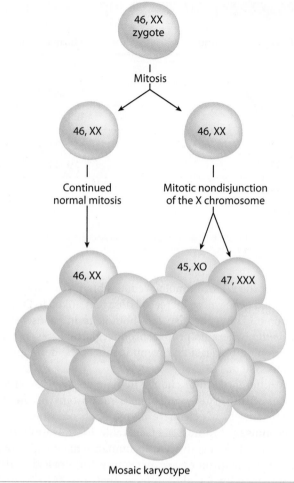

Turner syndrome mosaic females contain 46, XX and 45, XO cells, and they may also have cells with 47, XXX.

**Figure 13.5  Chromosome mosaicism.** Mosaicism usually begins with a normal diploid zygote. Mitotic nondisjunction produces one or more aneuploid cell lines that persist and are found in the newborn.

Uniparental disomy has two mechanisms of origin. The rarer mechanism requires nondisjunction of the same chromosome in both the sperm and egg. For example, uniparental disomy for chromosome 15 is produced when one gamete contributes two copies of the chromosome and the other does not contribute a copy of the chromosome. The second mechanism is more common. It involves nondisjunction in one parent that results in an aneuploid gamete contributing two copies of chromosome 15. The other gamete is normal and contributes a single copy of chromosome 15. Gamete union results in trisomy 15 in the fertilized egg, a condition that is invariably incompatible with survival. By a process known as **trisomy rescue,** one of the extra copies of chromosome 15 is randomly ejected in one of the first mitotic divisions following fertilization. One result of trisomy rescue can be a cell with one chromosome from each parent. Zygotes with this result have normal

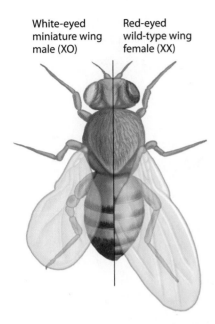

White-eyed miniature wing male (XO)    Red-eyed wild-type wing female (XX)

**Figure 13.6 Gynandromorphy in *Drosophila*.** White eye and miniature wing are X-linked recessive traits present in the hemizygous (XO) male half of the fly. Heterozygous genotypes for both genes are present in the wild-type (XX) female half of the fly.

chromosome content. Alternatively, trisomy rescue could result in a zygote that retains two copies of chromosome 15 from the same parent (i.e., uniparental disomy).

Uniparental disomy produces 10–20% of cases of AS and PWS. When uniparental disomy causes AS, both copies of chromosome 15 are from the father, and there is no maternal copy of chromosome 15. In uniparental disomy cases of PWS, both copies of chromosome 15 are from the mother, and there is no copy of the paternal chromosome 15. The discovery of uniparental disomy forms of AS and PWS stems from a mammalian process of regulating gene expression known as *genomic imprinting* (see Chapter 15).

---

**Genetic Insight** Mosaic organisms contain two or more distinct cell lineages that have different chromosomes resulting from mitotic nondisjunction. Uniparental disomy is a rare consequence of nondisjunction in which resulting cells have two copies of a chromosome from a single parent.

---

## 13.2 Changes in Euploidy Result in Various Kinds of Polyploidy

**Polyploidy** is the presence of three or more sets of chromosomes in the nucleus of an organism. Polyploidy in nature can result either from the duplication of euploid chromosome sets from a single species or from the combining of chromosome sets from different species. Many types of polyploidy are possible—triploids ($3n$), tetraploids ($4n$), pentaploids ($5n$), hexaploids ($6n$), octaploids ($8n$), and so on. Polyploids whose karyotype is

**(a)** Fertilization by multiple pollen grains

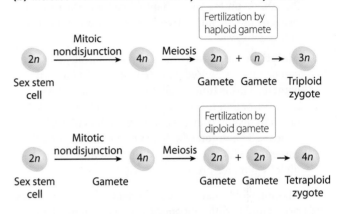

**(b)** Increase in chromosome number by mitotic nondisjunction

**(c)** Increase in chromosome number by meiotic nondisjunction

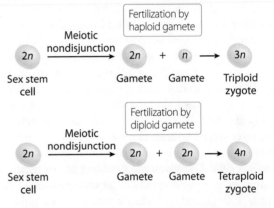

**Figure 13.7 Mechanisms creating triploid and tetraploid zygotes in plants.**

comprised of chromosomes derived from a single species are designated **autopolyploids** (*auto* = "self"), and polyploids with chromosome sets from two or more species are called **allopolyploids** (*allo* = "different"). Terms such as *autotetraploid* ($4n$ chromosomes that all derive from a single species), and *allohexaploid* ($6n$ with chromosomes from two or more species) are used to describe a polyploid organism's genomic content. Polyploidy is common in plant species but is extremely rare in animals.

### Autopolyploidy and Allopolyploidy

Three mechanisms lead to autopolyploidy (**Figure 13.7**).

1. **Multiple fertilizations.** Fertilization of an egg by more than one haploid pollen grain results in a zygote that is triploid ($3n$) or higher. This is generally

a rare event, because most sexually reproducing plants have elaborate mechanisms to prevent fertilization of an egg by more than a single pollen grain.

2. **Mitotic nondisjunction.** Mitotic nondisjunction in sex stem cells can result in chromosome doubling. These cells divide by mitosis before entering meiosis, and mitotic nondisjunction doubles the number of chromosomes from $2n$ to $4n$. The gametes that result from meiotic division of $4n$ sex stem cells are $2n$. If a $2n$ gamete unites with a haploid gamete, the resulting progeny are $3n$; and if two $2n$ gametes unite, the result is a $4n$ zygote.

3. **Meiotic nondisjunction.** Meiotic nondisjunction can produce a diploid gamete instead of a haploid gamete. This is functionally similar to the mitotic nondisjunction mechanism, except that the breakdown in disjunction affects the products of a single meiosis. The union of a $2n$ gamete and a haploid gamete produces a $3n$ zygote, and the union of two diploid gametes produces $4n$ zygotes.

Allopolyploids carry multiple sets of chromosomes that originated in different species. The union of a haploid set of chromosomes from species 1 ($n_1$) and a haploid gamete from species 2 ($n_2$) produces a hybrid organism that may have either an even number or odd number of chromosomes, since related species may have different diploid numbers (**Figure 13.8a**). For example, if species 1 has $2n = 10$ and species 2 has $2n = 14$, the $n_1 + n_2$ hybrid will have 12 chromosomes. Interspecies hybrids of this sort are infertile, because the two chromosome sets are not homologous and fail to pair during meiosis. The fertility of these interspecies hybrids can be rescued if chromosome doubling takes place by nondisjunction in cells that give rise to sex cells. Chromosome doubling in our hypothetical interspecies hybrid results in 24 chromosomes in sex cells ($2n_1 + 2n_2$), and each chromosome has a homologous mate for proper meiotic pairing (**Figure 13.8b**). Completion of meiosis produces gametes containing $n_1 + n_2 = 12$ chromosomes. These gametes can unite with like gametes to reproduce the interspecific hybrid and maintain the $2n_1 + 2n_2 = 24$ chromosome number.

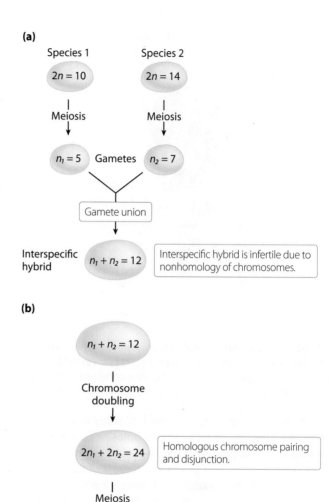

**Figure 13.8 The production of allopolyploidy.** **(a)** The union of two haploid gametes, $n_1 = 5$ plus $n_2 = 7$, from diploid species produces an interspecies hybrid with 12 chromosomes. **(b)** Chromosome doubling produces a fertile allopolyploid with 24 chromosomes that produces gametes with 12 homologous chromosome pairs.

## Consequences of Polyploidy

Allopolyploids occur in nature and are also produced by human manipulation. When used for commercial purposes, polyploidy generates three main consequences. First, fruit and flower size is increased. The nuclei and cells of polyploidy strains are larger than those of diploid strains, and many familiar fruit and vegetable varieties benefit from this effect. Apples ($3n = 51$), bananas ($3n = 33$), strawberries ($8n = 56$), peanuts ($4n = 40$), and potatoes ($4n = 48$) are just a few examples.

Increased fruit and flower size in polyploidy plants comes at the cost of the second effect—fertility. The problem is particularly acute for odd-numbered polyploids ($3n$, $5n$, etc.), in which the odd number of chromosomes cannot be evenly divided at the first meiotic division. The result is an unequal distribution of chromosomes that makes almost all of the resulting gametes nonviable. This reproductive disadvantage is turned into commercial advantage in cultivated plants with odd-numbered polyploidy. Certain "seedless" fruits and vegetables in the produce aisle of your local grocery store are odd-numbered polyploids.

Allopolyploids exhibit a third characteristic of commercial importance—increase in heterozygosity relative to diploids that comes about when inbred lines are crossed and is the basis of additional growth vigor. This phenomenon is known as **hybrid vigor,** and it consists of more rapid growth, increased production of fruits and flowers, and improved resistance to disease among the heterozygous (hybrid) progeny of inbred lines.

## Reduced Recessive Homozygosity

The pattern of single-gene inheritance in polyploids differs from that in diploids with respect to the proportions of dominant and recessive phenotypes from certain crosses. This difference is tied directly to the additional number of gene copies in polyploid genomes. A dominant phenotype is produced by any genotype containing one or more copies of the dominant allele, and the recessive phenotype is produced only by the homozygous recessive genotype. In the case of a phenotype decided by a single gene with a dominant and a recessive allele, the likelihood of producing the recessive phenotype in a tetraploid strain is decreased compared to the likelihood of producing it in a diploid.

Taking an autotetraploid with the genotype $AAaa$ as an example, let's determine the probability that progeny produced by self-fertilization would have the genotype $aaaa$. We'll use the designations $A_1$, $A_2$, $a_3$, and $a_4$ for alleles of the gene. The ratio of dominant to recessive alleles is 2:2 in the tetraploid, and six diploid gamete genotypes are produced by homologous disjunction: $A_1A_2$, $A_1a_3$, $A_1a_4$, $A_2a_3$, $A_2a_4$, and $a_3a_4$. Among these gametes, only one (one-sixth of the total) contains two recessive alleles. The probability of union of two fully recessive gametes is therefore $(1/6)(1/6) = 1/36$, much less than the 1/4 probability of producing homozygous recessive offspring from heterozygous diploids with the genotype $Aa$. **Genetic Analysis 13.1** guides you through an analysis of a genetic cross involving polyploids.

## Polyploidy and Evolution

The disadvantages in growth and reproduction experienced by polyploid organisms can be outweighed by the evolutionary advantages of polyploidy. More than half of all contemporary flowering plant species are derived from ancestors that evolved by polyploidy.

Evolution by polyploidy is a sudden, dramatic event that can lead to the development of a new species over a span of just one or two generations. Species that have had a quiescent genetic history can experience a sudden burst of evolutionary change through the development of polyploidy by two mechanisms. First, allopolyploidy can result in the evolution of a new species, owing to the fact that the newly polyploid progeny are reproductively isolated from their nonpolyploid progenitor by chromosomal differences that make hybridization between the progenitor and the new species unlikely. Second, polyploidy produces

gene duplication that relaxes natural selection constraints on duplicated copies of genes. (We discuss these ideas in Chapter 22.)

No common plant species embodies the evolutionary impact of polyploidy more dramatically than *Triticum aestivum*, common bread wheat. *Triticum aestivum* is a naturally occurring allohexaploid ($6n$) that developed through the union of diploid genomes of three ancestral species in two hybridization events (**Figure 13.9**). Modern members of the genus *Triticum* have 14 ($2n$), 28 ($4n$), and 42 ($6n$) chromosomes. The evolutionary history of modern wheat begins with the hybridization of two diploid species that contain 14 chromosomes each. Einkorn wheat (*T. monococcum*) is a cultivated variety of wheat that can still be found around the world. Represented by the genomic formula $AA$, *T. monococcum* hybridized with a wild grass, *T. searsii* (genomic formula $BB$), to form an allotetraploid variety called Emmer wheat (*T. turgidum*). Emmer wheat (genomic formula $AABB$) was being cultivated approximately 10,000 years ago when it underwent a second hybridization event with another wild diploid grass species, *T. tauschii* (genomic formula $DD$), to form

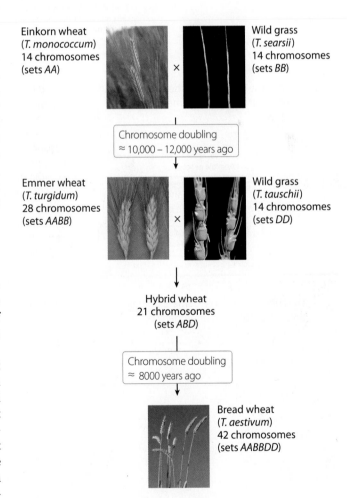

**Figure 13.9** The evolution of modern wheat (*Triticum aestivum*).

Flower color in an autotetraploid plant is a single-gene character with two alleles, $R_1$ and $R_2$, at the gene locus. The $R_1$ allele produces color, but the $R_2$ allele does not. As a consequence, flower-color intensity is determined by the number of $R_1$ alleles in the genotype. The genotype–phenotype correspondence is as follows:

| Genotype | Phenotype |
|---|---|
| $R_1R_1R_1R_1$ | Dark red |
| $R_1R_1R_1R_2$ | Light red |
| $R_1R_1R_2R_2$ | Pink |
| $R_1R_2R_2R_2$ | Light pink |
| $R_2R_2R_2R_2$ | White |

A pink-flowered plant is self-fertilized. What are the expected flower-color phenotypes, and in what proportions are they expected?

| Solution Strategies | Solution Steps |
|---|---|

### Evaluate

1. Identify the topic this problem addresses, and explain the nature of the required answer.

2. Identify the critical information given in the problem.

1. This problem concerns self-fertilization of an autotetraploid. The answer requires determination of the phenotypes of progeny and the expected frequency of each phenotype.

2. The plant is identified as an autotetraploid, and the specific genotype–phenotype relationships are given.

TIP: Autotetraploids are $4n$ and carry four homologous chromosomes derived from a single species.

### Deduce

3. Identify the genotype of the self-fertilized plant and the possible gametes it produces.

TIP: The gametes of an autotetraploid are diploid. Each gamete contains two of the chromosomes in the tetraploid genotype.

3. The genotype of the pink-flowered plant is $R_1R_1R_2R_2$. The gametes will be diploid. Six random combinations of chromosomes can form during gametogenesis. The first $R_1$ chromosome can occur in a gamete with the second $R_1$ or with either of the $R_2$ chromosomes, forming three of the gametes. The second $R_1$ can occur with either of the $R_2$ chromosomes, forming two more gametes, or the two $R_2$ chromosomes can form a gamete, making the sixth combination.

4. Determine the genotype and expected frequency of each possible gamete.

TIP: Add the predicted frequencies of the gametes to be sure their sum is 1.0.

4. Each combination of chromosomes in the gametes will form with equal frequency, meaning that the expected frequency of each gamete is 1/6. One combination contains both of the $R_1$ chromosomes, and one contains both of the $R_2$ chromosomes. The remaining gametes are different combinations with the genotype $R_1R_2$, for a combined frequency of 4/6.

### Solve

5. Describe the possible gamete unions and the production of progeny by fertilization.

TIP: Use a Punnett square to display gamete unions.

5. The results of union of the three gamete genotypes are as follows:

| | $R_1R_1$ $\left(\frac{1}{6}\right)$ | $R_1R_2$ $\left(\frac{4}{6}\right)$ | $R_2R_2$ $\left(\frac{1}{6}\right)$ |
|---|---|---|---|
| $R_1R_1$ $\left(\frac{1}{6}\right)$ | $R_1R_1R_1R_1$ $\left(\frac{1}{36}\right)$ | $R_1R_1R_1R_2$ $\left(\frac{4}{36}\right)$ | $R_1R_1R_2R_2$ $\left(\frac{1}{36}\right)$ |
| $R_1R_2$ $\left(\frac{4}{6}\right)$ | $R_1R_1R_1R_2$ $\left(\frac{4}{36}\right)$ | $R_1R_1R_2R_2$ $\left(\frac{16}{36}\right)$ | $R_1R_2R_2R_2$ $\left(\frac{4}{36}\right)$ |
| $R_2R_2$ $\left(\frac{1}{6}\right)$ | $R_1R_1R_2R_2$ $\left(\frac{1}{36}\right)$ | $R_1R_2R_2R_2$ $\left(\frac{4}{36}\right)$ | $R_2R_2R_2R_2$ $\left(\frac{1}{36}\right)$ |

6. Summarize the genotypes, phenotypes, and frequencies expected from this cross.

> TIP: Add the predicted frequencies to be sure their sum is 1.0.

6. Self-fertilization of a pink plant with the $R_1R_1R_2R_2$ genotype is expected to produce the following outcome:

| Genotype | Phenotype | Frequency |
|---|---|---|
| $R_1R_1R_1R_1$ | Dark red | 1/36 |
| $R_1R_1R_1R_2$ | Light red | 8/36 |
| $R_1R_1R_2R_2$ | Pink | 18/36 |
| $R_1R_2R_2R_2$ | Light pink | 8/36 |
| $R_2R_2R_2R_2$ | White | 1/36 |

For more practice, see Problems 1, 2, and 11.

*T. aestivum* (genomic formula *AABBDD*), the modern allohexaploid species. The latter event may have resulted from human intervention during the process of domestication of wheat.

---

Genetic Insight  Polyploidy increases fruit and flower size, causes changes in fertility, and alters growth vigor. Despite having possible growth and reproductive disadvantages, polyploidy offers multiple evolutionary advantages.

---

## 13.3 Chromosome Breakage Causes Mutation by Loss, Gain, and Rearrangement of Chromosomes

We have seen that the proper balance of gene dosage is important for promoting normal growth and development and that changes in gene dosage can have substantial phenotypic consequences. For this reason, mutations that result in the loss or gain of chromosome segments also have the potential to produce severe abnormalities. Such changes may be large enough to be detected by microscopy and can affect extensive regions of chromosomes containing potentially scores to hundreds of genes. Alternatively, the changes may be so small that their detection is visually impossible by conventional microscopy. These chromosome alterations can sometimes be detected with molecular methods. In this section, we examine alterations to chromosome structure by chromosome breakage and other events that lead to the loss or gain of chromosomal segments.

### Partial Chromosome Deletion

When a chromosome breaks, both strands of DNA are severed at a location called a **chromosome break point.** The broken chromosome ends at a break point retain their chromatin structure, and they can adhere to other broken chromosome ends or to the ends of intact chromosomes. Alternatively, since part of the broken chromosome is **acentric** (without a centromere), it can be lost during cell division.

Chromosome breakage can result in **partial chromosome deletion,** by the loss of a portion of a chromosome. The size of the deletion and the specific genes deleted are significant factors in the degree of ensuing phenotypic abnormality. Small deletions known as **microdeletions** affect one or a few genes and are too small to be readily seen by conventional microscopy. Microdeletions may behave as recessive mutations if the remaining single copies of genes on the homologous unbroken chromosomes are haplosufficient. If one or more of the deleted genes is dosage sensitive, however, the mutation will act as a dominant mutation and may have severe consequences.

Larger chromosome deletions are detected by microscopy through the observation of altered chromosome banding patterns. In these larger deletions, more genes are affected, and the likelihood of substantial phenotypic consequences is very high. A chromosome break that detaches one arm of a chromosome leads to a **terminal deletion** (Figure 13.10a). The chromosome fragment broken off in terminal deletion contains one of the chromosome ends, or *termini*, consisting of a telomere and additional genetic material. Without a centromere, the acentric fragment lacks a kinetochore. It is unable to attach spindle fibers and cannot migrate to a pole of the cell during division. Acentric chromosome fragments are lost during cell division. Organisms carrying one wild-type chromosome and a homolog with a terminal deletion are called **partial deletion heterozygotes.** A human condition known as cri-du-chat syndrome (OMIM 123450) is an example of a chromosome syndrome caused by terminal deletion of 5p15.2–5p15.3 (Figure 13.10b). The syndrome is named for the distinctive cat-cry-like sound emitted by infants with the condition.

In contrast to a terminal deletion, which results from a single break at one end of a chromosome, an **interstitial**

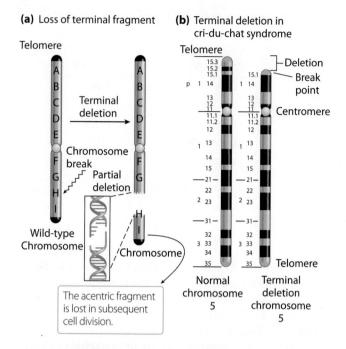

**(a)** Loss of terminal fragment

**(b)** Terminal deletion in cri-du-chat syndrome

**Figure 13.10 Chromosome terminal deletion. (a)** A double-stranded DNA break at a chromosome break point in region H leads to terminal deletion of the acentric fragment. **(b)** Terminal deletion of chromosome 5 in cri-du-chat syndrome.

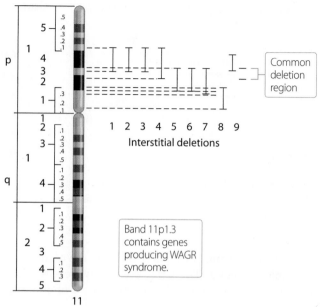

**Figure 13.11 Interstitial deletions of chromosome 11p1.3 in WAGR syndrome.** Deletions 1 through 7 result in WAGR, but deletions 8 and 9 do not. The smallest common deletion region is band 11p1.3.

**deletion** is the loss of an internal segment of a chromosome that results from two chromosome breaks. Interstitial deletions can be seen in many organisms, including humans. WAGR syndrome, an interstitial deletion condition seen in humans, is actually a series of conditions caused by deletion of multiple contiguous genes in the vicinity of 11p1.3 (**Figure 13.11**). The initials WAGR stand for **W**ilms tumor (a type of hereditary kidney cancer), **a**niridia (the absence of the iris in the eye), **g**enitourinary abnormalities, and mental **r**etardation. Patients with the largest deletions of 11p1.3 have all four conditions, whereas patients with smaller deletions may have just one or two of the disorders.

## Unequal Crossover

The processes of reciprocal recombination achieves the recombination of alleles on homologous chromosomes without causing a gain or loss of chromosomal material that would result in mutation (see Chapters 5 and 12). Not every recombination event between homologs is reciprocal, however, and occasionally an **unequal crossover** between homologs takes place. This can result in the **partial duplication** and **partial deletion** of genetic material on the resulting recombinant chromosomes. An organism carrying one homolog with duplicated material is a **partial duplication heterozygote,** whereas one with material deleted from one chromosome is a **partial deletion heterozygote.** We introduced unequal crossover in an earlier chapter to explain the most common mutation

causing X-linked recessive color blindness (see Chapter 4 Case Study).

Unequal crossover is rare and occurs most commonly when repetitive regions of homologous chromosomes misalign. The human condition known as Williams-Beuren syndrome (WBS; OMIM 194050) is frequently found in partial deletion heterozygotes for a segment of chromosome 7. In wild-type chromosome 7, this region contains duplicate copies of the gene *PMS*, designated $PMS_A$ and $PMS_B$, that are located near one another and have 17 genes located in between (**Figure 13.12a**). Misalignment of the homologous chromosomes results in mispairing of $PMS_A$ on one chromosome with $PMS_B$ on the homologous chromosome. A copy of *PMS* on each chromosome is looped out from each homolog during misalignment (**Figure 13.12b**). Unequal crossing over between the misaligned chromosomes results in one recombinant chromosome that has a partial deletion chromosome 7 that results in WBS. This chromosome contains a nonfunctional hybrid $PMS_A$-$PMS_B$ gene that is missing intact $PMS_A$ and $PMS_B$ genes as well as the 17 genes normally found between $PMS_A$ and $PMS_B$ (**Figure 13.12c**). The partial duplication chromosome (containing duplicated copies of the hybrid $PMS_A$-$PMS_B$ gene and the 17 intervening genes) does not cause readily identifiable phenotypic abnormalities.

## Detecting Duplication and Deletion

Large deletions or duplications of chromosome segments can be detected by microscopic examination that reveals

**(a)** Normal chromosome 7 structure

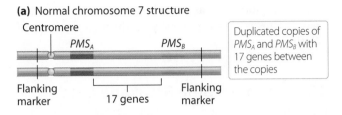

Duplicated copies of $PMS_A$ and $PMS_B$ with 17 genes between the copies

**(b)** Homologous chromosome misalignment and unequal crossover

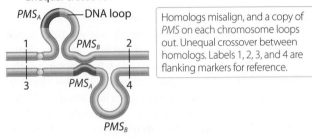

Homologs misalign, and a copy of *PMS* on each chromosome loops out. Unequal crossover between homologs. Labels 1, 2, 3, and 4 are flanking markers for reference.

**(c)** Deletion and duplication recombinant chromosomes

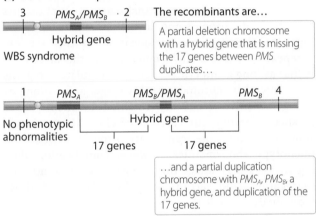

The recombinants are…

A partial deletion chromosome with a hybrid gene that is missing the 17 genes between *PMS* duplicates…

…and a partial duplication chromosome with $PMS_A$, $PMS_B$, a hybrid gene, and duplication of the 17 genes.

**Figure 13.12** **Unequal crossover in creation of Williams-Beuren syndrome.**

**(a)** Wild-type chromosome

FISH probes *A*    *B C*

**(b)** Microinterstitial deletion

*A*    *B C*

No fluoresence detected from probe B.

**(c)** Microduplication

*A*    *B C*

Two fluorescent spots indicate the target of probe B is duplicated.

**Figure 13.13** **Detection of chromosome microdeletion and microduplication by FISH.** **(a)** Three FISH probes identify genes *A*, *B*, and *C*. **(b)** Microdeletion of a chromosome segment containing *B* prevents probe hybridization. **(c)** Microduplication results in hybridization of probe *B* to duplicated genes.

altered chromosome banding patterns resulting from the structural change to the chromosome. Such deletions and duplications are generally quite large. In human chromosomes, duplications and deletions of about 100,000 to 200,000 base pairs are at the lower limit of chromosome banding visualization. Microdeletions and microduplications are considerably smaller and are generally not easily detected by chromosome banding analysis. Instead, molecular techniques such as FISH (fluorescent in situ hybridization) can be used to detect the absence or duplication of a particular gene or chromosome sequence (**Figure 13.13**; also see Chapter 11).

An alternative microscopic observation indicating partial chromosome deletion or duplication can be made at prophase I homologous chromosome synapsis during meiosis. Homologous pairs that are mismatched because one contains a large duplication or deletion will form an **unpaired loop** in synapsis (**Figure 13.14**). Along most of the length of the homologous pair, normal synaptic

pairing occurs. But in regions of structural difference, the extra material present on one chromosome bulges out to allow synaptic pairing on either side. The material in the loop is normal genetic material if one chromosome carries a deletion, and it is duplicated genetic material if one homolog carries a duplication.

**Genetic Insight** Partial deletion and partial duplication create abnormalities by affecting genes that are dosage sensitive. Larger deletions and duplications alter chromosome banding patterns and affect synapsis between homologs, whereas microdeletions and microduplications are detected by molecular analysis.

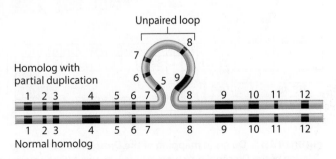

**Figure 13.14** **An unpaired loop at synapsis.** The partial duplication heterozygote shown here has duplicated genetic material of bands 6 through 9. The extra material forms an unpaired loop at synapsis to allow homologous regions to align correctly.

## Deletion Mapping

**Pseudodominance** is a genetic phenomenon that occurs when a normally recessive allele is "unmasked" and expressed in the phenotype because the dominant allele on the homologous chromosome has been deleted. Pseudodominance is used to map genes in deleted chromosome regions by a method known as **deletion mapping.**

**Figure 13.15** shows deletion mapping using pseudodominance to map the *Notch* gene (*n*) in *Drosophila*. The *Notch* gene resides on the X chromosome, and its location is revealed by the correlation of pseudodominance in partial X-chromosome deletion strains. Pseudodominance appears in females that are heterozygous for the partial deletion, carry the recessive allele on the intact X chromosome, and have lost the dominant allele by partial X-chromosome deletion. In the figure, the gray segments represent chromosome segments present and color identifies segments that have been deleted. The first two partial deletions (rJ1 and 258-42) do not lead to pseudodominance (in other words, the dominant wild-type phenotype is observed), indicating that the regions deleted do not contain the *Notch* gene. The other two partial deletions, 62d18 and N71a, result in pseudodominance, indicating that the *Notch* gene locus containing the dominant allele is in the region 3C5 to 3C9. To home in on the location of *Notch*, progressively smaller partial deletions are used to identify the smallest deletion segment common to all deletions resulting in pseudodominance. In this instance, and using additional data not shown, the smallest partial deletion common to genomes expressing pseudodominance for

Notch is region 3C-7, and this is where the gene resides. Genetic Analysis 13.2 guides you through analysis of deletion mapping.

## 13.4 Chromosome Breakage Leads to Inversion and Translocation of Chromosomes

Chromosome breakage involves double-strand DNA breaks that sever a chromosome. Breakage that is not followed by reattachment of the broken segment leads to partial chromosome deletion; but what happens if the broken chromosome reassembles but the broken segment reattaches in the wrong orientation or if the broken segment reattaches to a nonhomologous chromosome? The answers are that reattachment in the wrong orientation produces a **chromosome inversion,** whereas attachment to a nonhomologous chromosome results in **chromosome translocation.** We discuss two types of chromosome inversion events and two types of chromosome translocation in this section. A repeating theme that will emerge from this discussion is that as long as no critical genes or regulatory regions are mutated by chromosome breakage, and as long as dosage-sensitive genes are retained in their proper balance, heterozygous carriers of chromosome inversion or chromosome translocation may experience no phenotypic abnormalities. However, complications during meiosis may affect the efficiency of chromosome segregation, and fertility may be affected in those individuals.

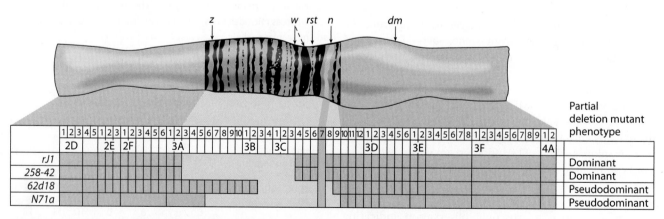

**Figure 13.15** **Deletion mapping of the *Drosophila Notch* (*n*) gene.** The extent of each partial deletion of the *Drosophila* X chromosome is shown by the colored bars for four partial deletion mutants. The retention of the dominant character or the emergence of notch by pseudodominance is indicated. The smallest X-chromosome segment missing from all pseudodominant mutants is region 3C-7, indicating this as the location of the gene.

In *Drosophila*, the X-linked recessive mutant traits singed bristle, lozenge eye, and cut wing are encoded at linked genes. Five strains of *Drosophila* produced by the cross of pure-breeding wild-type and pure-breeding mutant flies (*SLC/SLC × slc/slc*) are expected to have the trihybrid genotype *SLC/slc* and express the wild-type phenotypes. Females of each strain exhibit pseudodominance for one or more of the traits, however, due to partial deletion of the X chromosome.

Comparative X-chromosome maps showing the extent of deletions in each pseudodominant strain (indicated by dashed lines) are given below along with the pseudodominant phenotypes found in each strain. Use this information to locate each gene as accurately as possible along the X chromosome.

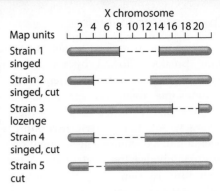

| Solution Strategies | Solution Steps |
|---|---|
| **Evaluate** | |
| 1. Identify the topic this problem addresses, and explain the nature of the required answer. | 1. This problem addresses deletion mapping using pseudodominance to locate the position of each gene. The answer requires construction of a map of gene locations. |
| 2. Identify the critical information given in the problem. | 2. The deletion regions on chromosomes and the corresponding pseudodominant phenotypes are given. |
| **Deduce** | |
| 3. Review the meaning of pseudodominance and the connection between chromosome deletion and pseudodominance. | 3. Pseudodominance is the appearance of a recessive trait in a presumed heterozygous organism due to deletion of a chromosome segment carrying the dominant allele. In deletion mapping using pseudodominance, the location of a gene maps to the smallest common deletion region shared by all organisms expressing the pseudodominant trait. |
| **Solve** | |
| 4. Interpret the meaning of the pseudodominant phenotype in strain 1. | 4. Strain 1 is missing chromosome material from the 8th to the 14th map unit. The appearance of the pseudodominant phenotype singed indicates that the *singed* gene maps to this interval. |
| 5. Compare strain 2 to strain 1, and interpret the meaning of the new pseudodominant phenotype cut. <br><br> **TIP:** Compare deletion mutants that share pseudodominance phenotypes to see where their deletions overlap. | 5. Strain 2 has a deletion from map units 4 to 13 that includes both *singed* and *cut*. <br><br> This narrows the location of *singed* to the interval between 8 and 13 map units. The *cut* location is between the 4th and 8th map unit, based on its appearance with the deletion of this interval. |
| 6. Assess pseudodominance of strain 3. | 6. Co-occurrence of the deletion between map units 16 and 20 and the appearance of the pseudodominant lozenge phenotype map the *lozenge* gene to this location. |
| 7. Assess strains 4 and 5, and refine the locations of the genes further where possible. <br><br> **TIP:** Again, compare deletion mutants that share pseudodominance phenotypes to see where their deletions overlap. | 7. Strain 4 contains a deletion between map units 4 and 12 and confines the location of *singed* to the interval between 8 and 12. This strain provides no additional information about the location of *cut*. <br><br> The deletion between map units 3 and 6 in strain 5 includes *cut* and refines its location to between map units 4 and 6. |
| 8. Identify gene locations based on the deletion mapping analysis. | 8. Based on the data for pseudodominance in these five strains, *cut* resides in the interval between units 4 and 6, *singed* lies between 8 and 12, and *lozenge* is between 16 and 20. |

For more practice, see Problems 4, 10, and 26.

## Chromosome Inversion

Chromosome inversions occur as a result of chromosome breaks followed by reattachment of the free segment in the reverse orientation. Two kinds of chromosome inversion are observed, depending on whether the centromere is part of the inverted segment (**Figure 13.16**). **Paracentric inversion** results from the inversion of a chromosome segment on a single arm and *does not* involve the centromere, whereas **pericentric inversion** reorients a chromosome segment that *includes* the centromere.

Inversion most commonly affects just one member of a homologous pair, and such organisms are either paracentric or pericentric **inversion heterozygotes** in which one chromosome has normal structure and the homolog contains an inversion. Inversion heterozygotes may experience no genetic or phenotypic abnormalities, as long as no critical genes or regulatory DNA sequences are disrupted by chromosome breaks. In such cases, the 180-degree reorientation of inverted

segments does not change the genetic content or gene expression of the affected chromosome.

Chromosome inversion does, however, cause a difference in linear order of genes between the homologs; thus, to bring the homologs of an inversion heterozygote into synaptic alignment during meiosis requires the formation of an unusual **inversion loop** at synapsis. Note, however, that an organism that is homozygous for an inversion carries the same order of genes and chromosome regions on both homologs and therefore will experience normal chromosome synapsis without the need for inversion loop formation.

In inversion heterozygotes, inversion loop formation readily occurs and does not affect subsequent chromosome segregation. Crossing over takes place between the homologs; but whereas crossing over that occurs *outside* the region spanned by the inversion loop takes place in the normal manner, crossing over *inside* the region of the inversion loop results in duplications and deletions among the recombinant chromosomes. **Figure 13.17** illustrates crossover within the inversion loop between chromosome regions B and C in a paracentric inversion heterozygote. Following crossover, one normal-order chromosome (1·ABCDEFGHI-1') and one inverted-order chromosome (3·ADCBEFGHI-3') are unchanged by recombination (the dot represents the centromere). The recombinant chromosomes, however, are abnormal: One is a **dicentric chromosome** with two centromeres (2·ABCDA·4), and the other is an acentric fragment that has no centromere (2'-IHGFEDCBEFGHI-4'). At anaphase I, when centromeres on homologous chromosomes normally migrate toward opposite poles, a **dicentric bridge** forms as the dicentric chromosome is pulled toward both poles of the cell. Eventually the bridge snaps under the tension, at a random break point. Both products of the break have a centromere, but both are also missing genetic material. In contrast, the acentric fragment, lacking a centromere, has no mechanism by which to migrate to a pole of the cell and will be lost during meiosis. The completion of meiosis of this paracentric inversion heterozygote results in two viable gametes, one with the normal-order chromosome (1·ABCDEFGHI-1') and one with the inverted-order chromosome (3·ADCBEFGHI-3'), and two nonviable gametes with partial deletion chromosomes.

Crossover in the inversion loop in a pericentric inversion heterozygote yields two viable gametes and two nonviable gametes. One viable gamete contains the normal-order chromosome (1 ABCDE·FGHI-1') and one containing the inversion-order chromosome (3 ABCHGF·EDI-3'). Crossover also results in two nonviable gametes, each having a combination of deletions and duplications (2 ABCDE·FGHCBA-4 and (4'-IDE·FGHI-2') (**Figure 13.18**).

Three observations about recombination in inversion heterozygotes have important genetic implications:

1. **The probability of crossover within the inversion loop is linked to the size of the inversion loop.**

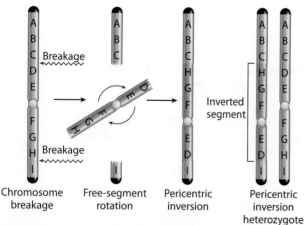

**(a)** Paracentric inversion

Centromere

Breakage

Breakage

Chromosome breakage · Free-segment rotation · Paracentric inversion · Paracentric inversion heterozygote

Inverted segment

**(b)** Pericentric inversion

Breakage

Breakage

Chromosome breakage · Free-segment rotation · Pericentric inversion · Pericentric inversion heterozygote

Inverted segment

**Figure 13.16 Paracentric and pericentric chromosome inversion.**

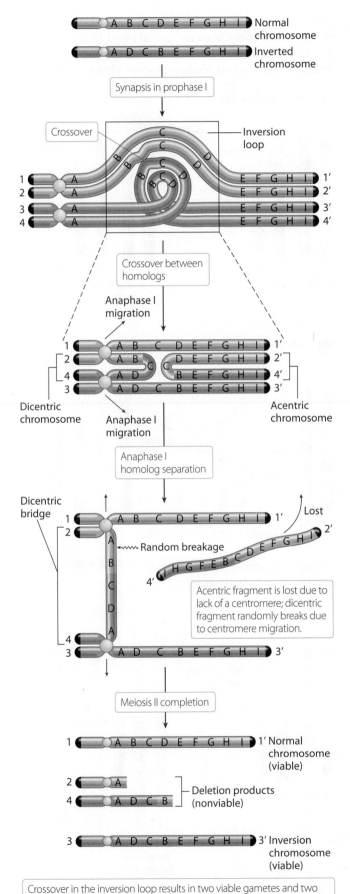

**Figure 13.17** The consequences of crossover in the inversion loop in paracentric inversion heterozygotes.

Small inversions produce small inversion loops that have a low frequency of crossover. On the other hand, larger inversions produce loops that span more of the chromosome and correlate with a higher probability of crossover.

2. **Inversion suppresses the production of recombinant chromosomes.**   The viable gametes produced by inversion heterozygotes contain either the normal-order chromosome or the inversion-order chromosome, but no recombinant chromosomes are viable, due to duplications and deletions of chromosome segments. The absence of recombinant chromosomes in progeny is identified as **crossover suppression.** Actually, crossovers do occur between homologous chromosomes carried by inversion heterozygotes, but because the recombinant chromosomes contain duplications and deletions, there is little possibility of viability for any progeny formed from the gametes that contain them. Geneticists take advantage of crossover suppression to mark homologous chromosomes with dominant alleles that aid in the interpretation of genetic crosses. **Experimental Insight 13.1** describes research by Hermann Müller, who used the so-called ClB chromosome to identify and later investigate lethal X-linked mutations induced in *Drosophila* by X-ray exposure.

3. **Fertility may be altered if an inversion heterozygote carries a very large inversion.**   When an inversion spans all or nearly all the length of a chromosome, at least one crossover will occur. Since meiosis produces two viable and two nonviable gametes, approximately half the gametes will be lost. No such loss of fertility is expected for organisms with small inversions.

---

Genetic Insight   Inversions rearrange chromosome structure by a 180-degree rotation of chromosome segments. Inversion heterozygotes experience crossover suppression through nonviability of any gametes carrying chromosomes resulting from crossover within the inversion loop.

---

## Chromosome Translocation

Chromosome translocation takes place following chromosome breakage and the reattachment of a broken segment to a *nonhomologous* chromosome. If no critical genes are severed or have their regulation disrupted by the breakage or translocation events, **translocation heterozygotes,**

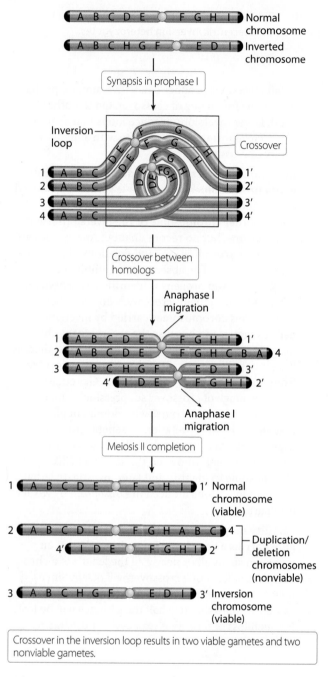

Figure 13.18 **The consequences of crossover in the inversion loop in pericentric inversion heterozygotes.**

chromosome is translocated to a nonhomologous chromosome and there is no reciprocal event (**Figure 13.19a**). **Reciprocal balanced translocation** is produced when breaks occur on two nonhomologous chromosomes and

**(a)** Unbalanced translocation

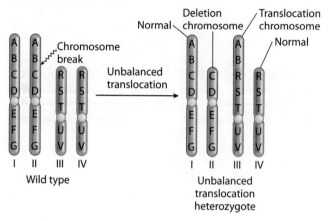

**(b)** Reciprocal balanced translocation

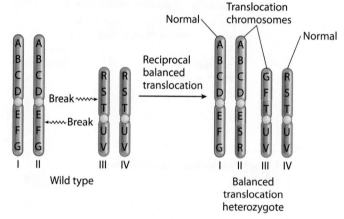

**(c)** Robertsonian translocation (chromosome fusion)

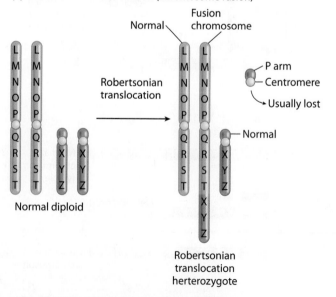

Figure 13.19 **Unbalanced, reciprocal balanced, and Robertsonian chromosome translocations.**

with one normal chromosome and one altered chromosome in each homologous pair, have a normal outward phenotype and a normal pattern of gene expression. Even if no phenotypic abnormalities are detected, however, certain translocation heterozygotes can experience semisterility as a result of abnormalities of chromosome segregation, as we describe below.

Three principal types of translocation are observed. **Unbalanced translocation** arises from a chromosome break and subsequent reattachment to a nonhomologous chromosome in a one-way event; that is, a piece of one

## Experimental Insight 13.1

### Hermann Müller and the *Drosophila* C/B Chromosome Method

Hermann Müller, a student of Thomas Hunt Morgan, made numerous important contributions to genetics. Among Müller's accomplishments were his discovery that X-rays induce mutations by chromosome breakage and his development of a genetic method to identify lethal X-ray–induced mutations of the X chromosome in *Drosophila*.

To identify these mutations, Müller created an X chromosome called the C/B chromosome: C for **c**rossover suppression, *l* for presence of a recessive **l**ethal mutation, and B for a dominant mutation producing an abnormal **b**ar-shaped eye. Crossover suppression results from the presence of multiple inversions that prevent the appearance of recombinants between inverted and wild-type X chromosomes in females. Bar eye is a dominant mutant phenotype that permanently marks the inversion chromosome, since it cannot be reshuffled by recombination. Potentially lethal recessive mutations (*m?*) are generated on male X chromosomes by X-ray exposure.

*Drosophila* males that are hemizygous for C/B (C/B/Y) die as a result of the lethal mutation (*l*) on the X chromosome. Female carriers of C/B (C/B/+) survive and preserve the chromosome. Müller began his search for lethal X-ray–induced mutations by exposing male fruit flies to X-rays to induce mutations in germ-line cells. X-ray–exposed males were then crossed to a bar-eyed female (C/B/+), in Cross I. Next, bar-eyed female progeny from Cross I were individually mated to wild-type males, in Cross II. Cross II would be expected to produce a 2:1 ratio of females to males if X-ray exposure *did not* induce a lethal mutation on the X chromosome. In this case, only males inheriting the C/B chromosome would die. If on the other hand a lethal mutation was induced, only female progeny would be produced by Cross II. Males inheriting the C/B chromosome would die, but so would males inheriting the X chromosome with the induced lethal mutation.

Identifying X-ray–induced lethal mutations using the C/B method is highly accurate: It requires only a determination of whether males are produced by Cross II. Müller recognized that when X-ray exposure induced a lethal mutation, he could study it by means of the Cross II females with normal eyes, which are heterozygous carriers of the induced lethal mutation. Müller used the C/B method to demonstrate that X-ray exposure induces mutations at a rate more than 150 times greater than the spontaneous mutation rate in *Drosophila*. His work led to the characterization of numerous mutations and to the identification of the linear relationship between the level of X-ray exposure and the frequency of induced lethal mutations.

**Müller's C/B method.** X-ray–exposed males are mated to bar-eyed females carrying the C/B chromosome in Cross I. Progeny bar-eyed females that potentially carry a lethal X-linked mutation [*m(?)*] are crossed to wild-type males in Cross II. The absence of male progeny from Cross II identifies the occurrence of an induced lethal mutation.

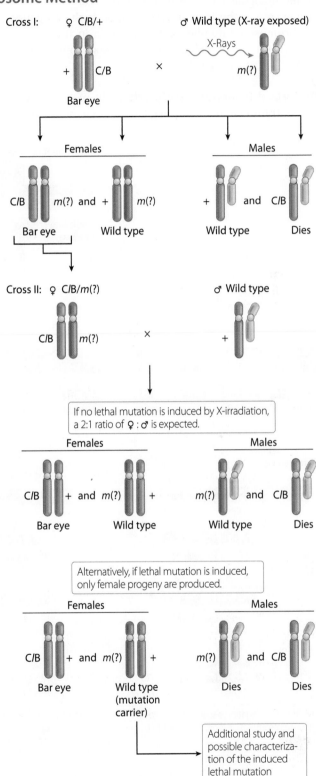

the resulting fragments switch places when they are reattached (Figure 13.19b). **Robertsonian translocation,** also known as **chromosome fusion,** involves the fusion of two nonhomologous chromosomes (Figure 13.19c). One consequence of Robertsonian translocation is the reduction of chromosome number. Our discussion in this section focuses on reciprocal balanced translocations and on Robertsonian translocations.

**Reciprocal Balanced Translocation** In reciprocal balanced translocation, one member of each homologous pair is altered by translocation, and none of the four chromosomes has a fully homologous partner. Instead, the translocated chromosome segments homologous to the normal member of each pair are dispersed on two other chromosomes. The absence of complete homology between chromosome pairs requires formation of an unusual tetravalent synaptic structure, a cross-like configuration made up of the four chromosomes related by the translocation, to enable homologous regions to synapse during metaphase I, as shown in **Figure 13.20**. The chromosomes in the figure are labeled I, II, III, and IV so that

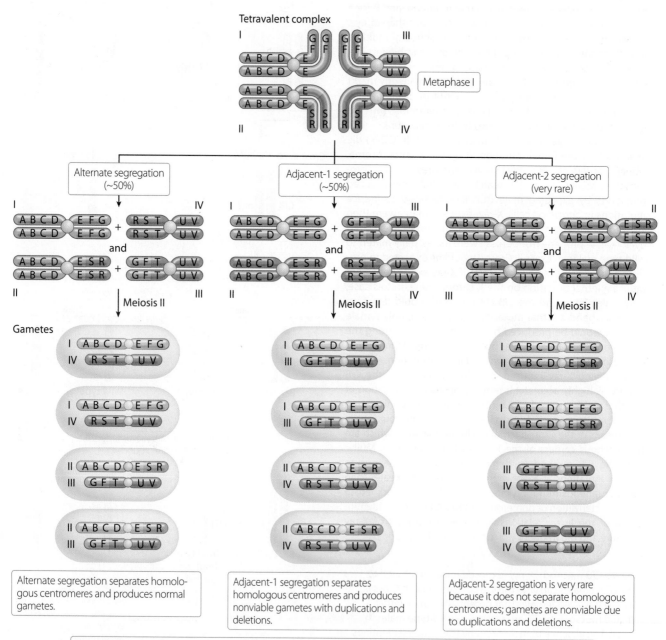

**Figure 13.20** **The tetravalent synaptic structure and alternate and adjacent chromosome segregation in reciprocal balanced translocation heterozygotes.**

we may more easily follow their progress in meiosis and meiotic outcomes.

Two patterns of chromosome segregation emerge from the tetravalent structures found in translocation heterozygotes. *Alternate segregation* and *adjacent-1 segregation* each occur in approximately 50% of meiotic divisions, although the actual proportions vary somewhat among different species. At anaphase I in **alternate segregation,** chromosomes I and IV move to one cell pole and chromosomes II and III move to the opposite pole. At the completion of meiosis, all gametes are viable because each contains a complete set of genetic information for the two chromosomes. Fertilization of a gamete containing chromosomes I and IV will produce a normal zygote, whereas fertilization of a gamete containing chromosomes II and III will produce a zygote with reciprocal balanced translocation heterozygosity, like the parent illustrated in the figure.

In anaphase I of **adjacent-1 segregation,** chromosomes I and III are moved to one cell pole and chromosomes II and IV go to the opposite pole. None of the gametes formed by this pattern of segregation is viable, because of duplications and deletions of genetic information. Gametes containing chromosomes I and III have a duplication of the F and G regions, along with deletion of the R and S regions. Conversely, gametes containing chromosomes II and IV have a duplication of the R and S regions and a deletion of regions F and G.

Occasionally, an unusual pattern of segregation known as adjacent-2 segregation takes place. It is rare because it requires that chromosomes I and II, which share homologous centromeres, move to the same pole of the cell at anaphase I. Correspondingly, chromosomes III and IV, which also share homologous centromeres, also move to the same cell pole (opposite chromosomes I and II). This is atypical of the usual pattern at anaphase I, in which homologous chromosomes (that carry homologous centromeres) are separated in the reduction division. None of the gametes or progeny resulting from adjacent-2 segregation are viable.

In summary, cell biologists conclude that in balanced translocation heterozygotes, only alternate segregation produces viable gametes and viable progeny. This pattern accounts for just one-half of all meiotic events in these individuals; thus, the semisterility of translocation heterozygotes is due to reduction by about one-half in the number of viable gametes that can be produced.

**Robertsonian Translocation** In organisms with a Robertsonian translocation, also known as chromosome fusion, two nonhomologous chromosomes fuse to form a single, larger chromosome, resulting in a reduction in chromosome number. If two pairs of chromosomes fuse by Robertsonian translocation, the number of chromosomes in a genome is reduced to $2n - 2$. This is a frequently observed mechanism by which chromosome number evolves

in related organisms. This mechanism accounts for the difference in chromosome number between human ($2n = 46$) and chimpanzee ($2n = 48$), as discussed in the Case Study. If multiple chromosomes undergo Robertsonian translocation, as was the case with mice on Madeira, larger reductions in chromosome number occur.

Carriers of a single Robertsonian translocation have one chromosome fusion. The homologs of the fused chromosomes remain separate chromosomes. **Figure 13.21** illustrates this pattern of Robertsonian translocation in humans in a condition called *familial Down syndrome* that is the cause of 5–10% of Down syndrome (trisomy 21) cases. Familial Down syndrome occurs when one parent is a carrier of a Robertsonian translocation of chromosome 21 to another autosome, most often chromosome 14. The translocation-heterozygous parent has a normal diploid

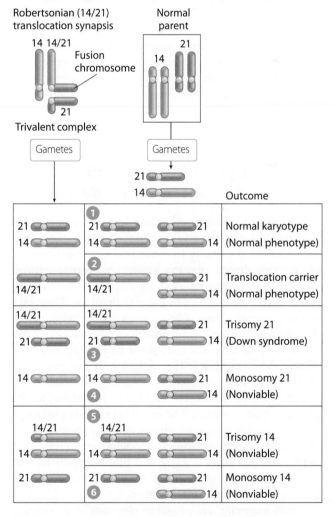

**Figure 13.21  Familial Down syndrome due to Robertsonian translocation.** For reproduction between a 14/21 Robertsonian translocation carrier and an individual with a normal karyotype, three nonviable zygotes (categories 4, 5, and 6) and three viable zygotes (categories 1, 2, and 3) are possible. Approximately one-third of the children from such unions (category 3) have trisomy 21 (Down syndrome).

genotype produced by a complete copy of chromosome 14, a complete copy of 21, and a 14/21 fusion chromosome. The fusion chromosome has lost the short arms of chromosome 14 and chromosome 21; but these contain no critical genetic information, and so the Robertsonian translocation carriers have a normal phenotype. Three possible patterns of segregation of the three chromosomes are equally likely following formation of the trivalent complex. Six possible gametes produced by these patterns are shown in the left column of the figure. When united with a normal gamete, three of the six possible gamete types result in nonviable zygotes (categories 4, 5, and 6 in the figure). The other three types of gametes produce viable zygotes (categories 1, 2, and 3). Two have normal phenotype and one, category 3, has Down syndrome. This form of Robertsonian translocation heterozygosity leads to about a 1 in 3 chance of producing a child with trisomy 21, and this high risk is present each time a child is conceived.

---

Genetic Insight   Chromosome translocation transfers all or part of a chromosome to a nonhomologous chromosome. In translocation carriers, abnormal synaptic structures must form, and meiotic chromosome segregation errors produce semisterility.

---

## 13.5 Transposable Genetic Elements Move throughout the Genome

**Transposable genetic elements** are DNA sequences of various lengths and sequence composition that have evolved the ability to move within the genome by an enzyme-driven process known as **transposition.** Transposable elements exist in dozens of forms that range in size from 50 bp to more than 10 kb. They vary in copy number from a few copies up to hundreds of thousands of copies. Transposable elements can create mutations, usually by their insertion into wild-type alleles. One example is the mutation of the pea plant *SBE1* gene that generates wrinkled peas studied by Mendel (see Table 2.6). The *SBE1* gene is mutated in pea plants by transposition. The recessive allele is generated by the insertion of approximately 800 bp of transposable DNA into what was formerly the wild-type allele. This inactivates the mutant allele and blocks production of the starch branching enzyme. This mutational process is known as **insertional inactivation.**

Evolutionarily, transposable elements can increase genome size. Many transposable elements seem to have the sole function of increasing their own copy number. As a consequence, organisms carrying certain transposable elements derive no useful benefit from their presence. Alternatively, some transposable elements contain expressed genes that may benefit the organism. In this and the following two sections, we discuss transposable elements in bacterial and eukaryotic genomes, and their evolutionary relationships.

## The Discovery of Transposition

Barbara McClintock discovered transposition in a series of studies of a mutant phenotype of kernel color in maize (*Zea mays*). The *C* gene for kernel color is located on chromosome 9, where a dominant wild-type allele *C* produces purple kernels and a mutant $c_1$ allele produces colorless kernels. One gene linked to *C* produces plump (*Sh*) or shrunken (*sh*) kernels, and a second linked gene produces shiny (*Wx*) or waxy (*wx*) kernels (**Figure 13.22a**). In experiments with several trihybrid strains of maize with the genotype *C Sh Wx/$c_1$ sh wx*, McClintock found a few unusual kernels that were mostly purple but had colorless sectors that varied among different kernels. Invariably,

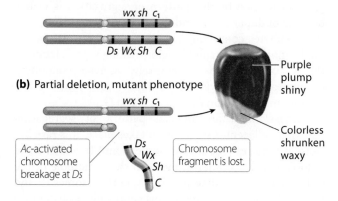

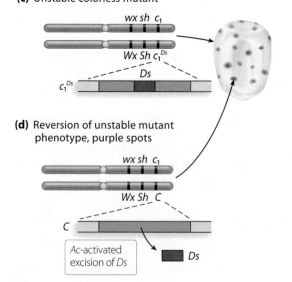

**(a)** Trihybrid, wild-type phenotype

**(b)** Partial deletion, mutant phenotype

*Ac*-activated chromosome breakage at *Ds*

Chromosome fragment is lost.

Purple plump shiny

Colorless shrunken waxy

**(c)** Unstable colorless mutant

**(d)** Reversion of unstable mutant phenotype, purple spots

*Ac*-activated excision of *Ds*

**Figure 13.22 Mutation of colorless sectors and reversion of the unstable colorless mutation in maize by the transposable genetic elements *Ds* and *Ac*.**

however, the purple regions were plump and shiny, but the colorless sectors were shrunken and waxy.

Looking at chromosome 9 in nuclei of cells from the colorless sectors of kernels, McClintock noticed a terminal deletion of one chromosome 9 homolog. In contrast, both chromosome 9 homologs were intact in cells from purple sectors. McClintock concluded that the simultaneous appearance of colorless, shrunken, and waxy resulted from pseudodominance due to deletion of the dominant alleles from one homolog (**Figure 13.22b**). Mitotic division of an original cell containing the chromosome deletion produced the abnormal sectors. The frequency of sectored kernels was too frequent to be a result of spontaneous chromosome mutation and, more importantly, McClintock saw that break points of chromosome 9 occurred in the same place in all affected kernels of a given strain. Based on these observations she concluded that a genetic element, later named a **dissociation (*Ds*) element,** was located at the site of chromosome breakage. What puzzled McClintock, however, was why *Ds* generated chromosome breakage in some cells but not in others. To explain this, she suggested that *Ds* alone could not generate chromosome breakage. Instead, chromosome breakage at *Ds* was activated by an unlinked genetic element she called an **activator (*Ac*) element.**

McClintock's *Ds/Ac* proposal proved to be the explanation for another highly unusual observation she made in maize. She found occasional colorless maize mutants that had an **unstable mutant phenotype.** These unstable mutants had kernels that were mostly colorless but also had purple spots. The patterns of purple spotting differed from kernel to kernel on the same maize ear, indicating that it developed by some sort of reversion in somatic cells that was perpetuated by subsequent mitotic division (**Figure 13.22c**). Her investigation led McClintock to conclude that the unstable mutant alleles were produced by the insertion of *Ds* into the *C* allele to form the mutant $c_1^{Ds}$ allele. This allele is mutated by the insertional inactivation process and as a result it produces no kernel color. The $c_1^{Ds}$ allele is reverted through the action of *Ac* that activates the excision of *Ds* in individual somatic cells of developing kernels. The reversion of $c_1^{Ds}$ to *C* in these and descendant cells leads to pigment production and purple spots.

McClintock's transposable genetic element hypothesis was that the unstable mutant phenotype was the result of a transposable genetic element (*Ds*) that created a mutation when it inserted into *C* and led to reversion when the expression of *Ac* led to its removal (**Figure 13.22d**). McClintock's hypothesis came at a time in biological history when the molecular structure of DNA and genes were first being described. It was difficult for many biologists to understand how genetic elements could be mobile, and so the transposition hypothesis was much debated for years. Eventually, however, more examples of transposition emerged in maize, in other plant species, in

animals, and in bacteria. Since McClintock's discovery of transposition in maize, the process has been identified in virtually all organisms. For her discovery of transposition, McClintock was awarded the 1983 Nobel Prize in Physiology or Medicine.

## Characteristics of Transposable Genetic Elements

Transposase is the enzyme responsible for excising or copying transposable genetic elements from chromosomes and reinserting them into new locations. A transposase gene is carried by some transposable elements, but not by all. In addition, some transposable elements carry one or more genes that impart specific characteristics to cells or organisms that contain them. On the other hand, some transposable elements do not encode any specific genes and are composed primarily of repetitive sequence.

These and other differences in genetic content explain two types of transposable elements. **Autonomous transposable elements** carry the transposase gene and possess DNA sequences that allow them to mobilize their own transposition without the aid of other genes or transposable elements. Conversely, **non-autonomous transposable elements** do not carry the transposase gene and may lack other necessary DNA sequences; this condition leaves them unable to move unless an autonomous element is present elsewhere in the genome. The *Ac* element described by McClintock is an example of an autonomous transposable element that encodes transposase and can drive its own transposition and that of the non-autonomous transposable element *Ds.*

## 13.6 Transposition Modifies Bacterial Genomes

Bacterial genomes contain two categories of transposable elements. Insertion sequence (IS) elements are simple transposable elements that contain only the sequences and genes necessary for autonomous transposition. In a previous chapter, we introduced insertion sequences in connection with the integration of F plasmids into bacterial chromosomes to form Hfr chromosomes (see Chapter 6). The second category contains two types of **transposons** that each carry multiple genes and confer new traits on the bacteria that contain them. Both types of transposable element can produce mutations if they insert into a wild-type gene, as is the case for the $c^{Ds}$ allele or in Mendel's peas.

### Insertion Sequences

Numerous insertion sequences have been identified in bacterial and plasmid DNA (**Table 13.4**). IS elements

| Table 13.4 | Characteristics of Insertion Sequence Elements in *E. coli* | | | | |
|---|---|---|---|---|---|
| Element | Length (bp) | Inverted Repeat Length (bp) | Direct Repeat Length (bp) | Number in *E. coli* | Integration Target Sequence[a] |
| IS1 | 768 | 23 | 9 | 5–8 | Random |
| IS2 | 1327 | 41 | 5 | 5 | Hotspots |
| IS4 | 1428 | 18 | 11 | 1–2 | $AAAN_{20}TTT$ |
| IS5 | 1195 | 16 | 4 | Variable | Hotspots |
| IS10R | 1329 | 23 | 9 | Variable | NGCTNAGCN |
| IS50R | 1531 | 9 | 9 | Variable | Hotspots |
| IS903 | 1057 | 18 | 9 | Variable | Random |

[a] N indicates any nucleotide.

vary somewhat in length, but all contain about 1000 bp of DNA. Each IS element contains a copy of the transposase gene that is bracketed by a short **inverted repeat (IR) sequence.** Different insertion sequences have different IR sequences. The IR sequences found on each end of a given IS element have the identical or nearly identical 5′-to-3′ sequence on *opposite* strands of DNA. **Figure 13.23** gives examples of inverted repeat sequences for two such IS elements.

The inverted repeats are integral to transposase-mediated transposition, which begins with cleavage at the ends of IS elements (**Figure 13.24**). Transposase also recognizes a specific sequence at the new integration site on the bacterial DNA and there catalyzes a double-stranded, staggered cleavage. Ligation of the IS into the new site first joins one end of each strand of the IS to the bacterial

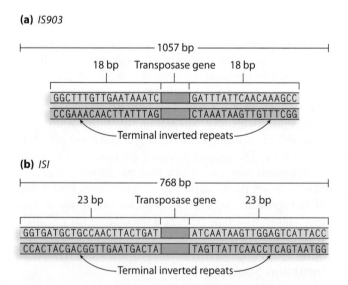

**(a)** *IS903*

Terminal inverted repeats

**(b)** *IS1*

Terminal inverted repeats

**Figure 13.23** *E. coli* insertion sequences IS903 and IS1.

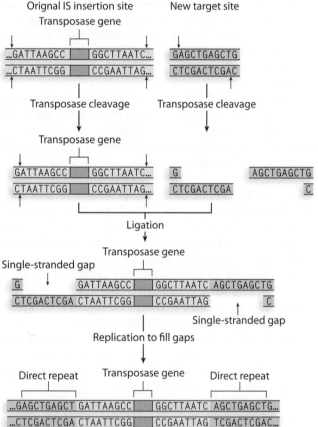

**Figure 13.24 Transposition of an IS element.** The IS element is removed from its original insertion site by transposase cleavage at the end of each inverted repeat. The new target site undergoes double-stranded, staggered cleavage by transposase. Ligation joins the IS element to the new target site at one end of each strand. Remaining single-stranded gaps are filled by DNA replication to create direct repeats that flank inserted IS elements.

The following DNA sequences occur on the same strand of DNA and are separated by a large number of nucleotides. Which of these sequences might be found flanking an insertion sequence? Explain your answer, and identify the relevant parts of your selected sequences.

**a.** 5'-TTAGCAC ... CAGGATT-3'

**b.** 5'-GGCCAAT ... ATTGGCC-3'

**c.** 5'-CCGACCGTA ... CCGACCGTA-3'

**d.** 5'-AGTATACCGC ... GCGGTATGGC-3'

| Solution Strategies | Solution Steps |
|---|---|
| **Evaluate** | |
| 1. Identify the topic this problem addresses, and explain the nature of the required answer. | 1. This problem requires you to recognize DNA sequences that might flank a bacterial insertion sequence. You must identify one or more of the given sequences as a candidate flanking sequence. |
| 2. Identify the critical information given in the problem. | 2. We are given single-stranded sequences from the same strand of DNA on opposite sides of potential insertion sequences. |
| **Deduce** | |
| 3. Determine the double-stranded sequences for each of the single-stranded sequences listed. | 3. The double-stranded sequences are<br><br>a. 5'-TTAGCAC ... CAGGATT-3'<br>   3'-AATCGTG ... GTCCTAA-5'<br>b. 5'-GGCCAAT ... ATTGGCC-3'<br>   3'-CCGGTTA ... TAACCGG-5'<br>c. 5'-CCGACCGTA ... CCGACCGTA-3'<br>   3'-GGCTGGCAT ... GGCTGGCAT-5'<br>d. 5'-AGTATACCGC ... GCGGTATGGC-3'<br>   3'-TCATATGGCG ... CGCCATACCG-5' |
| 4. Review what you know about the sequences flanking insertion elements. | 4. The sequences flanking insertion elements are inverted repeat sequences. |
| **Solve** | |
| 5. Identify any sequence that might be found flanking an insertion sequence. | 5. Sequences b and d in step 3 are the ones most likely to be found flanking insertion sequences because each sequence forms an inverted repeat sequence in double-stranded DNA. |

For more practice, see Problems 30 and 32.

---

DNA. DNA replication then fills the short single-stranded gaps at each end of the integrated IS to form a pair of duplicate sequences called **direct repeats.** Upon completion of transposition of an IS, the integrated IS contains the transposase gene flanked by inverted repeats and direct repeats. This outcome is known as target site duplication, and it is a virtually universal occurrence in bacterial and eukaryotic transposition, suggesting that basic transpositional mechanisms are shared.

Target site duplication and direct repeats are signs of transposition that can be read in the DNA sequence. Thus, searching for direct repeats can lead researchers to transposable elements and can facilitate the identification of mutations created by transposition. Searches of this kind have been instrumental in detecting numerous null mutations created in *E. coli* by insertional inactivation of a wild-type gene. **Genetic Analysis 13.3** guides you through an assessment of potential IR sequences.

---

Genetic Insight Insertion sequences are bacterial transposable genetic elements carrying the transposase gene that drives transposition and is flanked by inverted repeat sequences. Transposition is a mutational event that can inactivate or alter normal gene expression.

---

## Transposons

Like insertion sequences, transposons (Tn) are elements capable of transposition to other parts of the bacterial genome. Two types of transposons are found, *composite* and *simple*; they are distinguished by the genes they contain and the sequences that flank the transposon. Antibiotic-resistant genes are the most common genes carried by transposons, and transposition is a prominent mechanism of dissemination of antibiotic resistance in bacterial populations (Table 13.5). Other transposon genes include those that resist the toxic consequences of heavy metal exposure. In addition, transposons can carry genes that confer growth advantages in stressful or extreme environments.

**Composite transposons** have a central region of several thousand base pairs that usually contains one or more functional genes, such as those conferring antibiotic resistance. The gene-containing central region is flanked by complete IS elements that have opposite orientations and thus form inverted repeats. At least one of the IS elements of a composite transposon, and sometimes both IS elements, contains a copy of the transposase gene. Other genes are contained in the central element of composite transposons.

The composite transposon *Tn10* has a structure typical of many. It has a copy of *IS10R* on the right (R) side and a copy of *IS10L* on the left (L) side of a large central DNA segment containing the tetracycline resistance (*Tet$^R$*) gene (Figure 13.25). Each IS element encodes transposase. Each copy of *IS10* is 1329 bp, and the central segment containing *Tet$^R$* is about 6600 bp, giving this transposon a total length of approximately 9300 bp. The *Tn10* transposon readily inserts into plasmid DNA, allowing rapid dissemination of tetracycline resistance among bacterial strains that carry the plasmid.

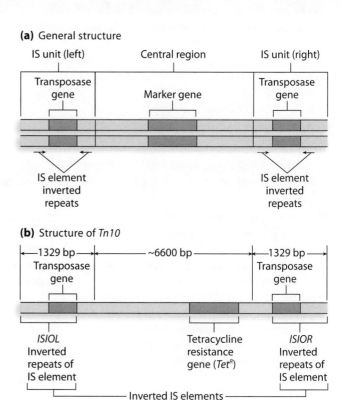

**(a)** General structure

**(b)** Structure of *Tn10*

**Figure 13.25  Structure of a composite transposon.**

**Simple transposons** are flanked by very short IR sequences of less than 50 bp. The IR sequences do not encode transposase. Instead, the transposase gene is carried by the simple transposon itself, along with other genes in the transposon central element. For example, the simple transposon *Tn3* contains a 4957-bp central element that encodes three genes: transposase (*TnpA*) and resolvase (*TnpB*), both of which are required for transposition, and β-lactamase (*bla*), which provides resistance to ampicillin and penicillin. The central element is flanked by two 38-bp inverted repeat sequences.

## Transposition Mechanisms

Target site duplication, and the creation of direct repeats of DNA sequences flanking at transposition sites, is a common outcome of bacterial and eukaryotic transposition. Two mechanisms of transposition that lead to target site duplication are also shared. **Conservative transposition** is a mechanism that excises a transposable element from one location and inserts it into a new location in a process that is driven by transposase. Conservative transposition is similar to the kind of cut-and-paste process you might use in altering a text file. The transposable element is cut from its original location and is subsequently pasted into a new location. This process moves transposable elements around the genome, but it does not increase the number of transposable elements.

| Table 13.5 | Characteristics of Bacterial Transposons | | | |
|---|---|---|---|---|
| Transposon | Insertion Sequences | Sequence Difference between IS Elements | Transposon Length (bp) | Marker Gene[a] |
| Tn5 | IS50L | 1-bp difference | 5700 | Kan$^R$ |
|  | IS50R |  |  |  |
| Tn9 | IS1 | None | 2500 | Cam$^R$ |
| Tn10 | IS10L | 2.5% difference | 9300 | Tet$^R$ |
|  | IS10R |  |  |  |
| Tn903 | IS903 | None | 3100 | Kan$^R$ |

[a] *Cam* = chloramphenicol, *Kan* = kanamycin, *Tet* = tetracycline.

**Replicative transposition,** as its name implies, involves replication of transposable elements that leads to increases in the number of transposable elements in a genome. **Figure 13.26** illustrates replicative transposition between two plasmids. Following the initiation of replication of the transposable element, transposase facilitates the formation of a **cointegrate** between the two plasmids. The cointegrate is a temporary fusion of the plasmids that is resolved by a recombination-like process that cuts and rejoins DNA strands to separate the plasmids. The outcome of replicative transposition is that both plasmids contain a copy of the transposable element.

**①** A donor plasmid carrying *Tn3* pairs with a recipient plasmid carrying a target sequence.

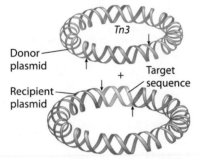

Donor plasmid · *Tn3* · Target sequence · Recipient plasmid

**②** Transposase cuts single strands of DNA adjacent to *Tn3* and at the target sequence; strands cross over.

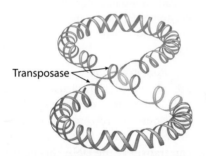

Transposase

**③** A cointegrate forms following new DNA synthesis to fill single gaps.

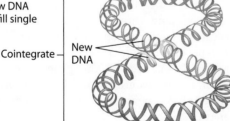

Cointegrate · New DNA

**④** Cointegrate resolution separates the plasmids; the donor plasmid retains *Tn3*, and the modified recipient plasmid gains a replicate copy of *Tn3*.

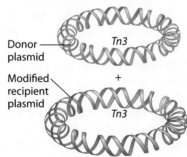

Donor plasmid · *Tn3* · Modified recipient plasmid · *Tn3*

**Figure 13.26  Replicative transposition.**

# 13.7 Transposition Modifies Eukaryotic Genomes

Transposable genetic elements are plentiful and highly varied in eukaryotic genomes; and the basic mechanisms that accomplish transposition in eukaryotes, and the potential mutational outcomes, are similar to those described earlier for bacteria. Eukaryotic transposable elements are divided into two groups: **DNA transposons,** which are transposed throughout eukaryotic genomes by conservative or replicative transposition, and **retrotransposons,** which use the enzyme reverse transcriptase to synthesize double-stranded DNA from a single-stranded RNA transcript. The *Ds* and *Ac* transposable elements first described by McClintock are DNA transposons, and numerous other examples are found in eukaryotic genomes. DNA transposons in *Drosophila* were fully characterized, however, with the identification and analysis of *P* elements, and we begin our discussion with that analysis.

## *Drosophila P* Elements

The genome of *Drosophila melanogaster* carries several dozen copies of a transposable genetic element called a *P* **element.** These DNA transposons were not part of the genome of *D. melanogaster* collected from the wild before about 1960. Today, however, all *D. melanogaster* collected in the wild carry *P* elements in their genome, suggesting that *P* elements were introduced into *D. melanogaster* about 1960, perhaps by cross-species transfer from a distantly related species. Since their introduction to the genome, *P* elements have quickly proliferated. The *Drosophila* life cycle can produce 20 to 25 generations per year; thus, *P* elements have been evolving for about 1000 generations or so in *D. melanogaster* since first being introduced into the genome.

The *P* elements exist in multiple forms. Full-length *P* elements encode transposase and are capable of autonomous transposition. These *P* elements are approximately 2900 bp in length, and they have a central region containing a gene for transposase that is encoded in four exons and three introns flanked by 31-bp inverted repeats. Transcription and translation of the transposase gene in full-length *P* elements produces an 87-kD transposase enzyme that activates *P* element transposition in germ-line cells. Several types of nonfunctional *P* elements are also found in the *D. melanogaster* genome, and none produce functional transposase and all are shorter than 2900 bp.

The *P* elements were discovered in *D. melanogaster* because they produce **hybrid dysgenesis,** a phenomenon in which sterility occurs in the $F_1$ progeny of a specific cross of laboratory-bred flies. Hybrid dysgenesis occurs when $F_1$ hybrids are formed by mating a female laboratory fly of the so-called M cytotype (*M* is for "maternal") to a wild-type male fly having what is known as the P ("paternal") cytotype. The P-cytotype male has three to four dozen *P* elements scattered throughout its genome and also possesses a repressor to block transposition in the genome. The M-cytotype female comes from a stock that has no *P* elements and does not produce the transposition repressor. The progeny of this cross between laboratory and wild flies are hybrids with a normal external appearance, but they are sterile because their gametes fail to develop.

The current model for hybrid dysgenesis explains why the phenotype occurs only when males have the P cytotype and females the M cytotype (**Figure 13.27**). In P-cytotype males, sperm, which contain chromosomes only and virtually no cytoplasmic material, do not possess a transposition repressor protein that is found in the cytoplasm. The eggs of M-cytotype females contain abundant cytoplasmic material but carry no transposition repressor protein, because the M cytotype is free of *P* elements. At fertilization, sperm add *P* element–laden chromosomes into an egg lacking transposition-repressing protein. Extensive transposition takes place, creating multiple mutations by insertion of *P* elements into functional genes or by inducing chromosome breaks similar to those observed by McClintock in the maize genome. Following embryonic development, the consequence of this widespread transpositional activity is widespread mutation by insertional inactivation that results in hybrid dysgenesis. In contrast, hybrid dysgenesis does not occur in the reciprocal cross between females with the P cytotype and males of either the M cytotype or the P cytotype. In these crosses, the cytoplasm of ova contains the transposition-repressing protein, which blocks *P* element transposition, and the germ line of the resulting F$_1$ hybrid progeny remains stable.

The *P* elements of *Drosophila* are readily manipulated, and they have become an important tool for genetic manipulation of the *Drosophila* genome (see Chapter 17 for further discussion).

## Retrotransposons

Retroviruses are viruses that infect eukaryotic cells and have a genome consisting of single-stranded RNA. On infection, the RNA is transcribed into double-stranded DNA by reverse transcriptase and the DNA is then integrates into a chromosome of the infected plant or animal cell. Integrated retroviral DNA encodes three genes, called *gag*, *env*, and *pol*. Gag and env encode proteins that form the retroviral particle. New retroviral particles are produced within infected cells and perpetuate the infection by invading new cells. The *pol* gene encodes the enzyme *reverse transcriptase* that directs the synthesis of double-stranded DNA from single-stranded RNA.

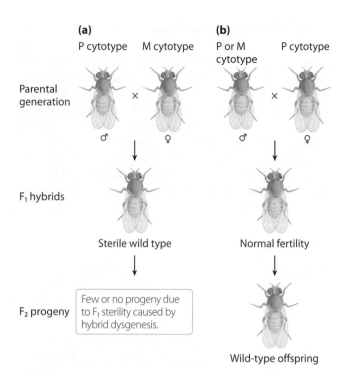

**Figure 13.27  Hybrid dysgenesis in *Drosophila*.** **(a)** Male *Drosophila* of the P cytotype crossed to females of the M cytotype produce F$_1$ progeny that are largely infertile due to mutations resulting from *P* element transposition. **(b)** Crosses of P-cytotype females to males with either the P or the M cytotype yield F$_1$ progeny normal fertility.

Retrotransposons are transposable elements that are related to retroviruses. All retrotransposons carry *pol*, and some contain *gag*, but none have *env*. As a consequence, retrotransposons are able to synthesize double-stranded DNA from single-stranded RNA and can transpose the DNA throughout the genome, but they do not possess the ability to produce protein particles.

**Figure 13.28** illustrates comparative structures of a retrovirus and three retrotransposons. Two constant features of retrotransposons are seen. First, all retrotransposons encode reverse transcriptase to catalyze transposition. Second, the gene or genes carried by retrotransposons are flanked by **long terminal repeats (LTRs)** that may be up to several hundred base pairs in length.

Multiple types of retrotransposons have been identified in eukaryotic genomes, including *Ty* elements in yeast, *copia* elements in *Drosophila*, and *L1* elements in humans. The various retrotransposons have been identified largely through analysis of mutations that have been traced to retrotransposon insertion.

***Ty* Elements of Yeast**    Many different forms of *Ty* retrotransposons of yeast are found, all sharing the common features of retrotransposons. In *Ty* elements, the central element is approximately 6 kb, flanked by LTRs that are

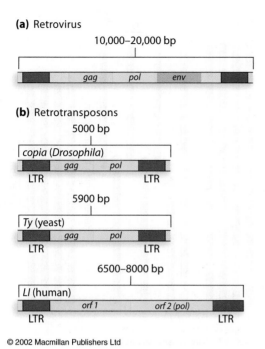

**(a)** Retrovirus

10,000–20,000 bp

gag    pol    env

**(b)** Retrotransposons

5000 bp

*copia (Drosophila)*

gag    pol

LTR              LTR

5900 bp

*Ty* (yeast)

gag    pol

LTR              LTR

6500–8000 bp

*LI* (human)

orf 1        orf 2 (pol)

LTR                    LTR

© 2002 Macmillan Publishers Ltd

**Figure 13.28  Retrovirus structure and selected eukaryotic retrotransposons.**

each about 330 bp in length. Both LTRs contain promoters that direct the transcription of different genes in the central region. Approximately 50 to 100 copies of *Ty* elements are present in the typical *Saccharomyces cerevisiae* genome. The *Ty* elements cause mutation in yeast genes by insertion.

**Copia Elements of Drosophila**  Multiple forms of the retrotransposon *copia* are found in the *Drosophila* genome. *Copia* elements have a central element of 5 to 8.5 kb that contains *pol* and *gag* genes and is flanked by LTRs of 250 to 600 bp each. The word *copia* comes from the Latin for "abundance," and befitting this designation, more than 5% of the *Drosophila* genome is composed of *copia* retrotransposons. This abundance leads to many mutations throughout the genome that are usually the result of insertion of *copia* into a wild-type gene.

**LINE and SINE Elements of Humans**  More than 45% of the human genome is composed of repetitive DNA sequences, many of which are derived from former transposable genetic elements that no longer transpose within the genome. The presence of these sequences, however, suggests that human and mammalian genomic evolution has been substantially influenced by transposable genetic elements. Among the remaining functional transposable genetic elements in the human genome, LINE (long interspersed elements) and SINE (short interspersed elements) families of elements stand out because of their relative abundance and their ability to cause spontaneous human gene mutations.

Human *L1* elements are retrotransposons that are particularly common members of the LINE family of elements in the human genome. The *L1* elements vary in length from about 6500 bp to 8000 bp. Full-length *L1* elements encode a protein with nuclease and reverse transcriptase function and may also encode a second RNA-binding protein, but shortening of the element affects its ability to transpose. The human genome contains approximately 600,000 copies of *L1*, constituting more than 17% of the total genome. Evidence that *L1* elements actively transpose in the human genome is drawn in part from studies of spontaneous mutation traced to *L1* transposition. For example, mutations of the *F8* gene, an X-linked gene whose mutation causes an X-linked recessive version of the blood-clotting disorder hemophilia, are traced to *L1* insertion into the gene.

SINE elements, too, are common in the human genome. One form, the *Alu* element, is the most common of the SINE elements. The *Alu* elements vary in length from 100 to 300 bp and are each flanked by direct repeats of 7 to 20 bp. They are so named because each element can be cleaved into two segments by the restriction endonuclease *Alu*I (*Al-LOO-one*) that recognizes the 4-bp restriction enzyme target sequence `5'-AGCT-3'`. The human genome contains approximately 1.2 million *Alu* elements, constituting nearly 11% of its total. Genomic analysis indicates that *Alu* elements are relatively recent additions to the human genome. They are present in the genomes of our closest primate relatives—chimpanzee, gorilla, and orangutan—but not in the genomes of other mammals.

## CASE STUDY

### Human Chromosome Evolution

Researchers can trace the evolution of human chromosomes by comparing chromosome structure and genetic composition of humans to those of other species that share a common ancestor. We describe two such comparative approaches here: One compares syntenic clusters of genes (genes on the same chromosome) in distantly related species, and the second

compares banding patterns of chromosomes in closely related species.

Figure 13.29 compares syntenic clusters of genes on 20 chromosomes (19 autosomes and the X chromosome) in the mouse genome and their relation to the same sequences on the 23 chromosomes (22 autosomes and the X chromosome) in

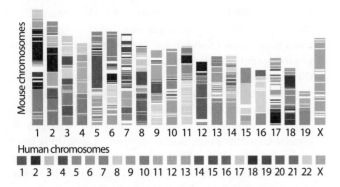

**Figure 13.29 Evolutionary conservation of chromosome synteny between mouse and human chromosomes.** Each of 23 human chromosomes is uniquely colored and its segments superimposed on 20 mouse chromosomes.

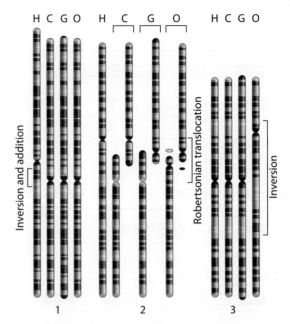

**Figure 13.30 Human and great ape chromosome evolution.** Chromosomes 1, 2, and 3 of human (H), chimpanzee (C), gorilla (G), and orangutan (O) are compared to determine the events leading to different chromosome numbers and structures.

humans. Published in 2002 by a large research group known as the Mouse Genome Sequencing Consortium, this study compares 342 syntenic chromosome segments. The average size of the syntenic segments is a little less than 10 million base pairs. Syntenic groups of genes found in the human genome are dispersed among several chromosomes in the mouse genome. Interestingly, human chromosomes 17 and 20 each correspond entirely to a portion of mouse chromosomes 11 and 2, respectively. In both cases, the human chromosome corresponds to a long cluster of contiguous syntenic groups in the respective mouse chromosome. Comparison of X chromosomes of human and mouse reveal very strong sequence and genetic similarity.

This comparison leads to two salient evolutionary conclusions. First, mouse and human share similar syntenic clusters because their common ancestor carried these clusters. Human and mouse chromosomes have diverged from those of their common ancestor by numerous rearrangements, including chromosome translocation, chromosome fusion, and chromosome inversion, that have changed many attributes of chromosome structure; but they also retain large segments of genes and sequences as syntenic clusters. Second, for X-linked genes specifically, the strong syntenic relationship has been maintained by natural selection driven by the requirements of embryonic development and the necessity to maintain a balance in dosage of X-linked genes by random X-inactivation.

Figure 13.30 illustrates the banding patterns of chromosomes 1, 2, and 3 of human (H), chimpanzee (C), gorilla (G), and orangutan (O). These four closely related primate species last shared a common ancestor between 30 and 35 million years ago. In each of the three chromosomes, strong similarity of banding patterns directly reflects the strong genetic

similarity between the species. Structural and numerical differences between the chromosomes allow reconstruction of the evolutionary events that shaped the contemporary chromosomes of each species. Taking the events from the perspective of human chromosomes, we can reconstruct the evolution for each chromosome as follows.

- Chromosome 1 is very similar in the four primate species, with the exception of a pericentric inversion and the addition of a small segment near the centromere of the human chromosome (1q1.2 to 1q2.1).

- Chromosome 2 holds the explanation for the difference in diploid number between humans ($2n = 46$) and our close relatives ($2n = 48$). The reduction in human diploid number is the result of a Robertsonian translocation fusing two small acrocentric chromosomes that belong to separate chromosome pairs in chimp, gorilla, and orangutan.

- Chromosome 3 shows strong similarity of banding pattern in the four species with the exception of the orangutan chromosome, which has undergone a pericentric inversion that changed the relative arm lengths and altered the position of the centromere in comparison to the other primate chromosomes.

---

SUMMARY          Mastering**GENETICS**™          *For activities, animations, and review quizzes, go to the study area at www.masteringgenetics.com.*

---

## 13.1 Nondisjunction Leads to Changes in Chromosome Number

- In euploid nuclei, the number of chromosomes is equal to a multiple of the haploid number (*n*), whereas aneuploid nuclei have additional or missing chromosomes.

- Chromosome nondisjunction is the failure of homologous chromosomes or sister chromatids to separate and is a common cause of aneuploid gametes.

- Aneuploidy alters the phenotype of an organism by changing the balance of gene dosage of critical genes.

- Human aneuploidy manifests as trisomy of particular autosomes and as trisomy or monosomy of sex chromosomes.
- Chromosomal mosaics are organisms containing cells with two or more distinct karyotypes.
- Uniparental disomy occurs when both homologous copies of a chromosome originate in a single parent.

## 13.2 Changes in Euploidy Result in Various Kinds of Polyploidy

- Polyploids carry three or more haploid sets of chromosomes.
- Allopolyploids carry chromosome sets from different species, whereas autopolyploids have multiple chromosome sets from a single species.
- Polyploidy is common in plant species, where increases in fruit and flower size alter fertility and can produce hybrid vigor.
- Polyploids have a reduced frequency of recessive homozygosity compared to diploid species.

## 13.3 Chromosome Breakage Causes Mutation by Loss, Gain, and Rearrangement of Chromosomes

- Chromosome breakage can result in terminal deletion or in interstitial deletion and may alter chromosome banding patterns.
- Heterozygosity for partial deletion or partial duplication produces phenotypic abnormalities through disturbances of gene dosage balance.
- Homologous chromosome synapsis involving a partial deletion or partial duplication chromosome produces a characteristic unpaired loop.
- Microdeletions and microduplications too small to be seen by banding changes are detected by molecular methods.
- The detection of pseudodominance provides important positional indicators for deletion mapping of genes.

## 13.4 Chromosome Breakage Leads to Inversion and Translocation of Chromosomes

- Chromosome breakage can lead to inversion or translocation of chromosome segments.
- Chromosome inversion heterozygotes have one chromosome with the normal order but have an inversion in the homolog. Homologs in these organisms form an inversion loop at synapsis.
- Paracentric inversions have two break points on one arm only, and the inversion does not include the centromeric region. Pericentric inversions have break points on each arm, and the centromeric region is included in the inverted region.
- Chromosome inversion is a crossover-suppression mechanism.
- A tetravalent synaptic structure containing chromosomes involved in reciprocal translocation leads to two patterns of chromosome segregation in meiosis.
- The reduction in the number of viable gametes produced by reciprocal balanced translocation heterozygotes results in semisterility.
- Robertsonian translocation occurs by the fusion of nonhomologous chromosomes.

## 13.5 Transposable Genetic Elements Move throughout the Genome

- Transposition is the process that moves transposable genetic elements in genomes and was first discovered in maize.
- Transposase is the enzyme responsible for transposition, and it is encoded by many transposable genetic elements.
- Transposition produces mutations through insertional inactivation.
- Autonomous transposable elements encode transposase to drive their movement, but non-autonomous elements must rely on other transposable elements for transposase production.

## 13.6 Transposition Modifies Bacterial Genomes

- Bacterial insertion sequences encode transposase and are flanked by inverted repeat sequences unique to each insertion sequence.
- Conservative and simple bacterial transposons carry the gene for transposase, and perhaps genes for antibiotic or heavy metal resistance.
- Conservative transposition occurs by the excision and reinsertion of a transposable element, whereas replicative transposition produces new copies of the transposable element.

## 13.7 Transposition Modifies Eukaryotic Genomes

- *Drosophila P* elements are common, transpose actively, and cause hybrid dysgenesis in certain crosses.
- Retrotransposons are transcribed into RNA and then reverse transcribed into double-stranded DNA, and the new DNA copy is transposed to a new location.

## KEYWORDS

acentric fragment (acentric chromosome) *(p. 433)*
activator (*Ac*) element *(p. 445)*
adjacent-1 segregation *(p. 443)*
alternate segregation *(p. 443)*
aneuploid *(p. 424)*
autonomous transposable element *(p. 445)*
chromosome aberration *(p. 423)*
chromosome break point *(p. 433)*
chromosome (paracentric, pericentric) inversion (inversion heterozygote) *(pp. 436, 438)*

## PROBLEMS

*For instructor-assigned tutorials and problems, go to www.masteringgenetics.com.*

## Chapter Concepts

*For answers to selected even-numbered problems, see Appendix: Answers.*

1.  Consider synapsis in prophase I of meiosis for two plant species that each carry 36 chromosomes. Species A is diploid and species B is triploid. What characteristics of homologous chromosome synapsis can be used to distinguish these two species?

2.  For one set of chromosomes carried by a triploid plant species, assume the chromosome pair as one bivalent involving chromosomes C1 and C2, and as one univalent with chromosome C3. Show the gametes that result from this synaptic pattern, and identify the frequency and content of the genetically different gametes produced by the species. In your answer, use the designations $C_1$, $C_2$, and $C_3$, $C_4$ to identify the individual chromosomes.

3.  If the haploid number for a plant species is 4, how many chromosomes are found in a member of the species that has one of the following characteristics? Explain your reasoning in each case.

    a.  diploid
    b.  pentaploid
    c.  octaploid
    d.  trisomic
    e.  triploid
    f.  monosomic
    g.  tetraploid
    h.  hexaploid

    In the list above, which plants are likely to be infertile or to have reduced fertility?

4.  From the following list, identify the types of chromosome changes you expect to show phenotypic consequences.

    a.  pericentric inversion
    b.  interstitial deletion
    c.  duplication
    d.  terminal deletion
    e.  trisomy

    f.  reciprocal balanced translocation
    g.  paracentric inversion
    h.  monosomy
    i.  polyploidy

5.  Mating between a male donkey ($2n = 62$) and a female horse ($2n = 64$) produces sterile mules. Recently, however, a very rare event occurred—a female mule gave birth to an offspring by mating with a horse.

    a.  Determine how many chromosomes are in the mule karyotype, and explain why mules are generally sterile.
    b.  How many chromosomes does the mule–horse offspring carry?
    c.  Why is it very unlikely that the offspring will have fully horse-like genetic characteristics?

6.  Studies of hybrid dysgenesis in *Drosophila* indicate that the transposition repressor protein produced by *P* elements is part of a process that limits the number of *P* elements present in a genome. Why is it advantageous to limit the number of *P* elements in a genome?

7.  What evidence suggests that *copia* elements of fruit flies and *Ty* elements of yeast are related to RNA-containing viruses?

8.  What can we conclude about a mutational event that renders *IS1* unable to transpose?

9.  In terms of the chromosome content of nuclei, what is meant by the term *mosaic*?

10. In *Drosophila*, an X-linked recessive allele produces yellow body color. The cross of a yellow female and a male with wild-type body color usually produces wild-type females and yellow males. Occasionally however, a yellow female is produced. Explain how the unusual female is produced.

## Application and Integration

11. The plants in this problem are the same as those described in Genetic Analysis 13.1, where flower color in the autotetraploid is a single-gene character determined by alleles $R_1$ and $R_2$ that have an additive relationship. The genotype–phenotype correspondence is as follows:

| Genotype | Phenotype |
|---|---|
| $R_1R_1R_1R_1$ | Dark red |
| $R_1R_1R_1R_2$ | Light red |
| $R_1R_1R_2R_2$ | Pink |
| $R_1R_2R_2R_2$ | Light pink |
| $R_2R_2R_2R_2$ | White |

  a. Predict the phenotypes and frequencies of progeny produced by self-fertilization of a light red plant.

  b. A light pink and a light red plant are crossed. Predict the frequencies of phenotypes among the progeny.

12. A normal chromosome and its homolog carrying a paracentric inversion are given. The dot (•) represents the centromere.

  Normal       `ABC•DEFGHIJK`

  Inversion   `abc•djihgfek`

  a. Diagram the alignment of chromosomes during prophase I.

  b. Assume a crossover takes place in the region between F and G. Identify the gametes that are formed following this crossover, and indicate which gametes are viable.

  c. Assume a crossover takes place in the region between A and B. Identify the gametes that are formed by this crossover event, and indicate which gametes are viable.

13. A pair of homologous chromosomes in *Drosophila* has the following content (single letters represent genes):

  Chromosome 1   `RNMDHBGKWU`

  Chromosome 2   `RNMDHBDHBGKWU`

  a. What term best describes this situation?

  b. Diagram the pairing of these homologous chromosomes in prophase I.

  c. What term best describes the unusual structure that forms during pairing of these chromosomes?

  d. How does the pairing diagrammed in part (b) differ from the pairing of chromosomes in an inversion heterozygote?

14. An animal heterozygous for a reciprocal balanced translocation has the following chromosomes:

  `MN•OPQRST`

  `MN•OPQRjkl`

  `cdef•ghijkl`

  `cdef•ghiST`

  a. Diagram the pairing of these chromosomes in prophase I.

  b. Identify the gametes produced by alternate segregation. Which of these gametes are viable?

  c. Identify the gametes produced by adjacent-1 segregation. Which of these gametes are viable?

  d. Identify the gametes produced by adjacent-2 segregation. Which of these gametes are viable?

  e. Among the three segregation patterns, which is least likely to occur? Why?

15. Dr. Ara B. Dopsis has an idea he thinks will be a boon to agriculture. He wants to create the "pomato," a hybrid between a tomato (*Lycopersicon esculentum*) that has 12 chromosomes and a potato (*Solanum tuberosum*) that has 48 chromosomes. Dr. Dopsis is hoping that his new pomato will have tuber growth like a potato and the fruit production of a tomato. He joins a haploid gamete from each species to form a hybrid and then induces doubling of chromosome number.

  a. How many chromosomes will the hybrid have before chromosome doubling?

  b. Will this hybrid be infertile?

  c. How many chromosomes will the polyploid have after chromosome doubling?

  d. Can Dr. Dopsis be sure the polyploid will have the characteristics he wants? Why or why not?

16. Suppose polymerase chain reaction (PCR) is used to amplify a single DNA marker on human chromosome 21. Further suppose that a couple who have a child with Down syndrome (trisomy 21) is examined for this marker. The mother has marker alleles of 310 and 380 bp. Her mate has marker alleles of 290 and 340 bp. What PCR bands are present in their child with Down syndrome if nondisjunction occurred in

  a. maternal meiosis I

  b. maternal meiosis II

  c. paternal meiosis I

  d. paternal meiosis II

17. Chromosome IV in *Drosophila* is a very small chromosome and carries a tiny amount of genetic material. Fruit flies that are trisomic for chromosome IV have no apparent phenotypic abnormalities, and they retain their fertility. Among the genes on chromosome IV is one for which a recessive allele *ey* produces the "eyeless" phenotype. A male that is trisomic for chromosome IV and has the genotype ++*ey* is crossed to a diploid eyeless female with the genotype *eyey*.

  a. Assuming random segregation of chromosomes takes place during spermatogenesis and that all sperm are viable, what sperm genotypes are expected and in what proportions?

  b. If these sperm are united with eggs from the eyeless female, what is the expected ratio of eyeless to normal-eyed flies among the progeny?

18. A healthy couple with a history of three previous spontaneous abortions has just had a child with cri-du-chat syndrome, a disorder caused by a terminal deletion of chromosome 5. Their physician orders karyotype analysis of both parents and of the child. The karyotype results for chromosomes 5 and 12 are shown here.

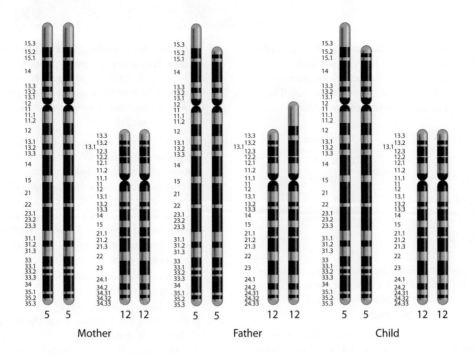

| 5 5 | 12 12 | 5 5 | 12 12 | 5 5 | 12 12 |
|---|---|---|---|---|---|
| Mother | | Father | | Child | |

a. Are the chromosomes in the child consistent with those expected in a case of cri-du-chat syndrome? Explain your reasoning.

b. Which parent has an abnormal karyotype? How can you tell? What is the nature of the abnormality?

c. Why does this parent have a normal phenotype?

d. Diagram the pairing of the abnormal chromosomes.

e. What segregation pattern occurred to produce the gamete involved in fertilization of the child with cri-du-chat syndrome?

f. What is the approximate probability that the next child of this couple will have cri-du-chat syndrome?

g. Do the karyotypes of the parents help explain the occurrence of the three previous spontaneous abortions? Explain.

19. A boy with Down syndrome (trisomy 21) has 46 chromosomes. His parents and his two older sisters have a normal phenotype, but each has 45 chromosomes.

a. Explain how this is possible.

b. How many chromosomes do you expect to see in karyotypes of the parents?

c. What term best describes this kind of chromosome abnormality?

d. What is the probability the next child of this couple will have a normal phenotype and have 46 chromosomes? Explain your answer.

20. Human chromosome 5 and the corresponding chromosomes from chimpanzee, gorilla, and orangutan are shown below. Describe any structural differences you see in the other primate chromosomes in relation to the human chromosome, and propose a mechanism to explain each difference.

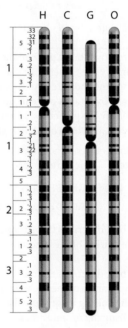

21. A small population of deer living on an isolated island are separated for many generations from a mainland deer population. The populations retain the same number of chromosomes and are interfertile, but that one chromosome has a different banding pattern.

a. Describe how the banding pattern of the island population chromosome most likely evolved from the mainland chromosome. What term or terms describes the difference between these chromosomes?

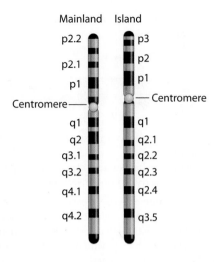

Mainland Island

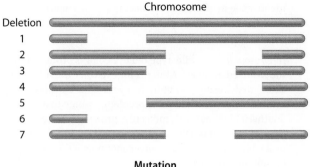

Chromosome

| Deletion | a | b | c | d | e | f | g |
|---|---|---|---|---|---|---|---|
| | | | | **Mutation** | | | |
| 1 | + | m | + | m | + | + | + |
| 2 | m | + | + | + | + | m | + |
| 3 | m | + | + | + | + | + | m |
| 4 | m | + | + | m | + | m | m |
| 5 | + | m | + | m | m | + | + |
| 6 | m | m | m | m | + | m | m |
| 7 | m | + | + | + | + | + | + |

b. Draw the synapsis of these homologs during prophase I in hybrids produced from the cross of mainland with island deer.

c. In a mainland–island hybrid deer, recombination takes place in band q1 of the homologous chromosomes. Draw the gametes that result from this event.

d. Suppose that 40% of all meioses in mainland–island hybrids involve recombination somewhere in the chromosome region between q2.1 and p2. What proportion of the gametes of hybrid deer are viable? What is the cause of the decreased proportion of viable gametes in hybrids relative to the parental populations?

22. In humans that are XX/XO mosaics, the phenotype is highly variable, ranging from females who have classic Turner syndrome symptoms to females who are essentially normal. Likewise, XY/XO mosaics have phenotypes that range from Turner syndrome females to essentially normal males. How can the wide range of phenotypes be explained for these sex-chromosome mosaics?

23. A plant breeder would like to develop a seedless variety of cucumber from two existing lines. Line A is a tetraploid line, and line B is a diploid line. Describe the breeding strategy that will produce a seedless line, and support your strategy by describing the results of crosses.

24. In *Drosophila*, seven partial deletions (1 to 7) shown as gaps in the diagram below have been mapped on a chromosome. This region of the chromosome contains genes that express seven recessive mutant phenotypes, identified in the following table as *a* through *g*. A researcher wants to determine the location and order of genes on the chromosome, so he sets up a series of crosses in which flies homozygous for a mutant allele are crossed with flies that are homozygous for a partial deletion. The progeny are scored to determine whether they have the mutant phenotype ("m" in the table), or the wild-type phenotype ("+" in the table). Use the partial deletion map and the table of progeny phenotypes to determine the order of genes on the chromosome.

25. Two experimental varieties of strawberry are produced by crossing a hexaploid line that contains 48 chromosomes and a tetraploid line that contains 32 chromosomes. Experimental variety 1 contains 40 chromosomes, and experimental variety 2 contains 56 chromosomes.

a. Do you expect both experimental lines to be fertile? Why or why not?

b. How many chromosomes from the hexaploid line are contributed to experimental variety 1? To experimental variety 2?

c. How many chromosomes from the tetraploid lines are contributed to experimental variety 1? To experimental variety 2?

26. In the tomato, *Lycopersicon esculentum*, tall (*D_*) is dominant to dwarf (*dd*) plant height, smooth fruit (*P_*) is dominant to peach fruit (*pp*), and round fruit shape (*O_*) is dominant to oblate fruit shape (*oo*). These three genes are linked on chromosome 1 of tomato in the order *dwarf-peach-oblate*. There are 12 map units between *dwarf* and *peach* and 17 map units between *peach* and *oblate*. A trihybrid plant (*DPO/dpo*) is test crossed to a plant that is homozygous recessive at the three loci (*dpo/dpo*). Progeny plants are grown with the results shown below.

| Progeny Phenotype | Number |
|---|---|
| Tall, smooth, round | 473 |
| Dwarf, peach, oblate | 476 |
| Tall, smooth, oblate | 12 |
| Dwarf, peach, round | 8 |
| Tall, peach, oblate | 17 |
| Dwarf, smooth, round | 13 |
| Tall, peach, round | 0 |
| Dwarf, smooth, oblate | 1 |
| | 1000 |

Identify the mechanism responsible for the resulting data that do not agree with the established genetic map.

27. In *Drosophila*, the wild-type red eye color is produced by the X-linked allele $w^+$. Mutants for eye color often lack the ability to deposit pigment in the eye and have white eye color. For the purpose of this problem, assume that in Southern blot analysis a molecular probe hybridizes to a 5.0-kb fragment of DNA from the eye-color locus. The probe binds to DNA fragments containing either wild-type or mutant sequence.

    a. If a male *Drosophila* has white eye color as a result of in-activation of $w^+$ by movement of a 3-kb *P* element into the wild-type allele, diagram the expected Southern blot pattern of DNA fragments from wild-type males and white-eyed males and females that carry the mutant allele. Explain your reasoning.

    b. Several male progeny of a female carrier of the mutant allele have red sectors on their eyes. The number and size of the sectors vary among the males. Explain the origin of these red sectors, and account for the variation in number and size.

    c. If Southern blotting is used to compare DNA isolated from a white sector and a red sector of the same eye, is a difference in DNA fragment size expected? Explain.

28. A *Drosophila P* element 2.5 kb in length is modified by adding a 1.0-kb intron sequence to one of its exons. A *copia* element of 6.0 kb is modified by adding the same 1.0-kb intron to its central region.

    a. A *Drosophila* genome carrying both transposable elements is induced to undergo transposition. What is the length of the newly transposed *P* element?

    b. What is the length of the newly transposed *copia* element?

    c. Explain the results for each case of transposition.

29. A biologist studying flight mechanisms in insects wants to introduce a dominant mutant allele producing oversized wings, called *flapper*, into the *Drosophila* genome. The biologist chooses a strain of fruit fly homozygous for a recessive mutant producing miniature wings. How will the biologist design the experiment using a *P* element to deliver the mutant allele to the genome?

30. One part of an inverted repeat sequence marking a transposon integration site is given. The transposon central region is 868 bp in length. Transposase creates a staggered cut in the DNA, breaking the strands at the location indicated by the lines.

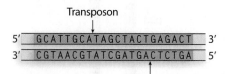

Transposon

```
5'  GCATTGCATAGCTACTGAGACT  3'
3'  CGTAACGTATCGATGACTCTGA  5'
```

a. Write the new sequence that surrounds the newly inserted transposon.

b. Describe the mechanism that creates the sequence surrounding the transposon insertion site.

31. Two *Not*I restriction enzymes cleave DNA on opposite sides of the *Dbm* gene in a species of yeast. A molecular probe for *Dbm* detects a DNA restriction fragment of 8.5 kb in organisms that are wild type at *Dbm*. In a strain of yeast, a *Ty1* transposable genetic element mutates *dbm*. *Ty1* is 5.6 kb in length.

    a. In haploid yeast with this *dbm* mutation, what is the length of the restriction fragment detected by the probe following *Not*I digestion?

    b. What DNA-fragment sizes are detected in a diploid yeast strain that is heterozygous for wild-type and mutant alleles at *dbm*?

    c. Insertion of *Ty1* into *dbm* causes a loss-of-function mutation. Explain why this is the case.

32. For the following crosses, determine as accurately as possible the genotypes of each parent, the parent in whom nondisjunction occurs, and whether nondisjunction takes place in the first or second meiotic division. Both color blindness and hemophilia, a blood-clotting disorder, are X-linked recessive traits. In each case, assume the parents have normal karyotypes (see Table 13.2).

    a. A man and a woman who each have wild-type phenotypes have a son with Klinefelter syndrome (XXY) who has hemophilia.

    b. A man who is color blind and a woman who is wild type have a son with Jacob syndrome (XYY) who has hemophilia.

    c. A color-blind man and a woman who is wild type have a daughter with Turner syndrome (XO) who has normal color vision and blood clotting.

    d. A man who is color blind and has hemophilia and a woman who is wild type have a daughter with triple X syndrome (XXX) who has hemophilia and normal color vision.

# Regulation of Gene Expression in Bacteria and Bacteriophage

# 14

Jacques Monod (right), André L'woff (middle), and François Jacob (left) on October 14, 1965, following the announcement of the award of the Nobel Prize in Physiology or Medicine for their work describing the lactose (*lac*) operon in *E. coli*.

## ESSENTIAL IDEAS

■ Gene expression in bacteria is controlled primarily through transcriptional regulation, often by regulating groups of genes known as operons.

■ Transcription of lactose (*lac*) operon genes is induced by lactose and is repressed in the absence of lactose.

■ Transcription of the repressible tryptophan (*trp*) operon adjusts to the level of available tryptophan.

■ Specialized regulatory processes control transcriptional response to environmental stress and regulate translation.

■ Bacteriophage use transcriptional regulation to express the genes responsible for infecting their hosts.

■ Competition between regulatory proteins determines the course of bacteriophage lambda infection in bacteria.

Take a moment to think about the ever-changing environment endured by the billions of *Escherichia coli* that populate your intestinal tract. These bacteria have evolved genetic control mechanisms that rapidly alter key patterns of gene expression, enabling the bacteria to adjust to diverse and constantly shifting environmental factors and nutritional conditions as well as to compete with the many other bacterial species in your gut. In your intestines, and across the wide array of other environmental niches that bacteria inhabit in nature, variable and even extreme conditions are the norm. In all these environments, bacterial survival depends on the ability to regulate gene expression to accommodate the immediate needs of the cell. In addition, these mechanisms contribute to the ability of bacterial cells to be almost entirely self-reliant when it comes to producing the proteins and enzymes necessary to generate the compounds that keep them alive.

Several of the mechanisms and processes that carry out transcriptional regulation of gene expression in bacteria and in bacterial viruses (bacteriophage) are the subject of this chapter. From each of these discussions, a central operational theme emerges—transcription is regulated by interactions between DNA-binding regulatory proteins and specific regulatory DNA sequences. This theme is common among all organisms, and we will see it repeated in discussions of the regulation of gene expression in eukaryotic genomes (Chapter 15).

In this chapter the regulatory systems we discuss are principally found in *E. coli*, the most widely used model bacterium. We begin with a general introduction to regulated gene expression and the most common mechanism of interaction between DNA-binding regulatory proteins and the DNA sequences that regulate transcription. Next we explore the organization, function, and regulation of the *E. coli* lactose (*lac*) operon system, whose gene transcription is induced (turned on) by the presence of the sugar lactose in the growth medium. This is followed by a discussion of mutational analysis and molecular explanation of the transcriptional control of *lac* operon genes. We then turn our attention to the genetic structure and molecular control of transcription of the tryptophan (*trp*) operon that contains the genes needed to synthesize the amino acid tryptophan. Lastly, after a discussion of post-transcriptional regulation of bacterial genes, we examine the genetically controlled process that controls infection of bacterial cells by bacteriophage λ (lambda).

## 14.1 Transcriptional Control of Gene Expression Requires DNA–Protein Interaction

Certain bacterial genes—specifically, those whose products are needed continuously to perform routine tasks—undergo **constitutive transcription,** a term identifying the genes as being transcribed continuously with no regulatory control. In contrast, the need for agile and calibrated responses to changing environmental conditions requires the **regulated transcription** of many bacterial genes.

Regulation of the transcription of bacterial genes takes place at two levels, and both levels of control result from interactions between DNA-binding proteins and specific regulatory sequences of DNA. The first level of control regulates the *initiation of transcription,* determining whether a particular gene or group of genes is transcribed at all. The second transcriptional control level determines the *amount of transcription,* regulating either the duration of transcription or the amount of mRNA transcript produced from the gene.

Most of the regulation of gene expression in bacterial genomes takes place through transcriptional regulation, which is the main focus of this chapter. Post-transcriptional regulatory mechanisms are also important for controlling the level of translation of mRNA or the activity of proteins and enzymes (see Chapter 14). **Figure 14.1** provides an overview of bacterial regulatory mechanisms.

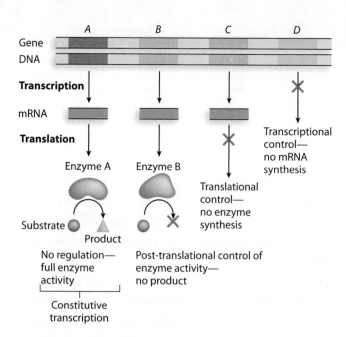

**Figure 14.1** **An overview of gene expression in bacteria.** Unregulated (constitutive) transcription and three patterns of regulated gene expression occur in bacteria.

## Negative and Positive Control of Transcription

Mechanisms of transcription control are described as negative or positive. **Negative control** of transcription involves the binding of a *repressor protein* to a regulatory DNA sequence, with the consequence of *preventing* transcription of a gene or a cluster of genes. On the other hand, **positive control** of transcription involves the binding of an *activator protein* to regulatory DNA, with the result of *initiating* gene transcription.

**Repressor proteins** are a broad category of regulatory proteins that exert negative control of transcription. In their active form, repressor proteins bind to regulatory DNA sequences, including those called **operators,** as we describe for the lactose operon. Repressor protein binding blocks transcription initiation by RNA polymerase. The repressor protein acts by occupying the space on regulatory DNA where the polymerase would otherwise bind or by preventing formation of the open promoter complex necessary for transcription initiation. Repressor proteins can be activated or inactivated by interactions with other compounds.

Repressor proteins commonly contain two active sites through which their functional role is performed. The **DNA-binding domain** is responsible for locating and binding operator DNA sequence or other target regulatory sequences. The **allosteric domain** binds a molecule or protein and, in so doing, causes a change in the conformation of the DNA-binding site. The property belonging to some enzymes of changing conformation at the active site as a result of binding a substance at a different site is known as **allostery**.

Allosteric domains operate in two modes. Certain repressor proteins undergo inactivation of their DNA-binding domain because of allosteric changes brought about by an **inducer** compound binding to the allosteric site (**Figure 14.2a**). If the inducer is removed from the allosteric site, the repressor's conformation is switched, the DNA-binding site is reactivated, and the protein can repress transcription. On the other hand, some repressor proteins require binding of a **corepressor** molecule at the allosteric site to activate the DNA-binding site (**Figure 14.2b**). In this case, transcriptional repression is reversed when the corepressor is removed from the allosteric site.

Positive control of transcription is accomplished by **activator proteins** that bind to regulatory DNA sequences called **activator binding sites.** Activator protein binding facilitates RNA polymerase binding at promoters and helps initiate transcription. Activator proteins have a DNA-binding domain that binds the activator binding site of DNA. In one mode of action for activator proteins, the DNA-binding domain remains inactive until the allosteric domain is bound by an **allosteric effector compound.** The induced allosteric change leads to the formation of a functional DNA-binding domain, allowing

the activator protein to bind to DNA (**Figure 14.3a**). Alternatively, certain activator proteins have a functional DNA-binding domain that is converted to an inactive conformation by binding of an **inhibitor** compound in the allosteric binding domain (**Figure 14.3b**).

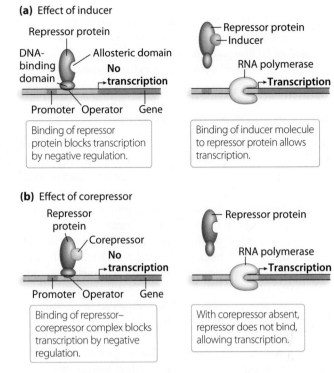

**(a)** Effect of inducer

Binding of repressor protein blocks transcription by negative regulation.

Binding of inducer molecule to repressor protein allows transcription.

**(b)** Effect of corepressor

Binding of repressor–corepressor complex blocks transcription by negative regulation.

With corepressor absent, repressor does not bind, allowing transcription.

**Figure 14.2  Mechanisms of negative control of transcription.**

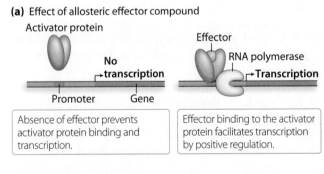

**(a)** Effect of allosteric effector compound

Absence of effector prevents activator protein binding and transcription.

Effector binding to the activator protein facilitates transcription by positive regulation.

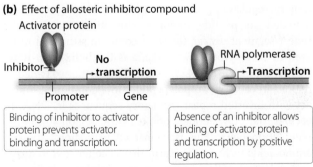

**(b)** Effect of allosteric inhibitor compound

Binding of inhibitor to activator protein prevents activator binding and transcription.

Absence of an inhibitor allows binding of activator protein and transcription by positive regulation.

**Figure 14.3  Mechanisms of positive control of transcription.**

## Regulatory DNA-Binding Proteins

Most DNA-binding proteins that exert regulatory control bind DNA at specific sequences to accomplish their regulatory activity. These interactions occur by association of the amino acid side chains of the proteins with the specific nucleotide bases and the sugar-phosphate backbone of DNA. The proteins make their contact with specific base pairs located in the major groove and the minor groove of the DNA helix using a unique pattern of hydrogen, nitrogen, and oxygen atoms that characterize each base pair.

To achieve protein–DNA specificity in these interactions, the protein must simultaneously contact multiple nucleotides. A common motif in the structures of DNA-binding regulatory proteins is the formation of protein secondary structures, most commonly α helices, that contain the amino acids that contact regulatory nucleotides. Frequently, two protein segments contact the DNA target sequence. The paired DNA-binding regions of a regulatory protein form in two ways. In one type of interaction, a single polypeptide folds to form two domains that bind specific DNA sequences. In the other type, the regulatory protein consists of two or more polypeptides joined to form a multimeric complex of two (dimeric), three (trimeric), or four (tetrameric) polypeptides. When identical polypeptides join together, the prefix *homo–* is used. A "homodimer" contains two identical polypeptides in the functional protein. When different polypeptides join together, the complex is identified by the prefix *hetero–*, as in "heterodimer."

Extensive studies of transcription-regulating proteins in bacteria have identified the characteristic structural features of DNA-binding regulatory proteins and the DNA sequence they bind. Bacterial regulatory DNA sequences frequently contain inverted repeats or direct repeats. In both kinds of interaction, each polypeptide of a homodimeric regulatory protein, or each of the binding regions of a folded polypeptide, interacts with one of the inverted repeat segment. By far, the most common structural motif seen in these proteins in bacteria is the **helix-turn-helix (HTH) motif** (Figure 14.4). In the HTH motif, two α-helical regions in each of two polypeptides in a homodimer interact with inverted repeat regulatory sequences in DNA. In each of the polypeptides, one α-helical region forms the recognition helix. This is connected to a short string of amino acids forming the "turn" that is connected to a second α-helical region known as the stabilizing helix that functions to hold the two polypeptides together. Hundreds of different DNA-binding regulatory proteins with the HTH

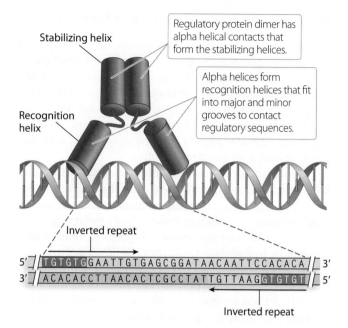

**Figure 14.4  The helix-turn-helix regulatory protein motif.** A polypeptide forming two alpha helices separated by a "turn" of amino acids unites with an identical polypeptide to form a homodimeric regulatory protein with stabilizing and recognition helices that bind inverted repeat DNA.

motif have been identified in bacteria. We will see some examples in later sections of this chapter and in discussions of the HTH and other regulatory protein motifs in eukaryotes (Chapter 15).

## 14.2 The *lac* Operon Is an Inducible Operon System under Negative and Positive Control

In comparing the genomes of different forms of life, one conclusion is that evolution has operated to restrict the total size of bacterial genomes compared to most others and to limit the percentage of repetitive (noncoding) DNA in them as well. These limitations are imposed by various factors, including the dependence of bacteria on their abilities to reproduce rapidly and respond quickly to environmental changes. Possession of a relatively small genome and small percentage of DNA that does not code for polypeptides speeds the DNA replication process and shortens the reproduction time. The need for rapid responsiveness to environmental change and for restricted genome size dictates another evolutionary adaptation in bacteria: the clustering and coordinated transcriptional regulation of genes involved in the same metabolic processes.

Clusters of genes undergoing coordinated transcriptional regulation by a shared regulatory region are called **operons.** Operons are common in bacterial genomes, and the genes that are part of a given operon almost always participate in the same metabolic or biosynthesis pathway.

Besides having a single promoter, shared by the operon genes, operons contain additional regulatory DNA sequences that interact with promoters to share transcriptional control.

In this discussion, we focus on the **lactose (*lac*) operon** of *E. coli*. This operon is responsible for the production of three polypeptides that permit *E. coli* to utilize the sugar lactose as a carbon source for growth and metabolic energy. In this section, we explain how the *lac* operon works, describe the circumstances under which its genes are transcribed, and identify the regulatory mechanisms that control operon gene transcription. In the following section, we turn our attention to mutational and molecular analyses of the *lac* operon to understand the function of operon genes and to explore the molecular interactions that regulate operon gene transcription.

## Lactose Metabolism

The monosaccharide sugar glucose is the preferred energy source of *E. coli*, just as it is for your cells. It is metabolized by the biochemical pathway called glycolysis, a sequence of biochemical reactions that oxidizes glucose to produce pyruvate and ATP (adenosine triphosphate, the compound used universally by cells to store and produce energy). Glycolysis occurs in virtually all cells as part of fermentation and cellular respiration, and *E. coli* will consume all available glucose before a genetic switch is flipped that changes their metabolic pathway to one that uses an alternative sugar. Lactose is one of many sugars that can serve as alternative carbon sources for bacteria. The genetic switch to lactose utilization requires that lactose be present in the cell, but it occurs only after glucose has been depleted. Lactose utilization is controlled by genes and regulatory sequences that form the *lac* operon, which is an **inducible operon** system—that is, under the specific circumstances that lactose is present in the growth medium and glucose is absent, transcription of the operon genes is activated, or induced. The inducible nature of the *lac* operon and other inducible operons means that expression of operon genes is limited to the circumstance in which the *inducer compound* is available. Other nutritional requirements may have to be met as well for transcription induction to occur.

Lactose is a disaccharide consisting of two monosaccharides, glucose and galactose, that are joined by a covalent β-galactoside linkage (**Figure 14.5**). Bacteria that have a ***lac*⁺ phenotype** ("lack plus") are able to grow on a medium containing lactose as the only sugar. *lac*⁺ strains accomplish this growth by producing a gated channel at the cell membrane that allows lactose to enter the cell and producing the enzyme β-galactosidase that breaks the β-galactoside linkage to release glucose and galactose. Glucose produced by lactose breakdown can immediately enter glycolysis. The molecule of galactose can be further processed to produce glucose. In addition to

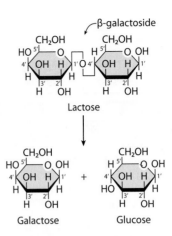

**Figure 14.5 Lactose metabolism.** The enzyme β-galactosidase cleaves the β-galactoside linkage of the disaccharide lactose and yields monosaccharides of glucose and galactose.

producing glucose and galactose, the breakdown of lactose produces a small amount of **allolactose,** a modified form of lactose. Allolactose plays a critical role in regulating the transcription of *lac* operon genes by acting as the inducer compound. Bacteria that are unable to grow on a lactose-containing medium are identified as having a ***lac*⁻ phenotype** ("lack minus"). These strains are either unable to import lactose to the cell, break it down once it is in the cell, or both.

## *Lac* Operon Structure

The *lac* operon consists of a multipart regulatory region and a structural gene region containing three genes (**Figure 14.6a**). The regulatory region contains three protein-binding regulatory sequences. One is the promoter that binds RNA polymerase, the second is the operator (*lacO*) sequence that binds the *lac* repressor protein, and the third is the CAP–cAMP region. These three regions partially overlap and are immediately upstream of the start of transcription of *lac* operon genes.

The three structural genes of the *lac* operon are identified as ***lacZ*,** a gene encoding the enzyme β-galactosidase; ***lacY*,** which encodes the enzyme permease; and ***lacA*,** which encodes transacetylase. These three genes are transcribed as a **polycistronic mRNA,** an mRNA molecule that is the transcript of all the genes in the operon. Each gene transcript that is part of a polycistronic mRNA contains a start and a stop codon sequence. The translation of a polycistronic mRNA generates a distinct polypeptide for each gene. The β-galactosidase produced by the *lacZ* gene is responsible for cleaving the β-galactoside linkage of lactose to release molecules of glucose and galactose. The enzyme also converts a small amount of lactose into allolactose, which has a chemical structure very similar to that of lactose. The permease enzyme of *lacY* functions at the cell membrane to facilitate the entry of lactose into the cell. Transacetylase, the product of

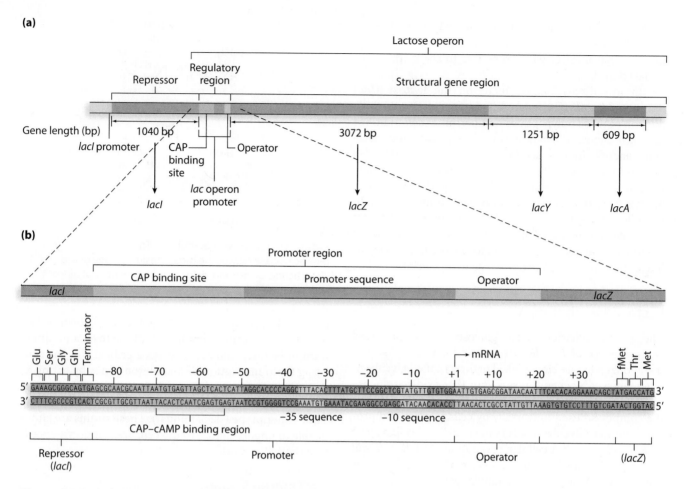

**Figure 14.6** **The lactose (*lac*) operon of *E. coli*.** **(a)** The repressor protein (*lacI*) is encoded by a 1040-bp segment under separate transcriptional regulation. The transcription regulatory region consists of a CAP binding site, a promoter consensus sequence region, and an operator sequence. The three structural genes of the *lac* operon encode the enzymes β-galactosidase (*lacZ*), permease (*lacY*), and transacetylase (*lacA*). **(b)** The DNA sequence of the regulatory region of the *lac* operon including the −10 and −35 consensus sequences, the operator, and the CAP binding site.

*lacA*, is not essential for lactose utilization, although in bacteria it protects against potentially damaging by-products of lactose metabolism. Our discussion focuses only on transcription of *lacZ* and *lacY*, and on the action of β-galactosidase and permease, since transacetylase is not essential for lactose utilization.

Adjacent to, but not part of the *lac* operon, is the regulatory gene, *lacI*, that produces the *lac* repressor protein. The *lacI* gene has its own promoter that is not regulated and drives constitutive transcription. The *lac* repressor protein is a homotetramer that has two functional domains. The first is a DNA-binding domain that binds the operator regions, and the second is an allosteric domain that binds the inducer substance allolactose.

**Figure 14.6b** shows the DNA sequence composition of the *lac* operon promoter, *lacP*, and the *lac* operator, *lacO*, which only spans about 80 base pairs. Notice that the promoter and the operator sequences overlap, and

that the operator sequence overlaps the +1 nucleotide that starts transcription. *LacP* contains the −10 and −35 consensus sequence sites that are critical for RNA polymerase binding (see Chapter 8 for a review). *LacO*, which binds the repressor protein produced by *lacI*, overlaps *lacP* near the start of transcription. Notice also that the CAP binding site is near the −35 and −10 regions of the promoter. We discuss this relationship in the next section.

## *Lac* Operon Function

The *lac* operon is transcriptionally silent when no lactose is available or when glucose is available to the cell (**Figure 14.7a**). In the absence of production of β-galactosidase, there is no allolactose in the cell and the constitutively produced *lac* repressor protein binds to *lacO*, using its DNA-binding domain. By its presence at the operator, *lac* repressor blocks RNA polymerase from

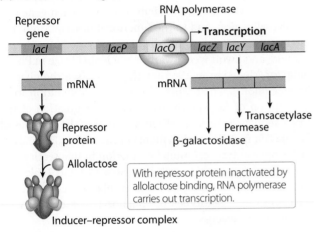

**(a)** Lactose unavailable (glucose available)

*lac* repressor protein binds to the operator (*lacO*) sequence and inhibits transcription.

**(b)** Lactose available (glucose unavailable)

With repressor protein inactivated by allolactose binding, RNA polymerase carries out transcription.

**Figure 14.7** *lac* **operon transcription regulation.** **(a)** When glucose and lactose are unavailable, *lac* operon genes are not transcribed. **(b)** Lactose availability in the absence of glucose induces operon gene transcription.

binding to *lacP* and prevents transcription initiation. This transcriptional regulatory interaction is an example of negative control of transcription that is achieved through the binding of repressor protein to the transcription-regulating operator sequence.

In contrast, the availability of lactose in the growth medium and the unavailability of glucose lead to the induction of transcription of the *lac* operon structural genes (Figure 14.7b). On this basis, the *lac* operon is identified as an inducible operon. With synthesis of β-galactosidase, the production of allolactose occurs. By binding to the allosteric domain of the repressor protein, allolactose forms the **inducer–repressor complex.** The formation of this complex induces an allosteric change that alters the conformation of the DNA-binding domain of the repressor protein to a form that does not recognize or bind the operator. The open operator allows RNA polymerase access to the promoter. Transcription of the *lac* operon genes is induced by this event. Synthesis of the polycistronic mRNA follows, and translation produces β-galactosidase, permease, and transacetylase.

The transcription driven by RNA polymerase that gains access to the *lac* promoter through the inducer–repressor complex mechanism alone is insufficient to generate enough copies of the polycistronic mRNA to drive active lactose metabolism. A second regulatory process featuring positive control of transcription is required to fully activate *lac* operon gene transcription. Positive control of *lac* operon transcription lies in a DNA–protein interaction that occurs at the **CAP binding site** of the *lac* operon promoter. This site is located at approximately −60 of *lacP* (see Figure 14.6b). The CAP binding site is the sequence that attracts the **CAP–cAMP complex,** a small molecular complex composed of a protein known as the catabolite activator protein (CAP) and the nucleotide cyclic adenosine monophosphate (cAMP). Binding of the CAP–cAMP complex to its binding site causes DNA to bend around the complex, and it increases the ability of RNA polymerase to transcribe *lac* operon genes (Figure 14.8). This positive regulatory effect leads to a high level of transcription of *lac* operon genes that allows the cell to metabolize lactose and grow on a lactose-containing medium.

The positive regulatory process is itself regulated indirectly by the level of glucose, which modulates the availability of cAMP. Cyclic AMP is synthesized from ATP (adenosine triphosphate) by the enzyme adenylate cyclase. During glycolysis, the availability of adenylate cyclase is limited and cAMP synthesis is reduced. Thus, when glucose is available, cAMP is very low in concentration, almost no CAP–cAMP can form, and *lac* operon gene transcription is highly inefficient. This effect of glucose in blocking *lac* operon gene transcription is known as **catabolite repression,** during which the presence of the preferred catabolite (glucose) represses the transcription of genes for an alternative catabolite (lactose, in this case).

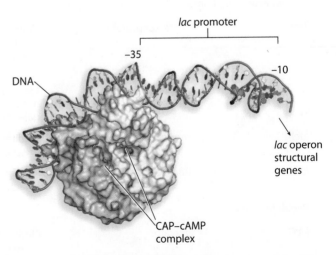

**Figure 14.8** **CAP–cAMP complex binding to the CAP binding site.** DNA bends at an approximate 90° angle around the CAP–cAMP complex.

Genetic Insight The *lac* operon is an inducible operon encoding three genes whose transcription is induced by allolactose, a derivative of lactose. When lactose is present and glucose is absent, the CAP–cAMP complex exerts positive regulation of transcription by helping RNA polymerase to efficiently bind the promoter.

With your budding understanding of *lac* operon gene transcription, perhaps the following question—a kind of chicken-and-egg conundrum—has occurred to you. Lactose must enter the cell so that allolactose can be produced to act as an inducer. Lactose cannot enter the cell without the aid of permease that helps bring lactose into the cell. But since the *lacY* gene that produces permease is part of the *lac* operon, and transcription is not induced until lactose is present inside the cell, how does lactose enter the cell in the first place? The answer lies in the reversibility of the interaction between the repressor protein and the *lac* operator. In the presence of glucose and the absence of lactose, the repressor protein is almost always bound to the operator sequence. Occasionally and spontaneously, however, the repressor protein loses contact with the operator sequence. While short-lived, this spontaneous release is just enough to allow momentary transcription of the operon and production of a few molecules of β-galactosidase and permease. This small amount of permease and β-galactosidase, amounting to no more than a few molecules per cell, is sufficient to bring the first molecules of lactose to cross the cell membrane and to generate allolactose. This trickle of lactose quickly induces more transcription, launching a transcriptional cascade that soon causes the cell to switch its metabolism to lactose utilization.

## 14.3 Mutational Analysis Deciphers Genetic Regulation of the *lac* Operon

The identification and description of the *lac* operon is traced to a series of publications in the early 1960s by François Jacob, Jacques Monod, André Lwoff, and several other colleagues. Their genetic analysis of numerous *lac* operon mutants led to the identification of each gene and regulatory region, and to the functional description of the operon we provided in the previous section. Jacob, Monod, and Lwoff were awarded the Nobel Prize in Physiology or Medicine in 1965 for this work (see the chapter opener photo). Their work also laid the foundation for a description of *lac* operon transcription regulation at the DNA sequence level. We discuss several of the analyses of *lac* operon mutants and elements of the molecular analysis of *lac* operon transcriptional regulation in this section. As you read this discussion, refer to Table 14.1 and Table 14.2 for a list of *lac* operon genes and regulatory sequences, as well as example genotypes and phenotypes associated with mutations we discuss.

### Analysis of Structural Gene Mutations

The genetic analysis of the *lac* operon by Jacob, Monod, and colleagues was made possible by the induction of operon mutations. Several dozen *lac*⁻ mutants were

| Table 14.1 | *lac* Operon Genes and Regulatory Sequences | | |
|---|---|---|---|
| Gene/Sequence | Product/Sequence Type | Function | Important Mutants |
| **Protein-Producing Genes** | | | |
| *lacI* | Repressor protein | Contains two binding sites, one for the operator and one for lactose, the inducer | $I^-$—Unable to bind to operator $I^S$—Unable to bind the inducer (allolactose) |
| *lacZ* | β-galactosidase | Cleaves lactose into two monosaccharides (glucose and galactose) | $Z^-$—No functional β-galactosidase |
| *lacY* | Permease | Facilitates lactose transport across the cell membrane | $Y^-$—No functional permease |
| *lacA* | Transacetylase | Protects against harmful by-products of lactose metabolism | $A^-$—No transacetylase |
| **Regulatory Sequences** | | | |
| *lacO* | Operator | Binds repressor protein to block transcription of operon genes | $O^C$—Fails to bind repressor protein |
| *lacP* | Promoter | Binds RNA polymerase | $P^-$—Fails to bind RNA polymerase or does so weakly |

**Table 14.2**    Synthesis of β-Galactosidase and Permease by Haploids and Partial Diploids with Structural Gene Mutations

| Genotype | β-Galactosidase[a] | | Permease[a] | | Description |
|---|---|---|---|---|---|
| | Lactose | No Lactose | Lactose | No Lactose | |
| 1. $I^+ P^+ O^+ Z^+ Y^+$ | + | − | + | − | Wild-type (*lac*$^+$) |
| 2. $I^+ P^+ O^+ Z^- Y^+$ | − | − | + | − | No functional β-galactosidase (*lac*$^-$) |
| 3. $I^+ P^+ O^+ Z^+ Y^-$ | + | − | − | − | No functional permease (*lac*$^-$) |
| 4. $I^+ P^+ O^+ Z^+ Y^- / I^+ P^+ O^+ Z^- Y^+$ | + | − | + | − | Wild-type response by complementation (*lac*$^+$) |

[a] Symbols + and − indicate production and no production, respectively, of functional enzymes.

generated by treatment of *E. coli* with mutagens. The mutants were first subjected to genetic complementation experiments to determine whether the *lac*$^-$ phenotypes of different mutants resulted from mutation of the same gene or from mutations of different genes. Investigations showed that *lac*$^-$ mutants formed two complementation groups, indicating that two genes are responsible for the *lac*$^-$ phenotype. The two complementation groups are today known to correspond to *lacZ* (β-galactosidase) and *lacY* (permease).

The complementation analysis was carried out using partial diploid bacterial strains that were produced by conjugation between F′ (*lac*) and F$^-$ bacteria. Recall that exconjugants produced by F′ × F$^+$ conjugation have two copies of a portion of the genome and are thus partially diploid. In the case of *lac* operon partial diploids, one copy of the *lac* operon information resides on the recipient bacterial chromosome, and the second copy of the operon is acquired on the F′ plasmid (see Chapter 6). The genotype of partial diploids is written with the F′ segment on the left and the recipient chromosome on the right. The homologous chromosomes are separated by a slash (/). For example, the genotype of a partial diploid demonstrating complementation of *lac* gene mutations can be written as follows:

$$F'\ I^+ P^+ O^+ Z^+ Y^- A^+ / I^+ P^+ O^+ Z^- Y^+ A^+$$

Analyzed as haploid genotypes, each portion of the partial diploid genotype above would produce the *lac*$^-$ phenotype. The F′ haploid lacks the ability to produce permease (*lacY*$^-$), and the bacterial haploid is unable to produce β-galactosidase (*lacZ*$^-$). Genetic complementation occurs in this partial diploid, however, and the resulting phenotype is *lac*$^+$ (see Table 14.2). The molecular basis of genetic complementation in this case is that the F′ portion of the partial diploid provides β-galactosidase by its *lacZ*$^+$ gene, and the recipient portion of the partial diploid provides permease by its *lacY*$^+$ gene. Based on the analysis of structural gene mutations, Jacob, Monod, and colleagues concluded that there are two protein-producing genes required for lac+ growth behavior and that *lacZ* and *lacY*

wild-type alleles are usually dominant to mutant alleles. Recombination mapping analysis revealed close genetic linkage of the three structural genes of the *lac* operon, but the order of these structural genes (*lacZ-lacY-lacA*) was ultimately determined by mutational analysis.

**Genetic Insight**  Genetic analysis of *lac* operon partial diploids identified genetic complementation between certain *lac*$^-$ strains and also identified two structural genes, *lacZ* (β-galactosidase) and *lacY* (permease), as essential to lactose metabolism. The order of the structural genes is *lacZ–lacY–lacA*.

## *Lac* Operon Regulatory Mutations

Mutations of regulatory components of the *lac* operon alter the inducible response of the operon to the presence of lactose and allolactose in the cell. Certain mutations of the *lac* operon lead to **constitutive mutants,** which are unresponsive to the presence or absence of lactose in the growth medium. These mutants continuously transcribe the operon genes, rather than transcribing the genes in an inducible manner. Other regulatory mutations block all response to lactose and render the cell *lac*$^-$. Genetic mapping of constitutive mutations identified two distinct sites of constitutive mutations of the *lac* operon: *lacO* and *lacI*. Constitutive mutations of *lacO* render the operator DNA sequence unrecognizable to the wild-type DNA-binding portion of the repressor protein. On the other hand, constitutive mutations of *lacI* result from production of a repressor protein with a mutated DNA-binding region that is unable to recognize and bind wild-type operator sequence. Both mutations prevent negative regulation of *lac* operon transcription.

The discovery that there are two sites of *lac* operon constitutive mutations suggested to Jacob and Monod that a negative regulatory system with two components exercises transcriptional control of the structural genes. They postulated that one constitutive mutation site is the gene producing a regulatory protein and the second is the target DNA-binding site for the regulatory protein binding.

**Operator Mutations** The genetic evidence indicating that the operator is the DNA sequence binding the repressor protein comes from the finding that *lac* operator (*lacO*) mutations are exclusively **cis-acting;** that is, they influence the transcription of genes only *on the same chromosome.* In the wild-type organism, *lacI*$^+$ produces repressor protein that has an allosteric (allolactose) binding domain and a functional operator binding domain. Repressor protein uses its operator binding domain to bind the regulatory sequence and block transcription (**Figure 14.9a**). Bacteria with operator mutations are constitutive for transcription of *lac* operon genes and have the genotype $I^+ P^+ O^C Z^+ Y^+$ (**Figure 14.9b**). The $O^C$ allele designation signifies an "operator-constitutive mutation." In $O^C$ mutants, the nucleotide sequence of the operator region is altered and is no longer recognized by wild-type repressor protein. In the absence of repressor protein bound to the operator sequence, constitutive transcription of the operon genes takes place and β-galactosidase and permease are produced continuously.

The crucial experiments revealing the cis-acting nature of *lacO* were performed with partial diploids. First it was shown that creation of partial diploids by conjugation of a constitutive *lac*$^+$ strain ($I^+ P^+ O^C Z^+ Y^+$) with a *lac*$^-$ strain producing defective β-galactosidase ($I^+ P^+ O^+ Z^- Y^+$) does not alter the constitutive transcription of β-galactosidase. Note that *lacO*$^c$ in the partial diploid appears dominant to *lacO*$^+$. Dominance on the part of *lacO*$^c$ arises because transcription of the wild-type *lacZ*$^+$ allele is exclusively controlled by the *lacO*$^c$ mutation, since these two alleles are on the same chromosome. The wild-type operator has no effect on the *lacZ*$^+$ allele because operator DNA is a cis-acting element, not a trans-acting element.

In a second experiment, the *lacZ* alleles were on different chromosomes, and the partial diploid genotype F′ $I^+ P^+ O^C Z^- Y^+ / I^+ P^+ O^+ Z^+ Y^-$ was produced using two *lac*$^-$ strains. In this case, the F′ strain is constitutive for permease production but does not produce functional β-galactosidase due to a *lacZ* mutation. The bacterial recipient strain produces β-galactosidase by the wild-type inducible mechanism, but it does not produce functional permease, due to mutation of *lacY*. The partial diploid produces permease constitutively, but β-galactosidase is produced only when transcription is induced by lactose. This result could occur only if the operator is a cis-acting element. In this case, the operator allele in cis to $Z^+$ is wild type, so β-galactosidase production falls under the inducible control of the wild-type operator sequence. Notice that in this partial diploid, the wild-type operator appears to be dominant to the $O^C$ mutant.

The apparent difference in the dominance relationship of $O^+$ and $O^C$ alleles is understandable if the *lac* operator is a cis-acting element that only controls the transcription of genes on the same DNA molecule. Taken together, the two experiments reveal the *lac* operator to

**(a)** *I*$^+$ (wild type)

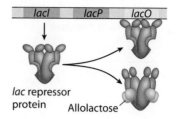

Repressor binds operator when the inducer is absent and forms an inducer–repressor complex when inducer is present.

**(b)** *O*$^c$ (operator constitutive mutation)

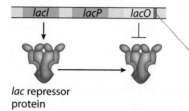

Operator-site mutation prevents repressor protein binding and leads to constituitive synthesis of the *lac* operon.

**(c)** *I*$^-$ (repressor mutation)

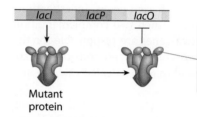

Repressor protein mutation prevents repressor binding to the operator and produces constitutive synthesis of the *lac* operon.

**(d)** *I*$^s$ (super-repressor mutation)

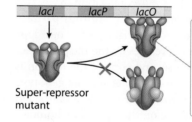

Repressor protein mutation blocks binding to the inducer, preventing formation of the inducer–repressor complex. Mutant repressor protein binds to the operator, preventing transcription.

**Figure 14.9** **Regulatory mutations of *lacI* and *lacO*.** (a) Wild-type *lacI* and *lacO*. (b) Operator-constitutive (*lacO*$^C$) mutation. (c) *lacI*$^-$ (operator-binding domain) mutation. (d) *lacI*$^S$ (super-repressor) mutation of the allosteric binding domain.

be **cis-dominant.** The principle of cis-dominance reflects the ability of the operator to only influence the transcription of downstream genes. For the *lac* operon, the "dominant" operator allele can differ, depending on the alleles carried by structural genes on each chromosome. If both wild-type structural genes are in cis to *lacO*$^c$, the mutant operator is dominant because it constitutively transcribes both genes. This is the case in the first experiment. On the other hand, if wild-type structural genes are on different chromosomes, as in the second experiment, then the *lacO*$^+$ allele is dominant because it exerts inducible transcriptional control on one of the two genes required for lactose metabolism (**Table 14.3**).

**Table 14.3**    Synthesis of β-Galactosidase and Permease by Haploids and Partial Diploids with Regulatory Mutations

| Genotype | β-Galactosidase | | Permease | | Description |
|---|---|---|---|---|---|
| | Lactose | No Lactose | Lactose | No Lactose | |
| 1. $I^- P^+ O^+ Z^+ Y^+$ | + | + | + | + | Constitutive transcription due to *lacI⁻* mutation. |
| 2. $I^+ P^+ O^C Z^+ Y^+$ | + | + | + | + | Constitutive transcription due to *lacO^C* mutation. |
| 3. $I^S P^+ O^+ Z^+ Y^+$ | − | − | − | − | Transcription is not inducible, due to *lacI^S* mutation. |
| 4. $I^+ P^- O^+ Z^+ Y^+$ | − | − | − | − | No effective transcription, due to *lacP⁻* mutation. |

**Constitutive Repressor Protein Mutations**    Experimental evidence supporting the hypothesis that the repressor gene produces a regulatory protein comes from the analysis of mutants that constitutively transcribe *lac* operon genes where the mutant allele is recessive to wild-type allele.

To see the dominance relationship of these alleles, let's first consider a haploid cell with the *lac* operon genotype $I^- P^+ O^+ Z^+ Y^+$. This cell constitutively transcribes and produces both β-galactosidase and permease (**Figure 14.9c**). Similarly, a haploid strain with the genotype $I^- P^+ O^+ Z^+ Y^-$ produces β-galactosidase constitutively, but no permease is produced, and bacteria with the genotype $I^- P^+ O^+ Z^- Y^+$ constitutively produce permease but do not produce β-galactosidase.

In contrast, a partial diploid with the genotype F′ $I^+ P^+ O^+ Z^- Y^+ / I^- P^+ O^+ Z^+ Y^-$ expresses both enzymes to their normal inducible manner. The $I^+$ allele can be on either the F′ plasmid or the recipient chromosome and have the same effect, inevitably resulting in the dominance of $I^+$ over $I^-$. This outcome indicates that *lacI* produces a regulatory protein that is **trans-acting.** In this context, *trans* refers to a protein capable of diffusing through the cell and binding to a cis-acting target sequence.

The molecular explanation of the trans-acting ability of the *lac* repressor protein is that a *lacI⁻* mutant alters the DNA-binding domain of the protein, rendering it incapable of binding the operator sequence. In the absence of negative control, transcription is constitutive. In partial diploids that are $I^+/I^-$, however, repressor protein with a functional DNA-binding domain is present in the cell and responds normally to the addition or removal of lactose from the cell.

**Super-Repressor Protein Mutations**    A second set of repressor protein mutations produces a different consequence for *lac* operon transcription. These mutants produce mutant repressor protein with an altered allosteric domain. The mutant proteins are unable to bind allolactose and are unresponsive to lactose addition or removal from cells. The DNA-binding domain is unaffected by the

allosteric domain mutation; but as a result of the nonfunctional allosteric domain, mutant repressor proteins cannot release the operator even in the presence of allolactose.

Haploids and partial diploids with mutations of the allosteric domain of the repressor protein are identified as $I^S$ mutants and are designated super-repressors. These mutants are **noninducible,** meaning that operon gene transcription cannot be induced (**Figure 14.9d** and Table 14.3). Haploids with the genotype $I^S P^+ O^+ Z^+ Y^+$ produce a repressor protein that binds normally to operator sequence; but lacking a functional allosteric domain, the protein is not removed from the operator by lactose in the cell. Such mutants are *lac⁻* and cannot be induced to metabolize lactose. Cultures of partial diploid bacteria with the genotype F′ $I^S P^+ O^+ Z^+ Y^+ / I^+ P^+ O^+ Z^+ Y^+$ may initially have some inducible responsiveness to lactose, but this ability is lost as mutant repressor protein binds to operator sequences. This partial diploid reveals the dominance of $I^S$ over $I^+$.

**Promoter Mutations**    Mutations of promoter consensus sequences significantly reduce transcription or may eliminate it entirely (see Figure 8.11). To know the specific effect of a promoter mutation usually requires direct testing of transcription in the mutant organism. Promoters, like operators, are cis-acting regulatory sequences, and most mutations of *lacP* significantly reduce, and may entirely eliminate, transcription of *lacZ* and *lacY* genes, which are located in cis. This reduces β-galactosidase and permease production to such a low point that haploid bacteria with the genotype $I^+ P^- O^+ Z^+ Y^+$ are *lac⁻*.

**Table 14.4** summarizes the conditions for *lac* operon gene transcription given the presence or absence of glucose and lactose. Active transcription of operon genes takes place only when glucose is depleted from the cell and lactose is present. Under these conditions, the following events occur:

1. Cyclic AMP level rises as a result of the availability of adenylcyclase.

| Table 14.4 | Transcription Conditions for the *lac* Operon | | | | | |
|---|---|---|---|---|---|---|
| **Glucose** | **Lactose** | **cAMP** | **Allolactose** | ***lac* Operon Transcription** | **Explanation** | |
| Present | Absent | Absent | Absent | Basal | Glucose is present to provide energy. | |
| Present | Present | Absent | Present | Basal | Glucose is present to provide energy; absence of cAMP prevents positive transcription regulation. | |
| Absent | Absent | Present | Absent | Basal | CAP-cAMP forms, but no inducer is present to block repressor binding at operator. | |
| Absent | Present | Present | Present | High | Inducer and CAP-cAMP available to induce and positively regulate transcription. | |

2. CAP–cAMP complex forms and binds to the CAP site of the *lac* promoter.

3. Allolactose is produced by a side reaction of the metabolism of lactose by β-galactosidase.

4. Repressor protein conformation is modified by interaction with allolactose, causing the protein to release from the operator, thus allowing operon gene transcription.

To test your understanding of the *lac* operon, see **Genetic Analysis 14.1** which guides you through analysis of some *lac* operon mutants.

---

**Genetic Insight** *Lac* operator mutants, unable to bind repressor protein, exhibit constitutive transcription by cis activity. Analysis of *lac* repressor protein mutations reveals two functional domains in the trans-acting protein. *Lac* promoter mutations effectively prevent transcription by cis action.

---

## Molecular Analysis of the *Lac* Operon

In the 50 years since Jacob, Monod, and colleagues described their genetic analysis of the *lac* operon, molecular analysis has identified the DNA sequences of its components (see Figure 14.6b). This and other accumulated molecular information weaves a virtually complete picture of *lac* operon transcription regulation, revealing it to be somewhat more complex, but wholly consistent, with the description presented above.

**Experimental Insight 14.1** discusses two important pieces of experimental molecular evidence derived from DNA footprint protection analyses that pertain to transcriptional regulation of the *lac* operon. The first observation is that the repressor protein binding location at the *lac* operator overlaps with the promoter binding location of RNA polymerase. This observation supports the hypothesis that repressor protein binding blocks RNA polymerase binding and transcription initiation and, conversely, that when the repressor protein is not bound to the operator, RNA polymerase can access and initiate

transcription at the promoter. The second observation identifies three distinct segments of operator DNA sequence. These operator segments, designated $O_1$, $O_2$, and $O_3$, interact differently with the repressor protein, and the result of the interactions provides a mechanism by which repressor protein binding can block RNA polymerase access to the promoter.

Additional molecular analysis reveals that the repressor protein is a homotetrameric protein formed by the union of four identical 360–amino acid polypeptides (**Figure 14.10**). The four polypeptides are joined together at their C-terminal ends and are arranged as two identical bundles. One end of each bundle forms an operator DNA-binding domain, and the other end forms the allosteric domain. The three operator DNA segments that are the targets of repressor protein binding share a conserved, 21-bp inverted repeat sequence. In each sequence,

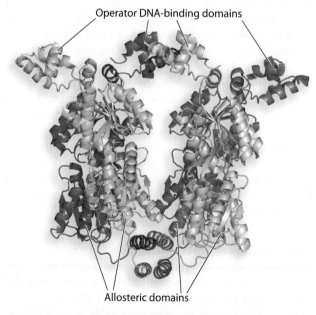

Operator DNA-binding domains

Allosteric domains

**Figure 14.10 The homotetrameric structure of the *lac* repressor protein.** Operator binding and allosteric domains are formed on opposite sides of the protein.

Evaluate the following *lac* operon partial diploids. Indicate whether the production of functional β-galactosidase from *lacZ* and of permease from *lacY* is "inducible," "constitutive," or "noninducible" for each partial diploid.

**a.** $I^- P^+ O^+ Z^+ Y^+ / I^+ P^+ O^+ Z^- Y^-$

**b.** $I^+ P^+ O^C Z^+ Y^- / I^+ P^+ O^+ Z^- Y^+$

**c.** $I^+ P^+ O^C Z^- Y^+ / I^S P^+ O^+ Z^+ Y^+$

| Solution Strategies | Solution Steps |
|---|---|
| **Evaluate** | |
| 1. Identify the topic this problem addresses, and describe the information required in the answer. | 1. This problem concerns an analysis of patterns of transcriptional regulation and the production of functional β-galactosidase and permease by operon genotypes. The answer requires a determination of whether the enzymes are produced inducibly, constitutively, or not at all. |
| 2. Identify the critical information given in the problem. | 2. The *lac* operon genotypes of three partial diploids are given. |
| **Deduce** | |
| 3. Describe the consequences of any mutations in genotype *a*. <br><br> **TIP:** Assess regulatory mutations first; then consider the consequences for structural gene transcription. | 3. The $I^-$ mutation produces a repressor protein that is unable to bind operator sequence. The $Z^-$ mutation will not produce functional β-galactosidase, and the $Y^-$ mutation will not produce functional permease. |
| 4. Describe the consequences of any mutations in genotype *b*. | 4. The $O^C$ mutation alters the operator sequence and prevents binding and transcriptional repression by repressor protein. The $Z^-$ and $Y^-$ mutations block production of functional β-galactosidase and permease. |
| 5. Describe the consequences of any mutations in genotype *c*. | 5. The $I^S$ mutation produces a super-repressor protein that has an altered allosteric domain and will not interact with allolactose. The $O^C$ and $Z^-$ alter function as described above. |
| **Solve** | **Answer a** |
| 6. Determine the expression pattern of functional enzymes for partial diploid *a*. | 6. Wild-type repressor protein is trans-active and binds the wild-type operator. This cis-acting operator blocks transcription of $Z^+$ and $Y^+$ when lactose is not in the cell, but permits transcription when lactose is present. Therefore, both enzymes are produced inducibly. |
| | **Answer b** |
| 7. Determine the expression pattern of functional enzymes for partial diploid *b*. | 7. $O^C$ is cis-active on $Z^+$, resulting in constitutive transcription. $Y^+$ is under the cis-active transcriptional control of $O^+$. Therefore, β-galactosidase is produced constitutively, and permease is produced inducibly. |
| | **Answer c** |
| 8. Determine the expression pattern of functional enzymes for partial diploid *c*. | 8. The $O^C$ sequence is not recognized by either the wild-type repressor or the super-repressor. Both repressors have wild-type DNA-binding sequences. Cis-active $O^C$ constitutively transcribes $Y^+$. The super-repressor binds $O^+$, and its cis activity renders $Z^+$ and $Y^+$ noninducible. Therefore, β-galactosidase is noninducible, and permease production is constitutive. |

For more practice, see Problems 5, 16, 17, and 18.

## Experimental Insight 14.1

### Regulatory Proteins Binding to *lac* Operon Regulatory Sequences

DNase I footprint protection analysis of the kind described in Research Technique 8.1 (see page 271) has been used to precisely identify the binding locations of *lac* repressor protein relative to the location of RNA polymerase binding in the regulatory region of the *lac* operon. Recall from the earlier description of this technique that identical control and experimental DNA fragments containing regulatory sequences are end-labeled with $^{32}$P. The experimental fragments are then exposed to DNA-binding proteins, but the control fragments are not. All fragments are then exposed to DNase I that randomly digests those segments not protected by bound proteins. The resulting DNase I-digested DNA fragments are separated by gel electrophoresis to reveal the "footprint" of protein protection.

The figure here shows the results of footprint analysis of a 123-bp segment of the *lac* operon regulatory region from position +39 to −84. Control DNA in first lane ❶ is not protein protected. The gel shows that the promoter regions protected by ❷ RNA polymerase and ❸ *lac* repressor protein partially overlap one another. The relative positions of these protein protected regions are consistent with the model that repressor protein binding can interfere with RNA polymerase binding.

Separate DNase I footprint analysis of the *lac* operator region detects three segments of DNA sequence that are protected by *lac* repressor protein : $O_1$, $O_2$, $O_3$. Lane a of the gel shown is control DNA not bound by protein, and is therefore unprotected DNA. The experimental analysis identifies one protected segment, designated $O_1$, as the principal operator sequence. The two other regions of protein-protected operator DNA sequence are designated $O_2$ and $O_3$. Lanes d through g of the DNA footprint-protection gel are protected by repressor protein, and show the footprint gaps corresponding to these operator elements.

Lanes of the gel also identify two regions, designated $C_1$ and $C_2$, that are protected from DNase I digestion by the CAP-cAMP complex. This segment contains the consensus sequences for the CAP binding site that partially overlaps operator regions $O_1$ and $O_3$. The relative positions of these protein-binding sites indicate two kinds of interactions between proteins binding the *lac* promoter and operator. First, when CAP-cAMP is bound to the CAP binding site, RNA polymerase gains enhanced access to the promoter, establishing conditions for efficient transcription of *lac* operon genes. Second, the overlap of the CAP binding region with $O_1$ suggests that when repressor protein is bound to DNA, the CAP-cAMP complex is unable to bind, thus preventing positive regulation of transcription.

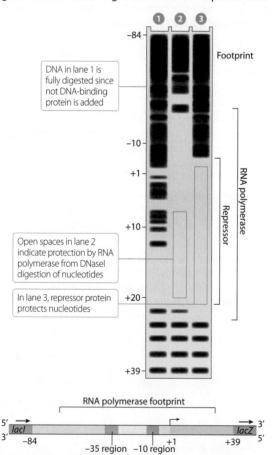

**DNaseI footprint protection analysis of the *lacP* and *lacO* regions and model.**

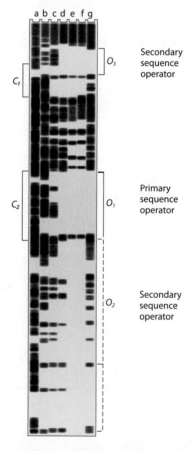

***lac* repressor protein footprint protection and DNA binding.**

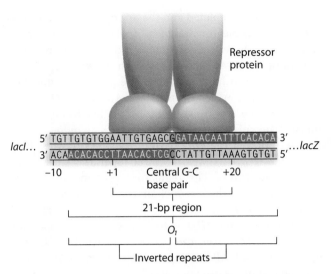

lacI... 5′ TGTTGTGTGGAATTGTGAGCGGATAACAATTTCACACA 3′ ...lacZ
3′ ACAACACACCTTAACACTCGCCTATTGTTAAAGTGTGT 5′

−10        +1    Central G-C    +20
base pair

21-bp region

$O_1$

Inverted repeats

**Figure 14.11  The *lacO* region $O_1$ contains an inverted repeat sequence.** The central G - C base pair is the pivot point of this region of twofold nucleotide symmetry of an inverted repeat sequence.

a central G - C base pair is at the midpoint of a twofold axis of symmetry (**Figure 14.11**). On either side of the central G - C base pair are inverted repeat sequences of 10 bp each that are the specific binding location for polypeptides in each half of the repressor protein. Mitchell Lewis and his colleagues examined the crystal structure of DNA-bound repressor protein in a 1996 study and determined that the tetrameric repressor protein binds to $O_1$ and $O_3$ and induces **DNA loop** formation that draws the $O_1$ and $O_3$ regions closer together (**Figure 14.12**). This DNA loop structure contains part of the *lac* promoter and prevents transcription by blocking access of RNA polymerase.

Parallel experiments examining mutated operator DNA sequences reveal how constitutive operator mutations are caused by alterations of the DNA sequence in region $O_1$. **Figure 14.13** shows several base-pair substitutions that cause constitutive operator mutations. Each of these changes disrupts the twofold symmetry of $O_1$, masking the sequence from recognition by repressor protein. Since $O_1$ is the primary binding target of the repressor protein, and $O_1$ must be bound before binding to $O_3$ can occur, $O_1$ mutation also disrupts binding to $O_3$. The inability of repressor protein to bind to mutant operator sequence means that the transcription-repressing DNA loop cannot form. This in turn leaves the promoter available for binding by RNA polymerase and opens the door to continuous transcription and constitutive expression of the *lac* operon genes.

## Positive and Negative Regulation of the Arabinose Operon

As a final illustration of the role of regulatory proteins in positive and negative control of transcription, we turn to

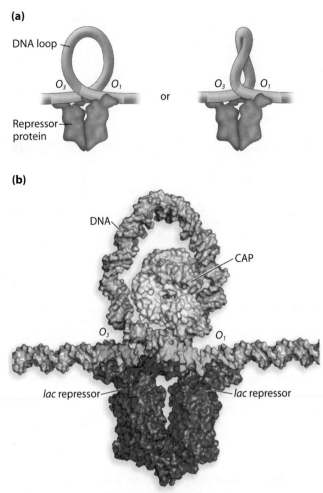

**Figure 14.12  *Lac* repressor protein binding. (a)** Repressor protein binding to *lacO* regions $O_1$ and $O_3$ forms a loop of DNA that inhibits RNA polymerase binding at *lacP*. **(b)** The crystal structural model of *lac* repressor binding at *lacO*.

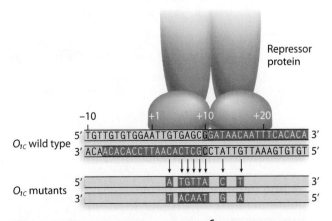

**Figure 14.13  Constitutive operator ($O^C$) mutations.** Eight base-substitution mutations in *lacO* region $O_1$ producing operator-comstitutive mutations. Each mutation disrupts the twofold symmetry of the operator inverted repeat sequences and prevents *lac* repressor protein binding.

**(a)**

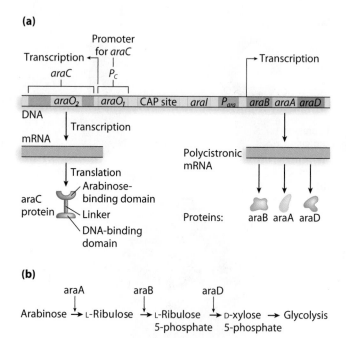

**(b)**

araA       araB       araD

Arabinose → L-Ribulose → L-Ribulose → D-xylose → Glycolysis
                 5-phosphate   5-phosphate

**Figure 14.14 The arabinose (ara) operon and arabinose metabolism. (a)** The regulatory protein araC is encoded by the gene *araC* under the control of promoter $P_C$. AraC binds three regulatory regions: *araO₁*, *araO₂*, and *araI*. The three operon genes, *araB*, *araA*, and *araD*, are transcribed as a polycistronic mRNA from promoter $P_{ara}$. **(b)** The proteins araA, araB, and araD catalyze separate steps of arabinose metabolism.

the **arabinose (*ara*) operon** system, where a single regulatory protein carries out both positive and negative transcriptional regulation. Arabinose is another of the alternative sugars that bacteria use as a source of carbon

when glucose is depleted and the *ara* operon contains the genes responsible for arabinose metabolism. The transcription of these genes is under inducible control, and the *ara* operon is an inducible operon system.

Arabinose is metabolized by the protein products of three *ara* operon genes, *araB*, *araA*, and *araD*. These genes produce a polycistronic mRNA that has a unique mechanism of transcriptional control (**Figure 14.14**). Transcription of the *ara* operon genes is controlled by a promoter ($P_{ara}$) as well as three operator sites: *araI*, *araO₁*, and *araO₂*. The regulatory protein for the *ara* operon is araC, the product of the *araC* gene, within which the *araO₂* operator is located. The *araC* gene has its own promoter ($P_C$), which overlaps the operator sequence *araO₁*. The *araC* gene is adjacent to the *ara* operon, with a CAP binding site separating $P_C$ from *araI* and the rest of the *ara* operon.

In 1989, Robert Schleif and colleagues proposed a DNA loop model to explain the regulation of the *ara* operon. Schleif's model proposes a structure that is reminiscent of the DNA loops formed when repressor protein binds to the *lac* operator (see Figure 14.12a). A distinctive feature of Schleif's model is the proposed dual role araC protein plays in both positive and negative control of transcription. In the absence of arabinose, Schleif's DNA loop model predicts araC protein will bind to operator sequences *araI*, *araO₁*, and *araO₂* (**Figure 14.15a**). A homodimer of araC protein binds at *araO₁*, and monomers of araC bind to *araI* and *araO₂*. The araC proteins bound at *araI* and *araO₂* then link to one another, inducing the formation of a DNA loop. The DNA loop prevents RNA polymerase from binding to $P_{ara}$, thus blocking transcription of the *ara* operon

**(a)** No arabinose

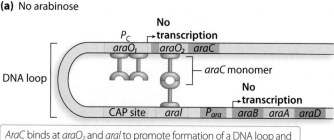

AraC binds at *araO₂* and *araI* to promote formation of a DNA loop and inhibit transcription of operon genes. An *araC* dimer binds at *araO₁* to inhibit further transcription of *araC*.

**Figure 14.15 Negative and positive transcriptional regulation of the *ara* operon by araC protein. (a)** No arabinose. **(b)** Arabinose present.

**(b)** Arabinose present

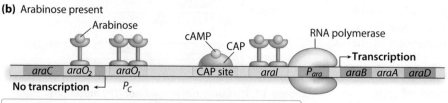

Arabinose binds to araC protein and breaks the DNA loop between *araO₂* and *araI*. An additional araC protein molecule binds at *araI*. CAP–cAMP binds at the CAP site, and RNA polymerase transcribes the operon genes.

genes. In addition to blocking operon gene transcription, the araC dimer bound to $araO_1$ occupies the RNA polymerase binding site of $P_C$ and prevents additional transcription of *araC*. When arabinose is absent from the growth medium, araC acts as a negative control protein to suppress gene transcription. The combination of the transcription-suppressing DNA loop and *araC* repression means that in the absence of arabinose, the *ara* operon is turned off by negative transcriptional control of araC protein.

When arabinose is present, it binds to araC proteins at all three operator sites (**Figure 14.15b**). This breaks the connection between araC proteins at *araI* and $araO_2$, and destroys the DNA loop that suppresses *ara* operon transcription. After the DNA loop is broken, a second araC protein–arabinose complex binds at *araI*, producing araC–arabinose dimers at *araI* and $araO_1$ that flank the CAP binding site. As with transcriptional activation of the *lac* operon, CAP must complex with cAMP and bind at the CAP binding site to activate *ara* operon gene transcription. The formation of the CAP–cAMP complex occurs when glucose is absent. The large complex comprised of dimers of araC–arabinose must bind at *araI* and $araO_1$, and the CAP–cAMP complex must bind at the CAP binding site, for RNA polymerase to bind efficiently and initiate transcription at $P_{ara}$. With these complexes in place, the *ara* operon is transcribed, and the B, A, and D proteins necessary to metabolize arabinose are produced. In this instance, araC exerts positive transcriptional control on *ara* gene transcription. In the presence of arabinose and the absence of glucose from the growth medium, araC operates in conjunction with CAP–cAMP to transcribe *araB, araA,* and *araD* in the *ara* operon.

# 14.4 Transcription from the Tryptophan Operon Is Repressible and Attenuated

The *lac* and *ara* operons are examples of inducible operons that produce proteins responsible for the breakdown of sugars that are alternative energy sources to glucose. Operons like *lac* and *ara* that are involved in catabolism of alternative energy sources are typically inducible, since they are called upon only when glucose is depleted and the alternative sugar is available. In contrast, operons involved in anabolic pathways (pathways that synthesize compounds needed by the cell) can be regulated by negative feedback mechanisms that operate through activity of the end product of the pathway to block operon gene transcription. Operons of this kind are **repressible operons.**

In addition to the negative feedback mechanism, certain repressible operons have a second regulatory capability known as **attenuation** that has the ability to fine-tune

transcription to match the momentary requirements of the cell, achieving a more-or-less steady state of compound availability. The difference between attenuation and inducibility can be clarified by an analogy. Inducible operons, such as *lac* or *ara,* are akin to light switches that provide illumination in one setting ("on") and no illumination in the alternative setting ("off"). Inducible operons are turned on and off by molecular switches controlled by DNA-binding proteins. Attenuation, on the other hand, works more like a dimmer switch that allows illumination to be incrementally adjusted up or down. For several amino acid operons, the regulation of gene expression has evolved to maintain steady amino acid levels in cells. In such systems, feedback inhibition turns off operon gene transcription when the amino acid is readily available and attenuation fine-tunes the amino acid level to maintain a steady-state concentration.

## Feedback Inhibition of Tryptophan Synthesis

The tryptophan (*trp*) operon ("trip operon") in the *E. coli* genome contains five structural genes that share a regulatory region containing a promoter (*trpP*), an operator (*trpO*), and a **leader region (*trpL*)** that contains the **attenuator region** (**Figure 14.16**). The regulatory region spans 312 base pairs and the five structural genes span approximately 6800 base pairs. The five structural genes transcribed in the operon are, in order, *trpE, trpD, trpC, trpB,* and *trpA*. Together, the protein products of these genes are responsible for synthesis of the amino acid tryptophan. Outside the operon, a sixth gene, *trpR*, encodes the repressor protein that is not activated until it pairs with tryptophan.

Transcription of *trp* operon genes is regulated by a feedback inhibition system that responds to free tryptophan in the cell. In this system, tryptophan acts as a corepressor by binding to and activating the *trp* repressor protein that is not active without its bound corepressor. Feedback inhibition is the principal mechanism turning on and turning off *trp* operon gene transcription (**Figure 14.17**). In the absence of tryptophan, the inactive repressor is unable to bind *trpO*, and operon gene transcription takes place. When tryptophan is present, however, it binds the repressor to activate it, and the repressor–corepressor complex binds the operator to block transcription.

Based on this description, and knowing about the feedback inhibition of gene transcription, one might expect that $trpR^-$ bacteria that are mutant for the repressor protein to show constitutive transcription of operon genes regardless of whether tryptophan is present. Surprisingly, however, this is not the case. In wild-type bacteria ($trpR^+$), tryptophan synthesis is very low when tryptophan is present in the cell; but, while tryptophan synthesis by $trpR^-$ strains is higher under the same

**Figure 14.16  The tryptophan (*trp*) operon.** Transcription is initiated from the promoter $P_{trp}$ and progresses through the tryptophan leader (*trpL*) region to transcribe the five operon genes (*trpE* to *trpA*) in a polycistronic mRNA. The protein products of the operon genes catalyze successive steps of tryptophan synthesis.

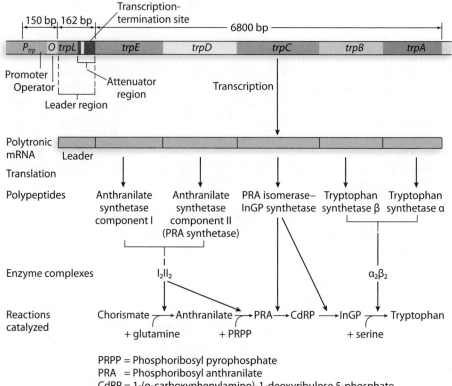

PRPP = Phosphoribosyl pyrophosphate
PRA  = Phosphoribosyl anthranilate
CdRP = 1-(o-carboxyphenylamino)-1-deoxyribulose 5-phosphate
InGP = Indole-3-glycerol phosphate

conditions, it is not at 100% capacity (**Table 14.5**). Both *trpR⁺* and *trpR⁻* strains synthesize tryptophan at 100% of capacity when tryptophan is absent. This suggests that a second regulatory mechanism is also affecting transcription of *trp* operon genes.

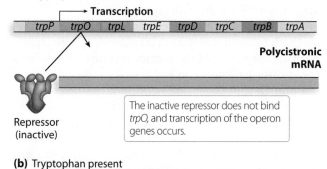

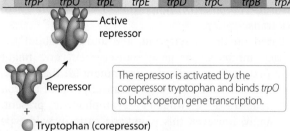

**Figure 14.17  *Trp* operon transcription regulation by the repressor.** **(a)** Tryptophan absent. **(b)** Tryptophan present.

## Attenuation of the *trp* Operon

The second mechanism regulating *trp* operon gene transcription is attenuation that is controlled by alternative folding undertaken by mRNA synthesized from the 162-bp *trpL* region. RNA polymerase binds to *trpP* and initiates transcription of *trpL*. The *trpL* region contains four repeat DNA sequences, and the mRNA transcript contains complementary repeats as well as start codon and stop codon sequences for a short polypeptide of 14 amino acids whose translation plays a pivotal role in attenuation (**Figure 14.18a**). Two features of the *trpL* region are critical to its attenuation function. First, the four repeat sequences, designated 1, 2, 3, and 4, can form different stem-loop structures (**Figure 14.18b**). (Stem-loop structures are discussed in Chapter 8 in connection with intrinsic transcription termination in bacteria; see Figure 8.7) Second, among the codons for the 14 amino acids encoded by *trpL* mRNA, there are two back-to-back tryptophan codons (UGG) that function to sense the availability of tryptophan.

| Table 14.5 | Percentage of Full Tryptophan Expression for *trpR⁺* and *trpR⁻* Strains | |
|---|---|---|
| | **Tryptophan Present** | **Tryptophan Absent** |
| *trpR⁺* | 8% | 100% |
| *trpR⁻* | 33% | 100% |

**(a)** *TrpL* attenuator region

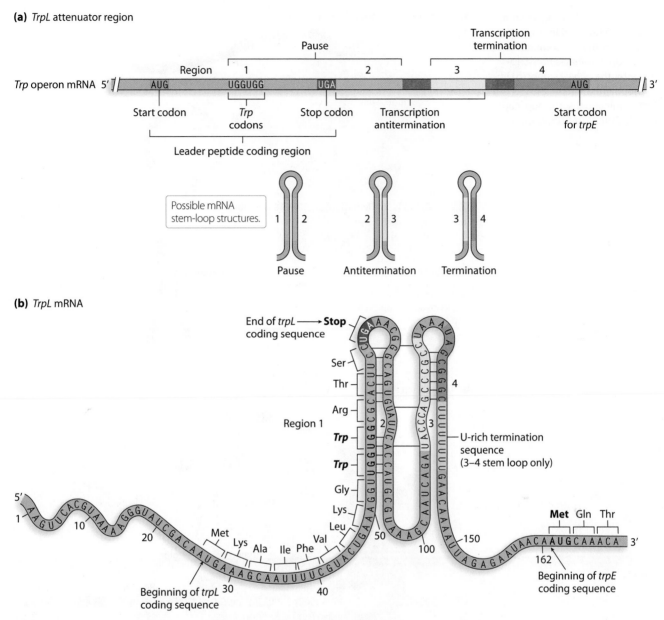

**(b)** *TrpL* mRNA

**Figure 14.18  The *trpL* attenuator region and mRNA. (a)** The *trpL* attenuator contains four inverted repeat sequences that encode regions 1 through 4 in *trpL* mRNA. Three alternative stem loops can form in mRNA. **(b)** The *trpL* mRNA attenuation sequence contains 162 nucleotides. Possible base pairing between attenuator regions and the coding sequence of the trp leader polypeptide are shown.

The formation of stem loops of *trpL* mRNA is directly tied to the continuation or termination of transcription of the five *trp* operon genes. In the *trpL* region mRNA, region 1 is complementary to region 2, region 2 is complementary to region 3, and region 3 is complementary to region 4. Two of these stem-loop structures, the *3–4 stem loop* and the *2–3 stem loop*, are central to attenuation. The third type of stem loop, the *1–2 stem loop*, plays a minor role in attenuation.

The **3–4 stem loop** of mRNA, which is the **termination stem loop,** signals transcription termination. This is identified as the transcription termination site in

Figure 14.16a. Formation of the 3–4 stem-loop halts RNA polymerase progress along the DNA, terminating transcription in the leader region before it reaches the structural genes of the operon (**Figure 14.19a**). Notice that region 4 is followed immediately by a poly-uracil sequence (a poly-U tail). This configuration—an mRNA stem loop followed by a uracil string—is the same as one described in connection with intrinsic termination of transcription in bacteria (see Chapter 8). Formation of a 3–4 stem loop may be preceded by formation of a 1–2 stem loop, which can induce a pause in the attenuation process. Formation of the 1–2 stem loop leaves only region 4 for

**Figure 14.19** *TrpL* mRNA stem-loop formation. **(a)** In tryptophan abundance, the 3-4 (termination) stem loop terminates transcription after the poly-U string. **(b)** In tryptophan starvation, the 2-3 (antitermination) stem loop leads to polycistronic mRNA synthesis.

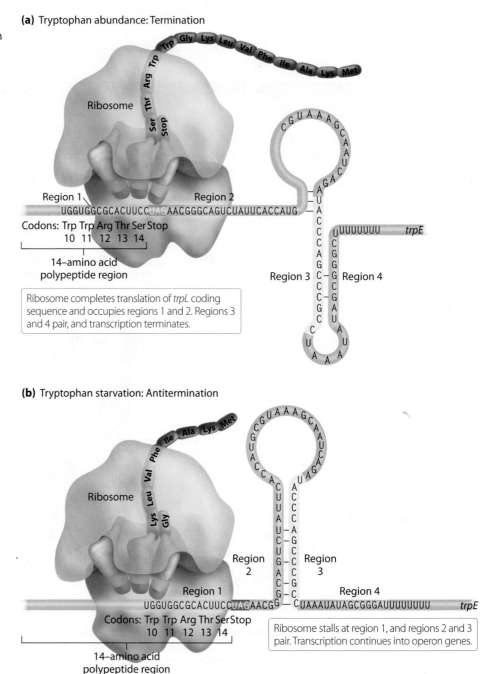

**(a)** Tryptophan abundance: Termination

Ribosome

Region 1      Region 2
UGGUGGCGCACUUCCUAGAACGGGCAGUCUAUUCACCAUG

Codons: Trp Trp Arg Thr Ser Stop
      10   11   12   13   14

14–amino acid
polypeptide region

Region 3      Region 4      *trpE*

Ribosome completes translation of *trpL* coding sequence and occupies regions 1 and 2. Regions 3 and 4 pair, and transcription terminates.

**(b)** Tryptophan starvation: Antitermination

Ribosome

Region 2      Region 3

Region 1
UGGUGGCGCACUUCCUAGAACGG—CUAAAUAUAGCGGGAUUUUUUUU    *trpE*

Codons: Trp Trp Arg Thr Ser Stop
      10   11   12   13   14

Region 4

14–amino acid
polypeptide region

Ribosome stalls at region 1, and regions 2 and 3 pair. Transcription continues into operon genes.

pairing with region 3. Alternatively, a **2–3 stem loop** of mRNA, which is the **antitermination stem loop,** forms when region 1 is unavailable for immediate pairing with region 2; formation of this stem loop precludes the formation of a 3–4 stem loop (**Figure 14.19b**). The antitermination stem loop allows RNA polymerase to continue transcription through the leader region and into the structural genes of the *trp* operon, beginning with the transcription of *trpE*. If transcription progresses past region 4, a polycistronic mRNA spanning the five *trp* genes is produced. Translation of the five enzymes required for tryptophan synthesis follows.

Each mRNA transcribed from the *trpL* operon eventually forms either a 2–3 stem loop or a 3–4 stem loop;

but what determines the type of stem loop an mRNA will form? The coupling of transcription and translation that is a prominent feature of bacterial gene expression plays a critical role in deciding this outcome. Transcription of the *trpL* region begins at the +1 nucleotide after RNA polymerase initiates transcription. Transcription across repeat regions 1 and 2 can lead to formation of a 1–2 stem loop that temporarily pauses the progress of RNA polymerase. The pause is only momentary, however; it lasts just long enough for a ribosome to bind at the start codon in *trpL* and begin translation of the 14–amino acid polypeptide starting with the AUG codon identified in Figure 14.18. Translation initiation breaks the 1–2 stem loop, RNA polymerase resumes transcription, and the ribosome and RNA polymerase begin their coupled progression.

Notice three features of the leader mRNA depicted in Figure 14.19b: (1) The polypeptide-coding sequence overlaps the entirety of leader region 1, and the stop codon is immediately adjacent to region 2; (2) codons 10 and 11 of the mRNA specify tryptophan making completion of translation dependent on tryptophan availability; and (3) region 4 is followed immediately by a poly-U string, a feature associated with intrinsic termination of transcription. As coupled transcription and translation proceed, the relative positions of RNA polymerase and the ribosome are determined by how efficiently the ribosome can progress along the mRNA. This process, in turn, is tied directly to the availability of tryptophan and the rapidity with which tryptophan is inserted into the nascent polypeptide chain. When the cell has an adequate supply of tryptophan, the ribosome makes steady progress along *trpL* mRNA, arriving at the stop codon where it partially overlays region 1 and region 2. Simultaneously, RNA polymerase is transcribing region 3, followed by region 4. With a portion of region 2 occupied by the ribosome and unavailable for pairing in a stem loop, region 3 forms a stem loop with region 4, the only available complementary segment of the mRNA. The 3–4 stem loop, being immediately followed by a poly-U string, causes transcription to spontaneously terminate at the end of region 4 by the intrinsic process. Formation of the 3–4 stem loop (the termination stem loop) stops transcription of the *trp* operon in the leader sequence before RNA polymerase reaches the beginning of the *trpE* gene. Transcription thus ceases only when the system senses that no additional tryptophan is needed to supply translation.

When the cell is starved for tryptophan, the supply of charged tRNA$_{trp}$ is low. The ribosome is forced to pause momentarily at codons 10 and 11 to await the arrival of a charged tryptophan tRNA that will incorporate tryptophan into the nascent polypeptide. As the ribosome pauses, its mass covers region 1. Meanwhile, RNA polymerase continues to transcribe *trpL*. As RNA polymerase transcribes region 3, the region finds a complementary partner in region 2, leading to 2–3 stem-loop formation. Region 3 is not followed by a poly-U string, making intrinsic termination impossible. Transcription continues through region 4 and on into the structural gene region of the operon to produce the polycistronic mRNA transcript of the operon. Formation of a 2–3 stem loop (the antitermination stem loop) thus permits transcription and translation of the enzymes necessary to synthesize tryptophan when the system senses that the available supply of tryptophan is insufficient to support translation.

Each *trpL* mRNA makes a molecularly based "decision" about whether to form a 3–4 or a 2–3 stem loop, depending on the availability of charged tRNA$_{trp}$ at the moment tRNA$_{trp}$ is needed by ribosomes. It is likely that

at any given moment in time, a single bacterial cell contains a mixture of *trpL* mRNAs with 2–3 stem loops and *trpL* mRNAs with 3–4 stem loops. The balance shifts in the direction of more 3–4 stem loops and fewer 2–3 stem loops at higher levels of tryptophan concentration and shifts in the opposite direction—more 2–3 stem loops and fewer 3–4 stem loops—as tryptophan concentration falls. The resulting fine-tuning allows each cell to maintain a relatively steady concentration of tryptophan by turning tryptophan synthesis up or down to meet the needs of the cell.

---

Genetic Insight  Attenuation is a mechanism that adjusts the rate of transcription of operon genes to meet the needs of a cell. Attenuation of the *trp* operon is controlled by which type of stem loop *trpL* mRNA forms in direct response to tryptophan availability in the cell.

---

## Attenuation Mutations

The attenuation model is supported by mutagenesis experiments. For example, experiments in which one of the two adjacent tryptophan codons (in positions 10 and 11 of the *trpL* mRNA) has been altered by missense mutation to specify another amino acid have provided evidence of the importance of the back-to-back tryptophan codons in the *trpL* transcript. Mutation of one tryptophan UGG codon affects the attenuator responsiveness to tryptophan. If both tryptophan codons are altered by missense mutation, the attenuator no longer senses tryptophan concentration and instead senses the availability of the amino acid encoded by the mutated codons. Mutagenesis experiments have also targeted regions 3 and 4 of the leader sequence (**Figure 14.20**). Base substitutions that reduce the percentage of complementary base pairs binding these two regions

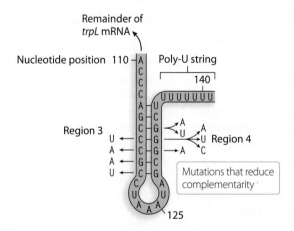

**Figure 14.20 Mutations of *trpL*.** Mutational analyses identify several base substitutions in *trpL* that decrease the efficiency of transcriptional regulation in the attenuator region by disrupting formation of the 3-4 stem loop.

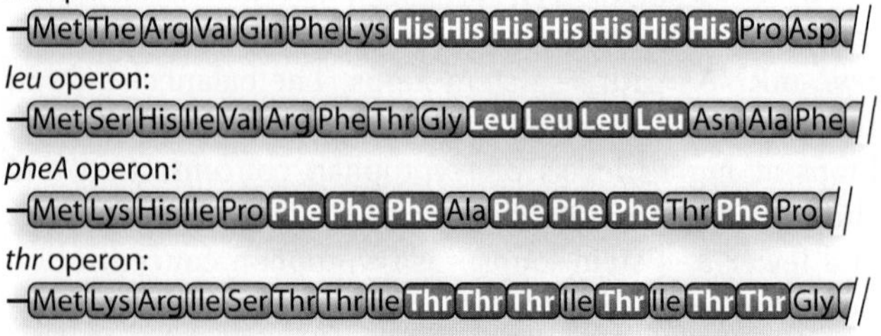

*his* operon:

—|Met|Thr|Arg|Val|Gln|Phe|Lys|**His**|**His**|**His**|**His**|**His**|**His**|**His**|Pro|Asp| //

*leu* operon:

—|Met|Ser|His|Ile|Val|Arg|Phe|Thr|Gly|**Leu**|**Leu**|**Leu**|**Leu**|Asn|Ala|Phe| //

*pheA* operon:

—|Met|Lys|His|Ile|Pro|**Phe**|**Phe**|**Phe**|Ala|**Phe**|**Phe**|**Phe**|Thr|**Phe**|Pro| //

*thr* operon:

—|Met|Lys|Arg|Ile|Ser|Thr|Thr|Ile|**Thr**|**Thr**|**Thr**|Ile|**Thr**|Ile|**Thr**|**Thr**|Gly| //

**Figure 14.21**   **Four bacterial amino acid operons with attenuator control of transcription.**   The regulatory amino acid for each operon is shown in bold.

destabilize the termination stem loop and reduce the efficiency of the mutated operon system in repressing structural gene transcription.

## Attenuation in Other Amino Acid Operon Systems

Attenuation represses transcription of structural genes in several amino acid operon systems in bacteria such as *E. coli* and *Salmonella typhimurium*. Like the *trp* operon, these other amino acid operons also contain multiple codons for the target amino acid in their leader transcripts (**Figure 14.21**). For example, the leader polypeptide of the *E. coli* histidine operon contains a run of seven consecutive histidine residues in the attenuator. Similarly, the phenylalanine leader polypeptide contains seven phenylalanine residues in a span of nine amino acids in the attenuator region. Like the *trp* operon, these operons use attenuation to form antitermination stem loops to regulate operon gene transcription. **Genetic Analysis 14.2** examines mutations of the *trp* operon.

## 14.5 Bacteria Regulate the Transcription of Stress Response Genes and Also Regulate Translation

The need on the part of bacteria to respond rapidly to changing environmental conditions suggests that transcriptional regulation must accommodate both common and rare circumstances, and also that the regulation of translation must be available under certain circumstances. This section presents examples of transcriptional regulation in bacteria under rarely encountered conditions and concludes with examples of how bacteria regulate translation.

## Alternative Sigma Factors and Stress Response

The operon mechanisms described to this point are examples of the regulatory strategies employed by bacterial cells under conditions they encounter routinely. In response to rare or unusual environmental circumstances, however,

bacteria switch gene transcription patterns to use genes that are not normally expressed. The response of *E. coli* to heat stress illustrates how expression of an *alternative sigma* ($\sigma$) *factor* alters gene transcription by activating the transcription of specialized heat stress response genes.

*Escherichia coli* grow vigorously at 37°C and can tolerate only narrow temperature variation. At low temperatures, their growth slows—an important reason refrigeration is used to preserve foods. At the other extreme, high temperatures kill the bacteria. This is the reason cooking is so efficient at reducing bacterial contamination of food. At the less dramatically elevated temperatures of 45°C, *E. coli* change their pattern of transcription by activating the expression of genes that are part of the heat shock response by the cell. The heat shock response protects *E. coli* cells from certain kinds of heat-induced damage. Similar mechanisms are common in other microorganisms as well as in fruit flies, plants, and animals, including humans.

Heat shock response in bacteria involves expression of an **alternative sigma ($\sigma$) factor** that changes the promoter-recognition capacity of the RNA polymerase core enzyme. Recall that the RNA polymerase core enzyme is bound by a sigma factor to form the holoenzyme (see Chapter 8). Under normal growth conditions, the RNA polymerase holoenzyme recognizes bacterial promoters containing an AT-rich Pribnow box at the −10 site. The common sigma factor, identified as $\sigma^{70}$, forms this holoenzyme that transcribes a wide array of bacterial genes under normal physiological conditions.

Bacteria grown at 45°C undergo several changes, including initiation of the expression of heat shock proteins, which are expressed only at high temperature, and of chaperon proteins, a class of proteins that either refold or degrade other proteins damaged by high heat. At these higher temperatures, $\sigma^{70}$ is unstable, and RNA polymerase containing it functions very poorly. To explain the transcription of heat shock proteins in the presence of poorly functioning $\sigma^{70}$-containing RNA polymerase, researchers proposed and quickly found genetic evidence pointing to an alternative, high-temperature $\sigma$ factor.

The evidence came from studies of mutant, temperature-sensitive *E. coli* that grow normally at 37°C but fail to grow at 45°C. This temperature sensitivity is a conditional lethal mutation affecting a gene called *rpoH*, which encodes an alternative sigma factor known as $\sigma^{32}$. When $\sigma^{32}$ binds an RNA polymerase core enzyme, the holoenzyme recognizes different promoter sequences than are recognized by holoenzymes containing $\sigma^{70}$ (**Figure 14.22**). In contrast to the AT richness that characterizes the Pribnow box sequence of bacterial promoters, the −10 region of promoters recognized by $\sigma^{32}$-containing RNA polymerase is rich in G-C base pairs.

The promoter for *rpoH* is recognized by $\sigma^{70}$-containing RNA polymerase when the temperature is elevated. The polypeptide translated from *rpoH* mRNA is very active in

**(a)**

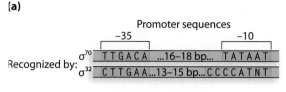

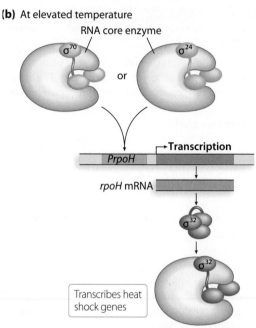

**Figure 14.22  Alternative sigma factors for heat shock genes. (a)** Promoter sequences recognized by $\sigma^{70}$- and $\sigma^{32}$-containing RNA polymerase. **(b)** At elevated temperature, $\sigma^{70}$ and $\sigma^{24}$ transcribe *rpoH*, which encodes $\sigma^{32}$ that in turn joins the RNA core enzyme to transcribe heat shock genes.

stimulating transcription of heat shock genes. In addition, transcription of a third sigma factor known as $\sigma^{24}$, which is normally present in *E. coli* cells at a very low level, is greatly elevated. The RNA polymerase holoenzyme containing $\sigma^{24}$ also recognizes the *rpoH* promoter and transcribes the gene at elevated temperatures that inactivate $\sigma^{70}$.

A second transcriptional change that occurs as a consequence of high heat is a change in the chaperon proteins. At normal growth temperatures, several chaperon proteins bind the small amount of $\sigma^{32}$ present in the cell

to inhibit its ability to form holoenzyme. At high temperatures, chaperone proteins release $\sigma^{32}$, leaving it free to join an RNA polymerase core enzyme and form a holoenzyme. Free chaperon proteins are redirected to bind heat-damaged cellular proteins instead. In this role, chaperon proteins either degrade the proteins they bind or assist in refolding the proteins.

Several additional examples of the use of alternative sigma factors in bacteria have been described. For example, *Bacillus subtilis* is a bacterium that normally propagates by vegetative growth, but poor growth conditions switch the growth mode to sporulation by activating the expression of alternative sigma factors. The gene transcription evidence shows that as growth conditions deteriorate, transcription of the common sigma factor is replaced by the transcription of two alternative sigma factors. The new sigma factors recognize the unique promoters and transcribe genes used in sporulation. A broad array of evidence shows that switching transcription from the normal sigma factor to alternative sigma factors induces a genome-wide change in the pattern of gene expression that silences previously active genes and initiates transcription of specialized genes that are used only under restrictive or extreme growth conditions. Table 14.6 compares and contrasts the mechanisms of gene regulation in bacterial systems.

---

**Genetic Insight**   Alternative sigma factors cause genome-wide changes in gene expression by permitting transcription of genes whose products are necessary only under extraordinary circumstances.

---

### Translational Regulation in Bacteria

Transcriptional regulation is far and away the predominant mode of controlling gene expression in bacteria, but bacteria are also capable of translational regulation. Translational regulation takes place by two mechanisms, one that binds protein to an mRNA to prevent its translation and another that pairs complementary *antisense RNA* with the mRNA to block its translation.

**Translation repressor proteins** regulate translation by binding mRNA in the vicinity of the Shine-Dalgarno

| Table 14.6 | Mechanisms of Transcription Regulation in Bacteria |
|---|---|
| **Mechanism** | **Actions and Outcomes** |
| 1. Operon-specific control | Inducer substances, such as lactose, and negative feedback mechanisms, such as tryptophan availability, regulate gene transcription in coordinately controlled operons. |
| 2. CAP-cAMP control | CAP-cAMP is utilized as a positive regulator of transcription for genes in several different operons, including the *lac* and *ara* operons. |
| 3. Alternative sigma factors | Extreme growth conditions, such as heat stress and starvation, induce transcription of alternative sigma factors that exclusively transcribe specialized genes by inducing a global change in transcription, affecting the entire genome. |

Describe the effects on attenuation and on tryptophan synthesis of the following mutations of the tryptophan codons (UGG) in the attenuator region of the operon.

a. The tryptophan codons are mutated to UAGUGG.

b. The tryptophan codons are mutated to UUGUUG.

| Solution Strategies | Solution Steps |
| --- | --- |
| **Evaluate** | |
| 1. Identify the topic this problem addresses, and describe the information required in the answer. | 1. This problem concerns the consequences of mutations to the UGG (tryptophan) codons in the attenuator region of the *trp* operon. The answer requires a description of mutational consequences for tryptophan regulation and synthesis. |
| 2. Identify the critical information given in the problem. | 2. The mutant codon sequences are given. |
| **Deduce** | |
| 3. Examine the nature of the mutation in part (a). | 3. The base substitution in mutant (a) creates a stop codon in place of the first tryptophan codon. |
| 4. Examine the nature of the mutation in part (b). | 4. Two base substitutions are seen in mutant (b). Each creates a leucine codon in place of a tryptophan codon. |
| **Solve** | |
| 5. Describe the consequence of the mutation in part (a). | 5. UAG is a stop codon that halts translation of the polypeptide. The location of this stop codon will prevent the ribosome from covering repeat region 2. The 2–3 stem loop is the only regulatory configuration that can form, and it will lead to constitutive tryptophan synthesis. |
| TIP: Compare the transcription of the wild-type operon to that of this mutant operon (see Figures 14.18 to 14.20). | |
| 6. Describe the consequence of the mutation in part (b). | 6. Both mutant codons in this case encode leucine. These mutational changes will prevent attenuation of the *trp* operon in response to tryptophan level. Instead, tryptophan synthesis will attenuate in response to the level of leucine since the availability of leucine to add to the polypeptide will determine which stem loop will form. |

For more practice, see Problems 7, 15, and 25.

sequence. Protein binding in this location interferes with recognition of the Shine-Dalgarno sequence by the 16S rRNA in the small ribosomal subunit and so blocks translation initiation. One of the clearest examples of this kind of regulatory protein–mRNA interaction is seen in the translational regulation of ribosomal proteins in *E. coli*. The ribosomal proteins are encoded in a series of operons that produce polycistronic mRNAs. These operons are under a certain degree of transcriptional regulation, but the most prominent control of production of ribosomal proteins is at the translational level. One of the protein products from each ribosomal protein operon can bind that operon's polycistronic mRNA near the 5′-most Shine-Dalgarno sequence, thus preventing binding of the small ribosomal subunit to the polycistronic mRNA and inhibiting synthesis of the proteins encoded by the operon.

Bacterial translation can also be inhibited by the activity of **antisense RNA,** an RNA molecule that is complementary to a portion of a specific mRNA. The binding of an mRNA by an antisense RNA prevents ribosome attachment to the mRNA and blocks translation. Several examples of bacterial translational regulation by antisense RNA have been described. One of the first mechanisms of antisense control of translation comes from the regulation of transposase production by the bacterial insertion sequence *IS10*. Transposase is the enzyme that drives the movement of transposable genetic elements in genomes (see Chapter 13). Transposase cuts DNA for transposable element removal and insertion. A low level of transposition can be tolerated by bacterial genomes and may even be advantageous (see Chapter 12). Excessive transposase expression, however, leads to excessive transposition,

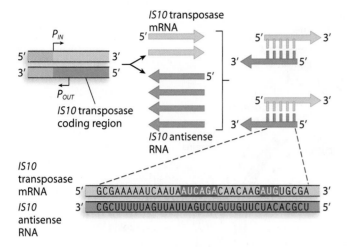

**Figure 14.23** **Antisense RNA control of the expression of *IS10* transposase.**

which may cause lethal mutations due to transposon insertion into critical genes.

The *IS10* insertion sequence contains two promoters. One, called $P_{IN}$, is relatively weak and controls transcription of the DNA strand coding for active transposase. The second promoter, $P_{OUT}$, is much stronger. This promoter is embedded in the transposase gene and directs transcription of the noncoding strand of the gene, producing an antisense RNA that is complementary to the 5′end of transposase mRNA and covers up the Shine-Dalgarno sequence of the mRNA, preventing its recognition by the small ribosomal subunit (**Figure 14.23**). As a consequence of the stronger $P_{OUT}$ promoter, *IS10* antisense RNA is more abundant than transposase mRNA. This results in most of the transposase mRNA being bound by antisense RNA and effectively prevents translation of nearly all transposase mRNA. Nevertheless, an occasional transposase mRNA escapes antisense binding and undergoes translation. This generates a low level of transposase that initiates the rare event of *IS10* transposition within the bacterial genome.

Genetic Insight Translation repressor proteins and antisense RNAs regulate translation in bacteria by targeting specific transcripts for regulatory inhibition.

# 14.6 Antiterminators and Repressors Control Lambda Phage Infection of *E. coli*

Bacteriophage (or phage, for short) are viruses that infect bacterial cells. Like all viruses, they must infect host cells to reproduce. Their tiny genomes do not contain all the genes necessary for replication, transcription, and translation, so phage are obligate parasites that use an ingenious array of tricks to accomplish these molecular processes.

The secret to their reproductive success lies in their ability to commandeer bacterial proteins and enzymes to preferentially express phage genes over bacterial genes.

Given the limited content of phage genomes, some of the most important genes for phage reproduction are those that redirect the activity of bacterial host genes to serve phage requirements. Successful phage infection requires (1) that genetic regulatory switches be controlled through phage gene expression to redirect the action of host genes and (2) that phage gene expression initiate a sequence of events leading the bacterium to participate in the expression of phage genetic information. In no bacteriophage is there a clearer picture of the processes that control regulatory genetic switching than in lambda (λ) phage.

Recall that all bacteriophage are capable of infecting and reproducing within the host bacterial cell. The infection ends with the lysis of the host cell, in a process called the lytic cycle (see Figure 6.16). But certain bacteriophage known as temperate phage, of which λ phage is an example, are also capable of a lysogenic cycle, or lysogeny. The lysogenic cycle is characterized by integration of the phage into the host chromosome, converting the host into a lysogen. Lysogenic integration is site specific, meaning it occurs at a sequence shared by the phage and the bacterial host (see Figure 6.20). The phage enzyme integrase is responsible for lysogenic integration. In this section, we discuss the two life cycles of λ phage, examining the regulatory proteins that control which life cycle a particular infection will undertake, as well as the actions of the proteins that control each life cycle.

## The Lambda Phage Genome

The λ phage genome is composed of approximately 48 kb of linear, double-stranded DNA that encodes nearly 60 genes (**Figure 14.24**). Its injection into a host bacterial cell leads to an immediate circularization inside the host cell that is accomplished by the joining of two single-stranded **cohesive (*cos*) ends** that are each 12 nucleotides in length. A host DNA ligase seals the two gaps that are left when the cohesive ends join and produces a circularized λ phage that is ready to begin gene expression.

The λ phage genome is organized as a series of operons. The genes in each operon are expressed in a well-defined sequence. Expression of genes in certain operons begins immediately after circularization. The specific order of gene expression is critical to the ability of λ phage to carry out successful infection of its bacterial host. Consequently, certain **early genes** are expressed shortly after circularization, while **late genes** are expressed later in the infection cycle. The transcription-regulating segments of the operons are also organized into early and late regions.

Immediately following circularization of the λ phage chromosome, **early promoters** and **early operators** control transcription of genes whose protein products in-

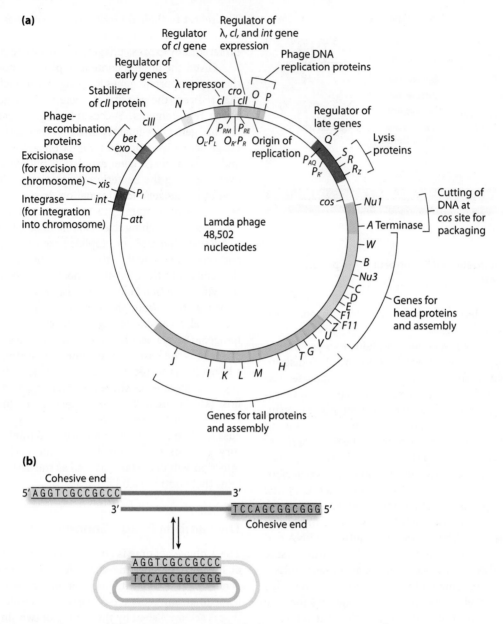

**Figure 14.24 The genome map of λ (lambda) phage. (a)** The λ phage genome is organized into operons that function at defined times during infection of a host cell. **(b)** The cohesive (*cos*) site is the region that enables the linear phage chromosome to circularize when it enters the host bacterial cell.

teract to determine whether the phage undergoes the lytic cycle or the lysogenic cycle (see Chapter 6). The lytic cycle results in a rapidly progressing infection leading to lysis (rupture) of the host cell and release of scores of progeny phage. In the lysogenic life cycle, on the other hand, the phage chromosome integrates into the host chromosome, as noted above. Expression of genes in the integrated phage chromosome (the prophage) is minimal; only the genes necessary to maintain lysogeny are expressed. Replication of the bacterial chromosome produces daughter cells that carry a copy of the prophage. Lysogeny continues until the prophage excises itself from its integration site, reactivating phage gene expression and the lytic cycle.

## Early Gene Transcription

Upon circularization of the phage chromosome, the protein products of early λ phage genes engage in a molecular tug-of-war for control of a genetic switch that determines whether the infection will result in the lytic cycle or the lysogenic cycle. At the beginning of each infection, transcription begins at two λ early promoters. The early promoter $P_R$ controls rightward transcription of early genes, beginning with the *cro* gene (for *c*ontrol of *r*epressor and *o*thers) that produces the cro protein (**Foundation Figure 14.25, ❶**) The early promoter $P_L$ controls leftward transcription beginning with the *N* gene, which produces an antiterminator

## Regulation of Bacteriophage Entry into the Lytic or Lysogenic Cycle

**1** N is produced by transcription from $P_L$.

**1** Transcription from $P_R$ produces cro

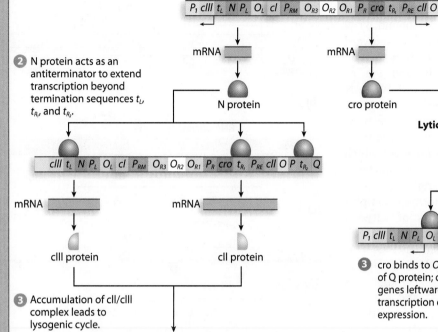

**2** N protein acts as an antiterminator to extend transcription beyond termination sequences $t_L$, $t_{R_1}$, and $t_{R_2}$.

**2** Accumulation of cro protein leads to lytic cycle.

**Lytic cycle development**

cro protein

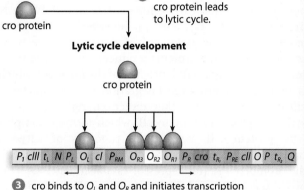

**3** cro binds to $O_L$ and $O_R$ and initiates transcription of Q protein; cro blocks *cl* transcription. Transcription of genes leftward of $O_L$ lead to phage particle assembly, and transcription of genes rightward of $O_R$ lead to late gene expression.

**3** Accumulation of cII/cIII complex leads to lysogenic cycle.

**Lysogenic cycle development**

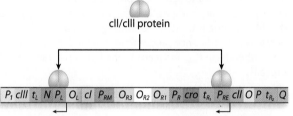

cII/cIII protein

**4** cII/cIII binding to $P_L$ leads to expression of integrase that stimulates prophage integration

**5** cII/cIII binding to $P_{RE}$ leads to expression of *cl*, the λ repressor protein, blocking synthesis of cro and permitting its own synthesis.

protein that plays a pivotal role in the establishment of lysogeny ❶.

The genetic switch triggering the lysogenic pathway begins with the binding of antitermination protein N to three transcription-terminating DNA sequences: $t_L$, $t_{R1}$, and $t_{R2}$ (see Figure 14.25, ❷). When not bound by N protein, termination sequence $t_L$ acts to block leftward transcription beyond *N*. In the other direction, $t_{R1}$ and $t_{R2}$ prevent rightward transcription beyond *cro* or beyond three other early genes—*cII*, *O*, and *P*. When N protein binds $t_L$, $t_{R1}$, and $t_{R2}$, however, several genes leftward of $t_L$ and rightward of $t_{R1}$ and $t_{R2}$ are transcribed. One of the proteins produced by leftward transcription is integrase (the product of the *int* gene), which is required for prophage integration into the bacterial chromosome. In the other direction, rightward transcription produces protein cII, which forms a complex with protein cIII, one of the products of leftward transcription. Together, the cII/cIII complex binds to the promoter $P_{RE}$ (for repressor *establishment*) ❸. This promoter initiates leftward transcription of the *cI* gene, producing the cI protein, which is also known as the λ repressor protein (Figure 14.25, ❹ and ❺). The cII/cIII complex also binds to $P_L$ to activate additional synthesis of integrase.

## Cro Protein and the Lytic Cycle

Entry into the lytic cycle requires the transcription of late genes that are regulated by **late promoters** and **late operators.** These genes are rightward of $P_R$, and are involved in the synthesis of head and tail proteins, as well as products that lyse the host cell. The genetic switch governing whether λ phage enters the lytic or the lysogenic cycle hinges on the binding of cro protein and λ repressor protein, respectively ❷. Both cro protein and λ repressor protein have affinity for operator sequences $O_{R1}$, $O_{R2}$, and $O_{R3}$, located between $P_R$ and $P_{RM}$, but have opposite binding affinities and transcription effects (see Figure 14.25 ❸). These three operator sequences each have a 17-bp target for binding of either cro protein or λ repressor protein. The $O_{R1}$ sequence lies fully within $P_R$, and $O_{R3}$ lies fully within $P_{RM}$; $O_{R2}$ is split between the two promoters (Figure 14.26).

The cro protein product is a 66-amino acid monomer that forms a globular structure. Functional cro protein is a homodimer that precisely spans the 17 bp of DNA that are its target binding sequence on the operators. Dimerized cro protein has very strong and equal binding affinity for $O_{R3}$ and $O_{R2}$, but lower affinity for $O_{R1}$. As cro protein concentration increases, however, it binds, in order, to $O_{R3}$, $O_{R2}$, and $O_{R1}$.

The presence of cro protein at the operator sequences blocks the access of RNA polymerase to $P_{RM}$, exerting negative control of *cI* gene transcription and preventing

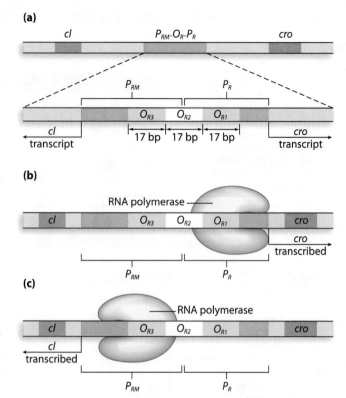

**Figure 14.26** **Transcription of λ phage genes *cro* and *cl*.** (a) Promoters $P_R$ and $P_{RM}$ overlap three operator sites—$O_{R1}$, $O_{R2}$, and $O_{R3}$—that are competitively bound by regulatory proteins. (b) The *cro* gene is transcribed from $P_R$ and leads to the lytic pathway if its influence predominates. (c) The *cl* gene is transcribed from $P_{RM}$ and leads to the lysogenic pathway if it prevails.

production of λ repressor protein. At the same time, cro protein binding exerts positive control on $P_R$, leading to enhanced transcription of *cro* and other genes that are rightward of $P_R$. Among these rightward genes is *Q*, a gene producing Q protein, which is a positive regulator of transcription of late genes that are rightward of the late promoter $P_{R'}$. These late genes include genes encoding proteins of the phage head and tail as well as genes required for lysis of the host cell.

## The λ Repressor Protein and Lysogeny

The λ repressor protein is the product of the *cI* gene. This protein is a 236-amino acid polypeptide containing 92 amino acids in the C-terminal domain (amino acids 1–92), 105 amino acids in the N-terminal domain (amino acids 132–236), and the remaining 39 amino acids (93–131) linking the two domains. Functional λ repressor protein is dimeric, and monomers are linked at their C-terminal ends. The resulting dimers have a dimension that spans 17 bp of DNA, precisely the size of each operator sequence (Figure 14.27a).

The structure of λ repressor protein is similar to that of cro protein, and the two proteins target the same operator segments, but with opposite patterns of affinity. The binding affinity of λ repressor protein is greatest for $O_{R1}$ and $O_{R2}$ and is lower for $O_{R3}$. Therefore, binding by λ repressor protein to the operators takes place in the order $O_{R1}$, $O_{R2}$, $O_{R3}$. Binding to $O_{R1}$ and $O_{R2}$ exerts negative control on $P_R$, blocking access of RNA polymerase to *cro*, *Q*, and the other genes that are rightward of $P_R$. This negative regulation blocks the transcription required to initiate the lytic cycle. Repressor protein binding to the operator sequences initiates positive control of transcription at $P_{RM}$ and leads to continuous transcription of *cI* that acts to maintain repression. It also activates the action of integrase, causing insertion of the phage chromosome into the host chromosome, at which point the λ phage genome is relatively quiescent. Continued positive regulation of *cI* gene expression by λ repressor serves to maintain the lysogenic state and is the only genetic activity in the prophage.

## Lysogen Induction

Lysogeny is a semipermanent state that can be maintained for an extended period of time by the ongoing binding of λ repressor protein to $O_{R1}$, $O_{R2}$, and $O_{R3}$. The persistence over long periods of the lysogenic state raises two questions. First, what makes lysogeny come to an end, and second, how does the phage resume the lytic cycle and produce progeny phage?

**Induction** is the process that brings lysogeny to an end and reinitiates the lytic cycle. You might think of induction as the flip of a switch, triggered by DNA damage done by extracellular forces. The principal force causing injury to DNA is ultraviolet light, whose effects on DNA activate DNA repair. Among the numerous proteins activated in the DNA repair cascade is the protein RecA, whose role in mutation repair is to activate recombination.

When bacterial DNA is damaged by UV light, however, the protease (protein-destroying) activity of RecA protein is also activated. The protease activity targets the amino acid segment of λ repressor monomers that join the N- and C-terminal regions of each protein (**Figure 14.27b**). The C terminus is clipped off each monomer, effectively breaking apart repressor dimers and causing the N-terminal ends to fall off DNA. With λ repressor no longer bound to DNA, the $O_{R1}$, $O_{R2}$, and $O_{R3}$ sequences are exposed, positive regulation of *cI* transcription ends, and negative regulation of *cro* transcription ends, allowing production of cro protein (**Figure 14.27c**). The cro protein binds to the operators no longer occupied by repressor protein. This leads to the expression of *Xis* that produces the enzyme excisionase that removes the lysogen from its integrated location. This event triggers

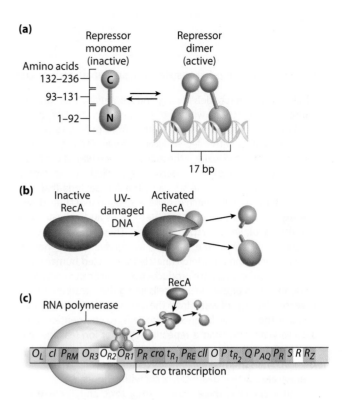

**(a)** 

**(b)** 

**(c)** 

**Figure 14.27  Lysogeny maintenance and termination.** **(a)** A homodimeric λ repressor protein binds to 17-bp operator sequences to regulate its own transcription and maintain lysogeny. **(b)** UV light and other DNA-damaging agents activate RecA, which cleaves λ repressor monomers to inactivate repressor protein. **(c)** Lysogeny ends with the removal of λ repressor protein from operator sequences and the initiation of transcription of *cro*.

the resumption of the lytic cycle and ultimately results in host cell lysis and the release of progeny phage.

In summary, λ phage offers an elegant example of a regulatory system that controls which way a genetic switch is flipped. The crucial interaction is between the protein products of the early genes *cro* and *cI* that compete for binding to operator sequences $O_{R1}$, $O_{R2}$, and $O_{R3}$. If cro protein prevails by successfully binding to $O_{R2}$ and $O_{R3}$, expression of *cI* is repressed, and the synthesis of late genes leading to completion of the lytic cycle is assured. On the other hand, if λ repressor protein prevails, its early occupation of $O_{R1}$ and $O_{R2}$ prevents transcription of late genes, ensuring that the lysogenic cycle will proceed.

---

Genetic Insight  In bacteriophage λ, an ordered series of promoters and terminator sequences control expression of genes required early and late in the lytic cycle. Competitive interaction between the λ repressor protein and the cro protein for the same DNA binding sites decides whether a cell enters the lytic cycle or the lysogenic cycle.

---

## *Vibrio cholerae*—Stress Response Leads to Serious Infection

Cholera is a severely debilitating and potentially fatal disease caused by infection with the intestinal bacterium *Vibrio cholerae*. The bacterium is transmitted from person to person through contact with infected fecal material. The ingestion of fecal-contaminated water is the most common way of contracting cholera. Most ingested bacteria are killed by the highly acidic environment of the stomach, but the bacteria that survive undergo a rapid switch in gene regulation that shuts down the expression of routinely activated genes and activates the expression of stress response genes. These genetic switch mechanisms enable the bacteria to survive in the hostile stomach environment. Unfortunately for infected humans, the *V. cholerae* stress response produces toxins that can rapidly lead to degradation of the mucosal cells lining the intestines and to excessive leakage of water from the damaged cells. The leakage disturbs the osmotic balance of the cells; to compensate, they secrete water, initiating a repeating cycle of ion leakage and water release that produces watery diarrhea and severe dehydration. Unless immediate antibiotic treatment and rehydration therapy are started, death can occur within hours.

In *V. cholerae*, three genes—*ToxS*, *ToxR*, and *ToxT*—exert positive control over the transcription of genes producing virulence (active bacterial growth that causes disease). The expression of *ToxS* and *ToxR* genes is stimulated by the environmental cues encountered by *V. cholerae* in the hostile environment of the stomach. A protein complex formed by the products of these genes activates transcription of *ToxT*. The polypeptide product of *ToxT* is a transcription-activating protein that binds to the promoter $P_{ctx}$ that controls transcription of two genes, *CtxA* and *CtxB* (abbreviations for "cholera toxin A" and "cholera toxin B") that are part of an operon. The polypeptide products of *CtxA* and *CtxB* are the cholera toxins that initiate the series of actions that lead to cholera symptoms.

Preventing cholera is an obvious public health priority that has the potential to save thousands of lives annually. Equally important is gaining understanding of how the ToxS-ToxR complex and ToxT operate in promoter recognition, and identifying the other genes they regulate. Similarly, gathering information about the stress response and virulence genes in *V. cholerae* will help medical practitioners and microbiologists understand how the bacterium produces its lethal effects and how care providers might intervene early in the infection process to prevent the most serious consequences of infection.

 Mastering **GENETICS**™    *For activities, animations, and review quizzes, go to the study area at www.masteringgenetics.com.*

### 14.1 Transcriptional Control of Gene Expression Requires DNA–Protein Interaction

▌ Regulated genes are under transcriptional control, whereas constitutive genes are not regulated.

▌ In negative control of transcription, regulatory proteins bound to DNA reduce or eliminate transcription.

▌ Regulatory proteins, also called repressors, have a DNA-binding domain to bind regulatory DNA sequences and an allosteric domain to bind a regulatory molecule.

▌ An inducer molecule binds to the repressor molecule at an allosteric site to inhibit its action.

▌ In positive regulatory control activator proteins bind DNA at promoters and other regulatory sequences and initiate or increase transcriptional efficiency.

### 14.2 The *lac* Operon Is an Inducible Operon System under Negative and Positive Control

▌ Bacterial operons transcribe two or more genes under the coordinated regulatory control of shared promoters, operators, and other regulatory elements.

▌ The lactose (*lac*) operon is an inducible operon system that produces three proteins—β-galactosidase (*lacZ*), permease (*lacY*), and transacetylase (*lacA*) that are required to metabolize lactose and its by-products. Its regulatory control center contains a promoter and an operator sequence (*lacO*).

▌ Negative control of *lac* operon gene transcription is exerted by a repressor protein (*lacI*) that binds to the *lacO* region to block transcription. Allolactose inactivates the repressor protein by changing its conformation and preventing it from binding to the operator.

▌ Positive control of transcription of *lac* operon genes is exerted by the CAP-cAMP complex that forms in the absence of glucose and binds to the CAP site of the *lac* promoter.

### 14.3 Mutational Analysis Deciphers Genetic Regulation of the *lac* Operon

▌ Mutation studies determined the order of *lac* operon genes as *lacZ-lacY-lacA*.

▌ The analysis of mutant haploid and partial diploid bacteria identified the trans-acting repressor protein that binds the operator sequence.

▌ *lac* operator mutation analysis indicates that the operator is a cis-acting element that controls transcription of immediately adjacent genes on the chromosome.

▌ The *lac* repressor binding site overlaps the RNA polymerase binding location in the *lac* promoter.

▌ *lac* repressor protein binding induces DNA loop formation that prevents RNA polymerase binding at the promoter.

- The CAP-cAMP complex binds to the CAP binding site of the *lac* promoter and facilitates RNA polymerase binding.
- Transcription of arabinose (*ara*) operon genes is regulated by the araC protein that exercises both negative and positive control.

## 14.4 Transcription from the Tryptophan Operon Is Repressible and Attenuated

- The tryptophan (*trp*) operon is a repressible operon that produces five polypeptides that participate in tryptophan synthesis.
- *Trp* operon transcription is inhibited by a feedback mechanism involving tryptophan as a corepressor.
- *Trp* operon gene expression is attenuated to maintain the cellular concentration of tryptophan at a steady state. Multiple amino acid operons are regulated by an attenuation mechanism.
- The *trpL* (leader) region contains an attenuator sequence of four DNA repeats that form one of two alternative mRNA stem loops.
- The 2–3 (antitermination) stem loop formed by mRNA permits transcription of five *trp* operon structural genes in a polycistronic mRNA.
- The 3–4 (termination) stem loop of mRNA terminates transcription before RNA polymerase binds to the structural genes of the operon.

## 14.5 Bacteria Regulate the Transcription of Stress Response Genes and Also Regulate Translation

- Alternative sigma factors are used to generate RNA polymerases that recognize promoters of genes not transcribed by the common bacterial RNA polymerase.
- Genes transcribed using alternative sigma factors are required only under specialized circumstances, such as in response to heat shock.
- The translation of bacterial mRNA can be blocked by RNA-binding translation repressor proteins or by antisense RNA that binds to mRNA from specific genes.

## 14.6 Antiterminators and Repressors Control Lambda Phage Infection of *E. coli*

- Early genes of the bacteriophage λ genome produce proteins that compete to bind at the same regulatory region. The protein that prevails determines whether the phage infection will follow the lytic cycle or the lysogenic cycle.
- Completion of the lytic cycle requires the expression of late λ phage genes.
- Lysogen integration and maintenance requires ongoing expression of the λ repressor protein, which regulates its own transcription.
- Lysogen integration is reversed by environmental changes that lead to induction and to resumption of the lytic cycle.

## KEYWORDS

activator binding site (*p. 461*)
activator protein (*p. 461*)
allolactose (*p. 463*)
allosteric domain (allostery) (*p. 461*)
allosteric effector compound (*p. 461*)
alternative sigma (σ) factor (*p. 480*)
antisense RNA (*p. 482*)
attenuation (attenuator region) (*p. 475*)
CAP binding site (CAP–cAMP complex) (*p. 465*)
catabolite repression (*p. 465*)
cis-acting (*p. 468*)
cis-dominant (*p. 468*)
cohesive (*cos*) ends (*p. 483*)
constitutive transcription (constitutive mutants) (*pp. 460, 467*)

corepressor (*p. 461*)
DNA-binding domain (*p. 461*)
DNA loop (*p. 473*)
early genes (early operators, early promoters) (*p. 483*)
helix-turn-helix (HTH) motif (*p. 462*)
inducer (*p. 461*)
inducer–repressor complex (*p. 465*)
inducible operon (*p. 463*)
induction (*p. 487*)
inhibitor (*p. 461*)
*lac*⁺ phenotype (*p. 463*)
*lac*⁻ phenotype (*p. 463*)
*lacA* gene (*lacY* gene, *lacZ* gene) (*p. 463*)
late genes (late operators, late promoters) (*pp. 483, 486*)

leader region (*trpL*) (*p. 475*)
negative control (of transcription) (*p. 461*)
noninducible (*p. 469*)
operator (*p. 461*)
operon [lactose (*lac*), tryptophan (*trp*), arabinose (*ara*)] (*pp. 462, 463, 474*)
polycistronic mRNA (*p. 463*)
positive control (of transcription) (*p. 461*)
regulated transcription (*p. 460*)
repressible operon (*p. 475*)
repressor protein (*p. 461*)
stem loop [3–4 (termination), 2–3 (antitermination)] (*p. 477, 478*)
trans-acting (*p. 469*)
translation repressor protein (*p. 481*)

## PROBLEMS

 Mastering**GENETICS**™

*For instructor-assigned tutorials and problems, go to www.masteringgenetics.com.*

### Chapter Concepts

*For answers to selected even-numbered problems, see Appendix: Answers.*

1. Bacterial genomes frequently contain groups of genes organized into operons. What is the biological advantage of operons to bacteria? Identify the regulatory components you would expect to find in an operon. How are the expressed genes of an operon usually arranged?

2. Transcriptional regulation of operon gene expression involves the interaction of molecules with one another and of regulatory molecules with segments of DNA. In this context, define and give an example of each of the following:

   a. operator

   b. repressor

c. inducer
d. corepressor
e. promoter
f. positive regulation
g. allostery
h. negative regulation
i. attenuation

3. Why is it essential that bacterial cells be able to regulate the expression of their genes? What are the energetic and evolutionary advantages of regulated gene expression? Is the expression of all bacterial genes subject to regulated expression? Compare and contrast the difference between regulated gene expression and constitutive gene expression.

4. Identify similarities and differences between an inducible operon and a repressible operon in terms of
   a. the transcription-regulating DNA sequences.
   b. the presence and action of allosteric regulatory molecules.
   c. the organization of structural genes of the operon.

5. The transcription of β-galactosidase and permease is inducible in $lac^+$ bacteria with a wild-type $lac$ operon. Explain the mechanism by which lactose gains access to the cell to induce transcription of the genes.

6. Is attenuation the product of an allosteric effect? Is attenuation the result of a transcriptional or a translational activity? Explain your answers.

7. The $trpL$ region contains four repeated DNA sequences that lead to the formation of stem-loop structures in mRNA.

What are these stem-loop structures, and how do they affect transcription of the structural genes of the $trp$ operon?

8. The CAP binding site in the $lac$ promoter is the location of positive regulation of gene expression for the operon. Identify what binds at this site to produce positive regulation, under what circumstances binding occurs, and how binding exerts a positive effect.

9. What role does cAMP play in transcription of $lac$ operon genes? What role does CAP play in transcription of $lac$ operon genes?

10. How would a $cap^-$ mutation that produces an inactive CAP protein affect transcriptional control of the $lac$ operon?

11. Explain the circumstances under which attenuation of operon gene expression is advantageous to a bacterial organism. Would you expect attenuation to be found in a single-celled eukaryote? In a multicelled eukaryote?

12. Describe how araC protein functions as both a positive and a negative regulator of $ara$ (arabinose) operon transcription.

13. Describe the lytic and lysogenic life cycles of λ bacteriophage. What roles do λ repressor and cro protein play in controlling transcription from $P_R$ and $P_{RM}$, and how are these roles linked to lysis and lysogeny?

14. Define $antisense$ $RNA$, and describe how it affects the translation of a complementary mRNA. Why is it more advantageous to the organism to stop translation initiation than to inactivate or destroy the gene product after it is produced?

## Application and Integration

For answers to selected even-numbered problems, see Appendix: Answers.

15. Attenuation of $trp$ operon transcription is controlled by the formation of stem-loop structures in mRNA. The attenuation function can be disrupted by mutations that alter the sequence of repeat DNA regions 1 to 4 and prevent the formation of mRNA stem loops. Describe the likely effects on attenuation of each of the following mutations under the conditions specified.

| Mutated Region | Tryptophan Level |
| --- | --- |
| a. Region 1 | Low |
| b. Region 1 | High |
| c. Region 2 | Low |
| d. Region 2 | High |
| e. Region 3 | Low |
| f. Region 3 | High |
| g. Region 4 | Low |
| h. Region 4 | High |

16. In the $lac$ operon, what are the likely effects on operon gene transcription of the mutations identified below?
    a. Mutation of consensus sequence in the $lac$ promoter

b. Mutation of the repressor binding site on the operator sequence
c. Mutation of the $lacI$ gene affecting the allosteric site of the protein
d. Mutation of the $lacI$ gene affecting the DNA-binding site of the protein
e. Mutation of the CAP binding site of the $lac$ promoter

17. Identify which of the following $lac$ operon haploid genotypes transcribe operon genes inducibly and which transcribe genes constitutively. Indicate whether the strain is $lac^+$ or $lac^-$.
    a. $I^+ P^+ O^+ Z^+ Y^-$
    b. $I^+ P^+ O^C Z^- Y^+$
    c. $I^- P^+ O^+ Z^+ Y^+$
    d. $I^+ P^- O^+ Z^+ Y^+$
    e. $I^+ P^+ O^+ Z^- Y^+$
    f. $I^+ P^+ O^C Z^+ Y^-$
    g. $I^+ P^+ O^C Z^+ Y^+$

18. Complete the following table, indicating whether functionally active β-galactosidase and permease are produced in the presence and absence of lactose. Use "+" to indicate the presence of a functional enzyme and "−" to indicate its absence. Indicate whether the partial diploid strain is $lac^+$ (able to grow on lactose only medium) or $lac^-$ (cannot grow on lactose medium).

| Genotype | β-Galactosidase | | Permease | | Phenotype |
|---|---|---|---|---|---|
| | Lactose | No Lactose | Lactose | No Lactose | |
| Example: $I^+ P^+ O^+ Z^+ Y^+$ | − | + | − | + | $lac^+$ |
| a. $I^S P^+ O^+ Z^+ Y^+ / I^- P^+ O^+ Z^+ Y^+$ | | | | | |
| b. $I^- P^+ O^+ Z^- Y^+ / I^+ P^+ O^C Z^+ Y^-$ | | | | | |
| c. $I^+ P^+ O^+ Z^- Y^+ / I^+ P^- O^+ Z^+ Y^-$ | | | | | |
| d. $I^- P^+ O^C Z^+ Y^+ / I^+ P^- O^+ Z^+ Y^+$ | | | | | |
| e. $I^+ P^+ O^C Z^+ Y^- / I^+ P^+ O^+ Z^+ Y^-$ | | | | | |
| f. $I^+ P^+ O^+ Z^- Y^+ / I^S P^+ O^+ Z^+ Y^-$ | | | | | |
| g. $I^S P^+ O^+ Z^- Y^+ / I^+ P^+ O^C Z^+ Y^-$ | | | | | |

19. List possible genotypes for *lac* operon haploids that have the following phenotypic characteristics:

    a. The operon genes are constitutively transcribed, but the strain is unable to grow on a lactose medium. List two possible genotypes for this phenotype.

    b. The operon genes are never transcribed above a basal level, and the strain is unable to grow on a lactose medium. List two possible genotypes for this phenotype.

    c. The operon genes are inducibly transcribed, but the strain is unable to grow on a lactose medium. List one possible genotype for this phenotype.

    d. The operon genes are constitutively transcribed, and the strain grows on lactose medium. List two possible genotypes for this phenotype.

20. Suppose each of the genotypes you listed in parts (a) and (b) in Problem 19 are placed in a partial diploid genotype along with a chromosome that has a fully wild-type *lac* operon.

    a. Will the transcription of operon genes in each partial diploid be inducible or constitutive?

    b. Which partial diploids will be able to grow on a lactose medium?

21. Four independent *lac⁻* mutants (mutants A to D) are isolated in haploid strains of *E. coli*. The strains have the following phenotypic characteristics:

    Mutant A is *lac⁻*, but transcription of operon genes is induced by lactose.

    Mutant B is *lac⁻* and has uninducible transcription of operon genes.

    Mutant C is *lac⁺* and has constitutive transcription of operon genes.

    Mutant D is *lac⁺* and has constitutive transcription of operon genes.

    A microbiologist develops donor and recipient varieties of each mutant strain and crosses them with the results shown below. The table indicates whether inducible, constitutive, or noninducible transcription occurs, along with *lac⁺* and *lac⁻* growth habit for each partial diploid. Assume each strain has a single mutation.

| Mating | Transcription and Growth |
|---|---|
| A × B | *lac⁺*, inducible |
| A × C | *lac⁺*, inducible |
| A × D | *lac⁺*, constitutive |
| B × C | *lac⁺*, inducible |
| B × D | *lac⁺*, constitutive |
| C × D | *lac⁺*, inducible |

Use this information to identify which *lac* operon gene is mutated in each strain.

22. Suppose the *lac* operon partial diploid $cap^- I^+ P^+ O^+ Z^- Y^+ / cap^+ I^- P^+ O^+ Z^+ Y^-$ is grown.

    a. Will this partial diploid strain grow on a lactose medium?

    b. Is transcription of β-galactosidase and permease inducible, constitutive, or noninducible?

    c. Explain how genetic complementation contributes to the growth habit of this strain.

23. A bacterial inducible operon, similar to the *lac* operon, contains three genes—*R*, *T*, and *S*—that are involved in coordinated regulation of transcription. One of these genes is an operator region, one is a regulatory protein, and the third produces a structural enzyme. In the table below, "+" indicates that the structural enzyme is synthesized and "−" indicates that it is not produced. Use the information provided to determine which gene is the operator, which produces the regulatory protein, and which produces the enzyme.

| Genotype | Enzyme Synthesis With . . . | |
|---|---|---|
| | Inducer Present | Inducer Absent |
| $R^+ S^+ T^+$ | + | − |
| $R^- S^+ T^+$ | − | − |
| $R^+ S^- T^+$ | + | + |
| $R^+ S^+ T^-$ | + | + |
| $R^- S^+ T^+ / R^+ S^- T^-$ | + | + |
| $R^+ S^- T^+ / R^- S^+ T^-$ | + | + |
| $R^+ S^+ T^- / R^- S^- T^+$ | + | − |

24. A repressible operon system, like the *trp* operon, contains three genes, *G*, *Z*, and *W*. Operon genes are synthesized when the end product of the operon synthesis pathway is absent, but there is no synthesis when the end product is present. One of these genes is an operator, one is a regulatory protein, and the other is a structural enzyme involved in synthesis of the end product. In the table below, "+" indicates that the enzyme is synthesized by the operon, and "−" means that no enzyme synthesis occurs. Use this information to determine which gene corresponds to each operon function.

| Genotype | Enzyme Synthesis | |
|---|---|---|
| | With End Product Present | With End Product Absent |
| $G^+ Z^+ W^+$ | − | + |
| $G^- Z^+ W^+$ | + | + |
| $G^+ Z^- W^+$ | − | − |
| $G^+ Z^+ W^-$ | + | + |
| $G^- Z^+ W^+/G^+ Z^- W^-$ | + | + |
| $G^+ Z^- W^+/G^- Z^+ W^-$ | + | + |
| $G^- Z^- W^-/G^+ Z^+ W^+$ | − | + |
| $G^+ Z^+ W^-/G^- Z^- W^+$ | − | + |

25. What is the likely effect of each of the following mutations of the *trpL* region on attenuation control of *trp* operon gene transcription? Explain your reasoning.

a. Region 3 is deleted.

b. Region 4 is deleted.

c. The entire *trpL* region is deleted.

d. The start (AUG) codon of the *trpL* polypeptide is deleted.

e. Two nucleotides are inserted into the *trpL* region immediately after the polypeptide stop codon.

f. Twenty nucleotides are inserted into the *trpL* region immediately after the polypeptide stop codon.

g. Ten nucleotides are inserted between regions 2 and 3 of *trpL*.

h. Two nucleotides are inserted immediately following the polypeptide start codon.

i. The entire polypeptide coding sequence of *trpL* is deleted.

j. The eight uracil nucleotides immediately following region 4 are deleted.

26. If mutations of each region listed below prevent the binding of araC protein, how will transcriptional regulation of the *ara* operon be affected?

a. *araI*

b. *araO₁*

c. *araO₃*

d. *araO₁* and *araO₃*

27. Two different mutations affect $P_{RE}$. Mutant 1 decreases transcription from the promoter to 10% of normal. Mutant 2 increases transcription from the promoter to tenfold greater than the wild type. How will each mutation affect the determination of the lytic or lysogenic life cycle in mutant λ phage strains? Explain your answers.

28. How would mutations that inactivate each of the following genes affect the determination of the lytic or lysogenic life cycle in mutated λ phage strains? Explain your answers.

a. *cI*

b. *cII*

c. *cro*

d. *int*

e. *cII* and *cro*

f. *N*

29. The bacterial insertion sequence *IS10* uses antisense RNA to regulate translation of the mRNA that produces the enzyme transposase, which is required for insertion sequence transposition. Transcription of the antisense RNA gene is controlled by $P_{OUT}$, which is over 10 times more efficient at transcription than the $P_{IN}$ promoter that controls transposase gene transcription.

a. If a mutation reduced the transcriptional efficiency of $P_{OUT}$ so as to be equal to that of $P_{IN}$, what is the likely effect on the transposition of *IS10*?

b. If a mutation of $P_{IN}$ eliminates its ability to function in transcription, what is the likely effect on the transposition of *IS10*?

30. Northern blot analysis is performed on cellular mRNA isolated from *E. coli*. The probe used in the northern blot analysis hybridizes to a portion of the *lacY* sequence. Below is an example of the autoradiograph from northern blot analysis for a wild-type *lac⁺* bacterial strain. In this gel, lane 1 is from bacteria grown in a medium containing only glucose (minimal medium). Lane 2 is from bacteria in a medium containing only lactose. Following the style of this diagram, draw the autoradiograph appearance for northern blots of the bacteria listed below. In each case, lane 1 is for mRNA isolated after growth in a glucose-containing (minimal) medium, and lane 2 is for mRNA isolated after growth in a lactose-only medium.

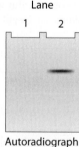

Lane

1    2

Autoradiograph
of northern blot

a. *lac⁺* bacteria with the genotype $I^+ P^+ O^C Z^+ Y^+$

b. *lac⁻* bacteria with the genotype $I^+ P^+ O^+ Z^- Y^+$

c. *lac⁻* bacteria with the genotype $I^+ P^- O^C Z^+ Y^+$

d. *lac⁺* bacteria with the genotype $I^- P^+ O^C Z^+ Y^+$

e. *lac⁻* bacteria with the genotype $I^+ P^+ O^+ Z^- Y^+$ that has a polar mutation affecting the *lacZ* gene

f. *lac⁻* bacteria with the genotype $I^+ P^+ O^C Z^- Y^-$

g. *lac⁻* bacteria with the genotype $I^+ P^+ O^+ Z^+ Y^+$ and a mutation that prevents CAP-cAMP binding to the CAP site

31. The electrophoresis gel shown below in part (a) is from a DNase I footprint analysis of an operon transcription control region. DNA sequence analysis of a 35-bp region is shown in part (b). The control region, labeled with $^{32}$P at one end, is shown in a map in part (c). Separate samples of control-region DNA are exposed to DNase I, and the resulting DNase I-digested DNA is run in separate lanes of the electrophoresis gel. Unprotected DNA is in lane 1, DNA protected by repressor protein is in lane 2, and RNA polymerase-protected DNA is in lane 3. The numbers along the electrophoresis gel correspond to the 35-bp sequence labeled on the map in part (c). Use the information provided to solve the following problems.

    a. Determine the DNA sequence of the 35-bp region examined.
    b. Locate the regions of the sequence protected by repressor protein and by RNA polymerase.

32. For the following *lac* operon partial diploids, determine whether the synthesis of *lacZ* mRNA is "constitutive," "inducible," or "uninducible," and indicate whether the merodiploid is *lac⁺* or *lac⁻* (able or not able to utilize lactose).

| Genotype | LacZ mRNA Synthesis | Lac Phenotype |
|---|---|---|
| a. $I^- P^+ O^+ Z^+ Y^+ / I^+ P^+ O^+ Z^+ Y^+$ | | |
| b. $I^+ P^+ O^C Z^+ Y^+ / I^+ P^+ O^+ Z^- Y^+$ | | |
| c. $I^S P^+ O^+ Z^+ Y^+ / I^+ P^+ O^+ Z^+ Y^+$ | | |
| d. $I^+ P^+ O^+ Z^- Y^+ / I^+ P^- O^+ Z^+ Y^+$ | | |
| e. $I^+ P^+ O^+ Z^+ Y^- / I^+ P^+ O^+ Z^+ Y^-$ | | |

33. The following hypothetical genotypes have genes *A*, *B*, and *C* corresponding to *lacI*, *lacO*, and *lacZ*, but not necessarily in that order. Data in the table indicate whether β-galactosidase is produced in the presence and absence of the inducer for each genotype. Use this data to identify the correspondence between *A*, *B*, and *C* and the *lacI*, *lacO*, and *lacZ* genes. Carefully explain your reasoning for identifying each gene.

| Genotype | β-Galactosidase Production | |
|---|---|---|
| | Inducer Present | Inducer Absent |
| 1. $A^- B^+ C^+$ | + | + |
| 2. $A^+ B^+ C^-$ | + | + |
| 3. $A^- B^+ C^+ / A^+ B^+ C^+$ | + | + |
| 4. $A^+ B^+ C^- / A^+ B^+ C^+$ | + | − |

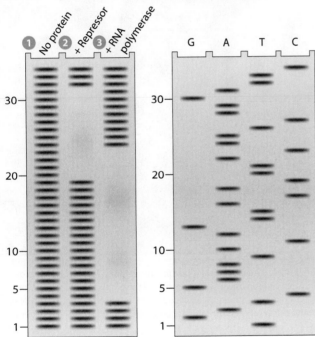

**(a)** Phase I treatment    **(b)** DNA sequencing

**(c)**

1    35

# 15

# Regulation of Gene Expression in Eukaryotes

## ESSENTIAL IDEAS

▌ Regulatory DNA sequences bind regulatory proteins to control the initiation or silencing of transcription in eukaryotes.

▌ Chromatin remodeling regulates gene transcription by means of modifications affecting nucleosomes.

▌ The structure of chromatin varies among different types of cells and sets the gene-expression program for distinct cell types.

▌ RNA-mediated mechanisms regulate eukaryotic gene expression by post-transcriptional interactions with mature mRNA.

Wild-type petunia flowers have solid color due to expression of a chromosomal pigment gene. Transgenic petunias with an extra copy of the pigment gene have colorless (white) regions due to a process called cosuppression caused by regulatory RNAs that inactivate the chromosomal copy and the transgenic copy of the pigment gene

I f the 46 chromosomes in a single nucleus from any cell in your body were stripped of their associated proteins and laid end to end, they would span almost 2 meters. Yet in their normal compacted state, these chromosomes can fit inside a nucleus that is about 5 microns (5 millionths of a meter) in diameter and still leave room for DNA replication, transcription, pre-mRNA processing, and numerous other activities to take place. This efficient packaging and access to DNA is made possible by the chromatin structure of your genome and the dynamic changes of which chromatin is capable throughout the cell cycle.

The genomes of eukaryotic organisms—yours included— are considerably larger on average than those of bacterial and archaeal species, and they are packaged much differently as well. One major packaging difference is the localization of chromosomes in a nucleus in eukaryotic cells. Nuclear localization sequesters the chromosomes and encapsulates DNA replication, transcription, and the various RNA-processing activities. A second difference is the incorporation of DNA into chromatin.

The process of chromatin condensation initiates at the beginning of prophase and culminates in fully condensed chromosomes in metaphase. This is an essential predecessor of efficient chromosome separation in anaphase. Chromatin condensation also plays a pivotal role in permitting or blocking transcription. No cell in your body expresses all 22,000 or so genes of the human genome. Instead, most human cell types express only a few thousand genes, while the other genes are transcriptionally silent. In recent decades, cell biologists studying the close connection between structural changes in chromatin and the transcription of eukaryotic genes have succeeded in uncovering many crucial details.

The processes that regulate gene expression in eukaryotes (see Chapters 8 and 9) are more varied and multifaceted than those governing gene expression in bacterial genomes (**Figure 15.1**). In the present chapter, we focus on five additional elements that are central to the regulation of transcription and gene expression in eukaryotes: (1) regulatory sequences other than promoters that contribute to the regulation of transcription; (2) mechanisms that remodel chromatin or reconfigure the association between nucleosomes and DNA to regulate transcription; (3) epigenetic mechanisms that exert transcriptional regulatory control in cell lineages over the long term; (4) the transmission of epigenetic states from one generation of cells to another to exercise long-term control of differential gene expression; and (5) RNA-based mechanisms operating post-transcriptionally to regulate the availability of mature mRNA for translation and therefore the ability to produce polypeptides.

# 15.1 Cis-Acting Regulatory Sequences Bind Trans-Acting Regulatory Proteins to Control Eukaryotic Transcription

Despite the considerable differences between eukaryotes and bacteria, the basic mechanisms controlling transcription are similar in a broad sense in both groups of organisms. The DNA–protein interactions in eukaryotes follow a scheme familiar from bacterial processes. *Activator proteins* bind regulatory sequences to stimulate transcription (positive regulation of transcription), and *repressor proteins* bind other sequences to hinder transcription (negative regulation of transcription). Unlike their counterparts in bacteria, however, eukaryotic transcription activators and repressors are found in large complexes composed of hundreds of distinct regulatory proteins that bind a wide and diverse array of regulatory sequences. These proteins aggregate in diverse combinations that activate or repress transcription of different patterns of genes in different tissues and at different times in the life cycle.

## Structural Motifs of DNA-Binding Proteins

Hundreds of DNA-binding regulatory proteins have been identified in eukaryotes. These proteins are the primary sources of regulated gene expression in eukaryotes. These proteins are categorized on the basis of their structure, called the structural motif of the protein, and many distinct **structural motifs** exist. The most common elements of the structural motifs are alpha helices and other protein secondary structures that form the **functional domains** of regulatory proteins. Certain functional domains allow the proteins to identify and bind specific regulatory DNA sequences in the major groove, or minor groove, of DNA. These DNA binding domains generally consist of a few dozen amino acids, among which several make direct contact with or interact with nucleotide base pairs in regulatory sequences. The amino acids in functional domains of regulatory proteins are able to recognize specific DNA sequences by means of interactions with the unique arrays of nitrogen, oxygen, and hydrogen atoms extending from each base pair. The interactions of two common structural motifs with regulatory DNA are shown in **Figure 15.2**. The characteristics of these and other common motifs are described in **Table 15.1**.

## Transcriptional Regulatory Interactions

Three sets of regulatory DNA sequences are commonly involved in eukaryotic regulation of transcription of specific genes. The first set of regulatory sequences is the *core promoter region* containing the TATA box and other sequences; it is immediately adjacent to the start of transcription and is the sequence to which RNA polymerase II and its associated

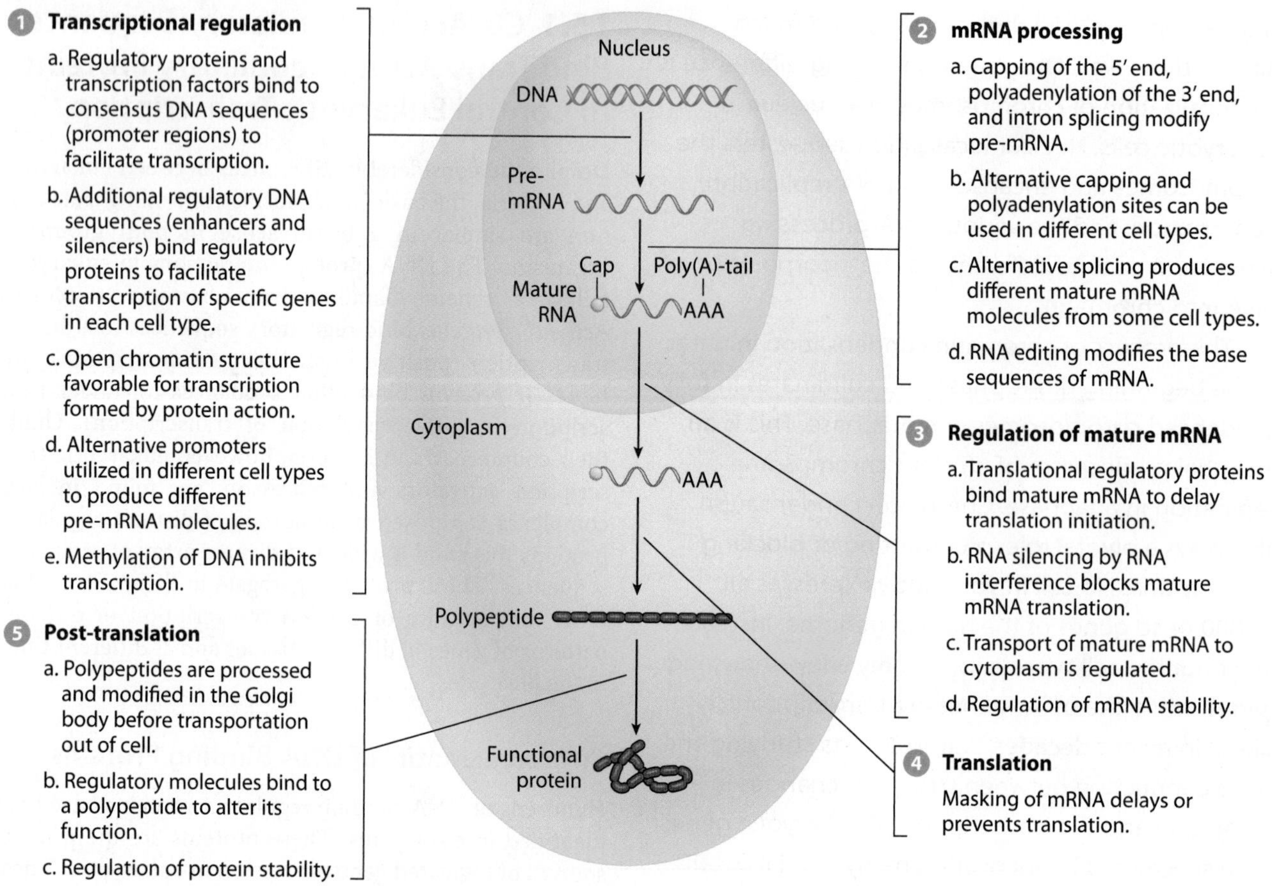

**1** **Transcriptional regulation**

a. Regulatory proteins and transcription factors bind to consensus DNA sequences (promoter regions) to facilitate transcription.

b. Additional regulatory DNA sequences (enhancers and silencers) bind regulatory proteins to facilitate transcription of specific genes in each cell type.

c. Open chromatin structure favorable for transcription formed by protein action.

d. Alternative promoters utilized in different cell types to produce different pre-mRNA molecules.

e. Methylation of DNA inhibits transcription.

**5** **Post-translation**

a. Polypeptides are processed and modified in the Golgi body before transportation out of cell.

b. Regulatory molecules bind to a polypeptide to alter its function.

c. Regulation of protein stability.

**2** **mRNA processing**

a. Capping of the 5′ end, polyadenylation of the 3′ end, and intron splicing modify pre-mRNA.

b. Alternative capping and polyadenylation sites can be used in different cell types.

c. Alternative splicing produces different mature mRNA molecules from some cell types.

d. RNA editing modifies the base sequences of mRNA.

**3** **Regulation of mature mRNA**

a. Translational regulatory proteins bind mature mRNA to delay translation initiation.

b. RNA silencing by RNA interference blocks mature mRNA translation.

c. Transport of mature mRNA to cytoplasm is regulated.

d. Regulation of mRNA stability.

**4** **Translation**

Masking of mRNA delays or prevents translation.

Nucleus

DNA

Pre-mRNA

Cap | Poly(A)-tail
Mature RNA | AAA

Cytoplasm

AAA

Polypeptide

Functional protein

**Figure 15.1** **An overview of gene regulation mechanisms in eukaryotes.**

**Figure 15.2** **Structural motifs and DNA binding. (a)** Alpha helices of dimeric leucine zipper proteins bind regulatory DNA. **(b)** Helix-loop-helix proteins use one alpha helix from each polypeptide to bind regulatory DNA.

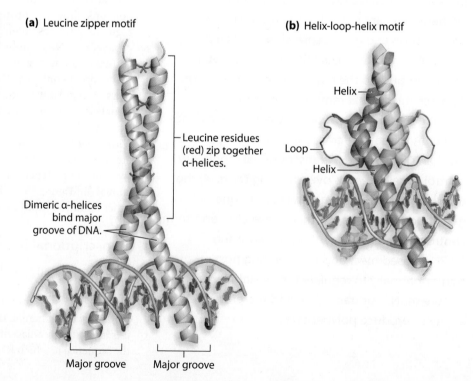

**(a)** Leucine zipper motif

Leucine residues (red) zip together α-helices.

Dimeric α-helices bind major groove of DNA.

Major groove     Major groove

**(b)** Helix-loop-helix motif

Helix

Loop

Helix

| Table 15.1 | Common Structural Motifs of Eukaryotic DNA-Binding Regulatory Proteins |
| --- | --- |
| **Structural Motif** | **Characteristics** |
| Helix-loop-helix | Two alpha helices separated by a loop in each polypeptide. Two polypeptides join to form a DNA-binding dimeric protein. One alpha helix from each dimer binds regulatory DNA. |
| Leucine zipper | Two alpha helices, one containing multiple leucine amino acids, in each polypeptide. Two polypeptides form a functional dimeric protein. The leucine-containing alpha helices face one another and interdigitate to form the "zipper," and the other alpha helix of each polypeptide binds DNA. |
| Homeodomain | Three alpha helices form in single functional polypeptides, and the longest helix interacts with regulatory DNA sequences. |
| Zinc finger | Protruding loops or "fingers" containing about 24 amino acids each contain a central molecule of zinc bound to two cysteine and two histidine amino acids. Two or three zinc fingers form in each polypeptide, and each finger interacts with regulatory DNA. |

transcription factors bind (**Figure 15.3**). Upstream of the core promoter are various *proximal elements* that are the second set of regulatory sequences. These sequences bind regulatory proteins and are essential to promoter recognition and eventual transcription. At greater distances from the core promoter are **enhancer sequences** (or **enhancers**), the third set of regulatory sequences, which bind regulatory proteins and interact with proteins bound to other promoter segments. Unlike core promoter and proximal promoter elements, which are invariably located upstream of and close to the genes they regulate, enhancers can be upstream or downstream and are occasionally found *within* genes. Moreover, although some enhancer sequences are close to the genes they regulate, others are great distances away from their gene targets; some are thousands to tens of thousands of nucleotides away from the genes they regulate.

All three of these regulatory regions contain **cis-acting regulatory sequences,** which means they regulate trans-cription of genes located *on the same chromosome* as the sequences. Cis-acting regulatory elements are able to regulate only transcription of genes on the *same* chromosome and do not influence the transcription of genes on other chromosomes.

RNA polymerase II (Pol II) and various general transcription factors (GTFs) are attracted to and bind the core promoter (see Section 8.3). Transcriptional activator proteins or transcriptional repressor proteins bind to proximal promoter elements and to enhancers. All these proteins are **trans-acting regulatory proteins:** They are able to identify and bind target regulatory sequences on *any* chromosome. RNA polymerase II, for example, is able to bind any core promoter region if the right general transcription factors are also present. Similarly, transcription activator and repressor proteins can bind any target regulatory sequence and can influence transcription with equal efficiency no matter where the sequence occurs.

Besides the regulatory proteins that bind regulatory DNA sequences, many additional proteins also associate with regulatory regions of DNA by protein–protein interactions that form larger complexes. At enhancers, for example, aggregation of multiple proteins, a few binding enhancer sequences and the others binding other proteins, form a large protein complex known as an **enhanceosome.** Enhanceosomes direct DNA bending into loops that bring the enhanceosome into contact with RNA polymerase and transcription factors bound at the core promoter or with protein complexes bound to proximal promoter elements (see Figure 8.12). The DNA loops can be small or large, in keeping with the observation that enhancers may be close to or quite distant from the genes they regulate.

In a broad sense, enhancer activity controls the timing and location of eukaryotic gene transcription to help ensure the proper function and development of organisms (for example, by making specific polypeptides available at crucial times). Often, genes whose transcription is controlled in a time-dependent manner produce polypeptides required for particular metabolic or developmental

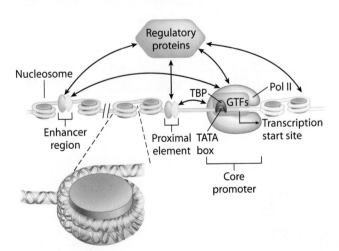

**Figure 15.3 Regulatory interactions in eukaryotic transcription.** TATA-binding protein (TBP), other general transcription factors (GTFs), and RNA polymerase II (Pol II) bind the core promoter. Other regulatory proteins bind proximal promoter and enhancer regions and interact with nucleosomes to activate transcription.

processes. Enhancers also control the expression of genes in specific tissues or cell types, producing tissue-specific patterns of polypeptide production.

---

---

## Enhancer-Sequence Conservation

Despite the diversity of the combinations through which regulatory sequences and proteins control transcription in eukaryotes, scientists have identified much useful information about those regulatory components. For example, studies of enhancer-sequence composition in the eukaryotic virus SV40 (simian virus 40) revealed modular sequences that have since been found to be similar to those of enhancers of other eukaryotes. The SV40 enhancer consists of adjacent regions of conserved sequences located about 200 bp upstream of the transcription start point of regulated genes. Each of seven segments of conserved sequence binds specific regulatory proteins (Figure 15.4a).

Comparisons among species reveal DNA-sequence conservation in enhancers. This implies that natural selection is operating to retain enhancer function, that is, to retain the capacity to bind specific regulatory proteins by conserving sequence composition. Figure 15.4b shows enhancer sequences for the β-interferon gene in several mammals; the abbreviations represent the enhancer-binding proteins whose binding relies on certain sequences. The species listed in the figure share a common ancestor from which their different lineages diverged approximately 100 million years ago.

Genomic sequence analysis points to the likelihood of strong evolutionary constraint on enhancer-sequence diversification. In these studies, sequence analysis first identifies potential enhancer sequences. Next, the conserved sequences that bind regulatory proteins are identified. Interestingly, comparisons of evolutionarily conserved sequences in mouse and human genomes find that fewer than half of the conserved sequences bind proteins. It is suggested that those conserved sequences that do bind proteins may have their diversification constrained by the requirement that they be efficiently recognized by regulatory proteins. From this perspective, the evolution of enhancer could be slowed by the necessity that the sequences be recognized by regulatory proteins.

## Yeast Upstream Activator Sequences

The regulation of transcription by enhancer sequences is well understood in the yeast *Saccharomyces cerevisiae*, where transcription of genes involved in the galactose utilization pathway, among others, are carefully regulated by enhancer-like sequences. When the monosaccharide galactose is the only sugar in the growth medium, strains of *gal*+ yeast will induce the transcription of four enzyme-producing genes, *GAL1*, *GAL2*, *GAL7*, and *GAL10*, that together import extracellular galactose (*GAL2*) and then, through a short series of biochemical reactions, break down intercellular galactose into glucose-1-phosphate for glycolysis (*GAL1*, *GAL7*, and *GAL10*; Figure 15.5). Each of the four genes has its own promoter, but transcription of the genes is regulated by another gene, *GAL4*, which produces a regulatory protein. Gal4 protein is a transcription activator protein that binds to an enhancer-like element called **upstream activator sequence (UAS)** (in this case

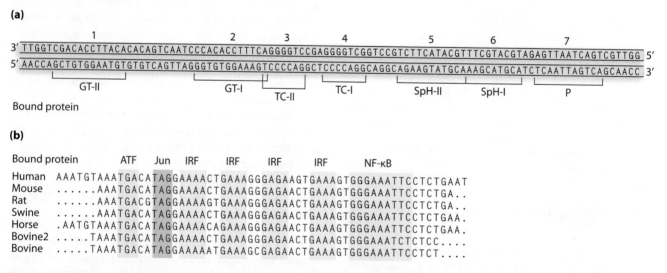

**(a)**

**(b)**

Figure 15.4 **Enhancer sequences and the regulatory proteins that bind them. (a)** The SV40 enhancer sequence contains seven short sequence segments targeted by specific regulatory proteins. **(b)** The enhancer sequence of β-interferon contains multiple sequences (colored boxes) conserved among mammalian species. Highlighted sequences are crucial to binding of specific regulatory proteins.

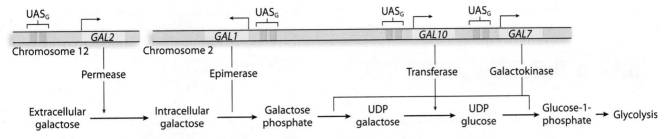

**Figure 15.5 Galactose utilization in *S. cerevisiae.*** Galactose utilization requires the action of products of each of four galactose-utilization (*GAL*) genes.

designated UAS$_G$) located upstream of each of the four *GAL* genes. The Gal4 regulatory protein is continuously available in yeast cells and interacts with Gal80, the product of the *GAL80* gene. When Gal80 protein binds to Gal4 protein, it inactivates Gal4 and blocks its ability to activate transcription.

The UAS$_G$ sequences are cis-acting regulatory elements, and Gal4 protein is a trans-acting regulatory protein. Each UAS$_G$ element contains two 17-bp repeat sequences that are the binding sites for Gal4 protein. In its active, DNA-binding form, Gal4 is a homodimeric protein composed of two identical polypeptides that form two active domains. The DNA-binding domain, at one end of the Gal4 dimer, targets the 17-bp repeats of UAS$_G$. The activation domain, at the opposite end, is a target for binding by the protein Gal80. Since Gal4 and Gal80 are each constitutively produced, they are normally bound to one another at the activation domain of Gal4. In this configuration, the DNA-binding domain of Gal4 is inactive, and the dimer is unable to bind UAS$_G$. Without Gal4 binding to UAS$_G$, transcription of *GAL* genes is blocked (Figure 15.6a). Conversely, when galactose is present, galactose and Gal3, the protein product of another *GAL* gene, bind to Gal80. Binding of the galactose–Gal3 complex alters Gal80 and causes it to release Gal4. The free Gal4 dimer then binds UAS$_G$ and activates *GAL* gene transcription (Figure 15.6b).

In the *GAL* gene system, Gal4 acts as an activator protein through its transcription-initiating effect. Its target DNA sequence is UAS$_G$, which acts like an enhancer sequence and is separated from *GAL* gene promoters by a large number of nucleotides. Gal4 binding leads to the formation of a multiprotein complex known as **Mediator,** which is an enhanceosome that forms after the activator protein Gal4 binds UAS$_G$. When inducing the formation of a DNA loop, Mediator makes contact with the general transcription apparatus—including TFIID (transcription factor II D) and RNA polymerase II (Pol II)—at a *GAL* gene promoter (see Figure 8.18). Thus, the transcription of *GAL* genes by RNA polymerase II is dependent on transcription activation by Gal4 binding to UAS$_G$ elements and causing the formation of Mediator. Distant enhancers and silencers use the same mechanism of DNA loop formation to regulate transcription of targeted genes.

## Locus Control Regions

The human β-globin gene was the focus of our attention in an earlier chapter (see Chapter 10). Recall that this gene produces the β-globin polypeptide, two copies of which join with two α-globin polypeptides produced by the α-globin gene to form the heterotetrameric hemoglobin molecule. The β-globin gene is, however, only one of six very closely related globin genes forming the β-globin complex on human chromosome 11 (Figure 15.7a). Located close to the β-globin complex is a regulatory region known as a **locus control region (LCR).** LCRs are highly specialized enhancer elements that regulate the transcription of multiple genes packaged in complexes of closely related genes. The LCR regulating transcription of genes in the β-globin complex contains four distinct cis-acting regulatory sequences, designated HS1 to HS4. Together these elements orchestrate the sequential developmental expression of the

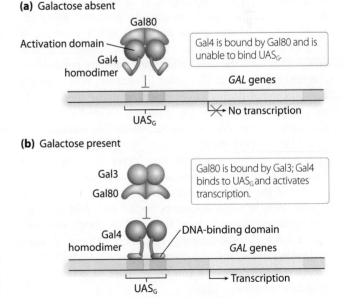

**Figure 15.6 Regulation of *GAL* gene transcription.** **(a)** When galactose is absent, Gal80 protein binds the activation domain to inactivate Gal4 protein and block *GAL* gene transcription. **(b)** When galactose is present, Gal3 protein binds Gal80 protein to prevent it from binding Gal4 protein. Gal4 protein, in turn, uses its DNA binding domain to bind the two 17-bp segments of UAS$_G$ to help initiate *GAL* gene transcription.

**(a)** β-globin–gene complex

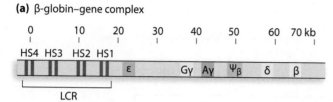

**(b)** Developmental expression of β-globin–complex genes

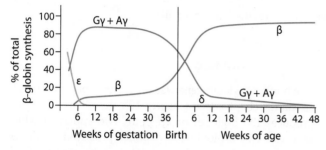

**Figure 15.7 Locus control and developmental expression of human β-globin–complex genes. (a)** The locus control region (LCR) of the human β-globin complex contains four regulatory segments (HS1 to HS4). **(b)** The LCR regulates the expression of five genes (ψβ is an unexpressed pseudogene) in a developmental pattern matched to gestational age.

β-globin–complex genes as a fetus develops during gestation. The LCR and the six genes it regulates occupy just over 70 kb.

Each gene of the β-globin complex produces a distinct globin polypeptide that imparts a different oxygen-carrying capacity to hemoglobin. During gestation, the oxygen requirements of the developing fetus change as its size increases and its organs develop. As gestation proceeds, transcription of the genes of the β-globin complex is switched from one to the next to produce hemoglobin molecules that have the oxygen-carrying capacity required by the developing fetus. The order of expression of β-globin complex genes during development matches the order in which they occur on the chromosome. **Figure 15.7b** shows the expression profile of these genes during development. The HS1 to HS4 components of the β-globin–complex LCR bind regulatory proteins that direct the formation of small DNA loops, and these serve as a bridge to the promoters of the β-globin–complex genes (**Figure 15.8**). The composition of enhanceosomes bound to the LCR varies during development to vary the resulting loops and thus produce the developmentally regulated pattern of gene expression from the β-globin complex. A similar LCR drives transcription of a smaller number of genes in the α-globin complex.

Mechanism of transcriptional activation by LCR

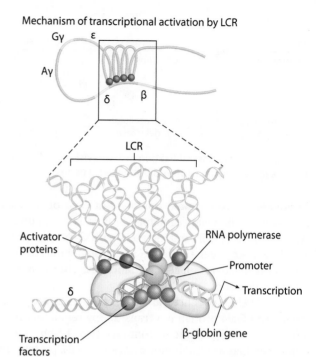

**Figure 15.8 Human β-globin–complex locus control region.** In combination with regulatory proteins that vary with developmental stage, the LCR forms DNA loops that also vary with developmental stage, allowing it to activate transcription of specific genes of the complex.

## Repressor Proteins and Silencer Sequences

A common mode by which repressor proteins inhibit transcription in bacteria is to bind to operator sequences that overlap promoters, blocking the binding of RNA polymerase (see Chapter 14). In eukaryotes, this mechanism of transcription inhibition is not seen. Instead, eukaryotic repressors inhibit transcription through other mechanisms. One of these is the binding of eukaryotic repressors to **silencer sequences,** cis-acting regulatory sequences that block transcription by directly preventing enhancer-mediated transcription.

The galactose-utilization genes in yeast offer an example of this direct mechanism of transcription repression. When glucose is present in the yeast growth medium, the protein Mig1 is produced. Mig1 binds a silencer sequence located between $UAS_G$ and the *GAL1* promoter (**Figure 15.9**). Mig1 in turn attracts the protein

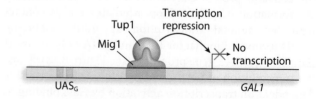

**Figure 15.9 Transcription repression of the yeast *GAL1* gene.** The proteins Mig1 and Tup1 bind to the Mig1 site to repress transcription when glucose is available in the growth medium.

---

**Genetic Insight** Enhancer sequences are cis-acting regulatory sequences that bind complexes containing regulatory proteins; these in turn interact with specific gene promoters to direct transcription.

Tup1, and together these proteins form a repressor complex that prevents UAS$_G$ from directing the initiation of transcription. Other mechanisms for transcription repression are discussed elsewhere in the chapter.

## Transcription Regulation by Enhancers and Silencers

The enhancers and silencers controlling transcription of a gene can be nearby or far from the gene they regulate although, DNA loop formation can bring even very distant sequences together. Some evidence suggests that occasionally the same DNA sequence can act as both an enhancer and as a silencer, depending on the regulatory proteins that bind the sequence. In yeast, enhancers and silencers are situated relatively close to the genes they regulate. The β-globin–complex LCR in humans is also very close to the genes it regulates. Often, however, the distance between an enhancer or silencer sequence and the gene it targets for regulation is vast. An example is provided by the *SHH* (*Sonic hedgehog*) gene, which in humans and other mammals directs the development of limbs and in its wild-type form produces five digits (fingers and toes) on each appendage. *SHH* is expressed in a tissue-specific manner in limbs under the direction of an enhancer that is 1 million base pairs (1 megabase) away from the gene. Genomic sequencing analysis reveals that the *SHH* enhancer is actually located in an intron of a neighboring gene.

A general model for eukaryotic transcription regulation must depict the characteristic action of enhancers and silencers while taking the variability of their locations and their tissue-specific patterns of regulation into account. The model depicted in **Figure 15.10**, for *SHH*, shows two distant enhancers controlling transcription of the same gene in a tissue-specific manner. In this example, *SHH* gene is shown expressed in the brain and in limbs. Transcription in these tissues is controlled by different regulatory proteins and transcription factors produced in each type of cell. One combination of regulatory proteins binds one enhancer in brain cells, but a different combination of regulatory proteins binds an alternative enhancer in limb cells. The different regulatory proteins present in different types of cells leads to tissue-specific patterns of expression of the target gene, producing a different polypeptide in each case. Similar models depicting the binding of repressor proteins to silencer sequences describe how distant silencers can inhibit transcription of targeted genes.

This model illustrates an important aspect of eukaryotic transcription regulation. Only when all of the necessary transcription factors and regulatory proteins are present in a cell can the assembly of protein complexes required for the tissue-specific or development-stage-specific pattern of transcription take place. The protein complexes assembled at regulatory sequences

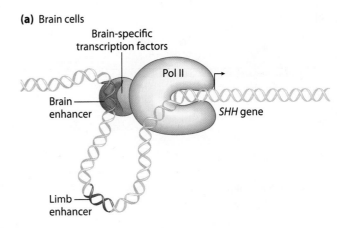

**(a)** Brain cells

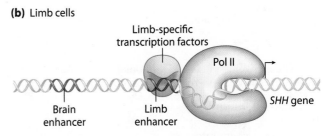

**(b)** Limb cells

**Figure 15.10 Tissue-specific enhancer action. (a)** The brain-specific enhancer is bound by brain-specific transcription factors and activates *SHH* transcription in brain cells. **(b)** A different, limb-specific enhancer binds different, limb-specific transcription factors to express *SHH* differently in limb cells.

direct particular patterns of expression by activating transcription of certain genes while blocking transcription of other genes. The polypeptides that are ultimately produced in each cell or at each stage of development drive the processes that make cells distinctive and lead to the observed developmental changes.

## Insulator Sequences

Considering that enhancers can be located far from the genes they regulate, what mechanisms direct enhancer action toward the intended gene and away from other nearby genes that are not regulated by the same enhancer? The answer lies in **insulator sequences,** cis-acting sequences located between enhancers and promoters of genes that are to be insulated from the effects of the enhancer. Insulators are protein-binding sequences that direct enhancers to interact with the intended promoter and that block communication between enhancers and other promoters (**Figure 15.11**). The mechanism of this activity may consist of allowing the formation of DNA loops containing enhancers and their intended promoter targets while preventing the formation of DNA loops containing an enhancer and a promoter that is not its intended target.

Certain insulator sequences have a different function, namely, to stop the spread of heterochromatin. For example, position effect variegation (PEV) in fruit flies

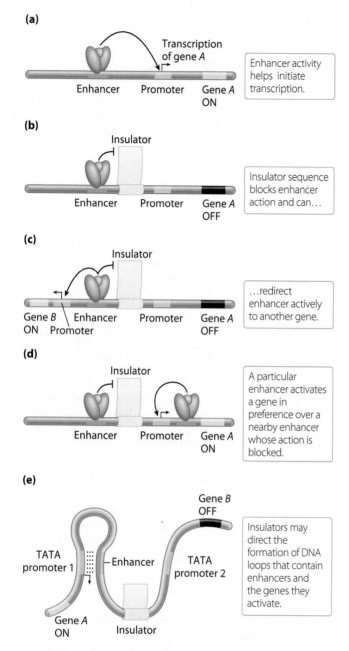

**Figure 15.11 Insulator and enhancer interactions.**

## Enhancer Mutations

Our previous discussions of mutations have described numerous ways in which changes in DNA can result in abnormal polypeptides or abnormal levels of polypeptide production. Here, we take a moment to consider examples of enhancer mutations that are the cause of hereditary disorders in humans.

The term *thalassemia* is used to describe certain hereditary anemias in which mutation leads to an imbalance of production of α-globin and β-globin polypeptides. This imbalance reduces the amount of functional hemoglobin, since each molecule needs an equal number of both polypeptides. Many distinct types of thalassemia result from different mutations of the α-globin or β-globin genes. In some thalassemia patients, however, no mutations of either globin gene were detected. Furthermore, the promoters of both genes were wild type, so the search for the source of the mutations in this group of patients had to be expanded. In several cases, the thalassemia mutations are due to deletion or chromosome-rearrangement mutations that alter the LCR of one of the globin gene complexes. These deletions result in enhancer mutations that alter the level of transcription of affected genes and lead to an imbalance of polypeptide production.

Base-substitution mutations in enhancers are another source of enhancer dysfunction. The *SHH* enhancer, located 1 megabase from the *SHH* gene it regulates, is mutated in certain cases of a condition called polydactyly, in which extra fingers and toes can form during development. The extra digits result from abnormal expression of the *SHH* gene. In studies of certain human families with polydactyly, single-base substitutions in the *SHH* enhancer have been identified. In addition, studies in mice in which a deletion of the *SHH* enhancer has occurred reveal significant abnormalities of limb development.

---

Genetic Insight Silencers and insulators direct the action of enhancers toward certain promoters and away from others. Enhancer mutations can alter normal patterns of gene expression.

---

## 15.2 Chromatin Remodeling Regulates Eukaryotic Transcription

Up to this point in the chapter, we have described transcription regulation in terms of interactions between regulatory proteins and regulatory sequences. In a sense, this pattern of regulation is analogous to, albeit more complex than, the processes described for regulating transcription in bacteria (see Chapter 14). However, the defining feature of eukaryotic DNA is its packaging into chromatin. How, then, do the activator and repressor transcription factors bind to regulatory DNA that is packaged into chromatin? There are three basic mechanisms by which trans-acting proteins access specific regulatory DNA sequences in eukaryotic chromosomes.

is seen in X-chromosome inversions that bring the *white* gene close to the centromere (see Figure 11.16). The spread of centromeric heterochromatin that blocks gene transcription is controlled by insulators. In this case, their function is to prevent the uncontrolled spread of heterochromatin following replication, thus keeping the expression of critical genes from being blocked.

Ultimately, what matters about enhancers, silencers, and other regulatory sequences are their effects on transcription. Enhancers and silencers contain sequences that attract particular proteins and may control the transcription of genes in particular tissues or at specific times during development (see Chapter 20).

First, some regulatory sequences are not tightly bound by histones, which thus allow more or less direct entry to the regulatory DNA. These sequences include the "linker" sequences between nucleosomes and sequences with specific characteristics that prevent histones from binding efficiently.

Second, proteins called *chromatin remodelers* can enzymatically change the distribution or composition of histone octamers (nucleosomes). These remodeler enzymes have three modes of operation. One type of chromatin-remodeling enzyme reorganizes nucleosomes by inducing nucleosome movement. These enzymes usually repress transcription by moving nucleosomes. A second type of chromatin-remodeling enzyme changes nucleosome organization by either sliding them along the chromosome or removing them from the DNA. These enzymes usually work by uncovering enhancers or promoters, and thus are associated with gene activation. The third type of chromatin-remodeling enzyme changes the composition of histone octamers, replacing specific histone proteins with variant proteins. These changes are associated with gene activation. Chromatin-remodeling enzymes are recruited to specific sites in the chromatin by trans-acting factors that bind to specific DNA sequences.

Finally, proteins called *chromatin modifiers* can enzymatically modify histones by adding or removing methyl or acetyl groups at specific amino acid residues, most commonly lysines, of histone proteins. The addition of acetyl groups is associated with gene activation and is typically found in euchromatin. In contrast, removal of acetyl groups and addition of methyl groups to specific lysine residues are associated with gene repression and typically found in heterochromatin. As with chromatin-remodeling enzymes, chromatin-modifying enzymes are recruited to specific sites in chromatin by trans-acting factors that bind to specific DNA sequences.

This combination of activities determines the relative access of trans-acting transcription factors to cis-acting DNA sequences in particular cells, at different times of organismal development, and under certain physiological conditions. Thus chromatin remodelers and chromatin modifiers mediate the reversible transition from inactive heterochromatic DNA to active euchromatic DNA.

## Open and Covered Promoters

Two contrasting states of nucleosome association with promoter sequences, known as *open promoters* and *covered promoters*, are at opposite ends of a continuum of nucleosome association with regulatory DNA sequence. Most promoters fall somewhere between these extremes with respect to their association with nucleosomes, but an examination of open promoters and covered promoters can help us understand how chromatin structure contributes to transcription regulation.

**Open promoters** cause genes to be constitutively transcribed. These promoters have a **nucleosome-depleted region (NDR),** which is a 150- to 100-bp region containing

few nucleosomes that lies immediately upstream of the start of transcription. These promoters do not generally contain a TATA box. Instead, a region rich in adenine and thymine, known as a poly A/T tract, is located in the NDR, near the transcription start site (Figure 15.12a). The poly A/T tract contains binding sequences (BS) that attract transcription activators (ACT). This binding region is usually flanked by sequences that help position two nucleosomes, one upstream and one downstream, of the NDR. The downstream nucleosome, identified as the +1 nucleosome, is placed at the transcription start site. This +1 nucleosome contains a variant histone 2A protein known as H2AZ that is readily modified for removal from the transcription start site at transcription initiation, allowing RNA polymerase II to bind and access the transcription start sequence.

**Covered promoters,** on the other hand, characterize genes whose transcription is regulated. Transcription of these genes is blocked until nucleosomes are displaced or removed from the promoter to allow transcription activators to bind to the necessary sequences, an event that leads in turn to RNA polymerase II binding and transcription initiation (Figure 15.12b). These promoters generally contain

**(a)** Open promoter

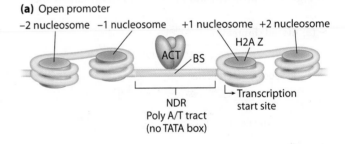

**(b)** Covered promoter

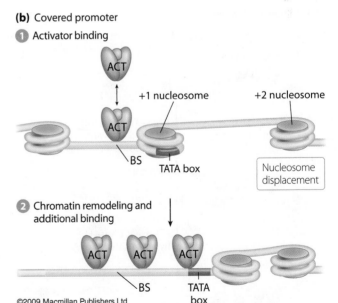

©2009 Macmillan Publishers Ltd

**Figure 15.12 Transcription of open and covered promoters. (a)** Open promoters have a nucleosome-depleted region (NDR) and no TATA box. Activator proteins (ACT) are attracted to binding sequences (BS) to recruit RNA polymerase II for transcription. **(b)** With covered promoters, transcription is activated by activator-protein binding and displacement of nucleosomes.

TATA boxes and other transcription-factor binding sequences. At covered promoters, there is active competition between nucleosomes and transcription-activating factors for binding. As a result, regulatory mechanisms are required that remodel chromatin to give activator proteins access to binding sequences in order to initiate transcription.

## Chromatin Remodeling by Nucleosome Modification

**Chromatin remodeling** refers to chromatin modifications that reposition nucleosomes in such a way as to open or close promoters and other regulatory sequences. Moving nucleosomes off regulatory sequences improves access to them by transcription-activating regulatory proteins. **Open chromatin** is chromatin in which the association of DNA with nucleosomes is relaxed in regions containing regulatory sequences, allowing access by regulatory proteins. Modifications that cause regulatory DNA to be covered by nucleosomes, thus restricting the access of regulatory proteins to the sequences, produce **closed chromatin.** In closed chromatin, regulatory sequences cannot be efficiently accessed by regulatory proteins, and genes are transcriptionally silent.

Molecular biologists can determine experimentally whether a region of DNA contains closed chromatin or open chromatin by assessing the sensitivity of the region to the DNA-digesting enzyme DNase I. This enzyme randomly cuts DNA in open chromatin regions but is not able to do so where chromatin is closed. Regions of open chromatin, sensitive to DNase I digestion, are known as **DNase I hypersensitive sites.** Where DNase I hypersensitivity is detected, genes are potentially transcribable. The experimental analysis of DNA for DNase I hypersensitivity is much like DNA footprint protection analysis described in Research Technique 8.1 (page 271). Fragments of DNA created by exposure to DNase I are separated and analyzed by gel electrophoresis.

DNase I hypersensitivity occurs in the immediate vicinity of transcribed genes and can also appear 1000 base pairs or more upstream or occasionally downstream of actively transcribed genes. Hypersensitive regions surround promoters, enhancers, and other transcription-regulating sequences. The open chromatin complexes detected by DNase I hypersensitivity are the sites for binding by transcription-activating proteins and for transcription (**Figure 15.13**). **Genetic Analysis 15.1** guides you through an analysis for the presence of DNase I hypersensitivity in a region of DNA.

**Chromatin remodelers** are the protein complexes that carry out chromatin remodeling by moving nucleosomes in two principal ways (**Figure 15.14**). First, chromatin remodelers can cause nucleosomes to *slide* along DNA. In this process, DNA spools around the nucleosome until promoter or enhancer sequences are free from

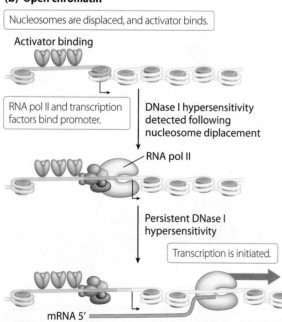

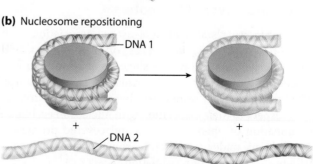

**Figure 15.13 Closed and open chromatin structure.**
**(a)** Closed chromatin is inaccessible to transcriptional proteins and insensitive to DNase I digestion. **(b)** Open chromatin binds transcriptional proteins and is DNase I hypersensitive.

**Figure 15.14 Nucleosome displacement to expose regulatory sequences.** **(a)** Nucleosome can be displaced by sliding, or **(b)** can be repositioned on other DNA regions.

The tissue enzyme TE2 is expressed in various mouse tissues at different times during the life cycle. Identical DNA fragments were isolated from a region immediately upstream of *TE2* and analyzed for DNase I hypersensitivity. The DNA was collected from embryonic (E) and adult (A) mouse heart (H), kidney (K), and thymus gland (T). In the analysis, a radioactive label was attached to one end of each DNA fragment, and the samples from each tissue were exposed to DNase I to determine if the regions upstream of *TE2* were DNase I hypersensitive. The DNA from each sample was then separated by gel electrophoresis, and the results are as shown below.

a. Based on the gel results, is there evidence that chromatin remodeling plays a role in the expression of *TE2*? Explain your reasoning.

b. In which tissue(s) and at what times during development do the results indicate the expression of *TE2* was most likely taking place?

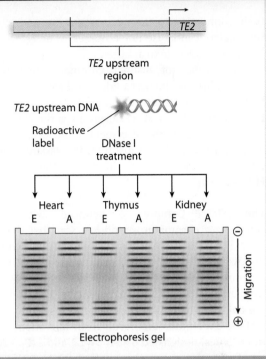

| Solution Strategies | Solution Steps |
|---|---|

### Evaluate

1. Identify the topic this problem addresses, and explain the nature of the required answer.

1. This problem concerns an experimental analysis for DNase I hypersensitivity in the region upstream (i.e., the promoter region) of *TE2*. The answers require interpretation of experimental results with respect to chromatin structure and gene expression.

2. Identify the critical information given in the problem.

   TIP: DNase I hypersensitivity is detected when chromatin structure is open and potentially accessible to transcription-activating proteins. Closed chromatin is not hypersensitive to DNase I.

2. Gel electrophoresis results are given for identical DNA fragments from embryonic and adult heart, thymus, and kidney. All DNA fragments were exposed to DNase I.

### Deduce

3. Compare and contrast the meaning of a continuous series of bands in some lanes of the gel versus lanes in which gaps are seen between bands.

3. A continuous series of DNase I–digested bands indicates DNase I hypersensitivity. Hypersensitivity correlates with open chromatin that is accessible to transcription. Gaps between gel bands indicate that certain fragment lengths of DNA are not created by DNase I treatment. This result signals the absence of DNase I hypersensitivity in those regions and suggests closed chromatin structure and no transcription.

4. Evaluate the gel, and describe the patterns of DNase I–digestion bands for each sample.

4. Discontinuous band patterns are observed in adult heart and embryonic thymus gland DNA. This absence of DNase I hypersensitivity suggests closed chromatin structure. Each of the other DNA samples indicates hypersensitivity to DNase I.

### Solve

5. Determine whether the gel data indicates chromatin modification near *TE2*.

Answer a

5. The DNase I hypersensitivity results indicate differential patterns of *TE2* expression in different tissues and at different times of development due to chromatin modifications. DNase I hypersensitivity resulting from open chromatin appears in embryonic and adult kidney, in embryonic heart, and in adult thymus DNA. Hypersensitivity is not seen in adult heart or in embryonic thymus DNA indicating closed chromatin.

6. Name the tissues in which *TE2* is expressed, and describe the developmental timing.

Answer b

6. *TE2* expression is likely to occur at embryonic and adult stages in the kidney, in the embryonic heart, and in the adult thymus gland. *TE2* expression is unlikely to occur in adult heart or in embryonic thymus gland.

For more practice, see Problems 3, 17, and 18.

their connection to the nucleosome. Second, chromatin remodelers can cause nucleosomes to be *relocated* from one DNA molecule to another.

A number of distinct chromatin remodelers are known. Three of the best-understood categories, classified by their main functions, are the *SWI/SNF complex*, which both slides and relocates nucleosomes; the *ISWI complex*, which helps direct the placement of nucleosomes; and the *SWR1 complex*, which substitutes the variant histone protein H2AZ in nucleosomes in place of the more common H2A protein.

**The SWI/SNF Complex** Pronounced "switch sniff," this category of chromatin remodelers was first described in yeast and is now known to operate in all eukaryotes. It was discovered through analysis of mutations that affect two unconnected activities of yeast. One set of yeast mutants were unable to switch (SWI) mating type, a process tied to the ability of haploid yeast strains to fuse to form diploid strains. SWI mutations result from alterations of any of three genes, designated *SWI1*, *SWI2*, and *SWI3*. A second set of mutants were sucrose nonfermenting (SNF) mutants. SNF mutants lose the ability to grow on medium containing the sugar sucrose owing to mutation of any of three genes designated *SNF2*, *SNF5*, and *SNF6*. The discovery that *SWI2* and *SNF2* are the same gene indicated that the activity blocked in *SWI* and *SNF* mutants was broader than just mating-type switching or the ability to initiate the transcription of genes needed for sucrose fermentation.

Research reveals that the **SWI/SNF complex** is composed of 8 to 12 proteins. The specific proteins and other details of SWI/SNF composition vary somewhat among eukaryotic species, but in each species the complex functions to open chromatin structure by displacing or ejecting nucleosomes. These actions expose promoter and other regulatory sequences to allow binding of transcription factors or activators that help initiate transcription (Figure 15.15). Demethylation and acetylation of certain amino acids of nucleosomes may also occur during transcription activation, as we discuss shortly.

**The ISWI Complex** Chromatin remodelers of the **ISWI (imitation switch) complex** function primarily to control the placement of nucleosomes into an arrangement that causes the region to be transcriptionally silent. These proteins have the ability to "measure" the length of linker DNA between bound nucleosomes in order to place the nucleosomes at regular intervals where they will cover promoters, thus preventing regulatory proteins from having access to the TATA box and other regulatory sequences. There is some evidence that certain nucleosome modifications can block ISWI activity, by a process that could be related to the opening of promoter and chromatin structure (Figure 15.16).

**The SWR1 Complex** The switch remodeling 1, or **SWR1 complex**, is responsible for replacing the common histone

**(a) Closed chromatin**

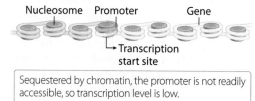

Sequestered by chromatin, the promoter is not readily accessible, so transcription level is low.

**(b) Open chromatin**

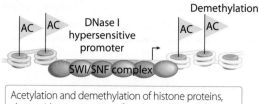

Acetylation and demethylation of histone proteins, along with repositioning of nucleosomes by SWI/SNF, cause chromatin structure to open.

**(c) Transcription initiation**

Transcription factors and RNA polymerase gain access to promoters in open chromatin and initiate transcription.

**Figure 15.15** Function of the SWI/SNF protein complex.

2A protein of nucleosomes with a variant form known as H2AZ that differs from the more common form by amino acid differences internal to the protein and in the amino terminal (N-terminal) protein tail. The differences found in H2AZ alter its pairing with other H2A proteins and its interactions with H3/H4 tetramers in the nucleosome.

H2AZ is found primarily at the so-called +1 nucleosome that is affiliated with the start of transcription. These promoters often contain NDRs and tend to be constitutively transcribed. In some species, H2AZ is found at other nucleosomes surrounding the transcription start. Functional analyses in several species suggest that the role of H2AZ is in the creation of unstable nucleosomes that might then be displaced, ejected from DNA, or modified to regulate transcription (see Figure 15.16).

Genetic Insight Chromatin remodelers are protein complexes that alter chromatin architecture at promoters to regulate the access of transcription-activating proteins to regulatory sequences.

## Chemical Modifications of Chromatin

The proteins called **chromatin modifiers** chemically modify histone proteins in the nucleosomes by adding or removing specific chemical groups. These modifications alter the

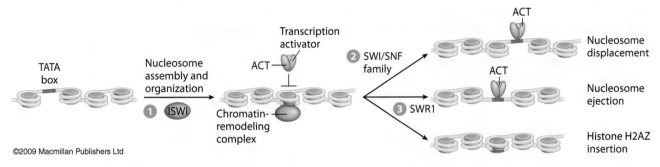

©2009 Macmillan Publishers Ltd

**Figure 15.16  The actions of chromatin-remodeling complexes.**  ❶ ISWI assembles and organizes nucleosomes in a regular pattern and contributes to transcription repression.  ❷ The SWI/SNF family opens chromatin structure and helps initiate transcription by either relocating nucleosomes away from regulatory sequences or ejecting nucleosomes.  ❸ SWR1 inserts the modified histone protein H2AZ into nucleosomes to help facilitate displacement.

strength of association between nucleosomes and DNA. The changes can cause chromatin structure to relax, leading to open promoters and to transcription activation, or they can lead to closed promoter structures that inhibit transcription. The principal chemical modifications to nucleosomes take place through the addition and removal of acetyl and methyl groups at specific amino acids in the N-terminal (amino terminal) region of histones. Less often, the addition or removal of phosphate groups modifies the N-terminal domain of histones.

The most common chemical modification associated with the opening of chromosome structure is the addition of acetyl groups ($COCH_3$) by **histone acetyltransferases (HATs).** The HATs are chromatin-modifying enzymes that add acetyl groups to specific positively charged amino acids in the N-terminal tails of histones. In their unacetylated form, these positively charged amino acids promote nucleosome adherence to negatively charged DNA. Acetylation neutralizes the positive charge and relaxes the tight hold the nucleosomes have on DNA. The amino acid lysine is common in N-terminal tails and is a frequent target of HATs. The acetyl groups on lysine and other N-terminal amino acids are removed by **histone deacetylases (HDACs),** a modification associated with the closing of chromatin structure (**Figure 15.17**).

A second common chemical modification of amino acids in N-terminal tails of histone proteins is methylation, the addition of methyl ($CH_3$) groups by chromatin-modifying **histone methyltransferases (HMTs).** Lysine is frequently targeted for methylation, but arginine is a target as well. Methylation plays a role in converting open chromatin to closed chromatin in conjunction with

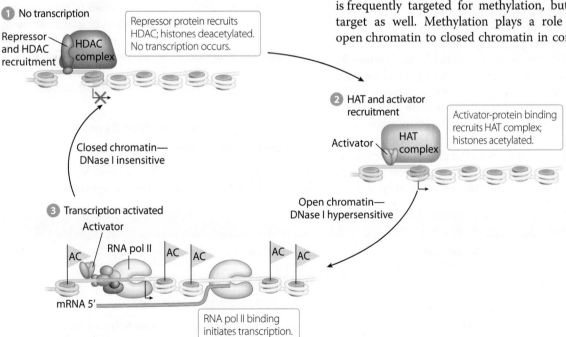

**Figure 15.17  Acetylation and deacetylation in open and closed chromatin structure.**  Histone deacetylases (HDACs) deacetylate amino acids in N-terminal histone protein tails and close the chromatin structure. Histone acetyltransferases (HATs) acetylate N-terminal amino acids and help open the chromatin structure to activate transcription.

deacetylation. Conversely, demethylation, in conjunction with acetylation, forms open chromatin. Demethylation is carried out by **histone demethylases (HDMTs).** Occasionally, phosphorylation and dephosphorylation, the addition and removal of phosphate groups, respectively, are also associated with modification of chromatin structure.

---

**Genetic Insight**    Chromatin modifiers alter chromatin structure by adding and removing chemical groups from N-terminal tails of histone proteins, thus influencing transcription.

---

Lysine (single-letter abbreviation K) is the N-terminal amino acid most frequently targeted for acetylation and can also be targeted for methylation (**Figure 15.18**). Arginine (single-letter abbreviation R) is a common target for methylation, and the amino acids threonine (T) and serine (S) are phosphorylation targets. In most instances, these amino acids are targets for a single kind of alteration, but certain lysine residues are the objects of competition between HATs and HMTs. For example, in H3 (histone 3 protein), the lysine (K) amino acids at positions 4, 9, and 27 can be either acetylated or methylated. Acetylation of these lysine residues occurs in open chromatin, whereas their methylation occurs in closed chromatin.

Multiple chemical modifications of N-terminal amino acids are required to remodel chromatin from a closed to an open structure, and vice versa. No single acetylation or methylation event determines chromatin structure. Instead, more than 150 identified chemical modifications that make chromatin structure dynamic are tied to the transcriptional activity or silencing of genes.

Because different patterns of modifications of histone tails lead to greater or lesser amounts of transcription by contributing to the opening and closing of chromatin structures, molecular biologists Thomas Jenuwein and C. Davis Allis suggested that a "histone code" exists. This speculative code would consist of different combinations of chemical modifications in histone N-terminal tails, resulting in different changes to the chromatin structure. More recently, two studies examining different aspects of chromatin complexity in two evolutionarily distant

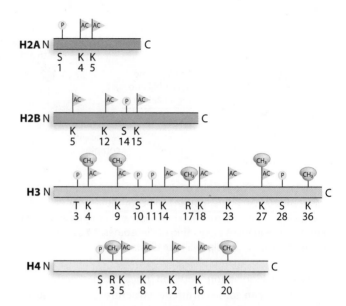

**Figure 15.18    Amino acid targets for acetylation, methylation, and phosphorylation in N-terminal histone tails.** The numbers are the amino acid positions in each histone. Three lysine amino acid positions in H3—K4, K9, and K27—are competitive targets for acetylation and methylation.

eukaryotes suggest chromatin exists in only a limited number of distinct states (**Table 15.2**). Examining the combinatorial complexity of chromatin modifications in *Drosophila* cells in 2010, Guillaume Filion and colleagues identified five principal types of chromatin, each designated by color (the Greek word *chroma* means "color"). A similar study of chromatin in *Arabidopsis* by Francois Roudier and colleagues in 2011 examined histone modifications and DNA methylation to identify four prominent chromatin states (CS) that roughly correspond to those in *Drosophila*. These studies have refined our earlier view of heterochromatin and euchromatin by defining specific molecular characteristics of different chromatin states.

## An Example of Transcription Regulation in *S. cerevisiae*

To illustrate the role of chromatin modifications in transcription initiation, we turn to transcription regulation of

| Table 15.2 | Principal Chromatin States in *Drosophila* and *Arabidopsis* | |
|---|---|---|
| **Drosophila** | **Arabidopsis** | **Function of Chromatin State** |
| Yellow | CS1 | Active gene transcription (euchromatin) |
| Red | CS1 | Active gene transcription (euchromatin) |
| Blue | CS2 | Polycomb repressed genes (facultative heterochromatin) |
| Green | CS3 | Repressed repetitive sequences (constitutive heterochromatin) |
| Black | CS4 | Repressed transcription (distinct from other heterochromatin) |

Data from Filion, G.J., et al. 2010 and Roudier, F., et al. 2011.

the *PHO5* gene in the yeast species *S. cerevisiae*. Our discussion of this particular example is based on numerous studies that collectively paint a comprehensive picture of the actions associated with chromatin modification in *PHO5* transcription initiation and regulation.

*PHO5* is a repressible gene encoding an acid phosphatase enzyme that removes phosphate groups from other proteins. In yeast, *PHO5* transcription is activated by phosphate starvation, but it is repressed when phosphate level is high. In the repressed state, access of transcription factors and RNA polymerase II to the promoter's TATA box is blocked by a nucleosome labeled −1 in Figure 15.19a. Similarly, access of transcription activator proteins to a UAS element, labeled UASp2, is blocked by a nucleosome labeled −2. In the repressed state, the transcription activator protein Pho2 and the acetylase protein NuA4 are present upstream of the promoter at a UAS element labeled UASp1. Upstream of these are nucleosomes labeled −3 and −4. There is a low level of acetylation of nucleosomes −1 to −4 in the repressed state. Together, the presence of the nucleosomes −1 to −4 blocks access of activator protein and transcription factors to *PHO5* regulatory sequences.

Transcription of *PHO5* occurs when phosphate level falls. The Pho4 protein attaches to Pho2, forming a protein complex that begins transcription activation. Additional acetylation of the −1 to −4 nucleosomes takes place under the direction of NuA4. The Pho4–Pho2 complex then initiates chromatin modification by displacing nucleosome −2 (Figure 15.19b), making UASp2 available for binding by the Pho4 protein. The SWI/SNF protein complex assembles, and additional chromatin modification displaces nucleosomes −1 (that previously covered the TATA box), −3, and −4. With chromatin opened by nucleosome displacement, general transcription protein and RNA polymerase II are able to bind the promoter and initiate transcription of the *PHO5* gene.

## Epigenetic Heritability

The transcription of eukaryotic genes is an intricate process in which multiple regulatory proteins interact to control gene expression on two different levels. One level consists of the interactions necessary to transcribe individual genes. The second level consists of interactions that regulate the developmental patterns of gene expression by creating and maintaining specific chromatin states for certain genes.

Activating the transcription of an individual gene requires a confluence of regulatory proteins that remodel or modify chromatin to provide enhancer and promoter access to transcription factors that initiate and carry out transcript synthesis, as we saw in the detailed description of *PHO5* transcription. Mechanisms controlling differential chromatin state formation and maintenance produce patterns of gene expression in different types of cells that are required for the growth and development of complex organisms. In a broad sense, these regulatory processes are the reason a single fertilized egg can develop and produce many distinct types of cells (liver cells, muscle cells, brain cells, and so on) that look and act differently even though they carry the same genetic information.

Among the trillions of somatic cells in your body are scores of different cell types, and yet all these cells contain the same genetic information. The differences of morphology and function between cell types are genetically controlled, as evidenced by the fact that daughter cells have the same structures and functions as parental cells; but DNA sequence variability *is not* the reason for those differences. Instead, the differences between somatic cells are **epigenetic,** resulting from the distinct chromatin states affecting gene transcription in specific types of cells.

Certain epigenetic patterns are heritable from one generation of cells to the next, causing daughter cells to have the same patterns of gene expression as their parent and sibling cells. There is also evidence that epigenetic differences are heritable through reproduction, that is, from one generation of the organism to the next. On the other hand, some epigenetic changes occur in the course of normal growth and

**(a)** High phosphate                                    ©2009 Macmillan Publishers Ltd

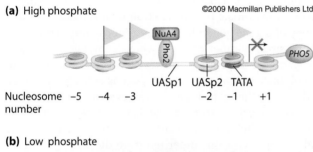

Nucleosome   −5   −4   −3      −2   −1   +1
number

**(b)** Low phosphate

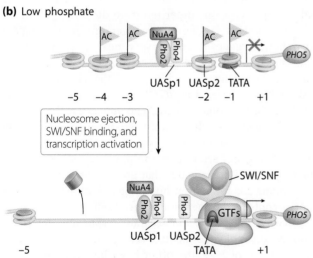

**Figure 15.19** **Transcription control of *PHO5* in** *Saccharomyces cerevisiae.* **(a)** Transcription is repressed in high-phosphate conditions. **(b)** In low-phosphate conditions Pho4 joins Pho2 at UASp1 and NuA4 directs acetylation of nearby nucleosomes. The SWI/SNF complex attaches, leading to the ejection of nucleosomes −1 to −4. RNA polymerase II and general transcription factors initiate *PHO5* transcription.

development, in some cases resulting from different physiological conditions. These changes are potentially reversible and variable during the life cycle of an organism, during which the transcription of certain genes is turned on and later off again, or vice versa. Changes of this latter type are not dependent on passing through meiosis.

We have previously encountered examples of heritable variation of gene expression that has an epigenetic basis. For instance, position effect variegation (PEV) in *Drosophila*, results from the movement of the transcriptionally active $w^+$ allele into the centromeric region of the fruit-fly X chromosome (see Figure 11.16). The DNA sequence of the gene is not altered. Instead, the spread of heterochromatin closes chromatin structure and blocks gene transcription by an epigenetic mechanism. The repressed transcriptional state is then maintained in daughter cells through mitotic division. The result is patches of cells descendant from original progenitor cells that share the same pattern of inactivation of $w^+$ expression. These cells form patches of white in the eye of the fly. In vertebrates and plants, deacetylation and methylation of N-terminal tails of histones lead to heterochromatin formation that closes chromatin structure, whereas demethylation and acetylation lead to open chromatin structure and are associated with euchromatic regions of genomes.

How is epigenetic control maintained in cells? In general, any acetyl and methyl groups that are present on histones before DNA replication are maintained in the same patterns after DNA replication. The specific molecular mechanics of this process are not entirely clear, but the partial disassembly and subsequent reassembly of nucleosomes is an essential component (see Figure 11.8). Recall that chromatin structure is broken down as the replication fork passes (see Chapter 11). Nucleosomes are separated from the parental DNA strands so the latter can serve as templates for the synthesis of daughter strands. The nucleosomes partially break apart, and old nucleosome segments along with newly synthesized nucleosome segments are reassembled on both new duplexes.

Following DNA replication, the newly formed nucleosomes carry only part of their previous epigenetic information. The original epigenetic state must be quickly reestablished by epigenetic marking of the newly synthesized histones. Old histones are able to modify new histones to have the same pattern of epigenetic marks. This process takes place among adjacent nucleosomes, thus preserving local epigenetic control of gene transcription. The interaction must also occur over long distances so as to maintain higher order chromatin structure, such as that characterizing inactivated X chromosomes.

---

**Genetic Insight**  Epigenetic inheritance occurs through heritable modifications of chromatin that control transcription without changing DNA sequence. Epigenetic states can be maintained through cell division or can vary during the life cycle.

---

## Genomic Imprinting

Different regions of eukaryotic genomes characterized by specific chromatin states that maintain epigenetically controlled transcriptional activation and silencing through successive mitotic cell divisions. The maintenance of open and closed chromatin structure is essential to normal growth and development, which requires distinctive patterns of gene expression in different types of cells. While specific epigenetically controlled patterns of gene expression can be passed along in mitosis, the expression patterns for certain genes may have to be reset in meiosis. A specialized example of this kind of resetting of epigenetic patterns in meiosis occurs in certain mammalian genes in a mechanism known as **genomic imprinting.** For the small number of mammalian genes subject to genomic imprinting, both copies of the gene are functional but just one is expressed.

You have probably become accustomed, by this point in your genetic studies, to the idea that in mammals, two copies of each autosomal gene are inherited and expressed—one copy is on the chromosome is inherited from the mother, and the other copy is on the chromosome comes from the father. One reason we know that expression of both gene copies is usually required for normal development and normal phenotype in mammals is that deletion or inactivation of one copy of a gene generally causes abnormalities to appear. For a small number of genes whose expression is subject to genomic imprinting, however, this pattern does not hold. Instead, one copy of the gene is actively expressed while the other copy is silent. The expressed gene copy is always inherited from a particular parent (for some genes it is the mother, for others it is the father), and the silent copy is the one inherited from the other parent.

The best-studied examples of genomic imprinting are two human genes encoded very near one another on chromosome 15. The insulin growth factor 2 (*IGF2*) gene on the paternally derived copy of the chromosome is expressed, whereas the *IGF2* gene on the maternally derived chromosome is silent. The opposite is the case for the *H19* gene, which is expressed from the maternally derived chromosome 15 but is silent on the paternal copy. These two genes are in a region of chromosome 15 containing several other genes that are also imprinted. They are among the few dozen human genes whose transcription is controlled by genomic imprinting.

Two regulatory sequences are responsible for these two instances of genomic imprinting. One is an enhancer downstream of *H19*; the other is an insulator sequence, called the **imprinting control region (ICR),** located between *H19* and *IGF2* (**Figure 15.20**). In the maternal chromosome, activator proteins bind the enhancer sequence and direct transcription of *H19* by interacting with transcription factors and RNA polymerase II at the promoter. The ICR in the maternal chromosome is bound by an insulator protein that blocks the enhancer from

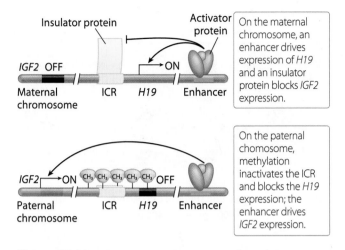

**Figure 15.20  Differential genomic imprinting of chromosome 15 in humans.**

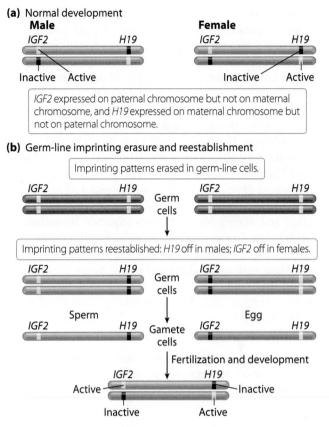

**Figure 15.21  Inheritance of genomic imprinting.** The genomic imprinting patterns on chromosome 15 are erased and reestablished in sex-specific forms early in gametogenesis to ensure reproductive success.

affecting *IGF2*. On the paternal chromosome, on the other hand, extensive methylation of the ICR and *H19* prevents insulator protein binding and blocks transcriptional protein binding at the *H19* promoter. In the absence of the insulator protein, the enhancer stimulates transcription of *IGF2*.

Genomic imprinting silences expression of paternal *H19* and maternal *IGF2* and directs transcription of paternal *IGF2* and maternal *H19* in all somatic cells. This pattern is essential for normal development, and any other pattern produces profound abnormalities. A genetic condition called Prader-Willi syndrome (OMIM 176270) most often results from partial deletion of the portion of the paternal copy of chromosome 15 containing *H19* and *IGF2*. The condition can also occur if the paternal chromosome 15 is not properly imprinted. A different condition called Angelman syndrome (OMIM 105830) is most often produced by partial deletion of the same portion of the maternal chromosome 15. Angelman syndrome also occurs if the maternal chromosome is not properly imprinted.

Given the importance of imprinting for certain genes, and considering the different imprinting patterns of gene expression in maternally derived versus paternally derived chromosomes, how does the inheritance of correctly imprinted chromosomes occur? The answer is that in primordial germ-line cells, the inherited imprinting patterns are first erased and then are reestablished in the sex-specific pattern of the germ line early in gametogenesis (Figure 15.21). In the female germ line, methylation of the paternal chromosome is reversed by demethylase activity, and the insulator protein is removed from the ICR on the maternal chromosome. Both chromosomes are then re-imprinted with the female-specific pattern. In the male germ line, both chromosomes have their imprinting erased and then reestablished in the male-specific pattern. These processes ensure that each parent passes a properly imprinted chromosome during reproduction.

## Silencing by Nucleotide Methylation

The methylation pattern identified in genomic imprinting of the ICR and *H19* gene is a limited type of methylation that can silence gene expression in many vertebrates, particularly mammals, that differs from methylation of amino acids in N-terminal histone protein tails. In this case, methyl (CH₃) groups are attached to specific DNA *nucleotides*, not to amino acids in histone protein tails, to silence genes. Nucleotide methylation is performed by specialized DNA methyltransferases that add methyl-groups primarily to cytosines located in **CpG dinucleotides,** side-by-side cytosine and guanine nucleotides in the same DNA strand. The *p* in CpG represents the single phosphoryl group in the phosphodiester bond connecting the nucleotides. Complementary strands of DNA containing CpG dinucleotides each have 5'-CG-3'.

Most of the cytosines in CpG dinucleotides of vertebrate genomes are methylated, but the pattern is not random. Instead, sequences rich in CpG, called **CpG islands,** are targeted for methylation. In mammalian genomes, CpG islands are clustered at promoters. When CpG islands are unmethylated, chromatin in the promoter region is open, allowing access by transcription factors and RNA polymerase. Thus active transcription is linked to low levels of methylation of CpG islands (**Figure 15.22a**). On the other hand, if CpG islands are methylated, promoter regions are closed, and transcription is repressed.

In the laboratory, the detection of methylated versus unmethylated CpG islands is performed by isolating DNA and treating it with certain restriction enzymes that have CG-rich restriction sequences (**Figure 15.22b**). The restriction enzymes *Msp*I and *Hpa*II each have the restriction sequence 5′-CCGG-3′. The *Hpa*II enzyme is "methyl sensitive," however, meaning that if its restriction sequence is methylated (5′-CC^meGC-3′), the sequence *will not* be digested. On the other hand, *Msp*I is not sensitive to methylation and digests its restriction sequence regardless of methylation. In the example illustrated in the figure, different-sized restriction fragments are produced by *Hpa*II and *Msp*I restriction digestion of the same CpG island region depending on whether the region is methylated. **Genetic Analysis 15.2** demonstrates an analysis of CpG island methylation.

---

**Genetic Insight**  Methylation of nucleotides, primarily cytosine, closes chromatin structure at CpG islands to silence transcription and also operates in genomic imprinting.

---

## 15.3 RNA-Mediated Mechanisms Control Gene Expression

In the past several years, RNA has emerged as a key component in the regulatory control of eukaryotic gene expression. Largely unknown before the mid-1990s, RNA-mediated regulatory mechanisms have rapidly become a major focus of research in plants and animals. This important area of inquiry emerged unexpectedly from experiments designed to produce a more colorful petunia.

In the early 1990s, Richard Jorgensen and his colleagues were attempting to deepen the color of petunias by introducing into the petunia genome a pigment-producing gene under the control of an active promoter. The researchers hoped that active transcription of this recombinant gene would dramatically deepen flower color. To Jorgensen's surprise, however, rather than exhibiting more intense color overall, many of the resulting flowers were variegated (see the chapter opener photo). Some flowers had stripes of deep pigment and stripes lacking pigment, and some flowers were almost entirely white. The researchers called this phenomenon **cosuppression,** because expression of both the introduced pigment gene and the petunia's natural pigment-producing gene were suppressed.

By 1995, similar gene-silencing phenomena had been documented in numerous plant species, in the fungus *Neurospora crassa*, in the nematode worm *Caenorhabditis elegans*, and in the fruit fly *Drosophila*. The fundamental

**(a)** Chromatin structure at CpG islands

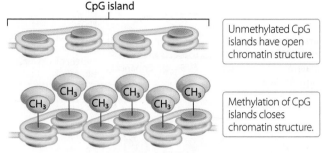

CpG island

Unmethylated CpG islands have open chromatin structure.

Methylation of CpG islands closes chromatin structure.

**(b)** Detecting methylation status of CpG islands

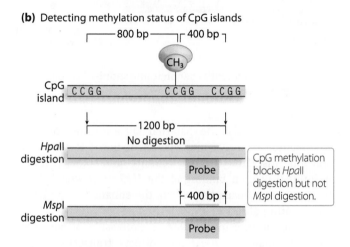

CpG methylation blocks *Hpa*II digestion but not *Msp*I digestion.

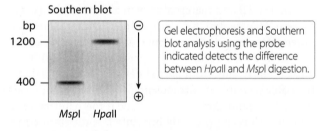

Gel electrophoresis and Southern blot analysis using the probe indicated detects the difference between *Hpa*II and *Msp*I digestion.

**Figure 15.22  Methylation of cytosine nucleotides in mammalian CpG islands. (a)** Methylation closes chromatin, whereas demethylation opens chromatin. **(b)** Comparative restriction digestion by *Msp*I and methyl-sensitive *Hpa*II results in differential restriction digestion if a restriction sequence is methylated. **(c)** Southern blot analysis detects CpG methylation.

mechanism behind this form of regulation was identified in 1998 by a research team led by Andrew Fire and Craig Mello. Fire and Mello found that double-stranded RNA (dsRNA) molecules were taking part in a post-transcriptional regulatory mechanism now known universally as **RNA interference (RNAi).** Fire and Mello received the Nobel Prize in Physiology or Medicine in 2006 for their work.

### Gene Silencing by Double-Stranded RNA

RNA interference silences gene expression either by blocking transcription of targeted genes or by blocking gene expression posttranscriptionally. Posttranscriptional silencing occurs following binding of small regulatory

M. Ethyl Layton, a biologist studying the effects of CpG island methylation on gene expression in eukaryotes, constructs two recombinant genetic systems. Both systems contain the bacterial *lacZ* gene. One recombinant joins *lacZ* to a 4.0-kb segment of a human gene promoter region containing numerous CpG dinucleotides. The second recombinant joins *lacZ* to a 4.0-kb segment of a *Drosophila* gene promoter that also contains numerous CpGs. The table lists the number of methylated CpG dinucleotides and the relative percentage of expression of *lacZ* in the recombinant gene system.

| | Number of Methylated CpGs | Relative percentage of *lacZ* Transcription |
|---|---|---|
| Human | 0 | 100 |
| | 9 | 50 |
| | 20 | 25 |
| | 35 | 2 |
| Drosophila | 0 | 100 |
| | 7 | 99 |
| | 14 | 97 |
| | 26 | 95 |

a. What general conclusion can you draw about the role of methylation as a gene-silencing mechanism in humans?

b. Based on these data, what do you conclude about the role of methylation in transcription in humans as compared to *Drosophila*?

| Solution Strategies | Solution Steps |
|---|---|
| **Evaluate** | |
| 1. Identify the topic addressed by this question, and describe the information required in the answer. | 1. This problem addresses the influence that methylation of CpG dinucleotides has on transcription of the *lacZ* gene in a recombinant system in which transcription is driven by either a human promoter or a fruit-fly promoter. The answer requires interpretation of the experimental results and a determination of the effect of methylation on human and *Drosophila* promoters. |
| 2. Identify the critical information given in the problem. | 2. The two recombinant DNA molecules are described as combining the bacterial *lacZ* gene with either a human promoter or a *Drosophila* promoter that each contain numerous CpG dinucleotides. Data are provided on the levels of *lacZ* transcription associated with different numbers of methylated CpGs in each recombinant. |
| **Deduce** | |
| 3. Describe any correlation between the number of methylated CpG dinucleotides and *lacZ* expression. | 3. Transcription of the *lacZ* gene decreases as the number of methylated CpG dinucleotides in the human gene promoter increases, but the transcription of *lacZ* is essentially unchanged by increasing numbers of methylated CpGs in the *Drosophila* promoter. |
| 4. Compare and contrast the differences between the recombinants in terms of your expectation based on chapter reading. | 4. When there is no methylation, 100% transcription occurs in both recombinants. Transcription driven by the *Drosophila* promoter is not reduced by methylation, but transcription driven by the human promoter gradually decreases as the number of methylated nucleotides increases, and it falls nearly to zero for the largest number of methylation events. Chapter reading indicates that methylation of CpG islands reduces transcription in mammalian genomes. |
| **Solve** | Answer a |
| 5. Describe the meaning of the experimental results for human gene expression. | 5. Methylation of CpG islands reduces transcription from human promoters. |
| | Answer b |
| 6. Describe how methylation of CpGs affects human gene expression versus gene expression in *Drosophila*. | 6. Transcription percentage gradually decreases with additional methylation of CpGs in human promoters, but no level of methylation reduces transcription from the *Drosophila* promoter. |

For more practice, see Problems 5 and 16.

RNAs to mRNA targets by complementary base pairing. The binding of these regulatory RNAs can either lead to the destruction of the target mRNAs or can block their translation. Alternatively, some regulatory RNAs enter the nucleus where they bind DNA to block transcription of targeted genes. Any of these regulatory processes first require that small regulatory RNA molecules use complementary base pairing to bind their targets. The regulatory RNAs in RNAi are derived from various sources that produce double-stranded RNAs by folding. An enzyme known as **Dicer** (Figure 15.23) cuts the double-stranded

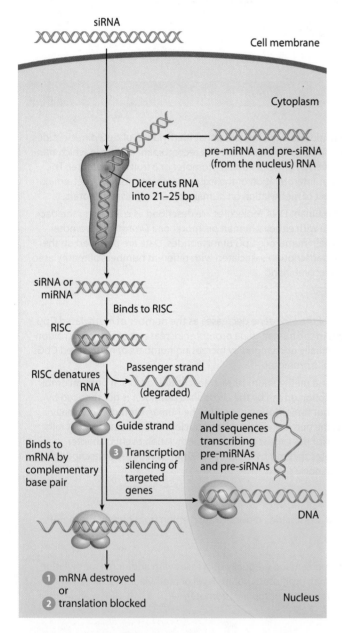

**Figure 15.23 Gene silencing by RNAi.** Dicer cuts dsRNA into 21- to 25-bp siRNA or miRNA segments that are then denatured by RISC. RISC–guide strand complexes can degrade targeted mRNAs, block translation of target mRNAs, or enter the nucleus to modify chromatin.

RNA into 21- to 25-bp fragments. These fragments are then bound by a protein complex called the **RNA-induced silencing complex (RISC)** that denatures the double-stranded RNAs into single strands of 21 to 25 nucleotides. The RNA single strands produced by RISC are identified as the **guide strand,** which is biologically active; and the passenger strand, which is usually degraded. The guide strand remains bound to RISC, and the complex directs one of three gene-silencing processes (numbers 1 through 3 in the figure): ❶ The complex uses complementary base pairing to attach the guide strand to mRNA, and the mRNA is destroyed; ❷ the RISC–guide RNA binds to complementary mRNAs and blocks their translation; or ❸ the complex directs chromatin-modifying enzymes to the nucleus, where they silence transcription of selected genes.

What is the origin of the dsRNA? It can be produced from endogenous genes, the transcription of other endogenous sequences, or it can come from exogenous sources. In many eukaryotes, genes encode precursors of dsRNA that are processed into 21- to 24-nt **microRNAs (miRNAs)** at a Dicer complex. Most genes encoding miRNAs are transcribed by RNA polymerase II, and the resulting transcript folds back on itself into a dsRNA. The targets of miRNAs are mRNAs that are either destroyed or have their translation blocked subsequent to activity mediated through RISC.

In contrast to miRNAs, **small interfering RNAs (siRNAs)** are usually not derived from genes, but rather are complementary RNAs that come from exogenous sources or from other endogenous transcription. For example, if both strands of a genomic region happen to be transcribed, dsRNA can form. Transcription from opposite strands of repetitive elements, such as transposons, can also lead to dsRNA production. In the latter case, the two strands do not have to be derived from the same genomic location. Some eukaryotes possess RNA-dependent RNA polymerases, which can produce dsRNA using single-stranded RNA as a template. The exogenous sources of dsRNAs can direct either posttranscriptional silencing through the destruction of target mRNAs, or via transcriptional silencing of target genes that takes place by chromatin modifying processes. Finally, exogenous sources of dsRNA, such as those produced by some RNA viruses, can trigger virus-induced gene silencing.

**Cleaving dsRNA** Dicer activity cleaves dsRNA molecules into 21- to 25-bp fragments. The general mechanism of action for dsRNA cleavage was identified in 2006 when Jennifer Doudna and her colleagues determined the crystal structure of Dicer in the intestinal parasite *Giardia intestinalis*. Doudna's research group used the crystal structure to determine that the dsRNA binding site on Dicer, called PAZ, is separated by 65-Å from the sites of two RNase domains that cut the RNA. The 65-Å space between PAZ and the RNase domains corresponds to the 24-bp length of dsRNA (Figure 15.24). Dicer repeats this action, each time

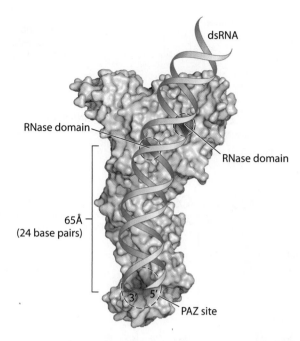

**Figure 15.24  Dicer structure and interaction with dsRNA.** The distance between the PAZ binding site and the location of RNases determines the length of siRNA.

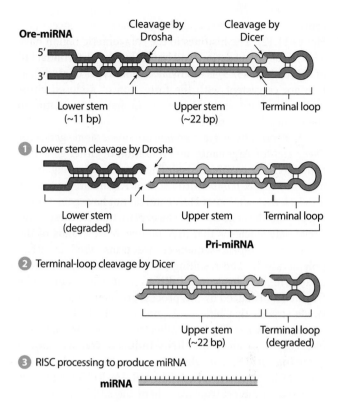

**Figure 15.25  Stepwise Drosha processing of pri-miRNA to produce miRNA.**

acting as a molecular ruler measuring off precisely sized dsRNAs. The spacing between the PAZ site and RNase domains vary among species and appear to correlate with difference in the lengths of siRNAs produced by subsequent RISC processing of dsRNAs in each species.

The creation of miRNA is similar to the creation of siRNAs. Precursor transcripts of miRNAs are synthesized in the nucleus of a cell and are processed into miRNAs by Dicer activity. The transcript that will form miRNA is called **primary microRNA (pri-miRNA).** The pri-miRNA folds to form a double-stranded stem typically containing 65 to 70 nucleotides and having free ends on one side and a single-stranded loop on the other side (**Figure 15.25**). In animals, the **Drosha** enzyme complex cuts pri-miRNA near the middle of the stem and produces two segments, one of which, now called **precursor microRNA (pre-miRNA),** contains the remainder of the upper stem, which is approximately 21 to 25 bp, and the terminal loop. The pre-miRNA is transported to the cytoplasm, where Dicer removes the terminal loop, leaving dsRNA of approximately 21 to 25 bp. RISC then binds the dsRNA and separates the strands to create miRNAs. In contrast to animals, plants use a single Dicer enzyme to perform all the miRNA processing activities.

**RISC and Argonaute**    After production, both siRNA and miRNA are bound by RISC, a multiprotein complex within which a protein of the **Argonaute** gene family plays a central role in how the RISC–guide strand silences gene expression. Many species encode multiple Argonaute

proteins—humans encode eight, for example—and each seems to direct a somewhat different activity by RISC–guide strand.

The best-understood mechanism of gene silencing by RISC–guide strand involves complementary binding of the guide strand to a target mRNA. If the percentage of base-pair complementation is high enough, this binding forms a structure that allows an RNase domain of Argonaute to cut the targeted mRNA strand near the middle of the guide strand–mRNA duplex, thus causing cleavage of the mRNA. When the guide strand–mRNA base pairing is less well matched—that is, when only a core of complementary base pairs are present in the guide strand–mRNA duplex—the RNase domain of Argonaute is unable to cut the duplex. Instead, the duplex retains its double-stranded form, causing translation to be blocked.

## Chromatin Modification by RNAi

The third mechanism by which the RISC–guide strand complex silences gene expression is through chromatin modification, and it is best studied in yeast (*Schizosaccharomyces pombe*). The first evidence of a role for RNAi in chromatin modification came from the study of centromeric heterochromatin in *S. pombe*. The centromeres of *S. pombe*, like those of other complex eukaryotes, contain a central element surrounded by repeat sequences (see Figure 11.12). The histones in the centromeric region have a low level of acetylation, and lysine 9 of the N-terminal tail of H3 (that is, H3K9) is methylated. Both types of modification are consistent with the formation of a closed chromatin structure and the spread of heterochromatin to silence nearby genes.

*Schizosaccharomyces pombe* possesses single genes for Dicer and for Argonaute, and mutation of either gene disrupts RNAi activity in the cell. The surprising finding, however, was that *S. pombe* with Dicer or Argonaute mutations also lack methylation of H3K9 and do not have gene silencing around the centromere. The explanation for these additional deficiencies is that in *S. pombe*, both strands of the centromeric repeat sequences are transcribed by RNA polymerase II. The resulting mRNAs are complementary and form double-stranded RNAs that Dicer cuts. The fragments produced by this process are then separated into single strands that bind to Argonaute, which then joins a protein known as Chp1 and other proteins to form a RISC-like complex called the **RNA-induced transcriptional silencing (RITS) complex** (Figure 15.26), which carries the siRNA into the nucleus. The siRNA–RITS complex is attracted to the centromere, where the siRNA appears to use complementary base pairing to form a duplex with transcripts of the centromeric repeat sequences. This pairing attracts other proteins that promote the deacetylation of histones and the methylation of H3K9 to close the chromatin structure and spread heterochromatin outward from the centromere.

## The Evolution of RNAi

RNAi is widespread in eukaryotes, and the mechanism of transcriptional silencing in *S. pombe* is thought to be related to RNAi-mediated transcriptional silencing in other eukaryotic species. But how did RNAi evolve? The answer is still under investigation, but the operating hypothesis is that RNAi evolved by helping organisms protect their genomes against the mutational effects of transposable genetic elements (described in Chapter 13).

Transposable elements are diverse but make up large percentages of the genomes of complex eukaryotes. For example, almost half the human genome is composed of transposable elements. In the human genome and in other eukaryotic genomes, most of these transposons are located in heterochromatin and are silent. Researchers have discovered that mutations in the RNAi machinery of an organism can reactivate normally quiescent transposons by reversing transcriptional silencing. This can lead to the movement of some transposable elements around the genome and potentially to the production of new mutations. The evidence suggests that RNAi plays a role in silencing the transcription of transposons. RNAi also plays a protective role in response to viral infection. In plants, the infection of one

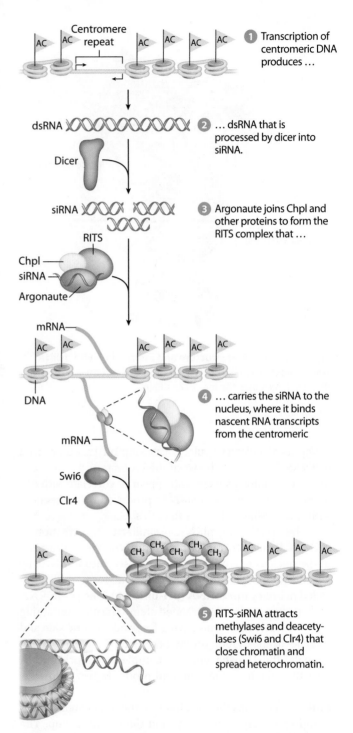

**Figure 15.26  RNA-induced transcriptional silencing (RITS) in yeast.**

leaf by a virus can generate an RNAi response that blocks viral replication and prevents the infection from spreading throughout the plant. In support of this observation, plants with Dicer or Argonaute mutations are much more susceptible to the spread of viral infections than are plants without Dicer or Argonaute mutations. These findings are consistent with the idea that RNAi evolved as a genome-protection mechanism against transposable genetic elements and viral infection. To return to Jorgensen's petunias and their cosuppression for a moment, biologists now know that RNAi is responsible for blocking expression of the chromosomal pigment-producing gene as well as the introduced copy of the pigment-producing gene.

Both plants and animal genomes encode miRNAs, but the mode of action of miRNAs differs slightly between the two taxa. In plants, miRNAs display near-complete sequence complementarity with their mRNA targets and usually cleave the target and, less frequently, repress translation. In contrast, miRNAs in animals are usually only complementary to their targets at one end of the miRNA, and usually repress translation rather than cleave the target. These differences suggest that miRNAs may have evolved independently in the two lineages.

The study of RNAi is emerging as a powerful research tool that can be used in a multitude of ways. One frequent application of RNAi in research is the use of interfering RNAs to "knock out" the expression of selected genes. This is a way of discovering the gene's effect on the phenotype by examining how phenotype is altered in the absence of expression of the gene. A second approach to the use of RNAi is in medicine, where biomedical researchers are exploring the possible uses of RNAi to control the expression of genes that produce too much transcript or produce abnormal transcripts in disease. In certain cancers, for example, the disease process is driven in part by overexpression of certain genes. RNAi therapy would involve designing and constructing small RNA molecules that specifically bind and block the translation of the transcripts of disease-causing genes while not affecting the transcripts of other genes. (We discuss other experimental applications of RNAi in Chapter 16).

## X-Inactivation and Regulatory RNA

Lastly, we take up a different kind of RNA-mediated mechanism of gene silencing—this one occurs exclusively in mammalian females to balance the dosage of X-linked genes. To achieve the correct balance of X-linked gene expression between mammalian males and females, the dosage compensation mechanism known as X-inactivation randomly inactivates one of the X chromosomes in the nucleus of each cell in a female (see Figure 11.18). The X-inactivation center (XIC) is the source of X-inactivation-specific transcript (*Xist*) RNA that spreads along the length of the inactive X chromosome to silence virtually all its genes (see Chapter 11). X-inactivation is an epigenetic phenomenon in which *Xist* RNA induces formation of a closed chromatin structure by attracting deacetylase and methylase proteins that actively deacetylate and methylate amino acids in the N-terminal tails of histone proteins. These events cause the entire inactivated X chromosome to form heterochromatin that appears in the nucleus as a Barr body. The available evidence indicates that transcription and spreading of *Xist* RNA is essential for establishing X-inactivation. Other events maintain the epigenetic modifications.

X-inactivation is reversible in mammalian female germ-line cells. The reversal events remodel closed chromatin to a more open form that restores euchromatic regions in the chromosome. Demethylation and acetylation accompany this reversal process.

## CASE STUDY

### Environmental Epigenetics

Here's a simple question: How are traits passed from one generation to the next? The first answer that came to your mind was probably (and not incorrectly) that traits are passed by the transmission of genes from parents to offspring. But over the past decade or so, the answer to that question has expanded in an unexpected direction. Emerging evidence suggests that in certain cases, parental nutrition and diet may lead to epigenetically controlled modifications of gene expression and that in a few select instances, the affected genes can be transmitted to offspring in their epigenetically modified form. More surprisingly, the data also indicate that the epigenetically modified state of the genes may persist in later generations. In other words, it may be possible for the nutritional experience of grandparents to affect gene expression in their grandchildren!

Three lines of evidence suggest a role for nutrition and dietary history in the epigenetic modification of gene expression. The first comes from studies in honeybees, where it has been shown that genetically identical larvae can develop into either fertile queens or sterile worker bees

following differential feeding with royal jelly, the compound fed to larvae that become queens. Experimental analysis led by Ryszard Maleszka in 2008 reveals that silencing the expression of the DNA methyltransferase Dnmt3 by knocking out translation of the *Dnmt3* transcript by RNA interference leads to the development of fertile queens. In other words, blocking a major histone methylation pathway led to the expression of genes that are typically expressed only when a larva is fed royal jelly. The implication is that methylation is an important epigenetic mechanism for repressing gene expression and directing the development of worker bees. Methylation and the resulting transcriptional repression are subverted by feeding royal jelly to produce the development of fertile queen bees.

The second line of evidence comes from multiple studies of the connection between environmentally generated methylation of genes and variation in gene expression in rats and mice. In one study, genetically identical mice carry a modified agouti gene that produces yellow coat color and extreme obesity when the gene is expressed, whereas the normal brown coat color and normal body weight are produced if the modified gene is not expressed. The coat color and body weight of genetically identical mouse pups carrying this modified gene are determined by the diet of the mother in the weeks before impregnation and during pregnancy and lactation.

In controlled experiments, mothers that will transmit the modified agouti gene to their pups are fed either a diet enriched with three compounds that each act as donors of methyl groups to DNA—folic acid (vitamin B$_{12}$), choline chloride, and anhydrous betaine—or a diet without these compounds. The controlled dietary period begins 2 weeks before mating and continues through pregnancy and lactation. The pups produced are genetically identical, and after they are weaned, they are all fed the same diet. At 3 weeks of age, however, the appearance of the pups is dramatically different. Mice produced by mothers who were fed the enriched diet have brown coat color and normal body weight, whereas genetically identical mice produced by mothers not fed the enriched diet have yellow coat color and are obese. The difference indicates that the modified agouti gene is expressed when it is transmitted from mothers that were not fed the diet enriched with methyl donors. If the modified gene is transmitted from mothers receiving the enriched diet, however, the modified agouti gene is methylated and silenced.

The third line of evidence comes from an unfortunate event during World War II. A severe famine occurred in German-occupied Netherlands between November 1944 and May 1945. The famine reduced daily caloric intake to 500 to 800 calories per day, much less than the body needs to fuel its normal metabolic activities. Long-term studies have been performed on Dutch people who were conceived or born during the famine and on their descendants. Studies of the health effects of the famine find that so-called famine babies were often born severely underweight. As the famine babies grew into adults and aged, they suffered increased risk of cardiovascular disease, diabetes, and obesity compared to peers who had not been affected by the famine. The proposed explanation is that the restricted nutritional conditions in the womb caused alterations of gene expression, producing an energetically "thrifty" metabolism. More surprising, however, was that among the children of the famine babies, there is also an elevated risk of cardiovascular and other diseases. The explanation proposed for this second-generation effect is epigenetic modification of gene expression that is transmitted through multiple generations.

A 2008 study by Bastiaan Heijmans on the methylation pattern of the *IGF2* gene on chromosome 15 confirms the epigenetic control mechanism that we discussed previously in connection with genomic imprinting, Prader-Willi syndrome, and Angelman syndrome. Heijmans and colleagues found that *IGF2* in certain famine babies (now in their sixties) still bears the marks of famine. The *IGF2* genes of those exposed to famine during the first 10 weeks of gestation are marked by significantly fewer methyl groups than are the genes of their same-sex siblings not exposed to famine conditions. These results support the idea that prenatal conditions can impart specific epigenetic patterns to genes, and that environmental factors contributing to epigenetic patterns may play an important role in modifying gene expression over multiple generations.

---

SUMMARY                    *For activities, animations, and review quizzes, go to the study area at www.masteringgenetics.com.*

## 15.1 Cis-Acting Regulatory Sequences Bind Trans-Acting Regulatory Proteins to Control Eukaryotic Transcription

- Regulatory proteins in eukaryotes bind to specific nucleotides exposed in major and minor grooves of DNA.

- Promoters, proximal elements, and enhancers are cis-acting DNA sequences that bind trans-acting regulatory proteins to regulate transcription.

- Enhancer sequences are strongly conserved, indicating they perform essential functions.

- Upstream activator sequences (UAS) in yeast are enhancer-like elements that regulate the expression of genes such as those involved in galactose utilization.

- Locus control regions (LCRs) are specialized enhancers that control the sequential expression of sets of genes such as those in the developmentally regulated human β-globin gene complex.

- Silencer sequences bind repressor proteins to block transcription of targeted genes.

- Insulators block enhancer influence on certain genes and direct that influence to other genes.

## 15.2 Chromatin Remodeling Regulates Eukaryotic Transcription

- Open promoters are constitutively transcribed, whereas transcription from covered promoters is regulated.

- In regions of closed chromatin structure, the DNA is wound tightly around nucleosomes. These regions are transcriptionally silent.

- In regions of open chromatin structure, the association of DNA and nucleosomes is looser, allowing genes to be expressed.

- Chromatin-remodeling complexes displace nucleosomes to allow transcription initiation by RNA pol II and general transcription factors.

- Chromatin modification by the addition and removal of acetyl, methyl, and phosphate groups at specific amino acids in the N-terminal tails of the histone proteins opens and closes chromatin.

- Epigenetic states of chromatin are heritable in somatic cells that divide by mitosis and may be reset in germ-line cells that divide by meiosis.

▌ Genomic imprinting in mammalian genomes involves nucleotide methylation and the action of enhancer and insulator sequences.

▌ Methylation of cytosine nucleotide bases that are part of CpG dinucleotides in CpG islands is a prominent mechanism for silencing mammalian promoters.

## 15.3 RNA-Mediated Mechanisms Control Gene Expression

▌ RNA interference (RNAi) is an RNA-mediated mechanism for regulating gene expression in eukaryotes.

▌ Small interfering RNAs (siRNAs) and microRNAs (miRNAs) are principal regulatory RNA molecules.

▌ The Dicer protein complex processes dsRNAs into their regulatory form.

▌ The RISC complex carries regulatory RNAs to RNAs targeted for destruction or for blockage of translation.

▌ A specific form of regulatory RNA directs mammalian X-inactivation.

## KEYWORDS

Argonaute *(p. 514)*
chromatin modifier *(p. 506)*
chromatin remodeler (SWI/SNF, ISWI, SWR1) *(pp. 504, 506)*
chromatin remodeling *(p. 504)*
cis-acting regulatory sequence *(p. 497)*
closed chromatin *(p. 504)*
cosuppression *(p. 512)*
covered promoter *(p. 503)*
CpG dinucleotide (CpG island) *(p. 511)*
Dicer *(p. 514)*
DNase I hypersensitive site *(p. 504)*
enhanceosome *(p. 497)*
enhancer (enhancer sequence) *(p. 497)*
epigenetic modification *(p. 509)*

functional domain *(p. 495)*
genomic imprinting *(p. 510)*
guide strand *(p. 514)*
histone acetyltransferase (HAT) *(p. 507)*
histone deacetylase (HDAC) *(p. 507)*
histone demethylase (HDMT) *(p. 508)*
histone methyltransferase (HMT) *(p. 507)*
imprinting control region (ICR) *(p. 510)*
insulator sequence *(p. 501)*
ISWI complex *(p. 506)*
locus control region (LCR) *(p. 499)*
mediator *(p. 499)*
microRNA (miRNA) *(p. 514)*
nucleosome-depleted region (NDR) *(p. 503)*

open chromatin *(p. 504)*
open promoter *(p. 503)*
RNA-induced silencing complex (RISC) *(p. 514)*
RNA-induced transcriptional silencing (RITS) complex *(p. 515)*
RNA interference (RNAi) *(p. 512)*
silencer sequence *(p. 500)*
small interfering RNA (siRNA) *(p. 514)*
structural motif *(p. 495)*
SWI/SNF complex *(p. 506)*
SWR1 complex *(p. 506)*
trans-acting regulatory protein *(p. 497)*
upstream activator sequence (UAS) *(p. 498)*

## PROBLEMS

*For instructor-assigned tutorials and problems, go to www.masteringgenetics.com.*

### Chapter Concepts

*For answers to selected even-numbered problems, see Appendix: Answers.*

1. Devoting a few sentences to each, describe the following structures or complexes and their effects on eukaryotic gene expression:
   a. promoter
   b. enhancer
   c. silencer
   d. RISC
   e. Dicer

2. Describe and give an example (real or hypothetical) of each of the following:
   a. upstream activator sequence (UAS)
   b. insulator sequence action
   c. silencer sequence action
   d. enhanceosome action
   e. RNA interference

3. What is meant by the term *chromatin remodeling*? Describe the importance of this process to transcription.

4. What general role does acetylation of histone protein amino acids play in the transcription of eukaryotic genes?

5. What roles does methylation play in transcription? Compare and contrast the relative importance of methylation to transcription in a mouse and in the nematode worm *C. elegans*.

6. Why do DNA-binding proteins share common three-dimensional motifs? How do these common motifs allow proteins to locate specific DNA consensus sequences?

7. A CpG island is a region of DNA that contains many CpG dinucleotides. CpG islands are often found near the 5′ ends of promoters in mammalian genomes. Speculate about the likely role CpG islands play in these locations.

8. Most biologists argue that the regulation of gene expression is considerably more complex in eukaryotes than in bacteria. List and describe the four factors that in your view make the largest contribution to this perception.

9. Compare and contrast the transcriptional regulation of *GAL* genes in yeast with that of the *lac* genes in bacteria.

10. The term *heterochromatin* refers to heavily condensed regions of chromosomes that are largely devoid of genes.

Since few genes exist in heterochromatic regions, they almost never decondense for transcription. At what point during the cell cycle would you expect to observe the decondensation of heterochromatic regions? Why?

## Application and Integration

*For answers to selected even-numbered problems, see Appendix: Answers.*

12. The consequences of four deletions from the region upstream of the yeast gene *DBM1* are studied to determine the effect on transcription. The normal rate of transcription, determined from study of transcription of genes that do not have upstream deletions, is defined as 100%. The location of each deletion and the effects of deletions on *DBM1* transcription are shown below.

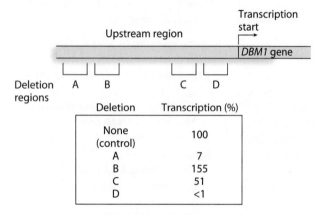

| Deletion | Transcription (%) |
|---|---|
| None (control) | 100 |
| A | 7 |
| B | 155 |
| C | 51 |
| D | <1 |

a. Which mutations(s) affect an enhancer sequence? Explain your reasoning.
b. Which mutation(s) affect a silencer sequence? Explain your reasoning.
c. Which mutation(s) affect the promoter? Explain your reasoning.

13. The *UG4* gene is expressed in stem tissue and leaf tissue of the plant *Arabidopsis thaliana*. To study mechanisms regulating *UG4* expression, six small deletions of DNA sequence upstream of the gene-coding sequence are made. The locations of deletions and their effect on *UG4* expression are shown below.

a. Explain the differential effects of deletions B and F on expression in the two tissues.
b. Why does deletion D raise *UG4* expression in leaf tissue but not in stem tissue?

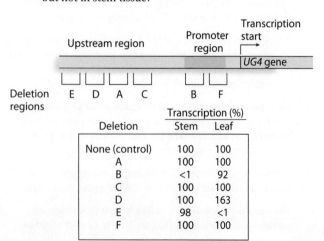

| Deletion | Transcription (%) Stem | Transcription (%) Leaf |
|---|---|---|
| None (control) | 100 | 100 |
| A | 100 | 100 |
| B | <1 | 92 |
| C | 100 | 100 |
| D | 100 | 163 |
| E | 98 | <1 |
| F | 100 | 100 |

c. Why does deletion E lower expression of *UG4* in leaf tissue but not in stem tissue?

14. A gene expressed in long muscle of the mouse is identified, and the regulatory region upstream of the gene is isolated. Various segments of the upstream sequence are fused to the *lacZ* gene, and each fusion is assayed to determine how efficiently it transcribes the gene. In the accompanying diagram, the dark bars indicate the upstream segments that are present in each of six different fusion genes. The transcriptional efficiency of each fusion is measured against the control fusion, that is, the full-length upstream segment fused to the *lacZ* gene.

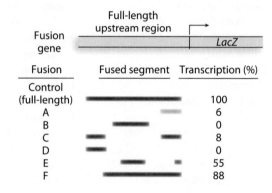

| Fusion | Fused segment | Transcription (%) |
|---|---|---|
| Control (full-length) | | 100 |
| A | | 6 |
| B | | 0 |
| C | | 8 |
| D | | 0 |
| E | | 55 |
| F | | 88 |

a. Identify the upstream region that contains the enhancer.
b. Identify the upstream region containing the promoter.
c. Speculate about the reason for the different transcription rates detected in fusions E and F.

15. A hereditary disease is inherited as an autosomal recessive trait. The wild-type allele of the disease gene produces a mature mRNA that is 1250 nucleotides (nt) long. Molecular analysis shows that the mRNA consists of a 5′ UTR (untranslated region) of 150 nt; a 3′ UTR of 100 nt; and four exons that measure 250 nt (exon 1), 320 nt (exon 2), 230 nt (exon 3), and 200 nt (exon 4). A mother and father with two healthy children and two children with the disease have northern blot analysis performed in a medical genetics laboratory. The results of the northern blot for each family member are shown below.

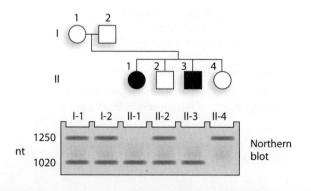

a. Identify the genotype of each family member using the sizes of mRNAs to indicate each allele. (For example, a person who is homozygous wild type is indicated as "1250/1250.")

b. Based on your analysis, what is the most likely molecular abnormality causing the disease allele?

16. The region of human DNA illustrated in the diagram below contains three restriction sequences (5'-CCGG-3') for the restriction enzymes *Hpa*II and *Msp*I. (Recall that *Hpa*II is methylation-sensitive, but that *Msp*I is not). The illustration identifies the distance between the restriction sequences and indicates the binding location of a molecular probe.

```
      ┌──── 1.7 kb ────┐┌──── 2.3 kb ────┐
5'   CCGG            CCGG            CCGG   3'
3'   GGCC            GGCC            GGCC   5'
              Molecular probe
```

a. If methylation occurs at any of the restriction sequences, which nucleotide is most likely to acquire a methyl group? Why?

b. Draw two Southern blots, each containing one lane with *Hpa*II-digested DNA and a second lane with *Msp*I-digested DNA. In Southern blot A, no methylation of the restriction sequences has occurred. In Southern blot B, the middle restriction sequence is methylated, but the other two sequences are not.

17. A muscle protein in mouse is produced through the use of alternative promoters in heart and skeletal muscle. A diagram of the gene region is shown below. The gene contains a total of six exons, and there are three restriction sites recognized by the restriction enzyme *Hin*dIII in the vicinity of the gene. In the diagram, the locations of heart ($P_H$) and skeletal ($P_S$) promoters are indicated, as are two molecular probes. Probe *a* hybridizes to exon 2, and probe *b* hybridizes to exon 4. Transcription of the gene in heart and skeletal muscle terminates after exon 6. The protein produced by the gene is recognized in heart and muscle samples by the same antibody.

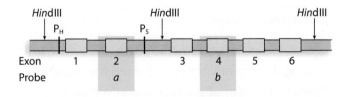

Diagram the expected results of *Hin*dIII digestion of DNA followed by the use of probes *a* and *b* in Southern blots. Messenger RNA and probes *a* and *b* are used for northern blots. The protein is probed with the antibody in western blots.

a. Southern blot analysis of DNA from heart muscle and skeletal muscle

b. Northern blot analysis of mature mRNA extracted from heart muscle and skeletal muscle

c. Western blot analysis of the protein from heart muscle and skeletal muscle

18. A muscle enzyme called ME1 is produced by transcription and translation of the *ME1* gene in several muscles during mouse development, including heart muscle, in a highly regulated manner. Production of ME1 appears to be turned on and turned off at different times during development. To test the possible role of enhancers and silencers in *ME1* transcription, a biologist creates a recombinant genetic system that fuses the *ME1* promoter, along with DNA that is upstream of the promoter, to the bacterial *lacZ* (β-galactosidase) gene. The *lacZ* gene is chosen for the ease and simplicity of assaying production of the encoded enzyme. The diagram shows the structure of the recombinant, as well as bars that indicate the extent of six deletions the biologist makes to the *ME1* promoter and upstream sequences. The table displays the percentage of β-galactosidase activity in each deletion mutant in comparison to the recombinant gene system without any deletions.

| Deletion | *LacZ* activity (%) |
|---|---|
| None (control) | 100 |
| A | 100 |
| B | 100 |
| C | 4 |
| D | <1 |
| E | 170 |
| F | 5 |

a. Does this information indicate the presence of enhancer and/or silencer sequences in the *ME1* upstream sequence? If so, where is/are the sequences located?

b. Why does deletion D effectively eliminate transcription of *lacZ*?

c. Given the information available from deletion analysis, can you give a molecular explanation for the observation that *ME1* expression appears to turn on and turn off at various times during normal mouse development?

# 16

# Forward Genetics and Recombinant DNA Technology

Thomas Hunt Morgan's (far right back row) fly room was the site of the original mutagenesis. The first screens relied on spontaneous mutants, but with the discovery by Hermann Muller (second from right back row) that X-rays are mutagenic, genetic screens became routine and powerful tools to uncover gene function. Also visible in this photo are Calvin Bridges (third from left back row) who used nondisjunction to prove the chromosome theory of heredity, and Alfred Sturtevant (middle front row) who constructed the first genetic map.

## ESSENTIAL IDEAS

▪ Forward genetic screens induce mutations to identify genes involved in a biological process; subsequent cloning sheds light on their molecular function.

▪ DNA can be amplified by either molecular cloning or the polymerase chain reaction.

▪ In molecular cloning, DNA fragments are ligated into a cloning vector, which in turn is replicated in a live host.

▪ Libraries are collections of clones of DNA fragments, derived from the DNA or mRNA isolated from cells or an organism.

▪ The precise sequence of DNA can be determined in an automated process.

▪ DNA sequences of specific genes can be discovered using recombinant DNA technology.

A central goal of biology is to understand the molecular and genetic bases of physiology and development. Beginning with Mendel and resuming in the first part of the 20th century, geneticists attempted to dissect the rules of heredity by connecting phenotypes to genetic loci. The discovery of DNA as the hereditary material indicated that genes are specific DNA sequences and that allelic differences reflect differences in those sequences. In the 1970s, discoveries stemming from the study of bacteria and their phages led to the development of tools to manipulate DNA in vitro. With these tools, collectively referred to as recombinant DNA technology, geneticists could for the first time obtain the precise DNA sequences of specific genes and alleles, thus identifying the molecular basis of phenotypic differences.

In this chapter we illustrate how a gene's existence can be inferred from mutant phenotypes and how the DNA sequences and functions of specific genes can be revealed using techniques of recombinant DNA technology. These discussions provide a foundation for Chapters 17 and 18, where we explore applications of recombinant DNA technology in the dissection of gene function and sequencing of genomes.

The exploration of how genes control physiological and developmental processes is approached in two ways that attack the problem from diametrically opposite directions. These opposite approaches are known as *forward genetic analysis* and *reverse genetic analysis*. The goals of forward and reverse analysis are the same: to identify the genes responsible for hereditary variation, to determine the structure and function of wild-type alleles controlling traits, and to describe how mutant alleles generate abnormal phenotypes. However, the two strategies begin at different ends of the process of gene identification.

**Forward genetic analysis** starts with a **genetic screen** that identifies specific phenotypic abnormalities in a population of organisms that have been mutagenized. The abnormal phenotype is then studied to identify the nature of the hereditary abnormality and, by inference, the normal functions of an associated gene. Ultimately, the sequence of the gene responsible for the abnormality is determined and may suggest the molecular function of the corresponding gene product (**Figure 16.1a**). In this chapter we discuss forward genetic screens and how recombinant DNA technology is used to identify the DNA sequence of specific genes. In contrast to forward genetic analysis, **reverse genetic analysis** begins with the sequence of a gene of unknown function that is then connected with a mutant phenotype (**Figure 16.1b**). This strategy is more fully explored in Chapter 17.

## 16.1 Forward Genetic Screens Identify Genes by Their Mutant Phenotypes

With the discovery by Hermann Muller that ionizing radiation induces mutations (see Section 12.3), geneticists realized that mutant organisms could be generated at will and systematically screened for phenotypes of interest. Mutant phenotypes provide information on the function of the wild-type allele and insight into biological processes. In an early example, in 1908 Archibald Garrod connected a hereditary condition, alkaptonuria, with the lack of a specific biochemical activity, the metabolism of benzene rings in homogentisic acid. He suggested that the wild-type version of the gene encodes the enzyme responsible for this biochemical activity. With the ability to generate mutations at will, geneticists began to employ systematic genetic screens to dissect other biological processes, and the genetic bases for entire biochemical pathways were elucidated.

The design of genetic screens to identify genes involved in specific biological processes is limited only by the imagination of the geneticist. An example is the research by Seymour Benzer that led to the field of behavioral genetics in the 1970s. Benzer believed mutations could be identified that specifically affect behavioral processes, such as one you are using now, the process of learning and memory. At the time, behavior was thought by many to be too complex to be dissected genetically. However, Chip Quinn, a graduate student in Benzer's lab, built on previous ideas and designed an ingenious screen to identify learning- and memory-deficient mutants in *Drosophila*. Wild-type flies could be taught that a pulse of odor would be followed by a shock; later, when the flies smelled the odor, they would take evasive action. When Quinn and Benzer subjected a mutagenized population of *Drosophila* to this genetic screen, they identified mutant strains of flies that could perceive the odor but seemed unable to associate the odor with the stimulus; either they did not learn or could not remember.

Two mutant genes identified in the study, *dunce* and *rutabaga*, were later shown to encode proteins involved in the production or degradation of the small signaling molecule cyclic adenosine monophosphate (cAMP). At the time, signaling via a cAMP pathway was known to be required for learning in the sea hare, *Aplysia*. Since both *Drosophila* mutants were defective in cAMP physiology, other genes that encoded proteins involved in cAMP signaling and response were also investigated for roles in learning. Ultimately, a transcription factor called *creb* (cAMP-*response* *element* *binding* protein), which activates or represses genes in response to cAMP signaling, was shown to be critical for storing memories in flies. Remarkably, *creb* is widely conserved in animal species, and mouse mutants lacking *creb* activity also fail to remember. A similar gene is found in our genome.

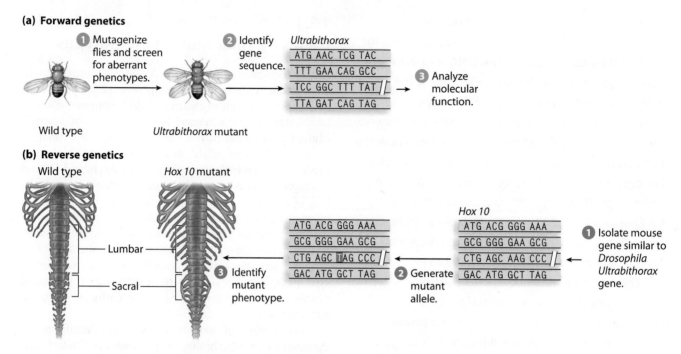

**Figure 16.1** **General strategies of forward and reverse genetics.**

A great strength of forward genetic screens is that they are unbiased; no prior knowledge of the molecular function of the encoded gene product is required. In a sense, by performing a mutagenesis, the geneticist is allowing the organism to reveal how its biological processes operate. Once genes in particular physiological or developmental processes have been identified by mutation, clues to the molecular function of the gene product can be obtained using recombinant DNA technology.

## Designing a Genetic Screen

Forward genetic screens begin with a **mutagenesis:** An organism is treated with a mutagen to create mutations randomly throughout the genome. A typical goal is to induce mutations in every gene in a population of mutagenized individuals, by an approach called **saturation mutagenesis.** The mutagenized population is then screened for phenotypic defects in whatever biological process is being studied, and the mutants are collected and propagated for further analysis. Strategies for mutagenesis depend on the biological process of interest, which dictates the experimental organism to use, the choice of mutagen, and the screening procedure to identify mutations.

**Choosing an Organism** The attributes that make an organism a good genetic model also make it a good choice for a mutagenesis experiment (see endsheets): An organism must be able to progress through its entire life cycle in the laboratory, have a short generation time (the time it takes to produce sexually mature progeny and complete the sexual life cycle), produce a reasonable number of

progeny, and be amenable to crossing. Organisms that are diploid must have a starting genotype (the genotype to be mutagenized) that is inbred—in other words, homozygous at all loci. This genotype allows newly induced mutations to be readily identified, without interference from the confounding effects of polymorphisms. Finally, it is advantageous to use the simplest organism possible for the biological process under study. Because *Saccharomyces cerevisiae* has a rapid life cycle and is easily manipulated in the laboratory, it is often used to investigate biological processes common to all eukaryotes. The principles elucidated in *S. cerevisiae* can often be extended to other eukaryotes, including humans.

**Choosing a Mutagen** The choice of mutagen is dictated by both the organism and the type of mutant alleles desired; different mutagens have different advantages and disadvantages (**Table 16.1**). Mutagens inducing different types of changes in DNA sequences were described in Section 12.4. Treatment with chemical mutagens can induce hundreds of mutations in a single individual, allowing saturation to be reached with only a few thousand mutagenized individuals. However, cloning genes identified by chemical mutagenesis may be laborious. In contrast, mutagens that result specifically in insertions of DNA, such as transposons, result in far fewer mutations per individual, making saturation difficult. But these mutagens have the advantage of being able to provide a DNA "tag" that facilitates finding and cloning the mutated genes. In all mutageneses, care must be taken to outbreed mutants of interest by crossing them with the wild-type progenitor strain, thus ensuring that the collected mutant

| Table 16.1 | Common Mutagens Used for Mutagenesis | | |
|---|---|---|---|
| Mutagen | Mutation Spectrum | Per Locus Mutation Rate | Allele Spectrum |
| *Chemical*<br>EMS | GC→AT conversion<br>Stop codons created: TGG → TAG<br>Splice sites destroyed: AG → AA | High | Null, hypomorph, hypermorph |
| *Radiation*<br>Fast neutron<br><br>X-ray<br>Gamma ray | Rearrangements (deletions, inversions, translocations) | Moderate | Usually loss-of-function (often null), but can be gain-of-function |
| *Insertional*<br>Transfer DNA<br>Transposons | Insertions | Low | Usually loss-of-function (often null) |

lines have only the mutation of interest and not others that were also induced during the mutagenesis.

**Strategy for Identifying Dominant and Recessive Mutations** The overall goal of mutagenesis is to identify multiple independent mutant alleles of each gene involved in the biological process of interest. Let us consider the identification of dominant and recessive mutations in a typical animal example. *Drosophila*, like most animals, spend most of their life cycle in the diploid state. Their germ cells are set aside early in development and do not contribute to

the somatic development of the remainder of the animal body. When *Drosophila* are treated with a mutagen—for example, by feeding males ethyl methanesulfonate (EMS) (see Table 16.1)—only the mutations induced in the germ cells are heritable and will be passed to the progeny of the mutagenized flies.

Newly induced dominant mutations can be identified in the $F_1$ generation that is produced by breeding the mutagenized males with wild-type females (Figure 16.2a). However, only a small fraction of the $F_1$ progeny will exhibit a mutant phenotype, since dominant mutations

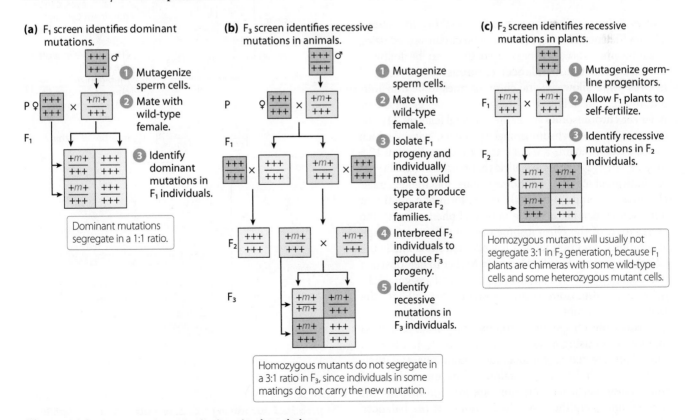

**Figure 16.2 Mutagenesis strategies in animals and plants.**

are rare. This rarity is due to the low probability that any change in the DNA sequence of a gene will produce a gain in function for the encoded gene product, either qualitatively or quantitatively.

Mutations that result in a loss of function are more common, but loss-of-function mutations are usually recessive and do not result in an observable phenotype in the $F_1$ generation. Therefore, further breeding must be performed to produce homozygous loss-of-function mutants. In *Drosophila*, recessive mutations are identified in an $F_3$ screen (**Figure 16.2b**). In this screen, $F_1$ individuals derived from the mating of mutagenized males with wild-type females are then crossed with wild-type females, producing an $F_2$ generation. The $F_2$ siblings are interbred, producing an $F_3$ population segregating for individuals that are homozygous for the induced mutation. The interbreeding of the $F_2$ to produce homozygous mutant $F_3$ is inefficient, since only half of the $F_2$ are heterozygous for the induced mutation. Nonetheless, such mutagenesis strategies are employed with many species, such as mice and zebrafish.

Identification of recessive mutations is somewhat simpler in organisms that self-fertilize, such as *Caenorhabditis elegans* and many plants (e.g., *Arabidopsis* and maize). In these organisms, $F_1$ individuals are self-fertilized to produce an $F_2$ generation from which recessive mutations can be identified. An example of an $F_2$ screen is shown in **Figure 16.2c**. In either an $F_2$ or $F_3$ screen, mutations resulting in homozygous lethality can be maintained in heterozygous siblings.

## Use of Balancer Chromosomes for Tracking Mutations

The inefficiency of an $F_3$ screen can be circumvented using chromosomes that are marked so they can be followed through generations. **Balancer chromosomes** developed in *Drosophila* allow specific chromosomes to be transmitted intact and followed through multiple generations. Balancer chromosomes have three general features: (1) one or more inverted chromosomal segments, within which meiotic recombinants are not transmitted (see Section 13.5 for a review); (2) a recessive allele that results in lethality, so an individual cannot be homozygous for the balancer chromosome; and (3) a "mark" in the form of a dominant mutation conferring a visible nonlethal phenotype, so the segregation of the chromosome can be followed through generations. An example of a balancer chromosome is the *ClB* chromosome used by Hermann Muller to demonstrate that X-rays induce mutations (Experimental Insight 13.1). Balancer chromosomes are available for all of the *Drosophila* chromosomes.

Mutations on specific chromosomes can be identified in *Drosophila* using balancer chromosomes (**Figure 16.3**). Male flies are fed EMS to induce mutations and then are mated with females containing a balancer chromosome. Note that while mutations are induced throughout the genome, only those on the homolog of the balancer chromosome are analyzed. Male $F_1$ progeny are selected

that inherit a mutagenized chromosome from their father and the balancer chromosome from their mother. Next, the selected males are mated to females of the balancer stock, producing $F_2$ progeny. The $F_2$ generation consists of both males and females heterozygous for the induced mutation and can be interbred to produce $F_3$ progeny. In the $F_3$ generation, 25% should be homozygous for the induced mutation and will not carry the dominant allele of the balancer chromosome; 50% will be heterozygous for the newly induced mutation and also carry the dominant allele; and the remaining 25% will die due to homozygosity for the balancer chromosome. The homozygous

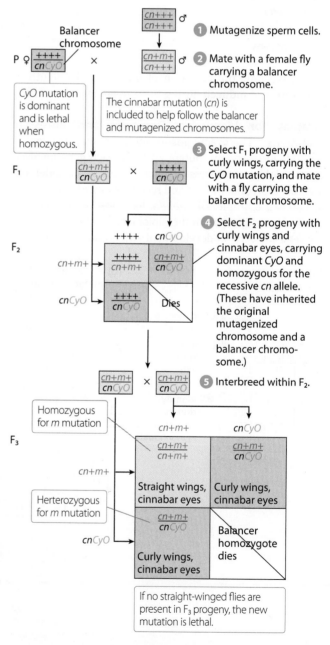

**Figure 16.3  Identifying recessive mutations in *Drosophila* using a balancer chromosome.**

progeny lacking the dominant allele from the balancer chromosome can be screened for an aberrant phenotype.

What happens if the new mutation results in lethality when it is homozygous? In that case, all surviving $F_3$ individuals will carry the dominant allele, located on the balancer chromosome. When a lethal mutation is identified in this way, the mutant allele can be propagated from the heterozygous siblings for propagation. This mutagenesis strategy was used by Eric Wieschaus, Christiane Nüsslein-Volhard, and colleagues in a screen to identify *Drosophila* mutations that disrupt pattern formation during embryogenesis. The research is described in detail in Section 20.2.

### Screening for Conditional Alleles in Haploid Organisms

One advantage of using haploid organisms in a forward genetic screen is that both recessive loss-of-function mutations and dominant mutations can be identified directly. With single-celled organisms, a population of mitotically active cells can be mutagenized, and mutants with an altered phenotype can be selected directly in the colonies derived from the mutagenized cells. A disadvantage is that mutations disrupting essential processes in growth and physiology are often lethal, interfering with the propagation of alleles and thus complicating genetic screening. Fortunately, it is often feasible to design a screen to identify conditional mutant alleles of essential genes. In conditional mutants, the encoded gene product is either functional or not needed under one environmental condition—the **permissive condition**—but is required and either inactive or absent under another—the **restrictive condition** (see Section 4.1).

With some lethal mutations, the mutant phenotype can be rescued by addition of a needed substance to the growth medium. For example, histidine auxotrophic mutants can grow only when histidine is present in the growth medium. In a screen for conditional mutants of this type, the mutagenized population is initially grown under permissive conditions—in this case, in a medium containing histidine—so that both mutant and wild type will grow. This mutagenized population is then replica plated, and the population is screened for phenotypic defects (e.g., lethality) when grown under the restrictive condition (e.g., a lack of histidine). Such genetic screens were performed by Beadle and Tatum to identify auxotrophs in *Neurospora* in the research that established biochemical genetics and produced the one gene–one enzyme theory (see Section 4.3).

If a missing substance cannot be supplied to the medium, other kinds of environmental conditions can be altered instead. In temperature-sensitive mutants, the stability of the mutant polypeptide product of an allele differs with temperature (see Section 4.1), often as a result of a missense mutation. This type of conditional lethal allele in the yeasts *S. cerevisiae* and *Schizosaccharomyces pombe* led to a molecular genetic understanding of the cell cycle, a

biological process shared by all eukaryotes. Mutagenized yeast were grown at a permissive temperature to allow propagation and then the mutant lines were exposed to a restrictive temperature, causing an arrest in growth of some of the mutant strains (**Figure 16.4a**). Surprisingly, in some mutant lines, growth was arrested at specific stages of the cell cycle, rather than randomly along the continuous spectrum of growth (the latter would be expected if the mutation had disrupted a metabolic pathway). These yeast mutants fell into discrete phenotypic categories defined by the stage of the cell cycle at which they were arrested. One possible explanation was the existence of specific checkpoints in the cell cycle (**Figure 16.4b**), and indeed, some of the genes identified by these mutations were

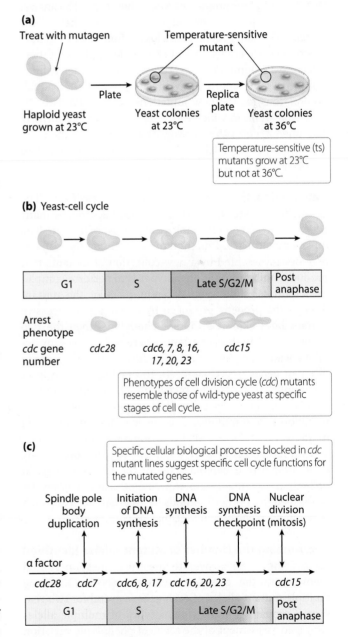

**Figure 16.4 Identification and analysis of conditional alleles.**

found to regulate the cell's progression through various stages of the cell cycle (Figure 16.4c). The studies in yeast provided the foundation for understanding the role of cell cycle regulation in cancer (see Section 3.1).

## Analysis of Mutageneses

Several important questions must be addressed in the analysis of mutants obtained by mutagenesis: (1) Are mutant alleles dominant or recessive with respect to the wild-type allele? (2) How many different genes have been identified in the mutagenesis? (3) How many different mutant alleles of each gene have been identified?

**Determining Dominance or Recessiveness**   The answer to the first question provides insight into whether the mutant allele likely represents a loss of function or a gain of function (see Sections 4.1 and 12.2 for descriptions of these categories). Dominance or recessiveness is determined using the same approach Mendel employed. Individuals homozygous for the new mutations are crossed with the wild-type strain in which the mutagenesis was performed. The phenotype in the $F_1$ progeny derived from the cross allows the mutant allele to be designated as dominant or recessive.

**Determining the Numbers of Genes Identified**   The answer to the second question provides clues to how many genes are involved in the biological process of interest. The most straightforward method of determining the number of genes represented by a new collection of mutants that produce similar mutant phenotypes is to perform complementation tests between different pairs of the mutant lines. If the progeny produced by crossing two recessive mutant lines exhibit a mutant phenotype, then the two mutations are in the same gene, whereas if the progeny exhibit a wild-type phenotype, then the two mutations are in different genes (see Chapter 4). In practice, we can limit the number of crosses by recognizing that complementation is communicative; that is, if mutation A is allelic to mutation B, and mutation B is allelic to mutation C, then mutations A and C are allelic. In some special cases, such as with mutations that are dominant or gametophytically lethal (lethal in a haploid stage of the life cycle, e.g., in pollen), complementation experiments cannot easily be performed, and other methods to ascertain allelism, such as mapping (see Section 5.2), may be employed.

**Determining the Number of Mutant Alleles Identified for a Gene**   The answer to the third question should follow from the complementation analysis. Obtaining multiple mutant alleles of each gene is useful for two reasons. Comparing mutant phenotypes of multiple alleles allows an assessment of the range of phenotypic variation that can be obtained by mutation of the gene in question

(see Section 4.1). The recovery of multiple alleles for each gene also provides information on the saturation of the genetic screen; in other words, it suggests what percentage of the genes that could be identified, have in fact been identified. When a mutagenesis experiment is shown to have produced multiple independent mutations in each gene identified, most genes in the process of interest likely have been mutated.

---

Genetic Insight   Forward genetic screens are an unbiased approach to identifying genes involved in a specific biological process; no prior knowledge of molecular function of the genes is required. Complementation tests reveal the number of genes and alleles identified in genetic screens.

---

Forward genetic screens often require the mutagenesis of thousands of individuals and the screening of large numbers of their progeny for mutant phenotypes. Each progeny may contain multiple mutations, but only a small fraction of the progeny will have a mutant phenotype of interest. For example, in their screens to identify auxotrophs, Beadle, Tatum, and colleagues screened many thousands of individual mutant lines to find the few arginine auxotrophs that were produced. While some screens necessitate the visual inspection of all progeny, others are specifically designed to highlight certain mutants of interest against the background of all other mutants. The designing of such screens is an art. Perhaps the most dramatic screen is one in which application of a simple selection technique allows mutants of interest to survive while those not of interest die. Examples include the isolation of bacteria resistant to antibiotics, insects resistant to insecticides, and plants resistant to herbicides. Similarly, isolation of mutants resistant to analogs of cellular chemicals or to high levels of naturally occurring hormones has proven useful in genetic screens. Often in such cases, mutations identify genes encoding proteins involved in the metabolism or signaling pathways of the respective chemicals.

Even when strong selection criteria cannot be applied, knowledge of the biological process of interest can influence the design of the screen. For example, when Wieschaus and Nüsslein-Volhard performed their screen for *Drosophila* embryogenesis mutants, they assumed that the mutations of interest were all likely to be lethal to the larva (see Section 20.2). Thus they could limit their intensive analysis to mutant lines in which larval lethality was evident. Genetic Analysis 16.1 challenges you to design a screen that identifies genes involved in a particular biological process.

## Identifying Interacting and Redundant Genes

Generally, mutant phenotypes reflect the response of the organism to a loss, or change, of a particular gene product.

In all eukaryotic organisms, proteins to be secreted from the cell or embedded in the plasma membrane are translated at the endoplasmic reticulum and travel via the Golgi apparatus to reach the plasma membrane. Outline a genetic screen for identifying genes involved in protein secretion.

| Solution Strategies | Solution Steps |
|---|---|
| **Evaluate** | |
| 1. Identify the topic this problem addresses, and explain the nature of the required answer. | 1. This problem is concerned with the design of a genetic screen. |
| 2. Identify the critical information given in the problem. | 2. The goal is to identify mutations in genes that function in protein secretion, a universal process among eukaryotes. |
| **Deduce** | |
| 3. Consider any information made available to us about genes involved in the secretory process. | 3. Since we have not been given any information about the genes involved in protein secretion, a forward genetic screen would be a good approach because forward genetic mutageneses are not biased by preconceptions about biochemical functions or gene sequences. |
| 4. Choose an organism for a forward genetic approach. | 4. Since secretory systems in all eukaryotes are similar, they are likely to be homologous, that is, inherited from a common ancestor. Thus we can choose any eukaryote amenable to genetic analysis. *Saccharomyces cerevisiae* would be a good choice since many genetic tools already exist for this model genetic organism. |
| 5. Consider the phenotypic consequence of a loss of protein secretion. | 5. Because loss of a functioning secretory system is likely to be lethal to any organism, we should use a strategy to identify conditional mutant alleles. |
| **Solve** | |
| 6. Design an approach for a genetic screen based on solution steps 3–5. | 6. A good design would be one similar to the procedure used to identify temperature-sensitive mutant alleles in genes of the cell cycle in *S. cerevisiae*. Mutagenesis of haploid cells could be performed at a permissive temperature (e.g., 25–30°C), followed by screening for mutant phenotypes at a restrictive temperature (e.g., 39°C). |
| 7. Describe how you would identify mutations specifically affecting secretion. | 7. A method to monitor secretion is required. One approach would be to follow the fate of a protein known to be secreted into the growth media and look for mutants in which the protein fails to be detected there. |

For more practice, see Problems 21, 22, 23, 27, 28, and 29.

However, individual genes do not act in isolation. The activity of other genes may modify, by either enhancing or suppressing, the phenotypic defects caused by the loss of a gene product. One approach to discovering genetic interactions is to carry out a genetic **modifier screen** to see if mutations in a second gene can enhance or suppress the phenotype of the first mutation. In one variation of a modifier screen, called an **enhancer screen,** mutations in a second site enhance the phenotype of the initial mutant. A complementary type of genetic screen, called a **suppressor screen,** seeks to identify second-site mutations that suppress the phenotype of the initial genotype. Note that both types of screen can be performed simultaneously.

Enhancer–suppressor screening strategies are almost limitless in number and sophistication and have the potential to identify genes that function in interacting genetic pathways.

Modifier screens can identify double mutants that display an unexpected phenotype, one that is not simply the combination of the phenotypes of the two single mutants. In perhaps the most dramatic form of enhancement, termed **synthetic lethality,** the two single mutants are viable but the double mutant is inviable. Synthetic lethality was first noted by *Drosophila* geneticists who observed that some pairwise combinations of mutant alleles were inviable. For example, when Alfred Sturtevant crossed *prune* (*pn*) mutant females (*pn* is on the X chromosome) with males from a stock of separate origin called S/E-S, he noted that the progeny consisted solely of *pn*⁺ females and no viable males (Figure 16.5a). Sturtevant determined that the S/E-S males carried an autosomal dominant mutation, which he called *Prune-killer* (*K-pn*), that in combination with *pn* results in lethality; but he noted that flies homozygous for *K-pn* mutation alone did not have an aberrant phenotype. In his cross, all male progeny inherited a *pn* allele from their mother and a *K-pn* allele from their father, and therefore these progeny

**(a)** Sturtevant's cross identifying synthetic lethality

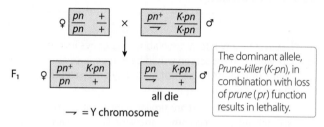

**(b)** Possible mechanisms for synthetic enhancement

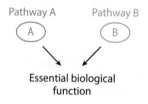

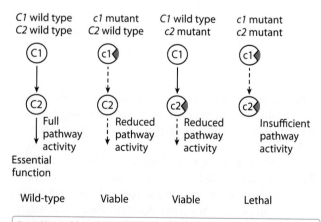

©2007 Macmillan Publishers Ltd

**Figure 16.5 Synthetic enhancement.**

died. In contrast, the female progeny were viable, since despite inheriting a *K-pn* allele from their father, they also inherited a *pn*⁺ allele from their father. In this example, both *pn* and *K-pn* mutants are viable, but the *pn*, *K-pn* double mutant results in lethality.

Figure 16.5b shows two possible mechanisms to explain synthetic lethality, or enhancement. In one mechanism, the two genes in question act in parallel complementary pathways. In this scenario, mutations resulting in the loss of either pathway can be compensated for by the activity of the remaining pathway. However, when both pathways are disrupted, a dramatic enhancement in mutant phenotype is observed. An alternative mechanism is possible when both genes are acting in the same pathway: A reduction in function of one component of the pathway results in a mild phenotype, but when two components are disrupted, the pathway no longer functions effectively. Note that in the latter scenario, hypomorphic alleles can result in synthetic enhancement, but null alleles cannot.

The first scenario, where two genes act in parallel, is an example of **genetic redundancy,** where the loss of the function of either gene alone is compensated for by the activity of the other nonmutant gene. Only when both genes are mutant would a conspicuous mutant phenotype be evident. In such a case, a 15:1 segregation ratio might be expected in the $F_2$ of a cross between the two recessive single mutants (see Section 4.2). In the most obvious case of genetic redundancy, two genes encode very similar proteins that can function interchangeably. In many instances the activities of the two genes do not fully compensate for one another, such that single mutations in either gene alone result in a mild phenotype while a severe phenotype is seen when both genes are mutant. Genetic redundancy caused by the presence of duplicate genes can arise through small-scale duplications or through whole-genome duplications. As we explore in detail in Chapter 18, genome sequences of eukaryotes show such duplications to be very common.

Genetic redundancy can also arise from the compensatory action of genes with little or no sequence similarity, but encoding biochemically different activities. With this type of genetic redundancy it is difficult to predict the redundancy of such genes based on their DNA sequence, but it can be uncovered by enhancer–suppressor screens. Enhancer–suppressor screens have been performed on many organisms including *Drosophila, C. elegans, Arabidopsis,* and mice (see Section 18.3) and are extremely successful at identifying interacting genetic pathways (see Section 20.3).

---

**Genetic Insight** Enhancer–suppressor screens can reveal that two genes are functioning in interacting genetic pathways.

# 16.2 Specific DNA Sequences Are Recognized Using Recombinant DNA Technology

We have seen that the use of cleverly designed genetic screens enables researchers to isolate mutants affecting almost any biological process. To discover the molecular basis for the resulting phenotypes, though, researchers must also be able to clone the genes they have identified. Recombinant DNA technology provides tools for proceeding from mutant phenotype to DNA sequence, so that allelic differences can be examined at the molecular level.

**Recombinant DNA technology** is the set of techniques developed for amplifying, maintaining, and manipulating specific DNA sequences in vitro as well as in vivo. This technology, which is based on advances in microbiology—particularly in understanding the life cycles of bacteria and their viruses, the bacteriophages—has revolutionized the study of genetics. With the ultimate goal of studying specific genes and their functions, biologists use recombinant DNA techniques to (1) fragment DNA into easily managed pieces and then separate and purify these fragments; (2) create many copies of DNA molecules of identical sequence; (3) combine DNA fragments to construct chimeric, or recombinant, DNA molecules; (4) determine the exact sequence of specific DNA molecules; (5) identify fragments of DNA containing complementary sequences; (6) introduce specific DNA molecules into living organisms; and (7) assay the phenotypic effects of the introduced DNA.

The major challenges of recombinant DNA technology are the identification of specific DNA sequences and their manipulation in vitro. To see these challenges in perspective, consider that each of your cells contains two copies each of 22 autosomes and 2 sex chromosomes. Collectively, a haploid set of 23 chromosomes contain 3 billion base pairs and carry some 25,000 or so genes. A typical gene encodes an mRNA transcript consisting of a few thousand bases, although the mRNA may be transcribed from a region that spans millions of base pairs. Molecular analysis of genes and of allelic variation is possible only by distinguishing a gene of interest from others in the genome.

Recombinant DNA technology allows researchers to divide the genome into smaller segments that can then be analyzed and reassembled to provide a molecular view of genes and the genome. In the following sections we describe the development of recombinant DNA technology tools and their application to identify specific DNA sequences, culminating on approaches to identify genes originally identified by only a mutant phenotype.

## Restriction Enzymes

Restriction enzymes cut DNA at specific sequences (see Chapter 10). Each restriction enzyme recognizes a specific restriction sequence at which it cuts both strands of the sugar-phosphate backbone of the DNA, cleaving the restriction sequence in the same way each time it is encountered. Restriction enzymes were originally discovered in bacterial cells, where they protect the bacteria from invasions of nucleic acids, such as the injected genomes of bacteriophages, by digesting foreign DNA. They were given the name restriction enzymes because they restrict the growth of the bacteriophages. Bacterial cells also contain **restriction-modification systems,** which modify the restriction enzyme recognition sequences in the bacterial DNA by the addition of methyl groups and thus protect the bacteria's own DNA from being digested by endogenous restriction enzymes. **Experimental Insight 16.1** explains how restriction-modification systems were identified and how they became an indispensable part of molecular biology.

Restriction enzymes are common in bacteria. The names given these enzymes are generally derived from the first letter of the bacterial genus and first two letters of the species moniker, followed by a Roman numeral. For example, *Eco*RI is derived from *Escherichia coli*; the letter R denotes the strain from which the enzyme was obtained (RY13), and the numeral (I) indicates it was the first enzyme identified. *Eco*RI recognizes the palindromic sequence:

$$5'-GAATTC-3'$$
$$3'-CTTAAG-5'$$

Recall that a palindrome has the same 5'-to-3' base sequence in both of its antiparallel DNA strands. Most restriction enzymes recognize palindromic sequences. For example, *Eco*RI cuts the sugar–phosphate bond between the G and the adjacent A residues in both strands, and the staggered cut results in two products, each ending with a four-base, single-stranded sequence:

$$5'-G \qquad AATTC-3'$$
$$3'-CTTAA \qquad G-5'$$

The single-stranded segments at the ends of each *Eco*RI fragment are referred to as **sticky ends** because they can "stick" to a complementary base-pair sequence by hydrogen bonding. Production of sticky ends facilitates the combining of DNA fragments generated with restriction enzymes, and complementary base pairing plays a role in almost all recombinant DNA techniques. The principle is that if two DNA molecules produced by restriction enzyme digestion have complementary sticky ends, they can be combined by complementary base pairing. The *E. coli* genome itself is protected from digestion by the *Eco*RI endonuclease by the enzyme *Eco*RI methylase, which adds a methyl group to the A adjacent to the T in both strands of the DNA. This is the "modification" in the *Eco*RI restriction-modification system.

Hundreds of restriction enzymes have been isolated from bacteria and are commercially available (see Table 10.1). While many restriction enzymes produce sticky ends, either with 5' overhangs (as in *Eco*RI) or with 3'

## Experimental Insight 16.1

### From Bacteriophage to Restriction Enzymes: Basic Research Spawned a Biological Revolution

Basic biological research aims to discover and understand phenomena from every part of the spectrum of life. Thousands of biologists engage in this research every day, and most have specialties that may seem obscure or trivial to non-scientists. Nevertheless, their discoveries can not only revolutionize research but contribute to changing how we view the world.

In the mid-1960s, Werner Arber was studying a bacterial phenomenon called *host-controlled restriction and modification*, which acts as a simple immune system for bacteria invaded by bacteriophages. He showed that *E. coli* produces two enzymes that affect the same short palindromic DNA sequences. One enzyme, called a *restriction endonuclease*, cleaves DNA like a pair of molecular scissors. The second enzyme, called a *modification enzyme*, adds methyl groups ($CH_3$) to DNA, thereby preventing restriction endonucleases from binding to the DNA.

In 1970, Hamilton Smith extended Arber's work by studying a restriction endonuclease from *Haemophilus influenzae*. Smith isolated *Hind*II and determined that it cleaves at the sequence

$$5'\text{-GTPyPuAC-}3' \quad \longrightarrow \quad 5'\text{-GTPy PuAC-}3'$$
$$3'\text{-CAPuPyTG-}5' \qquad\qquad 3'\text{-CAPu PyTG-}5'$$

*Hind*II generates blunt ends by cleaving its target sequence between the central purine (Pu = A or G) and pyrimidine (Py = T or C). Smith's work on *Hind*II identified some important characteristics of restriction enzymes. First, *Hind*II cleaves foreign DNA into large fragments, but it does not affect *H. influenzae* DNA. This confirmed Arber's idea that bacterial DNA is protected from the action of the bacteria's own restriction enzymes. Second, each resulting DNA fragment has the same three base pairs at its ends, indicating that cleavage occurs only at the target sequence. Smith also discovered that restriction enzymes cleave every copy they encounter of their target sequence.

In 1971, Daniel Nathans pioneered the use of restriction endonucleases to address genetic and genomic questions. Nathans used *Hind*II to digest the small genome of the Simian virus SV40 and found that 11 DNA fragments were formed. In 1973, Nathans digested SV40 with two newly discovered restriction endonucleases. He then used the three sets of restriction fragments to create the first *restriction map* of the SV40 genome, by determining the number and order of restriction sites for each enzyme and assembling these into a map.

By the time Nathans completed his SV40 genome map, biologists were already looking for other restriction enzymes. Within 5 years, over 100 more restriction enzymes were discovered. Many formed sticky ends on digested DNA, and Paul Berg realized that DNA fragments from different organisms could be joined together if they had complementary sticky ends. This finding led to his creating the first recombinant DNA molecule, in 1975.

Arber, Smith, and Nathans shared the Nobel Prize in Physiology or Medicine in 1978 for their work on restriction enzymes, and Berg won the prize in 1980 for the development of recombinant DNA. Since then, restriction enzymes have become a ubiquitous tool in genetic and genomic research. Arber's initial study of an obscure event in bacteria had spawned a revolution as momentous as Watson and Crick's description of DNA structure or Mendel's description of the laws of heredity.

---

overhangs, some restriction enzymes leave **blunt ends** that lack a single-stranded segment. Blunt-ended DNA molecules can also be recombined.

Some restriction enzymes recognize 4-bp sequences, others recognize sequences of 5 base pairs, or 6, or 8. The length of the recognition sequence influences how frequently a given enzyme will cut DNA. If the DNA of an organism were to consist of 25% A, 25% T, 25% G, and 25% C and the bases were randomly distributed, then a restriction enzyme that had a 4-bp recognition sequence would be expected to cut the DNA once every 256 bp ($1/4 \times 1/4 \times 1/4 \times 1/4 = 1/256$). Likewise, a restriction enzyme that recognized a 6-bp sequence would cut the DNA once every 4096 bp ($1/4^6$) on average, and a restriction enzyme that recognized an 8-bp sequence would cut the DNA once every 65,536 bp ($1/4^8$) on average. In reality, genomes of most organisms do not consist of equal amounts of each of the four bases. For example, most genomes of multicellular eukaryotes are AT-rich (that is, their genomes have a higher content of A and T than of G and C), and so restriction enzymes that recognize a GC-rich sequence would cut less frequently on average than would enzymes that recognize an AT-rich sequence.

Scientists use data from restriction experiments, including the number of restriction sites and the number of base pairs between the sites, to create maps of specific DNA sequences. These **restriction maps** provide a foundation for further manipulation of the cloned DNA fragments—for example, by suggesting where to further subdivide cloned fragments in order to clone still smaller fragments, in a process known as **subcloning.**

Let's use the genome of *E. coli* lambda phage in an example of the restriction mapping process. The DNA of the phage genome can be isolated by purifying the phage and removing its protein coat. If this is done gently, the isolated nucleic acid will be the entire lambda chromosome, which is a linear molecule 48,502 bp in length. Electrophoresis of the chromosome in an agarose gel including a fluorescent stain for DNA (see Chapter 10) would reveal a single fluorescent 48.5-kb band (first lane in **Figure 16.6**). If the purified lambda chromosome is first digested with *Apa*I, two fragments, one measuring 10.1 kb and the other 38.4 kb, are generated, indicating that *Apa*I must cut the genome once. This allows us to begin drawing the restriction map as shown below.

If we digest the purified lambda chromosome with *Xho*I, two fragments, one 33.5 kb and one 15 kb, are generated, indicating that *Xho*I must also cut the genome once:

However, two orientations are possible for the *Xho*I restriction map relative to the *Apa*I restriction map drawn above. It could also be drawn as shown below.

To determine which order is correct, we need to perform a double digest, in which both enzymes are used simultaneously to cut the lambda genome. This experiment generates three pieces: 10.1 kb, 15 kb, and 23.4 kb. Since the 15-kb *Xho*I fragment remained intact but the 33.5-kb *Xho*I fragment was cut into two fragments (10.1 kb and 23.4 kb) by *Apa*I, we conclude that the map must be:

The other possible map can be eliminated as incorrect since it would generate fragments of 4.9 kb, 10.1 kb, and 33.5 kb:

**Genetic Analysis 16.2** provides additional practice at constructing a restriction map.

To analyze DNA from organisms with large genomes, researchers must fragment the genomes into more manageable pieces. For example, the *Physcomitrella patens* genome consists of 400 million base pairs, and when digested with a restriction enzyme like *Eco*RI that cuts on average every 4096 bp, approximately 100,000 different DNA fragments are produced. When this digested DNA is electrophoresed through an agarose gel, the fragments making up the resulting "smear" range from 20+ kb down to smaller than 100 bp (**Figure 16.7**). A smear results because, although the

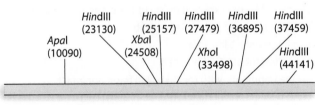

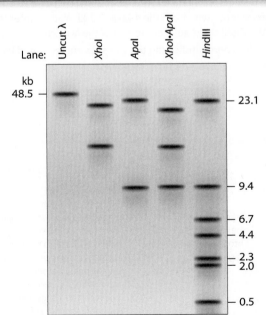

**Figure 16.6  Restriction mapping of lambda phage.**

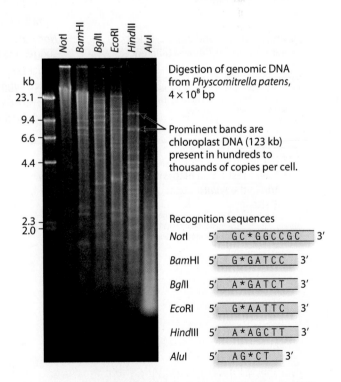

**Figure 16.7  Restriction enzyme digestion of genomic DNA.**

You isolate a plasmid from *E. coli* and wish to begin your analysis of it by making a restriction map. Using three restriction enzymes, ❶ *Bam*H1, ❷ *Eco*RI, ❸ *Not*I you perform six different digestions: single digests using each enzyme alone and double digests using each combination of two enzymes. Agarose gel electrophoresis of the resulting fragments produces the results shown here. Draw a restriction map of the plasmid.

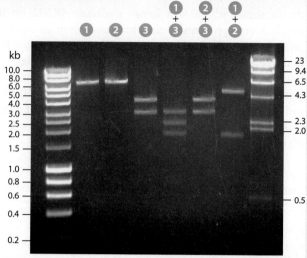

## Solution Strategies

### Evaluate

1. Identify the topic this problem addresses, and explain the nature of the required answer.

2. Identify the critical information given in the problem.

### Deduce

3. Identify the sizes of each of the fragments of the single digests, and determine how many times each enzyme cuts the plasmid.

4. Identify the sizes of each of the fragments of the double digests.

5. Compare single- and double-digest results for similarities and differences.

TIP: In analyzing double digests, the relative position of restriction sites can be determined by observing which fragments remain intact and which are cut into smaller fragments.

PITFALL: If two sites are very close to one another, there will be fewer fragments than expected in the double digest.

## Solution Steps

1. This problem asks you to construct a restriction map of a plasmid.

2. Electrophoresis results for three single digests and the three possible double-digest combinations are provided.

3. *Bam*HI—a single 7-kb fragment. Since plasmids are circular, *Bam*HI must cut the plasmid only once.
   *Eco*RI—a single 7-kb fragment. One site in the plasmid.
   *Not*I—two fragments, 3 kb and 4 kb. *Not*I must cut the plasmid at two sites.

4. *Not*I + *Bam*HI—three fragments: 3 kb, 2.3 kb, 1.7 kb
   *Not*I + *Eco*RI—two fragments: 4 kb, 3 kb
   *Bam*HI + *Eco*RI—two fragments: 5.3 kb, 1.7 kb

5. *Not*I + *Bam*HI—three fragments, with the 3-kb *Not*I fragment intact, suggesting the *Bam*HI site is within the 4-kb *Not*I fragment.
   *Not*I + *Eco*RI—two fragments, with both the 4-kb and 3-kb *Not*I fragments intact, suggesting the *Eco*RI site is adjacent to one of the *Not*I sites.
   *Bam*HI + *Eco*RI—two fragments, indicating the two sites are separated by 1.7 kb (or 5.3 kb the long way around the plasmid).

### Solve

6. (a) Draw a restriction map with *Not*I sites.
   (b) Add in the *Bam*HI site. (c) Add in the *Eco*RI site.

TIP: Drawing of the restriction map does not require the three enzymes to be examined in any particular order.

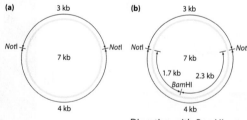

(b) Digestion with *Bam*HI cuts the 4-kb *Not*I fragment into 2.3-kb and 1.7-kb fragments.

(c) The *Eco*RI site must be adjacent to one of the *Not*I sites and is 1.7 kb from the *Bam*HI site. The relative order of the *Eco*RI and adjacent *Not*I sites cannot be determined, since the resolution of gel electrophoresis is not sufficient.

For more practice, see Problems 14, 16, 17, and 19.

enzyme cuts every 4096 bp on average, the distances between *Eco*RI sites will vary due to variation in the genome sequence, and the resolving power of agarose gel electrophoresis is not sufficient to separate all of the different fragments into discrete bands. This lack of resolution is compounded in larger genomes, such as ours, where digestion with *Eco*RI produces approximately 730,000 pieces (3,000,000,000/4096).

---

Genetic Insight   Restriction enzymes cut double-stranded DNA at specific recognition sequences and are used to fragment DNA into small pieces. Restriction maps of DNA allow specific fragments to be identified for further experimental manipulations.

---

## Molecular Cloning

After a genome under study has been reduced to smaller pieces by restriction enzymes, the individual pieces must be reproduced in large amounts—generally, either by molecular cloning or PCR—so that each of them can be analyzed in greater detail. Molecular cloning arose from discoveries in bacterial enzymology and utilizes bacteria and their plasmids or phages to amplify and propagate specific fragments of DNA.

In molecular cloning, isolated DNA fragments are inserted into a **vector,** a carrier fragment of DNA with attributes that will allow amplification (replication) in a biological system. Then the recombinant DNA molecule is introduced into a biological system that amplifies the DNA, making many identical copies called **DNA clones.** Molecular cloning produces a large quantity of identical DNA molecules that can be analyzed by a variety of techniques, including restriction enzyme analysis and DNA sequencing.

Molecular cloning has three general steps:

1. The joining together of the cloning vector and a donor DNA fragment to produce a **recombinant clone**

2. Selection of organisms containing copies of the cloned DNA segment of interest

3. Amplification of the recombinant clone in a biological system

In this section, we describe how DNA fragments are combined in vitro, the attributes of some common cloning vectors, and the means of their amplification. These discussions are followed in Section 16.4 with a description of how DNA libraries, collections of cloned fragments from DNA usually derived from the nucleic acids from a single source, are constructed.

**Creating Recombinant DNA Molecules**   One common method of producing recombinant DNA is to digest DNA

from the donor source and DNA of the cloning vector with the same restriction enzyme. The resulting linear fragments from the two DNA sources can then be annealed at their complementary sticky ends. **Figure 16.8** illustrates restriction digestion by *Eco*RI of both the vector DNA—a plasmid, in this case—and DNA from the human genome. Mixing the two DNAs in a test tube allows the sticky ends to hybridize to one another by complementary base pairing, after which the remaining single-stranded nicks are sealed with DNA ligase (see Chapter 7), resulting in a recombinant DNA molecule. In this case, a recombinant plasmid containing human DNA is formed.

While it is common to cut both source and vector DNA with the same enzyme, variations on this theme are frequently employed. For example, two different restriction enzymes that create complementary sticky ends are sometimes used. When different restriction enzymes are used to digest vector and donor DNA, complementary

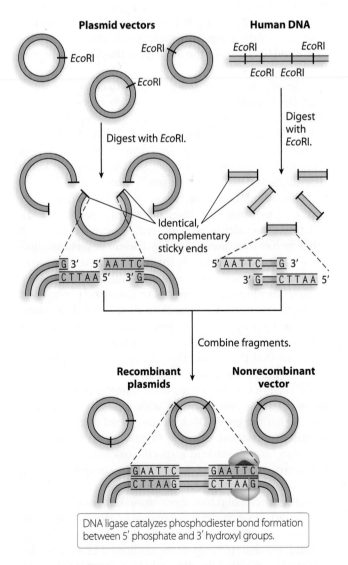

**Figure 16.8  Making recombinant DNA molecules.**

sticky ends are called **cohesive compatible ends.** For example, *Bam*HI recognizes the 6-bp sequence

5'-GGATCC-3'
3'-CCTAGG-5'

and leaves sticky ends

5'-G          GATCC-3'
3'-CCTAG          G-5'

*Sau*3A recognizes the 4-bp sequence

5'-GATC-3'
3'-CTAG-5'

and leaves a sticky end

5'-N          GATCN-3'
3'-NCTAG          N-5'

(where N represents any nucleotide). Since the sticky ends created by the two enzymes are the same (5'-GATC-3'), the ends of a *Bam*HI and a *Sau*3A fragment can combine to create recombinant DNA molecules. However, in this case, the resulting ligated products will often lack an intact *Bam*HI site, since the 5' Ns from the *Sau*3A site may not be Gs.

Usually the goal of this process is to create recombinant DNA molecules in which a single piece of source DNA is combined with a single cloning vector molecule. However, because digested DNA from both sources is mixed together in a test tube, a variety of recombinant molecules may arise. For example, some recombinants may have a single donor-DNA insert, whereas others may have two or more donor fragments that join together and then insert into the vector. In addition, the sticky ends of vectors can rejoin each other rather than incorporating a donor insert, producing a **nonrecombinant vector.** Because neither nonrecombinant vectors nor clones with multiple inserts are desired for cloning, techniques to favor the production of single-insert clones have been developed. For example, the occurrence of nonrecombinant clones can be reduced by removal of the 5' phosphates on the vector DNA, so that the vector DNA cannot ligate to itself to produce nonrecombinant clones.

A feature of experiments using a single restriction enzyme or using two enzymes with cohesive compatible ends is that the insert DNA can be ligated into the vector in either orientation. One way to ensure that insert DNA is cloned into a vector in a specific orientation is to use two restriction enzymes with different compatible ends, a process called **directional cloning** (**Figure 16.9**). Directional cloning has three desirable features. First, only insert DNA fragments possessing the two different compatible ends will be efficiently cloned into the vector. Second, the inserted fragments are ligated in a particular orientation dictated by the cohesive compatible ends. And third, due to the incompatibility of the two ends of

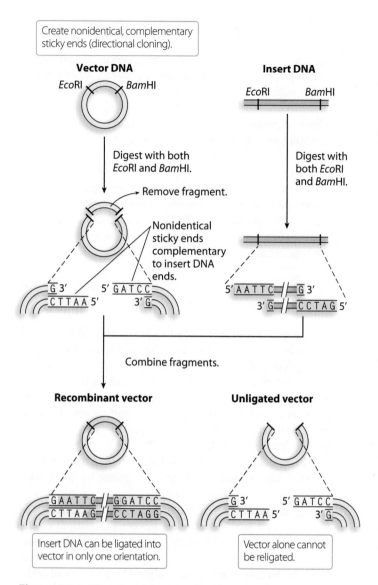

**Figure 16.9  Directional cloning of DNA molecules.**

the digested vector DNA, the vector cannot religate to itself, thus minimizing the creation of nonrecombinant vectors.

While hundreds of restriction enzymes are commercially available, cohesive compatible ends are not always possible to produce at the positions necessary for constructing certain recombinant DNA molecules. One approach to creating compatible ends in such a case is to generate blunt ends, ends without any overhang, and ligate the blunt-ended molecules.

Some restriction enzymes naturally create blunt ends, but any restriction enzyme site can be converted into a blunt end (**Figure 16.10**). For example, DNA polymerase (see Chapter 7) can use a 5' overhang as a template, and add dNTPs to the recessed 3' end until a blunt end has been produced. Alternatively, 3' overhangs can be made blunt by a DNA exonuclease (see Chapter 7) that

Create blunt ends by filling or trimming.

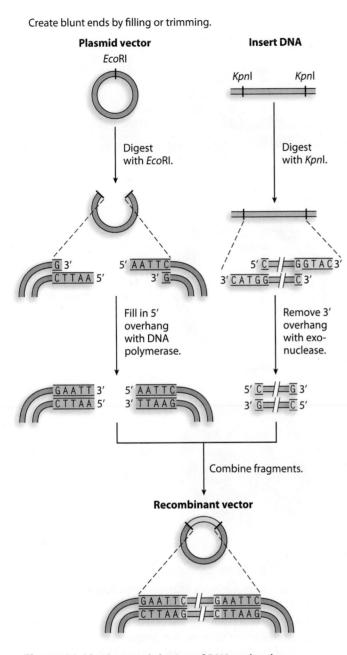

**Figure 16.10  Blunt-end cloning of DNA molecules.**

**Plasmids as Cloning Vectors**  Plasmids are circular DNA molecules that replicate autonomously in bacteria and usually carry nonessential genes. The F-factor involved in *E. coli* conjugation (see Chapter 6) is a plasmid. Plasmids used as cloning vectors replicate independently of the bacterial chromosome and, unlike the F-factor, which can recombine into the *E. coli* chromosome, always remain separate from it. Most plasmids used as cloning vectors have been modified in the laboratory to possess several features that facilitate the production of recombinant DNA molecules (**Figure 16.11**). For example, plasmids are equipped with an *origin of replication* that drives efficient replication of the plasmid within the bacterial host. They also contain a gene conferring a trait that permits bacteria harboring the plasmid to be selectively grown. Genes conferring resistance to antibiotics are commonly used as selectable markers.

Two types of plasmids, identified as pUC-based plasmids and pBR-based plasmids, are most frequently used in constructing recombinant plasmids capable of transforming competent bacteria. Both types have many different forms, developed through extensive genetic engineering in the laboratory. In these vectors, the *β-lactamase* gene, conferring resistance to ampicillin, is often used as the selectable marker. The *origin of replication (ori)* was derived from a naturally occurring *E. coli* plasmid called the ColE1 plasmid. The ColE1 *ori* allows these plasmids to be maintained at a high copy number of 100–200 plasmids per cell.

Both pUC and pBR plasmids also contain a **multiple cloning site (MCS)** that has several different restriction enzyme sites into which DNA can be cloned. These restriction enzyme sites occur only within the MCS and nowhere else in the plasmid. In pUC-based plasmid cloning vectors, the MCS is embedded in the *lacZ* gene, which encodes β-galactosidase, an arrangement that provides a colorimetric assay for determining which bacteria harbor vectors with an insertion of DNA into the MCS (see Figure 16.11). Although the normal substrate for β-galactosidase is lactose, the enzyme can also cleave lactose analogs, such as X-gal. When the colorless substrate X-gal is added to the growth medium, bacteria with a functional *lacZ* gene producing β-galactosidase will convert X-gal to a blue product. When a fragment of DNA is inserted into the MCS, the *lacZ* gene is disrupted and rendered nonfunctional. Bacteria then will appear as white colonies, whereas bacteria harboring a

degrades only single-stranded DNA, and "chews back" the 3′ overhang. In some procedures shearing force, rather than restriction enzymes, is used to produce random DNA fragments (as when DNA is passed through a fine needle), and the ends of the fragmented DNA can then be blunted by treatment with a DNA polymerase and exonuclease. Conversely, blunt ends can be converted into sticky ends by ligation of short oligonucleotides onto the blunt-ended DNA molecules. The oligonucleotides can be synthesized to have sequences for any restriction enzyme desired, thus adding any specific restriction site to the end of any DNA molecule. Oligonucleotides of this type are called **linkers.**

**Figure 16.11** A plasmid cloning vector.

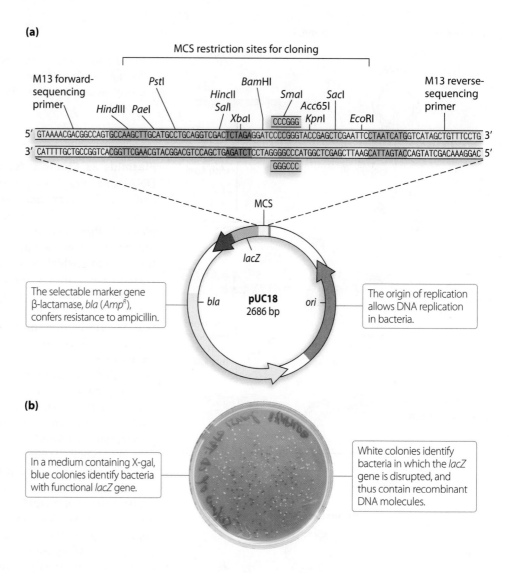

(a)

MCS restriction sites for cloning

M13 forward-sequencing primer

*Hind*III  *Pae*I  *Pst*I  *Hinc*II *Sal*I *Xba*I  *Bam*HI  *Sma*I CCCGGG *Acc*65I *Kpn*I *Sac*I *Eco*RI

M13 reverse-sequencing primer

5′ GTAAAACGACGGCCAGTGCCAAGCTTGCATGCCTGCAGGTCGACTCTAGAGGATCCCCGGGTACCGAGCTCGAATTCCTAATCATGGTCATAGCTGTTTCCTG 3′
3′ CATTTTGCTGCCGGTCACGGTTCGAACGTACGGACGTCCAGCTGAGATCTCCTAGGGGCCCATGGCTCGAGCTTAAGCATTAGTACCAGTATCGACAAAGGAC 5′
GGGCCC

MCS

*lacZ*

The selectable marker gene β-lactamase, *bla* (*Amp*^R), confers resistance to ampicillin.

*bla*

**pUC18**
2686 bp

*ori*

The origin of replication allows DNA replication in bacteria.

(b)

In a medium containing X-gal, blue colonies identify bacteria with functional *lacZ* gene.

White colonies identify bacteria in which the *lacZ* gene is disrupted, and thus contain recombinant DNA molecules.

cloning vector that does not contain a fragment of DNA inserted in the MCS are blue. This difference allows rapid identification of colonies harboring vectors with inserts in the MCS. Thus, selection based on antibiotic resistance allows identification of bacteria that have been transformed, and *blue versus white* screening allows identification of bacteria harboring plasmid vectors with an insertion of recombinant DNA.

**Amplifying Recombinant DNA Molecules** Recombinant DNA molecules are introduced into *E. coli* by transformation, the same process described by Griffiths and by Avery, MacLeod, and McCarty in their early investigations of the hereditary function of DNA (**Figure 16.12**; see Chapter 6). To amplify recombinant DNA molecules in modern laboratories, DNA is mixed with *E. coli* in a test tube. The bacteria are chemically treated with divalent cations (such as Ca$^{2+}$) or an electrical shock to open pores in their membranes, thus making the bacteria competent to take up exogenous DNA by transformation. The bacterial strains used in recombinant DNA experiments are

chosen for characteristics that do not allow them to survive well outside of the laboratory.

The concentrations of DNA used to transform competent bacteria are determined empirically and are chosen so that individual bacterial cells are likely to take up no more than one DNA molecule. After transformation, the bacteria are allowed to recover for a short period of time and are then plated on growth medium that selects for cells containing the selectable marker gene, conferring resistance to an antibiotic, encoded on the DNA vector. When the transformed bacteria are plated on media containing the antibiotic, only those bacteria harboring vector DNA will survive.

Recombinant DNA molecules introduced into microbial cells are amplified by repeated cycles of DNA replication. Since the recombinant vector has an origin of replication, it will amplify by autonomous replication using bacterial enzymes. Then, when the bacterium divides, each of its progeny will receive copies of the recombinant DNA molecule. Because a single bacterium with a recombinant DNA molecule can grow into a colony consisting of

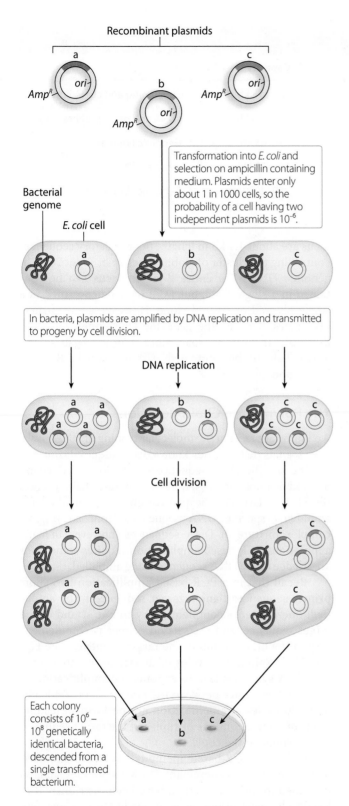

**Figure 16.12  Amplification of recombinant DNA molecules in bacteria.**

some $10^8$ bacteria, each with multiple copies of the recombinant DNA molecule, billions of identical copies of DNA molecules are made.

**Genetic Insight**  Plasmids are used as vectors to amplify DNA molecules. Biological systems are used to carry out the amplification. Antibiotic resistance produced by expression of a gene on the vector is used to identify transformed bacteria.

The use of plasmid vectors for cloning large DNA fragments is limited, mainly because large plasmids (over 20 kb) are not efficiently maintained in a high copy number. This limitation restricts the usefulness of plasmids in cloning eukaryotic genomic DNA, where genomes can be large (human genome is $3 \times 10^9$ bp) and individual genes are often much longer than 20 kb and therefore cannot be cloned in a single plasmid. To overcome these limitations of plasmids, vectors capable of handling larger clones have been developed (**Table 16.2**). Two general approaches have been employed to propagate larger DNA fragments. In one approach, vectors based on the life cycle of bacteriophages—in particular, bacteriophage lambda—accommodate larger fragments of DNA. The second approach harnesses single-copy origins of replication to efficiently propagate even larger recombinant DNA molecules in both bacteria and yeast.

**Bacteriophage Vectors**  Bacteriophage lambda is capable of both a lytic life cycle and a lysogenic life cycle (see Section 14.6). Phage propagation through the lysogenic life cycle requires the presence of all the genes of the lambda genome, but genes that are specifically involved in the lysogenic life cycle are dispensable for the lytic life cycle. If the genes required for lysogeny are removed, they can be replaced by up to 23 kb of DNA from another source, and the recombinant phage can then be propagated through a lytic life cycle (**Figure 16.13a**). In bacteriophage-based vectors, it is the replication of the phage within the bacterium that amplifies the recombinant DNA molecule.

The size of inserted DNA that can be accommodated is further increased by taking advantage of another feature of the lambda bacteriophage system: rolling circle replication (see Chapter 6). During the lytic life cycle, lambda DNA replicated by rolling circle replication results in successive concatenation of 50-kb genomes into long DNA molecules. A lambda-encoded nuclease then recognizes a specific sequence within the lambda genome and cleaves the concatenated genomes into single-genome units. Subsequently, specific sequences called *cohesive end sequence,* or *cos,* **sites** in the lambda genome will interact with lambda phage coat proteins to "package" the individual lambda genomes into discrete phage particles in vitro. The *cos* sites are the only lambda sequences required for DNA to be packaged, so when DNA from another source is concatenated with *cos* sequences derived from lambda, the ligated DNA can be packaged into phage particles. In this case, neither the genes for lysis nor the genes for lysogeny are in the phage

| Vector | Form of DNA | Host | Capacity | Uses |
|--------|-------------|------|----------|------|
| Plasmid | Circular | *E. coli* | <15 kb | Subcloning and cDNA libraries |
| Lambda | Linear phage chromosome | *E. coli* | <23 kb | cDNA and genomic libraries |
| Cosmid | Circular | *E. coli* | 30–45 kb | Genomic libraries |
| BAC | Bacterial chromosome | *E. coli* | 100–200 kb | Genomic libraries |
| YAC | Yeast chromosome | *S. cerevisiae* | 250–2000 kb | Genomic libraries |

**Table 16.2** Cloning Vectors

particles; thus, after infection of a host bacterium, the injected DNA does not enter the lambda life cycle. If an *origin of replication* and a selectable marker are included in the vector, however, the DNA can be replicated as a plasmid in the bacterium (**Figure 16.13b**). Vectors with these features are known as **cosmid vectors.** Since the lambda phage can hold up to 50 kb, cosmid vectors can carry up to 45 kb of insert sequence along with 5 kb of *cos, origin of replication,* and selectable marker sequence.

**Artificial Chromosomes**   While both lambda and cosmid vectors have been historically important, vectors called artificial chromosomes, which have the capacity to carry even larger DNA fragments, are now more frequently used. These were developed through accumulated knowledge of how chromosomes propagate in prokaryotes and eukaryotes, and of the functions of different chromosome regions in replication.

Yeast artificial chromosomes (YACs) were the first artificial chromosomes developed and are used as cloning vectors in *S. cerevisiae.* A YAC vector contains sequences corresponding to a centromere (see Section 11.3), telomeres, a selectable marker, and a cloning site, and it can accept an insert size of 200 kb to 2 megabases (Mb). YACs carrying an insert smaller than 200 kb are often unstable and do not properly segregate at mitosis.

Bacterial artificial chromosomes (BACs) were developed shortly after YACs. Although BACs have a smaller insert-size capacity (100–500 kb) than YACs, they are the preferred artificial chromosome cloning vector, largely due to the ease of using *E. coli* rather than yeast as a host. Like plasmids, BAC vectors contain an origin of replication, a selectable marker gene, and an MCS. However, the origin of replication in BAC vectors is derived from the F-factor plasmid. Unlike replication via the ColE1 origin, replication via the F-factor origin is strictly controlled, producing only one or two copies of the F-factor per cell. This difference allows large plasmids to be maintained, circumventing the problem encountered with plasmids that have high copy numbers.

The utility of BAC cloning vectors becomes apparent when we consider the typical sizes of eukaryotic genes. For example, while individual globin genes in the β-globin

locus are about 1.4 kb in length, the regulatory sequences controlling the cluster of globin genes span about 70 kb of genomic DNA. The entire β-globin locus can be contained in a single BAC, but would not be contained in a single plasmid or cosmid clone. However, some eukaryotic genes, such as the gene for Duchenne's muscular dystrophy in humans, span more than a megabase and are unlikely to be contained within a single BAC or YAC clone.

## The Polymerase Chain Reaction

An important facet of the molecular cloning of DNA, as we have seen, is that no prior knowledge is required concerning the specific DNA sequence to be cloned. A second method of amplifying DNA, the polymerase chain reaction (PCR), is a fast and inexpensive enzymatic method for amplifying specific DNA sequences from an original complex mixture of DNA sequences; however, in many applications PCR-based amplification requires some prior knowledge of the DNA sequence being amplified.

Recall that in a PCR-based amplification two short primer sequences flank a target sequence to be amplified; one primer is complementary to one strand of the target sequence, and the second primer is complementary to the other strand of the target sequence (see Figure 7.24). Reiterative cycles of denaturation, annealing, and DNA synthesis lead to exponential amplification of the target sequence between the two primers. Typically, 30 to 40 cycles of amplification are performed, leading to production of billions of copies of the target sequence from a single initial copy.

PCR and PCR-based methods have revolutionized many aspects of recombinant DNA technology and, as described in the next section, have been instrumental in the development of new generations of DNA sequencing technologies. The extreme sensitivity of PCR has facilitated the study of genes from organisms that cannot be propagated in pure cultures in the laboratory, and therefore their DNA is scarce. For example, PCR has allowed the study of the evolution of viral sequences during the infection of a single host organism, facilitated the study of human ancestry, and allows the monitoring of

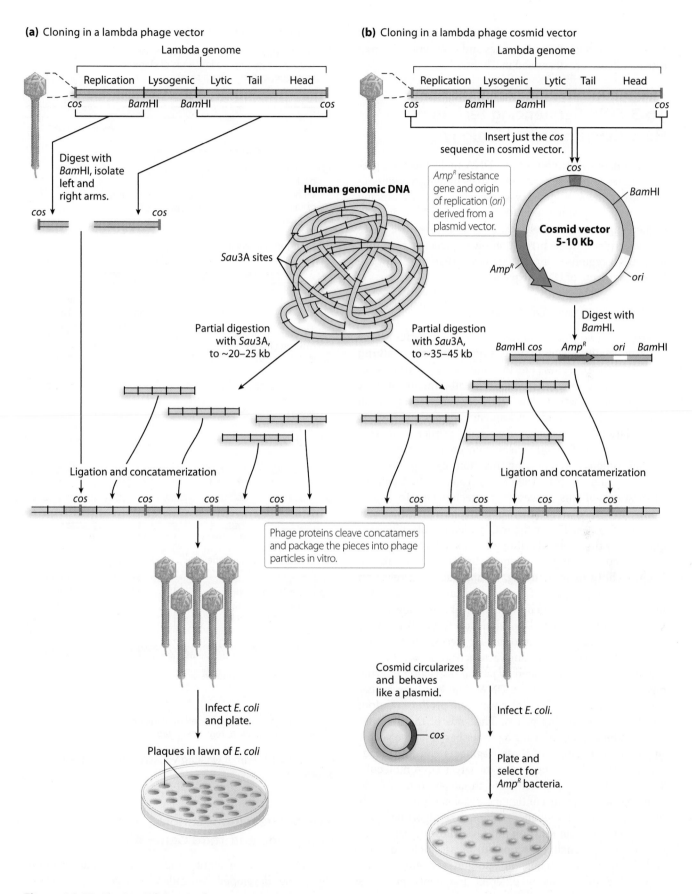

**(a)** Cloning in a lambda phage vector

**(b)** Cloning in a lambda phage cosmid vector

Lambda genome

| Replication | Lysogenic | Lytic | Tail | Head |

*cos*   *Bam*HI   *Bam*HI   *cos*

Digest with *Bam*HI, isolate left and right arms.

*cos*

*cos*

**Human genomic DNA**

*Sau*3A sites

Partial digestion with *Sau*3A, to ~20–25 kb

Partial digestion with *Sau*3A, to ~35–45 kb

Insert just the *cos* sequence in cosmid vector.

*cos*

*Amp^R* resistance gene and origin of replication (*ori*) derived from a plasmid vector.

*Bam*HI

**Cosmid vector 5-10 Kb**

*Amp^R*

*ori*

Digest with *Bam*HI.

*Bam*HI *cos*   *Amp^R*   *ori*   *Bam*HI

Ligation and concatamerization

*cos*   *cos*   *cos*   *cos*

Ligation and concatamerization

*cos*   *cos*   *cos*   *cos*

Phage proteins cleave concatamers and package the pieces into phage particles in vitro.

Infect *E. coli* and plate.

Plaques in lawn of *E. coli*

Cosmid circularizes and behaves like a plasmid.

*cos*

Infect *E. coli*.

Plate and select for *Amp^R* bacteria.

**Figure 16.13 Cloning in bacteriophage vectors.**

trafficking of endangered species. Other applications of PCR, including its use in forensics and genotyping human disease alleles, are discussed in Chapter 7.

## 16.3 DNA Sequencing Technologies Have Revolutionized Biology

The ultimate description of any DNA molecule is its precise sequence of bases. The principles of Sanger sequencing, also known as dideoxy sequencing, is a sequencing method developed in the 1970s (see Chapter 7). In dideoxy sequencing, approximately 800 to 1000 consecutive bases are determined in each reaction—a sequencing read, but most DNA regions of interest are larger than this. How are larger fragments of DNA sequenced?

### Sequencing Long DNA Molecules

There are two basic strategies for sequencing large DNA molecules. The first technique, **primer walking** (**Figure 16.14a**), relies on the successive synthesis of primers based on the progressive attainment of new sequence information. The DNA sequence information obtained in the first dideoxy sequencing reaction provides a foundation for the design of a second primer. If the second primer is 600 to 800 bases from the first primer, the second dideoxy sequencing reaction can extend the known sequence up to 1800 bases from the first primer. Reiterations of this process allow technicians to "walk" along a long DNA molecule, designing new primers every 600 to 800 bases. The speed with which a molecule is sequenced by this method is limited by its reiterative nature.

A second method for sequencing large molecules of DNA is **shotgun sequencing,** an approach that relies on redundant sequencing of fragmented target DNA in the hope that all regions will be sequenced at least a few times. In this technique, a large DNA molecule (e.g., a BAC clone of 100 kb) is fragmented into smaller pieces, and the fragments are ligated into cloning vectors (**Figure 16.14b**). The fragments may be generated by partial restriction enzyme digestion or by shearing the DNA. The key here is that fragmentation is done in such a way as to produce random and hence overlapping pieces. The ends of these clones can then be sequenced using a primer based on the vector sequence. The clones of these fragments can be considered a *library* of sequences from the larger DNA molecule. The strategy is to sequence enough clones to be able to assemble a complete contiguous sequence on the basis of overlaps in the sequences. Computer algorithms are available to perform much of this task, allowing data from millions of sequencing reactions to be assembled quickly (see Section 18.1). Thus, in shotgun sequencing, the sequencing of the many different fragments proceeds simultaneously ("in parallel"), allowing long DNA molecules to be sequenced rapidly.

**(a)** Sequencing by primer walking

**1** Primers (gray), initially based on vector sequences (orange), allow ends of clone to be sequenced from both sides.

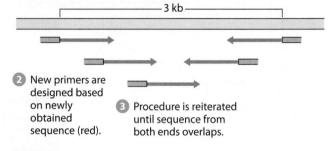

**2** New primers are designed based on newly obtained sequence (red).

**3** Procedure is reiterated until sequence from both ends overlaps.

**(b)** Shotgun sequencing

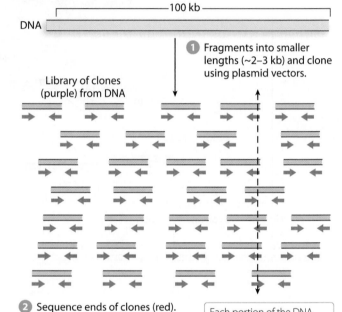

Library of clones (purple) from DNA

**1** Fragments into smaller lengths (~2–3 kb) and clone using plasmid vectors.

**2** Sequence ends of clones (red).

**3** Assemble sequences into a single contiguous sequence (green) by computer.

Each portion of the DNA should be sequenced in ~10 independent clones to facilitate assembly.

Contig

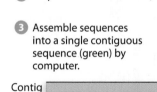

Primers ▪         ▪
PCR product ▬▬▬▬▬
Sequence ➡    ⬅

**4** Use PCR (with primers based on flanking sequences) to close remaining gaps.

**Figure 16.14  Primer walking versus shotgun sequencing approaches.**

### New Sequencing Technologies: Next Generation and Third Generation

New generations of sequencing technologies are continually being developed. So-called next-generation DNA-sequencing technology ascertains the sequence of a single strand of DNA by synthesizing the complementary strand

and detecting which base is added at each step. After fragmentation, individual DNA molecules are denatured and single stranded DNA molecules are captured and immobilized on beads. The beads, with each bead harboring a single DNA molecule, are then placed in wells. Linkers are added to the DNA fragments allowing amplification by PCR using primers complementary to the linkers. The amplified fragments are sequenced by flushing the wells sequentially with solutions containing the four bases, A, G, C, and T. Light is produced only when the solution contains the complementary nucleotide of the first unpaired base of the template. The light is detected by a sensor, and the order of the bases is revealed by the sequence of solutions that produce signals. Thus, next generation sequencing technologies are based on sequencing "by synthesis"—rather than by chain termination, as is the case with dideoxy sequencing.

One major advance of next-generation sequencing technologies is that thousands to millions of sequencing reactions are run simultaneously, producing orders of magnitude more sequence than dideoxy sequencing; thus they are also referred to as massively parallel or high throughput. Another major advance of next-generation sequencing technologies over dideoxy sequencing is that the DNA to be sequenced does not first need to be cloned but is instead directly amplified by PCR, and the resulting PCR products are then sequenced. For this reason, these techniques are called direct sequencing technologies. The elimination of the initial cloning step has two significant advantages. First, it facilitates the sequencing of DNA samples that are found in only trace amounts—for example, the small DNA samples obtained from extinct organisms, such as that from Neanderthal bones or from woolly mammoths preserved in the permafrost of Siberia. In fact, direct sequencing is so powerful that it enabled sequences representing environmental contamination of the woolly mammoth DNA to be identified, thus providing information about organisms that coexisted with the mammoth (e.g., grass species). Analysis of the Neanderthal genome is discussed in more detail in Section 18.2. Second, certain kinds of DNA that are difficult to clone, such as highly repetitive heterochromatin, can now be sequenced. On the other hand, these methods have the disadvantage of producing sequencing reads that are shorter (20–500 bases) than the 800- to 1000-base sequences obtained with Sanger dideoxy sequencing.

Third-generation sequencing technologies are now being developed in which millions of single DNA molecules are sequenced *directly* and *in parallel* without the need for either cloning or PCR amplification. It is worth noting that while sequencing technologies have changed, the process of determining DNA sequence is always based on the precise chemistry of complementary base pairing.

With the advent of the new technologies, the price of sequencing has plummeted, making new applications possible. The cost of sequencing a million base pairs by dideoxy technology was \$10,000 in 2001. In 2005 the price dropped to \$1000 for pyrosequencing, and to just \$1 using the third-generation sequencing technology available in 2010. Lower cost has resulted in an explosion of sequences available in public databases; their number went from less than 10 billion base pairs in 2000 to nearly 300 billion in 2010. A stated goal is to produce the genome sequence of any one person for less than \$1000, the so-called thousand-dollar genome, within the next decade. When this is feasible, it may be routine for your genome sequence to be part of your medical file. Now next-generation sequencing is being employed to determine whether parents may be carriers of mutations in the genes behind 448 childhood recessive diseases.

These new possibilities raise some unprecedented social issues. From the earliest days in the development of recombinant DNA technologies, the potential ethical problems and possible environmental and other risks have been the subject of intense debate. In 1975, following an initial self-imposed moratorium, scientists met at Asilomar to draw up a set of guidelines addressing many of the safety concerns. Potential ethical problems raised by the sequencing of individual genomes, including questions pertaining to confidentiality, will need to be addressed by similar public debates.

---

Genetic Insight  Overlaps in the sequences from multiple sequencing reactions allow contiguous sequences of millions of base pairs to be assembled. Through massively parallel sequencing technologies, entire genomes are sequenced at low cost.

---

## 16.4 Collections of Cloned DNA Fragments Are Called Libraries

A **DNA library** is a collection of cloned fragments of DNA, usually derived from the nucleic acids of a single source (recall our use of *library* in the preceding discussion of shotgun sequencing). DNA libraries come in two varieties; those derived from the genomic DNA of an organism are called **genomic libraries,** and those derived from mRNA are called **complementary DNA (cDNA) libraries.** Since the source of nucleic acids for each type of library differs, the kinds of sequences represented in each type also differ.

In theory, genomic libraries should contain all the sequences found in the genome of the source organism. For example, a human genomic library would contain all $3 \times 10^9$ bp in the haploid genome sequence. This would include the exons and introns of genes, the regulatory sequences controlling gene expression, intergenic sequences (noncoding sequences between genes), and repetitive sequences (centromeres, telomeres, ribosomal DNA, transposons, retroelements, etc.). By contrast, cDNA libraries are derived from mRNA and thus represent the DNA sequences that are transcribed in the tissue from which the mRNA is derived. Since only a fraction of the genes present in the genome are likely to be expressed in any particular tissue, and even those are expressed at different levels, only a fraction of the genes are represented, and in different amounts, in any cDNA library. Thus the number of times a specific sequence is represented in a library differs significantly between genomic and cDNA libraries.

## Constructing Genomic Libraries

Genomic libraries are collections of individual clones derived from the genomic DNA of an organism. To construct a genomic library, genomic DNA, usually from a single individual, is isolated and fragmented into smaller pieces that are then ligated into cloning vectors (**Figure 16.15**). The recombinant vectors are transformed into bacteria (in the case of plasmid and BAC vectors) or used to infect bacteria (in the case of phage vectors) that grow into colonies or plaques that collectively contain clones representing the entire genome. A genomic library contains each sequence in the genome at approximately the same frequency. Thus, sequences representing the exons and introns of genes, the regulatory sequences controlling their expression, and repetitive and intergenic sequences are all approximately equally represented in the genomic library. However, in practice, some sequences are not efficiently maintained in the host cells and will be underrepresented, so the entire genome is not fully represented in any typical genomic library. For

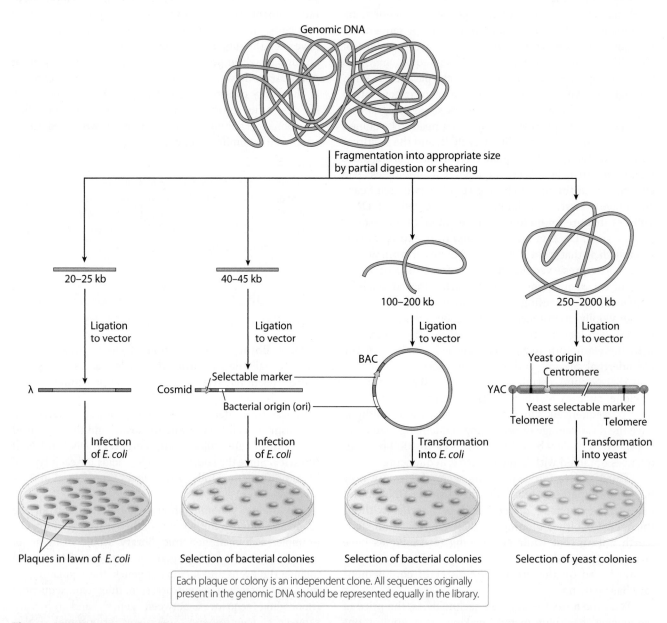

**Figure 16.15 Construction of genomic libraries.**

example, repetitive DNA tends to be underrepresented due to its propensity to undergo intragenic recombination that results in deletion of DNA sequences within clones.

Three desirable attributes for a genomic library are that (1) the genomic clones are broadly representative of DNA of the entire genome, (2) the genomic clones are large enough to be useful for sequencing and subcloning, and (3) the genomic clones are roughly similar in size. Let's look at how these attributes are achieved.

To ensure that a genomic library is broadly representative, care must be taken to fragment it into random pieces of an appropriate size for cloning into a vector. Random fragmentation is accomplished by two different methods. In one technique, the DNA is *partially digested* with an enzyme that cuts very frequently (e.g., a restriction enzyme that has a 4-bp recognition sequence). Partial digestion refers to the use of *less* restriction enzyme than would be needed to cut the DNA at every restriction sequence the enzyme recognizes, resulting in cuts at some of the restriction sequences but not all of them. Since a 4-bp recognition sequence should occur every 256 base pairs on average, partial digestion of DNA in which, on average, only one in 400 recognition sequences are cut should result in DNA fragments of approximately 100 kb. Thus, partial digestion with an enzyme that otherwise cuts frequently will generate random, large genomic DNA fragments with sticky ends, as desired. The second technique for obtaining random fragmentation of DNA is random shearing of genomic DNA with subsequent enzymatic treatment to create blunt ends. In theory, either technique should provide random representation of genomic DNA from the entire genome.

The size of DNA clones in genomic libraries results from technical choices that seek a balance between, on the one hand, the difficulty of isolating, cloning, and propagating large molecules of DNA and, on the other hand, the greater number of smaller fragments that would have to be cloned in order to span the entire genome. As we discuss in Chapter 18, however, a set of genomic libraries that each have a different-sized insertion can be useful for determining the sequence of an entire genome.

## Constructing cDNA Libraries

The starting material for a cDNA library is mRNA, often derived from a specific tissue or cell type. Messenger RNA cannot be cloned directly, because it is single stranded and is of course RNA, not DNA. Cloning of mRNA sequences can be accomplished by synthesizing a double-stranded cDNA copy of the mRNA and then ligating the cDNA into a vector. cDNA libraries are especially useful for working with eukaryotic organisms whose gene sequences are interrupted by many long introns.

The concept and development of cDNA libraries required advances in understanding the life cycle of retroviruses and the movement of retrotransposons (see Section 13.7). The availability of the enzyme **reverse transcriptase,** found in RNA-containing retroviruses, and of retrotransposons, which use single-stranded RNA as a template to produce a complementary strand of DNA, makes cloning from mRNA possible. Reverse transcriptase creates cDNA by first transcribing a single-stranded DNA molecule complementary to mRNA acting as a template. The poly-A tail added to RNA polymerase II transcripts of eukaryotes facilitates the construction of cDNA libraries from such mRNA, since the first strand of cDNA can be synthesized using an oligo dT primer (**Figure 16.16**). The mRNA template is then enzymatically removed, and the second strand of DNA is synthesized by DNA polymerase, using the first cDNA strand as a template.

The composition of a cDNA library reflects the level of expression of different genes active in the tissue from which the mRNA was extracted. Genes that are highly expressed are represented in the mRNA at a higher frequency than genes expressed at a lower level, and genes not expressed in the tissue of origin are not represented. In contrast to genomic libraries, which represent all genes at approximately equal frequency, the frequency with which any particular gene will be represented in a cDNA library is difficult to estimate, since it depends on the expression level of the gene in the original mRNA population (**Figure 16.17**).

Since cDNA libraries are usually made from mature cytoplasmic mRNA, the only sequences included in the cDNA clones are the 5'-UTR, the exons, and the 3'-UTR; the clones will lack any intronic and intergenic sequences. Since the genetic code is universal, cDNA clones derived from one organism can be expressed in any other organism as long as appropriate transcriptional (e.g., promoter) and translational signals are inserted to promote efficient gene expression in the host organism. A cDNA library constructed with such features is called an *expression library*. An example is described in Section 16.5.

## Screening Libraries

Once a library or other collection of clones is produced, how do biologists identify clones containing specific DNA sequences? As we have seen with PCR and the analysis of nucleic acids by Southern and northern blotting (see Research Technique 10.2 on page 344), all techniques to identify fragments containing specific nucleic acid sequences take advantage of the exquisite specificity of complementary base pairing between single-stranded molecular probes and single-stranded target-sequence regions of DNA or RNA. Recall that in Southern blotting, single-stranded DNA fixed to a solid support membrane (the Southern blot) is probed with a labeled, single-stranded DNA (the molecular probe). The labeled probe will hybridize to DNA fragments fixed on the membrane that contain complementary sequence. When the excess

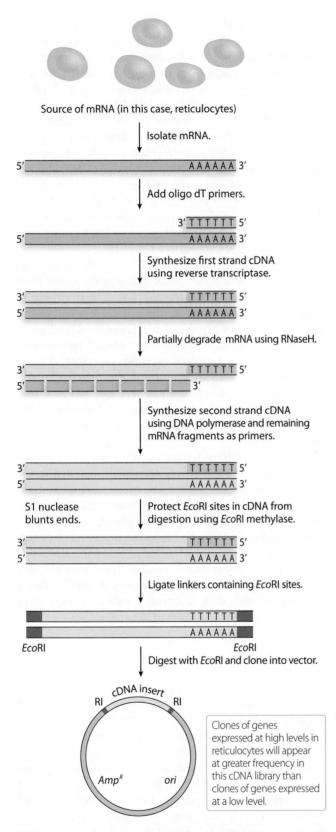

**Source of mRNA (in this case, reticulocytes)**

↓ Isolate mRNA.

5'  ▭AAAAAA 3'

↓ Add oligo dT primers.

3' TTTTTT 5'
5'  AAAAAA 3'

↓ Synthesize first strand cDNA using reverse transcriptase.

3'  TTTTTT 5'
5'  AAAAAA 3'

↓ Partially degrade mRNA using RNaseH.

3'  TTTTTT 5'
5'  ▯ ▯ ▯ ▯ ▯ ▯ 3'

↓ Synthesize second strand cDNA using DNA polymerase and remaining mRNA fragments as primers.

3'  TTTTTT 5'
5'  AAAAAA 3'

S1 nuclease blunts ends.     Protect *Eco*RI sites in cDNA from digestion using *Eco*RI methylase.

3'  TTTTTT 5'
5'  AAAAAA 3'

↓ Ligate linkers containing *Eco*RI sites.

TTTTTT
AAAAAA

*Eco*RI                          *Eco*RI

↓ Digest with *Eco*RI and clone into vector.

cDNA insert
RI          RI

Amp^R          ori

Clones of genes expressed at high levels in reticulocytes will appear at greater frequency in this cDNA library than clones of genes expressed at a low level.

**Figure 16.16  Construction of cDNA libraries.**

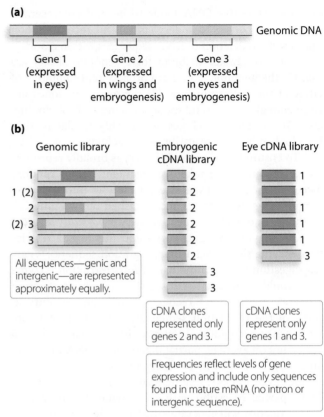

**(a)**

Genomic DNA

Gene 1 (expressed in eyes)    Gene 2 (expressed in wings and embryogenesis)    Gene 3 (expressed in eyes and embryogenesis)

**(b)**

Genomic library        Embryogenic cDNA library        Eye cDNA library

All sequences—genic and intergenic—are represented approximately equally.

cDNA clones represented only genes 2 and 3.

cDNA clones represent only genes 1 and 3.

Frequencies reflect levels of gene expression and include only sequences found in mature mRNA (no intron or intergenic sequence).

**Figure 16.17  Content of genomic versus cDNA libraries.**

(Figure 16.18). A membrane is laid on top of the bacterial colonies growing on a petri dish. Each colony contains clones of a different fragment from the library, and some of the bacteria in each colony stick to the membrane. The bacteria remaining on the petri dish serve as a resource for a later step in the procedure. The membrane-bound bacteria are lysed, and their DNA is denatured. The membrane can then be probed with a labeled single-stranded nucleic acid and treated as described for Southern blots. DNA that hybridizes with the probe is detected (e.g., by autoradiography), and the colonies it came from are identified by their position on the original petri dish. The same protocol is followed for phage, which form plaques in lawns of bacteria spread on petri dishes, and for yeast, which forms colonies similar to those of bacteria.

Genetic Insight  Genomic DNA sequences are represented equally in a genomic library, whereas transcribed sequences are represented in a cDNA library according to their abundance in the corresponding mRNA population.

## 16.5 Specific Genes Are Cloned Using Recombinant DNA Technology

To isolate a specific gene, an appropriate probe must be employed; but where does a gene-specific probe come from? The answer is something of a chicken-and-egg

probe is washed from the membrane the DNA fragments that have hybridized with the probe can be detected.

The same concept applies when a labeled probe is mixed with cloned DNA fragments from a library

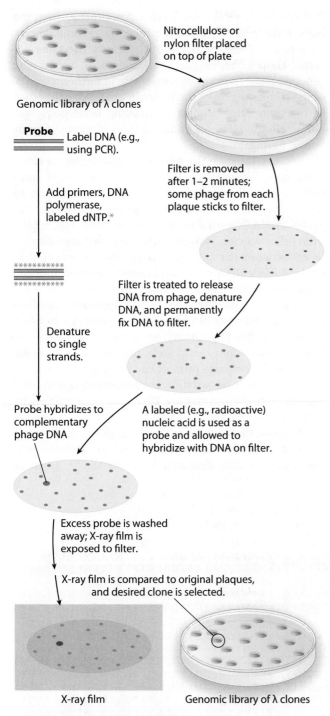

**Figure 16.18** **Screening libraries for specific sequences.**

## Cloning Highly Expressed Genes

Situations in which specific genes are expressed at high levels can be exploited to identify those genes. Consider a cDNA library made from mRNA isolated from human red blood cell precursors (reticulocytes). Up to 60% of the total protein in red blood cells is hemoglobin, so the genes for α- and β-globin are likely to be the most highly expressed genes in reticulocytes. If we isolate mRNA from reticulocytes, synthesize cDNA, and label the cDNA, we can use that labeled cDNA as a probe on a cDNA library made from red blood cells (Figure 16.19). It follows that the labeled cDNA representing the α- and β-globin genes will be among the most abundant of the labeled cDNA molecules. The mixture of labeled cDNAs is then hybridized with clones of the cDNA library. Since

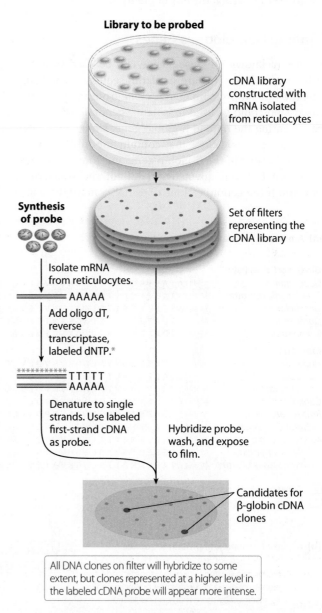

**Figure 16.19** **Cloning a highly expressed gene.** The expression of β-globin in reticulocytes is used as an example.

story, because a researcher must already have access to a similar sequence in order to identify a specific clone in a library. Fortunately, there are several ways to derive probes for seeking out specific DNA sequences or genes within a genomic or cDNA library. In the following discussion, we first consider approaches that take advantage of unique biological attributes of certain types of cells in order to clone specific genes. We then address the challenge of cloning genes for which only a mutant phenotype is known, such as genes identified in forward genetic screens.

the hybridization reaction is a bimolecular reaction, cDNA clones in the library that represent more highly expressed genes (e.g., β-globin) will hybridize more quickly and will exhibit a stronger signal than those less well represented in the labeled cDNA probe.

In our example, we might expect two different genes represented by strongly hybridizing clones, the β-globin gene and the α-globin gene. Since the β-globin and α-globin protein sequences are known, examination of the DNA sequences of strongly hybridizing clones could quickly identify the genes. However, this is a special situation, and most genes cannot be identified so easily. For example, in an experiment with *S. cerevisiae*, it was determined that cells can contain 15,000 mRNA molecules derived from about 2000 different genes, and the most commonly expressed genes are represented by 1% to 2% of the total mRNA population.

## Cloning Homologous Genes

Clones of known DNA sequence can be used to isolate other DNA clones that are similar in sequence. For example, the human β-globin cDNA clone from the preceding discussion could be used to isolate a genomic DNA clone representing the human β-globin gene. Unlike the cDNA clone, the genomic clone may contain not only the exons, but the introns of the β-globin gene and some sequences upstream (5′) and downstream (3′) of the transcribed portion. If the genomic clone does not contain the entire β-globin gene, we can use a reiterative process using the ends of the genomic clone to probe the genomic library once again. These probes should identify the same genomic clone as before plus some additional, overlapping genomic clones whose sequences extend farther 5′ and 3′ of the coding region of the β-globin gene. We will return to this reiterative process of isolating overlapping genomic clones when we consider the technique of positional cloning.

We can also use the human β-globin gene to clone other, homologous genes. If the degree of sequence similarity is enough to permit the homologous sequences to cross-hybridize, the human β-globin gene can be used to probe cDNA libraries made from reticulocyte mRNA from other species. However, when the phylogenetic distance between the species is large, the sequences may not be sufficiently similar to permit cross-hybridization, since stretches of 20 or so consecutive conserved bases are required. However, a PCR-type approach based on **reverse translation** of amino acid sequences can often circumvent these specificity problems. If sequences as short as 6 or 7 amino acids are conserved in the proteins of interest, degenerate primers can be designed using reverse translation of two different conserved regions of a protein, and the intervening DNA can be amplified by PCR using cDNA as a substrate.

Suppose that we wish to clone a β-globin gene from a wombat, an Australian marsupial, and we have the protein sequence data displayed in **Figure 16.20a**. In three

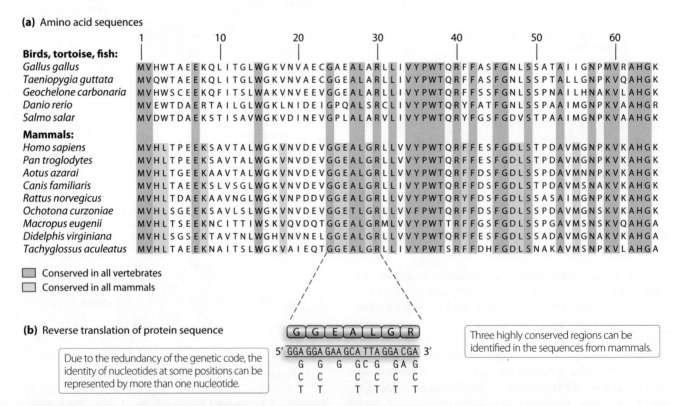

**Figure 16.20  Using reverse translation and PCR to clone homologous genes.**

regions, sequences of at least seven amino acids are conserved in all mammalian β-globin genes:

```
GGEALGR
LSELHCDKLHVDPENF
ALAHKYH
```

These sequences can be reverse translated (**Figure 16.20b**). Primers should be at least 20 nucleotides in length to be specific enough for use on a complex mixture of nucleic acids, such as genomic DNA or cDNA. The primers can be used to amplify cDNA, in a process called reverse transcription PCR (rtPCR) that starts with mRNA, which is reverse transcribed into cDNA, which is then used as the template for a PCR reaction. With cDNA as a template, either two gene-specific primers can be used or, alternatively, one gene-specific primer in combination with an oligo dT primer can be used. PCR-based techniques can be used to amplify any gene that shares enough sequence similarity to a known homologous gene. Since most genes belong to a family of related genes, this approach revolutionized the study of homologous genes both within and across species.

## Cloning Genes by Complementation

Another approach to identifying specific genes is to detect genetic complementation of a mutant phenotype by an introduced wild-type gene. This approach is restricted to cases in which transgenic organisms can be generated. Consider the yeast temperature-sensitive cell-cycle mutants described in Section 16.1. If clones of a yeast cDNA expression library are transformed into a yeast cell-cycle mutant, any clones that complement the mutant phenotype so that the cells grow normally should contain wild-type alleles of the mutated gene (**Figure 16.21**). In this case, the yeast strain would first be transformed and grown at the permissive temperature. The resulting yeast colonies would then be transferred to an environment maintained at the restrictive temperature. Only the yeast colonies receiving a clone encoding a wild-type version of the mutant gene in question would be able to continue growth at the restrictive temperature; in those colonies, the mutant phenotype would have been complemented by the added gene.

Complementation experiments can also be used to identify orthologous genes from other species, if there is sufficient conservation of protein function. For example, research in which a yeast cell-cycle mutant was transformed using a human cDNA expression library (one in which the human cDNA clones were first fused with sequences allowing for their transcription and translation in yeast) has led to identification of human genes orthologous to the mutated yeast genes. The fact that both human and plant genes can complement these yeast mutants demonstrates the universality of the cell cycle machinery and indicates that such proteins were present in the common ancestor of eukaryotes.

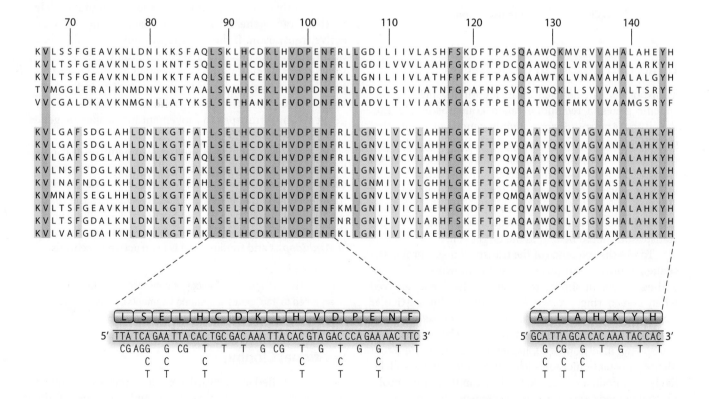

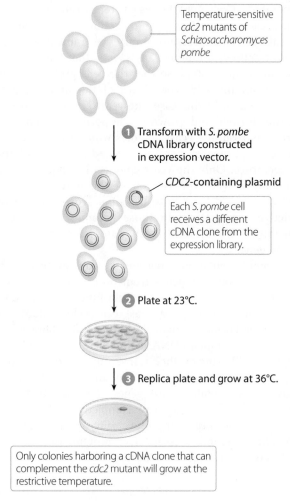

**Temperature-sensitive** *cdc2* mutants of *Schizosaccharomyces pombe*

① Transform with *S. pombe* cDNA library constructed in expression vector.

*CDC2*-containing plasmid

Each *S. pombe* cell receives a different cDNA clone from the expression library.

② Plate at 23°C.

③ Replica plate and grow at 36°C.

Only colonies harboring a cDNA clone that can complement the *cdc2* mutant will grow at the restrictive temperature.

**Figure 16.21 Cloning by complementation.**

## Using Transposons to Clone Genes

Transposons can be used as an identifying tag to clone specific genes, a technique called **transposon tagging.** Recall that transposons are mobile genetic elements that can integrate into the genome with little if any target-sequence specificity (see Chapter 13). If the sequence of a transposon is known, the transposon sequence can be used as a probe on a genomic library constructed from DNA of a strain in which the same transposon has been inserted into a target gene. Sequences adjacent to the transposon should belong to the target gene.

To use the sequence of the transposon as a probe, the sequence of the transposon must be known—a chicken-and-egg problem similar to others we have encountered when considering probes. A solution in this particular case is to "trap" the transposon in a gene whose sequence is already known. Allele instability is characteristic of transposon insertion (see Chapter 13). If researchers first identify unstable mutant alleles of a cloned gene—alleles likely to contain a transposon—they can then use a probe for that cloned gene to isolate the transposon sequence.

For transposon tagging to succeed in practice, the biology of transposons must be considered. Since transposons often occur in high copy numbers in the genome, techniques are necessary to distinguish the copy of the transposon in the target gene from all other copies of the transposon in the genome. The ideal situation is to begin with a genotype harboring only a single copy or a low copy number of a transposon of known sequence and then mobilize the transposon to create a mutant collection, which is then screened for phenotypes. Another consideration is that, since transposons are mobile, a transposon that has been inserted into the target gene may jump out again. To circumvent this problem, the transposon that is used as a tag is often separated into two components—the transposase and the inverted repeats whose sequence the transposase recognizes (see Section 13.5). The inverted repeats form the functional part of a nonautonomous element that cannot move on its own but can be mobilized if transposase is supplied in trans. Ideally, the transposase activity is produced from a mutant transposon that is not capable of moving because it lacks the inverted repeats. The new insertions of the nonautonomous transposon can be stabilized by removal of the transposase source through outcrossing.

A general protocol for using a transposon to tag a gene in a diploid eukaryote is shown in **Figure 16.22.** Two lines are initially crossed, one of which is homozygous for a stable mutant allele of the target gene and the second of which carries an active transposon system and is homozygous for the wild-type allele at the target gene. The $F_1$ of this cross is heterozygous for the mutant allele of the target gene in a genomic background with an active transposon. If the transposon moves into the gene of interest, thus creating a second mutant allele, the $F_1$ individual will display the homozygous recessive mutant phenotype. Screening of a large number of $F_1$ individuals is usually required to find any with mutations in the target gene, since transposon movement into a specific gene is a rare event.

Transposon tagging is limited to organisms that harbor active transposons or into which an active transposon can be introduced from another species. However, this limitation is not often a problem since, for example, the maize Ac/Ds transposon system (see Section 13.5), when introduced on a transgene, is active in other plant species, such as *Arabidopsis* and tomato, and is even active in zebrafish.

**Genetic Insight** The biology of transposons can be exploited to "tag" genes, permitting fragments containing the genes to be identified for molecular cloning.

## Positional Cloning

Genes identified in forward genetic screens are known only by their mutant phenotype, and so none of the

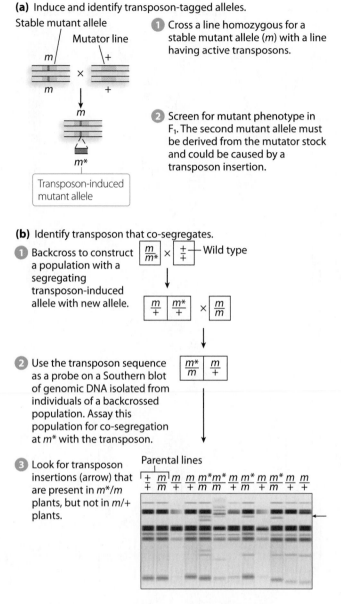

**(a)** Induce and identify transposon-tagged alleles.

Stable mutant allele

Mutator line

*m*

*m*

*+*

*+*

×

❶ Cross a line homozygous for a stable mutant allele (*m*) with a line having active transposons.

*m*

*m\**

Transposon-induced mutant allele

❷ Screen for mutant phenotype in F₁. The second mutant allele must be derived from the mutator stock and could be caused by a transposon insertion.

**(b)** Identify transposon that co-segregates.

❶ Backcross to construct a population with a segregating transposon-induced allele with new allele.

$\frac{m}{m*}$ × $\frac{+}{+}$ — Wild type

$\frac{m}{+}$ $\frac{m*}{+}$ × $\frac{m}{m}$

❷ Use the transposon sequence as a probe on a Southern blot of genomic DNA isolated from individuals of a backcrossed population. Assay this population for co-segregation at *m\** with the transposon.

$\frac{m*}{m}$ $\frac{m}{+}$

❸ Look for transposon insertions (arrow) that are present in *m\**/*m* plants, but not in *m*/+ plants.

Parental lines

$\frac{+}{+}$ $\frac{m}{m}$ $\frac{m}{+}$ $\frac{m}{+}$ $\frac{m*}{m}$ $\frac{m*}{m}$ $\frac{m}{+}$ $\frac{m*}{m}$ $\frac{m}{+}$ $\frac{m*}{m}$ $\frac{m}{+}$ $\frac{m}{+}$

**Figure 16.22** **Use of transposons for tagging genes.**

approaches to cloning genes we have thus far discussed are applicable. How do biologists find the DNA sequence for a gene when only a mutant phenotype is known? They do it by combining a genetic map made from recombination frequencies (Chapter 5) with a physical map of the genome based on DNA clones, or, when available, the genome sequence, in order to find the DNA sequence at a specific map position. **Figure 16.23** provides an overview of the relationships between genetic maps based on the segregation of genetic loci, physical maps based on sets of overlapping genomic clones, genes, and the DNA sequence of the genome. If two DNA markers are identified that flank the gene of interest, the gene must reside in the intervening DNA. The DNA for the gene can ultimately be identified by isolating a set of DNA clones that collectively span the region between the flanking markers. This approach is referred to as **positional cloning, or**

**chromosome walking,** since it consists of "walking" along the chromosome in sequential steps, from one flanking marker toward the other, by joining overlapping DNA clones.

Positional cloning is done in three steps. The first step is to construct a genetic map that shows the location of the gene of interest relative to mapped DNA markers. The second step is to identify DNA clones that span the markers flanking the gene of interest. The third step is to identify the gene of interest within the spanning DNA and determine its nucleotide sequence.

**Step 1: Constructing a Genetic Map Using DNA Markers** In 1980, a landmark paper by David Botstein and colleagues proposed an idea for creating a genetic map based on DNA markers, for the purpose of performing positional cloning of human genes. In the decades that followed, the cloning of many human "disease genes" was accomplished using this protocol. Even now that the human genome has been sequenced, a similar mapping protocol continues to be used for gene identification in humans, and the general approach to positional cloning, as described here, can be applied to any organism.

The key to positional cloning is to identify molecular markers flanking the gene you wish to clone. The flanking DNA markers define the two ends of the DNA sequence in which the gene of interest is encoded. Any DNA sequence that varies between individuals can potentially be a DNA marker, including single nucleotide polymorphisms (SNPs), restriction fragment length polymorphisms (RFLPs), small insertions or deletions, and various repeating DNA sequence variants (see Chapter 10). Once a collection of polymorphic DNA markers has been identified, detecting the segregation of these markers in a "mapping population" will allow placement of each marker at a particular location in the genome. Map construction with molecular markers follows the same procedure as with phenotypic markers (see Chapter 5); different DNA markers that co-segregate are physically linked, with a recombination frequency proportional to the distance in map units between them.

To examine how mapping works, let's take an example from *Arabidopsis* in which a gene is mapped in an F₂ population (**Figure 16.24**). The first step in the construction of a genetic map is to identify two strains that differ in DNA sequence, in this case, Landsberg (L) and Columbia (C). Each strain is highly homozygous due to inbreeding, yet they differ from each other at polymorphic loci throughout the genome. The mapping population is generated by crossing the two homozygous lines to produce an F₁ generation that is heterozygous at all loci that differ between the two inbred lines. These F₁ individuals are then interbred or allowed to self-fertilize. At each locus in the genome, individuals in the resulting

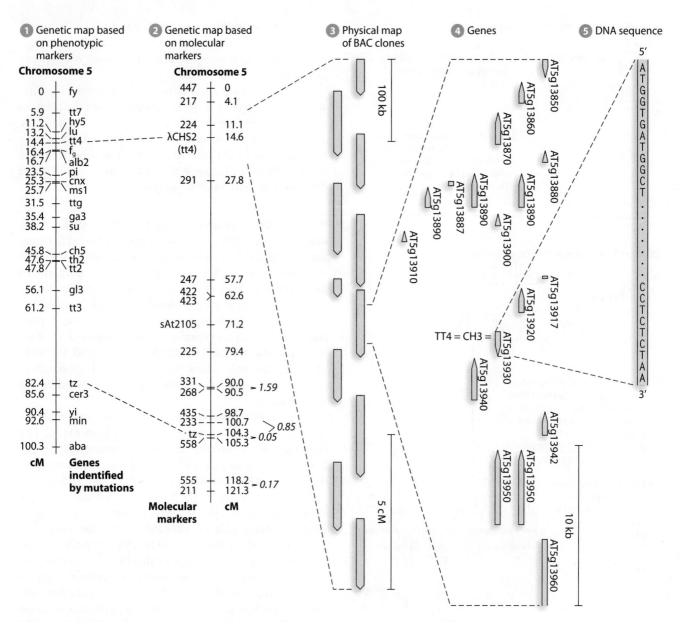

**Figure 16.23 Correlating genetic maps and physical maps of *Arabidopsis* to locate DNA sequences of genes.**

$F_2$ population can be either homozygous for alleles of one or the other of the original inbred lines, or they can be heterozygous. Alleles for the gene of interest, in this case *AP2*, are also segregating in the $F_2$ population, since one parent (L) was homozygous for a recessive *ap2* mutant allele while the other parent (C) was homozygous for the wild-type *AP2* allele.

The genotypes of each $F_2$ individual are determined for the DNA markers and the gene of interest. DNA markers that co-segregate with the mutation are linked to the gene of interest, where the distance from the gene is proportional to the recombination frequency; unlinked DNA markers should segregate independently of the mutation. In most cases, the alleles of DNA markers are codominant, so that examination of DNA allows direct determination of genotype. In contrast,

only $F_2$ individuals homozygous for a recessive mutation in the gene of interest can be accurately genotyped, and the genotype of phenotypically wild-type $F_2$ individuals has to be determined in the $F_3$ or by a test cross. While this example is one using a model genetic system, genetic mapping in humans with molecular markers follows a similar protocol (modifications described in Chapter 5).

Because the number of DNA sequence differences between two strains is likely to be greater than the total number of genes in the organism, genetic maps based on DNA markers are often dense enough for flanking markers closely linked to the gene of interest to be identified. Once DNA markers are found that flank the gene of interest, the next step of identifying the DNA between the markers can proceed.

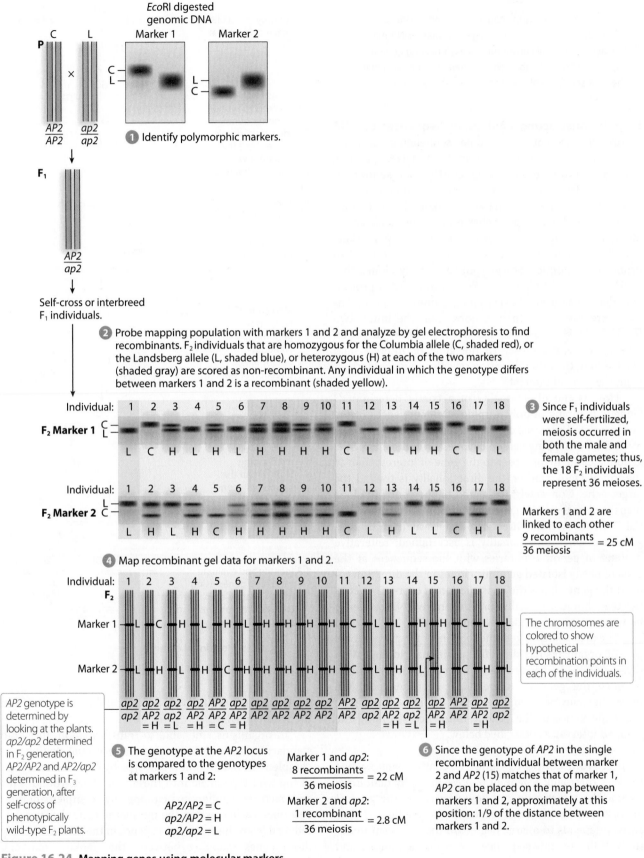

**Figure 16.24  Mapping genes using molecular markers.**

Genetic Insight  DNA sequence polymorphisms between individuals are used to construct genetic maps. DNA markers then act as points of correspondence connecting genetic maps (based on recombination frequency) to physical maps (the DNA sequence of the chromosomes) of a species.

## Step 2: Constructing Contiguous Sequences of DNA

Before the advent of genome sequencing projects, researchers were forced to assemble the DNA spanning two markers by constructing **contiguous sequences (contigs),** from sets of overlapping genomic clones (**Figure 16.25**). The DNA markers flanking the gene of interest ❶ can be used as probes on a genomic library to identify genomic clones that contain the DNA markers ❷. The ends of these genomic clones can be used to probe the genomic library again to identify clones that overlap the initial clones ❸. Reiteration of this process will identify additional overlapping genomic clones (contigs) extending in both directions from the initial two flanking DNA markers. Extension in one direction reveals sequences closer to the gene of interest, and extension in the other direction reveals sequences farther from the gene of interest.

How is the directionality of a contig determined? Genetic mapping of polymorphic DNA sequences in the newly isolated genomic clones can resolve the directionality of the overlapping genomic clones ❹. If the end of the genomic clone maps closer to the gene of interest than to the initial DNA marker, the directionality is toward the target gene. Conversely, if the end of the genomic clone maps farther from the gene of interest than from the initial DNA marker, the directionality is away from the target gene. Once directionality is ascertained, reiterative probing of genomic libraries with the sequences at the ends of newly isolated genomic clones in the direction toward the gene allows the construction of ever-larger contigs that should eventually span the entire DNA sequence between the two flanking DNA markers ❺.

The availability of genome sequences for many model genetic organisms has simplified positional cloning. With these species, the construction of a contig is not required, so once the gene of interest has been mapped, the researcher can proceed directly to the identification of candidate genes in the genome sequence spanning the mapped interval, as described below.

## Step 3: From Contig to Gene

A contig spanning two markers that flank a gene of interest must, by definition, include the target gene, but how do we find the gene of interest among the other genes in the contig ❻? The answer depends to a large degree on the organism under study. If the genome sequence is known, the number and identity of candidate genes—that is, sequences that could encode the gene of interest—are essentially known. In contrast, if the genome sequence is not known, experimental

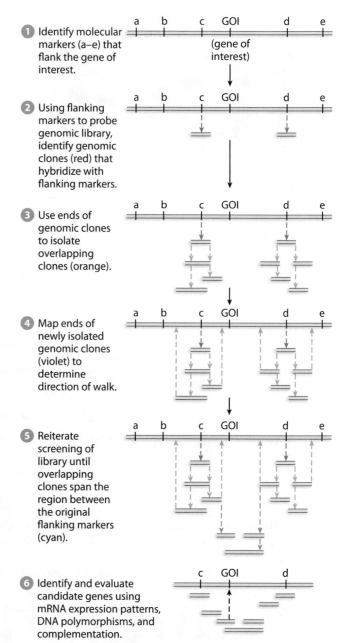

① Identify molecular markers (a–e) that flank the gene of interest.

(gene of interest)

② Using flanking markers to probe genomic library, identify genomic clones (red) that hybridize with flanking markers.

③ Use ends of genomic clones to isolate overlapping clones (orange).

④ Map ends of newly isolated genomic clones (violet) to determine direction of walk.

⑤ Reiterate screening of library until overlapping clones span the region between the original flanking markers (cyan).

⑥ Identify and evaluate candidate genes using mRNA expression patterns, DNA polymorphisms, and complementation.

**Figure 16.25  Positional cloning.**

approaches are required to identify candidate genes within the spanning DNA.

In organisms amenable to transformation, the "gold standard" of gene identification for positional cloning is to complement the mutant phenotype by introducing a copy of the wild-type allele into the mutant background. This approach is similar to cloning by complementation described earlier, except the number of candidate genes is reduced from the entire set of genes in the genome to only those genes mapping between the flanking markers. Transformation experiments are routine in many model genetic organisms and are described in more detail in Chapter 17.

In organisms not amenable to transformation (e.g., humans), other approaches can be used to identify and characterize candidate genes. First, direct sequencing of candidate genes and comparison of the sequences in wild-type and mutant individuals can reveal the gene of interest. Missense or nonsense mutations might be expected in each of the mutant alleles relative to the wild-type allele. However, because mutations outside of the coding region may be responsible for the altered gene expression, noncoding sequences may have to be surveyed as well. Note that for non-inbred species, if there is only a single mutant allele to examine, it may be difficult to tell whether differences in the DNA sequences of candidate genes are the cause of the mutant phenotype or simply polymorphisms existing in the population.

In a second approach to identifying the target gene, the nature of the phenotypic defect conferred by the mutant allele may provide clues to probable gene expression patterns, and candidate genes can be assayed for those expression patterns in specific cells and tissues. It

may also be possible to detect changes in RNA expression patterns, for example, mutations resulting in altered patterns of splicing, or those resulting in mRNA that is less stable than the wild-type mRNA. Genes can also be surveyed based on the type of protein they are thought to encode. If it is possible to predict the biochemical function of the target gene, some candidate genes may have features that make them appear more likely than others to be able to perform that function. However, in many cases this knowledge will be lacking.

This general approach to positional cloning has been applied to various model genetic systems. In the 1980s and 1990s, many genes in *Drosophila, C. elegans,* and *Arabidopsis* were identified by positional cloning protocols, long before their genomes were sequenced. Positional cloning strategies have been particularly successful in identifying genes associated with human diseases, despite the infeasibility of performing controlled crosses and complementation experiments in humans. The Case Study describes one such effort that led researchers to identify the gene responsible for Huntington disease.

## CASE STUDY

### Positional Cloning in Humans: The Huntington Disease Gene

Huntington disease, an inevitably fatal, late-onset neurodegenerative disorder, is named for George Huntington, the physician who published the classic description of the disease and its inheritance in 1872. His description specified the symptoms of movement disorder, personality change, and cognitive decline and, notably, outlined the autosomal dominant pattern of inheritance, a feature that went unappreciated until after the rediscovery of Mendel's work in 1900. Huntington recognized the pattern of inheritance thanks to the collective experience of his father and grandfather, both also physicians, who had the unique opportunity of observing several generations of the disease in a local family. He did not encounter the juvenile onset form of the disease, however, which presents additional symptoms, such as rigidity and seizures. In a form of inheritance termed *anticipation*, juvenile-onset Huntington is inherited through a paternal allele.

**MAPPING OF THE *HD* GENE** Researchers have compiled extensive pedigrees depicting the transmission of Huntington disease in a large family in Venezuela. The pedigrees span 10 generations and include nearly 20,000 individuals, three-quarters of whom are living. In the early 1980s, James Gusella and Susan Wexler and colleagues, studying this Venezuelan kindred as well as a large Ohio family, mapped the *HD* gene to the short arm of chromosome 4 (see Chapter 5 Case Study). Additional polymorphic markers linked to dominant mutant alleles of the *HD* gene further confined it to a region of 2.2 Mb on chromosome 4 (**Figure 16.26**). Mutant *HD* alleles were known from a large number of unrelated families from diverse genetic backgrounds, suggesting that dominant mutant *HD* alleles have arisen multiple times independently. Mapping data from 75 families identified a haplotype shared by about

one-third of the families, suggesting the *HD* gene was likely to reside within 500 kb of the shared haplotype (see Chapter 5 for a discussion of haplotypes).

**CANDIDATE GENE IDENTIFICATION** Before 2001, the year a draft of the human genome sequence was published, identification of genes in large stretches of human genomic sequence was an arduous task. In the case of cloning of the *HD* gene in the early 1990s, it required construction of a contig of cosmid and YAC clones spanning the *HD* locus, using the techniques described throughout this chapter for isolating overlapping genomic clones. To identify transcribed sequences within the 500-kb genomic region, a novel exon-trapping approach was used. Fragments of the genomic DNA were cloned into a vector, where they were flanked by two exons contained in the vector sequence. When assayed in human cells in culture, if the genomic DNA did not contain an exon, the two vector exons would become spliced together in post-transcriptional processing, generating a transcript of a defined size. However, if the cloned genomic DNA contained an exon, it would be spliced to the two flanking exons, creating a transcript of a larger size. This technique revealed the presence of four transcribed genes in the region.

Two approaches were undertaken to evaluate the candidate genes. First, the mRNA expression patterns of the genes were analyzed. However, no difference in expression patterns or levels for any of the four genes was detected between normal and HD individuals. Second, the candidate genes were examined for DNA polymorphisms. One of the candidate genes was polymorphic between individuals. A striking difference in the lengths of a trinucleotide repeat sequence in this gene was observed; normal individuals had 17 to 34

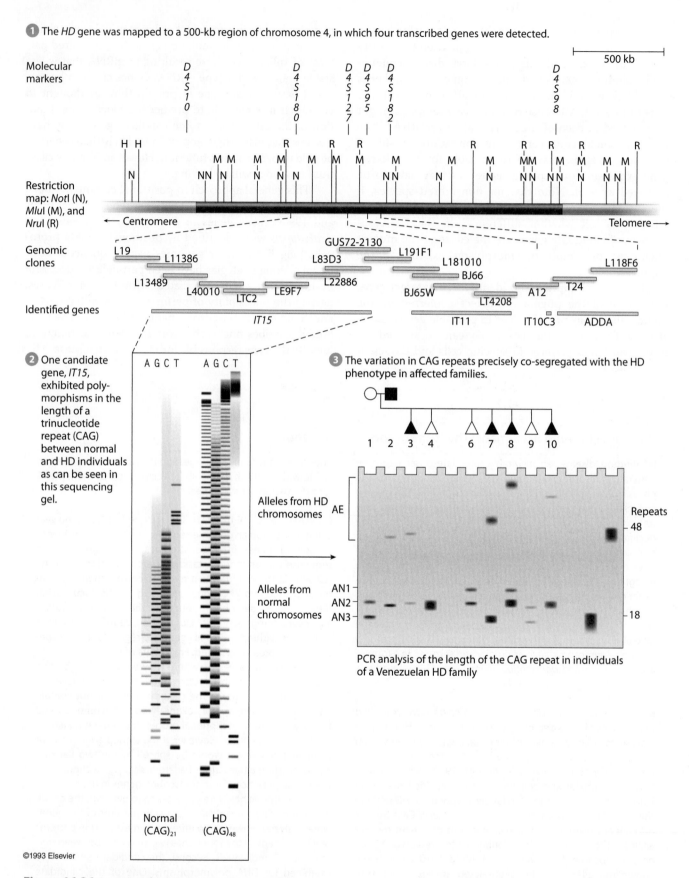

1  The *HD* gene was mapped to a 500-kb region of chromosome 4, in which four transcribed genes were detected.

2  One candidate gene, *IT15*, exhibited polymorphisms in the length of a trinucleotide repeat (CAG) between normal and HD individuals as can be seen in this sequencing gel.

3  The variation in CAG repeats precisely co-segregated with the HD phenotype in affected families.

PCR analysis of the length of the CAG repeat in individuals of a Venezuelan HD family

©1993 Elsevier

**Figure 16.26  Locating the Huntington disease gene.**

copies of a CAG repeat, and HD individuals had from 42 to more than 66 copies. The same correlation was seen in all 75 families, strongly suggesting that this was the *HD* gene. As further supporting evidence, the length of the repeats in the HD individuals also correlated with the age of onset of disease symptoms.

The *HD* gene spans 210 kb, encoding an mRNA of over 10 kb, and has an open reading frame of 9432 bases encoding a protein of 3144 amino acids. In this case, there is little in the protein sequence that provides a clue to function (and possible treatment). However, knowledge of the DNA sequence has provided a way of testing for the presence of the disease allele in families in which it is segregating. This information can be used in prenatal diagnostics to eliminate the allele from the next generation if therapeutic abortion is an option. While this test may seem to be a blessing, it introduces many ethical quandaries. Should one test a child for an adult-onset disease where there is no prospect for treatment or a cure at present? Might testing in a young adult inadvertently provide information about another individual, such as a parent, who does not wish to know his or her own genetic status?

Analysis of the polymorphic CAG repeat has also provided insight into the phenomenon of anticipation. The CAG alleles whose length approaches the high end of the normal range [(CAG)$_{27-35}$] are unstable during transmission and change size from one generation to the next. Instability occurs in both maternal and paternal inheritance, but large expansions have been noted only during male transmission; this explains why juvenile patients almost always inherit the mutant allele from their father. While the molecular basis of this gender bias has become apparent, the mechanistic basis is still unknown.

---

## SUMMARY

*For activities, animations, and review quizzes, go to the study area at www.masteringgenetics.com.*

### 16.1 Forward Genetic Screens Identify Genes by Their Mutant Phenotypes

▌ Forward genetic screens are designed to identify genes by creation of a mutant phenotype, often allowing researchers to infer the biological function of a gene.

▌ Complementation tests are used to discover the number of alleles and the number of genes affected in a forward genetic screen.

▌ Mutations resulting in lethality can be identified in genetic screens for conditional alleles.

▌ Enhancer and suppressor genetic screens identify genes that act in related or redundant pathways.

### 16.2 Specific DNA Sequences Are Recognized Using Recombinant DNA Technology

▌ Restriction enzymes, which cut at specific DNA sequences, are used to fragment large DNA molecules into defined smaller pieces.

▌ A restriction map of a DNA molecule can be constructed by analyzing patterns of DNA fragments after restriction enzyme digestion.

▌ DNA fragments can be ligated to create recombinant DNA molecules, usually composed of a vector that can be amplified in a biological system and a target DNA insert to be amplified.

▌ While cohesive compatible ends facilitate the creation of recombinant DNA molecules, any two DNA fragments can be ligated if their ends are made blunt.

▌ Use of two different restriction enzymes facilitates directional cloning of DNA molecules.

▌ Amplification of recombinant DNA molecules in a biological system allows the production of DNA clones.

▌ Bacteriophage and bacterial and yeast artificial chromosomes allow the cloning of large DNA molecules.

▌ The polymerase chain reaction is an enzymatic method of amplifying DNA, which can then be cloned into a vector or sequenced directly.

### 16.3 DNA Sequencing Technologies Have Revolutionized Biology

▌ Long DNA molecules can be sequenced using primer walking methods or by shotgun sequencing and reassembly via computer algorithms.

▌ Massively parallel sequencing technologies permit DNA to be sequenced without first cloning the DNA into a vector.

### 16.4 Collections of Cloned DNA Fragments Are Called Libraries

▌ Genomic libraries are collections of cloned DNA fragments that represent the entire genome of an organism.

▌ cDNA libraries are collections of cloned DNA fragments that represent the mRNA population of an organism or tissue.

▌ DNA hybridization, which depends on complementary base pairing, is a means of identifying similar sequences in a mixture of DNA sequences.

### 16.5 Specific Genes Are Cloned Using Recombinant DNA Technology

▌ Homologous genes can be identified by using similar DNA sequences as probes or by using conserved sequences to design PCR primers so as to amplify related sequences.

▌ Some genes can be cloned by complementation of a mutant phenotype.

▌ Transposons and other integrating elements can be used to tag genes, facilitating their subsequent cloning.

▌ Positional cloning, or chromosome walking, provides a means of identifying cloned genes known only from a mutant phenotype.

▌ In positional cloning approaches, mutations are first mapped and then contigs of DNA are isolated that span the target gene. The target gene can be identified by expression analyses, DNA sequence analyses, or complementation experiments.

## KEYWORDS

## PROBLEMS

*For instructor-assigned tutorials and problems, go to www.masteringgenetics.com.*

## Chapter Concepts

*For answers to selected even-numbered problems, see Appendix: Answers.*

1. What purpose do the *β-lactamase* and *lacZ* genes serve in the plasmid vector pUC18?

2. The human genome is $3 \times 10^9$ bp in length.

   a. How many fragments would be predicted to result from the complete digestion of the human genome with the following enzymes: *Sau*3A (˅GATC), *Bam*HI (G˅GATCC), *Eco*RI (G˅AATTC), and *Not*I (GC˅GGCCGC)?

   b. How would your initial answer change if you knew that the average GC content of the human genome was 40%?

3. Ligase catalyzes a reaction between the 5′-phosphate and the 3′-hydroxyl at the ends of DNA molecules. The enzyme calf intestinal phosphatase catalyzes the removal of the 5′-phosphate from DNA molecules. What would be the consequence of treating the vector, before ligation, with calf intestinal phosphatase?

4. You have constructed four different libraries: a genomic library made from DNA isolated from human brain tissue, a genomic library made from DNA isolated from human muscle tissue, a human brain cDNA library, and a human muscle cDNA library.

   a. Which of these would have the greatest diversity of sequences?

   b. Would the sequences contained in each library be expected to overlap completely, partially, or not at all?

5. You wish to clone the human gene encoding myostatin, which is expressed only in muscle cells.

   a. Assuming the human genome is $3 \times 10^9$ bp and that the average insert size in the genomic libraries is 100 kb, how frequently will a clone representing myostatin be found in the genomic library made from muscle?

   b. How frequently will a clone representing myostatin be found in the genomic library made from brain?

   c. How frequently will a clone representing myostatin be found in the cDNA library made from muscle?

   d. How frequently will a clone representing myostatin be found in the cDNA library made from brain?

6. The human genome is $3 \times 10^9$ bp. You wish to design a primer to amplify a specific gene in the genome. In general, what length of oligonucleotide would be sufficient to amplify a single unique sequence? To simplify your calculation, assume that all bases occur with an equal frequency.

7. Based on the sequences in Figure 16.21, how would you clone a β-globin gene from the California newt?

8. Genetic maps and physical maps are both representations of a genome.

   a. What are the similarities and differences between how genetic and physical maps are created?

   b. If genetic maps of a particular organism are independently constructed in two different laboratories, will they be identical? What about two independently constructed physical maps?

   c. How can the information in genetic and physical maps be combined?

9. It is often desirable to clone cDNAs into a vector in such a way that all the cDNA clones will have their 3′ end in one orientation in the plasmid and their 5′ end in the other orientation. This is referred to as directional cloning. Outline how you would directionally clone a cDNA library in the plasmid vector pUC18.

10. You sequence your genomic DNA using both Sanger sequencing and a massively parallel sequencing technology. Will there be differences between the types of sequences represented in the two data sets? If so, what types of differences?

11. Using the data on the inside front cover, calculate the average number of kilobase (kb) pairs per centimorgan in the

four multicellular eukaryotic organisms. How would this information influence strategies to positionally clone genes in these organisms?

12. A major advance in the 1980s was the development of technology to synthesize short oligonucleotides. This work both facilitated DNA sequencing and led to the advent of the development of PCR. Recently, rapid advances have occurred in the technology to chemically synthesize DNA, and sequences up to 10 kb are now readily produced. As this process becomes more economical, how will it affect the gene-cloning approaches outlined in this chapter? In other words, what types of techniques does this new technology have potential to supplant, and what techniques will not be affected by it?

13. Assuming that sequencing technologies will continue to advance and that everyone will soon have his or her own genome sequenced as part of a standard medical procedure, discuss some of the ethical ramifications of the knowledge generated.

## Application and Integration

*For answers to selected even-numbered problems, see Appendix: Answers.*

14. The bacteriophage lambda genome can exist in either a linear form (see Figures 16.6 and 16.13) or a circular form. The circular form occurs when the 20-bp cos sites (cohesive ends) anneal at their complementary base pairs and are ligated.
    a. How many fragments will be formed by restriction enzyme digestion with *Xho*I, *Xba*I, and both *Xho*I and *Xba*I in the linear and circular forms of the lambda genome?
    b. Diagram the resulting fragments as they would appear on an agarose gel after electrophoresis.

15. The restriction enzymes *Xho*I and *Sal*I cut their specific sequences as shown below:

    | | | |
    |---|---|---|
    | *Xho*I | 5'-C | TCGAG-3' |
    | | 3'-GAGCT | C-5' |
    | *Sal*I | 5'-G | TCGAC-3' |
    | | 3'-CAGCT | G-5' |

    Can the sticky ends created by *Xho*I and *Sal*I sites be ligated? If yes, can the resulting sequences be cleaved by either *Xho*I or *Sal*I?

16. The bacteriophage φX174 has a single-stranded DNA genome of 5386 bases. During DNA replication, double-stranded forms of the genome are generated. In an effort to create a restriction map of φX174, you digest the double-stranded form of the genome with several restriction enzymes and obtain the following results. Draw a map of the φX174 genome.

    | | | | |
    |---|---|---|---|
    | *Pst*I | 5386 | *Pst*I + *Psi*I | 3078, 2308 |
    | *Psi*I | 5386 | *Pst*I + *Dra*I | 331, 1079, 3976 |
    | *Dra*I | 4307, 1079 | *Psi*I + *Dra*I | 898, 1079, 3409 |

17. You have identified a 0.80-kb cDNA clone that contains the entire coding sequence of the *Arabidopsis* gene *CRABS CLAW*. In the construction of the cDNA library, linkers with *Eco*RI sites were added to each end of the cDNA, and the cDNA was cloned into the *Eco*RI site of the MCS of the vector shown below. You perform digests on the *CRABS CLAW* cDNA clone with restriction enzymes and obtain the following results. Can you determine the orientation of the cDNA clone with respect to the restriction enzyme sites in the vector? The enzymes listed in green region are found only in the MCS of the vector.

    | | |
    |---|---|
    | *Eco*RI | 0.8, 3.0 |
    | *Hind*III | 0.3, 3.5 |
    | *Eco*RI + *Hind*III | 0.3, 0.5, 3.0 |

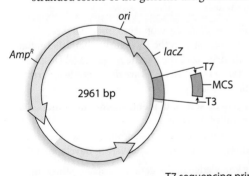

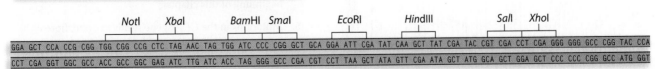

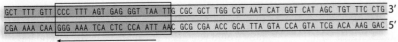

18. You have isolated a genomic clone with an *Eco*RI fragment of 11 kb that encompasses the *CRABS CLAW* gene (see Problem 17). You digest the genomic clone with *Hind*III and note that the 11-kb *Eco*RI fragment is split into three fragments of 9 kb, 1.5 kb, and 0.5 kb.

    a. Does this tell you anything about where the *CRABS CLAW* gene is located within the 11-kb genomic clone?

    b. Restriction enzyme sites within a cDNA clone are often also in the genomic sequence. Can you think of a reason why occasionally this is not the case? What about the converse: Are restriction enzyme sites in a genomic clone always in a cDNA clone of the same gene?

19. To further analyze the *CRABS CLAW* gene (see Problem 18), you create a map of the genomic clone. The 11-kb *Eco*RI fragment is cloned into the *Eco*RI site of the MCS of the vector shown in Problem 17.

    You digest the double-stranded form of the genome with several restriction enzymes and obtain the following results. Draw, as far as possible, a map of the genomic clone of *CRABS CLAW*.

    | *Eco*RI | 11.0, 3.0 | | |
    |---|---|---|---|
    | *Eco*RI + *Xba*I | 4.5, 6.5, 3.0 | *Xba*I | 4.5, 9.5 |
    | *Eco*RI + *Xho*I | 10.2, 3.0, 0.8 | *Xho*I | 13.2, 0.8 |
    | *Eco*RI + *Sal*I | 6.0, 5.0, 3.0 | *Sal*I | 6.0, 8.0 |
    | *Eco*RI + *Hind*III | 9.0, 3.0, 1.5, 0.5 | *Hind*III | 12.0, 1.5, 0.5 |

    What restriction digest would help resolve any ambiguity in the map?

20. You have isolated another cDNA clone of the *CRABS CLAW* gene from a cDNA library constructed in the vector shown in Problem 17. The cDNA was directionally cloned using the *Eco*RI and *Xho*I sites. You sequence the recombinant plasmid using primers complementary to the T7 and T3 promoter sites flanking the MCS. The first 30 to 60 bases of sequence are usually discarded since they tend to contain errors.

    a. Which sequence represents the 5′ end of the gene? Which sequence represents the 3′ end of the gene?

    b. Will the long stretch of T residues in the T3 sequence exist in the genomic sequence of the gene?

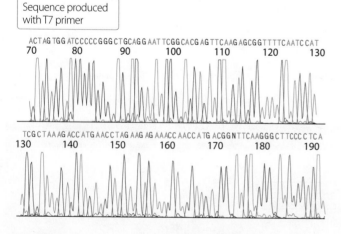

Sequence produced with T7 primer

Sequence produced with T3 primer

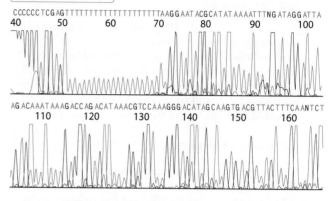

c. Can you identify which sequences are derived from the vector (specifically the MCS) and which sequences are derived from the cDNA clone?

d. Can you identify the start of the coding region in the 5′ end of the gene? What does the sequence preceding the start codon represent?

21. In our screen in Genetic Analysis 16.1, designed to identify conditional mutants of *S. cerevisiae* in which the secretory system was defective, we were successful in identifying 12 mutants.

    a. Describe the crosses you would perform to determine the number of different genes represented by the 12 mutations.

    b. Based on your knowledge of the genetic tools for studying baker's yeast, how would you clone the genes that are mutated in your respective yeast strains? What are two approaches to cloning the human orthologs of the yeast genes?

22. How would you design a genetic screen to find genes involved in meiosis?

23. The eyes of *Drosophila* develop from imaginal discs, groups of cells set aside in the fly embryo that differentiate into the adult structures during the pupal stage. Despite their importance in nature, eyes are dispensable for fruit-fly life in the laboratory.

    a. Devise a genetic screen to identify genes directing development of the fly eye.

    b. What complications might arise from genetic screens targeting an organ that differentiates late in development?

24. Given your knowledge of the genetic tools for studying *Drosophila*, outline two methods by which you could clone the *dunce* and *rutabaga* genes identified by Seymour Benzer's laboratory in the genetic screen described at the beginning of this chapter.

25. Mutations in the *CFTR* gene result in cystic fibrosis in humans, a condition in which abnormal secretions are present in the lungs, pancreas, and sweat glands. In the effort to positionally clone the *CFTR* gene, the gene was mapped to a region of 500 kb on chromosome 7 containing three candidate genes.

    a. Using your knowledge of the disease symptoms, how would you distinguish between the candidate

genes to decide which is more likely to encode the
*CFTR* gene?

b. How would you prove that your chosen candidate is the
*CFTR* gene?

26. You have cloned the cDNA for the *CFTR* gene (see
Problem 25). You have used the cDNA, which is 4.5 kb in
length, to identify a 250-kb BAC clone from a genomic
library that fully contains the *CFTR* gene.

a. Describe the strategies that you will use to sequence
each of these clones.

b. You assume that the vast majority of the disease-causing
mutations in this gene are within exons or at intron–exon
boundaries. How would you wish to identify mutations in
patients with a minimum amount of sequencing?

27. How would you devise a screen to identify recessive mutations in *Drosophila* that result in embryo lethality? How
would you propagate the recessive mutant alleles?

28. In land plants, there is an alternation of generations
between a haploid gametophyte generation and a diploid
sporophytic generation. Both generations are typically
multicellular and may be free-living. The male (pollen) and
female (embryo sac) gametophytes are the haploid generation of flowering plants.

a. How would you devise a screen to identify genes
required for female gametophyte development in
*Arabidopsis*?

b. How would you devise a screen to identify genes
required for male gametophyte development?

29. Most organisms display a circadian rhythm, in which
biological processes are synchronized with day length
(e.g., in humans, rapid movement between time zones
results in jet lag, in which established circadian rhythms
are out of synch with daylight hours). In *Drosophila*,
pupae eclose (emerge as adults after metamorphosis)
at dawn.

a. Using this knowledge, how would you screen for
*Drosophila* mutants that have an impaired circadian
rhythm?

b. In most plants, such as *Arabidopsis*, genes whose encoded
products have roles related to photosynthesis have
expression patterns that vary in a circadian manner.
Using this knowledge, how would you screen for
*Arabidopsis* mutants that have an impaired circadian
rhythm?

c. In each case, how would you clone the genes you identified by mutation?

# 17 Applications of Recombinant DNA Technology and Reverse Genetics

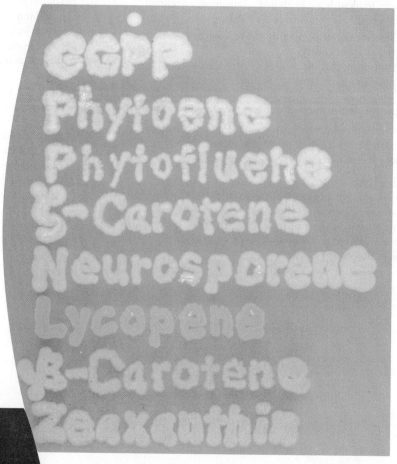

Transgenic *E. coli* expressing the genes for the carotenoid biosynthetic pathway derived from plants. Carotenoid pigments, responsible for the red and orange colors of tomatoes, peppers, and oranges, act as a buffer system to absorb excess electrons and radicals produced during photosynthesis.

## ESSENTIAL IDEAS

- Transgenic organisms are created by harnessing biological vectors to introduce genes into organisms.

- Phenotypes of transgenic organisms can provide information on gene function.

- Reverse genetics techniques start with a gene sequence and then proceed to the identification of a mutant phenotype.

- Recombinant DNA technology in humans is a pathway to the development of gene therapy.

- Cloning of plants and animals produces genetically identical individuals.

The advent of recombinant DNA technology opened the way to studying gene function at the molecular level. After introducing the basic strategies for the in vitro manipulation of DNA and for identifying the sequence of any given gene (see Chapter 16), an obvious next step is the precise manipulation of gene action in living organisms. One of the central technical developments propelling this advance is the ability to create *transgenic organisms,* which are the recipients of genes inserted into their genomes. The creation of transgenic organisms is a powerful tool for manipulating the activity of specific genes, observing the resultant phenotypes, and in this way acquiring new insight into biological processes. In addition, transgenic organisms can be fashioned for specific medical, agricultural, or industrial purposes.

The production of transgenic organisms is now routine in genetic analysis and can be adapted to an almost limitless number of experimental approaches. One of the most significant uses of recombinant DNA technology, however, is in the pioneering of techniques for *reverse genetics*. Whereas forward genetics approaches begin genetic investigation with a mutant phenotype and proceed toward the identification of a gene sequence, reverse genetics approaches begin with a gene sequence and seek to identify the corresponding mutant phenotype. For example, a reverse genetics experiment might generate loss-of-function alleles of specific genes by inserting transgenes into an organism and then observing how the transgenic phenotypes differ from the wild type. Reverse genetic analysis has risen to prominence as a result of the enormous quantity of DNA sequence data made available since the late 1990s.

Transgenic analysis is also used for exploring gene regulation. Regulatory sequences can be dissected by transgenic experiments that fuse all or part of the regulatory sequences for one gene with the expressed segment of another gene whose product is easily visualized. Regulatory sequences from different genes can be mixed and matched to alter the time, place, and amount of gene expression, providing new prospects for the study of gene function and the production of specific gene products.

In this chapter we discuss these applications of recombinant DNA technology, focusing on the methods used to create transgenic organisms and manipulate gene activity.

# 17.1 Introducing Foreign Genes into Genomes Creates Transgenic Organisms

The introduction of a gene from one organism into the genome of another organism creates a **transgenic organism**. The introduced gene is known as a **transgene;**

if the introduced gene comes from a different species, it is a heterologous transgene. The two principal challenges to creating a transgenic organism are (1) the need to introduce DNA into a cell in such a way that the DNA integrates into the genome and (2) the need to provide appropriate regulatory sequences so that the transgene will be properly expressed.

Because cells of different organisms differ in the ability to import DNA from their environment and in their propensity to recombine exogenous DNA into their genomes, protocols for introducing transgenes vary according to the organism. Nevertheless, the production of transgenic organisms is surprisingly straightforward, perhaps because naturally occurring mechanisms have evolved in most lineages of life for the uptake or delivery of DNA. Many organisms or cells will absorb DNA from their environment, and once inside the cell, one potential fate of the DNA is to recombine into the genome. Recall our discussion of certain naturally occurring versions of this process, including gene transfer by Hfr donors into recipient bacteria, transduction of genes from a bacterial donor to a recipient, and gene transfer between and within species by transformation (see Chapter 6).

The design of transgenes utilizes techniques of recombinant DNA technology (see Chapter 16). The expression of transgenes is like the expression of any gene: The gene sequence must first be transcribed into mRNA and then translated into a polypeptide. The universality of the genetic code permits the expression of coding sequences even when transferred between the most distantly related organisms—even when one is a prokaryote and the other a eukaryote. However, regulatory sequences and their molecular interactions with transcriptional and translational machinery vary significantly among organisms, and they are not interchangeable between distantly related organisms. Thus, for transgenes to be efficiently expressed, they must be combined with host regulatory sequences.

## Transgenes in *Escherichia coli*

Bacterial transformation by a recombinant plasmid is the primary method for generating transgenic bacteria. As we describe in Section 17.2, foreign DNA can be introduced into bacteria, such as *E. coli,* using a plasmid vector possessing sequences required for DNA replication and also possessing a selectable marker, such as antibiotic resistance, to facilitate the identification of transformants.

**Expression vectors** are vectors that have been furnished with sequences capable of directing efficient transcription and translation of transgenes (**Figure 17.1**). For transgenes to be properly expressed in *E. coli*, regulatory sequences compatible with the transcription and translation machinery in *E. coli* need to be present in the vector. Expression vectors for use in *E. coli* are constructed from plasmids that have been equipped with promoter

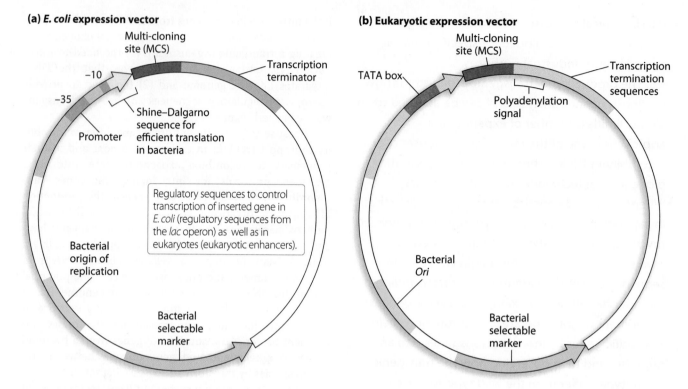

**(a)** *E. coli* **expression vector**

Multi-cloning site (MCS)

−10

−35

Promoter

Shine–Dalgarno sequence for efficient translation in bacteria

Transcription terminator

Regulatory sequences to control transcription of inserted gene in *E. coli* (regulatory sequences from the *lac* operon) as well as in eukaryotes (eukaryotic enhancers).

Bacterial origin of replication

Bacterial selectable marker

**(b) Eukaryotic expression vector**

Multi-cloning site (MCS)

TATA box

Polyadenylation signal

Transcription termination sequences

Bacterial *Ori*

Bacterial selectable marker

**Figure 17.1** **Expression vectors for *E. coli* and eukaryotes.**

sequences that bind RNA polymerase upstream of the multi-cloning site (MCS) of the plasmid. Recall that the MCS is a cluster of unique restriction sites into which the gene to be expressed is inserted in recombinant clones (see Chapter 16). Efficient translation of mRNA in *E. coli* also requires the presence of a Shine–Dalgarno sequence in the 5′ untranslated region of the mRNA, another feature that is built into *E. coli* expression vectors. In addition, since mRNA-splicing machinery does not exist in prokaryotes, eukaryotic transgenes must be free of introns if they are to be properly translated in prokaryotes. This requirement necessitates the use of cDNAs as eukaryotic transgenes in *E. coli* expression systems.

Expression of the heterologous gene carried by an expression vector can be either constitutive ("on" all the time) or regulated by the addition or removal of inducer compounds. An example of the latter approach is the use of the regulatory apparatus of the *lac* operon of *E. coli* to induce expression of transgenes: Fusion of the *lac* operator and CAP binding sites of the *lac* operon to the RNA polymerase binding site allows the transgene to be controlled in the same inducible manner as the genes of the *lac* operon (the *lac* operon is described in Chapter 14).

Two kinds of variation in the genetic mechanisms of living organisms can hamper the efficient production of functional transgenic products. The first complication affects the efficiency of translation. While the universal genetic code does indeed allow the expression of heterologous transgenes, organisms vary in the degree to which

they use specific codons when the genetic code contains more than one for a given amino acid or signal. In most species, synonymous codons are not used with equal frequency. For example, glycine is encoded by GGN, with N representing any nucleotide, but GGA and GGG are rarely used in *E. coli*, whereas these codons are commonly used in the other organisms listed in Table 17.1. The tRNAs corresponding to frequently used codons are expressed at higher levels than are the tRNAs for rarely used codons. This preferential use of codons is called **codon bias**. Thus, for efficient production of heterologous proteins in *E. coli*, the codon usage within the heterologous gene sequences may have to be altered to approximate the codon bias in *E. coli*. Note that such changes do not alter the amino acid sequence of the encoded protein; they only alter the efficiency with which translation occurs in *E. coli*. Codon bias can affect the expression of heterologous transgenes in any

| Table 17.1 | Preference in Different Organisms for Specific Glycine Codons | | | |
|---|---|---|---|---|
| Codon | E. coli | S. cerevisiae | H. sapiens | A. thaliana |
| GGA | 0 | 23% | 23% | 37% |
| GGG | 2% | 12% | 26% | 15% |
| GGC | 38% | 20% | 33% | 14% |
| GGT | 59% | 45% | 18% | 34% |

case where genes are being transferred between distantly related species.

A second possible obstruction to the production of functional heterologous proteins in *E. coli* is presented by the post-translational modifications many proteins must undergo in order to function. Post-translational modifications of proteins differ between species, in particular between eukaryotes and bacteria. For example, carbohydrate and lipid groups are added to many kinds of eukaryotic proteins. In addition, the functions of proteins may be modified by phosphorylation, acetylation, or methylation of amino acid residues; other post-translational polypeptide processing; and specific protein-folding activities. Most of these processes either do not occur, or they occur but with significant differences in bacterial cells. In such cases, eukaryotic cells, such as yeast or cells in tissue culture, and eukaryotic expression vectors must be used. **Eukaryotic expression vectors** have the eukaryotic features analogous to the features found in bacterial expression vectors, including sequences for the regulation of transcription (such as a TATA box for binding of RNA polymerase II), enhancer sequences for qualitative and quantitative control of transcription, and polyadenylation and transcription termination signals (see Figure 17.1).

---

Genetic Insight    The universality of the genetic code permits expression of heterologous genes controlled by regulatory sequences specific to the host organism.

---

## Production of Human Insulin in *E. coli*

A gene encoding insulin was among the first human genes to be expressed in *E. coli,* and human insulin was the first protein manufactured from recombinant DNA technology for therapeutic use in humans. Insulin, a protein hormone, regulates sugar metabolism in animals by stimulating liver and muscle cells to take in glucose, and fat cells to take in lipids, from the blood. Individuals who are unable to produce insulin, or whose cells cannot respond to it, have diabetes, an often debilitating disease that affects millions of people worldwide.

Insulin is cyclically produced in the pancreas by specialized cells in the islets of Langerhans and is released into circulating blood in response to the ingestion of sugar-containing carbohydrates. The pancreatic cells initially synthesize a 110–amino acid precursor protein called preproinsulin that is not secreted and does not have hormonal function until it is proteolytically processed. The "pre" amino acids—the 24 N-terminal amino acids of preproinsulin—are cleaved from the precursor to produce proinsulin, an event followed by the cleavage of an additional 35 amino acids—called the "pro" segment—from the middle of the protein. Further cleavage generates two amino acid chains, called the A chain and the B chain, that

are 21 and 30 amino acids, respectively, in length. The A chain is joined to the B chain by disulfide bonds between cysteine residues to produce insulin.

The amino acid sequence of insulin was determined by Fred Sanger in the early 1950s (**Figure 17.2, ❶**), but the human gene encoding insulin was not identified until the late 1970s. Even before the human insulin gene was cloned, however, molecular biologists began experiments designed to produce human insulin in *E. coli* by constructing recombinant plasmids containing chemically synthesized DNA encoding human insulin. An experimental strategy called the two-chain method utilized two synthetic genes, one encoding the A chain and the other encoding the B chain. Each synthetic gene was constructed from oligonucleotides whose sequence was based on the reverse translation of the amino acid sequences of the human insulin gene chains ❷.

The synthetic genes were cloned into separate plasmid vectors. In each case the chain was fused, in the same reading frame, to the 3′ terminus of the *lacZ* gene encoding β-galactosidase ❸. Genetic constructs like this, consisting of two or more genes or gene segments joined together to form a new artificial gene, are called **fusion genes.** Transcription and translation of a fusion gene produces a **fusion protein,** which in each of these cases contained the polypeptide of one insulin chain fused to the carboxyl terminus of β-galactosidase (the protein product of the *lacZ* gene). In the recombinant plasmid, transcription is under control of the *lac* operator regulatory sequences. Gene transcription is induced by lactose in the absence of glucose, as described in ❺ and ❻ (see also Chapter 14). Under appropriate growth conditions, up to 20% of the total protein produced by the recombinant *E. coli* strains is the fusion protein.

To separate the insulin peptides from β-galactosidase peptides, and to form functional insulin molecules, a methionine residue was engineered into the fusion protein at the junction between the N-terminal end of the insulin peptides and the C-terminal end of the β-galactosidase peptides to serve as a peptide cleavage site ❹. Treatment of proteins with cyanogen bromide (CNBr) cleaves peptide bonds at the carboxyl end of methionine residues ❼. There are no other methionine residues in the fusion protein, so CNBr treatment releases each of the insulin chains from the β-galactosidase peptides. When the A and B chains are purified from their recombinant host strains and mixed together under oxidizing conditions, disulfide bonds form to link the A and B chains and produce active insulin molecules ❽.

The original recombinant human insulin molecules produced by this method were identical to naturally occurring human insulin. Since the implementation of this synthetic process in the 1980s, however, more efficient methods for producing recombinant human insulin have been developed. Some of these methods have introduced amino acid changes in the recombinant human insulin, in

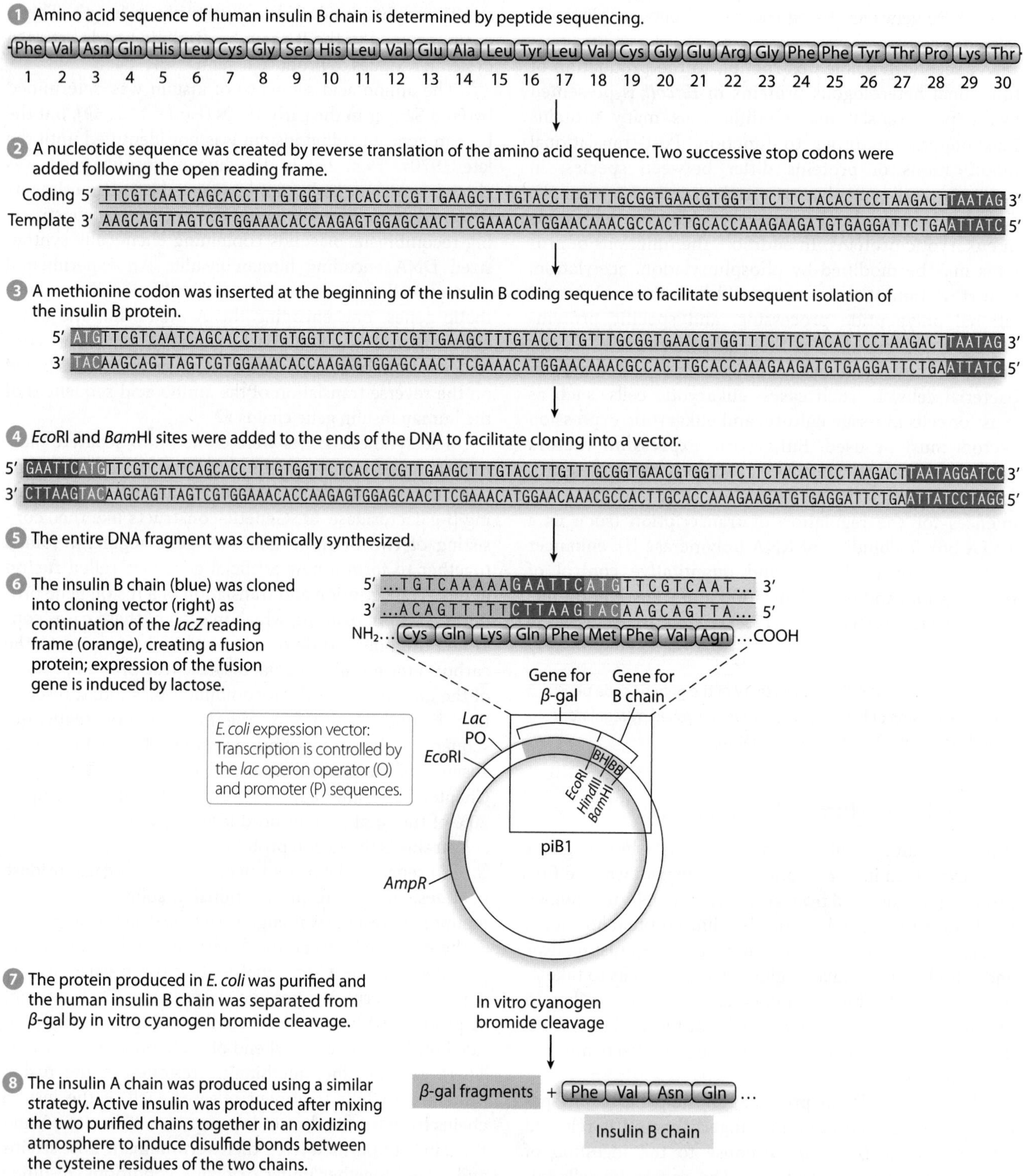

**①** Amino acid sequence of human insulin B chain is determined by peptide sequencing.

| Phe | Val | Asn | Gln | His | Leu | Cys | Gly | Ser | His | Leu | Val | Glu | Ala | Leu | Tyr | Leu | Val | Cys | Gly | Glu | Arg | Gly | Phe | Phe | Tyr | Thr | Pro | Lys | Thr |
|-----|-----|-----|-----|-----|-----|-----|-----|-----|-----|-----|-----|-----|-----|-----|-----|-----|-----|-----|-----|-----|-----|-----|-----|-----|-----|-----|-----|-----|-----|
| 1 | 2 | 3 | 4 | 5 | 6 | 7 | 8 | 9 | 10 | 11 | 12 | 13 | 14 | 15 | 16 | 17 | 18 | 19 | 20 | 21 | 22 | 23 | 24 | 25 | 26 | 27 | 28 | 29 | 30 |

**②** A nucleotide sequence was created by reverse translation of the amino acid sequence. Two successive stop codons were added following the open reading frame.

Coding 5' TTCGTCAATCAGCACCTTTGTGGTTCTCACCTCGTTGAAGCTTTGTACCTTGTTTGCGGTGAACGTGGTTTCTTCTACACTCCTAAGACTTAATAG 3'

Template 3' AAGCAGTTAGTCGTGGAAACACCAAGAGTGGAGCAACTTCGAAACATGGAACAAACGCCACTTGCACCAAAGAAGATGTGAGGATTCTGAATTATC 5'

**③** A methionine codon was inserted at the beginning of the insulin B coding sequence to facilitate subsequent isolation of the insulin B protein.

5' ATGTTCGTCAATCAGCACCTTTGTGGTTCTCACCTCGTTGAAGCTTTGTACCTTGTTTGCGGTGAACGTGGTTTCTTCTACACTCCTAAGACTTAATAG 3'

3' TACAAGCAGTTAGTCGTGGAAACACCAAGAGTGGAGCAACTTCGAAACATGGAACAAACGCCACTTGCACCAAAGAAGATGTGAGGATTCTGAATTATC 5'

**④** EcoRI and BamHI sites were added to the ends of the DNA to facilitate cloning into a vector.

5' GAATTCATGTTCGTCAATCAGCACCTTTGTGGTTCTCACCTCGTTGAAGCTTTGTACCTTGTTTGCGGTGAACGTGGTTTCTTCTACACTCCTAAGACTTAATAGGATCC 3'

3' CTTAAGTACAAGCAGTTAGTCGTGGAAACACCAAGAGTGGAGCAACTTCGAAACATGGAACAAACGCCACTTGCACCAAAGAAGATGTGAGGATTCTGAATTATCCTAGG 5'

**⑤** The entire DNA fragment was chemically synthesized.

**⑥** The insulin B chain (blue) was cloned into cloning vector (right) as continuation of the *lacZ* reading frame (orange), creating a fusion protein; expression of the fusion gene is induced by lactose.

5' ...TGTCAAAAAGAATTCATGTTCGTCAAT... 3'
3' ...ACAGTTTTTCTTAAGTACAAGCAGTTA... 5'

NH₂...| Cys | Gln | Lys | Gln | Phe | Met | Phe | Val | Agn |...COOH

*E. coli* expression vector: Transcription is controlled by the *lac* operon operator (O) and promoter (P) sequences.

Gene for β-gal    Gene for B chain

Lac PO
EcoRI
EcoRI
HindIII
BamHI

piB1

AmpR

In vitro cyanogen bromide cleavage

**⑦** The protein produced in *E. coli* was purified and the human insulin B chain was separated from β-gal by in vitro cyanogen bromide cleavage.

**⑧** The insulin A chain was produced using a similar strategy. Active insulin was produced after mixing the two purified chains together in an oxidizing atmosphere to induce disulfide bonds between the cysteine residues of the two chains.

β-gal fragments + | Phe | Val | Asn | Gln | ...

Insulin B chain

**Figure 17.2 Producing human insulin in *E. coli*.** This strategy was used in the late 1970s by the City of Hope National Medical Center and the biotechnology company Genentech to produce human insulin in *E. coli*.

order to create proteins that have different desired effects on the uptake of glucose by targeted cells. These various forms of recombinant human insulin are used by millions of insulin-dependent diabetics around the world every day.

The ease and economy of working with bacteria as compared to eukaryotes has made it practical to produce many eukaryotic proteins in bacteria for both medical and industrial applications. In addition to human insulin, proteins such as human growth hormone (HGH) and erythropoetin (which induces red blood cell formation) are produced in prokaryotic systems. The recombinant systems used to produce these and many other pharmaceutical and industrial agents are safe and effective sources of otherwise scarce material. For example, before the production of human insulin by recombinant DNA technology, insulin was extracted from pig and cow pancreases collected as a by-product of the meat industry. Pig and cow insulin are very similar to human insulin, but not identical to it; as a result, allergic reactions compromised their use by diabetics. Insulin extractions from animals also carry a risk of contamination from the source tissues. Likewise, HGH extracted from the pituitary glands of human cadavers carries a risk of transmitting neurological disease (e.g., Creutzfeldt-Jacob disease) due to the possible presence of contaminating proteins. Both recombinant human insulin and recombinant HGH have proven safe and effective over decades of use.

Many proteins used in industrial processes as well as in everyday household products are produced in bacteria. For example, proteases are protein-degrading enzymes added to laundry detergents to aid in removing stains from clothing. Isolation of genes encoding proteases from psychrophilic, or cold-loving, bacteria has allowed the industrial production of proteases that act in cold water, leading to substantial savings in energy costs stemming from household hot water usage. The genetic engineering of *E. coli* and other microbes to produce proteins or compounds used in industry, agriculture, and health care is an active field that will flourish in the coming years as more microbial systems are investigated at the genomic and physiological levels. An example of the transfer of an entire biochemical pathway into *E. coli* in order to produce a medically important compound is described in **Experimental Insight 17.1**.

---

Genetic Insight  Eukaryotic genes encoding proteins with pharmaceutical and industrial uses can be expressed by transgenic bacteria.

---

## Changing the Sequence of DNA Molecules

Sometimes, the wild-type version of a gene is the one that geneticists wish to express as a transgene. But we have seen that in some cases, it is desirable to express a modified version in which specific nucleotides have been changed. One reason it is sometimes desirable to alter the sequence of an encoded protein is to render the protein either more or less active. For example, changes in the identities of specific amino acids can sometimes cause an enzyme to be constitutively active or to be more stable at high or at low temperatures. A second reason to change a gene's nucleotide sequence is to improve its expression in a species with a different codon bias than that of the species from which the gene was derived (see "Transgenes in *Escherichia coli*").

Specific nucleotide changes can be accomplished through a technique called **site-directed mutagenesis,** outlined in **Figure 17.3**. The key to site-directed mutagenesis

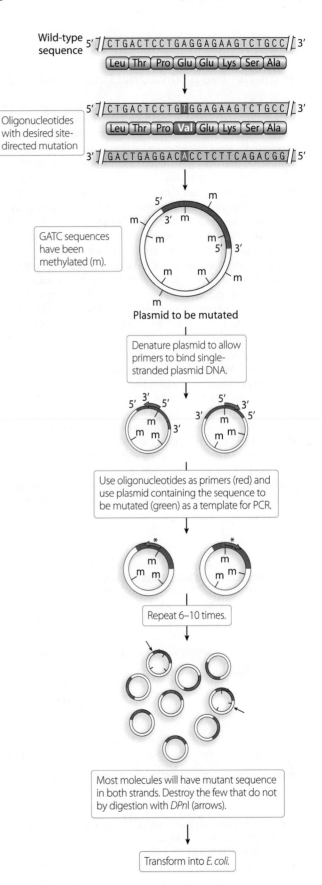

**Figure 17.3  Site-directed mutagenesis.**

## Experimental Insight 17.1

### Plant-Derived Antimalarial Drugs Produced in *E. coli*

The production of amorphadiene in *E. coli* exemplifies the use of genetic engineering to produce a high-value pharmaceutical product. Amorphadiene is the immediate precursor to artemisinin, a potent antimalarial drug. Artemisinin has been touted as the next-generation antimalarial drug since it is effective at treating multiple stages of malarial infection and exhibits no cross-resistance with existing antimalarial drugs, such as chloroquine and quinine. Chloroquine and quinine have been used to fight malarial infection for several decades, but their effectiveness is decreasing due to the evolution of resistant strains of *Plasmodium,* the malaria parasite.

Like many modern drugs, artemisinin was originally discovered in plant extracts. Currently the drug is extracted and purified from the sweet wormwood plant, *Artemisia annua*. The logistics of growing *Artemisia* are limiting factors, however, and the cost of producing large amounts of artemisinin from its natural source is also prohibitive. Production of artemisinin in a fermentable biological system such as *E. coli* could increase drug supply, conserve natural resources, and dramatically lower production costs.

Artemisinin is a complex terpene molecule produced in several biosynthetic steps. All plants produce the precursors of the terpene pathway, isopentenyl pyrophosphate (IPP) and dimethylallyl pyrophosphate (DMAPP), but the specific terpenes produced from them by each plant species vary. The final two steps in artemisinin biosynthesis, from farnesyl pyrophosphate (FPP) to artemisinin, are catalyzed by enzymes encoded by genes specific to *Artemisia*. While *E. coli* naturally produces IPP and DMAPP, the pathway is subject to feedback inhibition, preventing large quantities of these molecules from accumulating.

This obstacle to producing large quantities of amorphadiene in *E. coli* is circumvented by use of a combination of eight genes from *Saccharomyces cerevisiae* and *E. coli* to recreate the biosynthetic pathway leading to FPP production. ❶ A mutant *E. coli* strain is used in which the normal feedback inhibition of the FPP biosynthetic pathway is lacking. ❷ Expression of the eight *S. cerevisiae* genes is coordinated by distribution of the genes into two operons—one containing three genes and one containing five—controlled by *lac* operon regulatory sequences (see Chapter 14 for a review of the *lac* operon system). In this way, gene expression is induced in the presence of either lactose or the synthetic inducer isopropyl-β-D-thiogalactopyranoside (IPTG). Step ❸ is to clone the amorphadiene synthetase (ADS) gene from *Artemisia* and place it under the control of *lac* operon regulatory sequences.

Initial experiments with this system were disappointing, since only low levels of ADS protein were produced in *E. coli*. The reason was discovered to be differences in codon bias between *Artemisia* and *E. coli*. When codons preferred by *Artemisia* were replaced with synonymous codons preferred by *E. coli,* the production of ADS protein in *E. coli* became much more efficient. ❹ Now the bacteria produced a large quantity of amorphadiene, which could be converted into artemisinin either by chemical synthesis or in vivo by the introduction of the artemisinin synthetase gene from *Artemisia*.

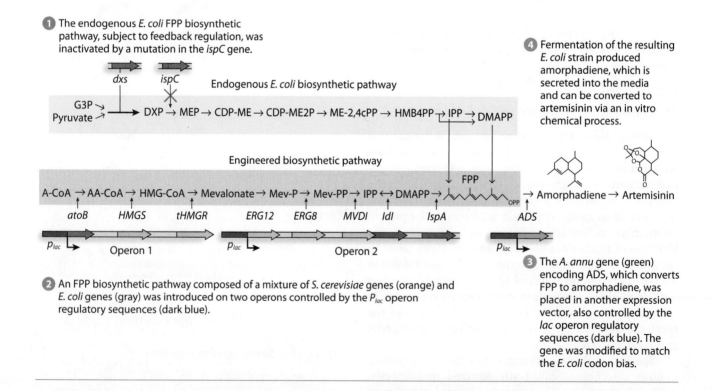

❶ The endogenous *E. coli* FPP biosynthetic pathway, subject to feedback regulation, was inactivated by a mutation in the *ispC* gene.

❹ Fermentation of the resulting *E. coli* strain produced amorphadiene, which is secreted into the media and can be converted to artemisinin via an in vitro chemical process.

❷ An FPP biosynthetic pathway composed of a mixture of *S. cerevisiae* genes (orange) and *E. coli* genes (gray) was introduced on two operons controlled by the $P_{lac}$ operon regulatory sequences (dark blue).

❸ The *A. annu* gene (green) encoding ADS, which converts FPP to amorphadiene, was placed in another expression vector, also controlled by the *lac* operon regulatory sequences (dark blue). The gene was modified to match the *E. coli* codon bias.

is to design oligonucleotide primers containing the desired nucleotide change in the middle of the primer sequence. The primers are then used in a polymerase chain reaction (PCR) to amplify the gene sequence to be mutated. In the PCR procedure, the gene sequence to be mutated is first cloned into a plasmid vector, after which the plasmid is denatured into single strands by being heated to 95°C and then quickly cooled. Next, addition of the primers (containing the mutation), a DNA polymerase, and a supply of dNTPs leads to in vitro DNA synthesis from the 3′ hydroxyl nucleotide of the primers. The result is a mixture of double-stranded plasmids, some of which contain the mutation of interest and others having the original sequence.

After the PCR is completed, the nonmutant plasmids can be selectively degraded through the use of methylation-sensitive restriction enzymes. For this, the original plasmid must first have been grown in an *E. coli* strain in which the DNA is methylated on the adenine residue in the sequence GATC, which naturally occurs in methyl-directed DNA repair (see Chapter 12). The methylated GATC sequence is cleaved by the restriction enzyme *Dpn*I, whereas the in vitro PCR-synthesized DNA is not methylated and is resistant to *Dpn*I-mediated cleavage. Thus, during transformation, only the mutant plasmid is introduced into *E. coli*.

As with DNA sequencing, the technology for chemically synthesizing DNA molecules has improved significantly in recent years, in terms of both accuracy and cost. Consider the example of human insulin genes. In the late 1970s, the construction of the B-chain gene from 18 chemically synthesized oligonucleotides 10 to 12 nucleotides long and of the A-chain gene from 12 oligonucleotides 10 to 15 nucleotides long was a monumental task. Today, however, oligonucleotides tens to hundreds of bases in length are inexpensive to construct, and they constitute one of the most widely used tools for manipulation of DNA molecules in vitro (e.g., in PCR and sequencing reactions).

More recently, chemical syntheses of DNA molecules up to 50,000 bases in length have become feasible. Geneticists are able to design a DNA molecule from scratch and synthesize it for subsequent use in living organisms. This ability is useful when multiple changes are required in a DNA molecule before its introduction into a transgenic organism. As with sequencing technologies, advances in chemical synthesis of large DNA molecules have the potential to transform biotechnology and biological research. In 2008, the entire 582,970-bp genome of *Mycoplasma genitalium* was chemically synthesized in vitro, cloned into a YAC vector, and propagated in *Saccharomyces cerevisiae*. The synthetic genome was then transplanted into a receptive *Mycoplasma* cytoplasm, generating a cell that would use the genetic information contained on the synthetic chromosome. This ability to synthesize genome-sized nucleic acid molecules could revolutionize experimental biology.

## Generation of Transgenic Fungi

Transgenes can be introduced into fungal cells in a manner similar to bacteria using a plasmid system developed for the fungus *Saccharomyces cerevisiae* (baker's yeast). In addition, DNA can be readily integrated into the genome of many fungi by homologous recombination, making direct manipulation of the fungal genome feasible.

**Yeast Plasmids**   Some strains of *S. cerevisiae* harbor a circular 6.3-kb plasmid that, because of its approximately 2-μm diameter, is known as the 2-micron plasmid. This plasmid can be modified into a recombinant plasmid by the insertion of transgenes. An *E. coli* origin of replication and appropriate selectable markers are also introduced into the 2-micron plasmid, which already contains the *S. cerevisiae* origin of replication (**Figure 17.4**). With these additions, the plasmid becomes a **shuttle vector,** a vector that can replicate in two species—in this case, both *E. coli* and *S. cerevisiae*—and thus can be used to shuttle DNA sequences between them. With a shuttle vector, DNA sequences can be manipulated in *E. coli*, where manipulation is easier, after which the modified plasmids can be shuttled into yeast for heterologous protein expression.

**Integrating DNA into the Genome of *S. cerevisiae***   If DNA that is introduced into an organism has no origin of replication, it undergoes one of two fates: enzymatic degradation or integration into the host genome. Enzymatic degradation, accomplished by nucleases that are common in cells, will eliminate the introduced DNA. Integration of DNA into the host genome, in contrast, allows the introduced nucleic acid to persist in the host cell. Integration is accomplished by either of two distinct mechanisms of recombination: illegitimate recombination or homologous recombination.

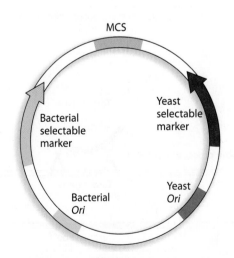

**Figure 17.4   Shuttle vector for *E. coli* and *Saccharomyces cerevisiae*.**

**Illegitimate recombination** integrates introduced DNA at a random, nonhomologous location. This form of recombination does not require any homology between the introduced DNA and the genomic DNA into which the former is integrated. In contrast, the second mechanism for integration of introduced DNA, **homologous recombination** between the introduced DNA and the host genomic sequence, requires a significant length of DNA sequence in common between the two recombining molecules. The relative frequencies with which these mechanisms occur depend on the species into which the DNA is introduced. In most plant and animal species, illegitimate recombination is the most common fate, although techniques exist to select for individuals in which homologous recombination has occurred (as described later in this chapter). In bacterial and fungal species, introduced DNA is often recombined in the genome in a homologous manner.

Segments of DNA introduced into *S. cerevisiae* have a propensity to undergo homologous recombination.

An introduced circular molecule of DNA can recombine by either a single crossover or a double crossover (**Figure 17.5a**). In a single crossover, the entire molecule of introduced circular DNA is integrated into the yeast genome with no loss of any genomic DNA. If recombination of a circular molecule occurs by double crossover, however, only DNA between the homologous flanking sequences is integrated into the recipient genome, and the integration is accompanied by a concomitant loss from the genome of the DNA between the homologous sequences. Thus, recombination with two crossovers results in replacement of the genomic DNA with the introduced DNA flanked by the homologous sequences.

Introducing a linear rather than circular molecule of DNA favors retrieval of recombinants produced by double crossover, since a single crossover will cause a deletion event resulting in recombinant molecules lacking a large portion of the original chromosome and therefore likely to be lethal (**Figure 17.5b**). Linearized DNA molecules

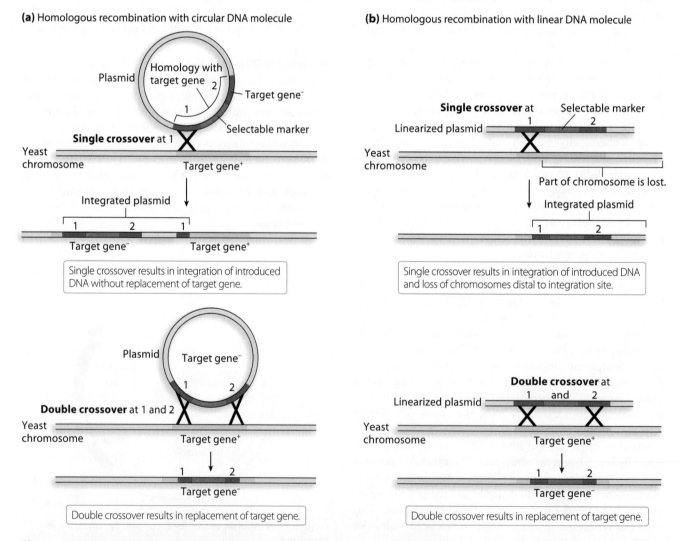

**(a)** Homologous recombination with circular DNA molecule

**(b)** Homologous recombination with linear DNA molecule

Single crossover results in integration of introduced DNA without replacement of target gene.

Single crossover results in integration of introduced DNA and loss of chromosomes distal to integration site.

Double crossover results in replacement of target gene.

Double crossover results in replacement of target gene.

**Figure 17.5  Homologous recombination in yeast: Single versus double crossovers.**

recombine at a higher frequency than circular ones, making the introduction of linear molecules the method of choice for homologous recombination experiments.

Taking advantage of this tendency for homologous recombination to occur in yeast, yeast geneticists create recombinant yeast both through gene insertion and gene replacement. Loss-of-function alleles are created by replacing the target gene with heterologous DNA, often a selectable marker gene, thus eliminating the production of functional wild-type protein by the target gene. Gene insertions that result in a deletion of the entire coding region of the gene create null alleles that produce no protein product. Such insertion alleles are often called **gene knockouts,** because the insertion "knocks out" the function of the gene, creating a recessive loss-of-function allele. Conversely, inserting a functional allele, often creating a gain-of-function allele, is called a knock-in.

The ease with which homologous recombinants are generated in *S. cerevisiae* has allowed the production of a large number of yeast strains for genetic analysis of biological processes in this organism. As described in greater detail in Section 17.4, loss-of-function alleles of every gene in the *S. cerevisiae* genome have been generated and can be ordered from a stock center. Such stocks have greatly facilitated genetic research by relieving scientists of the need to produce mutations in the genes of interest at the start of every new genetic experiment.

---

Genetic Insight  Baker's yeast, *S. cerevisiae,* exhibits a high level of homologous recombination, allowing researchers to create targeted gene knockouts and other tailored alleles.

---

## Transformation of Plant Genomes by *Agrobacterium*

Our food is mainly derived from plants, and humans have been genetically modifying plants since the beginning of agriculture, nearly 10,000 years ago. For most of this history, genetic improvement was limited to interbreeding wild and domesticated species to select for traits already present in nature. The recently developed techniques for introducing DNA from many sources into plants have added a new dimension to the genetic modification of plants for agricultural purposes. By these new means, the genetic variation available in plants has been extended to include not only genes from other plant species but also genes derived from animals, fungi, and bacteria.

The most widely used method of generating transgenic plants takes advantage of a natural plant transformation system that has evolved in the soil bacterium *Agrobacterium tumefaciens.* In nature, this bacterium is the cause of crown gall disease, an uncontrolled cell division in plant cells. This disease results in tumors (galls), typically at the crown (the base near the soil) of the plant.

Wild strains of *A. tumefaciens* harbor a large plasmid (200 kb) called the tumor-inducing plasmid, or **Ti plasmid** (**Figure 17.6a**). A portion of the Ti plasmid, a region referred to as the **transfer DNA (T-DNA)** is transferred from the bacterium into the nucleus of a plant cell. Mary-Dell Chilton and colleagues conclusively demonstrated the nature of this remarkable cross-kingdom transfer of DNA in the late 1970s by demonstrating that *Agrobacterium* Ti plasmid DNA can be detected inside plant cells. Once inside the plant cell, the T-DNA can recombine illegitimately with the plant nuclear genome, resulting in an insertion of the T-DNA at a random location in the plant genome (**Figure 17.6b**).

From the bacterial perspective, the outcome of this natural transformation event is the expression of genes in the T-DNA that encode proteins causing plant cells to (1) divide in an uncontrolled manner and (2) produce amino acids only the bacterium can utilize as an energy source. The bacteria essentially reprogram plant cells to become food factories for the bacteria. Bacterial genes encoding plant-hormone-biosynthesizing enzymes cause transformed plant cells to produce high levels of two plant hormones, auxin and cytokinin, which in turn cause uncontrolled division of plant cells, resulting in tumor formation (**Figure 17.6c**). The other genes on the T-DNA encode opine-biosynthesizing enzymes. Opines, such as nopaline and octopine, are amino acids that do not naturally occur in plants; therefore, plants do not produce any enzymes capable of metabolizing opines. *Agrobacterium* does have such enzymes, however; consequently, the opines produced by the plant cells can be used as carbon and nitrogen sources by the bacteria. Other genes on the Ti plasmid, but not located within the T-DNA region, encode enzymes required for the transfer of the T-DNA to the plant cell.

Sequence analysis has revealed that the genes involved in the transfer of T-DNA are evolutionarily related to those involved in the transfer of the F-factor in *E. coli* (see Chapter 6). Thus, *Agrobacterium* has evolved a mechanism to transfer DNA into plant cells by adapting genes usually involved in bacterial conjugation. A striking aspect of this cross-kingdom gene transfer is that the genes on the T-DNA have evolved to be transcribed and translated efficiently in plant cells instead of in bacterial cells. In nature, *Agrobacterium* normally transforms plants only; but in the laboratory, the bacterium has the ability to transfer DNA into almost any eukaryotic cell, including human cells.

**Creating Transgenic Plants**  Scientists can use *Agrobacterium* to transfer any gene of interest into plants by removing the opine- and tumor-producing genes normally found in the T-DNA and replacing them with DNA encoding the gene of interest. The T-DNA then transfers the gene of interest into the plant cell, where it becomes integrated into the genomic DNA of the plant.

**(a)**

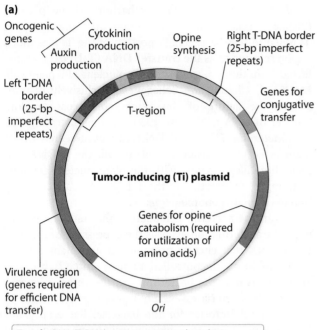

Oncogenic genes

Cytokinin production

Auxin production

Opine synthesis

Right T-DNA border (25-bp imperfect repeats)

Left T-DNA border (25-bp imperfect repeats)

T-region

Genes for conjugative transfer

**Tumor-inducing (Ti) plasmid**

Genes for opine catabolism (required for utilization of amino acids)

Virulence region (genes required for efficient DNA transfer)

*Ori*

Transfer DNA (T-DNA) contains auxin and cytokinin biosynthetic genes and genes for amino acid biosynthesis.

**(b)**  *Agrobacterium tumefaciens*  (1–2 microns wide)    **Plant cell** (5–50 microns wide)

Ti plasmid

T-DNA

T-strand

Virulence proteins

A single strand of T-DNA is transferred into the plant cell and is integrated into the plant nuclear genome.

**(c)**

Expression of auxin and cytokinin biosynthetic genes leads to uncontrolled cell division and gall formation; gall cells produce the unusual amino acids that *Agrobacterium* uses as carbon and nitrogen sources.

**Figure 17.6  Crown gall disease caused by *Agrobacterium* via plant transformation.**

**Figure 17.7a** depicts the manner in which the Ti plasmid is modified for transformation procedures. First, the tumor-inducing and opine genes are deleted from the Ti plasmid, producing what is called a "disarmed" Ti plasmid. Then, the gene of interest is inserted between the two ends of the T-DNA region, referred to as the left and right borders. These border regions contain sequences required for efficient transfer. Proteins encoded by genes of the Ti plasmid outside of the T-DNA recognize specific sequences in the left and right border and catalyze the transfer of a single strand of T-DNA from the bacterium to the plant cell; when this occurs, the gene of interest that has been inserted between the two border sequences will be transferred as well. As with any other protocol for constructing transgenic organisms, a selectable marker is included (between the left and right borders) in addition to the gene of interest to allow efficient selection of transformed plants. For experiments with plants, genes conferring resistance to either antibiotics (inhibiting translation in the chloroplast) or herbicides may be employed as selectable markers. The selectable marker genes are usually expressed using a promoter that confers constitutive expression, so that transgenic plants can be selected at any stage of their development.

Because the Ti plasmid is too large to be easily manipulated, most experimental protocols that use *Agrobacterium* construct a strain harboring two plasmids: One is a disarmed Ti plasmid, and the second is a plasmid that contains left and right border sequences flanking the DNA of interest (**Figure 17.7b**). This strategy, separating the functional elements of the Ti plasmid into two plasmids, is referred to as the binary approach. It results in the efficient transfer of the DNA of interest into the plant cell and its subsequent integration into the plant genome.

Unlike bacteria and yeast, which are single-celled organisms, transformed plant cells must be regenerated into an entire plant in order to reveal the effects of transgenes on the plant phenotype. Traditionally, scientists have taken advantage of a unique feature of plant development, the **totipotency** of most plant cells: Under the appropriate environmental and hormonal conditions, an entire normal plant can be regenerated from a single isolated plant cell. Thus, after infection of plant cells with the modified *Agrobacterium* strain and selection of transformed cells on the basis of the selectable marker gene, progeny plants can be regenerated from the individual transformed cells (**Figure 17.7c**). This technique has been successfully applied to a wide variety of flowering plant species, including crop species such as rice, maize, and tomatoes.

Plant researchers using *Arabidopsis* as a model system for studying basic biological processes sought an easier method of transformation that would not require regeneration from a single transformed cell. After several different techniques were attempted, they discovered that the simple technique of dipping *Arabidopsis* flowers into a culture of *Agrobacterium* works surprisingly well.

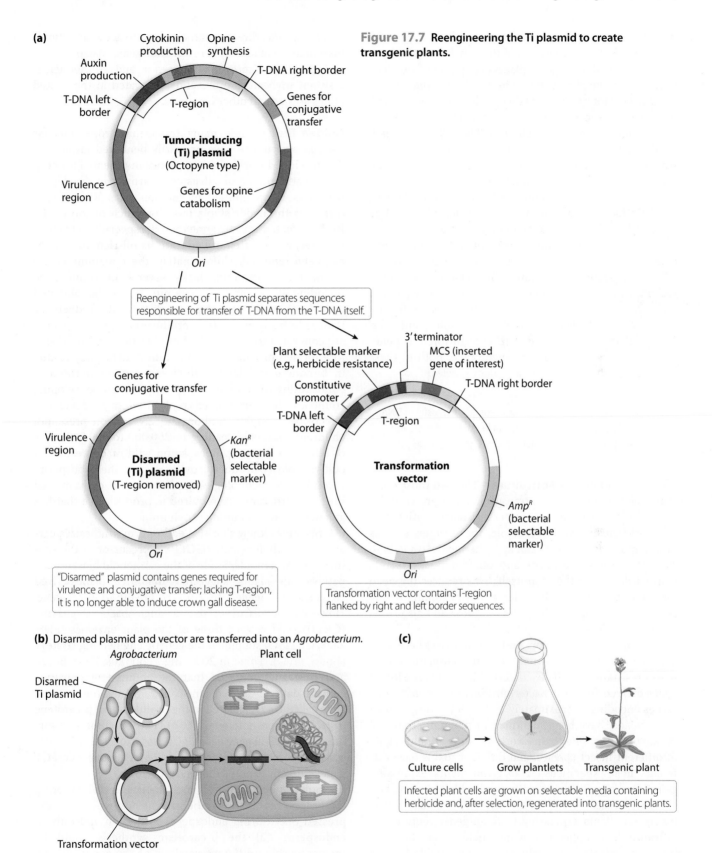

**(a)**

Cytokinin production

Opine synthesis

Auxin production

T-DNA right border

T-DNA left border

T-region

Genes for conjugative transfer

**Tumor-inducing (Ti) plasmid** (Octopyne type)

Virulence region

Genes for opine catabolism

*Ori*

Reengineering of Ti plasmid separates sequences responsible for transfer of T-DNA from the T-DNA itself.

Genes for conjugative transfer

Plant selectable marker (e.g., herbicide resistance)

3' terminator

MCS (inserted gene of interest)

Constitutive promoter

T-DNA right border

T-DNA left border

T-region

Virulence region

*Kan^R* (bacterial selectable marker)

**Disarmed (Ti) plasmid** (T-region removed)

**Transformation vector**

*Amp^R* (bacterial selectable marker)

*Ori*

*Ori*

"Disarmed" plasmid contains genes required for virulence and conjugative transfer; lacking T-region, it is no longer able to induce crown gall disease.

Transformation vector contains T-region flanked by right and left border sequences.

**(b)** Disarmed plasmid and vector are transferred into an *Agrobacterium*.

*Agrobacterium*

Plant cell

Disarmed Ti plasmid

Transformation vector

Genes on disarmed plasmid produce conjugative and virulence proteins that act in trans on T-DNA border sequences of transformation vector to effect transfer of T-DNA, which contains the inserted gene of interest, into plant cell.

**(c)**

Culture cells

Grow plantlets

Transgenic plant

Infected plant cells are grown on selectable media containing herbicide and, after selection, regenerated into transgenic plants.

**Figure 17.7 Reengineering the Ti plasmid to create transgenic plants.**

It allows the T-DNA to be transferred directly from *Agrobacterium* to the egg cell of the female gametophyte. In this protocol, transgenic plants are selected from seed produced by the plant exposed to *Agrobacterium*.

Many plant species are susceptible to *Agrobacterium*-mediated transformation. If they are not, DNA can be directly introduced into their cells. The cell walls of isolated plant cells are first removed enzymatically, after which the cells are mixed with heterologous DNA and given a heat or electrical shock to depolarize the membrane and facilitate the entry of DNA. Once in the cell, the DNA has the same fate as DNA transferred into fungi, as described above. In plants, homologous recombination is rare relative to illegitimate recombination, so the most common outcome is the insertion of the heterologous DNA into a random location in the genome. In another technique, DNA is introduced into plant cells by particle gun bombardment, the use of high pressure to fire microscopic particles coated with DNA into plant cells. The particles are propelled with enough force to penetrate the cell wall and plasma membrane. Both of these techniques can be applied to any plant species.

---

**Genetic Insight**    The natural cross-kingdom transfer of DNA between *Agrobacterium* and plants can be exploited to create transgenic plants expressing heterologous genes.

---

**Transgenic Plants in Agriculture**    The two most common traits engineered into transgenic crops grown today are herbicide resistance and insect resistance. With herbicide-resistant crops—for example, the varieties sold as Roundup Ready—farmers can apply herbicide to a field to clear the ground of weeds and other non-crop plants without damaging the crop itself. This reduces the need for tilling to plow weeds under at the beginning of the season. Less tilling results in less soil loss and also saves on the use of fossil fuels.

Cotton and maize crops resistant to insect herbivory are two of the most widely grown transgenic crops. Insect resistance is usually conferred by the expression of genes derived from the bacterium *Bacillus thuringiensis*. Genes encoding approximately 100 insect toxins, known as Bt toxins, have been identified in different strains of *B. thuringiensis*. The toxins work by perforating the guts of different insect species, and different toxins have different "host" specificity. Transgenic plants expressing genes encoding Bt toxins are less palatable to insects and exhibit reduced insect herbivory. As a consequence, transgenic plants expressing Bt toxin genes require significantly less application of insecticides than do non-transgenic plants, thus reducing the insecticide load in the environment.

While Bt toxins are clearly toxic to insects, other herbivores, such as humans, are impervious to the compounds. The properties of Bt toxins have been appreciated for some time. Organic farmers routinely spray *B. thuringiensis* directly on their crops to act as a "natural" insecticide. Millions of acres of transgenic maize, cotton, and potatoes expressing Bt genes and of herbicide-resistant soybeans are presently cultivated in the United States and several other countries.

**Golden Rice**    While many transgenic crops thus far used in agriculture have primarily benefited farmers in the developed world, the humanitarian potential for crop modification in aid of subsistence farmers in developing countries is exemplified by Golden Rice. Rice (*Oryza sativa*) is the major staple food for much of the world. Because oil tends to become rancid, especially in tropical climates, rice is often milled until its oil-rich outer layer has been removed. Unfortunately, the remaining edible grain, the endosperm, lacks several micronutrients, including provitamin A. Vitamin A can be obtained directly through consumption of animal products or indirectly from plants that produce carotenoids, which are converted to vitamin A after ingestion and are therefore termed provitamin A. Vitamin A deficiency results in blindness and increased disease susceptibility, thus contributing to childhood mortality in many developing countries. It is estimated that vitamin A deficiency affects between 140 million and 250 million preschool children worldwide, leading to 250,000 to 500,000 cases of blindness per year. Because no wild or domesticated cultivars of rice produce provitamin A in the endosperm, recombinant technologies, rather than a conventional breeding program, are required to produce rice that has an endosperm containing provitamin A.

Scientists knew that rice endosperm synthesizes geranylgeranyl diphosphate (GGPP), a precursor in the synthesis of carotenoids. Study of the carotenoid biosynthetic pathway in plants suggested that five plant-derived enzymes are needed to convert GGPP to β-carotene. However, the discovery that a single bacterial enzyme (CRTI) could replace three of the plant enzymes (PDS, ZDS, CRTISO) simplified the genetic engineering strategy (**Figure 17.8a**). Then, in 2000, Ingo Potrykus, Peter Beyer, and colleagues reported that the addition of only two genes, a daffodil-derived gene called *PSY* and the bacterial gene called *CRTI*, resulted in the production of β-carotene in rice endosperm (**Figure 17.8b**). This outcome was surprising because a gene called *LCY* was expected to be necessary as well, but apparently the endogenous rice *LCY* gene is already expressed in endosperm.

Subsequently, work has focused on tailoring the process so that (1) the transgenes would be expressed only during endosperm formation and only in endosperm, (2) the β-carotene synthesis could be increased using different versions of the genes, (3) the selectable marker could be removed from the transgenic lines, and (4) the transgenes could be introduced into rice cultivars that are typically used by subsistence farmers in southeast and south central Asia and Africa. These improvements have led to transgenic lines that

**(a)** Synthesis of GGPP

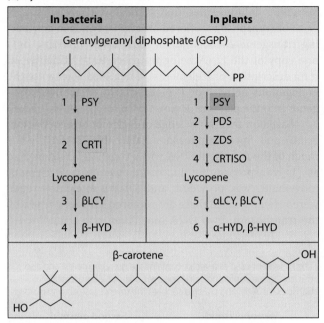

| In bacteria | In plants |
|---|---|
| Geranylgeranyl diphosphate (GGPP) | |

**(b)** Recombinant plasmids

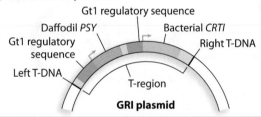

GRI plasmid

First-generation golden rice (GRI): Daffodil phytoene synthase (*PSY*) and bacterial gene (*CRTI*) from *Erwinia uredovora* are driven with rice glutelin-1 (Gt1) endosperm regulatory sequences.

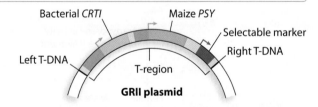

GRII plasmid

Second-generation golden rice (GRII): A maize *PSY* gene was exchanged for the daffodil *PSY* gene, boosting the production of β-carotene.

**(c)** Appearance of wild-type and transgenic rice

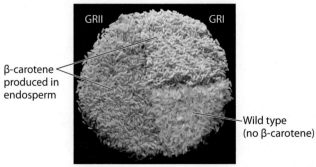

**Figure 17.8  The generation of golden rice.**

should provide a significant fraction of the required daily intake of provitamin A from a serving of Golden Rice (**Figure 17.8c**).

The funding for the research to produce Golden Rice was public, in part from the Rockefeller Foundation, but patents on many of the techniques and tools used to generate the transgenic rice are held by biotech companies. Fortunately, these companies agreed to license the inventors of Golden Rice to provide the technology free of charge for humanitarian use in developing countries. Golden Rice is an example of how customized crops can be developed to address specific nutritional needs and public health problems caused by dietary deficiencies.

Transgenic plants have been largely accepted in some parts of the world, but many concerns have been raised about their introduction. Some critics fear that transgenes could be adverse to human health—for example, that people may have allergic reactions to the protein product of a transgene. Another concern is that the transgenes may "escape" into the environment if transgenic crop plants interbreed with related species growing nearby. The likelihood of this occurrence can be reduced by not growing transgenic crops in environments harboring related species that have potential to interbreed. Transgenic crops must be tested to allay these concerns, but we must also recognize that, while the concerns about transgenic agricultural crops are valid, they are equally applicable to the cultivation of crops developed by traditional breeding methods.

---

Genetic Insight  The construction of transgenic plants introduces precise genetic modifications to meet particular agricultural requirements.

---

## Transgenic Animals

Protocols for the generation of transgenic animals are similar to those described for fungi, but as with plants, homologous recombination occurs much less frequently than illegitimate recombination (i.e., recombination not based on sequence homology). *Caenorhabditis elegans, Drosophila,* and *Mus musculus* (mice) are three of the most widely used genetic model animals and provide examples of the variety of methods available for creation of transgenic animals. Totipotency is not characteristic of most of their cells; thus, methods to produce transgenic animals rely on the injection of DNA into eggs, embryos, or cells that will give rise to gametes, with the hope that the injected DNA will be integrated into the genome either by homologous or illegitimate recombination.

Where injection directly into gametes is not feasible, DNA can be injected into isolated cells that are subsequently transplanted into an embryo. The embryo then develops as a **genetic chimera,** an organism in which some cells have a different genotype than others and will transmit transgenes to progeny only if the embryonic germ cells carry a copy of the transgene. As with the protocols utilized in fungi and plants, methods for the production of transgenic animals vary depending on the biological characteristics specific to each type of organism.

### C. elegans

In the nematode worm *C. elegans,* one protocol for creating transgenic animals is to inject DNA directly into the gonads of hermaphrodites during oocyte development (**Figure 17.9**). The gonads are syncytial, meaning that gonadal cells each contain many nuclei and a large amount of cytoplasm. Eventually, each nucleus gives rise to a germ cell. If the injected DNA is integrated into the genome of a germ cell, the mechanism of integration is almost always illegitimate recombination. The DNA is often, but not always, inserted as a concatemer, that is, as multiple tandem copies of the inserted DNA. Concatemers are undesirable because they result in abnormal levels of gene expression, either because of the additional copies producing too much gene product or because of RNA-mediated gene-silencing effects triggered by the repetitions in the concatemer (discussed in Chapter 15). Alternatively, the injected DNA may exist as extrachromosomal arrays that are not integrated into the genome, which may not be properly segregated during mitosis.

As with other systems for gene transfer, a selectable marker is built into the injected DNA to facilitate identification of cells that have been transformed. In *C. elegans,* a dominant mutant allele of the *roller-6* gene [specifically, *rol-6(su1006)*] can be used, causing animals with this mutant allele to exhibit a behavioral defect: Rather than moving in a serpentine pattern, mutants tend to roll in tight circles. Animals with several copies of this dominant mutant allele do not survive, so it serves as a "marker" that selects against concatemers of transgenes.

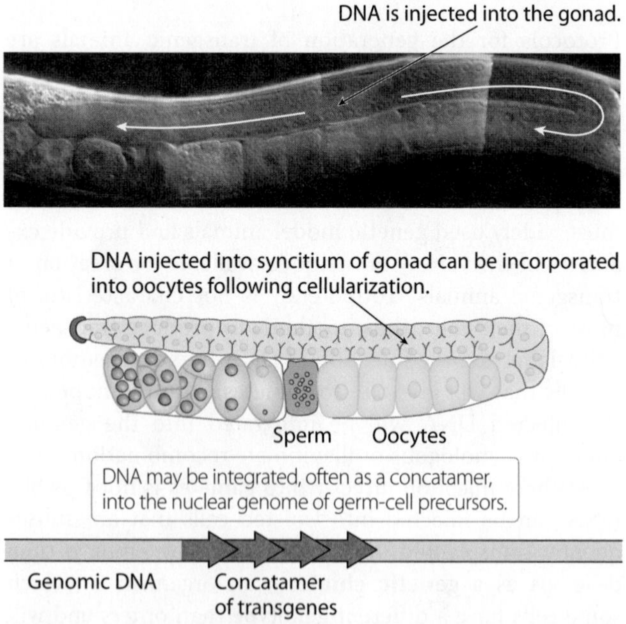

DNA is injected into the gonad.

DNA injected into syncytium of gonad can be incorporated into oocytes following cellularization.

Sperm      Oocytes

DNA may be integrated, often as a concatamer, into the nuclear genome of germ-cell precursors.

Genomic DNA      Concatamer of transgenes

**Figure 17.9  Transgenic *C. elegans*.**

### Drosophila

In the 1980s, Gerald Rubin and Allan Spradling demonstrated that *P* transposable elements, a class of transposons, offered an efficient means of creating transgenic *Drosophila,* in most cases inserting only one copy of the DNA being transferred (see Chapter 15 for a description of *P* elements). Their idea was to use the endogenous activity of ***P* elements** to transpose transgenes into the genome (**Figure 17.10**).

Based on their knowledge of *P* element transposition, Rubin and Spradling reasoned that they could replace much of the *P* element DNA with exogenous DNA as long as (1) transposase, the enzyme that controls *P* element movement, was provided; and (2) the *P* element ends were retained, since these are required for recognition by the transposase. Two DNA molecules, one a modified *P*

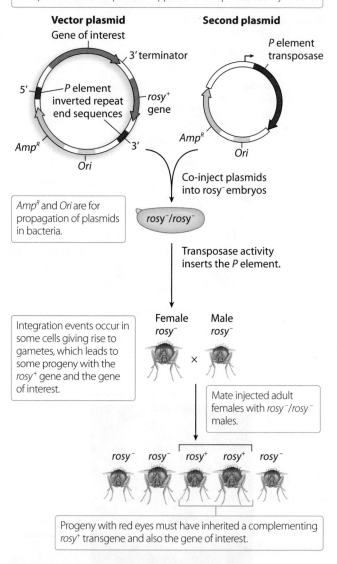

**Figure 17.10  *P* element–mediated transformation in *Drosophila*.**

element and the other a DNA molecule encoding the transposase but lacking the sequences required for transposition, are co-injected into a *Drosophila* embryo. The modified *P* elements are induced to insert into the genome at random positions by the action of the transposase. Typically, only a single *P* element is inserted, precluding the problems associated with the concatemeric arrays seen in transgenic *C. elegans*. This strategy resembles the use of *Agrobacterium* to transform plants in that it too utilizes a biological system that has evolved to recombine DNA into a host genome.

Since *P* elements transpose only in the germ-line cells of *Drosophila*, the injection is made into an early-stage embryo, targeting those cells that will give rise to the germ line. Early-stage *Drosophila* embryos are syncytial, and nuclei at the posterior end of the syncytium are most likely to give rise to the germ cells. The fly derived from the injected embryo is therefore a chimera in which most soma and some gametes are wild type, but some soma and gametes are transgenic. When the injected fly is mated with an uninjected fly of the same strain, gametes into whose genomes a *P* element was inserted will produce transgenic progeny.

A commonly used selectable marker in *Drosophila* is the *rosy* (*ry*) gene. In the procedure under discussion, the embryos to be injected are $ry^-/ry^-$ and have rosy eyes, rather than the wild-type red eyes. A wild-type, $ry^+$, copy of the gene is included in the modified *P* element, in addition to the DNA to be transformed into the fly. While flies derived from the injected embryos will have rosy eyes, some of the progeny derived from transgenic gametes of the injected fly will have red eyes due to the action of the $ry^+$ allele on the inserted *P* element. As is characteristic of transposons, *P* elements insert into the genome at random locations.

**Vertebrates**   A general approach to create transgenic vertebrates is to inject DNA directly into the nucleus of a fertilized egg cell, in a manner similar to that described earlier for *C. elegans* and *Drosophila*. The injected DNA can become integrated into the genome at random positions by illegitimate recombination. Because the DNA integrates randomly into the genome, the transgene becomes inserted at different locations in the genomes of different individual animals. In organisms such as salmon, each injected egg has the potential to develop into a transgenic individual (**Figure 17.11**).

Two features of this method lead to variability in the expression of the transgene. First, due to the integration of the transgenes as multicopy concatemers, gene expression levels can be affected as described for *C. elegans*. Second, the expression of the transgene can be abnormal because of the chromosomal environment in which it is located. For example, if the transgene is inserted into heterochromatin, gene expression may be altered as described for position effect variegation in *Drosophila* (see Chapter 15). Note that the problem of transgene position effects is shared by all transgenic organisms in which the transgene is integrated

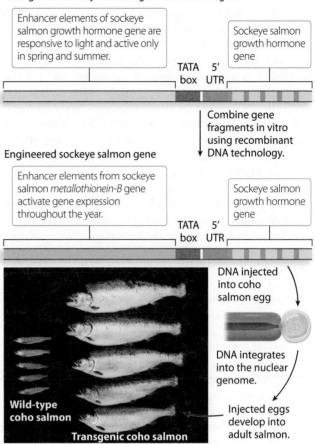

**Figure 17.11  Creation of transgenic salmon through injection of DNA into salmon eggs.**

into the genome by illegitimate recombination. While position effects can pose problems in *Drosophila*, *C. elegans*, and plants, they are exacerbated in mice due to the larger average size of vertebrate genes and the larger amount of heterochromatin in vertebrate genomes. To overcome this variability, methods to more precisely insert transgenes were developed for mice.

***Mus musculus***   Mice are important genetic models for human diseases and human physiology; thus, the ability to create transgenic mice enables scientists to dissect not only the genetic and molecular basis of mouse development and physiology but also, by proxy, many aspects of human development and physiology. Two methods are available to create transgenic mice, a targeted approach and a nontargeted approach. In the nontargeted approach, the transgene is randomly inserted into the genome through illegitimate recombination; in the targeted approach, the transgene is inserted into a specific locus in the genome through homologous recombination. The latter method transformed the study of mouse biology since it allows for the creation of mice with specific loss-of-function (or knockout) and gain-of-function alleles. In 2007, Mario Capecchi,

Martin Evans, and Oliver Smithies shared the Nobel Prize in Medicine or Physiology for their work leading to the development of knockout mice.

Problems associated with variable genomic positions and expression of transgenes led geneticists to explore the possibility of using homologous recombination for the integration of transgenes. Homologous recombination would provide more consistent transgenic mouse strains and would also facilitate the creation of mutations in specific mouse genes, which would be extremely useful for studying mammalian biology. Thus, methods were developed to identify mice in which exogenous DNA had been inserted into the genome by homologous recombination as opposed to the much more frequent illegitimate recombination (**Figure 17.12a**). The identification is accomplished by selecting for the homologous recombinant and at the same time selectively killing the transformants resulting from illegitimate recombination.

The overall strategy is similar to that described for homologous recombination in yeast. The transformation vector contains two regions of DNA homologous to the target locus flanking a positive selectable marker. An example of a positive selectable marker is the *Neomycin* (*Neo*) gene, whose product metabolizes the drug G418, which blocks translation and is lethal to mammalian cells. A vector containing these elements is capable of being integrated into the genome by homologous recombination, but more than 99% of integrations will occur by illegitimate recombination. To select against nonhomologous recombination events, a negative selectable marker is added to the vector outside one of the regions of homology to the target gene. A commonly used negative selectable marker is a *thymidine kinase* (*tk*) gene derived from a herpes simplex virus. Thymidine kinase catalyzes the addition of a phosphate to deoxythymidine, forming deoxythymidine monophosphate, which is eventually converted to deoxythymidine triphosphate, one of the substrates for DNA synthesis. In contrast to mammalian thymidine kinase, thymidine kinase from herpes simplex virus can also catalyze the addition of phosphate to thymidine analogs that cause chain termination when incorporated into DNA. Because the endogenous mammalian thymidine kinase does not recognize the thymidine analogs as substrates, only those cells expressing the herpes simplex virus *tk* gene are sensitive to the thymidine analogs. Thus, cells harboring the viral *tk* gene will be selected against when plated on media containing the thymidine analog ganciclovir. Such thymidine analogs are also used as potent antiviral medications, since only cells harboring the virus are sensitive to the analog.

For transformed mouse cells to survive, they must acquire the positive marker and must lose the negative marker. The occurrence of a homologous recombination event between the negative and positive markers is one possible way in which the introduced DNA can be integrated to produce a cell that possesses the positive and lacks the negative marker. Selection for this type of transformation is called **positive–negative selection.** A related protocol, negative–positive–negative selection, where negative selectable markers are positioned at each end of the introduced DNA, has been successfully used to identify homologous recombination events in plants, such as rice, and should be generally applicable to any species.

What types of mammalian cells are typically targeted for gene transfer? The blastocyst-stage mammalian embryo consists of an outer sphere of cells and a small pool of cells inside the sphere. At the blastocyst stage, the internal cells, known as embryonic stem (ES) cells, are totipotent. The production of a transgenic mouse starts with the isolation of ES cells from the mouse strain to be transformed (**Figure 17.12b**). The ES cells are grown in culture, and DNA is introduced into the cells, often by transiently depolarizing their membranes to make the cells permeable to DNA. The cells are then transferred to media containing the agents for positive and negative selection, and transformed cells in which homologous recombination occurred are selected.

The selected transformed ES cells are reintroduced into a blastocyst from a mouse of a genotype different from that of the transformed cells allowing the progeny derived from the transformed ES cells to be detected. For example, alleles conferring differences in coat color are often used. The blastocyst, now carrying transformed ES cells, is implanted into a surrogate female mouse. Because only some of the ES cells in the host blastocyst are transgenic, the mouse that develops from the embryo in which the transformed cells were introduced is a genetic chimera in which some tissues are derived from the transformed ES cells and other tissues are derived from host ES cells. Chimeric animals can be readily identified by their variegated coat color. It is hoped that at least some of the gametes of the chimeric offspring of the host mouse will be derived from the transformed ES cells, so that some mice in the subsequent generation will be heterozygous for the mutation caused by the homologous recombination event. If two heterozygous offspring of this generation are interbred, mice homozygous for the mutation can be produced. Technologies for the construction of other transgenic mammals, including sheep, cats, cows, horses, monkeys, and rats, follow a similar protocol.

---

**Genetic Insight** Transgenic mammals can be created by selecting for homologous recombination events in embryonic stem cells. The cells can then be reintroduced into an embryo from which a genetically chimeric animal develops.

---

## Site-Specific Recombination

In some cases it is desirable to manipulate transgenes after they have been introduced into an organism. For example, the ability to remove the positive selectable marker gene

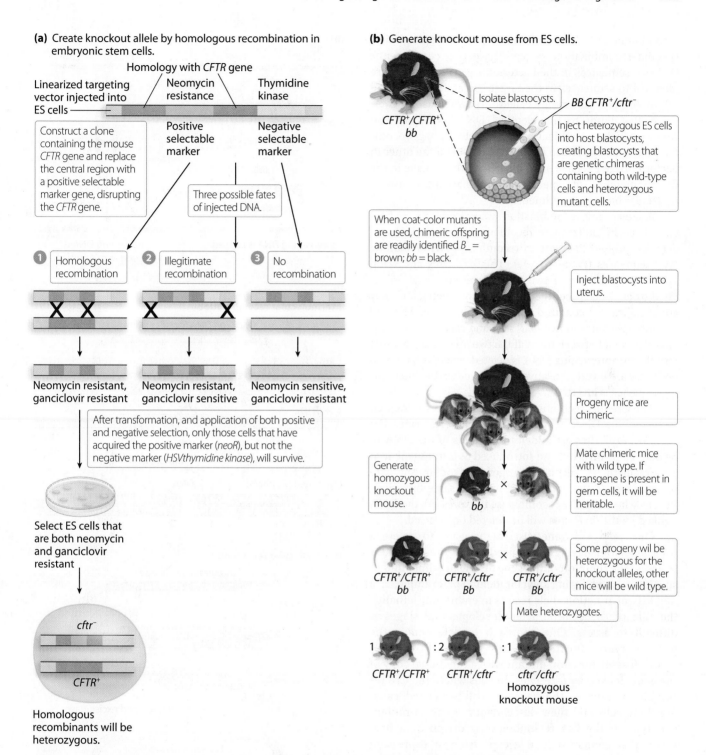

**(a)** Create knockout allele by homologous recombination in embryonic stem cells.

Homology with *CFTR* gene

Linearized targeting vector injected into ES cells

Neomycin resistance

Thymidine kinase

Positive selectable marker

Negative selectable marker

Construct a clone containing the mouse *CFTR* gene and replace the central region with a positive selectable marker gene, disrupting the *CFTR* gene.

Three possible fates of injected DNA.

① Homologous recombination

② Illegitimate recombination

③ No recombination

Neomycin resistant, ganciclovir resistant

Neomycin resistant, ganciclovir sensitive

Neomycin sensitive, ganciclovir resistant

After transformation, and application of both positive and negative selection, only those cells that have acquired the positive marker (*neoR*), but not the negative marker (*HSVthymidine kinase*), will survive.

Select ES cells that are both neomycin and ganciclovir resistant

*cftr⁻*

*CFTR⁺*

Homologous recombinants will be heterozygous.

**(b)** Generate knockout mouse from ES cells.

Isolate blastocysts.

*CFTR⁺/CFTR⁺ bb*

*BB CFTR⁺/cftr⁻*

Inject heterozygous ES cells into host blastocysts, creating blastocysts that are genetic chimeras containing both wild-type cells and heterozygous mutant cells.

When coat-color mutants are used, chimeric offspring are readily identified *B_* = brown; *bb* = black.

Inject blastocysts into uterus.

Progeny mice are chimeric.

Mate chimeric mice with wild type. If transgene is present in germ cells, it will be heritable.

Generate homozygous knockout mouse.

*bb*

×

*CFTR⁺/CFTR⁺ bb*   *CFTR⁺/cftr⁻ Bb*   ×   *CFTR⁺/cftr⁻ Bb*

Some progeny wil be heterozygous for the knockout alleles, other mice will be wild type.

Mate heterozygotes.

1   : 2   : 1

*CFTR⁺/CFTR⁺*   *CFTR⁺/cftr⁻*   *cftr⁻/cftr⁻* Homozygous knockout mouse

**Figure 17.12  Homologous recombination in mice creating a loss-of-function allele in CFTR (cystic fibrosis transmembrane conductance regulator).**  Mutations in the human ortholog are the cause of cystic fibrosis.

after selection of transformants mitigates one of the concerns raised by critics of transgenic plants. In addition, in vivo manipulation of transgenes facilitates the production of conditional alleles of genes whose null allele is lethal. The ability to specifically recombine DNA molecules makes in vivo manipulation of transgenes feasible.

Several bacteriophages use **site-specific recombination** systems during their life cycle, either for intramolecular recombination within the bacteriophage genome or for intermolecular recombination into host genomes. These recombination systems can be harnessed for producing recombinant DNA molecules in vitro and for recombining DNA molecules in vivo. Bacteriophage site-specific recombination systems have two components: (1) DNA sequences in the bacteriophage genome that are identical to sequences in the target bacterial genome and (2) an enzyme, commonly called a recombinase or integrase, that binds to the identical DNA sequences and catalyzes their recombination. Two bacteriophage recombination systems, one in bacteriophage λ and the other in bacteriophage P1, have proven particularly valuable in the development of site-specific recombination for use in molecular biology experiments.

A site-specific recombination system derived from bacteriophage P1 utilizes Cre recombinase, a bacteriophage-encoded protein that acts to recombine DNA containing *loxP* sequences (**Figure 17.13a**). The *loxP* sites are 34-bp sequences consisting of two 13-bp inverted repeats separated by an 8-bp spacer that provides asymmetry, and they are specifically recognized by Cre recombinase. The Cre recombinase binds to two *loxP* sites and catalyzes a recombination event between them. If the two *loxP* sites are direct repeats, the intervening DNA is deleted, whereas if the two *loxP* sites are inverted relative to one another, the intervening sequence is inverted.

The Cre–*lox* recombination system has been adapted to recombine DNA in vivo in transgenic organisms. For example, *loxP* sites are added to the ends of the DNA to be deleted or inverted and introduced as a transgene into an organism. A second transgene encoding the Cre recombinase is also introduced into the same organism. In cells where the Cre recombinase is expressed, the DNA flanked by the *loxP* sites will be deleted or inverted.

One reason a geneticist might want to delete a transgene after having introduced it into the genome is to assess the function of the gene at specific times and in specific tissues during development. For example, if a null loss-of-function allele results in embryonic lethality, the role of the gene at later developmental stages is difficult to assess. One approach to determining the post-embryonic function of such genes is to complement a loss-of-function mutant with a functional copy of the gene flanked by *loxP* sites. In cells where the Cre recombinase is active, the transgene will be deleted, causing these cells and their descendants to have a mutant genotype. If the Cre recombinase is driven by a promoter that confers inducible expression or expression that is temporally or spatially restricted, a genetic chimera can be created, allowing an assessment of gene function in specific tissues.

A second application is the removal of selectable markers in transgenic organisms. An objection to the use of transgenic organisms in agriculture is that some transgenic strains contain a selectable marker providing resistance to antibiotics, which might spread into the natural population. The antibiotic selectable marker genes were used to select the transgenic organism but are no longer needed once the transgenic organism has been identified.

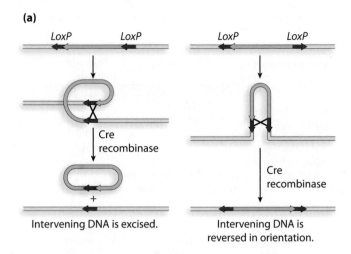

**(a)**

Intervening DNA is excised.

Intervening DNA is reversed in orientation.

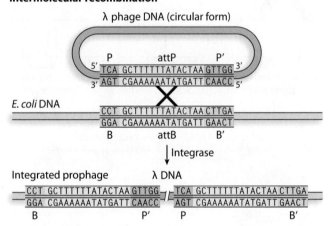

**(b)**

**Intermolecular recombination**

λ phage DNA (circular form)

*E. coli* DNA

Integrated prophage    λ DNA

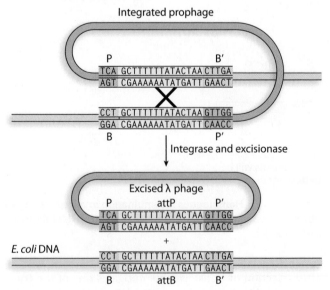

**Intramolecular recombination**

Integrated prophage

Excised λ phage

*E. coli* DNA

**Figure 17.13** Bacteriophage site-specific recombination systems.

One strategy for eliminating the selectable marker is to flank the unwanted transgene with *loxP* sites in a direct repeat orientation. A plant containing this transgene is then crossed with another transgenic plant expressing the Cre recombinase, and the unwanted transgene is deleted in the $F_1$. It is then possible to segregate the transgene encoding the Cre recombinase away from the desired transgene in subsequent generations.

In addition to in vivo recombination of DNA molecules, site-specific recombination systems can be used to manipulate DNA in vitro. A recombination system from bacteriophage λ has been adapted for this task. Recall that during the lysogenic life cycle, the genome of bacteriophage λ is integrated into the host *E. coli* genome (see Chapter 6). Integration is mediated by the bacteriophage λ-encoded protein integrase, which recognizes nearly identical DNA sequences in the bacteriophage λ genome (where the sequence is identified as attachment phage, or attP) and the *E. coli* genome (where the sequence is called attachment bacteria, or attB). Integrase mediates site-specific recombination between the sequences (**Figure 17.13b**). The attP and attB sequences share 15 identical nucleotides but differ in the three flanking nucleotides on either side, depending on whether the sequence is attP or attB. Thus, following recombination between the attP and attB sites, a hybrid site is formed (as shown in the top part of Figure 17.13b). An additional bacteriophage λ-encoded protein, excisionase, is required to act with integrase for the reverse recombination that excises bacteriophage λ from the *E. coli* genome. The ability to insert and excise from the bacterial genome allows bacteriophage λ to undergo either a lysogenic or lytic life cycle (see Chapter 14). Since the site-specific recombination system only requires bacteriophage λ-encoded proteins and short DNA sequences, the system can be adapted to recombine any DNA molecules. Recombination is accomplished by adding specific attB or attP sequences to the ends of the DNA molecules to be recombined and supplying purified integrase enzyme.

**Genetic Analysis 17.1** asks you to put some of these ideas to work by designing a mouse model of a human disease.

---

Genetic Insight   When sequences that are targets for recombination by bacteriophage enzymes are incorporated into transgenes, the transgenes can be manipulated in vivo.

---

## 17.2 Transgenes Provide a Means of Dissecting Gene Function

Most of the transgenes described in this chapter thus far have been heterologous genes transferred for the purpose of producing a protein product or conferring a favorable agronomic trait. However, we have also seen that another important use of transgenes is to study gene function, by creating loss-of-function and gain-of-function alleles, as well as to monitor gene expression patterns. This section describes in greater detail the ways transgenes can reveal genetic function.

While an almost limitless array of transgenes can be constructed for genetic analysis, many fall into two categories. One category consists of **reporter genes** to investigate gene regulation. Fusion of the regulatory sequences of a gene to a reporter gene provides information about where, when, and how much a gene is expressed. Some reporter genes facilitate live imaging and monitoring of gene expression in real time. The second category consists of gain-of-function alleles generated by placing coding regions from one gene under control of regulatory sequences derived from another gene. An allele constructed in this way can be expressed ectopically, meaning at times or in places where the gene is not normally expressed. The use of either or both of these types of transgenes can complement analyses of loss-of-function alleles by providing information on how genes are normally expressed and the phenotypic consequences of changing their normal expression pattern.

### Monitoring Gene Expression with Reporter Genes

A gene can act as a reporter if its product can be detected directly or is an enzyme that produces a detectable product. The regulatory sequences of the gene of interest are used to drive the expression of the reporter gene. Two types of reporter gene fusions can be constructed: transcriptional and translational (**Figure 17.14**).

In a *transcriptional fusion*, regulatory sequences directing transcription of the gene of interest are fused directly with the coding sequences of the reporter gene. In this case, the reporter gene will be transcribed in the pattern directed by the regulatory sequences to which it is fused. In *translational fusion*, not only the regulatory sequences but also the coding sequence of the gene of interest are fused to the reporter gene in such a way that the reading frame for translation is maintained between the gene of interest and the reporter gene. As a result, the reporter protein is translationally fused with the protein of interest, and the location of the reporter protein provides information not only on the spatial and temporal transcriptional expression pattern but also on the subcellular location of the fusion protein. In translational fusions, care must be taken to find out if the fusion protein is still functional, since the addition of the reporter protein could interfere with the proper folding or activity of the protein of interest.

Some frequently used reporter genes are represented in **Figure 17.15**. The choice of reporter gene depends on the biological question being addressed. With some reporter

**Figure 17.14 Transcriptional versus translational gene fusions.**

**Gene in eukaryotic genome**

5' upstream regulatory sequences

Transcription start site

Exon 1  Exon 2  Exon 3  3' downstream sequences

5'UTR | ATG  STOP | 3'UTR

Intron 1  Intron 2

**Transcriptional fusion**

5' upstream regulatory sequences

Transcription start site

3' downstream sequences

5'UTR | ATG  Reporter gene  STOP | 3'UTR

**Translational fusion**

5' upstream regulatory sequences

Transcription start site

Exon 1  Exon 2  Exon 3  3' downstream sequences

5'UTR | ATG  Reporter gene | STOP | 3'UTR

Intron 1  Intron 2

**(a)** *Lin-3* regulatory sequences driving *lacZ* reporter gene in *C. elegans*

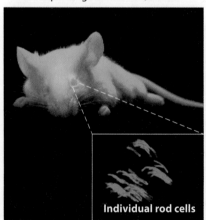

**(b)** *PHABULOSA* regulatory sequences driving β-*glucuronidase* reporter gene in *Arabidopsis*

**(c)** *CaMV 35S* regulatory sequences driving *luciferase* reporter gene in tobacco

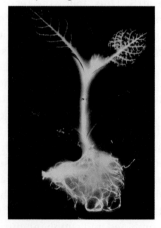

**(d)** *RHODOPSIN* regulatory sequences driving *GFP* reporter gene in *Mus musculus*

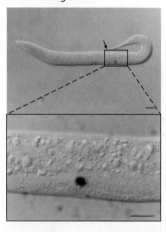

Individual rod cells

**(e)** *Mus musculus* neurons expressing three different fluorescent reporter genes, derived from modifying GFP

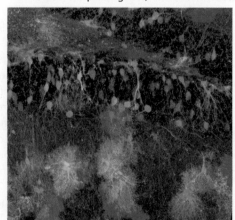

**Figure 17.15 Reporter genes.**

Mouse models of human diseases are valuable research tools that can be used to test therapies and drugs. How would you make a mouse model of Huntington disease, which is caused by an autosomal dominant mutation consisting of an expanded sequence of trinucleotide repeats?

| Solution Strategies | Solution Steps |
|---|---|
| **Evaluate** | |
| 1. Identify the topic this problem addresses, and explain the nature of the required answer. | 1. This problem asks how to construct a specific strain of transgenic mouse. |
| 2. Identify the critical information given in the problem. | 2. The desired disease model is of Huntington disease (HD), described as an autosomal dominant mutation that consists of an expanded sequence of trinucleotide repeats. |
| **Deduce** | |
| 3. Identify the inheritance pattern of the HD phenotype. | 3. Since HD is dominant, a phenotype should be evident if a mutant allele is introduced into the genome. |
| 4. Evaluate the ways in which the *HD* allele can be transferred into mice. | 4. Transgenic mice can be generated by random integration of a transgene or, alternatively, by homologous recombination that replaces the endogenous gene with a mutant version. |
| 5. Choose a method to generate a transgenic mouse. <br><br> **PITFALL:** Randomly integrated transgenes may have variation in expression patterns. | 5. Since we want the transgene to be expressed in the same pattern as the wild-type mouse *HD* gene, homologous recombination is the best approach, because the mutant *HD* gene will then be in the same genomic context and will be expressed in the same pattern as the wild-type gene. |
| **Solve** | |
| 6. Design a strategy to replace the wild-type mouse *HD* gene with a mutant version of the human *HD* gene. <br><br> **PITFALL:** Since a functional allele is desired, the positive selectable marker must not interfere with *HD* gene function. | 6. The positive–negative selection approach outlined in Figure 17.12 to produce a transgenic mouse by homologous recombination results in a loss-of-function allele. This approach must be modified to create a gain-of-function allele. <br> a. Construct a vector in which a human mutant *HD* gene is flanked by mouse *HD* regulatory sequences (5′ and 3′ of the *HD* gene). <br> b. The positive selective marker gene can be placed downstream of the *HD* gene, in a position not likely to interfere with *HD* gene expression, or could be removed using the Cre–*lox* approach outlined in Figure 17.13. <br> c. Another transgenic mouse expressing the wild-type human gene driven by the same regulatory sequences would provide a useful control to compare the specific phenotypic effects induced by the expression of the mutant allele. |

For more practice, see Problems 3, 4, 10, 16, and 22.

---

genes, the assay to monitor gene expression requires sacrificing the organism, whereas the expression of other reporter genes can be traced in a living organism. Reporter genes usually require substrates that must penetrate into the tissues or cells where the reporter genes are located. In addition, reporter genes vary in their sensitivity.

One of the first reporter genes to be developed emerged from research on the *lac* operon in *E. coli* (see Chapter 14).

To purify and study the activity of β-galactosidase, encoded by the *lacZ* gene, a number of β-galactosides were synthesized and tested as substrates. Two β-galactosides, abbreviated X-gal and ONPG, were found to be useful. β-galactosidase cleaves the colorless substrate, ONPG, into a yellow product. This assay is typically used for in vitro measurement of β-galactosidase activity. In contrast, X-gal, also colorless, is cleaved by β-galactosidase into a blue product. This assay can be used in bacteria in vivo, since bacterial cells can take up the X-gal substrate without a reduction in viability (see Chapter 16).

The *lacZ* gene can be used in conjunction with the substrate X-gal as a reporter gene in animal systems (Figure 17.15a). However, since plants have an endogenous β-galactosidase activity, *lacZ* is not suitable for studying plant systems. An alternative system is the *E. coli uidA* gene encoding β-glucuronidase, which enzymatically cleaves a colorless precursor, X-gluc, into a blue product (Figure 17.15b). Conversely, since animals have endogenous β-glucuronidase activity, the *uidA* gene cannot be used as a reporter in animals. A limitation of both of these reporter genes in organisms other than bacteria is that in order for the substrate to be taken up effectively into internal tissues, the tissue to be stained must be bathed in a solution that kills the cells.

Research into reactions that cause the natural emission of light in some animals has led to the development of reporter genes that cause light to be produced in living cells. For example, luciferase, the enzyme responsible for the glow of fireflies, catalyzes a reaction between the substrate luciferin and ATP that results in the emission of light. Transgenic plants expressing the luciferase gene will emit a yellow-green glow if supplied with the substrate (Figure 17.15c). However, luciferin is not delivered to all cells of the plant in equal measure, which in many cases limits its usefulness as a reporter gene.

The development of **green fluorescent protein (GFP)** led to great strides both in genetics and cell biology by providing a noninvasive means of visualizing gene and protein expression patterns in living organisms (Figure 17.15d). The *GFP* gene, derived from the jellyfish *Aequoria victoria*, is the source of the natural bioluminescence of this species. Its wild-type protein product, consisting of 238 amino acids, fluoresces green (a 509-nm wavelength) when illuminated with UV light (a 395-nm wavelength), which in this case is the "substrate," delivered by laser.

Because UV light, with its short wavelength, can be harmful to organisms (e.g., causing thymidine dimers to form in DNA, as described in Chapter 12), the wild-type *GFP* gene was mutated to produce variants that respond to lower-energy wavelengths. A major improvement was a mutation that shifted the excitation wavelength to 488 nm, corresponding to blue light and minimizing the potential damage to cells being illuminated. Subsequent modification of the GFP protein sequence has led to the production of variants that emit other colors (e.g., yellow, cyan, blue). Genes encoding fluorescent reporter proteins have also been isolated from marine corals and other jellyfish. The availability of multiple fluorescent reporter genes makes it possible to visualize the expression of several genes simultaneously in a single organism (Figure 17.15e). Osamu Shimomura, Martin Chalfie, and Roger Y. Tsien received the 2008 Nobel Prize in Chemistry for their discovery and development of GFP.

Reporter genes can be used to dissect regulatory DNA sequences and identify specific sequences required for particular aspects of gene regulation. The general approach is to start with a clone in which all the regulatory sequences required for proper gene expression are present, and then to assay the effects of deleting or changing specific portions of the clone. An example of such an analysis of the *Drosophila even-skipped* (*eve*) gene, which is expressed in seven stripes in the segmentation pattern of the embryo, is shown in Figure 17.16. Overlapping deletions spanning large regions are assayed first, with regions identified as important for gene regulation dissected with smaller deletions. The concept is similar to that described earlier for deletion mapping (see Chapter 7). When specific sequences required for proper gene expression are deleted, expression of the reporter gene will be correspondingly altered. If genomic sequence is available from two or more related species, regulatory elements may be predicted by searching for sequences that are conserved between the related species, using a method known as *phylogenetic footprinting* (discussed in Chapter 19). Such initial genomic sequence analyses can direct subsequent experimental tests that use reporter genes to analyze in transgenic organisms.

---

Genetic Insight   Reporter genes are used for dissecting gene regulatory sequences and for monitoring gene expression patterns in living organisms.

---

## Investigating Gene Function with Chimeric Genes

With chimeric gene fusions, scientists can cause any gene of interest to be driven by any regulatory sequences active in the organism being used as a host. Figure 17.17 shows one way experimenters can take advantage of this potential to obtain information on gene function. Recessive loss-of-function mutations in the *eyeless* gene of *Drosophila* result in a failure of eyes to develop. The *eyeless* gene is normally expressed only in the eye imaginal discs during *Drosophila* development. Imaginal discs are groups of precursor cells that are set aside during embryonic development. These grow by mitotic proliferation during larval life and later differentiate into adult body tissues during metamorphosis. If the *eyeless* gene is ectopically expressed in other imaginal discs, such as those that would normally give rise to the antennae or legs, the imaginal discs will differentiate as eye tissue instead. This outcome indicates that cells in any imaginal disc are capable of differentiating into eyes and that the *eyeless* gene product can promote the development of eyes from any imaginal disc. Thus, when the *eyeless* allele is ectopically expressed as a gain-of-function mutation in inappropriate imaginal discs, the resulting phenotype is the converse of the phenotype of the loss-of-function *eyeless* allele—ectopic eyes rather than an absence of eyes.

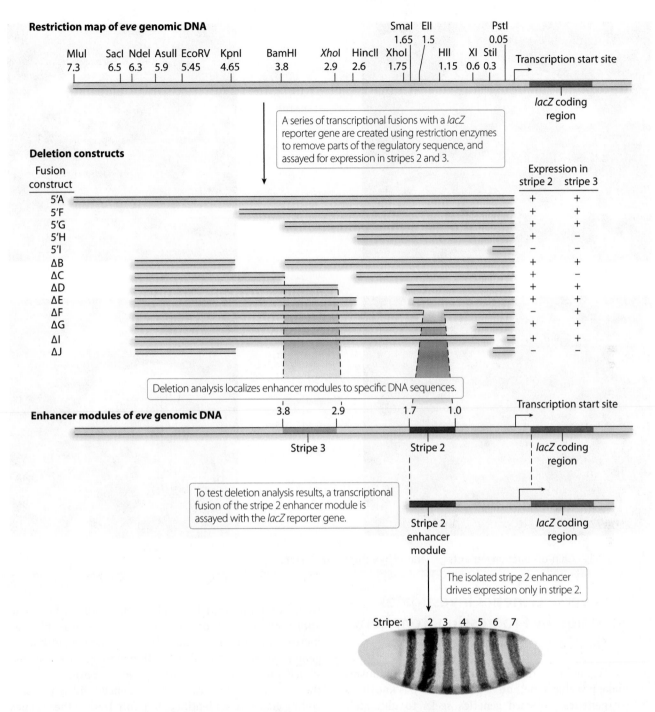

**Figure 17.16** **Use of reporter gene in promoter analysis of the *even-skipped* (*eve*) gene.**

In cases where the gain-of-function and loss-of-function phenotypes are complementary, interpretation of the effects of ectopic expression is straightforward. Thus, in the preceding example, *eyeless* is revealed to be a master control gene for the differentiation of eyes in *Drosophila*. However, ectopic expression of genes can also lead to enigmatic phenotypes that are more difficult to interpret. For example, ectopic expression of *eyeless* during embryogenesis leads to embryonic lethality, a phenotype that is not easily reconciled with the loss-of-function phenotype. Therefore, when considering gain-of-function alleles

generated by ectopic expression, we must remember that the phenotypes represent what the gene is capable of doing when expressed in particular contexts and may not reflect the normal function of the gene.

Genetic Insight  Construction of chimeric genes by fusing heterologous promoters with coding regions derived from other genes can create gain-of-function alleles that may shed light on gene function.

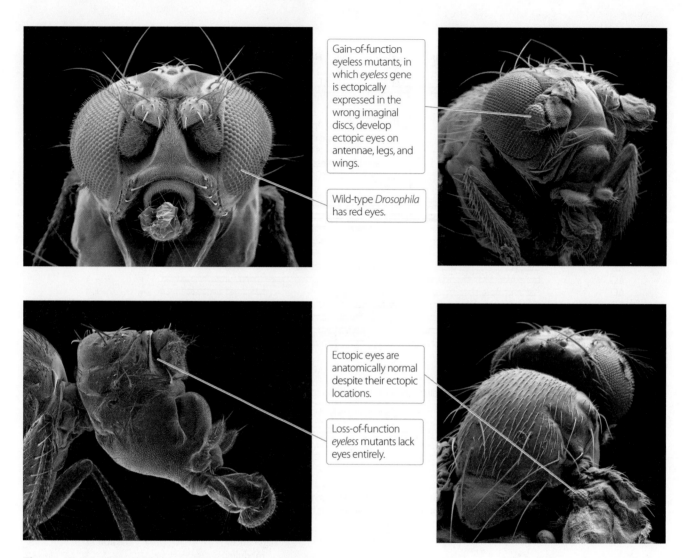

Gain-of-function eyeless mutants, in which *eyeless* gene is ectopically expressed in the wrong imaginal discs, develop ectopic eyes on antennae, legs, and wings.

Wild-type *Drosophila* has red eyes.

Ectopic eyes are anatomically normal despite their ectopic locations.

Loss-of-function *eyeless* mutants lack eyes entirely.

**Figure 17.17** **Gain-of-function mutants generated by ectopic expression.**

## 17.3 Reverse Genetics Investigates Gene Action by Progressing from Gene Identification to Phenotype

Genetic technologies that introduce DNA into organisms also make possible a method of genetic analysis known as reverse genetics. Forward genetics seeks to ultimately identify a gene sequence by beginning with a mutant phenotype and investigating the abnormal physiology or development of the phenotype as an indicator of the gene or genes involved in the abnormality (see Chapter 16). **Reverse genetics** works in the opposite direction, beginning with a gene sequence and seeking to identify an associated mutant phenotype (see Figure 16.1).

Forward genetics was for a long time the primary—and for much of the last century, the only—approach to uncovering gene function. The development of molecular methods for gene identification and advances in sequencing technologies are now making reverse genetic approaches increasingly valuable and common. The reasons for this shift in emphasis are twofold. First, the enormous amount of genomic sequence available has increased by orders of magnitude the number of known gene sequences, and only a fraction of them have been assigned a function by forward genetics. For example, when the *E. coli* genome was fully sequenced, 4288 protein-coding genes were identified, but only 1853 of these genes had been previously identified through forward genetic screens. Second, genomic sequencing and reverse genetic screens have also uncovered a degree of gene duplication not previously suspected. Gene duplications often result in genetic redundancy. In forward genetic screens, such duplicated genes would not be identified since mutation of only one of the genes would not usually result in a conspicuous mutant phenotype. However, reverse genetics approaches, where the functions of both duplicates are disrupted, are particularly suited in these situations to provide evidence of gene function.

Reverse genetics begins with the creation of a mutant allele for a gene that is identified only by its sequence. The selection of mutational tools is largely dependent on the biology of the experimental organism. In organisms in which homologous recombination readily occurs, targeted sequence changes, such as deletions, are the method of choice. In organisms amenable to transformation but in which homologous recombination is rare, two approaches are widely used. The first is to generate a large collection of random insertion mutations and then screen them for mutations in the gene of interest using PCR-based techniques. The second approach is to harness a gene-silencing phenomenon known as RNA interference (RNAi), which is induced by double-stranded RNA molecules. In species not amenable to large-scale transformation experiments, nontransgenic methods of mutagenesis can be used. These basic techniques for reverse genetics are described in the rest of this section. (Genomic aspects of reverse genetics are discussed in Chapter 18).

## Use of Insertion Mutants in Reverse Genetics

Reverse genetics approaches for most of the commonly used model genetic organisms utilize **knockout libraries,** collections of mutants in which most or all genes have been mutated by inactivating (or "knocking out") their expression. Most knockout mutants are produced by the insertion of exogenous pieces of DNA into the genome to generate loss-of-function alleles; thus, most alleles in the libraries are null alleles. *Saccharomyces cerevisiae* and *E. coli* geneticists have, for example, systematically generated loss-of-function alleles of all known *S. cerevisiae* and *E. coli* genes by homologous recombination. In these knockout library collections, each strain has a single mutation in a different gene.

In many model genetic organisms, it is not technically simple or economically feasible to systematically generate loss-of-function mutants for all genes. However, if an organism is easy to transform, populations of random mutants can be generated by transposon insertions or T-DNA insertions (in the case of plants) (Table 17.2). These populations can then be screened for mutations in specific genes. Screening is accomplished using PCR-based techniques with a primer that is specific to the gene of interest

and a primer that is specific to the insertional mutagen used. For model genetic systems such as *Drosophila*, where *P* elements have been used as an insertional mutagen, large populations of mutants generated by insertions have been characterized to such an extent that mutations in specific genes can be ordered directly from a stock center. Similar knockout libraries based on T-DNA and transposon insertions are available for *Arabidopsis*. Such knockout libraries are an invaluable resource for large-scale reverse genetics experiments that aim to elucidate the function of every gene in model genetic organisms (see Chapter 18). An example of an application of reverse genetics to determine the function of closely related genes in *Arabidopsis* is described in the Case Study at the end of this chapter.

---

**Genetic Insight**  Reverse genetics investigates gene function by beginning with a gene sequence and proceeding toward discovery of a mutant phenotype. Knockout libraries facilitate reverse genetics screening by providing null alleles for all or most genes in a genome.

---

## RNA Interference in Gene Activity

In the late 1980s, researchers introduced a chalcone synthase transgene into *Petunia* in an effort to increase the amount of floral pigment. To their surprise, some transgenic lines exhibited complete loss of pigment production (Chapter 15). Not only was the chalcone synthase transgene not expressed properly in these lines, but the endogenous chalcone synthase gene was also silenced, a phenomenon they termed co-suppression. A similar phenomenon was subsequently observed in both fungal and animal systems. This method of silencing genes was initially called quelling in *Neurospora* and **RNA interference (RNAi)** in animals. The phenomenon is now universally known as RNAi. In the 1990s, Andrew Fire and Craig Mello, who won the 2006 Nobel Prize in Physiology or Medicine for their work on RNAi, used a genetic approach to dissect and elucidate the biochemical mechanism for RNAi in *C. elegans*.

| Table 17.2 | Reverse Genetics Approaches in Model Genetic Organisms |
|---|---|
| **Species** | **Reverse Genetics Tools** |
| *Escherichia coli* | Knockouts by homologous recombination |
| *Saccharomyces cerevisiae* | Knockouts by homologous recombination |
| *Arabidopsis thaliana* | Random T-DNA and transposon insertions; TILLING; RNAi |
| *Drosophila melanogaster* | Random *P* element insertion lines; RNAi |
| *Caenorhabditis elegans* | RNAi loss-of-function alleles |
| *Mus musculus* | Knockouts by homologous recombination; RNAi |

Double-stranded RNA (dsRNA) can act as a trigger for the degradation not only of the double-stranded RNA itself but also of any RNA molecules that are complementary to the double-stranded RNA (see Chapter 15). A primary role of this gene-silencing system is to silence repetitive DNA. Transcription from several different copies of repetitive elements often generates double-stranded RNA molecules, since collectively both strands of the repetitive DNA are often transcribed. In addition, RNAi protects cells against double-stranded RNA viruses. Thus, dsRNA-mediated gene silencing acts as a genomic immune system to silence both repetitive DNA sequences and invading nucleic acids.

To take advantage of endogenous RNAi activity as a way of silencing genes, scientists introduce double-stranded RNA that is complementary in sequence to the target gene (**Figure 17.18**). The mRNA of the target gene will then be degraded through the action of Dicer and Argonaute enzymes (described in Chapter 15), causing a loss-of-function phenotype of the target gene. The efficiency of silencing can approach that of a null allele, although often the phenotypes induced represent a range of partial loss-of-function phenotypes.

The double-stranded RNA can be introduced directly into cells or organisms by injection of double-stranded RNA or indirectly by infection with a double-stranded RNA virus. Alternatively, a transgene can be designed that results in the production of double-stranded RNA, a method that has the added advantage of being heritable. In

animals, transient introduction of double-stranded RNA into cell cultures has been successful. One of the methods for introducing double-stranded RNA into *C. elegans* is surprisingly simple. *Caenorhabditis elegans* normally eats *E. coli* as food, and remarkably enough, when *C. elegans* is fed *E. coli* that is producing double-stranded RNA, the double-stranded RNA will be taken up into *C. elegans* and will silence genes in many organs of the *C. elegans* body. While in this case the RNAi phenotype is not indefinitely heritable, the phenotypic effects can be seen in several subsequent generations produced by self-fertilization of the worm that was fed the *E. coli*.

The advantages of the RNAi approach to reverse genetics include the ease and rapidity of applying the method. It allows large-scale reverse genetic screens to be conducted in cell cultures and whole organisms without the laborious preparatory task of creating mutagenized populations. In addition, transient RNAi-mediated gene silencing offers an alternative means of applying reverse genetics in species for which stable transformation protocols do not exist.

In a related approach, synthetic micro-RNAs have been created to target the degradation of specific mRNAs. Like RNAi-mediated gene silencing, synthetic micro-RNA–mediated gene silencing takes advantage of endogenous gene-silencing machinery (see Chapter 15). The synthetic microRNAs are designed according to principles derived from known microRNAs but are customized to direct the translational repression or mRNA cleavage of the gene of interest.

## Reverse Genetics by TILLING

Reverse genetics can also be performed on species that cannot be transformed easily as long as the species is amenable to standard genetic analyses. One approach to reverse genetics that can be applied to any genome is **targeted induced local lesions in genomes (TILLING)**. In a TILLING protocol, a population of organisms of an inbred strain is randomly mutagenized throughout the genome (**Figure 17.19**). Enough independent lines are produced to bring the level of mutagenesis to near saturation, at which, ideally, each gene is represented by multiple mutant alleles in the mutagenized population. Often, the mutagen employed in the development of the mutagenized lines is a chemical such as ethyl methanesulfonate (EMS), a potent mutagen that causes a spectrum of alleles (Chapter 16). DNA from the mutagenized lines is screened systematically using PCR-based methods to search for mutations in a particular gene of interest.

For each individual of the mutagenized population, both progeny and DNA are collected. The generation derived from the mutagenized population is often referred to as the $M_1$ generation (**Figure 17.19a**). DNA is isolated from $M_1$ individuals or from $M_2$ families of organisms. Any mutation induced in the mutagenesis will

Double-stranded RNA can be introduced by:

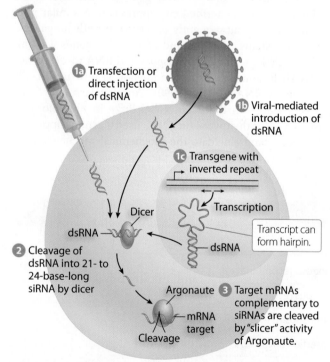

**1a** Transfection or direct injection of dsRNA

**1b** Viral-mediated introduction of dsRNA

**1c** Transgene with inverted repeat

Transcription

Dicer

Transcript can form hairpin.

dsRNA

dsRNA

**2** Cleavage of dsRNA into 21- to 24-base-long siRNA by dicer

Argonaute **3** Target mRNAs complementary to siRNAs are cleaved by "slicer" activity of Argonaute.

mRNA target

Cleavage

**Figure 17.18 Reverse genetics using RNAi.**

**(a)** Seeds are mutated to produce $M_1$ generation. Each $M_1$ plant is heterozygous for mutations in different genes (colors).

**$M_1$ individuals**

**(b)** Mutations in specific genes are identified by analyzing DNA isolated from each $M_2$ family. For example, one representative $M_2$ family with a mutant (red) segregating

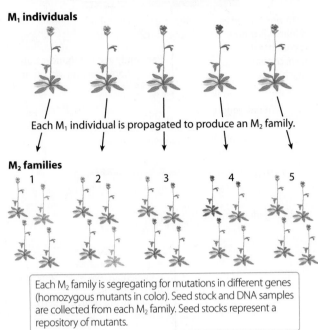

Each $M_1$ individual is propagated to produce an $M_2$ family.

**$M_2$ families**

1    2    3    4    5

Each $M_2$ family is segregating for mutations in different genes (homozygous mutants in color). Seed stock and DNA samples are collected from each $M_2$ family. Seed stocks represent a repository of mutants.

**Figure 17.19  Reverse genetics by TILLING.**

be either heterozygous (if the DNA was derived from an $M_1$ individual) or segregating (if the DNA was derived from an $M_2$ family). A region of the target gene is chosen for PCR-based amplification. The PCR products generated in this analysis are expected to contain both the wild-type sequence and mutant sequence. Those that consist solely of the wild-type allele can be distinguished from those consisting of a mixture of the wild-type allele and a mutant allele.

The PCR products are first denatured and allowed to reanneal, creating some homoduplex DNA, in which the strands are fully complementary if derived from the same allele, and some heteroduplex DNA (**Figure 17.19b**). Heteroduplex DNA is composed of strands that are largely complementary but contain one or more mismatched base pairs, indicating that the strands are derived from DNA containing different alleles. Heteroduplex DNA can be distinguished from homoduplex DNA by either a difference in migration of the products during electrophoresis or by differential susceptibility to an endonuclease that cleaves heteroduplex DNA at mismatched base pairs. Heteroduplex DNA forms only in DNA samples in which a mutation in the target gene is present. Screening progeny from several thousand mutagenized individuals often allows identification of multiple mutant alleles of the target gene. Individuals homozygous or heterozygous for the mutant allele can then be identified in the appropriate $M_2$ family.

When chemical mutagenesis is used to produce TILLING alleles, it results in both null alleles and partial

loss-of-function alleles. The spectrum of alleles of different phenotypes produced by TILLING approaches is often useful for dissecting gene function, even in organisms where gene knockouts are available. While TILLING was developed for studies in model genetic species, it is suitable for any organism that can be mutagenized and genetically analyzed. It is currently being applied to several crop plants.

---

Genetic Insight RNAi can be harnessed to rapidly create either transient or stable loss-of-function alleles by gene silencing. TILLING is a reverse genetics approach for generating and analyzing chemically induced mutations throughout a genome.

---

## Enhancer Trapping

**Enhancer trapping** uses a variation of an insertional library to identify genes based on expression patterns. This approach combines the generation of a large number of random insertion mutants with the expression of a reporter gene (Figure 17.20). In its simplest application, a population of transgenic organisms is generated by random insertion of a transposon (or T-DNA) containing the coding sequence of a reporter gene fused with a minimal promoter for RNA polymerase II transcription. If the insertion occurs near enhancer and silencer regulatory sequences that can act in conjunction with the minimal promoter of the reporter gene, the reporter gene can be expressed in a pattern that reflects the regulatory capability of the nearby genomic DNA sequences. The enhancer (or silencers) of the adjacent genomic DNA are co-opted, or "trapped," by the insertion to drive expression of the reporter gene. While reporter gene expression may not precisely reflect the expression of the adjacent gene, the expression of the reporter often at least partially reflects the normal gene expression pattern of the adjacent gene. Enhancer trapping techniques were first pioneered in *Drosophila* and have now been adapted to other systems. Because they identify genes by gene expression patterns, enhancer trapping techniques complement forward genetic screens.

Genetic Analysis 17.2 tests your understanding of the reverse genetics analytical techniques discussed in this section.

## 17.4 Gene Therapy Uses Recombinant DNA Technology

The ability to manipulate gene expression through the introduction of a transgene raises the possibility that human genetic diseases could be treated by the introduction of a functional copy of the mutant gene. The use of genes as therapeutic agents to cure or alleviate disease symptoms is termed **gene therapy.** From a genetic perspective, gene therapy is similar to a genetic complementation experiment in which the gene introduced by gene therapy compensates for a genetic abnormality in the

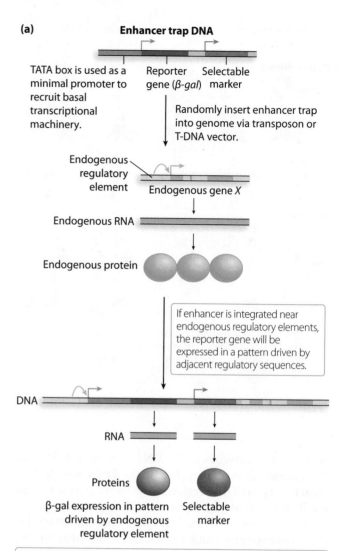

**(a)**

**Enhancer trap DNA**

TATA box is used as a minimal promoter to recruit basal transcriptional machinery.

Reporter gene (*β-gal*)  Selectable marker

Randomly insert enhancer trap into genome via transposon or T-DNA vector.

Endogenous regulatory element

Endogenous gene X

Endogenous RNA

Endogenous protein

If enhancer is integrated near endogenous regulatory elements, the reporter gene will be expressed in a pattern driven by adjacent regulatory sequences.

DNA

RNA

Proteins

β-gal expression in pattern driven by endogenous regulatory element

Selectable marker

If enhancer trap disrupts coding region of gene, a loss-of-function allele is created. However, insertion of vector may occur 5′ or 3′ to a gene and still "trap" enhancers without causing a loss-of-function mutation.

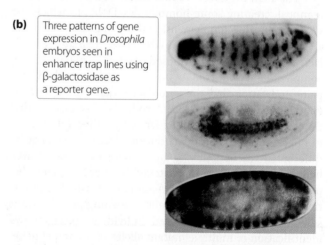

**(b)** Three patterns of gene expression in *Drosophila* embryos seen in enhancer trap lines using β-galactosidase as a reporter gene.

**Figure 17.20 Enhancer trapping to reveal expression patterns of endogenous genes.** (a) Strategy for generation of enhancer trap lines. (b) Examples of enhancer trap lines in *Drosophila*.

In searching the mouse genome, you identify the sequences of three genes orthologous to the single *hedgehog* gene of *Drosophila*: *Sonic hedgehog*, *Indian hedgehog*, and *Desert hedgehog*. Describe the research design you would use to learn the function of each of the genes and whether that gene function is unique or redundant in the mouse.

| Solution Strategies | Solution Steps |
|---|---|
| **Evaluate** | |
| 1. Identify the topic this problem addresses, and explain the nature of the required answer. | 1. This question asks about identifying the function of genes known only by sequence. |
| 2. Identify the critical information given in the problem. | 2. While only one *hedgehog* gene exists in *Drosophila*, three *hedgehog* genes exist in mouse, raising the question of whether the three mouse genes have different functions or whether there is any sharing of function. |
| **Deduce** | |
| 3. Consider possible approaches to discovering the functions of genes known only by sequence. | 3. Functions of genes known only by sequence can be determined by reverse genetics approaches. |
| 4. Consider possible approaches to reverse genetics available in mice. | 4. Homologous recombination approaches can be used to produce loss-of-function mutations in mice. As other possibilities, gain-of-function alleles can be constructed with chimeric genes, and gene expression patterns can be deduced with reporter genes. |
| **Solve** | |
| 5. Describe how you would create loss-of-function knockout alleles of each of the three genes by homologous recombination. | 5. The positive–negative selection approach outlined in Figure 17.12 can be used with a homologous recombination vector in which the coding sequence of the hedgehog genes is replaced with that of a positive selectable marker gene. |
| 6. Describe a genetic approach to determine whether the genes have unique or redundant functions. | 6. Homozygous mutant lines can be bred and the phenotypes of each of the three single knockouts examined. Interbreeding the single mutant lines will lead to the creation of strains in which combinations of two or more genes are inactive. Comparison of phenotypes of single mutants with those of multiple mutants allows an assessment of whether the genes exhibit unique or redundant functions. |
| 7. Describe how experiments using chimeric genes and reporter genes could support the conclusions you reached based on knockout alleles. <br><br> PITFALL: The reporter gene expression might not reflect the normal expression of the gene determined experimentally (e.g., by hybridization to mRNA in situ). | 7. Information on gene function can also be obtained by knowing where and when a gene is expressed, the subcellular location of the encoded gene product, and the phenotypic effects of ectopic expression. Since eukaryotic gene regulatory sequences may be located either in introns or at large distances upstream (5′) or downstream (3′) of the coding sequences, the best approach is to fuse the reporter gene in frame with the endogenous gene using homologous recombination. Comparisons of expression patterns of the three genes can also guide assays to assess mutant phenotypes. <br><br> The coding sequences of the three hedgehog genes could be fused with regulatory sequences derived from another gene that is expressed in a different pattern or at higher levels than endogenous *hedgehog* gene expression, and the phenotype assayed. Complementary loss- and gain-of-function phenotypes provide powerful evidence of gene function. |

For more practice, see Problems 11, 12, 16, and 18.

altered cell. Two types of gene therapy, classified as somatic gene therapy and germinal gene therapy, are feasible.

**Somatic gene therapy** targets somatic cells, whose descendants will not give rise to germ cells. Any genetic alterations induced in the targeted cells will be passed to daughter cells by mitosis, but the alteration will not be inherited by progeny of the individual undergoing somatic gene therapy. The specific somatic cells to be targeted depend on the disease in question. For example, in individuals with cystic fibrosis, the epithelial cells of the lungs represent a logical target, since lungs are severely affected in cystic fibrosis. On the other hand, for diseases of the blood, cells of the various hemopoietic lineages are the target cells; they can be removed from bone marrow, treated, and returned to the same individual. Somatic gene therapy turns the treated individual into a genetic chimera that has the transgene present in the target cells but not in other somatic cells or in germ cells. Somatic gene therapy can potentially be used to treat several genetic diseases whose phenotype becomes apparent early in childhood.

The alternative strategy for gene therapy, **germinal gene therapy,** targets cells of the germ line, which give rise to gametes. Because germinal gene therapy alters germ-line cells, the therapeutic transgene is transmitted to the progeny of the treated individual. Both types of gene therapy have been successful in animal systems, but for ethical reasons only somatic gene therapy has been attempted in humans. In this section, we discuss somatic gene therapy in humans and describe modifications of these protocols suggested by successful somatic gene therapy experiments in mice.

## Human Gene Therapy

The primary difficulties in human somatic gene therapy concern the delivery of the transgene to the somatic cells of interest and the proper expression of the transgene in targeted cells. The DNA encoding the gene must be delivered to the proper cells, pass through the cell membrane and into the nucleus, and once there, be expressed at a level that is sufficient to provide normal gene function. In some cases—for example, in hemopoietic diseases—the cells to be treated can simply be extracted from the body, treated in vitro, and then injected back into the body. However, in other cases—for example, cystic fibrosis, in which lung epithelial cells are the target—the cells must be treated in situ because they cannot be removed from the patient.

The choice of vector to deliver the DNA to the cells is pivotal. Gene therapy methods often take advantage of viruses that have evolved mechanisms to access specific cell types. Essentially, viruses are harnessed to transduce the transgene into the target cells the way the transduction of DNA between bacteria is performed by bacteriophage (see Chapter 7). The viruses can be "disarmed" so that they no longer have the ability to cause the diseases associated

with their wild-type relatives. Several types of viral vectors have been used, including gamma-retroviruses, lentiviruses, and adenoviruses (Table 17.3). Each has advantages and disadvantages for gene therapy protocols.

Many viral vectors deliver transgenes by integrating into the genome of the target cell. Integration provides a mechanism for stable gene transfer and thus permanent correction of the defect. Integration of the vector into the genome is not without risks, however; the insertion may cause a detrimental mutation, a problem that has plagued most human gene therapy experiments to date. The treatment of a serious immune system disease called severe combined immune deficiency syndrome (SCIDS) provides an example. SCIDS patients lack the ability to produce a category of blood cells called T cells that are critical to the body's defense against infection. One form of SCIDS is due to mutations in the gene encoding the gamma subunit ($\gamma$ chain) of the interleukin-2 receptor and is X-linked. In the mid-1990s, a gene therapy approach was designed using a retroviral vector carrying the $\gamma$ chain cDNA driven by viral regulatory sequences. The retrovirus carrying the cDNA was bounded by long terminal repeats (LTRs; see Chapter 12). In one study, 9 of 10 patients were successfully treated, and they exhibited functioning, adaptive immune systems following gene therapy. In three patients, however, an uncontrolled increase of mature T cells, termed T-acute lymphoblastic leukemia, developed in the years immediately following treatment. In each of these individuals, the retrovirus became inserted into the *LMO2* gene in such a way that the retroviral LTR promoter was able to cause unregulated expression of *LMO2*. This gene is known to be required for the differentiation of hemopoietic cells. Its overexpression is thought to be what led these patients to develop leukemia.

One of the concerns raised by the use of retroviruses as gene therapy vectors is that they may act as mutagens. Once this possibility was recognized, the SCIDS gene therapy trials just described were suspended. However, the high proportion of treated individuals whose immune defects were corrected in the study suggests that gene therapy can be a viable approach to treating such diseases.

In addition to concerns over safety and efficacy, the use of viral vectors presents technical challenges stemming from size limits on the amount of DNA that can be packaged in the viral capsid (similar limits were discussed

| Table 17.3 | Viruses Used as Vectors in Gene Therapy | | |
|---|---|---|---|
| **Virus Type** | **Integration into Genome** | **Target** | **Capacity** |
| Retrovirus | Integrates; insertional mutagen | Infects dividing cells | 8 kb |
| Lentivirus | Integrates; insertional mutagen | Infects nondividing cells | 8 kb |
| Adenovirus | Nonintegrating | Infects nondividing cells | 7.5 kb |
| Adeno-associated virus | Episomal, but can integrate | Infects nondividing cells | 4.5 kb |

in regard to bacterial transduction in Chapter 7). In most cases, the amount of DNA that can be packaged by a virus is much smaller than the size of a typical human gene. For example, the transcribed region of the *CFTR* gene spans approximately 170,000 bp and results in a 6132-bp processed mRNA encoding a 1480–amino acid protein. Since viral vectors can accept only 5 to 10 kb of DNA, only a cDNA of the *CFTR* gene lacking all endogenous gene-expression regulatory elements can be accommodated in a viral vector. In the absence of these endogenous regulatory sequences, the expression of the *CFTR* coding sequence is driven by viral regulatory sequences, which might not regulate the transgenes in a manner appropriate for proper gene function in the target cells.

Virus-based gene therapy continues to be employed in selected experimental cases despite past failures and continuing concerns over the safety of the procedures. Successes in treating patients with cystic fibrosis, SCIDS, and several other human hereditary conditions offer hope that continued research will identify effective vectors for delivering treatment that is sustained, targeted, and safe.

## Curing Sickle Cell Disease in Mice

The ideal somatic gene therapy would be one that corrects the specific mutation causing the genetic disease rather than just compensating for the mutant allele. Advances in understanding the biology of embryonic stem (ES) cells have brought new forms of somatic gene therapy that may approach the ideal for some genetic diseases. Embryonic stem cells are totipotent, meaning they have the potential to differentiate into any cell type in the body. In addition, as discussed in Section 17.1, the genome of an ES cell can be manipulated by homologous recombination. Thus, if ES cells can be isolated from an individual, gene mutations within the cells could perhaps be corrected, and the cells could then be induced to differentiate into the appropriate cell type to treat the genetic disease. As illustrated in the mouse experiment described below, the ability to create and manipulate ES cells provides a means of isolating cells from an individual, correcting mutations in the cells, and reintroducing the "corrected" cells into the body.

In many cases, the diagnosis of a genetic disease is not made until early childhood, when the body no longer possesses any ES cells, because they form only during early embryogenesis (see Section 17.1). How can ES cells be obtained from a person who has none? The answer is to create ES cells from other cells of the body.

In 2006 and 2007, a series of experiments demonstrated that mouse or human fibroblasts, a type of cell occurring in connective tissue, could be reprogrammed in vitro to behave like stem cells. These reprogrammed cells have been called induced pluripotent stem cells, or iPS. (The word *pluripotent* is used because scientists do not yet know if the iPS cells are totipotent.) This reprogramming of differentiated cells was accomplished by expressing a combination of three to four transcription factors, including Oct4, Sox2, c-Myc, and Klf4. The transcription factors that were used are normally expressed in ES cells and appear to be sufficient to induce reprogramming of the transcriptional networks of differentiated somatic cells into networks characteristic of ES cells. These advances set the stage for using iPS cells in gene therapy. Proof of principle (a phrase used by scientists to mean proof that the general idea is valid) was provided using a mouse model for sickle cell disease (**Figure 17.21**). The basic strategy being tested consisted of (1) harvesting adult cells, (2) reprogramming adult cells into iPS cells, (3) repairing the genetic defect through homologous recombination, (4) differentiating the iPS cells into hemopoietic precursors in vitro, and (5) transplanting the corrected cells into bone marrow of affected mice.

The starting point for this test of somatic gene therapy was the creation of a "humanized" mouse model for sickle cell anemia by substituting human $\alpha$-globin genes for the endogenous mouse $\alpha$-globin genes and substituting human $\beta^S$ (sickle) globin genes for the mouse $\beta$-globin genes. Mice homozygous for the $\beta^S$-globin allele ($h\beta^S/h\beta^S$) exhibited typical disease symptoms, including severe anemia and erythrocyte sickling. Fibroblasts isolated from the tail of $h\beta^S/h\beta^S$ mice were infected with retroviruses encoding the Oct4, Sox2, and Klf4 transcription factors and with a lentivirus encoding the c-Myc transcription factor. Expression of these four transcription factors resulted in the reprogramming of the fibroblast cells into iPS cells. On either side of the *c-Myc* gene on the lentivirus, *lox* sites had been placed, to allow the gene to be excised from the genome when the cells were infected with an adenovirus encoding Cre recombinase. This was important because continued expression of *c-Myc* predisposes cells to become cancerous. Although the other three transgenes were not removed in this experiment, their removal by a similar mechanism is also recommended.

To correct a $\beta^S$-globin allele, a transformation vector encoding the $\beta^A$-globin allele was introduced into the iPS cells, and hygromycin- and ganciclovir-resistant homologous recombinants were created using the procedure described in Section 17.1. The corrected iPS cells were now heterozygous at the $\beta$-globin locus ($h\beta^A/h\beta^S$). The $h\beta^A/h\beta^S$ iPS cells were then differentiated into hemopoietic progenitors (HPs, cells that have the potential to differentiate into any of the hemopoietic lineages) by infection with another retrovirus encoding the *HoxB4* gene, which induces the differentiation of ES cells into HPs when incubated with cytokines secreted from bone marrow cells. The $h\beta^A/h\beta^S$ HPs were then transplanted back into $h\beta^S/h\beta^S$ mice in which the endogenous $h\beta^S/h\beta^S$ bone marrow cells had been eliminated by irradiation, so that now the $h\beta^A/h\beta^S$ HPs constituted the primary source of hemopoietic cells. In this particular experiment, the *HoxB4* coding sequence was translationally fused with

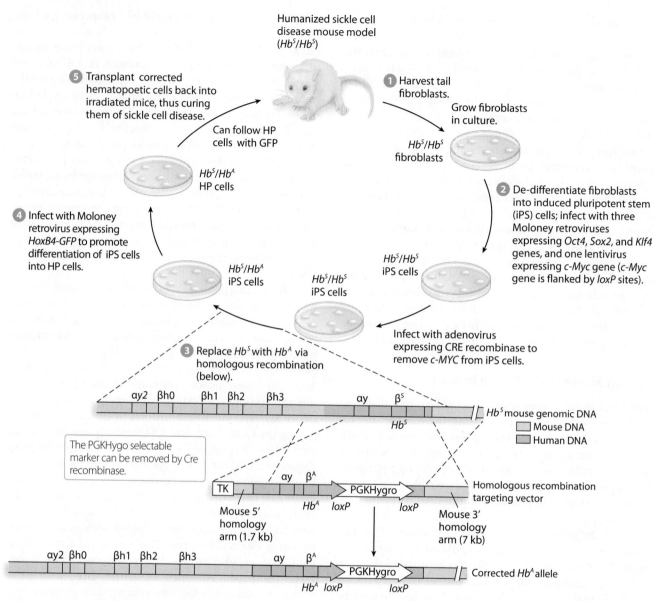

**Figure 17.21 Genetic therapy for mice with sickle cell disease.**

that of green fluorescent protein (GFP), so the activity of the $h\beta^A/h\beta^S$ HPs could be monitored by the presence of $GFP^+$ cells in the blood. Subsequently, by all physiological tests, the mice receiving the $h\beta^A/h\beta^S$ HPs were cured of sickle cell disease.

These experiments in mice provide proof of principle for the therapeutic potential of using ES or iPS cells in gene therapy to correct genetic defects, but at least two facets of gene therapy continue to cause concern. Problems associated with using retroviruses and oncogenes for reprogramming need to be resolved before implementing such a gene therapy protocol in humans. In addition, whether iPS cells are truly totipotent or still contain an epigenetic memory of their origin remains to be determined. Since an individual's own cells are used as the source for genetic modification, there are no impediments due to immune system incompatibility. However, this approach is limited to those diseases, such as blood disorders, in which cells can be isolated, genetically corrected, and reintroduced into the body.

---

Genetic Insight Somatic gene therapy is designed to correct genetic defects in somatic cells.

---

# 17.5 Cloning of Plants and Animals Produces Genetically Identical Individuals

Many plants have the capacity for vegetative (asexual) propagation in addition to sexual propagation. For example, poplar and aspen (*Populus* sp.) groves often consist of vegetatively propagated clones, all genetically identical. Some of these clonal groves are estimated to be at least 10,000 years old. Humans, taking advantage of the ability of plants to reproduce vegetatively, have been clonally propagating plants for centuries in agricultural practices. In these protocols, heterozygous genotypes of agriculturally desirable specimens are propagated intact, without the segregation of alleles that occurs during sexual reproduction. Heterozygous genotypes often exhibit hybrid vigor, resulting in high yields in comparison to inbred varieties.

Perhaps the most conspicuous example of agricultural vegetative propagation is the cultivation of grapes (*Vitus vinifera*), which were domesticated 6000 to 7000 years ago. Most grape cultivars are highly heterozygous; that is, they have two different alleles at many genomic loci. Thus, when they are self-fertilized or crossed with another cultivar, extensive segregation of genotypes and phenotypes is observed in the progeny. Because this presents an obstacle to controlling the properties of grape plants through breeding, cultivars that possess favorable phenotypes are propagated by cuttings (that is, additional plants are grown from pieces of source plants). In most vineyards, the vines are chimeric: The shoots are all genetically identical and chosen on the basis of their fruit phenotype, and the roots, also identical to one another, are of a different genotype that is chosen for being well adapted to local soil conditions.

Several wine grape cultivars can be traced back to the Middle Ages, and some are likely to be even older. For example, Pinot was first described in Roman times and is thought to be at least 2000 years old. While clonal propagation allows maintenance of specific genotypes, somatic mutations—due, for example, to errors in DNA replication and transposable element activity—can accumulate over time and lead to phenotypic variation. Thus, a mutation in a gene required for pigment synthesis led to the formation of Pinot blanc, a white-berry cultivar, from Pinot noir, the ancestral black-berry cultivar.

Unlike plants, most animals do not readily propagate clonally in nature—but there are exceptions. For example, some aphid species undergo multiple parthenogenetic (clonal) generations in the spring and summer, followed by sexual reproduction in the autumn. Since most animal cells are not totipotent, ES cells being the exception, animals do not readily regenerate from single cells. Thus techniques for cloning animals, and in particular mammals, from single differentiated cells are considerably more complicated than those for cloning plants.

Dolly, a sheep, was the first cloned mammal. In the protocol used to produce Dolly, a diploid nucleus is isolated from a differentiated cell of the animal to be cloned (**Figure 17.22**). This nucleus, containing all the nuclear genetic information of the animal from which it was taken, is injected into an egg cell that has had its own nucleus removed. The egg cell can be derived from the

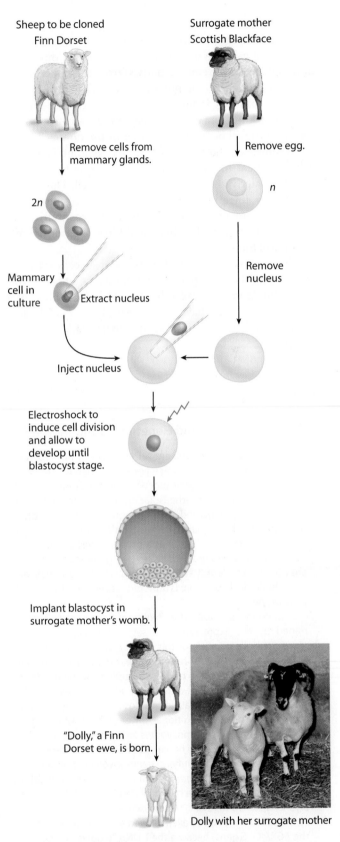

**Figure 17.22 Cloning animals by nuclear implantation.**

animal to be cloned (if it possesses egg cells) or from a different individual. If the nuclear transplantation is successful, the genome of the donor nucleus will direct the development of the embryo derived from the egg cell. The use of a diploid donor nucleus means that fertilization with a sperm cell is not required to produce a diploid nucleus in the embryo; thus, the genetic constitution of the embryo will be identical to that of the donor. Bear in mind, however, that while the nuclear genome is genetically identical to that of the donor, the mitochondrial genome is derived from the surrogate egg cell. The diploid egg cell is then induced to begin embryogenesis and implanted into a surrogate mother. If all goes well, it will develop into a normal embryo, and birth of a normal offspring will follow.

In most mammals, the frequency of success with this protocol has been quite low. Dolly's was the only one out of 270 implanted egg cells that resulted in the birth of a sheep. Donor cells have been derived from adult animals—Dolly's donor cell was a mammary gland cell—and

are therefore highly differentiated somatic cells rather than totipotent embryonic stem cells. In differentiated somatic cells, such as those of the mammary gland, the patterns of facultative heterochromatin (see Chapter 15) are vastly different from those of embryonic stem cells. In other words, although the sequences of nucleotides in the genomes of differentiated and embryonic stem cells are identical, the epigenetic modifications of the histones and DNA methylation patterns differ. The low frequency of success in the initial attempts to clone mammals was likely due to deficiencies in reprogramming the genetic material of the injected nucleus to mimic the epigenetic modifications characteristic of an embryonic stem cell. However, advances in knowledge of ES cell biology, and their application in the creation of iPS cells from differentiated cells, suggest that the cloning of mammals will increase over time. Despite the difficulties, many different mammals besides sheep have been successfully cloned, including mice, cows, horses, donkeys, cats, and dogs.

## CASE STUDY

### Reverse Genetics and Genetic Redundancy in Flower Development

In this case study, we present an example of how forward genetics and reverse genetics work together to provide a broader view of both gene function and evolution. We start with forward genetics—the isolation of a mutant that alters flower development and the subsequent identification of the mutant gene sequence using recombinant DNA technology. Using the cloned gene as a probe, genes of similar sequence are also cloned. Finally, reverse genetics approaches are applied to identify mutant alleles of related genes, and their biological function is inferred based on the mutant phenotypes.

In flowering plants, the identity of floral organs is determined by the expression of a set of transcription factors (For further description of this activity, see Chapter 20.) The identity of *Arabidopsis* reproductive organs (stamens and carpels) is determined in part by the activity of the *AGAMOUS* gene. Recessive null loss-of-function *agamous* alleles lead to the development of petals in the positions usually occupied by stamens and of an additional flower in the position usually occupied by carpels. Homozygotes are sterile and do not produce gametes (hence the name, *AGAMOUS*). In forward genetics screens aimed at identifying genes involved in *Arabidopsis* flower development, *agamous* mutant alleles induced by either EMS or T-DNA have been isolated (**Figure 17.23 ❶**; see also Chapter 16 to review genetic screens).

The T-DNA–induced allele proved a useful tool for cloning the *AGAMOUS* gene, because the T-DNA "tagged" the gene ❷ (see also Chapter 16 to review how transposons or T-DNA can be used to clone genes). First, a genomic library is constructed from DNA isolated from *agamous* mutants (see Chapter 16 for construction and screening of genomic libraries). Then the genomic library is screened with a probe consisting of T-DNA

sequence. The probe identifies genomic clones in the library that have T-DNA sequence. Since the T-DNA was inserted into the *AGAMOUS* gene, *Arabidopsis* DNA adjacent to the T-DNA sequences encodes the *AGAMOUS* gene.

The genomic clone encoding *AGAMOUS* can be used to identify an *AGAMOUS* cDNA clone from a library constructed with mRNA from wild-type flowers ❸ (see also Chapter 16 to review cDNA libraries). Sequencing of the *AGAMOUS* cDNA clones reveals that the encoded protein has a similarity to known eukaryotic transcription factors. This conclusion is based on the similarity between a 60–amino acid domain of the AGAMOUS protein and DNA-binding domains in yeast and mammalian transcription factors.

When the *AGAMOUS* cDNA is used to probe a Southern blot of restriction-enzyme-digested *Arabidopsis* genomic DNA, sequences related to the *AGAMOUS* gene sequence can be identified ❹ (see Chapter 10 to review Southern blotting). The same *AGAMOUS* cDNA can be used as a probe on the flower cDNA library to identify clones of related genes (see Chapter 16 to review the identifying of homologous genes). Genes related to *AGAMOUS* were called *AGAMOUS-LIKE*, or *AGL*, genes. These related genes possess the same highly conserved DNA-binding domain but differ in the rest of their protein sequences. To determine how the *AGL* genes are related to *AGAMOUS* and to each other, a phylogenetic tree can be constructed ❺ (see Chapter 1 to review phylogenetic trees).

Since the related genes are known by gene sequence only, a reverse genetics approach can be undertaken to determine gene function. Transposon- or T-DNA–induced mutant alleles of many of the *AGL* genes in *Arabidopsis* can be identified in available knockout libraries ❻ (see Section 17.3). Researchers were initially surprised that plants homozygous

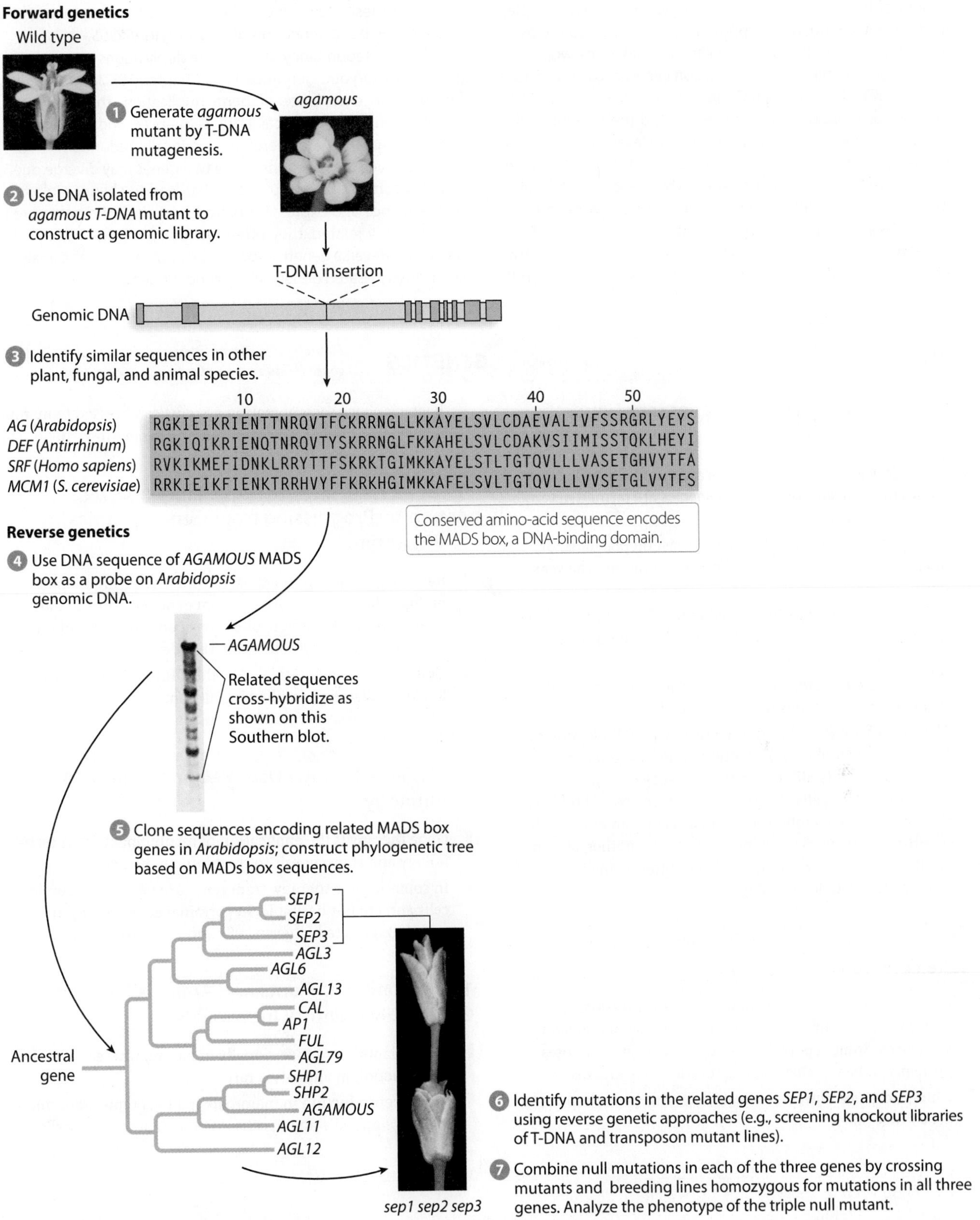

**Forward genetics**

Wild type

**1** Generate *agamous* mutant by T-DNA mutagenesis.

*agamous*

**2** Use DNA isolated from *agamous T-DNA* mutant to construct a genomic library.

T-DNA insertion

Genomic DNA

**3** Identify similar sequences in other plant, fungal, and animal species.

|   | 10 | 20 | 30 | 40 | 50 |
|---|---|---|---|---|---|

AG (*Arabidopsis*)     RGKIEIKRIENTTNRQVTFCKRRNGLLKKAYELSVLCDAEVALIVFSSRGRLYEYS
DEF (*Antirrhinum*)    RGKIQIKRIENQTNRQVTYSKRRNGLFKKAHELSVLCDAKVSIIMISSTQKLHEYI
SRF (*Homo sapiens*)   RVKIKMEFIDNKLRRYTTFSKRKTGIMKKAYELSTLTGTQVLLLVASETGHVYTFA
MCM1 (*S. cerevisiae*) RRKIEIKFIENKTRRHVYFFKRKHGIMKKAFELSVLTGTQVLLLVSETGLVYTFS

Conserved amino-acid sequence encodes the MADS box, a DNA-binding domain.

**Reverse genetics**

**4** Use DNA sequence of *AGAMOUS* MADS box as a probe on *Arabidopsis* genomic DNA.

— *AGAMOUS*

Related sequences cross-hybridize as shown on this Southern blot.

**5** Clone sequences encoding related MADS box genes in *Arabidopsis*; construct phylogenetic tree based on MADs box sequences.

SEP1
SEP2
SEP3
AGL3
AGL6
AGL13
CAL
AP1
FUL
AGL79
SHP1
SHP2
AGAMOUS
AGL11
AGL12

Ancestral gene

*sep1 sep2 sep3*

**6** Identify mutations in the related genes *SEP1*, *SEP2*, and *SEP3* using reverse genetic approaches (e.g., screening knockout libraries of T-DNA and transposon mutant lines).

**7** Combine null mutations in each of the three genes by crossing mutants and breeding lines homozygous for mutations in all three genes. Analyze the phenotype of the triple null mutant.

**Figure 17.23  Use of forward and reverse genetics to determine gene function.**

for loss-of-function alleles of many single genes did not display an aberrant phenotype. Hypothesizing that the more closely related the genes, the more similar their functions would be, researchers combined loss-of-function alleles of closely related genes ❼. The loss-of-function alleles were due either to T-DNA or transposon insertions in the genes and the mutant alleles identified by PCR-based genotyping. For example, *sep1* mutants—having mutations of the *SEPALLATA1* gene—were crossed with *sep2* mutants, after which *sep1 sep2* double mutants were identified in the F₂ generation. Disappointingly, the *sep1 sep2* double mutants did not differ significantly from wild-type plants. However, *sep1 sep2 sep3* triple mutant plants proved to have flowers consisting solely of sepals, which indi-cates that these genes have a function related to floral organ specification but distinct from the role of *AGAMOUS*.

Genetic redundancy due to gene duplications is prevalent in most eukaryotic genomes (see Chapter 18). Immediately following an occurrence of gene duplication, the duplicate genes often have identical DNA sequences and expression patterns, and they are therefore genetically redundant. Over time, however, the functions of the two genes may diverge due to the accumulation of mutations that lead to changes in protein sequence and expression pattern. Yet, since the genes are evolutionarily related, they often function in similar biological processes. Reverse genetics approaches can facilitate the analysis of closely related genetically redundant genes.

## SUMMARY

*For activities, animations, and review quizzes, go to the study area at www.masteringgenetics.com.*

### 17.1 Introducing Foreign Genes into Genomes Creates Transgenic Organisms

▐ Genes introduced into an organism are called transgenes. Genes introduced from another species are termed heterologous transgenes.

▐ Transgenes can be introduced into yeast on plasmids, or alternatively, by homologous recombination into the yeast chromosome.

▐ *Agrobacterium* and its tumor-inducing plasmid can be harnessed to create transgenic plants in which the transfer DNA carries the desired transgene.

▐ Transgenic *Drosophila* are created by injection into embryos of a *P* element transposon carrying the transgene.

▐ Transgenes are introduced into mice by direct injection of DNA into isolated cells. Detection of homologous recombination events is facilitated by positive–negative selection of embryonic stem cells. Transgenic mice are created by injection of transgenic embryonic stem cells into an embryo that is subsequently implanted into a surrogate mother, and the resulting progeny are chimeric. Non-chimeric mice are then selected in the following generation.

### 17.2 Transgenes Provide a Means of Dissecting Gene Function

▐ Reporter genes are used to monitor gene-expression patterns in transgenic organisms and for the dissection of regulatory sequences. Some reporter genes, such as the green fluorescent protein, can be visualized in real time in living organisms.

▐ Chimeric genes represent novel alleles that provide clues to gene function.

▐ Bacteriophage recombination systems can be used to manipulate DNA sequences in vitro and transgenes in vivo.

### 17.3 Reverse Genetics Investigates Gene Action by Progressing from Gene Identification to Phenotype

▐ Reverse genetics approaches, in which determination of biological function proceeds from gene sequence to mutant phenotype, make use of collections consisting of mutants that are each defective in a different defined gene.

▐ Collections of insertion alleles, the TILLING process, and RNAi-mediated gene silencing all contribute to the reverse genetics analysis of model organisms.

### 17.4 Gene Therapy Uses Recombinant DNA Technology

▐ Gene therapy is the application of recombinant DNA technology and transgenesis to treat human diseases.

▐ In somatic gene therapy, transgenes are targeted to somatic cells and are not heritable. In germinal gene therapy, transgenes are targeted to germ cells and are thus heritable.

### 17.5 Cloning of Plants and Animals Produces Genetically Identical Individuals

▐ Many plants reproduce clonally in nature, whereas clonal reproduction in animals is rare.

▐ Clonal reproduction in mammals requires reprogramming of differentiated somatic cells into pluripotent stem cells.

## KEYWORDS

positive–negative selection *(p. 578)*
reporter gene *(p. 581)*
reverse genetics *(p. 586)*
RNA interference (RNAi) *(p. 587)*
shuttle vector *(p. 569)*

site-directed mutagenesis *(p. 567)*
site-specific recombination *(p. 579)*
somatic gene therapy *(p. 591)*
targeted induced local lesions in genomes
   (TILLING) *(p. 588)*

Ti plasmid *(p. 571)*
totipotency *(p. 572)*
transfer DNA (T-DNA) *(p. 571)*
transgene *(p. 563)*
transgenic organism *(p. 563)*

## PROBLEMS

Mastering**GENETICS**

*For instructor-assigned tutorials and problems, go to www.masteringgenetics.com.*

### Chapter Concepts

*For answers to selected even-numbered problems, see Appendix: Answers.*

1. What are the advantages and disadvantages of using *GFP* versus *lacZ* as a reporter gene in mice, *C. elegans*, and *Drosophila*?

2. A transcriptional fusion of regulatory sequences of a particular gene with a reporter gene results in relatively uniform expression of the reporter gene in all cells of an organism, whereas a translational fusion with the same gene shows reporter gene expression only in the nucleus of a specific cell type. Discuss some biological causes for the difference in expression patterns of the two transgenes.

3. Using animal models of human diseases can lead to insights into the cellular and genetic basis of the diseases. Duchenne muscular dystrophy (DMD) is the consequence of an X-linked recessive allele.
   a. How would you make a mouse model of DMD?
   b. How would you make a *Drosophila* model of DMD?

4. Compare methods for constructing homologous recombinant transgenic mice and yeast.

5. Chimeric gene-fusion products can be used for medical or industrial purposes. One idea is to produce biological therapeutics for human medical use in animals from which the

products can be easily harvested—in the milk of sheep or cattle, for example. Outline how you would produce human insulin in the milk of sheep.

6. Why are diseases of the blood more likely targets for treatment by gene therapy than are many other genetic diseases?

7. Injection of double-stranded RNA can lead to gene silencing by degradation of RNA molecules complementary to either strand of the dsRNA. Could RNAi be used in gene therapy for a defect caused by a recessive allele? A dominant allele? If so, what might be the major obstacle to using RNAi as a therapeutic agent?

8. Compare and contrast methods for making transgenic plants and transgenic *Drosophila*.

9. What are the advantages and disadvantages of using insertion alleles versus alleles generated by chemicals (via TILLING) in reverse genetic studies?

10. You have cloned the mouse ortholog of the gene associated with human Huntington Disease (*HD*) and wish to examine its expression in mice. Outline the approaches you might take to examine the temporal and spatial expression pattern at the cellular level.

### Application and Integration

*For answers to selected even-numbered problems, see Appendix: Answers.*

11. The *CBF* genes of *Arabidopsis* are induced by exposure of the plants to low temperature.
   a. How would you examine the temporal and spatial patterns of expression after induction by low temperature?
   b. Can you design a method that would indicate these changes in gene expression in a way that a farmer could recognize them by observing plants growing in the field?

12. You have identified five genes in *S. cerevisiae* that are induced when the yeast are grown in a high-salt (NaCl) medium. To study the potential roles of these genes in acclimation to growth in high-salt conditions, you wish to examine the phenotypes of loss- and gain-of-function alleles of each.
   a. How will you do this?
   b. How would your answer differ if you were working with tomato plants instead of yeast?

13. You have generated three transgenic lines of maize that are resistant to the European corn borer, a significant pest in many regions of the world. The transgenic lines were created using *Agrobacterium*-mediated transformation with a

T-DNA having two genes, the first being a gene conferring resistance to the corn borer and the second being a gene conferring resistance to a herbicide that you used as a selectable marker to obtain your transgenic plants. You crossed each of the lines to a wild-type maize plant and also generated a $T_2$ population by self-fertilization of the $T_1$ plant. The following segregation results were observed (herbicide resistant/herbicide sensitive):

| Cross | Line 1 | Line 2 | Line 3 |
|---|---|---|---|
| Transgenic ($T_1$) × wild type | 1:1 | 3:1 | 5:1 |
| Self-cross ($T_2$) | 3:1 | 15:1 | 35:1 |

Explain these segregation ratios.

14. Bacterial *Pseudomonas* species often possess plasmids encoding genes involved in the catabolism of organic compounds. You have discovered a strain that can metabolize crude oil and wish to identify the gene(s) responsible. Outline an experimental protocol to find the gene or genes required for crude oil metabolism.

15. Two complaints about some transgenic plants presently in commercial use are that (1) the Bt toxin gene is constitutively expressed in them, leading to fears that selection pressures will cause insects to evolve resistance to the toxin, and (2) the selectable marker gene conferring kanamycin resistance remains in the plant, leading to concerns about increased antibiotic resistance in organisms in the wild. How would you generate transgenic plants that produce Bt only in response to being fed upon by insects and without the selectable marker?

16. In *Drosophila*, loss-of-function *Ultrabithorax* mutations result in the posterior thoracic segments differentiating with an identity normally found in the anterior thoracic segment. When the *Ultrabithorax* gene was cloned, it was shown to encode a transcription factor and to be expressed only in the posterior region of the thorax. Thus, *Ultrabithorax* acts to specify the identity of the posterior thoracic segments. Similar genes were soon discovered in other animals, including mice and men. You have found that mice possess two closely related genes, *Hoxa7* and *Hoxb7*, which are orthologous to *Ultrabithorax*. You wish to know whether the two mouse genes act to specify the identity of body segments in mice.

   a. How will you determine where and when the mouse genes are expressed?

   b. How will you create loss-of-function alleles of the mouse genes?

   c. How will you determine whether the mouse genes have redundant functions?

17. You have identified an enhancer trap line generated by *P* element transposition in *Drosophila* in which the marker gene from the enhancer trap is specifically expressed in the wing imaginal disc.

   a. How can you identify the gene adjacent to the insertion site of the enhancer trap?

   b. How would you show that the expression pattern of the enhancer trap line reflects the endogenous gene expression pattern of the adjacent gene?

18. When the *S. cerevisiae* genome was sequenced, only about 40% of predicted genes had been previously identified in forward genetics screens. This left about 60% of predicted genes with no known function, leading some to dub the genes *fun* (function unknown) genes.

   a. As an approach to understanding the function of a certain *fun* gene, you wish to create a loss-of-function allele. How will you accomplish this?

   b. You wish to know the physical location of the encoded protein product. How will you ascertain such information?

19. Translational fusions between a protein of interest and a reporter protein are used to determine the subcellular location of proteins in vivo. However, one complication is that fusion to a reporter protein sometimes renders the protein of interest nonfunctional because the addition of the reporter protein interferes with proper protein folding, enzymatic activity, or protein–protein interactions. You have constructed a fusion between your protein of interest and a reporter gene. How will you show that the fusion protein retains its normal biological function?

20. The highlighted sequence shown below is the one originally used to produce the B chain of human insulin in *E. coli*. The sequence of the human gene encoding the B chain of insulin was later determined from a cDNA isolated from a human pancreatic cDNA library and is shown below in black. Explain the differences between the two sequences.

```
ATGTTCGTCAATCAGCACCTTTGTGGTTCTCACCTCGTTGAAGC
TTTGTACCTTGTTTGCGGTGAACGTGGTTTCTTCTACACTCCT
                AAGACTTAA
GCCTTTGTGAACCAACACCTGTGCGGCTCACACCTGGTGGAAGC
TCTCTACCTAGTGTGCGGGGAACGAGGCTTCTTCTACACACCC
                AAGACCCGC
```

21. In enhancer trapping experiments, a minimal promoter and a reporter gene are placed adjacent to the end of a transposon so that genomic enhancers adjacent to the insertion site can act to drive expression of the reporter gene. In a modification of this approach, a series of enhancers and a promoter can be placed at the end of a transposon so that transcription is activated from the transposon into adjacent genomic DNA. What types of mutations do you expect to be induced by such a transposon in a mutagenesis experiment?

22. The *RAS* gene encodes a signaling protein that hydrolyzes GTP to GDP. When bound by GDP, the RAS protein is inactive, whereas when bound by GTP, RAS protein activates a target protein, resulting in stimulation of cells to actively grow and divide. A single base-pair mutation results in a mutant protein that is constitutively active, leading to continual promotion of cell proliferation. Such mutations play a role in the formation of cancer. You have cloned the wild-type version of the mouse *RAS* gene and wish to create a mutant form to study its biological activity in vitro and in transgenic mice. Outline how you would proceed.

```
                  Gly Ala Gly Gly Val Gly
Wild-type RAS DNA:  5'...GGC GCC GGC GGT GTG GGC...3'
                                 ↓
Mutant RAS DNA:                 GTC
                                Val
```

23. Vitamin E is the name for a set of chemically related tocopherols, which are lipid-soluble compounds with antioxidant properties. Such antioxidants protect cells against the effects of free radicals created as by-products of energy metabolism in the mitochondrion. Different tocopherols have different biological activities due to differences in their retention by binding to gut proteins during digestion. The one retained at the highest level is α-tocopherol, while γ-tocopherol is retained at less than 10% of that efficiency. In *Arabidopsis*, α-tocopherol is the most abundant tocopherol in leaves, while γ-tocopherol is the most abundant in seeds. An enzyme encoded by the *VTE4* gene can convert γ-tocopherol to α-tocopherol. How would you create an *Arabidopsis* plant that produces high levels of α-tocopherol in the seeds?

24. The *Drosophila even-skipped* (*eve*) gene is expressed in seven stripes in the segmentation pattern of the embryo. A sequence segment of 8 kb 5' to the transcription start site (shown as +1 in the figure) is required to drive

expression of a reporter gene (*lacZ*) in the same pattern as the endogenous *eve* gene. Remarkably, expression of each of the seven stripes appears to be specified independently, with stripe 2 expression directed by regulatory sequences in the region 1.7 kb 5′ to the transcription start site. To further examine stripe 2 regulatory sequences, you create a series of constructs, each containing different fragments of the 1.7 kb region of 5′ sequence. In the figure, the bars at left represent the sequences of DNA included in your reporter gene constructs, and the + and − signs at right indicate whether the corresponding *eve:lacZ* fusion gene directs stripe 2 expression in *P* element–mediated transformed *Drosophila* embryos. How would you interpret the results—that is, where do the regulatory sequences responsible for stripe 2 expression reside?

25. You have cloned a gene for an enzyme that degrades lipids in a bacterium that normally lives in cold temperatures. You wish to use this gene in *E. coli* to produce industrial amounts of enzyme for use in laundry detergent.

a. How would you accomplish this?

b. You have managed to produce transgenic *E. coli* expressing mRNA of your gene, but only a low level of protein is produced. Why might this be so? How could you overcome this problem?

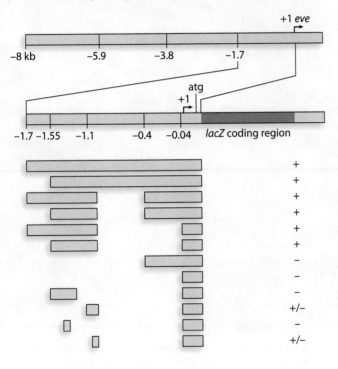

# 18 Genomics: Genetics from a Whole-Genome Perspective

Sequences of entire genomes of many species from Charles Darwin's "entangled bank" have clarified evolutionary relationships of life on Earth and provided the genetic blueprints of genes that define organisms, although the precise functions of most genes are presently unknown.

## ESSENTIAL IDEAS

- The goal of sequencing the human genome stimulated technological advances that enabled its realization. In addition to the human genome, researchers have now sequenced the genomes of hundreds of bacteria and archaea and scores of eukaryotes.

- The evolutionary history of a species is written in its genome and can be read both from its gene content and its chromosome architecture.

- Genome-wide analyses of gene expression, protein–protein interactions, protein–DNA interactions, and genetic interactions provide insights into the biological functions of the genes.

Genomics, the scientific study of biological processes from the perspective of the whole genome, originated in the Human Genome Project (HGP). This audacious project was initiated in the 1980s to sequence and analyze the human genome. At the time, neither the technologies for generating large amounts of DNA sequence nor the computing power to analyze such large amounts of data existed.

Although a primary goal of the HGP was to sequence the human genome, several model genetic organisms were also sequenced under its auspices, including those that have appeared most often in the pages of this book: *Escherichia coli*, *Saccharomyces cerevisiae*, *Caenorhabditis elegans*, *Drosophila melanogaster*, *Arabidopsis thaliana*, and *Mus musculus*. The genome sequences of these model organisms have contributed to our understanding of the organisms themselves as well as to interpretations of human genome structure, function, and evolution. Since then, the genomes of hundreds of other prokaryotes and scores of other eukaryotes have also been sequenced. Due to ever-decreasing costs and ever-improving technologies, genome sequencing is becoming increasingly affordable and routine. It is proving so useful that, in the future, species may be defined by characteristics of their genomic sequence.

In the initial analyses of the genomes of model organisms, two findings stand out. First, even in well-studied organisms, only a fraction of genes identified by genome sequencing had been previously identified by forward genetic analyses; this brings up the question as to the function of all these previously unknown genes. Second, genomic analyses have also revealed the highly dynamic nature of genomes, providing insights into the extent of differences between species, between individuals of a species, and the rates at which DNA sequences evolve.

This chapter provides an overview of genomics by describing three of its major subdivisions. **Structural genomics** is concerned with the sequencing of whole genomes and the cataloging, or annotation, of sequences within a given genome. It provides a parts list of the genetic tool kit of an organism. **Evolutionary genomics** is the comparison of genomes, both within and between species. It illuminates the genetic bases of similarities and differences between individuals or species. **Functional genomics** uses genomic sequences to understand gene function in an organism. Together, these three approaches contribute to the ultimate goal of understanding the role of every gene a given genome contains.

# 18.1 Structural Genomics Provides a Catalog of Genes in a Genome

Genomes vary enormously in size, from several hundred kilobases in some bacterial species to several thousand megabases in some vertebrate and plant species

(Table 18.1). Genomes may consist of a single DNA molecule, as in many bacterial and archaeal species, or of hundreds of chromosomes, as in some eukaryotic species. From a broad perspective, gene number generally increases with organismal complexity. However, genomes also vary in their proportions of coding versus noncoding DNA sequences; and in multicellular eukaryotes, genome size can increase much more than gene number due to a disproportionate increase in noncoding DNA.

These differences aside, even the smallest bacterial genomes are thousands of times longer than the 600 to 900 bp that can be sequenced in a traditional single dideoxy sequencing reaction (see Chapter 7). It is clear that to sequence any genome entirely would require many such reactions. For example, the human genome, with its $3 \times 10^9$ bp, would require at least $5 \times 10^9$ reactions. Were technicians to run these reactions sequentially, designing a new primer for each reaction based on the sequence obtained from the previous one, it would take decades to sequence the entire genome.

What, then, is an efficient way to sequence DNA molecules (i.e., chromosomes) millions of bases in length? The answer is to break them into smaller fragments and sequence all the fragments in parallel. After that, computer algorithms assemble the sequences of the fragments into a single contiguous sequence. Two basic approaches to this general mode of attack differ only in the starting DNA to be fragmented and sequenced. In one approach, often called **clone-by-clone sequencing,** each chromosome is first broken into overlapping clones that are then arranged in linear order to produce a physical map of the genome. Each clone in the map is then sequenced separately. In the second approach, called **whole-genome shotgun (WGS) sequencing,** DNA representing the entire genome is fragmented into smaller pieces, and a large number of fragments are chosen at random and sequenced.

The choice of approach is dictated by the genome to be sequenced. The clone-by-clone approach relies on the availability of specific genetic resources and thus is applicable only to some model organisms. In contrast, the whole-genome shotgun approach is applicable to any genome and is the approach in widespread use today. We describe both approaches here because they both played a role in the sequencing of the human genome.

## The Clone-by-Clone Sequencing Approach

The clone-by-clone approach begins with construction of a physical map. Genetic maps facilitate the construction of a physical map. For this reason, clone-by-clone sequencing is usually applied only to species that have a history of genetic analysis, and thus for which tools such as genetic maps are available. The physical map of a genome is a set of overlapping genomic clones assembled into contigs that, once

| Table 18.1 | Examples of Sequenced Genomes | | | |
|---|---|---|---|---|
| **Organism** | **Description** | **Genome Size (Mb)**[a] | **Predicted Number of Genes**[b] | **Predicted Genes/Mb** |
| *Escherichia coli* | Single-celled eubacterium | 4.64 | 4200 | 905 |
| *Agrobacterium tumefaciens* | Single-celled eubacterium | 5.67 | 5419 | 956 |
| *Rickettsia prowazekii* | Parasitic eubacterium | 1.11 | 834 | 751 |
| *Aeropyrum pernix* | Single-celled archaebacterium | 1.67 | 2694 | 1631 |
| *Arabidopsis thaliana* | Multicellular flowering plant | 130 | 28,775 | 238 |
| *Oryza sativa* | Multicellular flowering plant (rice) | 450 | 37,544 | 83 |
| *Saccharomyces cerevisiae* | Single-celled fungus (baker's yeast) | 12 | 6294 | 521 |
| *Neurospora crassa* | Multicellular fungus (bread mold) | 43 | 10,580 | 246 |
| *Caenorhabditis elegans* | Multicellular nematode worm | 100 | 19,100 | 91 |
| *Drosophila melanogaster* | Multicellular insect (fruit fly) | 180 | 14,100 | 78 |
| *Takifugu rubripes* | Multicellular fish (puffer fish) | 380 | 31,000 (est.) | 82 |
| *Ornithorhynchus anatinus* | Multicellular monotreme (platypus) | 437 | 18,500 | 41 |
| *Mus musculus* | Multicellular mammal (mouse) | 3000 | 24,502 | 8 |
| *Pan troglodytes* | Multicellular mammal (chimpanzee) | 3200 | 20,947 | 7 |
| *Homo sapiens* | Multicellular mammal (human) | 3200 | 22,763 | 7 |

[a] Genome sizes given for most multicellular eukaryotes are estimates since sequences of the heterochromatic regions of the genomes are not known.
[b] Gene number estimates are based on current annotations and could change with new experimental evidence.

assembled, cover the entire genome (see Chapter 16 for a description of contigs). The genome sequence is then determined by shotgun sequencing each of the clones (see Figure 16.14) after which the sequences of the clones are combined into larger contiguous sequences based on the physical map. The completeness of the resulting genome sequence depends on the quality and completeness of the sets of overlapping clones.

Comparison of the genetic map with the physical map can provide information that helps align the physical map with the known chromosomes of the organism. The genome sequencing of *S. cerevisiae, C. elegans, A. thaliana,* and, to some extent, humans relied on this use of physical maps. For these species, a direct correspondence has been drawn between the genome sequence and the chromosomes, each of which is represented by either a single contiguous sequence or a small number of contiguous sequences with gaps at the centromeres and at other highly repetitive sequences.

## Whole-Genome Shotgun Sequencing

The whole-genome shotgun (WGS) approach sequences genomic DNA by the shotgun method without prior construction of a physical map. For this reason, WGS can

be applied to any genome. The starting material for shotgun sequencing is DNA from the entire genome. The DNA is broken into fragments and sequenced, and the sequences are assembled into contigs based on sequence overlaps (**Figure 18.1**). To produce enough overlapping of sequences

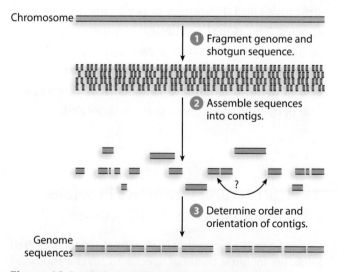

**Figure 18.1 Whole-genome shotgun sequencing.**

for this purpose, technicians commonly generate sequence totaling approximately 10 to 12 times the actual length of the genome (this degree of overlap is called $10–12 \times$ coverage); thus, any one sequence is contained in multiple reads to minimize the chance of sequencing errors. The ease with which sequences are assembled into contigs depends on the sequencing read lengths, which vary between technologies (see Chapter 16).

Repetitive DNA presents a major obstacle in WGS sequencing. Dispersed repetitive DNA sequences (for example, transposons and retrotranposons) interfere with genome assembly, as explained in **Figure 18.2**, because they can map to multiple locations within the genome. Consequently, the assembled sequence must often remain broken at repetitive sequences. One way of circumventing this problem is to use *paired-end sequence* data to bridge the gaps in the assembly due to repetitive DNA sequences. In **paired-end sequencing,** sequence is generated from both ends of a genomic DNA fragment of known size. If paired-end sequences flank a repetitive element, they allow the assembly of a **scaffold,** a set of contigs that are physically linked by paired-end sequences, containing the repetitive element. The relative orientations of paired-end

sequences and their distance from one another can be incorporated into assembly algorithms.

Let's examine how this works. Typically, three genomic libraries, each containing cloned DNA fragments of a different size—one of 2- to 3-kb clones, a second of 6- to 8-kb clones, and a third of larger clones (20–30 or more kilobases)—are generated (**Figure 18.3**). Paired-end sequence generated from clones in the different libraries will provide information on whether two particular sequences are physically linked and the approximate distance between the two sequences. Even if repetitive DNA occurs between the paired-end sequences, they can still be linked into a scaffold. Dispersed repetitive DNA in the genome often

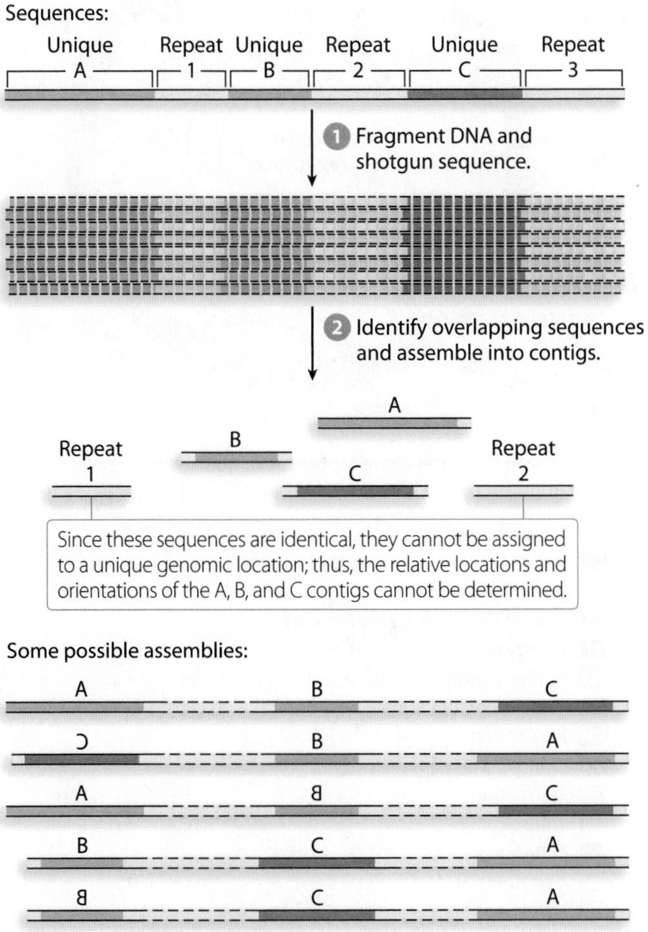

**Figure 18.2  The problem of repetitive DNA.**

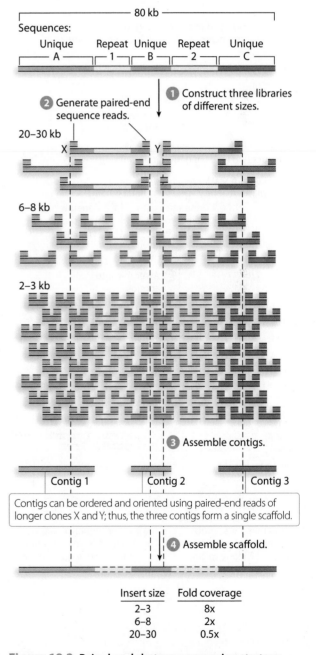

**Figure 18.3  Paired-end shotgun sequencing strategy.**

| Insert size | Fold coverage |
|---|---|
| 2–3 | 8x |
| 6–8 | 2x |
| 20–30 | 0.5x |

consists either of simple, short repeats (microsatellites or minisatellites) or transposable element sequences (up to 10,000 bp). Most repeat sequences are flanked by paired-end sequence from at least one of the different libraries. However, repetitive sequences longer than the largest available clones (for example, centromeric repeat sequences) cannot be spanned using this approach and thus cause gaps between contigs.

**WGS Sequencing of a Bacterial Genome**    For an idea of how the WGS approach works in practice, let's consider two examples, a small bacterial genome with little repetitive DNA and a large eukaryotic genome containing a significant proportion of repetitive DNA.

The first genome to be sequenced by a paired-end WGS approach (at the Institute for Genomic Research, or TIGR, in 1995) was that of *Haemophilus influenzae*, a Gram-negative bacterium whose natural host is humans (**Figure 18.4**). The *H. influenzae* genome is $1.8 \times 10^6$ bp and has relatively few dispersed repetitive elements. Paired-end sequence was generated from three genomic libraries. These sequence data were assembled into 140 contigs whose relative orders and orientations were unknown. The *H. influenzae* genome is a single circular chromosome, so the assembled sequence had 140 gaps for which sequence information was lacking. However, with information on the physical linkage of paired-end reads, the gaps could be divided into two categories: 98 were **sequence gaps** within

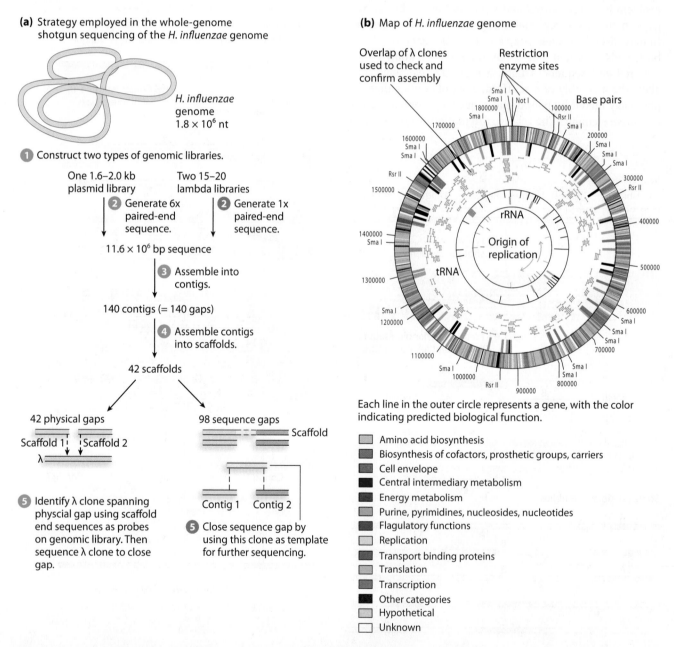

**(a)** Strategy employed in the whole-genome shotgun sequencing of the *H. influenzae* genome

*H. influenzae* genome $1.8 \times 10^6$ nt

1 Construct two types of genomic libraries.

One 1.6–2.0 kb plasmid library

Two 15–20 lambda libraries

2 Generate 6x paired-end sequence.

2 Generate 1x paired-end sequence.

$11.6 \times 10^6$ bp sequence

3 Assemble into contigs.

140 contigs (= 140 gaps)

4 Assemble contigs into scaffolds.

42 scaffolds

42 physical gaps

Scaffold 1    Scaffold 2

λ

5 Identify λ clone spanning physcial gap using scaffold end sequences as probes on genomic library. Then sequence λ clone to close gap.

98 sequence gaps

Scaffold

Contig 1    Contig 2

5 Close sequence gap by using this clone as template for further sequencing.

**(b)** Map of *H. influenzae* genome

Overlap of λ clones used to check and confirm assembly

Restriction enzyme sites

Base pairs

rRNA

Origin of replication

tRNA

Each line in the outer circle represents a gene, with the color indicating predicted biological function.

- Amino acid biosynthesis
- Biosynthesis of cofactors, prosthetic groups, carriers
- Cell envelope
- Central intermediary metabolism
- Energy metabolism
- Purine, pyrimidines, nucleosides, nucleotides
- Flagulatory functions
- Replication
- Transport binding proteins
- Translation
- Transcription
- Other categories
- Hypothetical
- Unknown

**Figure 18.4**    **Whole-genome shotgun sequencing of the *Haemophilus influenzae* genome.**

a scaffold, meaning gaps for which a clone was available for further sequencing that could close the gap, and 42 were **physical gaps** between scaffolds, meaning gaps for which there was no clone to supply the sequence.

Sequence gaps were closed by sequencing of spanning clones identified through paired-end sequencing. Two approaches were used to close the physical gaps. First, the lambda genomic libraries were probed with sequences derived from the ends of the scaffolds: If a single genomic clone hybridized with ends of two scaffolds, the clone should span the gap between the two scaffolds. Second, using primers specific to sequences at the ends of scaffolds, polymerase chain reaction (PCR) methodology was employed to amplify spanning sequence. With this combination of approaches, the entire 1,830,137-bp sequence of the *H. influenzae* genome was assembled into a single contig.

**WGS Sequencing of a Eukaryotic Genome**  The genome of *Drosophila* was the first large eukaryotic genome containing a significant fraction of repetitive DNA to be sequenced using a WGS approach. The *Drosophila* genome is approximately 180 Mb, of which 120 Mb is considered to be euchromatic and the remaining 60 Mb heterochromatic. Because centromeric heterochromatic DNA is not efficiently cloned, owing to its highly repetitive nature, only the euchromatic portion of the genome was sequenced.

Paired-end sequencing was accomplished using three genomic libraries of 2 kb, 10 kb, and 130 kb (**Figure 18.5**). The 10-kb clones were large enough to span most of the dispersed repetitive elements (such as transposons and retrotransposons) found in the *Drosophila* genome, while the 130-kb clones provided long-range linking information from which to infer overall structure in the sequence assembly. Most of the 12 ×-coverage sequence generated could be assembled into 50 scaffolds representing almost 115 Mb of the euchromatic portion of the genome. The remaining sequence was assembled into almost 800 additional scaffolds representing about 5 Mb; thus, the assembled *Drosophila* genome sequence had several hundred physical gaps. Genetic and physical maps of *Drosophila* were used to assign the 50 large scaffolds and an additional 84 scaffolds to specific regions of the four chromosomes, corresponding to most of the euchromatic regions of the chromosome arms.

The WGS sequencing of the *Drosophila* genome benefited from the genetic resources that *Drosophila* geneticists had constructed throughout the 20th century, such as genetic maps of morphological and molecular markers. These tools allowed sequences to be assigned to specific chromosomal locations. They also provided a benchmark for assessing the completeness of the assembled sequence: Of the 2783 previously known genes of *Drosophila*, 2778 could be found in the scaffolds, thus accounting for an estimated 97.5% of the euchromatic DNA.

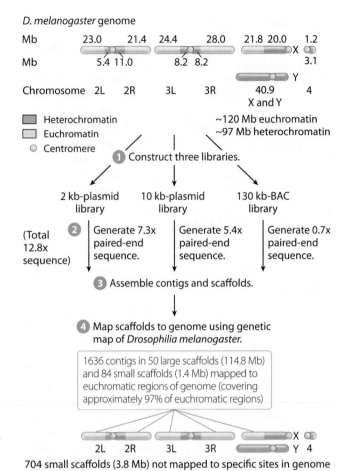

**Figure 18.5  Whole-genome shotgun sequencing of the *Drosophila melanogaster* genome.**

**The Future**  Rapid technological advances are continually changing how genomes are sequenced. Nearly all genome projects today employ massively parallel sequencing technologies (see Chapter 16) for more detail about the WGS technique. For most organisms whose genomes are being sequenced, researchers do not have extensive genetic maps, mutant collections, or other genetic resources. Thus the completeness of the genome sequences is not as easy to assess as it was for *Drosophila*, nor is the assignment of sequences to specific chromosomes straightforward. However, the ease with which genomes can now be sequenced, coupled with advances in forward and reverse genetic technologies (see Chapters 16 and 17), makes it feasible to develop into a genetic model almost any organism for which there is an interesting biological question.

Genetic Insight  Whole-genome shotgun (WGS) sequencing can be used to obtain the complete sequence of small genomes and reasonable coverage of the euchromatic DNA of complex genomes.

## The Human Genome

The U.S. Human Genome Project began officially in 1990 with a projected time scale of 15 years and a budget of $3 billion. This government-funded project took a clone-by-clone approach to sequencing the human genome; therefore, it started by developing tools to build a physical map. In 1998, however, the newly founded Celera Corporation announced that it would provide a human genome sequence in just 3 years by using a WGS sequencing approach. Competition from this private company increased the pace of the publicly funded project, so that the genome sequencing was completed 4 years ahead of schedule.

In 2000, then-President Bill Clinton, appearing at a press conference with J. Craig Venter (president of Celera) and Francis Collins (director of the Human Genome Sequencing Consortium), announced the completion of a "draft" of the human genome sequence. In fact, there were two draft sequences—one furnished by the HGP clone-by-clone approach and one by the Celera WGS approach—and both had numerous gaps. In subsequent years, a "complete" sequence of the human genome has been generated by targeted sequencing of specific regions of the genome to connect adjacent contigs and ensure that the error rate is less than 1/10,000. The gaps between the scaffolds and contigs were closed by the same approaches described earlier for the *H. influenzae* and *Drosophila* genomes, resulting in a genomic sequence consisting of approximately one contig for each chromosome arm.

## Metagenomics

In both the number of individual organisms and their total mass, microbial populations constitute the majority of life on earth. However, unlike model genetic organisms, which are convenient for scientists to study, only a small fraction of microbes can be cultivated in the laboratory. How can we begin to understand microbial diversity without being able to grow the necessary range of microorganisms in the lab? One approach is to apply WGS sequencing to DNA isolated from entire natural communities consisting of a range of organisms. The data derived from such sequencing projects is called a **metagenome.**

One of the first metagenomics projects provides an example. It was an environmental genomic shotgun sequencing of DNA isolated from microorganisms from the Sargasso Sea, a region of ocean bounded by the Gulf Stream off the southeast coast of the United States. In this study, approximately 265 Mb of sequence was generated and assembled into a large number of contigs, representing an estimated 1800 different genomes. However, none of the estimated 1800 genomes were complete, and many were represented by only one or a few contigs. This situation highlights a complication arising in metagenomic studies: Species in any given environmental sample are not equally represented, and so data from common species are

over-weighted relative to those of scarcer ones. Consequently, any complete genome sequences that are produced are likely to belong to very common species while genomes of rare species are represented by only a small, incomplete number of contigs.

Despite such limitations, metagenomic analyses provide information on species diversity and relative population levels in an environmental setting and also contribute to the identification of gene sequences of organisms living in a particular environment. Such analyses have been applied to ecological communities living in acidic mine tailings, contaminated groundwater, and drinking-water systems, and to more "natural" (less affected by humans) ecosystems such as soils, oceans, and hot springs. In addition, metagenomic analyses of several microbial biomes of humans, including the gut, mouth, and skin, have revealed that, collectively, our microbial biomes possess more genes than our own genome. The same sequencing strategy can be applied to any biological system from which purified DNA belonging to a single species is difficult to obtain. An application of metagenomics is presented in the Case Study at the end of this chapter.

---

Genetic Insight  Metagenomics provides insight into the composition of entire communities of microbes, which are likely to consist mostly of previously unknown species that cannot easily be cultured.

---

## Annotation to Describe Genes

The genome sequence can be considered the finest-scale physical map of the genome, and in it are encoded all the genes of the organism. But scientists must employ a variety of methods for searching through the sequence to find the genes. **Annotation** is the process of attaching biological functions to DNA sequences. Genome annotation is the process of identifying the location of genes and other functional sequences within the genome sequence, and gene annotation defines the biochemical, cellular, and biological function of each gene product the genome encodes. Until annotated, a genome sequence is nothing but a very long string of As, Ts, Cs, and Gs. The goal of annotation is to identify known genes, regulatory sequences, and so on as well as to identify sequences that are likely to be genes— but whose function, if they are genes, is as yet unknown. Annotations may be based on experimental evidence, the gold standard, or computational analysis, which then must be confirmed experimentally.

**Experimental Approaches to Annotation**  Experimental approaches to identifying transcribed sequences in a genome make use of cDNA. The sequences of cDNA clones are compared with genomic sequences to identify the parts of the genome that undergo transcription leading to production of RNA molecules (see Chapter 16 for a review of cDNA and genomic libraries). In theory, a complete set of cDNA clones representing all the genes from an organism would allow complete annotation of the transcribed regions of its genome. In practice, though, complete sets of cDNA clones are not available. Nevertheless, for many organisms, large numbers of cDNA clones have been either fully or partially sequenced. The partial cDNA sequence reads are called expressed sequence tags (ESTs). Comparing cDNA sequences with the genomic sequence allows accurate annotation of gene exons and introns, and it also provides clues to alternative splicing (**Figure 18.6**).

**Computational Approaches to Annotation** The genomes of multicellular eukaryotes often contain tens of thousands of genes, for many of which little or no experimental data have been collected. In the absence of experimental data concerning the existence or function of a gene, computational approaches are used to identify possible genes within genome sequences. The use of computational approaches to decipher DNA-sequence information is termed **bioinformatics.**

Bioinformatic annotation algorithms predict gene structure by identifying open reading frames (ORFs), sequences that have the potential to code for polypeptides. Most of these algorithms initially search for ORFs larger than a minimum size, such as 50 amino acids, since ORFs of that size are less likely to occur at random. Data derived from known cDNA sequences of the organism under analysis—for example, information on codon biases—can be used to fine-tune the algorithms employed for gene annotation. Even so, predictions are not infallible, especially in large eukaryotic genomes, where exons are often small relative to introns and are dispersed over large distances.

In comparison with experimental data, bioinformatic algorithms are usually only 50 to 90% successful in correctly predicting exons, but they can provide enough information to assist in the design of experimental approaches for clarifying gene structures. In contrast to protein-coding genes, genes that code for RNA molecules are not easy to predict computationally, since most computational methods begin with a search for ORFs. Thus, experimental or comparative genomic approaches are usually required for annotating genes whose products are noncoding RNA. The process by which genes are predicted is explored further in **Research Technique 18.1**.

Another bioinformatic method of gene annotation is to compare genome sequences of related species. As we discuss in a later section, this and other forms of comparative genomic analysis are becoming increasingly powerful as the genome sequences of more species become available. After genes are predicted computationally, either from algorithms or phylogenetic comparisons, they must then be confirmed experimentally.

## Annotation to Describe Biological Functions

In addition to pinpointing genes and their structural components, gene annotation should also describe biochemical and biological function. Let us consider the *lacI* gene, which encodes the lac repressor protein of *E. coli*. The biochemical function of the encoded protein is to bind to DNA and allolactose, and its cellular function is to regulate transcription of the *lac* operon (see Chapter 14). The biological function of the *lacI* gene is regulation of gene expression in response to sugar availability in the environment. In this particular example, the annotation we make can be quite detailed, since we know a great deal about the function of the *lacI* gene.

Genes that are similar to each other in sequence are assumed to encode gene products with similar biochemical functions. Genes similar in sequence to the *lacI* gene, for example, are likely to encode transcription factors that regulate gene expression; but the nature of the genes they regulate may not be easy to predict. Initial annotation of the eukaryote genomes represented in **Figure 18.7** categorized genes by their presumed biochemical or cellular function. Many genes have known biochemical and cellular functions, learned from previous experimental evidence; and about an equal number of genes have a presumed biochemical function based on sequence similarity to known proteins. While biochemical and cellular functions can sometimes be predicted, ascertainment of the biological functions of genes requires analyses of mutant phenotypes (see Chapters 16 and 17 for descriptions of approaches to mutant analysis).

Examination and comparison of whole-genome sequences has allowed researchers to recognize **gene families,** groups of genes that are evolutionarily related. Some gene families are prominent in certain species,

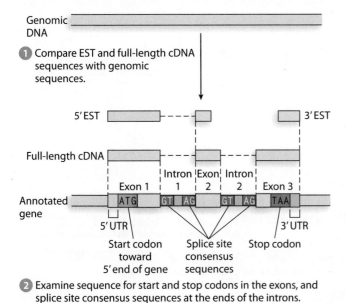

❶ Compare EST and full-length cDNA sequences with genomic sequences.

❷ Examine sequence for start and stop codons in the exons, and splice site consensus sequences at the ends of the introns.

**Figure 18.6 Experimentally acquired clues for gene annotation.**

## Bioinformatics

**PURPOSE** What do computer algorithms "look for" in a DNA sequence during annotation of a bacterial, archaeal, or eukaryotic genome? Often, the first step in annotation is the identification of open reading frames (ORFs). In bacteria and archaea, all ORFs that are translated into protein will have a start codon (ATG) and a stop codon (TAA, TAG, or TGA) with an uninterrupted open reading frame lying between. In eukaryotes, however, where genes may be separated into multiple exons, only the amino-terminal exon has a start codon, and only the last-coding exon has a stop codon; but all internal exons have the sequences that ensure proper splicing, as do the 3′ end of the first exon and the 5′ end of the last exon.

**PROCEDURE** Let's practice examining a nucleotide sequence to see if we can identify sequences that might encode biological information. However, as this example illustrates, the identification of ORFs quickly becomes a computational problem more suited to computers than to pencil and paper. To simplify the analysis, we'll assume we are looking at DNA sequence from a bacterium, so that we need not consider the requirements of exon–intron cutting and splicing.

In searching for potential ORFs, six reading frames must always be considered since proteins can be encoded in either strand of the double-stranded DNA molecule. There are three reading frames in the forward direction and three reading frames in the complementary strand, in the reverse direction. Consider the first 21 nucleotides of the below sequence.

5′ TTGCAGTATGGGCTAGACCAAAGAGAGAGTTGATAACTAGCCGAAACGAACCATGTTCGTCAATCAGCACCTTTGTGGTT
CTCACCTCGTTGAAGCTTTGTACCTTGTTTGCGGTGAACGTGGTTTCTTCTACACTCCTAAGACTTAAGCTAGCTAAGTA
TAGATGGCGAGGTGACACACACACACACAGGTAGATATTAA 3′

**1** Identify the three reading frames in the forward direction and in the complementary strand.

The three reading frames in the forward direction

rf1 5′ // TTG CAG TAT GGG CTA GAC CAA // 3′
rf2 5′ // T TGC AGT ATG GGC TAG ACC AA // 3′
rf3 5′ // TT GCA GTA TGG GCT AGA CCA A // 3′

The three reading frames in the complementary strand

rf4 3′ / AAC GTC ATA CCC GAT CTG GTT / 5′
rf5 3′ / AA CGT CAT ACC CGA TCT GGT T / 5′
rf6 3′ / A ACG TCA TAC CCG ATC TGG TT / 5′

**2** Highlight all potential start codons (ATG); note that these can occur in any of the six reading frames.

**3** Highlight any stop codons (TTA, TAG, TGA) that are in the same reading frame as the four identified start codons.

**The forward direction**

```
         rf2-1      rf2                              rf2              rf2-2
5′ TTGCAGTATGGGCTAGACCAAAGAGAGAGTTGATAACTAGCCGAAACGAACCATGTTCGTCAATCAGCACCTTTGTGGTT
   CTCACCTCGTTGAAGCTTTGTACCTTGTTTGCGGTGAACGTGGTTTCTTCTACACTCCTAAGACTTAAGCTAGCTAAGTA
                                                                   rf2            rf2
   TAGATGGCGAGGTGACACACACACACACAGGTAGATATTAA 3′
   rf2  rf2-3      rf2
```

**The reverse complementary sequence**

```
                                        rf4         rf4              rf4
5′ TTAATATCTACCTGTGTGTGTGTGTGTCACCTCGCCATCTATACTTAGCTAGCTTAAGTCTTAGGAGTGTAGAAGAAACC
   ACGTTCACCGCAAACAAGGTACAAAGCTTCAACGAGGTGAGAACCACAAAGGTGCTGATTGACGAACATGGTTCGTTTCG
                                          rf4              rf4         rf4-1

   GCTAGTTATCAACTCTCTCTTTGGTCTAGCCCATACTGCAA 3′
     rf4                     rf4
```

> We find that the rf2-1, rf2-3, and rf4-1 potential start codons are followed almost immediately by in-frame stop codons, preventing the open reading frames from encoding more than 2, 3, or 5 amino acids.

**4** Identify open reading frames and corresponding amino acid sequences.

The rf2-2 start codon is followed by an open reading frame of 93 nucleotides that could encode a protein of 31 amino acids:

```
5′ TTGCAGTATGGGCTAGACCAAAGAGAGAGTTGATAACTAGCCGAAACGAACCATGTTCGTCAATCAGCACCTTTGTGGTT
                                                     M  F  L  N  Q  H  L  C  G  S

   CTCACCTCGTTGAAGCTTTGTACCTTGTTTGCGGTGAACGTGGTTTCTTCTACACTCCTAAGACTTAAGCTAGCTAAGTA
   S  H  L  V  E  A  L  Y  L  V  C  G  E  R  G  F  F  Y  T  P  K  T  *

   TAGATGGCGAGGTGACACACACACACACAGGTAGATATTAA 3′
```

For more practice with bioinformatics concepts, see Problems 4, 5, 6, and 15.

**(a)** *Arabidopsis thaliana*

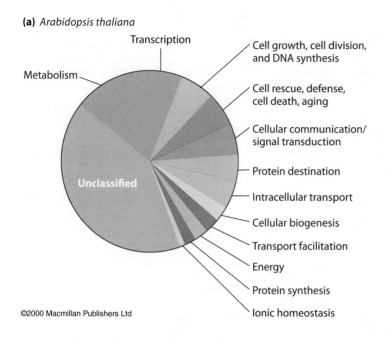

©2000 Macmillan Publishers Ltd

**(b)** *Homo sapiens*

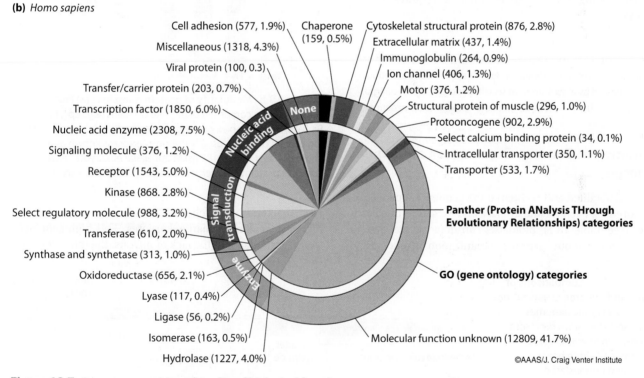

**Figure 18.7  Genome annotation of predicted biological function.**

while others may be entirely absent. The 23,000 genes of the human genome can be placed in about 10,000 gene families. While most mammals largely share this set of gene families, only 3000 to 4000 of these gene families are found throughout eukaryotes. Other lineages, such as fungi and plants, have their own sets of lineage-specific gene families.

Expansion and retention of particular gene families depends on the importance of their biological functions to the organism. For example, in mammals, the gene family encoding olfactory receptors is often the largest in the genome, frequently consisting of over 1000 members. However, the olfactory receptor gene family is much larger in organisms that rely heavily on this sense (a mouse has over 900 of these genes) than in species in which the sense of smell is diminished (humans have only 339). In humans, the largest gene family encodes proteins functioning in the immune system, but this family of genes is absent in both *Arabidopsis* and *Saccharomyces*, where the largest gene families encode protein kinases.

Conservation does not necessarily extend throughout entire genes, and evolutionary relationships between genes may also be recognized through conserved protein domains. Many eukaryotic proteins are modular, consisting of distinct protein domains joined together (**Figure 18.8**). Because many protein domains correlate with exon structure in genes—that is, one or more exons precisely encode a particular protein domain—a hypothesis has been advanced that composite genes (genes that encode multiple conserved protein domains) are generated by exon shuffling (see Chapter 18), through duplications, translocations, and inversions of chromosomal segments. The modular structure of proteins means that the number of genes is much larger than the number of unique functional protein domains. Exon shuffling creates new genes with novel arrangements of protein domains that can be appropriated to fulfill new biological roles. The available data indicate that the protein repertoires of multicellular eukaryotes are generally more complex, averaging more different domains per protein, than those of single-celled eukaryotes. Knowledge of conserved protein domains often provides insight into potential biochemical activities of proteins, but again, understanding the biological function requires mutant analysis.

---

**Genetic Insight**  Genome annotation is the process of attaching biological information to DNA sequences. Annotation is facilitated by algorithms that search for potentially meaningful sequences, but it must be confirmed by experimental evidence.

---

## Variation in Genome Organization among Species

Having obtained and compared the genome sequences for hundreds of bacteria and archaea and for scores of eukaryotes (see Table 18.1), biologists can draw several general conclusions about genome organization (**Figure 18.9**).

**(a)**

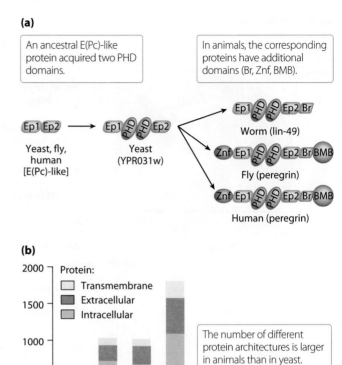

**(b)**

**Figure 18.8  Modularity of protein domains.  (a)** Proteins are often modular, composed of discrete domains (e.g., Ep1, Ep2, PHD, Br, BMB, Znf). Complex proteins can evolve by mixing and matching protein domains, usually through a process known as exon shuffling (see Chapter 18). **(b)** Multicellular eukaryotes have more complex protein architectures than single-celled eukaryotes.

First, bacteria and archaea have fewer genes and much higher gene density than eukaryotes. This high gene density is attributable to the lack of introns, the more compact

**Figure 18.9  Comparisons of gene and genome organization.** In the eukaryotic genomes depicted, thick lines represent exons, thin lines represent introns, and white boxes represent untranslated regions (UTRs).

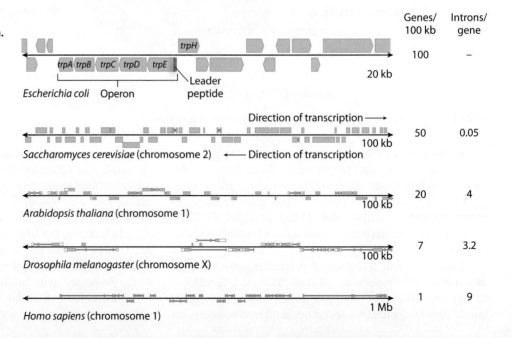

size of regulatory sequences, and the generally less complex structures of most encoded proteins in bacteria and archaea. Second, eukaryotes differ widely in both gene number and gene density, and the genomes of single-celled eukaryotes tend to encode fewer genes than those of multicellular eukaryotes. Third, species that have evolved to be obligate parasites often experience genome contraction. As parasites become dependent on their hosts for nutrients, they lose the genes they no longer need. This trait is reflected in the reduced genome size compared to the other eubacteria of *Rickettsia prowazekii*, the eubacterium responsible for typhus in humans (see Table 18.1).

Just as gene number and density vary among eukaryotes, so does the proportion of repetitive DNA in the genome. The human genome consists of over 50% repetitive DNA: Approximately 45% consists of transposable elements (transposons, retrotransposons, and retroelements); a further 3% consists of microsatellite sequence; and about 5% contains recent gene duplications. Additional repetitive DNA is present in the centromeric and telomeric sequences. The repetitive DNA that is not centromeric or telomeric is often called *dispersed repetitive DNA* since it is distributed throughout the genome. The proportion of repetitive DNA in a genome is a significant factor influencing gene density. Some features of genome organization can be seen in human chromosome 21, shown in **Figure 18.10**.

The annotated genome sequences of model genetic organisms can be browsed through the websites (see endsheets). The host site for the human genome (http://genome.ucsc.edu/) also acts as a portal to the annotated genomes of several additional species.

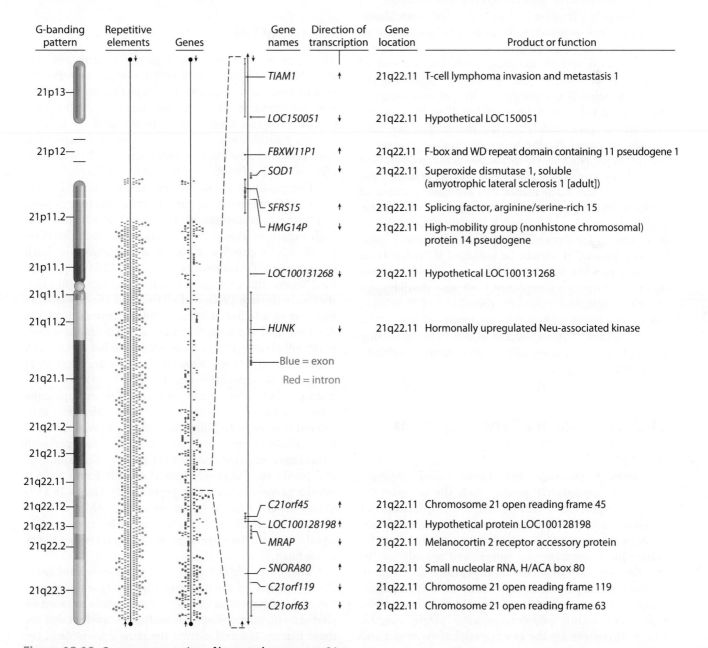

**Figure 18.10  Genome annotation of human chromosome 21.**

## Three Insights from Genome Sequences

Analyses of genome sequences from a range of bacteria, archaea, and eukaryotes have produced many insights into the nature of genomes, of which three are particularly important. First, genomic comparisons demonstrate that the genomes of all organisms are dynamic in nature. Transposable elements (see Chapter 13) are just one of the factors driving genome evolution; large- and small-scale chromosomal duplications as well as deletions and other rearrangements also contribute. Substantial genetic variation is seen even within species, thus providing raw material for natural selection and leading to the evolution of new species. Second, genome sequencing of model organisms reveals the limitations of forward genetic screens. Even in intensely studied species, such as *E. coli* and *S. cerevisiae*, forward genetic screens (see Chapter 17) identified only a fraction (1/3 to 1/2) of the genes identified by genome sequencing. What are the functions of all these previously unknown genes?

The third insight obtained from the analysis of genomes is the discovery that the number of genes in the human genome is comparable to that of various other multicellular eukaryotes. Over the last 25 to 30 years, the estimates of gene number in the human genome have steadily decreased. Having once estimated our genome to contain as many as 80,000 to 120,000 genes, we may find it humbling to discover that we and other animals have fewer genes than many plants. The estimated number of 20,000 to 25,000 genes in the human genome is typical for vertebrates, and it is not much higher than the 14,000 or so estimated for *Drosophila*. If some of us have "gene number anxiety," it should be assuaged by recognizing that gene number does not translate directly into protein number or organism complexity. Both exon shuffling and alternative splicing increase the complexity of proteins in eukaryotes, and these processes are much more prevalent in animals than in either fungi or plants. In the remaining pages of this chapter, we address these major insights in more detail.

## 18.2 Evolutionary Genomics Traces the History of Genomes

*Evolutionary genomics,* sometimes called phylogenomics or comparative genomics, is the comparative study of genomes. **Interspecific comparisons** of genomes—comparisons between species—identify sequences conserved over evolutionary time and thus facilitate the annotation of genomes and provide insight into the evolution of genes and organismal diversity. In contrast, **intraspecific comparisons** identify sequence polymorphisms that are responsible for the genetic differences within populations of a single species. These differences are the raw material of evolution and form the basis of population genetics and the evolution of species.

The evolutionary history of each organism can be traced in its genome and in the composition of its chromosomes. Evolutionary genomics has revealed the striking fact that a large number of genes are shared by phylogenetically distant species, reaffirming that all life on earth is related. Species that are more closely related to one another share a larger number of genes than species that are more distantly related. In closely related species, the similarities in sequence go beyond shared genes to conserved chromosomal segments. Evolutionary genomics has also brought to light important information concerning the highly dynamic nature of the genome. Changes, in the form of mutations, can be observed even in the time scale of a single generation.

### The Tree of Life

The large amount of DNA sequence information now available has revolutionized how biologists perceive the **tree of life,** the phylogenetic tree depicting the evolutionary relationships between organisms. Morphological and physiological traits were once the primary basis of species classification, but DNA sequence comparisons have provided new clarity concerning questions that the older methods of study were unable to resolve.

Comparisons of DNA sequences of the same gene from different species are particularly useful for assessing phylogenetic relationships. Due to their ubiquity and high degree of conservation, genes encoding the ribosomal RNAs provide a universal sequence for such comparisons. By comparing ribosomal RNA sequences, Carl Woese and colleagues revealed through pioneering studies in the late 1970s that all forms of life on earth fall into one or another of three distinct domains: Bacteria, Archaea, and Eukarya. Since then, relationships within many eukaryotic groups have been clarified using DNA sequence comparisons, allowing the basic architecture of the tree of life to be determined (**Figure 18.11**). Some surprising relationships have emerged. For example, the fungi and metazoans, which had traditionally been considered two separate "kingdoms" of life, were discovered to be relatively closely related and are now grouped with Amoebozoa in a clade called the Unikonts. Since animals and plants are the most conspicuous life-forms from a human perspective, the tree presented in Figure 18.11 is biased toward a focus on the interrelationships of those two groups. If all its branches could be presented in equal detail, the tree would more closely resemble a very dense bush.

The tree of life in Figure 18.11 was constructed using DNA sequence information (see Chapter 1) and comparison of the alignment of *homologous nucleotides* to ascertain phylogenetic relationships. **Homologous nucleotides** are those that are descended from the same nucleotide in the

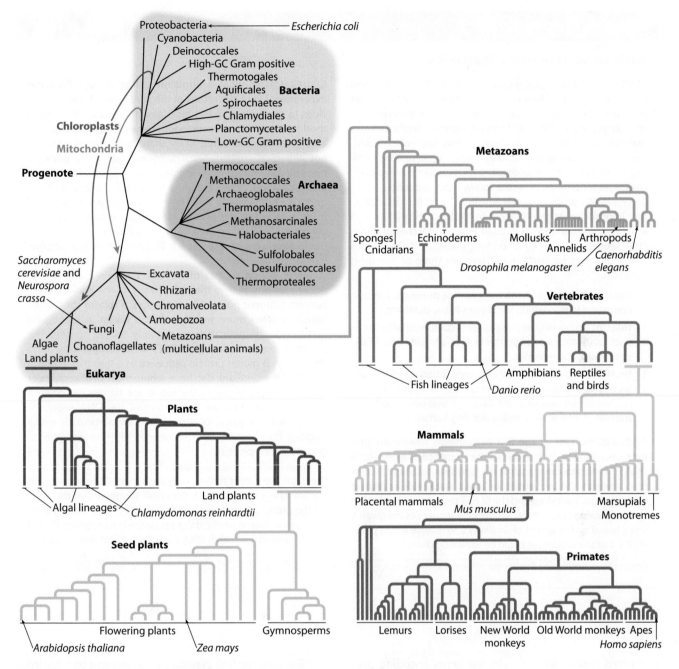

**Figure 18.11** **The tree of life, highlighting the phylogenetic relationships of model organisms discussed in this book.**

common ancestor of the two species being compared. Highly conserved protein-coding DNA sequences are analyzed to identify ancient evolutionary branch points, or **nodes.** Some protein-coding sequences have been conserved over time scales of more than a billion years. Conversely, noncoding sequences are compared to clarify recent nodes in species evolution. Intron and intergenic sequences, on which there may be little selective pressure to maintain a specific sequence, can accumulate mutations and change rapidly over time. A strategy developed to

search for homologous sequences is described in **Research Technique 18.2**.

## Interspecific Genome Comparisons: Gene Content

Genome sequencing indicates that certain genes are found in all organisms, whether bacteria, archaea, and eukaryotes, and suggests that these genes must have arisen early in the evolution of life on earth. Such highly

## Research Technique 18.2

### Basic Local Alignment Search Tool

**PURPOSE** Homologous genes are derived from a common ancestral gene and often have similar functions. A computer program called a **basic local alignment search tool (BLAST)** was developed in 1990 by Stephen Altschul, David Lipman, and colleagues to search for homologous sequences. BLAST, perhaps the most widely used and most important tool employed in bioinformatic endeavors, allows scientists to search databases for sequences similar to any input sequence.

The BLAST program at the National Center for Biotechnology Information at the National Institutes of Health (http://blast.ncbi.nlm.nih.gov/Blast.cgi) enables searches of either DNA sequence similarity or protein similarity. Various types of searches can be performed. Here are three of the most common.

▌ **nucleotide blast (blastn):** a nucleotide query sequence is compared to nucleotide sequences in the database.

▌ **tblastn:** a protein query sequence is compared with the nucleotide databases, hypothetically translated into all six potential reading frames.

▌ **tblastx:** a nucleotide query sequence is translated into all six possible reading frames and compared against the nucleotide sequences in the database that have also been translated into all six possible reading frames.

**PROCEDURE** One of the first experiments researchers perform once they have determined the sequence of a gene is to "BLAST" their sequence against the GenBank database. To perform a search, the user enters an "input" nucleotide or protein sequence into a window, and the BLAST program then searches chosen databases for similar sequences. Sequences are given a score based on the extent of similarity and relative to the probability that the sequences could be similar by chance.

For more practice, see Problems 14 and 15.

**CONCLUSION** What information can be derived from this experiment? First, the results of the BLAST search can provide clues to the biological and biochemical function of the gene used as a query. Since homologous genes are descended from a common ancestor, they likely share biochemical activity if not biological context. Second, knowledge of the phylogenetic distribution of homologous genes allows inferences to be made about when the gene evolved. For example, if the query is a human gene and if homologous genes are detected in all eukaryotes, the protein is likely to perform a function conserved in all eukaryotes. Conversely, if only mammals have homologous genes, the gene is likely to perform a function specific to mammals.

Since related species often have conserved amino acid sequences but, due to the redundancy of the genetic code, possess different nucleotide sequences, a tblastn (or tblastx) search is often more sensitive than a blastn in identifying homologous sequences from distantly related species. The β-globin sequences in Figure 16.23 were compiled by using the human β-globin protein sequence as a query in a tblastn search of the GenBank database, where most DNA sequences determined anywhere in the world are deposited. When a researcher has no prior knowledge of the DNA sequence being used as a query, tblastx searches are particularly useful because they identify DNA sequences with the potential to encode similar proteins.

What if a BLAST search fails to find any other sequences in the database similar to the query sequence? If the sequence is known to encode a protein, the result suggests that the gene for the protein is unlikely to be conserved in a broad phylogenetic sense. Alternatively, if the sequence is noncoding DNA, a lack of similarity to other DNA sequences is not unexpected.

---

conserved genes—for example, the genes encoding proteins needed for DNA synthesis—are involved in biological processes common to all species. Other genes have a more recent origin and define specific clades of species. For instance, genes encoding tubulin are found in all eukaryotes, implying that the tubulin gene evolved before the diversification of the eukaryotes. Still other genes are shared among more restricted clades of organisms, and some genes are confined to only closely related species. In this way, the phylogenetic distribution of gene families provides information on when specific genes evolved. Furthermore, the set of genes shared among any group of organisms can be considered to represent the minimum genomic content of the common ancestor of that group of organisms, thus providing information on the evolution of both genomes and organisms.

Because the first genomes to be sequenced were from phylogenetically diverse organisms, many genes appeared to be specific to particular taxa. As more genome sequences were determined, however, genes initially thought to be unique were found to have counterparts in the genomes of related species. Indeed, two closely related species may share almost their entire genome content; the genomic differences between sister taxa define the differences between the two species. For example, genome content is very similar in four closely related *Saccharomyces* species (*S. cerevisiae, S. paradoxus, S. mikatae,* and *S. bayanus*), all separated by 5 to 20 million years (**Figure 18.12**). Throughout the genomes of the four *Saccharomyces* species, just a handful of species-specific genes were detected, with an average of one unique gene for every 0.5 million years of evolutionary

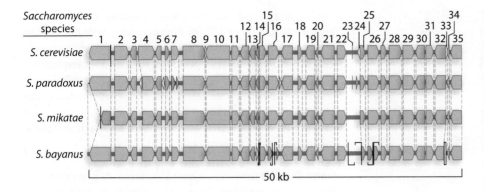

Figure 18.12 **Comparison of four *Saccharomyces* genomes.** Predicted open reading frames (ORFs) are depicted as arrows pointing in the direction of transcription. Orthologous ORFs are connected by dotted lines. ORFs with a one-to-one correspondence are shown in blue; ORFs with a one-to-two correspondence are in red (*S. paradoxus* has two genes in place of gene 7 of *S. cerevisiae*); ORFs that are unmatched (gene 24 in *S. cerevisiae*) are in white. Sequence gaps are indicated by vertical black lines.

distance. It is not yet clear whether this rate is typical for other organisms. But it does bring up the question, how do new genes form?

---

Genetic Insight  The architecture of the tree of life can be ascertained by comparing the DNA sequences of the same genes found in different species. The different rates of DNA-sequence evolution in functional and nonfunctional sequences can be exploited to resolve the evolutionary relationships of organisms that diverged in both ancient and recent times.

---

**The Births and Deaths of Genes**  In tracing the evolutionary history of genes by comparing genome sequences, geneticists obtain clues to the mechanisms through which new genes arise (**Figure 18.13**). These mechanisms include the following.

1. **Gene duplication by duplication of genomic DNA.** Duplication of genetic material can duplicate a part of a gene, a single gene, a chromosome or chromosome segment, or the entire genome (see Chapter 13).

2. **Gene duplication by unequal crossover.** In a special case of gene duplication, one or more genes can be duplicated by unequal crossover due to misalignment of homologous chromosomes at synapsis during prophase I of meiosis. Gene duplication by unequal crossover is indicated by the detection of tandem repeats, or back-to-back copies, of genetic material (see Chapter 13 and Chapter 3 Case Study).

3. **Exon shuffling.** During an exon-shuffling event, exons from two or more genes are combined in a new genomic context (see Figure 18.8a). The rearranging could occur through illegitimate recombination events or, alternatively, through retrotransposition events.

4. **Reverse transcription.** Reverse transcription of cellular RNAs using a retrotransposon-encoded reverse transcriptase, and their insertion into the genome, often leads to the formation of **pseudogenes,** sequences recognizable as mutated gene sequences, but can also produce new genes.

More than 10,000 pseudogenes have been recognized in the human genome, and many were derived from reverse transcription. In addition, the insertion of a retrotransposon into a new genomic location can alter the expression pattern of adjacent genes, potentially leading to new gene functions.

5. **Derivation of exons from transposons.** Transposons have sequences encoding a DNA-binding protein called transposase that is necessary for movement of the transposon. Transposase sequences can be made to perform a new function if fused with other exons derived from the genome. For example, the *RAG1* and *RAG2* genes of jawed vertebrates, whose protein products are involved in rearrangement of DNA sequences during the maturation of the immune system, were derived from sequences encoding a transposase.

6. **Lateral (horizontal) gene transfer.** The movement of genes from one species into the genome of another species is referred to as lateral gene transfer. Such events are common in prokaryotes, which may exchange genes with even distantly related organisms (see Chapter 6). Endosymbioses lead to large-scale lateral gene transfer events, such as in the case of the mitochondrion and chloroplast. While less common between eukaryotes, lateral gene transfer has been documented in some protists and plants.

7. **Gene fusion and gene fission.** Two genes can fuse into a single gene by deletion of the stop codon and transcription termination signals that normally separate genes. Alternatively, a single gene may be split into two genes, each with its own regulatory sequences.

8. **De novo derivation.** Exons can be derived de novo from previously intronic or intergenic sequences that are incorporated into exons of adjacent genes.

Comparisons between the genomes of several related *Drosophila* species have provided insights into the origins of new genes in a multicellular eukaryote. The major source of new genes, slightly less than 80% of the time, was gene duplication, in which the duplicates were either

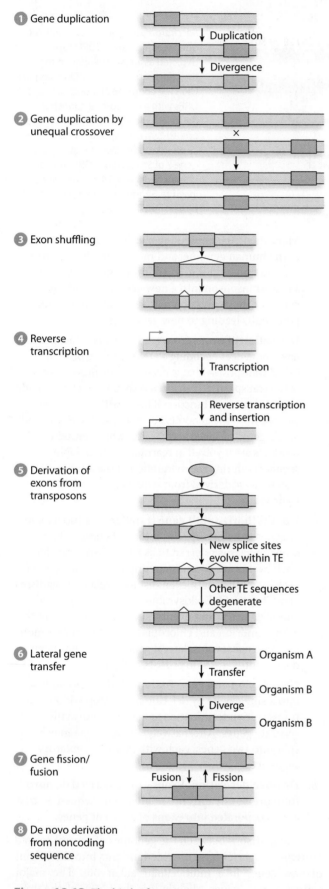

1 Gene duplication

↓ Duplication

↓ Divergence

2 Gene duplication by unequal crossover

×

3 Exon shuffling

4 Reverse transcription

↓ Transcription

↓ Reverse transcription and insertion

5 Derivation of exons from transposons

↓ New splice sites evolve within TE

↓ Other TE sequences degenerate

6 Lateral gene transfer

Organism A

↓ Transfer

Organism B

↓ Diverge

Organism B

7 Gene fission/ fusion

Fusion ↓ ↑ Fission

8 De novo derivation from noncoding sequence

**Figure 18.13 The birth of genes.**

tandemly arranged or dispersed at distant chromosomal locations. A further 10% of new genes were derived from retrotransposition events; and surprisingly, approximately 12% arose de novo, from previously noncoding sequences.

Two mechanisms—gene duplication and lateral gene transfer—stand out as being the major mechanisms responsible for generation of genes in eukaryotes and prokaryotes, respectively. Let's consider each of these mechanisms in greater detail.

**Gene Duplication**    The high frequency of gene duplication is one surprising discovery arising from evolutionary genomics. Most genomes contain a mosaic of gene families derived from both ancient and more recent duplication events, indicating that genomes are dynamic and continuously changing over time. A study in 2000 by Michael Lynch and John Conery counted the duplicated genes in nine eukaryotic species and estimated the duplication rate: approximately 0.01 genes per million years. Thus, for an average eukaryotic genome with 10,000 to 30,000 genes, this research suggests that one gene duplicates and is maintained in the genome every 3000 to 10,000 years.

The fate of duplicated genes depends on the molecular basis of the duplication. If the entire gene including regulatory sequences is duplicated, both copies will be able to produce a functional protein product in the correct amount, time, and place. In this case, the duplicate genes are genetically redundant and are free to evolve new functions, as long as the composite functions of the two duplicate genes retain the function of the original gene. Fully redundant genes are not maintained over long time periods, usually because the duplicate genes undergo one of three likely fates (**Figure 18.14**). First, the vast majority of new genes degenerate due to a lack of positive selection, without which mutations will slowly accumulate and render the genes nonfunctional. Pseudogenes form a significant fraction of the genomes of some organisms. Second, mutations in each of the two copies can result in the two genes having complementary activities such that their combined activity is the same as the activity of the gene before duplication, in a process called **subfunctionalization.** Third, in a process called **neofunctionalization,** a mutation in one of the duplicates could provide a function not performed by the original gene. In rare cases where the new function provides a selective advantage, the gene can be maintained and become fixed in the population. In the latter two cases, both copies remain functional, whereas in the first case, only a single copy retains activity.

Repeated duplication events produce families of related genes. Through gene duplications, gene losses, and speciation events, the relationships among these genes often become complex. Three terms describe different relationships of evolutionarily related genes. The broadest term is *homology*, which is defined as descent from a common ancestor. Thus **homologous genes,** or **homologs,** have descended from a common ancestral gene and are

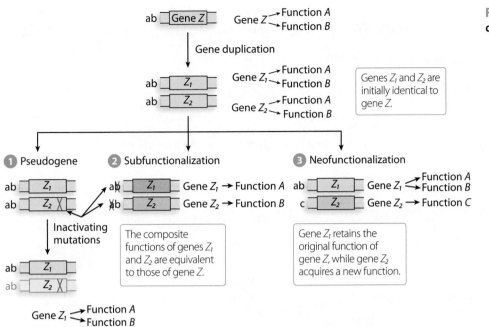

**Figure 18.14 The fates of duplicate genes.**

said to constitute a gene family (**Figure 18.15**). Two other terms define specific relationships between homologous genes. **Paralogous genes,** or **paralogs,** are genes whose origin lies in a gene duplication event. No indication of the age of the duplication event leading to the paralogs is implied. Generally, paralogs perform biologically distinct, but biochemically related, functions. **Orthologous genes,** or **orthologs,** are genes whose origin lies in a speciation event. They are genes in different species that are derived from a single ancestral gene in two species' last common ancestor. Orthologs most often, but not always, have equivalent functions in the two organisms being compared. The globin genes in Figure 18.15 illustrate these evolutionary relationships.

Gene duplication has been a key mechanism in generating new genes that over time have made possible the evolution of complex organisms. During globin gene evolution, gene duplication has permitted specialization, which in turn has allowed greater physiological complexity. Both subfunctionalization and neofunctionalization can be seen within the globin gene family. Neofunctionalization can be seen in the gene duplication event that produced the hemoglobin and myoglobin genes, where hemoglobin functions to carry oxygen in the blood and myoglobin functions to bind oxygen in muscles. Subfunctionalization has also occurred in the globin genes, if an assumption is made that the ancestral β-globin was active throughout the life cycle of the organism. If so, subfunctionalization is now evident between the ε-globin and β-globin paralogs, where the ε-globin is active in the embryo and the β-globin is active in the adult. Other examples include duplications of an ancestral gene leading to a family of genes that allow trichromatic vision in some primate species, including humans (see Chapter 3), and in the creation of another gene family that specifies identity along the anterior–posterior axis of animals (see Chapter 20).

**Lateral Gene Transfer**    Lateral gene transfer, also known as horizontal gene transfer, is the transfer of genetic material between two species. Lateral gene transfer may have been extensive early in the evolution of life, but as specialized genetic mechanisms evolved for control of gene expression, lateral gene transfer became less frequent within the eukaryotic lineage. A common lateral gene transfer event occurs through the sharing of plasmids among bacterial species (see Chapter 6), but other lateral gene transfer events between bacterial species and between bacterial and archaeal species also have been documented. Based on comparison of the sequenced bacterial and archaeal genomes, an estimated 1.5 to 14.5% of genes in any genome are the result of lateral gene transfer. This is likely to be an underestimate, since ancient transfer events may not be detectable. In an extreme example of lateral gene transfer, hyperthermophilic bacterial species (bacteria able to live in extremely hot environments) have acquired genes from archaeal species. Nearly a quarter of the genes in the bacterium *Thermotoga maritima* are most similar to archaeal genes, indicating an archaeal origin. One acquired archaeal gene encodes a reverse gyrase, a topoisomerase that induces positive supercoils in DNA and is required for adaptation to living at high temperatures.

While genes encoding proteins with metabolic functions appear to have been donated in lateral gene transfer events, those that encode proteins for information processing (e.g., replication, transcription, and translation) are not commonly transferred. One possible explanation for this bias is that proteins with information processing functions often act in large complexes and are not easily incorporated into existing complexes in other species.

Although lateral gene transfer is relatively common among bacteria and archaea, transfer between either bacteria or archaea and eukaryotes, or between eukaryotes, is rare. This is due in part to the different transcriptional

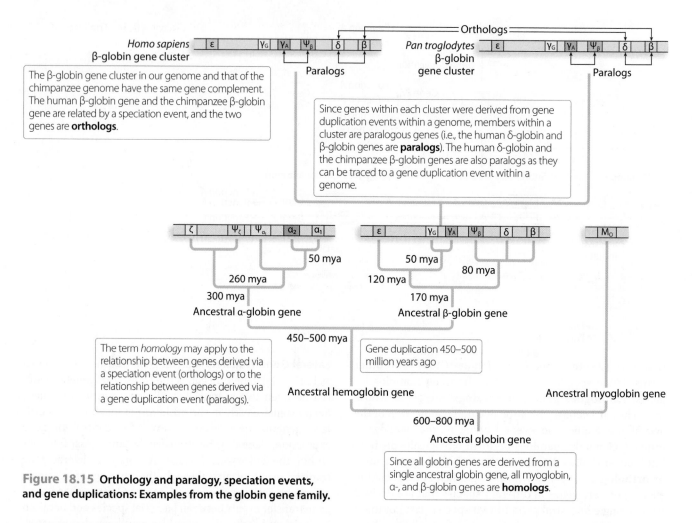

The β-globin gene cluster in our genome and that of the chimpanzee genome have the same gene complement. The human β-globin gene and the chimpanzee β-globin gene are related by a speciation event, and the two genes are **orthologs**.

Since genes within each cluster were derived from gene duplication events within a genome, members within a cluster are paralogous genes (i.e., the human δ-globin and β-globin genes are **paralogs**). The human δ-globin and the chimpanzee β-globin genes are also paralogs as they can be traced to a gene duplication event within a genome.

The term *homology* may apply to the relationship between genes derived via a speciation event (orthologs) or to the relationship between genes derived via a gene duplication event (paralogs).

Gene duplication 450–500 million years ago

Since all globin genes are derived from a single ancestral globin gene, all myoglobin, α-, and β-globin genes are **homologs**.

**Figure 18.15 Orthology and paralogy, speciation events, and gene duplications: Examples from the globin gene family.**

and translational control mechanisms in eukaryotes on the one hand and to bacteria and archaea on the other. Even though the bacterium *Agrobacterium tumefaciens* transfers genes to plant cells (see Chapter 17), there is little evidence that those genes have entered the germ line of the transformed plants. Conversely, there is no evidence of transfer of genes from transgenic plants to soil bacteria. However, there is one prominent exception to this generalization: the transfer of genetic material from endosymbionts to their hosts. The most conspicuous examples are the large-scale transfers of genes from mitochondria and chloroplasts to the nucleus in eukaryotic cells (explored in greater detail in Chapter 19). Finally, lateral gene transfer between eukaryotes is not thought to be common; but it has been documented, for example, between parasitic flowering plants and their flowering plant hosts as well as between fungi and aphids.

**Genetic Insight** The birth of new genes can occur by a variety of mechanisms. While most new genes quickly decay due to an accumulation of deleterious mutations, new genes can also begin to perform new functions, thus driving the evolution of new species.

## Interspecific Genome Comparisons: Genome Annotation

By comparing the genome sequences of closely related species, researchers are often able to refine their annotations of predicted genes whose existence has not been experimentally confirmed. If the predicted gene in fact functions as a gene, orthologous genes are likely to exist in related species.

**Conserved Coding Sequences**   Comparative genomic analyses can facilitate the discovery of previously unannotated genes. Sequences that are conserved in the genomes of two or more species are more likely to be functional (e.g., encode genes) than sequences that are not conserved. Due to the redundancy of the genetic code, amino acid sequences of proteins are often more conserved than the nucleotide sequences that encode them. Thus when searching for conserved coding sequences, the nucleotide sequences of each of the genomes are first translated into all six potential reading frames and the hypothetical amino acid sequences are then compared (see tblastx in Research Technique 18.2). Conserved sequences can then be used to direct experimental examination of the predicted genes, leading to refinement of the genome annotation.

Gene annotation can be hampered by a lack of homology to known genes, and genes or exons of a small size (e.g., encoding proteins of less than 100 amino acids) are particularly difficult to predict. Consider that stop codons occur, on average, about once in 21 codons (3/64) in a random sequence. Thus random ORFs of 63 amino acids occur frequently (approximately 5% of the time in any random 189-bp sequence). Furthermore, in multicellular eukaryotes, the coding sequences of genes are typically broken into small exons (often encoding fewer than 100 amino acids) dispersed over large distances, thus making their unambiguous identification a challenge. Annotation of such genes is feasible only with either experimental evidence or evidence of similar sequences in other genomes.

In the case of the *Saccharomyces* species (see Figure 18.12), comparisons between the four genomes led to prediction of over 40 previously unannotated genes encoding proteins between 50 and 100 amino acids in length. Comparisons of the human genome with the genomes of other vertebrates have aided in the identification of exons and significantly refined the annotation of

the human genome. This is one respect in which the genome sequencing of model genetic organisms has greatly increased our knowledge of our own genome.

**Conserved Noncoding Sequences** Besides helping to identify open reading frames, genome comparisons have also detected the presence of **conserved noncoding sequences (CNSs).** Noncoding DNA was once called "junk" DNA (a term originally coined by Sydney Brenner) since junk, as opposed to garbage, is something we tend to keep even though it serves no identifiable purpose. Today, however, we know that at least some of this noncoding DNA is functional; it contains regulatory, centromeric, and telomeric sequences and genes that produce functional noncoding RNAs, such as microRNA genes.

There are two methods for identifying conserved noncoding sequences, and they approach the task from opposite directions. In **phylogenetic footprinting,** conserved sequences are identified by searching for similar sequences in species separated by large evolutionary distances (**Figure 18.16**). Conversely, in **phylogenetic shadowing,** conserved sequences are identified by first

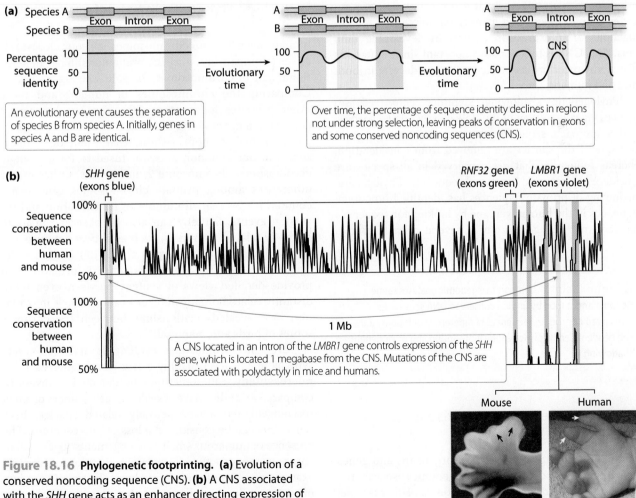

**Figure 18.16 Phylogenetic footprinting. (a)** Evolution of a conserved noncoding sequence (CNS). **(b)** A CNS associated with the *SHH* gene acts as an enhancer directing expression of the *SHH* gene in the developing limb bud.

eliminating sequences that are not conserved in closely related species. Comparative sequence analyses complement promoter analysis experiments (as described in Chapter 17), and are now often the first step to predicting regulatory sequences, which are then tested by experiment.

Regulatory sequences controlling expression of genes in most multicellular eukaryotes consist of enhancer modules spanning hundreds and potentially tens of thousands of base pairs (see Chapter 15). A large number of CNSs that correspond to regulatory sequences have been identified by phylogenetic footprinting using comparisons of mammalian or other vertebrate genomes (**Figure 18.16a**). Comparisons between mammals and fish have shown that enhancer modules can be conserved over large evolutionary distances (the lineages leading to fish and humans separated about 400 million years ago). Conserved noncoding sequences are often clustered in the genome, and they are often adjacent to evolutionarily conserved genes involved in basic developmental processes. For example, comparisons between the human, mouse, and fugu (pufferfish) genomes identified a CNS corresponding to an enhancer module approximately 1 megabase distant from the *Sonic hedgehog* (*SHH*) gene (**Figure 18.16b**). When this CNS was tested for regulatory activity, it drove expression of a reporter gene in mouse in a manner reminiscent of the endogenous *SHH* expression pattern in developing limb buds. This CNS is functionally important since mutations in this enhancer are associated with polydactyly in both mice and humans.

Phylogenetic shadowing identifies conserved sequences via comparison of multiple closely related species. In this approach, sequences that are not conserved in at least one of the species are removed from consideration, whereas sequences that *are* conserved in all species are considered as potential functional sequences. Phylogenetic shadowing of primate sequences has identified functional sequences in the human genome by looking for sequences that have not changed in any of several primate species (**Figure 18.17**).

---

Genetic Insight  Comparative genomic analyses refine genome annotation by finding conserved DNA sequences. Conserved noncoding sequences of different evolutionary ages can be identified by phylogenetic footprinting and phylogenetic shadowing.

---

## Interspecific Genome Comparisons: Gene Order

Just as the evolutionary history of organisms and genes can be traced by comparisons of genomes, so can the evolutionary histories of chromosomes. For example,

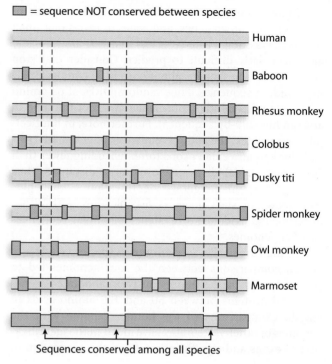

□ = sequence NOT conserved between species

Sequences conserved among all species

©2004 Macmillan Publishers Ltd

**Figure 18.17** **Phylogenetic shadowing of primate species.**

humans have $2n = 46$ chromosomes, but our closest relatives (chimpanzees, gorillas, orangutans) have an additional pair of chromosomes, $2n = 48$ (see Figure 13.29). Comparing the chromosomes of humans and these other primates for **synteny**—the conserved order of consecutive genes along the length of a chromosome or chromosomal segment —shows that a pair of chromosomes in our common ancestor fused to form a single chromosome, chromosome 2, in humans. Other minor differences among primate chromosomes can be accounted for by a small number of translocation and inversion events. Synteny can also be observed in more distantly related mammals, such as between mouse and human lineages that diverged about 100 million years ago (**Figure 18.18**). Genome sequence information can provide detailed views of synteny between even more distantly related organisms. For example, such information has revealed relationships between the chromosomes of birds and mammals.

Even if chromosome synteny is not conserved, synteny at the level of only a few genes, referred to as **microsynteny**, can sometimes be detected. Conversely, comparative studies have revealed large numbers of small rearrangements between closely related species. In a sense, this can be considered a loss of microsynteny. The presence of numerous small rearrangements suggests that chromosome structure is dynamic on a local scale. An example of a loss of microsynteny can be seen in the loss of strict colinearity between the mouse and human

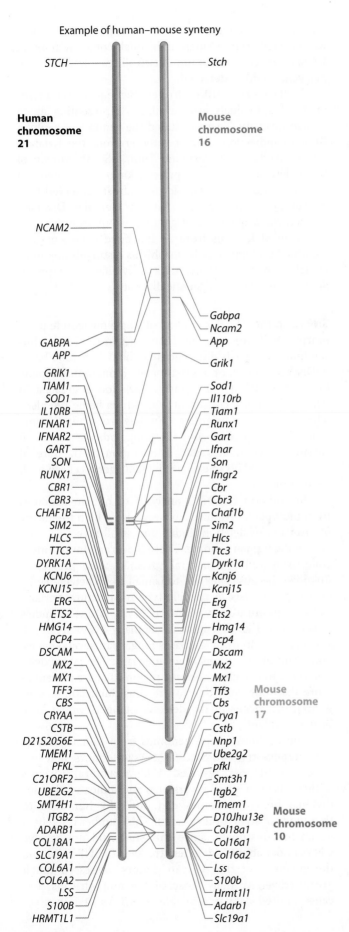

Example of human–mouse synteny

**Figure 18.18  Synteny between human and mouse chromosomes.**

chromosomes shown in Figure 18.18. As we discuss later in this chapter, small rearrangements are also found within individuals of a single species.

Another striking feature of most eukaryotic genomes examined to date is the evidence of past whole-genome duplications as well as smaller duplications involving only segments of chromosomes. Evidence for both whole-genome and smaller segmental duplications can be seen in the *Arabidopsis* genome in **Figure 18.19**. Whole-genome duplications (e.g., polyploidy) are particularly abundant in plants (see Chapter 13), although they are not limited to plants. Evidence of past genome duplications is seen in fungal (e.g., *S. cerevisiae*) as well as vertebrate genomes (e.g., *Dabio rerio*). Smaller, segmental duplications are common in all genomes. Both whole-genome and segmental duplications result in gene duplications on a massive scale and thus contribute significantly to both gene and genome evolution.

## Intraspecific Genome Comparisons

It is convenient to speak of "sequencing the genome of a species" as though one genome represents all members of that species, but logic tells us that this is not the case. Allelic differences, defined by polymorphisms in DNA sequences, are the ultimate cause of phenotypic differences between individuals of a species. And this genetic diversity, the raw material on which natural selection can act, is seen in intraspecific comparisons of the genomes of any two individuals that are not clones.

The study of allelic distributions is the foundation of population genetics (the subject of Chapter 22). Just as the evolutionary history of life in general is written in the genomes of species, the evolutionary history of a species is reflected in the distribution of polymorphic alleles among populations. While population genetics has been an established field for many decades, we are just beginning to examine genetic diversity from a genomic perspective.

The sequences representing the genomes of model organisms were derived from either a haploid individual or an inbred (homozygous at most or all loci) laboratory strain of a diploid organism and thus lack polymorphisms. The DNA sequence of the individual or individuals used to construct the initial complete genome sequence is called the **reference genome sequence.** Polymorphisms in these species can be identified by comparing the genome sequences of different strains collected from different populations derived from the wild with the reference genome sequence. The reference genome sequence can be used to expedite the assembly

Chromosome number

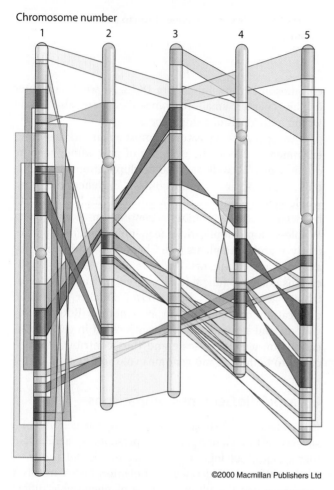

©2000 Macmillan Publishers Ltd

**Figure 18.19 Segmental duplications in the *Arabidopsis* genome.** Colored bands connect duplicated segments. Twisted bands connect duplicated segments having reversed orientations. Many segmental duplications are often relics of a past whole-genome duplication.

of whole-genome shotgun (WGS) sequence data from each new subject. Through the use of next generation sequencing technologies, this "resequencing" of genomes is inexpensive and is becoming increasingly common.

Two intriguing questions arise in the course of studying the genetic diversity of humans through genome analyses: (1) To what extent does genomic sequence vary from one person to another? and (2) What does it mean to be human in the genomic sense?

## Human Genetic Diversity

The first two human genome sequencing projects identified a limited set of polymorphisms of the human genome. The DNA sequenced in the publicly funded human genome project was isolated from sperm cells of a number of anonymous male donors and from white blood cells of anonymous female donors. Thus multiple alleles for a given site were sometimes revealed in the data from

this project. In contrast, since DNA sequenced by Celera was isolated from a single individual, company founder J. Craig Venter, a maximum of two alleles for any autosomal gene could be detected.

By the end of 2010, entire genome sequences for hundreds of individuals were available, representing much human diversity. These included the genome of !Gubi, a Khoisan indigenous hunter-gatherer from the Kalahari Desert; Archbishop Desmond Tutu, a South African of Bantu descent; and Inuk, a paleo-Eskimo from Greenland represented by 4000-year-old permafrost-preserved hair. In addition, through the Human Genome Diversity Project, the sequencing of genomic DNA from a broad spectrum of humans from all inhabited continents has identified millions of polymorphisms distinguishing individuals and populations, thus providing an unprecedented view of human genetic diversity.

**SNPs and Indels** A sampling of single nucleotide polymorphism (SNP) variation between two randomly chosen humans reveals differences at about 1 in 1000 bases in DNA sequence, or approximately 3 million base pairs in the $3 \times 10^9$ of the genome. Variation is greatest in African genomes, consistent with Africa being the place where our species originated. Studies producing genome sequences of parents and offspring indicate that SNP variation accumulates due to mutation at the rate of about 30 new SNPs in each individual's germ cells in each generation.

In addition to SNPs, analyses of human genomes from multiple individuals have revealed a high frequency of insertion or deletion (indel) and small inversion variants in the human genome. These indels and inversions—collectively called structural variants—were previously unknown because they are too small to be detectable by karyotype analysis. A specific type of structural variant, called **copy-number variants (CNVs),** is due to indels greater than 1 kb in length. While many CNVs are small, some are hundreds of kilobases long, span several genes, and result in alterations of gene dosage. The larger deletions are often in chromosomal regions that are present in more than one copy due to previous duplications, suggesting that genes in the deleted segments would have been redundant.

Comparisons of the genomes from four donors of African ancestry, two Southeast Asian donors, and two Caucasian donors revealed 1565 structural variants (either CNVs or small inversions) and indicated that individual human genomes can differ by hundreds to thousands of structural variants. This number is likely to be an underestimate, since this study did not resolve CNVs under about 5 kb in length. As with SNP variation, the genomes of the African donors possessed much greater diversity than those of the non-Africans. Because limited data is available on CNVs, it is not yet known how they contribute to the overall extent of genetic differences between individuals or how often variants arise.

**Human History**    The distribution of polymorphic alleles among populations has provided insight into the evolutionary history of humans as a species (**Figure 18.20**). For example, human diversity is higher in African populations than in non-African populations, indicating that African populations are the oldest human populations and all modern human populations are derived from an ancestral African population. Most allelic diversity is shared by multiple, or even most, human populations; but "private alleles" (alleles restricted to a single population) also exist for all populations.

The vast majority of SNPs and indels are neutral with respect to natural selection, but a small fraction of them determine the phenotypic differences between individuals, and a few are known to be selectively advantageous alleles associated with diet and adaptation to environmental conditions. For example, lactase persistence, the ability to digest fresh milk as an adult, is due to gain-of-function mutations in regulatory sequences of the *lactase* (*LCT*) gene. Such mutations arose and were selected for independently in northern European and multiple African populations that adopted a pastoral dairy lifestyle. Another conspicuous example concerns alleles at loci determining skin color: Alleles conferring darker skin color occur in environments with high levels of UV irradiation and are ancestral. Selection for alleles conferring lighter skin color occurs in environments with low UV irradiation (due to the need for UV irradiation for photosynthesis of vitamin $D_3$). Recent human migrations have sometimes resulted in the transfer of certain alleles to environments where they are maladaptive, as in the case of people of northern European descent living in the high-UV environment of Queensland, Australia, and consequently exhibiting the highest rates of skin cancer in the world.

Studies indicate that the level of genetic diversity among humans is relatively low compared to many other species. This is interpreted to mean that *Homo sapiens* is a comparatively young species and that populations have not had much time to diverge from one another. For example, compared to human SNP variation of approximately 1 in 1000 bases, two chimpanzees will differ by about 1 in 500 bases, and two inbred lines of maize will often differ by more than 1 in 100 DNA bases. It is important to stress that the extent of phenotypic differences between two organisms does not necessarily correlate with the extent of genetic differences. For example, dogs may exhibit marked phenotypic differences between individuals due to selection by humans, despite having minimal sequence variation of only about 1 in 670.

---

**Genetic Insight**    Polymorphisms identified by genome sequencing are used for studying allelic variation in populations. Human genetic variation is limited in comparison to many species and consists of SNPs, indels, and copy-number variations.

---

**What It Means to Be Human**    While comparisons between different human genomes provide insight into what makes each of us an individual, we must look to our closest primate relatives to understand what makes us human. Humans and chimpanzees diverged about 6 million years ago. Comparisons between their genomes have helped define unique attributes of the human genome that may contribute to the differences that distinguish the two species.

Approximately 4.0 to 4.5% of the euchromatic sequence in each species is lineage specific. Single-nucleotide substitutions differentiating the two genomes number about 30 million (a frequency of approximately 1%, or 10 times the number between any two humans). The vast majority of SNPs are in noncoding DNA, since orthologous proteins in humans and chimpanzees are very similar; 29% of proteins are identical in sequence, and an average pair of orthologous proteins differs by only two amino acids. Indels number about 5 million, but they account for about 3% of the difference in the euchromatic sequence between the human and chimpanzee genomes. The two genomes have also undergone numerous changes in terms of gene gains and losses since they diverged from a common ancestor; chimpanzee and human genomes differ in composition by several hundred genes.

Many of the indel differences between the two species are due to lineage-specific activity of transposable elements resulting in the amplification of element numbers. For example, short interspersed elements (SINEs; see Chapter 13 for a description) are more active in humans than in chimpanzees. There are approximately 7000 lineage-specific insertions in the human genome, compared to about 2300 in the chimpanzee genome. Conversely, the chimpanzee genome contains two retroviral elements that are unlike any other older retroviral elements in either genome and must have been introduced by infection of the chimpanzee germ line after its divergence from the lineage leading to humans.

How can we determine which changes were functionally important in the evolution of humans? A first step is to determine whether the changes occurred in the lineage leading to chimps or in the lineage leading to humans. Having the genome sequence of a third, more distantly related species, such as *Gorilla*, allows researchers to separate human and chimp alleles into those that are

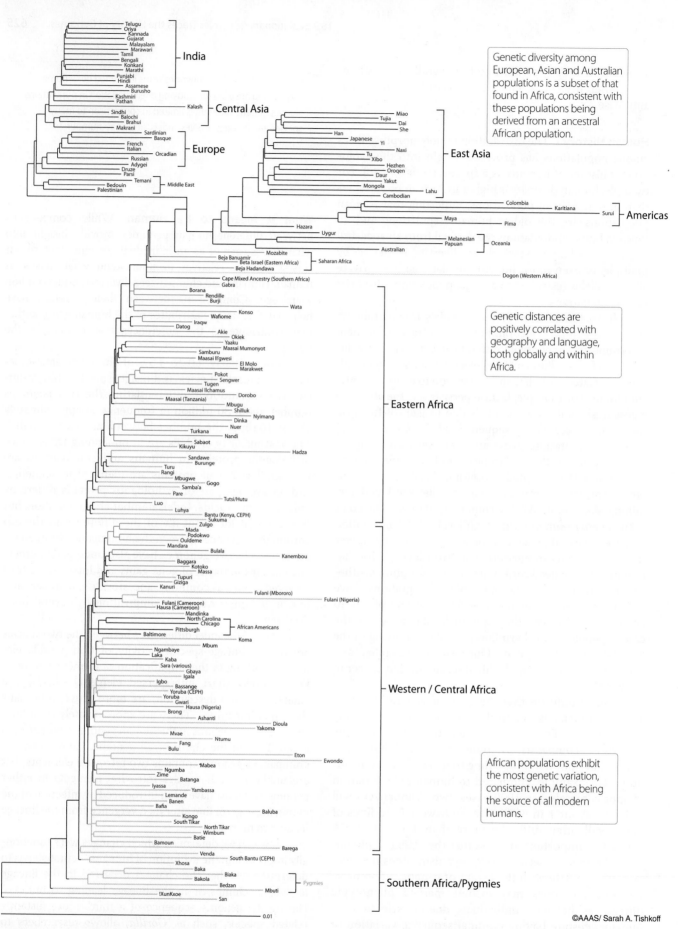

Genetic diversity among European, Asian and Australian populations is a subset of that found in Africa, consistent with these populations being derived from an ancestral African population.

Genetic distances are positively correlated with geography and language, both globally and within Africa.

African populations exhibit the most genetic variation, consistent with Africa being the source of all modern humans.

©AAAS/ Sarah A. Tishkoff

**Figure 18.20 Cladogram showing genetic distances between human populations.** Phylogenetic tree based on 1327 polymorphic markers: 848 microsatellites, 476 indels, 3 SNPs.

ancestral and those that are derived (**Figure 18.21**). Human alleles are considered *ancestral* if they are shared by humans and gorillas, but differ in chimps. Conversely, the human alleles are considered *derived* if they differ from alleles shared by chimps and gorillas. It is assumed that most genetic changes that make humans unique are likely to be derived alleles that have arisen in the human lineage after its separation from chimps.

Neandertals are descendants of a lineage that separated from humans about 400,000 years ago, left Africa, and lived in Eurasia until 30,000 years ago. Modern humans left Africa somewhat later, about 65,000 years ago, and coexisted with Neandertals in Eurasia. In mid-2010, a Neandertals genome sequence was determined, and comparisons of the Neandertals, human, and chimpanzee genomes have clarified the time of acquisition of the alleles in humans that are derived with respect to chimps. Neandertals share most of the derived alleles with humans, indicating that these changes in human evolution date to the time between the human–chimp divergence and human–Neandertals divergence. However, for some genes, Neandertals have ancestral alleles, thus pointing to changes in the human lineage after divergence from Neandertals.

Since modern humans and Neandertals co-inhabited Eurasia for millennia, questions have persisted about whether interbreeding occurred between those lineages. Comparisons between the Neandertals genome and a number of human genomes suggest that a small amount of gene flow (1–4%) may have occurred from Neandertals into the ancestors of non-Africans, but not of Africans. As more genomes of humans and our close relatives, both extant and extinct, are sequenced, we will be able to obtain an unprecedented view of our recent evolutionary history. (Methods to analyze and interpret such data are discussed in Chapter 22).

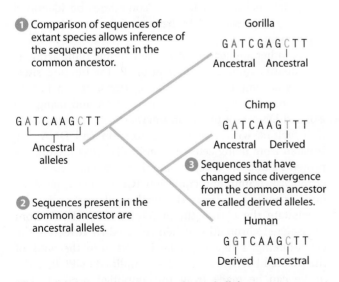

**Figure 18.21 Ancestral versus derived alleles.**

## 18.3 Functional Genomics Aids in Elucidating Gene Function

While the genome sequence supplies a catalog of genes for an organism, it does not directly provide an understanding of how the genes direct the organism's development and physiology. For this, we need to know when and where genes are expressed, the phenotypes of loss- and gain-of-function alleles, which other genes act in the same or redundant pathways, and which proteins each gene product interacts with. Functional genomics is the study of gene function from a whole-genome perspective. High-throughput technologies, in which a large number of genes are analyzed simultaneously, have enabled genome-wide examination of RNA- and protein-expression patterns, genetic interactions, and protein–DNA as well as protein–protein interactions. In addition, high-throughput technologies have facilitated the creation of mutant alleles of all genes in the genome of some model genetic species. In this section, we describe some high-throughput technologies of functional genomics and consider what we have learned by applying them to model organisms.

### Transcriptomics

One important clue to the function of a gene is when and where the gene is expressed. The study of gene expression from a genomic perspective is called **transcriptomics,** and the set of transcripts present in a cell or organism is called the **transcriptome.** Northern blotting is used to analyze gene expression (see Chapter 10). However, northern analysis is not amenable to a high-throughput design. Two high-throughput techniques used to analyze the transcriptome are DNA microarrays and high-throughput sequencing of cDNA. DNA microarrays are in widespread use today, but high-throughput sequencing will likely become the dominant method in the coming years.

**Expression and Tiling Arrays**  **DNA microarrays** consist of collections of synthesized DNA fragments, attached to a solid support and representing sequences present in a genome (**Figure 18.22**). The DNA fragments are of a fixed length, usually 25 to 70 bases. The specific DNA sequences are chemically synthesized on a silicon substrate, called a chip, at high density—tens of thousands to millions of oligonucleotide sequences per array, each sequence located on a different spot. Following hybridization with a fluorescent probe representing cDNA, the intensity of the signal from each of the spots reflects the concentration of the sequence complementary to the probe. For sequenced genomes, two types of microarrays have been produced: *expression arrays* and *tiling arrays.*

An **expression array** carries unique sequences from every annotated gene of the genome. Hybridization of an expression array with labeled cDNA probes produces quantitative information about the relative expression

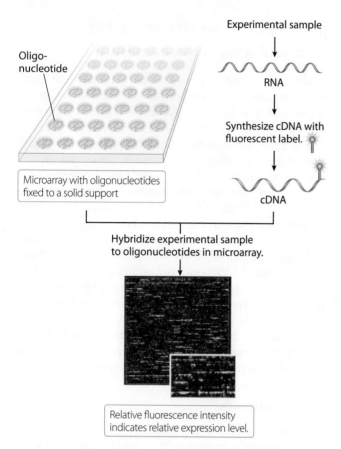

**Figure 18.22 Transcriptome analysis using oligonucleotide arrays.**

levels of the genes represented on the array. The power to examine gene expression patterns through the use of expression arrays is limited only by the degree to which mRNA can be extracted from specific cells or tissues and converted to cDNA before labeling.

An example from the budding yeast *S. cerevisiae* illustrates how microarray data can provide insight into the function of genes not previously identified by forward genetic approaches. Diploid yeast cells of *S. cerevisiae* produce haploid cells through the developmental process of sporulation, which consists of meiosis and spore morphogenesis. From forward genetic studies, approximately 150 genes were known to be involved in sporulation, and these could be classified into four groups defined by expression patterns and mutant phenotypes.

To examine genome-wide expression patterns during sporulation, diploid yeast cells were induced to sporulate, RNA samples were taken at seven time points spanning 11 hours, and their expression levels were compared to identify genes whose expression was either induced or repressed (Figure 18.23). More than 1000 genes exhibited significant changes at some point during the sporulation process: In about half of these cases the genes became induced, and in the other half the genes became repressed. In other words, more than six times as many

genes as had been identified previously were likely to play some role during sporulation.

The researchers categorized the induced genes by their expression patterns, expanding the four previously described patterns to at least seven. Genes with expression patterns similar to those of known genes could be hypothesized to have biological roles similar to those of the known genes. For example, some "Early I" genes (see Figure 18.23) are known to function in the synapsis of homologous chromosomes. By extrapolation, other Early I genes whose functions are unknown may also have roles in synapsis of chromosomes, suggesting areas for experimental study to support or refute the predicted roles. Similarly, comparisons of sequences upstream of coordinately regulated genes can provide information on gene regulation. For example, more than 40% of the Early I genes have a consensus URS1 sequence to which the transcription factor UME6 binds, suggesting that this set of genes is coordinately regulated by the same transcription factor. The temporal gene expression patterns during sporulation provide clues to the functions of hundreds of previously uncharacterized genes, some with homologs in humans.

Microarray technology is now being applied to the study of human cancers, allowing precise characterization of gene expression in morphologically similar but molecularly different cancers.

The second type of array, the **whole-genome tiling array,** contains all sequences of the genome or a genomic interval, including introns, exons, untranslated regions (UTRs), and intergenic regions. One of many applications of a whole-genome tiling array is to precisely map transcription patterns on the genomic DNA sequence (Figure 18.24). Labeled cDNA is used to probe genomic tiling arrays to identify sequences being transcribed into mRNA or other types of RNA. Such experiments facilitate gene annotation by providing data on the boundaries of exons and length of UTRs, in addition to identifying novel transcripts. Genes that have not yet been annotated using computational approaches can sometimes be identified by using expression data from a genomic tiling array experiment.

Another application of whole-genome tiling arrays is the identification of transcription factor binding sites. This is accomplished by isolating transcription factors and chromatin bound to genomic DNA and using the isolated DNA as a probe on microarrays. The transcription factors, with associated chromatin and DNA, are isolated from living cells by chemically cross-linking the proteins and DNA together and then, via a process called chromatin immunoprecipitation (ChIP), using an antibody specific to a transcriptional regulatory protein to precipitate the chromatin of interest. The DNA from the isolated chromatin is then released by reversing the cross-linking. Linkers can be ligated onto the ends of the isolated DNA, which is then amplified by PCR, and a probe can be made from the amplified product. The

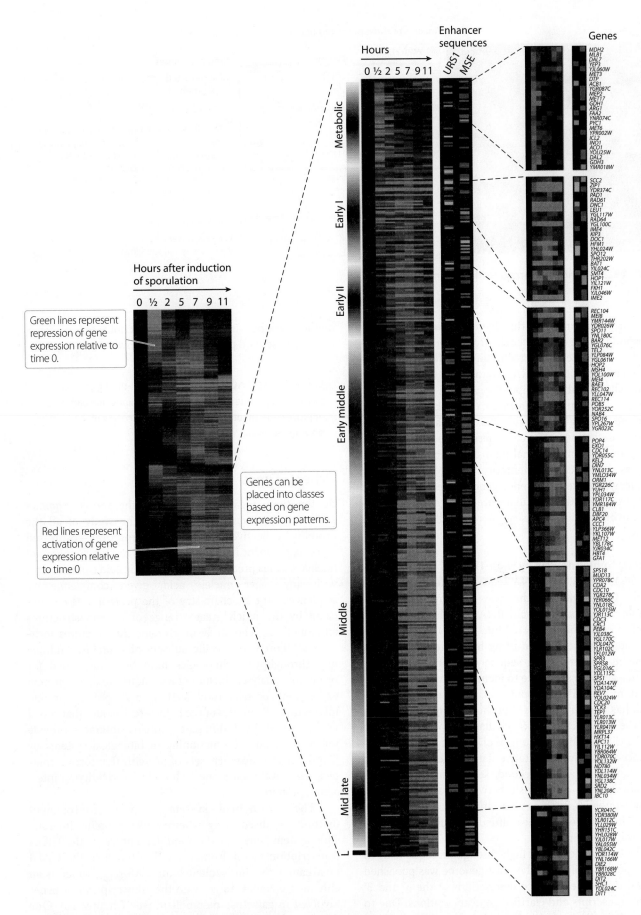

**Figure 18.23 Analysis of yeast transcription patterns using microarrays.**

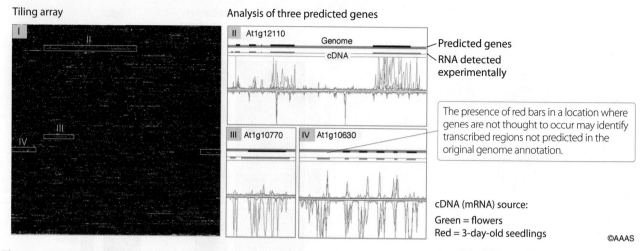

**Figure 18.24** **Using tiling arrays to identify transcription.**

spots on the microarray that hybridize with the probe correspond to the sequences the transcription factor was bound to in the cell. This technique provides a genome-wide view of protein–DNA interactions and is known colloquially as ChIP on chip.

Whole-genome tiling arrays can also be used to discover and analyze polymorphisms. Microarrays are very sensitive, detecting sequence differences of a single base pair by a reduction in signal intensity relative to a probe that has 100% complementarity to its target. Since microarrays utilize a reference genomic sequence, allelic differences from that sequence are detected as a reduction in hybridization intensity where the allelic difference resides.

**Transcriptome Analysis by High-Throughput Sequencing** High-throughput DNA sequencing techniques (see Chapter 16) provide another way of assaying the transcriptome. In this approach, RNA is isolated from the cells of interest and converted into cDNA, which is then fragmented and sequenced using high-throughput DNA sequencing. The resulting sequence is compared to the reference genome sequence to identify sequences that are present in the cDNA population. The sequencing approach has two advantages over those using microarrays. First, the sequencing approach has the potential to be more sensitive. Since millions of cDNA fragments can be sequenced, precise quantitative data on gene expression levels can be obtained. Second, sequencing approaches can more easily distinguish between transcripts with similar sequences, such as alternative splice variants and SNPs, which are sometimes difficult to distinguish with hybridization techniques.

The first application of high-throughput sequencing to transcriptome analysis of the yeast genome was published in 2008. It provided precise descriptions of the 5′ and 3′ ends of transcripts and clarified gene annotations. Due to the decreasing cost of high-throughput sequencing, these methods are becoming the preferred choice for transcriptome analyses and polymorphism discovery.

**Genetic Insight** Microarrays and high-throughput sequencing provide information on gene expression patterns, improve gene annotation, and facilitate discovery of polymorphisms.

## Other "–Omes" and "–Omics"

By the same logic that produced the terms *genomics* and *transcriptomics*, **proteomics** is the study of all the proteins—collectively known as the **proteome**—expressed in a cell, tissue, or individual. Whereas the biochemistry of nucleic acids is predictable—any nucleic acid can base-pair with any other nucleic acid, given complementary sequences—the biochemistry of the proteome is complicated by the much greater range of protein structures and functions. The study of proteins thus requires techniques tailored to specific subsets of proteins. Multiple high-throughput technologies have been developed for proteomic analyses, including techniques to study protein expression, protein modification, and protein–protein interactions. Examples of the latter—techniques that reveal whether and how different proteins interact—provide information on the functioning of biological systems by identifying, for instance, sets of proteins that form a complex. Here we discuss one technique for identifying interacting proteins.

The **two-hybrid system** is a high-throughput method for discovering whether two proteins interact. This system relies on the modular nature of the GAL4 transcription factor from yeast that binds to the GAL4 upstream activation sequence (or $UAS_{GAL4}$), which is an enhancer element, to activate the transcription of genes involved in galactose metabolism (see Chapter 15). One domain of the GAL4 protein, the DNA-binding domain,

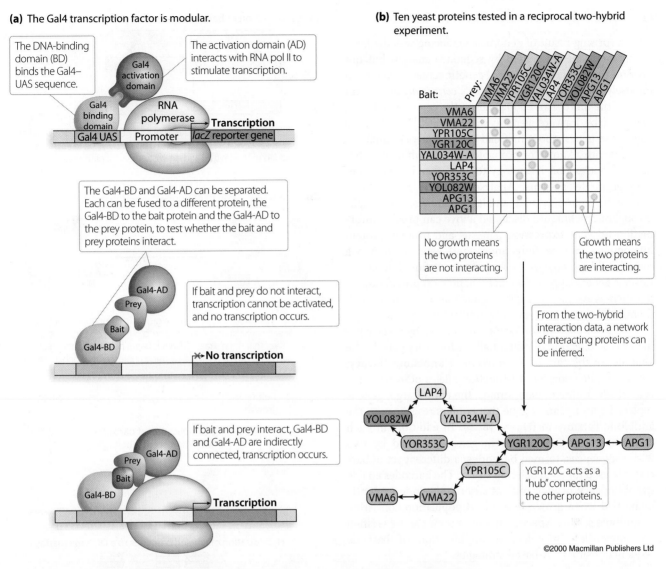

**(a)** The Gal4 transcription factor is modular.

The DNA-binding domain (BD) binds the Gal4–UAS sequence.

The activation domain (AD) interacts with RNA pol II to stimulate transcription.

Gal4 activation domain

Gal4 binding domain

RNA polymerase

Gal4 UAS | Promoter | *lacZ* reporter gene

**Transcription**

The Gal4-BD and Gal4-AD can be separated. Each can be fused to a different protein, the Gal4-BD to the bait protein and the Gal4-AD to the prey protein, to test whether the bait and prey proteins interact.

Gal4-AD

Prey

Bait

Gal4-BD

If bait and prey do not interact, transcription cannot be activated, and no transcription occurs.

**No transcription**

Prey

Gal4-AD

Bait

Gal4-BD

If bait and prey interact, Gal4-BD and Gal4-AD are indirectly connected, transcription occurs.

**Transcription**

**(b)** Ten yeast proteins tested in a reciprocal two-hybrid experiment.

No growth means the two proteins are not interacting.

Growth means the two proteins are interacting.

From the two-hybrid interaction data, a network of interacting proteins can be inferred.

LAP4 → YAL034W-A
YOL082W → YOR353C → YGR120C → APG13 → APG1
YPR105C
VMA6 → VMA22

YGR120C acts as a "hub" connecting the other proteins.

**Figure 18.25** **Identifying protein–protein interaction networks.** **(a)** The two-hybrid system identifies interacting proteins. **(b)** Application of the two-hybrid system identifies networks of interacting proteins.

binds to the $UAS_{GAL4}$ sequence; a second domain, the activation domain, activates transcription by interacting with RNA polymerase II as well as other chromatin factors (**Figure 18.25a**). The two domains can be physically separated.

To test whether two proteins interact, one of the proteins to be tested is translationally fused (see Chapter 17) to the GAL4 DNA-binding domain (BD), and the other protein to be tested is translationally fused to the GAL4 activation domain (AD). Both of these chimeric genes are then transformed into a single yeast strain. If the two proteins interact, the GAL4-BD and GAL4-AD will be brought together, and GAL4-activated genes will be transcribed. Conversely, if the two proteins do not interact, no transcription of the GAL4-activated reporter gene will be observed. To facilitate the screening process, an auxotrophic yeast strain is often used in which $UAS_{GAL4}$ drives

expression of a gene that will complement the auxotrophic defect. For example, a *histidine* auxotroph with a $UAS_{GAL4}$ : HIS transgene will not grow on media lacking histidine unless GAL4-mediated transcription is active (**Figure 18.25b**). However, certain interactions cannot be detected with the standard two-hybrid system, including those in which the interacting proteins are not efficiently transported into the nucleus and those in which proteins require a third partner for interaction.

Two-hybrid approaches have been applied successfully to many model systems, providing information on their protein-interaction networks. In *S. cerevisiae*, all pairwise combinations of the more than 6000 proteins encoded in the genome have been tested, providing an overview of protein-interaction networks in the living yeast cell (see Figure 18.25b). The sum of all of the protein–protein interactions in an organism is known as the **interactome**.

## Genomic Approaches to Reverse Genetics

One surprising result of genome sequencing was the large number of genes identified by sequence analysis but not previously identified by forward genetic screens. Even in an intensely studied organism such as *S. cerevisiae,* only about 1000 of the more than 6000 genes in the genome had been identified by forward genetic screens. Of the remaining 5000 or so genes, about half had some sequence similarity to genes with a known or probable function while the other half did not exhibit homology to any other known genes in other model systems. Analyses of other multicellular eukaryotic genomes had similar outcomes. The high-throughput techniques discussed above can provide information on gene expression patterns and protein–protein interactions, but to fully understand gene function, we must be able to analyze loss- and gain-of-function alleles. Reverse genetic approaches (see Chapter 17) provide an experimental avenue for exploring such alleles, and through them, the function of previously unidentified genes.

An essential tool for a genomic analysis by reverse genetics is a collection of mutant alleles for every gene in the genome. In the case of *S. cerevisiae,* a **knockout library,** containing deletion loss-of-function alleles of every gene, is available. In the mutant strains, the entire target gene is replaced with a marker gene that confers resistance to the antibiotic kanamycin (**Figure 18.26**). In addition, in each deletion strain, the kanamycin gene is flanked by two 20-bp sequences, termed **barcodes;** a different set of barcodes is used for each deletion strain. The barcodes enable the abundance of each mutant strain to be independently quantified when grown in a mixed population consisting of multiple strains. Specific mutant strains can be verified and quantified by selective amplification of barcode sequences using PCR-based strategies.

## Use of Yeast Mutants to Categorize Genes

A challenge for the future is to determine more precisely the molecular and biological roles of all genes to illuminate why they are maintained in the genome. As an initial step in this direction, yeast deletion strains have been analyzed to categorize *S. cerevisiae* genes as either essential for life or nonessential. The deletion strains are first constructed in diploid yeast. The heterozygous diploid deletion strain is then induced to undergo meiosis, allowing the phenotypes of deletion alleles to be analyzed in the haploid progeny. When mutations in each of the 6300 genes of *S. cerevisiae* were examined, deletion alleles of 1102 genes were not recoverable in haploid progeny (**Figure 18.27**). These genes, about 20% of the yeast genome, define the essential gene set of *S. cerevisiae,* meaning that they are required for survival of the organism. In addition, 186 of the deletion mutants had a reduced-growth phenotype as heterozygotes before induction of meiosis, thus indicating haploinsufficiency of these genes. (Recall that haploinsufficiency is a dominant

**(a)** Construction of barcoded yeast deletion mutants

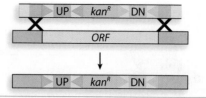

The coding regions of each gene were replaced by a selectable marker gene (e.g., kanamycin resistance), and barcodes unique to each gene were added upstream (UP) and downstream (DN) of the marker gene.

**(b)** Competitive growth of pools of deletion mutant strains

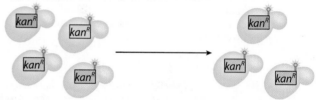

The barcoded mutant strains can be grown in competition with wild type or each other. In this example, the "blue" strain does not grow as well as the other three strains. DNA is isolated before and after growth, and each gene can be analyzed by using fluorescently labeled barcode primers.

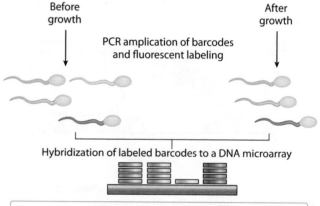

The relative proportion of growth of each strain can be examined by hybridizing the products derived from a DNA microarray.

©2007 Macmillan Publishers Ltd

**Figure 18.26  Barcoded knockout libraries for phenotypic analyses of mutants.**

phenotype in diploid organisms that are heterozygous for a loss-of-function allele.) For the remaining 5000 genes, both haploid deletion mutants and homozygous diploid mutants were obtained. However, 891 of these mutant strains exhibited a slow-growth defect in rich media under optimal conditions, which indicates that the genes are required for vital biological processes in optimal growth conditions. This leaves about 4000 genes for which no obvious mutant phenotype is detected under optimal growth conditions. These genes are referred to as nonessential, but that classification is dependent on environment; in other words, the genes are nonessential under optimal laboratory growth conditions.

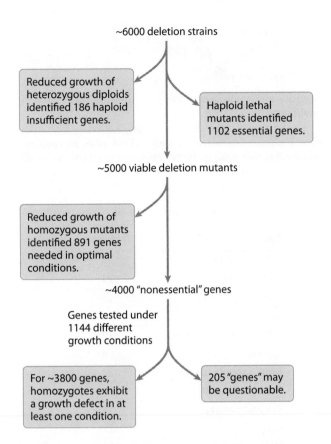

**Figure 18.27  Global analysis of yeast deletion mutants.**

One possible explanation for the lack of conspicuous mutant phenotypes associated with 4000 nonessential *S. cerevisiae* genes is that the genes are required only under specific growth conditions. To test this hypothesis, each mutant strain was grown under a variety of environmental conditions, including variations in temperature, media composition, and the presence of antifungal compounds, salts, and other chemicals known to perturb specific biological processes. As a result, yeast geneticists discovered measurable growth defects under at least one environmental condition for nearly all of the 4000 genes previously identified as nonessential. Thus these genes are required for efficient growth in at least one tested environmental condition; they are not really "nonessential" from an evolutionary perspective, since their presence is likely to provide a selective advantage. Growth defects were not found for only about 200 deletions, suggesting that (1) these genes are authentically nonessential, (2) the conditions to test their importance were not met, or (3) their annotation as genes is incorrect.

To further analyze the essential genes, conditional alleles are required. Traditionally, temperature-sensitive alleles isolated in forward genetic screens have been used to study functions of essential genes. Libraries of engineered conditional alleles of *S. cerevisiae* essential genes have also been constructed for this purpose. In one

approach, each essential gene is placed under the control of a tetracycline-repressible promoter. In the absence of tetracycline, the gene is expressed; but upon addition of tetracycline, gene expression is repressed, creating a loss-of-function phenotype. In another approach, a short peptide tag that confers heat-inducible protein degradation is added to the coding regions of essential genes. Under the normal growth temperature of 30°C, the protein is stable; but at 37°C, the tagged proteins are degraded and lose the ability to function.

Other types of libraries that have been constructed provide additional tools for identifying potential gene functions in *S. cerevisiae*. For example, a library in which every gene is a translational fusion with green fluorescent protein (GFP) permits visual determination of the subcellular location of proteins.

**Experimental Insight 18.1** describes the outcome of some attempts to identify the essential gene set and minimal genome in certain bacteria.

---

**Genetic Insight**  Knockout libraries of loss-of-function mutations permit the detection of essential genes and nonessential genes in standard growth conditions. Such libraries can aid in defining growth conditions in which otherwise nonessential genes demonstrate essential functions.

---

## Genetic Networks

Identification of genetic interactions can provide clues to gene function by revealing that two genes act in the same pathway or redundant pathways (see Chapter 16). Data derived from double mutants identify sets of interacting genes that define **genetic networks.**

An extreme example of a genetic interaction is *synthetic lethality,* where the mutation of either gene alone is not lethal but mutation of both genes together results in lethality (see Figure 16.5). An estimate of the number of synthetic lethal interactions in *S. cerevisiae* was obtained by using mutants representing 132 genes and analyzing their genetic interactions. For genes whose single-mutant phenotype is inviability, conditional alleles were used; for nonessential genes, null alleles were used. Each of the 132 mutants was crossed with 4700 viable deletion mutants, and the double-mutant phenotypes were examined. Approximately 4000 different synthetic lethal interactions were identified, involving about 1000 different genes. The number of interactions per gene ranged from 1 to 146, with an average of 34. One striking feature of this genetic interaction study is that essential genes exhibited about five times as many interactions as did "nonessential" genes. These results suggest that genetic networks consist of a small number of essential genes participating in many interactions and a larger number of nonessential genes participating in fewer interactions (**Figure 18.28**).

## Experimental Insight 18.1

### Prokaryotic Phylogenomics Hints at the Minimal Genome for Life

Genomic analyses of hundreds of bacterial species have provided a wealth of information about bacterial physiology and pathogenesis. Because the genomes of bacteria vary enormously in size, one question that arises from comparative genomics is, what essential components constitute the minimal genome for a free-living organism?

Defining the minimal genome has been approached from two directions. One approach compares the genomes of distantly related bacteria to identify genes that are shared, assuming that the set of shared genes should reflect those essential for biological functions. The first such comparison, between the genomes of *Haemophilus influenzae* and *Mycoplasma genitalium,* identified 256 shared genes. Subsequent comparisons between other bacterial and archaeal species identified about the same number of genes. Such studies lay a foundation for establishing the minimal genome, but the numbers they provide are likely to be an underestimate due to the inability of this method to identify genes divergent in sequence but coding for proteins with similar functions.

The second approach is to discover which genes are essential by monitoring the effects of mutations in each gene of a genome. This can be accomplished by random mutagenesis. Genes whose mutant phenotype is lethality are considered to be essential. A study of *Bacillus subtilis* systematically disrupted each of its approximately 4100 genes by homologous recombination. In these experiments, only 271 genes were shown to be essential. This number is likely to be an underestimate since genes were only inactivated singly, and genetic redundancy would obscure the identification of some essential genes. Note, moreover, that the description of genes as *essential* and *nonessential* is relative to the extent that classification is dependent on growth conditions. The bacteria were grown in rich media containing most amino acids and vitamins required for life, so the conditions were quite benign compared to their normal growth environments.

Most essential genes encode proteins necessary for fundamental biochemical activities, such as DNA, RNA, and protein metabolism; synthesis of the cell envelope (lipid and cell wall); cell division; and cell energetic functions.

**Essential Genes in *Bacillus subtilis***

| Function | Number |
| --- | --- |
| DNA metabolism | 27 |
| RNA metabolism | 14 |
| Protein synthesis | 95 |
| Cell envelope | 44 |
| Cell shape and division | 10 |
| Glycolysis | 8 |
| Respiratory pathways | 22 |
| Nucleotide synthesis | 10 |
| Enzyme cofactors | 15 |
| Other | 15 |
| Unknown | 11 |
| Total | 271 |

More than half of the essential genes had homologs in eukaryotes or archaea, attesting to the fundamental importance of the gene products for maintaining life. Other essential genes, such as those for cell-wall synthesis, were more phylogenetically restricted since they pertained to physiological processes that differ considerably among organisms.

If the same level of synthetic lethality exists for the remaining genes in the yeast genome, it is estimated that 200,000 different synthetic lethal interactions will occur among all yeast genes and that 1% of all double mutants will result in synthetic lethality. Thus, while only 1000 genes are essential under optimal laboratory growth conditions as defined by single-mutant phenotypes, additional genes become essential when organisms are compromised by a mutation in another gene. One explanation for the observed levels of synthetic lethality is that where there are multiple genetic pathways, some of the pathways buffer one another, creating stable genetic systems that are better able to withstand environmental and genetic perturbations.

Genetic networks defined by genetic interactions often identify groups of genes having similar molecular functions, such as translation, lipid metabolism, or DNA repair (see Figure 18.28). If a gene of unknown function belongs to a genetic network in which many genes have known roles—say, in lipid metabolism—experiments to identify the molecular function of the unknown gene might begin by investigating whether the gene in question also plays a role in lipid metabolism.

Genetic networks constructed on the basis of genetic interactions can be examined in comparison with groupings based on other gene attributes, such as their annotations, expression patterns, or interactomes (discerned from protein–protein interaction data). Prediction of

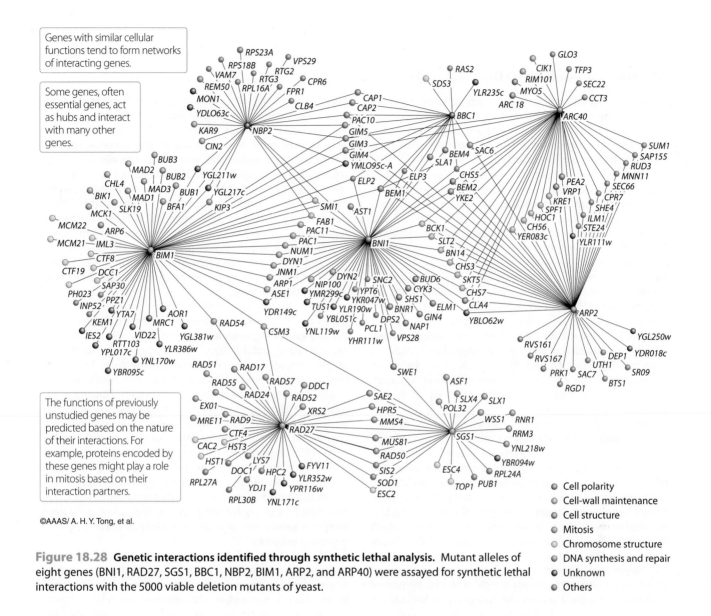

Genes with similar cellular functions tend to form networks of interacting genes.

Some genes, often essential genes, act as hubs and interact with many other genes.

The functions of previously unstudied genes may be predicted based on the nature of their interactions. For example, proteins encoded by these genes might play a role in mitosis based on their interaction partners.

©AAAS/ A. H. Y. Tong, et al.

- Cell polarity
- Cell-wall maintenance
- Cell structure
- Mitosis
- Chromosome structure
- DNA synthesis and repair
- Unknown
- Others

**Figure 18.28** **Genetic interactions identified through synthetic lethal analysis.** Mutant alleles of eight genes (BNI1, RAD27, SGS1, BBC1, NBP2, BIM1, ARP2, and ARP40) were assayed for synthetic lethal interactions with the 5000 viable deletion mutants of yeast.

biological functions of genes based on correlations between different data sets is referred to as **systems biology.**

Genetic interaction data often correlate well with gene expression data since genes that compensate for one another in function often exhibit similar expression patterns. In contrast, genetic interactions and protein–protein interactions overlap less often. One reason is that physically interacting proteins are likely to act in the same protein complex, whereas in genetic interactions, the proteins they encode often act in compensating pathways that would normally be composed of different protein complexes with related functions. Note that the latter holds true primarily when null alleles are used to test genetics interactions; however, when hypomorphic alleles are used, genetic interactions are often seen with genes

encoding proteins that act in the same complex or pathway (see Chapter 16).

The ultimate objective of functional genomics studies is to define the molecular function of every gene in an organism by compiling genomic data and searching for correlations that suggest hypotheses for further experimentation. This section focused on studies in *S. cerevisiae*, but similar approaches are being taken in other organisms.

Genetic Insight   Genetic networks derived from systematic genetic interaction analyses can provide clues to gene functions.

## CASE STUDY

### Genomic Analysis of Insect Guts May Fuel the World

In metagenomic analysis, biologists study genomes collected from the multiple organisms that together inhabit a single environment. Two recent studies suggest that metagenomic analysis of insect digestive tracts could potentially have a significant impact on production of biofuels.

Much of the current supply of ethanol for fuel is produced from cellulose that comes from the lignocellulose component of corn. Lignocellulose is a mixture of cellulose (a complex carbohydrate composed of glucose molecules) and lignin (the rigid structural material that protects cellulose). The production of corn ethanol requires high temperature, high heat, and the use of toxic chemicals to break down the lignin and hydrolyze the cellulose. This step is followed by microbial fermentation of the sugar and distillation of ethanol. Obtaining ethanol from corn in this way has adverse effects on the environment, consumes a great deal of energy, and may not be economically viable. These are principal reasons why the investigation of lignocellulose digestion in insects is attractive. Identification and characterization of the genes responsible for lignocellulose digestion may allow the development of new, biologically based methods of biofuel production.

In 2007, the microbiologist Falk Warnecke and his colleagues conducted a metagenomic study of the microbes in the gut of the wood-eating termite species *Nasutitermes*. Termites are wood-digesting creatures whose ancestors have inhabited cellulose-rich environments for more than 100 million years. *Nasutitermes* has a bacteria-laden gut that acts like a tiny bioreactor for digesting the lignocellulose in wood. Lignocellulose provides energy for these microorganisms, which first break down lignin to liberate cellulose and then break down cellulose via hydrolysis driven by hydrolase enzymes.

*Nasutitermes* has a three-part stomach, the main part of which, designated P3, contains a rich microbial mixture of hundreds of bacterial species that are primarily responsible for wood digestion. Warnecke and his colleagues collected *Nasutitermes* in Costa Rica. Then, in the laboratory, they isolated and emptied P3 and found that its total volume in each insect is just 1 microliter (μL). They isolated and performed shotgun sequencing on the DNA from the P3 microbial mass.

Warnecke estimates that the DNA in this metagenomic analysis may come from as many as 300 bacterial species whose symbiotic relationship with the termite allows the termite to derive energy from wood. Gene-identification analyses indicate that many of the most frequently found genes in these bacteria produce glycoside hydrolases (GH) that hydrolyze lignocellulose. More than 700 different GH genes representing more than 45 different gene families were found in this study. A large group of previously unidentified genes was also found, and Warnecke speculates that these genes might be involved in various kinds of lignocellulose binding and digestion reactions.

While Warnecke's study detected numerous bacterial genes that may carry out cellulose digestion, it did not identify any genes responsible for lignin digestion. However, a second study, published in 2008 by Scott Geib and colleagues, examined lignin digestion in the Asian longhorn beetle (*Anoplophora glabripennis*) and the Pacific dampwood termite (*Zootermopsis angusticollis*). Biochemical analysis of the digestive tracts and digestive products of both insects shows significant evidence of lignin digestion, suggesting that either the genomes of these organisms encode lignin-digesting enzymes or that the organisms carry symbiotic microbes whose genomes encode the enzymes. The researchers did not perform metagenomic analyses of the insect genomes or digestive system contents, but they did identify a single fungal species in the gut of the Asian longhorn beetle whose genome is likely to encode lignin-digesting enzymes.

A great deal of additional "bioprospecting" research will be necessary to characterize the genes that encode the enzymes driving lignin and cellulose digestion in insect guts. In the process, further genomic and metagenomic analyses may suggest ways these genes can be cloned and used to replace the costly current methods of lignocellulose-based ethanol production.

---

## SUMMARY

 Mastering**GENETICS**™

*For activities, animations, and review quizzes, go to the study area at www.masteringgenetics.com.*

### 18.1 Structural Genomics Provides a Catalog of Genes in a Genome

▎ Genomes can be sequenced in either a clone-by-clone approach or a whole-genome shotgun approach.

▎ Paired-end sequencing facilitates assembly of scaffolds consisting of sequence fragments generated by shotgun sequencing.

▎ Metagenomics studies the genetic sequences of communities of organisms whose member species may be difficult to cultivate individually.

▎ Genome annotation indicates the locations of genes and other functional sequences in a genomic sequence. It aims to ascribe biological function to sequence data.

▎ Functions of some annotated genes may be predicted based on sequence similarities with known genes as analyzed through computational approaches and bioinformatics, but experimental verification is required.

### 18.2 Evolutionary Genomics Traces the History of Genomes

▎ A phylogenetic tree of life can be constructed by comparing sequences of orthologous genes.

- New genes can be produced by gene duplication due to unequal crossing over or by larger-scale duplications of DNA, retrotransposition, and other mechanisms.

- Most new genes degenerate rapidly; but some are retained and may acquire new functions, driving the evolution of new species.

- Gene duplication has been a key feature in the evolution of complex organisms. Lateral gene transfer is a common mechanism of acquisition of new genes in bacteria and archaea, but it is less common in eukaryotes.

- By comparing genomes of related species, researchers can identify conserved genes and noncoding sequences and refine gene annotation. Conserved noncoding sequences often consist of gene regulatory sequences.

- Intraspecific genome comparisons identify genetic variation within a species and provide information about its evolutionary history and population dynamics. Both intra- and interspecific comparisons reveal that genomes are dynamic and can change rapidly on evolutionary timescales.

## 18.3 Functional Genomics Aids in Elucidating Gene Function

- DNA microarrays and high-throughput sequencing can reveal polymorphisms, global transcription patterns, and transcription-factor binding sites.

- Protein–protein interactions can be determined by using genetic tools developed from the study of yeast.

- Knockout libraries are used to perform genome-wide forward genetic screens that elucidate gene function. They have allowed classification of all yeast genes as essential or nonessential under specific growth conditions.

- Genes classified as essential under optimal growth conditions have on average more genetic interactions than those classified as nonessential.

- Genome-wide analyses of synthetic lethal interactions in yeast reveal large numbers of genes that are essential in genetically compromised organisms.

- Systems biology is a research approach that correlates data sets derived from functional genomics in order to define and elucidate gene function.

## KEYWORDS

annotation (gene annotation, genome annotation) *(p. 608)*
barcode *(p. 632)*
basic local alignment search tool (BLAST) *(p. 616)*
bioinformatics *(p. 609)*
clone-by-clone sequencing *(p. 603)*
conserved noncoding sequence (CNS) *(p. 621)*
copy-number variant (CNV) *(p. 624)*
DNA microarray *(p. 627)*
evolutionary genomics (phylogenomics, comparative genomics) *(p. 603)*
expression array *(p. 627)*
functional genomics *(p. 603)*
gene families *(p. 609)*

genetic networks *(p. 633)*
homologous genes (homologs) *(p. 618)*
homologous nucleotides *(p. 614)*
interactome *(p. 631)*
interspecific comparison (intraspecific comparison) *(p. 614)*
knockout library *(p. 632)*
metagenome *(p. 608)*
microsynteny *(p. 622)*
neofunctionalization *(p. 618)*
node *(p. 615)*
orthologous genes (orthologs) *(p. 619)*
paired-end sequencing *(p. 605)*
paralogous genes (paralogs) *(p. 619)*
phylogenetic footprinting (phylogenetic shadowing) *(p. 621)*

physical gap *(p. 607)*
proteome (proteomics) *(p. 630)*
pseudogene *(p. 617)*
reference genome sequence *(p. 623)*
scaffold *(p. 605)*
sequence gap *(p. 606)*
structural genomics *(p. 603)*
subfunctionalization *(p. 618)*
synteny *(p. 622)*
systems biology *(p. 635)*
transcriptomics (transcriptome) *(p. 627)*
tree of life *(p. 614)*
two-hybrid system *(p. 630)*
whole-genome shotgun (WGS) sequencing *(p. 603)*
whole-genome tiling array *(p. 628)*

## PROBLEMS

*For instructor-assigned tutorials and problems, go to www.masteringgenetics.com.*

### Chapter Concepts

*For answers to selected even-numbered problems, see Appendix: Answers.*

1. You have discovered a new species of Archaea from a hot spring in Yellowstone National Park.
   a. After growing a pure culture of this prokaryote, what strategy might you employ to sequence its genome?
   b. How would your strategy change if you were unable to grow the strain in culture?

2. Repetitive DNA poses problems for genome sequencing.
   a. Why is this so?
   b. What types of repetitive DNA are most problematic?
   c. What strategies can be employed to overcome these problems?

3. When the whole-genome shotgun sequence of the *Drosophila* genome was assembled, it comprised 134 scaffolds made up of 1636 contigs.
   a. Why were there so many more contigs than scaffolds?
   b. What is the difference between physical and sequence gaps?
   c. How can physical gaps be closed?
   d. How can sequence gaps be closed?

4. How do cDNA sequences facilitate gene annotation? Describe how the use of full-length cDNAs facilitates discovery of alternative splicing.

5. How do comparisons between genomes of related species help refine gene annotation?

6. You are designing algorithms for the bioinformatic prediction of gene sequences. How might algorithms differ for predicting genes in bacterial versus eukaryotic genomic sequence?

7. You have sequenced a 100-kb region of the *Bacillus anthracis* genome (the bacterium that causes anthrax) and a 100-kb region from the *Gorilla gorilla* genome. What differences and similarities might you expect to see in the annotation of the sequences—for example, in number of genes, gene structure, regulatory sequences, repetitive DNA?

8. You have just obtained 100 kb of human genomic sequence. What are three methods you might use to identify potential genes in the 100 kb? What are the advantages and limitations of each method?

9. The human genome contains a large number of pseudogenes. How would you distinguish whether a particular sequence encodes a gene or a pseudogene? How do pseudogenes arise?

10. Based on the tree of life in Figure 18.11, would you expect human proteins to be more similar to fungal proteins or to plant proteins? Would you expect plant proteins to be more similar to fungal proteins or human proteins?

11. When comparing genes from two sequenced genomes, how does one determine whether two genes are orthologous? What pitfalls arise when one or both of the genomes are not sequenced?

12. What are the differences between expression arrays and genome tiling arrays? What types of data can be obtained using microarrays? Can high-throughput sequencing supplant most applications of microarrays?

13. The two-hybrid method facilitates the discovery of protein–protein interactions. How does this technique work? Can you think of reasons for obtaining a false-positive result, that is, where the proteins encoded by two clones interact in the two-hybrid system but do not interact in the organism in which they naturally occur? Can you think of reasons you might obtain a false-negative result, in which the two proteins interact in vivo but fail to interact in the two-hybrid system?

## Application and Integration

*For answers to selected even-numbered problems, see Appendix: Answers.*

14. Go to http://blast.ncbi.nlm.nih.gov/Blast.cgi and follow the links to *nucleotide blast*. Type in the sequence below; it is broken up into codons to make it easier to copy.

```
5′ ATG TTC GTC AAT CAG CAC CTT TGT GGT TCT
CAC CTC GTT GAA GCTTTG TAC CTT GTT TGC
GGT GAA CGT GGT TTC TTC TAC ACT CCT AAG
          ACT TAA 3′
```

As you will note on the BLAST page, there are several options for tailoring your query to obtain the most relevant information. Some are related to which sequences to search in the database. For example, the search can be limited taxonomically (e.g., restricted to mammals) or by the type of sequences in the database (e.g., cDNA or genomic). For our search, we will use the broadest database, the "nucleotide collection (nr/nt)." This is the nonredundant (nr) database of all nucleotide data (nt) in GenBank and can be selected in the "database" dialogue box. Other parameters can also be adjusted to make the search more or less sensitive to mismatches or gaps. For our purposes, we will use the default setting, which is automatically presented. Press "search." What can you say about the DNA sequence?

15. In the course of the *Drosophila melanogaster* genome project, the following genomic DNA sequences were obtained. Can they be assembled into a contig?

```
5′ TTCCAGAACCGGCGAATGAAGCTGAAGAAG 3′
5′ GAGCGGCAGATCAAGATCTGGTTCCAGAAC 3′
5′ TGATCTGCCGCTCCGTCAGGCATAGCGCGT 3′
5′ GGAGAATCGAGATGGCGCACGCGCTATGCC 3′
5′ CCATCTCGATTCTCCGTCTGCGGGTCAGAT 3′
```

Using the assembled sequence, perform a blastn search using the "nucleotide collection (nr/nt)" database. Does the search produce sequences similar to your assembled sequence, and if so, what are they? Can you tell if your sequence is transcribed, and if it represents protein-coding sequence? Perform a tblastx search, first choosing the "nucleotide collection (nr/nt)" database and then limiting the search to human sequences by typing *Homo sapiens* in the organism box. Are homologous sequences found in the human genome? Annotate the assembled sequence.

16. Consider a phylogenetic tree below with three related species (A, B, C) that share a common ancestor (last common ancestor, or LCA). The lineage leading to species A diverges before the divergence of species B and C.

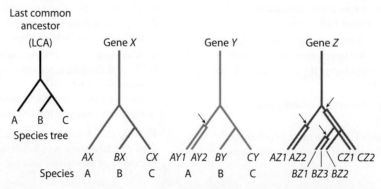

a. For gene *X*, no gene duplications have occurred in any lineage, and each gene *X* is derived from the ancestral gene *X* via speciation events. Are genes *AX*, *BX*, and *CX* orthologous, paralogous, or homologous?

b. For gene *Y*, a gene duplication occurred in the lineage leading to A after it diverged from that leading to B and C. Are genes *AY1* and *AY2* orthologous or paralogous? Are genes *AY1* and *BY* orthologous or paralogous? Are genes *BY* and *CY* orthologous or paralogous?

c. For gene *Z*, gene duplications have occurred in all species. Define orthology and paralogy relationships for the different *Z* genes.

17. You have isolated a gene that is important for the production of milk and wish to examine its regulation. You examine the genomes of human, mouse, dog, chicken, pufferfish, and yeast and note that all genomes except yeast have an orthologous gene.

a. How would you identify the regulatory elements important for the expression of your isolated gene in mammary glands?

b. What does the existence of orthologous genes in chicken and pufferfish tell you about the function of this gene?

18. When the human genome is examined, the chromosomes appear to have undergone only minimal rearrangement in the 100 million years since the last common ancestor of eutherian mammals. However, when individual humans are examined or when the human genome is compared to that of chimpanzees, a large number of small indels and SNPs can be detected. How are these observations reconciled?

19. Why do you think the *Rickettsia prowazekii* genome is so much smaller than the other bacterial genomes listed in Table 18.1?

20. Substantial fractions of the genomes of many plants consist of segmental duplications; for example, approximately 40% of genes in the *Arabidopsis* genome are duplicated. How might you approach the functional characterization of such genes using reverse genetics?

21. A modification of the two-hybrid system, called the one-hybrid system, is used for identifying proteins that can bind specific DNA sequences. In this method, the DNA sequence to be tested, the bait, is fused to a TATA box to drive expression of a reporter gene. The reporter gene is often chosen to complement a mutant phenotype; for example, a *HIS* gene may be used in a *his⁻* mutant yeast strain. A cDNA library is constructed with the cDNA sequences translationally fused to the GAL4 activation domain and transformed into this yeast strain. Diagram how trans-acting proteins that bind to cis-acting regulatory sequences can be identified using a one-hybrid screen.

22. A substantial fraction of almost every genome sequenced consists of genes that have no known function and that do not have sequence similarity to any genes with known function.

a. Describe two approaches to ascertaining the biological role of these genes in *S. cerevisiae*.

b. How would your approach change if the genes of unknown function were in the human genome?

23. In the globin gene family shown in Figure 18.16, which pair of genes would exhibit a higher level of sequence similarity, the human δ-globin and human β-globin genes or the human β-globin and chimpanzee β-globin genes? Can you explain your answer in terms of timing of gene duplications?

24. You are studying similarities and differences in how organisms respond to high salt concentrations and high temperatures. You begin your investigation by using microarrays to compare gene expression patterns of *S. cerevisiae* in normal growth conditions, in high-salt concentrations, and at high temperatures. The results are shown below with the values of red and green representing the extent of increase and decrease, respectively, of expression for genes *a–s* in the experimental conditions versus the control (normal growth) conditions. What is the first step you will take to analyze your data?

25. In conducting the study described in Problem 24, you have noted that a set of *S. cerevisiae* genes are repressed when yeast are grown under high-salt conditions.

a. How might you determine whether this set of genes is regulated by a common transcription factor?

b. How might you approach this question if genome sequences for the related *Saccharomyces* species, *S. paradoxus*, *S. mikatae*, and *S. bayanus*, were also available?

26. Using the two-hybrid system to detect interactions between proteins, you obtain the following results: A clone encoding gene *A* gave positive results with clones B and C; clone B gave positive results with clones A, D, and E but not C; and clone E gave positive results only with clone B. Another clone, F, gave positive results with clone G but not with any of A–E. Can you explain these results?

27. To follow up your two-hybrid results of Problem 26, you isolate null loss-of-function mutations in each of the genes *A–G*. Mutants of genes *A*, *B*, *C*, *D*, and *E* grow at only 80% of the rate of the wild type, while mutants of genes *F* and *G* are phenotypically indistinguishable from the wild type. You construct several double-mutant strains: The *ab*, *ac*, *ad*, and *ae* double mutants all grow at about 80% of the rate of the wild type, but *af* and *ag* double mutants exhibit lethality. Explain these results.

28. *PEG10* (paternally expressed gene 10) is a paternally expressed gene that has an essential role in the formation of the placenta of the mouse. In the mouse genome, the *PEG10* gene is flanked by the *SGCE* and *PPP1R9A* genes. To study the origin of *PEG10*, you examine syntenous regions spanning the *SGCE* and *PPP1R9A* loci in the genomes of several vertebrates, and you note that the *PEG10* gene is present in the genomes of placental and marsupial mammals but not in the platypus, chicken, or fugu genomes.

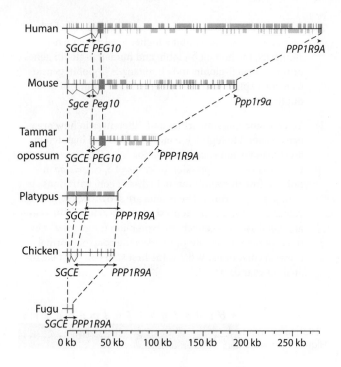

The green bars indicate the exons of each gene. The gray bars represent LINEs and SINEs, and the blue bars represent long terminal repeat (LTR) elements of retrotransposons. Solid black lines link introns, and dashed black lines connect orthologous exons. Arrowheads indicate direction of transcription.

Using the predicted protein sequence of *PEG10*, you perform a tblastn search for homologous genes and find that the most similar sequences are in a class of retrotransposons (the sushi-ichi retrotransposons).

Propose an evolutionary scenario for the origin of the *PEG10* gene, and relate its origin to its biological function.

# Cytoplasmic Inheritance and the Evolution of Organelle Genomes

# 19

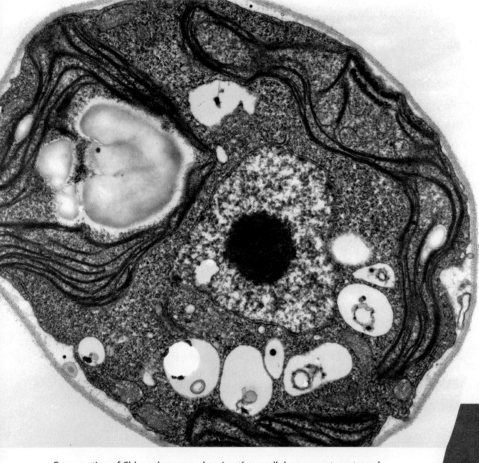

Cross section of *Chlamydomomas* showing three cellular compartments each with their own genetic material: nucleus (blue), mitochondrion (red), and chloroplast (green).

ESSENTIAL IDEAS

▪ Mitochondria and chloroplasts possess their own genomes encoding a small number of genes.

▪ Eukaryotic cells may have many copies of organelle DNA; therefore, multiple genotypes may coexist in a single cell.

▪ The inheritance of organelle genomes can be uniparental, as in maternal inheritance in mammals, or biparental.

▪ The organization and expression of organellar genomes reflect their evolutionary origins as symbiotic bacteria.

▪ Genes have been, and continue to be, transferred from the organellar genomes to the genome of the host cell.

Soon after the rediscovery of Mendel's laws in the early 1900s, Carl Correns and Erwin Baur, working independently, each noted a pattern of inheritance that was distinctly non-Mendelian. Both Correns and Baur were studying the inheritance in plants of a variegated phenotype in which individual branches had either white, green, or variegated leaves. Reciprocal crosses between flowers growing on white or green branches produced progeny that invariably exhibited the phenotype of the female parent in the cross.

The green coloration in land plants and green algae is due to the presence of the green pigment chlorophyll, which harvests light for photosynthesis. In plants, chlorophyll is found in **chloroplasts,** which are the organelles where photosynthetic reactions convert light energy and $CO_2$ into fixed organic carbon. The variegated and white phenotypes studied by Correns and Baur are caused by a failure of chloroplast development in some cells, which as a consequence remain colorless (white).

In the 1950s, studies demonstrated that chloroplasts contain their own genome. In combination with the observation that chloroplasts are strictly maternally inherited in many plants, this discovery suggested an explanation for the maternal inheritance seen by Correns and Baur: The mutations they were studying must reside on the chloroplast genome. As we will see, the cell's energy-producing and energy-capturing organelles—mitochondria and chloroplasts, respectively—each possess their own genome and may be either uniparentally or biparentally inherited depending on the species. Furthermore, uniparental inheritance may be maternal, paternal, or genetically determined. In this chapter, we explore the genetic transmission of the organelle genomes, the remarkable evolutionary events that led to the development of organelles, and the surprisingly dynamic interactions between the organelle and nuclear genomes of eukaryotes.

## 19.1 Cytoplasmic Inheritance Transmits Genes Carried on Organelle Chromosomes

**Cytoplasmic inheritance** refers to the transmission of genes on mitochondrial and chloroplast chromosomes—genes that are located in the cytoplasmic organelles as opposed to the nucleus. As with nuclear genes, expression of mitochondrial and chloroplast genes produces proteins and RNAs that perform specific functions in cells. However, genetic analysis of cytoplasmic inheritance differs from that of nuclear gene inheritance because the cytoplasm contained within a fertilized egg is not usually derived from equal contributions of both parental gametes.

In many eukaryotic species, the mitochondria and chloroplasts in fertilized eggs are **uniparental** in their origin. This means that just one parental gamete—often

the maternal gamete—contributes all of the cytoplasm and cytoplasmic organelles. In some species, organelles are inherited in a uniparental manner even though equal amounts of cytoplasm are inherited from both parental gametes. In such cases, the organelles derived from one of the gametes are selectively destroyed. In still other species, both parental gametes make contributions of cytoplasm and cytoplasmic organelles to the zygote; this pattern is termed **biparental.** Biparental cytoplasmic contributions are often *unequal* because one gamete contributes more of the cytoplasm and the other gamete makes a smaller contribution. Additional reasons that the study of cytoplasmic inheritance differs from the study of nuclear inheritance may be summarized as follows:

1. Multiple organelles may be present in eukaryotic cells.

2. Each mitochondrion or chloroplast may contain multiple copies of its chromosome. The potential presence of tens to hundreds of copies of organelle chromosomes in each cell stands in contrast to the two copies of nuclear genes present in the cells of diploid organisms.

3. The sizes (six to hundreds of kilobases), numbers of genes (few to hundreds), and identity of the specific genes contained in the organelle genomes are highly variable from one species to another.

4. Traits controlled by cytoplasmic inheritance can also be influenced by nuclear genes. Most biological functions ascribed to mitochondrial or chloroplast genes are produced through the joint action of nuclear genes and organelle genes.

### The Discovery of Cytoplasmic Inheritance

Erwin Baur and Carl Correns were working independently of one another in 1908—Baur on *Pelargonium* (geraniums) and Correns on *Mirabilis jalapa* (the four o'clock plant)—when each made his discovery of non-Mendelian inheritance. Correns was studying leaf-color inheritance in the four o'clock plant, using ovules and pollen from flowers located on branches with different leaf colors. He began his investigation by doing self-fertilization experiments and found that seeds derived from self-fertilization of flowers on green branches produced plants that contained only green leaves. Seeds derived from self-fertilization of flowers on white branches produced seedlings that had only white leaves. These latter seedlings grew poorly and never produced mature plants. The self-fertilization of flowers from branches with variegated leaves produced a mixture of progeny that were either variegated, had all white leaves, or had all green leaves.

In his next experimental step, Correns made reciprocal crosses between branches. Using pollen from a flower located on a branch with one leaf color, he fertilized ovules from a flower located on a branch with a different leaf color. The results, as shown in Figure 19.1, were progeny that invariably exhibited the phenotype of the female parent in the cross. This is *not* the result predicted by Mendelian genetics (which predicts no difference in the results of reciprocal crosses), nor is it the result expected if leaf color were inherited on a sex chromosome. Instead, the outcome suggested that transmission of leaf color occurs through **maternal inheritance**—that is, through genes transmitted in the ovule. Leaf color in the four o'clock flower is controlled exclusively by maternal inheritance,

and pollen (the male gamete) makes no contribution to the phenotype.

White leaves are produced when leaf cells contain mutated chloroplasts that lack the ability to produce chlorophyll. Variegated leaves are produced by plants whose cells contain a mixture of normal and mutant chloroplasts. The green patches of variegated leaves are composed of cells containing chloroplasts that can produce chlorophyll, whereas the white leaf patches are composed of cells containing mutated chloroplasts that are unable to produce chlorophyll. Modern-day plant biology explains these results as a consequence of cytoplasmic inheritance and states that the allelic differences reside in a gene in the chloroplast genome.

In the 1950s, several decades after Correns and Baur described their observations of non-Mendelian inheritance in plants, Yasutane Chiba and colleagues suggested that mitochondria and chloroplasts contain their own genomes. This assertion was based on the results of staining with the compound Feulgen, which specifically stains DNA. In studying mitochondria and chloroplasts from a variety of plants and animals, Chiba detected Feulgen-positive spots in the cytoplasm of virtually all cells examined, and determined that the Feulgen-stained cytoplasmic DNA was contained within the organelles. This result is consistent with the presence of chromosomes in mitochondria and chloroplasts.

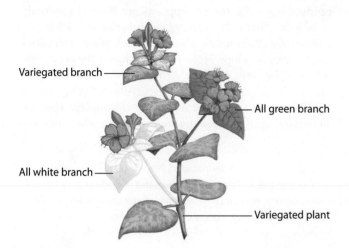

Variegated branch

All green branch

All white branch

Variegated plant

> **Genetic Insight** Cytoplasmic inheritance is a consequence of the presence of genes in the mitochondria and chloroplasts of eukaryotes.

## Homoplasmy and Heteroplasmy

Figure 19.1 illustrates that if an ovule is obtained from a flower on a branch with all green leaves, then it contains chloroplasts that produce chlorophyll, and its progeny plants will have only green leaves regardless of the leaf color of the pollen-producing plant. Similarly, an ovule obtained from an all-white-leafed branch contains mutated chloroplasts, and all progeny will have only white leaves due to the transmission of defective chloroplasts from the ovule. Ovules from variegated plants can produce progeny with green, white, or variegated leaves. This apparent departure from the maternal inheritance pattern for green and white leaves can be reconciled by the observation that each plant cell contains many copies of chloroplast genes.

The amount of nuclear genetic material is constant: haploid cells have a single copy of each chromosome, and diploid cells have two copies of each chromosome. In contrast, the number of copies of organelle genes in each cell is much higher and varies significantly with both organism and cell type. Copy number variation occurs at two levels.

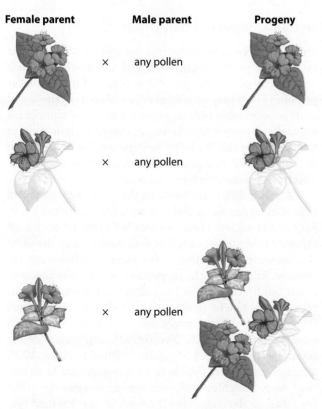

| Female parent | Male parent | Progeny |
|---|---|---|
| | × any pollen | |
| | × any pollen | |
| | × any pollen | |

**Figure 19.1** **Reciprocal crosses demonstrating maternal inheritance of chloroplasts in four o'clock plants.**

First, the number of organelles per cell can vary from one to hundreds, and second, the number of copies of the organelle genome per organelle also varies from one to many. Thus the terms *homozygous* and *heterozygous* are not applicable to alleles of genes on organelle genomes. Rather, a cell or organism in which all copies of a cytoplasmic organelle gene are the same is identified as **homoplasmic** and is said to exhibit **homoplasmy** for that gene (Figure 19.2a). On the other hand, if variation exists among the copies of an organelle gene, the cell or organism is **heteroplasmic** and exhibits **heteroplasmy,** carrying a mixture of alleles of an organellar gene. Note that in a heteroplasmic organism, some cells can be homoplasmic wild type, other cells homoplasmic mutant, and still others heteroplasmic.

Homoplasmic and heteroplasmic genotypes for chloroplast genes explain the maternal inheritance of variegation observed by Correns in four o'clock flowers (Figure 19.2b). Ovules derived from flowers on branches that contain green leaves are homoplasmic for wild-type chloroplast genes and transmit only wild-type chloroplasts to their progeny. In contrast, ovules derived from flowers on branches with white leaves are homoplasmic for a chloroplast mutation, and only mutant chloroplasts are passed to progeny.

The progeny phenotypes derived from flowers on variegated branches illustrate the complexity of organellar genetics. Consider an ovule produced on a variegated branch that consists of a mixture of cells: Some of them are heteroplasmic, inheriting a cytoplasm containing many chloroplasts; some of them are wild type; and others harbor the mutant allele. During mitoses and meiosis that produce egg cells, the chloroplasts are divided randomly among daughter cells. If an egg cell inherits both wild-type and mutant chloroplasts, a heteroplasmic plant with variegated leaves develops. However, if by chance the organelles inherited by an egg cell are all wild type, the branches of the resulting plant will be green. Alternatively, chance might result in an egg cell inheriting chloroplasts that are all mutant, in which case the plant will have white leaves.

---

**Genetic Insight** Individual cells contain multiple copies of mitochondrial and chloroplast genes. In a homoplasmic cell or organism, all copies are the same; a heteroplasmic cell or organism contains a mixture of alleles of a gene.

---

## Genome Replication in Organelles

Organelle DNA is packaged into protein–DNA complexes in an area of the organelle called the **nucleoid.** Each nucleoid usually contains multiple copies of the organellar genome. There may be several nucleoids per organelle and multiple organelles per cell, resulting in a copy number for organelle genomes that is in the range of hundreds to thousands per cell. To better understand the transmission of mutations in organellar genomes, and their phenotypic effects, let us examine how organellar DNA is replicated.

A major difference between the nuclear genome and that of an organelle is their relationship to the cell cycle. Each of the nuclear chromosomes is duplicated once each mitotic cycle, so that daughter cells have exactly the same chromosome constitution as the parent cell following cell division. In contrast, the replication of organelle genomes is not tightly coupled to the cell cycle. Rather, the replication of organelle genomes depends on three factors (Figure 19.3). First, organelle transmission genetics depends on the growth, division, and segregation of the organelles themselves ("organelle division" in Figure 19.3), although there does appear to be a mechanism to ensure that each daughter cell receives approximately equal amounts of the organelles present in the mother cell. Second, the segregation of genes encoded in the organelle genome is connected to the division and segregation of

**(a)** Homoplasmic and heteroplasmic cells

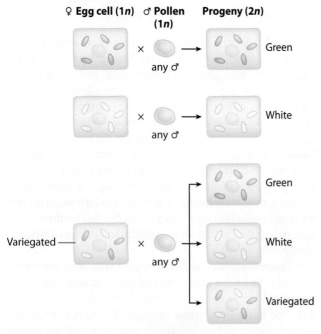

Nucleus

Chloroplasts mutant

Wild type

Green | White | Variegated

Homoplasmic cells have organelles with the same genotype.

Heteroplasmic cells contain a mixture of alleles.

**(b)** Phenotpe of progeny depends only on the genotype of the maternal parent.

♀ Egg cell (1*n*)  ♂ Pollen (1*n*)  Progeny (2*n*)

× any ♂ → Green

× any ♂ → White

Variegated × any ♂ → Green / White / Variegated

**Figure 19.2 Homoplasmy and heteroplasmy in cells.**

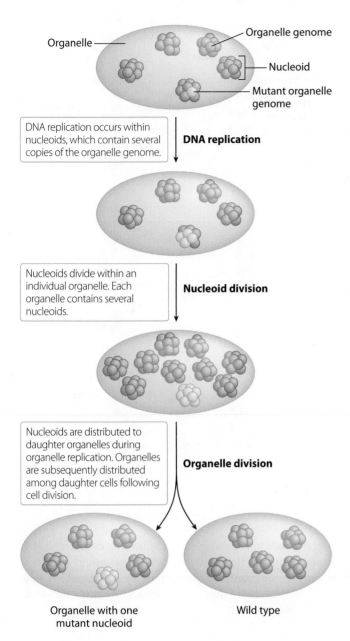

Organelle

Organelle genome

Nucleoid

Mutant organelle genome

DNA replication occurs within nucleoids, which contain several copies of the organelle genome.

**DNA replication**

Nucleoids divide within an individual organelle. Each organelle contains several nucleoids.

**Nucleoid division**

Nucleoids are distributed to daughter organelles during organelle replication. Organelles are subsequently distributed among daughter cells following cell division.

**Organelle division**

Organelle with one mutant nucleoid

Wild type

**Figure 19.3  Factors in replication of organelle genomes.**

nucleoids within an organelle ("nucleoid division" in Figure 19.3). Details of this process are still being discovered, but differences in the replication rate of nucleoids have been observed both between cells and between organelles. Third, organelle transmission genetics depends on the replication of the individual organelle genomes ("DNA replication" in Figure 19.3). There is evidence that DNA molecules within a nucleoid are related to each other; they are sometimes physically linked, which would suggest that they are products of DNA replication.

## Variable Segregation of Organelle Genomes

The variation in the numbers of organelles and of their genomes in different somatic cells and tissues can significantly influence the phenotypic effects of mutations in organelle genes. Consider again the case of the variegated leaves, keeping in mind the presence of multiple organelle chromosomes in each cell. If a cell is homoplasmic with regard to this trait, cells descended from this cell by division will also be homoplasmic. However, cells that are heteroplasmic can produce both heteroplasmic and homoplasmic descendants.

To see how this happens, imagine a plant cell in which a mutation occurs in a chloroplast genome. Through segregation of nucleoids during chloroplast division, chloroplasts in which all copies of the genome harbor the mutations can arise. During cell division, the chloroplasts are randomly distributed to the daughter cells. If by chance all the organelles inherited by a daughter cell are of a single genotype, homoplasmic cells can be generated from a heteroplasmic ancestral cell (see the cells at the bottom of the far-right columns in Figure 19.4). This random segregation of organelles during replication is termed **replicative segregation.** Replicative segregation is of great importance since it affects the proportion of mutant organelle genomes in a cell, thus influencing the severity (penetrance and expressivity) of phenotypes produced by mutations in organellar genomes. It can lead to genetically mosaic organisms with both "mutant" cells and "wild type" cells; and, as we see with the variegated plants, it can influence transmission of mutant alleles to subsequent generations depending on the organellar genotype of the germ cells.

In heteroplastic individuals, penetrance and expressivity will depend on the ratio between mutant and wild-type organelle alleles, which can vary among cells and tissues. In some cases, wild-type alleles can complement mutant alleles within an organelle, so a heteroplasmic individual can often tolerate a high frequency of mutant alleles without a mutant phenotype being evident or becoming severe. For organelle inheritance between generations, the number of chloroplast or mitochondrial genomes present in the germ cells is important. In heteroplastic individuals, transmission will depend on what fraction of organelle genomes present in the gametes contain mutant versus wild-type alleles. Due to replicative segregation, gametes can be produced that are homoplastic wild type, homoplastic mutant, or heteroplastic, and they can have varying ratios of mutant and wild-type alleles. Thus replicative segregation can explain both variation in penetrance and expressivity between individuals and also variable transmission, where green, white, and variegated seedlings can all be derived from variegated plants.

The observation that mitochondria undergo frequent fusion and fission has implications for the segregation of mitochondrial DNA and creates the potential for genotypes within a cell's mitochondria to become mixed and homogenized. In contrast to mitochondria, chloroplasts do not undergo fusion events with any frequency. This influences the effect of replicative segregation in chloroplast DNA transmission.

**Figure 19.4 Development of homoplasmy from heteroplasmy by replicative segregation.**

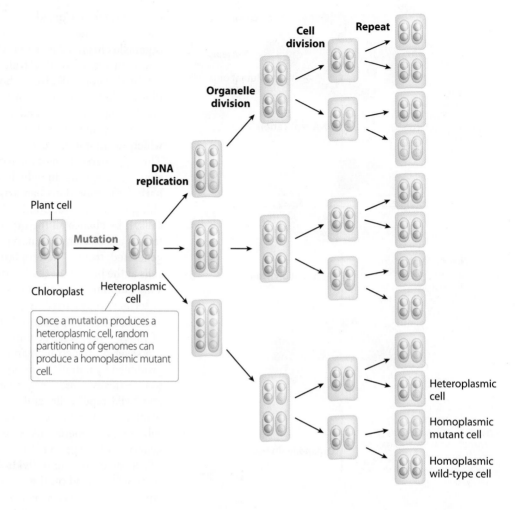

Once a mutation produces a heteroplasmic cell, random partitioning of genomes can produce a homoplasmic mutant cell.

Heteroplasmic cell

Homoplasmic mutant cell

Homoplasmic wild-type cell

Now that we have described some of the complexities of transmission of the organelle genomes, for the remainder of the chapter we will assume that individuals are homoplasmic, unless there is evidence that heteroplasmy exists.

---

**Genetic Insight** Random distribution of organelles to daughter cells, coupled with the dissociation of organelle replication from the cell cycle, leads to replicative segregation, through which heteroplasmic cells may give rise to homoplasmic daughter cells.

---

## 19.2 Modes of Cytoplasmic Inheritance Depend on the Organism

The inheritance of organelle genomes can be determined through two basic mechanisms. In many organisms, the transmission is biased to whichever gamete contributes the bulk of the cytoplasm to the zygote. In this case transmission can be either uniparental (maternal or paternal) or biparental. Alternatively, inheritance is genetically

determined: one gamete's organelles are destined to be transmitted to the progeny while the other gamete's organelle contributions are selectively destroyed. In this section, we explore three cases illustrating three different inheritance patterns.

### Mitochondrial Inheritance in Mammals

Maternal inheritance of mitochondria is the norm in mammals because the egg contributes all of the cytoplasm and the sperm contributes only a nucleus during fertilization. Maternal inheritance of the mitochondrial genome in mammals has three important consequences that we examine in this section:

1. Predictions of inheritance of mitochondrial mutations can be made based solely on the genotype of the mother.

2. Maternal inheritance allows the maternal lineage of organisms to be examined specifically.

3. Since there is no paternal contribution, phylogenetic trees constructed using mitochondrial DNA sequences can be interpreted as maternal genealogies reflecting the maternal history of species.

**Mother–Child Identity of Mitochondrial DNA**   Mothers and their children of both sexes share identical mitochondrial DNA. These identical genetic matches are put to many practical uses. One of the most dramatic examples in humans is the use of mitochondrial DNA to find matches between grandmothers and grandchildren who were separated during political unrest in Argentina during the 1970s. An Argentinean military dictatorship undertook a campaign of kidnapping and murder of political dissidents in the early 1970s. Among those kidnapped were pregnant women, who were allowed to give birth before they were murdered. The children of these women were adopted by unrelated families, and their identities were hidden from their real families.

As the political environment in Argentina became less repressive, a group known as Las Abuelas de la Plaza de Mayo (Grandmothers of the Plaza de Mayo) demanded an accounting of the murder of the dissidents and the return of the adopted children to their real families. Part of the process used to identify adopted children took advantage of the maternal inheritance of mitochondrial DNA—specifically, of the fact that each grandmother had transmitted her mitochondria to her biological children, all of whom, as a result, inherited identical mitochondrial genes (Figure 19.5). Her daughters in turn passed the same mitochondrial DNA to their biological children. By this hereditary transmission mechanism, grandmothers and the children of their daughters carry identical mitochondrial DNA. Comparisons of mitochondrial DNA revealed exact matches between individual abuelas and specific children of the murdered women, allowing many abuelas to be reunited with their grandchildren, whose mothers had been "disappeared."

**Mitochondrial DNA Sequences and Species Evolution**
Mitochondrial DNA sequences are used as a tool for deciphering the genealogical history and evolutionary relationships of mammalian species. Mitochondrial DNA is particularly well suited to such studies for two reasons. First, since mitochondria are strictly maternally inherited in mammals, there is no recombination of alleles, as there is with the nuclear genome. Second, some noncoding regions of mitochondrial genomes in general evolve quickly, with the result that many differences in mitochondrial DNA sequence are present even in closely related populations. This is particularly true for mammals, where the rate of mutation in the mitochondrial genome is about 10 times that of the nuclear genome, reflecting decreased levels of DNA mutation repair in mitochondria versus repair of nuclear DNA. Since there is little selective pressure to maintain a specific sequence in noncoding regions, mutations in these regions accumulate at a relatively steady rate.

Once a mitochondrial mutation occurs in the germ cells of an individual female, the mutation is transmitted to all her progeny. Therefore, maternal lineages can be traced by following the mutational changes back in time. The mitochondrial DNA sequences in the present population reflect the maternal genealogy of the population as a whole, and construction of a phylogenetic tree based on these sequences should allow the identification of the common ancestor(s) of the species.

**Mitochondrial Eve**   Analyses of mitochondrial DNA variation in human populations have refined our view of our early human ancestors' journey out of Africa. The regions around the Great Rift Valley of East Africa have been home to humans and our hominid ancestors for at least 4 million years. Based on the fossil record, dispersals from Africa have also been a regular feature throughout hominid evolution. *Homo erectus* reached as far as Java 1.7 million years ago, and *Homo neanderthalensis* was present in Europe as much as 500,000 years ago. Populations of *H. erectus* and *H. neanderthalensis* persisted in Eastern Asia and Europe, respectively, until approximately 30,000 years ago; meanwhile, hominid evolution continued in Africa.

Study of the age and locations of fossils belonging to the genus *Homo* produced two competing models of the evolution of *H. sapiens* (Figure 19.6). One model, called the *multiregional (MRE) model*, postulates that modern humans emerged gradually and simultaneously from *H. erectus* on different continents, and gene flow continued between the different populations. An alternative theory, the *recent African origin (RAO) model*, proposes that modern humans evolved from a small African population that migrated out of Africa, displacing other hominid species.

The RAO model postulates that modern humans arose approximately 120,000 to 200,000 years ago, whereas the MRE model posits a much older age for our species—up to 2 million years ago. The two models also make very different predictions about the genetic diversity of humans inhabiting the planet today. The MRE model predicts uniform genetic diversity among populations from most of the world, whereas the RAO model suggests genetic diversity should be greatest in Africa

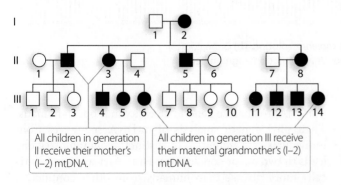

All children in generation II receive their mother's (I–2) mtDNA.

All children in generation III receive their maternal grandmother's (I–2) mtDNA.

**Figure 19.5**  **Maternal inheritance of mitochondrial genes in mammals.**

**(a)**

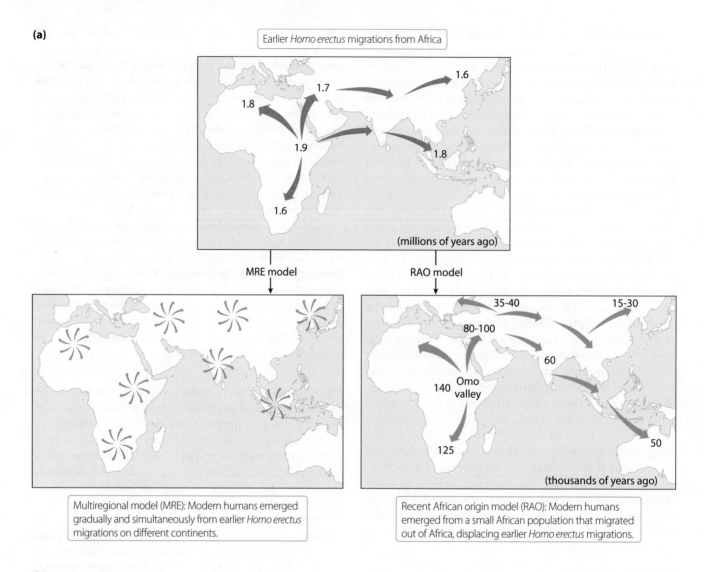

Earlier *Homo erectus* migrations from Africa

(millions of years ago)

MRE model          RAO model

(thousands of years ago)

Multiregional model (MRE): Modern humans emerged gradually and simultaneously from earlier *Homo erectus* migrations on different continents.

Recent African origin model (RAO): Modern humans emerged from a small African population that migrated out of Africa, displacing earlier *Homo erectus* migrations.

©1987 Macmillan Publishers Ltd

**(b)**

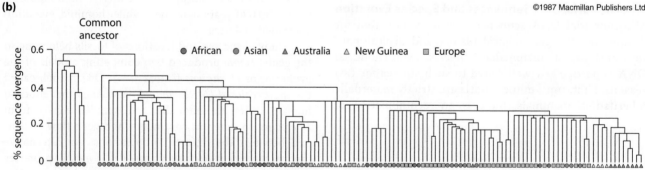

Figure 19.6 **Human migrations and evolution.** **(a)** Models of *H. sapiens* evolution. **(b)** Genealogical tree of modern humans based on phylogenetic analyses of mitochondrial restriction fragment length polymorphisms (RFLPs) strongly supports the RAO model.

since humans would have diversified there before migrating outward. In the RAO scenario, the genetic diversity outside of Africa would be a subset of that found in Africa and so would reflect the subpopulation of humans who migrated from Africa.

To distinguish between these hypotheses, Allan Wilson and colleagues used the mitochondrial genome (mtDNA) to analyze genetic diversity in modern humans. Their phylogenetic analysis of mtDNA sequences from individuals representative of distinct geographic regions leads to two major conclusions: First, Africans are genetically more diverse than humans from other continents, and second, the genetic diversity of non-Africans is a subset of that found in Africans. In addition, comparison of human sequences with those of chimpanzees allowed the researchers to estimate when the divergence of humans occurred. This is calculated by first working out the rate of sequence evolution in terms of base-pair changes per million years. The researchers divided the number of sequence differences between humans and chimpanzees by 5 to 7 million years (the divergence time of the two species) and then calculated the minimum divergence time of humans by applying the rate of sequence evolution to the two most divergent human sequences. Such calculations led to the estimate that modern humans first appeared about 200,000 years ago.

These conclusions agree with the RAO theory. They suggest that modern humans evolved in Africa and subsequently migrated around the world, largely displacing rather than interbreeding with other hominid populations (see Chapter 18). The mtDNA of all humans living today is descended from a female or group of females living in East Africa 120,000 to 200,000 years ago. The carrier of this ancestral mtDNA has been called our "mitochondrial Eve." See **Genetic Analysis 19.1** for practice interpreting data from another research project that analyzed mitochondrial DNA.

## Mitochondrial Mutations and Human Genetic Disease

Human biology is highly dependent on the cellular energy derived from oxidative phosphorylation reactions in our mitochondria. It is therefore not surprising that mitochondrial mutations can result in human genetic diseases (**Figure 19.7a**). The phenotypes of mitochondrial diseases are often highly pleiotropic, a reflection of the ubiquitous dependency of cells on mitochondrial function. A hallmark of such diseases is their strictly maternal transmission.

Leber's hereditary optic neuropathy (LHON) is a mitochondrial genetic disease in which degeneration of the central optic nerve results in blindness, usually in late adolescence to early adulthood (**Figure 19.7b**). Like most diseases caused by mitochondrial mutations, the LHON syndrome is accompanied by pleiotropic defects, primarily a range of heart abnormalities. LHON can be caused by mutations in a number of different genes that encode proteins of the NADH dehydrogenase subunit involved in electron transport. In the pedigree shown in Figure 19.7b, affected individuals have a single base-pair change, resulting in a missense (arginine to histidine) mutation in the subunit 4 gene, *ND4*.

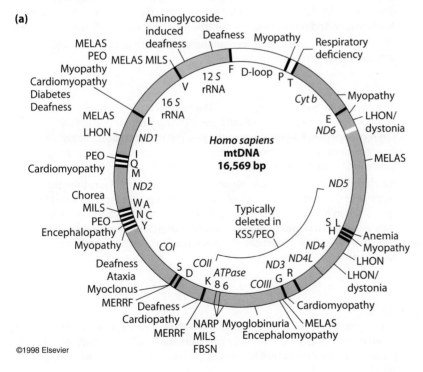

**Figure 19.7  Mutations in human mitochondrial genes leading to disease syndromes.  (a)** Muscle functioning, hearing, and vision all require high levels of energy produced by mitochondria. **(b)** Pedigree showing maternal inheritance with incomplete penetrance of LHON.

Although North American bison (*Bison bison*) and domestic cattle (*Bos taurus* and *Bos indicus*) descended from a common ancestor, they do not readily interbreed. However, because they still share the same chromosome number and structure, the production of fertile interspecific hybrids is possible. Male bison have been known to breed with female cattle, but not the converse. Twelve North American bison herds (numbered 1 through 12 below) were examined for evidence of such interbreeding by a comparison of their mtDNA sequences with those of several cattle breeds and related species. A phylogenetic tree constructed from the comparisons is presented here. The numbers represent confidence values for the particular relationships (100 is the maximum).

a. Explain why mtDNA but not nuclear DNA is used to detect bison--domestic cattle interspecific hybrids.

b. Based on this phylogeny, identify which bison herds show evidence of interspecific breeding with domestic cattle.

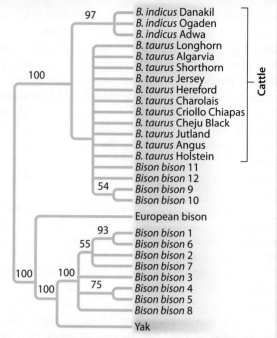

| Solution Strategies | Solution Steps |
|---|---|
| **Evaluate** | |
| 1. Identify the topic this problem addresses, and describe the nature of the requested answer. | 1. This problem presents a phylogenetic analysis of an mtDNA sequence in domestic cattle and in bison. We must explain why mtDNA was used rather than nuclear DNA, and then we must examine the phylogeny to identify bison herds that do and do not have bison–cattle hybridization in their lineage. |
| 2. Identify the critical information given in the problem. | 2. The phylogenetic tree depicts evolutionary relationships between cattle mtDNA and mtDNA samples from bison. |
| **Deduce** | |
| 3. Examine the major clades in the phylogenetic tree, and describe the membership of each clade. | 3. The phylogeny has two major clades. The bottom clade contains eight North American bison herds (*Bison bison* 1 through 8) and two outside reference species, European bison and yak. The upper clade contains fourteen domestic cattle breeds (*Bos taurus* and *Bos indicus*) and four North American bison herds (*Bison bison* 9 through 12). |
| 4. Identify the kind of phylogenetic evidence (based on mtDNA) that would be consistent with interspecific hybridization and also the kind that would be inconsistent with it.  TIP: In interspecies hybridization, bison mtDNA sequences would be more closely related to cattle sequences than they are to other bison sequences. | 4. If a clade consists only of domesticated breeds or only of bison, then the animals in the clade are more closely related to one another than they are to animals in other clades and do not have interspecific hybridization in their lineage. If a clade contains bison *and* domesticated cattle breeds, then there is a close relationship between the bison and the cattle in that clade. |

**Solve**

5. Explain why mtDNA but not nuclear DNA sequences were used in this phylogenetic analysis  TIP: In mammals, all mitochondrial DNA is maternally inherited.

*Answer a*

5. We are told that female cattle interbreed with male bison, but not the reverse. Since mtDNA is inherited maternally, the resulting hybrids would possess solely cattle mtDNA but would contain equal mixtures of cattle and bison nuclear DNA.

*Answer b*

6. Determine which bison are interspecies hybrids.  TIP: Bison of hybrid origin will harbor mtDNA more closely related to that of cattle than of bison.

6. Bison herds 9 to 12 are in the same clade as a number of breeds of domestic cattle, signifying that their mtDNA sequences are more closely related to domesticated cattle than to the wild bison and yak species. Thus these four herds have cattle mtDNA from interspecific hybridization in previous generations.

For more practice, see Problem 26.

Close inspection of the pedigree in Figure 19.7b reveals that, while all affected individuals have an affected mother, not all children of an affected mother exhibit LHON. If we assume strict maternal inheritance of the mitochondrial mutations, then the phenotype is not fully penetrant. There are three possible reasons for incomplete penetrance: the effects of heteroplasmy, the effects of genetic interactions with nuclear genes, and the effect of environmental factors interacting with mitochondrial gene mutations to produce a mutant phenotype. A discussion of mitochondrial gene–environment interactions appears in the Case Study at the end of this chapter, and an example of mitochondrial–nuclear interactions appears in **Experimental Insight 19.1**.

Heteroplasmy can lead to incomplete penetrance of a human hereditary disease because, as discussed earlier, each cell contains multiple mitochondria and each mitochondrion contains multiple copies of the mitochondrial genome. The numbers of organelles within a cell can influence expressivity, penetrance, and transmission of mutant alleles in various ways. There is no fixed number of copies of organelle genomes in cells. In mammals the numbers of copies of mitochondrial genomes in human cells vary from hundreds to hundreds of thousands, depending on the cell type and physiological state. In cells with both wild-type and mutant mitochondrial genotypes, the wild-type allele can complement the mutant allele.

In human pedigrees, heteroplasmic mothers can produce wild-type homoplasmic progeny, mutant homoplasmic offspring, or heteroplasmic offspring (**Figure 19.8a**). For mitochondrial transmission in mammals, the number of mitochondria present in the egg cell is what matters. Human oocytes typically have a small number (e.g., 10) of large mitochondria that are subsequently divided into many smaller mitochondria in the zygote. In humans, an egg cell contains up to 2000 mitochondrial genomes. In heteroplasmic individuals, replicative segregation can lead to variable penetrance, in which the ratio of mutant : wild-type mitochondrial genomes varies significantly between different progeny (**Figure 19.8b**).

Furthermore, replicative segregation of mitochondrial mutations over the lifetime of an individual can lead to variable ratios of mutant : wild-type mitochondrial genomes in different cells and tissues of the same heteroplasmic individual; and this too results in variable phenotypic penetrance. Disease symptoms will develop only when vulnerable cells contain a high proportion of mutant mitochondria. For example, in the case of another mitochondrial disease, called MERRF (myoclonic epilepsy with ragged red fibers), an individual who displayed the mutant genotype in 85% of his mitochondrial DNA did not exhibit a phenotype defect, whereas a cousin with 96% mutant mitochondria displayed a severe

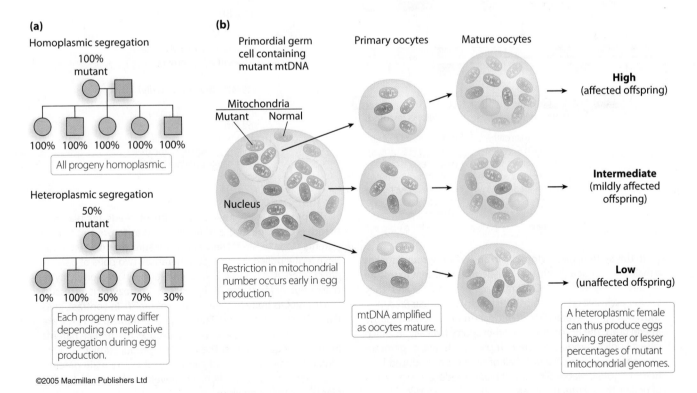

**(a)**

Homoplasmic segregation

100% mutant

100%  100%  100%  100%  100%

All progeny homoplasmic.

Heteroplasmic segregation

50% mutant

10%  100%  50%  70%  30%

Each progeny may differ depending on replicative segregation during egg production.

©2005 Macmillan Publishers Ltd

**(b)**

Primordial germ cell containing mutant mtDNA

Primary oocytes

Mature oocytes

Mitochondria
Mutant    Normal

Nucleus

Restriction in mitochondrial number occurs early in egg production.

mtDNA amplified as oocytes mature.

**High**
(affected offspring)

**Intermediate**
(mildly affected offspring)

**Low**
(unaffected offspring)

A heteroplasmic female can thus produce eggs having greater or lesser percentages of mutant mitochondrial genomes.

**Figure 19.8  Variable penetrance of mitochondrial mutations.**

# Cytoplasmic Male Sterility in Flowering Plants

You probably do not think of sterility as a useful trait in a crop plant; however, male sterility in one parent plant provides an efficient mechanism for producing hybrid seed. This is possible because the male sterile plant can act as the female parent in a cross with a second variety. Hybrid seed is desirable in many crop species because plants that are the progeny of crosses between two different varieties often exhibit higher yield than do either of the parents. This phenomenon is called hybrid vigor. Here we describe how hybrid seed can be produced by taking advantage of genetic interactions between specific nuclear and chloroplast genes.

In plants, male sterility is a failure to produce viable pollen. Some cases, called cytoplasmic male sterility (CMS), are maternally inherited and are due to mutations in the mitochondrial genome. However, the phenotypic defects of these mitochondrial mutations can often be suppressed by the presence of dominant alleles of one or more nuclear genes, called *restorer of fertility*, or *RF*, genes. The interaction between typical *CMS* and *RF* genes provides an example of how genetic interactions between nuclear and mitochondrial genotypes can influence phenotypes. It can be outlined as follows:

| Female Parent | × | Pollen Parent | Progeny Genotype | Progeny Phenotype |
|---|---|---|---|---|
| CMS *rf/rf* | | N *rf/rf* | CMS *rf/rf* | Male sterile |
| CMS *rf/rf* | | N *Rf/Rf* | CMS *Rf/rf* | Male fertile |

CMS = male sterile cytoplasm; N = wild-type cytoplasm; *Rf* = dominant nuclear RF allele; *rf* = recessive nuclear RF allele.

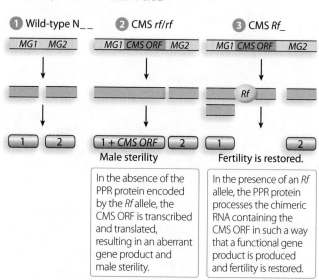

In this system, CMS cytoplasm in an *rf/rf* background makes a male sterile, but a dominant RF allele, *Rf*, is sufficient to restore fertility. Many different CMS mutants are known, and they exhibit exclusive relationships with particular nuclear *RF* genes, thus indicating several distinct nuclear–mitochondrial genome interactions. The RF loci may act either sporophytically, in which case all pollen produced from *Rf/rf* plants is fertile, or gametophytically, in which case only half of the pollen produced by a heterozygote is viable. Since most plants produce a vast excess of pollen, these latter plants are considered male fertile.

CMS mitochondrial genes usually have novel open reading frames (ORFs) that combine sequences of unknown origin fused with mitochondrial gene-coding sequences. Expression of the novel ORFs is driven by adjacent mitochondrial promoter sequences. Since most plants harboring CMS-causing ORFs have a full complement of normal mitochondrial genes, the CMS ORFs can be considered gain-of-function mutations.

Several RF genes encode proteins of the pentatricopeptide repeat (PPR) family. The functions of characterized PPR proteins include RNA processing, such as cleavage of RNA precursors and RNA editing. This discovery is consistent with the effects of *RF* genes on *CMS* genes, since in the presence of a restorer allele, transcripts of *CMS* ORFs fail to accumulate. One current hypothesis is that PPR proteins encoded by *Rf* alleles process transcripts produced by the *CMS* genes, rendering them nonfunctional.

CMS–RF systems have been harnessed to facilitate the production of hybrid seeds. For example, in maize, a double-cross hybrid scheme has been used that utilizes four breeding lines as parents.

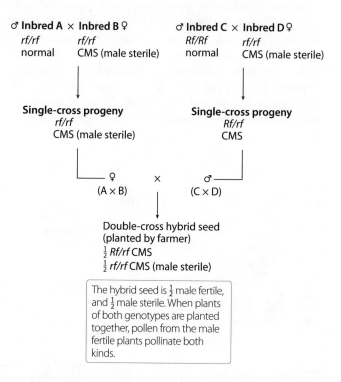

To produce each new generation of seeds for planting, breeders combine *CMS* and *RF* alleles so as to prevent female parents from self-fertilizing and to ensure that male parents have fertile pollen. In the first generation, two pairs of parents are crossed, A × B and C × D. Both A and C have normal cytoplasm but differ at the RF locus: A is homozygous recessive (*rf/rf*), and C is homozygous dominant (*Rf/Rf*). In contrast, lines B and D are CMS and *rf/rf*. The progeny produced by A × B are CMS *rf/rf*, male sterile, and can be used as the female parents in the subsequent cross. The progeny produced by C × D are CMS *Rf/rf*, male fertile, and can be used as the male parents. The seeds that ultimately result develop into larger, hardier plants due to hybrid vigor.

phenotype. See Genetic Analysis 19.2 for practice in analyzing a pedigree for evidence of various forms of nuclear and mitochondrial inheritance.

---

Genetic Insight  Mitochondrial genomes are maternally inherited in humans. Heteroplasmy may cause variable penetrance for the traits they produce. Mutations in human mitochondrial genes lead to diseases that often have pleiotropic phenotypic symptoms.

---

## Mating Type and Chloroplast Segregation in *Chlamydomonas*

*Chlamydomonas reinhardii* is a single-celled green alga with a haploid nuclear genome that harbors a single, large chloroplast containing 50 to 100 genomes divided among 5 to 15 nucleoids. Not surprisingly, the pattern of cytoplasmic inheritance in this alga differs from the mammalian pattern.

Matings between *Chlamydomonas* cells of different mating types produce diploid algae that undergo meiosis to produce haploid progeny. Mating compatibility is determined by the genotype at the $mt$ locus, and $mt^+$ individuals mate only with $mt^-$ individuals. Both mating types appear to contribute equally to the cytoplasmic content of the diploid zygote, but in approximately 95% of matings, the chloroplast genome is contributed by the $mt^+$ mating type. In the remaining 5% of matings, chloroplast inheritance is biparental. The first mutation in a chloroplast gene discovered in *Chlamydomonas* was isolated by Ruth Sager in 1954 and confers resistance to the antibiotic streptomycin ($str^R$). Analogous to reciprocal crosses between four o'clock flowers of different leaf types, reciprocal crosses between streptomycin-resistant and streptomycin-sensitive *Chlamydomonas* strains of different mating types give different results; the chloroplast genotype is contributed primarily by the $mt^+$ parent (Figure 19.9).

During the mating process in *Chlamydomonas*, the two cells of opposite mating type fuse, after which the single chloroplasts of each of the two mating types fuse to form a single chloroplast. The mechanism by which the $mt^-$ cell's chloroplast genome is eliminated is not known, but it is likely to involve degradation of that genome at some point in the mating process. Perhaps the degradation of organelles or their genomes provides a possible source of organelle DNA that may be transferred between genomes—into the nuclear genome, for example. (We will return to this topic later in the chapter, when we discuss the evolution of the organelles and their genomes.) For the cases in which biparental inheritance occurs, the presence of the two types of chloroplast genomes in the same organelle allows the genomes to undergo recombination

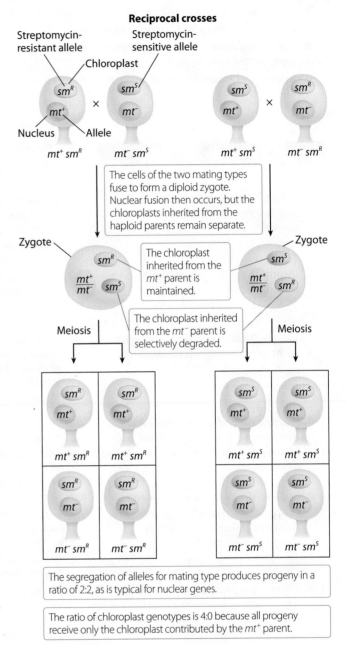

**Figure 19.9  Organelle segregation determined by mating type in *Chlamydomonas*.**

that may result in the segregation of recombinant and parental chloroplast genomes.

Haploid cells of *Chlamydomonas* typically have about 50 copies of the mitochondrial genome distributed among a small number of mitochondria in the germ cells and a larger number of mitochondria at other stages of the life cycle. Remarkably, though the chloroplast genome is preferentially transmitted by the $mt^+$ mating type, mitochondria are preferentially transmitted by the $mt^-$ mating type. The genetic mechanisms by which the different mating types preferentially transmit the different organellar genomes are presently unknown.

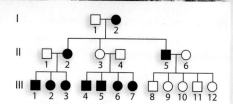

The pedigree shows transmission of a rare human hereditary disorder.

a. Determine the most likely mode of inheritance.

b. Identify any individuals in the pedigree whose phenotype is inconsistent with the expected phenotype.

c. Justify your proposed mode of inheritance by explaining the inconsistencies.

| Solution Strategies | Solution Steps |
|---|---|
| **Evaluate** | |
| 1. Identify the topic this problem addresses, and describe the nature of the requested answer. | 1. This problem concerns the mode of inheritance of a hereditary abnormality in a human pedigree. The answer requires proposing a mode of inheritance, identifying family members whose phenotypes are inconsistent with the proposed mode, and explaining those inconsistencies in a manner that justifies the proposed mode. |
| 2. Identify the critical information given in the problem. | 2. The pedigree gives the phenotype of each family member in three generations. |
| **Deduce** | |
| 3. Identify the possible modes of inheritance of the gene causing this abnormality.<br><br>TIP: Human cells contain maternally inherited mitochondria in addition to nuclear chromosomes. | 3. The possibilities are that the trait might be caused by the mutation of either a nuclear gene or a mitochondrial gene. If the mutated gene is nuclear, it might be either recessive or dominant and either autosomal or X-linked. If the mutation is mitochondrial, the transmission pattern will be maternal inheritance. |
| 4. Examine the pedigree to see whether the pattern is generally consistent with autosomal recessive or X-linked recessive inheritance. | 4. The pattern is inconsistent with X-linked recessive inheritance, in which many more males than females have the recessive phenotype. Here, the ratio of six females to four males is close to 1:1, so X-linked recessive inheritance is highly unlikely. Autosomal inheritance is unlikely, since siblings in generation III are either all affected or none affected within families. |
| 5. Examine the pedigree to see whether the pattern is generally consistent with X-linked dominant or autosomal dominant inheritance. | 5. In X-linked dominant inheritance, all daughters of males with the dominant mutation are also expected to have the trait. II-5 does not transmit the trait to any of his three daughters, thus making X-linked dominant inheritance highly unlikely. Autosomal dominant inheritance is possible, where II-3 is nonpenetrant; but there is only a 1/32 chance ($1/2^5$) that II-5 would have five children who do not have the trait. |
| 6. Examine the pedigree to see whether the pattern is consistent with maternal inheritance. | 6. The pedigree pattern is consistent with maternal (mitochondrial) inheritance. Affected individuals are all offspring of affected mothers (I-2, II-2) or of female II-3 (who may harbor the mutant allele but does not exhibit the phenotype). |
| **Solve** | Answers a and b |
| 7. Determine the mode of transmission that is consistent with the pedigree data. | 7. Maternal inheritance best explains the observed segregation pattern, but there is one inconsistency. Individual II-3 does not show the phenotype as expected under strict application of the rules of maternal inheritance. |
| | Answer c |
| 8. Explain the presence of the anomalous individuals whose phenotypes are inconsistent with maternal inheritance.<br><br>TIP: Heteroplasmy may occur among the multiple copies of mitochondrial chromosomes present in each cell.<br><br>TIP: Proteins produced by mitochondrial genes interact with proteins produced by nuclear genes. | 8. Lack of penetrance of the phenotype (as in II-3) may result from (1) variable penetrance owing to some individuals being heteroplasmic, since some could have a greater proportion of mutant mitochondria than others; (2) other genetic risk factors, such as alleles of nuclear genes (since both males and females show variable penetrance, alleles of autosomal genes may be influencing the penetrance of the mitochondrial mutation, although common alleles of X chromosome genes cannot be ruled out); (3) environmental factors that influence the penetrance of the phenotype. |

For more practice, see Problems 12, 14, 17, 18, 19, 20, and 22.

## Biparental Inheritance in *Saccharomyces cerevisiae*

*Saccharomyces cerevisiae* is a single-celled yeast that can grow either aerobically (with oxygen) or anaerobically (without oxygen). Mitochondria are not able to produce energy (ATP) when oxygen is unavailable; so under anaerobic growth conditions, yeast obtain their energy from fermentation, which does not require mitochondria. Under aerobic conditions, however, mitochondria-mediated aerobic respiration allows yeast to grow faster than they grow by fermentation. Thus mutations that eliminate mitochondrial function in yeast do not prevent growth, but they do cause the mutant yeast to grow at a slower pace than do wild-type yeast. This dual growth capacity makes *Saccharomyces* a versatile system for studying the genetics of mitochondrial biology.

In the mid-1950s, Boris Ephrussi noted that when grown on media that allow fermentative growth, some mutant colonies of yeast were much smaller relative to wild-type yeast colonies. He named these mutants *petite* and referred to the wild-type colonies as *grande*. Biochemical analyses revealed that the *petite* mutants are deficient in mitochondrial cytochrome activity and for this reason are unable to carry out respiratory growth. Therefore *petite* mutants are able to grow only by fermentation, and they grow more slowly than wild-type yeast growing by respiration. When *petite* mutants are transferred to media that permit only respiratory growth, they are unable to grow, and the mutations are lethal. Therefore *petite* mutants can be classified as conditional lethal mutations.

Recall that yeast grow as haploid cells (see Chapter 3). Their mating involves the fusion of two cells of different mating types, called **a** and **α**, to produce a diploid zygote. The diploid zygote can divide by mitosis for several generations, during which time its phenotype (petite or wild type) can be identified. When the zygote undergoes meiosis, four haploid progeny (ascospores) contained within an ascus are produced. Tetrad analysis can be performed on the ascospores to determine the segregation of alleles (see Section 5.7). Mutations in nuclear genes will segregate in a 2:2 ratio (mutant : wild type) when mutant lines are mated with wild type (**Figure 19.10a**). Both **a** and **α** gametes contribute mitochondrial genomes to the zygote.

Genetic analysis of *petite* mutants reveals that they fall into three distinct classes. One class, called *nuclear*, or *segregational*, *petites* (designated *pet⁻*), segregate 2:2 when crossed with the wild type (**Figure 19.10b**). In other words, *pet⁻* are mutations in nuclear genes. The existence of nuclear *petites* demonstrates that the functioning of the mitochondria depends not only on its own genome but also on genes contained in the nuclear genome. Both genomes encode genes whose products function in the organelle, as we discuss in a later section.

The other two classes of *petite* mutations—*neutral petites* and *suppressive petites*—do not show Mendelian inheritance and are the result of mutations in the

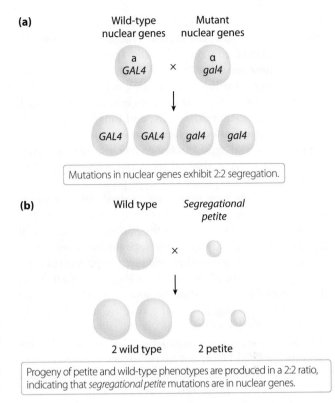

(a)

Mutations in nuclear genes exhibit 2:2 segregation.

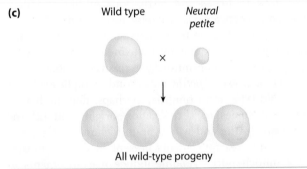

(b)

2 wild type    2 petite

Progeny of petite and wild-type phenotypes are produced in a 2:2 ratio, indicating that *segregational petite* mutations are in nuclear genes.

(c) Wild type    *Neutral petite*

All wild-type progeny

Progeny do not exhibit the petite phenotype, indicating that *neutral petite* mutants are not transmitted. Examination of *neutral petite* mutants indicates that they lack most or all mitochondrial DNA.

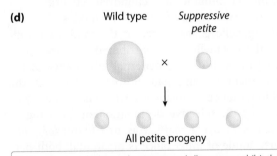

(d) Wild type    *Suppressive petite*

All petite progeny

Petite mitochondrial DNA dominates, and all progeny exhibit the petite phenotype. Examination of *suppressive petite* mutants indicates that they have deletions of only portions of their mitochondrial DNA.

**Figure 19.10** Transmission of *petite* phenotypes in *Saccharomyces cerevisiae*.

mitochondrial genome. When *neutral petites* are crossed with wild-type yeast, the diploid zygote grows normally, and the tetrads contain only wild-type spores (**Figure 19.10c**). These are called "neutral" because the *petite* phenotype is lost after the initial mating with wild type. Examination of *neutral petite* mutants reveals that they lack virtually all mitochondrial DNA, and thus they obviously lack proper mitochondrial function. When *neutral petites* are mated to wild-type *Saccharomyces*, essentially all mitochondrial DNA is derived from the wild-type parent, resulting in phenotypically wild-type progeny.

When *suppressive petites* are crossed with wild-type yeast, the diploid zygote has respiratory properties intermediate between those of the *petite* and wild type. If the diploid zygotes are grown mitotically for several divisions, the diploids tend to become *petite* in phenotype, and the tetrads contain all *petite* spores (**Figure 19.10d**). Thus the *suppressive petite* phenotype suppresses the wild-type phenotype, resulting in progeny that are all deficient in respiration. Analysis of the mitochondrial genome reveals that initially, most *suppressive petites* have small deletions of mitochondrial DNA; but upon further growth, the remaining mitochondrial DNA tends to become rearranged and duplicated. These gross defects in mitochondrial DNA lead to losses and disruptions of mitochondrial genes and to deficiencies in aerobic respiration.

Why do the mitochondria inherited from the *suppressive petite* parent overwhelm those of the wild-type parent? Two non-mutually exclusive possibilities are that (1) *suppressive petite* mitochondria replicate faster than wild-type mitochondria, perhaps due to having additional copies of a replication origin, and (2) the *suppressive petite* and wild-type mitochondria fuse, and the genomic rearrangements present in the *suppressive petite* mitochondrial genome induce rearrangements in the mitochondrial genomes inherited from the wild-type parent. The latter hypothesis has gained support from the observation that mitochondria within a cell often interact and fuse into a continuous mitochondrial network.

In summary, there are four primary modes of inheritance of organelle genes. Three of the modes are uniparental; the organelles are contributed primarily by a single parent, as in (1) the maternal inheritance of organelles in mammals and many flowering plants; (2) the paternal inheritance of organelles, which is seen in gymnosperms; or (3) selective degradation or silencing of organelle DNA during mating, as in *Chlamydomonas*. The fourth mode of inheritance is biparental; both parents contribute organelles and their genomes to the progeny, as in *Saccharomyces*.

As we learned in Section 19.1, mitochondria and chloroplasts contain their own genomes, composed of genes that are unique to the organelles and are expressed and replicated by mechanisms independent of those working on nuclear genes. The discussions that follow explore the structure, replication, function, and evolution of mitochondrial and chloroplast genomes.

# 19.3 Mitochondria Are the Energy Factories of Eukaryotic Cells

Enzymatically driven phosphorylation that transfers phosphates from adenosine triphosphate (ATP) to other molecules provides energy used by cells for many processes and functions. In most eukaryotes, mitochondria are the sites of energy production, where electron transport is coupled to oxidative phosphorylation to generate ATP. In many species, mitochondrial genes also participate in other metabolic processes and biochemical reactions, including ion homeostasis and biosynthetic pathways. The protein complexes that produce ATP are composed of gene products encoded by both the mitochondrial and nuclear genomes, thus necessitating coordination in the expression of genes from the two genomes.

The general structure of a **mitochondrion** can be described as two membranes surrounding the mitochondrial matrix (**Figure 19.11**). The enzyme complexes responsible for oxidative phosphorylation are found on the inner membrane. The mitochondrial matrix is the site of mitochondrial genome transcription, translation, and DNA replication. The mitochondrial genome is responsible for only a fraction of the genes needed to carry out these processes, however, and most of the proteins active in mitochondrial DNA replication, transcription, and translation are encoded in the nuclear genome.

Following their translation, nuclear-encoded mitochondrial proteins are transported into mitochondria. Examination of the mitochondrial genomes of different species reveals enormous diversity as to whether specific proteins are mitochondrial- or nuclear-encoded; only a few proteins are consistently encoded by the mitochondrial genome. Thus the synthesis and regulation of the protein complexes responsible for oxidative phosphorylation and other mitochondrial processes require coordination between the mitochondrial and nuclear genomes.

## Mitochondrial Genome Structure and Gene Content

Genetic mapping studies and direct observation of mitochondrial chromosomes by electron microscopy indicate the chromosomes often have a circular structure (**Figure 19.12a–b**). There is evidence, however, that circular mitochondrial genomes can assume a linear form and that the mitochondrial genomes of certain species are primarily linear. In the vast majority of species, the mitochondrial genome is a single molecule; but in a few species, the genome consists of more than one molecule. Thus, in some species, the mitochondrial genome consists of one (*Tetrahymena*) or more (*Amoebidium*) linear molecules that have terminal repeat sequences, which are reminiscent of telomeres.

Unlike the DNA in the nucleus, mitochondrial DNA is not packaged in chromatin composed of histones.

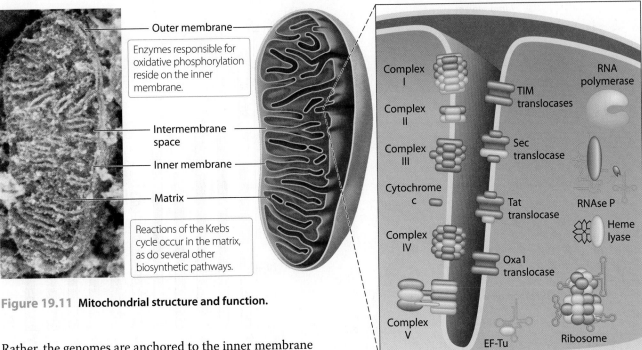

- Outer membrane

Enzymes responsible for oxidative phosphorylation reside on the inner membrane.

- Intermembrane space
- Inner membrane
- Matrix

Reactions of the Krebs cycle occur in the matrix, as do several other biosynthetic pathways.

**Figure 19.11  Mitochondrial structure and function.**

Complex I
Complex II
Complex III
Cytochrome c
Complex IV
Complex V

TIM translocases
Sec translocase
Tat translocase
Oxa1 translocase
EF-Tu

RNA polymerase
RNAse P
Heme lyase
Ribosome

Ribosomal RNA and a few proteins (blue) are always encoded by the mitochondrial genome, other products (purple) are always encoded by the nuclear genome, and still others (orange) may be encoded by either genome depending on the species.

Rather, the genomes are anchored to the inner membrane of the mitochondria, in a manner similar to that of bacterial chromosomes. These and other features described below give clues to the evolutionary origin of mitochondria, as we discuss further in a later part of this chapter.

**(a)**

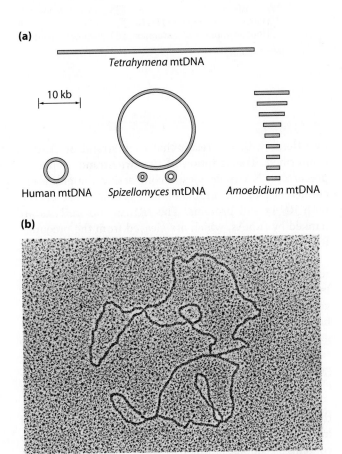

Tetrahymena mtDNA

10 kb

Human mtDNA   Spizellomyces mtDNA   Amoebidium mtDNA

**(b)**

**Figure 19.12  Genome structure of mitochondrial genomes.**

The gene content and size of mitochondrial genomes vary substantially among eukaryotes (Figure 19.13a). Genome sizes range from a low of 6 kb in the malarial parasite *Plasmodium* to hundreds or thousands of kilobases in flowering plants. However, as with nuclear genomes, the size in kilobases does not necessarily correlate with the number of genes. For example, the *Saccharomyces* mitochondrial genome is approximately five times as large as the human mitochondrial genome, but it contains only a few more genes. This is because much of the extra DNA, including some introns, is noncoding. In contrast to their nuclear genomes, mammalian mitochondrial genomes are particularly compact and have no introns and little noncoding DNA. Known gene numbers in mitochondrial genomes vary from a low of 6 in *Plasmodium* to a high of nearly 100 genes in certain jakobid flagellates such as *Reclinomonas americana* (Figure 19.13b).

As we discuss in a later section, all mitochondrial genomes are descended from a common bacterial ancestral genome that likely possessed thousands of genes. The differences between mitochondrial genomes reflect differential losses of genes from the ancestral genome in the different lineages. Gene losses in parasites—such as *Plasmodium*, which obtains its energy from its hosts—are often extreme, owing to loss of the genes encoding proteins required for oxidative phosphorylation.

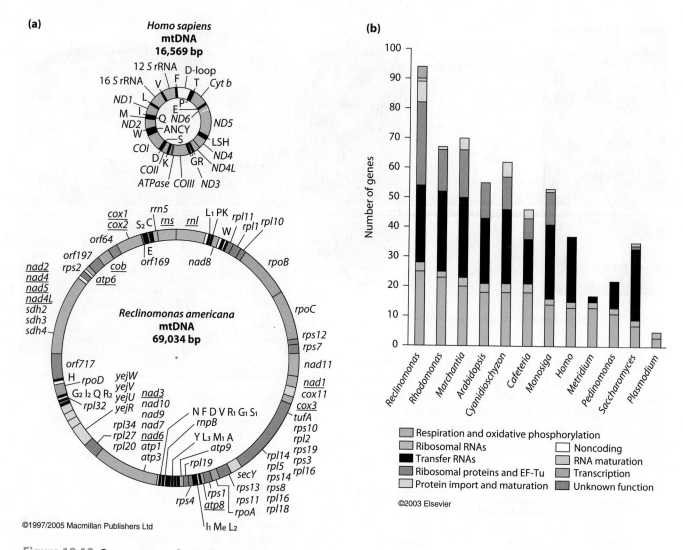

**Figure 19.13 Gene content of mitochondrial genomes.**

---

Genetic Insight Mitochondrial genomes encode genes whose products interact with nuclear gene products to carry out oxidative phosphorylation, which produces ATP to meet cellular energy requirements. The mitochondrial genomes of eukaryotic species vary widely in genetic content and structure.

---

## Mitochondrial Transcription and Translation

The mitochondrial genome is transcribed by an RNA polymerase similar to that found in bacteria. In some species, the mitochondrial RNA polymerase is encoded by a mitochondrial gene; in other species, it is encoded by a nuclear gene. Transcriptional regulation of mitochondrial gene expression also varies among species but in most cases has features reminiscent of bacterial operons. For example, transcription of the mammalian mitochondrial genome involves the production of just three polycistronic mRNA transcripts from only three promoters (Figure 19.14). All promoters are within the mitochondrial control region and they promote transcription in opposite directions, with the result that each strand of DNA is transcribed. Transcription of the two strands generates precursor RNA molecules encompassing the entire circumference of the mitochondrial genome that encode both RNAs and proteins. The rRNAs and mRNAs are flanked by tRNAs, which are cleaved from the precursor RNAs, thus releasing the rRNA and mRNA molecules.

Mitochondrial translation occurs on ribosomes that resemble bacterial ribosomes. The rRNAs utilized in mitochondria are always encoded by the mitochondrial genome, but the ribosomal proteins in mitochondria may be encoded by either the mitochondrial or nuclear genome. In *Reclinomonas americana*, Shine–Dalgarno sequences are present upstream of most protein-coding genes, but such sequences are not evident in the mitochondrial genes of most eukaryotes.

Most mitochondrial genomes encode many fewer than the 61 different tRNA genes that are theoretically required for translation of all codons. Recall that the genetic code contains 64 codons, of which 61 encode amino acids during translation. Each codon can be uniquely recognized by a

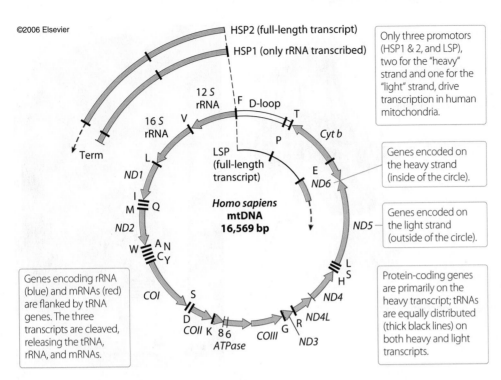

©2006 Elsevier

Only three promotors (HSP1 & 2, and LSP), two for the "heavy" strand and one for the "light" strand, drive transcription in human mitochondria.

Genes encoded on the heavy strand (inside of the circle).

Genes encoded on the light strand (outside of the circle).

Genes encoding rRNA (blue) and mRNAs (red) are flanked by tRNA genes. The three transcripts are cleaved, releasing the tRNA, rRNA, and mRNAs.

Protein-coding genes are primarily on the heavy transcript; tRNAs are equally distributed (thick black lines) on both heavy and light transcripts.

**Figure 19.14 Human mitochondrial transcription.**

complementary anticodon sequence in tRNA, but third-base wobble and the redundancy of the genetic code permit genomes to carry fewer than 61 unique tRNA genes. Consequently, only 32 different tRNA anticodon sequences (i.e., 32 different tRNA genes) are required to recognize the 61 codons.

The substantially lower number of unique tRNA genes in mitochondrial genomes compared to the number of codons is accommodated in different ways in the mitochondria of different species. In mammalian mitochondria, the rules of third-base wobble are more lenient than they are for nuclear genes. Certain mammalian tRNAs can read codons with any of the four bases in the third position, a system that reduces the number of different tRNA genes needed in mammalian mitochondria to 22.

In some mammalian species, not all mitochondrial tRNAs are encoded in the mitochondrial genome; instead, some nuclear-encoded tRNAs are imported into mitochondria. In extreme cases, such as *Plasmodium*, all tRNAs have to be imported since none are encoded in the mitochondrial genome. In addition to mechanisms that reduce the total number of different tRNA genes encoded in mitochondria, there are differences between the mitochondrial genetic codes of certain animals, plants, and fungi (Table 19.1). In many species, the mitochondrial genetic code is the same as the universal code, thus supporting the hypothesis that most of the changes listed in Table 19.1 occurred relatively late in the evolution of the major branches of eukaryotes. Some of the same differences have apparently evolved independently in multiple mitochondrial lineages, suggesting that certain changes may confer a selective advantage. It may be that the reduction in tRNA gene number in the mitochondrial genome allowed this evolution of the mitochondrial genetic code.

| Table 19.1 | Examples of Differences in the Mitochondrial Genetic Code | | | | | | |
|---|---|---|---|---|---|---|---|
| **Codon** | **Universal** | **Mitochondrial** | | | | | |
| | | Vertebrate | Echinoderms | *Saccharomyces* (Yeast) | *Chondrus* (Red Algae) | Land Plants | Ciliates |
| UGA | Stop | Trp | Trp | Trp | Trp | — | Trp |
| AUA | Ile | Met | — | Met | — | — | — |
| CUN | Leu | — | — | Thr | — | — | — |
| AGG, AGA | Arg | Ser/Stop/? | Ser | — | — | — | — |
| CGN | Arg | — | — | ? | — | — | — |

N, any of the four bases A, G, U, C; —, no change from the universal code.

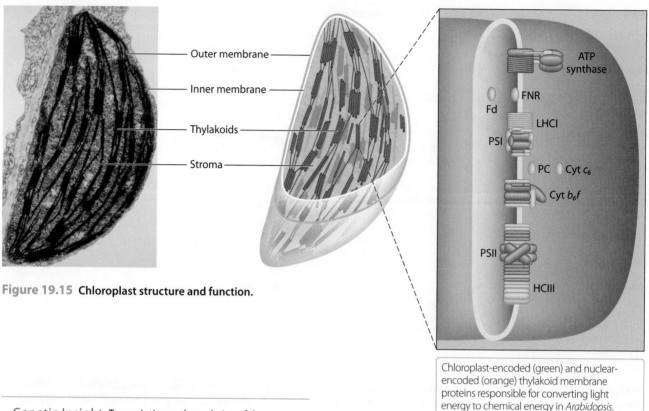

**Figure 19.15  Chloroplast structure and function.**

Chloroplast-encoded (green) and nuclear-encoded (orange) thylakoid membrane proteins responsible for converting light energy to chemical energy in *Arabidopsis*.

---

**Genetic Insight**  Transcription and translation of the mitochondrial genome are similar to those of bacteria. In some species, the genetic code of mitochondria deviates slightly from the universal code.

---

## 19.4  Chloroplasts Are the Sites of Photosynthesis

Chloroplasts—present in green plants, their green-algal relatives, and many other taxa that carry out photosynthesis—are only the most familiar of various organelles derived from a precursor organelle called a **plastid.** In the green tissues of plants, plastids differentiate into chloroplasts in response to light; but in nongreen tissues, plastids may differentiate into other types of specialized organelles. For example, tomatoes get their red color from pigments in a plastid derivative called a chromoplast. Regardless of type, all plastids and their derivatives in a given individual possess the same genome.

Chloroplasts resemble mitochondria in being enclosed by a double-membrane system (**Figure 19.15**). However, chloroplasts also possess a third membrane system, the thylakoid membranes. These membranes reside in the stroma, the region equivalent to the matrix of the mitochondrion. The protein complexes that carry out photosynthetic reactions are embedded in the thylakoid membranes. As with mitochondria, most chloroplast proteins are encoded in the nuclear genome but are produced and regulated through interactions between the two genomes (plastid and nuclear).

## Chloroplast Genome Structure and Gene Content

Many structural features of chloroplast genomes are similar to those of bacterial and mitochondrial genomes. For example, the chloroplast genome is anchored to the inner chloroplast membrane, and chloroplast genomes are not packaged in chromatin composed of histones. Like mitochondrial genomes, chloroplast genomes are generally found to be circular, on the basis of genetic and molecular mapping as well as direct observation with the electron microscope. However, there is evidence that linear chloroplast genomes may also occur. The similarity of chloroplast genomes and bacterial genomes reflects the ancestral evolutionary relationship that we explore in a later section.

Compared to mitochondrial genomes, chloroplast genomes are structurally less diverse. Chloroplast genomes range in size from 120 to 200 kb and usually encode 100 to 250 genes; the precise gene content varies between species. The chloroplast genome of *Marchantia polymorpha* is typical of many (**Figure 19.16a–b**). While chloroplast ribosomal proteins may be encoded by either the chloroplast or nuclear genomes, the rRNA is always encoded by the chloroplast genome, and the tRNA molecules are usually encoded by the chloroplast genome. Most of the remaining chloroplast genes with known functions encode proteins involved in photosynthesis.

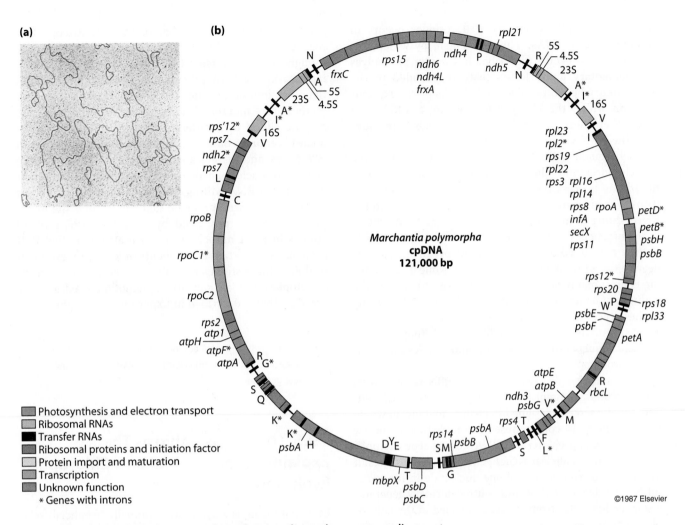

**Figure 19.16 Chloroplast genome of *Marchantia polymorpha*, a common liverwort.**

One of the photosynthetic genes in the chloroplast genome encodes the large subunit of ribulose-1, 5-bisphosphate carboxylase oxygenase, the enzyme responsible for the fixation of carbon from $CO_2$. The enzyme, often abbreviated RuBisCo, represents up to 50% of the protein content of green plants and is thus possibly the most abundant protein on the planet. RuBisCo is composed of two protein subunits, abbreviated rbcL and rbcS, for the large and small subunit, respectively. While rbcL is encoded in the chloroplast genome, rbcS is encoded in the nuclear genome, providing another example of the extensive coordination between the two genomes, which in this case must cooperate to produce appropriate quantities of the two subunits.

## Chloroplast Transcription and Translation

Transcription and translation of chloroplast genes are similar to those of bacteria. Many chloroplast genes are arranged in operons and as a result are coordinately transcribed. The RNA polymerase resembles that found in bacteria and, as in bacteria, recognizes consensus sequences (similar to those of bacterial promoters) at −10 and −35 of chloroplast gene promoters. Like bacterial mRNAs, chloroplast mRNAs are neither capped at their 5′ end nor polyadenylated at their 3′ end. However, some RNA processing occurs, such as the removal of introns from a few genes and RNA editing in most land plants (a process described in more detail later).

The ribosomes of chloroplasts are also similar to those of bacteria. For example, ribosome function is disrupted by aminoglycoside antibiotics, which also inhibit bacterial ribosome function. From 30 to 35 different tRNAs are usually encoded by the chloroplast genome, and as a result all codons can be translated without the additional wobble found in mitochondria. The kinds of deviations from the universal genetic code that are seen in mitochondrial genes are not observed in chloroplasts.

## Editing of Chloroplast mRNA

RNA editing is the process of altering the sequence of an RNA molecule after transcription from the DNA genome (Section 8.4). RNA editing was first discovered in the mitochondria of trypanosomes, where insertion (or, less frequently, deletion) of U residues occurs in mitochondrial mRNAs. The mechanism by which this editing process occurs (described in Chapter 8) involves complementary guide RNAs that are encoded in the mitochondrial genome. The guide RNAs provide a template on which the changes to the target mRNA are made; there, enzymes either add or delete U residues from the mRNA.

Subsequently, RNA editing has been noted in the mitochondria and chloroplasts of land plants, where the editing process results in C-to-U (or, less frequently, U-to-C) changes in organellar mRNAs. In contrast to the RNA editing to insert and delete bases, the RNA editing in the organelles of plants does not utilize a guide RNA. Rather, C-to-U editing is performed by an enzyme, C deaminase, which converts the C to a U, while U-to-C editing is presumably performed by the reverse reaction, the addition of an amine group to the U. Proper RNA editing in these cases requires the presence of certain sequences adjacent to the sites to be edited, suggesting that the adjacent sequences may represent binding sites for trans-acting proteins.

Not surprisingly, given that the mRNAs of several genes encoding proteins involved in photosynthesis are edited, genetic screens designed to identify mutants in which photosynthesis is compromised have identified nuclear genes controlling chloroplast RNA editing. For example, mutations in the nuclear *CCR4* gene of *Arabidopsis* result in a loss of U-to-C editing of one nucleotide in the *ndhD* mRNA within chloroplasts; this editing normally generates a start codon, AUG, from the ACG encoded in the chloroplast genome (**Figure 19.17**).

*CCR4* encodes a member of the pentatricopeptide repeat (PPR) family of proteins. These proteins are thought to play diverse roles in RNA processing, including cleavage of RNA precursor molecules. Surprisingly, the other four edited sites in *ndhD* RNA are edited correctly in *ccr4* mutants, leading to the idea that each site may be edited by a different trans-acting protein. The nuclear genomes of land plants encode large numbers of *PPR* genes, and there is a strong correlation between the number of nuclear-encoded PPR proteins and the extent of organellar RNA editing. While the mechanistic details are not yet known, it may be that each edited site in organellar RNA is processed by a distinct PPR protein! Studies in plant mitochondria have also identified PPR proteins as important components of RNA processing; in so doing, these studies have illuminated the mechanism of cytoplasmic male sterility, a phenotype used in plant breeding that is described in Experimental Insight 19.1.

---

**Genetic Insight** RNA editing of plant organelle mRNAs involves complex interactions with proteins encoded in the nuclear genome.

---

## 19.5 The Endosymbiosis Theory Explains Mitochondrial and Chloroplast Evolution

**Endosymbiosis** is a symbiotic (mutually beneficial) relationship between organisms in which one organism inhabits the body of the other. Several lines of evidence indicate that the mitochondria and chloroplasts inhabiting modern animal and plant cells are the descendants of formerly free-living bacteria that took part in ancient infections of eukaryotic cells. These ancient invaders established endosymbiotic relationships with their hosts and have evolved along with their hosts to produce the diversity we observe in organelles today. In this discussion we explore the principal lines of evidence supporting the **endosymbiosis theory** of mitochondria and chloroplast evolution, including the following evidence:

- The double-membrane system found in both organelles is derived from a similar membrane system found in bacteria.
- The organelles are similar in size to extant bacteria.
- Organelle DNA is packaged in a manner similar to the packaging of chromosomes in bacteria and dissimilar to that of DNA in the nuclear genome.
- The transcriptional and translational machinery of the organelles closely resembles that of bacteria.
- The protein-coding sequences of organelle genes are more like those of bacteria than like either the nuclear genes of eukaryotes or the sequences of archaea.

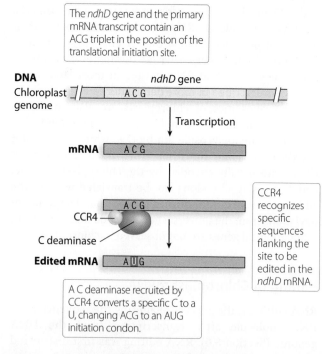

**Figure 19.17 A model for C-to-U RNA editing.**

## Separate Evolution of Mitochondria and Chloroplasts

The available genetic evidence indicates that mitochondria are monophyletic; that is, all mitochondria descended from a single common ancestor. Coupled with evidence that mitochondria bear strong similarities to bacteria, this finding suggests that the point of origin of all mitochondria was a single endosymbiotic event (Figure 19.18).

Based on the fossil record, the minimum age of the eukaryotes is approximately 1.5 billion years. One hypothesis concerning the origin of eukaryotes is that

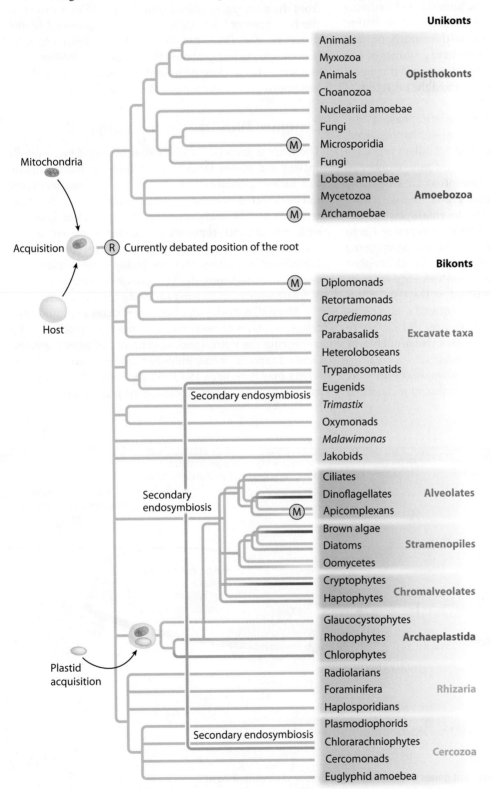

**Figure 19.18** The evolutionary history of the mitochondrion and the chloroplast.

they evolved from an anaerobic ancestor that acquired an aerobic **endosymbiont** (the mitochondrial ancestor). This event was perhaps linked with the global rise in atmospheric oxygen that began about 2 billion years ago and that could have provided a selective environment for aerobic organisms. Based on similarity in gene sequences, the closest extant relatives of mitochondria are free-living **α-proteobacteria,** although the precise lineage within the α-proteobacteria that gave rise to the mitochondria is unresolved. Extant α-proteobacteria have genomes of 4 to 9 Mb of DNA encoding 4000 to 9000 genes, so it appears that extensive gene loss has characterized the evolution of mitochondrial genomes.

Chloroplasts are also monophyletic, having descended from a single endosymbiotic event that occurred, according to the fossil record, at least 1.2 billion years ago (see Figure 19.18). Based on similarity of gene sequences, the closest extant relatives of chloroplasts are free-living **cyanobacteria,** although again the precise lineage within the cyanobacteria that gave rise to the chloroplasts is unresolved. Existing cyanobacteria have genomes of 1.6 to 9.0 Mb of DNA encoding 1900 to 7400 genes, implying extensive gene loss in the evolution of the chloroplast genome as well. Phylogenetic evidence also suggests multiple secondary symbioses (discussed at the end of this section) in which some eukaryotes acquired a photosynthetic eukaryotic symbiont (see Figure 19.18). These events resulted in the horizontal transmission of chloroplasts among unrelated eukaryotic lineages.

Two fundamental questions arise when we consider the genomes of the organelles. First, given that mitochondrial and chloroplast genomes contain from 6 to 100 and from 20 to 200 genes, respectively, what happened to all the other genes of the ancestral symbiont? Second, given that the organelles contain many more organellar proteins than genes, what is the origin of the nuclear genes that encode so many organellar proteins? Are those nuclear genes derived from the ancestral symbiont genome, or did they evolve in the host genome? A possible answer was provided by the discovery that DNA can be transferred from organelle genomes to nuclear genomes; this led to the hypothesis that genes have been relocated from the ancestral endosymbiont genome to the nuclear genome during evolution.

## Continual DNA Transfer from Organelles

The nuclear genomes of eukaryotes bear evidence of both ancient and recent DNA transfer between the organellar and nuclear genomes (**Figure 19.19**). Ancient transfer events can be detected by comparative genomics of mitochondrial genomes and by comparing eukaryotic nuclear genomes with prokaryotic genomes. Sequencing of eukaryotic genomes has also revealed evidence of recent transfers. Transferred sequences that are highly similar must have been transferred recently. The extent of similarity would decrease with the increasing age of the transfer event.

Since all mitochondrial genomes share a common ancestor, transfers of organelle genomes can be detected by comparing the mitochondrial genomes of extant species. Genes found in some mitochondrial genomes but not others may have been lost from the organelle genome by transfer to the nucleus. Ancient gene transfers can also be

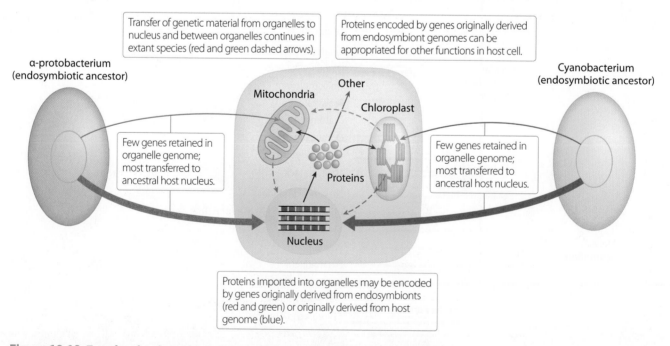

**Figure 19.19  Transfer of endosymbiont genes to the nuclear genome and destinations of encoded protein products.**

identified in comparisons between nuclear genomes of eukaryotes and the genomes of extant α-proteobacteria and cyanobacteria. Nuclear genes that are most similar to the genes of the living bacterial species are likely to have been derived from the bacterial endosymbiont. For example, comparisons between the *Arabidopsis* nuclear genome and genomes of three cyanobacteria suggest approximately 4300 *Arabidopsis* nuclear genes have a cyanobacterial origin. Thus a reasonable fraction of the *Arabidopsis* nuclear genome represents an acquisition of genetic information originally residing in the genome of the chloroplast (Figure 19.20). Similarly, comparisons between several eukaryotic nuclear genomes and those of α-proteobacteria detected at least 630 nuclear genes derived from the α-proteobacteria endosymbiont that gave rise to the mitochondrion. Thus, concomitant with the reduction in the organellar genomes is an increase in gene content in the nuclear genome. The importance of

this enormous amount of additional genetic information in the evolution of the eukaryotic lineage is difficult to overestimate (see Figure 19.19).

One surprise discovered through the analysis of eukaryotic genome sequences is that recent transfers of mitochondrial and chloroplast sequences seem to be included in all nuclear genomes. Mitochondrial DNA sequences found in the nucleus have been termed **nuclear mitochondrial sequences (NUMTS)**, while nuclear sequences derived from plastid genomes are called **nuclear plastid sequences (NUPTS)**. A surprising amount of organellar DNA sequence has been found in the nuclear genome of every organism examined. NUMTS and NUPTS are common in many plant species; the *Arabidopsis* genome contains 17 NUPTS, totaling 11 kb, and 14 NUMTS, one of which is 620 kb and represents almost two entire mitochondrial genomes. The human genome contains hundreds of NUMTS, ranging from 106 to 14,654 bp long (the latter being 90% of the length of the mitochondrial genome).

Three conclusions can be made from the observation of NUMTS and NUPTS. First, given the level of sequence similarity between NUMTS or NUPTS and the respective organelle genome sequences, most are thought to represent evolutionarily recent transfers of organelle DNA to the nuclear genome. Second, entire organelle genomes likely were transferred to the nuclear genome multiple times in evolutionary history. Third, the process is ongoing; DNA continues to move between the organelles and to the nucleus. While the rate of transfer is not known in most organisms, experiments to directly measure the rate of DNA transfer from chloroplast to nuclear genome in plants revealed a new integration of chloroplast DNA in the nuclear genome at a rate of 1 in 16,000 plants. This surprisingly high rate of DNA transfer between the organellar and nuclear genomes can account for the large numbers of evolutionarily recent insertions of organellar DNA (NUMTS and NUPTS) found in the nuclear genome of most organisms.

Although organelle genes are readily transferred into the nuclear genome, several criteria must be met for the transferred genes to be functional. Recall from Chapters 14 and 15 that the details of gene regulation differ between prokaryotes and eukaryotes. Since gene regulation in the organelles resembles that in bacteria, transferred genes must acquire sequences for proper transcriptional regulation in the nucleus. Researchers using an experimental system similar to the one for monitoring DNA transfer from chloroplast to nuclear genome in plants have demonstrated that transferred chloroplast genes can become functional nuclear genes at a reasonable frequency. In addition, as described in more detail later, the protein encoded by the transferred gene may be transported back to the organelle from which the gene was derived; or, alternatively, the protein may be transported to another cellular compartment. For the protein to be

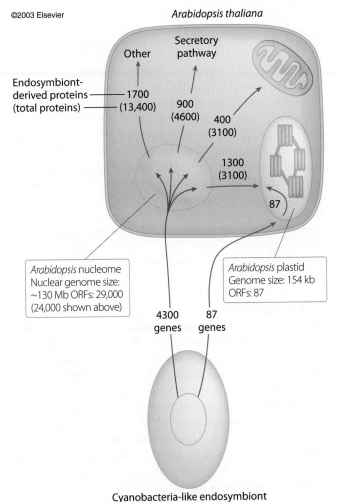

©2003 Elsevier

*Arabidopsis thaliana*

Other

Secretory pathway

Endosymbiont-derived proteins (total proteins) — 1700 (13,400)

900 (4600)

400 (3100)

1300 (3100)

87

*Arabidopsis* nucleome
Nuclear genome size: ~130 Mb ORFs: 29,000 (24,000 shown above)

*Arabidopsis* plastid
Genome size: 154 kb
ORFs: 87

4300 genes     87 genes

Cyanobacteria-like endosymbiont

| | Prochlorococcus | Synechocystis | Nostoc |
|---|---|---|---|
| Genome size (Mb): | 1.66 | 6.3 | 9.02 |
| ORFs: | 1694 | 3168 | 7281 |

**Figure 19.20 Origins of genes for organellar proteins.**

transported back to the organelle, an amino terminal signal sequence must be attached to it. Since signal sequences need only to have certain general structural features in order to function properly, the acquisition of functional signal sequences likely occurs at an appreciable frequency.

---

Genetic Insight  The mitochondrion and chloroplast are derived from two independent ancient endosymbiotic events. Most of the genes of the ancestral endosymbionts have been transferred to the nuclear genomes.

---

## Encoding of Organellar Proteins

Organelles contain many more proteins than they encode in their genomes; this is an indication that most organellar proteins are encoded in the nuclear genome. For example, the yeast mitochondrion contains approximately 400 proteins, but only 16 proteins are encoded in its mitochondrial genome. The nuclear-encoded organellar proteins are translated in the cytoplasm and then imported into the organelles. These organellar proteins are targeted to their final location by signal sequences, ranging from 15 to 25 amino acids in length, at the amino terminal end of the proteins. Different signal sequences label proteins for transport to different organelles and other locations, such as the outer membrane, intermembrane space, inner membrane, matrix, and stroma and thylakoid membrane systems.

When the endosymbiotic theory of the origin of mitochondria and chloroplasts was first proposed, its framers predicted that proteins were always targeted to the cell compartment from which the genes encoding them were originally derived. In other words, if a protein was encoded by a nuclear gene that had originally been derived from the endosymbiont that gave rise to the mitochondrion, the protein would be targeted back to the mitochondrion. Contrary to expectations, however, the relationships between the endosymbiont origins of genes and the final destination of gene products are complex and difficult to predict. For example, in *Arabidopsis*, less than half the proteins identified as coming from the cyanobacterial endosymbiont are found to be targeted to the chloroplast (see Figure 19.20). Conversely, a number of proteins targeted to the chloroplast were not acquired from the cyanobacterial symbiont, but rather are descended from the original eukaryotic host genome. Similar observations have been made concerning the mitochondrion. Thus the proteins encoded by nuclear genes originally derived from endosymbiont genomes may be targeted to any location in the cell.

While the diversity in the direction of protein transport was initially unexpected, perhaps consideration of the early stages of endosymbioses should have led scientists to expect it. When an endosymbiotic relationship was initially established, the genome of the ancestral

mitochondrion would have been similar in size to that of its prokaryotic ancestors. If the rate of DNA transfer was similar to that measured today, the nuclear genome must have experienced a bombardment of DNA from the endosymbiont. Before the evolution of the mitochondrial protein-import machinery, proteins produced by genes transferred to the nuclear genome had to remain in the cytoplasm or be transported to the plasma membrane. Reduction in the endosymbiont genome could occur only after the evolution of systems able to import proteins into the endosymbiont. Such systems were composed of proteins encoded by genes originally derived from both the nuclear and endosymbiont genomes.

---

Genetic Insight  Both the organellar and nuclear genomes encode organellar proteins. The host and endosymbiont ancestral genomes alike were the original sources of both organellar and nonorganellar proteins.

---

## The Origin of the Eukaryotic Lineage

The tree of life is often depicted as having three major branches—the Bacteria, the Archaea, and the Eukarya—based on comparison of sequences of the rRNA genes (see Chapter 1). The extensive gene flow from the bacterial endosymbiont to the nucleus, however, has resulted in the presence of significant numbers of "bacterial" genes in the nuclear genomes of eukaryotes. Given this situation, a simple tripartite view of life, in which three branches diverge from a single common ancestor, is overly simplistic. A fraction of the nuclear genome of every eukaryote is derived from bacterial endosymbionts, but where were all the remaining genes derived from? In other words, what was the original host of the α-proteobacterium that gave rise to the eukaryotes?

Two models have been proposed to answer this question. In one model, the original host is a cell described as having a nucleus but no mitochondria and as subsequently acquiring an α-proteobacterium as an endosymbiont. In this model, "eukaryotic" cells (cells having nuclei) existed before the endosymbiotic event, suggesting that such organisms lacking mitochondria might still exist. In the second model, the original host is a prokaryotic cell that acquires an α-proteobacterium as an endosymbiont; and subsequently, this host–endosymbiont system evolves other eukaryotic features, such as a nuclear membrane. If the latter model is correct, no intermediate eukaryotes lacking mitochondria should be found.

Two recent discoveries have contributed new fuel to this discussion. First, eukaryotic organisms that were originally thought to lack mitochondria, such as *Giardia intestinalis* (which causes diarrhea when it infects the human intestine), are now known to have mitochondria. In the case of *Giardia*, the mitochondria are reduced to double-membrane-bound structures called **mitosomes.** Mitosomes lack a genome, but proteins requiring an anaerobic environment to function are imported into them (see Figure 19.18). The nuclear genome of *Giardia* harbors genes of mitochondrial origin; this finding indicates that all of the mitochondrial genome was either transferred to the nucleus or lost. The extreme reduction of the mitochondrion to nothing but an anaerobic compartment allowing the cell to carry out specific reactions is likely a consequence of *Giardia's* parasitic lifestyle, in which all of its energy is derived from a host organism. This finding means that all known existing eukaryotes harbor mitochondria.

The second discovery concerns the nature of the genes in the nuclear genomes of eukaryotic organisms. Comparison of the complete genome sequences of the eukaryote *Saccharomyces cerevisiae* with two bacteria (*Escherichia coli* and *Synechocystis 6803*) and an archaea (*Methanococcus jannaschii*) revealed two general functional and evolutionary categories into which the yeast genes could be divided. One category of genes, called **informational genes,** encodes protein products that perform informational processes in the cell such as DNA replication, packaging of chromosomes, transcription, and translation. The informational genes of yeast resemble those found in *Methanococcus,* and this resemblance includes a similarity between the histones of the yeast and the histone-like chromatin proteins present in Archaea. The second category of genes, called **operational genes,**

encode proteins involved in cellular metabolic processes, such as amino acid biosynthesis, biosynthesis of cofactors, fatty acid and phospholipid biosynthesis, intermediary metabolism, energy metabolism, nucleotide biosynthesis, and some of the regulatory functions. While most yeast operational genes resemble those of Bacteria, some are more closely related to archaeal genes.

One scenario consistent with the apparent origins of informational and operational genes in yeast is that the original host cell of the α-proteobacterial endosymbiont was related to an archaeal cell (**Figure 19.21**). The original host genome would have contained both informational and operational genes, as would the α-proteobacterial endosymbiont. Over time, while both genomes retained their own informational genes, many endosymbiont operational genes were transferred to the nuclear genome and often replaced their functional equivalents. Unlike the cases of the mitochondria and chloroplasts, where the endosymbionts can be traced to specific lineages of Bacteria, the putative host may have been unrelated to any specific lineage of extant Archaea.

## Secondary and Tertiary Endosymbioses

The melding together of genomes did not happen only during the endosymbioses that formed mitochondria and chloroplasts. **Secondary** and even **tertiary endosymbiotic events** have occurred between different lineages of eukaryotes, resulting in the dispersal of plastids into eukaryotic

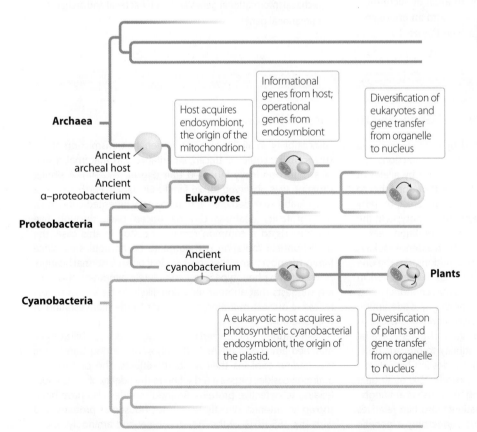

**Figure 19.21  One hypothesis for the evolution of the eukaryotes.**

Archaea

Ancient archeal host

Ancient α–proteobacterium

Proteobacteria

Cyanobacteria

Ancient cyanobacterium

**Eukaryotes**

**Plants**

Host acquires endosymbiont, the origin of the mitochondrion.

Informational genes from host; operational genes from endosymbiont

Diversification of eukaryotes and gene transfer from organelle to nucleus

A eukaryotic host acquires a photosynthetic cyanobacterial endosymbiont, the origin of the plastid.

Diversification of plants and gene transfer from organelle to nucleus lineages that are distantly related (see Figure 19.18). Typically, a non-photosynthetic eukaryote acquires a red or green algal endosymbiont. What happens to the nuclear genome of the secondary endosymbiont when one eukaryote envelops another eukaryote? Genes of the nuclear genome of the eukaryotic endosymbiont (the alga), whose products were targeted to the plastid, are translocated to the host nucleus in an occurrence analogous to the movement of genes from the organelle genomes to the primary endosymbiont host nuclear genome. Thus the nuclear genome of the algal endosymbiont, termed the **nucleomorph,** undergoes reduction to the extent that it encodes only some genes for products targeted to the plastid as well as some genes required for the maintenance of the nucleomorph genome. Thus the plastid is serviced by three different genomes (nuclear, nucleomorph, and plastid), and the nuclear genome of photosynthetic secondary endosymbionts is a mixture of four genomes (mitochondrial, chloroplast, and two nuclear genomes). Because secondary and tertiary endosymbioses have occurred many times during the evolution of eukaryotes, the mixing and coevolution of genomes seems to have been instrumental in shaping the evolution of life.

The mixing and melding of genomes can sometimes result in biological anomalies. For example, the discovery of a reduced chloroplast (or apicoplast) in *Plasmodium falciparum*, the malarial parasite, came as quite a surprise because this is clearly not a photosynthetic organism. *Plasmodium* resides within the phylum Apicomplexa, which would make it a descendant of an ancient secondary endosymbiosis involving a host eukaryote and an endosymbiotic chloroplast-containing red alga (see Figure 19.18). Is there a reason that *Plasmodium,* with its parasitic lifestyle, might have retained the apicoplast and its accompanying genome, albeit without any genes encoding proteins involved in photosynthesis?

One hypothesis explaining retention of the apicoplast in *Plasmodium* is based on differences in translation of organellar-encoded compared to nuclear-encoded genes. The initiator tRNA used in mitochondrial translation is a formylmethionyl-tRNA (tRNA$^{fMET}$), the same as used in bacteria. This special tRNA cannot be imported from the cytoplasm, since cytosolic translation in eukaryotes uses an initiator methionyl-tRNA that is not formylated. During the evolutionary history of *Plasmodium,* the gene encoding the enzyme that adds a formyl group to the methionyl-tRNA has been lost from the mitochondrial genome. It is thought that the tRNA$^{fMET}$ used in mitochondria might be imported from the apicoplast since the only methionyl-tRNA formyl transferase gene in *Plasmodium* is in the nuclear genome and the protein product of this gene is transported to the apicoplast. According to this hypothesis, the apicoplast may be maintained for the sole purpose of synthesizing tRNA$^{fMET}$ to be imported into the mitochondrion—a quirk of the evolutionary history of *Plasmodium.*

---

**Genetic Insight**  Eukaryotes may have descended from a symbiosis between an archaeal cell and a bacterial endosymbiont. Consequently, the nuclear genome contains a mixture of archaeal informational genes and both bacterial and archaeal operational genes.

---

## CASE STUDY

### Ototoxic Deafness: A Mitochondrial Gene–Environment Interaction

Phenotypic penetrance can be affected by both genetic and environmental factors. In the case of genetic interactions, the phenotypic effects of a mutation are influenced by alleles at other loci. The gene products of other loci are thought either to exacerbate or compensate for the mutational defect, thereby altering the expressivity or penetrance of the phenotype. In the case of environmental interactions, certain conditions either mitigate or enhance the phenotypic effects, in essence making the mutation a conditional allele. Some mutations, like the one described here, are subject to both these kinds of interaction. In this case, the locus of the key mutation is a mitochondrial gene.

A rare complication of the use of aminoglycoside antibiotics, such as streptomycin, gentamicin, and kanamycin, is irreversible loss of hearing, termed *ototoxic deafness*. Several observations point to a genetic susceptibility to ototoxic deafness. Due to pervasive use of aminoglycosides in China, it was reported that in a district of Shanghai, nearly 25% of all deaf individuals can trace their loss of hearing to the use of aminoglycosides. Nearly one-fourth of these patients also had relatives suffering from ototoxic deafness, thus suggesting a genetic

susceptibility. In all 22 cases where genetic transmission of the susceptibility could be traced, inheritance was maternal, a sign of a mitochondrially inherited trait (**Figure 19.22a**). A similar situation was observed for 26 families in Japan. Furthermore, a large Arab-Israeli pedigree with *maternally* inherited congenital (not ototoxic) deafness can be traced back through five generations to a common female ancestor. In this case, the mitochondrial mutation is thought to be homoplasmic since family members are either severely deaf or have normal hearing. However, the phenotype is not completely penetrant; this finding suggests that another mutation, likely to be an autosomal recessive nuclear mutation, contributes to the manifestation of the condition.

In studies on bacteria, aminoglycosides stabilize mismatched aminoacyl-tRNAs in the ribosome during translation; this finding explains their antibiotic effects. The presence of aminoglycosides causes a reduction in the fidelity of translation, leading to defective proteins. Aminoglycosides also have been shown to interact directly with both ribosomal proteins and with the 16S rRNA of the 70S ribosome; and aminoglycoside-

**(a)**

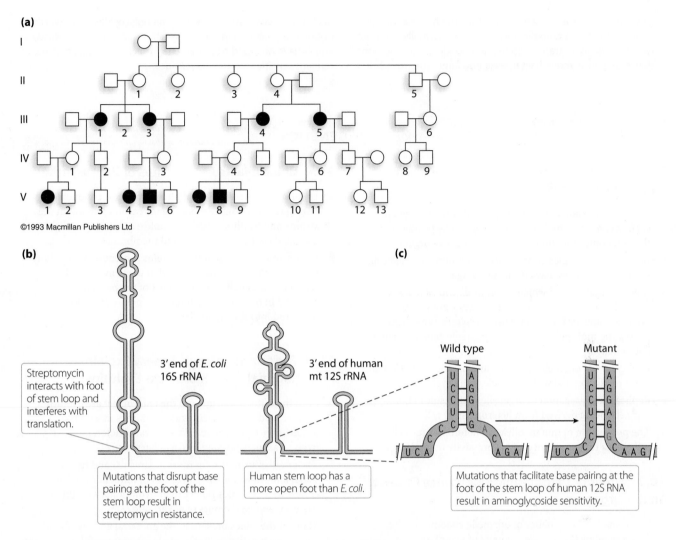

**(b)**

Streptomycin interacts with foot of stem loop and interferes with translation.

3' end of *E. coli* 16S rRNA

Mutations that disrupt base pairing at the foot of the stem loop result in streptomycin resistance.

3' end of human mt 12S rRNA

Human stem loop has a more open foot than *E. coli.*

**(c)**

Wild type

Mutant

Mutations that facilitate base pairing at the foot of the stem loop of human 12S RNA result in aminoglycoside sensitivity.

**Figure 19.22  Genetic and environmental interactions in ototoxic deafness.**

resistant bacteria have been shown to have point mutations in their 16S rRNA gene. Since the normal target of aminoglycosides is the evolutionarily related bacterial ribosome, the likely target of aminoglycoside ototoxicity in humans is mitochondrial ribosomes, and perhaps specifically the 12S rRNA that is homologous to the 16S rRNA of bacteria.

Sequencing of the mitochondrial 12S rRNA gene in individuals with congenital deafness in the Arab-Israeli family and in other individuals with ototoxic deafness revealed that they shared a single A-to-G mutation in their 12S rRNA genes (Figure 19.22b–c). The mutation lies at the foot of a stem loop conserved in bacteria, plants, and mammals. Studies on bacterial ribosomes have shown that this region of the 16S rRNA forms part of the aminoacyl site where mRNAs are decoded. Furthermore, aminoglycosides bind to this domain of the 16S rRNA, and bacterial mutants resistant to aminoglycosides map to this region of the 16S rRNA gene.

Thus the cause of the aminoglycoside-induced deafness is a mutation in the mitochondrial 12S rRNA gene, but three intriguing questions remain. First, why is deafness the primary, and perhaps only, phenotypic defect? A characteristic of many

mitochondrial diseases is pleiotropy due to a general loss of oxidative phosphorylation activity. However, in these cases of maternally inherited deafness or susceptibility to aminoglycosides, no obvious pleiotropic phenotypes are associated with the deafness. Is the cochlea especially susceptible to a loss of mitochondrial function? Are the cochlear mitochondria especially sensitive to aminoglycosides? Second, what is the nature of the autosomal recessive mutation that acts to enhance the effect of the 12S rRNA mutation in the Arab-Israeli family? Could it be a nuclear-encoded ribosomal protein gene that interacts with the mitochondrial 12S rRNA? And third, if our mitochondrial ribosomes are evolutionarily related to bacterial ribosomes, why are humans able to utilize aminoglycosides as antibiotics in the first place?

Clues to the answer of the third question have come from comparative studies of mitochondrial ribosome function. The mutation causing deafness creates an extension of base pairing by one base, in effect making the mitochondrial 12S rRNA more closely resemble the structure of the aminoglycoside-binding site of the bacterial 16S rRNA. Thus, in the 2 or so billion years since the separation of bacteria and mitochondria, the structure of the mitochondrial ribosome has changed just enough so that aminoglycosides do not normally interfere with the fidelity of translation in mitochondria; but mutations that result in a more bacteria-like ribosome structure bring back the ancient sensitivity to aminoglycosides. It is worth noting that—at least in this sense—translation in chloroplasts, which have diverged from bacteria for about 1.2 billion years, remains sensitive to aminoglycosides.

---

## SUMMARY

*For activities, animations, and review quizzes, go to the study area at www.masteringgenetics.com.*

### 19.1 Cytoplasmic Inheritance Transmits Genes Carried on Organelle Chromosomes

▌ Mitochondria and chloroplasts possess their own genomes, each encoding a small number of genes. The products of these genomes function within the respective organelle.

▌ Because many copies of organellar DNA occur in each cell, multiple genotypes may coexist in a single cell.

▌ Cells or organisms in which all genomic copies of an organelle gene have an identical sequence are said to be homoplasmic for that gene, whereas cells or organisms possessing multiple alleles for an organelle gene are called heteroplasmic.

▌ Replication of organelle genomes and organelle division are not directly coupled with the nuclear cell cycle.

▌ Replicative segregation of organelles can result in homoplasmic cells being derived from heteroplasmic cells.

▌ The proportion of mutant alleles in heteroplasmic cells influences the penetrance and expressivity of phenotypes.

### 19.2 Modes of Cytoplasmic Inheritance Depend on the Organism

▌ The transmission genetics of organelle genomes is often determined by the relative amounts of cytoplasm contributed by the parental gametes.

▌ Organelles are maternally inherited in mammals and many plant species, whereas in fungal species, mitochondria are often biparentally inherited. In some species, organelle inheritance is determined by alleles of a nuclear gene.

### 19.3 Mitochondria Are the Energy Factories of Eukaryotic Cells

▌ Mitochondria are the sites of energy production; the enzymes of oxidative phosphorylation are on the inner membrane.

▌ Mitochondrial mutations often have pleiotropic effects that reflect the role of mitochondria in energy production.

### 19.4 Chloroplasts Are the Sites of Photosynthesis

▌ Chloroplasts are the sites of photosynthesis, conducted by enzymatic reactions responsible for carbon fixation in the stroma and by photosystem complexes that convert light to chemical energy in the thylakoid membranes.

▌ Only a small fraction of the proteins present in a mitochondrion or chloroplast are encoded in the genome of the respective organelle; instead, most of the proteins are encoded in the nuclear genome and post-translationally imported into the organelles.

### 19.5 The Endosymbiosis Theory Explains Mitochondrial and Chloroplast Evolution

▌ Both the mitochondrion and the chloroplast are evolutionarily derived from ancient endosymbioses in which a bacterium (α-proteobacteria and cyanobacteria, respectively) was incorporated into a eukaryotic cell.

▌ The circular structure (in most organisms) and transcriptional and translational expression of mitochondrial and chloroplast genomes reflect their evolutionary origins as bacterial endosymbionts of eukaryotic cells.

▌ Many of the genes present in the ancestral endosymbiont have been transferred to the nuclear genome of the host cell and have contributed extensively to eukaryotic nuclear genome content.

▌ The process of DNA transfer from organelle genomes to the nuclear genome is ongoing, and recent transfers of organelle DNA into the nucleus can be detected in most, if not all, organisms.

▌ Genes transferred from the ancient endosymbiont genome to the host nuclear genome encode proteins that may be targeted to any compartments of the eukaryotic cell.

▌ Eukaryotic informational genes are related to archeal genes, thus suggesting that eukaryotes might be descended from an archaea-like cell that acquired a bacterial endosymbiont.

---

## KEYWORDS

α-proteobacteria *(p. 664)*
biparental inheritance *(p. 642)*
chloroplast *(p. 642)*
cyanobacteria *(p. 664)*

cytoplasmic inheritance *(p. 642)*
endosymbiosis *(p. 662)*
endosymbiosis theory *(p. 662)*
endosymbiont *(p. 664)*

heteroplasmic cell or organism (heteroplasmy) *(p. 644)*
homoplasmic cell or organism (homoplasmy) *(p. 644)*

PROBLEMS           Mastering**GENETICS**™          *For instructor-assigned tutorials and problems, go to www.masteringgenetics.com.*

## Chapter Concepts

*For answers to selected even-numbered problems, see Appendix: Answers.*

1. Reciprocal crosses of experimental animals or plants sometimes give different results in the F1. What are two possible genetic explanations? How would you distinguish between these two possibilities (i.e., what crosses would you perform, and what would the results tell you)?

2. How are some of the characteristics of the organelles (the mitochondria and chloroplasts) explained by their origin as ancient bacterial endosymbionts?

3. The human mitochondrial genome encodes only 22 tRNAs, but at least 32 tRNAs are needed for cytoplasmic translation. How are all codons in mitochondrial transcripts accommodated by only 22 tRNAs? The *Plasmodium* mitochondrial genome does not encode any tRNAs; how are genes of the *Plasmodium* mitochondrial genome translated?

4. What is the evidence that transfer of DNA from the organelles to the nucleus continues to occur?

5. Draw a graph depicting the relative amounts of nuclear DNA present in the different stages of the cell cycle (G1, S, G2, M). On the same graph, plot the amount of mitochondrial DNA present at each stage of the cell cycle.

6. What are the differences between the universal code and that found in the mitochondria of some species? Given that some changes ($UGA = stop \rightarrow TRP$) have occurred multiple independent times in evolution, can you think of any selective advantage to the mitochondrial code?

7. What is the evidence that the ancient mitochondrial and chloroplast endosymbionts are related to the α-proteobacteria and cyanobacteria, respectively?

8. Outline the steps required for a gene originally present in the endosymbiont genome to be transferred to the nuclear genome and be expressed, and for its product to be targeted back to the organelle of origin.

9. Consider the phylogenetic tree presented in Figure 19.18. How were the origins of secondary endosymbiosis in the brown algae determined?

10. Most large protein complexes in mitochondria and chloroplasts are composed both of proteins encoded in the organelle genome and proteins encoded in the nuclear genome. What complexities does this introduce for gene regulation (i.e., for ensuring that the appropriate relative numbers of the proteins in a complex are produced)?

11. What insights have analyses of human mitochondrial DNA provided into our recent evolutionary past?

## Application and Integration

*For answers to selected even-numbered problems, see Appendix: Answers.*

12. You are a genetic counselor, and several members of the family whose pedigree for an inherited disorder is depicted in Genetic Analysis 19.2 consult with you about the probability that their progeny may be afflicted. What advice would you give individuals III-1, III-2, and III-3?

13. A mutation in *Arabidopsis immutans* results in the necrosis (death) of tissues in a mosaic configuration. Examination of the mitochondrial DNA detects deletions of various regions of the mitochondrial genome in the tissues that are necrotic. When *immutans* plants are crossed with wild-type plants, the F₁ are wild type, and the F₂ are wild type and *immutans* in a 3:1 ratio. Explain the inheritance of the *immutans* mutation and a possible origin of the mitochondrial DNA deletions.

14. What type or types of inheritance are consistent with the following pedigree?

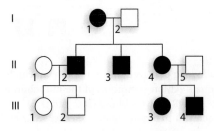

15. You have isolated (1) a streptomycin-resistant mutant ($str^R$) of *Chlamydomonas* that maps to the chloroplast genome and (2) a hygromycin-resistant mutant ($hyg^R$) of *Chlamydomonas* that maps to the mitochondrial genome. What types of progeny do you expect from the following reciprocal crosses?

$$mt^+ \ str^R \ hyg^S \times mt^- \ str^S \ hyg^R$$

$$mt^+ \ str^S \ hyg^R \times mt^- \ str^R \ hyg^S$$

16. You have isolated two *petite* mutants, *pet1* and *pet2*, in *Saccharomyces cerevisiae*. When *pet1* is mated with wild-type yeast, the haploid products following meiosis segregate 2:2 (wild type : petite). In contrast, when *pet2* is mated with wild type, all haploid products following meiosis are wild type. To what class of *petite* mutations does each of these *petite* mutants belong? What types of progeny do you expect from a *pet1* × *pet2* mating?

17. Consider this human pedigree for a vision defect.

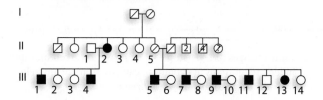

What is the most probable mode of inheritance of the disease? Identify any discrepancies between the pedigree and your proposed mode of transmission, and provide possible explanations for these exceptions.

18. A 50-year-old man has been diagnosed with MELAS syndrome (see Figure 19.7). His wife is phenotypically normal, and there is no history of MELAS syndrome in either of their families. The couple is concerned about whether their children will develop the disease. As a genetic counselor, what will you tell them? Would your answer change if it were the mother who exhibited disease symptoms rather than the father?

19. The first person in a family to exhibit Leber's hereditary optic neuropathy (LHON) was II-3 in the pedigree shown below, and all of her children also exhibited the disease. Provide two explanations as to why II-3's mother (I-1) did not exhibit symptoms of LHON.

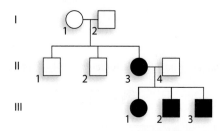

20. The following pedigree shows a family in which several individuals exhibit symptoms of the mitochondrial disease MERRF. Two siblings (II-2 and II-5) approach you to inquire about whether their children will also be afflicted with MERRF. What do you tell them?

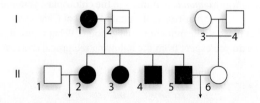

21. A 9-bp deletion in the mitochondrial genome between the gene for cytochrome oxidase subunit II and the gene for tRNA$^{LYS}$ is a common polymorphism among Polynesians and also in a population of Taiwanese natives. The frequency of the polymorphism varies between populations: the highest frequency is seen in the Maoris of New Zealand (98%), lower levels are seen in eastern Polynesia (80%) and western Polynesia (89%), and the lowest level is seen in the Taiwanese population. What do these frequencies tell us about the settlement of the Pacific by the ancestors of the present-day Polynesians?

22. What is the most likely mode of inheritance for the trait depicted in the following human pedigree?

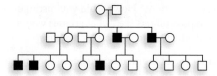

23. In 1918, the Russian Tsar Nicholas II was deposed, and he and his family were reportedly executed and buried in a shallow grave. During this chaotic time, rumors abounded that the youngest daughter, Anastasia, had escaped. In 1920, a woman in Germany claimed to be Anastasia. In 1979, remains were recovered for the tsar, his wife (the Tsarina Alexandra), and three of their children, but not Anastasia. How would you evaluate the claim of the woman in Germany?

24. The dodo bird (*Raphus cucullatus*) lived on the Mauritius Islands until the arrival of European sailors, who quickly hunted the large, placid, flightless bird to extinction. Rapid morphological evolution such as often accompanies island isolation had caused the bird's huge size and obscured its physical resemblance to any near relatives. However, sequencing of mitochondrial DNA from dodo bones reveals that they were pigeons, closely related to the Nicobar pigeon from other islands in the Indian Ocean. Why was mitochondrial DNA suited to the study of this extinct species?

25. Cytoplasmic male sterility (CMS) in plants has been exploited to produce hybrid seeds (see Experimental Insight 19.1). Specific CMS alleles in the mitochondrial genome can be suppressed by specific dominant alleles in the nuclear genome, called *Restorer of fertility* alleles, *Rf*. Consider the following cross:

$$♀ CMS1\ rf1/rf1\ rf2/rf2 × ♂ CMS2\ rf1/rf1\ Rf2/Rf2$$

What genotypes and phenotypes do you expect in the F$_1$? If some of the F$_1$ plants are male fertile, what genotypes and phenotypes do you expect in the F$_2$?

26. Wolves and coyotes can interbreed in captivity; and now, because of changes in their habitat distribution, they may have the opportunity to interbreed in the wild. To examine this possibility, mitochondrial DNA from wolf and coyote populations throughout North America—including habitats where the two species both reside—was analyzed,

and a phylogenetic tree was constructed from the resulting data (see Section 1.5 for details on how this is accomplished). Sequence from a jackal was used as an outgroup and a sequence from a domestic dog was included, demonstrating wolves as the origin of domestic dogs.

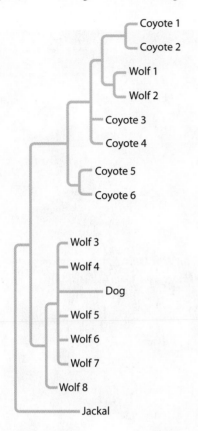

What do you conclude about potential interspecific hybridization between wolves and coyotes on the basis of this phylogenetic tree?

27. Considering the phylogenetic assignment of *Plasmodium falciparum*, the malarial parasite, to the phylum Apicomplexa (see Figure 19.18), what might you speculate as to whether the parasite is susceptible to aminoglycoside antibiotics?

28. *Elysia chlorotica* is a sea slug that acquires chloroplasts by consuming an algal food source, *Vaucheria litorea*. The ingested chloroplasts are sequestered in the sea slug's digestive epithelium, where they actively photosynthesize for months after ingestion. In the algae, chloroplast metabolism depends on the algal nuclear genome for over 90% of the required proteins. Thus it is suspected that the sea slug actively maintains ingested chloroplasts, supplying them with photosynthetic proteins encoded in the sea slug genome. How would you determine whether the sea slug has acquired photosynthetic genes by horizontal gene transfer from its algal food source?

# 20

# Developmental Genetics

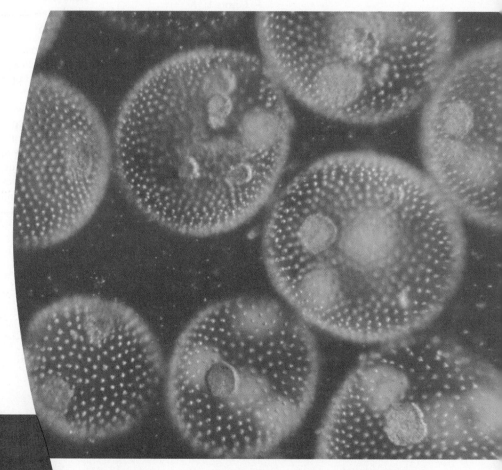

## ESSENTIAL IDEAS

- The formation of specialized cell types is directed by genes encoding transcription factors or signaling molecules.

- *Drosophila* embryos are subdivided into segments with unique identities by the sequential action of batteries of transcription factors.

- *Hox* genes specify the identity of body segments of *Drosophila* and are largely conserved throughout metazoans.

- Cells signal to either induce or inhibit neighboring cells from adopting particular developmental pathways.

- Morphological evolution can be the result of changes in gene expression patterns of a common genetic toolkit.

- Plant developmental genetics shares similarities with that of animals despite multicellularity evolving independently.

Multicellularity has evolved multiple times within the eukaryotes, as exemplified by *Volvox*, a chlorophyte green alga and member of a multicellular lineage independent of land plants and animals. In *Volvox*, the outer cells of cells are somatic while the germ cells will be derived from the inner cells.

The development of a multicellular organism from a single fertilized egg cell is one of the wonders of evolution. The fertilized egg undergoes an initial mitotic division to produce two genetically identical daughter cells. Those two cells divide to produce four identical cells, which divide to produce eight cells, and so on. Yet, while all cells in the growing embryo continue to carry the same genetic information, many of them acquire different identities as the embryo develops different body parts, organs, and tissues. This development is a genetically programmed process, occurring in the same way in all members of a species. Different species exhibit similarities in development, because of shared evolutionary ancestry, as well as differences in development, due to species-specific adaptations.

Geneticists rely on defects in development to reveal the mechanisms of normal development. As early as 1790, the German scientist and philosopher Johann Wolfgang von Goethe recognized the potential of this approach:

> From our acquaintance with . . . abnormal metamorphosis, we are enabled to unveil the secrets that normal metamorphosis conceals from us, and to see distinctly what, from the regular course of development, we can only infer.

Even so, the connections between developmental abnormalities, gene mutations, and the mechanisms that control normal development could not be understood in any detail until scientists began to apply the basic principles of genetics to the study of development. This process began around 1900, when the young embryologist Thomas Hunt Morgan decided to shift his research to focus on the nascent field of genetics, using the fruit fly *Drosophila* as his experimental organism. While Morgan never returned to the study of embryology, his students and his student's students blazed new trails by exploiting *Drosophila* genetics to illuminate many of the secrets of development in all metazoans (multicellular animals) and in plants as well.

In this chapter, we discuss the genetic processes that control development in complex multicellular organisms and the experimental approaches that lead to their discovery.

## 20.1 Development Is the Building of a Multicellular Organism

An animal begins its life as a single cell, the zygote, from which all the cell types, characterized by specific gene expression patterns, of the adult animal ultimately are derived. The key to understanding the molecular genetic basis of development is to understand how different patterns of gene expression are established and maintained as cells differentiate and specialize.

In 1915, Calvin Bridges (a student of Thomas Hunt Morgan) identified a *Drosophila* mutation in which the small hind wings, the halteres, developed into structures resembling the forewings (**Figure 20.1a**). A mutant in which an apparently normal organ or body part develops in the wrong place are called **homeotic mutations** (from the Greek *homeos*, meaning "the same" or "similar"), and they have been central to the progress geneticists have made in understanding how complex organisms develop and evolve. Ed Lewis (a student of Morgan's student Alfred Sturtevant) later identified the *bithorax* complex of genes as being responsible for the homeotic mutation observed by Bridges. As we discuss in this chapter, mutations in *bithorax* genes change the developmental program of a portion of the fruit-fly body, transforming the halteres into a second set of forewings. Another example is the dominant *Antennapedia* mutant, in which relatively normal fly legs develop in the positions that should be occupied by the antennae (**Figure 20.1b**). To understand the cascades of events responsible for such developments, we must first examine the phenomenon of cell differentiation and pattern formation.

**(a)** In a *bithorax* mutation, halteres seen in wild-type *Drosophila* (left) develop instead into a second set of wings (right).

Halteres

A second set of wings in the position normally occupied by halteres

**(b)** In an *Antennapedia* mutation, antennae in wild-type *Drosophila* (left) develop instead into legs (right).

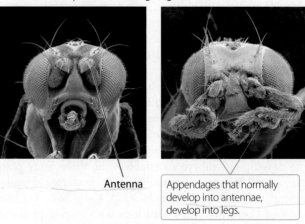

Antenna

Appendages that normally develop into antennae, develop into legs.

**Figure 20.1** **Inappropriate positions of organs and body structures in homeotic mutants.**

## Cell Differentiation

In an animal, fertilization of an egg cell by a sperm cell forms a single-celled zygote, which undergoes several mitotic divisions to form a small cluster of embryonic cells that are genetically identical. These embryonic cells are **totipotent,** which means they have the potential to differentiate into any tissue or cell type the animal can produce. In vertebrates, totipotent cells of early embryos are called **embryonic stem cells.** In totipotent cells, all genes have the potential to be expressed given the appropriate cues. As development proceeds, however, cells become **differentiated,** taking on different morphologies and undertaking different physiological activities.

Differentiation is characterized by changes in the patterns of gene expression that progressively limit which genes continue to be expressed by each cell type. At a certain stage in development, cells retain the potential to give rise to many different types of descendants, but not to all types—at this stage, the cells are said to be **pluripotent.** As development progresses further, however, most cells ultimately become specialized: They have characteristic shapes and carry out specific functions. These fully differentiated and specialized cells express only a subset of genes in the genome, and each cell type has its own characteristic pattern of gene expression. Thus development is a progressive process during which totipotent cells differentiate into specialized cell types through a series of genetically controlled steps that place ever more restrictive limits on their developmental potential.

While most cells of adult animals are fully differentiated and locked into a specific cell fate, there are some exceptions. In our bodies, various types of pluripotent stem cells—such as muscle, epidermal, epithelial, and hematopoietic (blood) cells—retain the capacity to develop into a range of further-specialized cells to replenish cells that are lost.

## Pattern Formation

How do genetically identical cells acquire different fates? Two mechanisms have been identified: Cells can inherit some definitive molecule that specifies cell fate, or the fate of cells can be determined by their neighbors through the action of signaling molecules. Inheritance of a fate-determining molecule depends on the identity of progenitor cells, whereas development through the influence of neighboring cells depends on the identity of those neighbors.

The term *pattern formation* describes the intricately interacting events that organize the differentiating cells in the developing embryo to establish the three body-plan axes of the mature organism: anterior–posterior, dorsal–ventral, and left–right (**Figure 20.2**). Cells have various ways of "knowing" their locations with regard to these axes. The combination of internal and external signals that a cell perceives during development provides information on the cell's location within an organism.

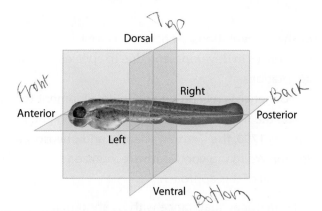

**Figure 20.2 The three embryonic axes of a zebrafish.**

To understand the role that such **positional information** plays in development, consider the French flag, which has a simple pattern of three vertical stripes in the order blue, white, and red, along a single (anterior–posterior) axis (**Figure 20.3a**). While French flags may come in various sizes, the proportions of the stripes within each flag remain generally constant, dividing the flag into thirds. Imagine the entire flag to be comprised of cells descended from a single parent cell. How do daughter cells know whether they are to differentiate as blue, white, or red? The cells could interpret their position by one or more of various mechanisms, but the simplest to envision is based on the concentration gradient of a molecule that is highly concentrated at one end of the embryonic flag and much less concentrated at the opposite end. The position of each cell on the flag's anterior–posterior axis is defined by the concentration of this molecule, in which threshold values define boundaries between discrete fates: Above a certain concentration, the result is blue cell identity; below this threshold concentration, white cells develop; and below an even lower threshold, red cells develop. Substances whose presence in different concentrations directs developmental fates are referred to as **morphogens.** If activation or repression of gene expression is dependent upon threshold concentrations of a morphogen (e.g., concentrations above which a gene is active and below which a gene is inactive), discrete boundaries of gene expression can be established.

Once a cell has acquired a specific identity, it may induce its neighbors to acquire a certain fate; this process is termed **induction.** A classic case of induction was first noted more than a century ago, when transplantation of cells from one region of a developing frog embryo to another region of a second embryo induced the surrounding cells to form a second body axis (**Figure 20.3b**). The region from which the transplanted cells were derived was called the **organizer,** because the cells of that region possess the ability to organize cells in the surrounding tissue. Alternatively, a cell that acquires a specific fate may then produce an inhibitory substance that prevents its neighbors from acquiring a certain fate, and this process is called **inhibition** (**Figure 20.3c**). Inhibition can be used to produce patterns of regularly spaced cells of a particular

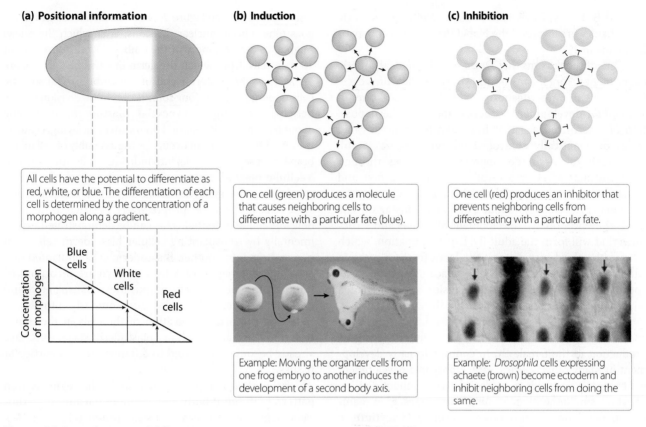

**(a) Positional information**

All cells have the potential to differentiate as red, white, or blue. The differentiation of each cell is determined by the concentration of a morphogen along a gradient.

Blue cells
White cells
Red cells

Concentration of morphogen

**(b) Induction**

One cell (green) produces a molecule that causes neighboring cells to differentiate with a particular fate (blue).

Example: Moving the organizer cells from one frog embryo to another induces the development of a second body axis.

**(c) Inhibition**

One cell (red) produces an inhibitor that prevents neighboring cells from differentiating with a particular fate.

Example: *Drosophila* cells expressing achaete (brown) become ectoderm and inhibit neighboring cells from doing the same.

**Figure 20.3  Mechanisms of differentiation.**

fate within a field of cells that would otherwise all differentiate in the same manner, such as in the example of *Drosophila* shown in Figure 20.3c. Other examples of such regularly spaced tissues include many epidermal features, such as bristles, feathers, hairs, and scales.

The developmental histories of cells can affect how the cells respond to cues from their neighbors. For example, for a cell to be able to respond to an inductive or inhibitory signal from neighboring cells, it must express the appropriate receptor. In addition, cells able to respond to a signal may behave differently depending on what other factors are present in the cell. When a cell divides, the daughter cells usually inherit the same set of transcription factors and chromatin states that existed in the cell they were derived from (the importance of chromatin states is discussed in Section 20.2). However, occasional asymmetric cell divisions in which the two daughter cells inherit different cellular constituents and acquire different fates underlie early developmental patterning events in some species.

Induction, inhibition, and asymmetric cell divisions are common mechanisms directing cell differentiation and pattern formation in multicellular organisms. When employed sequentially and reiteratively during embryogenesis, these processes enable a single-celled zygote to develop into a complex organism having a multitude of cell types. Each cell division in the embryo brings about changes in the

relative positional relationships between the cells, so new opportunities for cell–cell communication are constantly created. In keeping with the importance of positional information, induction, and inhibition in development, most genes identified as having prominent roles in developmental processes encode proteins that act as either transcription factors or signaling molecules.

---

**Genetic Insight**  Cells use positional information to interpret their location within a field of cells, and then differentiate accordingly. Cells that have attained a certain identity can influence fates of neighboring cells through induction or inhibition.

---

## 20.2 *Drosophila* Development Is a Paradigm for Animal Development

Discoveries about the developmental processes of *Drosophila* have made it ontogenetically one of the best-understood animals on the planet. These insights have in turn profoundly influenced how geneticists perceive the development and evolution of all other animals, ourselves included. For their work in unraveling some of the mechanisms underlying pattern formation in *Drosophila*,

Edward B. Lewis, Christiane Nüsslein-Volhard, and Eric Wieschaus were awarded the Nobel Prize in Physiology or Medicine in 1995.

One of the reasons that *Drosophila* is an ideal genetic experimental organism is its short, 9-day life cycle (Figure 20.4a). Embryogenesis spans the first 24 hours of *Drosophila* development, commencing with the deposition of a fertilized egg that immediately begins a rapid series of genetically controlled changes (Figure 20.4b). After embryogenesis, development progresses through three distinct larval stages, called instars. The first and second instar stages last about 24 hours each, and the third instar develops for about 2 days. Each instar stage is marked by progressive development of tissues and structures that will form the adult fly. During pupation, which follows the third instar stage, the larva forms a pupa in which metamorphosis will take place. Pupation lasts approximately 4 days; at its conclusion a fully formed adult fruit fly emerges, ready to begin the cycle anew.

The *Drosophila* egg has conspicuous anterior–posterior and dorsal–ventral polarities that are acquired during its production in the female fly. In contrast to early development in many other species, early embryonic development in *Drosophila* proceeds by nuclear division without division of cytoplasm. Rather than forming blastomeres, as in mammalian development, this process forms a **syncytium,** a multinucleate cell in which the nuclei are not separated by cell membranes (see Figure 20.4b). The fertilized egg undergoes nine mitotic nuclear divisions, after which the nuclei migrate to the periphery of the embryo. At this time, about 10 pole cells, from which the germ line will be derived, are set aside at the posterior end of the embryo. The somatic cells undergo another four rounds of mitotic divisions at the periphery, forming a **syncytial blastoderm** containing about 6000 nuclei. By about 3 hours after egg laying, cellularization of the syncytium occurs by the assembly of cell membranes that separate nuclei into individual cells, thus forming a **cellular blastoderm.**

During the syncytial blastoderm and cellularization stages, cells become progressively restricted in their developmental potential. This can be demonstrated experimentally by transplanting cellular blastoderm cells from one embryo into another. Blastoderm cells implanted into an equivalent region of a host embryo are incorporated normally into host structures, but those transplanted into different regions will develop autonomously into tissues reflecting the original position of the cells in the donor embryo. Thus, at the cellular blastoderm stage, cells have already become committed to differentiate into particular tissues.

*Drosophila* is typical of insects in the segmentation pattern of its adult body. Eight abdominal and three thoracic segments are easily distinguished (Figure 20.4c). While the head is not so clearly divided into sections, it

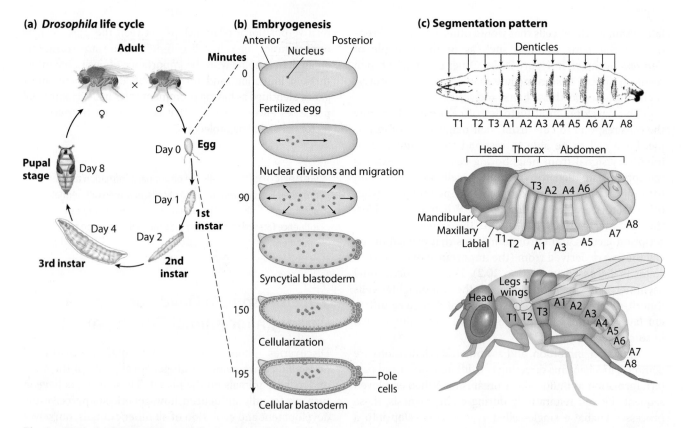

**Figure 20.4** Overview of *Drosophila* development.

too consists of at least three distinct developmental segments. The characteristic segments of the insect body are first visible during embryogenesis, where they are indicated by the pattern of denticles (small hooks for gripping during larval movement) on the ventral epidermis. The body plan established during embryogenesis determines the organization of tissues and organs in the adult fly.

## The Developmental Toolkit of *Drosophila*

Large-scale genetic screens (see Section 16.1) were commenced by Christiane Nüsslein-Volhard, Eric Wieschaus, and others in the late 1970s and early 1980s to identify and describe the function of genes directing pattern formation in *Drosophila* embryos. It is estimated that mutations in about 5000 of the 14,000 genes in *Drosophila* will result in a lethal phenotype. Most mutations resulting in lethality affect genes that have essential cellular functions, and these genes are sometimes described as **housekeeping genes.** However, several hundred genes producing lethal phenotypes are involved directly in developmental programs of pattern formation during embryogenesis.

Nüsslein-Volhard and Wieschaus faced a significant challenge when designing genetic screens for mutations in pattern formation, because flies in which segmental pattern formation is severely disrupted rarely survive beyond the larval stage. Their solution was to focus on embryos and larvae. They reasoned that mutations affecting embryonic pattern formation would not be lethal until larval formation, leaving a short window of time for observation of the effects of such mutations. From the types of spatial defect exhibited by the mutant phenotypes, mutants were grouped into four gene classes, with a fifth class identified earlier by Ed Lewis:

1. **Coordinate genes:** Defects affect an entire pole of the larva (**Figure 20.5a**).
2. **Gap genes:** Mutants are missing large, contiguous groups of segments (**Figure 20.5b**).
3. **Pair-rule genes:** Mutants are missing parts of adjacent segment pairs in alternating segments (**Figure 20.5c**).
4. **Segment polarity genes:** Defects affect patterning within each of the 14 segments (**Figure 20.5d**).
5. **Homeotic genes:** Defects affect the identity of one or more segments.

These five gene classes are expressed in sequence during embryogenesis: The coordinate genes act first, followed by gap genes, pair-rule genes, segment polarity genes, and finally homeotic genes. The cascade of gene expression subdivides the embryo in successive steps, first into broad regions and then into progressively smaller

domains, and each of the 14 resulting segments acquires a specific identity. The patterns of mRNA and protein expression of each gene correspond, both in space and in time, to its mutant phenotype (see Figure 20.5). For example, expression of the gap gene *knirps* spans a contiguous embryonic domain that is destined to become abdominal segments. These abdominal segments are missing in *knirps* mutants, as is evident in the early larva (see Figure 20.5c).

Expression of the pair-rule genes follows that of gap genes and produces seven stripes in the embryo. Curiously, the stripes of gene expression of pair-rule genes do not correspond to the segments of the adult insect, but rather straddle the boundaries between segments, thus occupying the posterior part of one segment and the anterior part of its neighbor. The domains of gene expression controlled by the pair-rule genes are therefore called **parasegments.** Expression of the segment polarity genes occurs in 14 polar stripes (i.e., each stripe has anterior and posterior "poles"), one for each segment of the embryo. The homeotic genes are the last to be expressed and affect broad domains of contiguous parasegments along the anterior–posterior axis. The anterior expression boundaries of the homeotic genes correspond to parasegment boundaries defined by the pair-rule genes. Thus, the sequential activation of different classes of genes during early development is reflected in the sequential subdivision of the organism, from a single-celled zygote into a segmented embryo.

When the expression pattern of a gene in a wild-type embryo corresponds precisely to the cell fates that are disrupted when the gene is mutated, the activity of the gene is said to be cell autonomous. A gene whose action is cell autonomous affects only the cells in which the gene is transcribed and expressed. Four of the five classes of genes act largely cell autonomously, an observation consistent with the identity of these genes as transcription factors. The exception is the segment polarity class of genes, which often encode signaling molecules that can act non-autonomously, that is in cells other than where the gene is expressed. In the following sections, we examine how the embryo is successively subdivided by the activity of these sets of genes.

---

Genetic Insight The coordinate, gap, and pair-rule genes act in succession to divide the *Drosophila* embryo into segments. Segment polarity genes and homeotic genes then direct the differentiation of segments that form the adult fly.

---

## Maternal Effects on Pattern Formation

In animals, the mother often supplies critical gene products to the egg that subsequently direct embryo development. These genes are called **maternal effect genes.** Note

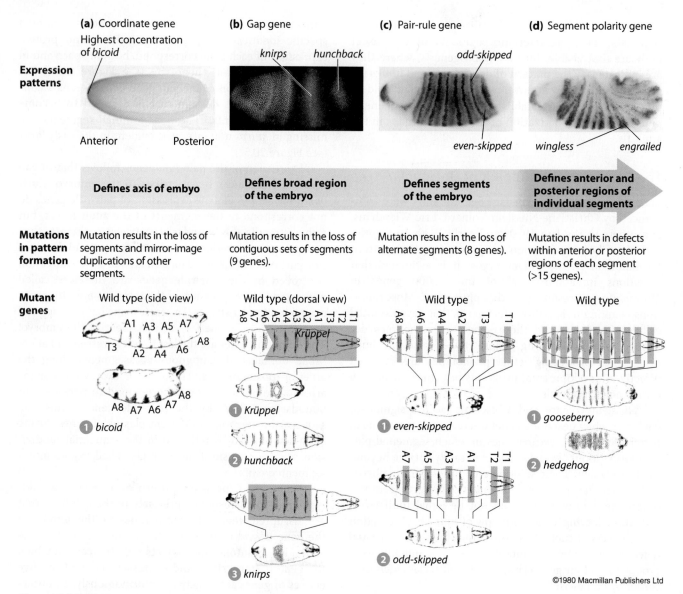

**Figure 20.5 Mutations causing defects in pattern formation in *Drosophila*.**

that maternal effects are different from maternal inheritance (introduced in Chapter 19), in that maternal effects entail the maternal deposition of protein or mRNA in the egg cell, whereas maternal inheritance refers to maternal transmission of genetic material (e.g., organelle genomes).

How can the maternal effect genes that influence development be identified in mutant screens, given that for these genes, the embryonic phenotype is determined by the genotype of the mother rather than that of the embryo? An answer becomes apparent when we compare the inheritance patterns observed with maternal effect genes against those observed with **zygotic genes,** genes that are active only in the zygote or embryo. For zygotic genes, *the genotype of the embryo determines the phenotype.* The

following cross illustrates this principle for an autosomal recessive mutation (*m*).

**Zygotic Genes**

| Parents | Offspring | Phenotype |
|---|---|---|
| *m/+* × *m/+* | *m/+, +/+* | Normal (3) |
| | *m/m* | Mutant (1) |

With maternal effect genes, where *the genotype of the mother determines the phenotype of the zygote,* the same cross as above, involving an autosomal recessive mutation (*m*), would give the following outcomes:

**Maternal Effect Genes**

| Parents (female × male) | Offspring | Phenotype |
|---|---|---|
| m/+ × m/+ | m/m, m/+, +/+ | All normal |
| m/+ × m/m | m/m, m/+ | All normal |
| m/m × +/+ or m/+ or m/m | m/m, m/+ | All mutant |

These divergent patterns allow discrimination between maternal effect genes and zygotic genes. Crosses can be performed to determine whether the genes are active maternally, zygotically, or both. When such crosses were performed to test the five classes of mutants described above, the coordinate genes were found to be maternally active; their expression *in the mother* rather than in the embryo provides positional information to the egg. Most gap genes are active zygotically; but at least one, *hunchback*, also exhibits maternal activity. All pair-rule, segment polarity, and homeotic genes act strictly zygotically. These findings make sense given the developmental stage at which the different classes of gene are active and the observation that zygotic gene expression commences only in the syncytial blastoderm stage of embryogenesis.

---

**Genetic Insight** Maternal effects can be distinguished from the actions of zygotic genes by examining the parental genotypes of mutant offspring. In *Drosophila,* the mother's genotype determines coordinate gene expression, whereas subsequent classes of developmental genes act for the most part zygotically.

---

## Coordinate Gene Patterning of the Anterior–Posterior Axis

The genetic control of development is essentially a process of regulating gene expression in three-dimensional space over time. It is not surprising, then, that most of the early-acting genes establishing the anterior–posterior axis of *Drosophila* encode transcription factors. The interaction of transcription factors with cis-acting regulatory elements of target genes provides spatial control of gene expression. This spatial control is coordinated over time by continual inputs from neighboring cells. In this section, we describe examples of the spatial and temporal regulation of gene expression that results in subdivision of a developing *Drosophila* embryo into the characteristic segments.

The *bicoid* gene plays a major role in the establishment of the anterior–posterior axis in *Drosophila.* Loss-of-function *bicoid* alleles result in a loss of anterior portions of the embryo; the anterior portions are replaced instead by a mirror-image duplication of posterior regions

(Figure 20.6a). *Bicoid* mRNA is anchored to the anterior region of the egg during oogenesis in the mother (Figure 20.6b). After translation, bicoid protein (Bicoid) diffuses from its site of synthesis at the anterior pole of the embryo throughout the syncytial embryo, owing to the absence of cell membranes to impede protein diffusion. The diffusion results in a gradient of Bicoid in which the highest concentration is at the anterior end and very little Bicoid is detected beyond the middle of the embryo.

Cytoplasmic transplantation experiments elegantly demonstrate how Bicoid specifies anterior identity. Anterior cytoplasm extracted from one wild-type embryo and then injected into a *bicoid* mutant embryo causes anterior structures to develop at the site of injection (see Figure 20.6a). When cloned *bicoid* was available, similar experiments were carried out with purified *bicoid* mRNA, which produced the same result. These findings indicate that the concentration gradient of Bicoid provides positional information along the anterior–posterior axis of the embryo, presumably by differentially regulating

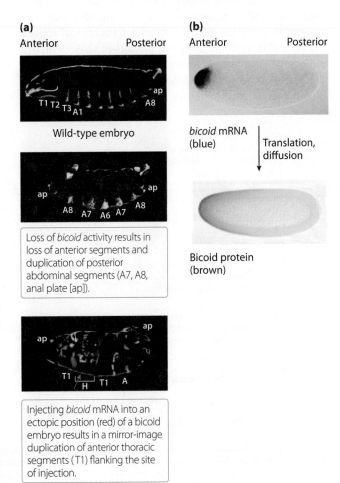

**(a)**

Anterior — Posterior

ap
T1 T2 T3 A1 — A8

Wild-type embryo

ap — ap
A8 A7 A6 A7 A8

Loss of *bicoid* activity results in loss of anterior segments and duplication of posterior abdominal segments (A7, A8, anal plate [ap]).

ap — ap
T1 H T1 A

Injecting *bicoid* mRNA into an ectopic position (red) of a bicoid embryo results in a mirror-image duplication of anterior thoracic segments (T1) flanking the site of injection.

**(b)**

Anterior — Posterior

*bicoid* mRNA (blue)

Translation, diffusion

Bicoid protein (brown)

**Figure 20.6** **Maternal *bicoid* patterning of the embryo along the anterior–posterior axis.**

several genes that respond to different concentrations of Bicoid. Among the known zygotic genes whose transcription is directly regulated by Bicoid is the gap gene *hunchback*.

Surprisingly, examination of the distribution of *hunchback* mRNA revealed that *hunchback* is also maternally expressed, and that its maternal expression is uniform throughout the egg (**Figure 20.7a**). The hunchback protein (Hunchback), on the other hand, is found only at the anterior end of the early embryo, implying that posterior *hunchback* mRNA is not translated. This seeming contradiction was explained by the discovery of another maternally expressed coordinate gene, *nanos*. The posterior end of the embryo is patterned by *nanos*, whose protein forms a gradient with the highest concentration at the posterior end. Rather than encoding a transcription factor, *nanos* encodes a protein that represses translation of *hunchback* mRNA. Thus, Hunchback is restricted to the anterior end of the embryo by posterior translational repression of maternal *hunchback* mRNA. In addition, zygotic *hunchback* expression in the anterior end is transcriptionally activated by anteriorly localized Bicoid.

Patterning of the posterior end of the embryo is governed by similar interactions. In addition to acting as a transcription factor, Bicoid acts as a translational repressor of the maternally supplied *caudal* mRNA, which is uniformly distributed throughout the egg. Translational repression of *caudal* mRNA by the anterior gradient of Bicoid results in a posterior gradient of *caudal* protein

(Caudal). The end result is an embryo with graded distributions of three transcription factors: Bicoid and Hunchback, in which the highest concentration is at the anterior end; and Caudal, in which the highest concentration is at the posterior end. The relative concentrations of these three proteins provide positional information along the length of the embryo, which is interpreted by the subsequently acting gap genes.

## Domains of Gap Gene Expression

The broad gradients of maternally supplied coordinate gene products are transformed into domains of gap gene expression with discrete boundaries. This occurs through a combination of cooperative binding of transcription factors—similar to the activation of the lambda repressor described in Chapter 14—and cross-regulatory interactions among the gap genes themselves. Let's first consider further how the gradual concentration gradient of Bicoid is translated into the more discrete pattern of *hunchback* mRNA expression.

As noted earlier, zygotic expression of the gap gene *hunchback* is confined to the anterior region of the embryo. Unlike Bicoid, which exhibits a gradual concentration gradient, the concentration of *hunchback* mRNA declines precipitously at a particular point along the anterior–posterior axis. Transcription of *hunchback* is activated by the binding of Bicoid to cis-regulatory elements 5′ to the *hunchback* coding region (**Figure 20.7b**). In this location,

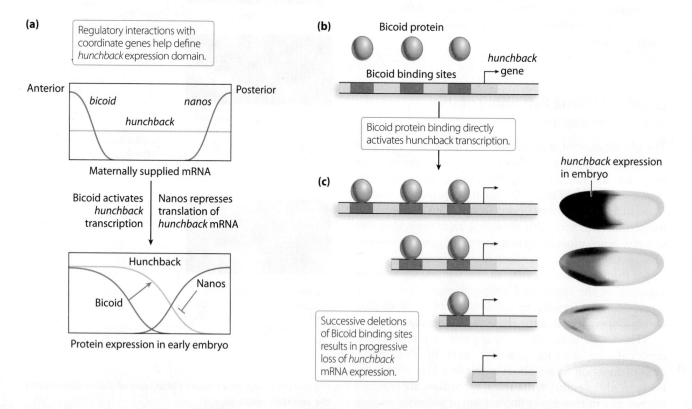

**Figure 20.7** Gap gene expression patterns are activated by coordinate genes.

there are multiple cis-acting sites to which Bicoid can bind, and these sites are bound in a cooperative manner, meaning that the binding of one Bicoid molecule to one site facilitates the binding of a second Bicoid molecule to a second nearby site, and so on. Mutation of the Bicoid binding sites alters the responsiveness of *hunchback* expression to Bicoid, and removal of all binding sites abolishes *hunchback* expression in the embryo (Figure 20.7c).

A threshold level of Bicoid must be present in order for *hunchback* expression to be activated. Consequently, *hunchback* expression occurs on one side of a threshold concentration with no expression on the other, and a sharp boundary is produced. In this manner, the gradual anterior concentration gradient of Bicoid is translated into a distinct anterior region of *hunchback* mRNA expression, which, after translation, produces a sharp gradient of Hunchback (see Figure 20.7a).

The gradient of hunchback protein is critical for the regulation of other gap genes, such as *Krüppel* (Figure 20.8), which is repressed by high levels of Hunchback but activated in the central region of the embryo where Bicoid levels are moderate. These interactions establish the anterior margin of *Krüppel* expression toward the posterior end of the Hunchback protein gradient. The posterior margin of *Krüppel* expression appears to be determined through negative regulation by other gap genes, *knirps* and *giant*. Similar regulatory interactions between other gap genes help establish the rest of the partially overlapping patterns of gap gene expression that subdivide the developing embryo into discrete domains.

## Regulation of Pair-Rule Genes

From the domains of gap gene expression emerge 14 narrower stripes of gene expression that represent the first manifestation of segmentation of the anterior–posterior body plan. Analysis of the regulation of the pair-rule gene *even-skipped* (*eve*) revealed that each stripe is established by independent enhancer modules of cis-acting regulatory sequences of *eve*. Each enhancer module from a pair-rule gene responds to specific combinations of gap genes (Figure 20.9a). Thus, the formation of stripes of gene expression is the result of combinatorial control of gene expression through multiple cis-acting regulatory elements of the pair-rule genes. This situation is conceptually similar to the regulation of the gap genes, as described earlier for *hunchback*.

Stripe 2 of *eve* provides an example of modularity in gene regulation. Gene expression within stripe 2 is controlled by a cis-regulatory element—the stripe 2 enhancer module—located about 1700 bp to 1000 bp upstream of the transcription initiation site of *eve* (see Figure 20.9a). When this regulatory element is isolated and used to drive a reporter gene (see Section 17.2) in transgenic *Drosophila* embryos, expression is observed only in stripe 2, indicating that these regulatory sequences are sufficient for stripe 2 expression. Detailed sequence analysis of this module identified binding sites for the gap proteins Hunchback, Krüppel, and Giant, as well as binding sites for Bicoid. Mutational analysis of different combinations of binding sites demonstrates that both Hunchback and Bicoid act as activators of *even-skipped* stripe 2 gene expression, while both Giant and Krüppel act as repressors.

Stripe 2 lies entirely within the *hunchback* expression domain of the embryo and is flanked on the anterior side by the *giant* expression domain and on the posterior side by the *Krüppel* expression domain (Figure 20.9b). It contains an intermediate level of Bicoid remaining from the maternally established gradient. Thus the position of *eve* stripe 2 along the anterior–posterior axis is a zone with a high concentration of Hunchback, low concentrations of Giant and Krüppel, and an intermediate concentration of Bicoid. Only in parasegment 3, which is the location of stripe 2, are both positive regulators present and both negative regulators absent (Figure 20.9c). This combination of gap and coordinate protein concentrations does not occur anywhere else along the axis of the embryo and uniquely defines the *eve* stripe 2 position. The integration of positive and negative regulators results in the precise limiting of *even-skipped* stripe 2 to a region only a few cells in width along the anterior–posterior axis. Similar combinatorial mechanisms are thought to control the

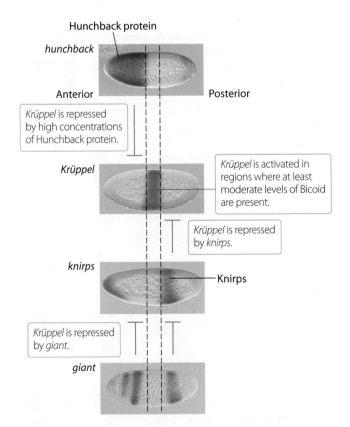

**Figure 20.8 Cross-regulatory interactions among gap genes define their expression patterns.**

Hunchback protein

*hunchback*

Anterior          Posterior

*Krüppel* is repressed by high concentrations of Hunchback protein.

*Krüppel*

*Krüppel* is activated in regions where at least moderate levels of Bicoid are present.

*Krüppel* is repressed by *knirps*.

*knirps*                              Knirps

*Krüppel* is repressed by *giant*.

*giant*

**Figure 20.9** Stripes of gene expression, established by combinatorial coordinate and gap gene activities.

expression patterns of all of the pair-rule and segment polarity genes.

The discovery that in multicellular organisms *the control of gene expression is modular* provided important insight into the evolution of organisms. Modularity of gene regulation allows changes in specific domains of expression without catastrophic disruption of global expression patterns.

---

Genetic Insight   The *Drosophila* embryo is successively subdivided by sequential action of maternal coordinate and largely zygotic gap genes. Specific concentrations of gap and coordinate proteins uniquely define anterior–posterior positions. Combinatorial activity of these proteins leads to the activation of pair-rule genes in precise, narrow stripes of embryonic tissue.

---

## Specification of Parasegments by *Hox* Genes

Having explored the mechanisms by which gap and pair-rule genes subdivide the *Drosophila* embryo into 14 segments, we can now consider how each segment acquires a unique identity through the action of the homeotic genes. Once again, the key discoveries were made through the study of mutations, pioneered by Edward B. Lewis starting in the 1950s.

As we saw at the beginning of the chapter, a remarkable aspect of homeotic mutant phenotypes is the development of relatively normal structures in inappropriate positions. Another general feature of homeotic mutations is that they cause identity transformations of serially repeated structures. Legs, for example, are appendages that are normally limited to the three thoracic segments in *Drosophila*, whereas antennae are appendages that normally develop only on the third cephalic (head) segment. In the case of *Antennapedia* mutants, however, a leg appears in a segment ordinarily reserved for an antenna (see Figure 20.1), suggesting that *Antennapedia* normally specifies the identity of one or more of the thoracic segments. Analyses of homeotic genes in *Drosophila* demonstrate that in fact they act in combination to specify the identity of each of the 14 body segments.

The homeotic genes of animals are also remarkable for being clustered in gene complexes. In *Drosophila* there are two homeotic clusters on the third chromosome: the ***Antennapedia* complex,** consisting of five genes, and the ***bithorax* complex,** consisting of three genes. In other

**(a) The *even-skipped* (*eve*) gene and and enhancer modules**

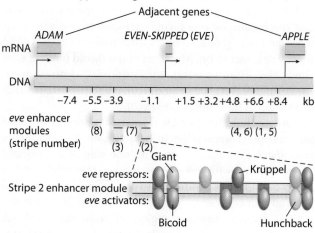

**(b) Distribution of gap gene expression**

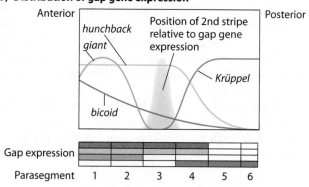

**(c) Occupancy of regulatory sites on *eve* stripe 2 enhancer module in different parasegments**

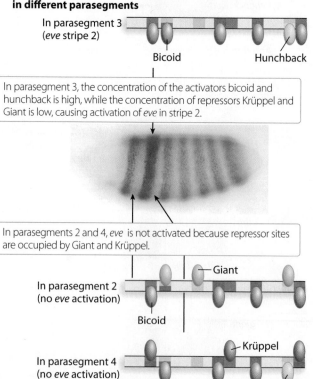

organisms, the homeotic genes are usually in a single cluster. Amazingly, the order of the genes within the complexes reflects the positions along the anterior–posterior axis that are influenced by each gene (Figure 20.10).

The cloning of the homeotic genes revealed another surprise: All eight genes encode closely related proteins, suggesting that all members of the complex were derived from a common ancestor through a series of gene duplications. All of the genes share a conserved sequence of DNA of 180 nucleotides that was dubbed the **homeobox,** which encodes a 60–amino acid protein domain, termed the **homeodomain,** with a helix-turn-helix motif. Such motifs had previously been recognized in bacterial and phage transcription factors, such as the *Lac* repressor and the lambda repressor proteins. They function to bind cis-regulatory DNA sequences of target genes. Since the homeobox genes of the *Antennapedia* and *bithorax* complexes share both molecular and functional similarity as well as having a common evolutionary origin, they are known collectively as *Hox* **genes.**

The patterns of *Hox* gene expression correlate with the regions affected in the corresponding mutants. Each of the *Hox* genes has a well-defined anterior boundary of expression but in most cases a more diffuse boundary on the posterior, resulting in overlapping domains of *Hox* gene expression. One surprising observation is that the anterior boundaries of *Hox* gene expression do not correspond to segmental boundaries but rather occur within the segments, at positions corresponding to boundaries of segment polarity gene expression. The result is that *Hox* gene expression is out of register with the groups of cells that give rise to segments in the adult fly and instead marks the boundaries of parasegments.

Because of the parasegmental pattern of *Hox* gene expression, mutations of those genes affect cellular identity in a parasegmental manner. Each parasegment of the embryo expresses a unique combination of *Hox* gene products, and it is this specific combination that gives each parasegment a unique identity. The activation of *Hox* genes is controlled by the earlier-acting gap and

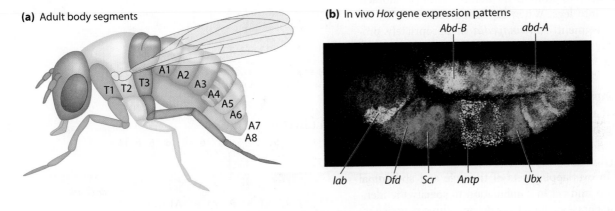

**(a)** Adult body segments

**(b)** In vivo *Hox* gene expression patterns

**(c)** *Hox* complexes on chromosome 3, and expression patterns in embryo

Figure 20.10 *Hox* **genes of the *Antennapedia* and *bithorax* complexes.**

pair-rule genes in a combinatorial manner similar to that described for the activation of pair-rule genes by the gap and coordinate genes. In the absence of all *Hox* gene activity, segments are formed; but they all differentiate into a "default" state that resembles a head segment. This outcome indicates that *Hox* genes are not required for the formation of the segments but rather for the specification of their identity.

**The *Antennapedia* Complex** The *Antennapedia* complex consists of five *Hox* genes—*labial, Deformed, Sex combs reduced, proboscipedia,* and *Antennapedia*—that act in combination to specify the cephalic and thoracic parasegments (see Figure 20.9b). The original *Antennapedia* mutant (see Figure 20.1) was dominant and was found to be the result of a gain-of-function allele (see Section 4.1). The *Antennapedia* gene is normally expressed only in parasegments 4 and 5, which give rise to thoracic segments that each produce a pair of legs. In flies carrying the dominant *Antennapedia* mutation, however, *Antennapedia* is expressed ectopically—meaning it is expressed at an inappropriate time or place or both. One of the normal roles of *Antennapedia* expression in the thoracic segments is to promote the differentiation of thoracic appendages into legs. When expressed ectopically in the third head segment, *Antennapedia* inappropriately promotes differentiation of head appendages (antennae) into legs instead.

**The *bithorax* Complex** In contrast to *Antennapedia* mutations that affect anterior body segments, mutations in the three genes of the *bithorax* complex—*Ultrabithorax, abdominal-A,* and *Abdominal-B*—affect more-posterior segments (**Figure 20.11a**). The *bithorax* complex genes are expressed in overlapping sets of thoracic and abdominal parasegments and act in combination to specify the identity of those parasegments. How do only three genes specify the identity of nine segments, one being thoracic and eight abdominal? The three genes vary not only in their spatial patterns of expression but also in expression levels between segments. Each has a sharp anterior border of expression and a more diffuse posterior boundary of expression. Thus each segment expresses a unique qualitative and quantitative pattern of *Hox* gene expression.

Loss of *Ultrabithorax* activity results in parasegments 5 and 6 having a combination of *Hox* gene products resembling that normally found in parasegment 4. This causes transformations of the identity of thoracic segment T3 and abdominal segment A1 into thoracic segment T2 (**Figure 20.11b**). Loss of the entire *bithorax* complex causes most abdominal segments to develop as T2, so each has legs as appendages (**Figure 20.11c**). This observation suggests that expression of *Antennapedia*, which promotes leg identity in appendages, extends posteriorly in such mutants, and that genes of the *bithorax* complex

**(a)** Wild type

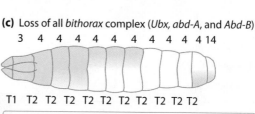

Both *Ultrabithorax* (*Ubx*) and *abdominal-A* (*abd-A*) have a diffuse posterior boundary of expression due to negative regulatory interactions between genes.

**(b)** Loss of *Ubx*

T3 and A1 are incorrectly specified as T2 due to a failure to repress *Antennapedia* in these segments.

**(c)** Loss of all *bithorax* complex (*Ubx, abd-A,* and *Abd-B*)

All segments posterior to T1 differentiate with T2 due to a failure to repress *Antennapedia* in all posterior segments.

**(d)** Loss of *abd-A* and *Abd-B*

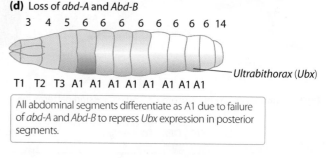

All abdominal segments differentiate as A1 due to failure of *abd-A* and *Abd-B* to repress *Ubx* expression in posterior segments.

**(e)** Loss of *Abd-B*

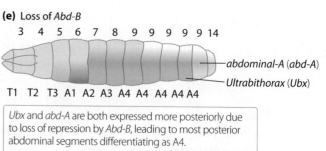

*Ubx* and *abd-A* are both expressed more posteriorly due to loss of repression by *Abd-B*, leading to most posterior abdominal segments differentiating as A4.

**Figure 20.11 Cross-regulatory interactions between *bithorax* complex genes, specifying thoracic and abdominal segment fates.**

normally repress posterior expression of *Antennapedia*. Such cross-regulatory interactions between *Hox* genes, whereby more posteriorly expressed *Hox* genes repress the expression of *Hox* genes normally expressed in more-anterior positions, is a common although not universal feature in the regulation of *Hox* genes (**Figure 20.11d–e**).

As you have probably noticed, there is no single *Hox* gene called *bithorax*; so what became of the original *bithorax (bx)* mutation that was isolated by Calvin Bridges? When Ed Lewis recognized that mutations such as *bithorax* could provide valuable insights into the genetic mechanisms of development, he began collecting mutations with similar but distinct phenotypic defects, some of which he called *postbithorax (pbx)*, *Contrabithorax*, *Ultrabithorax*, and *bithoraxoid (bxd)*. Each of these mutations mapped to a different position in the same chromosomal region, so that they were separable by recombination events, and double-mutant combinations could be constructed. At the time Lewis performed these studies, molecular cloning was unknown, and he assumed that each mutant he identified represented a different gene. When the *bithorax* complex was eventually cloned in 1983, however, many of the mutant phenotypes were found to result from mutations in different enhancer modules controlling the expression of a single

coding region that is now called the *Ultrabithorax* gene (**Figure 20.12a**).

Mutations of the regulatory elements can be either recessive, if an enhancer module acts to positively regulate gene expression, or dominant, if a silencer module acts to negatively regulate gene expression. While null loss-of-function alleles of *Ultrabithorax* result in embryo lethality, disruption of single enhancer modules results in milder defects. For example, recessive *Ultrabithorax*$^{bithorax}$ mutations *(bx)* result in the transformation of the anterior part of T3 into T2, causing the anterior portion of the haltere to develop as a wing (**Figure 20.12b**). Conversely, recessive *Ultrabithorax*$^{postbithorax}$ mutations *(pbx)* result in the transformation of the posterior region of T3 into T2 identity, and the posterior portion of the haltere develops as a wing. Only in the *Ultrabithorax*$^{bithorax}$ *Ultrabithorax*$^{postbithorax}$ double mutant is the identity of the entire T3 segment transformed into a T2 identity, causing a four-winged fly to develop (see Figure 20.1).

The cis-regulatory elements of *Ultrabithorax* span over 120 kb (see Figure 20.12a), and their modularity allows the evolution of changes in gene expression without catastrophic disruption of *Ultrabithorax* function, such as those caused by nonsense mutations within the coding

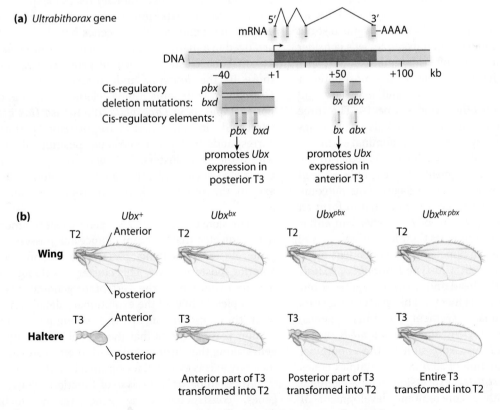

**Figure 20.12** Mutations in cis-regulatory elements of *Ultrabithorax* cause homeotic transformations.

region. Thus, *Ultrabithorax^{bithorax} Ultrabithorax^{postbithorax}* double mutants survive to adulthood because the remainder of the cis-regulatory elements controlling *Ultrabithorax* expression are intact. Genetic Analysis 20.1 asks you to evaluate cross-regulatory interactions among *Hox* genes.

---

Genetic Insight The eight (in total) *Hox* genes of the *bithorax* and *Antennapedia* complexes act in a combinatorial manner to specify *Drosophila* parasegmental identity. *Hox* genes have modular cis-regulatory regions containing separate enhancer modules that direct expression in discrete parasegments.

---

## Downstream Targets of *Hox* Genes

Given that combinatorial action of the *Hox* genes specifies parasegment identity and that *Hox* genes encode transcription factors, it follows that the downstream target genes activated by the *Hox* genes must differ between segments. These *Hox* target genes have been called **realizator genes,** and their expression contributes to the characteristic morphology of each segment. The transcriptional regulation of realizator genes by *Hox* genes is thought to occur in a manner similar to the regulatory systems described above, in which enhancer modules are composed of multiple binding sites for several distinct transcription factor inputs. As an example, let's consider the formation of appendages on each segment.

Wild-type flies have antennae on the most-anterior head segment and have mandibles and maxillary and labial sense organs on other head segments. The three thoracic segments have legs; T2 and T3 also have wings and halteres, respectively. The eight abdominal segments lack appendages. Loss of all *Hox* activity is lethal to the embryo and causes all segments to resemble a head segment having antennae as appendages. This outcome indicates that all segments have the potential to form an appendage, but that expression of *Hox* genes can either specify their identity or represses their formation.

The formation of an appendage is dependent upon a gene called *Distal-less*. In wild-type *Drosophila*, *Distal-less* is expressed in the head and thoracic segments but not in any abdominal segments. This pattern suggests that the abdominal segment identity genes, *Ultrabithorax*, *abdominal-A*, and *Abdominal-B*, negatively regulate *Distal-less* expression in the abdominal segments. Loss of function of all *bithorax* complex genes results in ectopic *Distal-less* expression in all abdominal segments, along with a concomitant development of appendages (legs) on all abdominal segments. Conversely, if *Ultrabithorax* is ectopically expressed at high levels throughout the embryo, *Distal-less* is not activated in any segment and no appendages are formed. Thus, action of specific *bithorax* complex Hox proteins on *Distal-less* cis-regulatory sequences represses *Distal-less* gene expression in the abdominal segments. The identity of the appendages is determined by the combinatorial activity of the *Hox* genes in conjunction with *Distal-less*. For example, the identity of the T1 leg is specified by *Distal-less* and *Sex combs reduced*, whereas the identity of the T2 leg is specified by *Distal-less* and *Antennapedia*.

## *Hox* Genes in Metazoans

Soon after the discovery of *Hox* gene clusters in *Drosophila*, researchers began to inquire whether *Hox* genes are a peculiarity of *Drosophila* development, or whether they are found in a broader range of species. Many developmental biologists did not expect to find *Hox* genes in other animals, since *Drosophila* is a segmented insect and there was no reason to expect that other animals would use the same genes to direct very different developmental programs. However, cross-hybridization studies using *Drosophila Hox* sequences as molecular probes revealed *Hox* gene sequences in the genomes of all animals, including insects, spiders, molluscs, and vertebrates (such as humans). This revelation suggested a common developmental mechanism among animals and affirmed their close evolutionary relationships.

Subsequent experiments showed not only that most animals have clusters of *Hox* genes but also that they are arranged in a manner similar to that in *Drosophila* (Figure 20.13). Each cluster consists of genes corresponding to those in the *bithorax* and *Antennapedia* clusters of *Drosophila*, with some minor deletions and duplications. For example, as in *Drosophila*, the mouse *Hox* genes are expressed in an anterior-to-posterior pattern that corresponds to the chromosomal position of the genes within the *Hox* clusters. This pattern suggests that *Hox* genes also specify identity along the anterior–posterior axis of the mouse and, by extension, of mammals in general.

The conservation of *Hox* gene clusters among animals indicates that a common ancestor possessed a *Hox* gene cluster specifying pattern formation along its anterior–posterior axis. This cluster was duplicated during the evolution of the vertebrate genome, which has four copies. While all the evolutionary details are not yet clear, the conservation of the *Hox* complexes over 500 million years suggests that the spatial colinearity of *Hox* genes along the chromosome with their expression along the body axis is essential for optimal functionality.

Mice embryos with loss-of-function alleles of *Hox* genes, constructed using gene-targeting techniques described in Chapter 17, exhibit defects in the identity of serially repeated structures. For example, loss of *Hox*

Why do loss-of-function mutations in *bithorax* complex genes result in homeotic transformations of parasegments into identities that correspond to more-anterior parasegments, whereas gain-of-function mutations (see Section 4.1) tend to result in identities corresponding to more-posterior parasegments?

| Solution Strategies | Solution Steps |
|---|---|
| **Evaluate** | |
| 1. Identify the topic this problem addresses, and explain the nature of the required answer. | 1. The subject of this question is the effect of mutations in the *bithorax* complex on segment pattern formation. The answer requires descriptions of why loss-of-function mutations lead to segments that resemble more-anterior segments, whereas gain-of-function mutations lead to the formation of segments that resemble more-posterior segments. |
| 2. Identify the critical information given in the problem. | 2. The question suggests there is a key difference between the effects of loss-of-function mutations and gain-of-function mutations of the *bithorax* complex. |
| **Deduce** | |
| 3. Describe the general patterns of expression and segmental pattern formation resulting from the normal expression of homeotic genes. <br><br> TIP: Use *Hox* genes as an example of a set of developmental genes. | 3. Homeotic genes, such as the *Hox* genes, specify segment identity in a combinatorial manner through overlapping expression domains in parasegments. Each gene has a well-defined anterior boundary but a more diffuse posterior boundary. Cross-regulatory interactions refine *Hox* gene expression domains, so that more-posterior genes repress more anteriorly expressed genes. |
| 4. Describe the general pattern of expression and the normal segmental pattern formation of *bithorax* genes. | 4. The *bithorax* complex consists of three genes, *Ubx*, *abd-A*, and *Abd-B*. *Ubx* is expressed in the anterior abdominal segments and posterior thoracic segments, *abd-A* is expressed in the middle abdominal segments, and *Abd-B* is expressed in the posterior abdominal segments. Segment identity is specified by the combination of *Hox* gene products and their levels of expression. |
| **Solve** | |
| 5. Explain why loss-of-function mutations of *bithorax* genes lead parasegments to take on a more-anterior identity. <br><br> TIP: Consider the cross-regulatory interactions of the *Hox* genes. | 5. The loss of function of a posterior gene leads to both the absence of expression of the mutant gene and posterior expansion in the expression domains of more-anterior genes. For example, the posterior gene *Abd-B* acts to repress *abd-A* in the most-posterior segments. Loss-of-function mutations in *Abd-B* result in a posterior expansion of *abd-A* expression into more-posterior abdominal segments. The result is that both middle and posterior abdominal segments acquire an identity that is similar to that of the middle abdominal segments—a homeotic transformation to more-anterior identity. |
| 6. Explain why gain-of-function mutations of *bithorax* genes lead parasegments to take on a more-posterior identity. <br><br> TIP: Gain-of-function *Antennapedia* mutations cause legs (a posterior structure) to develop in the position normally occupied by antennae (an anterior structure). | 6. Gain-of-function mutations cause gene expression at inappropriate times and locations. Gain-of-function alleles often, but not always, result in *Hox* gene expression in a more-anterior domain than in wild-type animals, thus resulting in homeotic transformations to a more-posterior identity. |

For more practice, see Problems 6, 7, 22, and 26.

function results in a homeotic transformation of the lumbar and sacral vertebrae, which do not normally bear ribs, into structures resembling more-anterior thoracic vertebrae that do carry ribs (see Figure 16.1). These and additional *Hox* gene mutations suggest *Hox* genes direct the development of body plans in chordates as well as in annelids, arthropods, molluscs, nematodes, and other animals.

Studies of *Hox* complexes in other metazoans reveal that gene duplication took place before the divergence of

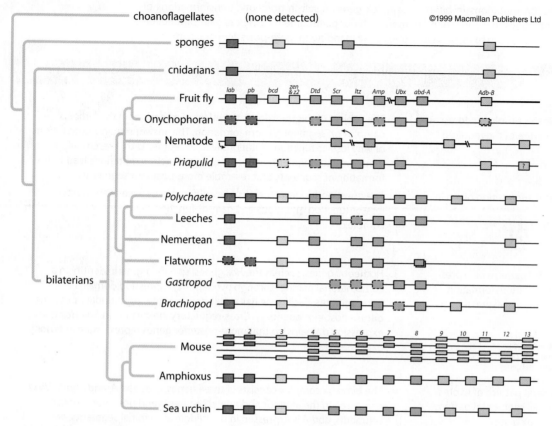

©1999 Macmillan Publishers Ltd

**Figure 20.13 Occurrence and arrangement of *Hox* complexes in metazoans.** *Hox* genes have not been detected in choanoflagellates, single-celled organisms that represent the sister clade to metazoans, but they are present in all metazoans. In the vertebrate lineage (exemplified by the mouse), the entire complex has been duplicated twice, resulting in four *Hox* complexes. Such events have produced duplicated genes that were later co-opted to new developmental functions.

bilaterian animals (animals that have bilateral symmetry). Thus, all bilaterian animals have essentially the same homeotic gene toolkit to pattern their anterior–posterior axis. This homology indicates that the differences between animals reflect how the toolkit is employed rather than differences in the component parts. Indeed, large-scale sequencing of cnidarian (jellyfish, sea anemone) genomes DNA suggests that other components of the genetic toolkit are also largely shared by all metazoans. Given that all animals share fundamental developmental patterning processes and genes, much of what we learn from the study of model animals such as *Drosophila*, *Caenorhabditis elegans*, and mice can be extended to other members of the animal kingdom, including ourselves.

Genetic Analysis 20.2 asks you to design an experimental strategy to genetically dissect development in another group of eukaryotes.

---

**Genetic Insight**  Metazoans share a common developmental genetic toolkit, thus suggesting that gene regulation carried out by *Hox* genes largely directs the development of the body plan in diverse metazoan species.

## Stabilization of Cellular Memory by Chromatin Architecture

The preceding sections describe how the basic body plan of *Drosophila* is established in early embryogenesis by the action of coordinate, gap, and segmentation genes and through spatially restricted patterns of *Hox* gene expression that specify segmental identity. The patterns of *Hox* gene expression are then faithfully propagated throughout the remainder of embryonic development. The proteins that activate *Hox* gene expression have an ephemeral pattern of expression; it disappears soon after *Hox* expression patterns are initiated. Thus, one challenge cells face during embryonic development is for specific lineages to maintain their identity as they proliferate.

Genetic screens for homeotic genes reveal that mutations at loci other than those encoding the *Hox* genes can also produce homeotic mutant phenotypes. In general, mutations at these other loci fall into two classes. The first class, exemplified by *trithorax* mutations, produces phenotypes reminiscent of multiple *Hox* loss-of-function mutations. In contrast, phenotypes of mutants of the second class, exemplified by *Polycomb* mutations, often resemble multiple gain-of-function homeotic gene mutants.

You are interested in the development of the body plan of kelp, a common brown alga found along many coastlines. Would reverse or forward genetics approaches—as represented, respectively, by looking for gene homology and by the use of mutagenesis—be more suited to identifying the genes required for early kelp development?

| Solution Strategies | Solution Steps |
|---|---|
| **Evaluate** | |
| 1. Identify the topic this problem addresses, and explain the nature of the required answer. | 1. This problem concerns the investigation of genes determining development of kelp. Devising an answer requires evaluating the relative potential of reverse genetic analysis versus forward genetic analysis (see Chapters 16 and 17). |
| 2. Identify the critical information given in the problem. | 2. Kelp is identified as brown algae, a form of life distinct from land plants and animals. |
| **Deduce** | |
| ③. Determine if looking for gene homology (a reverse genetic approach) has a high probability of successfully identifying developmental genes in kelp. | 3. Examination of Figure 18.12 indicates that kelp is only distantly related to either land plants or animals. Therefore, searching for brown algal genes based on the sequences of plant or animal developmental genes is something of a fishing expedition. |

**TIP:** Review Figure 18.12, to find the relationship of kelp to the other organisms you have been studying.

**PITFALL:** Distantly related organisms are likely to have evolved substantially since they last shared a common ancestor, and the extent of gene homology decreases as evolutionary distance between species increases.

| | |
|---|---|
| **Solve** | |
| 4. Determine whether the use of mutagenesis (a forward genetic approach) is likely to help identify kelp developmental genes. | 4. A good approach to finding developmental genes is to perform a mutagenesis experiment that will identify mutants in which pattern formation is perturbed. Mutagenesis can potentially affect any gene; thus, the forward genetic approach is not biased or restricted to genes that share homology with genes in other species. Mutants displaying abnormalities of wild-type pattern formation are likely to carry mutations of pattern-forming genes. |

**TIP:** How were genes that regulate development in *Drosophila* originally identified?

For more practice, see Problems 17, 19, 23, 25, and 28.

---

At the molecular level, expression of multiple *Hox* genes is found to be ectopic in *Polycomb* mutants and reduced in *trithorax* mutants. While *Hox* gene expression is established normally in both *Polycomb* and *trithorax* mutants, the expression either fails to be maintained (*trithorax* mutants) or is later activated in inappropriate locations (*Polycomb* mutants). Thus, rather than "remembering" what type of tissue they are destined to form, mutant *trithorax* and *Polycomb* cell lineages appear to "forget" their identity.

Both *trithorax* and *Polycomb* encode proteins that act in large protein complexes whose function is to modulate chromatin structure. Components of the complexes are encoded by genes known, respectively, as the trithorax group (trxG) genes and the Polycomb group (PcG) genes. Both the trxG and PcG protein complexes bind to specific DNA sequences, and each complex possesses a distinct type of histone-3-methyltransferase activity (see Section 15.3) in which the activity of the trxG complex is opposite to the activity of the PcG complex. The PcG complexes repress target gene expression by recruiting histone-modifying protein complexes capable of histone deacetylation. In contrast, trxG complexes recruit protein complexes that acetylate histone, leading to maintenance of active gene expression. These two types of modification are associated with transcriptionally inactive heterochromatin and transcriptionally active euchromatin, respectively (see Chapter 15). It is believed that trxG and PcG complexes are recruited to the cis-acting regulatory sequences of *Hox* genes to "lock" the chromatin into a particular form, allowing maintenance of either active or silent states of gene expression. In this way, these proteins provide a type of epigenetic cellular memory that is propagated through cell divisions occurring long after the initial activators of *Hox* gene expression patterns have disappeared.

Study of *trithorax* and *Polycomb* mutants has helped clarify that the establishment of euchromatic or heterochromatic chromatin at specific developmental genes is a primary mechanism by which the potential fates of cells become restricted as development proceeds from totipotent zygote to differentiated cell types. The relative rigidity or plasticity of these different chromatin states is directly responsible for a cell's ability to express some genes and not express others, thus influencing the developmental potential of particular cell types.

---

**Genetic Insight** Changes in chromatin structure provide an epigenetic mechanism of cellular memory during development. The formation of euchromatin and heterochromatin at specific developmental genes contributes to the restriction of potential developmental fates as cells differentiate.

---

## 20.3 Cellular Interactions Specify Cell Fate

We noted earlier in this chapter that cell fate is often triggered by cell signaling. The development of the *Caenorhabditis elegans* vulva provides an example of how inductive and inhibitory signals between cells direct the differentiation of distinct developmental fates in a group of pluripotent cells. While the adult *C. elegans* only contains about 1000 cells, its development provides a model of organogenesis. John Sulston, Sydney Brenner, and Robert Horvitz shared the Nobel Prize in Physiology or Medicine in 2002 for their research on the genetic regulation of organ development and programmed cell death in *C. elegans*.

### Inductive Signaling between Cells

*Caenorhabditis elegans* is a hermaphrodite nematode worm in which external genitalia, the vulva, forms a portal to the uterus through which eggs are laid. Early in their development, hermaphroditic worms produce sperm, which they store for later use. Eggs are subsequently produced in the gonads, fertilized with the stored sperm, and then extruded through the vulva. The vulva forms during the last larval stage, from six precursor cells called vulval precursor cells (VPCs); see **Figure 20.14a–b**. Three of these larval cells give rise to structures of the vulva itself: One is called the primary (1°) cell, and the other two are called secondary (2°) cells of the vulva. The other three cells differentiate into hypodermis and are called tertiary (3°) cells. The VPC closest to a specific gonadal cell, the

**(a)** Six cells, P3.p to P8.p, have potential to develop into vulva.

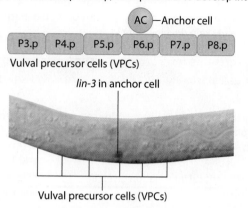

**(b)** The three cells closest to anchor cell P5.p to P7.p, form the vulva; the other cells develop into hypodermis.

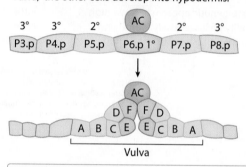

One cell has 1° identity and forms the central part; two flanking cells adopt 2° fate and form peripheral parts.

**(c)** Loss of the anchor cell results in loss of vulval development; all cells adopt hypodermal fate.

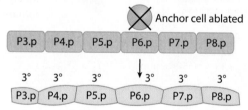

**(d)** Inductive signal from anchor cell induces vulval cell differentiation.

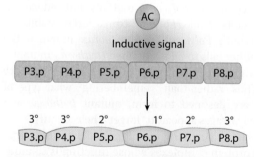

**Figure 20.14** **Inductive signaling during vulval development in** *C. elegans.*

anchor cell, differentiates as a 1° cell and forms the central part of the vulva. The two cells flanking the 1° cell differentiate as 2° cells and form the peripheral regions of the vulva. The 1° and 2° fates can be easily distinguished by their distinct cell-division patterns.

Initially, each of the six VPCs has the potential to differentiate along any of the pathways—1°, 2°, or 3°. This flexible cell-fate potential is demonstrated by laser-ablation experiments that destroy the anchor cell or one or more VPCs (**Figure 20.14c**). If the anchor cell is destroyed, no vulva will form, because all six VPCs differentiate with a 3° fate and become hypodermis. This suggests that the anchor cell must be present to induce VPCs to differentiate with 1° or 2° fates and thus form the vulva. Alternatively, if the VPC closest to the anchor cell is ablated, one of the cells that would normally differentiate with a 2° fate develops with a 1° fate instead, and the two cells flanking this new 1° cell differentiate as 2° cells, thus suggesting that any of the VPCs can differentiate with a 1° or 2° fate. Because all the VPCs are capable of the same kinds of differentiation, they are said to form an equivalence group.

What limits the number of VPCs destined to form the vulva to three? Given the loss of both the 1° and 2° fates when the anchor cell is removed, researchers hypothesized that the anchor cell might provide an **inductive signal** to induce vulval cell differentiation (**Figure 20.14d**). If this inductive signal is disseminated in a gradient, the cell closest to the anchor cell could acquire a different fate than cells that are more distant. Indeed, genetic analysis supports this inductive interaction model of vulval development.

As predicted by the inductive interaction model, mutations that eliminate either the inductive signal or the ability of cells to respond to the inductive signal result in a loss of vulval development, and all VPCs differentiate as hypodermis (**Figure 20.15a**). This mutant phenotype is called the vulva-less phenotype. In contrast, mutations that disseminate the inductive signal to all VPCs cause all VPCs to differentiate into vulval cells, producing a multi-vulva phenotype. Multi-vulva mutants lay eggs similarly to normal worms; however, the fertilized eggs of vulva-less worms cannot be laid and instead develop and hatch inside the mother's uterus. This abnormality is called the "bag of worms" phenotype. Progeny developing in the uterus eventually consume their mother from the inside and then hatch out of the carcass.

Recessive loss-of-function alleles at several loci produce a vulva-less phenotype. These genes encode proteins that act either in the production of the inductive signal from the anchor cell or that facilitate cell response to the inductive signal (**Figure 20.15b**). For example, the *lin-3* gene encodes a small, secreted protein expressed only in the anchor cell and is the inductive signaling molecule (see Figure 20.14a). Mutations that result in a loss

**(a)** Wild type

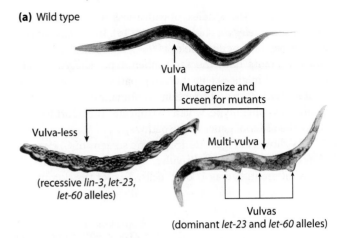

Vulva

Mutagenize and screen for mutants

Vulva-less

Multi-vulva

(recessive *lin-3*, *let-23*, *let-60* alleles)

Vulvas
(dominant *let-23* and *let-60* alleles)

**(b)**

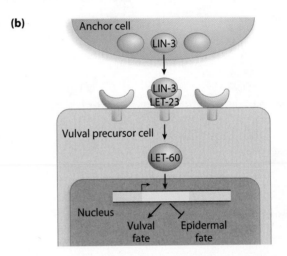

Anchor cell

LIN-3

LIN-3
LET-23

Vulval precursor cell

LET-60

Nucleus

Vulval fate     Epidermal fate

**Figure 20.15 Genetic analysis of vulval development in *C. elegans*.**

of active lin-3 protein result in the loss of the inductive signal from the anchor cell. In contrast, the *let-23* and *let-60* genes are expressed in the VPCs and act as the receptor (let-23) for the *lin-3*–encoded signal and as a signal transduction molecule (let-60) that communicates the signal from the plasma membrane to the nucleus, where changes in gene expression are induced. The absence of a receptor for lin-3, or the inability to transmit receipt of the signal, blocks the normal developmental fate of VPCs.

Epistatic analysis of developmental pathways, conducted by studying recessive and dominant mutations of the same gene, is used to identify a group of genes that interact to control a particular cellular process or pathway and to establish an order-of-function map for the genes in the pathway (see Section 4.3). Genetic analysis of developmental pathways can be more complicated than analysis of biochemical pathways because often there is no way of assaying intermediate steps in the developmental pathway. The analysis of double mutants and the availability of gain-of-function alleles can be crucial in these endeavors, as the studies of vulva-less and multi-vulva mutants in *C. elegans* show (**Figure 20.16**). In the case of recessive loss-of-function alleles of *lin-3*, *let-23*, and *let-60*, all single mutants have the same phenotype, suggesting all these genes might act in the same pathway. However, all double-mutant loss-of-function combinations also exhibit a vulva-less phenotype, which complicates the effort to discover the order of genes in the pathway.

As shown in Figure 20.15, genetic screens of *C. elegans* identified dominant multi-vulva mutations in which all VPCs differentiated as 1° or 2° cells. Two of the dominant mutations mapped to the same positions as *let-23* and *let-60*, suggesting that they might be gain-of-function alleles of these genes, and both dominant mutant alleles proved to be epistatic to recessive loss-of-function alleles of *lin-3* (i.e., the double mutants have a multi-vulva phenotype like the *let-23* and *let-60* gain-of-function single mutants), as outlined in Figure 20.16. The double-mutant phenotype indicates that the gain-of-function alleles of either *let-23* or *let-60* do not require the function of *lin-3* to exert their phenotypic effects, thus placing both *let-23* and *let-60* downstream of *lin-3*.

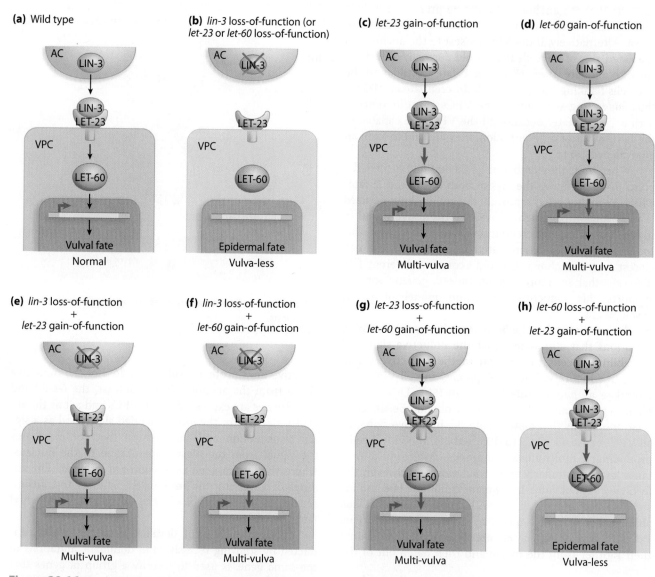

**Figure 20.16 Analysis of double-mutant phenotypes to find order of genes in developmental pathways. (a)** In wild-type worms, the vulva developmental pathway is active only in the presence of the signal (lin-3). **(b)** In *lin-3* mutants, no signal is present, and worms develop with a vulva-less phenotype. **(c)** and **(d)**, In either *let-23* or *let-60* gain-of-function alleles, the pathway is constitutively active, and worms develop with a multi-vulva phenotype. **(e)** and **(f)**, Gain-of-function alleles of *let-23* and *let-60* are epistatic to loss-of-function *lin-3* alleles. The pathway is constitutively active regardless of whether the *lin-3* signal is present. **(g)** and **(h)**, Gain-of-function alleles of *let-60* are epistatic to loss-of-function alleles of *let-23*. Conversely, loss-of-function alleles of *let-60* are epistatic to gain-of-function alleles of *let-23*. This places *let-60* downstream of *let-23*.

Similar analysis enables the ordering of the *let-23* and *let-60* genes in the pathway (see Figure 20.16). Dominant *let-60* alleles are epistatic to recessive *let-23* alleles, indicating that *let-60* can function in the absence of functional *let-23*, a finding that places *let-60* downstream of *let-23*. This conclusion is supported by the converse experiment, where recessive *let-60* alleles are epistatic to dominant *let-23* alleles, which indicates that *let-23* requires the function of *let-60* to exert a phenotypic effect. The genetic pathway was determined before the nature of the proteins had been analyzed. However, now that we know the molecular identities of lin-3 (signal), let-23 (receptor), and let-60 (signal transduction molecule), these epistatic relationships make sense. For example, dominant gain-of-function mutations of *let-60* result in constitutive activity of this protein, allowing it to transduce a signal independent of the state of the let-23 receptor. Likewise, gain-of-function alleles of *let-23* act as if they are receiving a signal all the time, whether or not *lin-3* is functional, and thus activate the downstream signal-transduction cascade, which in turn depends on having a functional allele of *let-60*.

---

**Genetic Insight**  Mutant analysis is used to reconstruct genetic pathways by which cell-fate decisions occur.

---

## Lateral Inhibition

Given that they are both induced by the *lin-3*–encoded signal, how are the 1° and 2° fates specified? One possibility is a differential response of the VPCs to a graded *lin-3* signal, where the highest concentration of signal produces a 1° fate and a lower concentration of signal produces 2° cells. However, when the cell that would normally be a 1° cell is ablated, a cell that would normally have been a 2° cell differentiates into a 1° cell instead. It is thus unlikely that the absolute concentration of signal perceived is solely responsible for directing cell fate.

A more likely explanation is that after reception of the *lin-3* signal, a second signal is sent from the 1° cell that inhibits the neighboring cells from becoming 1° cells (**Figure 20.17a**). This process is termed **lateral inhibition,** where an initial asymmetry is reinforced by signalling between adjacent cells (**Figure 20.17b**). All VPCs initially have the potential to express a lateral signal, encoded by the *lag-2* gene, and to express the receptor for the lag-2 signal, encoded by the *lin-12* gene. The *lag-2* gene is activated in response to the lin-3 signal, so it is expressed at higher levels in the 1° cell. Reception of lag-2 results in down-regulation of the *lag-2* gene in the receiving cells and up-regulation of its receptor, *lin-12* (Figure 20.17b). This creates a feedback loop that reinforces the initial asymmetry between the 1° and 2° cells. Continued feedback between the signal and its perception amplifies the differences between the two cells, causing them to acquire distinct developmental fates.

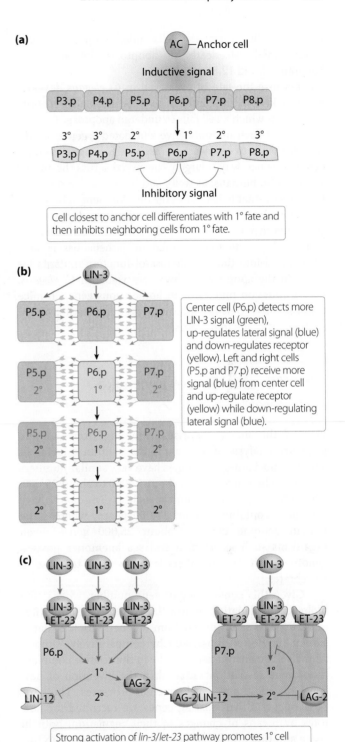

(a) Cell closest to anchor cell differentiates with 1° fate and then inhibits neighboring cells from 1° fate.

(b) Center cell (P6.p) detects more LIN-3 signal (green), up-regulates lateral signal (blue) and down-regulates receptor (yellow). Left and right cells (P5.p and P7.p) receive more signal (blue) from center cell and up-regulate receptor (yellow) while down-regulating lateral signal (blue).

(c) Strong activation of *lin-3/let-23* pathway promotes 1° cell fate, in turn activating the *lag-2/lin-12* pathway, which promotes a 2° cell fate in neighboring cells.

**Figure 20.17** Lateral inhibition in *C. elegans* vulval differentiation.

## Cell Death during Development

One of the striking observations made when Sulston, Brenner, and Horvitz tracked the fate of every cell during *C. elegans* development is that many cells are fated to die. Of the 1090 cells produced during the development of a hermaphrodite worm, 131 cells undergo a process called programmed cell death, or apoptosis (introduced in Sections 3.1 and 12.5).

Because the fate of every cell in *C. elegans* development is known, researchers have been able to identify mutants in which a cell fails to undergo apoptosis. Genetic analyses of such mutants have elucidated a genetic pathway that leads to cell death in response to a signaling molecule. This pathway is largely conserved across the animal kingdom (in humans, as well), and is a natural and important process that helps sculpt the development of tissues as well as maintain tissues in adult organisms. Indeed, it is estimated that $10^{11}$ cells are programmed to die every day in an adult human, many of them in epithelial tissues such as skin and intestine. While loss-of-function mutants for genes in the apoptosis pathway are viable in *C. elegans*, loss-of-function mutations in homologous genes in mice result in embryo death, indicating that cell death is an essential part of life in mammals.

## 20.4 "Evolution Behaves Like a Tinkerer"

One of the major surprises emerging from genome sequence analysis of animals is that, within a factor of about 2, most animal genomes have very similar numbers of genes. The range is from about 12,000 to about 25,000. Thus relatively simple animals such as *Drosophila* have a genome containing about 14,000 genes, whereas the human genome contains about 25,000 genes. Even organisms such as jellyfish and sea anemones possess genomes with gene numbers largely similar to those of vertebrates.

Given this consistency of gene number, what is the biological explanation of how the presumed "complexity" of vertebrates is produced from a genetic toolkit that is similar to the one possessed by comparatively "simple" animals? The answer seems to lie in the relative complexity of gene regulation rather than the invention of new genes for additional developmental processes. This proposal suggests that existing genes are recruited for new roles by means of changes in their regulation, both in space and time. Biologist Francois Jacob summed up this view of evolution when he said, "Evolution behaves like a tinkerer. . . . [It] does not produce novelties from scratch. It works on what already exists, either transforming a system to give it new functions or combining several systems to produce a more elaborate one."

A common theme in the evolutionary history of all genes, and particularly those influencing development, is the **co-option** of genes and genetic modules to direct the patterning or growth of novel organs. In this section, we consider an example of the co-option of genes by evolutionary "tinkering" to form newly evolved structures: digits (fingers and toes) on tetrapod limb appendages such as hands and feet. The study of the evolution of development is often referred to as **evo-devo.**

## Evolution through Co-option

Limb positioning in tetrapods (four-legged vertebrates) results in large measure from the expression of *Hox* genes that direct the anterior–posterior organization of the body. Work on both chickens and mice demonstrates that expression of *Hox* genes along the anterior–posterior body axis defines the position at which a limb will develop. The anterior limit of the expression domains of two *Hox* genes, *Hoxc8* and *Hoxc6*, demarcates the position of the forelimb, and the posterior limit of expression marks the position of the hindlimbs (**Figure 20.18a**). The expression of these two genes specifies the thoracic region of vertebrates, which is characterized by the formation of ribs from the vertebral column.

Once limb positions are specified, cells of the mesenchyme (loosely connected sub-ectodermal cells) send a signal to the overlying ectodermal cells. This signal promotes changes within a narrow band of cells that then forms the apical ectodermal ridge (AER), whose primary function is to direct limb-bud outgrowth by responding to signals produced in a group of mesenchymal cells toward the posterior side of the limb bud called the **zone of polarizing activity (ZPA;** Figure 20.18b). The ZPA acts as an organizer that promotes digit formation at the distal ends of limb buds through the production of a morphogen, now known to be a small secreted signaling protein called Sonic hedgehog (Shh). The *Sonic hedgehog (Shh)* gene is orthologous to the *Drosophila* segment polarity gene *hedgehog*. *Sonic hedgehog* is expressed principally in the neural tube, where it helps organize the brain, eyes, and other structures through patterning of a group of cells known as the floor plate, and in developing limbs, where it directs the development of digits. The Case Study in this chapter discusses the consequences of different *Shh* mutations on mammal development and morphology.

All extant tetrapods are characterized by five or fewer digits in each set, and each digit in the set has a unique identity. Tetrapod digits arise along the anterior–posterior axis of the limb bud. If you allow your arms to hang straight down, you will see that your thumb (digit 1) is in the anterior position on your hand, while your pinky (digit 5) is in the posterior position. *Sonic hedgehog* expressed in the ZPA plays an important role in initiating digit formation, and loss-of-function alleles of *Shh* result in a loss of digits 2–5; only digit 1 forms independently of *Shh* function. A second role of *Shh* in limb patterning is in the specification of digit identity. Experiments where a second ZPA is transplanted to an anterior position result in a mirror-image duplication of digits, suggesting that the ZPA instructs those digits closer to the ZPA to differentiate with posterior identity (see Figure 20.18b).

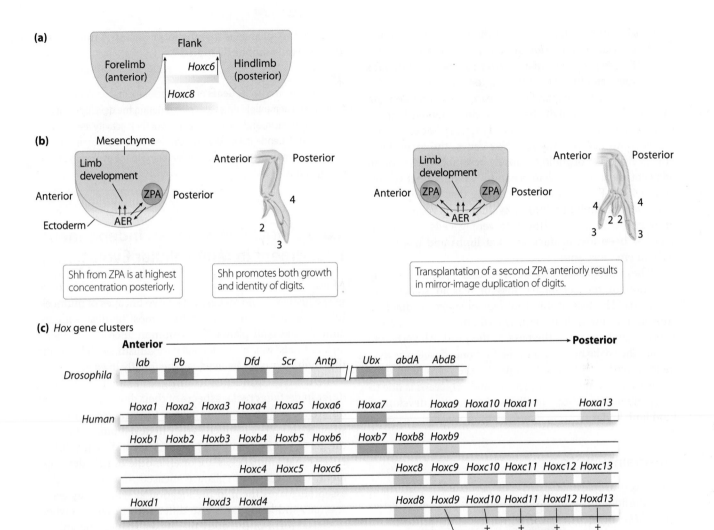

**Figure 20.18 Limb-position and digit determination.**

The *Hox* genes that play a conserved role in patterning the anterior–posterior axis in animals were considered likely to be the genes acting downstream of *Shh* to specify the patterning events in digits. In mice, five *Hox* genes are expressed in the limb bud at the time and place where the digits are developing: *Hoxd9, Hoxd10, Hoxd11, Hoxd12,* and *Hoxd13* (**Figure 20.18c**). These genes are also expressed in the posteriormost regions of the mouse embryo, where they contribute to patterning along the anterior–posterior body axis, and later in the developing nervous system. Despite the difference in position of hindlimb and forelimb along the body axis, the same five *Hox* genes are expressed in the developing digits of each limb. Their expression in the limb bud follows a precise temporal and spatial pattern and is dependent on *Shh* activity. The first gene to be expressed is *Hoxd9,* followed by *Hoxd10,* then *Hoxd11,* and so on through *Hoxd13.* Spatially, all genes share the same posterior boundary, but the anterior boundary of expression is different for each gene. Consequently, the five *Hoxd* genes subdivide the limb bud into five zones, each specified by a different combination of *Hoxd* gene expression. Analogous to patterning along the anterior–posterior axis, ectopic expression of different *Hoxd* genes within the developing limb bud results in transformations of digit identity. A similar combinatorial code of *Hox* gene expression also appears to specify the proximal–distal patterning of the limb buds (e.g., upper arm, forearm, hand, digits).

Mutations that expand or increase *Shh* expression and thus result in extra digits have been documented in mice, chickens, dogs, cats, and humans. However, because identity is controlled by only five *Hox* genes, the extra digits always have a morphology closely resembling that of an adjacent digit, rather than having a unique identity (see Figure 4.13). Finally, it is worth noting that the separation of the human limb bud into individual digits requires programmed cell death of the intervening cells—a process that has been lost in duck and bat limbs and has led to webbing in those animals.

These programs have been further modified during evolution in the secondary loss of legs in snakes and cetaceans. The loss of the front legs of snakes is due to a shift in anterior shift in both *Hoxc6* and *Hoxc8* gene expression all the way to the base of the head. All vertebrae behind the snake head, except the first one, develop as thoracic vertebrae with ribs. In contrast, the convergent evolution of loss of hind legs in snakes and cetaceans is due to independent alterations in *Shh* activity in the developing hind limb bud.

### Constraints on Co-option

The ancestral roles of *Hoxd* genes pertained to patterning along the anterior–posterior axis of the body. Therefore, the role of *Hoxd* genes in specifying digit identity represents a co-option of function of already existing genes. These same genes also acquired roles in the later differentiation of the nervous system. Likewise, the presence of the floor plate in all vertebrates is an indication that the floor plate evolved before limbs during vertebrate evolution. Limbs developed later within the tetrapod lineage; and in the course of limb evolution, *Shh* was co-opted to pattern digits, structures that did not previously exist. By what process are genes co-opted for new functions during evolution?

In the case of limb evolution, genes of the *Hoxd* cluster could have come under control of regulatory factors that caused the *Hoxd* genes to be expressed in developing limbs via the addition of limb-specific enhancer modules. As long as these changes in regulation did not disrupt the expression of the *Hoxd* genes during anterior–posterior patterning of the body axis, the changes would not result in defects of this earlier process. The change in gene expression pattern in the developing limb could be thought of as a gain-of-function mutation. The modularity of enhancers and silencers facilitates evolution by co-option because individual enhancer modules are free to evolve independently. Thus the patterning of a novel tetrapod organ, the limb, involved the co-option of, or tinkering with, preexisting genetic programs that already had developmental roles elsewhere. As noted above, the constraint on this type of evolutionary change is that the more ancestral functions of the gene must not be disrupted.

---

Genetic Insight  Genes of the developmental toolkit are engaged at multiple times and places during the development of an organism, and the consequences of their activity are context dependent. Co-option of existing genetic programs to pattern new structures is a common theme in the evolution of organisms.

---

## 20.5 Plants Represent an Independent Experiment in Multicellular Evolution

Multicellularity has evolved independently more than once in the history of life on Earth. The two lineages of multicellular organisms you are likely to be most familiar with are animals and land plants. The common ancestor of plants and animals was a single-celled organism, which means that multicellularity evolved independently in each lineage.

Due to their independent origins, animals and plants differ in certain crucial aspects of their development. One difference is that germ-line cells in animals separate from somatic (body) cells much earlier in development than do the germ-line cells in land plants. A second major difference is that animal cells are often motile during development, whereas plant cells are encased in a cell wall that essentially fixes them in the location at which they arise. Animals and land plants also differ with respect to when the basic form of the body plan takes shape. The animal body plan is established during embryogenesis, and subsequent development consists primarily of growth in size but without the addition of new organs. In contrast, throughout their lifetimes plants add new organs that are produced from pluripotent stem-cell populations. Finally, because plants often grow in a fixed location and are unable to migrate to different environments as many animals can, a plant must be able to alter its developmental program in response to changing environmental conditions throughout its lifetime. Thus, while identical twins in animals are nearly indistinguishable, genotypically identical plants may develop to look very different depending upon their growth environment. Despite such differences, the processes occurring during development in plants are remarkably similar to those in animals, especially in their reliance upon the coordinated action of transcription factors and signaling molecules.

### Development at Meristems

Plant growth occurs at organized groups of pluripotent cells called **meristems.** The two functions of meristems are generation of organs and self-maintenance (to ensure that a pool of stem cells is always present). The shoot meristem, which is the type of meristem that generates leaves, is divided into three functional domains—a peripheral zone from which leaves are formed, a rib zone from which part of the stem is derived, and a central zone that acts as a stem-cell reservoir to replenish cells lost to the developing leaves and stem (**Figure 20.19**). Meristems are generally indeterminate—that is, they can remain active for years, or in some cases the entire life of the plant—and through this time, the sizes of the central and peripheral domains remain remarkably constant. For example, the shoot meristem at the top of a pine tree can be active for centuries, continually producing leaves and side branches. It is the continual production of new organs from meristems throughout the life of a plant that allows plants to adjust and adapt to changing local environmental conditions.

The identity of the meristem determines what types of organs are produced from its periphery. Early in the life of a flowering plant, leaves are produced from the flanks of the shoot meristem, and roots are produced from the root meristem. In each leaf axil (meaning the upper side of the attachment point of the leaf to the stem) an axillary meristem is formed, from which a branch can arise. This reiterative formation of meristems that produce leaves that produce branches containing meristems forms the basis of most aboveground development of flowering plants. In response to appropriate environmental conditions, the identity of meristems can change. For example, shoot meristems, which have been producing leaves, are converted in response to seasonal changes into reproductive meristems. Reproductive meristems may either develop directly into a flower meristem, or alternatively produce leaves with flower meristems in their axils. In turn, flower meristems produce floral organs from their peripheral zones. Unlike the other meristems, flower meristems are determinate: No more stem cells are available after the flower meristem has produced a fixed number of organs.

Because each type of meristem is characterized by a specific pattern of gene expression, mutations in key genes can result in homeotic transformations of meristem types. We have all eaten one such mutant, cauliflower, in which meristems that would normally be specified as flowers behave instead as inflorescence meristems (see Figure 20.19, lower right). The genetic basis of this phenotype has been identified in *Arabidopsis* as loss-of-function alleles of two closely related genes encoding transcription factors.

---

**Genetic Insight** Meristems are organized groups of pluripotent stem cells that control plant development. The ability to produce new organs from meristems throughout life allows plants to adjust and adapt to changing local environmental conditions.

---

**Shoot meristem**

Central zone (stem-cell reservoir)

Peripheral zone (leaf formation)

Rib zone (stem development)

*Arabidopsis thaliana*

**Inflorescence meristem (im) producing flower meristems (fm)**

fm

im

*apetala 1 cauliflower* double mutant: homeotic conversion of flower meristems into inflorescence meristems

**Figure 20.19  Shoot meristems in plant growth.**

## Combinatorial Homeotic Activity in Floral-Organ Identity

Several flowering plant species have been adopted as models for the study of plant genetics. For example, peas (*Pisum sativum*), with which Mendel performed his experiments, and maize (*Zea mays*), in which transposons were discovered, were introduced in earlier chapters. Due to its small size, short generation time, and fully sequenced genome, the most widely used model plant is *Arabidopsis thaliana*. Since the 1980s, study of homeotic mutants in *Arabidopsis* and another plant species, *Antirrhinum* (snapdragon), has led to insights into the genetic basis of flower development and revealed developmental parallels with animals.

*Arabidopsis* flowers are composed of four concentric whorls of organs (**Figure 20.20**). The outermost whorl is occupied by sepals, organs that protect the flower bud during development. The second whorl is occupied by petals, which in many species attract pollinators. Stamens, the male organs that produce pollen, are located in the third whorl, and the female organs—carpels, containing the ovules, occupy the central whorl.

**Homeotic Floral Mutants of *Arabidopsis*** Recessive floral homeotic mutants of *Arabidopsis* fall into three classes, each having defects in two adjacent whorls (see Figure 20.20). One class, named the A class, exhibits homeotic transformations in the outer two whorls, where carpels develop in the positions normally occupied by sepals and stamens replace petals, so that the four floral whorls consist of carpels, stamens, stamens, and carpels (see Figure 20.20). A second class, the B-class mutants, exhibit homeotic transformations in the middle two whorls, where sepals replace petals and with carpels replace stamens, so that the four whorls consist of sepals, sepals, carpels, and carpels. In C-class mutants, homeotic transformations in the third and fourth whorls result in flowers where petals develop in the positions normally occupied by stamens, and the cells that would normally give rise to the carpels behave as if they were another flower meristem that reiterates the developmental cycle.

In *Arabidopsis*, A-class activity is promoted by two genes, *APETALA2* and *APETALA1*, B-class activity by the *APETALA3* and *PISTILLATA* genes, and C-class activity by the *AGAMOUS* gene. Double mutants either display an additive phenotype (e.g., *apetala3 agamous* flowers consisting of only sepals) or exhibit novel phenotypes (e.g., *apetala2 agamous* flowers with novel floral organs that do not exist in wild-type flowers). Additive double-mutant phenotypes suggest that the two genes do not interact, whereas nonadditive double-mutant phenotypes suggest that the two genes interact to influence a common developmental pathway. For example, in *apetala2 agamous* flowers, the first and fourth whorls have leaf-like carpels while the second and third whorls are occupied by organs with features of both petals and stamens. The

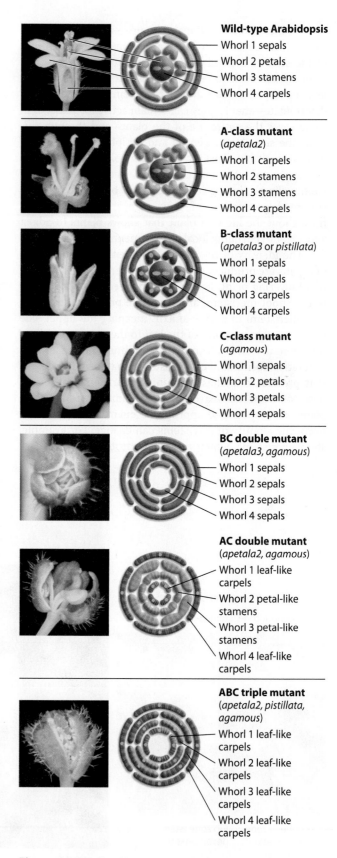

**Wild-type Arabidopsis**
- Whorl 1 sepals
- Whorl 2 petals
- Whorl 3 stamens
- Whorl 4 carpels

**A-class mutant** (*apetala2*)
- Whorl 1 carpels
- Whorl 2 stamens
- Whorl 3 stamens
- Whorl 4 carpels

**B-class mutant** (*apetala3* or *pistillata*)
- Whorl 1 sepals
- Whorl 2 sepals
- Whorl 3 carpels
- Whorl 4 carpels

**C-class mutant** (*agamous*)
- Whorl 1 sepals
- Whorl 2 petals
- Whorl 3 petals
- Whorl 4 sepals

**BC double mutant** (*apetala3, agamous*)
- Whorl 1 sepals
- Whorl 2 sepals
- Whorl 3 sepals
- Whorl 4 sepals

**AC double mutant** (*apetala2, agamous*)
- Whorl 1 leaf-like carpels
- Whorl 2 petal-like stamens
- Whorl 3 petal-like stamens
- Whorl 4 leaf-like carpels

**ABC triple mutant** (*apetala2, pistillata, agamous*)
- Whorl 1 leaf-like carpels
- Whorl 2 leaf-like carpels
- Whorl 3 leaf-like carpels
- Whorl 4 leaf-like carpels

**Figure 20.20 Floral homeotic mutations in *Arabidopsis*.**

*agamous* mutation has a phenotype effect in the first and second whorls in an *apetala2* background (compare the identities of these whorls in an *apetala2* single mutant to a *apetala2 agamous* double mutant), an effect not observed in a wild-type background, where phenotypic defects of *agamous* are limited to the third and fourth whorls. This indicates that *AGAMOUS* is ectopically active in first and second whorls in *apetala2* mutants. Likewise, based on the double mutant phenotype *APETALA2* is active in the inner whorls of *agamous* mutants.

On the basis of single and multiple mutant phenotypes, a model was formulated in which the identity of organs developing in any whorl is determined by the combination of homeotic genes active in that whorl (Figure 20.21). It was presumed that each class of gene is active in those whorls affected in the respective mutants: *APETALA2* and *APETALA1* in the outer two whorls, *APETALA3* and *PISTILLATA* in the middle two whorls, and *AGAMOUS* in the inner two whorls. Thus, each whorl is characterized by a different combination of homeotic gene activity that specifies floral organ identity. The A-class activity by itself in the first whorl specifies sepals, A- + B-class in the second whorl specifies petals, B- + C-class in the third whorl specifies stamens, and C-class by itself in the fourth whorl specifies carpels. To account for the mutant phenotypes (such as the *apetala2*

*agamous* mutant described above), a second postulate of the model is that the A- and C-class activities are mutually antagonistic, so that in an A-class mutant background, C-class activity is found in all four whorls; and conversely, in a C-class mutant background, A-class activity is in all four whorls. The specification of identity by combinations of homeotic gene activities and cross-regulatory interactions between the floral homeotic genes is reminiscent of specification of segmental identity in *Drosophila* by *Hox* genes.

The model successfully predicts the phenotypes of multiple mutants. For example, in a double mutant in which both B- and C-class activities are absent, only A-class genes are expressed in all four whorls, and a flower with only sepals develops (see Figure 20.20). In ABC triple mutants, in which all floral-organ-identity gene activity is compromised, leaf-like organs are found in all whorls. These observations suggest that since floral organs are evolutionarily derived from leaves, one role of the floral homeotic genes is to modify a leaf into a specialized floral organ.

**Homeotic MADS Box Transcription Factors**  As do animal homeotic genes, many floral homeotic genes encode closely related transcription factors. However, rather than encoding homeobox genes, the floral

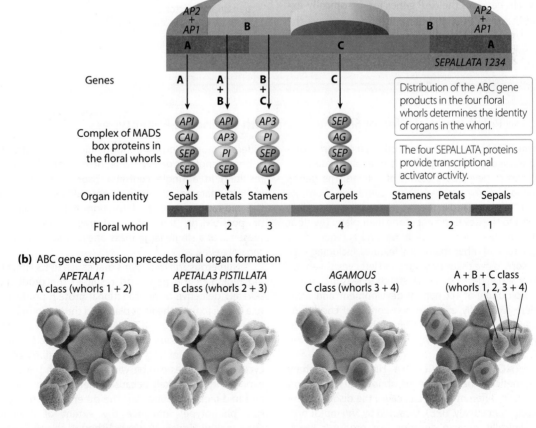

(a) The ABC model

(b) ABC gene expression precedes floral organ formation

Figure 20.21  **The ABC model of flower development.**

homeotic genes encode **MADS box** genes, named after the DNA-binding domain of the transcription factors. The name MADS box is derived from four members of the gene family: *MCM1* of *Saccharomyces cerevisiae*, *AGAMOUS* of *Arabidopsis*, *DEFICIENS* of *Antirrhinum*, and SRF of humans. All of the B- and C-class genes, as well as *APETALA1*, contain MADS boxes. Consistent with the model described above, the B-class genes are expressed in whorls two and three, and the C-class gene, *AGAMOUS*, is expressed in the third and fourth whorls (see Figure 20.21).

Subsequent studies have shown that the ABC classes of MADS box proteins interact with another class of MADS box protein encoded by the *SEPALLATA* (*SEP*) genes (see Chapter 17 Case Study). The SEP proteins together with the A-, B-, and C-class proteins form higher-order complexes that regulate transcription (see Figure 20.21). The SEP proteins provide a transcriptional activation activity to the complexes, an activity that the B and C proteins lack. Conversely, the A, B, and C proteins provide specificity to the complexes, an activity the SEP proteins lack. When A-, B-, or C-class genes are ectopically expressed throughout the flower meristem, they cause homeotic transformations of floral organ identity. For example, if B-class genes are ectopically expressed throughout the flower, the result is a flower with organ identities of petal, petal, stamen, stamen, from the first to the fourth whorls. In contrast, ectopic expression of the A-, B-, and C-class genes alone is not sufficient to convert the leaves of the *Arabidopsis* plant into floral organs. However, if the *SEP* genes are ectopically expressed in addition to, for example, the A and B genes, the combination is sufficient to convert leaves into petals. In this manner, the identities of leaves and floral organs are interconvertible by the absence or presence of the expression of the floral homeotic genes, consistent with floral organs evolving by modification of an ancestral leaf.

Studies of B- and C-class genes from flowering plants and gymnosperms (e.g., conifers) suggest that for all seed plants, C-class genes alone promote female reproductive development and that B + C gene activity promotes male reproductive development. However, unlike the *Hox* genes, which appear to have evolved at the base of the animal lineage and which control patterning in all known animals, the B- and C-class genes are unknown in earlier-diverging lineages of plants, such as ferns, lycophytes, and bryophytes, whose reproductive structures differ substantially in morphology and development and whose leaf-like organs evolved independently.

We have seen that the specification of serially repeated structures in both *Drosophila* and *Arabidopsis* is controlled in a similar manner via the combinatorial action of closely related transcription factors. Although the mechanism of developmental patterning in plants and animals is similar, the genes involved in development in the two kingdoms are not related; this is consistent with the independent evolution of multicellularity in plants and animals.

---

Genetic Insight  Floral organ identity is specified by the combinatorial action of a group of related MADS box transcription factors.

---

## CASE STUDY

### Cyclopia and Polydactyly—Different *Shh* Mutations with Distinctive Phenotypes

*Sonic hedgehog* (*Shh*), introduced in Section 20.4, is an evolutionarily conserved gene that performs multiple related but distinctive roles in developing tissues of animals. The gene's best-understood developmental roles, stemming from its expression in limb buds and in the neural tube, pertain to digit formation and to the development of the floor plate. The floor plate divides the brain into hemispheres, and is required for midline separation of other anatomical features, including separating developing eye tissue into right and left eyes. Given the central role of *Shh* in development, it stands to reason that *Shh* mutations profoundly affect normal development and morphology. Here we briefly examine two abnormal conditions that are caused by changes in *Shh* activity: holoprosencephaly/cyclopia and polydactyly.

**HOLOPROSENCEPHALY/CYCLOPIA** Holoprosencephaly (HPE) is a genetically heterogeneous abnormality, meaning that mutations in different genes can cause the disorder. One form of holoprosencephaly, HPE3, is caused by *Shh* mutations. HPE3 is a clinically variable disorder that produces many different morphological abnormalities in patients. The most subtle phenotypic defect is a slight loss of midline separation, resulting in a single central incisor. More severe defects include characteristic brain abnormalities; abnormalities of the mid-face, such as the formation of a proboscis-like nose; or possibly, in the most extreme cases, cyclopia, the presence of a single large mass of eye tissue rather than two separate eyes.

Numerous *Shh* mutations that cause HPE3 affect the coding region of the gene and result in the production of a severely defective or nonfunctional protein product, leading to a failure to form the floor plate and thus to form brain hemispheres (Figure 20.22a). To date, there are no specific genotype–phenotype correlations that tie specific *Shh* mutations to more severe or less severe manifestations of HPE3 or cyclopia. Pedigrees exhibit variation in both penetrance and expressivity, most likely because other genes involved in brain and mid-face formation (i.e., the other genes that cause the HPE phenotype) influence the extent of morphological abnormality (Figure 20.22b). Although the HPE3 mutations

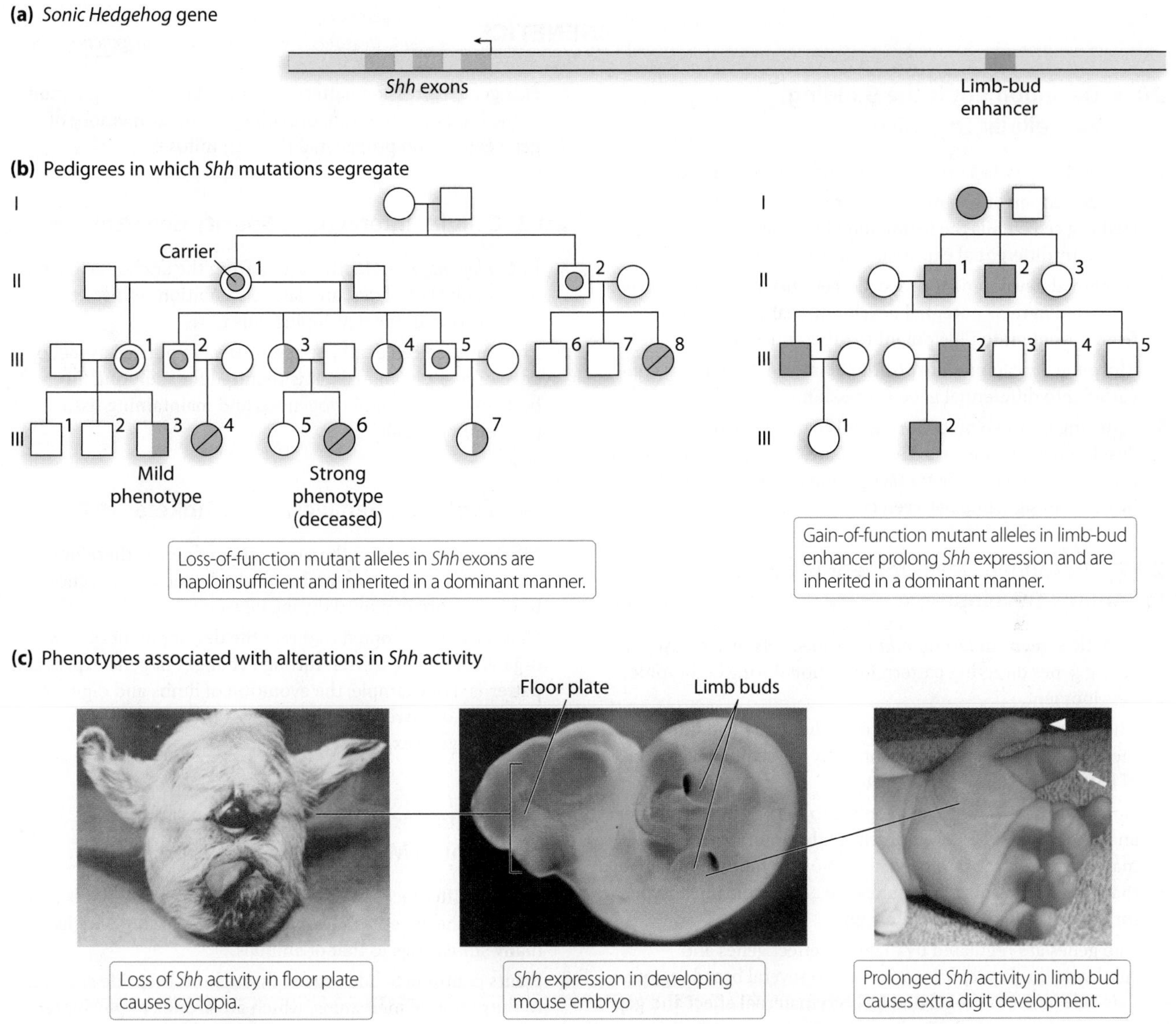

**(a)** *Sonic Hedgehog* gene

*Shh* exons

Limb-bud
enhancer

**(b)** Pedigrees in which *Shh* mutations segregate

Mild
phenotype

Strong
phenotype
(deceased)

Loss-of-function mutant alleles in *Shh* exons are
haploinsufficient and inherited in a dominant manner.

Gain-of-function mutant alleles in limb-bud
enhancer prolong *Shh* expression and are
inherited in a dominant manner.

**(c)** Phenotypes associated with alterations in *Shh* activity

Floor plate        Limb buds

Loss of *Shh* activity in floor plate
causes cyclopia.

*Shh* expression in developing
mouse embryo

Prolonged *Shh* activity in limb bud
causes extra digit development.

**Figure 20.22  Effects of alterations in Shh morphogen activity in the floor plate and the limb bud.**

in *Shh* are missense, nonsense, and frameshift loss-of-function alleles, familial cases of HPE3 are inherited in an autosomal dominant manner. This indicates that the *Shh* mutations are haploinsufficient: The presence of a single copy of a wild-type allele is not sufficient for normal activity. Thus, as with most genetic disorders that have been characterized in humans, both penetrance and expressivity of abnormal phenotypes are modified significantly by genetic background.

During the 1950s, an epidemic of cyclopia was reported among sheep in the Western United States (**Figure 20.22c**). The compound cyclopamine, found in the plant *Veratrum californicum*, was implicated as an environmental cause of the abnormalities. Evidence indicated that ingestion of *V. californicum* during gestation caused the production of lambs with cyclopia. In 2002, Philip Beachy and colleagues looked at the mechanism by which cyclopamine caused cyclopia and discovered that the compound binds directly to cells in the floor plate and blocks their response to Shh protein. This study illustrates that the action of normal proteins can be inhibited under certain environmental circumstances

to produce effects similar to those seen with gene mutation. When an environmental condition induces a phenotype similar to that caused by mutation, the environmental condition is said to induce a phenocopy of the mutant phenotype.

POLYDACTYLY If *Shh* expression is eliminated from the developing limb bud by loss-of-function mutations inactivating the Shh protein, limb patterning is perturbed and digits do not form. However, if *Shh* expression is altered by mutation in the cis-regulatory region of the gene, changes in the Shh protein concentration gradient can result in polydactyly, the presence of extra digits (see Figure 20.22c). The extra digits develop because Shh protein is present in high concentration in parts of the limb bud where it is not normally found. Polydactyly in humans (discussed in Section 4.2) is an autosomal dominant disorder. Its inheritance is dominant because the ectopic expression resulting from the mutation is a gain of function. The enhancer element responsible for appropriate *Shh* expression in the developing limb buds was identified using a phylogenetic footprinting approach (see Figure 18.17).

## 20.1 Development Is the Building of a Multicellular Organism

▌ Multicellularity has evolved independently multiple times.

▌ The development of a multicellular organism from a fertilized egg cell entails the formation of specialized cell types, driven by differential expression of genes.

▌ As animal development proceeds, cells become progressively restricted in their potential developmental fates, changing from totipotent to pluripotent to differentiated.

▌ Morphogens can provide positional information that is converted into differential gene expression.

▌ Signaling between neighboring cells can induce or inhibit developmental pathways. Genes controlling developmental processes often encode transcription factors or molecules involved in signaling between cells.

## 20.2 *Drosophila* Development Is a Paradigm for Animal Development

▌ Genetic screens in *Drosophila* identified sets of successively acting genes directing pattern formation during embryonic development.

▌ The *Drosophila* embryo is successively subdivided into segments, each with a unique identity, by the sequential action of batteries of transcription factors.

▌ Genes whose products are supplied to the egg by the mother and act to guide the development of the embryo are called maternal effect genes. The genotype of the mother, rather than that of the embryo, dictates the embryonic phenotype for the traits these genes determine.

▌ Gap genes are regulated by maternal effect genes and subdivide the *Drosophila* embryo into several broad regions. Pair-rule genes are regulated by both maternal effect and gap genes, and they subdivide the embryo into parasegments.

▌ Homeotic genes known as the *Hox* genes act in combination to specify the parasegments of *Drosophila*. *Hox* genes are largely conserved throughout the metazoan kingdom.

▌ Downstream targets of the *Hox* genes contribute to the morphogenesis of body segments.

▌ *Hox* gene expression patterns are maintained by regulation at the level of chromatin, providing a cellular memory of gene expression propagated through mitoses.

## 20.3 Cellular Interactions Specify Cell Fate

▌ In *C. elegans,* an inductive signal from the anchor cell determines vulval cell fates, and lateral inhibition refines cell specification in the developing vulva.

▌ Programmed cell death, or apoptosis, is a normal aspect of development in animals. It is required for sculpting the body plan during embryogenesis and maintaining tissues post-embryonically.

## 20.4 "Evolution Behaves Like a Tinkerer"

▌ Most animals possess the same types of genes; therefore, the differences between animals are largely due to differences in how genes are deployed during development.

▌ Genes can be co-opted to direct the development of new organs and tissues, often through changes in gene expression patterns. For example, the evolution of limbs and digits in tetrapods occurred through changes in *Hox* and *Sonic hedgehog* gene expression.

## 20.5 Plants Represent an Independent Experiment in Multicellular Evolution

▌ Despite differences in cellular behavior between plants and animals, the genetic control of development in plants has many similarities to that of animals.

▌ Plants continue to add organs throughout their life span due to the action of meristems, which are groups of pluripotent stem cells.

▌ Combinatorial action of homeotic genes specifies the identity of floral organs in flowering plants; the homeotic genes in plants encode MADS box transcription factors, analogous to the transcription factors encoded by the homeobox in animals.

## KEYWORDS

## Chapter Concepts

*For answers to selected even-numbered problems, see Appendix: Answers.*

1. Explain why many developmental genes encode either transcription factors or signaling molecules.

2. Bird beaks develop from an embryonic group of cells called neural crest cells that are part of the neural tube that gives rise to the spinal column and related structures. Amazingly, neural crest cells can be surgically transplanted from one embryo to another, even between embryos of different species. When quail neural crest cells were transplanted into duck embryos, the beak of the host embryo developed into a shape similar to that found in quails, creating the "quck." Duck cells were recruited in addition to the quail cells to form part of the quck beak. Conversely, when duck neural crest cells were transplanted into quail embryos, the beak of the embryo resembled that of a duck, creating a "duail," and quail cells were recruited to form part of the beak. What do these experiments tell you about the autonomy or non-autonomy of the transplanted and host cells during beak development?

3. How is positional information provided along the anterior–posterior axis in *Drosophila*? What are the functions of *bicoid* and *nanos*?

4. Early development in *Drosophila* is atypical in that pattern formation takes place in a syncytial blastoderm, allowing free diffusion of transcription factors between nuclei. In many other animal species, the fertilized egg is divided by cellular cleavages into a larger and larger number of smaller and smaller cells.
   a. What constraints does this impose on the mechanisms of pattern formation?
   b. How must the model that describes *Drosophila* development be modified for describing other animal species whose early development is not syncytial?

5. Consider the *even-skipped* promoter in Figure 20.9.
   a. How are the sharp boundaries of expression of *eve* stripe 2 formed?
   b. Consider the binding sites for gap proteins and Bicoid in the stripe 2 enhancer module. What sites are occupied in parasegments 2, 3, and 4, and how does this result in expression or no expression?
   c. Explain what you expect to see happen to *even-skipped* stripe 2 if it is expressed in a *Krüppel* mutant background. A *hunchback* mutant background? A *giant* mutant background? A *bicoid* mutant background?

6. What is the difference between a parasegment and segment in *Drosophila* development? Why do developmental biologists think of parasegments as the subdivisions that are produced during development of flies?

7. Why do loss-of-function mutations in *Hox* genes usually result in embryo lethality, whereas gain-of-function mutants can be viable? Why are flies homozygous for the recessive loss-of-function alleles *Ultrabithorax*^*bithorax* and *Ultrabithorax*^*postbithorax* viable?

8. Compare and contrast the specification of segmental identity in *Drosophila* with that of floral organ specification in *Arabidopsis*. What is the same in this process, and what is different?

9. Actinomycin D is a drug that inhibits the activity of RNA polymerase II. In the presence of actinomycin D, early development in many vertebrate species, such as frogs, can proceed past the formation of a blastula, a hollow ball of cells that forms after early cleavage divisions; but development ceases before gastrulation. What does this tell you about maternal versus zygotic gene activity in early frog development?

10. Ablation of the anchor cell in wild-type *C. elegans* results in a vulva-less phenotype.
    a. What phenotype is to be expected if the anchor cell is ablated in a *let-23* loss-of-function mutant?
    b. What about if the anchor cell is ablated in a *let-23* gain-of-function mutant?

11. In gain-of-function *let-23* and *let-60 C. elegans* mutants, all of the vulval precursor cells differentiate with 1° or 2° fates. Do you expect adjacent cells to differentiate with 1° fates or with 2° fates? Explain.

12. In mammals, identical twins arise when an embryo derived from a single fertilized egg splits into two independent embryos, producing two genetically identical individuals.
    a. What limits might there be, from a developmental genetic viewpoint, as to when this can occur?
    b. The converse phenotype, fusion of two genetically distinct embryos into a single individual, is also known. What are the genetic implications of such an event?

## Application and Integration

*For answers to selected even-numbered problems, see Appendix: Answers.*

13. *bicoid* is a coordinate gene with maternal effects.
    a. A female *Drosophila* heterozygous for a loss-of-function *bicoid* allele is mated to a male that is heterozygous for the same allele. What are the phenotypes of their progeny?
    b. A female that is homozygous for a loss-of-function *bicoid* allele is mated to a wild-type male. What are the phenotypes of their progeny?
    c. If loss of *bicoid* function in the egg leads to lethality during embryogenesis, how are females homozygous for *bicoid* produced? What is the phenotype of a male homozygous for *bicoid* loss-of-function alleles?

14. Given that maternal *bicoid* activates the expression of *hunchback* (see Figure 20.7), what would be the consequence of adding extra copies of the *bicoid* gene by

transgenic means, thus creating a female fly with two (the wild-type condition), three, or four copies of the bicoid gene? How would *hunchback* expression be altered? What about the expression of other gap genes and pair-rule genes?

15. What phenotypes do you expect in flies homozygous for loss-of-function mutations in the following genes: *Krüppel, odd-skipped, hedgehog, Ultrabithorax*?

16. The pair-rule gene *fushi tarazu* is expressed in the seven even-numbered parasegments during *Drosophila* embryogenesis. In contrast, the segment polarity gene *engrailed* is expressed in the anterior part of each of the 14 parasegments. Since both genes are active at similar times and places during development, it is possible that the expression of one gene is required for the expression of the other. This can be tested by examining expression of the genes in a mutant background—for example, looking at *fushi tarazu* expression in an *engrailed* mutant background, and vice versa.
    a. Given the hierarchy of gene action during *Drosophila* embryogenesis, what might you predict to be the result of these experiments?
    b. Based on your prediction, can you predict the phenotype of the *fushi tarazu* and *engrailed* double mutant?

17. In contrast to *Drosophila*, some insects (e.g., centipedes) have legs on almost every segment posterior to the head. Based on your knowledge of *Drosophila*, propose a genetic explanation for this phenotype, and describe the expected expression patterns of genes of the *Antennapedia* and *bithorax* complexes.

18. The bristles that develop from the epidermis in *Drosophila* are evenly spaced, so that two bristles never occur immediately adjacent to each other. How might this pattern be established during development?

19. You are traveling in the Netherlands and overhear a tulip breeder describe a puzzling event. Tulips normally have two outer whorls of brightly colored petal-like organs, a third whorl of stamens, and an inner (fourth) whorl of carpels. However, the breeder found a recessive mutant in his field in which the outer two whorls were green and sepal-like, while the third and fourth whorls both contained carpels. What can you speculate about the nature of the gene that was mutated?

20. A powerful approach to identifying genes of a developmental pathway is to screen for mutations that suppress or enhance the phenotype of interest. This approach was undertaken to elucidate the genetic pathway controlling *C. elegans* vulval development.
    a. A *lin-3* loss-of-function mutant with a vulva-less phenotype was mutagenized. Based on your knowledge of the genetic pathway, what types of mutations will suppress the vulva-less phenotype?
    b. In a complementary experiment, a gain-of-function *let-23* mutant with a multi-vulva phenotype was also mutagenized. What types of mutations will suppress the multi-vulva phenotype?

21. *Zea mays* (maize, or corn) was originally domesticated in central Mexico at least 7000 years ago from an endemic grass called teosinte. Teosinte is generally unbranched, has male and female flowers on the same branch, and has few kernels per "cob," each encased in a hard, leaf-like organ called a glume. In contrast, maize is highly branched, with a male inflorescence (tassel) on its central branch and female inflorescences (cobs) on axillary branches. In addition, maize cobs have many rows of kernels and soft glumes. George Beadle crossed cultivated maize and wild teosinte, which resulted in fully fertile $F_1$ plants. When the $F_1$ plants were self-fertilized, about 1 plant in every 1000 of the $F_2$ progeny resembled either a modern maize plant or a wild teosinte plant. What did Beadle conclude about whether the different architectures of maize and teosinte were caused by changes with a small effect in many genes or changes with a large effect in just a few genes?

22. The *Hoxd9–13* genes are thought to specify digit identity (see Figure 20.18).
    a. What would be the consequence of ectopically expressing *Hoxd10* throughout the developing mouse limb bud? What about *Hoxd11*? What about both *Hoxd10* and *Hoxd11*?
    b. You wish to examine the effect of loss-of-function alleles in developing limbs. How would you construct a mouse in which the function of *Hoxd9–13* is retained during anterior–posterior embryonic patterning but is absent from developing limbs?

23. Three-spined stickleback fish live in lakes formed when the last ice age ended 10,000 to 15,000 years ago. In lakes where the sticklebacks are prey for larger fish, they develop 35 bony plates along their body as armor. In contrast, sticklebacks in lakes where there are no predators develop only a few or no bony plates.
    a. In crosses between fish of the two different morphologies, the lack of bony armor segregates as a recessive trait that maps to the *ectodermal dysplasin* (*Eda*) gene. Comparisons between the *Eda*-coding regions of the armored and non-armored fish revealed no differences. How can you explain this result?
    b. Loss-of-function mutations in the coding region of the homologous gene in humans result in loss of hair, teeth, and sweat glands, as in the toothless men of Sind (India). What does this suggest about hair, teeth, and sweat glands in humans?

24. In *C. elegans* there are two sexes: hermaphrodite and male. Sex is determined by the ratio of X chromosomes to haploid sets of autosomes (X/A). An X/A ratio of 1.0 produces a hermaphrodite (XX), and an X/A ratio of 0.5 results in a male (XO). In the 1970s, Jonathan Hodgkin and Sydney Brenner carried out genetic screens to identify mutations in three genes that result in either XX males (*tra-1, tra-2*) or XO hermaphrodites (*her-1*). Double-mutant strains were constructed to assess for epistatic interactions between the genes (see table). Propose a genetic model of how the *her* and *tra* genes control sex determination.

| Genotype[a] | XX Phenotype | XO Phenotype |
| --- | --- | --- |
| Wild-type | Hermaphrodite | Male |
| $tra\text{-}1^{rec}$ | Male | Male |
| $tra\text{-}2^{rec}$ | Male | Male |
| $her\text{-}1^{rec}$ | Hermaphrodite | Hermaphrodite |
| $tra\text{-}1^{dom}/+$ | Hermaphrodite | Hermaphrodite |
| $tra\text{-}^{rec}1\ tra\text{-}2^{rec}$ | Male | Male |
| $tra\text{-}1^{rec}\ her\text{-}1^{rec}$ | Male | Male |
| $tra\text{-}2^{rec}\ her\text{-}1^{rec}$ | Male | Male |
| $tra\text{-}1^{rec}\ tra\text{-}1^{dom}/+$ | Hermaphrodite | Hermaphrodite |
| $tra\text{-}2^{rec}tra\text{-}1^{dom}/+$ | Hermaphrodite | Hermaphrodite |

[a] rec = recessive mutation; dom = dominant mutation.

25. The flowering jungle plant *Lacandonia schismatica*, discovered in southern Mexico, has a unique floral structure. Petal-like organs are in the outer whorls surrounding a number of carpels, and stamens are in the center of the flower. Closely related species are dioecious; female plants bear flowers that resemble those of *Lacandonia*, but without the central stamens. What type of mutation could have resulted in the evolution of *Lacandonia* flowers?

26. Homeotic genes are thought to regulate each other.
    a. What aspect of the phenotype of *apetala2 agamous* double mutants indicates that these two genes act antagonistically?
    b. Are similar interactions observed between *Hox* genes? Compare the action of the floral homeotic genes in specifying floral organ identity and the *Hox* genes in specifying identity along the anterior–posterior axis of animals.

27. Dipterans (two-winged insects) are thought to have evolved from a four-winged ancestor that had wings on both T2 and T3 thoracic segments, as in extant butterflies and dragonflies. Describe an evolutionary scenario for the evolution of dipterans from four-winged ancestors. What types of mutations could lead to a butterfly developing with only two wings?

28. Basidiomycota is a monophyletic group of fungi that includes most of the common mushrooms. You are interested in the development of the body plan of mushrooms. How would you identify the genes required for patterning during mushroom development?

29. In *Drosophila*, recessive mutations in the *fruitless* gene (*fru*) result in males courting other males; and recessive mutations in the *Antennapedia* gene (*Ant⁻*) lead to defects in the body plan, specifically in the thoracic region of the body, where mutants fail to develop legs. The two genes map 15 cM apart on chromosome 3. You have isolated a new dominant $Ant^d$ mutant allele that you induced by treating your flies with X-rays. Your new mutant has legs developing instead of antennae on the head of the fly. You cross your newly induced dominant $Ant^d$ mutant (a pure-breeding line) with a homozygous recessive *fru* mutant (which is homozygous wild type at the $Ant^+$ locus), as diagrammed below:

$$Ant^d\ fru^+ \times Ant^+\ fru \to \mathrm{F_1}\ Ant^d\ fru^+$$
$$Ant^d\ fru^+\ Ant^+\ fru \qquad Ant^+\ fru$$

a. What phenotypes, and in what proportions, do you expect in the $\mathrm{F_2}$ obtained by interbreeding $\mathrm{F_1}$ animals?
b. Your cross results in the following phenotypic proportions:

| | |
| --- | --- |
| Legs on head, normal courting behavior | 75 |
| Normal head, abnormal courting behavior | 25 |
| Legs on head, abnormal courting behavior | 0 |
| Normal head, normal courting behavior | 0 |

Provide a genetic explanation for these results and describe a test for your hypothesis.

c. Provide a molecular explanation for the reason your new $Ant^d$ mutant is dominant and for its novel phenotype.

# 21

# Genetic Analysis of Quantitative Traits

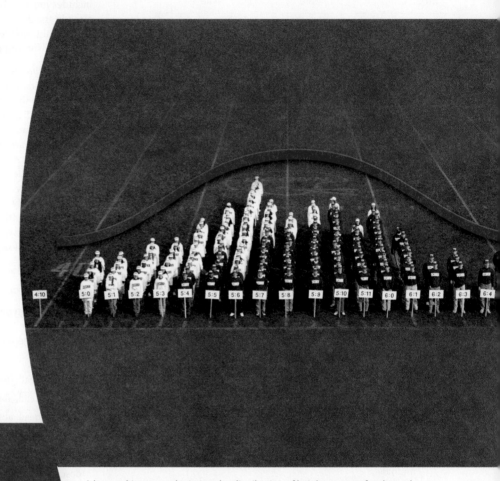

A human histogram depicting the distribution of heights among faculty and students of the University of Connecticut. The women are in white shirts and the men are in blue shirts. A line depicting the continuous distribution of adult height overlays the photo.

## ESSENTIAL IDEAS

▌ Quantitative traits are influenced by multiple genes and may also be influenced by the environment. They are continuously distributed along a phenotypic scale. Some quantitative traits are separated into distinct phenotypes by a threshold.

▌ The phenotypic distributions of quantitative traits are described by statistical measures that also estimate the genetic and environmental contributions to phenotype.

▌ The extent to which genetic variation contributes to phenotype variability can be estimated for quantitative traits and provides an indication of how traits may respond to artificial selection.

▌ The genes that influence quantitative traits are identified and mapped using genetic crosses and molecular and statistical techniques.

Explaining the connection between phenotypes and genotypes is simplest when the phenotypic variation in a trait is determined by variation in a single gene. For example, alleles of a single gene determine whether Mendel's peas are round or wrinkled. Other genes are not involved, and there is no evidence of gene interaction (i.e., epistasis) or of interaction between the gene and specific environmental factors in Mendel's observations. Variation of other genes determines whether the pods were inflated or constricted, whether the flower is purple or white, and so on. Similarly, your blood type—either A, B, AB, or O—is determined exclusively by inherited variation in a single gene;

and the environment in which you were raised had no effect on that outcome.

In reality, however, such direct correlations between phenotypes and genotypes are not common. For example, many traits display variation resulting from epistatic gene interactions (see Chapter 4). In addition, numerous traits, known as **polygenic traits,** result from the influence of multiple genes, which assort independently to produce a large number of genotypes and multiple phenotypes. The inheritance of polygenic traits is identified as **polygenic inheritance.** Further complicating the correlations between genotypes and phenotypes is the finding that many traits whose inheritance is influenced by multiple genes are also influenced by environmental factors. Thus both genetic variation and environmental variation contribute to the phenotypic variation of certain traits, which are therefore referred to as **multifactorial traits.**

An important measure of the influence of multiple genes and of environmental factors on phenotypes is the assessment of variation for traits in *quantitative* rather than *qualitative* terms. "Round seeds" versus "wrinkled seeds" or "blood type A" versus "blood type B" are examples of qualitative phenotypic differences. Qualitative phenotypes fall into discrete categories that correspond to particular genotypes and that are often distinctly different from one another. In contrast, quantitative phenotypic variation is usually continuous variation along a phenotypic scale and must be described using some unit of measure. For example, one might use kilograms to measure quantitative variation in the weight of cattle or centimeters to measure quantitative variation in the length of ears of corn. Traits of the latter kind are called **quantitative traits.** This term also applies to traits that vary over a phenotypic range. Some quantitative trait phenotypes are measured in values such as grams or centimeters, whereas others are identified in non-numeric ways such as a range of color phenotypes (i.e., from "black" to "white").

The genetic study and analysis of quantitative traits is the focus of the field of inquiry known as

**quantitative genetics.** In this chapter, we explore how quantitative genetics examines the hereditary variation of polygenic and multifactorial traits. In the process, we address some of the ways geneticists attempt to disentangle the genetic and environmental influences on trait variation and describe genetic approaches to interpreting the relative effects of those factors on quantitative trait phenotypes.

## 21.1 Quantitative Traits Display Continuous Phenotype Variation

For most of the traits we discuss in earlier chapters, phenotypic variation is controlled by allelic variation at single genes. The phenotypes of these single-gene traits commonly display **discontinuous variation,** meaning differences that allow organisms to be assigned to discrete, sharply distinguishable phenotypic categories. The discontinuous patterns of variation lead to the specification of consistent phenotype ratios, such as a 3:1 ratio among the $F_2$ progeny of self-fertilized $F_1$ organisms. Even when two genes take part in epistatic interactions that affect phenotypic expression, the phenotypes are discrete and occur in predictable ratios (see Section 4.3).

In contrast, polygenic and multifactorial traits are controlled by many genes and usually display **continuous variation,** which is phenotypic variation distributed across a range of values in an uninterrupted continuum. This section explores the genetic factors contributing to traits displaying continuous variation.

### Genetic Potential

Human adult height is an example of a multifactorial trait that varies continuously along a scale of measurement usually marked off in centimeters or inches. This continuous variation is demonstrated in the photograph at the start of the chapter, in which approximately 150 University of Connecticut students and faculty are arranged according to height. The height distribution of this sample, divided into 1-inch increments, ranges from 60 inches (5 feet) to 77 inches (6 feet 5 inches). The length of each line of individuals behind the height markers represents the frequency of each incremental category.

Adult height is influenced by multiple genes. For example, a 2011 study by Matthew Lanktree and many colleagues used the analysis of human genomic variation and statistical methods to suggest that more than 60 genes may influence adult height. While the actual number of genes influencing human height continues to be investigated, your own personal experiences, as well as population studies, tells you that taller parents tend to have taller children and shorter parents tend to have shorter children. In addition to this genetic influence, however, environmental and developmental factors can have a significant effect. If your genetics class is typical of most, a survey of your classmates would likely find that many of the men are taller than their fathers and grandfathers, and that many of the women are taller than their mothers and grandmothers. These differences are due almost exclusively to improved maternal and child health and nutrition, and only minimally to differences in the genetic makeup influencing adult height. Longitudinal studies confirm that much of the world's population is getting taller. During the 20th century, the height of the average American woman increased from approximately 5'2" in 1900 to almost 5'5" in 1990. An even more dramatic increase in average adult height can be detected by examining the doors of houses and other structures built during the Middle Ages. Most modern-day visitors have to stoop to enter! Such observations lead to the clear conclusion that adult height is a multifactorial trait.

To understand the role of genetics in such a trait, you might think of parents as transmitting to their children a "genetic potential" for reaching a certain maximum adult height; the genetic potential will be attained if the child grows and develops under ideal conditions. Not all of the children of a particular pair of parents will have the same genetic potential, since segregation and independent assortment of the contributing genes can produce many different genotypes. These processes produce offspring with different genotypes conveying genetic potential for a range of heights, including heights that are greater or lesser than those of their parents. On average, however, progeny genetic potential for height will be at approximately the midpoint of the two parents' genetic potential.

---

Genetic Insight The phenotypes of quantitative traits can be measured on quantitative scales. Quantitative traits produced by the cumulative influence of multiple genes are polygenic. Quantitative traits that are influenced by polygenic as well as environmental or developmental factors are multifactorial.

---

## Major Genes and Additive Gene Effects

The continuous phenotypic variation of polygenic traits results from the effects of multiple genes that may exert different amounts of influence. For example, the human *OCA2* gene has several alleles that strongly influence eye color. The color of the adult eye is further influenced by other genes that act less strongly than *OCA2*. A gene like *OCA2* is classified as a **major gene,** since it has a strong effect on the phenotype. Genes that have lesser effect on

the phenotype but that also make contributions are classified as **modifier genes.**

On the other hand, in many polygenic traits, the continuous phenotypic distribution results from incremental contributions by multiple genes. When the genes contributing to polygenic traits each contribute about an equal share to total phenotypic variation they are known as **additive genes.** Each allele of additive genes can be assigned a quantitative value that indicates its contribution to a polygenic trait known as an **additive trait** because phenotypes can be predicted by adding the values of the alleles together. For certain traits, each of the additive genes has an approximately equal effect. With flower color, for example, a single copy of an allele of an additive gene may contribute one unit of color, two alleles contribute two units of color, and so on. If the genes controlling hereditary variation of a trait are additive, no single gene has a major effect on phenotypic variation, so incremental differences in phenotype are observed.

To grasp the notion of additive genes requires a different way of thinking about genotypes and phenotypes than we have discussed previously. Since traits controlled by additive genes have a phenotype that is the sum of the allele values, it is possible for more than one genotype to correspond to certain phenotypes. Segregation and independent assortment of additive alleles produces the various genotypes, but the phenotype corresponding to each is based on the sum of the values of the alleles at all the contributing loci.

In the early 1900s, coinciding with the verification and expansion of the then recently rediscovered hereditary principles of Mendel, geneticists began to explore the hypothesis that the segregation of alleles of multiple genes played a role in phenotypic variation of particular traits. Known as the **multiple-gene hypothesis,** the proposal was that alleles at each of the contributing genes obeyed the principles of segregation and independent assortment and had an additive effect in the production of phenotypic variation.

The multiple-gene hypothesis was the foundation of quantitative genetics, and the plant geneticist Hermann Nilsson-Ehle was one of the first to use the hypothesis in his 1909 description of genetic control of kernel color in wheat. **Figure 21.1** illustrates one of Nilsson-Ehle's genetic models, describing the determination of wheat kernel color by additive alleles of two genes. In this model, only genetic effects on phenotype are being considered. The model predicts that kernel color spans a spectrum from deep red to white. Gene *A* and gene *B* each have two alleles. Alleles $A_1$ and $B_1$ are equivalent to one another, each adding an equal unit of color to the phenotype. Alleles $A_2$ and $B_2$ are also equivalent, neither adding any units of color to the phenotype. Under the additive genetic model, the more "number 1" alleles, either $A_1$ or $B_1$, the genotype contains, the darker the color of wheat kernels. Conversely, the fewer number 1 alleles (or the

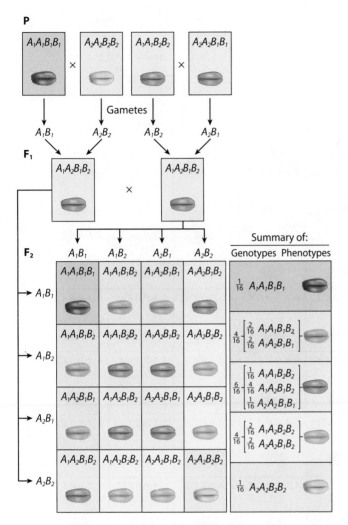

**Figure 21.1** **Polygenic inheritance of wheat kernel color controlled by two additive genes.** Each 1 allele (either $A_1$ or $B_1$) adds a unit of color, but 2 alleles ($A_2$ or $B_2$) add no units of color. Two distinct crosses of pure-breeding parents produce dihybrid $F_1$ with pink kernel color. Five phenotype classes are predicted among $F_2$ progeny in a ratio determined by the total number of $A_1$ plus $B_1$ alleles in the genotype.

more "number 2" alleles) there are in the genotype, the lighter the kernel color. The deepest red color is present when four number 1 alleles are present ($A_1A_1B_1B_1$). Conversely, white kernels are produced when no copies of number 1 alleles are in the genotype ($A_2A_2B_2B_2$).

Figure 21.1 shows a cross between pure-breeding red and pure-breeding white plants that have different phenotypes because they have different genotypes. The cross produces $F_1$ plants that are dihybrid ($A_1A_2B_1B_2$) and have an intermediate pink kernel color as a consequence of carrying just two number 1 alleles. Crossing the $F_1$ plants produces an $F_2$ generation with five different kernel colors, each dependent on the total number of number 1 alleles in the genotype. For these two loci, genotypes can have a maximum of four number 1 alleles and a minimum

of zero number 1 alleles. The five different totals of number 1 alleles produce the five different phenotypes in the $F_2$ generation, in proportions determined by independent assortment. Among the $F_2$, 1/16 carry four number 1 alleles and produce red kernels like the parental plant, 4/16 carry three number 1 alleles and have light red kernels, 6/16 have two number 1 alleles and have pink kernels, 4/16 carry a single number 1 allele and have light pink kernels, and the final 1/16 have no number 1 alleles and have white kernels like the parental plant.

As the number of additive genes contributing to a phenotypic trait increases, the number of phenotype categories increases as well. **Figure 21.2** illustrates an additive genetic model in which wheat kernel color is determined by three genes. In this example, genes *A*, *B*, and *C* each have two alleles whose additive effect is computed in the same way as for the two-gene system of Figure 21.1: Phenotype categories are determined by the number of "1" alleles contained in a genotype. A cross of pure-breeding dark red and pure-breeding white parental plants produces an $F_1$ of an intermediate (dark pink) color as a result of its trihybrid genotype ($A_1A_2B_1B_2C_1C_2$). Independent assortment produces an $F_2$ that falls into seven phenotypic categories that are determined by genotypes that have a maximum of six 1 alleles and a minimum of zero 1 alleles.

---

**Genetic Insight** Quantitative traits that are additive are determined by the sum of the relatively equal contributions of alleles of multiple genes.

---

## Continuous Phenotypic Variation from Multiple Additive Genes

The more phenotypes that occur along a limited scale of measurement, the narrower is the slice of the distribution each category occupies, and the less obvious the demarcation between categories may become. **Figure 21.3** shows five histograms illustrating the distribution of $F_2$ phenotypes produced by different numbers of additive genes that each have two alleles. As in the preceding examples, each number 1 allele adds a unit of color to the phenotype, but number 2 alleles do not. The proportions for each phenotype are determined using Pascal's triangle (see Figure 2.15). Notice the increase in the number of phenotype classes as the number of genes contributing to the phenotype increases from one to five. Moreover, the adjacent phenotype classes resemble one another more closely as the number of classes increases, blending into a continuous phenotypic distribution.

The number of distinct phenotype categories for a polygenic trait produced by the segregation of additive alleles of a given number of genes (*n*) is calculated as $2n + 1$. For example, for three additive genes contributing to a polygenic trait, $n = 3$, and the number of distinct

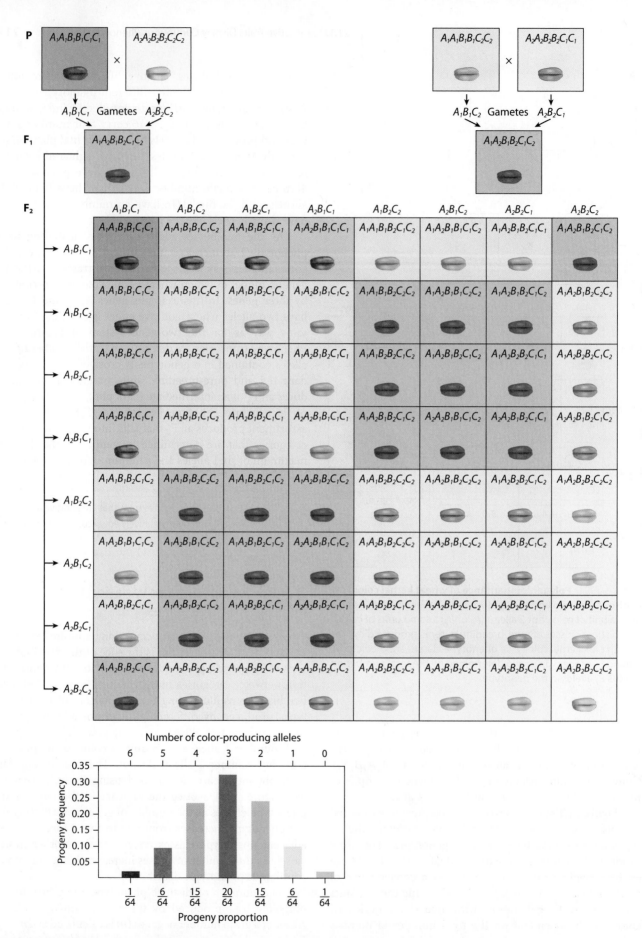

**Figure 21.2 A three-gene additive model for wheat kernel color.** Color is determined by total number of 1 alleles ($A_1$, $B_1$, and $C_1$) in the genotype. The $F_2$ have seven phenotypic classes in proportions generated by independent assortment at three loci.

**(a)** One locus: $A_1A_2 \times A_1A_2$

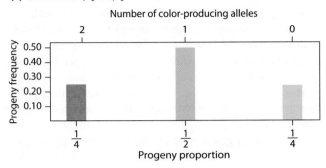

**(b)** Two loci: $A_1A_2B_1B_2 \times A_1A_2B_1B_2$

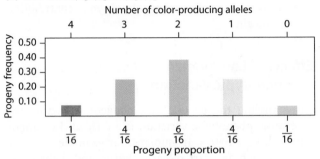

**(c)** Three loci: $A_1A_2B_1B_2C_1C_2 \times A_1A_2B_1B_2C_1C_2$

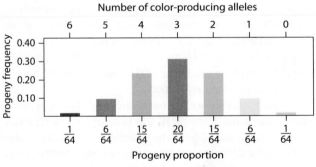

**(d)** Four loci: $A_1A_2B_1B_2C_1C_2D_1D_2 \times A_1A_2B_1B_2C_1C_2D_1D_2$

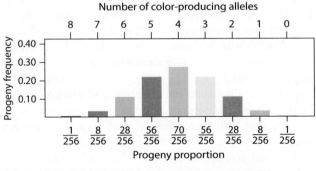

**(e)** Five loci: $A_1A_2B_1B_2C_1C_2D_1D_2E_1E_2 \times A_1A_2B_1B_2C_1C_2D_1D_2E_1E_2$

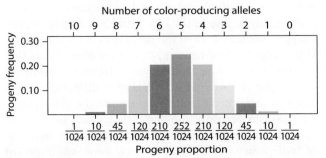

**Figure 21.3** **Phenotype distributions with additive genes.** The parents producing progeny in each example are heterozygous for each gene. The color-contributing alleles are designated as 1 for each gene. The number of $F_2$ phenotype categories increases with the number of additive genes.

phenotypic categories is $2(3) + 1 = 7$. **Table 21.1** lists the numbers of phenotypic categories for different numbers of contributing genes and gives the frequency of the most extreme phenotypes in each distribution. If more than two alleles occur for the contributing genes, the number of phenotypes can increase.

## Allele Segregation in Quantitative Trait Production

In 1916, plant geneticist Edward East undertook a comprehensive examination of the multiple-gene hypothesis by testing its ability to explain patterns of inherited variation that he produced in the length of the corolla (the petal-producing part of the flower) in *Nicotiana longiflora*. In this long-flower species of tobacco, the corolla is a tube-shaped structure whose length can be measured and compared with corollas in other plants.

East began his experiments with pure-breeding parental lines, one having a short corolla approximately 40 millimeters long and the other producing a long corolla of approximately 90 millimeters (**Figure 21.4**). Note that there is a small amount of variation in corolla length in each pure-breeding line, suggesting that despite attempts to produce pure-breeding lines, gene–gene interaction or multifactorial effects produce some variability. The $F_1$ progeny of this cross had an average corolla length of

| Table 21.1 | The Effect of Polygenes on Phenotypic Variation | |
|---|---|---|
| Number of Genes (*n*) | Number of Phenotype Categories | Frequency of Most Extreme Phenotypes |
| 1 | 3 | 1/4 |
| 2 | 5 | 1/16 |
| 3 | 7 | 1/64 |
| 4 | 9 | 1/256 |
| 5 | 11 | 1/1024 |
| 6 | 13 | 1/4096 |
| 7 | 15 | 1/16,384 |
| 8 | 17 | 1/65,536 |
| 9 | 19 | 1/262,144 |
| 10 | 21 | 1/1,048,576 |

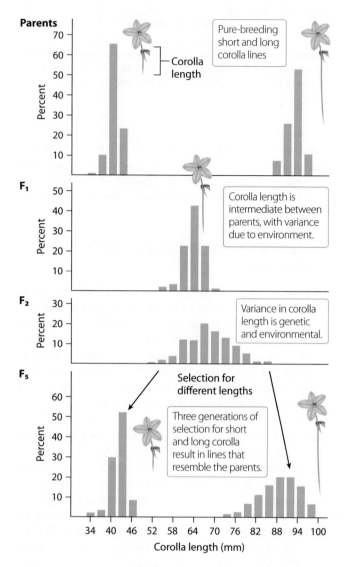

**Figure 21.4 Corolla length in tobacco.** Edward East determined that alleles of multiple genes control genetic variance in corolla length of tobacco (*Nicotiana longiflora*).

about 65 millimeters, approximately midway between the parental averages. These "mid-parental" values are an indication of strong genetic control of corolla length. Once again, there is some variability around the average corolla length value, but none of the $F_1$ have corolla lengths that are near the parental values.

East allowed $F_1$ plants to self-fertilize to produce about 450 $F_2$, among which he observed a wider distribution of corolla length than in the $F_1$, although the average length was about the same as that of the $F_1$. None of the $F_2$ East produced had corolla lengths equal to those of the pure-breeding parental lines. Then, over three additional generations beginning with $F_2$, East selectively bred plants to produce a line having a short corolla and a line having a long corolla, achieving new collections of plants with corolla lengths approximating those found in the original pure-breeding parents.

East reached two general conclusions based on his observations. Both conclusions are consistent with the models of continuous phenotypic variation of quantitative traits we have described. First, he concluded that corolla length in *Nicotiana longiflora*, particularly in the $F_2$, results from the segregation of alleles of multiple genes. Second, East concluded that the phenotypic expression of each genotype is influenced by nongenetic factors, that is, gene interaction or environmental factors that blur the connection between a given genotype and a specific phenotype. The nongenetic factors partially explain the variation around average corolla length. **Genetic Analysis 21.1** guides you through your own analysis of polygenic contributions to plant height.

## Effects of Environmental Factors on Phenotypic Variation

Disentangling the genetic and nongenetic factors that determine phenotypic variation is a difficult but important task in genetics. In humans, for example, common diseases such as heart disease, cancer, and diabetes are influenced by heredity, but nonhereditary factors are also critically important in disease development. Identifying the particular genes and the specific nonhereditary factors contributing to these diseases is the ultimate goal of research, but it must be approached in small, incremental steps that include modeling of the interactions of hereditary and nonhereditary factors.

Figure 21.5 shows a general approach taken by models of this kind. It displays the phenotypic ranges that would be associated with the genotypes $A_1A_1$, $A_1A_2$, and $A_2A_2$ under different assumptions of gene–environment interaction. In **Figure 21.5a**, no gene–environment interaction takes place, and each genotype corresponds to a distinct phenotype. Predictable correspondence of genotype and phenotype is seen in the $F_2$, where phenotypic distribution is discontinuous and a 1:2:1 phenotype ratio is found. **Figure 21.5b** shows the phenotypic ranges of parents and $F_1$ and $F_2$ progeny when moderate interaction occurs between the genotype and environmental factors. In each generation, a range of phenotypic values is associated with each genotype, and in the $F_2$, there is a small degree of overlap between the phenotypic ranges of different genotypes. In **Figure 21.5c**, substantial interaction between genes and environment takes place. A wide range of phenotypic values is associated with each genotype, and in the $F_2$ a significant degree of phenotypic overlap between the genotypes is seen, so that a large proportion of heterozygotes have phenotypes that overlap those of a homozygote. Gene–environment interaction of this kind is typical of multifactorial traits and can make it difficult to determine the genotype of an organism simply by looking at its phenotype. The use of a "phenotype scorecard" to predict the outcome of polygenic inheritance and of gene–environment

Dr. Ara B. Dopsis, a famous plant geneticist, develops several pure-breeding lines of daffodils. Under ideal growth conditions, line 1 plants are the tallest and grow to a height of 48 centimeters, whereas line 2 plants are the shortest and grow to 12 centimeters. Dr. Dopsis devises a genetic model with three additive genes to explain polygenic inheritance of plant height. He assumes that line 1 has the genotype $A_1A_1B_1B_1C_1C_1$ and that line 2 has the genotype $A_2A_2B_2B_2C_2C_2$. In answering the following questions, assume that genotype alone determines plant height under ideal growth conditions and that the alleles of the three genes are additive.

a. If these two pure-breeding parental plants are crossed, what will be the genotype and height of the $F_1$ progeny plants?

b. If $F_2$ are produced, what is the expected frequency of plants with different heights?

| Solution Strategies | Solution Steps |
| --- | --- |

### Evaluate

1. Identify the topic this problem addresses, and explain the nature of the requested answer.

2. Identify the critical information given in the problem.

1. This problem concerns assessment of a three-gene additive model for plant height, application of the model to crosses of pure-breeding parental plants of different heights, and evaluation of the $F_1$ and $F_2$ progeny.

2. The genotypes of the pure-breeding parents are given. In applying the polygenic additive model, we are to assume that genotype alone determines variation in plant height.

### Deduce

3. Deduce the contribution of each allele of the additive genes to height in line 1.

TIP: Assume that each allele makes an equal contribution in this additive genetic model.

4. Deduce the contribution of each allele of the additive genes to height in line 2.

5. Deduce the gametes produced by each pure-breeding line.

TIP: The laws of segregation and independent assortment apply to genes controlling polygenic traits.

3. The 48-cm height of line 1 plants is determined by six alleles of additive genes. Each "1" allele in the line 1 genotype contributes 48 cm/6 = 8 cm to plant height.

4. Six alleles also contribute equally to the 12-cm height of line 2 plants. Each "2" allele in the line 2 genotype contributes 12 cm/6 = 2 cm to plant height.

5. Line 1 has the genotype $A_1A_1B_1B_1C_1C_1$ and produces gametes with the genotype $A_1B_1C_1$. Line 2 has the genotype $A_2A_2B_2B_2C_2C_2$ and produces the gamete genotype $A_2B_2C_2$.

### Solve

6. Determine the genotype and height of $F_1$ plants.

7. Determine the frequency and height of each category of $F_2$ plants.

TIP: Either use Pascal's triangle (Figure 2.15) or determine the probability of genotypes containing different numbers of 1 and 2 alleles.

PITFALL: Remember that for most categories there are multiple genotypes with the corresponding number of 1 and 2 alleles.

Answer a

6. $F_1$ progeny of these pure-breeding parental plants will have the genotype $A_1A_2B_1B_2C_1C_2$. Based on the contribution of each 1 and 2 allele, the predicted $F_1$ plant height is [(3)(8 cm)] + [(3)(2 cm)] = 30 cm.

Answer b

7. The expected $F_2$ progeny are

| Number of Alleles | | Frequency | Height (cm) |
| --- | --- | --- | --- |
| 1 | 2 | | |
| 0 | 6 | 1/64 | 12 |
| 1 | 5 | 6/64 | 18 |
| 2 | 4 | 15/64 | 24 |
| 3 | 3 | 20/64 | 30 |
| 4 | 2 | 15/64 | 36 |
| 5 | 1 | 6/64 | 42 |
| 6 | 0 | 1/64 | 48 |

For more practice, see Problems 8, 9, and 20.

**Figure 21.5 The effect of gene–environment interaction.** The phenotype determined by a single gene with codominant alleles can be modified by the action of environmental factors.

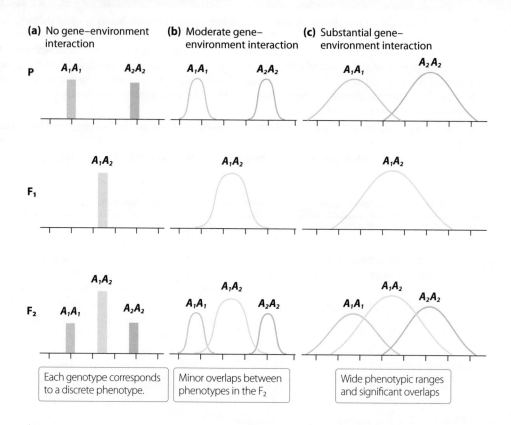

interaction in determining the multifactorial trait of height in a hypothetical plant is illustrated in **Experimental Insight 21.1**.

## Threshold Traits

Most polygenic and multifactorial traits exhibit a continuous phenotypic distribution, but certain of these traits, while having an underlying continuous distribution, can nevertheless be divided into distinct categories. Such traits are called **threshold traits.** Threshold traits are often encountered in medical contexts, where attempts are made, not always successfully, to identify two clinical categories—"unaffected" (or "normal") and "affected" (or "abnormal")—and thus to distinguish individuals who have an abnormality from those that do not. The vast majority of the members of a population will have phenotypes on the unaffected side of the threshold and will display the normal phenotype. A small proportion of the population, however, is found on the other side of the threshold and has an affected or abnormal phenotype. Cases that lie at the borderline between the two categories can be problematic to diagnose.

The genetic hypothesis explaining threshold traits proposes that the trait is polygenic or multifactorial, so that underlying the affected and unaffected phenotype categories is a continuous distribution of **genetic liability**—a term for the organism's risk of having the affected phenotype as the result of inheriting a particular genotype. Each member of a population has a specific

genetic liability determined by polygenic inheritance. **Figure 21.6** shows a continuous distribution of genetic liability for a population and the designation of a threshold that separates unaffected from affected individuals in the population. The portion of the population lying to the left of the **threshold of genetic liability,** by far the majority, are identified as unaffected or normal, and the small group to the right of the threshold are considered affected or abnormal.

Models are used to test the applicability of these concepts to real-world observations at the population level. In these models, the likelihood of crossing the threshold of liability increases when more "liability alleles" are present

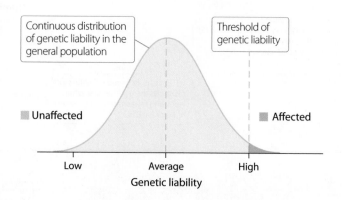

**Figure 21.6 Threshold traits.** A theoretical continuous phenotypic distribution and a threshold of genetic liability for a threshold trait.

## Experimental Insight 21.1

### Phenotype Scorecard: A Multifactorial Quantitative Phenotype Simulation

Here's a hands-on activity that illustrates an approach to modeling a multifactorial quantitative trait. In this hypothetical example, the mature height of a plant is under the control of five additive genes designated $A$ to $E$. Two alleles at each gene make different contributions to height. Each allele with the subscript 1 adds 5 centimeters to the genetic potential, and each allele with the subscript 2 adds 10 centimeters. Therefore, a plant homozygous for 1 alleles at each locus ($A_1A_1B_1C_1C_1D_1E_1E_1$) has genetic potential for a height of [(10 alleles)(5 cm)] = 50 cm, as compared to a plant carrying a genotype composed entirely of 2 alleles, which has a height potential of [(10 alleles)(10 cm)] = 100 cm. Plants carrying genotypes with different numbers of 1 and 2 alleles have different genetic potentials for height lying at 5-cm intervals along a continuum between 50 and 100 cm.

At this point, let's ask the following question: "How many 1 and 2 alleles must be present to give a height potential of 80 cm?" Each genotype contains a total of 10 alleles, two at each of the five loci. Therefore, any genotype with six 2 alleles and four 1 alleles will produce a height potential of [(6)(10) + (4)(5)] = 80 cm.

Here's a follow-up question: "What proportion of the progeny of two plants, each with a height potential of 75 cm, will have a height potential of 80 cm?" This problem is more complex. Plants with a height potential of 75 cm have five 2 alleles and five 1 alleles [(5)(10) + (5)(5) = 75]. Progeny genotypes that contain six 2 alleles and four 1 alleles will have a height potential of 80 cm. We can use the histogram in Figure 21.3e to predict the answer: 210 of the 1024 progeny (20.5%) have six copies of 2 alleles and four copies of 1 alleles.

Having examined the relationship between genotype and potential height in this model, let's examine the effect of five environmental factors on the attainment of height:

1. Amount of water
2. Amount of sunlight
3. Soil drainage
4. Nutrient content of soil
5. Temperature

Each environmental factor can vary from optimal to poor. If all factors are optimal, we'll assume that full potential height

is attained. However, if one or more of the environmental factors is less than optimal, then height is reduced. The state of each environmental factor has an effect on growth. In this exercise, we'll assume that the growth is affected according to the following scale:

| Environmental Factor State | Height Lost |
|---|---|
| Optimal (O) | 0 cm lost |
| Good (G) | 4 cm lost |
| Fair (F) | 8 cm lost |
| Marginal (M) | 12 cm lost |
| Poor (P) | 16 cm lost |

If, for example, one environmental condition is optimal, two are good, one is fair, and one is marginal, the loss of potential height is 28 cm.

The following table illustrates how the same genotype can produce different phenotypes under differing environmental conditions and how different genotypes can produce similar phenotypes under different conditions. Notice that the first two genotypes are identical but result in different phenotypes because of environmental differences. Also note that the third genotype has lower height potential than the other genotype but, in combination with a superior environment, results in the tallest plant. You can try your own combinations of genotypes and growth conditions to see different results.

| Genotype | Height Potential | Environmental Factor States 1 2 3 4 5 | Height Attained |
|---|---|---|---|
| $A_1A_2B_1B_2C_2C_2D_1D_2E_1E_2$ | 80 cm | G F O G M | 52 cm |
| $A_1A_2B_1B_2C_2C_2D_1D_2E_1E_2$ | 80 cm | F M G G F | 44 cm |
| $A_1A_1B_1B_2C_1C_2D_1D_2E_1E_2$ | 70 cm | O G G G G | 54 cm |

in the genotype. For example, **Figure 21.7a** depicts a hypothetical three-gene model in which alleles are designated as either 1 or 2 at each locus and in which genetic liability increases with a greater number of 1 alleles. In this model, the threshold of liability has arbitrarily been placed so that five 1 alleles must be present to exceed it. A greater number of 1 alleles increases the proportion of progeny that will lie to the right of the threshold of liability, and display an affected phenotype. The model can compare the risks represented by crosses between parents carrying different numbers of liability alleles.

Figure 21.7a illustrates Cross 1 between a parent with two 1 alleles and a parent with three 1 alleles. Both parents have the normal phenotype, because each is on the unaffected side of the threshold. Among the progeny of this cross, 1/32 (3%) are expected to carry five 1 alleles, but none can carry six 1 alleles. Thus 1/32 of the progeny lie to the right of the threshold of liability and have the affected phenotype. **Figure 21.7b** shows Cross 2 with different parents that produce a higher level of genetic liability in their progeny. In this cross, each parent carries three liability alleles, but neither is affected because the

**(a)** Cross 1: $A_1A_2B_1B_2C_2C_2$ × $A_1A_2B_1B_2C_1C_2$

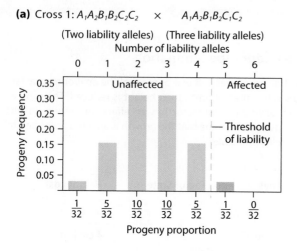

**(b)** Cross 2: $A_1A_2B_1B_2C_1C_2$ × $A_1A_2B_1B_2C_1C_2$

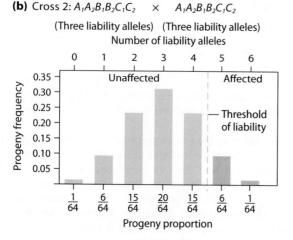

**Figure 21.7  A polygenic model for a threshold trait.** Any allele designated as 1 confers genetic liability, any allele designated as 2 confers no liability, and the alleles are additive. **(a)** In cross 1, the couple has a 1/32 chance of producing an affected child. **(b)** In cross 2, the couple has a 7/64 chance of producing an affected child.

liability threshold is 5 or more liability alleles. Among their progeny, however, independent assortment predicts that 7/64 (11%) will have genotypes that contain five or more 1 alleles. These progeny lie to the right of the threshold of liability and have the affected phenotype. The genotypes in the second cross confer almost a four-fold increased risk (3% versus 11%) of producing an affected offspring compared to the first cross. This difference is analogous to the difference we might see between a mating of individuals from a general population that has a low risk of producing a child with a threshold trait and a mating of parents that both come from families with a history of the trait.

The influence of environmental and developmental factors on phenotypes of threshold traits is an important additional component. These factors likely play a major role in determining whether individual organisms that have a genetic liability that places them near the threshold

of liability end up on one side or the other of the threshold. The threshold model envisions organisms possessing high genetic liability (i.e., possessing a genome with many liability alleles) as having the *potential* to develop the affected phenotype. Whether the affected phenotype develops may be due to the influence of other hereditary, developmental, or environmental factors. Less often, an organism may have a genetic liability slightly below the threshold but the influence of environmental factors could push the phenotype into the affected category.

Lastly, there is a caveat to consider with regard to defining the categories and classification of threshold traits, particularly in humans. Because these traits are quantitative and fall along a continuum, precise determination of categories and phenotypes can be inexact. For example, it is easy to classify a person's blood pressure as normal if it lies well within the normal range, or as abnormal if the blood pressure is very high. Many people, however, have "borderline" high pressures that are difficult to assign to either the normal or high blood pressure category.

---

**Genetic Insight**  Threshold traits are determined by polygenic inheritance and may also be influenced by environmental factors. They have a continuous distribution of genetic liability in a population. Most organisms carry a small number of liability alleles and have the unaffected, or normal, phenotype. A few organisms have genotypes with enough liability alleles to carry the organism across the threshold of liability and produce an abnormal (affected) phenotype.

---

## 21.2 Quantitative Trait Analysis Is Statistical

The statistical methods most often applied today to the study of quantitative traits are a direct extension of contributions made nearly a century ago by statistician and evolutionary biologist Sir Ronald Fisher. In 1918, Fisher used statistical analysis to show that quantitative traits result from the segregation of alleles of multiple genes displaying an additive effect. Fisher also showed that interactions between genes can be detected by these methods. In addition, he explored the role of gene–environment interaction and concluded that environmental factors contribute to continuous variation by blurring the lines between phenotypic classes. The tools and approaches described here and pioneered by Fisher allow scientists to identify genetic influences on phenotypes in terms of quantitative measurement rather than qualitative appearance. In the following description and illustrations of quantitative trait analysis, we revisit some concepts in statistics described in connection with chi-square analysis (see Section 2.5).

## Statistical Description of Phenotypic Variation

The first step in quantifying the phenotypic variation of a trait in a population is to construct a **frequency distribution** of values of the trait on a quantitative scale. A frequency distribution shows what proportion of the population exhibits each measured value of the trait or falls into each category defined for the trait. **Figure 21.8a** provides an example, showing the number and frequency of each designated height category in a sample of 1000 college-aged males. **Figure 21.8b** graphs the same data in a histogram that illustrates the continuous distribution of height among these subjects.

The individuals in this study are considered a random sample. They have not been selected for any attribute related to their height, and so their height distribution is assumed to resemble that of the general population of college-aged males. Random samples are used in quantitative trait analysis for two reasons. First, it is often impossible or impractical to collect data on every individual in a population; and second, random samples can be just as accurate in the statistical sense as "samples" consisting of whole populations. As an analogy, about 10 milliliters of blood—approximately two-tenths of 1% of a person's total blood volume—is usually drawn for most routine blood tests. The amount taken is not large enough to cause physiological problems, but it is representative enough to provide dependable information concerning a person's health status.

After the frequency distribution is constructed, the first piece of information obtained from it is the average, or mean, value ($\bar{x}$) for the distribution. Recall that this is calculated by summing all the values in the sample and dividing by the total number of individuals in the sample ($n$; see Section 2.5). For the sample of college-aged men in Figure 21.8, the mean height value is 175.33 cm (about 68.5 inches).

The shapes of frequency distributions vary depending on several factors, including the sample size and the number of classification categories for the trait. It is therefore necessary to provide a statistical description of the shape of the frequency distribution when comparing trait values. For example, it is important to report the **mode,** or **modal value,** that is, the most common value in a distribution. For the height data shown in Figure 21.8, the mode is the 173–175 cm category, containing 188 individual values. Each distribution also possesses a middle value, known as the **median,** or **median value.** In the height distribution, you can think of the median value as entry number 500 (in order of increasing height) of the 1000 entries in the distribution. This median value also resides in the 173–175 cm category.

Data in the real world are usually skewed—that is, unevenly distributed on either side of the mean, as Figure 21.8 illustrates. Therefore, to describe the frequency distribution, we must also have ways of measuring (and thus describing) the nature of the distribution around the mean. Two forms of measurement are commonly used.

**(a)** Number and frequency of heights in 3-cm intervals

| Height (cm) | Number | Frequency (%) |
|---|---|---|
| 155–157 | 4 | 0.4 |
| 158–160 | 8 | 0.8 |
| 161–163 | 26 | 2.6 |
| 164–166 | 53 | 5.3 |
| 167–169 | 89 | 8.9 |
| 170–172 | 146 | 14.6 |
| 173–175 | 188 | 18.8 |
| 176–178 | 181 | 18.1 |
| 179–181 | 125 | 12.5 |
| 182–184 | 92 | 9.2 |
| 185–187 | 60 | 6.0 |
| 188–190 | 22 | 2.2 |
| 191–193 | 4 | 0.4 |
| 194–196 | 1 | 0.1 |
| 197–199 | 1 | 0.1 |
| | 1000 | 100 |

**(b)** Graph displaying continuous variartion of height

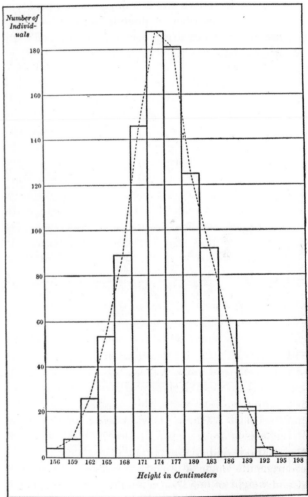

©1958 William Ernest Castle

**Figure 21.8 Adult male height.** The frequency distribution of height in 1000 college-aged males is shown in **(a)** tabular form and **(b)** in histogram form.

The first, called the **variance ($s^2$),** is a numerical measure of the spread of the distribution around the mean. This measure interprets how much variation exists among individuals in the sample. The variance value depends on the relationship between the width of the distribution and the number of observations in the sample. It will be small if all the observations are close to the mean, and it will be large if the observations are widely spread around the mean (**Figure 21.9**). The variance is determined by summing the square of the difference between each individual value and the sample mean and dividing that sum by the number of degrees of freedom (df) in the sample. The number of degrees of freedom is equal to the number of independent variables. Squaring the differences between individual values and the sample mean prevents positive and negative differences from canceling each other out. This is why the variance is expressed as squared units:

$$s^2 = \Sigma(x_i - \bar{x})^2/\text{df}$$

In our example of variation in a quantitative phenotype, the variance is described as **phenotypic variance ($V_P$).** Because we are measuring height in centimeters, the variance will be expressed in centimeters squared.

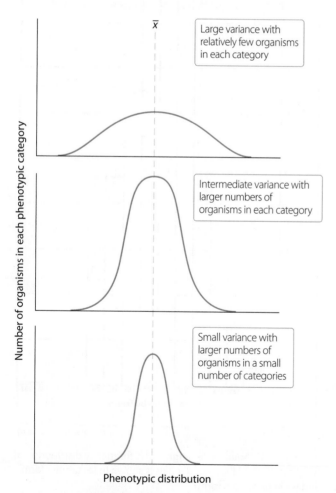

**Figure 21.9 Normal distributions.** The shape of curves depicting normal distributions is changed by the sample size and the number of outcome classes. Variance around the average is correspondingly large, intermediate, and small.

The second measure that describes the distribution of data is the **standard deviation ($s$),** a value expressing deviation from the mean in the same units as the scale of measurement for the sample. The standard deviation ($s$) is calculated as $s = \sqrt{s^2}$. In our sample of adult male height, $V_P = s^2 = 43.30$ cm$^2$, and the standard deviation is $s = 6.58$ cm.

---

**Genetic Insight** Quantitative traits are typically measured using a random sample. The mean is the average value in the sample, and deviation from the mean is measured as either the variance or the standard deviation.

---

## Partitioning Phenotypic Variance

A key part of analyzing quantitative trait variation is to analyze the factors thought to contribute to phenotypic variance, $V_P$. Quantitative phenotypes are the joint product of genes, environment, and gene interactions; consequently, phenotypic variance can be partitioned among those influences. As a first step, the phenotypic variance can be divided into two principal components: *genetic variance* ($V_G$) and *environmental variance* ($V_E$). Under this assumption, phenotypic variance can be expressed in terms of genetic variance plus environmental variance: $V_P = V_G + V_E$.

In this expression, **genetic variance ($V_G$)** is the proportion of phenotypic variance that is due to differences among genotypes. In highly inbred populations in which all individuals are homozygous for alleles controlling a quantitative phenotype, $V_G = 0$. Such populations are found only after strictly controlled laboratory inbreeding, however; they are almost never found in nature, due to the ubiquitous presence of genetic variation in natural populations. Genetic variation in natural populations generates individuals with different genotypes for quantitative traits and leads to phenotypic variability that is directly attributable to the genetic variability.

**Environmental variance ($V_E$)** is the portion of phenotypic variance that is due to variability of the environments inhabited by individual members of a population. Differences in sun exposure, in water and nutrient content of the soil, and in exposure to pests are examples of environmental variables that influence $V_E$ in plants. Carefully controlled laboratory experiments can sometimes control all of the environmental variables and produce a situation in which $V_E = 0$. In nature, however, such circumstances almost never occur. Individual members of natural populations are almost certain to experience variability in the environmental conditions they encounter. Some differences may be systematic and predictable. For example, members of a plant population growing below a natural spring will experience wetter growth conditions than plants living above the spring. Other environmental variables are sporadic or unpredictable. For example, a dry year might reduce the flow of water from a natural spring and affect the plants living below the spring more severely than those living above it.

Let's use an example to illustrate the dissection of $V_G$ and $V_E$ as components of $V_P$. Suppose that two different pure-breeding parental lines are established. Each line is genetically uniform, with $V_G = 0$; therefore, $V_P = V_E$ (Figure 21.10a). The pure-breeding lines are crossed to produce $F_1$ progeny that are genetically uniform. In the $F_1$, $V_G = 0$ because there is no genetic variation among the individuals, and $V_P = V_E$ (Figure 21.10b). Production of $F_2$ leads to genotypic variation and thus to the production of phenotypic variation that results from a combination of genetic variance and environmental variance (Figure 21.10c). Among the $F_2$, $V_P = V_E + V_G$. Since $V_E$ has been determined among the parents and in the $F_1$, genetic variance can be calculated by subtracting environmental variance from the phenotypic variance among the $F_2$. In other words, $V_G = V_P - V_E$. Genetic Analysis 21.2 provides practice in determining environmental and genetic variance.

## Partitioning Genetic Variance

Each allelic difference affecting a quantitative trait contributes to genetic variance in a population, but not necessarily each in the same way. Indeed, it can be difficult to measure the specific effect of each allelic variant. Nevertheless, genetic variance can theoretically be partitioned into three different *kinds* of allelic effects. **Additive variance ($V_A$)** derives from the additive effects of all alleles contributing to a trait. Additive variance is the result of incomplete dominance of alleles at a locus, which causes heterozygotes to have a phenotype intermediate between the homozygous phenotypes. **Dominance variance ($V_D$)** is variance resulting from dominance relationships in which alleles of a heterozygote produce a phenotype that is not intermediate between those of homozygotes (i.e., the nonadditive effects of alleles of contributing genes). Lastly, **interactive variance ($V_I$)** derives from epistatic interactions between the alleles of genes that influence a quantitative phenotype. Collectively these three components unite to produce the genetic variance in a model summarized by $V_G = V_A + V_D + V_I$. We use these values in the following section, to discuss *heritability*.

---

**Genetic Insight** Phenotypic variance of quantitative traits consists of the genetic variance ($V_G$), environmental variance ($V_E$). The genetic variance can be subdivided into separate components accounting for dominance, additivity, and epistatic interactions between genes.

---

# 21.3 Heritability Measures the Genetic Component of Phenotypic Variation

One goal of quantitative genetics is to estimate the extent to which genetic variation influences the phenotypic variation seen in a trait. This is a challenging task when a trait is determined by a combination of genetic variation, environmental variation, and gene–environment interaction. The concept of trait **heritability** was developed to help measure the proportion of phenotypic variation that is due to genetic variation.

Heritability differs from trait to trait. The phenotypic variation observed in a trait with high heritability is largely the result of genetic variation and thus can be strongly influenced by selection programs focused on changing the frequency of a phenotype in a population. Conversely, only a small proportion of the phenotypic variation of a trait with low heritability can be attributed to genetic variation, so the expression of the trait in a population is not effectively changed by selection processes. Heritability is an important measure of the potential responsiveness of a trait to natural selection or artificial selection. It is of special interest to evolutionary biologists and plant and animal breeders, who use it to assess the

**(a)**

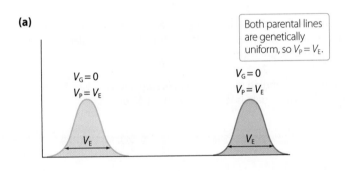

$V_G = 0$
$V_P = V_E$

$V_G = 0$
$V_P = V_E$

$V_E$                    $V_E$

> Both parental lines are genetically uniform, so $V_P = V_E$.

**(b)**

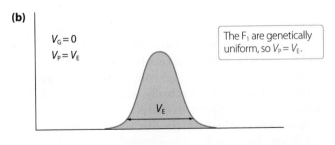

$V_G = 0$
$V_P = V_E$

$V_E$

> The $F_1$ are genetically uniform, so $V_P = V_E$.

**(c)**

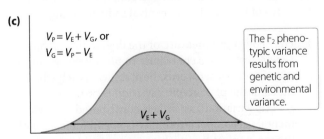

$V_P = V_E + V_G$, or
$V_G = V_P - V_E$

$V_E + V_G$

> The $F_2$ phenotypic variance results from genetic and environmental variance.

**Figure 21.10  Sources of phenotypic variance.**

Two pure-breeding lines of tomatoes, $P_1$ and $P_2$, producing fruit with different average weights, are crossed. The means and variances of their $F_1$ and $F_2$ progeny are shown in the table to the right.

a. What is the environmental variance ($V_E$) for this trait?
b. What is the genetic variance ($V_G$) determined from the $F_2$?

| Line | Average Fruit Weight (g) | $V_P$ |
|---|---|---|
| $P_1$ | 6.5 | 1.6 |
| $P_2$ | 14.2 | 3.5 |
| $F_1$ | 10.2 | 2.2 |
| $F_2$ | 9.8 | 4.0 |

| Solution Strategies | Solution Steps |
|---|---|

### Evaluate

1. Identify the topic this problem addresses, and explain the nature of the requested answer.

1. This problem concerns the determination of environmental variance and genetic variance for the tomato plant data given.

2. Identify the critical information given in the problem.

2. Fruit weight and phenotypic variance are given for the two pure-breeding parental lines and for the $F_1$ and $F_2$ progeny.

### Deduce

3. Describe the relationship between $V_P$, $V_G$, and $V_E$.

3. $V_P = V_G + V_E$

4. Identify the variance values that contribute to $V_P$ in each line and generation.

TIP: For organisms that are genetically identical, $V_P = V_E$.

4. Each of the pure-breeding parental lines ($P_1$ and $P_2$) and the $F_1$ progeny are genetically uniform. As a consequence, all phenotypic variance is due to environmental variance, and genetic variance makes no contribution. The $F_2$ contains genotypic variety, so both $V_G$ and $V_E$ contribute to $V_P$.

### Solve

5. Determine $V_E$ for this trait.

Answer a

5. In the genetically uniform $P_1$, $P_2$, and $F_1$, $V_G = 0$, and in each line $V_P = V_E$. The average environmental variance among these three lines is calculated as $(1.6 + 3.5 + 2.2)/3 = 2.43$ grams.

Answer b

6. Determine $V_G$ for this trait.

6. $V_G$ is calculated by rearranging the expression in step 3 to $V_G = V_P - V_E$. The genetic variance for these data is $V_G = 4.0 - 2.43 = 1.57$ grams.

For more practice, see Problems 4, 10, 12, and 14.

---

potential impact of selection on traits of agricultural or economic importance, as we describe in the case study that concludes this chapter.

Two widely used measures of heritability assess different components of the contribution of genetic variation to phenotypic variation. **Broad sense heritability ($H^2$)** estimates the proportion of phenotypic variation that is due to total genetic variation. This form of heritability is defined by the equality $H^2 = V_G/V_P$. **Narrow sense heritability ($h^2$)** estimates the proportion of phenotypic variation that is due to additive genetic variation. Narrow sense heritability is defined by the equality $h^2 = V_A/V_P$. Both measures of heritability are expressed as proportions that range in magnitude from 0.0 to 1.0. In all cases, greater heritability values indicate a larger role for genetic variation in phenotypic variation.

Heritability is easily misunderstood. An erroneous understanding can lead to the mistaken idea that genetic variation makes a much larger contribution to phenotypic variation than the data actually support. Heritability is difficult to apply to humans except under limited circumstances (described later in the discussion of twin studies), but it can be used for other organisms. The following attributes of heritability are central to its meaning:

1. Heritability is a measure of the degree to which *genetic differences* contribute to *phenotypic variation* of a trait. In other words, heritability is high when much of the phenotypic variation is produced by genetic variation and little is contributed by environmental variation. Heritability *is not* an indication of the mechanism by which genes control a trait, nor is

it a measure of how much of a trait is produced by gene action.

2. Heritability values are accurate only for the environment and population in which they are measured. Heritability values measured in one population cannot be transferred to another population, because both genetic and environmental factors may differ between populations.

3. Heritability for a given trait in a population can change if environmental factors change, and changes in the proportions of genotypes in a population can alter the effect of environmental factors on phenotypic variation, thus changing heritability.

4. High heritability does not mean that a trait is not influenced by environmental factors. Traits with high heritability can be very responsive to environmental changes.

## Broad Sense Heritability

We have seen that genetic variance ($V_G$) is a composite value that derives its magnitude from additive, dominance, and interaction variance. Unfortunately, genetic variance is not always easy to partition into these separate components. Fortunately, broad sense heritability ($H^2 = V_G/V_P$) can be used as a general measure of the magnitude of genetic influence over phenotypic variation of a trait, when $V_G$ cannot be partitioned.

In a 1988 study of the genetics and evolution of cave fish (*Astyanax fasciatus*), Horst Wilkens used broad sense heritability analysis to describe the genetic contribution to the evolution of the organism's eye tissue. Some populations of this species live in completely dark underground cave streams in Eastern Mexico and have a dramatically reduced amount of eye tissue in comparison to closely related fish living aboveground. In these populations, the eye tissue appears to be undergoing rapid evolutionary change. The eyes in sighted fish of this species are approximately 7 cm in diameter. In comparison, blind cave fish have less than 2 cm of eye tissue diameter.

Wilkens crossed sighted cave fish to blind cave fish, measured eye tissue mean and variance in the $F_1$, and then produced $F_2$ fish and measured their eye tissue as well. Since the $F_1$ fish are genetically uniform, the variance in the amount of eye tissue is due entirely to the environment. In these $F_1$, $V_E$ was 0.057 cm. Among the $F_2$, phenotypic variance ($V_P$) was 0.563 cm and was the result of both genetic and environmental variance ($V_G + V_E$). Broad sense heritability is derived by determining $V_G$ and dividing it by phenotypic variation. In this case,

$$V_G = V_P - V_E = 0.563 - 0.057 = 0.506$$
$$H^2 = V_G/V_P = 0.506/0.563 = 0.899$$

This broad sense heritability of approximately 0.90 means that approximately 90% of the phenotypic variation in eye size in these populations of cave fish is due to genetic variation.

## Twin Studies

Heritability can be quantified when both mating and environmental factors can be controlled. However, when mating and environmental variation are not among the controlled experimental parameters, heritability is far more difficult—some would say impossible—to measure accurately. This limitation applies to attempts to measure the heritability of traits in humans. Fortunately, studies of phenotypic variation in human twins can offer insights into broad sense heritability of human traits.

Identical twins, also known as monozygotic twins (MZ twins), are produced by a single fertilization event that is followed by a splitting of the fertilized embryo into two zygotes. Monozygotic twins share all of their alleles. Theoretically, broad sense heritability can be determined by assuming that phenotypic variance between them is fully attributable to environmental variance. Under this assumption, in MZ twin pairs, $V_P = V_E$.

Fraternal twins, on the other hand, are dizygotic (DZ twins), produced by two independent fertilization events that take place at the same time. Dizygotic twins are siblings that are born at the same time, but they are no more closely related than siblings born at different times. Like all full siblings, DZ twins have an average of 50% of their alleles in common. To control for differences between the sexes, only DZ twins of the same sex are used in twin studies. Phenotypic variance between DZ twins is the sum of environmental variance plus one-half of the genetic variance (the 50% of alleles not shared by the average DZ twin pair): In DZ twin pairs, $V_P = V_E + 1/2 V_G$. On the basis of these general formulas for calculation of $H^2$, broad sense heritability can be estimated for human traits by methods we do not discuss here (**Table 21.2**).

Studies of traits in human twins usually compare MZ twins to same-sex DZ twins to make heritability estimates more accurate. Even so, heritability studies of human twins are prone to several sources of error that lead to inaccurately high values. Following are the most common sources of error:

1. *Stronger shared maternal effects in identical twins than in fraternal twins.* These effects include the sharing of embryonic membranes and other aspects of the uterine environment that lead to more similar developmental conditions for identical twins than for fraternal twins.

2. *Greater similarity of treatment of identical twins than of fraternal twins.* Parents, other adults, and

| **Table 21.2** | Some Broad Sense Heritability ($H^2$) Values from Human Twin Studies |
|---|---|
| **Trait** | **Heritability ($H^2$), %** |
| *Biological Traits* | |
| Total fingerprint ridge count | 90 |
| Height | 85 |
| Maximum heart rate | 85 |
| Club foot | 80 |
| Amino acid excretion | 70 |
| Weight | 60 |
| Total serum cholesterol | 60 |
| Blood pressure | 60 |
| Body mass index (BMI) | 50 |
| Longevity | 29 |
| *Behavioral Traits* | |
| Verbal ability | 65 |
| Sociability index | 65 |
| Temperament index | 60 |
| Spelling aptitude | 50 |
| Memory | 50 |
| Mathematical aptitude | 30 |

peers have a tendency to treat identical twins more equally than they treat fraternal twins of the same sex. This gives identical twins a similar social and behavioral environmental experience, while fraternal twins more often are treated differently.

3. *Greater similarity of interactions between genes and environmental factors in identical twins than in fraternal twins.* Identical twins have the same genotype and are affected in similar, if not identical, ways by environmental factors. On the other hand, fraternal twins have genetic differences that can be influenced differently by environmental factors. This may result in greater variance between fraternal twins than between identical twins.

Because of the difficulties and the potential sources of error in making heritability estimates based on twin studies, the values in Table 21.2 are more likely to be too high than too low.

The study of identical twins reared together versus those reared apart is an alternative approach to estimating the influence of genes on phenotypic variation. Such studies measure the **concordance,** the percentage of twin pairs in which both have the same phenotype for a trait, versus the **discordance,** the percentage in which both twins have dissimilar phenotypes for a trait. Concordance and discordance frequencies give a general picture of the overall influence of genes on phenotypes. If phenotypic variation for a trait is 100% genetic, MZ twins should always be concordant for their phenotypes, whether reared together or apart. In this case, concor-

dance would be 100%. Dizygotic twins share an average of 50% of their genes in common and would have concordance of about 50% for a trait whose variation is completely genetic. When phenotypic variation of a trait is due entirely to nongenetic factors, on the other hand, concordance among MZ and DZ twins will be approximately equal, and both values will be significantly less than 100%. For traits with phenotypic variation that is determined to a significant extent by genetic variation, concordance among MZ twin pairs will be substantially less than 100% but significantly greater than for DZ twins. A number of human diseases, malformations, and other phenotypic variants fall into the latter category. Table 21.3 shows MZ and DZ twin concordance values for common malformations and other abnormalities that are determined to a large extent by genetic variation but are also a product of environmental triggers that play as yet undetermined roles.

---

**Genetic Insight** Broad sense heritability, which measures what proportion of phenotypic variation of a quantitative trait is due to genetic variation, is used to estimate the genetic contribution to phenotypic variation. Comparisons of concordance and discordance frequencies among MZ and DZ twins also give a broad measure of the degree to which genes influence a phenotype.

---

## Narrow Sense Heritability and Artificial Selection

Narrow sense heritability ($h^2 = V_A/V_P$) estimates the proportion of phenotypic variation that is due to additive

| **Table 21.3** | Concordance Values for Common Threshold Conditions in Humans | |
|---|---|---|
| **Trait** | **Percent Concordance** | |
| | MZ Twins | DZ Twins |
| Alzheimer disease | 60 | 25 |
| Autism | 70 | 10 |
| Cleft lip | 40 | 4 |
| Club foot | 30 | 2 |
| Congenital hip dislocation | 35 | 3 |
| Depression | 70 | 25 |
| Insulin-dependent diabetes | 50 | 10 |
| Pyloric stenosis | 25 | 3 |
| Reading disability | 70 | 45 |
| Schizophrenia | 60 | 20 |

genetic variance ($V_A$), variance resulting from the alleles of additive genes. These estimates are particularly useful in agriculture, where they predict the potential responsiveness of a trait in an animal or plant to artificial selection imposed through selective breeding programs or controlled growth conditions. High narrow sense heritability values are correlated with a greater degree of response to selection than low values, because additive genetic variance is responsive to selection.

Table 21.4 gives examples of $h^2$ values, covering a broad spectrum of magnitude, for several characteristics of plants and animals. Since higher $h^2$ values have the strongest correlation with selection response, biologists predict that traits such as body weight in cattle, back-fat thickness in pigs, and corn plant height will be most amenable to change through artificial selection schemes. On the other hand, litter size in pigs, egg production in poultry, and ear diameter in corn have low $h^2$ values and will be less responsive to selection.

Estimating the potential response to selection for a trait begins with calculation of a value known as the **selection differential (S)**, which measures the difference between the population mean value for a trait and the mean trait value for the mating portion of a population. Suppose, for example, that a goal of an artificial selection experiment is to increase plant height. Choosing taller-than-average plants to mate will be an effective way to increase the height of progeny if $h^2$ is high. If the population average height is 37.5 cm and the average height of plants selected for mating is 42 cm, then $S = 42$ cm $- 37.5$ cm $= 4.5$ cm.

The potential **response to selection (R)** depends on the extent to which the difference between the mating trait

mean value and the population mean value can be passed on to progeny. This probability is estimated using the formula $R = S(h^2)$. For this plant height example, let's assume we examine corn plant height, $h^2 = 0.70$ (see Table 21.4). In this case, $R = (4.5$ cm$)(0.70) = 3.15$ cm. Under stable growth conditions, the progeny plants could be expected to have a height equal to the population average plus the value of $R$, or 37.5 cm $+ 3.15$ cm $= 40.65$ cm. Narrow sense heritability can be measured by rearranging the terms in the response-to-selection equation to $h^2 = R/S$. For the plant-height example, $h^2 = 3.15$ cm$/4.5$ cm $= 0.70$. The case study for this chapter expands on the issue of selection response by describing an artificial selection experiment that is more than 120 years old.

Estimates of heritability have important practical applications for plant and animal breeders, and for evolutionary biologists. Whether traits are subjected to artificial selection by breeders or to natural selection, the extent to which the mean value of a trait changes in a population depends on its heritability. Breeders and evolutionary biologists predict substantial change in trait mean values (i.e., large values for $R$) when heritability is high, but little or no change in trait mean values when heritability is low. In other words, traits evolve when a substantial proportion of the phenotypic variation is due to genetic variation.

Figure 21.11a shows three examples in which the selection differentials are the same but the response to selection differs as a result of different degrees of heritability. This comparison illustrates that selection response is expected to be maximal when heritability is $h^2 = 1.0$. Selection response is substantially less when heritability is $h^2 = 0.2$, and there is no selection response when heritability is $h^2 = 0$. Selection also affects quantitative traits in natural populations. Figure 21.11b shows natural selection operating over many generations in three different modes that have different effects on phenotypic means and variances. In the mode known as **directional selection,** the mean phenotypic value is shifted in one direction because one extreme of the phenotype distribution is favored. This narrows the phenotypic range and reduces phenotypic variance. In contrast, natural selection favoring an intermediate phenotype over extreme phenotypes results in **stabilizing selection** that reduces the phenotypic variance. **Disruptive selection** occurs when both extreme phenotypes are favored over intermediate phenotypes. The result is an increase in the phenotypic variance and, potentially, a phenotypic split within the population.

| Table 21.4 | Some Narrow Sense Heritability ($h^2$) Values for Animals and Plants | |
|---|---|---|
| **Organism** | **Trait** | **Heritability ($h^2$)** |
| Cattle | Body weight | 0.65 |
| | Milk production | 0.40 |
| Corn | Plant height | 0.70 |
| | Ear length | 0.55 |
| | Ear diameter | 0.14 |
| Horse | Racing speed | 0.60 |
| | Trotting speed | 0.40 |
| Pig | Back-fat thickness | 0.70 |
| | Weight gain | 0.40 |
| | Litter size | 0.05 |
| Poultry | Body weight (8 weeks) | 0.50 |
| | Egg production | 0.20 |

Genetic Insight  Narrow sense heritability ($h^2$) measures the contribution of additive genetic variance to phenotypic variance and is used to determine the potential response of a quantitative trait to artificial selection or to natural selection.

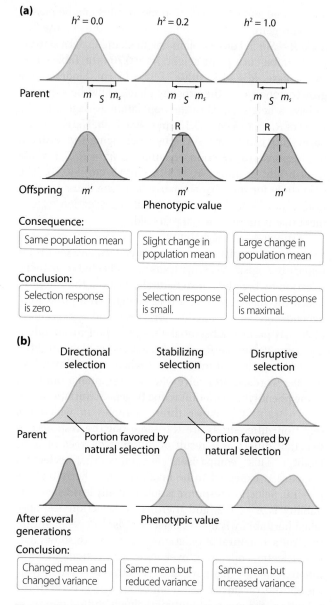

**Figure 21.11** **Response to artificial and natural selection.**
**(a)** Response to artificial selection after one generation depends on $h^2$. M is mean phenotype in parental generation; $M_S$ is mean phenotype of parents selected for reproduction; M' is mean phenotype of offspring after selection; selection differential is $S = M - M_S$. **(b)** Expected changes in phenotypic means and variances after several generations of natural selection.

## 21.4 Quantitative Trait Loci Are the Genes That Contribute to Quantitative Traits

The genes that contribute to the variation in a quantitative trait are collectively called **quantitative trait loci (QTLs)**. Individually, a gene that contributes to a quantitative trait is referred to as a **quantitative trait locus.** QTLs were initially of interest in agricultural plants such as tomatoes and corn, where they influence important

attributes including fruit sweetness, acidity, and color. QTL analysis has expanded greatly in recent decades through analysis of many distinct traits in plants and animals, including humans.

In one way, QTLs are no different from other genes we discuss. For example, they often produce polypeptides that operate in metabolic pathways producing compounds that give flavor or color to fruit. Identifying QTLs by experimental analysis is different from identifying other genes that control phenotypic variation, however, because many genes are influencing the trait and the presence or absence of a particular allele does not correlate well with distinct phenotypes. Specialized statistical methods have been developed to detect and map QTLs. This process is called **QTL mapping,** and it involves the identification of chromosome regions that are likely to contain QTLs.

The general process of QTL mapping is similar to the methods used to determine genetic linkage between genes. A chromosome region likely to contain a QTL is identified by the frequent co-occurrence of a specific genetic marker such as a single nucleotide polymorphism (SNP) in organisms with a particular phenotype. The SNP is not the QTL. Instead, it is genetically linked to the QTL. The connection between the genetic marker and the phenotype implies that a QTL exists near the genome location encoding the genetic marker.

### QTL Mapping Strategies

Contemporary QTL mapping uses DNA markers that have known chromosome locations to assist with the mapping and identification of genes. SNPs are particularly useful in these analyses, as are other DNA marker variants such as restriction fragment length polymorphisms (RFLPs) and variable number tandem repeats (VNTRs), in which different numbers of repeats of specific nucleotide base pairs occur in different chromosomes.

Multiple approaches can be taken in QTL mapping experiments. At its core, however, QTL mapping is a statistical process that seeks to identify regions of genomes containing genetic markers that are linked to QTLs. The statistical analysis for QTLs is closely related to the statistical analysis of genetic linkage using logarithm of the odds (Lod) score analysis (see Section 5.5). QTL analysis can lead to identification of the potential *chromosome location* of a QTL influencing phenotypic variation of a quantitative trait, but by itself it does not identify the actual QTL. Other genetic methods are available for QTL identification.

QTL mapping uses the parents and progeny produced by controlled crosses as the sources of DNA for genetic marker identification and as the source of data for the quantitative trait of interest. If, for example, a researcher wants to identify QTLs that influence large fruit size in tomatoes, he or she will cross two parental lines of tomatoes that differ in fruit size. The $F_1$ progeny of this cross could then be used to produce $F_2$ progeny or, as we illustrate here, the $F_1$ could be used in a backcross to one of the parental lines. Genetic markers will be determined in the original parental lines and in the backcross progeny. Tomato sizes produced by backcross progeny will be weighed and the results compared to genetic markers in the individual plants.

Figure 21.12a illustrates the structure of a backcross experiment designed to collect genetic marker and tomato-weight data for QTL mapping analysis. One parental tomato strain producing large fruit that averages 100 grams (g) contains genetic markers that are identified by the letter $L$. There are actually many markers linked to QTLs in the line, and for each marker gene tested, the large-tomato strain will have two copies of the large-strain marker allele genotype designated $LL$. Similarly, a small-tomato-producing strain, with an average

tomato weight of 10 grams, is characterized for the same genetic markers, and each of the loci tested in the small-strain genotype is designated $SS$. The $F_1$ progeny of the large × small cross is heterozygous for each marker locus and is designated $LS$. These plants in this example are shown to produce tomatoes that weigh 60 g. The backcross is made to the large-tomato strain, and the marker genotype will be either $LL$, if the $F_1$ transmits the large-strain allele, or $LS$, if the $F_1$ transmits the small-strain allele. The backcross progeny in this example produce tomatoes that vary in weight from 80 g to 88 g. Tomato weight from the backcross plants is greater than from the $F_1$ plants because the backcross plants are the result of a cross between the $F_1$ and the large-tomato strain.

Table 21.5 displays tomato-weight data for 10 backcross plants (1–10) and genetic marker data for two genes, marker A ($M_A$) and marker B ($M_B$), that are not linked to one another and are located in different parts of the genome. In an actual QTL backcross experiment several hundred backcross plants might be examined, and each plant might be genotyped for dozens of genetic markers that are ideally spaced about every 5 to 10 centimorgans (cM) in the genome. This number of genetic markers and their close proximity maximize the chance of identifying the location of QTLs detected by the analysis.

In Table 21.5, the average weight of tomatoes from backcross plants is 84 grams. Average tomato weight is compared for $LL$ plants versus $LS$ plants for each marker. There is almost no difference in average weight for $M_A$

**(a)** Parental cross and backcross

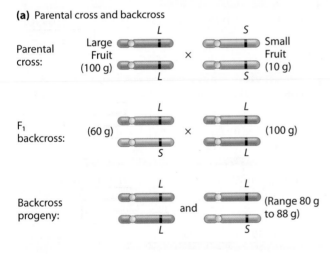

**(b)** Lod score profile

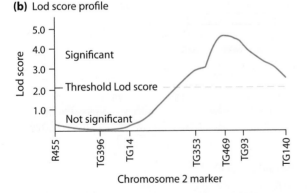

Chromosome 2 marker

**Figure 21.12 Quantitative trait locus (QTL) detection and mapping. (a)** Parental tomato plants producing large ($LL$) or small ($SS$) fruit are crossed to produce $F_1$ ($LS$). The $F_1$ are then backcrossed to the large-fruit line to yield backcross progeny that are either $LL$ or $LS$. **(b)** The significance of linkage between potential QTLs and genetic markers is tested among backcross progeny by Lod score analysis. A Lod score profile assessing fruit-weight QTLs reveals significant scores exceeding the threshold value on tomato chromosome 2.

| Table 21.5 | QTL Analysis of Tomato Weight in Backcross Progeny | | |
|---|---|---|---|
| **Backcross Plant** | **Average Fruit Weight (g)** | **Markers** | |
| | | $M_A$ | $M_B$ |
| 1 | 86 | $LS$ | $LL$ |
| 2 | 82 | $LL$ | $LS$ |
| 3 | 85 | $LL$ | $LL$ |
| 4 | 88 | $LL$ | $LL$ |
| 5 | 81 | $LS$ | $LS$ |
| 6 | 83 | $LS$ | $LS$ |
| 7 | 84 | $LL$ | $LL$ |
| 8 | 80 | $LL$ | $LS$ |
| 9 | 84 | $LS$ | $LS$ |
| 10 | 87 | $LS$ | $LL$ |
| Total average weight | 84 | | |
| $LL$ average weight | | 83.8 | 86.0 |
| $LS$ average weight | | 84.2 | 82.0 |

($LL$ = 83.8 g versus $LS$ = 84.2 g), but for $M_B$, $LL$ plants produce tomatoes that are 4 grams heavier on average than are the tomatoes from $LS$ plants ($LL$ = 86.0 g versus $LS$ = 82.0 g). These data may indicate that a QTL influencing tomato weight is located near $M_B$. Conversely, there is no evidence to indicate that a QTL is located near $M_A$.

To determine the statistical significance of the kind of information provided for genetic markers and tomato weight, a Lod score is calculated. The Lod score is an odds ratio of the probability of the data if a QTL is linked to the marker divided by the probability of the data if there is no QTL linked to the marker. The odds ratios for the backcross plants are added together, and the log (the *log* of the *odd*s) is taken to yield the Lod score. Like the analysis of Lod scores for genetic linkage, there is a threshold value for significance of the score (see Section 5.5). If the Lod score for a genetic marker is greater than the threshold value, the Lod score indicates a statistically significant probability that a QTL is linked to the marker.

In **Figure 21.12b**, a Lod score profile for several genetic markers located on chromosome 2 of tomato reveals significant evidence indicating genetic linkage to a QTL. Beginning at the marker designated TG353 and spanning to the right through marker TG140, the Lod score values are greater than the threshold value and give statistically significant evidence favoring linkage between these genetic markers and a QTL. On the other hand, the Lod scores falling below the threshold value in the figure give no statistical evidence of linkage to a QTL. For chromosome 2 in tomato, Lod scores for genetic markers to the left of TG353 are less than the threshold Lod score value. By using a large number of regularly spaced genetic markers distributed every few cM along each chromosome, QTL mapping analysis can potentially detect the location of any QTL influencing a quantitative trait phenotype. Commonly, multiple QTLs in a genome are identified.

Andrew Paterson and his colleagues published a 1988 study mapping 15 QTLs in the tomato genome that influence fruit weight, fruit acidity, and the amount of soluble solids in the fruit. Each trait has agricultural importance, and together they determine the quality and yield of tomato paste from the fruit. Paterson's study used 70 DNA markers spaced an average of 20 cM apart throughout the tomato genome. Collectively, these markers span about 95% of the 12 chromosomes that constitute the tomato genome.

The parental plants were two closely related and interfertile species: a domestic tomato (*Lycopersicon esculentum*) and a wild South American green-fruited tomato (*Lycopersicon chmielewskii*). The F₁ hybrids were backcrossed to *L. esculentum*, producing 237 backcross progeny plants for analysis. All backcross plants were grown under identical conditions to minimize the influence of environmental factors on the traits of interest. Individual fruits from backcross plants were assayed for fruit weight

(grams), soluble solids content (percentage), and acidity (pH). Lod score analysis was used to test whether genes influencing any of the three traits exhibited genetic linkage to genome markers. Significant Lod score values traced six genes influencing fruit weight, five influencing acidity, and four influencing soluble solids content to regions of nine chromosomes in the tomato genome (**Figure 21.13**).

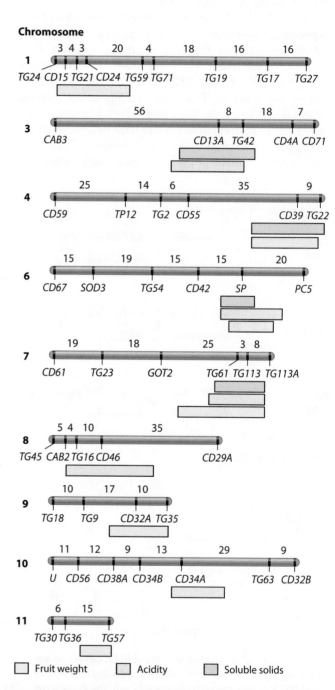

**Figure 21.13  QTL mapping in domestic tomato (*Solanum lycopersicon*).** Multiple QTLs influencing fruit weight, fruit acidity, and percentage of soluble solids of tomatoes are mapped. DNA marker designations are given below each chromosome and the distance (in cM) between markers is given as a number above the chromosomes.

## Identification of QTL Genes

Since QTL mapping identifies the location of genes influencing quantitative traits, but not the genes themselves; additional genetic analysis is required to identify the genes. To acquire information leading to gene identity, researchers use **near isogenic lines (NILs)**, also called **introgression lines (ILs).** These lines are derived from backcross progeny produced as described earlier. Different backcross progeny are self-fertilized over many generations to form highly inbred lines. The resulting lines are nearly isogenic, meaning they are genetically identical at almost all genes. The lines differ from one another, however, by carrying different crossovers that have introduced different alleles near the site of a QTL. The introduced differences are called introgressions, thus giving these lines their name.

**Figure 21.14a** illustrates six introgression lines (IL1 to IL6) descended from a cross between two original parental lines, one a domesticated species and the other a wild species. The chromosome colors illustrate crossovers that produce differences between the introgression lines. Crossover locations are identified by analysis of genetic markers, and each introgression line is characterized for a trait phenotype. In the figure, the bars to the right of each line indicate the percentage difference between the phenotype of the IL and the domesticated parental species. Two potential QTL regions, QTL-A and QTL-B, contain variations of the crossover segments. The greatest positive percentage difference relative to the domesticated species phenotype occurs in IL2 and IL3 that carry crossover chromosomes containing wild-species DNA in

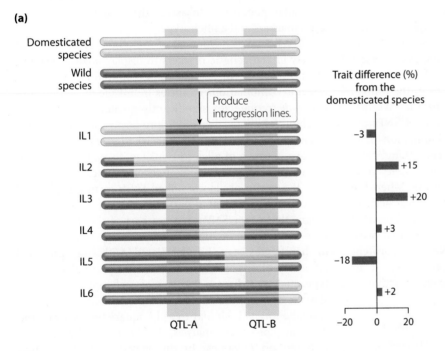

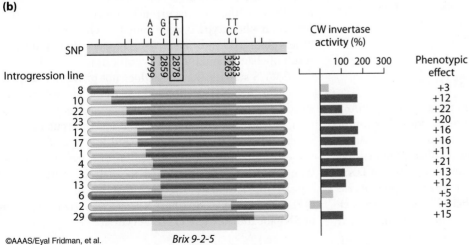

**Figure 21.14 QTL analysis in introgression lines. (a)** Six introgression lines (IL1 to IL6) formed by mating between a domesticated species and a wild species have different patterns of recombination in the region of two QTLs. The difference in trait expression between the trait in the domesticated species and each IL is given as a percentage. **(b)** Analysis of *Brix 9-2-5* in 13 introgression lines identifies SNPs that alter CW invertase activity. The SNP at position 2878 has a substantial influence on CW invertase function.

©AAAS/Eyal Fridman, et al.          *Brix 9-2-5*

the vicinity of QTL-A and domesticated species DNA near QTL-B.

To identify the genes responsible for QTL variation, "candidate genes," genes that are potentially responsible for the observed variation, must be identified and investigated. Genes in the QTL-A and QTL-B regions are located by examining DNA sequences, and sequence variants in candidate genes among introgression lines are identified. The sequence differences detected are studied to determine if they correlate with phenotypic variation.

**Figure 21.14b** illustrates the results of experimental analysis of tomato introgression lines by Eyal Fridman and colleagues in 2004 designed to identify genes contributing to Brix value in tomato. The Brix value of fruit refers to the total soluble solids content, of which sugars and acids are the primary constituents. Fridman and colleagues created a large number of ILs from an initial cross between the domesticated tomato species (*Solanum lycopersicum*) and a wild relative (*Solanum pennellii*).

The parental species and each of the ILs were studied for Brix value, and a QTL found to have a high Brix value, *Brix 9-2-5*, was intensively studied. DNA sequencing of the 484 nucleotides (positions 2799 to 3283) in *Brix 9-2-5* revealed the five SNP variants shown in the figure. The *Brix 9-2-5* QTL corresponds to a segment of the tomato *LIN5* gene that produces the cell wall enzyme invertase (CW invertase). In the figure, the positions of SNPs are shown relative to 13 ILs that carry recombination in or near *Brix 9-2-5*. The bar to the right of each IL indicates its percentage difference in CW invertase activity relative to *S. lycopersicum*. The results show that when the *S. pennellii* sequence is present, CW invertase activity is significantly greater than in *S. lycopersicum*. The SNP at position 2878 (boxed) was strongly correlated with increased CW invertase activity. DNA and protein sequence analysis revealed that this SNP produced an amino acid difference that altered CW invertase activity.

---

**Genetic Insight**  Quantitative trait loci (QTLs) are mapped to specific chromosome locations by Lod score analysis that identifies genetic linkage between QTLs and variable DNA markers whose chromosome locations are known.

---

## Genome-Wide Association Studies

The widespread availability of genome sequencing information has opened a new avenue to the identification of QTLs in numerous species, including humans. Known as **genome-wide association studies (GWAS),** the method seeks to tie the presence of a sequence variant of a DNA marker to a QTL influencing a specific phenotype. The relationship between an inherited genetic marker variant and the phenotype is by "association," which

means organisms that carry a particular variant are more likely to have a certain phenotype than are organisms that carry a different variant. The assessment of association is quantitative; that is, it expresses the percentage of organisms with a genetic marker that also display a certain phenotype versus the percentage that have the phenotype but not the genetic marker.

One advantage of GWAS over other QTL mapping approaches is that GWAS can scan the entire genome for QTLs by statistically testing for marker variants that are associated with phenotypic variation. Positive statistical results indicating association identifies chromosome regions that can be more closely inspected for genes that influence the trait. A second advantage of GWAS is that organisms in random mating populations can be analyzed. Rather than requiring controlled crosses and the formation of introgression lines as described earlier, GWAS uses "cases," or organisms with a particular phenotype, and compares them to "controls" that lack the particular phenotype to assess the association between QTL markers and a phenotype.

GWAS takes advantage of the tendency of alleles of closely linked genetic markers to display linkage disequilibrium (see the discussion in Section 5.6 for a review). Specific combinations of alleles in linkage disequilibrium occur at frequencies significantly greater than expected by chance. Linkage disequilibrium occurs because recombination has not reshuffled the alleles into random combinations. Groups of alleles in linkage disequilibrium form haplotypes along segments of chromosomes (see Section 5.6). If a group of closely linked SNPs form a haplotype, then identification of a particular SNP for one marker means that other SNPs that are part of the same haplotype are likely to be found nearby. The presence of SNPs in haplotypes can be correlated with the presence (affected) or absence (unaffected) of a particular phenotype, such as a disease that is genetically influenced. The statistical test of association between a SNP and the disease phenotype is similar to a chi-square test (see Section 2.5). Like chi-square analysis, significance of the outcome is based on $P$ values. In this statistical test, the null hypothesis is that the occurrence of a certain SNP and a particular phenotype is determined by chance. A $P$ value of less than 0.05 indicates that the association is not the result of chance, and the null hypothesis is rejected.

GWAS statistical analysis does not prove that a QTL at or near the SNP location influences the phenotype. Instead, it provides statistical evidence suggesting that a QTL *may be* located close by. Additional molecular analysis can identify candidate genes and link specific allelic variation directly to the production of a certain phenotype. GWAS has been applied to analysis of numerous human diseases and disorders that are influenced by QTLs. The GWA studies have uncovered numerous genes or SNP locations where QTLs may exist.

A 2001 study by Yasunori Ogura and colleagues used GWAS to identify several chromosome regions associated with Crohn's disease (CD), an inflammatory bowel disease that affects humans at a prevalence of 150 to 200 cases per 100,000 people. The severity of CD is highly variable, from relatively mild to potentially fatal. Clinicians describe CD severity using a scale that captures the quantitative nature of the trait, making CD a candidate disease for QTL analysis. The etiology of CD is unknown, but one prominent hypothesis proposed that it is an inflammatory response to intestinal bacteria and other microflora. CD clusters in families, and disease susceptibility is inherited, but susceptibility is influenced by multiple genes. In the study by Ogura and colleagues, the strongest statistical evidence of association of a genetic marker with a susceptibility gene came from chromosome region 16q12. A gene initially identified as *NOD2* and subsequently renamed *CARD15* (capase recruitment domain, member 15), is a candidate for a gene influencing susceptibility to CD.

*CARD15* encodes 12 exons that direct the production of a 1040–amino acid protein. Ogura and colleagues sequenced the exons and introns of *CARD15* in 12 CD patients from different families having multiple cases of CD. They performed the same gene sequencing on four healthy control individuals as well. The study identified an identical C-G base pair insertion at nucleotide 3020 of exon 11 in three of the 12 CD patients. The insertion, designated *3020insC,* induces a frameshift mutation that generates a premature stop codon, shortening the mutant protein by 1007 amino acids. Ogura and colleagues developed an allele-specific polymerase chain reaction (PCR) assay for 3020insC and tested 101 CD patients whose parents were heterozygous for the wild-type allele and the 3020insC allele. Of the 101 CD patients, 68 were homozygous for 3020insC (Figure 21.15a). Biochemical analysis shows mutant protein from the gene has only a small fraction of the activity of the wild-type protein. This diminished capacity reduces the sensitivity of the immune system to the microbial invader, and by a mechanism that remains to be elucidated, results in CD.

Mutations of *CARD15* are not the sole cause of CD; numerous CD patients do not carry 3020insC or any other known mutation of the gene. Research published in 2007 by the Wellcome Trust Case Control Consortium, a collaborative group composed of dozens of collaborators, confirmed the important role of *CARD15* in CD susceptibility and identified six additional genes that influence susceptibility to CD (Figure 21.15b). The same collaborative study examined GWAS of other human autoimmune disorders, including type 1 (insulin-dependent) diabetes and rheumatoid arthritis. GWAS analyses of rheumatoid arthritis and type I diabetes identify multiple genes influencing susceptibility. Interestingly, however, the analyses also identify two susceptibility genes, one on chromosome 1 and the other on chromosome 6, that are shared by both disorders, This valuable clinical insight may indicate that part of the mechanisms leading to development of these diseases are similar. Such information may prove valuable in the diagnosis and treatment of these diseases.

---

**Genetic Insight** Genome-wide association studies (GWAS) locate QTLs by identifying linkage disequilibrium between DNA marker variants and a QTL in randomly mating populations. Statistical analysis similar to chi-square analysis evaluates the association of genetic markers with QTLs.

---

**(a)**

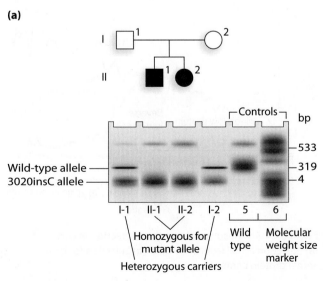

**(b)** SNPs significantly associated with Crohn's disease

| Chromosome | Gene |
| --- | --- |
| 1 | *IL23R* |
| 3 | *ATG16LI* |
| 5 | *IRGM* |
| 5 | *IBD5* |
| 10 | *NKX2-3* |
| 16 | *CARD15* |
| 18 | *PTPN2* |

**Figure 21.15 Detection of 3020insC in *CARD15* in a family with Crohn's disease. (a)** Gel electrophoresis of PCR products from four members of a family are shown in lanes 1 through 4. A wild-type control is in lane 5, and molecular weight size markers are in lane 6. **(b)** Seven QTLs influencing the expression of Crohn's disease, identified by GWAS.

## CASE STUDY

### Artificial Selection for Oil and Protein Content in Corn

An ongoing artificial selection experiment now 125 years in duration provides a clear example of the selection responsiveness of quantitative traits. The experiment, consisting of artificial selection for changes in the oil content and protein content of corn (*Zea mays*), was initiated in 1886 by Cyril Hopkins, then a professor of agronomy at the University of Illinois. It has been in continuous operation ever since.

The Hopkins experiment has two distinct parts, one focused on selecting separate lines of corn that have high and low oil content in kernels, and the other focused on raising and lowering the protein content of corn kernels. The oil content experiment began with a single strain of corn containing about 5% oil. More than 100 generations of selection have produced corn lines differing dramatically in oil content. At the extremes, one line designated as Illinois high oil (IHO) now contains more than 20% oil in kernels, whereas a line designated as Illinois low oil (ILO) contains virtually no oil at all (Figure 21.16a). The change in oil content of kernels has been continuous due to the artificial selection applied to each generation.

At generation 50, a reversal of artificial selection criteria was applied to a few plants in each line. Some IHO-line plants were redirected into an experiment to reverse the direction of change by lowering the oil content. When the reversal experiment began, oil content of IHO was about 12%; but over the next 50 generations, oil content of the reverse high oil (RHO) line was reduced to 5%. Also at generation 50, some plants from the ILO line were redirected into an experiment to reverse the direction of change in their oil content. The ILO line stood at 2% oil when the reverse experimental protocol began, but the oil content of the reverse low oil (RLO) line had been raised to 5% by generation 100. The results of this ongoing experiment indicate that continuous change in oil content can be achieved by the consistent application of selection criteria that favor high or low oil content of kernels. The reverse experiments indicate that genetic diversity is retained by many of the genes involved in oil production, even after many generations of intense selection.

The Illinois corn experiment also selected for the protein content of kernels in an experiment that almost precisely paralleled the design and results of the kernel oil content experiment. Figure 21.16b traces the divergence of two lines of plants, Illinois high protein (IHP) and Illinois low protein (ILP), that were established from the same founders as the oil experiment. As in the corn oil experiment, reversal experimental breeding was initiated at generation 50 to test the possibility of reversing the protein increase in the IHP line and reversing the loss of protein in the ILP line. The reverse lines, reverse high protein (RHP) and reverse low protein (RLP), showed reversal of some of the protein changes accrued over the first 50 generations.

In 2004, Cathy Laurie and her colleagues undertook a QTL analysis of the Illinois corn oil selection project in an attempt to identify QTLs associated with oil production. Using GWAS, they identified more than 50 potential QTLs influencing corn oil content that accounted for about one-half of the total genetic variation in oil production. None of the QTLs exerted more than 1% change in oil content, suggesting that a large number of genes, each with a very small influence on oil production, are responsible for the oil content of kernels. Laurie's analysis confirms that a large number of QTLs, acting additively, are responsible for oil content in corn. The smooth and sustained response to selection that has been observed throughout the Hopkins experiment is the effect of selection on many loci that each has a very small influence on oil content.

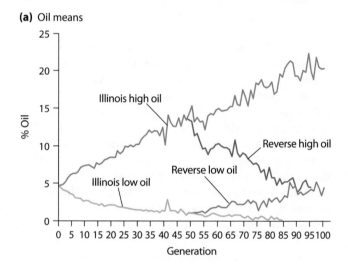

**(a)** Oil means

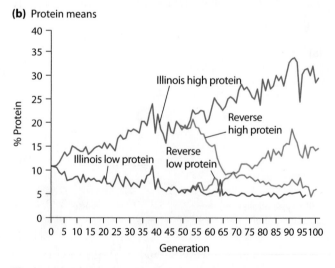

**(b)** Protein means

**Figure 21.16 Results of long-term selection in corn.**
**(a)** Selection for oil level in kernels. **(b)** Selection for corn kernel protein content.

---

SUMMARY                        *For activities, animations, and review quizzes,*
                                                    *go to the study area at www.masteringgenetics.com.*

---

### 21.1 Quantitative Traits Display Continuous Phenotype Variation

▎ Quantitative phenotypic traits are polygenic and are described by scales of measure that can be assigned values having a quantitative basis.

▎ The phenotypes of multifactorial traits result from polygenic inheritance and the influence of environmental factors.

▎ Most quantitative traits have a continuous phenotypic distribution. Those influenced by larger numbers of genes are more likely to display continuous variation. Discontinuous variation in phenotype is particularly likely with threshold traits.

▎ Threshold traits are explained by additive alleles and have a threshold of liability that separates one phenotypic category (unaffected) from another (affected). The threshold of liability is crossed when a sufficient number of additive alleles accumulate in the genotype.

### 21.2 Quantitative Trait Analysis Is Statistical

▎ Quantitative traits are analyzed using statistical methods that evaluate the mean, median, mode, and variance of quantitative trait phenotype distribution.

▎ The frequency distribution for the phenotype range is described by the variance or the standard deviation in sample values. In the case of quantitative trait phenotypes, the phenotypic variance ($V_P$) is a useful measure of the sample distribution.

▎ The phenotypic variance of a trait is the sum of genetic variance ($V_G$) and environmental variance ($V_E$).

▎ Genetic variance is partitioned into additive variance ($V_A$), dominance variance ($V_D$), and interactive variance ($V_I$), the latter resulting from the epistatic interaction of genes determining a phenotype.

### 21.3 Heritability Measures the Genetic Component of Phenotypic Variation

▎ Heritability is a measure of the extent to which genetic variation contributes to total phenotypic variation.

▎ Broad sense heritability ($H^2$) measures the ratio of genetic variance to phenotypic variance ($V_G/V_P$). One method of applying broad sense heritability analysis to humans is through twin studies that give a general estimate of heritability.

▎ Narrow sense heritability ($h^2$) measures the contribution of additive genetic variance to phenotypic variance ($V_A/V_P$).

▎ Narrow sense heritability is used to predict the selection response ($R$) of a trait to artificial selection.

### 21.4 Quantitative Trait Loci Are the Genes That Contribute to Quantitative Traits

▎ QTL mapping is used to determine the location of potential QTLs in genomes using methods that closely resemble recombination mapping.

▎ Controlled crosses and analysis of recombinant chromosomes are required for QTL mapping.

▎ Specific genes influencing quantitative trait phenotypes are identified and their variation characterized through QTL candidate locus analysis.

▎ Genome-wide association studies (GWAS) scan the entire genome of organisms in random mating populations for statistical evidence of QTLs.

---

KEYWORDS

additive genes (additive traits) *(p. 710)*
additive variance ($V_A$) *(p. 721)*
broad sense heritability ($H^2$) *(p. 722)*
concordance *(p. 724)*
continuous variation *(p. 709)*
directional selection *(p. 725)*
discontinuous variation *(p. 709)*
discordance *(p. 724)*
disruptive selection *(p. 725)*
dominance variance ($V_D$) *(p. 721)*
environmental variance ($V_E$) *(p. 720)*
frequency distribution *(p. 719)*
genetic liability *(p. 716)*
genetic variance ($V_G$) *(p. 720)*

genome-wide association studies
   (GWAS) *(p. 730)*
heritability *(p. 721)*
interactive variance ($V_I$) *(p. 721)*
introgression line (IL) (near isogenic line)
   *(p. 729)*
major gene *(p. 710)*
median (median value) *(p. 719)*
mode (modal value) *(p. 719)*
modifier gene *(p. 710)*
multifactorial trait *(p. 709)*
multiple-gene hypothesis *(p. 710)*
narrow sense heritability ($h^2$) *(p. 722)*
phenotypic variance ($V_P$) *(p. 720)*

polygenic inheritance *(p. 709)*
polygenic trait *(p. 709)*
quantitative genetics *(p. 709)*
quantitative trait *(p. 709)*
quantitative trait locus (QTL) *(p. 726)*
QTL mapping *(p. 726)*
response to selection ($R$) *(p. 725)*
selection differential ($S$) *(p. 725)*
stabilizing selection *(p. 725)*
standard deviation ($s$) *(p. 720)*
threshold of genetic liability *(p. 716)*
threshold trait *(p. 716)*
variance ($s^2$) *(p. 719)*

## Chapter Concepts

*For answers to selected even-numbered problems, see Appendix: Answers.*

1. Which of the following traits would you expect to be inherited as quantitative traits?
   a. body weight in chickens
   b. growth rate in sheep
   c. milk production in cattle
   d. fruit weight in tomatoes
   e. coat color in dogs

2. For the traits listed in the previous problem, which do you think are likely to be multifactorial traits with phenotypes that are influenced by genes and environment? Identify two environmental factors that might play a role in phenotypic variation of the traits you identified.

3. Compare and contrast broad sense heritability and narrow sense heritability, giving an example of each measurement and identifying how the measurement is used.

4. In a cross of two pure-breeding lines of tomatoes producing different fruit sizes, the variance in grams (g) of fruit weight in the $F_1$ is 2.25 g, and the variance among the $F_2$ is 5.40 g. Determine the genetic and environmental variance ($V_G$ and $V_E$) for the trait and the broad sense heritability of the trait.

5. Describe the difference between continuous phenotypic variation and discontinuous variation. Explain how polygenic inheritance could be the basis of a trait showing continuous phenotypic variation. Explain how polygenic inheritance can be the basis of a threshold trait.

6. Calculate the mean, variance, and standard deviation for a sample of turkeys weighed at 8 weeks of age that have the following weights in ounces: 161, 172, 155, 173, 149, 177, 156, 174, 158, 162, 171, 181.

7. Provide a definition and an example for each of the following terms:
   a. additive genes
   b. concordance of twin pairs
   c. multifactorial inheritance
   d. polygenic inheritance
   e. quantitative trait locus
   f. threshold trait

## Application and Integration

*For answers to selected even-numbered problems, see Appendix: Answers.*

8. Three pairs of genes with two alleles each ($A_1$ and $A_2$, $B_1$ and $B_2$, and $C_1$ and $C_2$) control the height of a plant. The alleles of these genes have an additive relationship: each copy of alleles $A_1$, $B_1$, and $C_1$ contributes 6 cm to plant height, and each copy of alleles $A_2$, $B_2$, and $C_2$ contributes 3 cm.
   a. What are the expected heights of plants with each of the homozygous genotypes $A_1A_1B_1B_1C_1C_1$ and $A_2A_2B_2B_2C_2C_2$?
   b. What height is expected in the $F_1$ progeny of a cross between $A_1A_1B_1B_1C_1C_1$ and $A_2A_2B_2B_2C_2C_2$?
   c. What is the expected height of a plant with the genotype $A_1A_2B_2B_2C_1C_2$?
   d. Identify all possible genotypes for plants with an expected height of 33 cm.
   e. Identify the number of different genotypes that are possible with these three genes.
   f. Identify the number of different phenotypes (expected plant heights) that are possible with these three genes.

9. For the three-gene system in the previous problem, suppose that instead of incomplete dominance among the additive alleles of each gene, the 1 allele is dominant in each case and the 2 allele is recessive. Under this revised scheme, the dominant phenotype contributes 10 cm to expected height and the recessive phenotype contributes 4 cm.
   a. What is the expected height of a plant that is homozygous for 1 alleles?
   b. What is the expected height of a plant that is homozygous for 2 alleles?
   c. What is the height of the $F_1$ progeny of these homozygous plants?
   d. What are the phenotypes and proportions of each phenotype among the $F_2$?

10. Two inbred lines of sunflowers ($P_1$ and $P_2$) produce different total weights of seeds per flower head. The mean weight of seeds (grams) and the variance of seed weights in different generations are as follows.

| Generation | Mean Weight/Head (g) | Variance |
|---|---|---|
| $P_1$ | 105 | 3.0 |
| $P_2$ | 135 | 3.8 |
| $F_1$ | 122 | 3.5 |
| $F_2$ | 125 | 7.4 |

   a. Use the information above to determine $V_G$, $V_E$, and $V_P$ for this trait.
   b. Determine $H^2$ for this trait.

11. A total of 20 men and 20 women volunteer to participate in a statistics project. The height and weight of each subject are given in the table.

| Subject | Men | | Women | |
|---|---|---|---|---|
| | Height (in.) | Weight (lb) | Height (in.) | Weight (lb) |
| 1 | 65 | 136 | 60 | 95 |
| 2 | 66 | 146 | 61 | 103 |
| 3 | 67 | 141 | 62 | 110 |
| 4 | 67 | 148 | 62 | 109 |
| 5 | 68 | 147 | 62 | 118 |
| 6 | 68 | 166 | 63 | 137 |
| 7 | 69 | 165 | 63 | 152 |
| 8 | 69 | 173 | 64 | 134 |
| 9 | 69 | 159 | 64 | 127 |
| 10 | 70 | 188 | 64 | 166 |
| 11 | 70 | 183 | 65 | 129 |
| 12 | 70 | 179 | 65 | 130 |
| 13 | 70 | 190 | 66 | 148 |
| 14 | 71 | 169 | 66 | 152 |
| 15 | 71 | 186 | 67 | 155 |
| 16 | 71 | 190 | 67 | 149 |
| 17 | 72 | 206 | 68 | 157 |
| 18 | 72 | 210 | 68 | 138 |
| 19 | 73 | 238 | 69 | 162 |
| 20 | 74 | 267 | 70 | 169 |

a. Draw one histogram for height of the subjects and a separate histogram for weight. Use different colors for men and women so that you can visually compare the distributions by sex and plot weights in 10-pound intervals (i.e., 90–99 lbs, 100–109 lbs, 110–119 lbs, etc.).

b. Calculate the mean, variance, and standard deviation for height and weight in men and women.

c. Compare the numerical values with the visual distribution of heights and weights you drew in the histograms and describe whether you think your visual impression matches the numerical values.

12. In *Nicotiana*, two inbred strains produce long ($P_L$) and short ($P_S$) corollas. These lines are crossed to produce $F_1$, and the $F_1$ are crossed to produce $F_2$ plants in which corolla length and variance are measured. The following table summarizes mean and variance of corolla length in each generation. Calculate $H^2$ for corolla length in *Nicotiana*.

| Generation | Mean Corolla Length (mm) | Variance |
|---|---|---|
| $P_L$ | 85.75 | 4.21 |
| $P_S$ | 43.15 | 2.89 |
| $F_1$ | 62.26 | 3.62 |
| $F_2$ | 67.37 | 38.10 |

13. Suppose the length of maize ears has narrow sense heritability ($h^2$) of 0.70. A population produces ears that have an average length of 28 cm, and from this population a breeder selects a plant producing 34-cm ears to cross by self-fertilization. Predict the selection differential ($S$) and the response to selection ($R$) for this cross.

14. In a line of cherry tomatoes, the average fruit weight is 16 grams. A plant producing tomatoes with an average weight of 12 grams is used in one self-fertilization cross to produce a line of smaller tomatoes, and a plant producing tomatoes of 24 grams is used in a second cross to produce larger tomatoes.

a. What is the selection differential ($S$) for fruit weight in each cross?

b. If narrow sense heritability ($h^2$) for this trait is 0.80, what are the expected responses to selection ($R$) for fruit weight in the crosses?

15. Two pure-breeding wheat strains, one producing dark red kernels and the other producing white kernels, are crossed to produce $F_1$ with pink kernel color. When an $F_1$ plant is self-fertilized and its seed collected and planted, the resulting $F_2$ consist of 160 plants with kernel colors as shown in the following table.

| Kernel Color | Number |
|---|---|
| White | 9 |
| Dark red | 12 |
| Red | 39 |
| Light pink | 41 |
| Pink | 59 |

a. Based on the $F_2$ progeny, how many genes are involved in kernel color determination?

b. How many additive alleles are required to explain the five phenotypes seen in the $F_2$?

c. Using clearly defined allele symbols of your choice, give genotypes for the parental strains and the $F_1$. Describe the genotypes that produce the different phenotypes in the $F_2$.

d. If an $F_1$ plant is crossed to a dark red plant, what are the expected progeny phenotypes and what is the expected proportion of each phenotype?

16. In studies of human MZ and DZ twin pairs of the same sex who are reared together, the following concordance values are identified for various traits. Based on the values shown, describe the relative importance of genes versus the influence of environmental factors for each trait.

| Trait | Concordance | |
|---|---|---|
| | MZ | DZ |
| Blood type | 100 | 65 |
| Chicken pox | 89 | 87 |
| Manic depression | 67 | 13 |
| Schizophrenia | 72 | 12 |
| Diabetes | 62 | 15 |
| Cleft lip | 51 | 6 |
| Club foot | 40 | 4 |

17. During a visit with your grandparents, they comment on how tall you are compared to them. You tell them that in your genetics class, you learned that height in humans has high heritability, although environmental factors also

influence adult height. You correctly explain the meaning of heritability, and your grandfather asks, "How can height be highly heritable and still be influenced by the environment?" What explanation do you give your grandfather?

18. An association of racehorse owners is seeking a new genetic strategy to improve the running speed of their horses. Traditional breeding of fast male and female horses has proven expensive and time-consuming, and the breeders are interested in an approach using quantitative trait loci as a basis for selecting breeding pairs of horses. Write a brief synopsis (~50 words) of QTL mapping to explain how genes influencing running speed might be identified in horses.

19. The case study in this chapter describes a long-term selection experiment designed to change the oil and protein content of corn kernels. Beginning with a single strain of corn, oil content has changed progressively in separate lines selected for high oil content and low oil content. After 50 generations of steady selection and change, however, reverse selection experiments were initiated that produced reversals in oil content.

   a. Describe the genetic basis for phenotype change that raised and lowered oil content in each line.
   b. Describe the genetic basis for the reversal of directional change of oil content in each line.

20. Suppose a polygenic system for producing color in kernels of a grain is controlled by three additive genes, $G$, $M$, and $T$. There are two alleles of each gene, $G_1$ and $G_2$, $M_1$ and $M_2$, and $T_1$ and $T_2$. The phenotypic effects of the three genotypes of the $G$ gene are $G_1G_1 = 6$ units of color, $G_1G_2 = 3$ units of color, and $G_2G_2 = 1$ unit of color. The phenotypic effects for genes $M$ and $T$ are similar, giving the phenotype of a plant with the genotype $G_1G_1M_1M_1T_1T_1$ a total of 18 units of color and a plant with the genotype $G_2G_2M_2M_2T_2T_2$ a total of 3 units of color.

   a. How many units of color are found in trihybrid plants?
   b. Two trihybrid plants are mated. What is the expected proportion of progeny plants displaying 9 units of color? Explain your answer.
   c. Suppose that instead of an additive genetic system, kernel-color determination in this organism is a threshold system. The appearance of color in kernels requires 9 or more units of color; otherwise, kernels have no color and appear white. In other words, plants whose phenotypes contain 8 or fewer units of color are white. Based on the threshold model, what proportion of the $F_2$ progeny produced by the trihybrid cross in part (b) will be white? Explain your answer.
   d. Assuming the threshold model applies to this kernel-color system, what proportion of the progeny of the cross $G_1G_2M_1M_2T_2T_2 \times G_1G_2M_1M_2T_1T_2$ do you expect to display colored kernels?

21. New Zealand lamb breeders measure the following variance values for their herd.

| Trait | $V_P$ | $V_G$ | $V_A$ |
|---|---|---|---|
| Body mass (kg) | 42.4 | 20.5 | 7.4 |
| Body fat (%) | 38.9 | 16.2 | 5.7 |
| Body length (cm) | 51.6 | 26.4 | 8.1 |

   a. Calculate the broad sense heritability ($H^2$) and the narrow sense heritability ($h^2$) for each trait in this lamb herd.
   b. Comparing one trait to another, how would you characterize the potential response to selection ($R$)?

22. Cattle breeders would like to improve the protein content and butterfat content of milk produced by a herd of cows. Narrow sense heritability values are 0.60 for protein content and 0.80 for butterfat content. The average percentages of these traits in the herd and the percentages of the traits in cows selected for breeding are as follows.

| Trait | Herd Average | Selected Cows |
|---|---|---|
| Protein content | 20.2% | 22.7% |
| Butterfat content | 6.5% | 7.4% |

   a. Determine the selection differential ($S$) for each trait in this herd.
   b. Which trait is likely to be the most responsive to artificial selection applied by the cattle breeders through selection of cows for mating?

23. In human gestational development, abnormalities of the closure of the lower part of the midface can result in cleft lip, if the lip alone is affected by the closure defect, or in cleft lip and palate (the roof of the mouth), if the closure defect is more extensive. Cleft lip and cleft lip with cleft palate are multifactorial disorders that are threshold traits. A family with a history of either condition has a significantly increased chance of a recurrence of midface cleft disorder in comparison to families without such a history. However, the recurrence risk of a midface cleft disorder is higher in families with a history of cleft lip with cleft palate than in families with a history of cleft lip alone.

   a. Suppose a friend of yours who has not taken genetics asks you to explain these observations. Construct a genetic explanation for the increased recurrence risk of midface clefting in families that have a history of cleft disorders versus families without a history of such disorders.
   b. Construct a similar explanation of why the recurrence risk of a cleft disorder is higher in families with a history of cleft lip with cleft palate than in families with a history of cleft lip alone.

24. The children of couples in which one partner has blood type O (genotype $ii$) and the other partner has blood type AB (genotype $I^AI^B$) are studied.

   a. What is the expected concordance rate for blood type of MZ twins in this study? Explain your answer.
   b. What is the expected concordance rate for blood type of DZ twins in this study? Explain why this answer is different from the answer to part (a).

25. Answer the following in regard to multifactorial traits in human twins.

   a. If the trait is substantially influenced by genes, would you expect the concordance rate to be higher in MZ twins or higher in DZ twins? Explain your reasoning.
   b. If the trait is produced with little contribution from genetic variation, what would you expect to see if you compared the concordance rates of MZ twins versus DZ twins? Explain your reasoning.

# Population Genetics and Evolution

# 22

The evolution of Hawaiian fruit flies is a model for species evolution in reproductive isolation.

## ESSENTIAL IDEAS

- The Hardy-Weinberg equilibrium predicts frequencies of genotypes in populations.
- The impact of natural selection on allele frequencies can be estimated.
- The effect of mutations on allele frequencies can be quantified.
- The effects of the migration on allele frequencies in populations can be determined.
- Chance events can lead to changes in allele frequency.
- Inbreeding and assortative mating are patterns of nonrandom mating that can alter genotype frequencies.
- Species evolve by processes that lead to the genetic isolation.

In 1970, Theodosius Dobzhansky, one of the most influential geneticists of the 20th century, wrote:

> In biology nothing makes sense except in the light of evolution.

Dobzhansky and the other architects of the *modern synthesis of evolution* (see Chapter 1) identified evolution and evolutionary analysis as central organizing principles of biology, necessary for understanding modern forms of life and their origins. Evolution shaped the living world we see today, just as it shaped life in the past and will continue to shape life into the future.

The modern synthesis focused on uniting two elements of evolutionary biology. One was the large-scale evolutionary change linked to speciation and to the divergence of taxonomic groups above the species level. The second element included what was known about Mendelian inheritance and the connection between inherited variation in DNA and evolutionary change. Among the four evolutionary processes—natural selection, mutation, migration, and genetic drift—natural selection is the principal process of large-scale change, although each of the processes plays a role (see Sections 1.4 and 22.8).

In the beginning of the chapter, however, we focus largely on evolution at the population level (Sections 22.1–22.7). At this level, all four of the evolutionary processes operate to drive the evolution of populations. Collectively, the impact of the evolutionary processes is most directly described in terms of their ability to change the frequencies of alleles over time.

The impact of evolutionary processes on populations has been a focus of population biologists, evolutionary biologists, and mathematicians since the beginning of the 20th century, several decades before DNA was identified as the hereditary molecule and its structure became known. Since those early days, the central predictions made about populations on the basis of evolutionary principles have been proven correct time and again in countless experiments and observations. The application of evolutionary principles to populations forms the foundation of the field of **population genetics,** which is where our discussion begins.

## 22.1 The Hardy-Weinberg Equilibrium Describes the Relationship of Allele and Genotype Frequencies in Populations

The origin of population genetics can be traced to the earliest years of the 1900s, shortly after the rediscovery of Mendel's laws of heredity, and to a time when George Udny Yule, William Castle, Karl Pearson, Godfrey Hardy, Wilhelm Weinberg, and others first debated the fate of genes in populations. In 1902, the inheritance of brachydactyly, an autosomal dominant condition characterized by shortening of fingers and toes, was described in humans as a trait paralleling a Mendelian pattern of heredity. In contemplating this observation, Yule proposed that since three-quarters of the progeny of a cross of heterozygous parents with brachydactyly will also display shortened digits, the frequency of the dominant allele might be expected to increase over time. William Castle thought Yule was wrong; and in 1903 he offered, as a partial refutation of Yule's contention, a mathematical demonstration that in the absence of natural selection, genotype frequencies remain stable in populations. Karl Pearson supported Castle's position by showing that if two alleles of a gene had equal frequency in a population, there would be a single, stable equilibrium frequency for their genotypes. Reginald Punnett (of Punnett square fame) also thought Yule was wrong; but unable to formulate a mathematical argument to refute Yule, he took the problem to his friend and regular cricket partner Godfrey Hardy.

Hardy, a mathematician rather than a biologist, quickly identified a "very simple" solution to the question of the fate of alleles in populations. He showed that with random mating and in the absence of evolutionary change in a population, the allele frequencies result in a stable equilibrium frequency. Hardy also showed that at equilibrium, genotypes occur in predictable frequencies derived directly from allele frequencies. In 1908, Hardy penned a letter to the editors of *Science* magazine that began with these self-effacing words:

> I am reluctant to intrude in a discussion concerning matters of which I have no expert knowledge, and I should have expected the very simple point which I wish to make to have been familiar to biologists. However, some remarks of Mr. Udny Yule, to which Mr. R. C. Punnett has called my attention, suggest it may be worth making.

In his letter, Hardy laid out the concept that has become known as the **Hardy-Weinberg (H-W) equilibrium.** The name recognizes Hardy's explanation of allele and genotype frequencies in populations as well as an independent explanation of the same principle by Wilhelm Weinberg (a German physician) that was also published in 1908. The H-W equilibrium is a cornerstone of population genetics and was the first of many developments in evolutionary genetics that culminated in the modern synthesis. Hardy may have been reluctant to intrude into matters of biology, but biologists for the last 100 years have been glad he did!

### Populations and Gene Pools

A **population** is a group of interbreeding organisms. The collection of genes and alleles found in the members of a population is known as a **gene pool.** The gene pool is the source of genetic information from which the next generation is produced. Each population member carries a portion of the gene pool in its genome, but typically, the amount of genetic variation in a gene pool is greater than the variation carried by individual members of the population. The pattern of mating between individuals and the effect of evolutionary processes on alleles determine (1) how alleles are dispersed into genotypes and (2) their frequencies in successive generations.

The H-W equilibrium serves as a model that calculates the frequencies of alleles and genotypes in a theoretical population that is infinite in size, practices random mating, and does not experience evolutionary change (Table 22.1). Under these conditions, the H-W equilibrium predicts that allele frequencies will be stable from generation to generation, that the frequencies of genotypes are predictable from their constituent allele frequencies, and that genotype frequencies too will remain the same in successive generations.

In nature, however, no real population meets all the criteria assumed by the H-W equilibrium. For example, all populations are finite in size and are subject to genetic drift as a consequence (a phenomenon we encounter in Section 22.6). In addition, natural selection, migration, and mutation each exert their influences on a population. Despite these circumstances, most populations adhere closely enough to the assumptions of the H-W equilibrium that alleles are distributed into genotypes in the proportions it predicts. The H-W equilibrium has proven to be a dependable arithmetic tool for assessing population genetic structure and detecting evolutionary change and nonrandom mating, and it is applied in numerous ways to the analysis of autosomal and X-linked genes in populations.

## The Hardy-Weinberg Equilibrium

The predictions of the H-W equilibrium can be modeled for any number of alleles of an autosomal gene, but the simplest model is for two alleles, here designated $A_1$ and $A_2$. For the general case, the alleles are given frequencies of

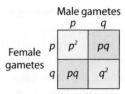

**Binomial expansion**
$$(p + q)(p + q) = p^2 + pq + pq + q^2 = p^2 + 2pq + q^2 = 1$$

**Figure 22.1  The Hardy-Weinberg equilibrium for autosomal genes.** The Punnett square method and the binomial expansion of alleles with frequencies $p$ and $q$ predict genotype frequencies under assumptions of the Hardy-Weinberg equilibrium.

$f(A_1) = p$ and $f(A_2) = q$, with the frequencies equal in males and females. Since $A_1$ and $A_2$ are the only alleles that occur at this gene, the sum of their frequencies is $p + q = 1.0$. Rearrangements of this equality allow the frequency of one allele to be used to determine the frequency of the other allele; thus, $p = 1 - q$ and $q = 1 - p$. Allelic segregation predicts the relationship between allele frequencies and genotype frequencies in populations. For the two alleles in our example, there are three genotypes: $A_1A_1$, $A_1A_2$, and $A_2A_2$. The genotype frequencies are computed using a binomial expansion $[(p + q)^2]$, where the two $(p + q)$ expressions represent male and female contributions to mating. Alternatively, a representation of random mating in the population that resembles a Punnett square can be used. Both methods make the same genotype frequency predictions of $f(A_1A_1) = p^2$, $f(A_1A_2) = 2pq$, and $f(A_2A_2) = q^2$ (Figure 22.1). The summation of these three genotype frequencies equals unity: $p^2 + 2pq + q^2 = 1.0$.

We can demonstrate the application of the H-W equilibrium by assigning frequencies to each allele in a hypothetical population: say, $f(A_1) = p = 0.6$ and $f(A_2) = q = 0.4$. As required, the sum of the two allele frequencies is $0.6 + 0.4 = 1.0$. In this hypothetical population example, 60 percent of gametes carry $A_1$ and 40 percent carry $A_2$ (Figure 22.2). If the population is in H-W equilibrium,

| Table 22.1 | The Hardy-Weinberg Equilibrium |
|---|---|

**Assumptions**

1. Population size is infinite.
2. Random mating occurs in the population.
3. Natural selection does not operate.
4. Migration does not introduce new alleles.
5. Mutation does not introduce new alleles.
6. Genetic drift does not take place.

**Predictions**

1. Allele frequencies remain stable over time.
2. Allele distribution into genotypes is predictable.
3. Stable equilibrium frequencies of alleles and genotypes are maintained.
4. Evolutionary and nonrandom mating effects are predictable.

|  | Male gametes | |
|---|---|---|
|  | $A_1$ 0.60 | $A_2$ 0.40 |
| Female gametes $A_1$ 0.60 | $A_1A_1$ 0.36 | $A_1A_2$ 0.24 |
| $A_2$ 0.40 | $A_1A_2$ 0.24 | $A_2A_2$ 0.16 |

**Binomial expansion:**
$(0.60 + 0.40)(0.60 + 0.40) = 0.36 + 0.24 + 0.24 + 0.16$
$A_1A_1 = 0.36$
$A_1A_2 = 0.48$
$A_2A_2 = 0.16$
Total = 1.00

**Figure 22.2  Application of the Hardy-Weinberg equilibrium.** The Punnett square method and the binomial expansion method applied to population in which $f(A_1) = 0.60$ and $f(A_2) = 0.40$.

probability predicts that an $A_1$-containing gamete from a male and an $A_1$-containing female gamete will unite to produce $A_1A_1$ progeny with a probability of $(0.6)(0.6) = 0.36$. Similarly, the production of $A_2A_2$ progeny, from the union of two $A_2$-containing gametes, has a probability of $(0.4)(0.4) = 0.16$. Heterozygous progeny are produced in two ways, with a combined frequency predicted as $(0.6)(0.4) + (0.6)(0.4) = 0.48$. The sum of frequencies of the three genotypes is $(0.36) + (0.48) + (0.16) = 1.00$. The binomial expansion method of calculating the genotype frequencies in progeny makes identical predictions.

In this example we see one of the predictions of the H-W equilibrium: Random mating for one generation produces genotype frequencies that can be predicted from allele frequencies. For any frequencies of $p$ and $q$ between 0.0 and 1.0, an expected equilibrium distribution of genotype frequencies can be derived (Figure 22.3). Notice that as the frequency of $p$ decreases and $q$ increases, the proportions of genotypes shift, altering the frequency of each homozygous class and the frequency of heterozygotes in the population. Heterozygous frequency has a maximum of 0.50 (50 percent) when the frequencies are $p = q = 0.50$.

This example also allows us to observe the second prediction of the H-W equilibrium: With random mating and no evolution, allele frequencies do not change from one generation to the next. We see this if we count the alleles in progeny genotypes, recognizing that *all* of the alleles in $A_1A_1$ are alleles of a single type, and all the alleles in $A_2A_2$ progeny are alleles of the other type. The $A_1A_1$ progeny are 36 percent of the new generation, and $A_2A_2$ are 16 percent. Among the 48 percent of the progeny that are heterozygotes, exactly *one-half* of the alleles are $A_1$ and *one-half* are $A_2$. Consequently, the frequency of $A_1$ among the progeny is 36 percent plus 24 percent, or 60 percent of

the alleles carried by progeny, which is the same frequency that was seen in the parental generation. The $A_2$ frequency is 16 percent plus 24 percent, or 40 percent of the progeny-generation alleles, also the same as the frequency found in the parental generation. Expressed as $p$ and $q$, the frequency of $A_1$ in the progeny generation is $f(A_1) = p^2 + pq$, and the frequency of $A_2$ is $f(A_2) = q^2 + pq$.

The evolutionary observation that random mating leads to predictable genotype frequencies and that allele frequencies are stable from one generation to the next can be portrayed in a mating-table format that shows the consequence of reproduction under the assumptions of the H-W equilibrium (Table 22.2). In the mating-table analysis, parental genotypes unite to reproduce at proportions predicted by their frequency. If parents have the same genotype, there is no reciprocal mating to account for; but if different genotypes occur in the parents, the reciprocal matings must be taken into account. The progeny of each mating are predicted according to Mendelian principles. The frequency or fraction of offspring with each genotype is summed once the table is filled. The term that is the sum of each genotype frequency can be simplified to show that offspring are produced in the genotype proportions $p^2$, $2pq$, and $q^2$, just as they occur in the parents. This analysis is compelling evidence that in the presence of random mating and the absence of evolutionary change, the allele frequencies in populations are stable over time.

In populations that meet the assumptions of the H-W equilibrium, a single generation of random mating will "reset" the genotype frequencies in the population into the predicted proportions $p^2$, $2pq$, and $q^2$. Moreover, if a population *is not* initially in H-W equilibrium, we can predict the consequence of one generation of random mating. As an example, Figure 22.4 illustrates the effect of uniting two previously separate populations with different frequencies of $A_1$ and $A_2$ to form a new population. Each of the contributing populations originally contained 500 individuals, and the new population contains 1000 individuals. Immediately after forming the new population, the genotypes *are not* in Hardy-Weinberg proportions, One generation of mating in the new population under Hardy-Weinberg assumptions, however, produces genotype frequencies in the next generation that *are* in H-W equilibrium The new population has new allele frequencies as a result of mixing the two populations.

## Determining Autosomal Allele Frequencies in Populations

Determination or estimation of allele frequencies are a commonly used measure of the genetic structure of populations and to assess Hardy-Weinberg equilibrium. Comparisons of allele frequencies between populations can identify relationships and diversification of populations, and documentation of allele frequency change over

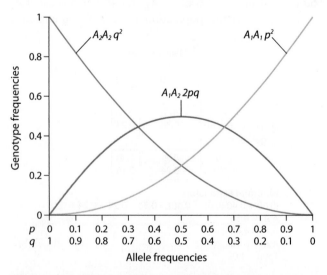

**Figure 22.3  The Hardy-Weinberg equilibrium for two autosomal alleles.** Each curve shows the frequency of the genotype for the indicated frequencies of the alleles $p$ and $q$.

| Table 22.2 | Hardy-Weinberg Mating Table for Two Alleles of an Autosomal Gene | | | |
|---|---|---|---|---|
| **Mating** | **Mating Frequency** | **Progeny Genotypes** | | |
| | | $A_1A_1$ | $A_1A_2$ | $A_2A_2$ |
| $A_1A_1 \times A_1A_1$ | $(p^2)(p^2) = p^4$ | $p^4$ | — | — |
| $A_1A_1 \times A_1A_2$ | $2[(p^2)(2pq)] = 4p^3q$ | $2p^3q$ | $2p^3q$ | — |
| $A_1A_1 \times A_2A_2$ | $2[(p^2)(q^2)] = 2p^2q^2$ | — | $2p^2q^2$ | — |
| $A_1A_2 \times A_1A_2$ | $(2pq)(2pq) = 4p^2q^2$ | $p^2q^2$ | $2p^2q^2$ | $p^2q^2$ |
| $A_1A_2 \times A_2A_2$ | $2[(2pq)(q^2)] = 4pq^3$ | — | $2pq^3$ | $2pq^3$ |
| $A_2A_2 \times A_2A_2$ | $(q^2)(q^2) = q^4$ | — | — | $q^4$ |
| Total | 1.0 | $p^2$ | $2pq$ | $q^2$ |

Among the progeny, a common term is factored out of each summation to produce the frequency of each genotype:

$$A_1A_1 = p^4 + 2p^3q + p^2q^2 = p^2(p^2 + 2pq + q^2) = p^2$$
$$A_1A_2 = 2p^3q + 2p^2q^2 + 2p^2q^2 + 2pq^3 = 2pq(p^2 + pq + pq + q^2) = 2pq$$
$$A_2A_2 = p^2q^2 + 2pq^3 + q^4 = q^2(p^2 + 2pq + q^2) = q^2$$

The sum of progeny genotype frequencies is $p^2 + 2pq + q^2 = 1.0$.

time is a hallmark of population evolution. Three equally valid methods can be used to determine the frequencies of alleles of autosomal genes in populations. The most appropriate method is usually dictated by the type of genotype or phenotype information available and the composition of the population or of the sample data.

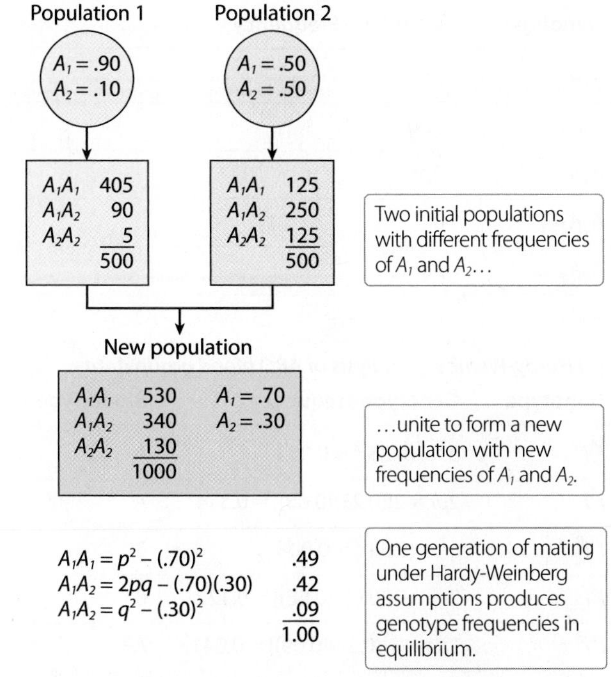

**The Genotype Proportion Method** The first method, called the **genotype proportion method,** can be used to determine allele frequencies whether or not the population is in H-W equilibrium, and requires that the genotype of each population member be known so the alleles can be accurately partitioned. This approach calculates allele frequencies by adding the frequency of the homozygotes for the allele and the frequency of one-half of the heterozygotes carrying the allele. As an example, suppose that a population has the following composition: $B_1B_1 = 0.64$, $B_1B_2 = 0.32$, $B_2B_2 = 0.04$. Applying the genotype proportion method, the frequency of $B_1$ is the sum of the frequency of $B_1B_1$ plus one-half the frequency of $B_1B_2$ heterozygotes. In this case, $f(B_1) = p = (0.64) + [(0.5)(0.32)] = 0.80$. Similarly, for $B_2$, the allele frequency is calculated by adding the frequency of $B_2B_2$ and one-half the frequency of $B_1B_2$, or $f(B_2) = q = (0.04) + [(0.5)(0.32)] = 0.20$. For this example, notice that $p + q = 0.80 + 0.20 = 1.0$.

**The Allele-Counting Method** A second approach to determining allele frequencies in populations, called the **allele-counting method,** can also be used whether or not the population is in H-W equilibrium. For alleles to be counted by this method requires that each one (in each organism) be identifiable through the phenotype. In practice, the allele-counting method is feasible when alleles are codominant, since there is then a distinct phenotype for each genotype. The human MN blood group system, for example, is a codominant system produced by two

**Figure 22.4 One generation of random mating produces Hardy-Weinberg equilibrium frequencies for genotypes of autosomal genes.**

alleles, *M* and *N*. Both alleles are present in all human populations and produce three blood group phenotypes: type M, type MN, and type N. Each blood group has a corresponding genotype. Individuals with blood type M or blood type N have homozygous genotypes *MM* and *NN*, respectively, and the blood type MN is produced by the *MN* genotype. MN blood group testing of 1482 members of a Japanese population produced the following results:

| Blood group | M | MN | N | |
|---|---|---|---|---|
| Number | 406 | 744 | 332 | = 1482 |

The allele frequency calculation recognizes that each of the 1482 people in the sample carries two alleles of the gene and that there are (2)(1482) = 2964 alleles represented in the sample. The frequency of each allele is determined by counting the two alleles of that type from each homozygote and the single allele of that type from each heterozygote. The allele frequencies are therefore $f(M) = [(2)(406) + (744)]/2964 = 0.525$ and $f(N) = [(2)(332) + (744)]/2964 = 0.475$.

**The Square Root Method** The final method for allele frequency determination in populations, the **square root method,** is used only when the population is in H-W equilibrium and when the two alleles of a gene are dominant and recessive, respectively. In the human autosomal recessive disorder cystic fibrosis, for example, one allele (*cf*) is recessive and therefore is evident only in the homozygous genotype. When the recessive allele is in a heterozygous genotype, it is "hidden" by the dominant allele (*CF*). In a circumstance like this, the dominant phenotype consists of two genotypes, *CFCF* and *CFcf*. In contrast, the recessive phenotype is produced only by the homozygous recessive genotype *cfcf*. The correspondence of the recessive phenotype and homozygous genotype allows use of the Hardy-Weinberg principle to estimate the frequency of the recessive allele by taking the square root of the recessive homozygous genotype frequency. In the U.S. population, the frequency of cystic fibrosis among newborn infants is approximately 1 in 2000. Where $f(CF) = p$ and $f(cf) = q$, $f(cfcf) = q^2 = 0.0005$. The frequency of $q$ is thus estimated as the square root of 0.0005, or $f(q) = 0.022$; that is, about 2.2 percent.

With $f(cf)$ determined, the frequency of *CF* is estimated as $f(CF) = p = 1 - q = 1.0 - 0.022 = 0.978$. The frequency of carriers of cystic fibrosis is of practical importance for determining the chance that a person is a carrier of cystic fibrosis. According to the Hardy-Weinberg principle, the population frequency of carriers is $f(CFcf) = 2pq = 2(0.978)(0.022) = 0.043$. In other words, approximately 4.3 percent of the population, or about 1 in 23 people, carry a recessive mutant allele for cystic fibrosis. Estimates like this can be particularly valuable in genetic

counseling situations, where it is desirable to know the probability that a person who has a dominant phenotype might be a heterozygous carrier of a recessive allele. Genetic Analysis 22.1 provides more practice in calculating allele frequencies and applying the H-W equilibrium.

---

**Genetic Insight** Allele frequencies are determined in a population by three equally valid methods: the genotype frequency method, the allele-counting method, and the square root method.

---

## The Hardy-Weinberg Equilibrium for More than Two Alleles

Having examined the application of the H-W equilibrium to genes with two alleles, we can now consider the more complex case of a gene that has more than two alleles. We shall limit our discussion to three alleles, whose frequencies are represented by the variables $p$, $q$, and $r$, where $p + q + r = 1.0$, and where the trinomial expansion $(p + q + r)^2$ represents random mating and predicts the distribution of alleles in genotypes. Six genotypes are predicted by application of H-W equilibrium for a gene with three alleles (Table 22.3a). The sum of genotype frequencies resulting from the trinomial expansion is $(p + q + r)^2 = p^2 + 2pq + q^2 + 2pr + r^2 + 2qr = 1.0$.

| Table 22.3 | Hardy-Weinberg Principle for Three Alleles of a Gene |
|---|---|

*(a) Genotype prediction for three alleles*

| Genotype | Genotype Frequency |
|---|---|
| $A_1A_1$ | $p^2$ |
| $A_1A_2$ | $2pq$ |
| $A_1A_3$ | $2pr$ |
| $A_2A_2$ | $q^2$ |
| $A_2A_3$ | $2qr$ |
| $A_3A_3$ | $r^2$ |

*(b) Hardy-Weinberg analysis of ABO blood group data*

| Genotype | Genotype Frequency[a] | Blood Type |
|---|---|---|
| $I^AI^A$ | $p^2 = (0.23)^2 = 0.053$ | A |
| $I^Ai$ | $2pr = 2[(0.23)(0.68)] = 0.314$ | A |
| $I^BI^B$ | $q^2 = (0.09)^2 = 0.008$ | B |
| $I^Bi$ | $2qr = 2[(0.09)(0.68)] = 0.122$ | B |
| $I^AI^B$ | $2pq = 2[(0.23)(0.09)] = 0.041$ | AB |
| $ii$ | $r^2 = (0.68)^2 = 0.462$ | O |

Where $f(A_1) = p$; $f(A_2) = q$; $f(A_3) = r$; and $p + q + r = 1.0$

A worldwide survey of genetic variation in human populations reported the autosomal codominant MN blood group types in a sample of 1029 Chinese from Hong Kong. The sample contained 342 people with blood type M, 500 with blood type MN, and 187 with blood type N.

**a.** Determine the frequencies of both alleles (*M* and *N*) using the genotype proportion method and the allele-counting method.

**b.** Determine the expected genotype frequencies under assumptions of the Hardy-Weinberg equilibrium.

| Solution Strategies | Solution Steps |
|---|---|
| **Evaluate** | |
| 1. Identify the topic this problem addresses, and describe the nature of the required answer. | 1. This problem addresses the determination of allele frequencies from population data and the determination of expected genotype frequencies under assumptions of the Hardy-Weinberg equilibrium. |
| 2. Identify the critical information given in the problem. | 2. The number of individuals with each blood type is given, and the blood type is identified as an autosomal codominant trait. |
| **Deduce** | |
| 3. Determine the genotype corresponding to each blood group. | 3. For this autosomal codominant trait, blood type M individuals have the genotype *MM*, those with blood type N are *NN*, and MN individuals are *MN*. |
| 4. Calculate the frequency of each blood type in the sample. <br><br> TIP: The frequency of each genotype is the number of people with the genotype over the total sample size. | 4. Blood type M is $342/1029 = 0.332$, MN is $500/1029 = 0.486$, and N is $187/1029 = 0.186$. |
| **Solve** | **Answer a** |
| 5. Calculate allele frequencies using the genotype proportion method. | 5. The frequencies are <br><br> $$f(M) = (0.332) + [(0.5)(0.486)] = 0.575 \text{ and}$$ $$f(N) = (0.186) + [(0.5)(0.486)] = 0.425.$$ |
| 6. Calculate the allele frequencies by the allele-counting method. <br><br> TIP: Each individual carries two alleles. | 6. For the sample of 1029 people, there are 2058 alleles. The allele frequencies are <br><br> $$f(M) = [(2)(342)] + (500)/2058 = 0.575 \text{ and}$$ $$f(N) = [(2)(187)] + (500)/2058 = 0.425.$$ |
| | **Answer b** |
| 7. Determine the expected genotype distribution under Hardy-Weinberg assumptions. <br><br> TIP: Assume $M = p$ and $N = q$, and expand the binomial equation $(p + q)^2 = p^2 + 2pq + q^2$. | 7. The expected genotype frequencies are <br><br> $$MM = (0.575)^2 = (0.33)(1029) = 339.57,$$ $$MN = 2[(0.575)(0.425)] = (0.49)(1029) = 504.21, \text{ and}$$ $$NN = (0.425)^2 = (0.18)(1,029) = 185.22.$$ |

For more practice, see Problems 17, 18, 21, and 25.

The human ABO blood group system provides an opportunity for the application of the H-W equilibrium to a gene with three alleles (Chapter 4). Recall that among the three alleles producing ABO blood types—$I^A$, $I^B$, and $i$—$I^A$ and $I^B$ exhibit dominance over $i$ but are codominant to one another. These allelic relationships result in four blood types from the six genotypes (see Figure 4.3). Using $f(I^A) = p$, $f(I^B) = q$, and $f(i) = r$, along with data reporting the frequencies of each blood type in a population, we can estimate the frequency of each allele by applying a version of the square root method. This approach provides an approximate estimate of *ABO* allele frequencies based on observed frequencies of each blood group in a population. The allele frequencies in the U.S. population, for example, are derived as follows:

Step 1. Blood type O is found with recessive homozygous genotypes, and the frequency of the blood type is $r^2 = 0.46$. The square root of $0.46 = r$; thus, the allele frequency is $f(i) = r = 0.68$.

Step 2. The combined frequency of blood types A and O is $p^2 + 2pr + r^2 = (p + r)^2$, so $f(I^A) = p$ is estimated by the square root of the combined frequency of A plus O minus $r$. The calculation is $f(I^A) = p = \sqrt{[0.37 + 0.46]} - r = 0.91 - 0.68 = 0.23$.

Step 3. Having estimated $p$ and $r$, we can solve for $q$ by $q = 1 - (p + r) = 1 - (0.23 + 0.68) = 0.09$

In this way, from the U.S. population frequencies we can estimate that the frequencies of the ABO alleles are $f(I^A) = 0.23$, $f(I^B) = 0.09$, and $f(i) = 0.68$. Based on these estimated allele frequencies, Table 22.3b calculates genotype frequencies for the ABO blood types in the U.S. population.

### The Chi-Square Test of Hardy-Weinberg Predictions

Strictly speaking, the assumptions of the H-W equilibrium are unattainable in real populations. From a statistical perspective, however, what matters is whether the observed genotype frequencies in populations deviate *significantly* from the predictions of the H-W equilibrium. The chi-square statistic is used to compare observed and expected results in order to evaluate the validity of an estimate based on the H-W equilibrium (see Section 2.5).

We can illustrate the application of chi-square analysis to Hardy-Weinberg predictions by examining the data on the MN blood group in the Japanese sample described in the previous section. Having counted the frequency of alleles in this population sample to be $f(M) = p = 0.525$ and $f(N) = q = 0.475$, we multiply the expected frequency of each genotype ($p^2$, $2pq$, or $q^2$) by the sample size of 1482 people to predict the number of people who should have each blood group, assuming the population is in Hardy-Weinberg equilibrium. The expected numbers are

Blood group M = $(0.525)^2(1482) = 408.48$

Blood group MN = $2[(0.525)(0.475)](1482) = 739.14$

Blood group N = $(0.475)^2(1482) = 334.38$

The observed and expected numbers for this sample are shown in Table 22.4, where the chi-square calculation is also shown. The chi-square value for this sample is 0.064. There is one degree of freedom in the analysis (the statistical explanation for this determination is beyond the scope of our coverage). The corresponding $P$ value for the analysis is $0.9 > P > 0.8$ (see Table 2.4). The chi-square analysis shows that there is no significant deviation between the observed and expected genotype values.

| Table 22.4 | Chi-Square Analysis of Hardy-Weinberg Predictions | |
|---|---|---|

*Chi-square analysis of MN blood group data (Japanese volunteers)*

| Blood Group | Number Observed | Number Expected |
|---|---|---|
| M | 406 | 408.48 |
| MN | 744 | 739.14 |
| N | 332 | 334.38 |
| | 1482 | 1482.0 |

Chi-square calculation:

$$\chi^2 = (406 - 408.48)^2/408.48 + (744 - 739.14)^2/739.14 + (332 - 334.38)^2/334.38 = 0.064$$

For df = 1, a chi-square value of 0.064 corresponds to a $P$ value between 0.8 and 0.9.[a] (See Table 2.4, page 49.)

[a] The number of degrees of freedom is $3 - 1 - 1 = 1$ as stated in the text.

## 22.2 Genotype Frequencies for X-Linked Genes Are Estimated Using the Hardy-Weinberg Equilibrium

The pattern of transmission of X-linked genes differs from that of autosomal genes principally because males are hemizygous. Recall that males inherit their X chromosome from their mother and that males transmit their X chromosome exclusively to daughters but not to sons (see Section 3.3). In contrast, females express genes on their X chromosomes and transmit their X chromosomes in patterns identical to those seen for autosomes. With respect to H-W equilibrium, the consequences of these differences in the transmission and expression of X-linked genes change the way genotype frequencies are calculated. There are also consequences for predicting the outcome of random mating.

### Determining Frequencies for X-Linked Alleles

To illustrate the effect of sex-dependent differences in transmission and expression of X-linked genes on Hardy-Weinberg predictions, we look first at how the model predicts allele frequencies for those genes and then at what the consequences are for the expression of an X-linked recessive trait. In Table 22.5a, a mating table shows random mating involving an X-linked gene carrying alleles $A_1$ and $A_2$. Females shown in the mating table have one of three genotypes ($A_1A_1$, $A_1A_2$, and $A_2A_2$) and males have one of two genotypes ($A_1Y$ and $A_2Y$). If $f(A_1) = p$ and $f(A_2) = q$, the predicted female genotype frequencies are $f(A_1A_1) = p^2$, $f(A_1A_2) = 2pq$, and $f(A_2A_2) = q^2$, just like the frequencies predicted for an autosomal genotype since females can be homozygous or heterozygous for X-linked genes. Males, on the other hand, are hemizygous, and they are expected to have each genotype in proportion to the respective allele frequency. In this case, the predicted frequencies are $f(A_1Y) = p$ and $f(A_2Y) = q$. The mating table shows progeny distribution and tabulates the sum of progeny in each category by sex of the offspring. After common terms are factored out, the female genotypes are distributed in the proportions $p^2$, $2pq$, $q^2$, and the male genotypes appear with frequencies $p$ and $q$, the latter frequencies are the same as those in the previous female generation since this is the source of male X chromosomes. These predicted genotype frequencies match the patterns of transmission of X-linked genes; namely, that female X-linked inheritance is identical to female autosomal inheritance and that male inheritance of X-linked traits reflects inheritance from mothers.

The X-linked recessive trait of red-green color blindness that affects about 9 percent of males in human populations serves as an example to illustrate the mating table calculations. According to the model, the frequency of colorblind males is equal to the frequency of the recessive allele in the population; thus 9 percent of X chromosomes carry the recessive allele ($c$), and 91 percent carry the dominant allele ($C$; Table 22.5b). Using the frequencies of $f(C) = p = 0.91$ and $f(c) = q = 0.09$, the genotype frequencies for males are $f(CY) = p = 0.91$ and $f(cY) = q = 0.09$, and for females are $f(CC) = p^2 = 0.828$, $f(Cc) = 2pq = 0.164$, and $f(cc) = q^2 = 0.008$. From these calculations we can estimate that the male frequency of color blindness is 9 percent, whereas the female rate is less than 1 percent.

---

**Table 22.5    Estimating Genotype Frequencies for X-Linked Genes**

*(a) Hardy-Weinberg mating table for X-linked genes*

| Mating | Mating Frequency | Progeny Frequency[a] | | | | |
|---|---|---|---|---|---|---|
| | | Female Progeny | | | Male Progeny | |
| | | $A_1A_1$ | $A_1A_2$ | $A_2A_2$ | $A_1Y$ | $A_2Y$ |
| $A_1A_1 \times A_1Y$ | $p^2 \times p = p^3$ | $p^3$ | — | — | $p^3$ | — |
| $A_1A_2 \times A_1Y$ | $2pq \times p = 2p^2q$ | $p^2q$ | $p^2q$ | — | $p^2q$ | $p^2q$ |
| $A_2A_2 \times A_1Y$ | $q^2 \times p = pq^2$ | — | $pq^2$ | — | — | $pq^2$ |
| $A_1A_1 \times A_2Y$ | $p^2 \times q = p^2q$ | — | $p^2q$ | — | $p^2q$ | — |
| $A_1A_2 \times A_2Y$ | $2pq \times q = 2pq^2$ | — | $pq^2$ | $pq^2$ | $pq^2$ | $pq^2$ |
| $A_2A_2 \times A_2Y$ | $q^2 \times q = q^3$ | — | — | $q^3$ | — | $q^3$ |
| | 1.00 | $p^2$ | $2pq$ | $q^2$ | $p$ | $q$ |

[a] Progeny values are summed and simplified by the removal of common terms. For example,

$$A_1Y = p^3 + p^2q + p^2q + pq^2 = p\,(p^2 + pq + pq + q^2) = p$$

*(b) Hardy-Weinberg mating table for color blindness[a]*

| Mating | Mating Frequency | Progeny Frequency | | | | |
|---|---|---|---|---|---|---|
| | | Female Progeny | | | Male Progeny | |
| | | CC | Cc | cc | CY | cY |
| $CC \times CY$ | $(0.828)(0.91) = 0.754$ | 0.754 | — | — | 0.754 | — |
| $Cc \times CY$ | $(0.164)(0.91) = 0.149$ | 0.075 | 0.075 | — | 0.075 | 0.075 |
| $cc \times CY$ | $(0.008)(0.91) = 0.007$ | — | 0.004 | 0.004 | 0.004 | 0.004 |
| $CC \times cY$ | $(0.828)(0.09) = 0.075$ | — | 0.075 | — | 0.075 | — |
| $Cc \times cY$ | $(0.164)(0.09) = 0.015$ | — | 0.007 | 0.007 | 0.007 | 0.007 |
| $cc \times cY$ | $(0.008)(0.09) = 0.0007$ | — | — | 0.0007 | — | 0.0007 |
| | 1.0007 | 0.829 | 0.161 | 0.012 | 0.915 | 0.087 |

[a] All values are rounded, accounting for slight variation from expected values.

## Hardy-Weinberg Equilibrium for X-Linked Genes

Populations in H-W equilibrium have the same frequencies of alleles in males and females, and these populations will remain in equilibrium as long as random mating takes place and significant evolutionary effects do not occur. We saw earlier in the chapter that a population that is initially not in H-W equilibrium for alleles of autosomal genes can achieve equilibrium with one generation of random mating. In contrast, the difference in X-chromosome transmission between males and females makes multiple generations of random mating necessary for H-W equilibrium to be achieved for X-linked genes in a population that is initially not in equilibrium.

**Figure 22.5** shows the results of six generations of mating in a new population consisting initially of males that come from one population and an equal number of females that come from a different population. The males come from a population in which X-linked gene $A$ carries alleles at frequencies of $f(A_1) = 1.00$ and $f(A_2) = 0.00$. The females are drawn from a population in which the alleles of gene $A$ have frequencies of $f(A_1) = 0.00$ and $f(A_2) = 1.00$. The frequency of X-linked alleles in the new population is determined by a calculation that accounts for the single copy of an allele in males and the two copies of X-linked alleles in females: With $f(A_1) = p$ and $f(A_2) = q$, and using $p_F$ and $q_F$ to represent X-linked allele frequencies in females, and $p_M$ and $q_M$ to represent X-linked allele frequencies in males, the allele frequencies are calculated as:

$$p = 2/3\ p_F + 1/3\ p_M = (2/3)(1.00) + (1/3)(0.00) = 0.633$$
$$q = 2/3\ q_F + 1/3\ q_M = (2/3)(0.00) + (1/3)(1.00) = 0.366$$

The values 2/3 and 1/3 are used here because XX females carry twice as many X chromosomes as XY males

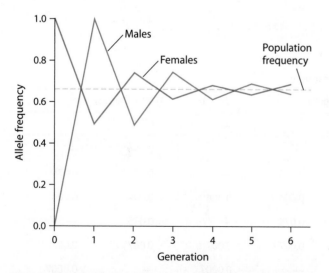

**Figure 22.5 The Hardy-Weinberg equilibrium for X-linked genes.** X-linked allele frequencies oscillate each generation as the allele frequencies in each sex approach equilibrium.

do. Due to the sex-dependent differences in transmission of X chromosomes, the male frequency of X-linked alleles is equal to the allele frequencies in females of the previous generation. Females, on the other hand, inherit one X-linked allele from each parent, and we estimate a female X-linked allele frequency to be the average of frequencies among males and females of the previous generation, $q_M + q_F/2$.

Tracing the oscillations of the $A_2$ allele in Figure 22.5, we see that the first generation of males has $q_M = 1.00$, identical to the frequency in females of the previous generation, whereas females have $q_F = 0.00 + 1.00/2 = 0.50$. These patterns are repeated in successive generations, causing the different male and female allele frequencies to oscillate around, and gradually approach, the equilibrium value. The oscillation in each generation is approximately half the magnitude of the previous oscillation. In this case, after six generations of random mating, the frequencies are close to equality in the sexes and are near to the population equilibrium allele frequency.

---

**Genetic Insight**  X-linked alleles require multiple generations of mating under H-W assumptions to reach H-W equilibrium frequencies. During the progression toward equilibrium, male and female allele frequencies oscillate around the eventual equilibrium allele frequency values of the population.

---

## 22.3 Natural Selection Operates through Differential Reproductive Fitness within a Population

Application of the H-W equilibrium to idealized populations provides insight into the mechanism that retains equilibrium when evolution does not occur. In the sense that the allele frequencies it describes do not change from generation to generation, the H-W equilibrium describes a static situation. But what happens to allele frequencies when evolution does occur? The simple answer is that allele frequencies change. The evolutionary impact can be quantified if the effect on allele frequencies can be estimated. In this section, we look at the effects of different mechanisms of natural selection on allele frequencies and H-W equilibrium. In later sections, we examine how the other evolutionary processes—mutation, migration (gene flow), and random genetic drift—affect allele frequencies and H-W equilibrium in populations (see Section 1.4 for an introduction to these processes).

### Differential Reproductive Fitness and Relative Fitness

Natural selection is an evolutionary mechanism that favors the reproductive success of certain members of a population over others as a result of differences in anatomical, physiological, or behavioral traits they possess. Individuals with the favored phenotype survive to reproductive age at higher rates than population members with other phenotypes, they reproduce at higher rates by virtue of the favorable phenotype, or both. This consequence of natural selection, which leads individuals with a particular phenotype to be more successful than individuals with other phenotypes at producing offspring for the next generation, is called **differential reproductive fitness.**

A common way to quantify differential reproductive fitness and measure the intensity of natural selection is to compare the relative abilities of organisms at producing offspring. The **relative fitness (*w*)** of organisms with the favored trait compared to those that do not have the trait. Organisms with the greatest reproductive success have the highest relative fitness and are assigned a relative fitness value of $w = 1.0$.

Since fitness values are measured relatively in a population, other organisms that reproduce less efficiently than the organisms with the highest relative fitness have their relative fitness reduced by a proportion called the **selection coefficient (*s*).** The selection coefficient identifies the proportionate difference between the fitnesses of organisms with different traits. For example, if an organism not having the favored trait reproduces 80 percent as well as the organism with the trait, the selection coefficient is $s = 0.2$, and the relative fitness of the organism is expressed as $w = 1 - s$, or $1 - 0.2 = 0.8$. If other organisms experience different levels of relative fitness, a second selection coefficient, designated *t*, is used. Where an organism with one genotype is most fit and organisms with either of two other genotypes experience reduced fitness, the relative fitness values for the two less fit genotypes are expressed as $w = 1 - s$ and $w = 1 - t$.

---

Genetic Insight  Natural selection is measured as relative fitness among organisms in a population with different traits. The intensity of natural selection is quantified by the selection coefficient that operates against one or more organisms.

---

## Directional Natural Selection

The pattern of natural selection called **directional natural selection** favors one phenotype that has higher relative fitness than other phenotypes in the population. Natural selection produces a directional change in allele frequencies when it operates to increase one allele frequency and decreases others.

In the directional selection example that follows, assume alleles $B_1$ and $B_2$ are codominant. The codominant relationship of the alleles will result in one genotype that occurs in organisms with the highest relative fitness and in reduced fitness in organisms with the other genotypes.

In this example, where the allele frequencies are $f(B_1) = 0.6$ and $f(B_2) = 0.4$, there are 1000 members of the population, the favored phenotype has a relative fitness of $w = 1.0$, and the other phenotypes have different relative fitness values of $w = 0.80$ and $w = 0.40$, the genetic profile of the population is as follows.

| Genotype | $B_1B_1$ | $B_1B_2$ | $B_2B_2$ |
|---|---|---|---|
| Frequency | 0.36 | 0.48 | 0.16 |
| Number | 360 | 480 | 160 |
| Relative fitness (w) | 1.0 | 0.80 | 0.40 |

In this example, the $B_1B_1$ organisms have the highest relative fitness ($w = 1.0$). In comparison, $B_1B_2$ organisms have $s = 0.20$ and $w = 1 - s = 0.80$, and organisms with the $B_2B_2$ genotype have a selection coefficient of $t = 0.60$ and a relative fitness of $w = 1 - t = 0.40$.

The impact of natural selection is computed in two steps. First, assuming natural selection has its effect before organisms reach reproductive age, the surviving number of organisms of each genotype is calculated by multiplying the original number of each genotype by the relative fitness value of the genotype. In this case the numbers of survivors of each genotype are $B_1B_1 = (1.0)(360) = 360$, $B_1B_2 = (0.80)(480) = 384$, and $B_2B_2 = (0.40)(160) = 64$. In this hypothetical population, 808 organisms of the original 1000 remain after natural selection.

The second step is determination of the allele frequencies after natural selection and of the genotype frequencies in the next generation. In this case, the frequencies are most readily calculated using the allele-counting method, since we can identify the genotype of each survivor. There are a total of 1616 alleles in the 808 survivors, and the allele frequencies after natural selection are $f(B_1) = [(2)(360) + (384)]/1616 = 1104/1616 = 0.683$, and $f(B_2) = [(2)(64) + (384)]/(2)(808) = 512/1616 = 0.317$. If we assume that random mating takes place among the survivors, the genotype frequencies in the next generation are $f(B_1B_1) = (0.683)^2 = 0.467$, $f(B_1B_2) = 2(0.683)(0.317) = 0.433$, and $f(B_2B_2) = (0.317)^2 = 0.100$.

The changes in allele frequencies are symbolized by the Greek delta ($\Delta$) and found by taking the absolute value of the difference between the original allele frequency and the allele frequency after selection. For this example, $\Delta B_1 = 0.60 - 0.683 = 0.083$, and $\Delta B_2 = 0.40 - 0.317 = 0.083$. Over successive generations in which natural selection continues to operate in favor of the survival of $B_1$ and against the survival of $B_2$, $f(B_1)$ increases while $f(B_2)$ decreases. If this pattern of natural selection continues for enough generations, the frequency of the $B_1$ allele will eventually become fixed at $f(B_1) = 1.0$, and the frequency of $B_2$ will be eliminated, so that its final frequency will be $f(B_2) = 0.0$. Once an allele frequency is either fixed ($f = 1.0$) or eliminated ($f = 0.0$), natural selection can no longer change the frequency. Population

**(a)**

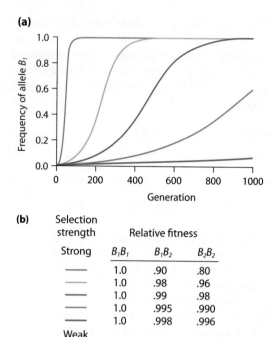

**(b)**

| Selection strength | Relative fitness | | |
|---|---|---|---|
| | $B_1B_1$ | $B_1B_2$ | $B_2B_2$ |
| Strong | | | |
| ——— | 1.0 | .90 | .80 |
| ——— | 1.0 | .98 | .96 |
| ——— | 1.0 | .99 | .98 |
| ——— | 1.0 | .995 | .990 |
| ——— | 1.0 | .998 | .996 |
| Weak | | | |

**Figure 22.6 The consequences of the intensity of natural selection on allele frequency.** (a) The curves illustrate the relationship between the rate of change in $f(B_1)$ and the intensity of natural selection. (b) Relative fitness values for natural selection of different intensities.

**Table 22.6** A Model of Directional Selection against a Recessive Lethal Allele

| | Genotype | | |
|---|---|---|---|
| | *BB* | *Bb* | *bb* |
| Frequency | 0.25 | 0.50 | 0.25 |
| Relative fitness (*w*) | 1.0 | 1.0 | 0.0 |
| Survivors after selection (total, 0.75) | 0.25 | 0.50 | 0.00 |
| Relative genotype frequencies | 0.25/0.75 = 0.333 | 0.50/0.75 = 0.667 | 0.00 |

Estimated allele frequencies after natural selection:

$$f(B) = (0.333) + (0.5)(0.667) = 0.667$$
$$f(b) = (0) + (0.5)(0.667) = 0.333$$

Estimated genotype frequencies after reproduction:

$$f(BB) = (0.667)^2 = 0.445$$
$$f(Bb) = 2(0.667)(0.333) = 0.444$$
$$f(bb) = (0.333)^2 = 0.111$$

allele frequencies of 0.0 or 1.0 can, however, be changed by migration and mutation.

Figure 22.6 illustrates that directional selection favoring $B_1$ increases the frequency of that allele at a pace determined by the intensity of natural selection. In the figure, $f(B_1)$ is 0.01 and $f(B_2)$ is 0.99 at generation 0, the starting point of natural selection. The progression of $f(B_1)$ toward fixation ($f = 1.0$) is most rapid with the strongest natural selection and is slowest when natural selection differences between the genotypes are relatively weaker. Under the strongest natural selection condition, $f(B_1)$ rises from 0.01 to 0.99 in less than 100 generations, but weaker natural selection dramatically slows the rate of increase of the allele frequency.

The concept of relative fitness values can be applied to populations in several ways, including the analysis of directional selection that operates against a recessive allele by selection against the homozygous recessive genotype. Table 22.6 illustrates a case in which frequencies $f(B) = 0.50$ and $f(b) = 0.50$ are subjected to natural selection against *bb*, where $w_{bb} = 0.0$ and $w_{Bb} = w_{BB} = 1.0$. No *bb* individuals survive to reproductive age, thus removing 25 percent of the population. When the relative genotype frequencies are determined using their new proportions in the surviving reproductive population, $f(B)$ and $f(b)$ are calculated to be $f(B) = 0.667$ and $f(b) = 0.333$. Among the progeny in generation 1, genotype frequencies are $f(BB) = 0.445$, $f(Bb) = 0.444$, and $f(bb) = 0.111$.

Directional natural selection against organisms with the recessive phenotype (and the homozygous recessive genotype) causes the frequency of the dominant allele to increase and the frequency of the recessive allele to decrease. Eventually, the recessive allele may be eliminated from the population gene pool. The recessive allele is not eliminated quickly, however, and its frequency changes slowly, especially as the allele gets less frequent. The slow pace of evolutionary change at low allele frequencies is due to the smaller number of recessive homozygotes in the population. For any particular generation, the frequency of the recessive allele ($q$) can be estimated by the equation $q_n = q_i/1 + nq_i$, where $n$ = the number of generations with selection, $q_i$ is the initial value of $q$, and $q_n$ is the value of $q$ after $n$ generations.

Numerous directional selection experiments, taking place over the last several decades of research, demonstrate that the theoretical predictions for populations are observed in the laboratory. A 1981 experiment by Douglas Cavener and Michael Clegg examined four subpopulations of *Drosophila melanogaster* for 50 generations to test the effectiveness of artificial directional selection at increasing the frequency of the allele $Adh^F$ of the *alcohol dehydrogenase* (*Adh*) gene. The enzyme product of $Adh^F$ rapidly breaks down ethanol, and flies that produce this enzyme have higher relative fitness in an ethanol-rich environment than do flies that do not produce the enzyme. An original population with an $Adh^F$ frequency of 0.38 was divided into four subpopulations of equal size. Two subpopulations reared on ethanol-rich food (population 1 and population 2) showed progressive increases in the

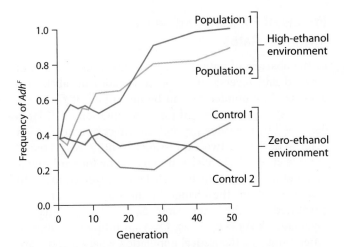

**Figure 22.7 Directional artificial selection favoring the Adh^F allele in experimental *Drosophila* populations.** The *Adh^F* allele increases in frequency in both experimental populations exposed to an ethanol-rich environment. Allele frequencies in two control populations (no natural selection) drift up and down around the starting allele frequency.

frequency of $Adh^F$ over 50 generations (**Figure 22.7**). In contrast, control populations (control 1 and control 2), which were reared on food without ethanol, showed an overall downward drift of $Adh^F$ frequency.

The rapidity with which allele frequencies change under directional selection is a function of the starting allele frequencies and the intensity of natural selection. Allele frequency change is slow when allele frequencies are low and can be more rapid when allele frequencies are higher. Similarly, the rate of allele frequency change is greatest when strong directional natural selection favors particular alleles. Two independent reports published in 2010, one by Xin Yi and colleagues and the other by Tatum Simonson and colleagues, identify rapid evolutionary changes that have occurred in the last 5000 years as native Tibetans have adapted to low oxygen conditions in the high-altitude environment of the Himalayan mountains. Strong directional natural selection has operated in favor of certain alleles of multiple genes that contribute to oxygen utilization and metabolism and has increased the frequencies of these alleles in comparison to populations of Han Chinese who inhabit the lowlands.

---

**Genetic Insight** Directional selection causes the frequency of a favored allele to rise toward fixation ($f = 1.0$) and decreases the frequency of disfavored alleles toward elimination ($f = 0.0$) at rates determined by the intensity of natural selection.

---

## Natural Selection Favoring Heterozygotes

A pattern of natural selection that can produce and maintain genetic diversity in populations is seen when the heterozygous genotype is favored. This type of natural selection results in a **balanced polymorphism,** in which alleles reach stable equilibrium frequencies that are maintained by the ongoing action of selection against the homozygous genotypes.

Natural selection pressure from malaria in human populations favored individuals who are heterozygous for $\beta^A$, $\beta^S$, and other alleles of the β-globin gene and lead to balanced polymorphisms in multiple populations (see Chapter 10). **Table 22.7** depicts a natural selection scheme similar to the one favoring heterozygous carriers of the $\beta^S$ allele. In this example, the relative fitness of the heterozygous genotype ($Cc$) is 1.0, the relative fitness of $CC$ is 0.80, and the fitness of $cc$ is 0.20, indicating that few of the recessive homozygotes survive to reproductive age. Beginning in generation 0 with $f(C) = f(c) = 0.50$, natural selection changes allele frequencies to $f(C) = 0.60$ and $f(c) = 0.40$ in the matings that produce generation 1.

Over successive generations, natural selection continues to favor the heterozygous genotype and select against both homozygous genotypes. Since natural selection against organisms with the homozygous recessive genotype is stronger ($s = 0.80$) than selection against organisms with the homozygous dominant genotype ($s = 0.20$), the frequency of $c$ will decrease relative to the frequency of $C$, but the recessive allele will not be eliminated. Instead, since natural selection operates in favor of heterozygotes, a balanced polymorphism is attained. Once attained, the equilibrium frequencies of the alleles will be maintained as a balanced polymorphism as long as natural selection remains steady. Population geneticists can predict the stable equilibrium frequencies of alleles in a balanced polymorphism using the relative intensity of natural selection

| Table 22.7 | A Model of Natural Selection Favoring the Heterozygous Genotype | | |
|---|---|---|---|

| | **Genotype** | | |
|---|---|---|---|
| | *CC* | *Cc* | *cc* |
| Frequency | 0.25 | 0.50 | 0.25 |
| Relative fitness | 0.80 | 1.0 | 0.20 |
| Survivors after selection | 0.20 | 0.50 | 0.05 |
| Relative genotype frequencies | 0.20/0.75 = 0.267 | 0.50/0.75 = 0.067 | 0.05/0.75 = 0.667 |

New allele frequencies after natural selection:

$$f(C) = 0.600$$
$$f(c) = 0.400$$

Genotype frequencies after reproduction:

$$f(CC) = (0.60)^2 = 0.36$$
$$f(Cc) = 2[(0.60)(0.40)] = 0.48$$
$$f(cc) = (0.40)^2 = 0.16$$

against the homozygous genotypes. Using the variables $s$ and $t$ to represent the natural selection coefficients operating against the homozygous genotypes, the relative fitness of $CC$ is $1 - s$ and the relative fitness of $cc$ is $1 - t$. Solving for the values of $s$ and $t$,

$$s = 1.0 - 0.80 = 0.20, \text{ and } t = 1.0 - 0.20 = 0.80$$

The stable equilibrium for $p$ and $q$, designated $p_E$ and $q_E$ in the balanced polymorphism, are calculated as ratios of selection coefficients operating against the homozygous genotypes. In this example, the equilibrium $p_E$ and $q_E$ values are

$$p_E = t/(s + t) = 0.80/0.20 + 0.80 = 0.80 \text{ and}$$
$$q_E = s/(s + t) = 0.20/0.20 + 0.80 = 0.20$$

---

**Genetic Insight** Natural selection can maintain balanced polymorphism in a population by heterozygous advantage, which confers higher relative fitness on heterozygotes than on homozygotes. The equilibrium allele frequencies are dependent on the intensity of selection against the homozygous genotypes.

---

## 22.4 Mutation Diversifies Gene Pools

**Mutation** is the ultimate source of all new genetic variation in populations, and as we have discussed, genetic variation is a necessary condition for evolution. On this basis, evolutionary biologists conclude that mutation is an indispensable component of evolution. By itself, however, gene mutation is a very slow evolutionary process because its effect on allele frequencies in populations is small and gradual. For example, if mutation converts one in every 10,000 $A_1$ alleles to $A_2$ alleles each generation, a population containing $f(A_1) = 0.90$ and $f(A_2) = 0.10$ in generation 0 will have frequencies $f(A_1) = 0.81$ and $f(A_2) = 0.19$ after 1000 generations, assuming no effects from the other evolutionary processes.

An additional reason that mutation alone is a slow evolutionary process has to do with the two directions in which mutation can affect any given allele. The **forward mutation rate ($\mu$)** pertains to $A_1$-to-$A_2$ mutations, or mutations that create a new $A_2$ allele by mutation of $A_1$, whereas the **reverse mutation rate ($v$)**, also known as the **reversion rate,** pertains to mutation of alleles in the opposite direction, $A_2$ to $A_1$. Forward and reverse mutation can create a balanced equilibrium, given a sufficient number of generations and the absence of other evolutionary processes. Keep in mind, however, as we explore this process, that the greater likelihood is for the effect of mutation to be balanced by the effects of natural selection, as described later under "Mutation–Selection Balance."

## Effects of Forward and Reverse Mutation Rates

In the absence of natural selection, the consequences of forward and reverse mutation (reversion) on allele frequencies in a population can be quantified. If $f(A_1) = p$ and $f(A_2) = q$, the effect of forward mutation on $f(A_1)$ is described by the value $\mu p$, and the effect of reversion on $f(A_2) = vq$. These two expressions identify, respectively, the rate at which $A_2$ alleles are created by forward mutation and the rate at which $A_2$ alleles are reverted to $A_1$. In each generation, the change in the frequency of $A_2$ is quantified by the expression $\Delta q$ ("delta $q$") that is calculated as $\Delta q = \mu p - vq$. Over an infinite number of generations in a theoretical population where $\mu$ and $v$ are constant and no other evolutionary processes are operating, allele frequency equilibrium is established.

Like the calculation in the previous section that determined the equilibrium frequencies of alleles when selection favors heterozygotes, the equilibrium frequencies of alleles subject only to mutation and reversion are a ratio of the frequencies of the respective events. If forward and reverse mutation alone are acting to change allele frequencies, the equilibrium values are purely a function of the ratios of the rate at which new copies of an allele are added and removed from the population gene pool. In this case, the equilibrium frequencies of $p$ and $q$ are $p_E = v/(\mu + v)$ and $q_E = \mu/(\mu + v)$. In a theoretical population where $f(A_1) = 0.99$, $f(A_2) = 0.01$, $\mu = 2 \times 10^{-6}$, and $v = 3 \times 10^{-8}$, $\Delta q$ is expressed as $\Delta q = [(2 \times 10^{-6})(0.99) - (3 \times 10^{-8})(0.01)] = 1.98 \times 10^{-6}$. This small change gradually increases $f(A_2)$ and decreases $f(A_1)$, leading eventually to equilibrium allele frequencies in this population that are

$$p_E = 3 \times 10^{-8}/(2 \times 10^{-6} + 3 \times 10^{-8}) = 0.015 \text{ and}$$
$$q_E = 2 \times 10^{-6}/(2 \times 10^{-6} + 3 \times 10^{-8}) = 0.985$$

### Mutation–Selection Balance

Unlike the theoretical population just described, mutations in the real world are commonly subject to natural selection. In some cases, natural selection operates against new dominant mutations that reduce relative fitness and thus leads to elimination of the new mutation. When a dominant mutation is lethal, or nearly so, mutant organisms have a relative fitness value that is effectively zero, and a high proportion of mutant organisms identified in each generation will have the mutation as a result of new mutations. If we consider such situations in humans, there will be no family history of the disorder before its appearance in the family. In such circumstances, the rate of occurrence of dominant mutations is approximately equal to the mutation frequency, since almost all individuals with the mutation are eliminated by natural selection each generation. In cases where the deleterious mutation is recessive, the mutant allele is masked by the wild-type dominant allele in heterozygous genotypes. Recessive mutant alleles are subjected to natural selection only when they occur in the homozygous recessive genotype. This results in the persistence of recessive mutant allele in most populations at a frequency somewhat greater than the mutation frequency.

Evolutionary biologists can quantify the frequency of mutant alleles in a population as a balance of the intensity of natural selection against the mutant and the frequency of mutation of the gene. This expression is called the **mutation–selection balance,** and it determines the equilibrium frequency of the mutant allele ($q_E$) by considering the rate of elimination of deleterious alleles by natural selection ($s$) and the rate at which new mutant alleles are generated ($\mu$). The following hypothetical examples illustrate calculation of the mutation–selection balance in the case of a lethal recessive mutation and a lethal dominant mutation.

For a recessive lethal mutation, the equilibrium frequency of the mutant allele is a ratio of mutation frequency and the selection coefficient. Consider the following relative fitness values.

| Genotype | $A_1A_1$ | $A_1A_2$ | $A_2A_2$ |
|---|---|---|---|
| Relative fitness | 1 | 1 | $1 - s$ |

The most widely used method for deriving the equilibrium frequency ($q_E$) identifies the balance between selection against a recessive genotype ($s$) and the rate of mutation ($\mu$) as

$$q_E = \sqrt{\mu/s}$$

This expression predicts that when selection against the recessive genotype is complete (i.e., $s = 1.0$), the equilibrium frequency of the mutant allele is approximately the square root of the mutation rate. When the selection coefficient is less than 1.0, the equilibrium frequency is greater than the square root of the mutation frequency.

In the case of complete selection against a lethal dominant mutant allele $B_2$, the relative fitness values of the genotypes are as follows.

| Genotype | $B_1B_1$ | $B_1B_2$ | $B_2B_2$ |
|---|---|---|---|
| Relative fitness | 1 | $1 - s$ | $1 - s$ |

In this case, $q_E = \mu$. In other words, when $s = 1.0$ against a lethal dominant mutation, the equilibrium frequency of the mutant allele is equal to the mutation frequency.

Numerous examples of mutation–selection balance have been investigated in organisms, including humans. Several studies of human hereditary disease alleles reveal that recessive mutant alleles are maintained in populations at frequencies predicted by calculating the mutation–selection balance.

**Genetic Insight** The genetic diversity that mutation adds to populations is essential for evolution. Forward and reverse mutation can establish an equilibrium between allele frequencies. Natural selection and mutation interact to produce a mutation–selection balance that maintains deleterious alleles in populations at low frequencies.

## 22.5 Migration Identifies Movement of Organisms and Genes between Populations

In population genetics, **migration** refers to the movement of organisms between populations and the reproduction of migrant organisms in their new populations. Allele frequency differences between the populations can lead to altered allele frequencies in the new mixed population. In evolutionary terms, migration is also known as **gene flow,** and the new mixed population is identified as an **admixed population.**

### Effects of Gene Flow

Gene flow has two principal effects on populations. First, in the short run, gene flow can change the frequency of alleles in the admixed population, particularly if the starting allele frequencies in one of the participating populations differ from those in the other and if the number of immigrants constitutes a large proportion of the admixed population. Second, in the long run, gene flow acts to equalize frequencies of alleles between populations that remain in genetic contact, often by the exchange of population members back and forth between the populations. This exchange can also slow genetic divergence of populations and can block speciation. Let's look at how both of these effects are explained.

The change in allele frequencies produced in an admixed population by gene flow from population 1 into population 2 can be described by the **island model** of migration that depicts a one-way process of gene flow, that is, from a mainland population to an island population. In the example illustrated in **Figure 22.8a**, gene flow changes allele frequencies by reducing $f(A_1)$ on the island from 1.0 to 0.60 and increasing $f(A_2)$ from 0.0 to 0.40. If the admixed population on the island consists of 1000 individuals, gene flow will have produced an almost instantaneous evolutionary change (**Figure 22.8b**). The admixed population has allele frequencies of $f(A_1) = 0.60$ and $f(A_2) = 0.40$, but the genotypes are not in H-W equilibrium immediately following migration. A single generation of random mating, however, will bring the genotype frequencies into ratios consistent with the H-W equilibrium: $A_1A_1 = 0.36$, $A_1A_2 = 0.48$, and $A_2A_2 = 0.16$.

The consequence of gene flow on allele frequencies in an admixed population is expressed by a formula that calculates $p_N$, the new value of $p$, as the weighted average of

**(a)** The island model of migration

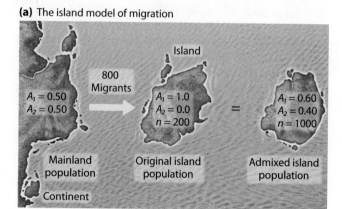

**(b)** Consequence of migration

| Island population | $A_1A_1$ | $A_1A_2$ | $A_2A_2$ |
|---|---|---|---|
| Original ($n = 200$) | 200 | 0 | 0 |
| Admixed ($n = 1000$) | 400 | 400 | 200 |
| Genotype frequencies in admixed population | .36 | .48 | .16 |

Genotype frequencies in admixed population

$f(A_1) = 0.60$
$f(A_2) = 0.40$

**Figure 22.8** **The island model of migration.**

the allele frequency among island residents and mainland immigrants. The expression uses $p_I$ and $p_C$ to represent $f(A_1)$ in the original island and mainland populations, respectively. The formula identifies the fraction of individuals or alleles from the mainland population as $m$, and the fraction contributed by island residents as $1 - m$. The value of $p_N$ as a result of gene flow is $p_N = (1 - m)(p_I) + (m)(p_C)$. Applying this formula to our example in Figure 22.8, we find $p_N$ $(0.20)(1.0) + (0.80)(0.50) = 0.60$.

## Allele Frequency Equilibrium and Equalization

We have just seen that gene flow can produce rapid evolutionary change in the allele frequencies of populations. In the short term, the effect of gene flow is determined by the change in the frequency of $p$ in the new gene pool of the island population. This value, $\Delta p_I$, is the difference in allele frequency before and after migration, and is defined as $\Delta p_I = p_N - p_I$. Substituting the formula for $p_N$ and simplifying gives $\Delta p_I = [(1 - m)(p_I) + (m)(p_C)] - p_I = m(p_C - p_I)$. Allele frequency equilibrium occurs when $\Delta p_I = 0$; thus, at equilibrium, $m(p_C - p_I) = 0$, indicating that $p$ remains constant either when there is no migration ($m = 0$) or when $p$ in the island gene pool equals the allele frequency in the mainland gene pool ($p_I = p_C$).

Population and evolutionary biologists use this analysis to conclude that gene flow has a homogenizing, or equalizing, effect on allele frequencies among participating

populations. By this mechanism, gene flow maintains the biological ability of populations to interbreed and can thus prevent evolutionary divergence of populations. In broader evolutionary terms, gene flow hinders the establishment of the reproductive isolation that is an important component of evolutionary divergence between populations and of potential speciation.

---

**Genetic Insight** Migration between populations allows gene flow that, through the influx of alleles capable of changing allele frequencies, can cause rapid population evolution. Over time, gene flow homogenizes allele frequencies among populations, reducing genetic diversity.

---

## 22.6 Genetic Drift Causes Allele Frequency Change by Sampling Error

The term **genetic drift** refers to chance fluctuations of allele frequencies that result from "sampling error," a statistical term signifying that a small sample taken from a larger population is not likely to contain all alleles in exactly the same frequencies as in the larger population. Genetic drift affects all populations, but it is especially prominent in small populations in which a small number of gametes unite to produce each subsequent generation.

To appreciate the consequences of genetic drift, picture a gene pool with alleles at frequencies $f(A_1) = f(A_2) = 0.50$ from which two separate samples are drawn. In sample one, 20 alleles are drawn at random, whereas in the second sample, 1000 alleles are drawn. These two separate draws represent the alleles that, in the two respective cases, unite to form the next generation. In the first sample, containing 20 alleles, each allele represents 5 percent (one allele out of 20) of the total for the next generation, whereas in the 1000-allele sample, each allele only represents 1/1000 of the alleles in the next generation. Any deviation from exactly 10 $A_1$ alleles and 10 $A_2$ alleles in the first sample will substantially change allele frequencies in the next generation. If, for example, the draw of 20 alleles contains 12 $A_1$ alleles and 8 $A_2$ alleles, the allele frequencies in the next generation will be $f(A_1) = 12/20 = 0.60$ and $f(A_2) = 8/20 = 0.40$. A change of such magnitude can easily occur by chance in the small sample, but it is very unlikely to occur in the larger sample of 1000 alleles.

Sampling errors of the kind described for the first sample can randomly raise or lower the frequency of an allele in a small population each generation. Once the allele frequencies are changed, the next generation, when it reproduces, has the new allele frequencies as a starting point. Over multiple generations, the frequency of an allele in a small population will randomly fluctuate, or "drift," sometimes increasing and sometimes decreasing, due to nothing more than "errors" in random sampling.

Allele frequency changes due to genetic drift are random. In the absence of any other evolutionary influence, one allele will ultimately reach fixation at a frequency of 1.0 and all other alleles will be eliminated. This can take a considerable number of generations, however, and the alleles fixed and eliminated may differ among experimental populations that are initially identical. Similarly, the number of generations required to fix or eliminate an allele can also differ. Figure 22.9 illustrates four different simulations of genetic drift of an allele in experimental populations and shows how the result of genetic drift for 30 generations can vary among populations that are initially identical. Each experimental population begins with 20 organisms and maintains that number throughout the 30 generations. The initial starting frequency of each allele is 0.50 in each population.

## The Founder Effect

Small population size provides the conditions under which sampling errors can produce significant genetic drift of allele frequency. Multiple mechanisms can generate small population size, including the establishment of a new population by a small number of founding organisms. A difference in allelic frequencies that is produced in this way in a newly established population is called the **founder effect.** The new population founders are usually drawn from a larger original population, and since the number of founders is small, the allele frequencies carried by the founders may differ from those in the original population, and some alleles may be missing altogether. These changes are due to sampling error. In this case, the founders may carry certain alleles at significantly higher frequencies or significantly lower frequencies than in the original population. As a result, the founder effect can create new populations with allele frequencies that differ substantially from those found in the original population.

Small human populations whose origins can be traced to religious, social, political, or other distinctions are usually established by a small number of individuals and contain few members of reproductive age. Often, the founders consist of several families, and the adults in the families are of reproductive age. Since the family members are related and share alleles, allele frequencies among the founders likely will differ from allele frequencies in the larger populations from which the founders emigrate, or may omit some alleles from the new population. If members of the newly founded community mate within the group, they are reproductively isolated from the surrounding populations and may be subject to genetic drift.

The consequences of founder effect and genetic drift can result in high frequencies of autosomal recessive disorders that are rare outside the affected populations. The Old Order Amish are a religious population established by about 200 founding members in Lancaster County, Pennsylvania, between 1720 and 1770. The founding population came from English and European populations and consisted of several extended families. Other Amish communities were established by different founders in Ohio, Indiana, and elsewhere in North America. These populations tend to be small, and mating within each population is common. Due to the founder effect and genetic drift, Amish populations exhibit high frequencies of several autosomal and X-linked recessive disorders that are rare in their populations of origin and in surrounding non-Amish communities.

One example of a disorder found in high frequency in an Amish community is Ellis–van Creveld syndrome (EvC) (OMIM 225500), an autosomal recessive disorder that produces short stature accompanied by short forearms and lower legs and by the frequent appearance of extra digits on hands or feet. In a survey of nearly 8000 Old Order Amish in Lancaster County completed several years ago, 43 cases of EvC were identified. From this information, the estimated frequency of the allele producing EvC in the population is estimated by taking the square root of the frequency of the recessive trait in the population. The calculation is $q = 43/8000 = 0.073$, or about 7.3 percent. Among other Amish populations, and in the general (non-Amish) population, the frequency of this recessive allele is $q < 0.001$.

The genealogical history of the Old Order Amish community in Lancaster County identifies founder effect followed by genetic drift as the probable cause of the high frequency of the allele causing EvC. All families with EvC trace their genealogies to Mr. and Mrs. Samuel King, who immigrated to Lancaster County in 1744. At the time, there were about 400 people in the Lancaster County population, and the evidence suggests that both Mr. King and Mrs. King were carriers of the recessive mutant allele. This information establishes the initial frequency of the mutant

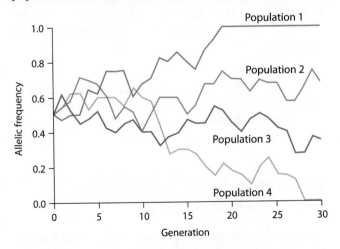

**Figure 22.9 Genetic drift of an allele frequency.** Four simulated populations each start with a frequency of 0.50 for a hypothetical allele whose frequency fluctuates randomly in each population over 30 generations. The allele eventually becomes fixed in population 1, is eliminated in population 4, and is still present in populations 2 and 3 at distinct frequencies.

allele in the founding population at $f(q) = 2/800 = 0.0025$, more than twice the frequency in the population of origin. Large family sizes and a tendency for the Amish to mate within the Lancaster County community subsequently contributed to the rise in the frequency of the allele in the population.

---

**Genetic Insight** Genetic drift is a random fluctuation of allele frequencies caused by variation in the sampling of gametes from generation to generation. Genetic drift can have large effects on allele frequencies in small populations. Founder effect is a form of genetic drift result from the establishment of a new population by founders who frequencies of alleles that differ from their population of origin.

---

## Genetic Bottlenecks

A second mechanism producing large allele frequency sampling errors in small populations is the **genetic bottleneck,** in which a relatively large population is substantially reduced in number by a catastrophic event independent of natural selection. The survivors of the bottleneck—a small, random sample of the original population—are likely to have a very low level of genetic diversity due to the loss of alleles from the gene pool. They are likely to carry alleles in frequencies that differ radically from those in the original population (**Figure 22.10**). The loss of genetic diversity can be quantified in two ways: first, by determining the percentage of polymorphic loci in the population, and second, by determining the percentage of loci in an average individual that are heterozygous.

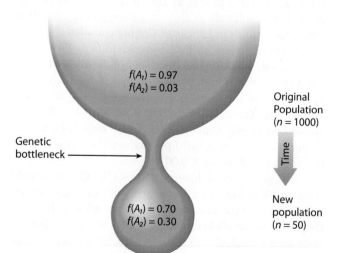

**Figure 22.10 A genetic bottleneck.** Catastrophic population reduction not due to natural selection can restrict or eliminate the alleles that pass through the bottleneck, and alter allele frequencies in the surviving population.

Genetic bottlenecks can affect single populations, or they can affect an entire species. An example of the latter case would be a near-extinction event such as the one that affected the Northern elephant seal (*Mirounga angustirostris*). This animal was historically distributed along the western coast of North America, in numbers that exceeded 150,000 in the mid-1800s. Extensive hunting devastated the rookeries where young elephant seals were raised, and by 1884 fewer than 100 elephant seals remained. Some biologists have estimated that the surviving population may have been as small as 20 individuals. The entire remaining population bred at an isolated rookery on Guadalupe Island, about 200 miles off the western shore of Baja California. Elephant seal protection measures put in place by the U.S. and Mexican governments in the early 1900s led to population growth and the reestablishment of additional rookeries. Today, the Northern elephant seal remains a protected species that has returned to its historic population size of approximately 150,000 individuals.

In 1974, Robert Selander and his colleagues collected blood samples from 159 Northern elephant seals from five populations and examined 24 blood protein and enzyme genes for evidence of genetic variation. All 24 genes were monomorphic, and the single allele of each gene was identical in all five populations! About 20 years later, A. Rus Hoelzel and colleagues expanded the genetic survey of Northern elephant seals to include 43 genes in 61 individuals from the five populations. They also found no genetic variation. Additionally, Hoelzel and colleagues examined variation of mitochondrial DNA in Northern elephant seals and found a low level of sequence variation in two distinctive mitochondrial DNA haplotypes that had frequencies of 0.725 and 0.275. The extremely limited genetic variation in Northern elephant seals is wholly consistent with the historical genetic bottleneck that left very little genetic variation in the surviving population members.

---

**Genetic Insight** Genetic bottlenecks are severe reductions in the number of individuals in a population that substantially reduce genetic diversity in the surviving populations.

---

## 22.7 Nonrandom Mating Alters Genotype Frequencies

Descriptions of population genetic structure based on the Hardy-Weinberg principle assume random mating within the population. If this assumption is not met, however—if mating in the population is nonrandom—the distribution of alleles into genotypes occurs in frequencies inconsistent with the chance predictions of the Hardy-Weinberg equilibrium.

Two forms of nonrandom mating in populations are most common. The first type is *inbreeding*, mating between related individuals. Inbreeding is nonrandom mating with respect to genotypes. The second form of nonrandom mating consists of two patterns of nonrandom mating with respect to specific phenotypes: *positive assortative mating* and *negative assortative mating*.

## The Coefficient of Inbreeding

**Inbreeding,** also known as **consanguineous ("with blood") mating,** is mating between related individuals who share a greater proportion of alleles with one another than with random members of a population. The principal genetic consequences of inbreeding are an increase in the frequency of homozygous genotypes in a population and a decrease in the frequency of heterozygous genotypes relative to the frequencies expected from random matings. The likelihood of homozygosity is increased because related organisms share alleles and are thus more likely to produce homozygotes, especially when the alleles involved are rare in the general population.

Inbreeding is a normal reproductive process for self-fertilizing plants and for some animals that reproduce by self-fertilization. The effect of self-fertilization on genotype proportions is shown in **Table 22.8**, where a heterozygous organism self-fertilizes and produces genotypes in generation 1 in a 1:2:1 ratio. Self-fertilization of generation 1 individuals produces a generation 2 that has an overall increase in the frequency of both homozygous genotypes and a decrease of one-half in the frequency of the heterozygous genotype. The decrease in heterozygous frequency of one-half occurs each generation. By generation 4, just over 6 percent of the progeny are heterozygous, and more than 93 percent are homozygous. Note, however, that the allele frequencies of $A_1$ and $A_2$ remain at $f(A_1) = f(A_2) = 0.50$ in each generation.

Among sexually reproducing organisms, the effect of inbreeding is similar; but it takes place over a larger number of generations since the proportion of organisms in a population participating in consanguineous matings is generally low. The population geneticist Sewall Wright

investigated the consequences of inbreeding in sexually reproducing populations and devised the **coefficient of inbreeding ($F$)** as an arithmetic measure of the probability of homozygosity for an allele obtained in identical copies from an ancestor. The coefficient of inbreeding quantifies the probability that two alleles in a homozygous individual are **identical by descent (IBD),** having descended from the same copy of the allele carried by a common ancestor of the inbred individual. A common ancestor is an ancestor shared by two inbreeding organisms, and potentially the source of identical alleles that could be carried by the inbreeding organisms. If inbreeding takes place, all genes in the genome are susceptible to the same inbreeding effects. Thus $F$ can also be used to estimate the proportion of loci that will be homozygous IBD.

The quantification of $F$ as a measure that a particular allele is IBD is most readily accomplished through pedigree analysis. The three key elements to determining $F$ from pedigrees are (1) the number of alleles of a gene carried by common ancestors, (2) the number of transmission events required to produce a genotype that is homozygous IBD, and (3) the probability of transmission for each event linking the allele in a common ancestor to the inbred individual. **Figures 22.11a** and **22.11b** show a mating between half-siblings having the same mother (I-2) as the common ancestor. The general solution for $F$ is $(1/2)^n$, where 1/2 is the probability of transmission of an allele and $n$ is the number of transmission events required to produce identity by descent. In this example, either the allele $A_1$ or $A_2$ of the mother could be transmitted to both II-1 and II-2 and then to their offspring III-1, so the general solution for $F$ is $(1/2)^n + (1/2)^n$. The arrows in the figure show the four transmission steps that are required for either allele to end up in III-1 IBD. Each required transmission event has a probability of 50 percent. Thus, the probability that either allele is found in III-1 in a homozygous IBD genotype is $(1/2)^4 = 1/16$. For this case, the inbreeding coefficient is the probability that *any* allele of a locus is homozygous IBD; thus, for each gene, $F = (1/2)^4 + (1/2)^4 = 1/8$. Notice that the arrows in the figure indicating transmission of alleles from I-2 to III-3 trace the two sides of a loop. This visual representation indicates the movement of the allele from generation to generation. If this loop were incomplete, identity by descent could not occur.

**Figure 22.11c** shows a first-cousin mating in a pedigree in which alleles from either I-1 or I-2 could make their way to IV-1. Here there are four alleles, any of which could be IBD in IV-1. Each allele must complete six transmission steps (indicated by arrows in the figure) to be identical by descent in IV-1, and the transmission probability for each step is 1/2. For each allele carried by I-1 and each allele carried by I-2, the probability the allele is IBD in IV-1 is $(1/2)^6 = 1/64$. For this pedigree, there are four alleles for each gene, two per common ancestor, and $F$ is determined by adding the probability of

| Table 22.8 | Consequences of Self-Fertilization to Genotype Frequencies |
|---|---|

P:  $A_1A_2$ (self-fertilization)

| Progeny Generation | Genotype | | |
|---|---|---|---|
| | $A_1A_1$ | $A_1A_2$ | $A_2A_2$ |
| 1 | 0.250 | 0.500 | 0.250 |
| 2 | 0.375 | 0.250 | 0.375 |
| 3 | 0.437 | 0.125 | 0.437 |
| 4 | 0.468 | 0.063 | 0.468 |

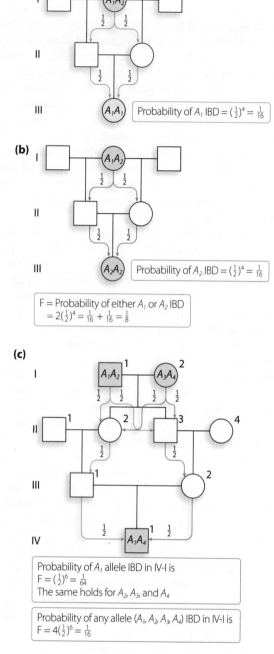

**Figure 22.11 Calculation of the inbreeding coefficient (F).**
(a) The probability of $A_1$ IBD equals the likelihood of four transmission events, each with a probability of 1/2. (b) The probability of $A_2$ IBD also requires four transmission events, each with a probability of 1/2. The likelihood of either allele IBD is $F = 2(1/2)^4$. (c) With two common ancestors, there are four alleles ($A_1$, $A_2$, $A_3$, and $A_4$) that can be IBD. For this first-cousin mating, the probability for each allele IBD is the same: $F = (1/2)^6$. For all shared alleles combined, $F = 4(1/2)^6 = 1/16$.

as $F = 4(1/2)^6 = 1/16$. **Genetic Analysis 22.2** demonstrates another computation of an inbreeding coefficient.

First-cousin mating is a form of inbreeding that is relatively common in many human societies and is common in mammals in general. Negative genetic outcomes in the form of infants with recessive conditions due to inbreeding occur when a recessive allele is rare in a population. In such cases there is a meaningful increase in the likelihood that a first-cousin mating will produce a child with a recessive phenotype. When the recessive allele frequency is $q = 0.01$, for example, the chance of producing a recessive homozygote from a first-cousin mating is only a few times more likely than the chance of producing a recessive homozygote by random mating.

## Inbreeding Depression

The genetic consequences of inbreeding for populations are an increase in the frequency of homozygous genotypes and a decrease in the frequency of heterozygous genotypes. One immediate impact of these consequences is seen in conservation genetics, where small, captive populations of individual organisms are bred to perpetuate a nearly extinct species. The increased frequency of homozygosity can lead to a phenomenon known as **inbreeding depression,** the reduction in fitness of inbred organisms, often as a result of the reduced level of genetic heterozygosity. The reduced fitness associated with inbreeding depression can be due either to an increase in the proportion of deleterious homozygous genotypes or to the higher fitness of heterozygotes.

The magnitude of inbreeding depression depends on the organism. Among plants that naturally reproduce by self-fertilization, the amount of inbreeding depression is small. Many bird species also experience only relatively minor inbreeding depression. This lack of negative consequence has been particularly beneficial in captive breeding programs that have bred bird species such as the California condor and then reintroduced the birds into their natural environment. In contrast to birds and plants, however, mammals experience severe inbreeding depression. Consequently, captive breeding programs for mammals must be managed far more carefully, from a genetic perspective. Multiple studies indicate that inbred mammals housed in zoos have higher mortality rates than do zoo mammals produced through programs designed to avoid inbreeding. The higher mortality of inbred mammals indicates the impact of inbreeding depression.

**Genetic Insight** Inbreeding is mating between organisms that are more closely related to one another than they are to random members of a population. The coefficient of inbreeding (F) quantifies the probability that two copies of an allele carried by a common ancestor will be found in a homozygous IBD genotype of a descendant.

the four complete loops (one for each allele) that could link an allele in a common ancestor to an inbred homozygous IBD descendant. In this case, $F = (1/2)^6 + (1/2)^6 + (1/2)^6 + (1/2)^6 = 1/16$. The value can also be determined

The pedigree shown below depicts crosses performed as part of an antelope captive-breeding program. Use the pedigree information to calculate the coefficient of inbreeding (*F*) for the mating of IV-1 and III-3 that produces the animal identified as V-1.

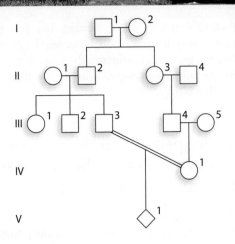

| Solution Strategies | Solution Steps |
|---|---|
| **Evaluate** | |
| 1. Identify the topic of this problem, and describe the nature of the required answer. | 1. This problem concerns determination of the coefficient of inbreeding (*F*) for a specific mating. |
| 2. Identify the critical information given in the problem. | 2. The pedigree depicting the common ancestry of the related animals is given. |
| **Deduce** | |
| 3. Count the number of transmission events that must occur for an allele to be identical by descent (IBD) in V-1. | 3. Counting from a common ancestor to individual V-1, there are seven transmission steps required to produce an allele that is IBD. |
| 4. Identify the transmission probability for each step of transmission. | 4. For an autosomal allele, the transmission probability is 1/2. |
| 5. Identify the number of alleles carried at an autosomal gene in the common ancestors of V-1. | 5. There are two common ancestors (I-1 and I-2) for the inbred individual (V-1). There are two alleles per gene in each common ancestor, for a total of four alleles at each locus. |
| **Solve** | |
| 6. Calculate the coefficient of inbreeding for this pedigree. | 6. The coefficient of inbreeding is $F = 4(1/2)^7 = 1/32$. |

For more practice, see Problems 33, 34, 35, and 36.

## Effects of Assortative Mating

When mates choose one another, they may do so because of some resemblance between them. This pattern of bias in mate selection is known as **positive assortative mating.** Conversely, some mates choose one another because they have dissimilar phenotypes. This biased pattern of mate selection is known as **negative assortative mating.** Although one or the other of these patterns may occur in a population, it never occurs with perfect consistency; thus, some matings may involve assortative mating, but others are random.

Preferential mating with respect to phenotype is most readily observed in humans and other animal species having traits for which a strong correlation of phenotypes is possible between mates. For example, humans often practice positive assortative mating for height and for skin color. It is important to understand that positive assortative mating for traits like height and skin color is different from inbreeding. Whereas inbreeding pertains to the entire genome, positive assortative mating is specific for only those genes contributing to a particular phenotype. In genetic terms, a consequence of positive assortative mating is the tendency to increase homozygosity and decrease heterozygosity for the genes that contribute to the phenotypes involved in mate selection. Negative assortative mating has the opposite effect on genotypes of genes associated with the mate-selection bias. Negative assortative mating increases the heterozygosity and decreases homozygosity. In either case, however, the effect is limited to just the genes that influence the phenotype, and its impact on the population overall is limited by the practice of mating that is otherwise random.

## 22.8 Species and Higher Taxonomic Groups Evolve by the Interplay of Four Evolutionary Processes

Our discussion to this point has focused on microevolution, that is, evolution operating at the population level. In this final section, we turn our attention to evolution at the species level and above.

This evolutionary change is driven by **reproductive isolation** that can result from any morphological, behavioral, or geographic condition or set of conditions that prevents one population from breeding with others. Reproductively isolated populations adapt separately to their particular circumstances, and divergence is a likely consequence. In each environment, differential reproductive success driven by natural selection allows the better-adapted organisms to leave more progeny. These processes are primary drivers of speciation.

### Processes of Speciation

Charles Darwin was the first to describe clearly the concept that existing species evolve from preexisting species. In his famous 1859 book, *On the Origin of Species by Means of Natural Selection*, he laid out two guiding principles of species formation that are still considered fundamental aspects of macroevolution. First, Darwin proposed that hereditary variation is present in all species and controls the phenotypic variability in each species. He suggested that this inherited variation allows parents to pass their phenotypic attributes to their offspring. Second, Darwin proposed that natural selection allows species members with favored phenotypic attributes to survive and reproduce in greater numbers than species members with other phenotypes. Darwin described his model combining these principles as "the theory of descent with modification through variation and natural selection." In other words, Darwin viewed inherited variation and the operation of natural selection as the elements essential to the transformation of one species into another.

Innumerable biological investigations in the last 150 years have verified and elaborated upon Darwin's original proposals as well as quantified the effects and the interplay of each of the four evolutionary processes (natural selection, mutation, migration, and random genetic drift) on speciation. The clear picture of speciation that emerges from these studies is that the evolutionary lineages leading from ancestral organisms to descendant forms are almost never simple, straight lines of descent.

Instead, the evolutionary history of modern species is filled with side branches that died out because a species, once developed, could not adapt to new environments or was displaced by competing species. It can be tempting to look backward into the evolutionary past and identify a linear step-by-step process leading to modern species, but this perspective minimizes the occurrence of adaptive changes that led to evolutionary "dead ends." More important, the backward-looking approach ignores a major reality of evolution: Evolutionary history is far more like a multibranched bush rather than like a tree with a long, straight trunk connecting past and present.

The evolutionary history of modern horses and their relatives zebras and donkeys (all three being members of the genus *Equus*) is an example of the typical complexity of evolutionary history (**Figure 22.12**). One can trace a lineage leading more or less directly from *Hyracotherium* in the early Eocene (about 54 million years ago) to modern *Equus*, but this would ignore the many other branches of the evolutionary tree that did not produce modern-day organisms.

The evolutionary tree of *Equus* also illustrates two patterns by which new species evolve from preexisting species. **Anagenesis** is the process by which an original species is transformed into a new species over an extended period of time that spans many generations. During anagenesis, a preexisting species experiences natural selection that leads to adaptive change of that species into a new species. In contrast, the pattern of species evolution known as **cladogenesis** is one of branching in which an ancestral species gives rise to two or more new species.

The evolution of *Equus* is an example of reconstruction of a phylogenetic tree through the use of fossil evidence. Increasingly, however, it is becoming possible to isolate small amounts of DNA from extinct species and to undertake sequence-based and gene-based comparisons of the extinct species with related living species. These analyses reinforce the idea that ancestral and descendant species share genetic information as a consequence of their common ancestry, and they make it possible to understand the capabilities of extinct species more fully than is possible by studying fossil remains alone.

One recent example comes from genomic analysis of DNA isolated from Neandertal bones excavated from a cave in Eastern Europe. The isolated Neandertal DNA was fragmentary, but provided enough material for a nearly complete sequence determination of the Neandertal genome. Interestingly, a comparison of Neandertal and modern human genomes published in 2010 detects the possible presence of a small percentage of Neandertal DNA sequences in the genomes of anatomically modern humans indigenous to Europe and Asia, but not in Africa (see Chapter 18). Estimates indicate that between 2 percent and 4 percent of the DNA of anatomically modern humans has a Neandertal origin. Neandertal DNA made its way into modern European and Asian genomes through mating between Neandertals and modern humans who lived in

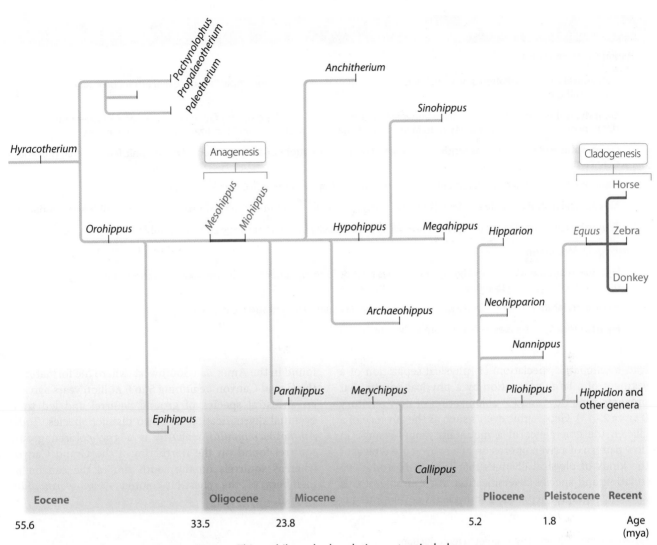

**Figure 22.12 Evolution of the genus *Equus*.** This multibranched evolutionary tree includes examples of anagenesis and cladogenesis as well as numerous examples of evolutionary branches that did not lead to modern species.

Europe and Asia at the same time and in some of the same locations. Temporal and geographic cohabitation is supported by the archeological record. Neandertals were not present in Africa, accounting for the absence of Neandertal DNA in indigenous African populations.

### Reproductive Isolation and Speciation

Although anagenesis and cladogenesis are distinct patterns of species formation, they share two essential features: (1) inherited variation controlling critical phenotypic variation and (2) adaptation through natural selection. Reproductive isolation is also an important component for both cladogenesis and anagenesis, although the precise mechanisms of isolation may differ.

The concept of cladogenesis and reproductive isolation of species derives from work by Theodosius Dobzhansky, Ernst Mayr, and other evolutionary biologists who recognized that new species can form when reproductive barriers prevent the exchange of genes (by

gene flow or migration) between populations. In describing the necessity of reproductive isolation in this context, two mechanisms are identified (**Table 22.9**). **Prezygotic mechanisms** of reproductive isolation are those that prevent mating between members of different species or prevent the formation of a zygote following interspecies mating. On the other hand, **postzygotic mechanisms** of reproductive isolation result in the failure of a fertilized zygote to survive, or result in sterile offspring of an interspecies mating. These mechanisms of reproductive and genetic isolation lead to patterns of speciation that most frequently occur in an allopatric pattern or in a sympatric pattern. Not all speciation events fit neatly into these two categories, and other, less common patterns are described in the literature of evolutionary biology.

**Allopatric Speciation**  In **allopatric speciation**, populations are separated by a physical barrier and thus develop in separate geographic locations or environments. Two principal mechanisms create the separations that

| **Table 22.9** | Mechanisms of Reproductive Isolation |
| --- | --- |

*Prezygotic Mechanisms*

**Behavioral isolation:** Patterns of sexual behavior in different species are incompatible, or sexual attraction is lacking between them.

**Gametic isolation:** Mating takes place between different species, but the gametes fail to unite with one another due to differences in gamete compatibility or to failure of male gametes to survive until fertilization of female gametes.

**Geographic isolation:** Species reside in separate geographic locations or are separated by geographic features that prevent their contact.

**Habitat isolation:** Species inhabit different ecosystems that prevent them from coming into contact.

**Mechanical isolation:** Male and female genitalia or reproductive structures of different species are anatomically incompatible.

**Temporal isolation:** Timing of reproductive ability or receptivity in different species is incompatible.

---

*Postzygotic Mechanisms*

**Hybrid breakdown:** Viable and fertile interspecies hybrids form, but after the $F_1$ generation the fitness of the progeny of hybrids is less than that of progeny from non-hybrids.

**Hybrid inviability:** The fertilized zygote of an interspecies mating fails to survive gestation.

**Hybrid sterility:** Interspecies hybrids are viable but infertile.

---

lead to allopatric speciation: (1) physical separation of a segment of a large population by a physical barrier that prevents gene flow and (2) colonization of new territory (**Figure 22.13**). Geographic events such as the advance of a glacier, the emergence of a mountain range, change in flow pattern of a river, or erosion of a canyon are typical of the kinds of physical changes that lead to reproductive isolation and species diversification. An example of this kind of geographic separation and species development is

found in the American Southwest, where the formation of the Grand Canyon beginning 5 to 6 million years ago split an ancestral species of ground squirrel and led to its eventual diversification into two distinct species. Today, *Ammospermophilus leucurus* is a gray-colored ground squirrel found on the north rim of the Grand Canyon, whereas squirrels on the south rim of the canyon are members of the chestnut-colored *Ammospermophilus harrisii*.

**(a)** Population bifurcation by a barrier to reproduction

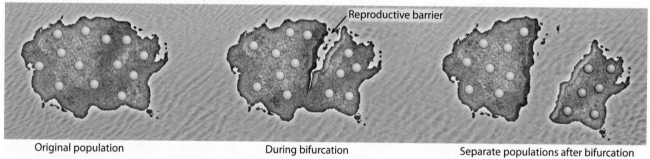

Original population                    During bifurcation                    Separate populations after bifurcation

**(b)** Colonization of new territory by migration

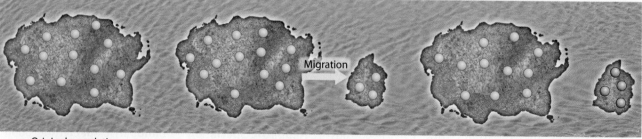

Original population                    During colonization                    After colonization

**Figure 22.13   Processes leading to allopatric speciation.**

The colonization model of allopatric speciation predicts that new species diversify following colonization of new habitats. The diversification of *Drosophila* species on the Hawaiian Islands is a case study of this mechanism (**Figure 22.14**). The Hawaiian Islands are part of a long chain of landmasses and submarine structures that stretch in a northwest-to-southeast direction and are produced by the movement of the Pacific tectonic plate over a volcanic hotspot that lies in the earth's mantle beneath it. As the plate slides toward the west, new islands are produced by volcanic activity of the hotspot. The oldest of the islands are Nihau and Kauai to the northwest; the youngest island is Hawaii, which is still growing by volcanic eruptions of Mauna Loa and Kilauea.

Hundreds of species of *Drosophila* inhabit the Hawaiian Islands, and evolutionary questions concerning how the species originated, how they are related to one another, and what evolutionary mechanism explains their diversification can be explored in light of the island model of allopatric speciation. In 2005, James Bonacum and his colleagues examined genetic and morphologic data in numerous Hawaiian *Drosophila* species to test the model. They found that the most closely related species occur on adjacent islands and that the phylogenetic pattern of species formation corresponds to the pattern of emergence of islands. These results provide support and documentation for the model of allopatric speciation by colonization.

**Sympatric Speciation**    In **sympatric speciation,** populations share a single habitat but are isolated by genetic or postzygotic mechanisms that prevent gene flow. Species that diverge while occupying the same geographic area are sympatric species, and their speciation usually involves either a genetic mechanism that prevents successful interbreeding during divergence or a behavioral, seasonal, or ecosystem-based process that limits the organisms available for reproduction. Populations diverging by a process of sympatric speciation live in the same geographic area but have very limited or episodic gene flow between them.

One clear example of sympatric speciation occurs in plant species that diversify from one another through the development of polyploidy. Mating between a polyploid species and one that is not polyploid can result in reduced fertility of hybrid individuals. We discuss the development of polyploidy through nondisjunction and highlight

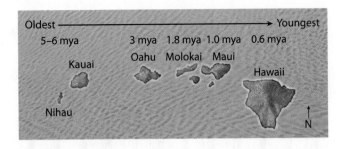

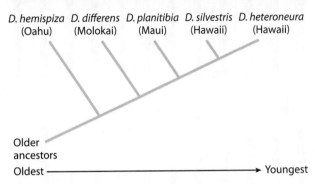

**Figure 22.14 Phylogenetic relationships among Hawaiian *Drosophila* species.** Evolutionary evidence supports the colonization of younger islands and the formation of new species following migration from older islands.

the evolution of the modern bread wheat species (*Triticum aestivum*) from a wild diploid grass to its contemporary allohexaploid form (see Figure 12.8). Animals that develop nocturnal or diurnal patterns of activity that make them more likely to encounter other members of the population that are active at the same time are another example of potential sympatric speciation. Similarly, changes in the seasonality of reproduction can limit organisms to the ability to reproduce only during certain times of the year. Organisms living in the same geographic area that do not have the same reproductive seasonality will be unable to mate.

Genetic Insight  A species is a reproductively isolated group of organisms that displays distinctive genetic and morphological characteristics. New species can descend from ancestral species by multiple processes.

# CASE STUDY

## CODIS—Using Population Genetics to Solve Crime and Identify Paternity

Each of us, with the exception of monozygotic multiple births, has a unique genome. Moreover, given the amount of genetic variability uncovered by human genome analysis, each of us may be genetically different from any other person who has

ever lived. The concept of genetic uniqueness has practical applications in individual identification through the identification of genotypes of DNA marker genes. This analysis is commonly known as DNA fingerprinting, or DNA profiling. In-

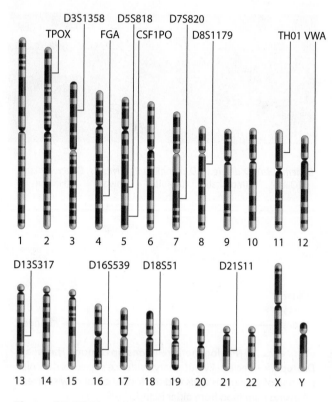

**Figure 22.15** The original 13 independently assorting CODIS STRP genetic markers.

dividual identification uses laboratory analyses of selected genetic markers in combination with statistical analysis based on the H-W equilibrium to determine the probability that a particular individual is the source of a specific DNA sample.

The genetic markers used in DNA profiling are simple tandem repeat polymorphisms (STRPs) that contain multiple copies of repeating DNA sequences. These markers are also known as variable number tandem repeat polymorphisms (VNTRs), and as minisatellite or microsatellite sequences. The DNA repeats for a particular STRP marker range in length from 2 base pairs (bp) to about 20. STRP analysis is carried out using polymerase chain reaction (PCR) amplification of targeted DNA sequences followed by gel electrophoresis. These methods are usually highly automated.

In 1997, the U.S. Federal Bureau of Investigation (FBI) selected 13 independently assorting human STRP markers to form the core of the bureau's Combined DNA Index System (CODIS; Figure 22.15). Extensive analysis determined the number and frequencies of alleles for the 13 original CODIS STRPs in most human populations. Studies also precisely defined laboratory methods for the analysis of CODIS STRPs. More STRPs were added to the original CODIS markers in later years, and more than 20 CODIS markers are in use today.

The statistical power of CODIS-based identification rests on the Hardy-Weinberg equilibrium and the product rule of probability for independently assorting genes. The STRP allele frequencies in each population are used to predict population genotype frequencies. Table 22.10 lists the frequencies of alleles for three example STRP loci, *D3S1358*, *VWA*, and *FGA*. Using these frequencies, let's determine the probability that a person selected at random from the population is homozygous for the 14 allele of D3S1358 (i.e., has the 14/14 genotype), is heterozygous 15/19 for *VMA*, and is heterozygous 20/25 for *FGA*. The frequency of the 14 allele of D3S1358 is $f(14) = 0.134$ and the homozygous frequency is $f(14/14) = (0.134)(0.134) = 0.018$ (1.8%). The genotype frequency for *VMA*, $f(15/19)$, is $2[(0.119)(0.088)] = 0.0209$ (2.09%) and, $f(20/25) = 2[(0.125)(0.094)] = 0.0235$ (2.35%). Based on independent assortment of the three markers, the joint probability of the genotypes is determined by the product rule: $(0.0180)(0.0209)(0.0235) = 8.84 \times 10^{-6}$, or approximately 1 in

| Table 22.10 | Allele Frequencies for Three STRP Loci Used in CODIS | | | | |
|---|---|---|---|---|---|
| **D3S1358** | | **VWA** | | **FGA** | |
| Allele | Frequency | Allele | Frequency | Allele | Frequency |
| 12 | 0.015 | 12 | 0.015 | 18 | 0.015 |
| 13 | 0.015 | 14 | 0.131 | 19 | 0.061 |
| 14 | 0.134 | 15 | 0.119 | 20 | 0.125 |
| 15 | 0.270 | 16 | 0.186 | 21 | 0.180 |
| 16 | 0.229 | 17 | 0.257 | 22 | 0.209 |
| 17 | 0.162 | 18 | 0.189 | 23 | 0.131 |
| 18 | 0.162 | 19 | 0.088 | 24 | 0.146 |
| 19 | 0.015 | 20 | 0.015 | 25 | 0.094 |
| | | | | 26 | 0.018 |
| | | | | 27 | 0.015 |

8.84 million. Most often, all 13 CODIS markers are used, and researchers estimate that the likelihood of two unrelated people having the same genotype is very small. According to some estimates, the theoretical probability of a random match of two unrelated people is about $10^{-15}$, or about one in a quadrillion!

CODIS markers have been used in countless criminal and paternity cases since 1997. In criminal cases where the DNA fingerprint of a suspect is compared to a sample from a crime scene, the genotypes for all the markers are compared to determine if any mismatches exist—that is, to see whether the suspect carries an allele not found in the crime scene sample, or vice versa. The detection of such a mismatch results in the *exclusion* of the suspect as the source of the genetic material from the crime scene. Figure 22.16 shows an example analysis of the genes *D3S1358, VWA*, and *FGA* from a crime scene sample, the crime victim, and three suspects. Mismatches between the crime scene sample and Suspects 1 and 3 exclude these two suspects as sources of the crime scene sample. On the other hand, Suspect 2 is not excluded on the basis of these three genes. Additional CODIS genes can be analyzed, and if they do not exclude Suspect 2 as the source of the crime scene sample, the

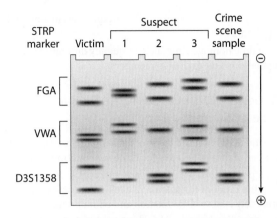

**Figure 22.16 DNA marker analysis.** CODIS STRP data produced by PCR analysis of three loci, *FGA, VWA*, and *D3S1358*. Suspects 1 and 3 are excluded as the source of crime scene DNA by mismatches at each gene. Suspect 2 is not excluded by this analysis.

probability that another person is the source of the crime scene genotype can then be calculated using the methods described.

---

SUMMARY · Mastering**GENETICS** · *For activities, animations, and review quizzes, go to the study area at www.masteringgenetics.com.*

## 22.1 The Hardy-Weinberg Equilibrium Describes the Relationship of Allele and Genotype Frequencies in Populations

- A population is a group of interbreeding organisms that share a collection of genes known as a gene pool.

- If there are two alleles at a locus, with frequencies represented by $p$ and $q$, the sum of allele frequencies $p + q = 1.0$.

- The Hardy-Weinberg equilibrium predicts that two alleles will be distributed into genotypes of frequencies $p^2$, $2pq$, and $q^2$. The sum of genotype frequencies $p^2 + 2pq + q^2 = 1.0$.

- The Hardy-Weinberg equilibrium assumes that the members of a population mate at random and that the population is not altered by any of the four evolutionary processes.

- Allele frequencies in populations can be determined by the genotype proportion method, allele counting, or the square root method.

- The Hardy-Weinberg equilibrium can be used even when more than two alleles occur for a gene.

- Chi-square analysis compares the number of observed genotypes to the number expected under assumptions of the Hardy-Weinberg equilibrium.

## 22.2 Genotype Frequencies for X-Linked Genes Are Estimated Using the Hardy-Weinberg Equilibrium

- The Hardy-Weinberg equilibrium can be used to predict X-linked genotypes and equilibrium frequencies of X-linked alleles.

## 22.3 Natural Selection Operates through Differential Reproductive Fitness within a Population

- Relative fitness is the comparative capacity of individuals with different phenotypes to make genetic contributions to the next generation due to influence of natural selection.

- A selection coefficient is the percentage decrease in reproductive success experienced by an organism possessing a relative fitness that is less than 1.0.

- Directional selection drives the frequency of the favored allele toward fixation in the population and the disfavored allele toward elimination.

- Balanced polymorphism is a stable allele frequency equilibrium resulting from natural selection that favors heterozygotes.

## 22.4 Mutation Diversifies Gene Pools

- Forward and reverse mutation slowly change the frequencies of alleles in populations.

- Deleterious mutations are removed by natural selection, striking an equilibrium frequency by balancing mutation and selection rates.

## 22.5 Migration Identifies Movement of Organisms and Genes between Populations

- Gene flow is the transfer of alleles by the migration of individuals between populations.

▮ Gene flow can produce new allele frequencies in admixed populations.

▮ Gene flow homogenizes allele frequency differences among populations that exchange members.

## 22.6 Genetic Drift Causes Allele Frequency Change by Sampling Error

▮ Genetic drift is caused by errors in sampling when a small number of individuals produce a new generation.

▮ Small populations are particularly susceptible to genetic drift, the random fluctuation of allele frequencies that will ultimately lead to allele fixation and elimination.

▮ Founder effect is a special form of genetic drift that occurs when a small number of individuals from a larger population establish a new, small population.

▮ A genetic bottleneck is a random and substantial reduction in population size that significantly changes allele frequencies among survivors.

## 22.7 Nonrandom Mating Alters Genotype Frequencies

▮ Inbreeding is nonrandom mating based on genotype that occurs between relatives who are more closely related to one another than to a random member of the population.

▮ The coefficient of inbreeding ($F$) is the probability that an allele is homozygous identical by descent in an inbred individual.

▮ Inbreeding increases the frequency of homozygosity and decreases heterozygosity.

▮ Inbreeding depression often develops in inbred populations due to the cumulative effects of numerous homozygous loci.

## 22.8 Species and Higher Taxonomic Groups Evolve by the Interplay of Four Evolutionary Processes

▮ New species emerge in reproductive isolation through adaptive change in response to conditions.

▮ Prezygotic reproductive isolation prevents mating between individuals in different populations. Postzygotic reproductive isolation reduces the ability of individuals from different populations to produce living and fertile offspring.

▮ In allopatric speciation, new species develop as a result of the physical separation of populations into different geographic areas.

▮ Sympatric speciation results from genetic differences that prevent reproduction among organisms that occupy the same habitat.

## KEYWORDS

admixed population *(p. 751)*
allele-counting method *(p. 741)*
allopatric speciation *(p. 759)*
anagenesis *(p. 758)*
balanced polymorphism *(p. 749)*
cladogenesis *(p. 758)*
coefficient of inbreeding (*F*) *(p. 755)*
differential reproductive fitness *(p. 747)*
directional natural selection *(p. 747)*
forward mutation rate (μ) *(p. 750)*
founder effect *(p. 753)*
gene pool *(p. 738)*

genetic bottleneck *(p. 754)*
genetic drift *(p. 752)*
genotype proportion method *(p. 741)*
Hardy-Weinberg (H-W) equilibrium *(p. 738)*
identical by descent (IBD) *(p. 755)*
inbreeding (consanguineous mating) *(p. 755)*
inbreeding depression *(p. 756)*
island model *(p. 751)*
migration (gene flow) *(p. 751)*
mutation *(p. 750)*
mutation–selection balance *(p. 751)*

negative assortative mating *(p. 757)*
population *(p. 738)*
population genetics *(p. 738)*
positive assortative mating *(p. 757)*
postzygotic mechanism *(p. 759)*
prezygotic mechanism *(p. 759)*
relative fitness (*w*) *(p. 747)*
reproductive isolation *(p. 758)*
reverse mutation rate (v; reversion rate) *(p. 750)*
selection coefficient (*s; t*) *(p. 747)*
square root method *(p. 742)*
sympatric speciation *(p. 761)*

## PROBLEMS

 Mastering**GENETICS**

*For instructor-assigned tutorials and problems, go to www.masteringgenetics.com.*

### Chapter Concepts

*For answers to selected even-numbered problems, see Appendix: Answers.*

1. Compare and contrast the terms in each of the following pairs:
   a. population and gene pool
   b. random mating and inbreeding
   c. natural selection and genetic drift
   d. a polymorphic trait and a polymorphic gene
   e. founder effect and genetic bottleneck

2. In a population, what is the consequence of inbreeding? Does inbreeding change allele frequencies? What is the effect of inbreeding with regard to rare recessive alleles in a population?

3. Identify and describe the evolutionary forces that can cause allele frequencies to change from one generation to the next.

4. Describe how natural selection can produce balanced polymorphism of allele frequencies through selection that favors heterozygotes.

5. Thinking creatively about evolutionary mechanisms, identify at least two schemes that could generate allelic polymorphism in a population. Do not include the processes described in the answer to Problem 4.

6. Genetic drift, an evolutionary factor affecting all populations, can have a significant effect in small populations, even though its effect is negligible in large populations. Explain why this is the case.

7. Over the course of many generations in a small population, what effect does random genetic drift have on allele frequencies?

8. Catastrophic events such as loss of habitat, famine, or overhunting can push species to the brink of extinction and result in a genetic bottleneck. What happens to allele frequencies in a species that experiences a near-extinction event, and what is expected to happen to allele frequencies if the species recovers from near extinction?

9. George Udny Yule was wrong in suggesting that an autosomal dominant trait like brachydactyly will increase in frequency in populations. Explain why Yule was incorrect.

10. The ability to taste the bitter compound phenylthiocarbamide (PTC) is an autosomal dominant trait. The inability to taste PTC is a recessive condition. In a sample of 500 people, 360 have the ability to taste PTC and 140 do not. Calculate the frequency of
    a. the recessive allele
    b. the dominant allele
    c. each genotype

11. In a population of 500 males, 40 are found to have color blindness, an X-linked recessive trait. Calculate the expected frequency of
    a. males with the recessive phenotype
    b. males with the dominant phenotype
    c. the frequencies of each female genotype

d. Use a mating table analysis to show that under Hardy-Weinberg conditions, the frequency of the dominant and recessive alleles for color blindness remain stable.

12. Biologists have proposed that the use of antibiotics to treat human infectious disease has played a role in the evolution of widespread antibiotic resistance in several bacterial species, including *Staphylococcus aureus* and the bacteria causing gonorrhea, tuberculosis, and other infectious diseases. Explain how the evolutionary mechanisms mutation and natural selection may have contributed to the development of antibiotic resistance.

13. Two populations of deer, one large one living in a mainland forest and a small one inhabiting a forest on an island, regularly exchange members who migrate across a land bridge that connects the island to the mainland.
    a. If you compared the allele frequencies in the two populations, what would you expect to find?
    b. An earthquake destroys the bridge between the island and the mainland, making migration impossible for the deer. What do you expect will happen to allele frequencies in the two populations over the following 10 generations?
    c. In which population do you expect to see the greatest allele frequency change? Why?

14. Directional selection presents an apparent paradox. By favoring one allele and disfavoring others, directional selection can lead to fixation (a frequency of 1.0) of the favored allele, after which there is no genetic variation at the locus, and its evolution stops. Explain why directional selection no longer operates in populations after the favored allele reaches fixation.

15. What is inbreeding depression? Why is inbreeding depression a serious concern for animal biologists involved in species-conservation breeding programs?

16. Certain animal species, such as the black-footed ferret, are nearly extinct and currently exist only in captive populations. Other species, such as the panda, are also threatened but exist in the wild thanks to intensive captive-breeding programs. What strategies would you suggest in the case of black-footed ferrets and in the case of pandas to monitor and minimize inbreeding depression?

## Application and Integration

*For answers to selected even-numbered problems, see Appendix: Answers.*

17. Genetic Analysis 22.1 (page 743) predicts the number of individuals expected to have the blood group genotypes *MM*, *MN*, and *NN*. Perform a chi-square analysis using the number of people observed and expected in each blood-type category, and state whether the sample is in Hardy-Weinberg equilibrium (See pages 48 and 49 for the chi-square formula and table).

18. In a population of rabbits, $f(C_1) = 0.70$ and $f(C_2) = 0.30$. The alleles exhibit an incomplete dominance relationship in which $C_1C_1$ produces black rabbits, $C_1C_2$ tan-colored rabbits, and $C_2C_2$ rabbits with white fur. If the assumptions of the Hardy-Weinberg principle apply to the rabbit population, what are the expected frequencies of black, tan, and white rabbits?

19. Sickle cell disease (SCD) is found in numerous populations whose ancestral homes are in the malaria belt of Africa and Asia. SCD is an autosomal recessive disorder that results from homozygosity for a mutant β-globin gene allele. Data on one affected population indicates that approximately 8 in 100 newborn infants have SCD.
    a. What are the frequencies of the wild-type ($\beta^A$) and mutant ($\beta^S$) alleles in this population?
    b. What is the frequency of carriers of SCD in the population?

20. Epidemiologic data on the population in the previous problem reveal that before the application of modern medical treatment, natural selection played a major role in shaping the frequencies of alleles. Heterozygous individuals have the highest relative fitness; and in comparison to

heterozygotes, those who are $\beta^A \beta^A$ have a relative fitness of 82 percent, but only about 32 percent of those with SCD survived to reproduce. What are the estimated equilibrium frequencies of $\beta^A$ and $\beta^S$ in this population?

21. The frequency of tasters and nontasters of PTC (see Problem 10) varies among populations. In population A, 64 percent of people are tasters (an autosomal dominant trait) and 36 percent are nontasters. In population B, tasters are 75 percent and nontasters 25 percent. In population C, tasters are 91 percent and nontasters are 9 percent.

    a. Calculate the frequency of the dominant ($T$) allele for PTC tasting and the recessive ($t$) allele for nontasting in each population.

    b. Assuming that Hardy-Weinberg conditions apply, determine the genotype frequencies in each population.

22. Tay-Sachs disease is an autosomal recessive neurological disorder that is fatal in infancy. Despite its invariably lethal effect, Tay-Sachs disease occurs at very high frequency in some Central and Eastern European (Ashkenazi) Jewish populations. In certain Ashkenazi populations, 1 in 750 infants has Tay-Sachs disease. Population biologists believe the high frequency is a consequence of genetic bottlenecks caused by pogroms (genocide) that have reduced the population multiple times in the last several hundred years.

    a. What is a genetic bottleneck?

    b. Explain how a genetic bottleneck and its aftermath could result in a population that carries a lethal allele in high frequency.

    c. In the population described, what is the frequency of the recessive allele that produces Tay-Sachs disease?

    d. Assuming mating occurs at random in this population, what is the probability a couple are both carriers of Tay-Sachs disease?

23. Cystic fibrosis (CF) is the most common autosomal recessive disorder in certain Caucasian populations. In some populations, approximately 1 in 2000 children have CF. Determine the frequency of CF carriers in this population.

24. Hemophilia A is a blood-clotting disorder caused by a loss-of-function mutation of the factor VIII gene located on the X chromosome. Males with a factor VIII mutation have hemophilia, as do females who are homozygous for a mutation. Worldwide, approximately 1 in 2000 males have hemophilia.

    a. Determine the frequency of the wild-type and mutant factor VIII alleles in humans.

    b. What is the frequency of female hemophiliacs? What is the frequency of female hemophilia carriers?

    c. Explain why the term *hemophilia carrier* is applied to females but not to males.

    d. Among 1 million people, what is the number of affected males? Affected females? Female carriers?

25. In a population of flowers growing in a meadow, $C_1$ and $C_2$ are autosomal codominant alleles that control flower color. The alleles are polymorphic in the population, with $f(C_1) = 0.80$ and $f(C_2) = 0.20$. Flowers that are $C_1C_1$ are yellow, orange flowers are $C_1C_2$, and $C_2C_2$ flowers are red. A storm blows a new species of hungry insects into the meadow, and they begin to eat yellow and orange flowers but not red flowers. The predation exerts strong natural selection on

the flower population, resulting in relative fitness values of $C_1C_1 = 0.30$, $C_1C_2 = 0.60$, and $C_2C_2 = 1.0$.

    a. Assuming the population begins in H-W equilibrium, what are the allele frequencies after one generation of natural selection?

    b. Assuming random mating takes place among survivors, what are the genotype frequencies in the second generation?

    c. If predation continues, what are the allele frequencies when the second generation mates?

    d. What are the equilibrium frequencies of $C_1$ and $C_2$ if predation continues?

26. Assume that the flower population described in the previous problem undergoes a different pattern of predation. Flower color determination and the starting frequencies of $C_1$ and $C_2$ are as described above, but the new insects attack yellow and red flowers, not orange flowers. As a result of the predation pattern, the relative fitness values are $C_1C_1 = 0.40$, $C_1C_2 = 1.0$, and $C_2C_2 = 0.80$.

    a. What are the allele frequencies after one generation of natural selection?

    b. What are the genotype frequencies among the progeny of predation survivors?

    c. What are the equilibrium allele frequencies in the predation environment?

27. ABO blood type is examined in a Taiwanese population, and allele frequencies are determined. In the population, $f(I^A) = 0.30$, $f(I^B) = 0.15$, and $f(i) = 0.55$. Assuming Hardy-Weinberg conditions apply, what are the frequencies of genotypes, and what are the blood group frequencies in this population?

28. A total of 1000 members of a Central American population are typed for the ABO blood group. In the sample, 421 have blood type A, 168 have blood type B, 336 have blood type O, and 75 have blood type AB. Use this information to determine the frequency of ABO blood group alleles in the sample.

29. A sample of 500 field mice contains 225 individuals that are $D_1D_1$, 175 that are $D_1D_2$, and 100 that are $D_2D_2$.

    a. What are the frequencies of $D_1$ and $D_2$ in this sample?

    b. Is this population in Hardy-Weinberg equilibrium? Use the chi-square test to justify your answer.

    c. Is inbreeding a possible genetic explanation for the observed distribution of genotypes? Why or why not?

30. In humans the presence of chin and cheek dimples is dominant to the absence of dimples, and the ability to taste the compound PTC is dominant to the inability to taste the compound. Both traits are autosomal, and they are unlinked. The frequencies of alleles for dimples are $D = 0.62$ and $d = 0.38$. For tasting, the allele frequencies are $T = 0.76$ and $t = 0.24$.

    a. Determine the frequency of genotypes for each gene and the frequency of each phenotype.

    b. What are the expected frequencies of the four possible phenotype combinations: dimpled tasters, undimpled tasters, dimpled nontasters, and undimpled nontasters?

31. Albinism, an autosomal recessive trait characterized by an absence of skin pigmentation, is found in 1 in 4000 people in populations at equilibrium. Brachydactyly, an

autosomal dominant trait producing shortened fingers and toes, is found in 1 in 6000 people in populations at equilibrium. For each of these traits, calculate the frequency of

a. the recessive allele at the locus
b. the dominant allele at the locus
c. heterozygotes in the population
d. For albinism only, what is the frequency of mating between heterozygotes?

32. Using the population data in Table 22.10,

a. Calculate the population frequency of individuals with the 16/18 genotype at *D3S1358*, the 14/18 genotype at *VWA*, and the 23/26 genotype at *FGA*.
b. Explain how the Hardy-Weinberg principles are used in the analysis of CODIS genotypes and other STRP-locus comparisons.

33. Evaluate the following pedigree, and answer the questions below for individual IV-1.

a. Is IV-1 an inbred individual? If so, who is/are the common ancestor(s)?
b. What is *F* for this individual?

34. Evaluate the following pedigree, and answer the questions below.

a. Which individual(s) in this family is/are inbred?
b. Who is/are the common ancestor(s) of the inbred individual(s)?
c. Calculate *F* for any inbred members of this family.

35. The following is a partial pedigree of the British royal family. The family contains several inbred individuals and

a number of inbreeding pathways. Carefully evaluate the pedigree, and identify the pathways and common ancestors that produce inbred individuals A (Alice in generation IV), B (George VI in generation VI), and C (Charles in generation VIII).

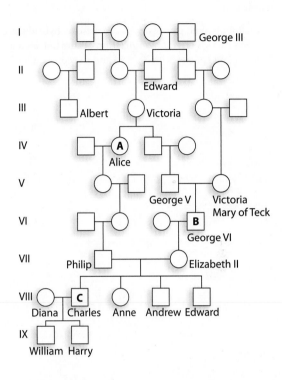

36. Draw a separate pedigree, identifying the inbred individuals and the inbreeding pathways, for each of the following inbreeding coefficients:

a. $F = 4(1/2)^6$
b. $F = 2(1/2)^5$
c. $F = 4(1/2)^8$
d. $F = 2(1/2)^7$

37. The human melanocortin 1 receptor (*MC1R*) gene plays a major role in producing eumelanin, a black-brown pigment that helps determine hair color and skin color. Jonathan Rees and several colleagues (J. L. Rees et al., *Am. J. Human Genet.* 66(2000): 1351–1361) studied multiple *MC1R* alleles in African and European populations. Although this research found several *MC1R* alleles in African populations, *MC1R* alleles that decrease the production of eumelanin were rare. In contrast, several alleles decreasing eumelanin production were found in European populations. How can these results be explained by natural selection?

38. Achromatopsia is a rare autosomal recessive form of complete color blindness that affects about 1 in 20,000 people in most populations. People with this disorder see only in black and white and have extreme sensitivity to light and poor visual acuity. On Pingelap Island, one of a cluster of coral atoll islands in the Federated States of Micronesia, approximately 10 percent of the 3000 indigenous Pingelapese inhabitants have achromatopsia.

Achromatopsia was first recorded on Pingelap in the mid-1800s, about four generations after a typhoon devastated Pingelap and reduced the island population to about 20 people. All Pingelapese with achromatopsia trace their ancestry to one male who was one of the 20 typhoon survivors. Provide a genetic explanation for the origin of achromatopsia on Pingelap, and explain the most likely evolutionary model for the high frequency there of achromatopsia.

39. New allopolyploid plant species can arise by hybridization between two species. If hybridization occurs between a diploid plant species with $2n = 14$ and a second diploid species with $2n = 22$, the new allopolyploid would have 36 chromosomes.

a. Is it likely that sexual reproduction between the allopolyploid species and either of its diploid ancestors would yield fertile progeny? Why or why not?

b. What type of isolation mechanism is most likely to prevent hybridization between the allopolyploid and the diploid species?

c. What pattern of speciation is illustrated by the development of the allopolyploid species?

# References and Additional Reading

## Chapter 1  The Molecular Basis of Heredity, Variation, and Evolution

Chargaff, E. 1951. Structure and function of nucleic acids as cell constituents. *Fed. Proc.* 10: 654–59.

Dronamraju, K. 1992. Profiles in genetics: Archibald E. Garrod. *Am. J. Hum. Genet.* 51: 216–19.

Dunn, L. C. 1965. *A Short History of Genetics*. New York: McGraw–Hill.

Garrod, A. E. 1902. The incidence of alkaptonuria: A study in chemical individuality. *Lancet* ii: 1616–20.

——— 1909. *Inborn Errors of Metabolism*. London: Frowde, Hodder and Stoughton.

Judson, H. F. 1978. *The Eighth Day of Creation: Makers of the Revolution in Biology*. Woodbury, NY: Cold Spring Harbor Press.

Lander, E. S., and R. A. Weinberg. 2000. Genomics: Journey to the center of biology. *Science* 287: 1777–82.

Organ, C. L., et al. 2007. Origin of avian genome size and structure in non-avian dinosaurs. *Nature* 446: 180–84.

Ridley, M. 1999. *Genome: The Autobiography of a Species in 23 Chapters*. New York: Perennial.

Schweitzer, M. H., et al. 2007. Analyses of soft tissue *Tyrannosaurus rex* suggests the presence of protein. *Science* 316: 277–80.

Sturtevant, A. H. 1965. *A History of Genetics*. New York: Harper and Row.

Watson, J. D. 1968. *The Double Helix*. New York: Atheneum.

Watson, J. D., and F. H. C. Crick. 1953. Genetical implications of the structure of deoxyribonucleic acid. *Nature* 171: 964–69.

——— 1953. Molecular structure of nucleic acids: A structure for deoxyribose nucleic acid. *Nature* 171: 737–38.

## Chapter 2  Transmission Genetics

Armstead, I. I., et al. 2007. Cross-species identification of Mendel's *I* locus. *Science* 315: 73.

Aubry, S., J. Mani, and S. Hörtensteiner. 2008. Stay-green protein, defective in Mendel's green cotyledon mutant, acts independently and upstream of pheophorbide *a* oxidase in the chlorophyll catabolic pathway. *Plant Mol. Biol.* 67: 243–56.

Bateson, W. 1913. *Mendel's Principles of Heredity*. Cambridge: Cambridge University Press.

Bhattacharyya, M., C. Martin, and A. Smith. 1993. The importance of starch biosynthesis in the wrinkled shape character of peas studied by Mendel. *Plant Mol. Biol.* 22: 525–31.

Bhattacharyya, M., et al. 1990. The wrinkled-seed character of pea described by Mendel is caused by a transposon-like insertion in a gene coding starch-branching enzyme. *Cell* 60: 115–22.

Bennett, R. L., et al. 1995. Recommendations for standardized human pedigree nomenclature. *Am. J. Hum. Genet.* 56: 745–52.

Detlefsen, J. A. 1918. Fluctuations of sampling in a Mendelian population. *Genetics* 3: 599–607.

Fisher, R. A. 1936. Has Mendel's work been rediscovered? *Ann. Science* 1: 115–37.

Hartl, D., and V. Orel. 1992. What did Gregor Mendel think he discovered? *Genetics* 131: 245–53.

Hellens, R. P., et al. 2010. Identification of Mendel's white flower character. *PLoS ONE* 5: 1–7.

Henig, R. M. 2001. *A Monk in the Garden: The Lost and Found Genius of Gregor Mendel, the Father of Genetics*. New York: Houghton-Mifflin.

Lester, D. R., et al. 1997. Mendel's stem length gene (*Le*) encodes a gibberellin 3β-hydroxylase. *Plant Cell* 9: 1435–43.

Martin, D. N., W. M. Proebsting, and P. Hedden. 1997. Mendel's dwarfing gene: cDNAs from the *Le* alleles and function of the expressed proteins. *Proc. Natl. Acad. Sci. USA* 94: 8907–11.

Mendel, G. (1866) 1966. Experiments in plant hybridization. In *The Origins of Genetics: A Mendel Source Book*, edited by C. Stern and E. Sherwood. Translated. San Francisco: W. H. Freeman.

Olby, R. C. 1985. *Origins of Mendelism*. London: Constable.

Orel, V. 1996. *Gregor Mendel: The First Geneticist*. Oxford: Oxford University Press.

Peters, J. (ed.). 1959. *Classic Papers in Genetics*. Englewood Cliffs NJ: Prentice-Hall.

Stern, C., and E. Sherwood (eds.). 1966. *The Origins of Genetics: A Mendel Source Book*. San Francisco: W. H. Freeman.

Stubbe, H. 1972. *History of Genetics: From Prehistoric Times to the Rediscovery of Mendel's Laws*. Cambridge, MA: MIT Press.

Tschermak-Seysenegg, E. 1951. The rediscovery of Mendel's work. *J. Hered.* 42: 162–72.

Welling, F. 1991. Historical study: Johann Gregor Mendel 1822–1884. *Am. J. Med. Genet.* 40: 1–25.

White, O. E. 1917. Studies of inheritance in *Pisum*. II. The present state of knowledge of heredity and variation in peas. *Proc. Am. Phil. Soc.* 56: 487–88.

## Chapter 3 Cell Division and Chromosome Heredity

Barr, M. L. 1960. Sexual dimorphism in interphase nuclei. *Am. J. Hum. Genet.* 12: 118–27.

Bridges, C. B. 1916. Nondisjunction as proof of the chromosome theory of heredity. *Genetics* 1: 1–52 and 107–63.

Gonzalez, A. N., et al. 2008. A shared enhancer controls a temporal switch between promoters during Drosophila primary sex determination. *Proc. Natl. Acad. Sci. USA* 105:18436–18441.

Gould, K. L., and P. Nurse. 1989. Tyrosine phosphorylation of the fission yeast cdc2+ protein kinase regulates entry into mitosis. *Nature* 342: 39–45.

Hartwell, L. H. 1991. Twenty-five years of cell cycle genetics. *Genetics* 129: 975–80.

Hartwell, L. H., et al. 1974. Genetic control of cell division cycle in yeast. *Science* 183: 46–51.

Hartwell, L. H., J. Culotti, and B. J. Reid. 1970. Genetic control of cell division in yeast. I. Detection of mutants. *Proc. Natl. Acad. Sci. USA* 66: 352–59.

Hartwell, L. H., and M. W. Unger. 1977. Unequal division in *Saccharomyces cerevisiae* and its implications for the control of cell division. *J. Biol. Chem.* 75: 422–35.

Hodgkin, J. 1989. *Drosophila* sex determination: Cascade of regulated splicing. *Cell* 56: 905–6.

Hunt, D. M., et al.1995. The chemistry of John Dalton's color blindness. *Science* 267: 984–88.

Hunt, T., and M. W. Kirschner. 1993. Cell manipulation. *Current Opinions in Cell Biol.* 5: 163–65.

Johnson, R. T., and P. N. Rao. 1970. Mammalian cell fusion: Induction of premature chromosome condensation in interphase nuclei. *Nature* 226: 717–22.

Koopman, P., et al. 1991. Male development of chromosomally female mice transgenic for *Sry. Nature* 351: 117–21.

Lyon, M. F. 1962. Sex chromatin and gene action in the mammalian X-chromosome. *Am. J. Hum. Genet.* 14: 135–48.

Masui, Y., and C. L. Markert. 1971. Cytoplasmic control of nuclear behavior during meiotic maturation of frog oocytes. *J. Exp. Zool.* 177: 129–45.

Morgan, T. H. 1910. Sex-limited inheritance in *Drosophila. Science* 32: 120–22.

Murray, A. W., and M. W. Kirschner. 1989. Cyclin synthesis drives the early embryonic cell cycle. *Nature* 339: 275–80.

Nathans, J., et al. 1986. Molecular genetics of inherited variation in human color vision. *Science* 232: 203–10.

Nathans, J., D. Thomas, and D. S. Hogness. 1986. Molecular genetics of human color vision: The genes encoding blue, green, and red pigments. *Science* 232: 193–202.

Page, D. C., A. de la Chapelle, and J. Weissenbach. 1985. Chromosome Y-specific DNA in human XX males. *Nature* 315: 224–26.

Page, D. C., et al. 1987. The sex-determining region of the Y chromosome encodes a finger protein. *Cell* 51: 1091–1104.

Rao, P. N., and R. T. Johnson. 1970. Mammalian cell fusion studies on the regulation of DNA synthesis and mitosis. *Nature* 225: 159–64.

Salz, H. K., and J. W. Erickson. 2010. Sex determination in Drosophila: The view from the top. *Fly* 4:60–70.

Stevens, N. M. 1905. Studies in spermatogenesis with especial reference to the "accessory chromosome." Washington, DC: Carnegie Institute of Washington, Publication No. 36.

Willard, H. F. 1996. X chromosome inactivation, *XIST*, and pursuit of the X-inactivation center. *Cell* 86: 5–7.

Wilson, E. B. 1895. *An Atlas of the Fertilization and Karyokinesis of the Ovum*. New York: Columbia University Press.

## Chapter 4 Gene Interaction

Beadle, G. W., and E. L. Tatum. 1941. Genetic control of biochemical reactions in *Neurospora*. *Proc. Natl. Acad. Sci. USA* 27: 499–506.

Bultman, S. J., E. J. Michaud, and R. P. Woychik. 1992. Molecular characterization of the mouse *agouti* locus. *Cell* 71: 1195–1204.

Duhl, D. M. J., et al. 1994. Pleiotropic effects of the mouse *lethal yellow* ($A^y$) mutation explained by detection of a maternally expressed gene and the simultaneous production of *agouti* fusion RNAs. *Development* 120: 1695–1708.

Flatt, T., M-P. Tu, and M. Tatar. 2005. Hormonal pleiotropy and the juvenile hormone regulation of *Drosophila* development and life history. *BioEssays* 27: 999–1010.

Garrod, A. E. 1902. The incidence of alkaptonuria: A study in chemical individuality. *Lancet* 2: 1616–20.

———. (1909) 1963. *Inborn Errors of Metabolism*. London: Oxford University Press.

Garrod, S. C. 1989. Family influences on A. E. Garrod's thinking. *J. Inherit. Metabol. Dis.* 12: 2–8.

Jackson, I. J. 1994. Molecular and developmental genetics of mouse coat color. *Ann Rev. Genet.* 28: 189–217.

Landsteiner, K., and P. Levine. 1927. Further observations on individual differences of blood group. *Proc. Soc. Exp. Biol. Med.* 24: 941–42.

Michaud, E. J., et al. 1994. A molecular model for the genetic and phenotypic characteristics of the mouse lethal yellow (Ay) mutation. *Proc. Natl. Acad. Sci. USA* 91:2562–2566.

Phillips, P. C. 1998. The language of gene interaction. *Genetics* 149: 1167–71.

Race, R. R., and R. Sanger. 1975. *Blood Groups in Man*. 6th ed. Cambridge: Oxford University Press.

Srb, A. M., and N. H. Horowitz. 1944. The ornithine cycle in *Neurospora* and its genetic control. *J. Biol. Chem.* 154: 129–39.

Siracusa, L. D. 1994. The *agouti* gene: Turned on to yellow. *Trends Genet.* 10: 423–28.

Yamamoto, F., et al. 1990. Molecular genetic basis of the histo-blood group ABO system. *Nature* 345: 229–33.

## Chapter 5 Genetic Linkage and Mapping in Eukaryotes

Bateson, W., E. R. Saunders, and R. C. Punnett. 1905. Experimental studies in the physiology of heredity. *Rep. Evol. Committee Royal Soc.* II: 1–55, 80–99.

Bregger, T. 1918. Linkage in maize: The *C* aleurone factor and waxy endosperm. *Am. Nat.* 52: 57–61.

Bridges, C. B., and T. M. Olbrycht. 1926. The multiple stock "Xple" and its use. *Genetics* 11: 41–56.

Creighton, H. B., and B. McClintock. 1931. A correlation of cytological and genetic crossing over in *Zea mays*. *Proc. Natl. Acad. Sci. USA* 17: 492–97.

The Huntington's Disease Collaborative Research Group (58 authors). 1993. A novel gene containing a trinucleotide repeat that is expanded and unstable on Huntington's disease chromosomes. *Cell* 72: 971–83.

Janssens, F. A. 1909. La theorie de la chiasmatypie. *La Cellule* 25: 389–411.

——— 1909. Spermatogenese dans la Batraciens. V. La theorie de la chiasmatypie. Nouvelles interpretation des cineses de maturation. *Cellule* 25: 387–411.

Green, M. M., and K. C. Green. 1949. Crossing-over between alleles at the Lozenge locus in *Drosophila melanogaster*. *Proc. Natl. Acad. Sci. USA* 35: 596–91.

Gusella, J. F., et al. 1983. A polymorphic DNA marker genetically linked to Huntington's disease. *Nature* 306: 234–38.

Hall, J. M., et al. 1990. Linkage of early-onset familial breast cancer to chromosome 17q21. *Science* 250: 1684–89.

——— 1992. Closing in on a breast cancer gene on chromosome 17q. *Am. J. Hum. Genet.* 50: 1235–42.

Houlahan, M. B., G. W. Beadle, and H. G. Calhoun. 1949. Linkage studies with biochemical mutants of *Neurospora crassa*. *Genetics* 34: 493–507.

Ikeda, Y., et al. 2006. Spectrin mutations cause spinocerebellar ataxia type 5. *Nat. Genet.* 38: 184–90.

Knudson, A. G. Jr. 1971. Mutation and cancer: Statistical study of retinoblastoma. *Proc. Natl. Acad. Sci. USA* 68: 820–23.

Lindegren, C. C. 1933. The genetics of *Neurospora*. III. Pure-bred stocks and crossing over in *N. crassa*. *Bull. Torrey Bot. Club* 60: 133–54.

Morgan, T. H. 1910. Sex-limited inheritance in *Drosophila*. *Science* 32: 120–22.

——— 1910. The method of inheritance of two sex-limited characters in the same animal. *Proc. Soc. Exp. Biol. Med.* 8: 17.

——— 1911. An attempt to analyze the constitution of the chromosomes on the basis of sex-limited inheritance in *Drosophila*. *J. Exp. Zool.* 11: 365–414.

——— 1911. Random segregation versus coupling in Mendelian inheritance. *Science* 34: 384.

Morgan, T. H., et al. 1915. *The Mechanism of Mendelian Heredity*. New York: Henry Holt.

Morton, N. E. 1955. Sequential tests for the detection of linkage. *Am. J. Hum. Genet.* 7: 277–318.

Stern, C. 1931. Zytologisch-genetische untersuchungen als beweise fur die Morgansche theorie des fakorenaustauchs. *Biol. Zentrabl.* 51: 547–87.

Stern, C., and D. Doan. 1936. A cytogenetic demonstration of crossing-over between X- and Y-chromosomes in the male of *Drosophila melanogaster*. *Proc. Natl. Acad. Sci. USA* 22: 649–54.

Strachan, T., and A. P. Read. 2004. *Human Molecular Genetics*. 3rd ed. London and New York: Garland Science.

Sturtevant, A. H. 1913. The linear arrangement of six sex-linked factors in *Drosophila* as shown by their mode of association. *J. Exp. Zool.*, 14: 43–59.

Weber, J. L., et al. 1993. Evidence for human meiotic recombination interference obtained through construction of a short tandem repeat polymorphism linkage map of chromosome 19. *Am. J. Hum. Genet.* 53:1079–1095.

Wexler, Alice. 1995. *Mapping Fate: A Memoir of Family, Risk, and Genetic Research*. New York: Times Books, Random House.

## Chapter 6 Gene Analysis and Mapping in Bacteria and Bacteriophage

Bachmann, B. J. 1990. Linkage map of *Escherichia coli* K-12, Edition 8. *Microbiol. Rev.* 54: 130–97.

Benzer, S. 1959. On the topology of the genetic fine structure. *Proc. Natl. Acad. Sci. USA* 45: 1607–20.

——— 1961. On the topology of the genetic fine structure. *Proc. Natl. Acad. Sci. USA*, 47: 403–16.

Blattner, F. R., G. Plunkett III, C. A. Bloch. 1997. The complete genome sequence of *Escherichia coli* K-12. *Science* 277: 1453–62.

Curtiss, R. 1969. Bacterial conjugation. *Ann. Rev. Microbiol.* 23: 69–123.

Davis, B. D. 1950. Nonfiltrability of the agents of recombination. *J. Bacteriol.*, 60: 507–8.

Hayes, W. 1953. Observations on a transmissible agent determining sexual differentiation in Bact. Coli. *J. Gen. Microbiol.* 8: 72–88.

Hotchkiss, R. D., and M. Gabor. 1970. Bacterial transformation with special reference to recombination processes. *Ann. Rev. Genet.* 4: 193–224.

Lederberg, J. 1986. Forty years of genetic recombination in bacteria: A fortieth anniversary reminiscence. *Genetics* 28: 491–511.

Lederberg, J., and E. L. Tatum. 1946. Gene recombination in *Escherichia coli*. *Nature* 158: 558–59.

Stent, G. S. 1963. *Molecular Biology of Bacterial Viruses*. San Francisco: W. H. Freeman.

Susman, M. 1970. General bacterial genetics. *Ann. Rev. Genet.* 4: 135–76.

Wollman, E. L., F. Jacob, and W. Hayes. 1962. Conjugation and genetic recombination in *E. coli* K-12. *Cold Spring Harbor Symp. Quant. Biol.* 21: 141–62.

Yanofsky, C., and E. S. Lennox. 1959. Transduction and recombination study of linkage relationships among the genes controlling tryptophan synthesis in *Escherichia coli*. *Virology* 8: 425–47.

Zinder, N. D. 1958. Transduction in bacteria. *Sci. Am.* 199: 38–46.

Zinder, N. D., and J. L. Lederberg. 1952. Genetic exchange in *Salmonella*. *J. Bacteriol.*, 64: 679–99.

## Chapter 7 DNA Structure and Replication

Avery, O. T., C. M. Macleod, and M. McCarty. 1944. Studies on the chemical nature of the substance inducing transformation of pneumococcal types: Induction of transformation by a desoxyribonucleic acid fraction isolated from pneumococcus type III. *J. Exp. Med.* 79: 137–58.

Blackburn, E. H. 1991. Structure and function of telomeres. *Nature* 350: 569–73.

Cairns, J. 1963. The bacterial chromosome and its manner of replication as seen by autoradiography. *J. Mol. Biol.* 6: 208–13.

DeLucia, P., and J. Cairns. 1969. Isolation of an *E. coli* strain with a mutation affecting DNA polymerase. *Nature* 224: 1164–66.

Georgescu, R. E., et al. 2007. Structure of a sliding clamp on DNA. *Cell* 132:43–54.

Greider, C. W., and E. H. Blackburn. 1987. The telomere terminal transferase of *Tetrahymena* is a ribonucleoprotein enzyme with two kinds of primer specificity. *Cell* 51: 887–98.

Griffith, F. 1928. The significance of pneumococcal types. *J. Hyg.* 27: 113–59.

Hanahan, D., and R. Weinberg. 2000. The hallmarks of cancer. *Cell* 100: 57–70.

Harley, C. B., A. B. Futcher, and C.W. Greider. 1990. Telomeres shorten during ageing of human fibroblasts. *Nature* 345: 458–60.

Hayflick, L., and P. S. Moorhead. 1961. The serial cultivation of human diploid cell strains. *Exp. Cell. Res.* 25: 585–621.

Hershey, A. D., and M. Chase. 1952. Independent function of viral protein and nucleic acid in growth of bacteriophage. *J. Genet. Phys.* 36: 39–56.

Huberman, J. A., and A. D. Riggs. 1968. On the mechanism of DNA replication in mammalian chromosomes. *J. Mol. Biol.* 32: 327–41.

Huberman, J. A., and A. Tsai. 1973. Direction of DNA replication in mammalian cells. *J. Mol. Biol.* 75: 5–12.

Huntington's Disease Collaborative Research Group. 1993. A novel gene containing a trinucleotide repeat that is expanded and unstable on Huntington's disease chromosomes. *Cell* 72: 971–83.

Kornberg, A. 1960. Biological synthesis of DNA. *Science* 131: 1503–8.

Lemon, K. P., and A. D. Grossman. 2000. Movement of replicating DNA through a stationary replisome. *Mol. Cell* 6: 1321–30.

Meselson, M., and F. W. Stahl. 1958. The replication of DNA in *Escherichia coli*. *Proc. Natl. Acad. Sci. USA* 44: 671–82.

O'Donnel, M., and J. Kuriyan. 2006. Clamp loaders and replication initiation. *Curr. Opin. Struct. Biol.* 16: 405–15.

Ogawa, T., and R. Okazaki. 1980. Discontinuous DNA replication. *Ann. Rev. Biochem.* 49: 421–57.

Rodriguez, R. L., M. S. Dalbey, and C. I. Davern. 1973. Autoradiographic evidence for bidirectional DNA replication in *Escherichia coli*. *J. Molec. Biol.* 74: 599–604.

Steitz, T. A. 1998 A mechanism for all polymerases. *Nature* 391: 231–32.

——— 2006. Visualizing polynucleotide polymerase machines at work. *EMBO J.* 25: 3458–68.

## Chapter 8  Molecular Biology of Transcription and RNA Processing

Berget, S. M., C. Moore, and P. Sharp. 1977. Spliced segments at the 5′ terminus of adenovirus 2 late mRNA. *Proc. Natl. Acad. Sci. USA* 74: 3171–75.

Bogenhagen, D. F., S. Sakonju, and D. D. Brown. 1980. A control region in the center of the 5S RNA gene directs specific initiation of transcription: II. the 3′ border of the region. *Cell* 19: 27–35.

Brenner, S., F. Jacob, and, M. Meselson. 1961. An unstable intermediate carrying information from genes to ribosomes for protein synthesis. *Nature* 190: 575–80.

Cech, T. 1987. The chemistry of self-splicing RNA and RNA enzymes. *Science* 236: 1532–39.

Chambon, P. 1981. Split genes. *Sci. Am.* 244: 60–71.

Cramer, P., et al. 2000. Architecture of RNA polymerase II and implications for the transcription mechanism. *Science* 288: 640–49.

Darnell, J. E. 1983. The processing of RNA. *Sci. Am.* 249: 90–100.

Dugaiczyk, A., et al. 1978. The natural ovalbumin gene contains seven intervening sequences. *Nature* 274: 328–33.

Hamkalo, B. 1985. Visualizing transcription in chromosomes. *Trends. Genet.* 1: 255–60.

Kersanach, R., et al. 1994. Five identical intron positions in ancient duplicated genes of eubacterial origin. *Nature* 367: 387–89.

Kim, M., et al. 2006. Distinct pathways for snoRNA and mRNA termination. *Mol. Cell* 24: 723–34.

Lees-Miller, J. P., L. O. Goodwin, and D. M. Helfman. 1990. Three novel brain tropomyosin isoforms are expressed from the rat α-tropomyosin gene through the use of alternative promoters and alternative RNA processing. *Mol. Cell. Biol.* 10: 1729–42.

Logsdon, J. M. Jr., et al. 1995. Seven newly discovered intron positions in the triose-phosphate isomerase gene: Evidence for the intron-late theory. *Proc. Natl. Acad. Sci. USA* 92: 8507–11.

Maniatis, T., S. Goodbourn, and J. A. Fischer. 1987. Regulation of inducible and tissue-specific gene expression. *Science* 236: 1237–45.

Maniatis, T., and R. Reed 2002. An extensive network of coupling among gene expression machines. *Nature* 416: 499–506.

Myers, R. M., K. Tilly, and T. Maniatis. 1986. Fine structure genetic analysis of the β-globin promoter. *Science* 232: 613–18.

Pribnow, D. 1975. Nucleotide sequence of an RNA binding site at an early T7 promoter. *Proc. Natl. Acad. Sci. USA* 72: 784–88.

Reed, R. 2003. Coupling transcription, splicing, and mRNA export. *Curr. Opin. Cell Biol.* 15: 326–31.

Reed, R., and T. Maniatis. 1985. Intron sequences involved in lariat formation during pre-mRNA splicing. *Cell* 41: 95–105.

Sakonju, S., D. F. Bogenhagen, and D. D. Brown. 1980. A control region in the center of the 5S RNA gene directs specific initiation of transcription: I. the 5′ border of the region. *Cell* 19: 13–25.

Sharp, P. 1994. Nobel Lecture: Split genes and RNA splicing. *Cell* 77: 805–15.

Sudhof, T. C., et al. 1985. Cassette of eight exons shared by genes for LDL and EGF precursors. *Science* 228: 893–95.

Tilghman, S., et al. 1978. The intervening sequence of a mouse beta-globin gene is traced to the 15S beta-globin mRNA precursor. *Proc. Natl. Acad. Sci. USA* 75: 1309–13.

## Chapter 9  The Molecular Biology of Translation

Blobel, G., and B. Dobberstein. 1975. Transfer of proteins across membranes. I. Presence of proteolytically processed and unprocessed nascent immunoglobulin light chains on membrane-bound ribosomes of murine myeloma. *J. Cell. Biol.* 67: 835–51.

Brenner, S., F. Jacob, and M. Meselson. 1969. An unstable intermediate carrying information from genes to ribosomes for protein synthesis. *Nature* 190: 576–81.

Carrell, R. W., and D. A. Lomas. 2002. Alpha-1 antitrypsin deficiency—a model conformational disease. *N. Engl. J. Med.*, 346: 45–53.

Cech, T. R. 2000. The ribosome is a ribozyme. *Science* 289: 878–79.

Chapeville, F. F., et al. 1962. On the role of soluble ribonucleic acid in coding for nucleic acids. *Proc. Natl. Acad. Sci. USA* 48: 1086–92.

Crick, F. 1966. Codon-anticodon pairing: The wobble hypothesis. *J. Mol. Biol.* 19: 548–55.

Crick, F., et al. 1961. General nature of the genetic code for proteins. *Nature* 192: 1227–32.

Fox, T. D. 1987. Natural variation in the genetic code. *Ann. Rev. Genet.* 21: 67–91.

Johnson, A. W., E. Lund, and J. Dahlberg. 2002. Nuclear export of ribosomal subunits. *Trends Biochem. Sci.* 27: 850–57.

Khorana, H. G., et al. 1967. Polynucleotide synthesis and the genetic code. *Cold Spring Harbor Symp.* 31: 39–49.

Kloc, M., N. R. Zearfoss, and L. D. Etkin. 2002. Mechanisms of subcellular mRNA localization. *Cell* 108: 533–44.

Kozak, M. 1978. How do eukaryotic ribosomes select initiation regions in mRNA? *Cell* 15: 1109–23.

——— 1987. An analysis of 5′-noncoding sequences from 699 vertebrate messenger RNAs. *Nuc. Acid Res.* 15: 8125–48.

Nirenberg, M. W., and P. Leder. 1964. RNA code words and protein synthesis. I. The effects of trinucleotides upon the binding of sRNA to ribosomes. *Science* 145: 1399–1407.

Nirenberg, M. W., and J. H. Matthaei. 1961. The dependence of cell-free protein synthesis in *E. coli* upon naturally occurring or synthetic polyribosomes. *Proc. Natl. Acad. Sci. USA* 47: 1588–1602.

Patil, C., and P. Walter. 2001. Intracellular signaling from the endoplasmic reticulum to the nucleus: The unfolding protein response in yeast and mammals. *Curr. Opin. Cell Biol.* 13: 349–55.

Sachs, A. B., P. Sarnow, and M. W. Hentz. 1997. Starting at the beginning, middle, and end: Translation initiation in eukaryotes. *Cell* 89: 831–38.

Shine, J., and L. Dalgarno. 1974. The 3′-terminal sequence of *Escherichia coli* 16S ribosomal RNA: Complementary to nonsense triplet and ribosome binding sites. *Proc. Natl. Acad. Sci. USA* 71: 1342–46.

Tsugita, A., et al. 1962. Demonstration of the messenger role of viral RNA. *Proc. Natl. Acad. Sci. USA* 48: 846–53.

Watson, J. D. 1963. Involvement of RNA in the synthesis of proteins. *Science* 140: 17–26.

Zheng, N., and L.M. Gierasch. 1996. Signal sequences: The same yet different. *Cell* 86: 849–52.

## Chapter 10  The Integration of Genetic Approaches: Understanding Sickle Cell Disease

Allison, A. C. 1954. Protection afforded by sickle cell trait against subtertian malarial infection. *Br. Med. J.* 1: 290–94.

Antonarakis, J. S., et al. 1984. Origin of sickle beta globin gene in blacks: The contribution of recurrent mutation or gene conversion or both. *Proc. Natl. Acad. Sci. USA* 81: 853–56.

Carlson, J., et al. 1994. Natural protection against severe *Plasmodium falciparum* malaria due to impaired rosette formation. *Blood* 84: 3909–14.

Cavalli-Sforza, L. L., and W. Bodmer. 1971. *The Genetics of Human Populations*. San Francisco: W. H. Freeman.

Cavalli-Sforza, L. L., and M. W. Feldman. 2003. The application of molecular genetic approaches to the study of human evolution. *Nat. Genet.* (suppl.), 33: 266–75.

Cholera, R., et al. 2008. Impaired cytoadherence of *Plasmodium falciparum*-infected erythrocytes containing sickle hemoglobin. *Proc. Natl. Acad. Sci. USA* 105: 991–96.

Cooke, G. S., and A.V. S. Hill. 2001. Genetics of susceptibility to human infectious disease. *Nat. Rev. Genet.* 2: 967–77.

Desai, D. V., and H. Dhanani. 2004. Sickle cell disease: History and origin. *Internet J. Hematology* 1: 2.

Herrick, J. B. 1910. Peculiar elongated and sickle shaped red blood corpuscles in a case of severe anemia. *Arch. Intern. Med.* 6: 517–21.

Ingram, V. E. 1957. Gene mutations in human haemoglobin: The chemical difference between normal and sickle cell haemoglobin. *Nature* 180: 326–28.

Kurnit, D. M. 1979. Evolution of sickle variant gene. *Lancet* 1: 104.

Livingstone, F. B. 1958. Anthropological implications of sickle cell gene distribution in West Africa. *Am. Anthropol.* 60: 533–62.

Neel, J. V. 1949. The inheritance of sickle cell anemia. *Science* 110: 64–66.

Pauling, L., et al. 1949. Sickle cell anemia, a molecular disease. *Science* 110: 543–48.

Savitt, T. L., and M. F. Goldberg. 1989. Herrick's 1910 case report of sickle cell anemia. The rest of the story. *JAMA* 261: 266–71.

Soloman, E., and W. F. Bodmer. 1979. Evolution of sickle variant gene. *Lancet* 1: 923.

Smithies, O. 1995. Early days of electrophoresis. *Genetics* 139: 1–4.

Southern, E. M. 1975. Detection of specific sequences among DNA fragments separated by gel electrophoresis. *J. Mol. Biol.* 98: 503–17.

Wiesenfeld, S. L. 1967. Sickle cell trait in human and biological evolution. *Science* 157: 1134–40.

## Chapter 11  Chromosome Structure

Bendich, A. J., and K. Drlica. 2000. Prokaryotic and eukaryotic chromosomes: What's the difference? *BioEssays* 22: 481–86.

Blackburn, E. H. 2000. Telomere states and cell fates. *Nature* 408: 53–56.

Carbon, J. 1984. Yeast centromeres: Structure and function. *Cell* 37: 352–53.

Clarke, L. 1990. Centromeres of budding and fission yeasts. *Trends Genet.* 6: 150–54.

Craig, J. M., W. C. Earnshaw, and P. Vagnarelli. 1999. Mammalian centromeres: DNA sequence, protein composition, and role in cell cycle progression. *Exp. Cell. Res.* 246: 249–62.

Cremer, T., and C. Cremer. 2001. Chromosome territories, nuclear architecture and gene regulation in mammalian cells. *Nat. Rev. Genet.* 2: 292–301.

Cross, I., and J. Wolstenholme. 2001. An introduction to human chromosomes and their analysis. In *Human Cytogenetics: Constitutional Analysis* (3rd ed.), edited by D. E. Rooney. Oxford: Oxford University Press.

Deal, R. B., J. Henikoff, and S. Henikoff. 2010. Genome-wide kinetics of nucleosome turnover determined by metabolic labeling of histones. *Science* 328: 1161–64.

DeLange, R. J., et al. 1969. Calf and pea histone IV. III. Complete amino acid sequence of pea seedling histone IV; comparison with the homologous calf thymus histone. *J. Biol. Chem.* 244: 5669–79.

Duan, Z., et al. 2010. A three-dimensional model of the yeast genome. *Nature* 465: 363–67.

Hayes, J. J., and J. C. Hansen. 2001. Nucleosomes and the chromatin fiber. *Curr. Opin. Genet. Dev.* 11: 124–29.

Heikanen, M., L. Peltonen, and A. Palotie. 1996. Visual mapping with high resolution FISH. *Trends Genet.* 12: 379–84.

Kornberg, R. D., and A. Klug. 1981. The nucleosome. *Sci. Am.* 244(2): 52–64.

Luger, K., et al. 1997. Crystal structure of the nucleosome core particle at 2.8 A resolution. *Nature* 389: 251–60.

Luger, K., and T. J. Richmond. 1998. DNA binding within the nucleosome core. *Curr. Opin. Struct. Biol.* 8: 33–40.

———— 1998. The histone tails of the nucleosome. *Curr. Opin. Genet. Dev.*, 8: 140–46.

Thanbichler, M., and L. Shapiro. 2006. Chromosome organization and segregation in bacteria. *J. Struct. Biol.* 156: 292–303.

Thanbichler, M., S. C. Wang, and L. Shapiro. 2005. The bacterial nucleoid: A highly organized and dynamic structure. *J. Cell. Biochem.* 96: 506–21.

Trask, B. J. 2002. Human chromosomes: 46 chromosomes, 46 years and counting. *Nat. Rev. Genet.* 3: 769–78.

Woodcock, C. L., and S. Dimitrov. 2001. Higher-order structure of chromatin and chromosomes. *Curr. Opin. Genet. Dev.* 11: 130–35.

Yunis, J. J. 1982. The origin of man: A chromosomal pictorial legacy. *Science* 215: 1525–30.

## Chapter 12  Gene Mutation, DNA Repair, and Homologous Recombination

Bzymek, M., et al. 2010. Double Holliday junctions are intermediates of DNA break repair. *Nature* 464: 937–41.

Cox, M. M. 2001. Recombinational DNA repair of damaged replication forks in *Escherichia coli*: Ques. *Ann Rev. Genet.* 35: 53–82.

De Laat, W. L., N. G. Jaspers, and J. H. Hoeijmakers. 1999. Molecular mechanism of excision nucleotide repair. *Genes Dev.* 13: 768–85.

Eichler, E. E., and D. Sankoff. 2003. Structural dynamics of eukaryotic chromosome evolution. *Science* 301: 793–97.

Holliday, R. 1964. A mechanism for gene conversion in fungi. *Genet. Res.* 5: 282–304.

Kowalczykowski, S. C. 2000. Initiation of genetic recombination and recombination-dependent replication. *Trends Biochem. Sci.* 25: 156–64.

Kunkel, T. A., and D. A. Erie. 2005. DNA mismatch repair. *Annu. Rev. Biochem.* 76: 681–710.

Lusetti, S. L., and M. M. Cox. 2002. The bacterial RecA protein and the recombinational DNA repair of stalled replication forks. *Annu. Rev. Biochem.* 71: 71–100.

Lynch, M. 2010. Evolution of mutation rate. *Trends Genet.* 26: 345–52.

McCann, J., and B. N. Ames. 1978. The *Salmonella*/microsome mutagenicity test: Predictive value for animal carcinogenicity. In *Advances in Modern Technology* (Vol. 5: Mutagenesis), edited by W. G. Flamm and M. A. Mehlman. Washington, DC: Hemisphere Publishing.

Ossowski, S., et al. 2010. The rate and molecular spectrum of spontaneous mutations in *Arabidopsis thaliana*. *Science* 327: 92–94.

Page, S. L., and R. S. Hawley. 2003. Chromosome choreography: The meiotic ballet. *Science* 301: 785–89.

Sekiguchi, J. M., and D. O. Freguson. 2006. DNA double-strand break repair: A relentless hunt uncovers new prey. *Cell* 124: 260–62.

Szostak, J. W., et al. 1983. The double-strand break repair model for recombination. *Cell* 33: 25–35.

## Chapter 13  Chromosome Aberrations and Transposition

Blakeslee, A. F. 1934. New jimson weeds from old chromosomes. *J. Hered.* 25: 80–108.

Drucker, R., and E. Whitelaw. 2004. Retrotransposon-derived elements in the mammalian genome: A potential source of disease. *J. Inherit. Metabol. Dis.* 27: 319–30.

Federoff, N. V. 1993. Barbara McClintock (June 16, 1902–September 2, 1992). *Genetics* 136: 1–10.

Feldman, M., and E. R. Sears. 1981. The wild gene resources of wheat. *Sci. Am.* 244: 102–12.

Gardner, R. J. M., and G. R. Sutherland. 2004. *Chromosome Abnormalities and Genetic Counseling.* 3rd ed. Oxford: Oxford University Press.

Gersh, M., et al. 1995. Evidence for a distinct region causing a catlike cry in patients with 5p deletions. *Am. J. Hum. Genet.* 56: 1404–10.

Grindley, N. D. F., and R. R. Reed. 1985. Transpositional recombination in prokaryotes. *Annu. Rev. Biochem.* 54: 863–96.

Hassold, T. J., and D. Chiu. 1985. Maternal age-specific rates of numerical chromosome abnormalities with special reference to trisomy. *Hum. Genet.* 70: 11–17.

Hassold, T. J., and P. Hunt. 2001. To err (meiotically) is human: The genesis of human aneuploidy. *Nat. Rev. Genet.* 2: 280–91.

Hassold, T. J., and P. A. Jacobs. 1984. Trisomy in man. *Annu. Rev. Genet.* 18: 69–97.

Hook, E. B., and A. Linsjo. 1978. Down syndrome by single year maternal age interval in a Swedish study: Comparison with results from a New York study. *Am. J. Hum. Genet.* 30: 19–27.

Hunt, P. A., and T. J. Hassold. 2008. Human female meiosis: What makes a good egg go bad? *Trends Genet.* 24: 86–93.

Kaiser, P. 1984. Pericentric inversions: Problems and significance for clinical genetics. *Hum. Genet.* 68: 1–47.

Madan, K. 1995. Paracentric inversions: A review. *Hum Genet.* 96: 503–15.

McClintock, B. 1939. The behavior in successive nuclear divisions of a chromosome broken at meiosis. *Proc. Natl. Acad. Sci. USA* 25: 406–16.

———— 1951. Chromosome organization and genic expression. *Cold Spring Harbor Symp. Quant. Biol.* 16: 13–47.

Medstrand, P., et al. 2005. Impact of transposable elements on the evolution of mammalian gene regulation. *Cytogenet. Genome Res.* 110: 342–52.

Miki, Y. 1998. Retrotransposal integration of mobile genetic elements in human disease. *J. Hum. Genet.* 43: 77–84.

Nelson D. L., and R. A. Gibbs. 2004. The critical region in trisomy 21. *Science* 306: 619–21.

O'Hare, K. 1985. The mechanism and control of *P* element transposition in *Drosophila*. *Trends Genet.* 1: 250–54.

Patterson, D., and A. Costa. 2005. Down syndrome and genetics—A case of linked histories. *Nat. Rev. Genet.* 6: 137–45.

Warburton, D. 2005. Biological aging and the etiology of aneuploidy. *Cytogenet. Genome Res.* 111: 266–72.

## Chapter 14  Regulation of Gene Expression in Bacteria and Bacteriophage

Bertrand, K., and C. Yanofsky. 1976. Regulation of transcription termination in the leader region of the tryptophan operon of *Escherichia coli*. *J. Mol. Biol.* 103: 339–49.

Dickson, R. C., et al. 1975. Genetic regulation: The *Lac* control region. *Science* 187: 27–33.

Fisher, R. F., et al. 1985. Analysis of the requirements for transcription pausing in the tryptophan operon. *J. Mol. Biol.* 182: 397–409.

Jacob, F., and J. Monod. 1961. Genetic regulatory mechanisms in the synthesis of proteins. *J. Mol. Biol.* 3: 318–56.

Fried, M. G. 1996. DNA looping and *Lac* repressor–CAP interaction. *Science* 274: 1930.

Lee, D., and R. Schleif. 1989. In vivo loops in *araCBAD*: Size limits and helical repeats. *Proc. Natl. Acad. Sci. USA* 86: 476–80.

Lewis, M., et al. 1996. Crystal structure of the lactose operon repressor and its complexes with DNA and inducer. *Science* 271: 1247–54.

Matthews, K. S. 1996. The whole lactose repressor. *Science* 271: 1245–46.

Pardee, A. B., F. Jacob, and J. Monod. 1959. The genetic control and cytoplasmic expression of inducibility in the synthesis of β-galactosidase by *E. coli*. *J. Mol. Biol.* 1: 165–78.

Ptashne, M. 2005. *A Genetic Switch: Phage Lambda Revisited*. 3rd ed. Cold Spring Harbor, NY: Cold Spring Harbor Laboratory Press.

Ptashne, M., and A. Gann. 2002. *Genes and Signals*. Cold Spring Harbor, NY: Cold Spring Harbor Laboratory Press.

Schlief, R. 2000. Regulation of the L-arabinose operon of *Escherichia coli*. *Trends Genet.* 16: 559–66.

———— 2003. AraC protein: A love-hate relationship. *BioEssays* 25: 274–82.

Schultz, S. C., G. C. Shields, and T. A. Steitz. 1991. Crystal structure of a CAP-DNA complex: The DNA is bent 90°. *Science* 253: 1001–7.

Tijan, R. 1995. Molecular machines that control genes. *Sci. Am.* 272: 55–61.

Ullman, A. 2003. *Origins of Molecular Biology: A Tribute to Jacques Monod*. rev. ed. Washington, DC: ASM Press.

## Chapter 15  Regulation of Gene Expression in Eukaryotes

Ambrose, V., and X. Chen. 2007. The regulation of genes and genomes by small RNAs. *Development* 134: 1635–41.

Black, D. L. 2000. Protein diversity from alternative splicing: A challenge for bioinformatics and postgenomic biology. *Cell* 103: 367–70.

Cairns, B. R. 2009. The logic of chromatin architecture and remodeling at promoters. *Nature* 461: 193–98.

Dillon, N., and P. Sabbattini. 2000. Functional gene expression domains: Defining the functional unit of eukaryotic gene regulation. *BioEssays* 22: 657–65.

Fedoriw, A. M., et al. 2004. Transgenic RNAi reveals essential function of CTCF in H19 gene imprinting. *Science* 303: 238–40.

Filion, G. J., et al. 2010. Systematic protein location mapping reveals five principal chromatin types in *Drosophila* cells. *Cell* 143:212–224.

Flintoft, L. 2010. Complex diseases: Adding epigenetics to the mix. *Nat. Rev. Genet.*, 11: 94–95.

Fuda, N. J., M. Behfar Ardehali, and J. T. Lis. 2009. Defining mechanisms that regulate RNA polymerase II transcription *in vivo*. *Nature* 461: 186–92.

Gregory, P. D., K. Wagner, and W. Hurz. 2001. Histone acetylation and chromatin remodeling. *Exp. Cell Res.* 265: 195–202.

Grewal, S. I. S., and S. Jia. 2008. Heterochromatin revisited. *Nat. Rev. Genet.* 8: 35–46.

Hassan, A. H., et al. 2001. Promoter targeting of chromatin-modifying complexes. *Frontiers in Bioscience* 6: 1054–64.

Horn, P. J., and C. L. Peterson. 2002. Chromatin higher order folding: Wrapping up transcription. *Science* 297: 1824–27.

Jenuwein, T., and C. D. Allis. 2001. Translating the histone code. *Science* 293: 1074–80.

Kappeler, L., and M. J. Meaney. 2010. Epigenetics and parental effects. *BioEssays* 32: 818–27.

Kucharski, R., et al. 2008. Nutritional control of reproductive status in honeybees via DNA methylation. *Science* 319: 1827–30.

Lawrence, P. A. 1992. *How to Make a Fly*. London: Blackwell Scientific.

Lusser, A., and J. T. Kadonaga. 2003. Chromatin remodeling by ATP-dependent molecular machines. *BioEssays* 25: 1192–1200.

MacRae, I. J.,et al. 2006. Structural basis for double-stranded RNA processing by Dicer. *Science* 311: 195–98.

Margueron, R., and D. Reinberg. 2010. Chromatin structure and the inheritance of epigenetic information. *Nat. Rev. Genet.* 11: 285–96.

Meister, G., and T. Tuschl. 2004. Mechanisms of gene silencing by double-stranded RNA. *Nature* 431: 343–49.

Moore, M. J. 2005. From birth to death: The complex lives of eukaryotic mRNAs. *Science* 309: 1514–18.

Roudier, F., et al. 2011. Integrative epigenomic mapping defines four main chromatin states in *Arabidopsis*. EMBO J., 1-11, doi:10.1038/emboj.2011.103.

Siomi, H., and M. C. Siomi. 2007. Expanding RNA physiology: MicroRNAs in unicellular organisms. *Genes Dev.* 21: 1153–56.

Small, S., et al. 1991. Transcriptional regulation of a pair-rule gene in *Drosophila*. *Genes Develop.* 5: 827–39.

Sudarsanam, P., and F. Winston. 2000. The SWI/SNF family of nucleosome-remodeling complexes and transcriptional control. *Trends Genet.* 16: 345–51.

Turner, B. M. 2000. Histone acetylation and the epigenetic code. *BioEssays* 22: 836–45.

Van Steensel, B. 2011. Chromatin: Constructing the big picture. EMBO J., 1-11,doi:10.1038/emboj.2011.135.

Visel, A., E. M. Rubin, and L. A. Pennacchio. 2009. Genomic views of distant-acting enhancers. *Nature* 461: 199–205.

Wang, H., et al. 2004. Using atomic force microscopy to study nucleosome remodeling on individual nucleosomal arrays in situ. *Biophys. J.* 87: 1964–71.

## Chapter 16  Forward Genetics and Recombinant DNA Technology

Bates, G. P. 2005. The molecular genetics of Huntington disease—a history. *Nat. Rev. Genet.* 6: 766–73.

Berg, P., et al. 1974. Potential biohazards of recombinant DNA molecules. *Proc. Natl. Acad. Sci. USA* 71: 2593–94.

———— 1975. Summary Statement of the Asilomar Conference on recombinant DNA molecules. *Proc. Natl. Acad. Sci. USA* 72: 1981–84.

Berg, P., and Singer, M. A. 1995. The recombinant DNA controversy: Twenty years later. *Proc. Natl. Acad. Sci. USA* 92: 9011–13.

Boone, C., H. Bussey, and B. J. Andrews. 2007. Exploring genetic interactions and networks with yeast. *Nat. Rev. Genet.* 8: 437–49.

Cohen, S. A., et al. 1973. Construction of biologically functional plasmids in vitro. *Proc. Natl. Acad. Sci. USA* 70: 3240–44.

Danna, K., and D. Nathans. 1971. Specific cleavage of simian virus 40 DNA by restriction endonuclease of *Hemophilus influenzae*. *Proc. Natl. Acad. Sci. USA* 68: 2913–17.

Forsburg, S. L. 2001. The art and design of genetic screens: Yeast. *Nat. Rev. Genet.* 2: 659–68.

Grunstein, M., and D. S. Hogness. 1975. Colony hybridization—Method for isolation of cloned DNAs that contain a specific gene. *Proc. Natl. Acad. Sci. USA* 72: 3961–65.

Hartwell, L. H., et al. 1973. Genetic control of the cell division cycle in yeast: V. genetic analysis of *cdc* mutants. *Genetics* 74: 267–86.

Hershfie, V., et al. 1974. Plasmid ColE1 as a molecular vehicle for cloning and amplification of DNA. *Proc. Natl. Acad. Sci. USA* 71: 3455–59.

The Huntington's Disease Collaborative Research Group. 1993. A novel gene containing a trinucleotide repeat that is expanded and unstable on Huntington's Disease chromosomes. *Cell* 72: 971–83.

Kelley, T. J. Jr., and H. O. Smith. 1970. A restriction enzyme from *Hemophilus influenzae*. II. Base sequence of the recognition site. *J. Molec. Biol.* 51: 393–409.

Jackson, D., R. Symons, and P. Berg. 1972. Biochemical methods for inserting new genetic information into DNA of simian virus 40: Circular SV40 DNA molecules containing lambda phage genes and the galactose operon of *Escherichia coli*. *Proc. Natl. Acad. Sci. USA* 69: 2904–9.

Kile, B. T., and D. J. Hilton. 2005. The art and design of genetic screens: Mouse. *Nat. Rev. Genet.* 6: 557–67.

Linn, S., and W. Arber. 1968. Host specificity of DNA produced by *Escherichia coli*. X. In vitro restriction of phage fd replicative form. *Proc. Natl. Acad. Sci. USA* 59: 1300–1306.

Margulies, M., et al. 2005. Genome sequencing in microfabricated high-density picolitrereactors. *Nature* 437: 376–80.

Page, D. R., and U. Grossniklaus. 2002. The art and design of genetic screens: *Arabidopsis thaliana*. *Nat. Rev. Genet.* 3: 124–36.

Poinar, H. N., et al. 2006. Metagenomics to paleogenomics: Large-scale sequencing of mammoth DNA. *Science* 311: 392–94.

Sambrook, J., E. F. Fitch, and T. Maniatis. 1989) *Molecular Cloning: A Laboratory Manual*. 2nd ed. Cold Spring Harbor, NY: Cold Spring Harbor Press.

Smith, H. O., and K. W. Wilcox. 1970. A restriction enzyme from *Hemophilus influenzae*. I. Purification and general properties. *J. Molec. Biol.*, 51: 379–91.

St Johnston, D. 2002. The art and design of genetic screens: *Drosophila melanogaster*. *Nat. Rev. Genet.* 3: 176–88.

Sturtevant, A. H. 1955. A highly specific complementary lethal system in *Drosophila melanogaster*. *Genetics* 40: 118–23.

Shuman, H. A., and T. J. Silhavy. 2003. The art and design of genetic screens: *Escherichia coli*. *Nat. Rev. Genet.* 4: 419–31.

Thomas, J. H. 1993. Thinking about genetic redundancy. *Trends Genet.* 9: 395–399.

## Chapter 17 Applications of Recombinant DNA Technology and Reverse Genetics

Austin, C. P., et al. 2004. The knockout mouse project. *Nat. Genet.* 36: 921–24.

Bellen, H. J., et al. 1989. P-element-mediated enhancer detection—A versatile method to study development in *Drosophila*. *Genes & Development* 3: 1288–300.

Chilton, M.-D., et al. 1977. Stable incorporation of plasmid DNA into higher plant cells: The molecular basis of crown gall tumorigenesis. *Cell* 11: 263–71.

Echeverri, C. J., and N. Perrimon. 2006. High-throughput RNAi screening in cultured cells: A user's guide. *Nat. Rev. Genet.* 7: 373–84.

Goeddel, D. V., et al. 1979. Expression in *Escherichia coli* of chemically synthesized genes for human insulin. *Proc. Natl. Acad. Sci.* 76: 106–10.

Halder, G., P. Callaerts, and W. J. Gehring. 1995. Induction of ectopic eyes by targeted expression of the eyeless gene in *Drosophila*. *Science* 267: 1788–92.

Hanna, J., et al. 2007. Treatment of sickle cell anemia mouse model with iPS cells generated from autologous skin. *Science* 318: 1920–23.

Hannon, G. J. 2002. RNA interference. *Nature* 418: 244–51.

Kimmelman, J. 2008. Science and society: The ethics of human gene transfer. *Nat. Rev. Genet.* 9: 239–44.

Martin, V. J. J., et al. 2003. Engineering a mevalonate pathway in *Escherichia coli* for production of terpenoids. *Nat. Biotechnol.* 21: 796–802.

McCallum, C. M., et al. 2000. Targeting induced local lesions in genomes (TILLING) for plant functional genomics. *Plant Physiol.* 123: 439–42.

Pelaz, S., et al. 2000. B and C floral organ identity functions require *SEPALLATA* MADS-box genes. *Nature* 405: 200–203.

Rubin, G. M., and A. C. Spradling. 1982. Genetic-transformation of *Drosophila* with transposable element vectors. *Science* 218: 348–53.

Small, S., A. Blair, and M. Levine. 1992. Regulation of *even-skipped* stripe-2 in the *Drosophila* embryo. *EMBO J.* 11: 4047–57.

Thomas, K. R., and M. R. Capecchi. 1987. Site-directed mutagenesis by gene targeting in mouse embryo-derived stem-cells. *Cell* 51: 503–12.

Waehler, R., S. J. Russell, and D. T. Curiel. 2007. Engineering targeted viral vectors for gene therapy. *Nat. Rev. Genet.* 8: 573–87.

Wilmut, I., et al. 1997. Viable offspring derived from fetal and adult mammalian cells. *Nature* 385: 810–13.

Wofenbarger, L. L., and P. R. Phifer. 2000. The ecological risks and benefits of genetically engineered plants. *Science* 290: 2088–93.

Ye, X., et al. 2000. Engineering the provitamin A (β-carotene) biosynthetic pathway into (carotenoid-free) rice endosperm. *Science* 287: 303–5.

## Chapter 18 Genomics: Genetics from a Whole-Genome Perspective

Adams, M. D., et al. 2000. The genome sequence of *Drosophila melanogaster*. *Science* 287: 2185–95.

The *Arabidopsis* Genome Initiative. 2000. Analysis of the genome sequence of the flowering plant *Arabidopsis thaliana*. *Nature* 408: 796–814.

Blattner, F. R., et al. 1997. The complete genome sequence of *Escherichia coli* K-12. *Science* 277: 1453–61.

Boffelli, D., M. A. Nobrega, and E. M. Rubin. 2004. Comparative genomics at the vertebrate extremes. *Nat. Rev. Genet.* 5: 456–65.

Boone, C., H. Bussey, and B. J. Andrews. 2007. Exploring genetic interactions and networks with yeast. *Nat. Rev. Gen.* 8: 437–49.

Brune, A. 2007. Woodworker's digest. *Nature* 450: 487–88.

Cavalli-Sforza, L. L. 2005. The Human Genome Diversity Project: Past, present, and future. *Nat. Rev. Gen.* 6: 333–40.

The Chimpanzee Sequencing and Analysis Consortium. 2005. Initial sequence of the chimpanzee genome and comparison with the human genome. *Nature* 437: 69–87.

Chu, S., et al. 1998. The transcriptional program of sporulation in budding yeast. *Science* 282: 699–705.

DeRisi, J. L., V. R. Iyer, and P. O. Brown. 1997. Exploring the metabolic and genetic control of gene expression on a genomic scale. *Science* 278: 680–86.

Feuk, L., A. R. Carson, and S. W. Scherer. 2006. Structural variation in the human genome. *Nat. Rev. Gen.* 7: 85–97.

Fleischmann, R. D., et al. 1995. Whole genome random sequencing and assembly of *Haemophilus influenzae* Rd. *Science* 269: 496–512.

Giaever, G., et al. 2002. Functional profiling of the *Saccharomyces cerevisiae* genome. *Nature* 418: 387–91.

Geib, S. M., et al. 2008. Lignin degradation in wood-feeding insects. *Proc. Natl. Acad. Sci. USA* 105: 12932–37.

Goffeau, A., et al. 1996. Life with 6000 genes. *Science* 274: 562–567.

Green, R. E., et al. 2010. A draft sequence of the Neandertal genome. *Science* 328: 710–22.

Hattori, M., et al. 2000. The DNA sequence of human chromosome 21. *Nature* 405: 311–19.

International Human Genome Sequencing Consortium. 2001. Initial sequencing and analysis of the human genome. *Nature* 409: 860–921.

Keeling, P. J., et al. 2005. The tree of eukaryotes. *Trends Ecol. and Evol.* 12: 670–76.

Kellis, M., et al. 2003. Sequencing and comparison of yeast species to identify genes and regulatory elements. *Nature* 423: 241–54.

Kobayashi, K., et al. 2003. Essential *Bacillus subtilis* genes. *Proc. Natl. Acad. Sci. USA* 100: 4678–83.

Lindblad-Toh, K., et al. 2005. Genome sequence, comparative analysis and haplotype structure of the domestic dog. *Nature* 438: 803–19.

Lockhart, D. J., and E. A. Winzeler. 2000. Genomics, gene expression and DNA arrays. *Nature* 405: 827–36.

Long, M., et al. 2003. The origin of new genes: Glimpses from the young and old. *Nat. Rev. Gen.* 4, 865–75.

Mazurkiewicz, P., et al. 2006. Signature-tagged mutagenesis: Barcoding mutants for genome-wide screens. *Nat. Rev. Genet.* 7: 929–39.

Mouse Genome Sequencing Consortium. 2002. Initial sequencing and comparative analysis of the mouse genome. *Nature* 420: 520–62.

Rosenberg, N. A., et al. 2002. Genetic structure of human populations. *Science* 298: 2381–85.

Tishkoff, S. A., et al. 2009. The genetic structure and history of Africans and African Americans. *Science* 324: 1035–44.

Turnbaugh, P. J., et al. 2007. The Human Microbiome Project. *Nature* 449: 804–10.

Uetz, P., et al. 2000. A comprehensive analysis of protein-protein interactions in *Saccharomyces cerevisiae*. *Nature* 403: 623–27.

Venter, J. C., et al. 2001. The sequence of the human genome. *Science* 291: 1304–51.

———— 2004. Environmental genome shotgun sequencing of the Sargasso Sea. *Science* 304: 66–74.

Warnecke, F., et al. 2007. Metagenomic and functional analysis of hindgut microbiota of a wood-feeding higher termite. *Nature* 450: 560–69.

Warren, W. C., et al. 2008. Genome analysis of the platypus reveals unique signatures of evolution. *Nature* 453: 175–84.

Wienberg, J. 2004. The evolution of eutherian chromosomes. *Curr. Opinion in Gen. Devel.* 14: 657–66.

Yamada, K., et al. 2003. Empirical analysis of transcriptional activity in the *Arabidopsis* genome. *Science* 302: 842–46.

## Chapter 19 Cytoplasmic Inheritance and the Evolution of Organelle Genomes

Brown, J. R. 2003. Ancient horizontal gene transfer. *Nat. Rev. Genet.* 4: 121–32.

Burger, G., M. W. Gray, and B. F. Lang. 2003. Mitochondrial genomes: Anything goes. *Trends Genet.* 19: 709–16.

Cann, R. L., M. Stoneking, and A. C. Wilson. 1987. Mitochondrial DNA and human evolution. *Nature* 325: 31–36.

Cavalli-Sforza, L. L., and M. W. Feldman. 2003. The application of molecular genetic approaches to the study of human evolution. *Nat. Genet.* 33(suppl.): 266–75.

Chase, C. D. 2007. Cytoplasmic male sterility: A window to the world of plant mitochondrial-nuclear interactions. *Trends Genet.* 23: 81–90.

Chen, X. J., and R. A. Butow. 2005. The organization and inheritance of the mitochondrial genome. *Nat. Rev. Genet.* 6: 815–25.

Embley, T. M., and W. Martin. 2006. Eukaryotic evolution, changes and challenges. *Nature* 440: 623–30.

Finlayson, C. 2005. Biogeography and evolution of the genus *Homo*. *Trends Ecol. Evol.* 20: 457–63.

Garrigan, D., and M. F. Hammer. 2006. Reconstructing human origins in the genomic era. *Nat. Rev. Genet.* 7: 669–80.

Gilson, P. R., et al. 2006. Complete nucleotide sequence of the chlorarachniophyte nucleomorph: Nature's smallest nucleus. *Proc. Natl. Acad. Sci. USA* 103: 9566–71.

Huang, C. Y., M. A. Ayliffe, and J. N. Timmis. 2003. Direct measurement of the transfer rate of chloroplast DNA into the nucleus. *Nature* 422: 72–76.

Kotera, E. M. Tasaka, and T. Shikanai. 2005. A pentatricopeptide repeat protein is essential for RNA editing in chloroplast. *Nature* 433: 326–30.

Lang, B. F., et al. 1997. An ancestral mitochondrial DNA resembling a eubacterial genome in miniature. *Nature* 387: 493–97.

Lehman, N., et al. 1991. Introgression of coyote mitochondrial DNA into sympatric North American gray wolf populations. *Evolution* 45: 104–19.

Leister, D. 2003. Chloroplast research in the genomic age. *Trends Genet.* 19: 47–56.

Martin, W. 2002. Evolutionary analysis of *Arabidopsis*, cyanobacterial, and chloroplast genomes reveals plastid phylogeny and thousands of cyanobacterial genes in the nucleus. *Proc. Natl. Acad. Sci. USA* 99: 12246–51.

Oda, K., et al. 1992. Gene organization deduced from the complete sequence of liverwort *Marchantia polymorpha* mitochondrial DNA. *J. Mol. Biol.* 223: 1–7.

Prezant, T. R., et al. 1993. Mitochondrial ribosomal RNA mutation associated with both antibiotic-induced and non-syndromic deafness. *Nat. Genet.* 4: 289–94.

Rivera, M. C., et al. 1998. Genomic evidence for two functionally distinct gene classes. *Proc. Natl. Acad. Sci. USA* 95: 6239–44.

Stegemann, S., and R. Bock. 2006. Experimental reconstruction of functional gene transfer from the tobacco plastid genome to the nucleus. *Plant Cell* 18: 2869–78.

Taylor, R. W., and D. M. Turnbull 2005. Mitochondrial DNA mutations in human disease. *Nat. Rev. Genet.* 6: 389–402.

Timmis, J. N., et al. 2004. Endosymbiotic gene transfer: Organelle genomes forge eukaryotic chromosomes. *Nat. Rev. Genet.* 5: 123–35.

Trifunovic, A., et al. 2004. Premature ageing in mice expressing defective mitochondrial DNA polymerase. *Nature* 429: 417–23.

Wallace, D. C., et al. 1988. Mitochondrial DNA mutation associated with Leber's hereditary optic neuropathy. *Science* 242: 1427–30.

Ward, T. J., et al. 1999. Identification of domestic cattle hybrids in wild cattle and bison species: A general approach using mtDNA markers and the parametric bootstrap. *Animal Conservation* 2: 51–57.

## Chapter 20  Developmental Genetics

Bender, W., et al. 1983. Molecular-genetics of the bithorax complex in *Drosophila melanogaster*. *Science* 221: 23–29.

Binns, W., L. F. James, and J. L. Shupe. 1964. Toxicosis of *Veratrum californicum* in ewes and its relationship to a congenital deformity in lambs. *Ann. N.Y. Acad. Sci.* 111: 571–76.

Bowman, J. L., D. R. Smyth, and E. M. Meyerowitz. 1991. Genetic interactions among floral homeotic genes of *Arabidopsis*. *Development* 112: 1–20.

Chen, J. K., et al. 2002. Inhibition of hedgehog signaling by direct binding of cyclopamine to smoothened. *Genes Dev.* 16: 2743–48.

Coen, E. S., and Meyerowitz, E. M. 1991. The war of the whorls: Genetic interactions controlling flower development. *Nature* 353: 31–37.

De Robertis, E. M., and H. Kuroda. 2004. Dorsal-ventral patterning and neural induction in *Xenopus* embryos. *Annu. Rev. Cell Dev. Biol.* 20: 285–308.

de Rosa R. J. K., et al. 1999. Hox genes in brachiopods and priapulids and protostome evolution. *Nature* 399: 772–76.

Driever, W., and C. Nusslein-Volhard. 1988. The Bicoid protein determines position in the *Drosophila* embryo in a concentration-dependent manner. *Cell* 54: 95–104.

Driever, W., V. Siegel, and C. Nusslein-Volhard. 1990. Autonomous determination of anterior structures in the early *Drosophila* embryo by the Bicoid morphogen. *Development* 109: 811–20.

Horvitz, H. R. 2003. Worms, life and death. *Bioscience Reports* 23: 239–69.

Lemons, D., and W. McGinnis. 2006. Genomic evolution of Hox gene clusters. *Science* 313: 1918–22.

Lettice, L. A., et al. 2002. Disruption of a long range *cis*-acting regulator for *Shh* causes preaxial polydactyly. *Proc. Natl. Acad. Sci. USA* 99: 7548–53.

Lewis, E. B. 1978. Gene complex controlling segmentation in *Drosophila*. *Nature* 276: 565–70.

Nanni, L., et al. 1999. The mutational spectrum of the *sonic hedgehog* gene in holoproencephaly: *Shh* mutations cause a significant proportion of autosomal dominant holoproencephaly. *Hum. Mol. Genet.* 8: 2479–88.

Nusslein-Volhard, C., and Wieschaus, E. 1980. Mutations affecting segment number and polarity in *Drosophila*. *Nature* 287: 795–801.

Shubin, N., C. Tabin, and S. Carroll. 1997. Fossils, genes, and the evolution of animal limbs. *Nature* 388: 639–48.

———— 2009. Deep homology and the origins of evolutionary novelty. *Nature* 457: 818–23.

Small, S., A. Blair, and M. Levine. 1992. Regulation of *even-skipped* stripe-2 in the *Drosophila* embryo. *EMBO J.* 11: 4047–57.

Sternberg, P. W., and M. Han. 1998. Genetics of RAS signaling in *C. elegans*. *Trends Genet.* 14: 466–72.

## Chapter 21  Genetic Analysis of Quantitative Traits

Castle, W. 1916. *Genetics and Eugenics*. Cambridge, MA: Harvard University Press.

deVicente, M. C., and S. D. Tanksley. 1993. QTL analysis of transgressive segregation in an interspecific tomato cross. *Genetics* 134: 585–96.

Dudley, J. W. 1977. 76 generations of selection for oil and protein percentage in maize. In Proc. Internat. Conf. on Quant. Gene., edited by E. Pollack, O. Kempthorne, and T. Bailey. Ames: Iowa State University Press.

East, E. M. 1910. A Mendelian interpretation of variation that is apparently continuous. *Am. Nat.* 44: 65–82.

———— 1916. Studies on size inheritance in *Nicotiana*. *Genetics* 1: 161–76.

Fridman, E., et al. 2004. Zooming in on a quantitative trait for tomato yield using interspecific introgressions. *Science* 305: 1786–89.

Lanktree, M. B., et al. 2011. Meta-analysis of dense gene centric association studies reveals common and uncommon variants associated with height. *Am. J. Hum. Genet.* 88: 6–18.

Laurie, C. C., et al. 2004. The genetic architecture of response to long-term selection for oil concentration in the maize kernel. *Genetics* 168: 2141–55.

Moose, S. D., J. W. Dudley, and T. R. Rocheford. 2004. Maize selection passes the century mark: A unique resource for 21st century genomics. *Trends Plant Sci.* 7: 358–64.

Nilsson-Ehle, H. 1909. Kreuzengsunter-su-chungen an hafer und weizen. Lunds Univ. Aarskrift, N.F. Atd., Ser. 2, 5: 1–122.

Ogura, Y., et al. 2001. A frameshift mutation in *NOD2* associated with susceptibility to Crohn's disease. *Nature* 411: 603–6.

Tanksley, S. D. 2004. The genetic, developmental, and molecular bases of fruit size and shape variation in tomato. *Plant Cell* 16: S181–89.

Weedon, M. N., et al. 2008. Genome-wide association analysis identifies 20 loci that influence human height. *Nat. Genet.* 40: 575–83.

Wellcome Trust Case Control Consortium. 2007. Genome wide association study of 14,000 cases of seven common diseases and 3,000 shared controls. *Nature* 447: 661–78.

## Chapter 22  Population Genetics and Evolution

Cavalli-Sforza, L. L., and W. F. Bodmer. 1971. *The Genetics of Human Populations*. San Francisco: W.H. Freeman and Co.

Cavener, D. R., and M. T. Clegg. 1981. Multigenic response to ethanol in *Drosophila melanogaster*. *Evolution* 35: 1–10.

Crow, J. F. 1986. *Basic Concepts in Population, Quantitative, and Evolutionary Genetics*. New York: W. H. Freeman.

Diamond, J. M., and J. I. Rotter. 1987. Observing the founder effect in human evolution. *Nature* 329:105–106.

Dobzhansky, T. 1970. *Genetics of the Evolutionary Process*. New York: Columbia University Press.

Elena, S. F., V. S. Cooper, and R. E. Lenski. 1996. Punctuated evolution caused by selection of rare beneficial mutations. *Science* 272: 1802–4.

Freeman, S., and J. C. Herron. 2001. *Evolutionary Analysis*. Upper Saddle River, NJ: Prentice Hall.

Green, R. E., et al. 2010. A draft sequence of the Neandertal genome. *Science* 328:710–722.

Hardy, G. H. 1908. Mendelian proportions in a mixed population. *Science* 28: 49–50.

McKusick, V. A. 2000. Ellis-van Crevald syndrome and the Amish. *Nature Genetics* 24:203–204.

National Institute of Standards (NIST). "Overview of STR Fact Sheets." Last updated May 20, 2011. www.cstl.nist.gov/strbase/str_fact.htm.

———— "Material Measurement Laboratory." Last updated May 25, 2011. www.cstl.nist.gov/strbase/fbicore.htm.

Ralls, K., K, Brugger, and J. Ballou. 1979. Inbreeding and juvenile mortality in small populations of ungulates. *Science* 206: 1101–3.

Simonson, T. S., et al. 2010. Genetic evidence for high-altitude adaptation in Tibet. *Science* 329: 72–75.

Slatkin, M. 1987. Gene flow and the geographic structure of natural populations. *Science* 236:787–792.

Smith, J. M. 1989. *Evolutionary Genetics*. New York: Oxford University Press.

Weinberg, W. 1908. Ueber den nachweis der vererbung beim menschen. *Jahreshefte des Vereins für Vaterländische Naturkunde in Württemburg* 64: 368–82. English translation in Boyer, S. H. 1963. *Papers on Human Genetics*. Englewood Cliffs, NJ: Prentice Hall.

Yi, X., et al. 2010. Sequencing of 50 human exomes reveals adaptation to high altitude. *Science* 329: 75–78.

Yule, G. U. 1902. Mendel's laws and their probable relations to intra-racial heredity. *New Phytologist* 1: 193–207.

# Answers

## Chapter 1

**2.** Protein, not DNA, was the focus of efforts to understand heredity before this discovery. Only after this discovery was serious attention turned to understanding DNA structure and function. Whereas the complexity of protein structure confounded thinking about mechanisms of inheritance, DNA structure provided profound insight into the mechanism of heredity, suggesting a simple, elegant mechanism for duplication (inheritance), change (mutation and evolution), and phenotype specification (coding). The finding that DNA is universal facilitated rapid progress because study results from all organisms were now directly related. This also fostered the development of recombinant DNA technologies in bacteria and bacteriophage, which led to the explosion of biological information that excites and confounds us today.

**4.** Evolution states that all life descended from a common ancestor, which passed its genes and the mechanisms by which those genes were used to its descendants. These mechanisms would include the structure of nucleotides, the structure of DNA, the enzymes that replicate and read DNA, the enzymes that translate mRNA into amino acid sequences, and many more. Any change to one of these components would be harmful, slowing or preventing reproduction, and would be removed by natural selection (mountains of experimental evidence demonstrate that mutations in genes encoding basic genetic machinery are lethal). Thus, once established as the genetic material, DNA would be maintained as the genetic material by natural selection and therefore would be expected to be found as the genetic material in all existing organisms.

**6.** *Genotype* refers to the genetic makeup of a cell or organism, whereas *phenotype* refers to observable characteristics of the cell or organism, such as appearance, physiology and behavior. The genotype of an organism is part of what determines the phenotype of the organism; however, environment also plays a role in the organism's phenotype. The genotype is heritable and, therefore, its contribution to phenotype will be inherited. The aspects of phenotype that are due to environment are not heritable.

**8.** The modern synthesis of evolution is the reconciliation of Darwin's evolutionary theory with the findings of modern genetics. Darwin's theory proposed that all evolution was adaptive. Genetic studies on mutation and genetic recombination initially argued against the importance of natural selection as an agent for change because mutation and recombination were nonadaptive. The modern synthesis stated that evolution is due to the combined action of adaptive and nonadaptive evolutionary forces. In particular, the modern synthesis explained how mutation and recombination could provide the raw material (new genotypes and phenotypes) on which natural selection acts.

**10a.** *Transcription* is the synthesis of RNA by RNA polymerase. The RNA is complementary to the strand of DNA that was used as the template for transcription.

**10b.** An *allele* is a specific form of a gene or genetic locus. ·

**10c.** The *central dogma of biology* originally stated that genetic information flows from DNA to RNA (by transcription) and from RNA to protein (by translation). A point of emphasis of this dogma was that information does not flow in the reverse direction and has been modified to account for reverse transcription.

**10d.** *Translation* is the synthesis of a polypeptide using the information in an mRNA. *Translation* and *protein synthesis* are synonyms.

**10e.** *DNA replication* is the process by which DNA is copied by DNA polymerase.

**10f.** A *gene* is a segment of DNA that contains all the information necessary for its proper transcription, including the promoter, transcribed region, and termination signals.

**10g.** A *chromosome* is a heritable molecule composed of DNA and protein that typically contains genes.

**10h.** The term *antiparallel* refers to the orientation of the two strands of nucleic acid in a double-stranded nucleic acid (RNA or DNA). Each end of the

double-stranded nucleic acid will contain the 5′ end of one strand and the 3′ end of the other.

**10i.** *Phenotype* refers to the observable characteristics of an organism, which include morphology, physiology, and molecular composition. The phenotype of an organism is a product of the interaction between its genotype and its environment.

**10j.** *Complementary base pair* refers to the two nucleotides on opposite, antiparallel strands of a double-stranded nucleic acid, which are hydrogen bonded to each other. One nucleotide contains a purine base that makes hydrogen bond to the pyrimidine base that is part of the other nucleotide.

**10k.** *Nucleic acid strand polarity* refers to the orientation of the nucleotides along a single strand of nucleic acid. One end of the strand terminates at the 3′ hydroxyl group of a ribose (or deoxyribose) sugar, whereas the other strand terminates at the 5′ phosphate group on the sugar. These are commonly referred to as the 3′ and 5′ ends of the strand.

**10l.** *Genotype* refers to the genetic makeup of an organism. The genotype can refer either to the organism's entire genetic makeup or to the genetic information at only one or a few loci.

**10m.** *Natural selection* is the process by which populations and species evolve and diverge through differential rates of survival and reproduction of members that are due to their inherited differences.

**10n.** *Mutation* is the process that generates new genetic variety through change to existing alleles.

**10o.** *Modern synthesis of evolution* is the term applied to the reconciliation of modern genetic analysis with Darwin's theory of evolution by natural selection.

**12.** The 5′ end is a phosphate group. The 3′ end is a hydroxyl group. A phosphodiester bond is a covalent bond that joins nucleotides in a strand of nucleic acid.

**14.** The central dogma is a description of the flow of genetic information. The flow is unidirectional, from DNA to RNA to protein. The process by which information flows from DNA to RNA is called transcription. Transcription is the synthesis of RNA by RNA polymerase. The RNA is complementary to the strand of DNA that was used as the template for transcription. The flow of information from RNA to protein is called translation. Translation is the synthesis of a polypeptide using the information in an mRNA.

**16a.** The mRNA sequence is 5′-UUCCAUGUC-3′.

**16b.** The amino acid sequence is Phe-His-Val.

**18a.** 6 clades

**18b.** A backbone (They are all vertebrates.)

**18c.** The mammalian and human clades share these characteristics: have backbones and four legs, have fur and produce milk.

**20.** The samples are (1) double-stranded DNA, (2) single-stranded RNA, (3) double-stranded RNA, and (4) single-stranded DNA.

*Evaluate:* DNA will contain thymine (T) but not uracil (U) whereas RNA will contain U but not T. Double-stranded DNA will contain equal proportions of adenine (A) and T and equal proportions of guanine (G) and cytosine (C) whereas single-stranded DNA typically does not. Double-stranded RNA will contain equal proportions of A and U and equal proportions of G and C whereas single-stranded RNA typically does not. *Deduce:* Sample 1 contains T but not U and the percentage of A is the same as T, therefore, it is probably double-stranded DNA. Sample 2 contains U but not T and the percentage of A is not the same as U; therefore, it is single-stranded RNA. Sample 3 contains U but not T and the percentage of A is the same as U; therefore, it is probably double-stranded RNA. Sample 4 contains T but not U and the percentage of A is not the same as T; therefore, it is single-stranded DNA. *Solve:* Samples 1 and 3 are double-stranded, whereas samples 2 and 4 are single-stranded.

**22.** Mammals

## Chapter 2

**2.** The genotype ratio will be 1/2 *BB* and 1/2 *Bb*. The phenotype will be all B.

**8a.** False. The expected *phenotype* ratio is 9/16 : 3/16 : 3/16 : 1/16, assuming simple dominance and independent assortment of genetic loci (*AaBb* × *AaBb*). There are nine different genotypes, and the genotype ratio is 1/16 (*AABB*) : 2/16 (*AABb*) : 1/16 (*AAbb*) : 2/16 (*AaBB*) : 4/16 (*AaBb*) : 2/16 (*Aabb*) : 1/16 (*aaBB*) : 2/16 (*aaBb*) : 1/16 (*aabb*).

**8b.** True

**8c.** True

**8d.** False. The law of *independent assortment* is of primary importance in predicting the outcome of dihybrid and trihybrid crosses. The law of segregation is also necessary, but not sufficient.

**8e.** False. Reciprocal crosses that produce identical results indicate that the traits being studied are autosomal.

**8f.** False. The law of segregation predicts that she will produce two gamete genotypes with respect to her albinism gene at equal frequency.

**8g.** True

**8h(1).** True

**8h(2).** False. There will be 1/16 *AABB*, 1/16 *AAbb*, 1/16 *aaBB*, and 1/16 *aabb*. All four genotypes will be true-breeding; therefore, 1/4 of the progeny will be true-breeding.

**8h(3).** False. Being "heterozygous at one or both loci" excludes only the progeny that are homozygous at both loci. In part (b) of this question, it was calculated that 1/4 will be homozygous at both loci; therefore, 3/4 will be heterozygous at one or both loci (2/16 *AaBB*, 2/16 *AaBB*, 4/16 *AaBb*, 2/16 *Aabb*, and 2/16 *aaBb*).

**10a.** Mottled is the dominant phenotype.

**10b.** The results are consistent with autosomal inheritance.

**10c.** 1/2 of the $F_2$ of both crosses are expected to be homozygous, and 1/2 are expected to be heterozygous.

**10d.** One cross would be a test cross, and the other would be a backcross. The test cross would be the mottled $F_2$ to a true-breeding leopard individual. The backcross would be the mottled $F_2$ to one of its parents (both are heterozygotes).

**12a.** The mode of fur color inheritance in these crosses is likely to be a single gene with two alleles controlling fur color, and black will be dominant to brown.

**12b.** Assign the letter *B* for the black allele and *b* for the brown allele. The brown male must be *bb*. The black female in the first cross must be *Bb*. The black female in the second cross must be *BB*.

**14a.** The results suggest that the inheritances of color and fin shape are each due to segregation of two alleles at a single gene.

**14b.** The results indicate that gold is dominant to black and that split fin is dominant to single fin.

**14c.** The chi-square value for color is 0.061, which corresponds to a *P* value between 0.7 and 0.9. The chi-square value for fin shape is 0.05, which corresponds to a *P* value between 0.7 and 0.9. Neither *P* value is less than 0.05, which indicates that the hypothesis of one gene with two alleles for color and fin shape cannot be rejected.

**16b.** Of the $F_2$ progeny, 3/4 will have yellow seeds, 1/4 will have green seeds, 3/4 will have round seeds, and 1/4 will have wrinkled seeds.

**16c.** 9/16 yellow, round; 3/16 yellow, wrinkled; 3/16 green, round; 1/16 green, wrinkled

**20.** The $\chi^2$ value is 4.81. There is one degree of freedom (df = 1) in this calculation and the *P* value is less than 0.05. Therefore, the hypothesis that bicolor corn-kernel color is the result of segregation of two alleles at a single genetic locus should be rejected for this experiment.

**22a.** 0.132

**22b.** 0.178

**22c.** 0.822

**24a.** 0.422

**24b.** 0.0313

**24c.** 0.4219

**24d.** 0.0469

**26.** 1800 full wings and gray bodies, 600 full wings and ebony bodies, 600 vestigial wings and gray bodies, 200 vestigial wings and ebony bodies

**28a.** *Evaluate:* The Blue Persian and Spanish Dwarf varieties are assumed to be true-breeding for seed color and plant height; therefore, the resulting $F_1$ generation will be heterozygotes. *Deduce:* Since tall and white are expressed in the heterozygotes, they are the dominant traits. The reappearance of short and blue in

the $F_2$ confirm that they are the recessive traits. *Solve:* Tall and white are the dominant traits, whereas short and blue are recessive.

**28b.** The expected phenotypic distribution in the $F_2$ is 9/16 tall, white; 3/16 tall, blue; 3/16 short, white; and 1/16 short, blue.

**28c.** The hypothesis being tested in this experiment is that the two pea plant varieties differ at two independently assorting genetic loci, and two alleles show simple dominance at each locus.

**28d.** The chi-square value is 0.78, which corresponds to a *P* value between 0.9 and 0.7, indicating that the hypothesis cannot be rejected (the results are consistent with the hypothesis).

**30a.** 9/16

**30b.** 7/16

**30c.** 1/2

**32a.** *Evaluate:* The total number of children in these families was 480. The number of children with CF was (52 × 1) + (32 × 2) + (18 × 3) + (2 × 4) = 178; therefore, 480 − 178 = 302 were normal. *Deduce:* The expected number of children with CF is 480 × ¼ = 120; therefore, 480 − 120 = 360 are expected to be normal. *Solve:* The chi-square value using these numbers is 37.4. The number of degrees of freedom is (2 classes) − 1 = 1. Using Table 3.4, the chi-square of 37.4 for 1 degree of freedom corresponds to a *P* value less than 0.001, which is much lower than 0.05, indicating that observed results are inconsistent with those expected and that you must reject the hypothesis that CF is inherited as an autosomal recessive trait based on these results. *Solve:* The observed results for the total number of children with CF in families in which both parents are carriers are inconsistent with expectations (*P* value < 0.001).

**32b.** One expects 38 families to have no CF children, 50.6 to have 1 child with CF, 25.3 to have 2 children with CF, 5.6 to have 3 children with CF, and 0.47 to have 4 children with CF.

**32c.** *Evaluate:* The observed values are given in the problem, and the expected values were calculated in the answer to part b. *Deduce:* The chi-square using these numbers is 47. The number of degrees of freedom is (5 classes) − 1 = 4. Using Table 3.4, the chi-square value of 47 for 4 degrees of freedom corresponds to a *P* value less than 0.001. This indicates that the difference between the observed and expected results is highly statistically significant and therefore is not consistent with that expected under binomial probability. *Solve:* The results are not consistent with expectations based on a binomial distribution (*P* value < 0.001).

**34.** 0.988

**36.** The probability of 5 unaffected children is 0.237, of 4 unaffected and 1 affected is 0.396, of 3 unaffected and 2 affected is 0.264, of 2 unaffected and 3 affected is 0.0879, of 1 unaffected and 4 affected is 0.0146, and of all 5 affected is 0.000977.

**38.** *Experiment One:* (1) Cross the true-breeding short, brown-furred and long, white-furred guinea pigs to create an $F_1$ population; (2) test that they are dihybrids by crossing male $F_1$ with female $F_1$ to produce $F_2$ guinea pigs; (3) cross both sets of pure-breeding parents to create 24 $F_1$ progeny and intercross all the $F_1$ (12 crosses) to produce 144 $F_2$; (4) compare observed phenotypic distribution with expected results. *Experiment Two:* (1) Cross the $F_1$ with their long, white-furred parent to produce backcrossed guinea pigs; (2) cross two of the $F_1$ males with their long, white-haired mother, thus producing 24 progeny; and (3) cross the long, white-haired male guinea pig with all of the short, brown-haired female $F_1$, which would produce 144 progeny.

**40.** Cross the parents to produce (*FfRrTt*) trihybrids, and self-fertilize the $F_1$ to create an $F_2$ population that will include all possible phenotypes and genotypes. Among the $F_2$, 3/64 will produce yellow, pear-shaped tomatoes and have axial flowers (*ffrr*). To determine which of these plants are *TT*, self-fertilize them and identify the plants that breed true for axial flower position.

**42a.** All four adults are heterozygous carriers of alkaptonuria (*Aa*).

**42b.** For Sarah and James, the chance that their next child will have alkaptonuria is 1/4, not *very low*; for Mary and Frank, the chance is 1/6000, not *0*.

**42c.** 1/4

**42d.** 3/4

**42e.** The probability that one of their children with alkaptonuria will have a child with alkaptonuria is dependent on the genotype of their child's mate. *Deduce:* If their child's mate has no family history of alkaptonuria, then there is a 1/1000 chance that they will be a carrier. If the mate is a carrier, then there will be a 1/2 chance that a grandchild will have alkaptonuria. The overall probability is therefore (1/1000) × 1/2 = 1/2000. The probability that their affected child will have a child with alkaptonuria will increase dramatically if their child's mate has a close relative with the disorder. For example, if their child's mate's grandmother had alkaptonuria, then the mate will have at least a 1/2 chance of being a carrier, and the overall probability that their first child will be affected is at least 1/2 × 1/2 = 1/4.

**44.** The genotypic ratios will be 4/9 *FFpp*, 4/9 *Ffpp*, and 1/9 *ffpp*. The phenotypic ratio will be 8/9 feathered legs and single comb, and 1/9 no leg feathers and single comb.

## Chapter 3

**2a.** 48 chromosomes

**2b.** 48 chromosomes

**2c.** 24 chromosomes (23 autosomes and 1 sex chromosome)

**2d.** 48 chromosomes

**2e.** 48 chromosomes

**2f.** 48 chromosomes

**4.** Cohesion opposes the pulling forces attempting to separate sister chromatids until all pairs of sister chromatids are attached to microtubules from opposite poles of the spindle (i.e., bipolar attachment). Premature, as well as delayed, sister chromatid separation can cause sister chromatids to partition together instead of separating during anaphase and can lead to errors in chromosome segregation. Sister chromatid cohesion is due to cohesin, a protein complex that binds to sister chromatids and attaches them to each other. When all pairs of sister chromatids are under the tension generated by bipolar attachment and sister chromatid cohesion, the protease separase is activated. Separase cleaves a component of cohesin, which simultaneously ends cohesion on all pairs of sister chromatids and allows sister chromatids to be pulled toward opposite poles of the spindle.

**8.** Anaphase II

**10.** A normal human female nucleus contains one Barr body, whereas a normal male nucleus contains no Barr bodies.

**12a.** *Dd*

**12b.** 50%

**12c.** There is no chance that her daughter will have OTD, but there is a 1/2 or 50% chance that her daughter will be a carrier for OTD.

**12d.** *d*Y

**12e.** 1/2 of the daughters and 1/2 of the sons will have OTD.

**14a.** The male parent was $M^+Y$ and $P^+p^-$ and the female was $M^+m^-$ and $P^+p^-$.

**14b.** Among the females with purple eyes, half are $M^+m^-$ and $p^-p^-$ and half are $M^+M^+$ and $p^-p^-$. Males with purple eyes and miniature wings are $m^-Y$ and $p^-p^-$.

**16.** The unusual number of Barr bodies seen in the rare male and female infants is the result of nondisjunction during meiosis, which creates abnormal gametes containing an additional X chromosome or lacking an X chromosome.

**18.** *Evaluate:* The syndrome is an X-linked, recessive trait. Therefore, males inheriting this mutation must express it (hemizygous), whereas it is possible for females to be heterozygous for the mutation (carrier females). *Deduce:* Because of random X inactivation, the tissue of heterozygous females will contain some cells that express the wild-type allele and some that express the mutant allele. The problem states that most female carriers do not show symptoms; therefore, the pattern of X-inactivation normally results in sufficient wild-type gene expression to promote normal development. *Solve:* In female carriers that show symptoms, X-inactivation must have occurred such that wild-type gene expression was insufficient for normal development. The symptoms are less severe in these symptomatic female carriers because they express some level of the wild-type allele.

**22.** The presence of horns is a sex-influenced trait controlled by a single, autosomal gene (designated *H*) in these cattle. Males need only one *H* allele, whereas females need two. *Cross I:* Male is *HH*, female is *hh*; $F_1$ are 1/2 *Hh* (horned males) and 1/2 *Hh* (hornless females); $F_2$ males are 3/4 horned (1/4 *HH* and 2/4 *Hh*) and 1/4 hornless (1/4 *hh*). $F_2$ females are 1/4 horned (1/4 *HH*) and 3/4 hornless (2/4 *Hh* and 1/4 *hh*).

**24a.** The reciprocal crosses produced different results, which is diagnostic for sex-linked traits.

**24b.** The female is the heterogametic sex (ZW), whereas males are homogametic (ZZ). Black spot is dominant, and nonspotted is recessive. Using Z-linked alleles designated *B* for black-spot and *b* for nonspotted, the parents of cross I are *Bb* male and *b*W female. Their progeny are 1/4 black-spot males (*Bb*), 1/4 nonspotted males (*bb*), 1/4 black-spot females (*B*W), and 1/4 nonspotted females (*b*W). For cross II, the parents are a *bb* male and a *B*W female. Their progeny are approximately 1/2 *Bb* males and 1/2 *b*W females.

**26.** Rare sex-reversed males carry an altered X chromosome that contains a fragment of the Y chromosome including the *SRY* gene. Thus they develop male sex characteristics yet lack a Y chromosome. Rare sex-reversed females carry an altered Y chromosome the lacks the *SRY* gene. Thus they develop female sex characteristics even though they have a Y chromosome.

**28a.** Female is *ECec; Vv* and the male is *ecY; Vv*.

**28b.** The parents are *ECec; Vv Ee* female and *ECY; Vv Ee* male.

**28c.** The parents are *ECec; Vv ee* female and *ecY; Vv Ee* male.

**30.** *Evaluate:* The children of the first mating, $Cc \times cY$, are expected to be 1/2 color blind and 1/2 normal, with equal proportions of male and female progeny in each class. The children of the second mating, $cc \times CY$, are expected to be 1/2 color blind and 1/2 normal, but all the color-blind individuals will be male and all the normal individuals will be female. *Deduce:* The reciprocal matings give different results, which is indicative of X-linked inheritance. *Solve:* The results suggest recessive inheritance because the mother in the first mating is not affected but has affected children. The results differ from those predicted for autosomal recessive inheritance because the second mating yields all males with one phenotype and all females with the other.

## Chapter 4

**2.** *Evaluate:* Epistasis indicates that two or more genes interact to contribute to a particular phenotype. Pleiotropy indicates that one mutant allele contributes to two or more mutant phenotypes. *Deduce and Solve:* Epistasis and pleiotropy can be distinguished by inheritance patterns in pedigrees or from crosses. If the inheritance pattern of one phenotype indicates that more than one gene is segregating, then epistasis is occurring. If two or more phenotypes are inherited together in a pattern that indicates segregation of alleles of a single gene, then pleiotropy is occurring.

**4a.** The bacteria in the 12 colonies that grew on minimal medium are prototrophs, whereas those in the 3 colonies that could not grow on minimal medium were auxotrophs.

**4b.** The bacteria in the three colonies that could not grow on minimal medium but could grow on minimal medium plus serine are serine-requiring auxotrophs. They carry mutations in one or more genes required for the biosynthesis of serine.

**4c.** Mutant 1 carries a mutation in the gene coding for a component of enzyme C, mutant 2 carries a mutation in the gene coding for a component of enzyme A, and mutant 3 carries a mutation in the gene coding for enzyme B.

**6.** Child b's parents must be 1. Child a's parents could be 1 or 3; however, since child b's parents are 1, child a's parents must be 3. Child c's parents could be 3 or 4; however, since child a's parents are 3, child c's parents must be 4. Child d's parents could be 2, 3, or 4; however, since child a's parents are 3 and child c's parents are 4, child d's parents must be 2.

**10a.** At the B locus, one parent was *Bb* and the other was *bb*. At the D locus, one parent was *Dd* and the other was *dd*.

**10b.** One parent was *BbDd* (brown) and the other was *bbDd* (yellow).

**10c.** At the B locus, one parent was *BB* and the other parent could have had any genotype (*BB*, *Bb*, or *bb*). At the D locus, one parent was *Dd* and the other was *dd*.

**12a.** One parent was *BBDdCc* and the other was *DdCc*. Any genotype is possible at the B locus.

**12b.** The parents were *BbddCc* × *bbddCc*.

**12c.** Both parents were *BbDdCc*.

**12d.** The parents were both *Bb* at the *B* locus. Between the two parents, there was a *DD* and a *CC* locus, but these need not have been in the same parent. The other *D* and *C* loci could have any genotype.

**14.** The $F_2$ will be 1/4 red, 1/2 pink, and 1/4 ivory.

**16.** 2/3 short stature and short limbs and 1/3 normal stature and limbs

**18a.** Pure-breeding red petunias are *aaBB*.

**18b.** Pure-breeding blue petunias are *AAbb*.

**18c.** The phenotypic distribution in the $F_2$ will be 9/16 purple, 3/16 red, 3/16 blue, and 1/16 white.

**20a.** *Evaluate:* Recall that variable expressivity indicates that individuals with the same mutant genotype vary in severity of mutant phenotype. *Deduce:* Evidence of variable expressivity would be the appearance of some individuals with two thumbs affect and others with only one thumb affected. The pedigree shows both types of affected individuals. *Solve:* Yes

**20b.** *Evaluate:* Recall that incomplete penetrance indicates that not all individuals with a mutant genotype show the mutant phenotype. *Deduce:* Evidence of reduced penetrance would be the appearance of an affected child that had unaffected parents. IV-4 and IV-5 are affected children of unaffected parents. The nonpenetrant individuals are III-5 and III-10. *Solve:* Yes

**22a.** Genetic heterogeneity

**22b.** There are five complementation groups: Group 1 is defined by mutations 1, 3, and 7; Group 2 is defined by mutations 4 and 8; Group 3 is defined by mutations 5, 6 and 10 and Group 4 is defined by mutation 2. Mutation 9 fails to complement any of the other mutants and may represent complementation group 5.

**24a.** The parental strains are $AAbb$ and $aaBB$.

**24b.** The phenotypic distribution will be 9/16 blue ($A\_B\_$), 6/16 purple ($A\_bb + aaB\_$), and 1/16 red ($aabb$).

**24c.** The progeny of the backcross of the $F_1$ to the $AAbb$ parent will be 1/2 blue ($AABb$ and $AaBb$) and 1/2 purple ($AAbb$ and $Aabb$) progeny. The progeny of the backcross of the $F_1$ to the $aaBB$ parent will be 1/2 blue ($AaBB$ and $AaBb$) and 1/2 purple ($aaBB$ and $aaBb$).

**26a.** $Y_1 \times G_1$ produces all yellow $F_1$, which produce 3/4 yellow and 1/4 green $F_2$. Green appears to be recessive and the 3:1 ratio suggests segregation of alleles of a single gene. $Y_2 \times G_1$ produces all green $F_1$, which produces 3/4 green and 1/4 yellow. Here, yellow appears recessive and, again, the 3:1 ratio suggest segregation of alleles at a single gene. $Y_1 \times Y_2$ produces all yellow $F_1$, which produces 13/16 yellow and 3/16 green. The sixteenths in the $F_2$ suggests alleles at 2 genes are segregating. The two genes segregating in the third cross correspond to the genes segregating in the first two crosses. **Solve:** The results of these crosses indicate that two genes control squash fruit color.

**26b.** In cross I, $Y_1$ is $AABB$, and $G_1$ is $aaBB$. Their yellow $F_1$ progeny are $AaBB$, and their $F_2$ are 3/4 yellow ($A\_BB$) and 1/4 green ($aaBB$). In cross II, $Y_2$ is $aabb$, $G_1$ is $aaBB$, their green $F_1$ are $aaBb$, and their $F_2$ are 3/4 green ($aaB\_$) and 1/4 yellow ($aabb$). The $F_1$ of $Y_1 \times Y_2$ resulting from cross III are $AaBb$ and their $F_2$ are 9/16 yellow ($A\_B\_$), 3/16 yellow ($A\_bb$), 3/16 green ($aaB\_$), and 1/16 yellow ($aabb$).

**26c.** The progeny will be 1/2 yellow ($AaBB$ and $AaBb$) and 1/2 green ($aaBB$ and $aaBb$).

**28a.** All the mutants are temperature-sensitive mutants, carrying a mutant allele of a gene required for normal growth at 37°C. The typical mechanistic explanation of this is that the gene is required for growth at all temperatures, and the mutant allele is functional at 25°C but not 37°C (an alternative explanation is that the gene is required only at 37°C). This explains the five mutants that cannot grow at 37°C. For the two mutants that grow slowly at 37°C, the alleles probably retain partial function at 37°C.

**28b.** This study identifies three complementation groups: Group 1 is defined by A, D and F, Group 2 is defined by B and G, and Group 3 is defined by C and E.

**30a.** The chi-square value is $\dfrac{(22-25)^2}{25} + \dfrac{(23-25)^2}{25} + \dfrac{(55-50)^2}{50} = 1.02$. There are three phenotypic classes; therefore, this calculation has two degrees of freedom. A chi-square value of 1.02 with 2 df gives a $P$ value between 0.5 and 0.7; therefore, the results are consistent with a 1:2:1 hypothesis, and that hypothesis cannot be rejected.

**30b.** The chi-square value is $\dfrac{(22-18.75)^2}{18.75} + \dfrac{(55-56)^2}{55} + \dfrac{(23-25)^2}{25} = 0.75$. There are three phenotypic classes; this calculation has two degrees of freedom. A chi-square value of 0.75 with 2 df gives a $P$ value between 0.5 and 0.7; therefore, the results are consistent with a 9:4:3 hypothesis, and that hypothesis cannot be rejected.

**30c.** Neither hypothesis can be rejected based on the chi-square analysis.

**30d.** Self-fertilize all the purple progeny and determine the proportion that breed true. If 1:2:1 is correct, all purple plants will breed true. If 9:4:3 is correct, then 2/3 of the purple plants will not breed true (i.e., they will produce some whites).

**32.** Strains 1 and 2 are homozygous for mutations in the same gene, $A$, that causes albinism. Strain 3 is homozygous for a mutation in a different gene, $B$, which causes albinism. Strains 1 and 2 are $aaBB$, as are the $F_1$ and $F_2$ of cross A. Strain 3 is $AAbb$. The $F_1$ of cross B and cross C are $AaBb$. The $F_2$ of cross B and cross C are 9/16 $A\_B\_$ (pigmented), 3/16 $A\_bb$ (albino), 3/16 $aaB\_$ (albino), and 1/16 $aabb$ (albino).

## Chapter 5

**2a.** Parental: 41% $DR$, 41% $dr$; recombinant: 9% $Dr$, 9% $dR$

**2b.** Parental: 41% $Dr$, 41% $dR$; recombinant: 9% $DR$, 9% $dr$

**4.** $E$ and $H$ are not genetically linked, because a single crossover in every meiosis results in production of equal proportions of $EH$, $Eh$, $eH$, and $eh$ gametes. $EH$ and $eh$ are parental gametes, whereas $Eh$ and $eH$ are recombinant gametes. The percentage of parental gametes is the same as the percentage of recombinant gametes, which is 50%.

**6a.** Yes, the $y$ and $w$ genes are expected to show linkage because they are less than 50 map units (m.u.) apart.

**6b.** Yes, $y$ is expected to assort independently of $f$ because $y$ and $f$ are more than 50 m.u. apart. The same applies to $w$ and $f$.

**6c.** There will be 24.265% of each of the following progeny types: gray with red eyes and forked bristles, gray with red eyes and normal bristles, yellow with white eyes and forked bristles, and yellow with white eyes and normal bristles. There will be 0.375% of each of the following types: gray with white eyes and forked bristles, gray with white eyes and normal bristles, yellow with red eyes and forked bristles, and yellow with red eyes and normal bristles.

**6d.** The female is heterozygous at $y$, $w$, and $f$, and the male is hemizygous recessive. Consider the linked $y$ and $w$ loci first. The $y$ and $w$ genes are 1.5 m.u. apart, so the females will make 0.4925 of each parental gamete type, which are $y+w+$ and $y-w-$. They will make 0.075 of each recombinant gamete type, which are $y+w-$ and $y-w+$. The $f$ gene is unlinked to $y$ and $w$, so its alleles assort independently of $y$ and $w$; 0.50 of each $y-w-$ genotype receives an $f+$ allele and 0.50 receives an $f$ allele. The fraction of $y+w+f+$ is $0.4925 \times 0.5 = 0.24625$. The same is true for $y+w+f$, $y-w-f+$, and $y-w-f-$. The fraction of $y+w-f+$ is $0.075 \times 0.50 = 0.0375$. The same is true for $y+w-f-$, $y-w+f+$, and $y-w+f-$.

**8a.** See figure.

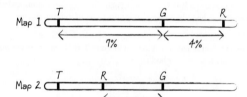

**8b.** A trihybrid organism with dominant alleles on one chromosome and recessive alleles on the homologous chromosome ($GRT/grt$) could be test crossed to a pure-breeding recessive ($grt/grt$). The number of testcross progeny in each outcome category can be used to determine which genetic map is correct.

**10.** Syntenic genes that are separated by 50 map units or more will assort independently because there will be one or more crossovers between them per meiosis.

**12a.** To measure distance between $Y$ and $L$, cross a $yl/y^+l^+$ female to a $yl/Y$ male. The progeny should be 36% yellow lozenge, 36% gray normal eyes, 14% yellow normal eyes, and 14% gray lozenge eyes. To measure the distance between $L$ and $F$, cross an $lf/l^+f^+$ female to an $lf/Y$ male. The progeny will be 34% lozenge forked bristles, 34% normal eyes and normal bristles, 16% normal eyes with forked bristles, and 16% with lozenge eyes and normal bristles.

**12b.** No cross can demonstrate genetic linkage between genes $L$ and $F$, because the recombination frequency between genes $Y$ and $L$ plus that between $L$ and $F$ is greater than 50%; therefore, the percent recombination between $Y$ and $F$ in any cross will be 50%.

**12c.** Syntenic genes (genes on the same chromosome) that are separated by more than 50 map units do not display genetic linkage and, therefore, assort independently.

**14a.** Nail–patella syndrome is a dominant trait.

**14b.** Yes, NPS appears to segregate with blood type A in this pedigree, indicating genetic linkage between these traits.

**14c.** I-1 is $I^On/I^On$, I-2 is $I^AN/I^On$, II-2 is $I^On/I^On$, II-4 is $I^AN/I^On$, II-6 is $I^AN/I^On$, II-7 is $I^On/I^On$, and II-9 is $I^AN/I^On$.

**14d.** III-6 is $I^ON/I^On$ and III-8 is $I^On/I^On$. Even though III-6 and III-8 are both $O$ blood type, III-6 has nail–patella syndrome because he inherited the recombinant chromosome, $I^ON$, from his mother, whereas III-8 inherited the nonrecombinant $I^On$ from her mother.

**14e.** The genotypes of III-11 and III-12 cannot be unambiguously determined, because they have type A blood and both of their parents are $I^AI^O$. Thus, either or both could be $I^AI^A$ or $I^AI^O$. For this reason, it is not clear whether III-11 and III-12 are parental-type or recombinant-type progeny.

**16a.** The order is $G$ $T$ $L$.

**16b.** The recombination frequencies are 0.139 between $G$ and $T$, 0.088 between $T$ and $L$, and 0.214 between $G$ and $L$.

**16c.** The recombination frequency for $G$ and $L$ is less than for $G$ and $T$ plus $T$ and $L$ because the double-crossover progeny do not appear to be recombinant for $G$ and $L$ and, therefore, are not counted.

**16d.** The interference value ($I$) is 0.49.

**16e.** The meaning of $I = 0.49$ is that only half of the expected number of double-crossover progeny were observed. This indicates that a meiotic cell undergoing a crossover between $G$ and $T$ is about half as likely to also have a crossover occur between $T$ and $L$. Similarly, a crossover between $T$ and $L$ reduces the likelihood of a crossover between $G$ and $T$.

**18a.** Yes, the data provides strong support for linkage between Rh and elliptocytosis because the maximum lod score supporting linkage is above 3 (it's about 5.5 for linkage at a θ value just over 0.1).

**18b.** The maximum lod score is about 5.5 for linkage at a θ value just over 0.1.

**18c.** The results support linkage at θ values from just under 0.05 to about 0.30.

**22.** The chi-square value is 168.3. For 3 degrees of freedom, this corresponds to a $P$ value of less than 0.001.

**24a.** The pure-breeding brown-eyed fly line is $ccd^+d^+$, the pure-breeding short-bristled line is $c^+c^+dd$, and the $F_1$ is $cc^+dd^+$.

**24b.** The cross should be a test cross of the $F_1$ dihybrid. The test-cross strain would be $ccdd$. The test cross will yield four progeny categories whose phenotypes will be determined by the dominant or recessive alleles contributed by the $F_1$ dihybrid.

**24c.** The progeny will be 36% $cd^+$, 36% $c^+d$, 14% $cd$, and 14% $c^+d^+$.

**24d.** The progeny will be 25% $cd^+$, 25% $c^+d$, 25% $cd$, and 25% $c^+d^+$.

**28a.** I-1 is either $N1/n2$ or $N2/n1$. I-2 is $n2/n2$. II-1 is $N1/n2$. II-2 is $n2/n2$. III-1 is $N1/n2$. III-2 is $N1/n2$. III-3 is $n2/n2$. III-4 is $N1/n2$. III-5 is $n2/n2$. III-6 is $N2/n2$. III-7 is $N1/n2$. III-8 is $n2/n2$.

**28b.** III-6 is a recombinant. Her genotype indicates that the marker allele 2 is on the same chromosome as the $NF1$ allele, unlike the allele arrangement in her mother (II-1).

**28c.** 1/8

**30a.** The gene order is scute, echinus, crossveinless. The allelic phase in the trihybrid is $+ e + / s + c$.

**30b.** The recombination frequency between scute and echinus is 0.067. The recombination frequency between echinus and crossveinless is 0.095. The recombination frequency between scute and crossveinless is 0.162.

**30c.** There are no discrepancies across this genetic interval.

**30d.** The chi-square value is 38,555, which corresponds to a $P$ value well below 0.01, which indicates that the results of this experiment are not due to independent assortment.

**32a.** The chi-square values for both sets of data correspond to P values well below 0.01 and therefore indicate that the results significantly deviate from expectation based on independent assortment. This supports linkage of the colorless and waxy genes.

**32b.** The recombination frequency from cross 1 was 0.27, and the recombination frequency from cross 2 was 0.24.

**32c.** Yes, both sets of data are compatible with the hypothesis of genetic linkage, although the recombination frequencies of the two sets differed slightly.

**32d.** The recombination frequency using combined data is 0.267

## Chapter 6

**2.** Link 1: Transfer of an entire $F^+$ plasmid from an $F^+$ cell converts an $F^-$ cell to an $F^+$ cell. Link 2: Integration of the F plasmid into the host chromosome converts an $F^+$ cell to an Hfr cell. Link 3: Precise excision of the F plasmid from the chromosome of an Hfr cell converts an Hfr cell to an $F^+$ cell. Link 4: Excision of the F plasmid plus some host DNA from the chromosome of an Hfr cell converts an Hfr cell into an $F'$ cell.

**4.** *Evaluate:* All three mechanisms can involve homologous recombination of the transferred DNA into the recipient chromosome. In all three mechanisms, if the DNA entering the cell is linear or does not contain an origin of replication, then recombination of the DNA into the recipient cell chromosome or episome is required for the DNA to be stably maintained. If the DNA entering the recipient is circular and contains sequences required for replication and maintenance, then recombination into the recipient cell chromosome or an episome is not required.

These three mechanisms differ in how DNA is transferred from one cell to another. Only conjugation requires genetic information for transfer in the donor cell (F plasmid DNA) and physical contact between the donor and recipient cell. Transduction is characterized by infection of the donor cell by a bacteriophage. On the other hand, transformation does not require particular genes in the donor or the help of a phage: DNA is released from the "donor" cell due to cell lysis and enters the "recipient" cell via DNA transporters.

**6.** Lysis of an infected bacterial host cell, and the release of progeny phage particles, is the end result of the lytic cycle of bacteriophage. Lysogeny involves the integration of the phage chromosome (known as a prophage once integrated) by site-specific recombination into a specific site (DNA sequence) in the bacterial chromosome. Once integrated, the prophage can replicate along with the rest of the bacterial chromosome until conditions induce excision of the prophage and resumption of the lytic cycle.

**8.** A *prophage* is a bacteriophage genome that is part of the host cell chromosome. It is formed by integration of a bacteriophage chromosome into the host cell chromosome by site-specific recombination.

**10.** In genetic complementation, bacterial lysis occurs because the two viruses have mutations in different genes. This is analogous to complementation analysis in eukaryotes. In recombination, bacterial lysis occurs because, although the viruses have mutations in the same gene, rare homologous recombination events produce recombinant wild-type viruses. Complementation and recombination can be differentiated by the frequency of bacterial lysis after simultaneous infection with two mutant viruses: lysis is frequent in the case of complementation (many plaques are formed), whereas it is rare in the case of recombination.

**18a.** Selection for $met^+$ was done on minimal medium containing glucose and phenylalanine. The $met^+$ transductants were assayed for cotransduction of $phe^+$ using minimal medium containing glucose. The $met^+$ transductants were assayed for cotransduction of $ara^+$ using minimal medium containing arabinose. Selection of $phe^+$ transductants was done on minimal medium containing glucose and methionine. The $phe^+$ transductants were assayed for cotransduction of $met^+$ on minimal medium containing glucose. The $phe^+$ transductants were assayed for cotransduction of $ara^+$ on minimal medium containing arabinose. The $met^+$ $phe^+$ transductants were selected on minimal medium containing glucose. The $met^+$ $phe^+$ transductants were assayed for cotransduction of $ara^+$ on minimal medium containing arabinose. The $ara^+$ transductants were selected on minimal medium containing arabinose, phenylalanine, and methionine. The $ara^+$ transductants were assayed for cotransduction of $met^+$ on minimal medium containing arabinose and phenylalanine. The $ara^+$ transductants were assayed for cotransduction of $phe^+$ on minimal medium containing arabinose and methionine.

**18b.** The gene order is *ara-phe-met*.

**20a.** 4 genes

**20b.** Mutations 1, 5, and 8 are in one gene. Mutation 2 is in a second gene. Mutations 3 and 7 are in a third gene. Mutations 4 and 6 are in a fourth gene.

**20c.** Complementation resulted in the formation of many plaques on each plate (the lysis of many different bacteria) due to coinfection by bacteriophage with mutations in different genes. The vast majority of these phage are mutants that cannot by themselves infect and lyse bacteria. Recombination between mutations in the same gene results in rare plaques (very few bacteria lyse); however, all the virus particles produced are wild type and can infect and lyse bacteria.

**20d.** Mutation 9 is a deletion that inactivates two genes. It overlaps mutations 1 and 7 but not mutation 3, 5, or 8.

**20e.** Mutation 10 is a deletion mutation that inactivates two genes. It overlaps mutations 4 and 8 but not mutation 1, 5, 6, or 9.

**20f.** The mutation order is 3, 7, 1, 5, 8, 4, 6, and 2.

## Chapter 7

**2.** *Evaluate:* The key results were those showing that enzymes that destroyed RNA and protein did not destroy the transforming principle, whereas enzymes that destroyed DNA did. *Solve:* The transforming principle was considered to be genetic material. The most reasonable interpretation of those results was that DNA was the only essential component of the transforming principle, and therefore, the genetic material.

**4.** *Evaluate:* Hershey and Chase prepared T2 particles whose protein was labeled with the radioactive sulfur, $S^{35}$, and whose DNA was labeled with radioactive phosphorous, $P^{32}$. They used the labeled T2 to infect bacteria and then separated the infected bacteria from the empty phage shells (phage ghosts) using a blender. They found that essentially all the $P^{32}$-labeled T2 DNA but little to none of the $S^{35}$-labeled T2 protein was in the infected bacterial cells. *Solve:* Since T2 genetic material must be inside the infected cells in order to direct new virus particle synthesis, these results pointed to DNA as the genetic material of phage T2.

**6. *Evaluate:*** The problem concerns the chemical bonds that form base pairs in double-stranded DNA. Recall that hydrogen bonds are weak, non-covalent bonds involving two atoms sharing a hydrogen nucleus. The distance between the atoms sharing the hydrogen nucleus is critical for hydrogen bonds to form. ***Deduce:*** The bases in the two complementary antiparallel DNA strands are aligned such that each of the atoms that share hydrogen nuclei (N and O or N and N) in each base are positioned next to each other at a distance that allows all possible hydrogen bonds to form. ***Solve:*** The bases in the complementary but parallel strands are not aligned in this manner; therefore, the atoms that could form hydrogen bonds do not align and are not close enough together to allow hydrogen bonding between all possible and necessary chemical groups.

**8a.** Phosphodiester

**8b.** Hydrogen bonds

**8c.** There are 24 phosphodiester bonds in the molecule.

**8d.** There are 33 hydrogen bonds in the DNA molecule.

**10.** DNA polymerase III determines which free nucleotide triphosphate is complementary to the base being copied. DNA polymerase III catalyzes phosphodiester bond formation between the $\alpha$-phosphate of the incoming nucleotide triphosphate and the 3′ hydroxyl group of the last nucleotide added to the strand.

**12.** RNA is synthesized and serves as a primer for elongation by DNA polymerase.

**14a.** DNA polymerase I is required to remove the RNA primer and fill in the gap with DNA. DNA polymerase III is responsible for the bulk of synthesis of DNA on the leading and lagging strands.

**14b.** The absence of DNA pol I will not prevent the bulk of DNA replication but will result in newly replicated DNA containing small segments of RNA and nicks at the junctions of polymerase III synthesized DNA and the 5′ end of the RNA primers.

**14c.** An *E. coli* mutant without a functional DNA polymerase III will be unable to replicate its DNA because it lacks the enzyme responsible for the bulk of DNA synthesis during replication.

**16a.** True

**16b.** False

**16c.** True

**16d.** False

**16e.** True

**18.** Helicase, SSB, primase, DNA pol III, DNA pol I, ligase

**20. *Evaluate:*** Recall the Meselson-Stahl experiment and consider how the results excluded the alternatives to the semiconservative model for DNA replication. Meselson and Stahl initially cultured *E. coli* in medium containing only N14 (heavy nitrogen) until all cells contained only N14/N14 DNA. They then cultured the N14/N14 *E. coli* in normal medium (N12) and collected samples after one, two, and three rounds of DNA replication. The results showed that before transfer to N12 medium, only N14/N14 DNA was present. After one round of replication in N12 medium, all of the DNA was N14/N12; after two rounds of replication, half the DNA was N14/N12 and half was N12/N12; and after the third round of replication, 1/4 of the DNA was N14/N12 and 3/4 was N12/N12. ***Deduce:*** The conservative model predicted that the original N14/N14 DNA would remain throughout, and therefore the results ruled out the conservative model. The dispersive model predicted that after each round of replication there would be only one form of DNA, which would become less and less dense. ***Solve:*** Although this model was not ruled out after one round of replication, the persistence of the N14/N12 DNA and the presence of two classes of DNA (N14/N12 and N12/N2) after rounds two and three ruled out the dispersive model.

**22. *Evaluate:*** Cells were incubated in medium containing 3H-thymine for a short period of time (a "pulse") and then transferred to medium containing an excess of unlabeled thymine (the "chase"). The cells were then collected and their DNA was prepared for electron microscopy, which can detect replication bubbles in DNA, and for autoradiography, which reveals the location of 3H-thymine incorporation into DNA. ***Deduce:*** Recall that bidirectional replication from a replication origin produces a replication bubble with DNA synthesis occurring at both ends, whereas unidirectional replication results in a replication bubble with DNA synthesis occurring at one end. The results showed that DNA replication bubbles contained regions of label on the sides of the midpoint of the bubble, which corresponds to the replication origin. ***Solve:*** If DNA synthesis was unidirectional, then label would be present on only one side of the origin.

**24. *Evaluate:*** DNA helicases unwind dsDNA during DNA replication and repair. Bloom syndrome is characterized by chromosome instability and an increased rate of cancer. Chromosome instability is evident when chromosomes are lost

from cells, typically because of a failure during mitosis. Mitosis fails to occur properly if chromosomes are not completely replicated. Cancer is a disease caused by accumulation of somatic mutations, which will accumulate at an elevated rate if DNA repair by DNA replication is defective. ***Solve:*** Based on the information provided, it is reasonable to speculate that lack of the DNA helicase encoded by the Bloom syndrome gene results in incomplete replication during S phase and during repair of DNA damage. Failure to completely replicate chromosomes could result in a failure to pass chromosomes on to progeny cells during mitosis, which would results in chromosome instability. Failure to repair DNA damage would also lead to an increased rate of somatic mutation, which would lead to cancer.

**26a.** Telomeric DNA is composed of a repetitive, short DNA sequence. In many organisms, the repeated sequence is 5′-TTAGGG-3′ or a variant thereof.

**26b.** Telomerase uses a segment of its RNA as the template to add multiple copies of a simple sequence to the 3′ end of each strand of DNA on a linear chromosome. This strand, which corresponds to the template for lagging strand synthesis, is copied by the normal mechanism of lagging strand synthesis after it is extended by telomerase.

**26c.** Telomeres are thought to provide two functions, one in chromosome replication and the other in chromosome protection. Telomeres provide a mechanism for replication of the ends of linear chromosomes. Without telomeres, lagging strand synthesis would fail to extend to the chromosome ends, leaving a gap at each end after each round of replication. This would shorten the chromosome and, after many rounds of replication, would result in loss of important DNA sequences (genes). Telomeres are repetitive DNA, which prevents loss of important DNA sequences if shortening occurs. Telomeres are also the binding site for telomerase, which extends the lagging strand template to compensate for sequences lost during incomplete lagging strand synthesis. Telomeres also provide a protective "cap" on the ends of linear chromosomes; this cap distinguishes normal chromosome ends from ends generated by double-stranded chromosome breaks (DNA damage). Without telomeric DNA and the proteins that bind telomeric DNA, the ends of chromosomes are recognized as broken chromosomes and are fused together by DNA repair enzymes. Such breakage can create chromosome end-to-end fusions, which then create dicentric chromosomes that can be broken during the next cell division, creating new breaks and new fusions in an endless cycle known as the bridge-break-fusion cycle.

**26d. *Evaluate:*** Germ-line cells divide many times, whereas many somatic cells are capable of a limited number of cell divisions (some are unable to divide at all). ***Solve:*** Telomerase is required to ensure complete chromosome replication in germ cells, ensuring that every mitosis produces two daughter cells with complete chromosomes. Telomerase is not required in somatic cells, because they cannot divide enough times to result in loss of important DNA at chromosome ends. It is also thought that the lack of telomerase in somatic cells prevents indefinite cell division, because loss of DNA at chromosome ends will activate DNA damage responses that stop division and lead to cell death. This response would help protect the organism from the spread of cancerous cells.

**28a.** The reaction would have equal concentrations of deoxycytidine triphosphate, deoxythymidine triphosphate, and deoxyguanidine triphosphate. It would also have a mixture of deoxyadenosine triphosphate and dideoxyadenosine triphosphate.

**28b.** Dideoxysequencing uses DNA synthesis to generate labeled DNA fragments of different lengths, which are then resolved by gel electrophoresis or column chromatography. To visualize the products of DNA synthesis in traditional dideoxysequencing, relatively high levels of template were necessary. The use of PCR allows detectable levels of DNA synthesis from much lower levels of template DNA.

**28c.** Dideoxynucleotides contain a hydrogen group instead of a hydroxyl group on their 3′ carbon. When a dideoxynucleotide is incorporated into a growing DNA strand, there is no 3′ hydroxyl group present to allow phosphodiester bond formation with the next nucleotide to be added; therefore, no additional nucleotides are added to this DNA strand, and thus synthesis of this strand is terminated.

**32.** Approximately 7500 origins of replication. Five minutes = 300 seconds. Working bidirectionally, each origin generates (300 sec.)(40 nt.)(2) = 24,000 nucleotides per second, requiring $1.8 \times 10^8 / 2.4 \times 10^4 = 0.75 \times 10^4$ origins.

## Chapter 8

**2.** The three major modifications of mRNA are 5′ capping, intron splicing, and 3′ polyadenylation. The process of 5′ capping involves addition of a guanosine monophosphate by guanylyl transferase to the 5′ end of a pre-mRNA via a 5′-to-5′ triphosphate linkage and the subsequent methylation of the guanine and sometimes additional nucleotides on the pre-mRNA. Intron splicing involves the removal of introns from the pre-mRNA and the joining of adjacent exons by the spliceosome. The process of 3' polyadenylation involves the cleavage of the pre-mRNA downstream of the polyadenylation sequence by cleavage factors and addition of 20 to 200 adenine nucleotides by polyadenylate polymerase.

**6.** DNA and RNA polymerases are similar in that both (1) catalyze phosphodiester bond formation to polymerize nucleotides into nucleic acids, (2) polymerize in a 5'-to-3' direction, and (3) are dependent on a DNA sequence template. DNA and RNA polymerases differ in that (1) RNA polymerase can initiate strand synthesis whereas DNA polymerase can only extend an existing strand, (2) most DNA polymerases can proofread using a 3'-to-5' exonuclease activity whereas RNA polymerases cannot, and (3) DNA polymerases use deoxyribonuclotide triphosphates as substrates whereas RNA polymerases use ribonucleotide triphosphates as substrates.

**8.** The primary transcripts of bacterial and eukaryotic genes differ in that bacterial transcripts often contain more than one coding sequence (they are polycistronic) whereas eukaryotic transcripts do not. Polycistronic mRNAs allow for coordinate regulation of production of several proteins by controlling initiation of transcription of only one gene. Eukaryotes accomplish this by coordinate regulation of transcription of multiple genes by gene-specific transcription factors. Prokaryotic and eukaryotic primary transcripts differ in that eukaryotic transcripts are extensively modified before translation whereas prokaryotic transcripts are not. The modification of eukaryotic transcripts includes 5' capping and 3' polyadenylation, which generate structures that are critical for regulation of the initiation of translation and for controlling the half-life of the mRNA. Mechanisms controlling translation initiation and mRNA half-life in bacteria do not involve these structures. The modification of eukaryotic transcripts also includes intron splicing, which is required for generating the complete open reading frame used in translation and allows for the generation of multiple, different (but related) mRNAs from a single primary transcript. This last mechanism increases the number of different proteins that are coded by a genome without increasing the number of genes present.

**10.** Recall that enhancers are DNA sequences that increase the level (rate) of transcription of genes in a position- and orientation-independent manner. Enhancers are binding sites for transcription factors that stimulate transcription of one or more genes. Since the expression of the transcription factors is often specific to the cell type or tissue, enhancers often provide for a mechanism to stimulate transcription of genes in a manner specific to the cell type or tissue. Possible rationales for the lack of enhancers in bacteria include (1) the lack of differentiated cell types in most bacteria; (2) little to no intergenic space on bacterial chromosomes, which makes long-range-acting enhancer sequences unnecessary; and (3) bacterial operons make coordinate regulation of protein synthesis by enhancers unnecessary.

**18. *Evaluate:*** This problem requires application of your understanding of the band shift assay to match each condition listed to the result shown on a gel. ***Deduce:*** The bands in lanes 2 and 4 have migrated the most rapidly; therefore, they correspond to naked DNA molecules. Lanes 1, 3, and 5 have migrated more slowly than naked DNA; therefore, they are DNA+ protein complexes. Lane 1 showed slightly higher mobility than lane 5, which showed higher mobility than lane 3. Higher mobility indicates less protein is bound to the DNA. ***Solve:*** Conditions c and d should result in naked DNA because c contains DNA only, and d contains DNA plus RNA pol II, which cannot bind to DNA in the absence of general transcription factors. Therefore, c and d correspond to lanes 2 and 4 (either lane is equally possible for either condition). Condition e contains the lowest number of transcription factors, followed by condition a, and condition b has the most transcription factors. Thus, lane 1 corresponds to condition e, lane 5 corresponds to condition a, and lane 3 corresponds to condition b.

**20a.** The organism transcribes as a wild type.

**20b.** The organism transcribes slowly (i.e., is leaky).

**20c.** The organism does not transcribe genes.

**20d.** Temperature-sensitive mutant

**22a. *Evaluate:*** Recall that AG is the consensus sequence found at the 3' end of introns. ***Deduce and Solve:*** Since this mutation is in an intron but causes a defect in β-globin, it must affect splicing efficiency. The mutation replaces A with U, changing the 3' splice site sequence AG to UG. This change is likely to affect the efficiency with which the spliceosome recognizes the end of intron 2 and leads to either inclusion of intron 2 in the mRNA—which results in an insertion or premature termination—or causes a change in the location of the 3' splice junction, which leads to an insertion, deletion, or frameshift mutation.

**22b.** This problem requires you to consider the structure of genes and identify important DNA sequences that are not part of exons. Non-exon-located mutations that could prevent gene function include mutations in the promoter or terminator sequences as well as in enhancer or silencer sequences. Mutations in the promoter would diminish or prevent transcription, which would reduce or

eliminate the mRNA. Mutations in the terminator could prevent or alter termination, which would elongate the mRNA. Mutations in an enhancer would diminish transcription, which would reduce mRNA abundance. Mutations in the silencer would enhance transcription, which would increase mRNA abundance.

**24a.** First, the eukaryotic promoter is unlikely to be recognized by bacterial RNA polymerase holoenzyme. Second, the introns will not be removed from the pre-mRNA, which will result in production of an abnormal protein. Third, sequences required for efficient translation initiation in bacteria are not present.

**24b.** First, I would make a cDNA copy of the gene. The cDNA is a DNA copy of the mRNA sequence, which lacks introns. Second, I would place the cDNA sequence downstream of a known bacterial promoter, which will ensure that the gene is transcribed. Third, I would modify the coding sequence upstream of the ATG start codon to contain a Shine–Dalgarno sequence, which is important for proper initiation of translation. Fourth, I would place an intrinsic or rho-dependent termination sequence downstream of the cDNA to ensure efficient transcription termination.

**26a. *Evaluate:*** This problem challenges you to interpret the results of a DNA footprinting experiment. Recall that DNA footprints correspond to sites where bound protein protects an end-labeled DNA fragment from digestion by the endonuclease, DNase. ***Deduce and Solve:*** The DNA-only and DNA + protein lanes differ because bands are missing from the DNA + protein lanes. This result indicates that the proteins are bound to the DNA. Since the proteins are transcription factors and RNA polymerase, which bind to promoters, it is reasonable to conclude that the DNA fragment contains a promoter sequence.

**26b.** 200 base pairs

**26c. *Evaluate:*** This problem challenges you to apply your understanding of promoter structure and function to design experiments to test a DNA fragment for promoter function. ***Tip:*** The function of a promoter is to provide all the DNA sequences required for binding of transcription factors and RNA polymerase and for start of transcription. ***Deduce and Solve:*** One reasonable experiment would be to clone this DNA sequence upstream of the coding sequence for a protein whose expression is easy to assay and then introduce that chimeric construct into cells and assay for protein expression. If the result is negative, then the orientation of the fragment should be inverted to check that it was not inserted backward in the first attempt. Also, a known, control promoter should be used to confirm that the protein-coding sequence is correct and that the protein can be detected in the cells used.

## Chapter 9

**2a.** Nirenberg and Matthaei developed an in vitro translation system that contained everything necessary for translation except for amino acids and mRNAs. Synthetic RNA comprised of only U (poly-U) was added to 20 separate reactions, each containing a different radioactive amino acid as well as the other 19 nonradioactive amino acids. Only the reaction with radioactive phenylalanine produced radioactive protein. Since the RNA sequence poly-U contains only UUU codons, then UUU codes for phenylalanine.

**2b.** Only reaction with radioactive proline produces radioactive protein. Since poly-C contains only CCC codons, CCC codes for proline.

**2c.** One type of polypeptide composed of alternating Arg and Glu amino acids was produced.

**2d.** No detectable polypeptides would be produced.

**4.** I. Preinitiation complex formation: the small ribosomal subunit and IF3 bind to the mRNA, the AUG start codon is identified by 16S rRNA base-pairing with the Shine–Dalgarno sequence, and the AUG codon is in the ribosomal P site.
II. Formation of the 30S preinitiation complex: fMet–tRNA$^{fMet}$ bound to IF2–GTP binds to start codon in the P site, and IF1 binds.
III. Formation of the 70S initiation complex: 50S ribosomal subunit binds, IF2 cleaves GTP to GDP + phosphate, and IF1, IF2–GDP, and IF3 leave the complex.

**6.** tRNAs that are charged with different amino acids have unique structural features that allow them to interact with their cognate aminoacyl tRNA synthetases. Unique features include the anticodon sequence as well as sequences and base modifications in the T-arm and D-arm.

**8a.** 5'-CUA-3' and 5'-CUG-3'

**8b.** 5'-UUU-3'

**8c.** 5'-GAG-3'

**8d.** 5'-CAU-3'

**8e.** 5'-AUC-3' and 5'-AUU-3'

**10.** See table.

| | Bacterial Ribosome | Eukaryotic Ribosome |
|---|---|---|
| **Similarities** | | |
| Composition | RNA and protein | RNA and protein |
| Number of subunits | two (small and large) | two (small and large) |
| tRNA binding sites | three (E, P, and A) | three (E, P, and A) |
| **Differences** | | |
| Number and size of rRNAs | three (16S, 23S, and 5 S) | four (18S, 28S, 5.8S, 5S) |
| Size of subunits | 30S and 50S | 40S and 60S |
| Numbers of proteins | 21 in the small subunit and 31 in the large subunit | ~35 in the small subunit and 45 to 50 in the large subunit |

**12a.** The errors in the diagram are (1) ribosome is moving in the wrong direction along mRNA, (2) mRNA contains T's, (3) amino terminal amino acid of the peptide is incorrect, (4) ribosomal subunit sizes are incorrect, (5) anticodon sequence of the tRNA in P site is incorrect, and (6) amino acids on tRNAs in P and A sites are incorrect.

**14.** 31

**16.** See table.

| DNA | Non-template (5′ to 3′) | | | | | | | | | |
|---|---|---|---|---|---|---|---|---|---|---|
| | | AAC | A̲T̲A | TGT | GAA | G̲G̲C̲ | GAG | AA̲T | GA̲A̲ | CGA̲ |
| Template (3′ to 5′) | | TTG | T̲A̲T | ACA | CT̲T̲ | CCG | CTC | TTA | CTT | G̲C̲T |
| mRNA (5′ to 3′) | | AA̲C | AUA | UGU | GAA | GGC | GAG | A̲A̲U | GAA | C̲GA |
| tRNA (3′ to 5′) | | UUG | UAU | ACA | CUU | CCG | CUC | UUA | CUU | GCU |
| Amino acid abbreviations | 3-letter | Asn | Ile | Cys | Glu | Gly | Glu | Asn | Glu | Arg |
| | 1-letter | N | I | C | E | G | E | N | E | R |

**20.** Soon after initiation of translation of an mRNA coding for a secretory protein, the amino terminus of the secretory protein is synthesized and is exposed on the surface of the ribosome. The amino terminus contains the signal sequence that marks this protein for cotranslational translocation into the endoplasmic reticulum (ER). The signal receptor particle (SRP) binds to the signal sequence and the ribosome and halts further translation. The SRP/ribosome/mRNA complex binds to the ER membrane—SRP binds to its receptor, and the ribosome binds to a protein translocation channel. SRP is released, translation resumes, and the growing polypeptide is extruded through the channel into the lumen of the ER. There, the signal sequence is cleaved by signal peptidase and the protein is glycosylated, folded with the help of chaperones, and packaged into transport vesicles destined for the Golgi apparatus. The carbohydrate on the protein is modified as the protein passes through the compartments of the Golgi, and the protein is packaged into vesicles destined for transport to the plasma membrane. Fusion of the transport vesicle membrane with the plasma membrane releases the secretory protein into the extracellular fluid.

**22a.** 5′-UGUGUGUGUGUGUGUG . . .-3′

**22b.** Cys-Val-Cys-Val-Cys-Val-Cys-Val . . .

**22c.** The experiment resulted in production of a polypeptide composed of alternating Val and Cys amino acids, which is what was predicted for a nonoverlapping triplet code.

**22d.** Two polypeptides, each composed of a single type of amino acid, would be produced if the code had been a doublet, nonoverlapping code.

**22e.** Translation of the RNA using an overlapping doublet or triplet code gives the same result—a single type of polypeptide with two alternating types of amino acids. This result does not differ from that predicted based on a nonoverlapping three-letter code but does differ from that predicted for a nonoverlapping two-letter code.

**24.** 438

**30a.** The start and stop codon are in bold print. 5′-*CAP*CCAAGCGUUAC**AUG** UAUGGAGAGAAUGAAACUGAGGCUUGCCACGUUUGU**UAA**GCACCUAUGCUACCG *AAAAAAAAAAAAAAAAAAAAAAAA*-3′

**30b.** Met-Tyr-Gly-Glu-Asn-Glu-Thr-Glu-Ala-Cys-His-Val-Cys (MYGENETEACHVC)

**32.** The consensus sequence is CCCGCCGCCACC**AUG**G. Also see table.

| Position | −12 | −11 | −10 | −9 | −8 | −7 | −6 | −5 | −4 | −3 | −2 | −1 | [start] | +4 |
|---|---|---|---|---|---|---|---|---|---|---|---|---|---|---|
| Percent A | 23 | 26 | 25 | 23 | 19 | 23 | 17 | 18 | 25 | 61 | 27 | 15 | [AUG] | 23 |
| Percent C | 35 | 35 | 35 | 26 | 39 | 37 | 19 | 39 | 53 | 2 | 49 | 55 | [AUG] | 16 |
| Percent G | 23 | 21 | 22 | 33 | 23 | 20 | 44 | 23 | 15 | 36 | 13 | 21 | [AUG] | 46 |
| Percent T | 19 | 18 | 18 | 18 | 19 | 20 | 20 | 20 | 7 | 1 | 11 | 9 | [AUG] | 15 |
| Consensus | C | C | C | G | C | C | G | C | C | A | C | C | **AUG** | G |

**34. Answer:** GCCACCAUGG

## Chapter 10

**2. Evaluate:** Recall that genetic diseases are those determined by genetic factors in the individual's own genome. On the other hand, infectious diseases are caused by foreign agents that have invaded the individual. **Deduce:** All genetic diseases are also molecular diseases since the genes involved encode molecules that are responsible for development of the disease. The only difference between the terms *genetic* and *molecular* is their point of emphasis: *genetic* emphasizes that a disease is not infectious, whereas *molecular* emphasizes that defects in one or more molecules are the ultimate cause of the disease. Sickle cell disease (SCD) is the first genetic disease for which the molecular mechanism was understood. SCD is seen in individuals homozygous for the $\beta^S$ allele, which encodes an altered β-globin polypeptide that leads to formation of an altered hemoglobin protein, polymerization in red blood cells under low oxygen conditions, cells with a sickle shape, blockage of capillaries in peripheral tissue, and ultimately pain and tissue damage.

**4. Evaluate:** Recall that electrophoretic mobility is the rate at which a molecule migrates during electrophoresis. The electrophoretic mobility of a protein is determined by its charge, size, and shape, which in turn are primarily a result of its amino acid sequence. **Deduce:** Two proteins with different electrophoretic mobility typically have some difference in amino acid sequence. If different alleles of a protein-coding gene code for polypeptides that differ in sequence, even if they differ at only one amino acid position, the charge or shape of the proteins can differ and result in differences in electrophoretic mobility.

**6.** See "Template Strand (DNA)" column of table.

| β-globin Form | Position | Amino Acid | Codon (mRNA) | Coding Strand (DNA) | Template Strand (DNA) |
|---|---|---|---|---|---|
| $\beta^A$ (wild type) | 7 | Glu | 5′-GAG-3′ | 5′-GAG-3′ | 5′-CTC-3′ |
| Siriraj | 7 | Lys | 5′-AAG-3′ | 5′-AAG-3′ | 5′-CTT-3′ |
| San Jose | 7 | Gly | 5′-GGG-3′ | 5′-GGG-3′ | 5′-CCC-3′ |
| $\beta^A$ (wild type) | 58 | Pro | 5′-CCU-3′ | 5′-CCT-3′ | 5′-AGG-3′ |
| ZIGUINCHOR | 58 | Arg | 5′-CGU-3′ | 5′-CGT-3′ | 5′-ACG-3′ |
| $\beta^A$ (wild type) | 145 | Tyr | 5′-UAU-3′ | 5′-TAT-3′ | 5′-ATA-3′ |
| Bethesda | 145 | His | 5′-CAU-3′ | 5′-CAT-3′ | 5′-ATG-3′ |
| Fort Gordon | 145 | Asp | 5′-GAU-3′ | 5′-GAT-3′ | 5′-ATC-3′ |

**8. Evaluate:** Recall that mutations that increase the length of a polypeptide either change the stop codon to a sense codon or shift the reading frame before, but close to, the stop codon. By comparing the sequences, we can determine that the wild-type and mutant sequences differ immediately after codon 144. **Deduce:** The continued change in sequence after codon 144 rules out base-substitution mutants, which would affect specific bases only, and points to a shift in the reading frame caused by the insertion of AG between codon 144 and codon 145. The two-nucleotide insertion shifts the reading frame to one that includes 13 sense codons (compared with two sense codons in the wild type) before a stop codon appears, which happens to be after codon 157.

**10.** The primary molecular parameter affecting the electrophoretic mobility of DNA and mRNA is size (typically expressed as length). For proteins, size (the number of amino acids), charge, and shape affect electrophoretic mobility.

**12.** Electrophoretic mobility is a measure of the size (length) of mRNA and DNA molecules; however, comparing them to each other is, in most cases, like comparing apples to oranges. The length of an mRNA is an inherent, biological property of the gene that encodes the mRNA and is not dependent on the method used to prepare the mRNA for gel electrophoresis. The size of the DNA molecule containing the gene is entirely dependent on the experimental method used to prepare it for electrophoresis. For example, the size of a DNA restriction fragment containing a gene is determined by which restriction enzyme is used. DNA fragment sizes will likely differ for different restriction enzymes and is unlikely to be the same length as the mRNA. The exception to this rule is the comparison of a cDNA molecule to an mRNA molecule. A cDNA molecule is a double-stranded DNA copy of the mRNA. If the DNA copy is perfect, then cDNA length in base pairs should equal the mRNA length in nucleotides.

**14.** *Evaluate:* The frequency of $\beta^S$ will be determined by the evolutionary forces acting on the phenotypes of $\beta^S\beta^S$, $\beta^S\beta^A$, and $\beta^A\beta^A$ individuals. The $\beta^S$ allele encodes an abnormal β-globin polypeptide that forms abnormal hemoglobin molecules that reduce the life span of red blood cells. This is most severe in $\beta^S\beta^S$ homozygotes and is detectable but less severe in $\beta^S\beta^A$ heterozygotes. The reduced life span of the red blood cells interrupts the life cycle of the malaria protists and results in a certain level of resistance to malaria. *Solve:* The severity of the anemia in $\beta^S\beta^S$ individuals decreases their reproductive fitness relative to $\beta^A\beta^A$, which selects against the $\beta^S$ allele. The less severe effect in $\beta^S\beta^A$ heterozygotes does not affect fitness, except in populations where malaria is endemic, where it increases reproductive fitness relative to $\beta^A\beta^A$. Based on this, a reasonable hypothesis would be that malaria is more prevalent in western Africa than southern Africa.

**16a.** Yes, the fetus has a 1/4 chance of having SCD.

**16b.** The fetus will develop SCD because it is homozygous for the $\beta^S$ allele.

**18.** Some restriction enzymes make a staggered double-stranded DNA cut at their recognition sequence, cutting the two DNA strands at different positions. This leaves single-stranded DNA ends, called "sticky ends" because they can form base pairs with the single-stranded ends of other restriction fragments and cause the fragments to stick together. Some restriction enzymes make a clean double-stranded DNA cut at their recognition sequence, cutting both strands at the same site. This leaves ends that are blunt in the sense that all of their nucleotides are base-paired. Blunt ends, therefore, are not sticky.

**20.** *Evaluate:* Restriction enzymes break two phosphodiester bonds between the same nucleotides in the same sequence on both strands of DNA. Two sites of action suggest two active sites in the enzyme. *Solve:* Restriction enzymes typically bind as dimers with each monomer binding to the recognition sequence on opposite strands and cutting that sequence in the same location. Thus, 5′GGTACC3′ / 3′CCATGG5′ is bound by a *Bam*HI dimer, where each monomer binds to and positions its active site to break the phosphodiester bond between the G's.

**24a.** One type of mutation would be a deletion of 3.5 kb within the gene but not including the region corresponding to the probe. A second type of mutation would be a point mutation that creates a new restriction site 4.0 kb in from the right end of the map.

**24b.** For the deletion mutant, I would expect either a smaller mRNA (3500 nucleotides shorter than the wild type) or less mRNA if the mutant mRNA is unstable. For the point mutation, I would not expect the mRNA to be different in length (unless the mutation alters pre-mRNA splicing), although the abundance of the mRNA may change if the mutation changes the stability of the mRNA.

**26.** *Evaluate:* Linear DNA molecules can be considered to move through an agarose gel matrix as extended (rod-like) molecules that have the same charge-to-mass ratio regardless of length. *Deduce:* Since their shape and charge-to-mass ratio is the same, the only factor affecting their relative electrophoretic mobility is their length. The gel will slow the longer molecule to a greater extent than the shorter molecule. The same rationale applies to mRNAs, assuming electrophoresis under conditions where mRNAs are linear.

**28.** *Evaluate:* Recall that restriction endonucleases are components of bacterial defense systems that protect against invasion by foreign DNA. *Deduce:* You can infer from the information provided that the restriction enzyme is able to cut foreign DNA, for example, an invading bacteriophage's genome. DNA from a bacteriophage is digested by the restriction enzyme, which decreases the likelihood that the bacteriophage will successfully infect the bacterium.

**30.** Yes, the father could be ♂1 but not ♂2, because puppy P3 has a band that is not in the mother or in ♂2 but is in ♂1.

# Chapter 11

**2.** Recall that the term *haploid* refers to one copy of genetic information. The terms do not conflict because a bacterium is haploid regardless of whether its genes are contained in one or more than one chromosome.

**4.** Approximately $2.9 \times 10^9 / 2 \times 10^2 = 1.45 \times 10^7$ nucleosomes.

**6.** Recall that the G-banding pattern of light and dark bands of chromosomes is characteristic for each chromosome. These distinctive band patterns allow a cytologist to unambiguously identify each chromosome in a human karyotype. This pattern allowed for the development of cytogenetics, which is the genetic analysis of an individual that is performed by microscopy. Genetic abnormalities associated with alterations in chromosome number or structure can be detected by cytogenetic analysis, allowing for the rapid diagnosis of some genetic diseases.

**8.** Interphase chromosomes will be less condensed than metaphase chromosomes. Chromosomes are difficult to resolve by microscopy in interphase whereas chromosomes are easily resolved in metaphase due to their high degree of condensation.

**10.** *Evaluate:* This problem asks you to consider DNA sequence elements that are essential, evolutionarily conserved components of bacterial or eukaryotic chromosomes. *Deduce:* Chromosomes must be replicated and passed on to progeny cells. Bacterial chromosomes must have all the genes essential for bacterial life, an origin of replication for initiation of replication, and a site for attachment to the bacterial membrane to ensure segregation of daughter chromosomes of each cell at cell division. Eukaryotic chromosomes must contain a centromere, a telomere at each end, and multiple origins of replication. Eukaryotic chromosomes must also contain genes essential for eukaryotic life, although these genes can be dispersed among the chromosomes. Natural selection will select against chromosomes that lack sequences required for replication or segregation and the cells that contain them because they will be less fit than those that contain these sequences. The same argument applies to chromosomes that lack essential genes.

**12.** Bacterial chromosomes are typically circular and are attached to the bacterial cell membrane. Circular chromosomes do not require telomeres; therefore, telomeres are not present on bacterial chromosomes and there would be no evolutionary advantage for a chromosome to have them. Bacteria do not have microtubules; therefore, bacterial chromosomes do not need a centromere to facilitate assembly of a microtubule binding site. The membrane attachment site of a bacterial chromosome serves to promote segregation of daughter bacterial chromosomes to opposite sides of a dividing cell, which is analogous to the function of eukaryotic chromosome centromeres.

**14.** Telomeres are composed of repetitive DNA in which the repeated sequence is a simple sequence (for example, TAAGGC repeated many times). Directly next to the telomere are telomere-associated sequences, which are also composed of repetitive DNA, but the repeated sequences are more complex and may include genes. Directly next to the telomere-associated sequences are "normal" chromosome sequences that contain genes and intergenic regions. Note that there is variation in this sequence organization among chromosomes in the same organism and between chromosomes in different organisms.

**16a.** *Evaluate:* Recall that nucleosomes are spaced 200 bp apart and that after S phase, there are two copies of each chromosome. The haploid genome size of *Arabidopsis* is $10^8$ bp. *Deduce:* The number of nucleosomes per genome in a diploid is given by $\frac{10^8}{200} \times 2 = 5 \times 10^5$. There are twice as many nucleosomes after completion of S phase. *Solve:* $2\left(\frac{10^8}{200} \times 2\right) = 10^6$.

**16b.** The histone proteins that were part of nucleosomes before S phase are recycled and used to form the new nucleosomes during S phase. The additional histone proteins required to double the nucleosome number are newly synthesized. Therefore, half of the histone protein present on chromosomes after S phase is newly synthesized, and the other half was already present.

**20a.** Recall that the *E. coli* chromosome is 1000 times longer than an *E. coli* cell, yet it fits into a small region of the cell called the nucleoid. As with other bacterial chromosomes, the *Methanococcus jannaschii* chromosome is compacted by supercoiling and the binding of proteins that fold the chromosome. The combination of supercoiling and folding (condensation) reduces the volume occupied by the chromosome, allowing it to fit into the region of the bacterial cell known as the nucleoid.

**20b.** The chromosome is likely compacted in two steps. First, it is bound by nucleoid-associated proteins and SMC proteins, which loop segments of the chromosome and condense them. Second, the DNA is supercoiled by the action of topoisomerases.

**20c.** Recall that most bacterial chromosomes and plasmids are negatively supercoiled. Negative supercoiling folds chromosomal DNA and promotes unwinding of regions of the chromosome. This promotes access to ssDNA for enzymes such as DNA polymerase and RNA polymerase. Supercoiling of the *M. jannaschii* chromosome folds the chromosome and promotes the function of DNA and RNA polymerases.

**22.** *Evaluate:* Recall that histones interact with DNA in a sequence-independent manner to form nucleosome core particles that are conserved in structure and function from yeast to man. Also recall that evolutionary change of a protein's sequence is under "functional constraint," which limits or prevents changes to sequences that perform essential functions. *Deduce and Solve:* Histone H4 is one of the core histones and is part of the nucleosome core particle. Most of the amino acid residues of H4 interact either with DNA or other histone proteins; therefore, most of the H4 amino acid sequence is under functional constraint (any change to the H4 amino acid sequence will likely be deleterious and therefore will be selected against by natural selection). Thus, little change occurs in the sequence of H4 over many millions of years that separate pea plants and cows from their common ancestor.

**24.** *Evaluate:* Recall that all four histones are synthesized at the beginning of S phase and that, after completion of S phase, half the histone proteins present will be new and half will be left over from previous cell cycles. *Tip:* $^{35}$S-containing methionine can be used in a pulse-chase experiment to radioactively mark all the histone proteins synthesized during one S phase. *Deduce and Solve:* Start with cells that contain unlabeled methionine and are in $G_1$ phase. Place them in medium containing $^{35}$S-methionine and allow them to complete S phase. Take a sample of cells for analysis and then transfer the remaining cells to medium containing unlabeled methionine and allow them to divide and go through multiple rounds of S phase and cell division, collecting samples after each S phase. Analyze the nucleosomes of each sample by microscopy and autoradiography. If nucleosomes are a mixture of old and new histones, then most or all nucleosomes will be radiolabeled after the first S phase, about half will be radiolabeled after the second round, and 1/4 after the third round. If all nucleosomes contain either old or new histones, then about half of the nucleosomes will be radiolabeled after the first S phase, 1/4 after the second, and 1/8 after the third.

**26a.** *Evaluate:* Recall that DNase I degrades DNA that is not protected by protein. "Large amounts" of DNase I would be expected to digest DNA completely, leaving only those sequences bound by nucleosomes protected. *Tip:* Review the DNase footprinting technique introduced in Chapter 9. *Deduce and Solve:* The only DNA sequences protected from DNase I would be those in the nucleosome core particle. There would be one band, approximately 145 bp in size.

**26b.** *Evaluate, Deduce, and Solve:* Each band would represent the DNA bound by histones in the nucleosome core particle. All non-nucleosomal DNA and all linker DNA between nucleosomes would be digested.

**26c.** *Evaluate:* Recall that the 10-nm fiber model for chromatin states that the 10-nm fiber is a linear array of nucleosomes and that the chromatin is not folded into higher order structures. *Deduce and Solve:* The only protection against digestion by DNase I was due to histones bond to DNA in nucleosomes, indicating that no other proteins and no higher order packing of the chromatin at this region occurred. Higher order structures would have generated fragments of DNA that are larger than the nucleosome core.

## Chapter 12

**2.** *Evaluate:* Recall that 5-BrdU is a base analog and that nitrous acid is a deaminating agent. *Deduce and Solve:* 5-BrdU is a thymidine analog that can base-pair like thymine or like cytosine. If 5-BrdU is incorporated in place of thymidine and then base-pairs like cytosine in the subsequent round of replication, it causes an A-T to G-C transition. If 5-BrdU is incorporated into DNA in place of cytidine and then base-pairs like thymine in the following round of replication, it causes a G-C to A-T transition. Nitrous acid converts cytosine to uracil, which will base pair with adenine in the next round of replication and cause a C-G to T-A transition. Nitrous acid can also convert adenine to hypoxanthine, which will base-pair with cytosine in the next round of replication and cause an A-T to G-C transition.

**4a.** The mutation is a frameshift mutation (insertion or deletion).

**4b.** TCT/G-TAC-ATA-TGC-GAG-ACA-AGN

**8.** *Evaluate:* This problem concerns the relationship between the coding sequence of a gene and the function of the encoded protein. Recall that the effect of a single-nucleotide substitution depends on whether the substitution changes the meaning of the codon and, if so, whether that change has a significant impact on the structure of the protein. *Deduce:* Nucleotide substitutions can result in silent, missense, or non-sense mutations. Silent mutations do not change the amino acid sequence of the protein and therefore have no effect on protein function. Missense mutations change one amino acid in a protein. *Solve:* Therefore, the effect on the function of the protein depends on the importance of the amino acid that was replaced and the functional similarity (or lack thereof) of the R-group on the substituted amino to that of the wild-type amino acid.

**10.** *Evaluate:* This problem tests your understanding of the effect of spontaneous mutations on gene function. Recall that spontaneous mutations are typically nucleotide substitutions and, if detected, are not silent. *Deduce and Solve:* (1) Recessive mutations are typically loss-of-function mutations. Wild-type gene sequences have been selected during evolution for optimum function; therefore, any change (mutation) to that sequence is likely to replace a nucleotide maintained by natural selection with one that reduces the function of the gene. (2) Forward mutations include all mutations in a gene that convert it from wild type to mutant, whereas reverse mutations are only those that precisely reverse a specific mutation to wild type. Thus, the number of possible nucleotide changes corresponding to a forward mutation is much greater than those that reverse a given mutation, making forward mutations far more frequent than reversion.

**12a.** *Evaluate:* Consider the close evolutionary relationship between mice and humans and the experimental utility of the mouse as a model research organism. *Deduce and Solve:* The mouse (*Mus musculus*) is a widely used model organism for genetic analysis of mammalian development and physiology, specifically in relation to human disease, because many of these processes in mice and humans are evolutionarily conserved. The other advantage is that with mice, researchers can perform experimental manipulations that are not possible when studying humans or even nonhuman primates.

**12b.** *Evaluate:* Consider the differences between mice and humans. *Deduce:* Although the mouse (*Mus musculus*) is a mammal, there are many developmental, behavioral, and physiological differences between mice and humans. In addition, not every human gene has a homolog in the mouse genome. *Solve:* Therefore, in cases where the physiology or genetics of mice and humans differ, mutations in a mouse homolog to a human disease gene may not provide useful information on the human disease process.

**14.** 1 mutation per 322,182 gametes

**16a.** 2 births with retinoblastoma, 8 births with achondroplasia, and 22 births with neurofibromatosis

**16b.** Two reasons that could explain why the neurofibromatosis (NF1) mutation rate is higher than retinoblastoma (RB1) mutation rate are (1) the *NF1* gene is larger than *RB1*, and (2) a higher percentage of mutations within *NF1* affect NF1 function as compared to *RB1*.

**18.** *Evaluate:* This problem tests your knowledge of the function of the *E. coli* *RecA* gene. *Deduce and Solve:* The *RecA* gene is required for recombination repair of DNA damage. *E. coli* with a null mutation in *RecA* would lack *RecA* function and would be deficient in recombination repair. Recombination repair is used to fill in a single-stranded DNA gap created by the lack of replication of a region due to DNA damage (for example, a UV photoproduct). Several steps in recombination repair are catalyzed by *RecA*, including two-stranded invasion events and a single-stranded DNA cleavage event.

**20.** *Evaluate:* This problem tests your understanding of mutation and homologous recombination. *Deduce and Solve:* Mutation is defined as "a change in DNA sequence." Gene conversion resulting from recombination changes the DNA sequence of a chromosome; therefore, it fits within the definition of mutation. Recombination also combines chromosome sequences in new ways, creating new stretches of DNA sequence; therefore, recombination also is arguably a form of mutation. However, *mutation* is typically reserved to describe DNA sequence changes that are due to processes other than homologous recombination and gene conversion.

**22.** Yes; heteroduplex DNA is always created during homologous recombination.

**24.** *Evaluate:* Recall that gene conversion is rare and results in the conversion of the genotype of one gamete out of four during meiosis. *Deduce and Solve:* The four products of meiosis in multicellular eukaryotes are not identifiable as such and instead are pooled with the four products or with those of hundreds if not thousands of other meioses. Furthermore, these gametes are detected only by mating individuals and observing the phenotype of the resulting progeny. The high numbers of gametes produced and the random sampling of gametes during zygote formation make statistically significant identification of aberrant 3:1 segregation impossible.

**30a. *Evaluate:*** This problem tests your ability to analyze yeast mutant phenotypes. ***Deduce and Solve:*** Prototrophic yeast are able to grow on minimal medium, whereas auxotrophic yeast cannot. Yeast in colonies 4 and 5 grew on complete medium at 25°C but not on minimal medium at either temperature, therefore, colonies 4 and 5 correspond to auxotrophic yeast mutants. The remainder can grow on minimal medium at 25°C and therefore are prototrophic yeast.

**30b. *Evaluate:*** This problem tests your ability to analyze yeast mutant phenotypes. ***Deduce and Solve:*** The yeast in colonies 1 and 2 can grow on all media at 25°C but not on any of the media at 37°C. These yeast mutants are temperature sensitive for growth. The yeast in colony 5 cannot grow on minimal medium at either temperature but can grow on minimal plus adenine at both temperatures. This yeast mutant is an adenine auxotroph.

**30c. *Evaluate:*** This problem tests your ability to analyze yeast mutant phenotypes. ***Deduce and Solve:*** The yeast in colony 4 have two separate mutant phenotypes. This mutant cannot grow on complete medium at 37°C and therefore has a temperature-sensitive growth phenotype. This mutant also cannot grow on minimal medium at 25°C but can grow on minimal plus adenine at 25°C; therefore, it is also an adenine auxotroph. The fact that the yeast mutant corresponding to colony 4 has two different mutant phenotypes is an indication that this mutant carries two separate mutations, one affecting growth independently of adenine metabolism and a second affecting only adenine metabolism.

**34a.** The ascus shows 6:2 segregation of $brp^+ : brp^-$.

**34b. *Evaluate:*** This problem tests your understanding of the process of gene conversion and its relationship to recombination during meiosis. ***Deduce and Solve:*** The aberrant ratio of 6 $brp^+$ : 2 $brp^-$ indicates that gene conversion in the $brp$ locus occurred during the meiosis producing this ascus. Gene conversion is associated with recombination, which indicates that a recombination event was initiated in the region between $ala$ and $cty$ during this meiosis.

**34c. *Evaluate:*** This problem tests your understanding of the genotype of asci that show evidence of gene conversion. ***Deduce and Solve:*** Recombination is accompanied by formation of heteroduplex DNA between the two Holliday junctions that form. A region of the heteroduplex can contain mismatched base pairs if that region in the homolog is not identical. In this case, the heteroduplex included the region containing the sequence difference that distinguishes $bry^-$ from $bry^+$ and therefore contained mismatched bases. These mismatches are repaired by mismatch repair; but the direction of the repair is not controlled, such that a homolog that should have $bry^-$ could be repaired to contain $bry^+$ information. The heteroduplex occupies only a portion of the region undergoing recombination, which in this case did not include the $ala$ or $cty$ loci. Therefore, there was no gene conversion at $ala$ or $cty$, and those alleles segregated normally—4:4.

## Chapter 13

**4.** b, interstitial deletion; c, duplication; d, terminal deletion; e, trisomy; f, reciprocal balanced translocation; g, paracentric inversion; h, monosomy; i, polyploidy

**6.** Since $P$ elements can cause mutations if they insert into genes, limiting their number reduces the likelihood of a new mutation.

**8.** Mutation of a single $IS1$ sequence in *E. coli* will prevent insertion of transposable DNA into the element. Copies of $IS1$ that are not mutated can undergo transposition.

**10.** Yellow-bodied females can be produced from this cross if nondisjunction in the female parent produces an egg with two X chromosomes, and the egg is fertilized by a sperm containing the Y chromosome. The XXY zygote will develop as female and will be homozygous for the recessive yellow-body allele.

**16c.** Two PCR marker combinations are possible: 290, 310, 340; and 290, 340, 380.

**16d.** Four PCR marker combinations are possible: 290, 310; 290, 380; 310, 340; and 340, 380.

**20.** The human and orangutan chromosome have identical banding patterns along their entire lengths, and all four species have the same chromosome 5 banding pattern from band 5q14.1 to the q arm telomere. In comparison to human chromosome 5, the chimpanzee chromosome has undergone a pericentric inversion with breakpoints at approximately 5p13.2 and 5q13.3. The gorilla chromosome differs from the human chromosome from 5q13.3 to the telomere of 5p. It may have undergone a balanced translocation with another chromosome.

**22.** Mosaicism refers to the condition in which the body has cells with more than one karyotype. The range of phenotypic effects observed in these sex chromosome mosaics is dependent on the relative percentages of cells with each karyotype. Greater percentages of XO cells correspond to phenotypes that are more similar to those with Turner syndrome.

**26.** In this analysis, both recombination frequencies are much lower than expected: Recombination between *peach* and *oblate* is 2.1% instead of 17%, and recombination

between *dwarf* and *peach* is 3.1% instead of 12%. This result is consistent with inversion of the chromosome region containing *peach* and the presence of heterozygosity for the inversion in the trihybrid line. Inversion heterozygosity is suppressing the appearance of most of the crossover chromosomes in this cross.

**28a.** 3.5 kb

**28b.** 7.0 kb

**28c.** The insertions of intron sequences into the $P$ element and into the *copia* element are likely to disrupt gene expression from each element.

## Chapter 14

**2a.** A DNA sequence that binds a regulatory protein, such as the *lac* operator sequence.

**2b.** A regulatory protein that binds DNA, such as the *lac* repressor protein.

**2c.** A compound that induces or activates transcription, such as lactose.

**2d.** A compound that interacts with another protein or compound to form an active repressor, such as the *trp* corepressor.

**2e.** A DNA sequence that binds RNA polymerase and regulates transcription, such as the *lac* promoter.

**2f.** A process of transcription regulation through which the binding of regulatory proteins to DNA activates transcription, such as the CAP binding site of the *lac* promoter.

**2g.** A process by which the stereochemistry of a protein is altered to change its interaction capabilities, such as the *lac* repressor protein.

**2h.** A process of transcriptional regulation through which binding of regulatory proteins to DNA blocks transcription, such as *lac* repressor protein binding to *lac* O.

**2i.** A mechanism of transcriptional regulation in which transcription level is modified (attenuated) to meet environmental requirements, such as *trp* operon attenuation.

**4a. *Similarities:*** Both have promoter and operator regulatory sequences. ***Differences:*** Inducible operons bind repressor protein to block transcription and may use positive control to help activate transcription. Inducible operons require an inducer substance to activate transcription. Repressible operons use a corepressor plus the pathway end product to repress transcription. Repressible operons often utilize attenuation.

**4b. *Similarities:*** Both types of regulatory systems utilize allostery in regulating transcription. ***Differences:*** The mechanism and consequences of allostery differ. In *lac* operon regulation, the repressor protein binds the operator, but allosteric change caused by allolactose prevents binding. In *trp* operon regulation, the corepressor protein cannot bind the operator until its allosteric shape is changed by binding to tryptophan.

**4c. *Similarities:*** Both types of operons contain multiple genes that share a single promoter and a single operator sequence. ***Differences:*** Repressible operons often use attenuation and contain a transcribed leader sequence that participates in determining structural gene transcription. This mechanism is not found in inducible operons.

**6.** Attenuation does not involve allosteric changes. Attenuation is the result of transcription of a leader sequence that undergoes translation. Coupling of transcription and translation dictates whether or not transcription continues past the leader sequence and into the structural genes.

**8.** The CAP binding site is part of the *lac* promoter and is located at approximately −60. It binds the CAP–cAMP complex and opens DNA slightly to allow efficient RNA polymerase binding at the *lac* promoter.

**10.** A $Cap^-$ mutation would alter the CAP binding site sequence and render it unrecognizable by CAP–cAMP. The required positive regulation of transcription would not occur, and *lac* operon transcription would be minimal. The strain would be $lac^-$.

**12.** In the absence of arabinose, araC protein negatively regulates transcription by binding at $araO_1$, $araI$, and $araO_2$, inducing the formation of a DNA loop that prevents RNA polymerase from binding to $P_{ara}$, thus blocking transcription of the *ara* operon genes. When arabinose is present, it binds to araC proteins at the *ara* operator sites. This prevents DNA loop formation and makes a CAP binding site available for binding. CAP–cAMP activates *ara* operon gene transcription.

**14.** Antisense RNAs are single-stranded RNAs that are complementary to a portion of specific mRNA transcripts. Bound to their mRNA targets, antisense RNAs can either block translation or lead to the destruction of mRNA. Blocking translation prevents the production of proteins that might initiate unnecessary or harmful actions.

**16a.** Blocks all transcription

**16b.** Produces constitutive transcription

**16c.** Blocks all transcription (this is an $I^S$ mutation)

**16d.** Produces constitutive transcription (this is an $I^-$ mutation)

**16e.** Only minimal transcription will occur.

**18.** See table.

| Genotype | β-Galactosidase | | Permease | | Phenotype |
|---|---|---|---|---|---|
| | Lactose | No Lactose | Lactose | No Lactose | |
| Example: $I^+ P^+ O^+ Z^+ Y^+$ | + | − | + | − | $lac^+$ |
| a. $I^S P^+ O^+ Z^+ Y^+ / I^- P^+ O^+ Z^+ Y^+$ | − | − | − | − | $lac^-$ |
| b. $I^- P^+ O^+ Z^- Y^+ / I^+ P^+ O^C Z^+ Y^-$ | + | + | + | − | $lac^+$ |
| c. $I^+ P^+ O^+ Z^- Y^+ / I^+ P^- O^+ Z^+ Y^-$ | − | − | + | − | $lac^-$ |
| d. $I^- P^+ O^C Z^+ Y^+ / I^+ P^- O^+ Z^+ Y^+$ | + | + | + | + | $lac^+$ |
| e. $I^+ P^+ O^C Z^+ Y^- / I^+ P^+ O^+ Z^+ Y^-$ | + | + | − | − | $lac^-$ |
| f. $I^+ P^+ O^+ Z^- Y^+ / I^S P^+ O^+ Z^+ Y^-$ | − | − | − | − | $lac^-$ |
| g. $I^S P^+ O^+ Z^- Y^+ / I^+ P^+ O^C Z^+ Y^-$ | + | + | − | − | $lac^-$ |

**20a.** $I^- P^+ O^+ Z^- Y^+ / I^+ P^+ O^+ Z^+ Y^+$ will have inducible transcription of both genes. $I^+ P^+ O^C Z^- Y^+ / I^+ P^+ O^+ Z^+ Y^+$ will have constitutive transcription of $lacY$ and inducible transcription of $lacZ$. $cap^+ I^+ P^- O^+ Z^+ Y^+ / I^+ P^+ O^+ Z^+ Y^+$ will have inducible transcription of both genes. $cap^- I^S P^+ O^+ Z^+ Y^+ / I^+ P^+ O^+ Z^+ Y^+$ will be noninducible.

**20b.** The first three partial diploids will be able to grow on a lactose medium, but the final partial diploid ($I^S P^+ O^+ Z^+ Y^+ / I^+ P^+ O^+ Z^+ Y^+$) will not.

**22a.** No; permease is not produced.

**22b.** Transcription of $lacZ$ is inducible from the $cap^+ I^- P^+ O^+ Z^+ Y^-$ chromosome. Only minimal transcription occurs from the other chromosome, so permease is noninducible.

**22c.** The $lacI$ gene has its own promoter and is not affected by $lac$ operon regulation of gene mutations. The $cap^-$ mutation minimizes transcription, but repressor protein produced from this chromosome is trans-active and binds $O^+$ on the other chromosome to induce $lacZ$ expression.

**24.** Gene Z is the enzyme, gene W is the repressor, and G is the operator.

**26a.** The mutant cannot activate $ara$ operon gene transcription. The araC dimer with $araO_2$ cannot form.

**26b.** The mutation may be defective in repressing $ara$ operon gene transcription by being unable to block RNA polymerase access to $P_C$.

**26c.** The mutant cannot repress $ara$ operon gene transcription. The araC dimer with $ara$ cannot form.

**26d.** The mutant cannot repress $ara$ operon gene transcription. The araC dimer cannot form, and $P_C$ cannot be blocked.

**28a.** The mutant is incapable of establishing lysogeny. Lytic gene transcription from the $O_R$ sites cannot be repressed.

**28b.** The mutant is incapable of establishing lysogeny.

**28c.** The mutant will be unable to carry out lysis. No transcription activation occurs from $O_L$ or $O_R$.

**28d.** The mutant will be unable to establish lysogeny. The mutant cannot undertake site-specific recombination to integrate the lysogen.

**28e.** The mutant will be unable to carry out lysis due to the $cro$ mutation, and it will be unable to establish lysogeny due to the $cII$ mutation.

**28f.** The mutant will be unable to carry out lysis. The expression of the late gene that takes place through the antiterminator activity of $N$ at $t_L$, $t_{R1}$, and $t_{R2}$ will not occur.

**30a.** The same band in both lanes

**30b.** No band in lane 1; band in lane 2

**30c.** No band in either lane

**30d.** The same band in both lanes

**30e.** No band in lane 1 and a band in lane 2.

**30f.** No band in either lane.

**32.** See table.

| Genotype | LacZ mRNA Synthesis | Lac Phenotype |
|---|---|---|
| a. $I^- P^+ O^+ Z^+ Y^+ / I^+ P^+ O^+ Z^+ Y^+$ | inducible | $lac^+$ |
| b. $I^+ P^+ O^C Z^+ Y^+ / I^+ P^+ O^+ Z^- Y^+$ | constitutive | $lac^+$ |
| c. $I^S P^+ O^+ Z^+ Y^+ / I^+ P^+ O^+ Z^+ Y^+$ | uninducible | $lac^-$ |
| d. $I^+ P^+ O^+ Z^- Y^+ / I^+ P^- O^+ Z^+ Y^+$ | uninducible | $lac^-$ |
| e. $I^+ P^+ O^+ Z^+ Y^- / I^+ P^+ O^+ Z^+ Y^-$ | inducible | $lac^-$ |

# Chapter 15

**2a.** UAS elements are found in the yeast genome, where they operate as enhancer-like regulatory sequences. Gal4 protein binds yeast UAS elements to activate transcription of galactose utilization genes.

**2b.** Insulator sequences shield genes from enhancer effects. The mechanism of action may be through the formation of specific DNA loops that protect particular genes from enhancers.

**2c.** Silencer sequences prevent transcription of particular genes. The mechanism of action may be through competitive protein binding at silencer sequences that overlap with enhancer sequences. The yeast Mig1 and Tup1 proteins bind a silencer sequence during glycolysis to prevent transcription of galactose utilization genes.

**2d.** The protein complexes that assemble at enhancers to facilitate transcription are known as enhanceosomes. The enhanceosome complex known as Mediator assembles at yeast enhancers. It contacts promoter-bound proteins to activate transcription.

**2e.** RNA interference describes the posttranscriptional regulation of mRNAs by regulatory RNA molecules. RNAi is a prominent feature of the regulation of gene expression in most eukaryotic genomes.

**4.** Acetylation occurs when acetyl groups are added to amino acids of the histone protein by acetylase enzymes. These acetylation events are most often associated with transcription activation, although there are many exceptions.

**6.** The retention of structural similarities among DNA-binding regulatory proteins is an evolutionary consequence of positive natural selection favoring the capability of these proteins to recognize and bind specific regulatory sequences.

**8.** Several factors can be cited, including the following: (1) the presence of a nucleus in eukaryotic cells, (2) the chromatin structure of eukaryotic genomes, (3) multicellularity that is frequent in eukaryotes, and (4) differential gene expression among different types of eukaryotic cells.

**10.** Heterochromatin regions will decondense for DNA replication during the S phase of the cell cycle to allow replisome access.

**12a.** Mutant A has an enhancer mutation. This deletion is located well upstream of the start of transcription and substantially reduces transcription.

**12b.** Mutant B affects a silencer sequence. This deletion results in a substantial increase in the level of transcription.

**12c.** Both mutants C and D are promoter mutations. Their location immediately upstream of the transcription start and the reduced levels of transcription from these mutants are consistent with promoter mutations.

**14a.** The enhancer is most likely in the region at the left-hand side that is present in mutant E and mutant F.

**14b.** The promoter region is most likely in the region at the right-hand side that is present in mutant E and in mutant F.

**14c.** Mutant E likely contains all or most of the enhancer and promoter sequences, but DNA between the sequences is missing. This leads to difficulty forming the correct DNA loop and appears to interfere with efficient transcription initiation. In contrast, mutant F contains additional DNA sequence, particularly between the enhancer and the promoter. Its higher level of transcription indicates greater efficiency in transcription initiation.

**18a.** Enhancer and silencer sequences are each detected in this analysis. The enhancer sequence is located in the deleted region that is common to mutant E and mutant F. The silencer sequence is located in the deletion region unique to mutant E.

**18b.** The deletion in mutant D deletes the *ME1* promoter sequence.

**18c.** It seems likely that regulation of *ME1* is developmentally controlled by the combined activity of an enhancer and a promoter that activate transcription, and a silencer sequence that represses transcription at particular times during development.

## Chapter 16

**2a.** *Sau*3A, $1.17 \times 10^7$; *Bam*HI, $7.32 \times 10^5$; *Eco*RI, $7.32 \times 10^5$; *Not*I, $4.58 \times 10^4$

**2b.** *Sau*3A, $1.08 \times 10^7$; *Bam*HI, $4.32 \times 10^5$; *Eco*RI, $9.72 \times 10^5$; *Not*I, $7.68 \times 10^3$

**4a.** The genomic libraries

**4b.** The two genomic libraries should completely overlap. The cDNA libraries should be a subset of the genomic libraries. The two cDNA libraries should only partially overlap.

**6.** A 16 base-pair (bp) sequence is predicted to occur randomly once in $4.3 \times 10^9$ bp; thus, oligonucleotides should be of at least this length to have a reasonable probability of being unique in the genome.

**8a.** Genetic maps are created by observing linkage between loci; the conclusion that two loci are linked is determined by an indirect measure of physical linkage. Physical maps are created by directly testing whether two pieces of DNA are physically linked.

**8b.** Genetic maps may differ slightly in the distances between loci due to stochastic variability in the frequency of crossovers. If the starting material for the physical maps is the same (i.e., clonal or inbred organisms), the physical maps will be identical.

**8c.** By inclusion of a molecular marker with a known position in a physical map into a genetic map

**10.** Sanger sequencing, based on chain termination, produces longer sequence reads than massively parallel sequencing, which is based on chain synthesis. Since Sanger sequencing often involves the cloning of DNA fragments, sequences that are difficult to clone and maintain in a bacterial host, such as repetitive DNA, will be underrepresented in the Sanger sequencing relative to next-generation sequencing, which does not involve cloning of individual DNA fragments.

**12.** Most recombinant DNA manipulations involving combining of DNA fragments of less than 10 kb, including changing specific base pairs in a known sequence, can be accomplished by synthesis. However, for instances where the exact sequence of the DNA in question is not known, standard recombinant DNA techniques will continue to be needed.

**18a.** Since the cDNA is also cut by *Hind*III, the cDNA may overlap one of the *Hind*III sites found in the genomic clone.

**18b.** The restriction site could be formed from sequences originating from two adjacent exons, such that the complete restriction site sequence is present in the cDNA after the spicing of the two exons together. Restriction sites in genomic regions not incorporated into mature mRNAs (e.g., introns, intergenic, spanning intron–exon boundaries) will not be found in the cDNA.

**20a.** The top sequence produced with the T7 primer is the 5′ end of the gene, and the bottom sequence produced by the T3 primer is the 3′ end of the gene. This is determined by assuming the cDNA is derived from a polyadenylated mRNA; thus, the stretch of T residues in the bottom sequence is the sequence complementary to the poly-A tail of the mRNA.

**20b.** No; the poly-A tail is added posttranscriptionally.

**20c.** The *Eco*RI site (GAATTC) can be found at position 95–100 in the 5′ sequence. Other sites including the *Sma*I, *Bam*HI, and *Xba*I sites can also be seen upstream of the *Eco*RI site. The *Xho*I site (CTCGAG) is found directly preceding the stretch of T residues in the 3′ sequence.

**20d.** The coding sequence should begin with ATG and usually follows the first ATG in the cDNA. The first ATG can be found at about position 143. The sequence preceding this ATG is the 5′ untranslated region (5′-UTR).

**22.** Since meiosis is not required for viability, a genetic screen searching for mutants that fail to undergo meiosis properly would work. However, the ability to cross the mutant for complementation tests would be useful, and thus a screen for conditional mutants would be desirable. Since chemical mutagenesis induces the broadest spectrum of alleles, it would be a better choice of mutagen than insertion of deletion alleles, which are often null. Finally, the simplest genetic system in which meiosis occurs would be the best system to examine this question. *S. cereviseae*, where many genetic tools are available, would be a good choice.

**24.** Because you have no a priori information on the nature of the gene product, homology-based techniques are not applicable. Positional cloning would work, and requires only a mutant phenotype to go from map position to gene. Transposon tagging would work by starting with an organism heterozygous for a mutation in one of the genes and mobilizing the *P* element. Since identifying the mutants is the most time consuming, positional cloning would likely be the best approach.

**26a.** The smaller 4.5-kb cDNA could be sequenced using a primer walking technique. For the 250-kb BAC clone, fragmentation and shotgun sequencing would be a good approach.

**26b.** I would focus on sequencing cDNA clones from the patients in order to identify both exonic and intron–exon boundary mutations. But this approach would not identify mutations in regulatory elements that are in non-transcribed regions.

**28a.** Since *Arabidopsis* is a flowering plant, the female gametophyte (egg) is retained on the female parent, on the placenta. Female gametogenesis can be directly observed within the ovules. Mutations resulting in female gametophytic mutations (e.g., lethality) can be observed as a 1:1 ratio.

**28b.** Since the male gametophyte (pollen) is produced in excess and is not retained on the plant, to observe male gametophytic mutations, I would observe the developing pollen directly.

## Chapter 17

**2.** Difference suggests posttranscriptional regulation. One possibility is that the protein is stable in only one cell type and is rapidly degraded in other cell types. Another possibility is that the mRNA is translated in only one cell type.

**4.** The principles are identical for both species, but the techniques differ because homologous recombination occurs frequently in yeast and rarely in mice. Thus positive–negative selection techniques are required in mice, while only positive selection is required in yeast.

**6.** Gene therapy often targets blood diseases because blood circulates throughout the body. Thus, replacement of mutant bone marrow cells with corrected ones allows the defect to be corrected throughout the body.

**8.** Both methods use "naturally" occurring biological entities. In plants, the Ti-plasmid is reengineered to have the gene of interest and then is reintroduced into *Agrobacterium*, which naturally transfers the T-DNA into the genome at random locations of plant cells. In *Drosophila*, the *P* element is reengineered to have the gene of interest and then injected into embryos, where it integrates into the genome at random locations.

**10.** While PCR or northern blotting approaches can give some perspective on expression patterns, observing them in situ provides more information. For this, either a transcriptional or translational fusion to a reporter gene (e.g., *lacZ* or *GFP*) would be best. The difficulty may be in initially identifying the sequences responsible for proper expression of the gene. These experiments are judged by the following standards: (1) How well does the observed expression pattern of the marker line match with all other data on expression patterns? and (2) Can a translational fusion gene complement a loss-of-function mutant phenotype?

**12a.** Loss-of-function alleles in *S. cerevisiae* can be produced by homologous recombination (see also Figure 17.5). Gain-of-function alleles, such as those that produce the gene product constitutively, can be constructed by making a gene fusion combining a promoter that drives transcription constitutively in *S. cerevisiae* with the coding region of the gene of interest (see also Figures 17.14 and 17.17).

**12b.** Since homologous recombination is not routine in tomato, loss-of-function alleles can be created by using RNAi-mediated mechanisms. In this case, a promoter that drives transcription constitutively can be transcriptionally fused with a sequence containing an inverted repeat, such that the mRNA produced can form a stem–loop including double-stranded RNA (see also Figure 17.18). The chimeric gene can be introduced into tomato using *Agrobacterium* (see also Figures 17.6 and 17.7). Gain-of-function alleles would be produced in a similar manner as described for *S. cerevisiae*, except the regulatory sequences need to be suited to tomato and the transgenic organisms produced by *Argobacterium*-mediated transformation.

**14.** There are two possible approaches: (1) Mutagenize the bacterial strain and screen for mutants that can no longer metabolize crude oil. Then clone the corresponding gene(s) using a complementation assay, as outlined in Chapter 16. (2) Alternatively, clone the gene(s) by transferring large genomic clones from the strain of interest into a related *Pseudomonas* strain that cannot metabolize crude oil. This approach often works in bacteria with specialized traits since the gene conferring the trait is often found within a single operon.

**16a.** There are two possible approaches: (1) perform in situ hybridization using probes made from each of the two genes, or (2) construct translational or transcriptional fusion genes with a reporter gene (*LacZ*, *GFP*) via homologous recombination methods. Translational fusions may retain their functionality as long as the marker gene fusion does not disrupt function of the protein of interest. If in a gene replacement, transcriptional fusions would result in a loss-of-function allele; but if in a recessive gene, transcription fusions would still provide information about gene expression in a phenotypically wild-type mouse.

**16b.** Use homologous recombination techniques to replace the coding region of the gene with a selectable marker. Alternatively, you can use an RNAi-based approach to create a loss-of-function phenotype. The former approach has the advantage of heritability.

**16c.** Create loss-of-function alleles via homologous recombination for each of the genes and examine the mutant phenotypes. Cross the two single mutants to create an F₁ population, interbreed the F₁s to produce an F₂, and identify a double-mutant strain and examine its phenotype. If the double-mutant strain exhibits phenotypic defects beyond what is expected by the addition of the single-mutant phenotypes, then the two genes have redundant functions. Alternatively, you can use an RNAi-based approach to create a loss-of-function phenotype, but again, this approach is not heritable.

**18a.** In yeast, I would create loss-of-function alleles by homologous recombination gene replacement.

**18b.** I would create a translational fusion with a reporter gene (e.g., *GFP*), preferably with all endogenous regulatory sequences.

**20.** Because the original sequence (highlighted) was from reverse translation of the protein sequence, in nucleotide positions where there is degeneracy in potential sequences, these may differ from the actual sequence encoded in the genome.

**22.** The mutant sequence can be created using site-directed mutagenesis. These clones could be used to create wild-type and mutant protein to be studied in vitro. To study in vivo consequences, it would be best to introduce the mutant version of the gene into its endogenous chromosomal location. Unlike the creation of loss-of-function alleles, where a selectable marker replaces the endogenous gene, the creation of gain-of-function mutations is slightly more complicated. One solution is to create the point mutation in the genome via homologous recombination and then remove the selectable marker using a Cre-lox-based system. However, care must be taken not to leave a "footprint" of non-endogenous sequences in any coding or regulatory sequences.

## Chapter 18

**2a.** Repetitive DNA can often be assembled in many different ways, making unambiguous assembly difficult. On a finer scale, repetitive DNA can also lead to polymerase slippage causing sequence errors.

**2b.** Dispersed, repetitive DNA that is longer than a single sequencing read and is found at many locations in the genome is particularly problematic.

**2c.** Paired-end sequencing is one approach to identify unique sequences flanking repetitive DNA.

**4.** cDNA sequences provide information on which genomic sequences are transcribed and processed into mature mRNAs. Different forms of full-length cDNAs from the same region of genomic DNA can indicate alternative splicing.

**6.** In eukaryotic genomes, one must account for the possible presence of introns; in prokaryotic genomes, open reading frames should be contiguous. Predictive algorithms must also take into account differences in promoter and enhancer elements/consensus sequences.

**8.** *Bioinformatic Method:* Use an algorithm to search for potential open-reading frames within the sequence. This method is only predictive and not very accurate, so experimental data is needed to confirm accuracy.

*Comparative Method:* BLAST the sequence against the database of known sequences. If sequences are conserved, they are likely to be functional. This method also needs experimental verification.

*Experimental Method:* Use the sequence as a probe against a cDNA library or other technique (e.g., microarray, rtPCR) to determine which sequences are transcribed. This is the best method, but it is also much more time- and labor-intensive than the others.

**10.** Human proteins are closer to fungal proteins than to plant proteins; plant proteins are equidistant from either human or fungal proteins.

**12.** Expression microarrays represent all the sequences of a genome that are transcribed, whereas tiling arrays represent all of the sequences in a genome. Expression or tiling microarrays can be used to obtain information on which sequences are transcribed in a specific tissue or cell type; tiling arrays can be used to identify binding sites of DNA binding proteins and to assay recombination events between polymorphic strains. For most applications, high-throughput sequencing can provide the same type of information that arrays can. The former provides more quantitative information and information on alternative splicing than the latter. At present, the cost for high-throughput sequencing is higher than for arrays, but this may change in the future.

**14.** This DNA sequence is the synthetic version (the nucleotide sequence inferred from reverse translation of the protein sequence) of the human insulin gene.

**16a.** All three genes are orthologs.

**16b.** *AY1* and *AY2* are paralogs; *AY1* and *BY* are orthologs; *BY* and *CY* are orthologs.

**16c.** *AZ1* and *AZ2* are paralogs; *BZ1* and *BZ2* are paralogs; *BZ2* and *BZ3* are paralogs; *BZ1* and *BZ3* are paralogs; *CZ1* and *CZ2* are paralogs; *AZ1* and *AZ2* are orthologous to *BZ1*, *BZ2*, and *BZ3*; *AZ1* and *AZ2* are orthologous to *CZ1* and *CZ2*; *CZ1* is orthologous to *BZ1* and *BZ2*; *CZ2* is orthologous to *BZ3*.

**18.** While large-scale chromosomal rearrangements appear to have been rare in primate evolution during mammalian evolution, small-scale rearrangements appear to be common and frequent.

**20.** Segmental duplications, resulting in large-scale gene duplication, can often lead to genetic redundancy, especially if the duplication is evolutionarily recent. Using reverse genetics (see Chapter 17), loss-of-function alleles can be created in the duplicate genes. Due to potential genetic redundancy, double mutants of the two paralogs might have to be constructed to observe an aberrant mutant phenotype.

**22a.** Any of these approaches are possible: (1) create a loss-of-function allele by gene replacement and examine for mutant phenotype, (2) create a gain-of-function allele by constitutive expression of the gene and examine for mutant phenotype, (3) create a reporter gene fusion allele by gene replacement to examine where and when in the cell the protein is expressed, (4) perform a synthetic enhancer screen in the loss-of-function background, (5) perform a two-hybrid screen to identify interacting proteins, (6) perform transcriptome analysis to examine the expression pattern of the gene.

**22b.** In a human genome, the possibilities are much more limited: only (5) and (6) from the answer to 22a would work.

**24.** The first step is to organize the data to identify genes that behave similarly and those that behave differently. For example *a, c, d, e, f, g, i, j, k, n, q,* and *r* all increase in expression with both high salt and high temperature; *b, p,* and *s* all increase in expression with both high salt and high temperature; *h* and *o* decrease in response to salt but increase in response to high temperature; *l* and *m* increase in response to salt but decrease in response to high temperature. This analysis provides information into possible roles of genes that may be involved in a general stress response versus genes that may have specific roles in response to salt or temperature stress.

**28.** The *PEG10* gene is likely derived form the insertion of a retrotransposon, and its protein-coding sequences have been co-opted to perform a role in placenta formation. Retrotransposons contain a gene encoding reverse transcriptase, a nucleic-acid-binding protein that could be co-opted to have a role in binding and regulating endogenous nucleic acid sequences. The presence of the gene in placental mammals only suggests that the insertion of the retrotransposon occurred in the common ancestor of therian mammals, after the divergence of the monotremes from the rest of the mammals.

## Chapter 19

**2.** Their membrane system, chromosomal organization, replication, transcription, and translation (ribosome structure) are all similar to those in bacteria.

**4.** Sequencing of eukaryotic genomes has revealed evidence of transfers that are recent and transfers that are ancient. Transferred sequences that are highly similar must have been transferred recently.

**6.** See Table 19.1 for variations in the genetic code in mitochondria. A consequence of tRNA gene number reductions and the change in the code minimizes errors such that the two closely related codons, UGA and UGG, are both Trp.

**8.** There are three steps in this process: (1) Transfer of organelle DNA from organelle to nucleus and integration into the nuclear genome. (2) Acquisition of nuclear transcriptional regulatory elements such that the organellar gene is transcribed in the nucleus. Translational regulatory elements are similar (except where there are changes in the genetic code or RNA editing, both of which might inhibit production of a functional protein). (3) For the protein to be targeted back to the organelle, protein sequences facilitating efficient subcellular targeting need to be acquired from adjacent genomic sequences.

**10.** Transcription and/or translational and/or posttranslational regulation of the multiple components of the complexes need to be coordinated such that appropriate stoichiometries of the subunits are produced.

**12.** The most appropriate advice would be the following: for III-1, none of your progeny will be afflicted; for III-2, all of your progeny will be afflicted, and the extent may vary between individuals depending on levels of homo- and heteroplasy; for III-3, all of your progeny will be afflicted, although the extent may vary between individuals depending on levels of homo- and heteroplasy.

**14.** Maternal inheritance

**16.** *pet1* is a segregational (nuclear) mutation; *pet2* is a neutral mutation. The expected progeny is 2 wild type : 2 *petite*.

**18.** Since inheritance of this syndrome is maternal and not paternal, there is no need to worry; but if the mother exhibits symptoms, then there is a probability that their children would be affected. The extent to which their children will be affected depends on whether the mother is homoplasmic (all offspring would be affected) or heteroplasmic (possibility that some might not be affected).

**20.** Sibling II-2's children will be afflicted, but II-5's children will not be afflicted because MERRF is maternally, not paternally, inherited.

**22.** Maternal inheritance is the most likely, but it is penetrant only in males.

**24.** mtDNA is found in many copies per cell and the probability of being preserved is higher than it is for nuclear DNA. It is also highly useful for elucidating evolutionary relationships.

**26.** If there was no interbreeding, all coyote sequences would be more closely related to each other than to wolf sequences; this is not the case, so there must have been interbreeding.

**28.** Examine the genome of the sea slug and search for genes that are closely related to nuclear genes of the algae. If they are present in the nuclear genome of the sea slug, horizontal gene transfer can be assumed to have occurred. Use sequencing to confirm that these genes are not present in relatives of the sea slug.

## Chapter 20

**2.** The neural crest cells differentiate autonomously with the identity of the species from which they are derived. However, they can recruit host cells to contribute to the beak, suggesting that the neural crest cells non-autonomously influence the developmental fate of neighboring host cells.

**4a.** In a syncitium proteins are free to diffuse, so mechanisms whereby factors are restricted by membranes are not functional.

**4b.** Gradients of morphogens such as *bicoid* and *nanos* either must not exist, or the gradients must be established in another manner, such as cell–cell communication.

**6.** Segments correspond to the clear morphological and anatomical divisions in the larva or adult organism. Parasegments are offset from the segments, spanning the posterior part of one segment and the anterior part of its neighbor, and correspond to domains of gene expression. Developmental biologists consider parasegments as the subdivisions that are produced during fly development because they correspond to the domains of gene expression that control pattern formation and identity in the organism.

**8.** *Similarities:* Both segmentation in *Drosophila* and floral organ whorls in *Arabidopsis* are serially repeated structures/segments with identities controlled by related sets of transcription factors that act combinatorially and exhibit cross-regulation. *Differences:* In *Drosophila* the genes are *Hox* genes and are encoded in complexes in the genome, whereas in *Arabidopsis* the genes are *MADS*-box genes that are dispersed throughout the genome.

**10a.** In a loss-of-function mutant, the phenotype is vulvaless.

**10b.** In a gain-of-function mutant, the phenotype is multi-vulval.

**12a.** This phenomenon can occur only when cells are totipotent. Once cells are only pluripotent, the identities of possible differentiation pathways are limited and may not be able to form a complete organism.

**12b.** The resulting individual will be a genetic mosaic, consisting of two distinct genotypes. This probably happens more than is acknowledged, but it is detected only when multiple parts of an individual are genotyped.

**14a.** Extra copies of *bicoid* would increase the amount of *bicoid* mRNA that the mother puts into her eggs, thus increasing the amount of bicoid protein. This would result in a posterior shift in threshold levels of *bicoid* required to activate downstream targets; hunchback expression would be increased; other gap genes and pair-rule genes are also likely to be affected, with a general shift of anterior gene expression patterns (and subsequent fates) to more posterior positions—shifting the other gap gene expression patterns to more posterior positions.

**16a.** Pair-rule genes might be expected to influence the expression of the segment polarity genes, which act at a later time in development.

**16b.** The *fushi tarazu* single mutant likely has a loss of the even-numbered parasegments (*fushi tarazu* is Japanese, meaning "too few segments"), and the engrailed single mutant likely has defects in the anterior part of each parasegment. Thus, one might predict that the double mutant would be a combination of these two single-mutant phenotypes.

**18.** This pattern could be established by lateral inhibition.

**20a.** Gain-of-function alleles in *let-23* and *let-60* would result in a vulva being produced in the *lin-3* loss-of-function background.

**20b.** Loss-of-function alleles of *let-60* would suppress the gain-of-function *let-23* multi-vulva phenotype and result in a vulvaless worm.

**22a.** The consequence of ectopically expressing *Hoxd10* throughout the developing mouse limb bud would be that the "thumb" would acquire an "index finger" identity; for *Hoxd11*, the "index finger" would acquire a "middle finger" identity,

and the "thumb" would have an altered identity promoted by *Hoxd9* + *Hoxd11* (but it is not clear what it might look like, since that combination is not found in any wild-type digit). For *Hoxd10* and *Hoxd11*, both the "thumb" and "index finger" would acquire a "middle finger" identity.

**22b.** To construct this model, you need to create a conditional allele. One approach would be to introduce *lox* sites flanking the *Hoxd9–13* cluster of genes via homologous recombination (see Chapter 17). The intervening DNA including the genes could then be excised by induction of the Cre recombinase protein, which could be controlled either by regulatory elements driving expression in the limb bud or perhaps by a heat shock.

**24.** *tra-1* and *tra-2* mutant alleles are epistatic to the *her-1* mutant allele, while the putative gain-of-function *tra-1* allele (*her-2*) is epistatic to recessive loss-of-function alleles of *tra-1* and *tra-2*. Since the wild-type allele of *tra-1* acts as a repressor of male development and the wild-type allele of *her-1* acts as a repressor of hermaphrodite development, the activities of the genes in a wild-type animal can be summarized as follows:

| genotype | *her-1* | *tra-2* | *tra-1* | phenotype |
|---|---|---|---|---|
| XX | off | on | on | hermaphrodite |
| XO | on | off | off | male |

Based on the observation that *tra-1* is epistatic to *her-1*, *tra-1* should be downstream of *her-1*. In addition, assuming that *her-2* is a gain-of-function allele of *tra-1* and is epistatic to *tra-2*, this places *tra-1* downstream of *tra-2*. Thus, a model for the control of sex determination can be constructed.

In one model, the X:A ratio influences the activity of a repressor, *her-1*, which represses *tra-2*, which in turn activates *tra-1*, which then promotes hermaphrodite (female) fate and represses male development:

| X:A ratio | ---] *her-1* | ---] *tra-2* | → *tra-1* | or | → | |
|---|---|---|---|---|---|---|
| high (XX/AA) | low | high | high | | → | hermaphrodite development |
| low (XO/AA) | high | low | low | | → | male development |

**26a.** In an otherwise wild-type background, the phenotypic effects of *agamous* mutations are confined to the third and fourth whorls; but in an *apetala2* mutant background, phenotypic effects of *agamous* mutations are also seen in the first and second whorls. This implies that in an *apetala2* mutant background, *AGAMOUS* is ectopically expressed in the first and second whorls, and the converse is true for the phenotypic effects of *apetela2* mutations.

**26b.** Yes; cross-regulatory interactions occur among *Hox* genes in animals. Posteriorly expressed *Hox* genes often repress the expression of *Hox* genes normally expressed in respective anterior positions. This is a common, although not universal, feature in the regulation of *Hox* genes.

**28.** Based on the phylogeny of eukaryotes (see Figure 18.11), the last common ancestor of Basidiomycota and animals (or plants) was likely a single-celled organism. Thus, as in the comparison of multicellular development of plants and animals, Basidiomycota are likely to utilize a unique set of genes to direct their development. While the genes might be expected to encode transcription factors and signaling molecules, they are not likely to be homologous to those directing development in plants and animals. Thus, a forward genetic screen to identify pattern formation mutants in mushrooms would likely be a more successful approach than any reverse genetic screens.

## Chapter 21

**2.** Traits 1a through 1d are likely to be multifactorial. Dietary nutrition and temperature are two environmental conditions likely to influence each trait.

**4.** $V_E = 2.25$, $V_G = 5.40 - 2.25 = 3.15$.

**6.** The mean is 165.75, $s^2 = 1137.22/11 = 103.38$, and $s = 10.17$.

**8a.** $A_1A_1B_1B_1C_1C_1 = 36$ cm and $A_2A_2B_2B_2C_2C_2 = 18$ cm.

**8b.** 27 cm

**8c.** 24 cm

**8d.** Any genotype with five "1" alleles and one "2" allele yields $[(5)(6 \text{ cm})] + (3 \text{ cm}) = 33$ cm.

**8e.** There are $(3)^3 = 27$ possible genotypes.

**8f.** Seven different phenotypes are possible.

**10a.** $V_E = 3.5$ and $V_G = 7.4 - 3.5 = 3.9$.

**10b.** $H^2 = 3.9/7.4 = 0.527$.

**12.** $H^2 = 34.48/38.10 = 0.905$.

**14a.** For the cross involving 12-gram tomatoes, $S = -4$ g; for the cross involving 24-gram tomatoes, $S = 8$ g.

**14b.** For the cross involving 12-gram tomatoes, $R = (-4)(0.8) = -3.2$ g; for the cross involving 24-gram tomatoes, $R = (8)(0.8) = 6.4$ g.

**16.** Blood type is known (see Chapter 4) to be the result of three alleles of a single gene, and the MZ results confirm the exclusive genetic determination of blood type. Chicken pox is an infectious disease, and there is no reason to suspect gene-dependent differences in infection as the equal concordance values indicate. For the five other conditions, MZ concordance is considerably higher than DZ concordance, suggesting that genes have a pronounced effect on the appearance of the conditions.

**18.** QTL analysis screens a large number of SNP and other DNA sequence variants and plots the results on the phenotype of interest, running speed in this case. Any DNA markers associated with faster running speed could potentially indicate the nearby location of a gene (quantitative trait locus) influencing running speed.

**22a.** For protein content, $S = 22.7 - 20.2 = 2.5\%$; for butterfat content, $S = 7.4 - 6.5 = 0.9\%$.

**22b.** Response to selection will be greater for protein content $[(2.5)(0.60) = 1.5\%]$ than for butterfat content $[(0.9)(0.80) = 0.72\%]$.

**24a.** 100%. Blood type is controlled by genotype alone, and MZ twins are genetically identical.

**24b.** 50%. Blood types A ($I^A i$) and B ($I^B i$) are both expected, with a probability of 1/2. The chance that DZ twins will have the same blood group is 1/4 for blood type A plus 1/4 for blood type B, or 1/2. It is also possible that since DZ twins are the result of independent fertilization events, one twin could be blood type A and the other blood type B.

## Chapter 22

**2.** Inbreeding is a genome-wide phenomenon that increases homozygosity and reduces heterozygosity. It does not change allele frequencies; rather, it nonrandomly distributes alleles into genotypes. Inbreeding can increase the probability that inbred organisms might be homozygous for a rare recessive allele.

**4.** Natural selection conferring the highest relative fitness on particular heterozygous organisms will eliminate the two alleles when they occur in homozygous genotypes. Equilibrium allele frequencies are established as a ratio of the selection coefficients operating against the alleles.

**6.** Among a very small number of population samples, each outcome contributes a large percentage to the total; in much larger samples, each individual outcome contributes a much smaller percentage to the total.

**8.** A genetic bottleneck substantially reduces the number of organisms in a population. Elimination and survival are random. Some alleles may be lost, and others may survive at much different frequencies than were present before the bottleneck. The overall result is less genetic diversity after the bottleneck. As populations increase in size after a bottleneck, they display a reduced level of genetic variability in comparison to diversity before the bottleneck.

**10a.** $\sqrt{0.28} = 0.53$.

**10b.** $1 - 0.53 = 0.47$.

**10c.** Using $p = 0.47$ for the dominant allele $T$ and $q = 0.53$ for the recessive allele $t$, $TT = (0.47)^2 = 0.22$, $Tt = 2(0.47)(0.53) = 0.50$, and $tt = (0.53)^2 = 0.28$.

**12.** Mutation can generate new antibiotic-resistant alleles that confer improved survival on bacteria carrying the mutant alleles that are exposed to an antibiotic.

**14.** Evolutionary processes, including directional selection, require inherited variability for their operation. In the absence of inherited variation, there is just a single allele of a gene, and selection has no alternative alleles on which to exert selective pressure.

**16.** Inbreeding may be reduced by carefully managing matings to ensure that the level of relationship is minimized to the extent possible when matings take place.

**18.** The expected frequencies of rabbits are black $(C_1 C_1) = (0.70)^2 = 0.49$, tan $(C_1 C_2) = 2(0.70)(0.30) = 0.42$, and white $(C_2 C_2) = (0.30)^2 = 0.09$.

**20.** In this problem, $s = 1 - 0.82 = 0.18$, and $t = 1 - 0.32 = 0.68$. The estimated equilibrium frequencies are $\beta^A = t/s + t = 0.68/0.86 = 0.791$, and $\beta^S = s/s + t = 0.18/0.86 = 0.209$.

**22a.** A genetic bottleneck is a substantial population reduction that eliminates population members at random.

**22b.** Since the loss of population members is random in a genetic bottleneck, the overall level of genetic diversity is reduced; certain alleles are eliminated and other alleles retained at frequencies that are higher or lower than before the bottleneck. In Ashkenazi populations, repeated bottlenecks followed by population growth generated a founder effect that brought alleles such as the recessive for Tay-Sachs disease to high frequency.

**22c.** In this population, the recessive allele frequency, $f(t)$, is $\sqrt{0.00133} = 0.0365$.

**22d.** The dominant allele frequency is $f(T) = 1 - 0.0365 = 0.9635$, and the carrier frequency is $2(0.9635)(0.0365) = 0.070$, or about 7 per 1000 people.

**26a.** Following one generation of natural selection, the relative genotype frequencies are $C_1 C_1 = 0.421$, $C_1 C_2 = 0.526$, and $C_2 C_2 = 0.032$. The approximate allele frequencies are $C_1 = 0.421 + (0.5)(0.526) = 0.684$, and $C_2 = 0.053 + (0.5)(0.526) = 0.316$.

**26b.** Following reproduction of the survivors of predation, the genotype frequencies are $C_1 C_1 = (0.684)^2 = 0.468$, $C_1 C_2 = 2(0.684)(0.316) = 0.432$, and $C_2 C_2 = (0.316)^2 = 0.010$.

**26c.** The equilibrium allele frequencies are predicted to be $C_1 = 0.2/0.6 + 0.2 = 0.25$, and $C_2 = 0.6/0.6 + 0.2 = 0.75$.

**28.** Assuming $f(I^A) = p$, $f(I^B) = q$, and $f(i) = r$, the frequency of $r = \sqrt{\dfrac{336}{1000}} = 0.579$, the frequency of $p$ is $\sqrt{0.421 + 0.336} - 0.579 = 0.290$, and the frequency of $q$ is $1 - (0.579 + 0.290) = 0.131$.

**30a.** For dimpling, the genotype frequencies are estimated to be $DD = (0.62)^2 = 0.3844$, $Dd = 2(0.62)(0.38) = 0.4712$, and $dd = (0.38)^2 = 0.1444$. For PTC tasting ability, the genotype frequencies are estimated to be $TT = (0.76)^2 = 0.5776$, $Tt = 2(0.76)(0.24) = 0.3648$, and $tt = (0.24)^2 = 0.0576$.

**30b.** The expected phenotypes are dimpled taster $(D\_T\_) = (0.8556)(0.9424) = 0.8063$; dimpled nontaster $(D\_tt) = (0.8556)(0.0576) = 0.0493$; undimpled taster $(ddT\_) = (0.1444)(0.9424) = 0.1361$; undimpled nontaster $(ddtt) = (0.1444)(0.0576) = 0.0083$.

**32a.** For *D3S1358*, the frequency of 16/18 heterozygotes is estimated to be $2(0.229)(9.162) = 0.0742$. For *VWA*, the estimated frequency of 14/18 heterozygotes is $2(0.131)(0.189) = 0.0495$. For *FGA*, the estimated frequency of 23/26 heterozygotes is $2(0.131)(0.018) = 0.0047$.

**32b.** Each heterozygous genotype of a CODIS gene is estimated by an expression similar to that used to determine the heterozygous class for a gene with two alleles (i.e., $2pq$). Homozygous genotypes are estimated as the square of the allele frequency (i.e., $p^2$). The joint probability for multiple genotype of independently assorting CODIS genes is estimated using the product rule.

**34a.** Individuals V-1, V-2, and V-3 are inbred.

**34b.** Common ancestors are I-1 and I-2.

**34c.** $F = 4(1/2)^8 = 0.015625$.

**38.** The recessive allele producing achromatopsia was present in the original (pre-typhoon) Pingelapese population, although the original allele frequency is unknown. The typhoon produced a genetic bottleneck that produced a frequency of approximately 1 copy in 40 alleles, or $q = 0.025$ for the recessive allele. Subsequent repopulation of the island was affected by genetic drift and inbreeding that may have acted to increase the frequency of the allele (genetic drift) and to increase the likelihood of individuals who are homozygous IBD (inbreeding).

# Glossary

**2-micron plasmid** A naturally occurring *Saccharomyces cerevisae* plasmid (circumference = 2 μm) that has been engineered to work as a vector in yeast.

**3′ polyadenylation (3′ poly-A tailing)** During eukaryotic pre-mRNA processing, an enzyme-driven modification that removes the 3′ end of the pre-mRNA and adds numerous adenines.

**3′ splice site** In eukaryotic pre-mRNA processing, the location of cleavage at the 3′ end of an intron. Contains an AG dinucleotide in a consensus sequence.

**3′ to 5′ exonuclease activity** DNA- and RNA-digesting activity that progresses in the 3′ to 5′ direction to remove nucleotides. See also *DNA proofreading*.

**3′ untranslated region (3′ UTR)** The untranslated segment of mRNA between the stop codon and the 3′ end of the transcript.

**5′ capping** In eukaryotic pre-mRNA processing, the addition of 7-methylguanosine to the nucleotide at the 5′ end of pre-mRNA by a triphosphate bridge. Methylation of adjacent nucleotides may also occur.

**5′ splice site** In mRNA processing, the location of cleavage at the 5′ end of an intron. Contains a GU dinucleotide in a consensus sequence.

**5′ to 3′ exonuclease activity** DNA- or RNA-digesting activity that progresses in the 5′ to 3′ direction to remove nucleotides.

**5′ to 3′ polymerase activity** DNA synthesizing activity of DNA polymerases that progresses in the 5′ to 3′ direction to add new nucleotides to a growing DNA strand. Requires a template strand.

**5′ untranslated region (5′ UTR)** The untranslated segment of mRNA between the 5′ end of the transcript and the start codon.

**6-4 photoproduct** A DNA lesion and potential mutagenic event caused by exposure to ultraviolet (UV) irradiation.

**−10 consensus sequence** See *Pribnow box*.

**10-nm fiber** The "beads-on-a-string" form of chromatin, in which DNA is wrapped around nucleosomes.

**30-nm fiber** A structure of chromatin in which histone 1 (H1) partially condenses chromatin fibers into a coiled form. Also known as *solenoid* or *solenoid structure*.

**30S initiation complex** In bacterial translation, the complex formed by a small ribosomal subunit, mRNA, and the tRNA carrying fMet.

**−35 consensus sequence** A specific consensus sequence of the bacterial promoter at which RNA polymerase is bound.

**70S initiation complex** The fully assembled bacterial ribosome that is prepared to initiate translation.

**300-nm fiber** A structural state of chromatin in which chromatin fibers are looped and condensed.

**α-globin gene** A gene belonging to a family of closely related genes that encode a globin polypeptide that is part of hemoglobin.

**α-proteobacteria** Lineage of bacteria that are the closest extant relatives of the lineage that gave rise to mitochondria.

**β^A allele** The common (wild-type) allele of the human β-globin gene.

**β-globin gene** A gene belonging to a family of closely related genes that encode a globin polypeptide that is part of hemoglobin.

**β^S allele** A specific mutant allele of the human β-globin gene that produces sickle cell disease in homozygous individuals.

**θ (theta) structure** In bacterial DNA replication, the name given to an intermediate structure of DNA replication of a circular molecule with a single origin of bidirectional replication.

**θ (theta) value** A variable indicating a recombination distance between genes. Used in lod score analysis.

**aberrant ratio** In fungi, a ratio of haploid spore genotypes within a single ascus indicating gene conversion.

**acentric fragment (acentric chromosome)** A chromosome fragment without a centromere.

**acrocentric chromosome** A eukaryotic chromosome in which the centromere is very near one end. Forms a chromosome with long and short arms of distinctly different lengths.

**activator (*Ac*) element** In transposition, a transposable genetic element containing a transposase gene.

**activator binding site** DNA sequence to which an activator protein binds to regulate gene expression. Term refers to regulatory sites in bacteria; in eukaryotes, the equivalent sequence would be called an enhancer element.

**activator protein** A transcription factor that binds to regulatory sequences associated with a gene and upregulates that gene's expression.

**addition rule** See *sum rule*.

**additive genes** Genes contributing to a polygenic trait and producing their effect by their cumulative contributions that are approximately equal for each gene.

**additive variance (*V_A*)** For quantitative traits, the component of genetic variance contributed by genes having an additive effect on phenotypic variance.

**adenine (A)** One of four nitrogenous nucleotide bases in DNA and RNA; one of the two types of purine nucleotides in DNA and RNA.

**adjacent-1 segregation** A pattern of chromosome segregation that can occur following reciprocal balanced translocation. Leads to gametes carrying gene duplications and deletions.

**admixed population** A population whose members are a blend of formerly distinct populations.

**agarose** An inert material derived from agar that is mixed with buffer and used to form gels for gel electrophoresis.

**allele** An alternative form of a gene.

**allele-counting method** A method for determining allele frequency in a sample by tabulating the number of alleles of each type.

**allelic phase** Describing the cis and trans arrangements of alleles of linked genes on homologous copies of a chromosome pair.

**allelic series** A group of alleles of a gene that display a hierarchy of dominance relationship among them.

**allolactose** A modified from of lactose that binds to the lac repressor protein, inducing an allosteric change that reduces the DNA binding ability of the complex.

**allopatric speciation** The development of new species in geographic isolation.

**allopolyploidy** A polyploidy organism arising through the union of chromosome sets from different species.

**allosteric domain** Domain of a protein that allows the protein to change shape when it binds to a specific molecule; the protein in the new shape is altered in its ability to bind to a second molecule (e.g., DNA). Also known as *allostery*.

**allosteric effector compound** Molecule that binds to the allosteric protein domain and subsequently induces a change in the bound protein.

**allostery** Reversible interactions of a small molecule with a protein that lead to changes in the shape of the protein and to a change in the interaction of the protein with a third molecule.

**alternate segregation** A pattern of chromosome segregation that can occur following reciprocal balanced translocation that leads to the production of viable gametes.

**alternative mRNA processing (alternative intron splicing, promoters, polyadenylation)** In eukaryotic pre-mRNA processing, alternative processes by which different mRNAs can be produced from the same gene using different promoters, polyadenylation sites, or by removal of different exon elements.

**alternative sigma (σ) factor** Different forms of the sigma subunit of bacterial RNA polymerase that induce distinct conformational changes to the RNA polymerase core and to the recognition of distinct promoters.

**Ames test** A laboratory method commonly used to determine whether a compound or one of its breakdown products is mutagenic.

**amino acid** An aminocarboxylic acid that is a component of a polypeptide or protein.

**aminoacyl site (A site)** The site on a ribosome at which incoming charged tRNAs match their anticodon sequence with mRNA codons.

**aminoacyl-tRNA synthetase (tRNA synthetase)** A group of enzymes whose specific functions are to identify particular tRNAs and catalyze the attachment of the appropriate amino acid at the 3′ terminus.

**amorphic mutation** See *null mutation*.

**anagenesis** Phylogenetic evolution of a new species from an ancestral species without branching.

**anaphase** The phase of mitosis during which sister chromatids separate (*anaphase A*) and move to opposite poles (*anaphase B*).

**aneuploid**  An uneven number of chromosomes. Usually the result of the gain or loss of a chromosome—that is, $2n + 1$ (trisomy) or $2n - 1$ (monosomy).

**annotation (gene annotation, genome annotation)**  The process of attaching biological functions to DNA sequences. Genome annotation is the process of identifying the location of genes and other functional sequences within the genome sequence; gene annotation defines the biochemical, cellular, and biological function of each gene product the genome encodes.

**Antennapedia complex**  One of two homeotic gene clusters in *Drosophila* consisting of five genes (*labial, Deformed, Sex combs reduced, proboscipedia,* and *Antennapedia*) that act in combination to specify the cephalic and thoracic parasegments.

**anticodon**  The nucleotide triplet sequence of transfer RNA that pairs with an mRNA codon sequence in translation.

**antiparallel**  Opposite 5′ and 3′ orientations of two complementary nucleic acid strands.

**antisense RNA**  An RNA molecule that is complementary to a portion of a specific mRNA.

**antitermination stem loop**  A stem loop that allows RNA polymerase to continue transcription through the leader region of bacterial attenuator controlled operons and into the structural genes of an operon (e.g., the 2-3 stem loop in *trp* operon regulation).

**apurinic site**  The location of a nucleotide that has lost its purine base.

**arabinose (*ara*) operon**  An inducible operon consisting of genes encoding enzymes allowing the use of arabinose as a carbon source. The operon is controlled by a single regulatory protein, which carries out both positive and negative transcriptional regulation.

**Archaea**  One of the three domains of life; separate from Bacteria and Eukarya.

**Argonaute**  Protein subunit of RISC (RNA-induced silencing complex) that binds small RNA molecules and provides either the catalytic "slicer" activity or the translational repressor activity.

**artificial cross-fertilization**  A controlled cross between plants made by an investigator who transfers pollen from one plant to fertilize the other plant.

**ascus**  The spore sac formed by fungi containing four (tetrad) or eight (octad) haploid spores. Also called *spores*.

**aster**  The structure forming during cell division that contains microtubules emanating from centrosomes.

**attachment site (att site)**  Identical or nearly identical sequences on the bacterial and bacteriophage chromosomes that are cut and used to integrate or to excise the bacteriophage chromosome from the bacterial chromosome.

**attenuation**  A gene regulatory mechanism that fine-tunes transcription to match the momentary requirements of the cell, achieving a more or less steady state of compound availability.

**attenuator region**  A regulatory region downstream of the promoter of repressible amino acid operons that exerts transcriptional control (in the form of *transcription termination*) based on the translation of a leader peptide, the efficiency of which is determined by the availability of specific amino acids.

**autonomous transposable element**  A transposable genetic element that encodes transposase and can initiate its own transposition.

**autopolyploidy**  A pattern of polyploidy produced by the duplication of chromosomes from a single genome.

**autoradiograph**  A photographic image obtained by exposure of X-ray film to the radioactive decay of isotopes attached to molecular probes. Used in the analysis of *gel electrophoresis*.

**autosomal dominant inheritance**  A pattern of hereditary transmission in which the dominant allele of an autosomal gene results in the appearance of the dominant phenotype.

**autosomal inheritance**  Hereditary transmission of genes carried on autosomes.

**autosomal recessive inheritance**  A pattern of hereditary transmission in which the recessive allele of an autosomal gene results in the appearance of the recessive phenotype.

**Bacteria**  One of the three domains of life; separate from Archaea and Eukarya.

**bacterial artificial chromosome (BAC)**  Cloning vector used in bacteria that utilizes the F plasmid origin of replication; can accept DNA inserts up to 500 kb.

**bacterial chromosome**  The main, usually singular, chromosome encoding the genome of a bacterium.

**bacteriophage (phage)**  A virus whose host is a bacterium.

**balanced polymorphism**  A genetic polymorphism maintained in a population because organisms with the heterozygous genotype have higher relative fitness than do organisms with either of the homozygous genotypes.

**balancer chromosome**  A chromosome with inversions used to maintain specific allele combinations (e.g., recessive lethal alleles) in genetic stocks.

**band (in electrophoresis gel)**  A region in an electrophoresis gel or in an autoradiograph where a protein of nucleic acid congregates. Usually visualized using a stain or molecular probe.

**barcode**  Short DNA sequences that identify specific strains in knockout libraries.

**Barr body**  The darkly staining inactive X chromosome visible in mammalian female nuclei. The result of random X inactivation.

**base analog**  A compound with structure similar to a naturally occurring nucleotide base that can substitute for the base in DNA.

**base-pair substitution mutation**  A DNA sequence change resulting in the substitution of one base pair for another.

**base stacking**  A phenomenon of DNA base-pair interaction that rotates the base pairs around a central axis of symmetry and imparts twisting to the double helix.

**basic local alignment search tool (BLAST)**  A computer program designed to search for homologous sequences in databases.

**bidirectional DNA replication**  The standard method of DNA replication that synthesizes new DNA in both directions from a replication origin.

**binomial probability**  A probability function using two coefficients, *a* and *b*, whose sum equals 1 and whose products predict the probability of events.

**bioinformatics**  The use of computational approaches to decipher DNA-sequence information.

**biosynthetic pathway**  A multistep biochemical pathway that synthesizes an end product or compound.

**biparental inheritance**  Condition in organellar inheritance where both parental gametes make contributions of cytoplasmic organelles to the zygote; contributions are often unequal because one gamete contributes more of the cytoplasm and the other gamete makes a smaller contribution.

**bithorax complex**  One of two homeotic gene clusters in *Drosophila* consisting of three genes (*Ultrabithorax, abdominal-A,* and *Abdominal-B*) that act in combination to specify the thoracic and abdominal parasegments.

**blending theory of heredity**  An obsolete theory of heredity proposing that the traits of offspring are the average of parental traits.

**blotting (in gel electrophoresis)**  The process of transferring proteins or nucleic acids from an electrophoresis gel to a permanent membrane or filter.

**blunt ends**  5′ or 3′ ends of double-stranded DNA lacking any single-stranded overhangs.

**branch point adenine**  In intron splicing, an adenine nucleotide near the 3′ splice site of an intron that joins with a guanine located at the 5′ splice site by a 2′-to-5′ phosphodiester bond to form a lariat intron.

**BRCA1-associated genome surveillance complex (BASC)**  A multiprotein complex that surveys the genome for mutations at the $G_1$- to S-phase cell cycle checkpoint.

**broad sense heritability ($H^2$)**  The proportion of total phenotypic variance that is contributed by total genetic variance.

**bulky adduct**  Large chemical groups added to nucleotides by alkylating agents.

**bypass polymerase**  A group of DNA polymerases that are unstable and synthesize short regions of DNA under conditions in which the main DNA polymerase is unable to function, such as when faced with DNA lesions that block replication. Also called *translesion DNA polymerase*.

**CAAT box**  A common consensus sequence component of eukaryotic promoters.

**CAP (catabolite activator protein)**  In bacterial transcription regulation, binds cAMP (cyclic AMP) at low glucose concentrations to positively regulate the transcription of operons that allow the use of alternative carbon sources.

**CAP binding site**  A bacterial DNA regulatory sequence to which the CAP–cAMP complex binds to positively regulate gene expression. See also *CAP–cAMP complex*.

**CAP–cAMP complex**  Formed by joining catabolite activator protein to cAMP, the complex binds to the CAP binding site of the bacterial *lac* promoter to regulate gene expression.

**catabolite repression**  Situation where the presence of the preferred catabolite (e.g., glucose) represses the transcription of genes for an alternative catabolite (e.g., lactose).

**Cdks (cyclin-dependent kinases)**  A group of multimeric proteins whose levels fluctuate during the cell cycle. Composed of cyclin proteins and protein kinases, Cdks control entry and progression through mitosis.

**cell cycle**  Consisting of interphase ($G_1$ phase, S phase, and $G_2$ phase) and M phase (mitosis or meiosis) in cells. The transition from one phase to the next is controlled by protein-based interactions.

**cellular blastoderm**  Stage of *Drosophila* embryogenesis in which the nuclei are located at the periphery of the embryo and are enclosed by cell membranes.

**central dogma of biology** The description of the functional relationship between DNA, RNA and proteins (DNA to RNA to protein).

**centromere** A specialized DNA sequence on eukaryotic chromosomes that is the site of kinetochore protein and microtubule binding.

**centrosome** A cytoplasmic region, containing a pair of centrioles in many eukaryotic species, from which the growth of microtubules forms the spindle apparatus during cell division.

**chaperone** A category of eukaryotic proteins that assist with the folding or movement of other polypeptides.

**Chargaff's rule** The observation that the percentage of adenine equals that of thymine and that guanine percentage equals cytosine percentage in DNA.

**charged tRNA** A tRNA to which the correct amino acid has been attached.

**chiasma (plural: chiasmata)** Points of contact between homologous chromosomes that are coincident with crossover locations between the homologs.

**chi-square test ($\chi^2$ test)** A statistical test to compare the observed results of an experiment with the results predicted by chance.

**chloroplast** An organelle, bounded by a double membrane, where photosynthetic reactions convert light energy and $CO_2$ into fixed organic carbon.

**chromatin** The complex of nucleic acids and proteins that compose eukaryotic chromosomes.

**chromatin modifier** Proteins that chemically modify histone proteins in the nucleosomes by adding or removing specific chemical groups, thereby modifying chromatin structure and regulating gene expression.

**chromatin remodeler** Proteins that reposition nucleosomes within chromatin in such a way as to open or close promoters and other regulatory sequences or that change the composition of nucleosomes, altering their biological activity (e.g., SWI/SNF, ISWI, SWR1).

**chromatin remodeling** Processes that modify the structure or composition of chromatin. Usually associated with alterations of nucleosome binding to DNA and affecting the regulation of gene transcription.

**chromatography** A technique for separating the components of a molecular mixture by their similarities and differences.

**chromosome** A structure composed of DNA and associated proteins that in total contain the genome of an organism.

**chromosome aberration** An abnormality of chromosome number or structure.

**chromosome arms [long arm (q arm), short arm (p arm)]** The segments of eukaryotic chromosomes between the centromere and the telomeres.

**chromosome banding** A group of laboratory methods that stain eukaryotic chromosomes to reveal distinctive patterns of light and dark bands. Chromosome banding by Giemsa staining produced standardized patterns for different chromosome of selected species. Also known as *Giemsa (G) banding*.

**chromosome break point** The location of a chromosome break.

**chromosome fusion** See *Robertsonian translocation*.

**chromosome inversion (paracentric, pericentric)** A structural alteration of a chromosome in which a segment breaks away from the chromosome and subsequently reattaches after 180° rotation. See also *inversion heterozygote*.

**chromosome scaffold** Composed of numerous nonhistone proteins, the superstructure of eukaryotic chromosomes.

**chromosome territory** The region within a nucleus occupied by a particular chromosome during interphase.

**chromosome theory of heredity** The theory developed in the early 20th century that genes are carried on chromosomes and that the meiotic behavior of chromosomes is the physical basis of Mendel's laws.

**chromosome translocation** The relocation of a chromosome or chromosome segment to a non-homologous chromosome.

**chromosome walking** See *positional cloning*.

**cis-acting** Acting on the same chromosome (e.g., DNA sequences that control expression of genes encoded on the same piece of DNA).

**cis-acting regulatory sequence** Sequences to which proteins bind to regulate transcription of genes located on the same chromosome as the sequences.

**cis-dominant** The principle that the operator can influence only the transcription of adjacent downstream genes.

**clade** In phylogenetics, a group of organisms defined by characteristics that are unique to the group and distinguish the group from others.

**cladistics** The classification of organisms by characteristics that are unique to the group and distinguish it from other groups. Involves branching of new species from ancestral species. See also *clade*.

**cladogenesis** Phylogenetic evolution by branching of descendant species from ancestral species.

**clamp loader** A multiprotein complex that pairs with DNA polymerase and the sliding clamp during replication.

**clone-by-clone sequencing** An approach to genome sequencing where each chromosome is first broken into overlapping clones that are then arranged in linear order to produce a physical map of the genome. Each clone in the map is then sequenced separately. Contrast with *whole-genome shotgun sequencing*.

**closed chromatin** Chromatin in which regulatory DNA is covered by nucleosomes, thus restricting the access of regulatory proteins to the sequences rendering genes in closed chromatin transcriptionally silent.

**closed promoter complex** The initial stage of transcription that forms when RNA polymerase loosely binds the promoter.

**coding strand** The nontemplate strand of DNA that has the same 5′-to-3′ polarity as its transcript and the same sequence, except for T in DNA and U in RNA.

**codominance** The equal and detectable expression of both alleles in a heterozygous organism.

**codon** The nucleotide triplet of mRNA that encodes a single amino acid.

**codon bias** The preferential use of specific codons where there is redundancy in encoding a specific amino acid.

**coefficient of coincidence (c)** The ratio of the observed number of double recombinants to the number of double recombinants expected to occur by chance.

**coefficient of inbreeding (F)** The probability that two alleles carried in an individual are homozygous identical by descent (IBD).

**cohesive (cos) ends** Short, single-stranded overhangs at 5′ and 3′ ends produced after digestion with certain restriction ends. The cohesive ends are termed compatible if they can base-pair with complementary single-stranded ends of another DNA molecule. Compare with **cohesive sites**.

**cohesive (cos) sites** The single-stranded ends of phage lambda that facilitate circularization or concatamerization of lambda phage genomes and that interact with coat proteins during packaging of phage particles. Compare with *cohesive ends*.

**cointegrate** In replicative transposition, the fusion of two circular transposable elements into a single, larger circular element.

**comparative genomics** See *evolutionary genomics*.

**compensatory mutation** A second mutation occurring at another site that fully or partially restores wild-type function lost when an initial mutation occurs.

**complementary base pairs** The specific pattern of purine-pyrimidine pairing of nucleic acid strands. In DNA, G with C and A with T; RNA uses U instead of T.

**complementary DNA (cDNA) library** Collection of DNA clones, originally derived via reverse transcription of mRNA molecules into DNA (cDNA) and cloned into a vector.

**complementary gene interaction (9:7 ratio)** A characteristic ratio of phenotypes produced by the interaction of two complementary genes that control a trait.

**complementation group** A group of mutations that affect the same gene.

**complete genetic linkage** The absence of crossing over between linked genes.

**complete initiation complex** The multisubunit complex that forms at the promoter immediately before the onset of transcription.

**composite transposon** In bacteria, a transposable element containing multiple genes located between terminal insertion sequences.

**concordance** In twin studies, the observation that both twins exhibit the trait.

**conditional probability** A probability prediction that is dependent on another previous event having taken place.

**conjugation** The short-term union of two bacterial cells for the unidirectional transfer of DNA from the "donor" to the "recipient." The transferred material may be plasmid DNA or donor bacterial chromosome DNA.

**conjugation pilus** The hollow filament extending from the donor bacterium to the recipient bacterium through which DNA is transferred. Also known as *conjugation tube*.

**conjugation tube** See *conjugation pilus*.

**consanguineous mating** See *inbreeding*.

**consensus sequences** A nucleotide sequence in a DNA segment derived by comparing sequences of similar segments from other genes or organisms.

The most commonly occurring nucleotides at each position comprise the sequence.

**conservative DNA replication** A disproven model of DNA replication positing that one duplex produced by replication contained the two original strands and the other two daughter strands.

**conservative transposition** In transposition, the removal of a transposable element from one location followed by insertion into a new location.

**conserved noncoding sequence (CNS)** Sequences that do not code for amino acids and are conserved across significant phylogenetic distances.

**constitutive heterochromatin** Chromosome regions containing chromatin that is always densely compacted. Usually containing highly repetitive DNA sequences.

**constitutive mutants** Mutants in which a gene is always expressed rather than being under regulatory control.

**constitutive transcription** State in which a gene is continuously transcribed.

**contiguous sequence (contig)** Overlapping DNA clones that together cover an uninterrupted continuous stretch of DNA sequence.

**continuous variation** In polygenic and multifactorial traits, the observation of phenotypic distribution over a continuous range.

**controlled genetic cross** Genetic crosses controlled by an investigator who usually knows the genotypes and/or phenotypes of the organisms being crossed.

**convergent evolution** Processes of independent evolution of similar structures in unrelated species. Also known as *homoplasmy*.

**co-option** A common theme in the evolutionary history of genes by which genes and genetic modules are reused in a new manner to direct the patterning or growth of novel organs.

**coordinate genes** Genes, often with maternal effects, that establish the major axes of the embryo, especially the anterior-posterior and dorsal-ventral axes; examples include *bicoid* and *nanos*.

**copy number variant (CNV)** A specific type of structural variant due to insertions or deletions (indels) greater than 1 kb in length.

**core DNA** The approximately 146 base pairs of eukaryotic DNA that wrap each nucleosome.

**core element** Consensus sequences in the active regions of promoters recognized by RNA polymerase I.

**core enzyme** The five-polypeptide component of bacterial RNA polymerase that actively carries out transcription.

**corepressor** An accessory molecule required for a repressor protein to exert its function.

**cosmid vector** Cloning vector used in bacteria that utilizes phage lambda *cos* sites for packaging of phage and a bacterial origin of replication for subsequent maintenance in bacteria; can accept DNA inserts of up to 40 kb.

**cosuppression** The silencing, via a small RNA mediated mechanism, of an endogenous gene due to the presence of a homologous transgene or virus. Cosuppression can occur at the transcriptional or post-transcriptional level.

**cotransduction** The simultaneous transduction of two or more genes contained on a donor DNA fragment into a recipient cell, where it undergoes homologous recombination to be spliced into the transductant chromosome.

**cotransduction frequency** The frequency with which two genes are transduced.

**cotransduction mapping** A method of mapping donor bacterial genes based on their frequency of cotransduction.

**cotransformation** Simultaneous transformation of two or more genes carried on a donor DNA fragment into a recipient.

**covered promoter** Promoter in which nucleosomes are found adjacent to the transcription start site, preventing efficient transcription initiation. This feature is common at highly regulated genes.

**CpG dinucleotide** See **CpG island.**

**CpG island** Region in which the frequency of CpG dinucleotides is higher than the average for the genome; commonly found near the transcription start sites of animal genes. The cytosines are often methylated when the gene is inactive and demethylated when the gene is transcriptionally active.

**crossing over** The breakage and reunion of homologous chromosomes that results in reciprocal recombination.

**crossover suppression** The significant reduction, or complete absence, of progeny with recombinant chromosomes due to duplications and deletions of genetic material following crossing over within the inversion loop in organisms that are heterozygous for an inversion.

**cryptic splice site** A 5′ or 3′ splice site that is not normally used except when a mutation either inactivates an authentic splice site or creates a new splice site at the cryptic site location. See also *splicing mutation*.

**cyanobacteria** Lineage of photosynthetic bacteria that are the closest extant relatives of the lineage that gave rise to the plastids.

**cyclin protein** A family of proteins whose levels fluctuate during the cell cycle. Cyclins pair with protein kinases to form cyclin-dependent kinases (Cdks) that help regulate the cell cycle.

**cytokinesis** Part of telophase, the process of cytoplasmic division between daughter cells.

**cytological markers** Structural differences between homologous chromosomes that serve to differentiate the chromosomes when they are visualized using microscopy.

**cytoplasmic inheritance (organellar inheritance)** The transmission of genes on mitochondrial and chloroplast chromosomes; genes that are located in the cytoplasmic organelles as opposed to the nucleus.

**cytosine (C)** One of four nitrogenous nucleotide bases in DNA and RNA; one of the two types of pyrimidine nucleotides in DNA and RNA.

**daughter cell** The genetically identical cells produced by mitotic cell division.

**daughter strand** A newly synthesized strand of DNA that is complementary to a template strand.

**deamination** A DNA lesion resulting in the loss of an amino group ($NH_2$) from a nucleotide base.

**degrees of freedom (df)** The number of independent variables in an experiment. In a chi-square test, most often the number of outcome class minus 1 ($n - 1$).

**delayed age of onset** The appearance of an abnormal phenotype that is not present at birth but appears later in life and is caused by an inherited mutation.

**deletion** The loss of genetic material (see also *interstitial, microdeletion, partial deletion, partial deletion heterozygote, terminal deletion*).

**deletion mapping** A method for mapping genes utilizing partial chromosome deletions with known locations to expose recessive mutants by pseudodominance.

**denaturation** In DNA, the separation of complementary strands of nucleic acids by hydrogen bond breakage. In polypeptides and proteins, the unfolding of tertiary or quaternary structures.

**densitometry** A technique for passing light through an electrophoresis gel to detect the presence of a stained band of protein or nucleic acid.

**deoxynucleotide monophosphates (dNMPs)** Monophosphate forms of deoxynucleotides.

**deoxynucleotide triphosphates (dNTPs)** Triphosphate forms of deoxynucleotides.

**deoxyribonucleic acid (DNA)** The hereditary molecule of organisms. Composed of two complementary strands of nucleotides with purine bases adenine (A) and guanine (G) and pyrimidine bases thymine (T) and cytosine (C).

**depurination** A DNA lesion occurring when a deoxyribose molecule loses its purine nucleotide base. See *apurinic site*.

**dicentric bridge** In a dicentric chromosome, the portion between the two centromeres that are drawn to opposite poles of the cell during division.

**dicentric chromosome** A chromosome with two centromeres.

**dicer** Ribonuclease that acts on double-stranded RNA responsible for the generation of small regulatory RNA molecules, such as microRNAs and small interfering RNAs; typically 21–30 nucleotides in length.

**dideoxy DNA sequencing** A method of DNA sequencing devised by Fred Sanger that uses a mixture of deoxynucleotide and dideoxynucleotide triphosphates to selectively block DNA replication, producing a ladder of partially synthesized DNA strands of different lengths. Also known as the *Sanger method*.

**dideoxynucleotide triphosphates (ddNTPs)** Rare DNA nucleotides absent oxygen molecules at the 2′ and the 3′ carbons that are most commonly used in dideoxynucleotide DNA sequencing.

**differential reproductive fitness.** See *relative fitness*.

**differentiation** Process by which cells become restricted in their developmental potential and take on specialized morphologies and physiological activities.

**dihybrid cross** A cross between organisms that are heterozygous for two loci.

**diploid number of chromosomes (2n)** The characteristic number of chromosomes in somatic cell nuclei of diploid species. Equal to twice the haploid (n) number of chromosomes found in the nuclei of gametes of sexually reproducing diploid species.

**direct repeat** Identical or nearly identical DNA sequences in the same orientation that are separated by intervening DNA.

**directional cloning** Technique whereby a DNA insert is cloned with a specific directionality with respect to sequences of the cloning vector; usually accomplished by using two different restriction enzymes.

**directional natural selection**  See *directional selection.*

**directional selection**  Natural or artificial selection that continuously changes the frequency of an allele in a direction toward fixation (frequency = 1.0) or toward elimination (frequency = 0.0).

**discontinuous variation**  A phenotype distribution containing discrete or separable categories.

**discordance**  In twin studies, the observation that the traits exhibited by the twins are different.

**disjunction**  The normal process separation of homologous chromosomes or of sister chromatids during cell division.

**dispersive DNA replication**  A disproven model of DNA replication positing that each strand of daughter duplexes is composed of segments of original DNA and segments of newly synthesized DNA.

**displacement loop (D loop)**  During DNA damage repair and homologous recombination, the displacement of a single strand of DNA by strand invasion.

**disruptive selection**  Natural or artificial selection of phenotypic extremes in a population, leading eventually to two strains with distinctive phenotypes.

**dissociation (*Ds*) element**  In transposition, a non-autonomous genetic element that is incapable of transposing on its own.

**DNA-binding protein**  A general term for a protein that binds to DNA; the interaction can be either DNA-sequence specific (most regulatory proteins) or DNA-sequence nonspecific (e.g., structural proteins such as histones).

**DNA clone**  A fragment of DNA that is inserted into a vector, such as a plasmid, cosmid, or artificial chromosome.

**DNA double helix (DNA duplex)**  The two complementary strands of DNA arranged in antiparallel orientation.

**DNA library**  Collection of DNA clones in which the DNA is usually derived from a single source.

**DNA ligase**  An enzyme active in DNA replication that joins together segments of a DNA strand by catalyzing formation of a phosphodiester bond.

**DNA loop**  In gene regulation, a condition where the DNA sequences between regulatory elements form an extended loop that allows distant regulatory sequences with associated DNA-binding proteins to interact.

**DNA microarray**  Collections of synthesized DNA fragments attached to a solid support and representing sequences present in a genome; can be used to assess transcription patterns, transcription factor binding sites, and recombination patterns, among other uses.

**DNA nucleotides**  DNA building blocks composed of deoxyribose sugar, a nitrogenous base, and one or more phosphate groups. See also *adenine (A), thymine (T), cytosine (C),* and *guanine (G).*

**DNA polymerase (pol I, pol II, pol III)**  The large multisubunit complex responsible for the synthesis of new strands of DNA during DNA replication or DNA repair.

**DNA proofreading**  The capacity of many types of DNA polymerase to utilize a 3′ to 5′ exonuclease activity to remove and replace mismatched or damaged nucleotides during replication. See also *3′ to 5′ exonuclease activity.*

**DNA replication**  The synthesis of new DNA strands by complementary base pairing of nucleotides in a daughter strand to those in a template strand.

**DNaseI hypersensitive site**  Regions of chromatin sensitive to cleavage by DNase I; these often represent open chromatin that is transcriptionally active.

**DNA transposon**  One type of transposable genetic element encoding a transposase and capable of transposition.

**DNA triplet**  Three DNA nucleotides corresponding to a codon of mRNA.

**dominance variance**  In polygenic and multifactorial inheritance, the portion of genetic variance attributed to the dominance effects of contributing genes.

**dominant epistasis (12:3:1 ratio)**  A characteristic ratio of phenotypes produced by the interaction of two genes that control a trait in which a dominant allele of one gene masks or reduces the expression of alleles of a second gene.

**dominant interaction (9:6:1 ratio)**  A characteristic ratio of phenotypes produced by the interaction of two genes that control a trait in which the presence of dominant alleles of both genes produces one phenotype, one dominant allele of either gene produces a second phenotype, and organisms with only recessive alleles for the interacting genes have a third phenotype.

**dominant negative mutation**  A dominant mutation that behaves as a loss-of-function, often due to blocking the formation or normal function of a multimeric protein complex.

**dominant phenotype**  The phenotype observed in a heterozygous organism that is identical to the phenotype observed in a homozygote. The phenotype produced when an organism is homozygous for the dominant allele or carries a single copy of the dominant allele in the heterozygous genotype. Compare with *recessive phenotype.*

**dominant suppression (13:3 ratio)**  A characteristic ratio of phenotypes produced by the interaction of two genes that control a trait in which the dominant allele of one gene suppresses the expression of the dominant allele of the second gene.

**donor cell (bacterial donor)**  The bacterial cell that is the source of DNA transferred to a recipient cell by either conjugation, transduction, or transformation.

**donor DNA**  DNA to be used in cloning or other recombinant DNA technologies.

**dosage compensation**  A mechanism for equalizing the expression of X-linked genes in males and females of a species.

**double Holliday junction (DHJ)**  An intermediate structure temporarily connecting chromatids of homologous chromosomes that forms during homologous recombination.

**double recombinant (double crossover)**  The occurrence of two crossovers between homologous chromosomes in a particular region. May involve two, three, or all four chromatids.

**double-strand break repair**  Following phosphodiester bond breakage on both strands of a DNA duplex, a mechanism of DNA damage repair. Related to the mechanism for homologous recombination.

**downstream**  Referring to a gene or sequence location that is toward the 3′ direction on the *coding strand.*

**drosha**  Ribonuclease that processes pri-microRNA molecules into pre-microRNA molecules in animals.

**duplicate gene action (15:1 ratio)**  A characteristic ratio of phenotypes produced by the interaction of

two genes that duplicate each other's action due to genetic redundancy.

**duplication**  The gain of genetic material by the inclusion of one or more additional copies of a chromosome segment. See also *microduplication, partial duplication,* and *partial duplication heterozygote.*

**early genes**  The first genes expressed following the infection of bacterial cells by bacteriophage. The fate of early gene expression determines whether the phage undergoes the lytic cycle or the lysogenic cycle.

**early promoters**  Regulatory sequences responsible for the activation of early genes or operons in bacteriophage.

**East-West (EW) resolution**  One of the possible patterns for resolving a Holliday junction to separate homologous chromosomes before meiotic anaphase.

**electrophoretic mobility**  A measurement of (1) the distance of migration or (2) the speed of migration of a nucleic acid or protein in gel electrophoresis.

**elongation factor (EF)**  A group of proteins associated with ribosomes that contribute to the elongation of the polypeptide product.

**embryonic stem cells**  In vertebrates, totipotent cells of early embryos that can give rise to any and all cell types of the organism.

**endosymbiont**  An organism that lives within the body or cell of another organism.

**endosymbiosis**  An (often) mutually beneficial relationship between organisms in which one organism, the endosymbiont, inhabits the body of the other.

**endosymbiosis theory**  Hypothesis that the mitochondrion and chloroplast are evolutionarily derived from bacterial endosymbionts related to extant α-proteobacteria and cyanobacteria, respectively.

**enhanceosome**  Protein complex that binds enhancer elements and directs DNA bending into loops that bring the protein complex into contact with RNA polymerase and transcription factors bound at the core promoter or with protein complexes bound to proximal promoter elements.

**enhancer**  A eukaryotic cis-acting DNA regulatory sequence to which trans-acting factors bind and stimulate transcription. See also *enhancer sequence.*

**enhancer screen**  A genetic screen designed to identify mutations in genes that worsen the phenotypic effects of mutations in another gene.

**enhancer sequence**  Sets of regulatory sequences that bind specific transcriptional proteins that can elevate transcription of targeted eukaryotic genes.

**enhancer trap**  A transgenic construct inserted randomly into the genome that allows identification of enhancer elements controlling specific patterns of gene expression.

**environmental variance ($V_E$)**  For quantitative traits, the proportion of the total phenotypic variance contributed by differences in the environment experienced by population members.

**epigenetic**  Heritable patterns or changes in gene expression that are not associated with any change in DNA sequence.

**epigenetic marks**  A collection of chemical marks and modifications, such as acteylation and methylation of histone proteins, that are functional in chromatin remodeling. Also known as *epigenetic modification.*

**epigenetic modification** Chemical modifications of DNA or associated histones, such as acetylation and methylation, that alter chromatin structure and influence gene transcription.

**epistasis** See *epistatic interaction*.

**epistatic interaction** A group of specific patterns of gene interaction in which an allele of one gene modifies or prevents the expression of alleles of another gene. Also known as *epistasis*.

**equilibrium frequency** The stable frequency of an allele in a population attained and maintained through the action of evolutionary processes.

**ethidium bromide (EtBr)** A compound used to stain DNA and RNA in electrophoresis gels.

**euchromatic region** See *euchromatin*.

**euchromatin** Chromosome regions containing chromatin that is not densely compacted. Most expressed genes are located within euchromatic regions of chromosomes. Also known as *euchromatic region*.

**Eukarya** One of the three domains of life; separate from Archaea and Bacteria. See also *eukaryote*.

**eukaryote** Referring to organisms belonging to the domain Eukarya.

**eukaryotic initiation factor (eIF)** A group of eukaryotic proteins that associate with ribosomal subunits and help initiate translation.

**euploid** A number of chromosomes that is an exact multiple of the haploid number.

**E(var) mutations** Mutations that enhance position effect variegation in *Drosophila*. Mutated genes produce proteins that are active in chromatin remodeling.

**evo-devo** The study of the evolution of development.

**evolution** (1) Any change in the genetic characteristics of a population, strain, or species over time. (2) The theory that all organisms are related by common ancestry and have diversified from common ancestors over time.

**evolutionary genetics** The study of evolution and evolutionary processes using genetic techniques and tools.

**evolutionary genomics** The comparison of genomes, both within and between species. It illuminates the genetic basis of similarities and differences between individuals or species.

**evolutionary processes** Four processes—natural selection, migration, mutation, and random genetic drift—that can cause changes in the genetic characteristics of a population or lineage.

**exconjugant cell** The cell that is the product of conjugation between a donor cell and a recipient cell.

**exit site (E site)** On the ribosome, the site through which an uncharged tRNA exits.

**exon** A nonintron segment of the coding sequence of a gene. Joined together following intron splicing, exons correspond to the mRNA sequence that is translated into a polypeptide.

**exonic splicing enhancers (ESEs)** Exon sequences that play a role in intron splicing.

**expression array** DNA microarray that carries unique sequences from every annotated gene of the genome and is used to monitor gene expression patterns.

**expression vector** Cloning vector possessing DNA sequences required for DNA fragments inserted into the vector to be transcribed and translated. Vectors with sequences facilitating expression in eukaryotes are called *eukaryotic expression vectors*.

**F (fertility) factor** The plasmid containing genes that confer the ability to act as a donor cell on a bacterium. May be either an extrachromosomal plasmid or may be incorporated into the donor bacterial chromosome. Also known as *F plasmid*.

**F′ cell (F′ donor)** An extrachromosomal fertility plasmid into which a portion of the donor bacterial chromosome has been incorporated.

**F plasmid.** See *F (fertility) factor*.

**F+ cell (F+ donor)** A donor bacterium containing an extrachromosomal fertility plasmid.

**F₁ generation (first filial generation)** The first generation of offspring. In genetic experiments, usually the offspring produced by crossing pure-breeding parents.

**F₂ generation (second filial generation)** The second generation, produced by crossing F₁ organisms.

**F₃ generation (third filial generation)** The third generation, produced by crossing F₂ organisms.

**facultative heterochromatin** Heterochromatic chromosome regions whose level of compaction can vary. Often contains repetitive DNA, but may also contain some expressed genes.

**first-division segregation** In *Neurospora* and other organisms forming an ascus, the separation of alleles at the first meiotic division due to no crossing over having occurred.

**fluorescent in situ hybridization (FISH)** A laboratory method for identifying genes or DNA sequences using molecular probes labeled with a compound that can emit fluorescent light upon excitation.

**forked-line diagram** A method for diagramming the probabilities of outcomes in a branching format.

**forward genetic analysis** The classical approach to genetic analysis whereby genes are first identified by mutant phenotypes caused by mutant alleles and the gene sequence is subsequently identified by recombinant DNA technologies. Also known as *forward genetics*.

**forward genetics** See *forward genetic analysis*.

**forward mutation** A mutation that alters a wild type and generates a mutant. Also known as *mutation*.

**forward mutation rate (μ)** The frequency of mutation from wild-type alleles to mutant alleles.

**founder effect** The random occurrence of allele and genotype frequency differences between a new population established by a small number of founders and the larger parental population.

**four-strand double crossover** Two crossover events between a pair of homologous chromosomes that involve all four chromatids.

**frameshift mutation** The insertion or deletion of DNA base pairs resulting in translation of mRNA in an incorrect reading frame.

**frequency distribution** A visual display or histogram of quantitative data.

**functional domain** A protein region with a specific function or interaction.

**functional genomics** Using genomic sequences and genome-wide patterns of transcripts and protein expression to understand gene function in an organism.

**functional RNAs** Various types of transcripts that are not translated and are functional as nucleic acids. See also *tRNA, rRNA, snRNA, miRNA, siRNA, ribozymes*.

**fusion gene** A recombinant gene comprised of DNA sequences from more than one source (e.g., the codon sequences derived from one gene and the sequences responsible for expression derived from a second gene).

**fusion protein** A recombinant protein encoded by DNA sequences from more than one source; made by combining the open reading frames of two unrelated genes.

**G₀** The "G zero" phase of the cell cycle, an alternative to G₁ of the cell cycle entered by mature cells that generally do not divide again until they die. Compare with $G_1$ and $G_2$ *phase*.

**G₁ phase** The "Gap 1" phase of the cell cycle during which genes are actively transcribed and translated and cells carry out their normal functions. Compare with $G_0$ and $G_2$ *phase*.

**G₂ phase** The "Gap 2" phase of the cell cycle during which the cell prepares to divide. Compare with $G_0$ and $G_1$.

**gain-of-function mutation** A mutation causing a gene to be overexpressed, to be expressed at the wrong time, or to encode a constitutively acting protein. Usually inherited as a dominant mutation.

**gametes** The reproductive cells produced by male and female reproductive structures; sperm or pollen in male animals and plants, and eggs in females.

**gap genes** In *Drosophila*, genes that control development in large contiguous regions along the anterior-posterior axis; examples include *hunchback*, *giant*, *krüppel*, and *knirps*.

**Gaussian distribution** See *normal distribution*.

**GC-rich box** An occasional upstream consensus sequence of eukaryotic promoters that is rich in guanine and cytosine.

**gel electrophoresis** A laboratory method for separating proteins or nucleic acid molecules or fragments using electrical current in a gel matrix.

**gene** The physical unit of heredity, composed of a DNA sequence that is transcribed and encodes a polypeptide or another functional molecule.

**gene conversion** Repair of mismatched (non-complementary) DNA nucleotides in heteroduplex DNA that forms during meiotic recombination. One allele is switched for another allele already in the genotype.

**gene dosage** The number of copies of a gene.

**gene–environment interaction** Interactions taking place between particular genes and specific environmental factors.

**gene family** A group of genes that is evolutionarily related via successive gene duplication events that are followed by diversification.

**gene interaction** Referring to genes that interact with one another due to their participation in the production of a particular product or trait.

**gene knockout** Loss-of-function allele of a gene usually obtained via a reverse genetic approach.

**gene pool** The total of all alleles present in breeding members of a population at a given moment.

**gene therapy** The use of genes as therapeutic agents to cure or alleviate symptoms of a genetic disease.

**generalized transducing phage** In transduction, a bacteriophage that carries a random segment of the chromosome of a donor cell to the recipient cell.

**generalized transduction** The transduction of a random segment of a donor chromosome into a recipient cell by a transducing phage. See also *generalized transducing phage*.

**genetic bottleneck** A period or event characterized by a substantial random reduction in population size. Loss of genetic diversity and allele frequency changes usually occur.

**genetic chimera** A tissue or organism comprised of cells of two or more distinct genotypes.

**genetic code** The universal set of correspondences of mRNA codons to amino acids. Used in translation to synthesize polypeptides.

**genetic complementation** (1) The observation of a wild-type phenotype in an organism or cell containing two different mutations. (2) The cross of two pure-breeding mutants that yields progeny that are exclusively wild type.

**genetic dissection** The use of mutations and recombinants in genetic analyses to identify and assemble the genetic components of a biological property or process.

**genetic drift** A process of evolution referring to random changes in allele frequencies that result from sampling errors. Occurs in all populations but is strongest in small populations.

**genetic fine structure** The method of high-resolution analysis of intragenic recombination to map genes at the nucleotide level.

**genetic heterogeneity** The observation of the same phenotype produced by mutation of any one of two or more different genes.

**genetic liability** See *threshold of genetic liability.*

**genetic linkage** The result of genes being located so near one another on a chromosome that their alleles do not assort independently. Identified by detecting certain pairs of alleles (parentals) that are transmitted together significantly more often than expected by chance and of other pairs of alleles (nonparentals or recombinants) that are transmitted together significantly less often than expected.

**genetic linkage mapping** Process for creating maps of genes based on their linkage relationships to other genes.

**genetic markers** Alleles of either expressed genes or non-coding chromosomal regions identifying a specific region of a chromosome. Can be used to trace or identify another gene, the chromosome, or a cell, organ, or individual.

**genetic network** Set of interacting genes identified from double mutants or other analyses indicating gene interaction.

**genetic redundancy** The situation where the functions of one gene are compensated for by the actions of another gene.

**genetic screen** A procedure whereby a population of organisms is mutagenized and their progeny are propagated and examined for mutant phenotypes. Also known as *mutagenesis.*

**genetic variance ($V_G$)** In polygenic and multifactorial inheritance, the proportion of total phenotypic variance contributed by genetic variation.

**genome** The entire complement of DNA sequences in a chromosome set of an organism.

**genome-wide association studies (GWAS)** Association analysis performed using genetic marker genes distributed throughout the genome. Designed to locate genes that may influence the variation of quantitative traits.

**genomic imprinting** Epigenetic phenomena that create differential expression of alleles depending on whether they were maternally or paternally inherited.

**genomic library** A set of clones consisting of the DNA representing the genome of an organism.

**genotype** (1) The genetic composition of an organism or a cell (i.e., all the alleles of all the genes). (2) The alleles of a single gene or a specified set of genes in a cell or organism.

**genotype proportion method** A method for estimating allele frequencies in a population by manipulation of genotype frequencies.

**genotypic ratio (1:2:1 ratio)** (1) A ratio or set of relative proportions between organisms with different genotypes. (2) The ratio of $1/4 : 1/2 : 1/4$ observed among the homozygous and heterozygous $F_2$ progeny of a monohybrid cross.

**germinal gene therapy** Gene therapy aimed at correcting the genetic defect in the germ cells, such that progeny would not inherit the genetic defect.

**germ-line cell** See *gametes.*

**Giemsa (G) banding** See *chromosome banding.*

**Goldberg-Hogness box** See *TATA box.*

**green fluorescent protein (GFP)** A gene, derived from the jellyfish *Aequoria victoria*, that is the source of the natural bioluminescence of this species, fluorescing green (a 509-nm wavelength) when illuminated with UV light (a 395-nm wavelength). When used as a reporter gene, GFP allows a noninvasive means of visualizing gene and protein expression patterns in living organisms.

**guanine (G)** One of four nitrogenous nucleotide bases in DNA and RNA; one of the two types of purine nucleotides in DNA and RNA.

**guide RNA (gRNA)** In RNA editing, the nucleic acid that directs the addition or removal of nucleotides from mRNA. Also known as *guide strand.*

**guide strand** See *guide RNA.*

**gynandromorphy** A condition in which the body of an organism is mosaic, appearing to contain both male and female features.

**hairpin structure** See *stem-loop.*

**haploid number of chromosomes (*n*)** One-half the diploid ($2n$) number. The number of chromosomes typically found in nuclei of gametes of diploid species.

**haploinsufficient** A wild-type allele that is unable to support wild-type function in a heterozygous genotype. Classified as a recessive wild-type allele. Compare with *haplosufficient.*

**haplosufficient** A wild-type allele that supports wild-type function in heterozygous organisms. Classified as a dominant wild-type allele. Compare with *haploinsufficient.*

**haplotype** The specific array of alleles encoded by linked genes in a segment of a single chromosome.

**Hardy-Weinberg equilibrium** The population genetic principle that in a population practicing random mating and in the absence of natural selection, mutation, migration, or random genetic drift, allele frequencies are stable at frequencies $p + q = 1.0$ for two alleles and are distributed into genotypes at frequencies $p2$, $2pq$, and $q2$.

**helicase** In DNA replication, the enzyme responsible for breaking hydrogen bonds between complementary nucleotides of a DNA duplex. Unwinding of the strands occurs ahead of the advancing replication fork.

**helix-turn-helix (HTH) motif** A DNA-binding protein domain consisting of two alpha helices: one helix binds to a specific DNA sequence, and the second helix stabilizes the interaction.

**hemizygous** Referring to the genotype of males that carry a single copy of each X-linked gene.

**hemoglobin (Hb)** A globin protein composed of four polypeptides (two α-globin and two β-globin) found in blood that transports oxygen.

**heritability** See *broad sense heritability* and *narrow sense heritability.*

**heterochromatin** A chromosome region containing densely compacted chromatin and few, if any, expressed genes. See *constitutive heterochromatin* and *facultative heterochromatin*. Also known as *heterochromatic region.*

**heteroduplex DNA** A DNA duplex created during homologous recombination by combining complementary strands of DNA from nonsister chromatids. Also known as *heteroduplex region.*

**heteroplasmic cell or organism** A cell or organism that harbors a mixture of alleles of an organellar gene. Also known as *heteroplasmy.*

**heteroplasmy** See *heteroplasmic cell or organism.*

**heterozygous advantage** In evolution, the greater relative fitness of heterozygous organisms compared to homozygous organisms in a population. May result in a *balanced polymorphism.*

**Hfr cell** An abbreviation for "high frequency recombination," pertaining to Hfr chromosomes or to Hfr donors in bacterial conjugation.

**Hfr chromosome** See *Hfr donor.*

**Hfr donor** A donor bacterial strain containing an F factor integrated into its chromosome. Also known as *Hfr chromosome.*

**histone acetyltransferase (HAT)** Chromatin-modifying enzyme that adds acetyl groups to specific positively charged amino acids (e.g., lysine) in the N-terminal tails of histones.

**histone deacetylase (HDAC)** Chromatin-modifying enzyme that removes acetyl groups to specific positively charged amino acids (e.g., lysine) in the N-terminal tails of histones.

**histone demethylase (HDMT)** Chromatin-modifying enzyme that removes methyl groups to specific positively charged amino acids (e.g., lysine) in the N-terminal tails of histones.

**histone methyltransferase (HMT)** Chromatin-modifying enzyme that adds methyl groups to specific positively charged amino acids (e.g., lysine) in the N-terminal tails of histones.

**histone proteins (H1, H2A, H2B, H3, H4)** Five proteins encoded by a gene family that form octameric nucleosomes (H2A, H2B, H3, and H4) and adhere to DNA to condense chromatin (H1).

**Holliday junction** A DNA structure that forms during meiotic recombination in which single strands are crossed over between nonsister chromatids of homologous chromosomes.

**Holliday model** Proposed originally by Robin Holliday; a model intended to explain meiotic recombination at a molecular level.

**holoenzyme** A fully functional multisubunit protein complex in bacteria, for example, the RNA polymerase holoenzyme.

**homeobox** A conserved sequence of DNA of 180 nucleotides encoding a *homeodomain* composed of three α-helices in a family of transcription factors found throughout eukaryotes; in metazoans some genes with homeobox genes are *homeotic genes.*

**homeodomain** A 60-amino acid DNA-binding domain.

**homeotic gene** Gene that controls the developmental fate of a region of the body of an organism; examples include the *Hox genes* in metazoans and the *MADS-box genes* in flowering plants.

**homeotic mutation** Mutation in which an apparently normal organ or body part develops in an inappropriate location.

**homologous chromosomes** Chromosomes that synapse (pair) during meiosis. Chromosomes with the same genes in the same order. Also known as *homologous pair, homologs*.

**homologous genes** Genes descended from a common ancestral gene. Also known as *homologs*.

**homologous nucleotides** Nucleotides descended from a common ancestral nucleotide.

**homologous recombination** Exchange of genetic information between homologous DNA molecules.

**homoplasmic cell or organism** A cell or organism in which all copies (alleles) of a cytoplasmic organelle gene are the same. Also known as *homoplasmy*.

**hotspot of mutation** A location within a gene or genome at which mutations occur much more often than average.

**housekeeping genes** Genes that have essential cellular or physiological functions.

***Hox* genes** Members of the homeobox gene clusters found throughout metazoans; the genes often pattern the anterior-posterior axis and are *homeotic genes*.

**hybrid dysgenesis** In *Drosophila*, the failure of $F_1$ progeny of P-cytotype males crossed with M-cytotype females to develop due to the presence of P elements.

**hybrid vigor** The greater growth, survival, and fertility of hybrids produced by crossing highly inbred lines.

**hybridization (of molecular probe)** In an electrophoresis gel or in gel blotting, the binding of a single-stranded nucleic acid probe to a single-stranded target nucleic acid by complementary base pairing.

**hydrogen bond** Weak electrostatic attraction formed by the sharing of a positively charged hydrogen atom by negatively charged oxygen and nitrogen atoms. Hydrogen bonds form between complementary nucleotides to hold nucleic acid strands together.

**hypermorphic mutation** A mutant whose phenotype is similar to, but greater than, the wild-type phenotype.

**hypomorphic mutation** See *leaky mutation*.

**identical by descent (IBD)** A homozygous genotype in an organism in which both copies of the allele in an individual can be traced back to a common ancestor.

**illegitimate recombination** Exchange of genetic information between non-homologous DNA molecules.

**imprinting control region (ICR)** Master regulatory cis-acting DNA sequences to which trans-acting factors bind to regulate genomic imprinting.

**inbreeding** Mating between relatives. Also known as *consanguineous mating*.

**inbreeding depression** A reduction in vigor, survival, or reproductive fitness of offspring due to inbreeding.

**incomplete dominance** The observation that the phenotype occurring in heterozygous organisms is intermediate between the phenotypes of homozygous organisms, but more similar to one homozygous phenotype than to the other. Also known as *partial dominance*.

**incomplete genetic linkage** The occurrence of crossing over between linked genes.

**incomplete penetrance** The occurrence of individual organisms that have a particular genotype or allele but not the corresponding phenotype.

**induced mutations** Mutations generated by exposure to physical, chemical, or biological mutagens.

**inducer** An accessory molecule that binds to a protein that leads to activation of gene expression. The inducer can bind to a repressor protein and prevent its function or bind to an activator protein and stimulate its function.

**inducer–repressor complex** A molecular complex consisting of a repressor protein and a bound inducer molecule.

**inducible operon** Operon that is not expressed under one set of environmental conditions, but whose transcription is activated under an alternative environmental condition (i.e., the *lac* operon).

**induction** Process by which one cell or tissue promotes a particular developmental fate in neighboring cells or tissues.

**inductive signal** A molecule that acts non-cell autonomously to influence cell fate; in *C. elegans* vulval development, the lin-3 protein secreted from the anchor cells acts as an inductive signal to influence the fate of vulval precursor cells.

**informational genes** Class of genes that encode protein products that perform informational processes in the cell such as DNA replication, packaging of chromosomes, transcription, and translation.

**ingroup** A species within a clade used to compare to other members of the clade.

**inhibition** Process by which one cell or tissue prevents a particular developmental fate in neighboring cells or tissues.

**inhibitor** An accessory molecule that converts activator proteins to an inactive conformation by binding to an allosteric binding domain of the activator protein.

**initial committed complex** In eukaryotic transcription, a partially completed multiprotein complex that is preparing to bind RNA polymerase II.

**initiation complex** In eukaryotic translation, the complex formed by the small ribosomal subunit, mRNA, and charged tRNA-carrying methionine.

**initiation factor (IF)** A group of proteins, associated with ribosomes, that contribute to ribosome assembly and translation initiation.

**initiator tRNA** The first charged tRNA associated with the ribosome.

**inosine (I)** A modified nucleotide found occasionally in anticodons that can base-pair with uracil, cytosine, or adenine.

**insertional inactivation** A process of mutation in which the insertion of DNA into a gene renders it nonfunctional.

**in situ hybridization** A laboratory method for hybridizing a molecular probe to a DNA sequence or a gene on an intact chromosome.

**insulator sequence** Cis-acting sequences that act to prevent cross-talk between regulatory elements of an adjacent gene and are located between enhancers and promoters of genes that are to be insulated from the effects of the enhancer.

**interactive variance ($V_i$)** In polygenic and multifactorial inheritance, the proportion of total phenotypic variance that is due to the interactions of genetic and environmental factors.

**interactome** The sum of all of the protein–protein interactions in an organism.

**interchromosomal domain** Open spaces between chromosome domains in the interphase nucleus.

**interference (I)** Measured on a zero to 1.0 scale, the measurement of the independence of crossovers. Expressed as 1.0 minus the *coefficient of coincidence*.

**internal control region (ICR)** Promoter consensus sequences of certain rRNA and tRNA genes that are downstream of the start of transcription (i.e., sequences that are internal to the transcriptional region of the gene).

**internal promoter element** Promoter consensus sequences of snRNA and tRNA genes that are downstream of the start of transcription (i.e., sequences that are internal to the transcriptional region of the gene).

**interphase** The multiphase period of the cell cycle between cell divisions. See also $G_1$ *phase, S phase,* and $G_2$ *phase*.

**interrupted mating** A technique used to map bacterial genes that stops conjugation at timed intervals to determine which genes have transferred from the donor cell to the recipient cell.

**interspecific comparison** Any comparison between different species. Compare with *intraspecific comparison*.

**interstitial deletion** The loss of a portion of a chromosome from within one arm.

**intragenic recombination** Crossing over within a gene.

**intragenic reversion** A reversion produced by a second site mutation within a single gene.

**intraspecific comparison** Any comparison between individuals of the same species. Compare with **interspecific comparison.**

**intrinsic termination** In bacterial transcription, the DNA sequence-dependent mechanism for *transcription termination*. Inverted repeat DNA sequences induce formation of 3′ mRNA stem-loop (hairpin) structures that are followed by multiple uracils (transcribed from adenines).

**introgression line** Lines of experimental organisms in which genome segments from two or more other lines are present due to repeated back crosses between hybrids and organisms of one parental line.

**intron** Intervening sequences between the exons of many eukaryotic genes. Present in DNA and pre-mRNA, but spliced out during pre-mRNA processing.

**intron self-splicing** The capacity of certain RNA transcripts to undergo self-generated splicing that does not require splicing enzymes of the spliceosome complex.

**intron splicing** The spliceosome complex-driven process that removes introns from eukaryotic pre-mRNA and ligates exons to form mature mRNA.

**inversion heterozygote** Organisms whose homologous chromosomes have different structural organization. Most commonly, one has normal structure whereas the homolog carries an inversion.

**inversion loop** At homologous chromosome synapsis in an *inversion heterozygote*, the structure that forms by the looping of one chromosome to align homologous regions.

**inverted repeat (IR) sequence** Identical or nearly identical DNA sequences located on the same molecule but with opposite orientations.

**IS (insertion sequence) element** Mobile DNA elements in bacteria that cause mutations by inactivating the expression of genes into which they insert.

**island model** In evolutionary genetics, a model of species evolution in which new species are reproductively isolated from an ancestral population.

**isoaccepting tRNA** The group of tRNAs that carry the same amino acid, but recognize synonymous codons.

**ISWI** Imitation switch complex that functions primarily to control the placement of nucleosomes into an arrangement that causes a region to be transcriptionally silent.

**joint probability** The likelihood of an outcome requiring the occurrence of two or more simultaneous or sequential events.

**karyokinesis** Part of telophase, the process of nuclear division between daughter cells.

**karyotype** A digital of analog photograph of chromosomes arranged by conventional chromosome numbering.

**kilobase (kb)** A length of nucleic acid containing 1000 nucleotides.

**kinetochore** The site of attachment of multiple proteins that connects a spindle fiber microtubule to the centromeric region of a chromosome. Forms during **M phase** of cell division.

**knockout library** Collections of mutants in which most or all genes of a particular organism have been mutated by inactivating (or "knocking out") their expression.

**Kozak sequence** A specific consensus sequence of eukaryotic mRNA that contains the authentic start codon (AUG) sequence.

**lac⁻ phenotype** Bacteria that are not able to grow on a medium containing lactose as the only sugar. Compare with *lac⁺ phenotype*.

**lac⁺ phenotype** Bacteria that are able to grow on a medium containing lactose as the only sugar. Compare with *lac⁻ phenotype*.

**lacA gene** A gene of the bacterial *lac* operon; encodes lac transacetylase. Compare with *lacY gene* and *lacZ gene*.

**lactose (lac) operon** An inducible operon consisting of genes (*lacA*, *lacY*, *lacZ*) encoding enzymes allowing the use of lactose as a carbon source. The operon is repressed by the lac repressor regulatory protein that binds to the lac operator sequence, and is activated by the CAP–cAMP complex that binds to sequences of the CAP binding site.

**lacY gene** A gene of the bacterial *lac* operon; encodes lac permease, which facilitates import of lactose into the cell. Compare with *lacZ gene* and *lacA gene*.

**lacZ gene** A gene of the bacterial *lac* operon; encodes β-galactosidase, which breaks down lactose into glucose and galactose. Compare with *lacY gene* and *lacA gene*.

**lagging strand** In DNA replication, the discontinuously synthesized strand whose Okazaki fragments are ligated to complete new strand synthesis. Compare with *leading strand*.

**large ribosomal subunit** The larger of two subunits of the ribosome.

**lariat intron structure** During intron splicing, the structure formed by covalent bonding of the 5′ guanine of an intron to the branch point adenine of the intron.

**late genes** Bacteriophage genes expressed late in the lytic cycle. Encode protein products required for packaging of phage particles and lysis of the host cell. *Late promoters* and *late operators* are the regulatory sequences responsible for late gene activation.

**lateral inhibition** Process by which one cell or tissue prevents neighboring cells or tissues from acquiring a developmental fate similar its own.

**law of independent assortment (Mendel's second law)** The random distribution of alleles of unlinked genes into gametes.

**law of segregation (Mendel's first law)** The separation of alleles of a gene during gamete formation.

**leader region** Transcribed region upstream of the major enzyme encoding genes of repressible amino acid biosynthesis operons (e.g., *trpL*). Region encodes a small peptide whose rate of translation reflects the concentration of the amino acid (e.g., tryptophan) in the cells and consequently regulates transcription of the operon.

**leader sequence** See *signal sequence*.

**leading strand** In DNA replication, the continuously synthesized strand. Compare with *lagging strand*.

**leaky mutation** A mutant whose phenotype is similar to, but less than, the wild-type phenotype. Also known as *hypomorphic mutation*.

**lethal allele** See *lethal mutation*.

**lethal mutation** An allele that results in the premature death of the organisms that carry it. Lethality most often affects homozygous organisms. Also known as *lethal allele*.

**linkage disequilibrium** The nonrandom distribution into gametes of alleles of linked genes.

**linkage equilibrium** The random distribution into gametes of alleles of linked genes achieved by crossing over between the genes.

**linkage group** A group of genes displaying genetic linkage.

**linker** A short, chemically synthesized oligonucleotide that can be ligated to DNA molecules.

**linker DNA** DNA between nucleosomes in the 10-nm fiber structure of chromatin.

**locus control region (LCR)** Specialized enhancer element that regulates the transcription of multiple genes, often complexes of closely related genes.

**lod score (log of the odds ratio)** Based on analysis of transmission in pedigrees, the statistic used to calculate the likelihood of genetic linkage between genes.

**long terminal repeats (LTRs)** Arrays of scores to hundreds of nucleotides that bracket the ends of retroviruses integrated into host chromosomes.

**loss-of-function mutation** A mutant that prevents the production of the wild-type protein or renders it inactive. Most commonly a recessive mutation.

**lysis** See *lytic cycle*.

**lysogenic cycle** The life cycle of a bacterium infected by a temperate bacteriophage that integrates into the host chromosome and replicates along with it.

**lysogeny** See *lysogenic cycle*.

**lytic cycle** The life cycle of a bacterium infected by a bacteriophage that replicates within the host cell and lyses the host to release progeny bacteriophage.

**M phase** The cell division phase of the cell cycle. Follows *interphase*.

**macroevolution** Evolutionary processes operating at the species level and higher.

**MADS-box** A conserved sequence of DNA of 168–180 nucleotides encoding a 56–60 amino acid DNA-binding domain in a family of transcription factors found throughout eukaryotes; in flowering plants, some MADS-box genes are *homeotic genes*.

**major gene** A gene that has a substantial effect on phenotypic variation.

**major groove** The larger of two grooves formed in the DNA sugar-phosphate backbone by the helical twist of the double helix and exposing certain base pairs.

**mapping function** Corrective calculations used to more accurately estimate recombination frequencies between linked genes. Mapping functions differ among certain species.

**map unit (m.u.), centiMorgan (cM)** A theoretical unit of distance between linked genes on a chromosome.

**maternal effect genes** Genes that act in the mother to impart gene products (RNA or protein) into the egg and subsequently the embryo. For maternal effect genes, the embryonic phenotype is determined by the genotype of the mother rather than that of the embryo.

**maternal inheritance** Transmission through the female (ovule or egg cell). Compare with paternal inheritance: transmission through the male (sperm).

**matrix attachment region (MAR)** Portions of the chromosome scaffold to which loops of chromatin are attached.

**mature mRNA** The fully processed product of eukaryotic transcription that moves to the cytoplasm for translation.

**mean (μ)** The average value of a group of values.

**median** In a sample distribution, the middle most values. Also known as *median value*.

**median value** See *median*.

**mediator** An enhanceosome complex that forms a bridge between activator proteins bound to enhancer elements and the basal transcriptional machinery bound to the promoter.

**meiosis** The process of cell division occurring in germ-line cells. Produces four haploid gametes or spores through two successive nuclear divisions in diploid species.

**meiosis I** First nuclear division characterized by homologous chromosomes separating. Compare with *meiosis II*.

**meiosis II** Second nuclear division characterized by sister chromatids separating. Compare with *meiosis I*.

**Mendelian genetics** Referring to genetic applications and analyses using the law of segregation and the law of independent assortment originally described through experiments and analysis by Gregor Mendel.

**meristem** Organized groups of pluripotent cells at the growing tips of plants that both generate organs and self-maintain to ensure that a pool of stem cells is always present.

**messenger RNA (mRNA)** A form of RNA transcribed from a gene and subsequently translated to produce a polypeptide or protein.

**metacentric chromosome** A chromosome with a centrally located centromere that produces long and short arms of approximately the same length.

**metagenome** Sequence derived from whole-genome shotgun sequencing of DNA from entire natural communities consisting of a range of organisms.

**metaphase** The stage of *M phase* during which chromosomes align in the middle of the cell.

**metaphase plate** The cell midline along which chromosomes align during metaphase.

**microdeletion** A small chromosome deletion detectable only by using molecular methods of analysis.

**microduplication** A small chromosome duplication detectable only by using molecular methods of analysis.

**microevolution** Evolutionary changes at the population level.

**micro RNA (miRNA)** Small (21–24 nuts) regulatory RNAs produced by Dicer and acting in a RISC complex to either repress translational or cleave target mRNA molecules. Compare with *RNA interference (RNAi)*.

**microsynteny** Conservation of the order of a small number of genes in the same order in related species.

**migration** A process of evolution referring to the movement of organisms and genes between populations. Also known as *gene flow*.

**minimal initiation complex** In eukaryotic transcription, a partially completed multiprotein complex that is preparing to bind RNA polymerase II.

**minor groove** The smaller of two grooves formed in the sugar-phosphate backbone by the helical twist of the double helix, exposing certain base pairs.

**missense mutation** A DNA base-pair substitution that leads to production of a polypeptide in which one amino acid substitutes for another.

**mitochondrion** An organelle, bounded by a double membrane, encoding polypeptides that interact with nuclear gene polypeptides in oxidative phosphorylation to generate ATP. In many species, mitochondria also participate in other metabolic processes and biochemical reactions, including ion homeostasis and biosynthetic pathways.

**mitosis** The process of cell division in somatic cells that produces genetically identical daughter cells through a single nuclear division.

**mitosome** Double-membrane-bound organelles that are evolutionarily derived from mitochondria but have lost all of the ancestral genome; proteins requiring an anaerobic environment to function are imported into them.

**mitotic crossover** Crossing over between homologous chromosomes during mitosis.

**modal value** See *mode*.

**mode** In a sample distribution, the most commonly occurring value. Also known as *modal value*.

**modern synthesis of evolution** Referring to the broad-based effort beginning in the middle of the 20th century to unite Mendelian genetics with Darwin's theory of evolution by natural selection.

**modifier gene** A gene that modifies the effect of a major gene.

**modifier screen** A genetic screen designed to identify mutations in genes that modify, either enhance or suppress, the phenotypic effects of mutations in another gene.

**molecular cloning** The process whereby a single DNA molecule is selectively cloned from a mixture of DNA molecules and then amplified to produce a large number of identical copies.

**molecular genetics** The subfield of genetics that studies hereditary transmission, variation, mutation, and evolution through the analysis of nucleic acids and proteins.

**molecular probe (probe)** A single-stranded nucleic acid or antibody protein labeled with a detectable marker that attaches to a specific target molecule, allowing target molecule detection in subsequent analysis. Single-stranded nucleic acid probes detect target nucleic acids, and antibody probes bind specific target proteins.

**monohybrid cross** A genetic cross between organisms that are heterozygous for one gene.

**monophyletic group** A group of organisms with a single common ancestor.

**monosomy** The presence of a single chromosome instead of a homologous pair, resulting in a chromosome number that is $2n - 1$.

**morphogen** Substance whose presence in different concentrations directs different developmental fates.

**multifactorial inheritance** The inheritance of traits whose phenotypic variation is the result of polygenic inheritance and environmental influences. See also *multifactorial trait*.

**multifactorial trait.** Traits whose phenotypic variation is the result of polygenic inheritance and environmental influences. See also *multifactorial inheritance*.

**multiple cloning site (MCS)** A vector DNA sequence containing several unique restriction enzyme target sequences facilitating cloning of inserted DNA fragments.

**multiple gene hypothesis** The hypothesis that alleles of multiple genes contribute to the production of certain traits.

**multiplication rule** See *product rule*.

**multipoint linkage analysis** A statistical method for testing and mapping alternative orders of multiple genes linked on a chromosome. Related to lod score analysis.

**mutagen** A chemical, physical, or biological agent capable of damaging DNA and creating a mutation.

**mutagenesis** A procedure whereby a population of organisms is mutagenized and their progeny are propagated and examined for mutant specific phenotypes. See also *genetic screen*.

**mutation** An inherited change in DNA.

**mutation frequency** The rate at which mutations occur per gene per unit of time. Most often expressed per gene per generation.

**mutation–selection balance** An arithmetic expression used to determine the equilibrium frequencies of alleles in populations as a result of allele elimination by natural selection and new allele creation by mutation.

**narrow sense heritability ($h^2$)** The proportion of total phenotypic variance that is contributed by additive genetic variance.

**natural selection** The evolutionary process operating through differences in survival, fecundity, and relative fitness of organisms with different genotypes and phenotypes.

**negative assortative mating** A pattern of preferential mating based on differences between the mating individuals.

**negative control (of transcription)** Condition where binding of a repressor protein to a regulatory DNA sequence prevents transcription of a gene or a cluster of genes.

**negative interference** Occurring when the *coefficient of coincidence* is greater than 1.0, the observation of more double crossovers than expected between a pair of genes.

**negative supercoiling** Twisting of the DNA duplex in the direction opposite to the turns of the double helix.

**neofunctionalization** The process, following gene duplication, whereby a mutation in one of the duplicates provides a function not performed by the original gene.

**neomorphic mutation** A mutant expressing a new or novel function not seen in the wild type.

**N-formylmethionine (fMet; tRNAfMet)** A modified methionine amino acid usually used as the amino acid that initiates bacterial translation. Carried by a specialized tRNA.

**node** An evolutionary branch point in a phylogenetic tree.

**non-autonomous transposable element** A transposable genetic element lacking a transposase gene and incapable of initiating transposition.

**nondisjunction** The failure of homolog or sister chromatid separation during cell division. Results in nuclei with the wrong number of chromosomes.

**nonhistone proteins** Numerous nuclear proteins that are not histones associated with chromosomes.

**non-homologous end joining (NHEJ)** An error-prone mechanism of double-stranded DNA break repair in eukaryotic genomes in which damaged nucleotides are removed and blunt ends of strands are joined.

**noninducible** Condition in which transcription of bacterial genes or operons cannot be activated.

**nonparental ditype (NPD)** In an ascus, the occurrence of four haploid spores that are each recombinant.

**nonpenetrant** An organism with a genotype corresponding to a mutant phenotype that instead displays the wild-type phenotype.

**nonrecombinant vector** Produced in a cloning experiment when the intended vector does not pick up a DNA insert.

**nonrevertible mutants** Mutations caused by partial deletion of DNA nucleotides that cannot be reverted to wild type.

**nonsense mutation** A type of point mutation producing a stop codon in mRNA.

**nonsister chromatid** A chromatid belonging to a homologous chromosome. Nonsister chromatids of homologs are involved in crossing over.

**non-template strand** See *coding strand*.

**normal distribution** The continuous distribution of outcomes predicted by chance. Also known as *Gaussian distribution*.

**northern blotting** A method for transferring mRNA from an electrophoresis gel to a permanent membrane or filter.

**north-south (NS) resolution** One possible pattern for resolving a Holliday junction to separate homologous chromosomes before meiotic anaphase.

**nuclear mitochondrial sequence (NUMTS)** Mitochondrial DNA sequences found in the nucleus as a result of recent transfer from the mitochondrial genome to the nuclear genome.

**nuclear plastid sequence (NUPTS)** Plastid DNA sequences found in the nucleus as a result of recent transfer from the plastid genome to the nuclear genome.

**nucleoid** The region of bacterial and archaeal cells (or mitochondria or chloroplasts) where the main chromosome resides.

**nucleolus** (plural: nucleoli) Nuclear organelle containing rRNA-encoding genes.

**nucleomorph** In a secondary endosymbiosis, the nuclear genome of the secondary endosymbiont.

**nucleosome** An octameric protein complex composed of two polypeptides each of histones H2A, H2B, H3, and H4, around which DNA wraps in chromatin.

**nucleosome-depleted region (NDR)** A 100- to 150-bp region containing few nucleosomes, which lies immediately upstream of the start of transcription.

**nucleotide excision repair** A mechanism of DNA damage repair in which a segment of one strand containing damaged nucleotides is excised and replaced.

**null mutation** A mutant that produces no functional product. Most commonly a recessive allele. Also known as *amorphic mutation*.

**Okazaki fragment** A short segment of newly synthesized DNA that is part of a lagging strand and is ligated to other Okazaki fragments to complete lagging strand synthesis.

**one gene–one enzyme hypothesis** Proposed by George Beadle and Edward Tatum in 1941, the hypothesis proposing that each gene encodes a specific protein product and controls a distinct function.

**open chromatin** Chromatin in which the association of DNA with nucleosomes is relaxed in regions containing regulatory sequences, allowing access by regulatory proteins and giving genes in open chromatin the potential to be transcriptionally active.

**open promoter** Promoters that reside in open chromatin, resulting in constitutive transcription. See also *open promoter complex*.

**open promoter complex** At transcription initiation, the stage at which RNA polymerase is bound and a short region of DNA opens to allow transcription from the template strand. See also *open promoter*.

**operational genes** Class of genes that encode proteins involved in cellular metabolic processes (e.g., amino acid biosynthesis, biosynthesis of cofactors, fatty acid and phospholipid biosynthesis, intermediary metabolism, energy metabolism, nucleotide biosynthesis).

**operator** Regulatory DNA sequences to which repressor or activator proteins bind. Term used in bacterial systems.

**operon** A set of adjacent genes that are transcribed in a polycistronic mRNA and are thus coordinately regulated; an operon is generally considered to include associated regulatory sequences (e.g., promoter, operator, etc.). Primarily found in bacteria and archaea.

**ordered ascus** The linear sequence in an ascus of haploid spores whose arrangement allows determination of the chromatids participating in crossing over.

**organizer** Groups of cells that possess the ability to influence the fates of cells in the surrounding tissues via non-autonomous signals.

**origin of migration** The starting point of nucleic acid or protein migration in gel electrophoresis.

**origin of replication** The specific sequence at which DNA replication begins.

**origin of transfer (oriT)** The site within the fertility (F) factor sequence where transfer to the recipient cell is initiated.

**orthologous genes** Genes in different species whose origin lies in a speciation event and that can be traced to a single gene in a common ancestor of the two species. Also known as *orthologs*.

**orthologs** See *orthologous genes*.

**outgroup** A species related to members of a *clade* but outside the clade; used to root the clade.

***P* element** A specific type of transposable genetic element prevalent in the *Drosophila* genome.

***P* value (probability value)** In the chi square test, the likelihood that a repeat experiment will produce a result as deviant or more deviant than expected in comparison to the experimental result being tested.

**paired-end sequencing** Sequence generated from both ends of a DNA clone; provides evidence of physical linkage of the two paired sequences.

**pair-rule genes** In *Drosophila*, genes that delimit parasegments along the anterior-posterior axis; examples include *even-skipped* and *odd-skipped*.

**paracentric inversion** A chromosome inversion in which the inverted segment does not include the region of the centromere.

**paralogous genes** Genes whose origin lies in a gene duplication event within an extant or ancestral species. Also known as *paralogs*.

**paralogs** See **paralogous genes.**

**paraphyletic group** A group of organisms which includes some but not all the members descended from a common ancestor.

**parasegment** In *Drosophila*, the posterior part of one segment and the anterior part of its neighbor. The stripes of gene expression of pair-rule genes correspond to parasegments, straddling the boundaries between segments.

**parental (nonrecombinant) chromosome** Chromosomes in gametes produced when crossing over does not take place between linked genes. Alleles marking each gene are retained in their initial (parental) configurations.

**parental ditype (PD)** In an ascus, the occurrence of four haploid spores that are each nonrecombinant.

**parental generation (P generation)** The parents of $F_1$ progeny. In controlled genetic crosses, the parents are pure-breeding.

**parental strand** The DNA strand acting as a template to direct the synthesis of a new ("daughter") strand of DNA.

**partial chromosome deletion** The loss of a segment of a chromosome.

**partial deletion heterozygote** An organism with one wild-type chromosome and a homolog that is missing a segment.

**partial diploid** An exconjugant bacterium that acquires a second copy of one or more genes by conjugation with an F′ donor cell.

**partial dominance** See *incomplete dominance*.

**partial duplication** The duplication of a segment of a chromosome.

**partial duplication heterozygote** An organism with one wild-type chromosome and a homologous chromosome with a duplicated segment.

**particle gun bombardment** Technique of using high pressure to fire microscopic particles coated with DNA into plant cells. The particles are propelled with enough force to penetrate the cell wall and plasma membrane.

**particulate inheritance** Mendel's theory that genetic information is transmitted from one generation to the next as discrete units or elements of heredity.

**Pascal's triangle** A diagram listing the coefficients of a given binomial expansion in which the binomial expression is expanded *n* number of times.

**PCR primers** In polymerase chain reaction, short single-stranded segments of nucleic acid that bind template DNA and serve as primers from which DNA polymerase begins strand synthesis.

**pedigree** A family tree composed of standard symbols that depicts relationships in successive generations and often displays individual phenotypes.

**penetrant** Expression of the phenotype corresponding to a particular genotype.

**peptide bond** A type of covalent bond that joins amino acids in polypeptide chains. Formed between the amino end of one amino acid and the carboxyl end of the adjoining amino acid.

**peptide fingerprint analysis** A form of chromatography in which polypeptide fragments are separated and distinctive patterns revealed.

**peptidyl site (P site)** The site on the ribosome where amino acids are joined by a peptide bond.

**pericentric inversion** A chromosome inversion in which the inverted segment includes the region of the centromere.

**permissive condition** Environmental condition in which environmentally sensitive (e.g., temperature sensitive) mutants exhibit the wild-type phenotype or can survive.

**phenocopy** A phenotype similar to a phenotype caused by mutation but that is produced instead by an environmental condition.

**phenotype** (1) The observable physical characteristics or traits of an organism. (2) The physical manifestation of a specific genotype.

**phenotypic ratio (3:1 ratio and 9:3:3:1 ratio)** A ratio or set of relative proportions between organisms with different phenotypes—e.g., The ratio of progeny produced by a monohybrid cross (3:1) or a dihybrid cross (9:3:3:1).

**phenotypic variance ($V_P$)** The total variance observed for a trait.

**phosphodiester bond** A type of covalent bond formed between two nucleotides in a nucleic acid strain. Formed between the 5′ phosphate group of one nucleotide and the 3′ OH of the adjacent nucleotide.

**photoproduct** A characteristic DNA lesion produced by exposure to ultraviolet light.

**photoreactive repair** A mechanism of DNA damage repair in bacteria that uses visible light energy to remove the damage done by ultraviolet irradiation.

**phylogenetic footprinting** Technique whereby conserved sequences are identified by searching for similar sequences in species separated by large evolutionary distances.

**phylogenetic shadowing**  Technique whereby conserved sequences are identified by first eliminating sequences that are not conserved in closely related species.

**phylogenetic tree**  A diagram of evolutionary relationships among organisms or genes based on morphological or molecular characteristics.

**phylogenomics**  Method for determining phylogenetic relationships of organisms using genomic DNA sequence information. See also *evolutionary genomics.*

**physical gap**  Sequence gap between scaffolds for which there is no clone to supply the sequence.

**plasmid**  One of multiple types of extrachromosomal circular DNA molecules that may be found in bacterial cells.

**plastid**  Organelle, bounded by a double membrane, descended from the cyanobacterial endosymbiont; specialized types of plastids include chloroplasts and chromoplasts.

**pleiotropy**  A single gene mutation that affects multiple and seemingly unconnected properties of an organism.

**pluripotent**  State of a cell when it can give rise to many but not all cell types of an organism.

**point mutation**  A DNA lesion at a defined location. Usually either a base pair substitution, or the insertion or deletion of one or a small number of base pairs

**polyacrylamide**  A synthetic compound mixed with buffer and used to form electrophoresis gels.

**polyadenylation signal sequence**  A hexanucleotide sequence of mRNA, usually AAUAAA, that identifies the location of 3′ pre-mRNA cleavage and polyadenylation.

**polycistronic mRNA**  In bacteria, an mRNA containing the transcripts of two or more genes.

**polygenic inheritance**  A quantitative trait dependent on the contributions of multiple genes. Also known as *polygenic trait.*

**polygenic trait**  See *polygenic inheritance.*

**polymerase chain reaction (PCR)**  A laboratory method for controlled replication of a specific target sequence of DNA in successive cycles. Using two short single-stranded primers that bind to sequences on opposite sides of the target sequence, exponential replication of the target sequence occurs.

**polypeptide**  A chain of amino acids joined by peptide bonds. Formed at ribosomes during translation.

**polyploidy**  The presence of more than two complete sets of chromosomes in a genome. See also allopolyploidy, autopolyploidy.

**polyribosome**  In translation, the simultaneous translational activity of multiple ribosomes on a single mRNA.

**population**  A group of organisms that mate with one another to establish the next generation.

**population genetics**  The subfield of genetics that studies the genetic structure and evolution of populations.

**positional cloning**  The process by which the DNA sequence of a gene identified only by mutant phenotype can be obtained by using genetic and physical maps. Also known as *chromosome walking.*

**positional information**  Process by which gene expression, or other chemical cues, establish geographical addresses along the axes of a developing embryo or organ primordium.

**position effect variegation (PEV)**  The observation in *Drosophila* of a specific type of mutation producing variegation of eye color due to the abnormal positioning of the *w* (white) gene for eye color.

**positive assortative mating**  A pattern of preferential mating based on similarity between the mating individuals.

**positive control (of transcription)**  Condition where binding of an activator protein to a regulatory DNA sequence stimulates transcription of a gene or a cluster of genes.

**positive-negative selection**  The use of both negative and positive selectable markers to follow the fate of introduced DNA to select for homologous recombination events.

**positive supercoiling**  Superhelical twisting of DNA.

**post-translational polypeptide processing**  In eukaryotes, modifications to polypeptides in the endoplasmic reticulum and Golgi apparatus after the completion of translation.

**postzygotic isolation mechanism**  Mechanisms operating after mating to reduce or prevent the possibility of producing hybrids between populations or species.

**precursor microRNA (pre-miRNA)**  The stem loop product derived from processing of pri-microRNAs. The pre-microRNA stem loop is further processed by Dicer to produce the mature single stranded microRNA from the double stranded region of the stem loop.

**precursor mRNA (pre-mRNA)**  The initial transcript of a eukaryotic gene requiring mRNA processing prior to translation.

**preinitiation complex**  In eukaryotic transcription, a large multiprotein complex containing several general transcription factors and RNA polymerase II.

**prezygotic isolation mechanism**  Mechanisms operating before mating to reduce or prevent the possibility of producing hybrids between populations or species.

**Pribnow box (−10 consensus sequence)**  A specific consensus sequence component of the bacterial promoter with a location centered at approximately −10 relative to the start of transcription.

**primary microRNA  (pri-miRNA)**  The primary transcript, with single stranded ends and a stem loop, from which pre-microRNAs are derived by processing. The single stranded ends of the pri-microRNA are removed, by Drosha in animals and Dicer in plants, to produce the pre-microRNA.

**primase**  The specialized RNA polymerase that synthesizes the RNA primer during DNA replication.

**primer annealing**  In PCR, the binding by complementary base pairing of a short single-stranded primer by complementary base pairing.

**primer extension**  In PCR, the synthesis of DNA by DNA polymerase beginning at the 3′ end of a short single-stranded primer.

**primer walking**  Technique for sequencing long DNA molecules where new sequencing primers are synthesized based on successive DNA sequence reads. Compare with **shotgun sequencing.**

**primosome**  In DNA replication, a multisubunit protein complex whose central component is primase.

**product rule**  The probability of an event requiring the sequential or simultaneous occurrence of two or more contributing events. The probabilities of contributing events are multiplied and their product

is the event in question. Also known as the *multiplication rule.*

**proliferating cell nuclear antigen (PCNA)**  In eukaryotic DNA replication, the functional equivalent of the bacterial sliding clamp that adheres DNA polymerase to the template strand and drives its progression.

**prometaphase**  In **M phase** of the cell cycle, sometimes identified as a stage between prophase and metaphase.

**promoter**  A regulatory sequence of DNA near the 5′ end of a gene that acts as the binding location of RNA polymerase and directs RNA polymerase to the start of transcription.

**promoter mutation**  A mutation altering promoter sequence and function.

**promoter-specific element**  A specific promoter consensus sequence located upstream of small nuclear RNA genes.

**prophage**  The designation for bacteriophage that has integrated into the host bacterial chromosome.

**prophase**  The stage of **M phase** during which chromosome condensation occurs.

**protein**  A string of amino acids encoded during translation of mRNA and linked together by peptide bonds. See also **polypeptide.**

**protein sorting**  In eukaryotes, the process using the polypeptide **leader sequence** to designate the destination of polypeptides.

**proteome**  Set of the proteins in a cell, tissue, or organism.

**proteomics**  The study of all the proteins, collectively known as the proteome, within a cell, tissue, or organism.

**proto-oncogene**  A broad category of normal genes producing protein whose generalized functions promote cell proliferation. It is often mutated in carcinogenesis.

**pseudoautosomal region (PAR)**  Homologous regions on the X and Y chromosomes that synapse and cross over.

**pseudodominance**  The phenotypic expression of a recessive allele on one chromosome due to deletion of a portion of the homologous chromosome containing the dominant allele.

**pseudogene**  Sequences recognizable as mutated gene sequences often derived from gene duplication or retrotransposition events.

**Punnett square**  Named in honor of early 20th-century geneticist Reginald Punnett, a checkerboard-like diagram that predicts the genotypes and genotype frequencies of progeny from a genetic cross.

**pure-breeding**  A group of genetically identical homozygous organisms that, when self-fertilized or intercrossed, only produce offspring that have a phenotype identical to the parents. Also known as *true-breeding.*

**pyrimidine dimer**  The specific type of lesion formed on DNA due to exposure to ultraviolet irradiation. Also known as *thymine dimer.*

**QTL locus analysis**  A method for characterizing the effects of quantitative trait loci on variation.

**QTL mapping**  A method for locating quantitative trait loci in a genome.

**quantitative genetics**  The subfield of genetics that studies quantitative traits.

**quantitative trait**  A trait exhibiting polygenic inheritance and displaying continuous phenotypic variation.

**quantitative trait locus (QTL)**  A gene contributing to the phenotypic variation of a quantitative trait.

**R (resistance) plasmid**  A type of bacterial plasmid conferring resistance to one or more antibiotic compounds.

**radial loop–scaffold model**  A model of chromatin structure that predicts rosettes of looped chromatin on a chromosome scaffold.

**random X-inactivation (Lyon hypothesis)**
Proposed by Mary Lyon in the mid-20th century, the process of randomly inactivating one copy of the X chromosome in each mammalian female nucleus early in zygotic development.

**reading frame**  The partitioning of sequential sets of mRNA trinucleotide segments (codons) that are used in translation to determine amino acid order of a polypeptide.

**realizator genes**  In *Drosophila*, the *Hox* target genes whose expression contributes to the characteristic morphology of each segment.

**recessive epistasis (9:3:4 ratio)**  A characteristic ratio of phenotypes produced by the interaction of two genes that control a trait in which alleles of one gene mask or reduce the expression of alleles of a second gene.

**recessive phenotype**  The phenotype observed in an organism that is homozygous for the recessive allele. Compare with *dominant phenotype.*

**recipient cell (F-cell)**  A bacterial cell that does not contain fertility factor DNA sequence and can conjugate with a donor bacterium.

**reciprocal cross**  Paired crosses involving distinct parental phenotypes in which the sexes are switched (i.e., if one cross is ♂ phenotype A × ♀ phenotype B, the reciprocal cross is ♂ phenotype B × ♀ phenotype A).

**reciprocal translocation (balanced, unbalanced)**
Exchange of chromosome segments between nonhomologous chromosomes. If all genes are present, the translocation is "balanced," but if genes are missing, the translocation is "unbalanced."

**recombinant (nonparental) chromosome**
Chromosomes in gametes produced by crossing over between linked genes. Alleles marking each gene are rearranged on chromatids by crossing over.

**recombinant clone**  A combination of DNA molecules from different sources (e.g., vector and insert DNA), that are joined together using recombinant DNA technology.

**recombinant DNA technology**  The set of laboratory techniques developed for amplifying, maintaining, and manipulating specific DNA sequences in vitro as well as in vivo.

**recombination frequency (r)**  The rate of occurrence of recombination between a pair of linked genes. Expressed as the number of recombinants divided by the total number of meioses.

**recombination nodule**  Protein aggregations along the synaptonemal complex that are thought to play a role in crossing over.

**recombination repair**  A DNA repair mechanism that reestablishes normal DNA molecules by exchanging a damaged strand segment for a normal strand segment from another chromatid.

**reference genome sequence**  The DNA sequence of the individual or individuals used to construct the initial complete genome sequence.

**regulated transcription**  Condition in which gene expression is controlled at the transcriptional level in response to changing environmental conditions.

**regulatory mutation**  A mutation altering a regulated attribute of gene expression.

**relative fitness (w)**  In evolutionary genetics, the measurement of the reproductive fitnesses of organisms in a population relative to one another. The organism class with greatest fitness has a relative fitness of $w = 1.0$.

**release factor (RF)**  Molecules that bind mRNA stop codons and contribute to translation termination.

**replicate cross**  Repeated crosses involving parents with the same genotypes and phenotypes.

**replication bubble**  A region of active bidirectional DNA replication containing replication forks on each end, an origin of replication in the middle, and leading and lagging strands in each half of the bubble.

**replication fork**  In DNA replication, the site of the replisome structure, and the site of synthesis of leading strand and lagging strand DNA.

**replicative segregation**  Random segregation of organelles during cell division.

**replicative transposition**  Transposition carried out by replicating a copy of a transposable element and inserting the copy in a new genome location.

**replisome**  The large molecular machine located at the replication fork that coordinates multiple reaction steps during DNA replication.

**reporter gene**  A gene whose expression is easy to assay phenotypically. Fusion of reporter genes with heterologous sequences allows both transcriptional and translational expression patterns to be visualized.

**repressible operon**  Operon that is expressed under one set of environmental conditions, but whose transcription is repressed under an alternative environmental condition (i.e., the *trp* operon).

**repressor protein**  A transcription factor that binds to regulatory sequences associated with a gene and represses that gene's expression.

**reproductive isolation**  The absence of interbreeding between populations or species; often involves geographic, physical, or behavioral mechanisms or conditions.

**response to selection (R)**  The amount of change in the phenotype of a trait between parental and offspring generations as a result of selection on the parents.

**restriction endonuclease**  One of a large number of DNA-digesting enzymes, usually of bacterial origin, that cut DNA at specific recognition sites called restriction sequences. Each enzyme has its own particular *restriction sequence* and generates double-stranded cleavage of DNA at the restriction sequence. Also known as *restriction enzyme.*

**restriction length polymorphism fragment**  A fragment of DNA generated by treatment with a restriction endonuclease.

**restriction map**  A map showing the numbers and relative positions of target sites for restriction enzymes of a DNA molecule.

**restriction sequence**  The specific base-pair sequence recognized by a particular restriction endonuclease.

**restriction-modification system**  System of a restriction enzyme with a specific recognition sequence and a modifying enzyme that adds methyl groups to bases of the recognition sequence. The system protects the bacteria's own DNA from being digested by endogenous restriction enzymes, but allows restriction of invading exogenous DNA.

**restrictive condition**  Environmental condition in which environmentally sensitive (e.g., temperature sensitive) mutants exhibit the mutant phenotype.

**retrotransposon**  A transposable element that uses reverse transcriptase to transpose through an RNA intermediate.

**reverse genetic analysis (reverse genetics)**  Genetic analysis that begins with a gene sequence, which is used to identify or introduce mutant alleles and subsequently to identify and evaluate the resulting mutant phenotype. It is the complementary approach to forward genetics.

**reverse mutation rate (v)**  The rate at which mutant alleles are reverted to wild-type alleles. Also known as *reversion rate.*

**reverse transcriptase**  Enzyme, derived from retroviruses or retrotransposons, that catalyzes the synthesis of a DNA strand (cDNA) from an RNA template.

**reverse transcription**  The process of DNA synthesis from an RNA template by the enzyme reverse transcriptase.

**reverse translation**  The process of using the genetic code to deduce the possible DNA sequences encoding a specific amino acid sequence.

**reversion mutation**  A mutation that alters a mutant to wild-type sequence and function. Also known as *reversion.*

**revertible mutant**  A point mutation caused by base-pair substitution or deletion of one or a few base pairs that can be reverted to wild type.

**rho-dependent termination (Rho protein)**  The process of bacterial *transcription termination* involving rho protein.

**rho utilization site (rut site)**  The site of attachment of rho protein that aids in rho-protein-driven bacterial *transcription termination.*

**ribonucleic acid (RNA)**  A family of polynucleotides that are transcribed from DNA. RNAs are composed of nucleotides containing the sugar ribose, one or more phosphate atoms and one of four nitrogenous bases (A, G, C, and U).

**ribonucleotides**  Composed of ribose, one or more phosphate groups, and one of four nitrogenous bases, the nucleotides that make up RNA. See also *adenine (A), uracil (U), guanine (G),* and *cytosine (C).*

**ribose**  The 5-carbon sugar molecule in ribonucleotides.

**ribosomal RNA (rRNA)**  A group of RNA molecules that compose part of the structure of ribosomes.

**ribosome**  Ribonucleoprotein particles, composed of rRNAs and numerous proteins, at which translation takes place.

**ribozymes**  Catalytically active RNAs.

**RNA editing**  The process of post-transcriptional addition or removal of nucleotide of certain mRNAs.

**RNA interference (RNAi)**  A regulatory gene-silencing mechanism based on double-stranded RNA, which can target complementary sequences for inactivation. The machinery can be harnessed to silence gene expression in a reverse genetic approach.

**GLOSSARY**

**RNA polymerase** The enzyme that catalyzes the synthesis of RNA. See also *RNA pol*, *RNA pol I*, *RNA pol II*, and *RNA pol III*.

**RNA polymerase I (RNA pol I)** In eukaryotic transcription, the enzyme that transcribes certain rRNA genes.

**RNA polymerase II (RNA pol II)** In eukaryotic transcription, the enzyme that transcribes protein-coding genes to produce mRNA.

**RNA polymerase III (RNA pol III)** In eukaryotic transcription, the enzyme that transcribes tRNA genes.

**RNA primer** In DNA replication, the short, single-stranded RNA segment synthesized by primase. The 3′ end of the RNA primer is used by DNA polymerase to begin synthesis of DNA.

**RNA-induced silencing complex (RISC)** Complex containing Argonaute protein that binds small RNA molecules and targets complementary RNA molecules for degradation or translational repression.

**RNA-induced transcription-silencing (RITS) complex** RISC-like complex that mediates small RNA-induced transcriptional gene silencing.

**Robertsonian translocation** The fusion of two non-homologous chromosomes, often with the deletion of a small amount of nonessential genetic material. Also known as *chromosome fusion*.

**rolling circle replication** A unidirectional mode of DNA replication used to replicate circular plasmid molecules in which the replicating circular molecule appears to reel off its nontemplate DNA strand, using the other as the template for replication.

**S phase** The middle phase of interphase, during which DNA replication takes place.

**Sanger method** See *dideoxy DNA sequencing*.

**saturation mutagenesis** Mutagenesis aimed at identifying multiple mutant alleles for all loci in the genome of an experimental organism.

**scaffold** A set of contigs that are physically linked.

**scanning** In eukaryotic translation, the process used by the small ribosomal subunit to locate the authentic start codon.

**secondary endosymbiosis (tertiary symbiosis)** Endosymbiotic event where one eukaryotic, usually photosynthetic, is an endosymbiont within another eukaryote resulting in an organism with genomes derived from at least two nuclear genomes and multiple organellar genomes.

**second-division segregation** Patterns of haploid spores in an ascus that indicate the alleles were separated at the second meiotic division as a result of crossing over between a gene and the centromere.

**second-site reversion** A specific type of reversion taking place at a location separate from the site altered to generate the original mutation.

**segment** Division of the body along the anterior-posterior axis into a series of morphological similar units.

**segment polarity genes** In *Drosophila*, genes that delimit the anterior and posterior regions of individual parasegments along the anterior-posterior axis; examples include *wingless*, *engrailed*, *hedgehog*, and *gooseberry*.

**selected marker screen** An experimental method used to detect microorganisms with a specific genotype.

**selection coefficient (s)** The value of the reduction in reproductive fitness for an organism (i.e., $w = 1.0 - s$).

**selection differential (S)** The difference between the population mean value for a phenotype and the phenotype value of population members selected as parents for the next generation.

**selective growth medium** The growth medium used in a selective marker screen.

**semiconservative replication** The established method of DNA replication in which each strand of a parental duplex acts as a template for daughter strand synthesis, and each daughter duplex is composed of one parental strand and a complementary daughter strand.

**semisterility** Reduced fertility, commonly the result of the occurrence of adjacent segregation during meiosis in balanced translocation heterozygotes.

**sequence gap** Gap between two contigs for which a clone is available for further sequencing that could close the gap.

**sex chromosome** Homologous chromosomes that differ between the sexes. Designated X and Y in species in which females are XX and males XY. Designated Z and W in species in which females are ZW and males are ZZ.

**sex determination** The genetically controlled processes that determine the sex of offspring.

**sex-influenced trait** A gene, usually autosomal, whose expression differs between males and females of a species. Also known as *sex-influenced expression*.

**sex-limited trait** A gene or trait expressed exclusively in one sex. Also known as *sex-limited gene*.

**sex-linked inheritance** The inheritance of genes on the sex chromosomes.

**shared derived characteristics** Characteristics or traits of organisms that evolve from more ancestral characteristics or traits found in ancestral organisms.

**Shine–Dalgarno sequence** In bacterial translation, the 5′ UTR mRNA consensus sequence that pairs with nucleotides near the 3′ end of 16S rRNA in the small ribosomal subunit to orient the start codon on the ribosome.

**shotgun sequencing** Method for sequencing large molecules of DNA that relies on redundant sequencing of fragmented target DNA in the hope that all regions will be sequenced at least a few times. Contrast with *primer walking*.

**shuttle vector** A vector that can replicate in two species and thus can be used to shuttle DNA sequences between them.

**sickle cell disease (SCD)** A human autosomal recessive disorder resulting from homozygosity; a specific mutant allele ($\beta^S$) of the β-globin gene that is part of hemoglobin protein.

**sigma (σ) subunit** Accessory protein that changes the promoter-recognition specificity of the bacterial RNA polymerase core enzyme.

**signal hypothesis** The accepted hypothesis proposing that the polypeptide *leader sequence* identify post-translational processing and transport.

**signal sequence** A string of amino acids at the N terminal and of certain eukaryotic polypeptides containing information directing post-translational processing and the extracellular destination of the polypeptide. Also known as *leader sequence*.

**signal-transduction pathway** Pathways through which cells receive external messages at cell surface receptors and transmit the messages into the cell to initiate action or response.

**silencer** A eukaryotic cis-acting DNA regulatory sequence to which trans-acting factors bind to repress transcription.

**silencer sequence** Regulatory DNA sequences that can repress transcription of specific genes that may be located distantly from the sequence. Also known as *silencers*.

**silent mutation** A base substitution mutation that changes one codon to a synonymous codon and does not alter the amino acid sequence of a polypeptide.

**simple transposon** In bacterial transposition, a transposon containing multiple genes between two inverted repeats.

**single nucleotide polymorphism (SNP)** A single base-pair difference in a specific genome location detected by comparing individual DNA sequences.

**single-stranded binding (SSB) protein** In DNA replication, a protein that adheres to each template strand following unwinding by helicase to prevent strand reannealing before the arrival of the replication fork.

**sister chromatid cohesion** The protein-based temporary attachment of sister chromatids facilitated by cohesin protein that resists the pulling forces of spindle fibers in metaphase.

**site-directed mutagenesis** Introduction of specific nucleotide changes in a DNA molecule in vitro.

**site-specific recombination** An exchange between two DNA molecules that requires specific sequences in common and that is catalyzed by an enzyme specific to that recombination (e.g., integration of phage lambda into the *E. coli* genome).

**sliding clamp** In bacterial DNA replication, the multisubunit protein complex that joins with DNA polymerase to hold polymerase on the template and helps drive polymerase along the template.

**small interfering RNA (siRNA)** Single-stranded 21- to 24-nucleotide RNA molecules derived from either endogenous or exogenous double-stranded RNA molecules that are incorporated in RISC to mediate RNAi. Endogenously produced siRNAs are most often from non-genic regions (e.g., repetitive RNA or products of an RNA-dependent RNA polymerase). Exogenously produced siRNAs are often derived from invading nucleic acids (e.g., transposons and viruses).

**small nuclear RNA (snRNA)** Regulatory RNAs operating in the nucleus.

**small nucleoid–associated proteins** In bacterial DNA, small proteins localized to the nucleoid and associated with the main chromosome.

**small ribosomal subunit** The smaller of two subunits of the ribosome.

**solenoid structure** See *30-nm fiber*.

**somatic cell** A body cell. All cells of a multicellular organism excepting those of the germ line.

**somatic gene therapy** Gene therapy aimed at correcting a genetic defect in the somatic cells.

**Southern blotting** A laboratory method devised by Edwin Southern for transferring DNA from an electrophoresis gel to a permanent membrane or filter.

**specialized transduction (specialized transducing phage)** Transduction from a donor cell to a recipient cell of a few select genes located near the site of bacteriophage integration.

**spindle fiber microtubule (kinetochore, polar, and astral microtubule)** Composed of tubulin proteins, the fibers emanating from centrosomes that attach to kinetochore regions (kinetochore), overlap to control cell shape (polar), or attach to the cell membrane to stabilize centrosomes (astral).

**spliceosome** The multiprotein complex that carries out intron splicing.

**splicing mutation** A mutation altering the normal splicing pattern of a pre-mRNA.

**spontaneous mutation** Mutations occurring due to spontaneous events or changes involving nucleotides or nucleotide bases.

**square root method** A method for estimating allele frequencies based on manipulation of the frequency of a homozygous genotype.

**SR proteins** Proteins rich in serine (S) and arginine (R); SR proteins are operative in intron splicing.

**SRY** The *sex-determining region of Y* gene that initiates male sex development in mammals.

**stabilizing selection** A pattern of natural or artificial selection that reduces population variation by removing organisms with extreme phenotypes.

**standard deviation (σ)** A statistical value that measures the scatter of outcome values around the mean or average outcome value. Expressed as the square root of the sum of squared deviations of each value from the mean value.

**start codon** Most commonly AUG, encoding methionine, the first codon translated in polypeptide synthesis.

**start of transcription** The DNA location at which transcription begins.

**stem-loop** Short double-stranded segments of RNA topped by a single-stranded loop containing unpaired nucleotides. Also known as a *hairpin structure*.

**sticky end** Short single-stranded overhangs created by the cleavage of DNA by specific restriction endonucleases, which can potentially base-pair with complementary single-stranded sequences.

**stop codon** One of three codons that bind a release factor instead of base-pairing with tRNA to initiate a series of events that stops translation.

**strand invasion** During synthesis-dependent strand annealing and meiotic recombination, the entry of the 3′ end of a displaced DNA into the intact sister chromatid.

**strand polarity (5′ and 3′)** The orientation of a nucleic acid strand indicating its 5′ phosphate and 3′ hydroxyl ends.

**strand slippage** During DNA replication, a mutational event leading to increased or decreased numbers of repeating nucleotides in newly synthesized DNA and caused by slippage of DNA polymerase on the template strand or slippage of the newly synthesized strand on DNA polymerase.

**structural genomics** The sequencing of whole genomes and the cataloging, or annotation, of sequences within a given genome.

**structural maintenance of chromosomes (SMC)** A category of bacterial proteins localized to the nucleoid and associated with the main chromosome.

**structural motif** The characteristic three dimensional structure of a protein, sometimes obtained from crystal structure data.

**Su(var) mutations** Mutations that suppress position effect variegation in *Drosophila*. Mutated genes produce proteins that are active in chromatin remodeling.

**subcloning** Process by which DNA clones are further subdivided in order to clone still smaller fragments for analyses.

**subfunctionalization** The process, following gene duplication, whereby mutations in each of the two copies can result in the two genes having complementary activities such that their combined activity is the same as the activity of the gene before duplication.

**submetacentric chromosome** A chromosome with a centromere located near the midpoint that produces long and short arms of different lengths.

**sugar-phosphate backbone** The alternating sugar (deoxyribose or ribose) and phosphate molecule pattern of nucleic acid strands formed by the formation of phosphodiester bonds linking nucleotides in the strand.

**sum rule** The probability of an event that can result from two or more equivalent outcomes. The probabilities of the contributing events are added, and their sum is the probability of the event in question. Also known as the *addition rule*.

**supercoiled DNA** The superhelical twisting of covalently closed circular DNA. See *positive supercoiling* and *negative supercoiling*.

**suppressor screen** A modifier genetic screen designed to identify mutations in genes that suppress the phenotypic effects of mutations in another gene.

**SWI/SNF (switch/sucrose nonfermentable)** A yeast chromatin-remodeling complex that modulates nucleosome positioning in an ATP-dependent manner.

**SWR1 (switch remodeling 1)** A chromatin-remodeling complex responsible for replacing the common histone 2A protein of nucleosomes with a variant form known as H2AZ.

**sympatric speciation** An evolutionary process in which new species form in overlapping regions. Reproductive isolation mechanisms accompanying speciation are usually behavioral or mechanical.

**synapsis** The close approach and contact between homologous chromosomes during early prophase I in meiosis.

**synaptonemal complex** A specialized three-layer protein complex, consisting of a central element and two lateral elements, that forms between homologous chromosomes at synapsis.

**syncitium** A multinucleated cell in which the nuclei are not separated by cell membranes.

**syncytial blastoderm** Stage of *Drosophila* embryogenesis in which the nuclei are located at the periphery of the embryo but are not separated by cell membranes.

**synonymous codon** The groups of codons that specify the same amino acid.

**syntenic genes** Genes located on the same chromosome.

**synteny** The conserved order of genes together on a chromosome in species that share a common ancestor.

**synthesis-dependent strand annealing (SDSA)** An error-free mechanism for repair of DNA double-strand breaks occurring after the completion of DNA replication and utilizing strand invasion to provide wild-type sequences for repair.

**synthetic lethality** The situation where a particular double mutant results in lethality but the two respective single mutants are viable.

**systems biology** Prediction of biological functions of genes based on correlations between different data sets.

**T strand** The DNA strand of the T-DNA cleaved to initiate the transfer of plasmid DNA during rolling circle replication.

**targeted induced local lesions in genomes (TILLING)** A reverse genetic approach in which a population of organisms of an inbred strain is randomly mutagenized throughout the genome, and this population is then screened to find mutations in a gene of interest for which the sequence is known.

**TATA-binding protein (TBP)** A general transcription factor protein that binds the TATA box and assists in binding other transcription factors and RNA polymerase II to promoters.

**TATA box** The thymine- and adenine-rich consensus sequence region found in most eukaryotic promoters. Also known as *Goldberg-Hogness box*.

**TBP-associated factors (TAF)** Specific general transcription factors that associate with TATA-binding protein.

**telocentric chromosome** A chromosome with a centromere located at one end, producing a long arm only.

**telomerase** The ribonucleoprotein complex whose RNA component provides a template used to synthesize repeating DNA segments that form chromosome telomeres.

**telomere** Repeating DNA sequences, synthesized by *telomerase*, at the ends of linear chromosomes in eukaryotes; contain dozens to hundreds of copies of specific short DNA sequence repeats that buffer the coding sequence of the chromosome from loss during successive cycles of DNA replication.

**telophase** The last stage of *M phase*, in which the nuclear contents are divided (*karyokinesis*) and the daughter cells are divided (*cytokinesis*).

**temperate phage** A bacteriophage, such as λ phage, that can integrate into the bacterial host chromosome and produce either the lytic or lysogenic life cycle.

**temperature-sensitive allele** A mutation evident only at or above a certain temperature due to an abnormality of the protein product that affects its stability.

**template strand** The DNA strand serving as a template for synthesis of a complementary nucleic acid strand.

**terminal deletion** The loss of a chromosome segment that includes the telomeric region.

**termination sequence** DNA sequences that serve to stop transcription. Also known as *transcription termination*.

**termination stem loop** Stem loop of an mRNA transcript that signals RNA polymerase to terminate transcription in the leader region of bacterial attenuator-controlled operons (e.g., *trp* operon).

**test cross** The cross of an organism with the dominant phenotype that may be heterozygous with an organism that is homozygous for a recessive allele. Also known as *test-cross analysis*.

**tetrad** An ascus containing four haploid spores.

**tetrad analysis** The analysis of genetic linkage by analysis of different tetrad segregation types.

**tetratype (TT)** In an ascus, the occurrence of both types of parentals and both types of recombinants among the spores.

**third-base wobble** The flexibility of purine-pyrimidine base pairing between the third base of a codon and the corresponding nucleotide of the anticodon.

**three-point test-cross analysis** A test cross designed to identify genetic linkage between three genes and to provide data for determination of recombination frequency between linked genes.

**three-strand double crossover** The occurrence of double crossover involving three of the four chromatids.

**threshold of genetic liability** In polygenic and multifactorial inheritance, a trait with different phenotypes (i.e., affected and unaffected) that are determined by whether individual organisms are above or below a particular critical value on the phenotypic scale. Also known as *threshold trait*.

**threshold trait** See *threshold of genetic liability*.

**thymine (T)** One of four nitrogenous nucleotide bases in DNA; one of the two types of pyrimidine nucleotides in DNA.

**thymine dimer** See *pyrimidine dimer*.

**tiling array** DNA array that contains all sequences of the genome or a genomic interval, including introns, exons, untranslated regions (UTRs), and intergenic regions.

**time-of-entry mapping** A method of donor gene mapping by conjugation that uses interrupted mating to determine the order and relative timing of gene transfer.

**Ti plasmid** A large (200 kb) circular plasmid of *Agrobacterium tumefaciens* that harbors genes for transfer of DNA into plants cells and genes that cause uncontrolled division of plant cells; hence, the tumor-inducing (Ti) plasmid. It has been engineered for the construction of transgenic plants.

**topoisomerases** Enzyme that relaxes DNA supercoiling by controlled strand nicking and rejoining.

**totipotency** State of a cell when it can give rise to any and all cell types of an organism.

**trans-acting** Acting between two molecules (e.g., DNA sequences that control expression of genes interacting with a diffusible protein product).

**trans-acting regulatory protein** Proteins that act in trans by binding to cis-acting regulatory sequences and consequently regulating nearby genes, either by activating or repressing transcription. Often referred to as *transcription factors*.

**transcription** The cellular process that synthesizes RNA strands from a DNA template strand.

**transcription factors (TFs)** Proteins that bind promoters and are functional in transcription.

**transcription-terminating factor I (TTFI)** A specific protein that binds a termination sequence to stop transcription.

**transcription termination** See *termination sequence*.

**transcriptome** Set of transcripts present in a cell, tissue, or organism.

**transcriptomics** The study of all the transcripts, collectively known as the transcriptome, within a cell, tissue, or organism.

**transductant** The bacterium that is the product of transduction.

**transduction** In bacterial systems, the process of transfer of DNA from a donor bacterial cell to a recipient cell using a bacteriophage as a vector. More generally can refer to the process by which foreign DNA is introduced into another cell via a viral vector.

**transfer DNA (T-DNA)** The portion of the Ti plasmid that is transferred from the bacterium into the nucleus of a plant cell.

**transfer RNA (tRNA)** A family of small RNA molecules that each bind a specific amino acid and convey it to the ribosome, where the anticodon sequence undertakes complementary base pairing with an mRNA codon during translation.

**transformant** The bacterium that is the product of transformation.

**transformation** 1) The bacterial process of gene transfer in which donated DNA fragments originating in a dead donor cell, or plasmid DNA, is taken up across the cell wall and membrane of a recipient cell and recombined into the transformant genome. 2) More generally refers to the process by which exogenous DNA is directly taken up by a cell resulting in a genetic alteration of the cell. 3) The conversion of animal cells to an abnormal unregulated state by an oncogenic virus or by transforming DNA.

**transgene** A gene that has been modified in vitro by recombinant DNA technology and introduced into the genome via transformation.

**transgenic organism** An organism harboring a transgene.

**transition mutation** A type of DNA base-pair substitution in which one purine replaces the other or one pyrimidine replaces the other.

**translation** The process taking place at ribosomes to synthesize polypeptides. Complementary base pairing between mRNA codons and tRNA anticodons determines the order of amino acids composing the polypeptide.

**translation repressor protein** In bacteria, proteins that regulate translation by binding mRNA in the vicinity of the Shine–Dalgarno sequence and thereby prevent ribosome binding.

**translesion DNA synthesis** Utilizing a bypass polymerase, a mechanism for replicating DNA in the presence of damage that blocks replication by the common polymerase.

**translocation heterozygote** An organism with chromosome translocation in which chromosome pairs consist of one normal chromosome and a homolog carrying a translocation.

**transmission genetics** The subfield of genetics concerned with assessment and analysis of gene transfer from parents to offspring. Synonymous with *Mendelian genetics*.

**transposable genetic element** A class of DNA sequences that can move from one chromosome location to another, either by excision and reinsertion or by replication and reinsertion of the replicated copy.

**transposition** The process by which mobile genetic elements move from one portion of a genome to another. See also *transposable genetic element*.

**transposon tagging** Technique used to identify and clone genes through insertion of a transposon into the target gene.

**transversion mutation** A type of DNA base substitution mutation in which a purine substitutes for a pyrimidine, or vice versa.

**tree of life** The phylogenetic tree depicting the evolutionary relationships between organisms.

**trihybrid cross** A genetic cross between organisms that are heterozygous for three genes.

**trinucleotide repeat disorder** A hereditary disorder caused by a mutant gene containing an increased number of repeats of a DNA trinucleotide sequence.

**trisomy** The presence in a genome of three copies of a chromosome rather than a homologous pair of chromosomes, and resulting in a number of chromosomes that is $2n - 1$.

**trisomy rescue** In a trisomic genome, the random loss of one extra chromosome to reduce the chromosome number to the diploid.

**trivalent synaptic structure** The specific synaptic arrangement of three homologous chromosomes in a trisomic genome.

**true-breeding** See *pure-breeding*.

**true reversion** A type of reversion that exactly reverses the original mutation.

**tumor suppressor gene** A broad category of normal genes whose generalized functions slow, pause, or stop cell proliferation. It is often mutated in carcinogenesis.

**two-hybrid system** A method for discovering whether two proteins interact using the GAL4 protein of yeast, which is separated into a DNA-binding domain and a transcriptional activation domain. The two GAL4 domains are fused with the two proteins of interest respectively, and the resultant fusion proteins are assayed for their ability to activate transcription, which indicates interaction of the two proteins of interest.

**two-point test-cross analysis** A test cross designed to identify genetic linkage between two genes and to provide data for determination of recombination frequency between linked genes.

**two-strand double crossover** The occurrence of a double crossover involving two of the four chromatids.

**ultraviolet (UV) repair** A multiprotein DNA damage repair system that corrects lesions caused by exposure to ultraviolet irradiation.

**uncharged tRNA** A tRNA not carrying an amino acid.

**unequal crossover** Resulting from the improper synaptic pairing of homologous chromosomes and crossing over between the mispaired chromosomes. A source of duplication and deletion of genetic material.

**uniparental disomy** In a genome, the presence of a pair of homologous chromosomes that originate from a single parent.

**uniparental inheritance** Condition in organellar inheritance whereby just one parental gamete—often the maternal gamete—contributes all of the cytoplasmic organelles.

**unordered tetrad** Haploid spores in an ascus that are arranged in random order.

**unpaired loop** At synapsis involving partial deletion or partial duplication of one chromosome of a homologous pair, the "extra" genetic material that does not have a homolog on the paired chromosome.

**unselected marker screen** An experimental technique used to screen microbial genotypes. Commonly used following selected marker screening.

**unstable mutant phenotype** A mutation with an unusually high frequency of reversion.

**upstream** Referring to a gene or sequence location that is toward the 5′ direction of a *coding strand*.

**upstream activator sequence (UAS)** An enhancer-like sequence in yeast, located just upstream of the genes they regulate.

**upstream control element**  An upstream consensus sequence found in certain eukaryotic gene promoters.

**uracil (U)**  One of four nitrogenous nucleotide bases in RNA; one of the two types of pyrimidine nucleotides in RNA.

**variable expressivity**  Variation in the degree, magnitude, or intensity of expression of a phenotype.

**variance ($S^2$)**  A statistical measurement of the variation of sample values around the mean value.

**vector**  A DNA fragment with attributes that will allow its amplification (origin of replication) in a biological system and serves as a carrier for foreign DNA inserted into it. Vectors usually also possess genes (e.g., encoding resistance to an antibiotic) that allow selection of hosts carrying the vector.

**western blotting**  A method for transferring protein from an electrophoresis gel to a permanent membrane or filter.

**whole-genome shotgun (WGS) sequencing**  An approach to genome sequencing whereby DNA representing the entire genome is fragmented into smaller pieces, and a large number of fragments are chosen at random and sequenced with the aim that all genomic regions will be sequenced multiple times. Compare with *clone-by-clone sequencing*.

**whole-genome tiling array**  A microarray on which sequences representing the entire genome are present.

**X/autosome ratio (X/A ratio)**  The ratio of X chromosomes to a pair of autosomes. Used in *Drosophila* as the mechanism of sex determination.

**X-linked dominant**  A pattern of inheritance consistent with the transmission of a dominant allele of a gene on the X chromosome. Compare with *X-linked recessive*.

**X-linked inheritance**  The pattern of inheritance characteristic of genes located on the X chromosome.

**X-linked recessive**  A pattern of inheritance consistent with the transmission of a recessive allele of a gene on the X chromosome. Compare with *X-linked dominant*.

**yeast artificial chromosome (YAC)**  Cloning vector used in yeast that utilizes an endogenous yeast origin of replication, centromere, and telomere; can accept DNA inserts in excess of 1 megabase.

**Y-linked inheritance**  The exclusively male-to-male transmission of genes on the Y chromosome.

**$Z_{max}$**  The most likely recombination distance (theta [$\theta$] value) between genes as determined by lod score analysis.

**zone of polarizing activity (ZPA)**  The posterior side of the limb bud that acts as an organizer, secreting Sonic hedgehog (Shh) protein that acts to pattern the developing limb.

**Z/W system**  The sex chromosome inheritance system in species in which the male is homogametic (ZZ) and the female is heterogametic (ZW).

**zygotic genes**  Genes that are active only in the zygote or embryo. For zygotic genes, the genotype of the embryo determines the phenotype.

# Credits

## Photo Credits

**Chapter 1** **p. 1, CO-01** Pearson Education/Eric Schrader; **p.3, 1.1a** Scala/Art Resource, NY; **p. 3, 1.1b** Dorling Kindersley Media Library; **p. 3, 1.2a** National Library of Medicine; **p. 3, 1.2b** Library of Congress; **p. 3, 1.2c** Curt Stern Papers - American Philosophical Society; **p. 6, 1.4a** Omikron/Photo Researchers, Inc. **p. 6, 1.4b** National Library of Medicine; **p. 7, 1.5** National Library of Medicine; **p. 14, 1.12** American Association for the Advancement of Science.

**Chapter 2** **p. 25, CO-02** Mendel Museum, Augustinian Abbey Brno; **p. 42, Experimental Insight 2.1** Shutterstock.

**Chapter 3** **pp. 66–67, 3.2 (all)** Jennifer Waters/Photo Researchers, Inc.; **p. 70, 3.5a** Don W. Fawcett/Photo Researchers, Inc; **p. 70, 3.5b** Louisa Howard Dartmouth College; **p. 84, 3.16** George Halder; **p. 94, 3.24a** AP Photo; **p. 97, 3.27** Shutterstock.

**Chapter 4** **p. 105, CO-04** iStockphoto; **p. 111, 4.3** Karen Petersen; **p. 116, 4.7** John Bowman; **p. 116, 4.8a** imagebroker RF/Photolibrary; **p. 116, 4.8b** Stanton K. Short, The Jackson Laboratory; **p. 119, 4.13** Biophoto Associates/Photo Researchers, Inc.

**Chapter 5** **p. 143, CO-06** Science Source/Photo Researchers, Inc.

**Chapter 6** **p. 184, CO-06** David Scharf/SPL/Photo Researchers, Inc.; **p. 186, 6.1** Huntington Potter, University of South Florida and David Dressler, Oxford University and Balliol College; **p. 188, 6.4** David Scharf/SPL/Photo Researchers, Inc.; **p. 203, 6.15 (top)** Eye of Science/Photo Researchers, Inc.; **p. 203, 6.22 (bottom)** Gopal Murti/SPL/Photo Researchers; **p. 210, 6.22** Lester V. Bergman/Corbis.

**Chapter 7** **p. 222, CO-07 (and inset)** John Bowman; **p. 224, 7.1** Centers for Disease Control; **p. 234, 7.12** From "The Chromosome of Escherichia coli" *Cold Spring Harb Symp Quant Biol.* 1963, 28: 43–46; **p. 236, 7.13** Ray Rodriguez; **p. 237, 7.15** From The Units of DNA Replication in Drosophila melanogaster Chromosomes. 1974. *Cold Spring Harb Symp Quant Biol.* 38: 205–223; **p. 253, 7.32b** Cancer Research Technology, Wellcome Images; **p. 255, 7.34** Dr. James Gusella.

**Chapter 8** **p. 260, CO-08** Ann L. Beyer; **p. 281, 8.19** Phillip A. Sharp.

**Chapter 9** **p. 304, Research Technique 9.1** Robert Traut; **p. 310, 9.11a** From Hamkalo and Miller. 1973. Electronmicroscopy of Genetic Material. *Annual Review of Biochemistry* 42:379, Figure 6a.

**Chapter 10** **p. 332, CO-10** SPL/Photo Researchers, Inc.; **p. 332, 10.1** Centers for Disease Control; **p. 342, 10.8a** John Bowman; **p. 342, 10.8b** Yikrazuul.

**Chapter 11** **p. 359, CO-11** Dr. Klaus Boller/Photo Researchers, Inc.; **p. 362, 11.2a,b** James C. Wang; **p. 364, 11.3b (top)** Victoria Foe; **p. 364, 11.3c (middle)** Barbara Hamkalo; **p. 366, 11.5** Ulrich K. Laemmli; **p. 369, 11.9** L. Willatt/Photo Researchers, Inc.; **p. 372, 11.12a** Wellcome Images; **p. 372, 11.12b** Hesed M. Padilla-Nash, Antonio Fargiano, and Thomas Ried. National Cancer Institute; **p. 373, 11.13b** From Cremer, T. and Cremer, C. 2001. Chromosome territories, nuclear architecture and gene regulation in mammalian cells. *Nature Rev Genet.* 2:292–301; **p. 379, 11.19a (1,2)** Joanne Davidson Staines and Paul Edwards.

**Chapter 12** **p. 383, CO-12** Shutterstock.

**Chapter 13** **p. 422, CO-13** SPL/Photo Researchers, Inc.; **p. 431, 13.9 (top left)** James King-Holmes/Photo Researchers, Inc.; **p. 431, 13.9 (top right)** Adina Breiman; **p. 431, 13.9 (middle left)** USDA Agricultural Research Service; **p. 431, 13.9 (middle right)** Jon Raupp; **p. 431, 13.9 (bottom)** Bjorn Svensson/Photo Researchers, Inc.

**Chapter 14** **p. 459, CO-14** Bettmann/Corbis.

**Chapter 15** **p. 494, CO-15 (left)** Jorgensen, R.A., Q. Que, and C.A. Napoli. 2002. Maternally controlled ovule abortion results from cosuppression of dihydroflavonol-4-reductase or flavonoid-3′, 5′-hydroxylase genes in Petunia hybrida. *Functional Plant Biology* 29:1501–1506; **p. 494, CO-15 (right)** Napoli, C., C. Lemieux, and R. Jorgensen. 1990. Introduction of a chimeric chalcone synthase gene into petunia results in reversible cosuppression of expression of homologous genes in trans. *Plant Cell* 2:279–289; **p. 494, CO-15 (bottom)** Richard Jorgensen.

**Chapter 16** **p. 522, CO-16** CalTech Archives; **p. 533, 16.7** John Bowman; **p. 534, Genetic Analysis 16.2** John Bowman; **p. 538, 16.11b** John Bowman.

**Chapter 17** **p. 562, CO-17** From F. X. Cunningham, Jr. and E. Gantt. 1998. Genes and enzymes of carotenoid biosynthesis in plants. *Annual Review of Plant Physiology and Plant Molecular Biology* 49: 557–583; **p. 572, 17.6c** John Bowman; **p. 575, 17.8c** Goldenrice.org; **p. 576, 17.9** Wormbase.org, used under a Creative Commons License: http://creativecommons.org/licenses/by/2.5/; **p. 577, 17.11** Robert Devlin, Fisheries and Oceans Canada; **p. 582, 17.15a** Russell J. Hill and Paul Sternberg; **p. 582, 17.15b** John Bowman; **p. 582, 17.15c** Keith V. Wood; **p. 582, 17.15d** John Wilson, Baylor College of Medicine; **p. 582, 17.15e** Jeff Lichtman, from Livet et al. 2007. Transgenic strategies for combinatorial expression of fluorescent proteins in the nervous system. *Nature* 450, 56–62, Figure 3d-right; **p. 585, 17.16** Stephen Small, From "Regulation of even-skipped stripe 2 in the Drosophila embryo." *EMBO Journal* volume 11 (1992), 4047–4057; **p. 586, 17.17 (top left and right)** Eye of Science/Photo Researchers, Inc.; **p. 586, 17.17 (bottom left)** David Scharf/Photo Researchers, Inc.; **p. 586, 17.17 (bottom right)** Eye of Science/Photo Researchers, Inc.; **p. 590, 17.20b** Clive Wilson; **p. 595, 17.22** Roslin Institute; **p. 597, 17.23 (all)** John Bowman.

**Chapter 18** **p. 602, CO-18** John Bowman; **p. 606, 18.4b** From Fleischmann et al. 1995. Whole genome random sequencing and assembly of Haemophilus influenzae Rd. *Science* 269, 496–512; **p. 621, 18.16 (right)** Robert Hill; **p. 621, 18.16 (left)** James Sharpe and Laura Lettice, From 1999. Identification of Sonic hedgehog as a candidate gene responsible for the polydactylous mouse mutant Sasquatch. *Current Biology* 9:97–100; **p. 628, 18.22** David J. Lockhart and Elizabeth A. Winzeler; **p. 629, 18.23** Patrick Brown, From 1998. The Transcriptional Program of Sporulation in Budding Yeast. *Science:* 282: 699–705; **p. 630, 18.24** Joseph R. Ecker, From 2003. Empirical Analysis of Transcriptional Activity in the Arabidopsis Genome. *Nature* 302: 842–846.

**Chapter 19** **p. 641, CO-19** Dartmouth Electron Microscope Facility; **p. 657, 19.11** Dr. David Furness, Keele University/Photo Researchers, Inc.; **p. 657, 19.12** Don W. Fawcett/Photo Researchers, Inc.; **p. 660, 19.15** Dr. Jeremy Burgess/Photo Researchers, Inc.; **p. 661, 19.16a** Richard D. Kolodnar.

**Chapter 20** **p. 674, CO-20** FLPA/Alamy; **p. 675, 20.1a (top left and right)** Edward B. Lewis/CalTech Archives; **p. 675, 20.1b (bottom left and right)** Eye of Science/SPL/Photo Researchers, Inc.; **p. 676, 20.2** Steve Baskauf; **p. 677, 20.3b (left)** Edward M. De Robertis; **p. 677, 20.3c (right)** James B. Skeath; **p. 680, 20.5a (left)** Daniel St. Johnston and Wolfgang Driever; **p. 680, 20.5b (2nd from left)** From J. Jaeger et al. 2004. Dynamic control of positional information in the early Drosophila embryo. *Nature* 430: 368–371; **p. 680, 20.5c (3rd from left)** from James B. Jaynes and Miki Fujioka. 2004. Drawing lines in the sand: even skipped et al. and parasegment boundaries. *Developmental Biology* 269: 609–622, Figure 2c; **p. 680, 20.5d (right)** from James B. Jaynes and Miki Fujioka. 2004. Drawing lines in the sand: even skipped et al. and parasegment boundaries. *Developmental Biology* 269: 609–622, Figure 3e; **p. 681, 20.6b (both)** Daniel St. Johnston; **p. 683, 20.8a–d** Matthias Mannervik; **p. 684, 20.9c** Stephen Small, From 1992. Regulation of even-skipped stripe 2 in the Drosophila embryo. *EMBO Journal* 11: 4047–4057; **p. 685, 20.10b** William McGinnis, From 2006. Genomic Evolution of Hox Gene Clusters. *Science* 313: 1918–1922; **p. 692, 20.14a** Paul W. Sternberg; **p. 693, 20.15a (top)** Erik Jorgensen; **p. 693, 20.15a (bottom left and right)** Paul W. Sternberg; **p. 697, 20.18b** Shutterstock; **p. 699, 20.19 (all)** John Bowman; **p. 700, 20.20 (all)** John Bowman; **p. 701, 20.21b (all)** John Bowman; **p. 703, 20.22c (left)** From Binns, W., James. L. F., Shupe, J. L. 1964. Toxicosis of veratrum californicum in ewes and its relationship to a congenital deformity in lambs. *Annals of the New York Academy of Sciences* 111: A2, 571–576; **p. 703, 20.22c (middle)** Robert Hill, from 2003. A long-range Shh enhancer regulates expression in the developing limb and fin and is associated with preaxial polydactly. *Human Molecular Genetics*, 12: 1725–1735; **p. 703, 20.22c (right)** Robert Hill, from 2003. A long-range Shh enhancer regulates expression in the developing limb and fin and is associated with preaxial polydactyly. *Human Molecular Genetics* 12: 1725–1735.

**Chapter 21** **p. 708, CO-21** Peter Morenus.

**Chapter 22** **p. 737, CO-22** Karl Magnacca.

## Text and Illustration Credits

**Chapter 1**  **p17, 1.13** Adapted by permission from N. A. Campbell and J. B. Reece, *Biology*, 8th ed., Fig 1.22, p. 17. © 2008.

**Chapter 3**  **p. 94, 3.24b** Adapted by permission from Macmillan Publishers Ltd: *Nature Genetics*, vol. 10, no. 2, Luis E. Figuera, Massimo Pandolfo, Patrick W. Dunne, Jose M. Cantú, and Pragna I. Patel, "Mapping of the congenital generalized hypertrichosis locus to chromosome Xq24–q27.1," copyright 1995. http://www.nature.com/ng/index.html.

**Chapter 4**  **p. 122, 4.16** Adapted from Monroe W. Strickberger, *Genetics*, 3e, p. 543. Macmillan, 1969. Used with permission of Pearson Education.

**Chapter 5**  **p. 163, 5.13** Data from Weber, J. L., et al. 1993. Evidence for human meiotic recombination interference obtained through construction of a short tandem repeat polymorphism linkage map of chromosome 19. *Am. J. Hum. Genet.* 53:1079–1095. **p. 176, 5.26a** Adapted by permission from Macmillan Publishers Ltd: *Nature*, v. 306, No 5940, James F. Gusella, et al., "A polymorphic DNA marker genetically linked to Huntington's disease," copyright 1983. http://www.nature.com/nature/index.html.

**Chapter 6**  **p. 194, 6.9a** Adapted from E.L. Wollman, F. Jacob, and W. Hayes, "Conjugation and Genetic Recombination in Escherichia coli K-12," *Cold Spring Harbor Symposia on Quantitative Biology*, 21, 1956, p. 141. Used with permission of the copyright holder, Cold Spring Harbor Laboratory Press and the Estate of Elie Wollman; **p. 197, 6.11b** Barbara J. Bachmann, "Linkage Map of Escherichia coli K-12, Edition 8+," *Microbiological Reviews*, v. 54, Issue 2, 1 June 1990, Copyright © 1990, American Society for Microbiology, p. 133. Reproduced with permission from American Society for Microbiology; **p. 213, 6.25** Adapted from Seymour Benzer, "Fine Structure of a Genetic Region in Bacteriophage," *Proceedings of the National Academy of Sciences of the United States of America*, Vol. 41, No. 6 (Jun. 15, 1955), pp. 344–354, Fig. 5A. Used with permission; **p. 214, 6.27** Adapted from Seymour Benzer, "On the Topography of the Genetic Fine Structure," *Proceedings of the National Academy of Sciences*, Vol. 47, No. 3 (1961), p. 410, Fig. 6. Used with permission; **p. 215, 6.28** Adapted from Seymour Benzer, "On the Topography of the Genetic Fine Structure," *Proceedings of the National Academy of Sciences*, Vol. 47, No. 3 (1961), p. 406, Fig. 3. Used with permission.

**Chapter 7**  **p. 237, 7.14** Reprinted from *Journal of Molecular Biology*, Vol. 32, J. A. Huberman and A. D. Riggs, "On the mechanism of DNA replication in mammalian chromosomes," pp. 327–341, Copyright 1968, with permission from Elsevier.

**Chapter 8**  **p. 271, Table 8.2** Adapted and reprinted from *Journal of Molecular Biology*, Vol. 304, R. H. Elbright, "RNA Polymerase: Structural Similarities Between Bacterial RNA Polymerase and Eukaryotic RNA Polymerase II," 687–698, Copyright 2000, with permission from Elsevier; **p. 274, 8.11** From R. M. Myers, K. Tilly, and T. Maniatis, 1986, "Fine structure genetic analysis of a beta-globin promoter," *Science*, 232: 613–618. Reprinted with permission from AAAS and the author. http://www.sciencemag.org; **p. 285, 8.23b** Reprinted from *Cell*, Vol. 103, D. L. Black, "Protein diversity from alternative splicing," pp. 367–370, Copyright 2000, with permission from Elsevier. **p. 286, 8.24** Adapted from J. P. Lees-Miller, L. O. Goodwin, and D. M. Helfman, "Three Novel Brain Tropomyosin Isoforms Are Expressed from the Rat cx-Tropomyosin Gene through the Use of Alternative Promoters and Alternative RNA Processing," *Molecular and Cellular Biology*, Apr. 1990, Fig. 8, p. 1739. Amended with permission from American Society of Microbiology.

**Chapter 10**  **p. 351, 10.15a and b** Adapted by permission from Macmillan Publishers Ltd: *Nature Communications*, v. 1. no. 8, Frédéric B. Piel, Anand P. Patil, Rosalind E. Howes, Oscar A. Nyangiri, Peter W. Gething, and Thomas N. Williams, "Global distribution of the sickle cell gene and geographical confirmation of the malaria hypothesis," copyright 2010. http://www.nature.com/ncomms/index.html; **p. 351, 10.15c** Adapted by permission from Macmillan Publishers Ltd: *Nature Reviews Genetics*, vol. 2, no. 12, Graham S. Cooke and Adrian V. S. Hill, "Genetics of susceptibility to human infectious disease," copyright 2001. http://www.nature.com/nrg/index.html. Data from World66.com.

**Chapter 11**  **p. 370, 11.10a** Figure adapted from J. J. Yunis and O. Prakash, "The origin of man: A chromosomal pictorial legacy," *Science*, 215:1527. Reprinted with permission from AAAS. http://www.sciencemag.org/content/215/4539/1525.abstract?sid=17df9c3a-287e-4f30-b241-c7b1c88b1126; **p. 374, 11.14** Adapted by permission from Macmillan Publishers Ltd: *Nature*, vol. 465, no. 7296, Zhijun Duan, Mirela Andronescu, Kevin Schutz, Sean McIlwain, Yoo Jung Kim, and Choli Lee, "A three-dimensional model of the yeast genome," copyright 2010. http://www.nature.com.

**Chapter 12**  **p. 402, 12.17** H. H. Watson, J. D. Winsten, J. A., eds. *Origins of Human Cancer*, 1977, 1431–1450. Cold Spring Harbor Laboratory Press, Cold Spring Harbor, N.Y. Used with permission. **p. 407, 12.22** Hardin, Jeff; Bertoni, Gregory Paul; Kleinsmith, Lewis J., *Becker's World of the Cell*, 8th, ©2012. Printed and Electronically reproduced by permission of Pearson Education, Inc., Upper Saddle River, New Jersey.

**Chapter 13**  **p. 451, 13.28** Adapted from Macmillan Publishers Ltd: *Nature*, vol. 420, no. 6915, Mouse genome sequencing consortium, "Initial sequencing and comparative analysis of the mouse genome," copyright 2002. http://www.nature.com.

**Chapter 14**  **p. 462, 14.4** Madigan, Michael T.; Martinko, John M.; Stahl, David A.; Clark, David P., *Brock Biology of Microorganisms*, 13th, ©2012. Printed and Electronically reproduced by permission of Pearson Education, Inc., Upper Saddle River, New Jersey.

**Chapter 15**  **p. 503, 15.12** Adapted by permission from Macmillan Publishers Ltd: *Nature*, vol. 461, no. 7261, Bradley R. Cairns, "The logic of chromatin architecture and remodelling at promoters," copyright 2009. http://www.nature.com; **p. 504, 15.14** Watson, James D.; Baker, Tania A.; Bell, Stephen P.; Gann, Alexander; Levine, Michael; Losick, Richard; *CSHLP, Inglis, Molecular Biology of the Gene*, 6th, ©2008. Printed and Electronically reproduced by permission of Pearson Education, Inc., Upper Saddle River, New Jersey; **p. 507, 15.16** Adapted by permission from Macmillan Publishers Ltd: *Nature*, vol. 461, no. 7261, Bradley R. Cairns, "The logic of chromatin architecture and remodelling at promoters," copyright 2009. http://www.nature.com; **p. 508, 15.18** Watson, James D.; Baker, Tania A.; Bell, Stephen P.; Gann, Alexander; Levine, Michael; Losick, Richard; *CSHLP, Inglis, Molecular Biology of the Gene*, 6th, ©2008. Printed and Electronically reproduced by permission of Pearson Education, Inc., Upper Saddle River, New Jersey. Adapted from Fig. 4.35, Copyright © 2007 from *Molecular Biology of the Cell*, 4e by Bruce Alberts et al. Reproduced by permission of Garland Science/Taylor & Francis Books, Inc. in the format Textbook via Copyright Clearance Center. Figure adapted from T. Jenuwein and C. D. Allis, "Translating the histone code," *Science*, 2001 Aug 10, 293(5532):1077–1078. Reprinted with permission from AAAS. http://www.sciencemag.org/content/293/5532/1074.abstract?sid=eaef9715-35ec-4d8b-b8d2-5183099ed927; **p. 509, 15.19** Adapted by permission from Macmillan Publishers Ltd: *Nature*, 461, p. 189, Nicholas J. Fuda, M. Behfar Ardehali, and John T. Lis, "Defining mechanisms that regulate RNA polymerase II transcription in vivo," copyright 2009. http://www.nature.com/nature/index.html. **p. 511, 15.20** Watson, James D.; Baker, Tania A.; Bell, Stephen P.; Gann, Alexander; Levine, Michael; Losick, Richard; *CSHLP, Inglis, Molecular Biology of the Gene*, 6th, ©2008. Printed and Electronically reproduced by permission of Pearson Education, Inc., Upper Saddle River, New Jersey.

**Chapter 16**  **p. 530, 16.5** Adapted by permission from Macmillan Publishers Ltd: *Nature Reviews Genetics*, vol. 8, no. 6. Charles Boone, Howard Bussey, and Brenda J. Andrews, "Exploring genetic interactions and networks with yeast," copyright 2007. http://www.nature.com/nrg/index.html. **p. 556, 16.26** Reprinted from *Cell*, 72, The Huntington's Disease Collaborative Research Group, Marcy E. MacDonald et al., "A novel gene containing a trinucleotide repeat that is expanded and unstable on Huntington's Disease chromosomes," p. 972, Copyright 1993, with permission from Elsevier. http://www.sciencedirect.com/science/journal/00928674.

**Chapter 18**  **p. 611, 18.7a** Adapted by permission from Macmillan Publishers Ltd: *Nature*, vol. 408, no. 6814. The Arabidopsis Genome Initiative, "Analysis of the genome sequence of the flowering plant Arabidopsis thaliana," copyright 2000. http://www.nature.com; **p. 611, 18.7b** Adapted from J. C. Venter, "The sequence of the human genome," *Science* 291: 1335. Reprinted with permission from AAAS. Used with permission from the J. Craig Venter Institute. http://www.sciencemag.org/content/291/5507/1304.abstract?sid=0ad9852f-d2bd-45bc-aa3b-68dd28966d53; **p. 621, 18.16** Adapted/reprinted from *Current Biology* 9, J. Sharpe et al., "Identification of Sonic hedgehog as a candidate gene responsible for the polydactylous mouse mutant Sasquatch," 97–100, Copyright (1999), with permission from Elsevier; **p. 622, 18.17** Adapted by permission from Macmillan Publishers Ltd: *Nature Reviews Genetics*, 5, Dario Boffelli, Marcelo A. Nobrega and Edward M. Rubin, "Comparative genomics at the vertebrate extremes," copyright 2004. http://www.nature.com/nrg/index.html; **p. 624, 18.19** Adapted by permission from Macmillan Publishers Ltd: *Nature*, vol. 408, no. 6814. The Arabidopsis Genome Initiative, "Analysis of the genome sequence of the flowering plant Arabidopsis thaliana," copyright 2000. http://www.nature.com; **p. 626, 18.20** Adapted from Sarah A. Tishkoff, et al., "The Genetic structure and history of Africans and African Americans," *Science*, 324, p. 1036, Figure 1. Reprinted with permission from AAAS and the author. http://www.sciencemag.org/content/324/5930/1035.abstract?sid=792ea3ba-1d34-45ce-aad6-622a1d4d6b00; **p. 630, 18.24** Adapted from K. Yamada et al., "Empirical analysis of transcriptional activity in the Arabidopsis genome," *Science*, 302, p. 843, Figure 2. Reprinted with permission from AAAS. http://www.sciencemag.org/

content/302/5646/842.abstract?sid=044b459a-d3fb-42e5-9c3d-1576af08980e#aff-1; **p. 631, 18.25b** Adapted by permission from Macmillan Publishers Ltd: *Nature*, vol. 403, no. 6770, Peter Uetz, Loic Giot, Gerard Cagney, Traci A. Mansfield, Richard S. Judson et al., "A comprehensive analysis of protein-protein interactions in : Saccharomyces cerevisiae," copyright 2000. http://www.nature.com; **p. 632, 18.26** Adapted by permission from Macmillan Publishers Ltd: *Nature Reviews Genetics*, vol. 8, no. 6. Charles Boone, Howard Bussey and Brenda J. Andrews, "Exploring genetic interactions and networks with yeast," copyright 2007. http://www.nature.com/nrg/index.html; **p. 635, 18.28** Adapted from A. H. Y. Tong, et al., Systematic genetic analysis with ordered arrays of yeast deletion mutants, *Science*, 294, 2364–2368, Fig. 3. Reprinted with permission from AAAS and the author. http://www.sciencemag.org/content/294/5550/2364.full.

**Chapter 19** **p. 648, 19.6b** Adapted by permission from Macmillan Publishers Ltd: *Nature*, 325, Rebecca L. Cann, Mark Stoneking, and Allan C. Wilson, "Mitochondrial DNA and human evolution," copyright 1987. http://www.nature.com/nature/index.html; **p. 649, 19.7a** Reprinted from *Biochimica et Biophysica Acta*, 1366, S. DiMauro et al., Mitochondria in neuromuscular disorders, p. 206, Copyright 1998, with permission from Elsevier. http://www.sciencedirect.com/science/journal/00052736; **p. 651, 19.8** Adapted by permission from Macmillan Publishers Ltd: *Nature Reviews Genetics*, vol. 6, no. 5, Robert W. Taylor and Doug M. Turnbull, "Mitochondrial DNA mutations in human disease," copyright 2005. http://www.nature.com/nrg/index.html; **p. 658, 19.13a (top)** Data from Macmillan Publishers Ltd: *Nature Reviews Genetics*, vol. 6, no. 5, Robert W. Taylor and Doug M. Turnbull, "Mitochondrial DNA mutations in human disease," copyright 2005. http://www.nature.com/nrg/index.html; **p. 658, 19.13a (bottom)** Data from Macmillan Publishers Ltd: *Nature*, v. 387, n. 6632, B. Franz Lang, Gertraud Burger, Charles J. O'Kelly, Robert Cedergren, G. Brian Golding, et al., "An ancestral mitochondrial DNA resembling a eubacterial genome in miniature," copyright 1997. http://www.nature.com/nature/index.html; **p. 658, 19.13b** Adapted from *Trends in Genetics*, 19, G. Burger and M. W. Gray, Mitochondrial genomes: anything goes, 709–716,

Copyright 2003, with permission from Elsevier. http://www.sciencedirect.com/science/journal/01689525; **p. 659, 19.14** Adapted from *Molecular Cell*, 24, Bonawitz et al., Initiation and beyond: multiple functions of the human mitochondrial transcription machinery, p. 814, Copyright 2006, with permission from Elsevier. http://www.sciencedirect.com/science/journal/10972765. **p. 661, 19.16b** Reprinted from *Trends in Genetics*, 3, K. Umesono and H. Ozeki, "Chloroplast gene organization in plants," p. 281, Copyright 1987, With permission from Elsevier. http://www.sciencedirect.com/science/journal/01689525. **p. 663, 19.18** Adapted by permission from Macmillan Publishers Ltd: *Nature*, v. 440, n. 7084, T. Martin Embley and William Martin, "Eukaryotic evolution, changes and challenges," copyright 2006. http://www.nature.com/nature/index.html; **p. 665, 19.20** Reprinted from *Trends in Genetics*, 19, D. Leister, "Chloroplast research in the genomic age," 47–56, Copyright 2003, with permission from Elsevier. http://www.sciencedirect.com/science/journal/01689525; **p. 669, 19.22a** Adapted by permission from Macmillan Publishers Ltd: *Nature Genetics*, 4, Toni R. Prezant, et al., "Mitochondrial ribosomal RNA mutation associated with both antibiotic–induced and non–syndromic deafness," copyright 1993. http://www.nature.com/ng/index.html; **p. 650, Genetic Analysis 19.1** Data from Macmillan Publishers Ltd: *Nature*, 351, Wayne RK, Jenks SM, "Mitochondrial DNA analysis implying extensive hybridization of the endangered red wolf Canis rufus" 565–568, Copyright 1991. http://www.nature.com/ng/index.html. Problem.

**Chapter 20** **p. 680, 20.5** Adapted by permission from Macmillan Publishers Ltd: *Nature*, 287, Christiane Nüsslein-Volhard and Eric Wieschaus, "Mutations affecting segment number and polarity in Drosophila," copyright 1980. Adapted from *Developmental Biology* 6th ed., Fig. 9.27, p. 286. Sunderland: Sinauer Associates, 2000. Used with permission. http://www.nature.com/nature/index.html; **p. 690, 20.13** Adapted by permission from Macmillan Publishers Ltd: *Nature*, v. 399, n. 6738, Renaud de Rosa, Jennifer K. Grenier, Tatiana Andreeva, Charles E. Cook, Andre Adoutte, et al., "Hox genes in brachiopods and priapulids and protostome evolution," copyright 1999. http://www.nature.com/nature/index.html.

**Chapter 21** **p. 714, 21.4** Adapted from *Biometrical Genetics*, ed. K. Mather, Fig. 2, p. 6. Dover Publications, 1949. Used with permission of the publisher; **p. 719, 21.8a** Data is from William Castle (1916) *Genetics and Eugenics*. Harvard University Press; **p. 719, 21.8b** Reprinted by permission of the publisher from *Genetics and Eugenics: A Text-Book for Students of Biology and a Reference Book for Animal and Plant Breeders*, Fourth Revised Edition, by W. E. Castle, p. 102, Cambridge, Mass.: Harvard University Press, Copyright © 1930 by the President and Fellows of Harvard College. Copyright © renewed 1958 by William Ernest Castle; **p. 724, Table 21.3** Data from R. Plomin et al. 1994. The genetic basis of complex human behaviors. *Science* 264: 1733–1739. **p. 729, 21.14b** Adapted from Eyal Fridman, et al., Zooming In on a Quantitative Trait for Tomato Yield Using Interspecific Introgressions, *Science*, Vol. 305, no. 5691, p. 1787, Figure 2. Reprinted with permission from AAAS and the author. http://www.sciencemag.org/content/305/5691/1786.abstract?sid=00f765da-7768-4583-b360-a5399b6ca820; **p. 731, 21.15a** Adapted by permission from Macmillan Publishers Ltd: *Nature*, 411, Yasunori Ogura, Denise K. Bonen, Naohiro Inohara, Dan L. Nicolae, and Felicia F. Chen, et al., "A frameshift mutation in *NOD2* associated with susceptibility to Crohn's disease," copyright 2001. http://www.nature.com/nature/index.html; **p. 731, 21.15b** Data from Paul R. Burton, David G. Clayton, Lon R. Cardon, Nick Craddock, and Panos Deloukas, et al., "Genome-wide association study of 14,000 cases of seven common diseases and 3,000 shared controls," *Nature*, 447, copyright 2007.

**Chapter 22** **p. 748, 22.6** Adapted from Freeman, Scott; Herron, Jon C., *Evolutionary Analysis*, 4th, ©2007. Printed and reproduced by permission of Pearson Education, Inc., Upper Saddle River, New Jersey; **p. 759, 22.12** Adapted from Campbell, Neil A.; Reece, Jane B., *Biology*, 8th, ©2008. Printed and Electronically reproduced by permission of Pearson Education, Inc., Upper Saddle River, New Jersey; **p. 761, 22.14** Adapted from Freeman, Scott; Herron, Jon C. *Evolutionary Analysis*, 4th, ©2007. Printed and Electronically reproduced by permission of Pearson Education, Inc., Upper Saddle River, New Jersey. **p. 762, Table 22.10** Data from cstl.nist.gov/strbase/str_fact.htm.

# Index

Note: A *b* following a page number indicates a box, an *f* indicates a figure, and a *t* indicates a table. Page numbers in **bold** indicate pages on which key terms are discussed.

Functional genomics    (*continued*)
    transcriptomics in, 627–630, 628*f*,
        629*f*, 630*f*
    yeast mutants to categorize genes in,
        632–633, 633*f*
Functional RNAs, **263**
Fungus, transgenic, generation of, 569–571,
    569*f*, 570*f*
Fusion genes, **565**
Fusion protein, **565**

**G**

G (Giemsa) banding, **370**
G₀ ("G zero"), of interphase, **65,** 65*f*
G₁ (Gap 1) phase, of interphase, **65,** 65*f*, 70*f*
G₂ (Gap 2) phase, of interphase, **65,** 65*f*, 66*f*
Gaertner, Carl Friedrich, genetic research of, 44*b*
Gain-of-function mutation, **107,** 108*f*, 109
Gamete(s), **2, 32, 64**
    frequencies of, determining from genetic maps,
        158–159, 158*f*
Gap genes, **679,** 680*f*
    expression of, domains of, 682–683, 682*f*, 683*f*
Garden pea. *See* Pea, garden *(Pisum sativum)*
Garrod, Archibald
    in documentation of human hereditary disor-
        der, 2, 124
    on gene-protein connection, 298
    genetic screens and, 523
Gaussian (normal) distribution, **48,** 48*f*
GC-rich box, **273**
Gel electrophoresis, 335–338, **336,** 336*f*, 337*f*
    in hemoglobin peptide fingerprint analysis, 338
    in SCD analysis, 346–348, 346*f*, 347*f*, 348f
    two-dimensional, in ribosomal protein
        identification, 304*b*
Gender. *See* Sex *entries*
Gene(s), **2**
    actions of, reverse genetics investigating,
        586–590, 587*t*, 588*f*, 589*f*, 590*f*
    additive, **710,** 711*f*, 712*f*
    alleles of, **3**. *See also* Allele(s)
    annotation describing, 608–609, 609*f*
    births and deaths of, 617–618, 618*f*
    carried on chromosomes, chromosome theory
        of heredity proposing, 81, 84–87
    chimeric, in gene function investigation,
        584–585, 586*f*
    cloning
        by complementation, recombinant DNA
            technology in, 549, 550*f*
        positional, 550–555, **551,** 552*f*, 553*f*,
            554*f*, 555–557*b*
        using transposons, recombinant DNA
            technology in, 550, 551*f*
    coordinate, **679,** 680*f*
        in patterning of anterior–posterior axis,
            681–682, 681*f*, 682*f*
    definition of, original, 260
    duplication of, 618–619, 619*f*, 620*f*
    expression of, 10–13
        sex-limited, **117**
        transcription in, 11–12
        translation in, 12–13, 13*f*
    fine structure of, discovery of, 210
    foreign, introducing into genomes to create
        transgenic organisms, 563–581
    function of
        dissecting, transgenes in, 581–585, 582*f*,
            583*b*, 585*f*, 586*f*
        investigating, chimeric genes in,
            584–585, 586*f*
    fusion, **565**
    gap, expression of, domains of, 682–683,
        682*f*, 683*f*

*globin,* **334,** 335*f*
    mutations, 335
*Hfr,* transfer of, 191–193, 192*f*
highly expressed, cloning, recombinant DNA
    technology in, 547–548, 547*f*
homeotic, **679,** 680*f*
    parasegmental pattern of expression of,
        684–688
homologous, cloning, recombinant DNA tech-
    nology in, 548–549, 548–549*f*
housekeeping, **679**
*hox,* **685,** 685–686, 685*f*
Huntington disease
    mapping (Case Study), 176–177*b*
    positional cloning in identification of
        (Case Study), 555–557*b*, 556*f*
identified in mutagenesis, number of, 528
informational, **667**
interacting, identifying, 528–530
*lac* operon, regulatory sequence for, 466*t*
lateral transfer of, 619–620
linkage groups of, **165**
linked, assortment of, 144–145
major, **710**
maternal effect, **679–680,** 679–681
messages of, RNA transcripts carrying,
    261–264
modifier, **710**
mutant alleles identified for, number of, 528, 529*b*
mutations of. *See* Mutation(s)
operational, **667**
order on chromosome, for three-point recom-
    bination mapping, 156–157
orthologous, **619**
pair-rule, **679,** 680*f*
    regulation of, 683–684, 684*f*
pairs of, recombination frequency of, **147**. *See
    also* Recombination frequency
    in three-point recombination
        mapping, 157
paralogous, **619**
pleiotropic, 121–122, 122*f*
producing variable phenotypes, 119–122
quantitative trait loci, identification of,
    729–730, 729*f*
realizator, **688**
recombination between, 152–163, 157–160, 162*b*
recombination within, 161, 161*f*
redundant, identifying, 530
regulation of, in evolution, 696–698
relative order of, finding, by three-point map-
    ping, 153–155, 154*f*
reporter, **581**
    in gene expression monitoring, 581–584,
        582*f*, 585*f*
segment polarity, **679,** 680*f*
silencing of
    by double-stranded DNA, 512, 514, 514*f*
    by nucleotide methylation, 511–512,
        512*f*, 513*b*
single, alternative transcripts of, 283, 285–287
syntenic, **144**
transfer of, in bacteria
    by conjugation, 185–193
    by transduction, 203–209
    by transformation, 201–203, 203*f*
tumor suppressor, **72**
two, dihybrid-cross analysis of, 35, 35*f*, 37–38
X-linked
    genotype frequencies for, estimating,
        Hardy-Weinberg equilibrium in,
        744–746, 745*t*, 746*f*
    Hardy-Weinberg equilibrium for, 746,
        746*f*
zygotic, **680,** 680–681

Gene content
    of chloroplast, 660–661
    of mitochondrial genome, structure and,
        656–658, 657*f*, 658*f*
Gene conversion, 411, **411,** 414–415, 414*f*, 415*f*
Gene dosage, **425**
Gene-environment interactions, **120,** 120–121
Gene expression
    in bacteria and bacteriophage, regulation of,
        459–489
        antiterminators and repressors in,
            483–487
        inducible operon system in, 462–466
            mutational analysis deciphering,
                466–475
        repression and attenuation in, 475–480
        stress response and, 480–481
        transcriptional, DNA–protein interac-
            tion in, 460–462, 460*f*, 461*f*
        translational, 481–483
    in eukaryotes, regulation of, 494–518
        chromatin remodeling in, 502–512
        cis-acting regulatory sequences in,
            495–502
        RNA-mediated mechanisms in, 512,
            514–517, 514*f*, 515*f*
    monitoring of, reporter genes in, 581–584,
        582*f*, 585*f*
Gene expression machine model, for coupling
    transcription with pre-mRNA pro-
    cessing, 283, 284*f*
Gene families, 609, **609,** 611
Gene flow, **751**
    effects of, 751–752, 752*f*
Gene interaction(s), 105–142, **122**
    allelic series as, 112–114, 114*f*, 115*f*
    codominance as, 110–112, 111*f*, 112*f*, 113*b*
    delayed age of onset of, 118, 118*f*
    epistatic, 127, **127,** 127*f*, 129–133, 134*b*
    incomplete dominance as, 109–110, 110f
    lethal mutations as, 114–117, 116*f*, 117*f*
    in Mendelian ratios modification, 122–133
    one gene-one enzyme hypothesis and,
        124–125*b*, 124–126
    in pathways, 123–124, 123*f*
    producing variable phenotypes, 119–122,
        119*f*, 120*f*, 122*f*
    sex-influenced traits as, 118, 118*f*
    sex-limited traits as, 117–118
    types of, 106
Gene knockouts, **571**
Gene order, 622–623, 623*f*
Gene pool, **738,** 738–739
    mutation diversifying, 750–751
Gene therapy, **590**
    in curing sickle cell disease in mice,
        593–594, 594*f*
    germinal, **592**
    human, 592–593, 592*t*
    somatic, **591,** 591–592
    using recombinant DNA technology, 590–594,
        592*t*, 594*f*
Gene transmission
    basic principles of, discovered by Mendel, 26–30
    in mitosis, 2
    in sexual reproduction, 2
Generalized transducing phages, **206,** 206*f*
Generalized transduction, **205,** 206–207, 206*f*
Genetic analysis
    in bacteria, 184–221
    of dinosaurs, 21–22*b*, 21*f*
    forward, **523**
    laying foundation of, 3
    of quantitative traits, 708–736
    reverse, **523**